FLOWERING PLANTS
OF JAMAICA

FLOWERING PLANTS
OF JAMAICA

BY

C. D. ADAMS

Reader in Botany, University of the West Indies

WITH CONTRIBUTIONS BY

G. R. PROCTOR, R. W. READ

AND OTHERS

ORIENS EX OCCIDENTE LUX

UNIVERSITY OF THE WEST INDIES
MONA · JAMAICA

Printed in Great Britain by
Robert MacLehose and Company Limited
The University Press, Glasgow

© C. D. ADAMS 1972

Set in Monotype Imprint
Preliminaries in 10 on 11 point
Text and Appendixes in 9 point
Keys and Indexes in 8 point

Dedicated to

ELSIE

'*Botanist by Marriage*'

Foreword

JAMAICA is a tropical island of great variety and contrast. This description applies both to the people and to the natural features. Within an area of approximately 4,400 square miles nearly every sort of tropical habitat can be found from desert conditions in parts of the south coast to mist forest high in the Blue Mountains only a few miles away. Visitors marvel at the scene and are full of questions about the rich and varied plant life which prevails. They, as well as our residents and especially our teachers, have so often lamented the absence of a book which would provide identification and other information about these plants, many of which are so beautiful and interesting and so unfamiliar, especially to those from temperate climates.

The British Museum of Natural History has amongst its oldest and most valued possessions the collections which Sir Hans Sloane made in Jamaica while serving as physician to the Duke of Albermarle, Governor of the island, in 1687–89. From the 17th century onwards, botanists have paid a great deal of attention to Jamaica's rich plant life. A vast number of plants from all parts of the tropical world, and also from temperate climates, have been introduced. Botanical gardens have been established at Bath, at Cinchona in the Blue Mountains, at Castleton and at Hope. The British Museum of Natural History began publication of the Flora of Jamaica in the year 1910. Five of the volumes of this work, which will probably run to 8 volumes, have appeared, but the earlier volumes are already out of print and out of date before the series has been finished. A comprehensive book on the flowering plants found in Jamaica will fill a long-standing need. Dr. Adams has been pursuing the task of preparing this book for some ten years and his meticulous work will, I am sure, be greatly appreciated by students of the Jamaican flora. He deserves our gratitude for his perseverance and achievement.

EDWARD SEAGA
Minister of Finance & Planning
Kingston, Jamaica

Preface

In 1959 a policy to promote the writing and publication of Caribbean Floras was implemented at the University College of the West Indies when the author was appointed to a lectureship in the Department of Botany at Mona. This post had been newly established to further the study of flowering plant systematics in the region. At that time a preliminary survey sufficed to indicate that the facilities and personnel then available were inadequate for the production of a comprehensive regional Flora. Moreover the estimated content of an Antillean Flora of some 10,000 species of vascular plants would have required more than one volume to encompass. It was decided that the initial project at Mona should be an attempt to put together in one volume data for the identification and naming of approximately 3,000 species of indigenous and naturalized or commonly cultivated flowering plants known to occur in Jamaica.

In view of the existence of several published parts of the definitive Fawcett & Rendle *Flora of Jamaica* and the continuing preparation of further volumes of that series at the Botany Department of the British Museum (Natural History) in London, it was decided that compression of the whole Jamaican angiosperm flora into one volume could be achieved in an independent publication by making the new book complementary in its content to the larger work. Fawcett & Rendle is distinguished by the provision of generic descriptions with at least one member of each genus illustrated by a line drawing, extensive synonymy and bibliography, as well as citations of all known exsiccata. The new work would, by omitting or minimizing these features and taking advantage of the existence of many new collections in the herbaria of the University and the Institute of Jamaica, besides the opportunities for daily and all the year round observations on living plants, be able to include useful field and ecological data which the other Flora did not have. It was also anticipated that a concise treatment at relatively low cost would be welcomed by many people in and outside of Jamaica as a comprehensive reference work of a kind which had never been available before.

The opportunity has been taken to carry out taxonomic revisions throughout the flora, to add taxa unknown to the earlier writers, to revise names according to the latest editions of the *International Code of Botanical Nomenclature* and to construct keys and modify descriptions in the light of new information.

The Introduction explains the methods of preparation of this Flora and its style and content. At this point it is only necessary to say that the richness and diversity of the Jamaican flora make certain minimal demands on the botanical knowledge and application of those who would aspire to some understanding of it. The use of some technical terms is unavoidable where considerations of accuracy and clarity are involved.

At the outset it was realized that several years of research would be necessary to bring our knowledge of the Jamaican flora up to a level from which acceptable publications in modern terms could result. This research has been made possible through financial assistance provided by the Scientific Research Council of Jamaica, the Ministry of Overseas Development of the United Kingdom (through the then Department of Technical Cooperation) and the University of the West Indies. Administrative arrangements were at first facilitated by the then Registrar of the University, Sir H. W. Springer and by the Director of the Institute of Jamaica, Mr. C. B. Lewis and then throughout by Professor A. D. Skelding, Head of the Department of Botany at Mona, to all of whom I am most grateful.

Publication of this volume has been enabled financially through the good offices of Sir George Taylor, Director of the Royal Botanic Gardens, Kew and by the late Mr. Stanley Smith of Nassau, Bahamas, acting through the Bentham–Moxon Trust. The Research and Publications Fund Committee of the University of the West Indies accepted guarantees.

It would be impossible for any one author to encompass with equal attention to detail and in a reasonable span of time complete revision of all the families and genera of this varied and difficult flora. I acknowledge gratefully the work of contributing authors and of all those who gave generously of their time and experience to discuss problems in their areas of special understanding and interest. The families *Bromeliaceae*, *Melastomataceae* and *Myrtaceae* were contributed by G. R. Proctor of the Institute of Jamaica; R. W. Read wrote *Palmae* and provided additional data on *Hohenbergia* and *Pitcairnia* (*Bromeliaceae*). The following genera were written or greatly assisted by: G. R. Proctor (*Rondeletia – Rubiaceae* and *Verbesina – Compositae*); B. D. Morley (*Columnea – Gesneriaceae*); H. A. Hespenheide (*Lepanthes – Orchidaceae*); R. E. Weaver (*Lisianthius – Gentianaceae*); Wm. T. Stearn (*Centaurium – Gentianaceae*). A special word of thanks is also due to Wm. T. Stearn of the Department of Botany, British Museum (Natural History), for his generosity in making available to me his unpublished text on the gamopetalous families prepared for volume VI of the *Flora of Jamaica* and for many hours of friendly discussion on nomenclatural and other matters.

Discussion or correspondence with the following provided valuable and greatly appreciated information and advice:

C. K. Allen (*Lauraceae*), W. R. Anderson (*Malpighiaceae*), M. R. Birdsey (*Araceae*), J. P. M. Brenan (*Caesalpiniaceae – Cassia, Commelinaceae*), D. Burch (*Euphorbiaceae – Euphorbia*, section *Chamaesyce*), J. F. M. Cannon (*Umbelliferae*), W. D. Clayton (*Gramineae*), J. E. Dandy (*Ruppiaceae* and nomenclature), W. G. D'Arcy (*Solanaceae – Solanum*), R. L. Dressler (*Orchidaceae*), A. W. Exell (*Combretaceae – Combretum*), F. R. Fosberg (*Amaranthaceae, Compositae*), L. A. Garay (*Orchidaceae*), the late H. B. Gilliland (*Gramineae*), S. A. Graham (*Lythraceae – Cuphea*), P. S. Green (*Oleaceae*), J. Haskett (*Pontederiaceae – Eichhornia*), D. Hillcoat (*Caesalpiniaceae, Mimosaceae*), C. E. Hubbard (*Gramineae*), D. R. Hunt (*Cactaceae*), P. F. Hunt (*Orchidaceae*), C. E. Jeffrey (*Compositae, Cucurbitaceae*), M. C. Johnston (*Rhamnaceae*), E. Launert (*Balsaminaceae, Labiatae*), J. Lewis (*Smilacaceae, Xyridaceae, Zingiberaceae* and other monocotyledons), G. Lucas (*Malvaceae*), M. E. Mathias (*Menispermaceae*), H. B. Moore (*Orchidaceae*), G. F. J. Pabst (*Orchidaceae*), L. L. Phillips (*Malvaceae – Gossypium*), R. M. Polhill (*Papilionaceae – Crotalaria*), D. M. Porter (*Zygophyllaceae*), P. H. Raven (*Onagraceae – Ludwigia*), M. Raymond (*Cyperaceae*), N. K. B. Robson (*Guttiferae*), V. Rudd (*Papilionaceae – Aeschynomene, Galactia*), the late N. Y. Sandwith (*Bignoniaceae*), B. G. Schubert (*Papilionaceae – Desmodium*), B. C. Stone (*Pandanaceae*), V. S. Summerhayes (*Orchidaceae*), P. Taylor (*Lentibulariaceae*), G. L. Webster (*Euphorbiaceae – Phyllanthus*), F. White (*Ebenaceae*).

Much of the research involving literature and typification was carried out at the British Museum in London and at the Royal Botanic Gardens, Kew; to the Directors and Keepers of those institutions I am deeply grateful for hospitality and the friendly cooperation of their staffs. In particular Miss Phyllis Edwards and Mr. R. G. C. Desmond, librarians, have been more than ordinarily helpful and I thank them. A short period of study was spent in the Harvard University Herbaria and the Arnold Arboretum at Jamaica Plain; I would like to record my gratitude to Dr. R. A. Howard, Director of the Arnold Arboretum, and to Dr. R. C. Rollins, Director of the Gray Herbarium, for facilities provided, and to Miss D. A. Powell for checking references. A long period was spent during 1964–65 in the Botany Department, Royal Holloway College at Englefield Green; Professor K. Wilson and his staff were especially friendly and hospitable and I have pleasant recollections of the time I spent with them.

Exploration in many parts of the Jamaican countryside involved numerous people whose names I never knew, but who are nonetheless deserving of appreciation. At Mona, where all the specimens were examined and curated and all new information was put on cards, I was willingly assisted by students and others among whom I must mention Mrs. Athelin Barnett, Miss Patricia Coke, Mrs Barbara Lee, Miss Lilly Lodenquai, Miss Ruby McKay, Miss Dorothy Mitchell, Miss Patricia Pearson, Mrs. Cicely Tobisch, Miss Carole Vickers and not least Mr. Rupert Fletcher and the late Mr. Leslie Wynter who grew many of the plants that required further study. Mrs. M. Allwood, Herbarium Assistant in the Botany Department, has been involved in many aspects of the preparatory work and has supervised all the curating. Among Mrs. Allwood's outstanding achievements has been the remounting of the whole Harris herbarium. Mrs. Valerie Been spent many hours extracting and sorting references and vernacular names and in preparing material for the indices. Mrs. Inge Judd typed the 1500 or so pages of manuscript in triplicate, a mammoth task carried out with care and efficiency. To these helpers the completion of this Flora is in good measure due.

Finally I am happy to acknowledge and thank those who were patient and interested enough to read and comment on manuscript and try out keys, in particular Robert D. Henry, Michael duQuesnay and my friend of many years Peter S. Green. J. E. Dandy, Peter Green, Brenda Hammerton and Wm. T. Stearn had the fortitude to read proofs. Wherever there is quality in this volume I would like to think that the credit can be shared; wherever there are weaknesses or errors the responsibility is wholly mine.

C. D. ADAMS

Mona, Jamaica
10th June 1970

Contents

Contents

Introduction

METHODS OF INVESTIGATION

ALTHOUGH ultimately information about Jamaican plants was obtained from many hundreds of sources including original descriptions and illustrations and accounts in local and other tropical Floras, the early part of the investigation in Jamaica was directed towards equating the specimens in the two herbaria, those of the University of the West Indies and Institute of Jamaica, with each other and with new observations of species in the field. Between November 1959 and early 1964 over 250 field excursions were made by the author and concurrently some 40,000 local herbarium specimens, mostly collected since 1950, were studied. Thus as an initial phase a taxonomic appraisal of the whole flowering plant flora was made. Standard procedure was to transfer all collectors' annotations verbatim or in summary to a species-card and in many instances to add descriptive data which were believed to be new or revisionary.

The names of many species were authenticated by reference to exsiccata in Jamaican herbaria which had been cited in various publications. The collections of Jamaican plants made by Wm. Harris had been distributed as duplicates to foreign herbaria. Many of these were cited in the Fawcett & Rendle Flora and some are the types of several hundred new species which he discovered and which were described by N. L. Britton, I. Urban and others. The University herbarium contains what is now believed to be the largest single collection of Harris's plants, comprising much syntype, isotype and even holotype material. The Institute of Jamaica herbarium contains numerous types of species named by or for G. R. Proctor, collector of the most comprehensive and representative specimens of Jamaican vascular plants which the author has been privileged to study.

Further verification of names was begun at the British Museum and Kew in 1961 with particular attention being given to monocotyledonous plants for which there are no modern accounts for Jamaica. Subsequently in 1964–65 and 1968 the nomenclatural review was continued and completed.

The first set of the author's collections, numbering some 8,000 gatherings, is housed in its entirety in the herbarium of the University of the West Indies at Mona, a duplicate set by species is at the British Museum (Natural History) at South Kensington, a third set is at Duke University, Durham, North Carolina and fourth and fifth sets are at the Botanische Staatssammlung, Munich, West Germany and the Missouri Botanical Garden, St. Louis, Missouri, respectively.

STYLE AND CONTENT OF THE FLORA

CLASSIFICATION

The placing of the Monocotyledons before the Dicotyledons is more in deference to traditional treatments in existing Floras for the region than to a conviction that one class is more primitive than the other. Even if some special

15

evidence, derived from study of Jamaican flowering plants, were to lend weight
to an opinion, a Flora such as this would hardly be the place to discuss it. The
arrangement is a modified version of Hutchinson's system set out in his *Families
of Flowering Plants*, 2nd edition, 1959.

The classification adopted for the Dicotyledons is patterned on that of Fawcett
& Rendle which in turn was based on the system of Engler & Prantl. Some of the
revisions embodied in the 1964 edition of Engler's *Syllabus der Pflanzenfamilien*
have been incorporated and others have not. The major variations from Fawcett
& Rendle are 1. *Picrodendron* moved from *Euphorbiaceae* to *Juglandales*; the
latest *Syllabus* has *Picrodendraceae* in *Rutales*, 2. *Guttiferales* have been moved
forward to a position near *Papaverales*, 3. *Cactales* have been moved to proximity
with *Centrospermae*, 4. *Lacistemataceae* have been shifted to *Violales*.

Variation from the latest *Syllabus* is mainly due to a wish not to depart from
the Fawcett & Rendle arrangement more than absolutely necessary. For example
Piperales remain in an early position and *Aristolochiaceae* remain near to
Balanophoraceae; *Bataceae* remain in *Centrospermae*; *Callitrichaceae* and
Buxaceae ally with *Euphorbiaceae* and *Balsaminaceae* is with *Tropaeolaceae* in
Geraniales.

The legume families and *Euphorbiaceae*, *Rubiaceae* and *Compositae* have
been accorded their own orders. Some first-hand study of *Buddleja americana*,
the type of the genus *Buddleja*, has convinced the author that *Loganiaceae* in
the broad sense is a satisfactory family and that no good purpose is served by
breaking it up.

DESCRIPTIONS OF FAMILIES

Because most keys to families are artificial and because almost all the important
tropical families occur in Jamaica, these descriptions are comprehensive and
provide a source of this generally useful information not otherwise available
outside specialized publications. The descriptions of the families are intended as
far as possible not to be interpretive nor to draw special attention to supposed
genetical relationships. For example in the numerous instances where unilocular
compound ovaries can be interpreted as being contributed to by 2 or more
carpels, the ovary has been described as " 1-locular " followed by such further
descriptive remarks as " placentas 3, parietal " or " styles 5, free " or " capsule
4-valved ", such information being provided more to enable accurate identification or definition of the taxon, less to establish its theoretical position in any
system.

GENERA

In order to save space, generic diagnoses are provided only where, in other
than monotypic families, one genus occurs in Jamaica and there is therefore no
key. Key-entries are written primarily to enable identification of the genus as it is
represented in Jamaica. The construction of the keys to the genera and species
is dichotomous but the statements in any couplet, except for the first entry, are
not always strictly equated. This is especially so where the homologous statement
is fully implied in an opposite sense and/or where the homologues may appear
again in a subsequent couplet with different values. For example:

1 Calyx 3–4 mm. long; leaf-blades 5 cm. or more long; style evident; petals blue   A

1 Calyx 1·5–2·5 mm. long; leaf-blades shorter; stigma sessile or stalked; petals not blue:

  2 Leaf-blades 2–4 cm. long; style evident; petals yellow   B

  2 Leaf-blades 1–2·5 cm. long; stigma sessile; petals pink   C

It is quite obvious that the words underlined are superfluous and can be omitted.

CIRCUMSCRIPTION OF GENERA

For the purposes of this Flora a broad concept of genus has usually been maintained. The decision to adopt such a convention is determined by several practical considerations. Firstly, because descriptions of genera are not provided and the key-entries are the only sources of genus diagnoses, the combinations of larger more obvious differences appear in the genus key. As a result smaller often unique differences are relegated to the species key. Identification is facilitated where any dichotomy provides several associated characters particularly if these refer to parts available at different seasons or phases of growth. Secondly, the adoption of fewer larger genera automatically means a reduction in the diversity of names and consequently less strain on the memories of those who wish to acquire some mastery of the vast array of flora in a tropical country. Once initial interests have been established then increasing specialization and depth of involvement can lead to more critical taxonomic limits. There is little doubt that in the course of time further research will tend on balance to strengthen the contentions of those who prefer the larger number of smaller genera and so support a general trend which has continued with fluctuations since the time of Linnaeus. Nevertheless, there would seem to be no firm philosophical foundation for this increase; one would expect that new resemblances would be discovered at about the same rate as new differences, but increasing specialist interest over smaller areas of study tends to give more weight to the latter.

That there are still fluctuations and reverse trends more than 200 years after the concept of genus received general acceptance, may be an indication that small segregate genera are less satisfactory and that their recognition at any time could be more a matter of fashion than anything else. Not very many groups in the Jamaican flora are much affected by these considerations but those that are at the present time show some astonishing anomalies and inconsistencies. The agrostologists, particularly in *Andropogon* and *Panicum*, are moving towards more genera; the cyperologists towards fewer; the characters which distinguish the " unacceptable " small genera within the *Cassia* complex are easier to define than the characters which distinguish " acceptable " genera in the *Mimosaceae*. These characters are often interpretable as homologous in these two groups, and the different patterns are set by the circumstance that in the *Caesalpiniaceae* genera can be based on the many different kinds of flowers, whereas in the *Mimosaceae* the flowers are very uniform and the burden of genus definition is carried by fruits and foliage. It is never the function of a Flora to make and defend taxonomic causes but some explanation is necessary to justify the arbitrary line that has been taken with some of the large genera that are nowadays often split up. Besides those already mentioned they include *Epidendrum* and *Euphorbia*. The provision of adequate synonymy is in part a measure to meet the critics and if by such devices the inevitable frustrations which may beset the student can be minimized then this Flora will have achieved much of its purpose.

SPECIES

The keys to the species and the descriptive remarks for each of them are integrated so that there is a minimum of repetition of information. This procedure has saved a great deal of print at the cost to the reader of having to refer to the key from time to time for some of the descriptive matter. As a rule the species-data have been derived entirely from Jamaican sources.

The citation of the currently accepted binomial is given in full. Synonyms are provided where there have been changes of name since the appearance of important publications on the flora of Jamaica, particularly where the currently accepted name is different from that used by Fawcett & Rendle[1] or, for example, Hitchcock in his *Manual of the Grasses of the West Indies*. Basionyms are given frequently in order to provide historical perspectives to common plants as well as cross references to other Floras and monographs where species may have been placed in different genera. Synonyms are cited by binomial, author and date of publication.

The method of citing genera, monographs and species-synonyms by author and date is to assist future workers in finding the full citation easily. As dates are not given consistently in the main volumes of *Index Kewensis*, then the provision of date enables the full citation to be reconstructed. If the reference sought is in one of the *Index Kewensis* Supplements, then having the date enables the researcher to refer immediately to the correct supplement. After some familiarity with authors and dates most of the important literature references come quickly to mind. A list of the references and monographs cited after genera is given in Appendix II. A further practical advantage of giving dates with synonyms is that the full stop is effectively displaced from those authors whose names are not abbreviated, e.g. *Scirpus micranthus* Vahl (1805). instead of *Scirpus micranthus* Vahl. The dates of publication of most of the older works important in establishing nomenclatural priority have now been determined with sufficient accuracy for working purposes. A number are however still subject to change as new information comes to hand. Where differences of opinion seem to exist about dates of publication Stafleu, F. A., *Taxonomic Literature*, Regnum Vegetabile **52** (1967), has usually been followed.

COMMON NAMES

These names are given where they are reliably reported and have some relevance to the identification of the species. Many names in current use are of mixed or doubtful application and an attempt has been made to particularize them where possible. The vernacular names are of great assistance to lay people and are also of interest to language students. Non-Jamaicans will be puzzled until they realize that many of the common names are adoptions from European sources applied to tropical plants which may or may not resemble the plants originally so named. Other names stem from every language source in Jamaica. Some of the names are clearly spontaneous reflections of superficial resemblances to cultivated plants, for example " Wild Lime " for *Adelia ricinella*, many others

---

[1] These authors are abbreviated in citations to " F. & R. "

relate to particular properties often expressed in anthropomorphic terms. A separate index of common names is provided. A valuable source of Jamaican names for plants is Cassidy, F. G. & Le Page, R. B., *Dictionary of Jamaican English* (1967).

DESCRIPTION OF SPECIES

These descriptions are rarely in any sense complete and are intended to provide only such additional information as will serve to confirm identity. Much new information has been gathered during field studies and data on ephemeral characters such as colour, odours, textures and shapes have been added. Many dimensions have been revised as a result of measurements made on living as opposed to herbarium specimens. Some of these revised measurements include data on juvenile phase or coppice regrowth where the leaves are often larger or, if compound, more complex. The herbarium collections at Mona are being enhanced by the addition of specimens of juvenile and seedling stages wherever these can be firmly identified. In *Eugenia* and perhaps other *Myrtaceae* the leaves of seedlings are considerably narrower than on mature plants of the same species. In melastomes foliage of juvenile branches is often more hairy than on flowering twigs; see also the note under *Myrica cerifera*. Such details can assist identification but may also confuse and so particularly aberrant measurements may be added to more usual ones in parentheses; for example in *Albizia lebbeck* one reads " pinnae 2–4 (–6) pairs " indicating that 5 or 6 pairs of pinnae occur on saplings and young trees and would be uncommon otherwise. Dimensions stated in this way usually indicate rare or unusual conditions or occasionally, for plants that are poorly known in Jamaica, extreme measurements reported in literature or from non-Jamaican materials.

FREQUENCY AND DISTRIBUTION IN JAMAICA

This information is based on parish records which can be verified from herbarium specimens and except for a very few common species visual records are not included in the statements made. Complete listing of parishes would only reflect our incomplete knowledge of the distribution of common and certainly widespread species. A feature of the Jamaican Flora is the very high incidence of local endemism. These facts are behind the decision to name up to three parishes only, or in a few instances where the distribution is discontinuous, four parishes, for rare and local species. Actual names of places where rare species are known to occur have been omitted purposely for reasons of conservation. Distribution of broader but significant localization have been indicated by compass points in preference to longer lists of parish names. The parishes are as follows:

Western Group (County of Cornwall)	Position in each County
Westmoreland (abbr. West.)	South-west
Hanover (abbr. Han.)	North-west
St. Elizabeth (abbr. St. Eliz.)	South-east
St. James	North
Trelawny (abbr. Trel.)	North-east

Central Group (County of Middlesex)   Position in each County
  Manchester (abbr. Manch.)   South-west
  St. Ann   North
  Clarendon (abbr. Clar.)   South-central
  St. Catherine (abbr. St. Cath.)   South-east
Eastern Group (County of Surrey)
  St. Andrew including Kingston (abbr. St. Andr.) South-west
  St. Mary   North-west
  Portland (abbr. Port.)   North-east
  St. Thomas (abbr. St. Thom.)   South-east

ECOLOGICAL DATA

Most of the herbarium material on which the Fawcett & Rendle Flora was based carried no field notes. The specimens of Wm. Harris mostly stated altitude. Since then much more information has been recorded, sufficient to enable a general statement of habitat to be made for most species. This information is largely original and may be of particular importance to tropical ecologists, especially in respect of the numerous species in Jamaica characterizing woodland on limestone – a formation that has received much less attention than most other tropical types. The descriptive statement of habitat is followed by recorded altitudinal range and flowering and fruiting months.

The classification of the vegetation of Jamaica has been insufficiently studied and the geographical units in which our species belong are not clearly defined. Comparisons of species distributions with relief, geological and rainfall maps suggest correlations which almost all remain to be tested. Except for certain outstandingly distinct habitats such as freshwater streams and sea-beaches, edaphic relationships cannot be taken for granted and statements implying strict dependences have been largely avoided. Thus the habitat notes are rarely exclusive; epiphytic may often be read to include epipetric; the several different types of limestone which exist have not been distinguished; the word shale is used freely in reference to several types of sedimentary non-calcareous rocks. These cautions are intended to emphasize the need for future observational and experimental studies, involving the cultural requirements of wild species and their detailed mapping.

Altitudinal data are often correlated with distribution and habitat. Most of the limestone formations occur below 3,000 feet in areas away from the main ridge of the Blue Mountains and Port Royal Mountains. There is very little limestone in these mountains above 3,000 feet and the higher parts are restricted to the easternmost parishes of St. Andrew, Portland and St. Thomas. Prominent topography is often associated with unusual distributions of plants and the John Crow Mountains, Hollymount and Dolphin Head are all examples of isolated high ground associated with high rates of endemism. Occasional discontinuous altitudinal ranges can be attributed to the peculiarities of certain species such as *Dodonaea viscosa* or *Borreria verticillata*, which have maritime and montane variants, and sometimes to the temporary occurrence of lowland weeds in the mountains carried up in road metal, building sand or gravel. Few species in Jamaica have total altitudinal ranges of greater than 5,000 feet. Heights above sea-level are stated in feet because these are the units used on the local topographical survey maps.

Flowering and fruiting times have a regular annual periodicity in some species and the winter flowering of members of *Compositae*, *Cyperaceae*, *Gramineae* and terrestrial members of the *Orchidaceae* is a conspicuous feature. In other groups the flowering is often less regular but patterns will probably emerge after more intensive study. A species recorded as flowering for all or most of the year need not do so continuously in the same locality. The distribution of rainfall in Jamaica is patchy and particular species often respond to local variations in rainfall and other environmental factors.

CITATION OF SPECIMENS

Where specimens exist three or four are cited at the end of each entry. These are exsiccata which are mainly housed in Jamaica, but of which duplicates have usually been distributed to herbaria in North America and Europe. The purpose of citing these specimens is primarily to enable the authoritative confirmation of identity by workers in Jamaica. It is also hoped that future monographers will see these specimens or their duplicates and use this work as a reference. The Jamaican herbaria will then be able to keep the nomenclature of their collections up to date. The citing of specimens is always a reminder to students, not only of botany or ecology but in applied related fields also, of the importance of preserving vouchers. Abbreviations have been used for the names of collectors as follows:

A = C. D. Adams; A & C = W. R. Anderson and M. Crosby; ACH = W. R. Anderson, M. Crosby and H. A. Hespenheide; A & D = C. D. Adams and M. duQuesnay; A & S = W. R. Anderson and D. Sternberg; H = Wm. Harris; H & B = Wm. Harris and N. L. Britton; HCR = H. A. and E. B. Hespenheide, J. S. Calver and R. E. Ricklefs; H & P = R. A. Howard and G. R. Proctor; HPS = R. A. Howard, G. R. Proctor and Wm. T. Stearn; J.P. = Jamaican Plants (Collectors of original Botanical Department, including from 1879 those made by J. H. Hart, D. Morris and others, now housed at Mona); P = G. R. Proctor.

DISTRIBUTION OUTSIDE JAMAICA

It has obviously not been possible to examine, and compare with Jamaican plants, specimens from all localities to which species have been attributed. The best that one can hope to do is what has been attempted, namely to collect as many literature records of distribution as possible and to check these by looking at specimens as opportunities present themselves. Particular care was taken to examine specimens where taxonomic limits were known to have been changed or were under critical review or where published statements seemed to be unrealistic or improbable.

COMPOSITION OF THE JAMAICAN FLORA

This account of the flora of Jamaica recognizes that most users of this book will be seeking to determine the names of relatively common widespread species many of which are introduced useful plants and weeds. The extent to which such plants have spread and established themselves is variable and difficult to judge and of course the process is still going on. The criteria that have been used to decide inclusion or exclusion have been arbitrary and each case has been decided on its merits. In general only those species restricted to horticultural cultivation and not known to have naturalized themselves or to exist in relict floras have been excluded. Over 350 plants which are common in cultivation have been included but these are frequently taken no further than to a key-entry.

Families described total 183. Of these 36 are monocotyledonous and include *Eriocaulaceae* not yet reported from Jamaica but with a fairly strong likelihood that at some time examples may be discovered. Dicotyledonous families number 147 and these likewise include *Callitrichaceae* although it also is not yet reported. Flowering plants belonging to 1150 genera and 3247 species are considered. A few others are known from the Cayman Is. but not yet from Jamaica; it has been thought useful to draw attention to some of those that might eventually turn up, either by including them in the keys or in notes.

The monocotyledons comprise 276 genera of which 36 are represented by cultivated species only, and 831 species of which 78 are casual or cultivated. The dicotyledons comprise 874 genera of which 119 are represented by cultivated species only, and 2416 species of which 282 are casual or cultivated. The estimated total known flowering plant flora of native and fully naturalized plants is 173 families, 996 genera and 2888 species.

Like most vegetation in the humid intertropical zone, the natural vegetation of Jamaica is predominantly woody. Of the native and fully naturalized species about half are woody and these in turn comprise about equal numbers of species of shrubs and trees. Of the herbaceous species there are about equal numbers of monocots and dicots. In many localities dominated by woody species shrublike growths have resulted from continual coppicing of plants that left to themselves would have grown into trees. It is this fact that largely accounts for the use of the phrase " shrub or tree " in many of the descriptions rather than to any property of the plants themselves.

Each full species-entry includes a statement of the distribution in Jamaica and outside Jamaica as well if any. From this it is clear that four genera (*Tetrasiphon, Jacaima, Salpixantha* and *Acanthodesmos*) and 784 species are in the state of our present knowledge thought to be endemic to the island. Most of these species are neo-endemics and are quite rare. They are often restricted to particular small or isolated areas. Some of them are, however, among our commonest and most widespread species including the orchids *Broughtonia sanguinea, Brassavola cordata* and *Oncidium tetrapetalum*, the palm *Thrinax parviflora, Senecio discolor* and *Vernonia acuminata (Compositae), Pilea grandifolia (Urticaceae), Galactia pendula (Papilionaceae), Cassia viminea (Caesalpiniaceae)* and dozens of others. The emphasis of distinctness which these species with restricted distribution gives to the Jamaican flora is slightly diminished when we learn that some of them have very close affinity with species in other islands of the Greater Antilles. For example *Galactia pendula* is very close to *Galactia longiflora* Arn. of the Lesser Antilles, *G. eggersii* Urb. of Puerto Rico and also to *G. sangsterae* Proctor recently described from Jamaica. Such affinities only serve to demonstrate a larger evolutionary pattern embracing the neighbouring islands in varying degrees. The flora of Jamaica has greatest affinity, as one would expect, with that of Cuba, and next with Hispaniola, then the common denominators become fewer with greater distance apart. Although endemic genera are few some large widespread genera as for example *Acalypha, Ilex, Pilea, Dendropanax, Columnea, Lisianthius, Rondeletia* or *Lepanthes* have groups of species which have undergone remarkable evolutionary radiations within this island. These groups are the truly native ones which have occupied and formed part of the natural woodland vegetation for many centuries – for far longer than man has been active here. The period during which man has occupied Jamaica and replaced much of the natural vegetation with introduced food plants, ornamentals and weeds has been infinitesimally short by comparison and most of these introductions have not had time to evolve very much in their new surroundings.

The two main indigenous families of monocotyledonous plants, comprising many epiphytes, are the *Bromeliaceae* and *Orchidaceae*. They number 267 species and their endemics number 82 or 30·7%. In contrast the *Gramineae* and *Cyperaceae* number 346 species but have only 4 or 1·2% of endemic species between them. Many of these sun-loving plants of open ground are widespread in tropical countries and very few even of the strictly West Indian grasses and sedges could have evolved in Jamaica. These facts lend weight to the contention that the climax vegetation of Jamaica was an almost entirely woody one. Forests existed in the wetter areas in which the native trees and epiphytes on them as well as shade-loving ferns, Peperomias, Pileas, Gesnerias and Psychotrias evolved together. An as yet not properly described type of thickety woodland occupied most of the centre and west of the island where more or less exposed rugged limestone was dominated by shrubs and small trees. This vegetation comprised in a substantial way, and with unique combinations of species, genera such as *Rondeletia*, *Coccoloba*, *Eugenia* and the woody sections of *Phyllanthus*, again carrying on their branches many locally evolved epiphytes. In the drier coastal areas, open woodlands gave way to scrub-thicket with characteristic succulent and other xeromorphic species still associated with a few endemic epiphytes. In none of these associations are grasses or sedges conspicuous components. The clues to this historical appreciation rest in recognition of the ancient indigenous genera. It is from studies of them that we shall learn about the natural ancestry and origins of the flora. It is in them that most of the improvements and refinements in the classification of Jamaican plants will be made in the future. Taxonomic considerations bedevil all statements about endemism and therefore influence our ideas about the history of the flora. Nevertheless, a high rate of endemism is usually taken to be correlated with long distance or long duration of isolation. The overall endemism rate for Jamaica, calculated on the indigenous content of this Flora and the most recent information on distribution is 27·2%.

Some species, often of useful or especially ornamental plants, were originally attributed to Jamaica but have not been found here since. These errors could be accounted for by the fact that from time to time Jamaica has been a centre of horticultural activity and a clearing house for useful plants which would then have been sent to Europe for description after the original data on sources were lost. Several of the plants attributed to Jamaica by Wm. Wright (1735–1819) have not been confirmed and may have come from islands of the Lesser Antilles; these include *Protium attenuatum* (*Burseraceae*), *Persea urbaniana* (*Lauraceae*), *Freziera undulata* (*Theaceae*), *Callicarpa reticulata* (*Verbenaceae*) and *Pilea lanceolata* and *Urera caracasana* (*Urticaceae*). Alternatively, errors could have been perpetrated intentionally in order to obscure the real natural localities of valuable species. For example, *Odontoglossum jamaicense* Griseb. is almost certainly of Central American origin and is of a group which has not been and undoubtedly never will be found wild in Jamaica. Some of the early collections of Purdie who explored both in Jamaica and Trinidad were confused as between localities in the parish of St. Ann in the former and St. Ann's in the latter; these records have now been sorted out as have certain other Purdie collections from Colombia which were at one time thought to have come from Jamaica. Grisebach reported *Pinguicula elongata* Benjam. for Jamaica, but this was really a Purdie gathering in Colombia. Conversely, strictly Jamaican plants were originally reported as from far distant places, such as *Brassavola cordata* from Brazil and *Oncidium pulchellum* from Guyana, although these are both endemic to Jamaica. Throughout this Flora notes have been inserted drawing attention to discrepancies between records of the past and what we now believe to be the truth. It is of some interest to know that some of the early records of species not now known to occur were undoubtedly authentic and were part of an as yet unexplained element in the West Indian flora which comprises rather widespread species of plants common in other islands even to-day but very rare or more usually altogether absent from Jamaica. These species include *Epidendrum ciliare* L. and *Guettarda scabra* (L.) Vent.

Some additions to the flora are inevitable in the future. There will be introductions of vigorous agricultural species as crops, fodder and weeds, some of which are bound to become established. Seymour Grass, *Andropogon pertusus*, has in less than a hundred years become perhaps the commonest grass in lowland Jamaica. Wynne Grass, *Melinis minutiflora*, was a very rare grass here in 1925 but is now one of the commonest species in the open parts of the St. Andrew hills. Other rare local species will unfortunately become extinct. In remote areas there are still species completely unknown to science waiting to be discovered and described. It can be safely stated however that the main phase of exploration is virtually closed. For many species we do not have complete descriptions; either flowers or fruits remain unknown and hence the classification is probably not as satisfactory as it could be; many distributions remain to be expanded or not as the case may be into areas where particular species might be expected to grow. There is much scope still for accumulating additional detailed information. But the stage is also set for a step forward into an experimental phase which will involve investigations into breeding systems by means of which the different patterns of variation can be explained. There is scope now for evaluating the biological potential of many of our wild species in purely scientific investigations or for horticultural or agricultural purposes. The basic classification is now available to enable systematic surveys to be carried out into the chemical, biochemical or other productive assets of our native plants. Classification is only a first step in stating the diversity and complexity of the remarkable plant world around us and not, as it can so easily become, an end in itself.

Sequence of Orders and Families

1. *Class* MONOCOTYLEDONES

Subclass CALYCIFERAE

Subclass COROLLIFERAE

Descriptions of Families

1. HYDROCHARITACEAE

Partly or wholly submerged herbs of fresh or salt water. Leaves radical or dispersed along elongated stems, alternate, opposite or whorled. Flowers actinomorphic, unisexual or bisexual, within a pair of bracts or a bifid spathe. Perianth-segments free in 1 or 2 series, 3 (–2) in each series, the outer usually green and valvate, the inner petaloid and imbricate. Stamens 1-many; anthers 2-locular, opening lengthwise. Ovary inferior, 1-locular with parietal placentas; styles as many as the placentas, entire or branched; ovules numerous on each placenta. Fruit dry or pulpy, rupturing irregularly. Seeds with straight embryo, without endosperm.

14 genera with about 100 species mostly in the warmer regions of the world.

1 Fresh-water plants with flowers pollinated at or above the surface of the water; pollen sphaeroid:
  2 Leaves spread along the stem, opposite or verticillate, short, thin; flowers arising in the axils of the leaves:
    3 Male spathes 2-several-flowered; petals much broader than the sepals   **1. Egeria**
    3 All spathes 1-flowered; petals narrower than the sepals   **2. Elodea**
  2 Leaves basal, alternate, elongated; male spathes many-flowered; female spathes 1-flowered   **3. Vallisneria**
1 Marine plants with flowers pollinated beneath the surface of the water; pollen united in strings:
  4 Leaves alternate, linear and undifferentiated into petiole and lamina; spathes of 2 connate bracts; styles 6–12, bifid   **4. Thalassia**
  4 Leaves opposite, oblong and differentiated into petiole and lamina; spathes of 2 free bracts; styles (2–) 3 (–5), entire   **5. Halophila**

1. EGERIA Planch. (1849)

1. E. densa Planch. in Ann. Sci. Nat. sér. 3, Bot. **11**: 80 (1849).—*Elodea densa* (Planch.) Casp. (1857). *Anacharis densa* (Planch.) Marie-Vict. (1931).
  Leaves mostly in whorls of 5, linear-oblong, thin, mostly somewhat crisped-undulate; sepals 3; petals 3, white.
  Rare (St. Andr., St. Cath.), in ponds, cultivated and naturalized; 600–1250 feet; fl. Nov–Jan, May. *P 24205! Robertson UCWI 20273! 20276! Sidrak UCWI 23350!* Native of subtropical S. Amer., introduced into Europe, S. Africa, Pacific Is. and in aquaria.

2. ELODEA Michx. (1803)

1. E. canadensis Michx., Fl. Bor. Amer. **1**: 20 (1803).—*Anacharis canadensis* Planch. (1849). Canadian Pondweed.
  Stems very brittle; leaves 6–12 mm. long, 3–4 mm. broad, translucent; petals white or tinged purplish.
  Cultivated (St. Andr.), in artificial ponds; 600 feet; fl. Nov. *Robertson UCWI 6766!* Native of Canada and the United States, introduced and naturalized in Europe.

3. VALLISNERIA L. (1753)

1. V. americana Michx., Fl. Bor. Amer. **2**: 220 (1803).
  Rhizomatous herb; leaves 6–12 mm. broad, very long and broadening towards the blunt tip; pedicels long and slender; flowers white.

Cultivated (Han.), in artificial ponds. *P 10002!* Canada to Florida and Mexico, Greater Antilles.

4. THALASSIA Konig (1805); Den Hartog (1970)

1. T. testudinum Konig in Konig & Sims, Ann. Bot. **2**: 96 (1805).—Turtle Grass.
Rhizomatous herb; sheaths brown; leaves green, linear, blunt-tipped, 8–15 mm. broad; flowers white.
Submerged in often large colonies in lagoons protected by coral reefs; SL down to 30 feet. *Goreau UCWI 2712! P15133! Stearn 329!* Bermuda, Bahamas, West Indies; Cayman Is.

5. HALOPHILA Thouars (1806); Den Hartog (1970)

1. H. decipiens Ostenf. in Bot. Tidsskr. **24**: 260, f. (1902).—*H. baillonis* Aschers. (1874), partly.
Slender-stemmed much-branched trailing herb; leaves simple, oblanceolate with conspicuous midrib, 1–2 cm. long, 3–6mm. broad.
Local (St. Andr., St. Eliz., Han., Port.), submerged in placid estuaries, creeks and brackish ponds. *Howe 930! Kirton! A. von der Porten!* Florida, West Indies, Indian and Pacific Oceans.

2. ALISMATACEAE

Perennial or annual herbs of ponds and marshes. Stems short. Leaves basal, sheathing but open at the base; blades flat. Flowers bracteate, often whorled in scapose racemes or panicles, actinomorphic. Perianth 2-seriate; sepals 3, imbricate, persistent; petals 3, imbricate, deciduous or rarely absent. Stamens hypogynous, 6 or more, free; anthers 2-locular, extrorse or opening by lateral slits. Carpels numerous, usually free; styles persistent; ovules solitary or few, basal or nearly so, anatropous or amphitropous. Fruits achenial or rarely follicular. Seeds curved; endosperm absent.
12 genera with about 70 species; widely distributed.

Flowers bisexual; achenes long-beaked **1. Echinodorus**
Flowers unisexual; achenes with short or obsolete beaks **2. Sagittaria**

1. ECHINODORUS L. C. Rich. (1815)

Leaves ovate-lanceolate to broad and cordate; scape, inflorescence-branches and petioles ribbed; inflorescence-branches in several whorls **1. berteroi**
Leaves linear-oblanceolate; scape slender; inflorescence umbellate or of few whorls **2. tenellus**

1. E. berteroi (Spreng.) Fassett in Rhodora **57**: 139 (1955).—*Alisma berteroi* Spreng. (1825). *E. cordifolius* Griseb (1857), partly, not *Alisma cordifolium* L. (1753).
Glabrous herb to 1 m. in flower, but sometimes much smaller in drier marginal habitats; largest leaves 7-veined with long petioles; petals white; achenes keeled and ribbed.
Very local (St. Cath.), along ditches and lagoon margins in fresh water; 40–50 feet; fl. Dec–Mar, June, fr. Dec–Mar. *A 10728! H12312! Proctor & Stearn 11570!* S. United States, Mexico, Ecuador, Bahamas, West Indies.

2. E. tenellus (Mart.) Buchen. in Abh. Nat. Bremen **2**: 21 (1869).—*Alisma tenellum* Mart. (1830).
Slender-stemmed herb up to about 20 cm. in flower; sometimes spreading by stolons; very variable in petiole-length and length/breadth ratio of blade; petals white.
Rare and local (St. Cath., St. Eliz.), in swamps; 50–1000 feet; fl. Jan, July, fr. Jan. *H12095! P6143! 21928!* Florida, Mexico, tropical C. and S. Amer., Cuba, Hispaniola.

2. SAGITTARIA L. (1753)

1 Leaves floating; achenes verrucose, irregularly shaped, with inconspicuous beak **1. guayanensis**

1 Leaves emersed; achenes smooth, laterally flattened with distinct short beak:
2 Blade narrowed to the petiole, elliptical to linear-lanceolate; achenes thick-margined; stamens more than 15   *2. lancifolia*
2 Blade sagittate; achenes thinly winged; stamens about 10   *3. intermedia*

1. S. guayanensis Kunth, Nov. Gen. **1**: 250 (1816).—*Lophotocarpus guayanensis* (Kunth) J. G. Sm. (1894).

Petioles tufted from a short submerged stem; leaf-blades rounded at the tip; petals white.

Very rare (St. Eliz.), in a small weedy pond; 100 feet; fl. Oct. *P 21448!* Tropical S. Amer., Hispaniola.

2. S. lancifolia L., Syst. Nat. ed. 10, **2**: 1270 (1759).

Erect herb up to 150 cm. in flower; flowers showy with white petals.

Locally common in swamps and riverside marshes; SL–100 feet; fl. and fr. all the year. *A 6024! H 6033! P 21835!* Subtropical and tropical Amer., S. United States, Greater Antilles; Grand Cayman.

3. S. intermedia M. Micheli in A. & C. DC., Monogr. Phan. **3**: 80 (1881).

Erect herb up to 60 cm. in flower; petals white; anthers orange; filaments green.

Rather local in central and western districts at pond-margins and along ditches; SL–1000 feet; fl. and fr. May, Aug–Nov, Jan. *A 10382! H 12156! P 18326!* Greater Antilles.

3. POTAMOGETONACEAE

Aquatic perennial herbs of fresh water. Leaves alternate or opposite, often polymorphic according to whether submerged or floating, sheathing at the base. Flowers bisexual in pedunculate axillary spikes; peduncle sheathed at the base; bracts absent. Perianth of 4 free shortly clawed valvate segments. Stamens 4 inserted on the claws of the segments; anthers sessile, 2-locular, opening extrorsely. Carpels 4, free, 1-locular; styles short or stigmas sessile; ovules solitary, campylotropous, attached to the adaxial angle of the carpel. Fruiting carpels sessile, indehiscent. Seeds without endosperm.

2 genera; *Potamogeton* widespread with about 90 species.

1. POTAMOGETON L. (1753); Ogden (1943)

Leaves alternate; carpels drupaceous in fruit.

1 Leaves 1–2 mm. broad, linear, sessile, margin smooth if sometimes very shallowly sinuate; inflorescence shortly pedunculate; spike less than 1 cm. long   *1. foliosus*
1 Leaves 1 cm. or more broad; peduncle and spike longer than above:
2 Leaf-blades lanceolate, almost always sessile and submerged, margin minutely denticulate   *2. illinoensis*
2 Leaf-blades linear-lanceolate when fully submerged, intergrading to elliptical and long-petiolate when on the surface, margin entire   *3. nodosus*

1. P. foliosus Raf. in Med. Repos. New York, ser. 2, **5**: 354 (1808).

Leaves 2–6 cm. long, acute, obscurely 3–5 veined; peduncles 0·5–1·8 mm. long; spikes subcapitate; fruit orbicular or flattened.

Local (St. Cath., Clar.), in flowing streams and rivers; 100–1250 feet; fl. and fr. Mar. *A 10853! J.P. 858! P 8268!* Canada to Mexico and Brazil, Greater Antilles, Hawaii.

2. P. illinoensis Morong in Bot. Gaz. **5**: 50 (1880).

Plant completely submerged except the inflorescence; leaves with 7–9 longitudinal veins.

Rather rare (St. Ann, St. Mary, St. Eliz.), in clear flowing rivers; SL–1000 feet; fl. Dec, fr. Dec–Mar. *Powell 838! Proctor & Mullings 21837!* United States, Canada, Mexico, C. Amer., West Indies.

3. P. nodosus Poir. in Lam., Encycl. Méth. Bot. Suppl. **4**: 535 (1816).

Inflorescences borne on the floating branches; spike 2–3 cm. long; submerged leaves translucent; floating leaves more opaque.

Common in shallow flowing streams; SL–1300 feet; fl. Nov–May. *A 10852! H 5982! 12883! P 8025! Stearn 547!* N. Amer., Mexico, West Indies, Africa, Eurasia.

4. RUPPIACEAE

Aquatic herbs of lagoons and marshes characteristically with seasonally variable salinity. Stems forking, long and slender. Leaves alternate or opposite, linear or setaceous, sheathing at the base. Flowers bisexual in short terminal spikes at first enclosed by the sheathing leaf-bases, at length exserted on the elongated peduncle; bracts absent. Perianth absent. Stamens 2; filaments very short and broad; anthers 2-locular, the loculi separated by the connective, opening extrorsely. Carpels 4, each with 1 pendulous campylotropous ovule; stigma peltate. Fruiting carpels at first sessile, at length long-stipitate, indehiscent. Seeds without endosperm.

1 genus with few species widely distributed.

1. RUPPIA L. (1753)

Peduncle becoming elongated in fruit and often spiral; leaf-tips usually obtuse or rounded
                                                                    1. *cirrhosa*
Peduncle not much longer in fruit than in flower; leaf-tips acute   **2.** *maritima*

1. R. cirrhosa (Petagna) Grande in Bull. Orto Bot. Univ. Nap. **5**: 58 (1918).—
    Buccaferrea cirrhosa Petagna (1787). *R. spiralis* L. ex Dumort (1827).

Slender-branched herb with submerged stems but usually some leaves floating at the surface; carpophores 1–2 cm. long.

Local (Clar.), in brackish streams; about SL; fl. Apr. *Stearn 753!* Cosmopolitan; Grand Cayman.

2. R. maritima L., Sp. Pl. **1**: 127 (1753).

Like the last.

Rather local (St. Andr., St. James), floating or submerged in masses in mangrove swamps and brackish lagoons; SL down to 3 feet; fl. Aug, fr. Dec. *H 12297!* Cosmopolitan.

5. ZANNICHELLIACEAE

Submerged aquatic herbs. Rhizome creeping, slender. Leaves opposite, alternate or crowded at nodes, linear, sheathing at the base; sheaths mostly ligulate at the apex. Flowers unisexual, monoecious or dioecious, axillary, solitary or in cymes. Perianth of 3 small free scales or absent. Stamens 1 (–3); anthers 1–2-locular, opening lengthwise; pollen globose or thread-like. Carpels 1 (–9), free; style simple or 2–4-lobed; ovules solitary, pendulous, orthotropous. Fruiting carpels sessile or stipitate, indehiscent. Seeds without endosperm.

7 genera with about 20 species mostly in tropical seas.

Styles simple; one anther inserted higher than the other   **1. Halodule**
Styles 2-fid; anthers inserted at the same height   **2. Syringodium**

1. HALODULE Endl. (1841); Den Hartog (1964), (1970)

1 Leaf-tips tridentate:
  2 Median tooth obtusely rounded; leaves 0·5–1 mm. wide with many secondary teeth and
    cilia on the median tooth; Panama   *H. ciliata* (Den Hartog) Den Hartog
  2 Median tooth acute; leaves linear 0·6–1·3 mm. wide; median tooth 1–10 times as long
    as the lateral teeth   **1.** *beaudettei*

1 Leaf-tips bicuspidate:
> 3 Leaves 0·3–1 mm. wide without secondary projections on the lateral teeth; inner side of the lateral teeth concave; Cayman Is.   *H. wrightii* Aschers.
> 3 Leaves 0·6–1·3 mm. wide, often with secondary projections on the lateral teeth; inner side of the lateral teeth convex; Bermuda   *H. bermudensis* Den Hartog

1. H. beaudettei (Den Hartog) Den Hartog in Blumea **12** (2): 303 (1964).—*Diplanthera beaudettei* Den Hartog (1960).

Erect shoots to 10 cm. high from a creeping rhizome; leaf-blades flat.

Occasional (St. Andr., Clar., St. Mary), in shallow lagoons on a sand and coral substrate; SL down to 1 foot. *P 7461! Stearn 762!* Atlantic coast of N. Amer. from N. Carolina southwards to the West Indies; Pacific coast of Panama, Guatemala and Nicaragua.

Den Hartog (1964), (1970) provides the tentative key, extracted and reproduced above, to four closely related species of *Halodule* occurring in the W. Atlantic and Caribbean region. *H. wrightii* when construed in his sense has been reported from Cayman Is., Bahamas, Virgin Islands, Cuba, Hispaniola, Puerto Rico, St. Vincent and the coasts of W. Africa and the Indian Ocean. The overlapping distribution of all these variants and the apparent absence of *H. wrightii* from Jamaica strongly suggest further collecting and field study.

2. SYRINGODIUM Kützing (1860); Den Hartog (1970)

1. S. filiforme Kützing in Hohenack., Alg. Marin. Sicc. **9**: n. 426 (1860); Dandy & Tandy in Journ. Bot. **77**: 116 (1939).—*Cymodocea manatorum* Aschers. (1868). Manatee Grass (Antigua), Eel Grass (Cayman Is.).

Rhizome creeping; leaf-blades more or less terete, 4–30 cm. long, 0·8–1·8 mm. broad, sheathed at the base; sheaths 1–5 cm. long; fruit ellipsoid, flattened, 5–6·5 mm. long, 2·5 mm. broad, beaked by the persistent style-base; seed 2·5–3 mm. long.

Probably generally distributed in lagoons and on reefs; mostly between SL and 2 feet down, but exceptionally in much deeper water down to 60 feet; fl. and fr. sporadically. *Howe 934! Skelding! Stearn 327!* Bermuda, Bahamas, S. United States, West Indies; Cayman Is.

6. NAJADACEAE

Submerged annual aquatic plants in fresh or brackish water. Stems slender, much branched. Leaves linear, subopposite or verticillate, sessile with a pair of minute scales within the sheath. Flowers unisexual, usually monoecious, solitary or clustered, sessile at the branch-bases. Male flowers with 2-lipped perianth; stamen 1; anther 1–4-locular, opening lengthwise. Female flower without perianth or perianth membranous and adhering to the carpel; ovary of 1 carpel, 1-locular; style solitary, stigmas 2–4; ovule solitary, anatropous, erect from the base. Fruit achenial, usually enclosed by a foliaceous sheath. Seed without endosperm; testa thin.

1 genus with about 40 species; widely distributed.

1. NAJAS L. (1753)

Stem and leaves not prickly-toothed; leaf-margin remotely serrulate to shallowly repand-dentate, especially towards the tip; in fresh water   1. *guadalupensis*
Stem and leaf-margins with coarse often blackish-purple prickly teeth; in fresh and brackish water   2. *marina*

1. N. guadalupensis (Spreng.) Magnus in Engl. & Prantl, Nat. Pflanzenf. **2** (1): 217 (1889).—*Caulinia guadalupensis* Spreng. (1825).

Plant trailing to 60 cm.; leaf-blades 10–25 mm. long, 0·5–1·5 mm. broad; fruit 2 mm. long; seeds narrowly ellipsoid, reticulate.

Common in shallow ponds and sluggish streams; 20–1750 feet; fl. Mar. *A 10784!*
H 8184! P 7834! Stearn 899! United States, continental tropical Amer., West
Indies.

2. N. marina L., Sp. Pl. **2**: 1015 (1753).
Plant rooting in mud or floating submerged; roots in water long, unbranched;
stem green, up to 40 cm. or more long; leaves 1–4·5 cm. long; flowers dioecious;
fruit 3–4 mm. long.
Rare (St. Andr., St. Cath., St. Eliz.), in lagoons, rivers and dams; SL–600 feet.
A 10741! Loveless! Widespread in temperate and tropical regions.

7. COMMELINACEAE

Annual or perennial herbs with leafy erect or trailing succulent stems. Leaves
alternate, often with rather broad blades, the sheathing base closed. Inflorescences
terminal, axillary or leaf-opposed, usually cymose; bracts spathaceous, leafy or
very small or absent. Flowers bisexual, less commonly some staminate only,
regular or zygomorphic. Perianth biseriate; sepals 3 (–2), free or fused, usually
green, more or less persistent; petals 3, free or fused, alternate with the sepals,
white or coloured, fugacious. Stamens 6 or fewer, hypogynous; filaments often
hairy; anthers 2-locular, opening lengthwise or by pores. Ovary superior, 3–2-
locular; style terminal, simple; stigma capitate or lobed; ovules 1–few in each
loculus, axile, orthotropous. Fruit usually a loculicidally dehiscent capsule, rarely
fleshy and indehiscent. Seeds often sculptured; embryo small; endosperm abundant,
mealy.
About 45 genera with some 550 species in warm regions of the world.

1 Inflorescences subtended by spathaceous bracts different in size, shape or texture from
   leaves or their bladeless sheathing bases:
  2 Inflorescence terminal, bracts paired; petals purple or pink; stamens 6:
    3 Sepals united into a tube at base, hyaline; petals connate; stems trailing; leaves rather
      thin, striped   **1. Zebrina**
    3 Sepals free, herbaceous; petals essentially free; stems ascending; leaves fleshy, not
      striped   **2. Setcreasea**
  2 Inflorescence axillary or leaf-opposed; petals white or blue, free:
    4 Peduncle usually perforating the leaf-sheath; calyx accrescent and forming a fleshy
      cover to the capsule; stamens 6; bracts usually paired   **3. Campelia**
    4 Peduncle not perforating the subtending leaf-sheath:
      5 Perfect stamens 6; bracts paired, rarely 3; perianth regular   **4. Rhoeo**
      5 Perfect stamens (2–) 3; bract solitary; perianth zygomorphic; lateral petals clawed
                 5. Commelina
1 Inflorescences without spathaceous bracts, terminal or in the axils of leaves or their
   bladeless sheathing bases:
  6 Inflorescences mostly axillary; stamens 1–3   **6. Callisia**
  6 Inflorescences terminal or also sparingly subterminal; stamens (5–) 6:
    7 Anthers opening by pores; seed with a red aril; perianth regular, but 2 posterior petals
      slightly larger; stamens subequal; large erect or scrambling herb; petals white or
      blue; cultivated ornamental; native of Brazil
             Dichorisandra hexandra (Aubl.) Standl.
    7 Anthers opening by slits; seeds exarillate:
      8 Perianth zygomorphic; flowers secund on their axes; erect herb; petals blue
                 7. Tinantia
      8 Perianth actinomorphic; flowers subumbellate; trailing herbs; petals white or tinged
      pink:
        9 Filaments hairy, unequal; peduncles not usually as much as three times longer
          than pedicels, densely pubescent   **8. Tripogandra**
        9 Filaments glabrous; peduncles usually five times or more longer than pedicels
                 9. **Leiandra**

1. ZEBRINA Schnizl. (1849)

1. Z. pendula Schnizl. in Bot. Zeit. **7**: 870 (1849).—Wandering Jew.
Trailing herb; leaves ovate, 3–7 cm. long, 1·5–3 cm. broad, purple-striped above,
purple beneath; outer bract with a hairy swelling on each side at the base; pedicels

2·5 mm. long; calyx 5 mm. long, bilabiate, ciliate-keeled; corolla-tube white, 6 mm. long; corolla-lobes ovate, bright pink within, whitish outside, about 5 mm. long, 2·5–3 mm. broad; stamens epipetalous, subequal with hairy filaments; anthers white; ovary 3-locular, with 2 ovules in each loculus; capsule dehiscent.

Common on rocky banks in usually rather shady places; 400–4000 feet; fl. June–July, Oct–Nov and during wet weather. *A 11393! H 12134! P 15424!* Native of C. Amer., introduced and naturalized in the West Indies and many other warm countries.

2. SETCREASEA K. Schum. & Sydow (1901); Hermann (1957)

1. S. purpurea Boom in Act. Bot. Neerl. **4** (2): 167–8 (1955).—Purple Heart, Purple Queen.

Perennial herb with suberect or ascending branches 30–50 cm. long; leaves oblong, acute, amplexicaul from sheaths 1–2·5 cm. long, ciliate-margined, 14–18 cm. long, 2–3·5 cm. broad, smaller above; pedicels up to 7 mm. long, glabrous below, white-pilose above; sepals oblong, glabrous above, 8–10 mm. long; petals 15–20 mm. long, cuneate; 3 filaments epipetalous and 3 adherent to petal-margins, variably few-haired or glabrous; anthers yellow.

Grown as a pot plant and planted out in gardens where it may become naturalized; fl. all the year. Native probably of Mexico but apparently now known only in cultivation.

3. CAMPELIA L. C. Rich. (1808)

1. C. zanonia (L.) Kunth, Nov. Gen. **1**: 264 (1816).—*Commelina zanonia* L. (1753).

Stem thick, erect or trailing, up to 1 m. long, exuding a glutinous mucilage when broken; leaves lanceolate, 10–18 cm. long, 3–5 cm. broad, or larger below; bracts 2, up to 3 cm. long; pedicels up to 6 mm. long; sepals 3·5 mm. long; petals longer than sepals; 3 of the stamens epipetalous with a few long moniliform hairs on lower half of filaments, 3 other stamens free and glabrous; anther-loculi subspherical, widely separated on deflexed connective-branches; stigma obconic, cupular; fruit obliquely cochleate, purplish-black with crimson juice; seeds black, smooth, shiny, rather flat.

Rare and local (St. Andr., St. Ann), among jagged and craggy limestone rocks in damp shady areas; 900–1700 feet; fl. and fr. May–Aug, Nov–Dec. *A 9455! J.P. 954! P 18416!* Mexico to Brazil, Greater Antilles.

4. RHOEO Hance (1853)

1. R. spathacea (Sw.) Stearn in Baileya **5**: 195 (1957).—*Tradescantia spathacea* Sw. (1788). *R. discolor* (L'Hérit.) Hance (1853). Moses-in-the-Bulrushes, Oyster Plant.

Plants bushy, 40–70 cm. high, often clonally gregarious; stems thick and branched near the base, covered with withered leaves; leaves linear-lanceolate up to 30 cm. long and 4·5 cm. broad, dark green above, purple along margin and beneath; peduncles up to 5 cm. long; spathaceous bracts purplish, acuminate; flowers numerous in umbelliform cymes; pedicels 1·6–1·7 cm. long; sepals ovate, 5–6 mm. long, 2·5–3 mm. broad; petals broadly ovate, 9 mm. long, 8 mm. broad; filaments 6–7 mm. long with long beaded hairs; anthers yellow with a red line over each loculus; seeds ellipsoid, rugose.

Common, on limestone banks and in rocky thickets and woodland margins; SL–2000 (–2900) feet; fl. and fr. most of the year. *H 12160! P 16620!* Native of C. Amer., probably Nicaragua, now cultivated and naturalized in many warm countries; Grand Cayman.

5. COMMELINA L. (1753)

1 Spathe open posteriorly, without included mucilage, margin ciliolate; staminodes 2; petals subequal; plant usually trailing and rooting   1. *diffusa*

1 Spathe closed posteriorly, with mucilage, margin membranous; staminodes 3; petals very
    unequal, the anterior small; plants tufted with more or less erect branches:
  2 Anterior petal subulate-lanceolate; capsule ideally 3-seeded; submargin of spathe
     glabrous or nearly so:
    3 Leaves ovate-lanceolate; sheath auriculate at junction with blade   2. *elegans*
    3 Leaves linear-lanceolate; sheath not auriculate   3a. *erecta* var. *angustifolia*
  2 Anterior petal ovate-lanceolate; capsule ideally 5-seeded; submargin of spathe ciliate;
    leaves broadly ovate to suborbicular   4. *benghalensis*

1. C. diffusa Burm. f., Fl. Ind.: 18, t. 7, f. 2 (1768).—*C. longicaulis* Jacq. (1791).
    C. cayennensis L. C. Rich. (1792). Water Grass.

Plants annual in seasonal climates, glabrous or nearly so, with long trailing
branches; leaves lanceolate to ovate, 2–6 cm. long, 1–2 cm. broad; spathe elongate-
acuminate, stalked in the axil of a leaf; flowers within the spathe in a pedunculate
cyme, exserted singly or in pairs, the first often longer-pedicellate staminate only
and on a posterior branch of the cyme; petals blue; staminodes yellow; capsule
3-locular, ideally 5-seeded, the posterior loculus indehiscent with one seed; seeds
reticulate, grey-black.

A common weed of cultivations, waste places and pastures; SL–3900 feet; fl.
and fr. most of the year. *A 5700! J.P. 1024! P 15806!* In the subtropics and tropics
of both hemispheres; Grand Cayman.

2. C. elegans Kunth, Nov. Gen. **1**: 259 (1816).

Branches tufted, more or less erect, 30–100 cm. long, more rarely decumbent
or trailing; leaves acute or acuminate, up to 10 cm. long and 3·5 cm. broad,
glabrous except on midrib beneath; auricles at upper sheath-margins with a few
long white hairs; peduncle 0·5–1·5 cm. long; spathe rather triangular in outline
with a few long white hairs on outer surface; flowers all hermaphrodite; lateral
petals with limb 1–1·2 cm. broad, light blue or white; staminodes yellow; posterior
lobe of capsule warted, indehiscent; seeds subglobose, rather smooth.

Common on rocky banks and in woodland margins; SL–2000 feet; fl. and fr. all
the year. *A 5653! H 12287! P 11852!* Bermuda, Bahamas, S.E. United States and
throughout tropical Amer.; Grand Cayman.

3. C. erecta L., Sp. Pl. **1**: 41 (1753).

3a. Var. angustifolia (Michx.) Fernald in Rhodora **42**: 439 (1940).—*C. angusti-*
    folia Michx. (1803).

Slender-stemmed perennial straggling herb with branches to 1 m. long; leaves
4–6 (–8) cm. long, 4–7 (–14) mm. broad; petals blue; posterior lobe of capsule
warted, smaller than anterior lobes.

Local (St. Cath., Clar., Manch.), in cactus scrub at margins of salinas and in
thickets on arid limestone; SL–30 feet; fl. and fr. Nov–Feb. *A 8815!* S. United
States, Mexico and Cuba.

4. C. benghalensis L., Sp. Pl. **1**: 41 (1753).

Plants more or less tufted with erect or trailing branches; some stems usually
decumbent with numerous short superficial or subterranean axillary branches, the
latter often producing cleistogamous flowers in reduced inflorescences; margin of
leaf-sheath declinate, with reddish or brownish setae, not auricled; spathe obtuse
on a short peduncle, often coarsely setulose on the outer surfaces; capsule oblong,
the anterior loculi ideally 2-seeded and loculicidally dehiscent, the posterior
loculus 1-seeded and indehiscent, not warted; seeds oblong or angular, reticulate,
greyish or grey-brown.

Locally common (St. Andr.), in cultivated ground and waste places; 500–700
feet; fl. and fr. Oct–Apr. *A 6180! 8169! Proctor & Powell 20543! Robertson UCWI
1358!* A weed widespread in tropical Asia and Africa, introduced into Barbados.
Although this plant was reported for the first time in Jamaica in October 1954,
several more or less distinct variants can be recognized within the still limited
range of its occurrence. The two most likely to be met with can be diagnosed as
follows:

Variant A. Leaves up 6 cm. long and 4 cm. broad, light green above; internodes,
sheaths and upper leaf-surface long-hispid; anterior petal subentire; lateral petals
blue.

Variant B. Leaves up to 4 cm. long and 2.5 cm. broad, dark green above; internodes, sheaths and upper leaf-surface shortly hispid; anterior petal notched;
lateral petals mauve. A further small-flowered variant has blue petals.

6. CALLISIA Loefl. (1758)

1. C. repens (Jacq.) L., Sp. Pl. ed. 2, **1**: 62 (1762).—*Hapalanthus repens* Jacq.
(1760).

Stem slender, rooting at the nodes, often purplish, glabrous except for a line of
small hairs; leaves ovate, 0·5–3 cm. long and up to about 1 cm. broad, cordate-
clasping at base, acuminate, margin ciliate; sheaths and bracteoles ciliate; flowers
in small axillary clusters, inconspicuous; pedicels angled; sepals (2–) 3, 3–4 mm.
long, ciliate-keeled; petals (2–) 3, oblong, shorter than the sepals, glabrous;
stamens (1–) 2–3; anthers with 2 elliptic-globose loculi widely separated by a
cordiform connective; filaments glabrous; ovary pilose at apex; style filiform;
stigma pilose; capsule oblong, 1·5–2 mm. long; seeds 1 mm. long, 0·8 mm. broad,
dark brown, wrinkled.

Locally common on sheltered rocky banks; 25–2900 feet; fl. Dec–Feb. *A 5810!*
H 11833! P 17421! Continental tropical Amer., West Indies; Grand Cayman.

7. TINANTIA Scheidw. (1839) nom. cons.

1. T. erecta (Jacq.) Schlecht. in Linnaea **25**: 185 (1852).—*Tradescantia erecta*
Jacq. (1790).

Stems up to 1 m. or more high, branched; leaves ovate-oblong, petioled,
acuminate, nearly glabrous, 5–10 cm. long, 2·5–5·5 cm. broad; sheaths loose;
inflorescence of 1–several small unilateral cymes, peduncle tomentellous and with
long glandular hairs, partial inflorescences bracteolate; sepals equal, hairy outside;
petals subequal, 9–15 mm. long, suborbicular; stamens 6, 3 longer, at least 3 with
hairy filaments; ovary 3-locular; capsule oblong, trigonous, obtuse, each loculus
with 2–5 seeds; seeds warty with linear hilum, rectangular, black.

Very local (St. Andr., St. Thom.), a weed along shady pathsides; 3800–4100
feet; fl. and fr. Sept. *H 12419! J.P. 1166! P 9624!* Native of tropical America,
Hispaniola, cultivated and naturalized in Asia and Africa.

8. TRIPOGANDRA Raf. (1837)

1. T. multiflora (Sw.) Raf., Fl. Tellur. **2**: 16 (1837).—*Tradescantia multiflora* Sw.
(1788).

Stem trailing and rooting, the flowering branches ascending; stems and leaves
often tinged purple; leaves ovate, subcordate to rounded at base, shortly acuminate,
up to 5 cm. long and 2·5 cm. broad, sheath with a line of hairs and a ciliate margin;
inflorescences terminal and axillary; flowers small and numerous in 1–4 stalked
umbels; peduncle up to 1·5 cm. long; pedicels slender 3–4 mm. long; sepals
ovate, pointed, about 2·5 mm. long; capsule enclosed by the calyx; seeds striate-
reticulate, bluish-grey, pitted on one side.

Locally common on sheltered stony banks, rarely in pastures; 1000–4800 feet;
fl. and fr. May–July, Oct–Jan. *A 5465! A. M. Barry! H 12124!* Also in Tobago and
from northern S. Amer. to Paraguay.

9. LEIANDRA Raf. (1837)

1. L. cordifolia (Sw.) Raf., Fl. Tellur. **2**: 17 (1837).—*Tradescantia cordifolia* Sw.
(1788).

Stems slender, trailing and rooting, glabrescent; leaves broadly ovate, acute,
cordate at base, glabrous except proximal margins minutely denticulate and sheath-
margins finely setose, rarely more than 2 cm. long and 1 cm. broad; peduncle
slender, up to 4·5 cm. long; pedicels and sepals puberulous; sepals 2–2·5 mm. long;
petals white.

Known in Jamaica only from the type; reported as from damp shady places in
the mountains. *Swartz!* Florida, Mexico to Panama and Peru, Cuba.

8. MAYACACEAE

Herbs creeping on wet ground or floating or partly submerged in fresh water. Leaves spirally arranged, numerous, narrow with a single vein, bidentate at the apex. Flowers bisexual, actinomorphic, axillary, solitary or several subterminally, subtended by membranous bracts; pedicels often reflexed after flowering. Sepals 3, subvalvate. Petals 3, imbricate, clawed. Stamens 3, hypogynous, opposite the sepals; filaments free; anthers 2–4-locular with porose dehiscence. Ovary superior, 1-locular; style simple; ovules several, biseriate on each of 3 parietal placentas, orthotropous. Fruit a capsule opening between the placentas. Seeds endospermous with a small embryo.

1 genus with 4 species; in tropical and subtropical America and West Africa (Angola).

1. MAYACA Aubl. (1775); Lourteig (1952)

1. M. fluviatilis Aubl., Hist. Pl. Guiane **1**: 42 (1775).

Leaves 3–15 mm. long, 0·3–1 mm. broad, subulate; pedicels 2–15 mm. long in flower, up to 2·5 cm. long in fruit; petals pinkish-mauve, blue or white; anthers yellow; capsule 3·5–5·5 mm. long, 2–3·5 mm. broad.

Uncommon (Clar., St. Eliz., Trel.), in sluggish streams, boggy ponds and swamps; 25–2300 feet; fl. and fr. Aug–Mar. *A 5945! H 9468! P 15889!* Throughout the American subtropics and tropics from S. United States to Paraguay, but rare in the West Indies.

9. XYRIDACEAE

Annual or more usually perennial herbs. Leaves mostly radical, tufted, linear, filiform, terete or flat, sheathing at the base. Inflorescence a bracteate globose to cylindrical long-pedunculate head; bracts closely imbricate, sometimes forming an involucre. Flowers bisexual, zygomorphic. Perianth 2-seriate; sepals (2–) 3, the 2 lateral keeled and chaffy, the inner membranous and forming a hood over the corolla before anthesis; corolla with a short or long tube and 3 equal spreading lobes. Stamens 3, opposite the corolla-lobes; anthers 2-locular, opening lengthwise; filaments epipetalous. Staminodes 3, alternating with the corolla-lobes or absent. Ovary superior, 1-locular with 3 parietal placentas or imperfectly 3-locular with basal or free central placentation; style simple or 3-lobed; ovules 1-many, orthotropous. Fruit a 3-valved loculicidal capsule enclosed in the persistent corolla-tube. Seeds small, endospermous with a small embryo.

4 genera with about 270 species mostly in the American tropics and subtropics but also in Africa, extreme S.E. Asia and Australasia.

1. XYRIS L. (1753)

Tufted, non-rhizomatous herbs; corolla deeply divided; staminodes conspicuous; pollen not spiny.

1. X. caroliniana Walt., Fl. Carol.: 69 (1788).—*X. jupicai* L. C. Rich. (1792).

Erect herb 15–40 cm. high in flower; leaves flat, 2·5–6 mm. broad; peduncle somewhat flattened above; bracteoles obtusely pointed or rounded at the tip; petals yellow.

Local (Clar., St. Eliz.), in damp open savannas, swamps and shallow pools; 50–2400 feet; fl. and fr. Nov–July. *A 7146! H 12093! P 23157!* S.E. United States, C. and S. Amer., Greater Antilles, Trinidad.

10. ERIOCAULACEAE

Annual or perennial herbs with narrow, sometimes membranous, tufted leaves. Flowers unisexual, actinomorphic, small and numerous, usually monoecious in a bracteate stalked head. Perianth 2-seriate, chaffy or membranous, the members of the outer series free or connate, the inner stipitate or cupular or rarely absent. Stamens as many or twice as many as the outer perianth-segments (rarely fewer), inserted on the corolla (when present); anthers small, 1–2-locular, opening lengthwise; filaments distinct. Staminate flowers often with a pistillode. Pistillate flowers rarely with staminodes. Ovary superior, 2–3-locular, each loculus with a single pendulous orthotropous ovule; style lobed. Fruit a membranous loculicidally dehiscent capsule. Seeds endospermous with a small embryo.

About 12 genera with some 1200 species mainly of wet sandy and swampy habitats in the tropics and subtropics, especially of the New World. *Eriocaulaceae* have not been reported from Jamaica to date but the family has been included because species of *Eriocaulon* have been reported from several neighbouring countries.

11. BROMELIACEAE

By G. R. Proctor with R. W. Read

Herbs, rarely perennial shrubs, mostly epiphytic. Roots, if present, serving merely as holdfasts in epiphytic species. Leaves spirally arranged, usually basal, dilatedsheathing below, simple, bearing minute scales at least when young, the margins entire or spinose-serrate. Inflorescence simple or paniculate, of spikes or racemes, more or less conspicuously bracteate. Flowers bisexual or functionally dioecious. Perianth biseriate, 6-merous; sepals and petals dissimilar, free or connate. Stamens 6 in 2 series; filaments free, or joined to the petals or to each other; anthers 2-locular, opening lengthwise; pollen smooth, porose or grooved. Ovary superior to inferior, 3-locular; ovules numerous, anatropous, on axile placentas; style short or slender; stigmas 3. Fruit a capsule or berry, or (in *Ananas*) fleshy and compound. Seeds often plumose; embryo small, located at the base of copious mealy endosperm.

About 45 genera and over 1000 species, with but one exception found exclusively in the tropical and subtropical parts of the Western Hemisphere. Commonly called Wild Pine in Jamaica. In many species the leaves fit together closely at the base, forming reservoirs which hold water. This water harbours an interesting fauna (A. M. Laessle, A microlimnological study of Jamaican Bromeliads, Ecology 42 (3): 499–517 (1961)).

1 Ovary wholly inferior; fruit a fleshy or leathery berry or compound; seeds naked; leaves usually spinose-serrate:
  2 Plants always terrestrial; leaves long-tapered to apex, not usually holding water:
    3 Inflorescence compound; ovaries always distinct, the fruits consisting of separate fleshy berries; inflorescence not bearing sterile leafy bracts at apex   1. **Bromelia**
    3 Inflorescence simple; ovaries fusing to form a compound fruit; inflorescence bearing a tuft of sterile leaflike bracts at the apex   2. **Ananas**
  2 Plants epiphytic or on rocks; wild species with hard inedible fruits; leaves mostly abruptly contracted at apex, overlapping tightly at base and holding water:
    4 Inflorescence densely cylindrical with numerous rigid 2–3-flowered branches, the flowers erect and separate; petals pink or mauve   3. **Aechmea**
    4 Inflorescence of a series of dense spherical, ellipsoid or cylindrical conelike spikes on a central axis, the flowers and fruits more or less compressed; petals white, green or light yellow   4. **Hohenbergia**
1 Ovary wholly or partly superior; fruit a capsule; seed variously appendaged:
  5 Leaves spinose-serrate, at least near base, long, flaccid, grasslike; flowers normally red, conspicuous, in racemes or openly paniculate; plants mostly on well drained rocks   5. **Pitcairnia**
  5 Leaves always entire; flowers usually not red, mostly inconspicuous; plants mostly epiphytic:

6 Appendage of seed apical, folded over at maturity; sepals strongly asymmetric; inflorescence of polystichous-flowered spikes   6. **Catopsis**
6 Appendage of seed basal, straight at maturity; sepals usually symmetric:
  7 Petals joined or closely agglutinated; inflorescence always of polystichous-flowered spikes   7. **Guzmania**
  7 Petals free:
    8 Petals naked, rarely with vertical folds; inflorescence of one or more distichous-flowered spikes, rarely simple and polystichous, the flowers usually erect, i.e. oriented towards apex of inflorescence or branch; leaves often more or less densely covered with evident minute scales   8. **Tillandsia**
    8 Petals each bearing one or two scales on inner surface; inflorescence simple and distichous, or a panicle of several distichous or polystichous-flowered spikes, the flowers usually spreading or secund; leaves nearly or apparently naked, the scales, if any, appressed and of microscopic size (except *V. incurva* has densely grey-green lepidote narrowly triangular leaves and strongly decurved or pendulous inflorescence)   9. **Vriesea**

1. BROMELIA L. (1753)

Inflorescence a dense headlike panicle, scapeless and sunk in the centre of the leaf-rosette
1. *superba*
Inflorescence an elongate panicle, terminating a definite scape   2. *pinguin*

1. B. superba Mez in Urb., Symb. Ant. **2**: 252 (1900).
Leaves up to 4 m. long.
Very rare (St. Andr., West.), in thickets; fl. June, fr. Nov. *H 5170! P 27712!* At one time cultivated at Hope Gardens. Endemic.

2. B. pinguin L., Sp. Pl. **1**: 285 (1753).—Pinguin, Ping Wing.
Inner leaves and bracts becoming bright pink or scarlet at anthesis; flowers rose-pink; fruit edible.
Frequent to common in coastal thickets or on dry brushy hillsides; SL–1750 feet; fl. Mar.–June, fr. Nov. *P 6536! Robertson & Wynter UCWI 3260!* C. and northern S. Amer., West Indies. In some places planted to mark boundaries or to form a living fence.

2. ANANAS Mill. (1754)

A single variable species, commonly cultivated for its edible fruit.

1. A. comosus (L.) Merr., Interpr. Rumph. Amb.: 133 (1917).—*Bromelia ananas* L. (1753). *B. comosa* L. (1754). *Ananassa sativa* Lindl. (1833). Pineapple, Sweet Pine.
Stiff terrestrial herb with a dense rosette of spinose-serrate leaves; inflorescence erect on a thick scape, densely conelike, subtended and crowned by spiny green bracts; flowers sessile; corolla violet or rose; ovaries coalescing with the floral bracts and axis to form a succulent compound fruit, sterile in the cultivated races.
Widely grown in Jamaica, both commercially and in private gardens; sometimes becoming naturalized. *H 5167! P 7798!* Native of the Guyanas and Brazil; cultivated throughout the tropics.

3. AECHMEA Ruiz & Pav. (1794) nom. cons.

1. A. paniculigera (Sw.) Griseb., Fl. Br. W.I.: 593 (1864).—*Bromelia paniculigera* Sw. (1788).
Leaves up to 1 m. long and 9 cm. broad, dark green, shiny; upper scape-bracts large, reflexed, pink, conspicuous; inflorescence bluish-purple, up to 1 m. long; sepals 7 mm. long; corolla pink, mauve or sometimes light blue.
Common and widespread in forested districts, low on trees or more often on shaded limestone rocks; 300–3100 feet; fl. ?Dec, Jan–Sept, fr. Mar, Oct, Nov. *A 10801! H5168! P 7797!* Endemic.

4. HOHENBERGIA J. A. & J. H. Schult. (1830); L. B. Smith (1960)

By R. W. Read

1 Lowermost spikes sessile:
  2 Floral bracts glabrous   **1.** *gnetacea*
  2 Floral bracts densely brown-lepidote   **2.** *polycephala*
1 Lowermost spikes stipitate:
  3 Floral bracts densely dark-brown-lepidote   **3.** *eriostachya*
  3 Floral bracts glabrous or light-coloured-lepidote and glabrescent:
    4 Floral bracts yellow or orange-yellow at anthesis, strongly striate when dry, 12 mm. or more long excluding mucro:
      5 Leaves except near base mostly entire; scrape-bracts slightly exceeding internodes; spikes remote, stipes of lowermost more than twice as long as lowermost floral bracts; sepals equalling or exceeding floral bracts   **4.** *inermis*
      5 Leaves regularly serrate; scape-bracts twice as long as internodes; spikes crowded, stipes of lowermost about equal to lowermost floral bracts; sepals completely covered by closely imbricate floral bracts:
        6 Wing of posterior sepals broadly rounded at apex; floral bracts rounded, appressed   **5.** *urbaniana*
        6 Wing of posterior sepals attenuate; floral bracts broadly acute, often divergent   **6.** *proctorii*
    4 Floral bracts not yellow or orange at anthesis, up to 10 mm. long excluding mucro:
      7 Floral bracts rounded or broadly acute, often mucronate:
        8 Floral bracts noticeably striate, straw-coloured when dry   **7.** *brittoniana*
        8 Floral bracts non-striate or nearly so, dark brown when dry:
          9 Leaf-blades conspicuously serrate throughout: mucro of floral bracts minute   **8.** *fawcettii*
          9 Leaf-blades entire or obscurely serrulate towards base; mucro of floral bracts to 4 mm. long   **9.** *jamaicana*
      7 Floral bracts attenuate:
        10 Lower spikes shorter than their stipes:
          11 Primary bracts much shorter than stipes; apical spike distinctly stipitate   **10.** *distans*
          11 Primary bracts about equalling or exceeding lower stipes; apical spike sessile or subsessile:
            12 Branches of inflorescence stiffly spreading, longer than primary bracts; scape-bracts barely exceeding internodes; fruiting spikes up to 2·5 cm. long   **11.** *abbreviata*
            12 Branches of inflorescence reflexed, shorter than lower primary bracts; scape-bracts to 4 times longer than internodes; fruiting spikes 3 cm. or more long   **12.** *laesslei*
        10 Lower spikes equal to or longer than their stipes:
          13 Mucro nearly or quite as long as base of sepal:
            14 Spikes subsessile; inflorescence normally very dense   **13.** *spinulosa*
            14 Spikes distinctly stalked; inflorescence rather lax; leaf about 8 cm. broad with spinulose margin; Grand Cayman   *H. caymanensis* Britton ex L. B. Sm.
          13 Mucro much shorter than base of sepal; at least lower spikes distinctly stalked; lower part of inflorescence open, the spikes more or less remote:
            15 Spikes subglobose to broadly ellipsoid, relatively short; lower primary bracts not much exceeding branches; widespread, except southern parishes   **14.** *penduliflora*
            15 Spikes cylindric, to 6 cm. long; lower primary bracts three times as long as branches; Westmoreland only   **15.** *negrilensis*

1. H. gnetacea Mez in Mart., Fl. Bras. **3** (3): 272, t. 60, f. 1 (1891).

This obscure species, known hitherto only from the type collection without locality or collector, was believed by Mez to be of Brazilian origin although it is unlike anything since collected in Brazil. L. B. Smith recognized its close affinity with *H. polycephala* and a recent collection from Trelawny (*Read 1858!*) adds evidence that this is probably another Jamaican endemic species.

2. H. polycephala (Bak.) Mez in C. DC., Monogr. Phan. **9**: 133 (1896).— *Aechmea polycephala* Bak. (1879).

Leaves grey-green, sometimes faintly mottled.

Common in the central part of the island, on trees or sometimes on shaded lime-stone rocks; near SL–2700 feet; fl. and fr. most of the year. *A 11745! H 5567! P 10155!* Endemic.

3. H. eriostachya Mez in Urb., Symb. Ant. **2**: 255 (1900).
Leaves noticeably glaucous, especially beneath; inflorescence dark brown throughout, the spikes on slender peduncles.
Uncommon (St. Andr., Port., St. Thom.), on trees in moist forests; 1000–3900 feet; fl. Mar, June, fr. Mar–May, Sept. *duQuesnay 351! Harris! P 5615!* Endemic.

4. H. inermis Mez in Fedde, Repert. Sp. Nov. **12**: 414 (1913).
Inflorescence decurved-spreading, the rachis straw-coloured or sometimes purple; floral bracts orange or yellow.
Locally common (West., St. James, Trel.), epiphytic; 400–2100 feet; fl. and fr. Mar, Aug–Nov. *ACH 756! H 9977! P 11239!* Endemic.

5. H. urbaniana Mez in Urb., Symb. Ant. **2**: 253 (1900).
Inflorescence erect to spreading; spikes yellowish, 9–18 cm. long, 1·8–2·5 cm. thick; floral bracts flat, abruptly acuminate, appressed, striate chiefly towards the apex.
Locally common (Clar., Manch.), on trees or sometimes on limestone rocks; 650–3100 feet; fl. and fr. Mar–Sept, Dec. *H 10884! P 7876!* Endemic.

6. H. proctorii L. B. Sm. in Phytologia **7** (5): 252, t. 1, f. 5, 6 (1960).
Inflorescence erect to spreading; spikes yellowish to brownish, 5–10 cm. long, 2–3 cm. thick; floral bracts convex, long-acuminate, not appressed, strongly striate throughout.
Locally frequent (Manch., St. Eliz., West.), on trees and shaded limestone rocks, sometimes associated with *H. spinulosa*; 750–2600 feet; fl. and fr. Dec–May, Sept. *A 13226! P 24000! Robertson UCWI 5402!* Endemic.

7. H. brittoniana L. B. Sm. in Contrib. Gray Herb **98**: 8, t. 1, f. 12–14 (1932).
Rare and local (Han.); 1300–1600 feet; *Britton 2313! H 9296!* Endemic. This species has not been collected since 1908.

8. H. fawcettii Mez in Urb., Symb. Ant. **2**: 254 (1900).
Very local (St. Andr., Port.); 3000–4500 feet; fl. Jan, July, fr. July, Sept. *H 5135! P 16611!* Endemic.

9. H. jamaicana L. B. Sm. & Proctor in Phytologia **7** (5): 251, t. 1, f. 3, 4 (1960).
Rare (Han., St. James), epiphytic; 650–1250 feet; fl. and fr. Feb, Apr, fr. July. *P 6543! 16452!* Endemic.

10. H. distans (Griseb.) Bak. in Saund., Refug. Bot. **4**: under t. 284 (1871).—
Aechmea distans Griseb. (1864).
Inflorescence lax, pendent, or, in the case of plants growing on rocks, the inflorescence resting on the ground; spikes 2–8 cm. long, on inflorescence-branches up to 15 cm. long.
Local (Han., West.), on trees and shaded limestone rocks; 300–1700 feet; fl. Aug, fr. Nov–May. *A 10954! H 9296!* Endemic.

11. H. abbreviata L. B. Sm. & Proctor in Phytologia **7** (5): 253, t. 1, f. 7, 8 (1960).
Inflorescence lax with slender branches up to 6 cm. long; primary and floral bracts light green; ripe fruits white.
Local (Trel., St. Ann), on limestone ledges and crags, or sometimes epiphytic; 1500–2200 feet; fl. June–July, fr. Apr, Oct. *P 11048! 15522!* Endemic.

12. H. laesslei L. B. Sm. in Bull. Bromel. Soc. **6**: 52 (1956).
Inflorescence spreading or decurved.
Very local (St. James), epiphytic; 1900–2100 feet; fl. July, fr. Aug. *Laessle IJ 9432! P 10393!* Endemic.

13. H. spinulosa Mez in Urb., Symb. Ant. **2**: 253 (1900).
Inflorescence erect, densely clavate, the individual spikes often difficult to distinguish; primary bracts pale- or grey-green; corolla cream or greenish-yellow.

Locally common (Manch., St. Eliz.), on trees or sometimes shaded limestone rocks; 400–3000 feet; fl. Dec–Mar, fr. Mar, Aug. *Britton 3771! P 7732! 16104!* Endemic.

Plants which have been identified as *H. spinulosa* from low elevations in western St. Elizabeth may not be the same; more collecting and study are required.

14. H. penduliflora (A. Rich.) Mez in C. DC., Monogr. Phan. **9**: 135 (1896).—*Pitcairnia penduliflora* A. Rich. (1850).

Scape curved-spreading, relatively slender, floccose; scape-bracts light green or sometimes blue-violet; spikes 1·5–4 cm. long; flowers sometimes fragrant; sepals 4–8 mm. long; corolla white. A variable species.

Common on trees or rocks, especially in the northern and eastern parishes; near SL–3500 feet; fl. Feb–Sept, fr. Mar, May–Sept, Dec. *A 6871! H 10205! P 6936!* Cuba.

15. H. negrilensis Britton ex L. B. Sm. in Proc. Amer. Acad. **70**: 151, t. 1, f. 7, 8 (1935).

A very large species with massive thyrsoid inflorescences up to 1 m. long, often tripinnate in the lower part; sepals about 5 mm. long.

Very local (West.), epiphytic; near SL–150 feet; fl. and fr. Nov. *P 11147!* Endemic.

Jamaica is the chief centre of speciation of this genus outside Brazil. The western part of the island is particularly rich in little-known species and it is certain that others than those dealt with here remain to be described. On the other hand some taxa, especially in the affinity of *H. jamaicana*, *H. caymanensis*, *H. spinulosa*, *H. penduliflora* and *H. negrilensis*, may not be sufficiently distinct to warrant species rank.

Gravisia aquilega (Salisb.) Mez resembles a species of *Hohenbergia* but the partial inflorescences are much less compact; the primary bracts of the inflorescence are conspicuous, bright wine-red and reflexed. The type of this species allegedly came from Jamaica but its existence there has not been confirmed. It is otherwise known from Costa Rica, northern S. Amer. and Trinidad and Tobago.

5. PITCAIRNIA L'Hérit. (1789) nom. cons.; Read (1969)

1 Leaves densely white- or silvery-lepidote over most of abaxial surface; scales symmetrical, fimbriate and concave when crowded   1a. *bromeliifolia* var. *bromeliifolia*
1 Leaves not as above:
  2 Leaves glabrous abaxially; teeth and margin of leaf-base mostly green or colourless
                                     1b. *bromeliifolia* var. *graminifolia*
  2 Leaves glabrous over most of abaxial surface but with a row of large asymmetrical scales along the often revolute margin and near apex; teeth and margin of leaf-base black and horny   1c. *bromeliifolia* var. *wynteri*

1. P. bromeliifolia L'Hérit., Sert. Angl.: 7, t. 11 (1789).

1a. Var. bromeliifolia.—*Hepetis angustifolia* Sw. (1788), not *P. angustifolia* Ait. (1789).

Tufted acaulescent herb with leaves up to about 1 m. long and 2 cm. broad; inflorescence simple or branched at base; flowers about 4 cm. long.

Frequent and locally abundant throughout the eastern half of the island, on rocky limestone banks and cliffs; SL–1250 feet; fl. and fr. mostly Jan–May. *A 10449! J.P. 1370! R 1792! 1969!* Endemic.

1b. P. bromeliifolia var. **graminifolia** Griseb., Fl. Br. W.I.: 594 (1864).

Like the last in habit; teeth on petiolar part of leaf rarely more than 2 mm. long; sepals (18–) 19–27 mm. long; petals red or rarely white.

Frequent (St. Andr., St. Cath., St. Mary, Port.), on stream and roadside banks primarily of shaly rocks; 250–2600 feet; fl. and fr. Dec–June. *A 6481! J.P. 606! R 1976! 1980!* Endemic.

1c. P. bromeliifolia var. **wynteri** Read in Brittonia **21** (1): 90 (1969).

Like the last; teeth on petiolar part of leaf 2–4 mm. long; sepals 16–18 (–19) mm. long.

Locally common (Clar., Han., Trel.), on limestone cliffs; 1500–2500 feet; fl. and fr. May. *H 11033! R 1922!* Endemic.

6. CATOPSIS Griseb. (1864)

1 Sepals 10–18 mm. long; leaves bearing chalky powder, especially towards base:
  2 Inflorescence slender, recurved, the flowers relatively few, 4–20; petals yellow, much
    longer than sepals   1. *nutans*
  2 Inflorescence stout, erect, the flowers relatively numerous, 20–75 or more; petals white,
    shorter than or equalling sepals   2. *berteroniana*
1 Sepals not more than 9 mm. long; leaves devoid of chalky powder:
  3 Leaves few, forming a cylindrical rosette; scape-bracts all much shorter than internodes
                                                       3. *nitida*
  3 Leaves many in a dense, spreading rosette; scape-bracts longer than internodes
                                                  4. *floribunda*

1. C. nutans (Sw.) Griseb., Fl. Br. W.I.: 599 (1864).—*Tillandsia nutans* Sw.
(1788).
Longest leaves usually less than 20 cm. long, rarely to 24 cm., spreading-recurved,
light green.
Uncommon epiphyte, especially on *Citrus* or other isolated trees in pastures;
400–2300 feet; fl. Feb, Sept–Oct, fr. Jan–Apr, Aug, Nov. *A 6383! H 5447!
P 11625!* C. and northern S. Amer., Cuba, Hispaniola.

2. C. berteroniana (J. A. & J. H. Schult.) Mez in C. DC., Monogr. Phan. **9**: 621
(1896).—*Tillandsia berteroniana* J. A. & J. H. Schult. (1830).
Leaves up to 40 cm. long, erect to slightly spreading, not recurved; floral bracts
yellow-green; capsule barely exceeding sepals.
Frequent, usually epiphytic on shrubs or small trees on wooded hilltops or in
low thickets; 250–4000 feet; fl. July, Oct–Mar, fr. Nov–Apr, June. *A 12155!
Harkness I 6539! P 15995!* Florida, C. and S. Amer., West Indies.

3. C. nitida (Hook.) Griseb., Fl. Br. W.I.: 599 (1864).—*Tillandsia nitida* Hook.
(1827).
Plants often forming a clump of several individuals attached by a common mass
of roots; leaves shining green, the blades ligulate; corolla white or cream.
Uncommon to frequent, epiphytic in rather dense moist forests and on wooded
hilltops; 2000–4500 feet; fl. Apr–Oct, fr. nearly all the year. *A 8372! H 5164!
P 23109!* C. Amer., Guianas, West Indies.

4. C. floribunda L. B. Sm. in Contrib. Gray Herb. **117**: 5 (1937).—*Pogospermum
floribundum* Brongn. (1864), provisional name. *Catopsis nutans* Griseb. (1864),
partly, not *Tillandsia nutans* Sw.
Leaves often speckled or minutely papillose, 15–30 (–40) cm. long, the blades
narrowly triangular, acuminate; corolla white.
Frequent, epiphytic chiefly on wooded hilltops; 1500–3000 feet; fl. June–July,
fr. nearly all the year. *A 12666! H 5161! P 7130!* Florida, C. and northern S. Amer.,
West Indies.

7. GUZMANIA Ruiz & Pav. (1802)

1 Inflorescence simple; corolla white:
  2 Inflorescence forming a rosette at top of axis, the outer bracts much enlarged
                                                  1. *lingulata*
  2 Inflorescence more or less elongate, the outer bracts relatively inconspicuous:
    3 Inflorescence elongate, linear, 2–3 cm. thick, sterile towards apex   2. *monostachia*
    3 Inflorescence clavate or ellipsoid, 4–7 cm. thick, fertile throughout   3. *erythrolepis*
1 Inflorescence compound, ample, laxly bipinnate, 30–50 cm. long; floral bracts shorter
    than or barely equalling sepals; corolla yellow   4. *fawcettii*

1. G. lingulata (L.) Mez in C. DC., Monogr. Phan. **9**: 899 (1896).—*Tillandsia
lingulata* L. (1753). *Caraguata lingulata* (L.) Lindl. (1827).
Leaves longitudinally red- or purple-striate on the basal sheaths; bracts of
inflorescence dark red or maroon at anthesis.
Common and widespread in moist forest areas, epiphytic or on shaded limestone
rocks; 400–3100 feet; fl. Mar, May–Aug, Dec, fr. Mar, Aug, Oct–Nov. *A 8439!
P 21424!* C. and S. Amer., West Indies.

2. G. monostachia (L.) Rusby ex Mez in C. DC., Monogr. Phan. **9**: 905 (1896).—
Renealmia monostachia L. (1753). *Tillandsia monostachia* (L.) L. (1762).
Fertile floral bracts light green with conspicuous purple-brown longitudinal
lines; sterile bracts at apex of inflorescence usually pale to bright red.
Common on trees, or occasionally on rocky banks or shaded limestone cliffs;
200–2900 feet; fl. Mar–July, Sept, Dec, fr. June–Mar. *A 11251! H 5153! P 6786!*
Florida, C. and S. Amer., West Indies. The sterile bracts at the apex of the inflorescence are normally light salmon-pink to scarlet, but occasionally a white-
bracted variant occurs. This is presumably *G. monostachia* var. *alba* Ariza-Julia,
originally described from the Dominican Republic (Bull. Bromel. Soc. **9**: 38
(1959)).

3. G. erythrolepis Brongn. ex Planch. in Fl. des Serres **11**: 25, t. 1089 (1856).—
G. harrisii Mez (1896).
Floral bracts bright red to rose-pink with pale margins, finely striate when dry.
Rare (St. Andr., St. Ann, St. Cath., Port.), on trees, rocks or cliffs; 2300–3500
feet; fl. Mar–May, fr. July–Sept, Dec. *H 5146! P 6493!* Cuba, Puerto Rico.

4. G. fawcettii Mez in C. DC., Monogr. Phan. **9**: 951 (1896).
Plant usually over 1 m. high, with the aspect of *Tillandsia excelsa*; bracts green
or often red abaxially with straw-coloured margins.
Local (St. Andr., Port.), epiphytic or on mossy ground in montane forest;
1500–4500 feet; fl. Dec–Apr, fr. May, Sept. *A 12115! H & B 10534! P 11559!*
Endemic.

8. TILLANDSIA L. (1753)

1 Stamens equal to or shorter than perianth:
  2 Stem evident and more or less elongated, often branched:
    3 Leaves minutely impressed-punctate-scaly, olive-green; inflorescence 4–10-flowered;
      sepals 10 mm. long   1. *tenuifolia*
    3 Leaves densely covered with minute spreading scales, usually grey, except when wet;
      inflorescence 1–2-flowered, rarely more; sepals less than 9 mm. long:
      4 Stems rather stiff, curved, less than 10 cm. long, concealed by the imbricate leaf-
        bases; inflorescence clearly terminal, often with more than 1 flower
        2. *recurvata*
      4 Stems lax, pendent, branched, up to 1 m. or more long, exposed between the more or
        less remote leaves; inflorescences apparently lateral, always 1-flowered
        3. *usneoides*
  2 Stem very short or obsolete, simple; flowers distichous in spikes:
    5 Inflorescences axillary, several to numerous:
      6 Inflorescences always simple, arching in flower, pendent in mature fruit; leaves
        ligulate, abruptly acute at apex, at least the lower half more than 3 cm. broad
        4. *complanata*
      6 Inflorescences compound, spreading or stiffly decurved; leaves narrowly deltate,
        long-attenuate at apex, except for the sheath mostly less than 2 cm. broad
        5. *antillana*
    5 Inflorescence terminal, solitary, compound:
      7 Sepals asymmetric, 5 mm. long; inflorescence laxly tripinnate; leaves flat, abruptly
        apiculate at apex, minutely and densely punctate-scaly   6. *fawcettii*
      7 Sepals symmetric or nearly so, more than 10 mm. long:
        8 Leaf-blades minutely and densely punctate-scaly, abruptly narrowed above the
          sheath, tapering to a long, narrow, involute apex 7a. *elongata* var. *subimbricata*
        8 Leaf-blades essentially glabrous, gradually narrowed above the sheath, flat, the
          apex acute to shortly acuminate:
          9 Sepals 25–45 mm. long; inflorescence usually bipinnate with lax, deflexed primary
            branches; spikes to 25 cm. long   8. *fendleri*
          9 Sepals 12–20 mm. long; inflorescence-branches usually stiff and erect; spikes
            3–7 cm. long:
            10 Inflorescence up to 1 m. or more high, usually tripinnate; floral bracts not
              keeled; sepals 15–20 mm. long; leaves up to about 5 cm. broad at middle
              9. *excelsa*
            10 Inflorescence 20–40 cm. high, bipinnate; floral bracts slightly keeled; sepals
              12–15 mm. long; leaves 1·5–3 cm. broad at middle   10. *selleana*

1 Stamens clearly longer than perianth; leaves always densely but often minutely scaly:
  11 Inflorescence open, floral bracts much less than twice as long as internodes:
    12 Leaf-blades not over 3 mm. broad and 9 cm. long; plant small; corolla bright rose-
      red   11. *argentea*
    12 Leaf-blades more than 1 cm. broad at least towards base, plants large:
      13 Flowers spreading or clearly divergent from rachis, rather few, inflorescence simple
        or laxly few-branched; sepals 20–30 mm. long; leaves spirally twisted from base,
        usually transversely banded   12. *flexuosa*
      13 Flowers appressed to rachis, usually numerous in an ample panicle; sepals 14–20
        mm. long; corolla greenish-white or white; leaves neither twisted nor banded:
        14 Sepals glabrous; floral bracts and rachillae green; leaf-scales closely appressed
            13. *utriculata*
        14 Sepals lepidote; floral bracts and rachillae red; leaf-scales subspreading
            14. *calcicola*
  11 Inflorescence dense, the floral bracts twice as long as internodes or more:
    15 Leaves densely beset with small but noticeable spreading silvery scales:
      16 Leaf-sheaths inflated, 2–4 cm. long; leaf-blades recurved or contorted; corolla
        violet   22. *pruinosa*
      16 Leaf-sheaths appressed tightly to the slender stem, about 1 cm. long; leaf-blades
        straight or nearly so, abruptly divergent or deflexed; corolla yellow
            26. *schiedeana*
    15 Leaves with minute inconspicuous appressed scales only:
      17 Leaf-blades linear-subulate to filiform; leaves closely fasciculate, the sheaths
        triangular with abruptly auricled bases:
        18 Spikes arching-recurved; floral bracts scarcely more than twice as long as inter-
          nodes; plants non-stoloniferous; sepals to 17 mm. long   15. *festucoides*
        18 Spikes straight, erect:
          19 Plants scaly-stoloniferous; sepals 15–20 mm. long   16. *juncea*
          19 Plants forming dense non-stoloniferous tufts; sepals 7–12 mm. long 17. *setacea*
      17 Leaf-blades broader, definitely if narrowly triangular; leaf-sheaths ovate or elliptic
        without auricled bases:
        20 Leaf-bases conspicuously inflated:
          21 Leaf-sheath gradually contracted into blade; floral bracts nearly smooth, faintly
            nerved towards apex, up to 22 mm. long, glabrous or nearly so 19. *balbisiana*
          21 Leaf-sheath abruptly contracted into blade; floral bracts distinctly nerved and
            keeled, not over 15 mm. long, scaly towards apex   23. *bulbosa*
        20 Leaf-bases flat, not inflated:
          22 Leaf-sheaths concolorous with the blades:
            23 Inflorescence densely compound, the primary bracts with long leafy tips;
              spikes mostly less than 7 cm. long, rarely more, the floral bracts coriaceous
              and closely imbricate, the axis concealed   18. *polystachia*
            23 Inflorescence simple or few-branched, the primary bracts without leafy tips;
              spikes up to 20 cm. long, rarely less than 8 cm., the floral bracts thin-papery
              and loosely imbricate, often exposing the axis   25. *valenzuelana*
          22 Leaf-sheaths distinctly brown or purple-brown at least near base, contrasting with
            the blades:
            24 Spikes 3–6 cm. broad; floral bracts much inflated and with strongly curved
              inner margins, more than 4 cm. long, about 3·5 cm. broad, diverging from
              rachis at a wide angle; inflorescence simple or few-branched   20. *compressa*
            24 Spikes less than 3 cm. broad, except sometimes in fruiting specimens; floral
              bracts not inflated and inner margins only slightly curved, up to 3 cm. long
              and 2 cm. broad, ascending at a narrow angle from rachis:
              25 Flowering spikes 1·5–3 cm. broad; inflorescence robust with main axis 3 mm.
                or more thick; branches usually numerous   21. *fasciculata*
              25 Flowering spikes less than 1·5 cm. broad; inflorescence slender; main axis up
                to 1·5 cm. thick:
                26 Leaves very minutely punctate-scaly, appearing green; floral bracts cori-
                  aceous, usually green; inflorescence compound, dense   18. *polystachia*
                26 Leaves densely cinereous-scaly, appearing grey; floral bracts thin-papery,
                  bright red at anthesis; inflorescence simple or few-branched
                  24. *canescens*

1. T. tenuifolia L., Sp. Pl. ed. 2, **1**: 410 (1762).—*T. pulchella* Hook. (1825). *T. subulata* Vell. (1825).

Stems long-lived, up to 30 cm. or more long, pendent-recurved; corolla white on Jamaican plants.

Local (St. Andr., St. Thom.), epiphytic, often forming large tufts or masses; 1500–4100 feet; fl. Sept, fr. Apr–Sept. *H 5138! P 24062!* Cuba, Hispaniola, south to Brazil and northern Argentina.

2. T. recurvata (L.) L., Sp. Pl. ed. 2, **1**: 410 (1762).—*Renealmia recurvata* L. (1753), excl. var. *β*. *T. uniflora* Kunth (1816). Old Man's Beard.

Plants forming loose to rather dense rounded tufts, not pendent.

Common epiphyte, also often growing on electric and telephone wires; SL–3500 feet; fl. and fr. all the year. *A 10755! H 5139! P 20671!* S. United States, Mexico to Argentina and Chile, West Indies, Cayman Is.

3. T. usneoides (L.) L., Sp. Pl. ed. 2, **1**: 411 (1762).—*Renealmia usneoides* L. (1753). Old Man's Beard, Spanish Moss.

Plants long-pendent, loose; flowers minute, inconspicuous, not often seen.

Locally abundant, hanging from trees, rarely cliffs, chiefly in or near south coastal areas; SL–1300 feet; fl. and fr. Jan–July. *H 5154! H & P 14521! Wynter UCWI 3095!* Virginia and Texas to Argentina and Chile, Greater Antilles, Antigua.

4. T. complanata Benth., Bot. Voy. Sulph.: 173 (1846).—*T. axillaris* Griseb. (1864).

Leaves often spotted red; floral bracts green with red tips; corolla pink.

Local (St. Andr., Port., St. Thom.), epiphytic in montane forest and mossy woodlands; 4000–5800 feet; fl. Jan–Feb, fr. Jan–Sept. *A 10507! Fawcett 7189! P 25698!* Costa Rica, S. Amer., Cuba.

5. T. antillana L. B. Sm. in Contrib. U.S. Nat. Herb. **29** (7): 282 (1949).

Inflorescences spreading or somewhat decurved, apparently produced in succession so that different stages of development can be seen at the same time; spikes to 45 cm. or more long, slender, with numerous sterile bracts in the basal two-thirds; floral bracts keeled; petals 4 cm. long.

Rare (West., St. Eliz.), epiphytic in dry woodlands; 250–2000 feet; fl. Nov, fr. Aug. *P 9269! 11182!* Guadeloupe.

6. T. fawcettii Mez in C. DC., Monogr. Phan. **9**: 752 (1896).

Plant 1 m. or more high; leaves about 4 cm. broad.

Very rare, only once collected in the Blue Mountains; fl. June. *H 5186!* Hispaniola.

7. T. elongata Kunth, Nov. Gen. **1**: 293 (1816).

7a. Var. subimbricata (Bak.) L. B. Sm. in Journ. Acad. Sci. Washington **43**: 68 (1953).—*T. subimbricata* Bak. (1887).

Leaves yellow-green; inflorescence yellowish with numerous spikes 15–30 cm. or more long; floral bracts 18–20 mm. long, strongly nerved; corolla violet-blue; capsules slender, long-exserted from calyx, 3–4 cm. long.

Frequent epiphyte on small trees in thorny woodlands, and beside ponds and swamps in the western third of the island; SL–400 feet; fl. July–Aug, fr. Sept–Mar. *H 10217! P 10553!* C. Amer., Colombia, Cuba, Trinidad.

8. T. fendleri Griseb. in Ges. Wiss. Gött. Nachr. **1864**: 17 (1865).

The largest Jamaican *Tillandsia*, the inflorescences up to 2 m. high; leaves numerous in a dense rosette, up to 1 m. long, glaucous near base, of young plants often mottled maroon; spikes long-stipitate; floral bracts glabrous, keeled towards apex, green to straw-coloured, often tinged with pink; petals white with blue margins.

Local (St. Andr., Port., St. Thom.), on trees, ledges or cliffs; 2300–4100 feet; fl. Jan–Apr, fr. Jan–Oct. *A 8155! H 5565! Proctor & Gentles 6670!* C. and northern S. Amer., Cuba, Hispaniola.

9. T. excelsa Griseb., Fl. Br. W.I.: 597 (1864).

Inflorescence sometimes mottled red becoming green; spikes sessile; corolla violet or white.

Widespread but not common, epiphytic; 1000–2600 feet; fl. Jan–Mar, fr. most of the year. *A 11751! H 10847! Proctor & Mullings 22063!* C. Amer., Cuba, Hispaniola.

10. T. selleana Harms in Notizbl. Bot. Gart. Berlin **10**: 799 (1929).

Inflorescence with reddish-orange scape and crimson bracts; spikes sessile; leaves reddish, 15–22 cm. long; corolla white; capsule about 2·5 cm. long.

Rare and local (Clar., Trel.), epiphytic or on limestone rocks; 2000–2500 feet; fl. Dec, fr. May. *A 12849! H 12762!* Hispaniola.

11. T. argentea Griseb., Cat. Pl. Cub.: 254 (1866).

Leaves minutely grey-scaly, almost whitish, forming a small very dense rosette 6–15 cm. across; corolla bright red, showy.

Rare (St. Eliz.), epiphyte on trees in pastures and dry thickets; 750–1000 feet; fl. Mar–May, fr. Mar. *Maxwell! P 11691!* Mexico, Cuba.

12. T. flexuosa Sw., Nov. Gen. & Sp. Pl.: 56 (1788).—*T. aloifolia* Hook. (1826).

The banded spirally twisted leaves are distinctive.

Widespread but uncommon, epiphytic especially in dry thorny woodlands; SL–1000 feet; fl. Sept–Oct, fr. Jan–Aug. *H 5671! P 8370! HPS 14733!* Florida, Panama and northern S. Amer., Bahamas?, West Indies; Grand Cayman.

13. T. utriculata L., Sp. Pl. **1**: 286 (1753).

Leaves forming a dense rosette; inflorescence erect, up to about 1 m. high, much-branched, the branches greenish or rarely bright red; floral bracts with broad membranous often dark purple margin; corolla white or sometimes pale greenish-cream.

Common epiphyte, especially in areas of low to moderate rainfall, rarely on cliffs; SL–2700 feet; fl. July–Dec, fr. all the year. *A 8135! H 5141! Proctor & Foster 9324!* S.E. United States, Mexico, Br. Honduras, Venezuela, West Indies.

14. T. calcicola L. B. Sm. & Proctor in Phytologia **16** (2): 77, t. 1, f. 21–22 (1968).

Plant in flower about 1 m. high; leaf-sheaths dark brown, blades up to 60 cm. long and 6 cm. broad at base; primary branches of inflorescence 25–35 cm. long; corolla 4 cm. long, zygomorphic; stamens exserted; capsule 3 cm. long.

Local (Trel.), on limestone cliffs; about 1500 feet; fl. Apr–June, fr. most of the year. *H & P 14433! R 1759! 1856!* Endemic.

15. T. festucoides Brongn. ex Mez in C. DC., Monogr. Phan. **9**: 678 (1896).

Inflorescence always compound, up to 17 cm. long, with up to 7 or more, distinct, more or less divergent spikes; scape and floral bracts often pink or red; corolla purple to mauve.

Rather rare (Han., West., St. Ann), epiphytic; 500–1500 feet; fl. Jan–Apr, fr. Feb, July. *A 6836! H 7049! C. B. Lewis & 6274!* Florida, Mexico to Costa Rica, Greater Antilles.

16. T. juncea (Ruiz & Pav.) Poir. in Lam., Encycl. Méth. Bot. Suppl. **5**: 309 (1817). —*Bonapartea juncea* Ruiz & Pav. (1802).

Leaves often reddish; bracts at top of scape crowded, overlapping and sub-involucrate especially in fruiting material; inflorescence simple or of a few densely crowded spikes, seldom more than 7 cm. long; corolla violet.

Common epiphyte; 250–3100 feet; fl. and fr. most of the year. *A 12511! H 5136! P 24127!* Mexico to Brazil and Peru, Cuba, Hispaniola, Trinidad.

17. T. setacea Sw., Fl. Ind. Occ. **1**: 593 (1797).—*Renealmia recurvata* var. β L. (1753). *T. tenuifolia* of authors, not L. (1762).

Leaves mostly 15–35 cm. long, rarely longer; scapes slender, about 1 mm. thick, excluding bracts; spikes few, mostly 1–3, turned edgewise to axis of inflorescence; corolla violet to purple.

Common epiphyte; 50–3000 feet; fl. July–Nov, fr. all the year. *A 12762! Britton 2642! P 22566! Yuncker 17934!* S.E. United States, Mexico to Salvador, Venezuela, N. Brazil, Greater Antilles; Grand Cayman.

18. T. polystachia (L.) L., Sp. Pl. ed. 2, **1**: 410 (1762).—*Renealmia polystachia* L. (1753). *T. angustifolia* Sw. (1788).

Leaves and bracts often purple-tinged along the margins, especially near base, usually 30–40 cm. long with long narrow flexuous tips; corolla mauve, fading whitish.

Common epiphyte, except at the extreme east and west of the island; 350–2900 feet; fl. Nov–Mar, fr. all the year. *A 10031! H 5144! P 22819!* Florida, C. and S. Amer., West Indies. A very variable species over its whole range and possibly taxonomically divisable.

19. T. balbisiana J. A. & J. H. Schult. in L., Syst. Veg. ed. nov. **7**: 1212 (1830).

Leaves dull olive-green, sometimes purple-margined; inflated sheaths forming an ovoid or ellipsoid pseudobulb up to 12 cm. long; bracts often crimson; corolla purple or violet, sometimes fading white.

Common except in areas of high rainfall, epiphytic in dry woodlands, on trees in pastures and beside roads and streams; SL–3000 feet; fl. June–Jan, fr. all the year. *A 8057! H 5526! P 23991!* Florida, Mexico to Panama, Venezuela, Bahamas, Cuba, Hispaniola, Cayman Is.

20. T. compressa Bert. in L., Syst. Veg. ed. nov. **7**: 1210 (1830).

Vegetatively apparently not distinct from *T. fasciculata*; scapes stout, usually rigidly erect; floral bracts pink to bright red, rarely green, tinged reddish or pale yellowish; corolla violet.

Common gregarious epiphyte, rarely on rocks; 400–4200 feet; fl. Apr, Oct, Dec, fr. all the year. *A 10130! H 5160! P 6063!* This species was assigned a rather wide range by Mez (as *T. fasciculata* var. *venosispica*), but without reference to Jamaica. The abundant Jamaican populations seem uniformly distinct from *T. fasciculata* where they occupy different ranges. *T. compressa* generally occupies cooler moister localities than *T. fasciculata*. Where they occur together putative hybrids possessing very low pollen viability and having intermediate characteristics have been found.

21. T. fasciculata Sw., Nov. Gen. & Sp. Pl.: 56 (1788).—*Renealmia polystachia* of Jacq. (1763), not L. (1753).

Leaves numerous, forming a stiff, rather dense rosette; sheaths brown near base; blades narrowly triangular, long-acuminate, rigid, not more than 3 cm. broad; primary and floral bracts red or bright pink at anthesis; corolla violet or mauve.

Common gregarious epiphyte at chiefly low elevations in areas of low to moderate rainfall; SL–2300 feet; fl. Jan–Nov, fr. Dec–Mar. *A 6870! H 5137! P 15370!* Florida, C. and northern S. Amer., West Indies, Trinidad. A variable species over much of its range, but in Jamaica quite uniform.

22. T. pruinosa Sw., Fl. Ind. Occ. **1**: 594 (1797).

Inner leaves closely enfolding base of inflorescence; leaf-blades usually longer than inflorescence, flexuous-spreading, involute-subulate, 2–4 mm. in diameter; inflorescence usually simple, subsessile in the upper leaves; floral bracts pink at anthesis, soon becoming green; corolla bluish-purple.

Frequent but often solitary epiphyte, commonest in west-central interior areas; 250–2900 feet; fl. sporadically all the year, fr. all the year. *A 7189! H 5527! P 23632!* Mexico to northern S. Amer. and Brazil, Cuba, Hispaniola.

23. T. bulbosa Hook., Exot. Fl.: t. 173 (1826).

Leaves covered with minute, appressed, silvery scales; scape erect, with leafy bracts; inflorescence usually green, simple or more often compound with a few (3–5) small divergent spikes; corolla violet.

Frequent epiphyte, rarely on rocks; 250–3000 feet; fl. Nov–Feb, fr. nearly all the year. *A 12154! H 5140! P 22851!* Mexico to northern S. Amer., Bahamas?, Cuba, Puerto Rico, Martinique, Trinidad and Tobago.

The Jamaican plants are noticeably larger than those from other countries. The inflated leaf-bases frequently house colonies of ants.

24. T. canescens Sw., Nov. Gen. & Sp. Pl.: 57 (1788).

Leaves numerous, mostly 10–20 cm. long, forming a dense rosette, the sheaths usually dark brown especially towards base; spikes 2–5 (–9) cm. long; corolla mauve to bluish.

Rare or uncommon on shaded limestone ledges, cliffs and crags, rarely on tree-trunks, confined to a small area along the west-central axis of the island; 2000–3000 feet; fl. Dec–May, fr. May–July. *A 8432! H 8831! P 21334!* Cuba; recorded doubtfully from Trinidad.

25. T. valenzuelana A. Rich. in Sagra, Hist. Cuba, Parte 2, **11**: 267 (1850).— *T. laxa* Griseb. (1864). *T. sublaxa* Bak. (1887).

Plants rather lax, with loose subflexuous leaves mostly 25–40 cm. long; floral bracts said to be pink or red by L. B. Smith (1938), nearly always greenish in Jamaican specimens; corolla usually violet or purple.

Frequent epiphyte, especially on trees overhanging streams, or sometimes on limestone ledges or crumbling shaly banks; 250–2600 feet; fl. Aug–Dec, fr. Jan, Mar, May, July, Dec. *A 6819! H 5555! P 25681!* Florida, southern Mexico to northern S. Amer., Greater Antilles.

26. T. schiedeana Steud., Nomencl. Bot. ed. 2, **2**: 688 (1841).

Stems often to 20 cm. or more long, the plants solitary or divergent in a loose cluster; leaf-blades somewhat flattened adaxially; bracts tipped light crimson; somewhat resembling an erect form of *T. recurvata* but differing especially in the much larger yellow flowers and capsules up to 4·5 cm. long.

Rather rare (St. Eliz., Manch., St. Andr.), epiphytic, chiefly on small orchard trees or in dense thickets; 700–2500 feet; fl. May–June, fr. Jan, Apr, June, Sept. *A 12515! H 5223! P 24066!* Mexico to northern S. Amer., Cuba, Hispaniola.

9. VRIESEA Lindl. (1843) nom. cons.

By R. W. Read

1 Leaves greyish-green, densely lepidote; inflorescence strongly decurved, simple
                                                               1. *incurva*
1 Leaves not greyish green, glabrous or glabrescent; inflorescence erect or ascending:
  2 Leaf-blade tessellated; flowers and fruits distichous, occurring singly in a usually simple spike   2. *platynema*
  2 Leaf-blade not tessellated; flowers and fruits lying at different angles or paired in the axils of bracts:
    3 Floral bracts exceeding sepals and longer than mature fruits; leaves often transversely red-banded   3. *ringens*
    3 Floral bracts shorter than sepals; leaves not transversely red-banded:
      4 Flowers in pairs or dense heads along the main axis:
        5 Flowers paired in the axils of primary bracts; leaves green or often suffused red beneath or at apex, the bracts red-tipped   4. *sintenisii*
        5 Flowers in dense corymbose clusters; leaves and bracts not marked with red
                                              5. *capituligera*
      4 Flowers several to many along several more or less elongate secondary branches:
        6 Spikes lax and open with flowers widely spaced and secund; floral bracts shorter than internodes   6. *swartzii*
        6 Spikes dense; floral bracts more than twice as long as internodes:
          7 Spikes stiffly ascending, the flowers erect; floral bracts acute; leaf-margin faintly purple; leaves of young plants often mottled maroon; corolla white
                                        7. *sanguinolenta*
          7 Spikes spreading, the flowers secund; floral bracts obtuse; leaf-margin not purple; plant glaucous grey-green; corolla green   8. *gibba*

1. V. incurva (Griseb.) Read in Phytologia **16** (6): 457–458 (1968).—*Tillandsia incurva* Griseb. (1865).

Leaves numerous, forming a thick subbulbous rosette; sheaths dark brown, usually with a deep purple margin, up to about 7 cm. broad; corolla yellowish-green.

Local (St. Andr., St. Ann, Port., St. Thom.), epiphytic in wet montane forests, at higher elevations than any other bromeliad in Jamaica; 2500–7000 feet; fl. Feb, Sept, Nov, fr. Apr, Aug, Dec. *A 10660! H 5131! P 8202! R 1762!* C. and northern S. Amer., Cuba. The leaf-bases secrete a mucilaginous gum said to be proteolytic (see Picado in Théses Fac. Sci. Paris 1913).

2. V. platynema Gaudich., Bot. Voy. Bonite: t. 66 (1843).
Leaf-tissue distinctly tessellated in living plants; corolla white or sometimes light green, only slightly exceeding the stamens.
Frequent on trees and shaded limestone crags of wooded hilltops, chiefly along the central axis of the island; 2000–3000 feet; fl. June, Aug, Dec, fr. May–July, Oct–Dec. *A 11253! H 5445! P 16536!* Mexico, Brazil, Argentina, Cuba.

3. V. ringens (Griseb.) Harms in Notizbl. Bot. Gart. Berlin **10**: 801 (1929).—
Tillandsia ringens Griseb. (1866).
Branches few, usually 1 or 2 plus terminal spike; corolla white, the petals flaccid-recurved; stamens exserted.
Frequent on trees or rarely shaded rocky banks in forested areas; 400–2750 feet; fl. Jan, Apr, fr. Feb, Apr, Aug, Oct, Dec. *P 10124! HPS 14637!* Panama, Colombia, Cuba, Hispaniola.

4. V. sintenisii (Bak.) L. B. Sm. & Pitt. in Journ. Acad. Sci. Washington **43**: 403 (1954).—*Caraguata sintenisii* Bak. (1889).
Primary bracts longer than flower; sepals coriaceous with a narrow papery margin; corolla yellow.
Frequent epiphyte in montane forest; 3000–5500 feet; fl. Dec–Jan, fr. Apr–Dec. *H 5134! P 6676! Wynter UCWI 3704!* Greater Antilles.

5. V. capituligera (Griseb.) L. B. Sm. & Pitt. in Journ. Acad. Sci. Washington **43**: 402 (1954)—*Tillandsia capituligera* Griseb. (1866).
Plants usually very large, massive, with numerous closely whorled leaves; bracts yellow-green; corolla whitish.
Local (St. Andr., St. Thom.), on trees, or sometimes on rocks or humus-rich ground in montane forests, an isolated record from swampy woodlands in northern Clarendon; 2300–5250 feet; fl. May–June, fr. May–Oct. *Harris! P 6823!* Cuba, Hispaniola, Colombia.

6. V. swartzii (Bak.) Mez in Engl., Pflanzenr. **4** (32): 400 (1935).—*Tillandsia swartzii* Bak. (1888).
Leaves yellow-green, faintly mottled; inflorescence to 1·5 m. high; corolla creamy-white.
On trees, or often on limestone rocks and ledges, on moist forested hills in the western half of the island; 1200–2500 feet; fl. Mar–June, fr. Apr–July. *H 10885! 11019! P 21336!* Endemic.

7. V. sanguinolenta Cogn. & Marchal, Pl. Ornem.: t. 52 (1874).
The only Jamaican species whose inflorescence-branches, flowers and fruits are always stiffly erect, not noticeably spreading or secund. In this respect our plant disagrees with published descriptions. Floral bracts with scarious margins; sepals 3–4 cm. long; corolla about 4 cm. long; capsule dark brown, apiculate, 4–5 cm. long.
Not common, on trees or rarely shaded rocks in areas of relatively high rainfall; 500–2750 feet; fl. July–Aug, fr. Jan, Mar–Apr, Aug–Sept, Nov. *A 11483! P 24406!* Costa Rica to Colombia, Cuba.

8. V. gibba L. B. Sm. in Contrib. U.S. Nat. Herb. **29**: 524, f. 80 (1954).
Leaf-blades light green beyond the brown basal sheaths, faintly mottled, powdery-glaucous beneath; sepals coriaceous, emerald-green and mucilaginous at anthesis, when dry dark brown with a broad thin pale margin.
Quite common, on large trees or occasionally on shaded limestone crags; 2000–2800 feet; fl. May–July, fr. Aug–Dec. *Dignum 1ʒ 2785! P 26421!* Endemic.

An apparently undescribed species of *Vriesea* resembling *V. sintenisii* has the inflorescence always simple and the flowers invariably occurring singly in the axils of much smaller primary bracts than those of *V. sintenisii* (*H 5442! P 22070!*). *V. macrostachya* Mez was wrongly attributed to Jamaica by Britton & Wilson in Sci. Surv. Porto Rico & Virg. Is. **5**: 142 (1923).

12. MUSACEAE

Massive herbs, monocarpic or with shortly branched rhizomes, the apparent stem above ground-level formed by tightly overlapping petiole-bases. Leaves spirally arranged, simple with a thick petiole and midrib; blade oblong with numerous veins extending to the margin. Inflorescence terminal, erect or pendulous. Flowers mostly unisexual in spirally arranged clusters, each cluster subtended by a large usually deciduous spathaceous bract. Perianth 2-seriate; calyx tubular, toothed, splitting; corolla zygomorphic, toothed or truncate. Perfect stamens 5; filaments slender; anthers linear, 2-locular, opening by slits; staminode 1. Ovary inferior, 3-locular with numerous anatropous ovules on axile placentas; style slender. Fruit fleshy, usually indehiscent. Seeds (not usually formed in cultivated varieties) endospermous with straight embryo and thick hard testa.

2 genera; *Musa* with perhaps 40 species indigenous in the Old World tropics is now distributed throughout the warmer parts of the world as numerous cultivars mostly derived from species by hybridization.

1. MUSA L. (1753); Cheesman (1947), (1948); Simmonds (1959), (1962)

The long-standing application of the botanical names *M. sapientum* L. (1759) for Banana and *M. paradisiaca* L. (1753) for Plantain, treated as two species, cannot be maintained. Both names refer to closely allied but superficially different triploid interspecifi hybrids. Most cultivated bananas and plantains are now known to have originated as hybrids between the species *M. acuminata* Colla (1820) and *M. balbisiana* Colla (1820) and their logical classification depends on the analysis of the contribution to any hybrid by each of these species respectively and the determination of ploidy. Simmonds (1959) has described a system of nomenclature based on the current code for cultivated plants.

13. STRELITZIACEAE

Shrubby herbs or trees with usually distichously arranged, sometimes very large leaves. Flowers bisexual, more or less zygomorphic in a cincinnus (unilateral cyme) in the axil of a conspicuous spathaceous bract. Perianth 2-seriate of 6 unequal segments; sepals free or adnate to the corolla; petals variously connate. Stamens 5, rarely 6; filaments free; anthers 2-locular, opening by vertical slits. Ovary inferior, 3-locular, each loculus with 1–many anatropous ovules; style filiform. Fruit a loculicidal capsule or a berry. Seeds endospermous with a straight embryo.

4 genera with about 150 species (mostly of *Heliconia*) discontinuously distributed in the tropics and subtropics of Africa, America and Melanesia.

1 Perianth-segments partly united; ovary with 1 basal ovule in each loculus; seeds not arillate; stamens 5; inflorescence terminal   **1. Heliconia**
1 Perianth-segments free; ovary with numerous axile ovules in each loculus; seeds arillate:
   2 Stamens 5; flowers strongly zygomorphic; herb or shrub with terminal inflorescence; cultivated; native of S. Africa; Bird of Paradise   *Strelitzia reginae* Ait.
   2 Stamens 6; flowers weakly zygomorphic; tree; inflorescences axillary; cultivated; native of Madagascar; Traveller's Tree   *Ravenala madagascariensis* Sonn.

1. HELICONIA L. (1771) nom. cons.

1 Inflorescence compact; spathes deeply concave, broadly ovate when opened out, distichous and overlapping tightly in flower; flowers included; rhizome short with approximated shoots; leaves very large with oblong blades overtopping the inflorescence   1. *caribaea*
1 Inflorescence open; spathes linear-lanceolate, hardly distichous and not overlapping; flowers laterally exserted; rhizome creeping and branching underground; leaves smaller with lanceolate blades; inflorescence distinctly pedunculate, overtopping the leaves:

2 Bracts mostly crimson; plant in flower 2–3 (–5) m. high   2. *swartziana*
2 Bracts orange or lemon-yellow, never crimson; plant in flower 80–120 cm. high
                                       3. *psittacorum*

1. H. caribaea Lam., Encycl. Méth. Bot. **1**: 426 (1785).—*Musa bihai* L. var. β L. (1753). Balisier (Lesser Antilles, Trinidad), Wild Plantain.

Plant up to 6 m. high to the tips of the leaves, often waxy-glaucous especially when young; bracts at flowering all yellow or yellow with crimson inside, with orange margins and green tips; perianth light green; fruits deep blue with 3 hard warted black seeds.

Local in clearings and margins of woodland, mostly on limestone in wet areas; 1000–2200 feet; fl. Jan–Feb, Aug–Sept, fr. Aug–Sept. *A 11589! Proctor & Mullings 22044! Robbins UCWI 2877!* West Indies.

2. H. swartziana Roem. & Schult. in L., Syst. Veg. ed. nov. **5**: 591 (1819).— *Bihai harrisiana* Griggs (1915). *H. harrisiana* (Griggs) L. B. Sm. (1939). *H. psittacorum* of Sw. (1791), not L. f. (1781).

Perennial herb with aromatic rhizome; leaf-blade up to 30 cm. long and 12 cm. broad, rounded at the base, caudate-acuminate, margin purplish-crimson; bracts green at the base merging to pinkish-yellow, merging to brilliant crimson at the tips; perianth orange-scarlet at base merging to light orange with a dark green spot at the tip of each member, lighter orange within; ovary red-tipped; fruit ripening blue.

Very local (Clar., St. Ann), in shady woodlands on rough limestone; (700–) 1800–2250 feet; fl. Feb–Sept, fr. Apr–Sept. *A 7301! Fawcett 8466! H 10841! P 7863!* Endemic. The closely related *H. hirsuta* L. f. extends through tropical Amer. from Nicaragua to Peru and southern Brazil.

3. H. psittacorum L. f., Suppl.: 158 (1781).

Plant in flower about 1 m. high; leaf-blade 30–45 cm. long, 2·5–6 cm. broad; flowers tipped with blackish-green, about 4·5 cm. long, triangular in section.

Common in gardens and occasionally naturalized; up to 700 feet; fl. mostly Sept–Dec. *H 10923! Stearn!* Tropical S. Amer., introduced and naturalized elsewhere.

Several other highly ornamental species have been introduced from C. and S. tropical Amer. into Jamaican gardens.

14. ZINGIBERACEAE

Perennial herbs with tuberous rhizomes. Leaves with distinctly sheathing bases, sometimes ligulate; blade with numerous closely parallel lateral veins, often sessile on the sheath. Flowers mostly bisexual, often asymmetric. Perianth 2-seriate, the 3 outer calyx-segments united into a tube; the 3 inner petaloid segments showy, more or less united, the posterior segment usually larger. Fertile stamen 1, with a large petaloid staminode (labellum) opposite to it; lateral staminodes petaloid or small or absent; anther 2-locular, opening lengthwise; filament distinct, usually slender. Ovary inferior, 3 (–2)-locular with axile placentas or 1-locular with parietal or rarely basal placentas; ovules usually numerous, anatropous or campylotropous; style simple, usually slender and often more or less enveloped in a channel of the filament whence it extends between the anther-loculi; stigma capitate, dentate or 2-lipped. Fruit a 3-valved loculicidal capsule or indehiscent. Seeds with endosperm. About 50 genera with 1500 species in the tropics and subtropics.

1 Leaves spirally arranged with tubular sheaths; plants not usually aromatic   **1. Costus**
1 Leaves mostly distichous with sheaths open on the side opposite the lamina; plants often aromatic:
  2 Lateral staminodes petaloid:
    3 Staminodes and labellum joined to form a single deeply 3-lobed organ; anther prolonged into a narrow crest with inflexed margins enfolding the style; inflorescence on a separate shoot; cultivated; Ginger   *Zingiber officinale* Roscoe
    3 Staminodes free from the labellum; anther-crest if present not enfolding the style:

4 Inflorescence on a separate shoot with the flowers borne at or near ground-level; bracts free; filament much shorter than half the length of the labellum; cultivated ornamental; Resurrection Lily　　　　　　　　　　*Kaempferia rotunda* L.
4 Inflorescence terminal:
 5 Bracts connate laterally for about half their length forming closed pouches each containing a cincinnus of a few flowers; anthers spurred at base; cultivated; native of Java; Madras Root, Turmeric　　　　*Curcuma domestica* Valeton
 5 Bracts not connate, each subtending 2–3 flowers; anthers more or less entire at base; filaments as long as or longer than labellum　　　　2. **Hedychium**
2 Lateral staminodes never petaloid, sometimes lacking, usually present as small teeth or short appendages at base of labellum:
 6 Inflorescence on a leafless peduncle from the rhizome or base of a leafy shoot, cone-like with imbricated primary bracts; native of tropical Asia; Cardamon
　　　　　　　　　　　　　　　　　Elettaria repens (Sonn.) Baill.
 6 Inflorescence terminal or if basal plant not cultivated:
  7 Labellum erect, stalked　　　　　　　　　　　　　　3. **Renealmia**
  7 Labellum horizontal or deflexed, sessile or nearly so:
   8 Primary bracts small or absent; sepals and petals white and pink; labellum bright yellow and red; cultivated ornamental; native of East Indies; Shell Ginger
　　　　　　　　　　　　　　　　　Zerumbet speciosum Wendl.
   8 Primary bracts conspicuous　　　　　　　　　　　　4. **Alpinia**

Besides the above, other spice plants and ornamental species belonging to the genera *Globba* and *Nicolaia* (*Phaeomeria*) may occur in gardens.

1. COSTUS L. (1753)

Stems and leaves at most puberulous; bracts shorter than ovary and calyx, crimson, the tip acute, erect; labellum white, 7–9 cm. broad　　　　　　　　1. *speciosus*
Stems and leaves covered with long hairs; bracts longer than ovary and calyx, green or becoming red basally in fruit, with a more or less distinct curled or reflexed tip; labellum bright yellow, about 5 cm. broad　　　　　　　　　　2. *villosissimus*

1. C. speciosus (Koenig) Sm. in Trans. Linn. Soc. **1**: 249 (1791).—*Banksia speciosa* Koenig (1783).

Rhizome thick, creeping and branching, forming large persistent colonies; shoots 1·5–2·5 m. high; leaf-sheaths overlapping, crimson, ciliate; leaves densely soft-puberulous beneath, glabrous above, up to 20 cm. long and 9 cm. broad; calyx divided halfway to base, crimson; corolla shortly tubular at base, lobes broadly oblong-elliptical, about 4 cm. long and 2 cm. broad, pink; labellum yellow and puberulous within near base; stamen boat-shaped, about 4 cm. long and 1·6 cm. broad, yellow on the recurved tip; ovary 3-locular with ovules in 2 rows in each loculus.

Locally common on roadside banks; 200–1000 feet; fl. June–Nov. *A 7331! P 11163! Wynter UCWI 3259!* Native of S.E. Asia, naturalized in northern S. Amer., Cuba and Trinidad.

2. C. villosissimus Jacq., Fragm.: 55, t. 80 (1804–07).

Rhizome shortly branched; shoots 1–2·4 m. high in flower, approximate; leaf-blades up to 30 cm. long and 10 cm. broad; calyx 1–1·5 cm. long, shortly lobed; corolla-tube 1·5–2 cm. long, lobes oblong, acute, yellow, up to 6 cm. long; stamen oblanceolate, obtuse; fruit obovoid, sulcate, white, 1·3–1·8 cm. long; seeds wrinkled, black with a white aril, 2 mm. long.

Rare and local (Port.), at woodland margins and in roadside hollows in wet areas; 100–700 feet; fl. most of the year, fr. Oct. *A 8114! P 16568!* Originally described from St. Vincent, also in Guyana and Peru.

2. HEDYCHIUM Koenig (1783)

1 Inflorescence compact with broad overlapping bracts spirally arranged:
 2 Labellum white; calyx glabrous　　　　　　　　　　　1. *coronarium*
 2 Labellum yellow; calyx pubescent　　　　　　　　　　　2. *flavum*
1 Inflorescence open with narrow bracts arranged in definite vertical rows and not overlapping:

3 Leaves linear-lanceolate; petals and staminodes vermilion; flowers faintly fragrant
3. *coccineum*

3 Leaves lanceolate to oblong-lanceolate; petals and staminodes yellow; flowers intensely fragrant
4. *gardneranum*

The species-status of some of these plants may be in doubt owing to the numerous hybrids which have been made from time to time.

1. H. coronarium Koenig in Retz., Obs. Bot. **3**: 73 (1783).—White Ginger Lily.

Shoots herbaceous, 1·2–2·5 m. high in flower; leaves usually with a crimson spot at top of sheath, blade up to 50 cm. long and 10 cm. broad; ligule conspicuous, 2–3 cm. long; bracts ovate, 4–5·5 cm. long, 2–3 cm. broad; flowers fragrant; calyx 4 cm. long, glabrous; corolla-tube 8 cm. long, lobes 3–3·5 cm. long; lateral staminodes oblong-lanceolate, 4·5 cm. long, variable in width from less than 1 cm. to more than 1·5 cm. broad; labellum emarginate, about 3·5 cm. long and broad, greenish below; capsule-valves separating widely, reddish-orange within.

Locally abundant especially in the eastern parishes, in wet areas, gregarious along ditches and streams; (100–) 1250–4000 feet; fl. June–Dec. *A 7766! H 11675! P 24032!* Native of S. Asia, introduced and cultivated elsewhere and often naturalized.

2. H. flavum Roxb., Fl. Ind. **1**: 81 (1820).—Yellow Ginger Lily.

Very similar to the last.

Rare and local (St. Andr., Port.), in damp sheltered places; 3600–4500 feet; fl. Sept–Feb. *H 11673!* Native of India.

3. H. coccineum Sm. in Rees, Cyclop. **17**: no. 5 (1811).

Shoots approximate from short rhizome-branches, 1·2–2 m. high in flower; leaves spirally arranged, 25–40 cm. long, 3–5 cm. broad; ligule 1·2–2·5 cm. long; bracteoles 1–1·5 cm. long; calyx tubular, hispid, 3-lobed above, 2–2·5 cm. long, green at base; corolla-tube 2·5 cm. long, lobes linear, 3·5 cm. long; lateral staminodes linear-oblanceolate, 2·7 cm. long, puberulous at base; labellum-limb 2-lobed, claw 1 cm. long, lobes 1·2 cm. long, 8 mm. broad; anther crimson; stigma top-shaped; ovary hispid.

Local (St. Andr., Port.), along roadsides; 750–4000 feet; fl. Mar–Apr, July–Oct. *A 8157! H 11676! P 23807!* Native of India and Burma.

4. H. gardneranum Sheppard ex Ker-Gawl. in Edw., Bot. Regist. **9**: t. 774 (1824).

Rhizome creeping and massive with circular shoot-scars; shoots up to 2 m. high; leaf-blades 20–45 cm. long, 10–15 cm. broad; ligule 1·5–2·5 cm. long; inflorescence erect, up to 45 cm. long; flowers strongly fragrant; calyx tubular, 3–3·5 cm. long, 3-toothed at apex, pilose; corolla-tube 5–5·5 cm. long, a little longer than the bracts, lobes narrowly linear, reflexed, 3·5–4·5 cm. long; lateral staminodes 2·5–3 cm. long, 1·5 cm. broad; labellum-limb bilobed, 2·5–3 cm. long; filament 6 cm. long, red or orange-red; capsule-valves 1·8 cm. long, red within.

Locally abundant (St. Andr., Port.), naturalized and gregarious on steep sheltered roadside banks; 3500–4500 feet; fl. July–Oct, fr. Dec–Jan. *A 7383! J.P. 1397! H 11669!* Native of the Himalayan region, introduced elsewhere.

3. RENEALMIA L. f. (1781) nom. cons.

1 Inflorescence racemose, the flowers solitary in the axils of bracts borne on a leafless scape separate from the leafy shoots; bracteoles 1·5 cm. long; calyx 1·5–2 cm. long
1. *sylvestris*

1 Inflorescence paniculate; bracts subtending helicoid several-flowered cymes (cincinni); bracteoles 8–13 mm. long; calyx 5–7 mm. long:

2 Inflorescence terminal on the leafy shoots, 6–12 cm. long; cincinni 2–4-flowered; bracts about 4 cm. long
2. *antillarum*

2 Inflorescence separate from the leafy shoots, 20–30 cm. long on a leafless or almost leafless scape 60–120 cm. long; cincinni up to 15-flowered; bracts 6–8 cm. long
3. *aromatica*

1. R. sylvestris (Sw.) Horan., Prodr. Monogr. Scit.: 32 (1862).—*Amomum sylvestre* Sw. (1788).

Leafy shoots 2–2·5 m. high; leaf-blade linear-oblanceolate, up to 35 (–65) cm. long and 8 (–14) cm. broad; bracts up to 4 cm. long, salmon-pink; corolla creamy-yellow to white, tube 1·3 cm. long, lobes 7 mm. long; labellum obovate; capsule oblong-globose, shiny black, sharply demarcated from the green pedicel, 20 mm. long and 18 mm. broad; seeds brown, shiny, 6 mm. long, 4·5 mm. broad.

Rare (Trel., Port., St. Thom.), in moist forest glades; 1000–2200 feet; fl. Jan–Mar, July–Sept, fr. Feb–May, July–Sept. *J.P. 1022! P 22111! 22707! R 1691! Stearn 971!* Endemic.

2. R. antillarum (Roem. & Schult.) Gagnepain in Bull. Soc. Bot. France **50**: 202 (1903).—*Alpinia antillarum* Roem. & Schult. (1817).

Shoots 1·5–2 m. high; leaf-blade linear-lanceolate, 10–32 cm. long, 3–7·5 cm. broad; calyx 7 mm. long, red; corolla white, tube 12 mm. long, lobes 7 mm. long; labellum obovate; capsule globose.

Rare (St. James, St. Thom.), in wet forest; about 1900 feet; fl. and fr. Feb. *H 9160! J.P. 899!* Greater Antilles, Martinique.

3. R. aromatica (Aubl.) Griseb. in Abh. Gött. Akad. **7**: 275 (1857).—*Alpinia aromatica* Aubl. (1775). *R. occidentalis* (Sw.) Sweet (1830).

Leafy shoots 2–2·5 m. high; leaf-blade lanceolate, 15–50 cm. long, 5–12 cm. broad; calyx 5–6 mm. long, red; corolla yellow; labellum reniform; capsule subglobose, red.

Uncommon (Trel., Port., St. Thom.), in woodland margins in wet areas; 400–1000 feet; fl. Mar, Aug–Sept, fr. Sept. *A 9295! H 6056! P 15624!* Northern S. Amer., Greater Antilles, Virgin Is., Guadeloupe.

4. ALPINIA Roxb. (1810) nom. cons.

1. A. allughas (Retz.) Roscoe in Trans. Linn. Soc. **8**: 346 (1807).—*Heritiera allughas* Retz. (1791).

Shoots in flower 1·2–3 m. high from a thick rhizome covered with overlapping scale-leaves; leaf-blade lanceolate or linear-lanceolate, up to 70 cm. long and 15 cm. broad, cuspidate; primary bracts about 5 mm. long distally, much larger basally and deciduous; axis and branches of inflorescence tomentose; calyx 9–11 mm. long, light green to pink, minutely pilose; corolla-tube 8 mm. long, lobes about 10 mm. long, obtuse, minutely pilose, greenish-white to pink; labellum obovate, bifid, 3 cm. long, light pink to rose-coloured; stamen 1·8 cm. long; anther bifid, 4–5 mm. broad; capsule globose, black, pericarp brittle, breaking irregularly, pulp light green drying white and mealy; seeds angled, black, 5–6 mm. in diameter.

Locally abundant in marshes and riverside swamps; 25–750 feet; fl. and fr. most of the year. *A 11504! H 11752! P 23805!* Native of S. Asia.

Guillainia purpurata Vieill., Red Ginger, is commonly grown in gardens in sufficiently wet areas. It is a native of the western Pacific, recently introduced into Jamaican gardens.

15. CANNACEAE

Perennial rhizomatous herbs. Leaf-blade large, broad, pinnately veined with distinct midrib; leaf-stalks tightly sheathing each other to form a false stem. Inflorescence terminal, racemose or paniculate. Flowers bracteate, bisexual, irregular. Perianth 2-seriate; sepals 3, imbricate, herbaceous, persistent in fruit; petals 3, connate at the base and adnate to the staminal column. Staminodes up to 5, petaloid; 3 outer imbricate, connate at base; 2 inner more or less connate; innermost abaxial (labellum). Fertile stamen free; anther solitary, 1-locular, lateral on a petaloid filament. Ovary inferior, 3-locular; style and stigma petaloid; ovules numerous on axile placentas, anatropous. Fruit a fibrous warted capsule. Seeds rounded, endospermous, very hard, usually black.

1 genus with 30 or more species probably all native in the New World tropics and subtropics, but several long established in the Old World.

1. CANNA L. (1753)

1 Petaloid staminodes light yellow or only very faintly spotted red; leaves lanceolate, long-
tapered at base; rhizome stoloniferous   1. *glauca*
1 Petaloid staminodes mostly red; leaf-blades oblong to elliptical, rather abruptly narrowed
at base; rhizome tuberous:
  2 Plants mostly 2·5 m. or more high; rhizome thick; flowers dark red, about 7·5 cm. long
                                                                2. *edulis*
  2 Plants mostly 1–2 m. high:
    3 Staminodes and labellum with deeply 2-lobed apices; flowers narrow; labellum
narrow, about 6 cm. long, strongly revolute   3. *sylvestris*
    3 Staminodes and labellum entire or merely emarginate, spreading:
      4 Petals green; 3 staminodes subequal, obovate-spathulate to oblanceolate, acute,
4–5 cm. long, 1–1·3 cm. broad; labellum narrow, reddish-yellow spotted red
                                                                     4. *indica*
      4 Petals light scarlet; 2 larger staminodes oblong-spathulate, emarginate; third stami-
node smaller, linear-lanceolate; labellum red spotted yellow   5. *coccinea*

1. C. glauca L., Sp. Pl. 1: 1 (1753).
Plant in flower to about 2 m. high, glabrous; leaves long-acuminate, up to 85
cm. long and 17 cm. broad, light green; sepals 13–15 mm. long; petals shortly
connate at base, about 6 cm. long, 10–12 mm. broad, yellow; staminodes 3, obovate,
7 cm. long, 20–24 mm. broad; labellum narrow, bifid; capsule angular-ellipsoid,
up to 4·5 cm. long and 2·5 cm. broad; seeds black, subspherical, 7 mm. long.
Rare (St. Cath., West.), at margins of swampy pastures and in low-lying damp
areas; 50–100 feet; fl. and fr. most of the year. *A 13025! Harris!* West Indies,
tropical Amer. southwards to Argentina; introduced elsewhere.

2. C. edulis Ker-Gawl. in Edw., Bot. Regist. 9: t. 775 (1824).—Spanish Arrowroot.
Plant in flower to about 3·5 m. high; leaf-blade oblong to ovate-oblong, shiny;
sepals 1·2 cm. long; petals oblong-lanceolate, 4 cm. long, 7–8 mm. broad; stami-
nodes 3, about 7 cm. long.
Cultivated (St. Andr.); fl. Dec. *H 6969!* Tropical and subtropical S. Amer.,
West Indies; introduced and cultivated in many warm countries.

3. C. sylvestris Roscoe, Monandr. Pl., Scit.: t. 10 (1828).
Plant in flower 75–200 cm. high; leaf-blade oblong to oblong-lanceolate, acuminate; sepals lanceolate, about 1 cm. long; petals lanceolate, crimson, to 4 cm. long;
staminodes 3, narrowly spathulate, 6–7 cm. long, scarlet or yellow with scarlet
blotches; capsule 3·5–4 cm. long, 1·8 cm. in diameter.
Occasional (St. Andr., St. Thom.), along pathsides and in open sheltered areas;
1250–5000 feet; fl. Jan–Apr, Aug, fr. Apr. *A 7760! H 10894!* Puerto Rico, Panama.

4. C. indica L., Sp. Pl. 1: 1 (1753).—Indian Shot, Wild Tapioca.
Plant in flower 1·2–2 m. high; leaf-blade oblong, acute; sepals about 1 cm. long,
free; petals lanceolate, acuminate, 3–4 cm. long, 4–5 mm. broad.
Cultivated and escaped; fl. Dec–Apr, fr. Apr–Sept. *H 6970! Stearn 784!* Widely
established in warm countries.

5. C. coccinea Mill., Gard. Dict. ed. 8 (1768).—Wild Canna.
Plant in flower 1·5–2 m. high; leaf-blade ovate, oblong to oblong-lanceolate,
acuminate and filiform-apiculate; sepals 10–15 mm. long, green suffused pink;
petals scarcely connate, lanceolate, acuminate, 3·5–4·5 cm. long; capsule 3 cm.
long, 2 cm. in diameter.
Mostly cultivated but occasionally escaping on to roadsides and waste places;
800–3500 feet; fl. Dec–May, fr. Apr. General throughout tropical Amer. and
established in many warm countries.

16. MARANTACEAE

Perennial herbs with rhizomes. Leaves with closely overlapping open-sheathed petioles, terete above and with a pulvinus (callus) below the blade; blade usually broad and often with one straight and one curved margin converging to an oblique acumen. Flowers bisexual, asymmetric, often in unequally pedicellate pairs in bracteate inflorescences. Perianth 2-seriate; outer perianth-segments free, equal; inner perianth-segments more or less connate into a tube with the lobes often unequal. Stamen 1; anther 1-locular. Staminodes variously petaloid. Ovary inferior 1–3-locular; style stout, simple; ovules solitary in each loculus, basal, erect, anatropous or campylotropous. Fruit a capsule or indehiscent. Seeds endospermous, often arillate.

About 30 genera with 350 species mostly in damp shady tropical and subtropical habitats.

1 Ovary 3-locular; bracts persistent; inflorescence spicate or capitate; outer staminode 1;
　　mostly cultivated ornamentals　　　　　　　　　　　　　　　　**1. Calathea**
1 Ovary 1-locular; bracts deciduous; inflorescence paniculate:
　2 Outer staminode 1; plants forming colonies in fresh-water swamps　　　**2. Thalia**
　2 Outer staminodes 2; cultivated and relict　　　　　　　　　　　**3. Maranta**

1. CALATHEA G. F. W. Meyer (1818)

1. C. allouia (Aubl.) Lindl. in Edw., Bot. Regist. **14**: under t. 1210 (1829).—
Maranta allouia Aubl. (1775).
Plant 1–1·5 m. high; leaves long-petiolate; callus 2 cm. long; blade lanceolate to oblong, up to 50 cm. long and 13 cm. broad, acuminate, acute to rounded at the base; peduncle about 10 cm. long; inflorescence ovoid, bracteate, about 10 cm. long; sepals lanceolate, 8 mm. long; corolla white, tube 2·5 cm. long, lobes 1 cm. long; outer staminode white, 1·2 cm. long.
Cultivated and occasionally escaped (St. Ann) into damp shady places; 2500 feet; fl. Aug–Sept. *Harris!* French Guiana, West Indies. Two species with spicate inflorescences occur in gardens and may also escape; *C. insignis* Peters., native of Panama and *C. lutea* (Aubl.) G. F. W. Meyer, native of Guyana, have the flowers subtended by two rows of showy overlapping bracts. The latter is naturalized in St. Thomas (*P 27842!*).

2. THALIA L. (1753)

1. T. geniculata L., Sp. Pl. **2**: 1193 (1753).
Leaves up to 2 m. high with very long petioles; callus about 1·5 cm. long; blade up to 60 cm. long and 20 cm. broad; plant in flower up to 4 m. high; inflorescence-branches markedly zig-zag; flowers purplish; sepals 2 mm. long; corolla tubular below, lobes 7 mm. long; outer staminode 1·4 mm. long; capsule indehiscent 1 cm. long; seed minutely tuberculate with a whitish aril.
Rare (St. Eliz., West.), gregarious at pond-margins and in riverside swamps; – 50–75 feet; fl. Dec–Jan. *H 11822! P 18455! Skelding UCWI 3592!* From Florida through the American subtropics and tropics to Argentina; Greater Antilles.

3. MARANTA L. (1753)

1. M. arundinacea L., Sp. Pl. **1**: 2 (1753).—Arrowroot.
Herb up to about 1 m. high with a fleshy rhizome; petiole about 7 cm. long; leaf-blade ovate-lanceolate to lanceolate, up to 22 cm. long and 8 cm. broad; sepals 1·5 cm. long; corolla-tube 1·3 cm. long, lobes 8–10 mm. long, white; outer staminodes 10 mm. long, obovate, emarginate; fruit 7 mm. long; seed rather rough with a yellowish aril.
Very occasionally cultivated (St. Andr., Port.) in the wetter areas; SL–700 feet; fl. June. *H 12523!* Native of Brazil now introduced and sporadically cultivated in the tropics and subtropics, a major crop in St. Vincent.

17. LILIACEAE

Herbs, shrubs or trees, the herbs with rhizomes, corms or bulbs; roots sometimes tuberous. Flowers usually bisexual, mostly actinomorphic, never in umbels. Perianth-members mostly 6, free or connate in 2 distinct but similar series. Stamens usually 6, free or adnate to the perianth-members and always opposite to them; anthers 2-locular. Ovary superior, usually 3-locular with axile placentation; ovules usually numerous and mostly 2-seriate in each loculus, rarely solitary, anatropous. Fruit a capsule or berry. Seeds with endosperm.

A heterogeneous family with about 240 genera and over 3500 species of world-wide distribution, but less numerous in tropical regions.

1 Plants climbing; perianth-segments free:
  2 Flowers showy, mostly yellow or red and yellow; leaves broad, tendril-tipped; style long, deflected to one side; fruit a capsule with red seeds; stems smooth
        1. Gloriosa
  2 Flowers very small, white or tinged pink; leaves obsolete, scaly, replaced by flat or setaceous clustered phylloclades; style short; fruit a berry; stems spiny
        2. Asparagus
1 Plants not climbing:
  3 Plants herbaceous; stem short or rhizomatous; leaves mostly basal and tufted:
    4 Leaves thick, succulent with prickly margins; fruit a capsule   **3. Aloe**
    4 Leaves thin, leathery or herbaceous with smooth margins:
      5 Perianth-segments free, reflexed; anthers basifixed, opening by terminal pores; filaments kinked; fruit a blue berry   **4. Dianella**
      5 Perianth-segments connate at least at the base; anthers medifixed; opening lengthwise:
        6 Leaves leathery; inflorescence racemose; flowers white to cream; fruit a dry berry with fleshy seeds   **5. Sansevieria**
        6 Leaves soft; inflorescence paniculate; flowers tawny; fruit a loculicidal capsule
        6. Hemerocallis
  3 Plants shrubby or arborescent with leaves spread along the stem:
    7 Perianth-segments free; leaves spine-tipped; fruit a large soft berry or capsular
        7. Yucca
    7 Perianth-segments connate at base:
      8 Ovules numerous; leaves elliptical   **8. Cordyline**
      8 Ovules solitary; leaves linear to lanceolate   **9. Dracaena**

1. GLORIOSA L. (1753)

Gloriosa superba L., with flowers changing colour from light green through yellow and orange to light red and *G. rothschildiana* O'Brien, with red and yellow flowers, are African plants sometimes grown in gardens. Some of these highly ornamental plants are hybrids of garden origin.

2. ASPARAGUS L. (1753)

Phylloclades setaceous, clustered, 3–6 mm. long, straight; flowers solitary   1. *setaceus*
Phylloclades flattened or angled, few together 12–18 mm. long, curved; flowers in racemes   2. *tetragonus*

1. A. setaceus (Kunth) Jessop in Bothalia **9**: 51 (1966).—*Asparagopsis setacea* Kunth (1850). *Asparagus plumosus* Bak. (1875).

Long twining stems to 4 or 5 m., tufted from a short rhizome; perianth greenish-white.

Cultivated and occasionally escaped (St. Andr., Manch.), in thickets and waste places; 500–1500 feet; fl. and fr. sporadically. *P 22938!* Native of S. Africa, cultivated in many warm countries.

2. A. tetragonus Bresler, Gen. Aspar., Diss. Berol.: 27 (1826).

Scrambling twiner to 3–5 m.; flowers fragrant; perianth-segments 6 (–8), white or tinged pink with a green mid-line, finally purplish; stamens 6 (–8); anthers purplish-black; pollen orange.

Uncommon (St. Andr., St. Eliz., West.), on fences and in thickets; 20–1500 feet; fl. Dec–Jan. *A 12050! Yuncker 17980!* Native of S. Africa.
Several other ornamental species of *Asparagus* are grown in gardens.

3. ALOE L. (1753)

1. A. vera (L.) Burm. f., Fl. Ind.: 83 (1768).—*A. perfoliata* var. *vera* L. (1753). *A. barbadensis* Mill. (1768). *A. vulgaris* Lam. (1783). Bitter Aloes, Sempervivum, Sinkle Bible.

A short-stemmed herb with tufted fleshy glaucous leaves, budding profusely to produce new plants from the short creeping rhizome; inflorescence 1–1·3 m. high, simple or sparingly branched; bracteoles persistent; perianth yellow; fruit not seen.

Occasional and locally gregarious (St. Cath., Clar.), and also in cultivation; where naturalized mostly on exposed limestone rocks and in pebbly soil in arid areas; SL–200 feet; fl. Jan–May. *A 6295! Stearn 812! Yuncker 18167!* Native probably of the Mediterranean region, cultivated and naturalized in many of the more arid parts of the Caribbean region; Grand Cayman.

Aloe is an Old World genus centred in southern Africa; there is no evidence that this species is indigenous in the New World tropics although it has been established there for a long time. Other species of *Aloe* are sometimes grown in gardens in Jamaica.

4. DIANELLA Lam. (1786)

1. D. ensifolia (L.) DC. in Redouté, Liliac.: t. 1 (1802).—*Dracaena ensifolia* L. (1767).

Perennial herb with branched rhizomes; roots orange; leaves linear, the midrib beneath and margins rough distally; inflorescence to 1 m. high, corymbose, bracteate; pedicels 5–10 mm. long; outer perianth-segments lanceolate, 8–9 mm. long, acute, olive-grey outside, bluish-grey inside, with narrow whitish margins; inner perianth-segments oblong, 7–8 mm. long, obtuse, olive-green with broad whitish margins; ovary subspherical, about 2 mm. in diameter; ovules pendulous; mature fruit a soft berry, lapis blue with juice of the same colour; seeds black, shiny.

Established locally (St. Andr., St. Mary), on roadside banks and rough grassy hillsides; 500–2500 feet; fl. and fr. sporadically. *A 8266! H 10652!* Native of East Indies and Pacific Islands.

5. SANSEVIERIA Thunb. (1794) nom. cons.

1 Leaves flat or nearly so, linear-oblanceolate, broadest at or above the middle, often
    blotched or mottled paler:
  2 Leaf-margin red   1. *metallica*
  2 Leaf-margin not red   2. *trifasciata*
1 Leaves terete, tapered towards the tip, not variegated   3. *cylindrica*

1. S. metallica Gérôme & Labroy in Bull. Mus. Hist. Nat. Par. **2**: 170, 173, f. 2 (1903).—Bowstring Hemp, Mother-in-Law's Tongue.

Stout herb with numerous branches from a creeping rhizome; leaves upright, very firm, 30–100 cm. long, 5–9 cm. broad at the middle, gradually narrowed to the base; inflorescence to 100–120 cm. long, with flowers in clusters of 2–3 along the axis; perianth white with greenish tips, the lobes about 1·5 cm. long; fragrant at night.

Gregarious and common in rocky ground around habitations, in sand near the sea and along roadsides; SL–1500 feet; fl. Jan–Apr, Sept–Oct. *A 6971! P 23221! Stearn 238!* Native probably of Africa now widespread in tropical countries; Cayman Islands.

2. S. trifasciata Prain, Bengal Pl. **2**: 1054 (1903).—Tiger Cat.

Like the last; leaves up to 60 cm. long, 3–5 cm. broad above the middle, transversely banded with light and dark green on both surfaces; berries globose, red, 8 mm. broad.

Local (Clar.), at roadside; 150 feet; fr. Apr. *Stearn 903!* Probably native of tropical Africa, now widely grown as an ornamental.

3. S. cylindrica Boj. ex Hook. in Curt., Bot. Mag. **85**: t. 5093 (1859).
Perennial herb with fleshy dark green leaves 60–80 cm. long; perianth white.
In gardens. Native of tropical Africa, cultivated and escaped, Antigua, Virgin Islands.

6. HEMEROCALLIS L. (1753)

1. H. fulva (L.) L., Sp. Pl. ed. 2, **1**: 462 (1762).—Tawny Daylily.
Roots tuberous; leaves basal, tufted, long and narrow; flowers brownish-orange in a branched inflorescence.
Locally escaped from gardens and established; 2300–4000 feet; fl. Jan–Feb. *H 12329!* Native of China and Himalayas. Several varieties are in cultivation in Jamaica.

7. YUCCA L. (1753)

1. Y. aloifolia L., Sp. Pl. **1**: 319 (1753).—Spanish Dagger.
Plant shrubby to 6 m. high, branching underground; leaves numerous in a tight spiral, 20–35 cm. long, 2·5–4 cm. broad at the middle; flowers in a large terminal panicle; perianth white; fruit 5–9 cm. long, 2·5–4 cm. broad, flat-ribbed, becoming a soft fleshy berry with coloured juice, perianth persistent at base; seeds black, shiny.
Extensively cultivated and locally naturalized in pasture-margins and around habitations; SL–2000 feet; fl. Apr, July, fr. Sept–Jan, not often formed. *Campbell 6610!* Native of southern N. Amer., introduced into the West Indies; Little Cayman.

8. CORDYLINE Commers. ex Juss. (1789) nom. cons.

Cordyline fruticosa (L.) A. Chev. (*C. terminalis* (L.) Kunth) is commonly grown in gardens and planted as a boundary marker or live fence. Plants have long slender sparingly branched stems and the leaves which are clustered in the new growth fall off to leave ring-scars. There seem to be several varieties, one of the commonest of which has broadly elliptic leaves variegated bright crimson; the flowers are rather small in branched terminal inflorescences; fruits ripen to red berries. Native of tropical Asia.

9. DRACAENA Vandelli ex L. (1767)

Several species of this shrubby or arborescent Old World genus have been grown in Jamaica for ornament or as live fence; none is naturalized.

18. PONTEDERIACEAE

Aquatic annual or perennial floating or rooted erect or prostrate herbs of fresh water. Stems short with usually opposite or verticillate leaves. Flowers bisexual, actinomorphic or zygomorphic in a spike, raceme or panicle, the inflorescence subtended by a spathaceous leaf-sheath; bracts small or absent. Perianth obscurely 2-seriate of 6 imbricated segments, free or basally connate. Stamens 6, 3 or 1, inserted on the perianth, sometimes unequal; filaments free; anthers 2-locular, opening lengthwise or rarely by pores. Ovary superior, 3-locular with axile placentas or 1-locular with 3 parietal placentas; style simple; stigma entire or lobed. Ovules anatropous, numerous or solitary. Fruit a capsule opening by 3 valves or indehiscent. Seeds longitudinally ribbed, endospermous, with straight embryo.

7 genera with about 30 species, mostly in the tropics and subtropics but extending into temperate America.

Stamens 6; anthers dorsifixed; ovary 3-locular; perianth blue or violet, showy
1. Eichhornia
Stamens 3; anthers basifixed; ovary 1-locular; perianth blue or white   **2. Heteranthera**

1. EICHHORNIA Kunth (1842) nom. cons.

Floating or stranded perennial rosette herbs, often with markedly inflated petioles, proliferating by offsets; inflorescence spicate; posterior perianth-lobe (inner series) larger, up
to 3·8 cm. long and 2·5 cm. broad; lower bract with a reduced blade; upper bract
bladeless   1. *crassipes*
Erect annual herb; petioles not inflated; inflorescence paniculate with unilateral racemose
branches; posterior perianth-lobe much smaller than above; lower bract with a large
leafy blade; upper bract spreading, linear-lanceolate   2. *paniculata*

1. E. crassipes (Mart.) Solms in A. & C. DC., Monogr. Phan. **4**: 527 (1883).—
  Pontederia crassipes Mart. (1823). Water Hyacinth.
 Petioles variously inflated from suborbicular to linear-fusiform, often apparently
according to the growth conditions; leaf-blade broad, cuspidate, shortly acuminate
or smoothly rounded; peduncle deflexed after flowering; anterior perianth-lobe
smaller, posterior larger with a yellow spot; stamens (in all Jamaican plants seen)
3 with filaments 5–7 mm. long and 3 with filaments 20–24 mm. long, all curved
upwards below the connective; style 2·3 cm. long, curved upwards placing the
stigma in a mid-styled position.
 Gregarious and locally abundant in still ponds and along the margins of slow-
moving rivers mainly in southern coastal areas; SL–600 (–1200) feet; fl. sporadically
throughout the year, fr. not seen. *A 9448! H 7085! P 8026! Stearn 427!* Probably
a native of Brazil, now established in many of the warmer parts of the world;
Grand Cayman.

2. E. paniculata (Spreng.) Solms in A & C. DC., Monogr. Phan. **4**: 530 (1883).—
  Pontederia paniculata Spreng. (1822). *Piaropus paniculatus* (Spreng.) Small
  (1913).
 Erect herb 40–100 cm. high; non-flowering shoots with a single terminal cordate
acutely acuminate leaf-blade; perianth at first violet-blue fading violet-rose, one
segment with 2 yellow streaks, more or less persistent in fruit.
 In ditches and pond-margins in several of the southern parishes; SL–1000 feet;
fl. mostly Oct–Mar. *A 10358! H 12071! P 18325!* S. United States, Cuba.

2. HETERANTHERA Ruiz & Pav. (1794) nom. cons.

Leaf-blade longer than broad, broadly ovate and subcordate to narrowly elliptic-lanceolate
and rounded at the base; plants mostly tufted with short stems; perianth blue, one
· segment with a yellow spot   1. *limosa*
Leaf-blade mostly broader than long, round-sided, distinctly cordate; plants creeping and
rooting; perianth white   2. *reniformis*

1. H. limosa (Sw.) Willd. in Ges. Naturf. Freunde, Neue Schr. **3**: 439 (1801).—
  Pontederia limosa Sw. (1788).
 Tufted herb 10–15 cm. high; leaf-blade about 2 cm. broad.
 Rare (St. Andr., St. Eliz.), in mud at margins of fresh-water ponds; about 1000
feet; fl. May, Nov. *H 6704! P 15369! Stearn 1026!* S. United States, Mexico to
Paraguay, Cuba, Hispaniola.

2. H. reniformis Ruiz & Pav., Fl. Peruv. & Chil. **1**: 43, t. 71, f.a (1798).
 Plant creeping, rhizomatous; stem and leaves softly succulent; leaf-blade 1·5–
4 cm. broad; peduncle deflexed in fruit; perianth about 4 mm. long, with outer
segments and pedicel glandular-pubescent.
 Locally common in ditches, swamps and sluggish streams; SL–1700 feet; fl.
Aug–Mar, fr. Mar. *A 8960! H 10589! H & P 14543!* S. United States to Paraguay,
Cuba, Hispaniola.

19. SMILACACEAE

Climbing or scrambling shrubs often with tendrils from the petioles and prickly stems. Rhizome sometimes massive and woody. Leaves alternate or opposite, mostly leathery, usually broad with 3 or more strong longitudinal veins. Flowers small, mostly unisexual and dioecious in axillary umbels, racemes or spikes. Perianth-segments 6, equal or subequal, usually free. Stamens usually 6; filaments free or united into a column; anther-loculi confluent. Ovary superior, 3-locular; ovules 1 or 2 in each loculus, pendulous; staminodes present in the female flowers. Fruit a berry. Seeds endospermous with a small embryo.

4 genera with about 300 species, mostly of *Smilax*, widely distributed in tropical and temperate regions.

1. SMILAX L. (1753); O. E. Schulz (1904)

Flowers dioecious; perianth-segments free; stamens 6 or more, free.

1 Sepals 3–6 mm. long; leaves rather thin and papery when dry:
  2 Peduncles of male inflorescences distinctly longer than the petioles at anthesis; larger leaves somewhat cordate at the base; male flowers 5·5 mm. long; stem sharply quadrangular, armed; cultivated   **1. regelii**
  2 Peduncles of male inflorescences much shorter than the petioles at anthesis; leaves rounded to acute at the base; male flowers 4·5–6 mm. long; stems mostly terete, not armed   **2. domingensis**
1 Sepals 1·5–2·5 mm. long; leaves papery or leathery; bases of larger leaves subcordate; stems armed or unarmed:
  3 Umbels in the axils of leaves only; leaves 7-nerved, leathery, apex truncate to emarginate, mucronate, often spine-margined; stem more or less quadrangular; Cayman Is.   *S. havanensis* Jacq.
  3 Umbels in the axils of bracts as well as leaves; leaves 3–5 (–7)-nerved, papery or rather leathery, mostly rounded at the tip, mucronate, entire; stem slightly angled   **3. balbisiana**

1. S. regelii Killip & Morton in Bot. Maya Area, Misc. Paper **12**: 272 (1936).— Jamaican Sarsaparilla.

Leaves ovate to oblong, variable in shape at the base, mostly 15–21 cm. long, 7·5–13·5 cm. broad on the main stem, smaller on the branches; tendrils arise well above the base of the petiole; peduncles 3–5 cm. long.

Cultivated (St. Andr., St. Thom., Trel.); 600–2000 feet; fl. Apr. *H 8952!* Native of Amazonian Brazil and the Guianas, now widely distributed.

2. S. domingensis Willd. in L., Sp. Pl. ed. 4, **4** (2): 783 (1806).

Leaves ovate to lanceolate, 5–13 cm. long, 3–7 cm. broad, acuminate to cuspidate at the tip; tendrils arise near the base of the petiole; peduncles rarely as much as 1 cm. long; pedicels thickened in fruit, 5–9 mm. long; flowers green, fragrant; berries reddish then finally black, 9 mm. in diameter, 1–2-seeded.

Occasional, twining on shrubs and trees in savanna thickets and wet woodlands, mostly on limestone; 800–4000 feet; fl. Aug, fr. Dec–May. *A 5920! 7927! H 11088!* Mexico, British Honduras, Guatemala, Greater Antilles.

3. S. balbisiana Kunth, Enum. Pl. **5**: 183 (1850).—*S. celastroides* Kunth (1850).
Briar Withe, Chainy Root (China Root).

Lower leaves up to 15 cm. long and 11 cm. broad or larger, broadly ovate, mucronate, upper leaves lanceolate to elliptical, much smaller and bractlike on the terminal branches; tendrils arise about one-third from the base to halfway on the petiole; peduncles very short to 6 (–10) mm. long; pedicels slender in fruit, 3–5 mm. long; flowers green; berries reddish-purple then black, 5–7 mm. in diameter, 1–3-seeded.

Very common in open woodlands and thickets; SL–5000 feet; fl. and fr. most of the year. *A 12576! H 5564! P 17473! H & P 14458!* Cuba, Hispaniola.

20. DIOSCOREACEAE

Climbers (rarely otherwise) with thick tuberous or woody mostly subterranean stems. Leaves broad, usually alternate, rarely opposite, often cordate with digitate primary venation, often with pellucid lines. Flowers unisexual or rarely bisexual, actinomorphic, small, in spicate, racemose or paniculate inflorescences. Perianth 2-seriate, 6-lobed. Male flowers with 6 or 3 epipetalous stamens, the latter with or without 3 staminodes; filaments free or shortly connate; anthers 2-locular; pistillode present or absent. Female flowers often with staminodes; ovary inferior, 3-locular; styles 3, free or connate; ovules 2 (–4) in each loculus, superposed, axile. Fruit a 3-valved capsule or a berry. Seeds often winged; embryo small; endosperm present.

6 (or more) genera with over 600 species, widely distributed.

Fruit a 3-winged capsule; seeds winged	1. **Dioscorea**
Fruit 1-winged, indehiscent; seeds not winged	2. **Rajania**

1. DIOSCOREA L. (1753); Coursey (1967)

1 Leaves 3-foliolate, alternate; stamens 6   1. *dumetorum*
1 Leaves simple, the blade cordate at base:
  2 Leaf-blade 3–5-lobed; stem 4-angled and often winged   2. *trifida*
  2 Leaf-blade entire:
    3 Fertile stamens 3; staminodes present; leaves glabrous, alternate; plant without bulbils; stem not winged; common native species lacking edible tubers
                                                       3. *polygonoides*
    3 Fertile stamens 6; staminodes wanting; mostly cultivated plants with edible tubers or bulbils, occasionally relict from abandoned cultivations:
      4 Stem 2- or 4-winged; leaves usually opposite; bulbils sometimes developed
                                                          4. *alata*
      4 Stem not winged; leaves alternate or opposite:
        5 Leaf-blades long-pilose on veins beneath; stems unarmed; axillary bulbils sometimes developed   5. *pilosiuscula*
        5 Leaf-blades essentially glabrous:
          6 Plant developing axillary bulbils; stems normally spineless; leaves mostly alternate   6. *bulbifera*
          6 Plant without bulbils; stems often spiny; leaves opposite or alternate:
            7 Flesh of tuber usually yellow, with starch grains small and roughly triangular; stock woody and cormous above the tubers; tubers of short dormancy with shoot production almost continuous   7. *cayenensis*
            7 Flesh of tuber usually white, with starch grains large and ovoid; stock not markedly woody; tubers of definite long dormancy with shoot production strongly seasonal   8. *rotundata*

1. D. dumetorum (Kunth) Pax in Engl. & Prantl, Nat. Pflanzenf. **2** (5): 134 (1887).—*Helmia dumetorum* Kunth (1850). Bitter Yam, Cluster Yam.
A rare introduction (Trel.); 500–2000 feet; fl. July–Aug. *H 8782!* Also in tropical Africa.

2. D. trifida L. f., Suppl.: 427 (1781).—Indian Yam, Yampie.
Tubers clustered, mostly about 15–20 cm. long.
Cultivated (St. Andr., Manch., Trel.); 700–2500 feet; fl. Aug–Nov, fr. Sept–Dec. *Harris! P 24072!* Native of S. Amer., introduced into many West Indian islands.

3. D. polygonoides Humb. & Bonpl. ex Willd. in L., Sp. Pl. ed. 4, **4** (2): 795 (1806).—Bitter Jessie, Wild Yam.
Vine twining often to the tops of trees; leaves glossy, light green with numerous pellucid lines about twice as long as broad; perianth white, light green or yellow; anther-loculi distant; female flowers with 6 staminodes of which 3 bear rudimentary anthers; capsule oval, 2·5–3 cm. long.

Very common in the margins of woodland, in damp open thickets and shaded gullies on limestone; 500–3000 (–4000) feet; fl. and fr. all the year. *A 6506! H 12358! P 24177! Stearn 580!* C. and S. tropical Amer., West Indies.

4. D. alata L., Sp. Pl. **2**: 1033 (1753).—Greater Yam, Water Yam.

Common in cultivation; up to about 2500 feet; fl. Oct–Nov, fr. Nov–Dec. *H 6632! Robertson UCWI 3452!* Native of S. Asia, introduced generally throughout the tropics; with many varieties.

5. D. pilosiuscula Bert. ex Spreng. in L., Syst. Veg. ed. 16, **2**: 152 (1825).

Slender vine; leaves alternate, cordate with broad sinus, acuminate, 8–14 cm. long, 4–7·5 cm. broad; female flowers in long simple spikes; capsule oblong-elliptical, 2 cm. long, 1 cm. broad; seeds 8 mm. long, winged at base.

Very rare (Port.), not recently collected. *McNab! Purdie!* Hispaniola, Puerto Rico, Antigua to Tobago, Guyana.

6. D. bulbifera L., Sp. Pl. **2**: 1033 (1753).—*D. lutea* of Griseb. (1864), partly. Potato Yam.

Tubers small or wanting, hard and bitter; bulbils edible.

Uncommon (Port.). *Deans!* Native of the Old World tropics, now widespread by introduction.

7. D. cayenensis Lam., Encycl. Méth. Bot. **3**: 233 (1789).—*D. occidentalis* R. Knuth (1928). Yellow Yam.

Common in cultivation; 700–2300 feet; fl. Aug–Nov, fr. Jan. *H 9415! P 24073!* Native of tropical Africa, introduced into all the Greater Antilles and other islands.

8. D. rotundata Poir. in Lam., Encycl. Méth. Bot. Suppl. **3**: 139 (1813).—*D. sativa* of Harris (1906), not L. Guinea Yam, White Yam.

Cultivated (St. Andr.); about 700 feet; fl. Aug–Dec, fr. Nov. *Harris 117! 124! 125! 128!* Native of tropical Africa; widespread in the tropics and especially suited to areas with strongly seasonal climates. Some authors do not distinguish *D. rotundata* from *D. cayenensis* but I. H. Burkill annotated the four specimens cited above, which were grown at Hope Gardens and bore the cultivar names ' Lucea ', ' Mosella ', ' Negro ', and ' White Yam ' respectively, as belonging to *D. rotundata* on account of the white tuber-flesh, cf. also Wm. Harris in Bull. Dept. Agric. Jamaica 4: 3–6 (1906).

2. RAJANIA L. (1753)

Leaf-blade rotundate, shortly cuspidate-acuminate, the acumen about 1 cm. long
　　　　　　　　　　　　　　　　　　　　　　　　　　1. *cyclophylla*
Leaf-blade ovate, acuminate　　　　　　　　　　　　　　2. *cordata*

1. R. cyclophylla (Urb.) R. Knuth in Notizbl. Bot. Gart. Berlin **7**: 218 (1917).— *Dioscorea cyclophylla* Urb. (1909).

A rare vine (Trel.); 2000 feet; fl. Sept. *H 9402!* Endemic.

2. R. cordata L., Sp. Pl. **2**: 1032 (1753).—Himber, Wild Yam.

Vine twining to 10 m.; leaves with rather few indistinct pellucid lines 5–6 times as long as broad; flowers in compound racemes fascicled in leaf-axils or terminal; perianth yellow.

Local (Trel.), in thickets and woodlands on limestone; 1200–2000 feet; fl. Aug–Sept, fr. July–Sept. *H 8780! 9398! A & C 1201!* West Indies.

21. ARACEAE

Herbaceous or shrubby terrestrial, epiphytic or climbing perennial plants, rarely aquatic, often with viscid, acrid or milky sap. Leaves alternate, simple or compound, usually broad; petiole sheathing at base. Flowers small, bisexual or unisexual and then mostly monoecious, sometimes with a foetid odour; arranged in a simple spadix subtended by one usually large caducous or persistent sometimes brightly coloured spathaceous bract. Perianth usually absent in unisexual flowers, of 4–6 free or connate segments in bisexual flowers. Stamens (1–) 2, 4 or 8, opposite the perianth-segments; anthers free or united, mostly 2-locular, opening by slits or pores. Ovary superior or embedded in the spadix (inferior), 1–many-locular; ovules 1–many. Fruit a berry. Seeds mostly with endosperm.

About 120 genera with 1800 species, widely distributed but most numerous in the subtropics and tropics.

1 Flowers all bisexual; spadix without a barren terminal appendix; leaves mostly simple; plants without laticiferous ducts:
  2 Perianth of distinct segments; spathe persistent, sometimes coloured; terrestrial, epiphytic or scandent; leaves entire; ovules anatropous   1. **Anthurium**
  2 Perianth absent; spathe deciduous; robust climbers; leaves cut or perforated; ovules anatropous or amphitropous; ripe fruit edible; cultivated; native of Mexico, Costa Rica and Panama   *Monstera deliciosa* Liebm.
1 Flowers unisexual, monoecious, the pistillate below the staminate in the spadix; perianth absent or a simple ring:
  3 Spadix adnate to spathe; aquatic floating herbs with ribbed spongy simple leaves; laticiferous ducts absent; pistillate flower solitary below a group of fused stamens; ovules orthotropous   5. **Pistia**
  3 Spadix free from or only basally adnate to spathe; terrestrial or climbing plants; latex or mucilage ducts present:
    4 Climbers; spadix not appendaged:
      5 Ovaries separate; stamens free; leaves simple or partite with fine venation   2. **Philodendron**
      5 Ovaries connate; stamens synandrous; leaves compound or at least deeply lobed with coarse venation   3. **Syngonium**
    4 Stems short or erect, not climbing; stamens synandrous or otherwise united; leaf-blade entire:
      6 Staminodes present in pistillate flowers; lower pistillate portion of spadix adnate to spathe; spadix not appendaged; stems erect, plant shrubby; leaves not peltate, not hastate, blade tapered to petiole   4. **Dieffenbachia**
      6 Staminodes absent from pistillate flowers; spadix free, terminated by a barren appendage; stems short and usually tuberous:
        7 Leaves not peltate, hastate, clear green or purplish; stigma discoid, 3–4-lobed; ovules anatropous; cultivated and naturalized in wet places; native of tropical Amer.; Coco   *Xanthosoma sagittifolium* (L.) Schott
        7 Leaves peltate; stigma capitate, 3–5-grooved; ovules orthotropous:
          8 Leaves green, not spotted or variegated; cultivated and naturalized in damp sheltered places; native of tropical Asia; Dasheen   *Colocasia esculenta* (L.) Schott
          8 Leaves spotted or multivariegated; cultivated ornamentals occasionally naturalized; native of south tropical Amer.; Caladium   *Caladium bicolor* (Ait.) Vent.

1. ANTHURIUM Schott (1829)

1 Stem elongated with leaves well separated; leaf-blades elliptical, narrowed to both ends, 5–12 cm. long; peduncle slender, up to 3 cm. long; spadix 1–3 cm. long   1. *scandens*
1 Stem short and thick with leaves clustered; leaf-blades broadly ovate to deltate, cordate to subtruncate at base, larger; peduncle and spadix mostly 6 cm. or more long:
  2 Spadix usually 25 cm. or more long, tapered and sometimes sterile towards tip, longer than the usually recurved spathe:
    3 Leaf-tip acuminate; blade-base usually with a narrow U-shaped sinus   2. *cordifolium*
    3 Leaf-tip acute or obtuse, not acuminate but often shortly cuspidate; blade-base usually openly cordate   3. *grandifolium*

2 Spadix up to about 16 cm. long, oblong, slightly tapered, usually not much longer than the spathe, mostly about as long, sometimes shorter:
4 Spathe rigidly erect, 6–16 cm. long, about as long as spadix; leaf-blades up to 55 cm. long and 35 cm. broad, mostly openly cordate to truncate, dull green
                                                                    4. *mancuniense*
4 Spathe spreading or laxly erect, 5–6 cm. long, mostly shorter than spadix; leaf-blades up to about 18 cm. long and broad, narrowly cordate, glossy dark green
                                                                       5. *venosum*

1. A. scandens (Aubl.) Engl. in Mart., Fl. Bras. **3** (2): 78 (1878).—*Dracontium scandens* Aubl. (1775). *A. violaceum* (Sw.) Schott (1832).

Stem wiry and trailing with long fleshy roots; petiole slender, 1–2 (–4) cm. long; spathe green or purple; flowers green; berries white, ripening light mauve.

Common in moist shaded woodlands, epiphytic or on rocks; 750–6000 feet; fl. and fr. most of the year. *A 6548! H 6992! P 23167!* Mexico to Peru and Guyana, Greater Antilles, Trinidad.

2. A. cordifolium Kunth, Enum. Pl. **3**: 76 (1841).

Leaf-blade up to 50 cm. long and 40 cm. broad or larger; peduncle slender to 50 cm. long; spathe light green, erect or reflexed; inflorescence green at first, light brown later, mostly rather slender and often sterile at tip; ripe berries mauve.

Rather local (St. Andr., Port., Han.), epiphytic or on banks and limestone rocks in sheltered wet areas; 300–1750 (–5000) feet; fl. and fr. sporadically. *A 7600! A. M. Barry! Fawcett!* Lesser Antilles, Trinidad.

3. A. grandifolium (Jacq.) Kunth, Enum. Pl. **3**: 77 (1841).—*Pothos grandifolius* Jacq. (1790). Junction Root, Wild Coco.

Leaf-blade up 75 cm. long and 60 cm. broad or larger, stout-petioled; peduncle 40–60 cm. long; spathe green, often reflexed; inflorescence dark purplish-brown at flowering, rarely sterile distally; ripe berries dark reddish-purple.

Very common in woodlands and on sheltered banks, sometimes epiphytic; 100–4000 feet; fl. and fr. all the year. *A 6867! H 6962! P 23076!* Venezuela.

4. A. mancuniense Adams in Phytologia **21** (2): 65 (1971).

Petiole 12–30 cm. long; peduncle 9–30 cm. long; spathe ovate, olive-green tinged pinkish; inflorescence light brown, 7–16 cm. long.

Rare and local (Manch., Trel.), on limestone ledges and cliffs in woodland; 700–2000 feet; fl. sporadically but mostly Nov–Apr, fr. Jan. *A 6095! H 8833! P 9952! 22975!* Endemic.

5. A. venosum Griseb., Cat. Pl. Cub.: 219 (1866).

Petiole 20–40 cm. long; peduncle 12–23 (–32) cm. long; spathe lanceolate, green; inflorescence 4–7 cm. long.

Very local (St. James), on moist shaded limestone crags and cliffs; about 2100 feet; fl. Mar–May, Aug. *P 7977! Proctor & Stearn 11748!* Cuba.

Several ornamental species and hybrids of *Anthurium* are grown in gardens.

2. PHILODENDRON Schott (1829) nom. cons.

1 Leaf-blade entire:
2 Leaf-blade narrowly lanceolate, 10–18 cm. long, 3–5 cm. broad, minutely cordate; petiole narrowly winged along its whole length   1. *schottii*
2 Leaf-blade broadly ovate, up to 35 cm. long and 25 cm. broad, deeply cordate; petiole winged in the lower part or wingless   2. *scandens*
1 Leaf-blade not entire; petiole winged in lower part or wingless:
3 Leaf-blade divided nearly to the base into 3 lanceolate, acuminate lobes 18–25 cm. long, 6–8 cm. broad   3. *tripartitum*
3 Leaf-blade cut or lobulate usually about halfway to the midrib into 5–10 divisions on each side   4. *lacerum*

1. P. schottii C. Koch in Ender, Ind. Aroid.: 63 (1864).—*P. lingulatum* Schott (1856), not (L.) C. Koch (1856).

Stem sparingly branched, climbing to 6 m. by means of short adventitious roots;

leaf-blade rounded and minutely cordate at the base, acuminate; spathe 6–8 cm. long, white or light green; flowers green or dull purple.

Local (Manch., Han., Port.), on tree trunks and mossy stumps in woodland on limestone in wet areas; 1200–3000 feet; fl. Nov–Feb, July–Aug, fr. July. *A 11576! H 7471! P 23882!* Endemic.

2. P. scandens C. Koch & H. Sello in Ind. Sem. Hort. Berol., App.: 14 (1853).—
    P. oxycardium Schott (1856).

Branched scrambler or climber by means of short clustered adventitious roots, sometimes trailing on the ground or pendulous from tree branches when the leaves are smaller and the internodes longer; leaves usually green but sometimes variegated with light yellow streaks; spathe and spadix 10–12 cm. long; spathe persistent.

Locally abundant in the wetter parishes and frequently cultivated for ornament; 100–2500 feet; fl. Apr–May. *H 8311! P 10136!* Mexico to Guyana, West Indies.

Bunting (1968) discusses vegetative differences within the broader circumscription of *P. scandens* and provides the following key to subspecies with tentative statements of their distribution:

1 Petioles of leaves of juvenile shoots very short, about one-fourth as long as the non-glossy
    broadly ovate blades; Guianas, Peru, Trinidad   ssp. *prieurianum* (Schott) Bunting
1 Petioles of leaves of juvenile shoots elongate, mostly one-half as long as blades or longer:
    2 Leaves of juvenile shoots with a silky lustre; Mexico to Colombia   ssp. *scandens*
    2 Leaves of juvenile shoots glossy:
        3 Leaves subcircular-ovate, with numerous prominent veinlets between the primary
            lateral veins; Hispaniola, Puerto Rico, Lesser Antilles
                                                    ssp. *isertianum* (Schott) Bunting
        3 Leaves ovate, with only the primary lateral veins prominent; Mexico, Cuba, Jamaica
                                                    ssp. *oxycardium* (Schott) Bunting

3. P. tripartitum (Jacq.) Schott in Schott & Endl., Melet. Bot.: 19 (1832).—
    Arum tripartitum Jacq. (1797).

Branched climber with stems up to 3–4 m. long; spathe and spadix about 10 cm. long; spathe greenish outside, yellowish towards apex and margins, light yellow with buff lines within, purplish at base inside.

Local on trees and rocky banks in wet areas; 100–2250 feet; fl. July–Aug. *A 9191! H 8432! P 16569!* Guatemala to Venezuela.

4. P. lacerum (Jacq.) Schott in Schott & Endl., Melet. Bot.: 19 (1832).—*Arum lacerum* Jacq. (1804).

Branched climber or scrambler with thick stems up to 30 m. long; leaves of mature plants as much as 1 m. long; peduncles mostly several together, longer than the spathes; spathes 10–15 cm. long, cream-coloured tinged purplish near the base outside.

Common climbing on rocks or epiphytic in trees, sometimes terrestrial or in swamps; 50–2500 feet; fl. most of the year. *A 9768! H 6961! P 23999!* Cuba, Hispaniola.

Several ornamental species of *Philodendron* are grown in Jamaican gardens. A climbing ornamental with very large usually variegated leaves, frequently planted on trees, is *Rhaphidophora aurea* (Linden & André) Birdsey, native of the Solomon Is., long known in cultivation as *Pothos aureus* Linden & André.

3. SYNGONIUM Schott (1829)

1. S. auritum (L.) Schott in Schott & Endl., Melet. Bot.: 19 (1832).—*Arum auritum* L. (1759). Five Finger.

Leaves of juvenile plants simple to 3-fid, eventually large and up to 9-fid, but flowering branches often again with 3-foliolate leaves; large climber with thinly milky sap; bracts winged; inflorescences often paired; spathe often reddish inside, light green to yellow outside; spathe-limb and male portion of spadix deciduous in fruit.

Very common on trees, rocks and sheltered banks, sometimes trailing on the ground in wooded areas; SL–4600 feet; fl. and fr. most of the year. *A 6563! P 22562! Robertson UCWI 2770!* Hispaniola.

4. DIEFFENBACHIA Schott (1832)

1. D. seguine (Jacq.) Schott in Schott & Endl., Melet. Bot.: 20 (1832).—*Arum seguine* Jacq. (1760). Dumb Cane.

Stem stout, usually shortly creeping then ascending; erect flowering shoots up to 1 m. high; leaf-blade ovate to elliptical, up to 40 cm. long and 15 cm. broad; spathe 18–25 cm. long, waisted near the middle and overlapping in the lower half which remains fleshy and encloses the pistillate flowers; berries orange or red.

Cultivated and completely naturalized around pond margins, and in ditches and wet bottom pastures; SL–900 feet; fl. Jan–Apr, Aug–Sept. *A 9403! H 8261! P 15586!* Generally distributed through tropical Amer., apparently absent from Cuba. The name Money Plant is given to a cultivated variant with white-spotted leaves. This is sometimes referred to a distinct species, *D. maculata* (Lodd.) G. Don (*D. picta* Schott).

5. PISTIA L. (1753)

1. P. stratiotes L., Sp. Pl. **2**: 963 (1753).—*P. occidentalis* Blume (1836). Water Lettuce.

A rosette herb growing to about 15 cm. or slightly more in diameter, propagating mostly by vegetative buds formed at the ends of short horizontal stolons; leaves suborbicular to obcuneate, up to about 10 cm. long and 6 cm. broad, shallowly emarginate, light grey-green; spathe axillary, about 10 mm. long, whitish with ciliate margins; seeds oblong, brown, verrucose, 2 mm. long.

Locally abundant floating in ponds and sluggish streams; 50–1100 feet; fl. and fr. sporadically. *H 11828! P 21514!* Generally distributed throughout the subtropics and tropics.

22. LEMNACEAE

Small to minute aquatic, mostly floating herbs with simple threadlike roots or roots wanting. Plant-body flat or subglobose, branching and budding. Flowers monoecious; perianth absent. Male flowers with 1 or 2 stamens. Female flowers with a 1-locular sessile ovary; style and stigma simple; ovules 1–7. Seeds with or without endosperm; embryo straight.

3 genera with about 25 species, cosmopolitan in fresh water; the smallest flowering plants.

Plants with roots; male flowers in pairs; filaments slender at least at the base; anthers 2-locular   **1. Lemna**
Plants without roots; male flowers solitary; filaments short and thick; anthers 1-locular; plants minute   **2. Wolffia**

1. LEMNA L. (1753)

Unit of plant-body 3–7 mm. long, 3–6 mm. broad; roots several in a cluster   1. *polyrhiza*
Unit of plant-body 1·5–3 mm. long, 1–2 mm. broad; roots usually solitary   2. *perpusilla*

1. L. polyrhiza L., Sp. Pl. **2**: 970 (1753).—*Spirodela polyrhiza* (L.) Schleiden (1839).

Local and sporadic (St. Cath., St. Eliz., West.), in ditches and stagnant shallow pools in low-lying areas; 20–50 feet. *H & B 10510! P 3166!* Widespread in warm countries.

2. L. perpusilla Torr., Fl. State New York **2**: 245 (1843).—*L. cyclostasa* (Ell.) Schleiden (1839) in synonymy. *L. minor* of Griseb. (1864), not L.

Occasional in shallow ponds and ditches, floating or on mud; 25–1000 feet. *H 9982! 10837! P 8696!* Widespread in warm countries.

2. WOLFFIA Horkel ex Schleiden (1844) nom. cons.

1. W. punctata Griseb., Fl. Br. W. I.: 512 (1864).
Rather uncommon, floating on still ponds; 25–2500 feet. *H & P 15188! P 8657!*
United States and Caribbean area.

23. TYPHACEAE

Perennial rhizomatous gregarious herbs of swamps and marshes. Leaves mostly radical, erect, linear, often very long, spongy and mostly with a plano-convex section. Flowers unisexual, monoecious, small and numerous, densely crowded in a long unbranched terminal spadix with the staminate flowers above and the pistillate flowers below. Scape long, terete. Perianth of scale-like bristles. Male flowers with 2 or more fertile stamens; filaments free or variously connate; anthers basifixed. Female flowers with a stipitate 1-locular ovary narrowed above to a simple style; stigma narrow or ligulate, unilateral; ovule solitary, pendulous. Fruit a nutlet with a persistent style. Seeds with mealy endosperm; testa striate.
1 genus with about 15 species, widely distributed.

1. TYPHA L. (1753)

1. T. domingensis Pers., Synops. Pl. **2**: 532 (1807).—Reedmace.
Plants in flower 2–3·3 m. high; leaves up to 2·5 m. long, 3–20 mm. broad; spadix in flower pale brown, about 40 cm. long.
Local along riverbanks and in ditches, swamps and boggy pastures, usually in fresh water but sometimes in brackish; SL–50 (–3500) feet; fl. and fr. Dec–May. *A 8910! H 10068! J.P. 823! Stearn 131!* Pantropical.

24. CYCLANTHACEAE

Perrenial rhizomatous terrestrial, climbing or epiphytic herbs or shrubs. Leaf-blades lobed, parted or entire; petiole sheathing. Inflorescence an axillary peduncled simple spadix protected when young by 2 or more spathaceous bracts. Flowers monoecious, densely crowded either in spirally arranged groups, each group of 1 pistillate flower surrounded by 4 staminate flowers or the sexes in alternate cycles. Male flowers mostly with a cupular lobed perianth; stamens usually numerous; anthers 2- or 4-locular, opening lengthwise; filaments connate and often swollen at the base. Female flowers with perianth free or those of several flowers more or less confluent; tepals when present 4, free or connate; staminodes 4. Ovary 4-carpellate, 1-locular, free or embedded in the spadix; styles 4 or absent; placentation parietal or apical; ovules numerous, anatropous. Fruit a fleshy syncarp of separate or united berries. Seeds small; endosperm copious; embryo straight.
11 genera with 180 species in C. and S. tropical Amer. and the Lesser Antilles.

1. CARLUDOVICA Ruiz & Pav. (1794); Harling (1958)

Terrestrial shrubs; leaf-blades flabelliform-parted, the (3–) 4 (–5) segments wedge-shaped, their apical parts regularly toothed or lobed; petiole 3–5 times longer than the blade; staminate and pistillate flowers in spirally arranged groups; 3 species native from Mexico to Bolivia, 1 introduced into Jamaica.

1. C. palmata Ruiz & Pav., Syst. Veg. Fl. Peruv. & Chil. **1**: 291 (1798).—*C. jamaicensis* Lodd. ex Fawcett & Harris (1902). Jippi Jappa.
Leaf-blade 40–80 cm. long; petiole 1–3·5 m. long; peduncle 30–50 cm. long in flower, longer in fruit; spathes 3–4, 10–20 (–30) cm. long, 3–5 (–10) cm. broad;

spadix 9–20 cm. long; stamens 40–55; staminodes 3–6 (–10) cm. long; seeds 2–3 mm. long, 1–1·5 mm. broad.

Locally common (St. Andr., St. Cath., St. Mary), on steep wooded or cleared hillsides in sheltered places; 400–1500 feet; fl. Dec–Apr. *A 10095! P 22141! Yuncker 18297!* Native in C. and northern S. Amer.; also introduced into several West Indian islands, East Africa and some Pacific islands.

25. PALMAE (ARECACEAE)

By R. W. Read

Trees, shrubs, undershrubs or vines with thick fibrous roots, perennial, sometimes monocarpic. Stems slender to massive, strongly fibrous, solitary, tufted or soboliferous, rarely branched, ringed with leaf-scars and smooth waxy internodes or rough with persistent petiole-base fibres or fissured. Leaves spirally arranged, with sheathing tubular armed or unarmed deciduous or persistent petioles; leaf-blade plicate in vernation, simple and linear-elliptic to orbicular with crenate, serrate or entire margins and pinnate or palmate venation, or pinnately or rarely bipinnately compound, or palmately or costapalmately divided; leaflets and segments induplicate or reduplicate in vernation, linear-elliptic, falcate, sigmoid or cuneiform, glabrous or scaly, hairy, prickly or waxy at least during development; stipules absent. Flowers bisexual or unisexual, and then plants monoecious, dioecious or polygamous, actinomorphic, in terminal or axillary infra-, inter- or suprafoliar bracteate sessile or pedunculate panicles, spikes or heads, individually pedicellate or sessile, solitary or clustered in triads of 2 male on either side of one female or in acervulae of lines of flowers with one to several females and several to many males. Perianth biseriate (or in *Thrinax* and allies a single usually irregularly 6-lobed series); sepals 3, imbricate or rarely valvate, free or connate; petals 3, imbricate or valvate, often shortly connate, rarely adnate to sepals; sepals and petals persistent in fruit. Stamens 3 to 200 or more; filaments free, connate or epipetalous; anthers (1–) 2-locular, opening lengthwise. Unisexual flowers usually with pistillodes or staminodes respectively. Ovary superior, of 3 or rarely more uniovulate carpels, 1-carpelled in *Thrinax* and allies; carpels free or variously united; ovules generally anatropous; styles as many as carpels, free or fused or stigmas sessile. Fruit small to very large, 1–3 or rarely more-seeded, dry and fibrous or baccate, with smooth, scaly, prickly, hairy or warty exocarp, fibrous or fleshy mesocarp and papery to bony endocarp. Seed with solid, hollow, perforated, homogeneous or ruminate mealy endosperm and a subapical, lateral or basal embryo.

About 236 genera with 2650 species, distributed in nearly all tropical and sub-tropical regions and extending to some warm temperate areas.

1 Plants armed either with needlelike prickles or toothlike thorns; leaves pinnate:
  2 Petiole with short toothlike thorns on the margin; inflorescence compact and nearly hidden among the leaf-bases; cultivated and escaped; native of tropical Africa; African Oil Palm   *Elaeis guineensis* Jacq.
  2 Petiole, sheath and leaf-rachis with scattered needlelike prickles:
    3 Trunk solitary, stout, often ventricose; abaxial surface of leaflets hirsute; large inflorescence-bract 1 m. or more long, with few large prickles   **1. Acrocomia**
    3 Trunks caespitose, slender, never ventricose; abaxial surface of leaflets glabrescent; large inflorescence-bract less than 50 cm. long, densely covered with large prickles
                                                 2. Bactris
1 Plants unarmed:
  4 Leaves pinnately divided:
    5 Leaflets cuneiform; leaf bipinnately divided; plants monocarpic; inflorescence shortly pedunculate; male flowers with numerous stamens; fruit several-seeded, containing an irritant; cultivated for ornament; native of Indo-Malaysia; Fish-tail Palm
                                               Caryota urens L.
    5 Leaflets linear-elliptic; leaf pinnately divided; plants polycarpic; fruit usually 1-seeded:
      6 Stems caespitose; leaf and leaflets arching strongly; inflorescence interfoliar, long-pedunculate; fruit ovoid, orange-yellow; cultivated ornamental; native of Madagascar; Ostrich Feather Palm   *Chrysalidocarpus lutescens* (Bory) H. Wendl.
      6 Stem solitary:

7 Leaflets with more than one large conspicuous nerve; inflorescence long-pedunculate; flowers sunk in pits; leaflets disposed in a single plane
4. **Calyptronoma**
7 Leaflets with a single large conspicuous nerve; inflorescence shortly pedunculate or
sessile; flowers not sunk in pits:
8 Leaf-blade with all leaflets lying in the same plane; inflorescence interfoliar; large
inner bract strongly grooved; seed very large; cultivated and naturalized; native
of the Old World tropics; Coconut   *Cocos nucifera* L.
8 Leaf-blade multifarious (leaflets disposed in different planes); inflorescences
infrafoliar at the base of the prominent green crownshaft; large inner bract not
grooved; seed small   3. **Roystonea**
4 Leaves palmate or costapalmate; fruit usually 1-seeded:
9 Leaves costapalmate, the costa continuous with the petiole and extending through the
length of the blade and strongly recurved; inflorescence large and much-branched
5. **Sabal**
9 Leaves palmate, lacking a prominent costa; inflorescence slender, subdivided into
5–15 pendent primary branches arising from an erect or arching rachis:
10 Leaf-sheath split below the insertion of the petiole; fruit white; seed smooth,
perforated   6. **Thrinax**
10 Leaf-sheath not split below the insertion of the petiole, the tubular portion having
the appearance of a finely woven net; fruit purple-black; seed cerebriform
7. **Coccothrinax**

1. ACROCOMIA Mart. (1824)

1. A. spinosa (Mill.) H. E. Moore in Gent. Herb. **9**: 238 (1963).—*Palma spinosa*
Mill. (1768). *Bactris globosa* Gaertn. (1788). *Cocos fusiformis* Sw. (1797). *A.
fusiformis* (Sw.) Sweet (1827). Maccafat.
Tree 5–10 m. high; trunk 40–50 cm. in diameter at 1 m. above ground, enlarged
irregularly above to 70–80 cm. in diameter, ringed with stiff dark prickles; leaves
multifarious, 2·5–3·5 m. long, with scattered prickles over the surface of the sheath,
petiole and rachis; inflorescence interfoliar, 1–1·5 m. long; larger bract tomentose
and somewhat prickly outside, smooth and light yellowish-green inside; flowers
light to golden yellow; fruit greenish-yellow to brown, 3·5–4·3 cm. in diameter;
seed 2·5–3 cm. in diameter, covered with a fibrous mucilaginous pulp.
Common as scattered individuals or in colonies along gullies and hedgerows
throughout the island; 50–1250 feet; fl. May–July, fr. Mar–May. *P 9322! R 1661!*
Endemic.

2. BACTRIS Jacq. (1763)

1. B. jamaicana L. H. Bailey in Gent. Herb. **4**: 177 (1938).—*Palma gracilis* Mill.
(1768), not *B. gracilis* J. B. Rodrigues (1875). Prickly Pole.
Stems several together; trunks 5–10 m. high, about 15 cm. in diameter, ringed
with leaf-scars and covered with sharp needlelike prickles up to 5 cm. long; leaves
2–3-farious; leaflet-apex oblique, midrib with small prickles, one margin with
conspicuously larger prickles than the other; inflorescence inter- to infrafoliar,
about 50 cm. long; flowers yellow; fruit depressed subtriangular-globose, hard, red,
12–13 mm. in diameter.
Very uncommon but widely distributed in damp savannas with an occasional
plant in moist forest at higher elevations; 1000–2250 feet; fl. July–Aug, fr. Sept.
P 10498! R 1692! 1703! Endemic.

3. ROYSTONEA O. F. Cook (1900)

Petals 3–3·2 mm. long, 1·5 mm. broad; persistent perianth at fruiting short, almost flat;
lowlands   1. *princeps*
Petals 4–5·5 mm. long, 2–2·5 mm. broad; persistent perianth at fruiting cupular, about
2 mm. deep; submontane   2. *altissima*

1. R. princeps (Beccari) Burret in Engl., Bot. Jahrb. **63**: 76 (1929).—*Oreodoxa
princeps* Beccari (1912). Morass Royal, Swamp Cabbage.
Tree 5–15 m. or more high; trunk smooth but banded with leaf-scars, irregularly
enlarged below, more slender above; leaves multifarious, 4–6 m. long; leaf-sheath

green, tubular, forming an erect crownshaft 1·5 m. or more long; flowers whitish; stamens 6, lavender to purple; fruit pyriform with the neck curved to one side, about 15 mm. long, greenish-yellow to purple.

Uncommon and rather local, restricted to the western parishes, in small colonies or as scattered individuals on poorly drained lowlands and in morass; SL–500 feet; fl. throughout the year but mainly Dec–Jan, fr. throughout the year. *P 18453! R 1570!* Endemic.

2. R. altissima (Mill.) H. E. Moore in Gent. Herb. **9**: 239 (1963).—*Palma altissima* Mill. (1768). *R. jamaicana* L. H. Bailey (1935). Mountain Cabbage.

Tree 6–20 m. or more high; trunk appearing slender, strongly and deeply ringed with leaf-scars; leaves multifarious, 4–5 m. long; leaf-sheath as above but much shorter in tall trees; flowers whitish; fruit as above, purple.

Common, on hill tops and mountain slopes in the central and eastern parishes; SL–2500 feet; fl. and fr. throughout the year. *H & P 13530! R 1678!* Endemic.

These two species are very similar and the possibility that they are ecological variants cannot be ruled out; further study is required.

4. CALYPTRONOMA Griseb. (1864)

1. C. occidentalis (Sw.) H. E. Moore in Gent. Herb. **9**: 252 (1963).—*Elaeis occidentalis* Sw. (1797). *Calyptrogyne occidentalis* (Sw.) Gómez Maza (1889). *Calyptronoma swartzii* Griseb. (1864). Long Thatch.

Tree up to 7 m. high; leaves 3–4 m. long, the youngest developing leaf green or red to maroon in colour; inflorescence interfoliar; lower bract ancipitous, upper bract fusiform, caducous, inserted near the base of the peduncle and after falling leaving a collar 6–9 cm. long; axes and petals of male flowers usually green but sometimes bright red; fruits about 13 mm. long, purple.

Common in damp woodlands and locally in lowland swamps; SL–2250 feet; fl. and fr. throughout the year. *H & P 14551! P 8581! R 1610! 1693!* Endemic.

5. SABAL Adans. (1763)

1. S. jamaicensis Beccari in Fedde, Repert. Sp. Nov. **6**: 94 (1909).—*S. umbraculifera* of Griseb. (1864), not Mart. (1837). Big Thatch, Bull Thatch.

Tree up to 10 m. or more high; trunk very heavy, 30–50 cm. in diameter, brown, fibrous; leaves 4 m. or more long, the blade about 2·5 m. long from hastula to apex, strongly plicate with the lateral halves folded downwards; petiole-base split; segments deeply bifid with a single long filament arising in each sinus; hastula about 15 cm. long, sagittate; flowers white; anthers yellow; fruit about 10 mm. in diameter, dark brown to black; seed smooth, shiny brown, depressed on one side.

Common, but populations disjunct or individuals scattered; SL–1700 feet; fl. Jan–July, fr. Jan–Apr. *P 11110! R 1649!* Endemic.

6. THRINAX Sw. (1788)

1 Leaves of mature plants densely covered abaxially with white or silvery overlapping and interlocking hyaline fimbriate scales, of younger plants similar scales less densely disposed; petioles persistently tan-velutinous abaxially; inflorescence-branches and flowers pink to purple at anthesis; primary bracts densely covered with rufous appressed scales   1. *excelsa*
1 Leaves of mature plants concolorous or very slightly lighter green beneath, scales when present widely dispersed; petioles soon glabrescent abaxially; inflorescence-branches and flowers white, ivory or yellowish at anthesis; primary bracts not rufous-scaly:
　2 Leaf-segments broadest well beyond the palman, of mature plants concolorous, glabrescent, scales when present minute (difficult to see with the aid of an 8 × lens); sheath long-linguiform; inflorescence-branches and flowers ivory to yellowish at anthesis, the axes densely granulose-puberulous (if not then distal pedicels very short)   2. *parviflora*
　2 Leaf-segments broadest at point of fusion, of mature plants concolorous or very slightly lighter green beneath, with scattered fimbriate scales each having a conspicuous translucent central portion; sheath deeply V-form; inflorescence and flowers white at anthesis, the axes glabrous   3. *multiflora*

1. T. excelsa Lodd. ex Griseb., Fl. Br. W.I.: 515 (1864).—*T. rex* Britton & Harris (1910). Broad Thatch.

Tree 3–11 m. high; trunk 12·5–16 (–20) cm. in diameter; leaf-blade (2–) 2·3–3·5 m. in diameter with the palman 45–100 cm. broad; hastula 3·4–3·9 cm. broad; drupe white, 8–11 mm. in diameter.

Locally common (Port., St. Thom.), in humid woodlands on craggy exposed limestone; 900–2500 feet; fl. Mar–Apr, fr. July–Oct. *H & B 10759! P 10483! R 1625! 1682!* Endemic.

2. T. parviflora Sw., Nov. Gen. & Sp. Pl.: 57 (1788).—*T. tessellata* Beccari (1907). *T. harrisiana* Beccari (1908). Broom Thatch, Thatch Pole.

Tree 1–10 (–13) m. high; trunk 5–14 (–15) cm. in diameter; leaf-blade 70–140 cm. in diameter; drupe white or ivory-coloured, often tessellated as a result of fungal activity, 6·5–7·5 mm. in diameter; fruit-pedicels (0·1–) 0·5–2·8 (–5·2) mm. long.

Very common in the central and western parishes, on well drained limestone; SL–3000 feet; fl. Apr–Sept at lower elevations, Dec–Feb at the higher, fr. sporadically. *H 3180! 10264! H & B 10524! P 11065! 25675! R 1676! 1681!* Endemic. At lower altitudes plants are often found with short thick trunks, straight leaf-segments and puberulous pedicels apparently contrasting with plants at higher altitudes having tall slender trunks, curled leaf-segments and glabrous pedicels. These combinations of characters are however not mutually exclusive and in some localities plants of the two types grow in close proximity; phenotypic differences are thus probably controlled by both genetical and ecological factors.

3. T. multiflora Mart., Hist. Nat. Palm. **3**: 255, t. 103 (1838).—*T. excelsa* of L. H. Bailey (1938), not Griseb. (1864). Bullhead Thatch, Sea Thatch.

Tree (1·5–) 2–10 (–12) m. high; trunk 8–13 cm. in diameter; leaf-blade 120–160 cm. in diameter; drupe white, smooth, about 7–8 mm. in diameter.

Common in thickets and exposed areas on coastal limestone; SL–250 feet; fl. Mar–June, fr. June–Aug. *H 9957! P 10568! R 1960!* Florida, Mexico, British Honduras, Bahamas, Cuba, Hispaniola.

7. COCCOTHRINAX Sarg. (1899)

1. C. jamaicensis Read in Principes **10** (4): 133–141 (1966).—*Thrinax argentea* of Griseb. (1864), partly. *C. fragrans* of L. H. Bailey (1939), not Burret (1929). Silver Thatch.

Tree 2–8 m. high; trunk 5–20 cm. in diameter; leaf-blade 80–140 cm. in diameter, the abaxial surface grey to silvery, covered with persistent hyaline hairs and scales; inflorescence 70–90 cm. long; flowers fragrant, white, on pedicels (1–) 2–6·2 mm. long; stamens 7–15; drupe purple-black, juicy, (7–) 7·5–9·5 mm. in diameter; seed brown with cerebriform surface markings.

Common mainly in limestone areas along the south coast; SL–1000 feet; fl. May, July, Nov. fr. Oct–Nov. *Proctor & Mullings 21975! R 1665! 1667!* Endemic.

26. PANDANACEAE

Trees or erect or climbing shrubs, often with stout prop roots. Leaves spirally arranged, linear, sessile, sheathing at the base, leathery, often spinulose on the margin and keel. Flowers dioecious in panicles or spadices, enclosed at first by spathaceous or leafy bracts. Perianth rudimentary or absent. Staminate flowers densely packed and scarcely distinguishable; stamens numerous; filaments free or connate; anthers erect, basifixed, 2-locular or again divided, opening lengthwise. Pistillate flowers with or without hypogynous staminodes or staminodes adnate to the ovary; ovary superior, 1-locular, free or confluent with adjacent ovaries into bundles with separate or united stigmas; style short or absent; ovules anatropous, solitary to many, basal or parietal. Fruit a syncarp, oblong or globose, the mature carpels woody, drupaceous or baccate. Seeds small with endosperm.

3 genera with about 900 species in the Old World tropics, particularly in the Pacific Is.

1. PANDANUS S. Parkinson (1773)

Inflorescence spicate or capitate with sessile flowers; pistillate flowers without staminodes; ovules solitary in each loculus; trees at maturity; Screw Pine.

Leaves up to 8 cm. broad with red marginal prickles; leaf-tip attenuated; sparingly cultivated as an ornamental; native of Madagascar   *P. utilis* Bory

Leaves 15 cm. or more broad with marginal prickles not red; leaf-tip abruptly narrowed to a slender cirrhus; plants mostly innovating profusely from the lower axils; cultivated and naturalized on beaches in wet areas; native of Indonesia   *P. dubius* Spreng.

27. AMARYLLIDACEAE

Perennial herbs with usually bulbous or rarely rhizomatous stocks. Leaves few, basal, entire, usually rather soft. Flowers bisexual, actinomorphic or subregular, often showy, solitary or umbellate and subtended by spathaceous bracts at the top of a naked simple scape. Perianth petaloid of 6 segments in 2 similar series often united below into a tube. Stamens 6, opposite the perianth-segments, hypogynous or epipetalous; filaments free or connate, often curved upwards in laterally directed flowers; anthers 2-locular, basi- or medifixed, opening lengthwise or rarely by terminal pores. Ovary superior or more usually inferior, usually 3-locular with axile placentas; style slender; ovules mostly numerous, anatropous. Fruit a capsule or fleshy berry. Seeds with fleshy endosperm and small embryo.

About 90 genera with over 1000 species widespread in warm temperate regions, less numerous in the tropics.

1 Ovary superior; fruit a loculicidal capsule:
  2 Rootstock rhizomatous; herbs cultivated in the mountains of eastern Jamaica with showy blue or rarely white flowers; native of S. Africa
                                      Agapanthus africanus (L.) Hoffmgg.
  2 Rootstock a tunicated bulb; flowers not blue:
    3 Perianth-segments free or nearly so; plants strongly odorous   **1. Allium**
    3 Perianth-segments distinctly united at base; plants not odorous   **2. Nothoscordum**
1 Ovary inferior; fruit a capsule or berry; rootstock usually bulbous:
  4 Corona absent:
    5 Ovules few; flowers red, small, 50 or more in a globose umbel; fruit a spherical berry ripening red; scape and petioles speckled purple; native of the savannas of tropical Africa, introduced as a garden ornamental; Powder-puff Lily
                                      Haemanthus multiflorus Martyn
    5 Ovules numerous; flowers in each inflorescence 1–few, rarely as many as 30, showy; perianth-tube distinct:
      6 Flowers several together in a showy umbel; stigma small, capitate; fruit a membranous capsule opening irregularly or by disintegration, with large fleshy greenish angular seeds   **3. Crinum**
      6 Flowers solitary or rarely paired; stigma lobed; fruit a 3-valved capsule with flat black seeds   **4. Zephyranthes**
  4 Corona present:
    7 Corona of small fimbriate scales independent of the stamen-filaments; perianth red; leaves linear; ovules numerous; fruit a 3-valved capsule with flat black seeds
                                         5. Hippeastrum
    7 Corona formed by the expanded filaments; perianth white:
      8 Leaves linear-oblong to elliptic-lanceolate; perianth-lobes linear; filaments long; fruit a membranous capsule with few large fleshy oval light green seeds; pedicels short or obsolete   **6. Hymenocallis**
      8 Leaves broader; filaments short; pedicels distinct:
        9 Corona notched but not toothed between the filaments; perianth-lobes ovate; lamina tapered to the petiole   **7. Eucharis**
        9 Corona toothed, deeply divided between the filaments; perianth-lobes elliptic; lamina broader than long, cordate at the junction with petiole, very strongly veined, cuspidate; cultivated ornamental; native of S.E. Asia and tropical Australia   *Eurycles amboinensis* (L.) Lindl.

1. ALLIUM L. (1753)

A. cepa L. (Onion), *A. fistulosum* L. (Escallion) and *A. sativum* L. (Garlic) are extensively cultivated. *A. schoenoprasum* L. (Chives) is also grown in gardens.

2. NOTHOSCORDUM Kunth (1843) nom. cons.

1. N. inodorum (Ait.) Nicholson, Ill. Dict. Gard. **2**: 457 (1885).—*Allium inodorum* Ait. (1789). *N. fragrans* (Vent.) Kunth (1843).

Leaves narrow, glaucous, often twisted; flowering scape to 60 cm. high; perianth-segments 9–12 mm. long, white with a purple mid-line and tip; fragrant; capsule up to about 1 cm. long; seeds angular, black.

Locally common (St. Andr.), a weed of open waste ground and roadside banks; 3000–5000 feet; fl. and fr. most of the year. *A 6434! H 12662! P 23686!* Native of southern subtropical Amer. and Mexico; introduced into Europe, N. Africa, Canary Is., Mauritius. *N. inodorum* is very close to *N. bivalve* (L.) Britton, the latter being usually a smaller plant with narrower leaves and shorter fruits. If taken together the name *N. bivalve* has priority and the range is extended to include the northern subtropical region of America.

3. CRINUM L. (1753)

1 Pedicels 3–5 cm. long; flowers 20 or more in each inflorescence; perianth-tube slightly
  curved, 7–10 (–12) cm. long, deep crimson; perianth-lobes 8–16 (–19) cm. long, about
  2·5 cm. broad, deep crimson outside, white with red flush inside   **1. amabile**
1 Pedicels very short or absent; flowers fewer than 12 in each inflorescence:
  2 Perianth-tube straight, 12–20 cm. long; perianth-lobes 6·5–8 cm. long, 12–15 mm.
    broad, white; filaments spreading uniformly; leaves straight   **2. americanum**
  2 Perianth-tube curved, 7–15 cm. long; perianth-lobes 7–9 cm. long, about 35 mm. broad,
    white with a broad crimson stripe; filaments declinate; leaves often undulate
                                            3. zeylanicum

1. C. amabile Donn, ex Ker-Gawl. in Curt., Bot. Mag. **39**: t. 1605 (1813).

A very large plant with leaves up to 150 cm. long and 19 cm. broad; scape to 120 cm. long, more or less two-edged; flowers fragrant.

Rare (St. Andr., St. Cath.), in sheltered waste places; 2000–3000 feet; fl. sporadically. *Fawcett!* Native of tropical Asia.

2. C. americanum L., Sp. Pl. **1**: 292 (1753).—*C. roozenianum* O'Brien (1891).
  Seven Sisters (Cayman).

Flowers sessile, (1–) 5–12 in each inflorescence; perianth-tube green or purplish; filaments purplish; seeds green.

Locally abundant (St. Eliz.), in wet grassy places at low elevations; fl. and fr. sporadically. *Distin! P 24528! Strickland! Robertson UCWI 25826!* Throughout subtropical and tropical Amer., West Indies; Grand Cayman.

3. C. zeylanicum (L.) L., Syst. Nat. ed. 12, **2**: 236 (1767).—*Amaryllis zeylanica* L. (1753).

Scape 40–80 cm. long, not sharply two-edged; flowers very shortly stalked, 3–10 (–12) in each inflorescence, open during one evening and night; style and filaments bright pink.

Locally common in undisturbed waste places, lawns and citrus plantations; 500–1000 feet; fl. Aug–Oct and sporadically. *Robertson UCWI 5530! Skelding UCWI 3324!* Native of tropical Asia, widely introduced and naturalized in Africa and Amer.

Several other species of *Crinum* have been introduced into Jamaican gardens.

4. ZEPHYRANTHES Herbert (1821) nom. cons.

1 Lobes of stigma short and thick, up to 1 mm. long; anthers 7–9 mm. long; perianth
  yellow, 3–4 cm. long   **1. citrina**
1 Lobes of stigma distinct, recurved, 2–3 mm. long:

2 Anthers 5–7 mm. long; perianth white, often tinged dull purple outside, 3·5–5 cm. long
2. *tubispatha*

2 Anthers 12–18 mm. long; perianth mostly pink, the tube green at base, up to 8 cm. long
3. *carinata*

1. Z. citrina Bak. in Curt., Bot. Mag. **108**: t. 6605 (1882).—Crocus.
Bulbous herb with very narrow leaves; capsules often ripening and the plant propagating by seed.
Occasionally escaping from gardens; 200–2300 feet; fl. sporadically. *P 23648!* Cultivated in many warm countries; Grand Cayman.

2. Z. tubispatha (L'Hérit.) Herbert in Edw., Bot. Regist. **7** App.: 36 (1821).—
Amaryllis tubispatha L'Hérit. (1789). Rain Flower, Wind Flower.
Bulbous herb with narrow often dark green purplish-tinged leaves.
Mostly in gardens but occasionally escaped and established in grassy places; fl. sporadically. *H 10810!* Cultivated in many warm countries.

3. Z. carinata Herbert in Curt., Bot. Mag. **52**: t. 2594 (1825).—*Atamosco carinata* (Herbert) P. Wilson (1924).
Like the last but usually larger in all its parts; stigma sometimes 4-lobed.
Escaped and naturalized especially on heavy soils in the central parishes, along roadside banks; 600–3000 feet; fl. sporadically. *H 9987! P 22897! Stearn 907! Wynter UCWI 3474!* C. Amer., West Indies. A small-flowered variant occurring in Grand Cayman and Cuba has been named *Z. rosea* Lindl. This and other introduced species have been reported from Jamaica and may have become naturalized.

5. HIPPEASTRUM Herbert (1821) nom. cons.

1. H. puniceum (Lam.) Kuntze, Revis. Gen. Pl. **2**: 703 (1891), (err. *purpureum*).—
Amaryllis punicea Lam. (1783). *H. equestre* (Ait.) Herbert (1821). Easter Lily.
Scape terete, 30–60 cm. long; umbel 2–4 flowered; pedicels up to 7 cm. long; perianth 8–10 cm. long, greenish at base, the segments light red.
Cultivated and naturalized along roadsides and in thickets near habitations; 1000–2450 (–3500) feet; fl. Mar–June, Dec. *P 11624! Wynter UCWI 3104!* C. and S. Amer., West Indies; Cayman Is.; introduced into most tropical countries.
Some ornamental varieties of *Hippeastrum*, usually with large dark red flowers, are cultivated in hill gardens.

6. HYMENOCALLIS Salisb. (1812); Sealy (1954)

1 Leaf-blade elliptical, narrowed to base into a more or less distinct petiole usually much less than one-sixth the greatest width of blade:
  2 Perianth-tube much longer than segments; flowers subsessile; leaf-tip cuspidate-acuminate; cultivated ornamental; native of Trinidad and northern S. Amer.
H. tubiflora Salisb.
  2 Perianth-tube as long as or shorter than segments; flowers shortly stalked; leaf-tip acute
1. *fragrans*
1 Leaves oblanceolate to linear-lanceolate, not distinctly petiolate, narrowest portion rarely less than one-fifth the greatest width of blade:
  3 Leaf-blade rounded at apex; perianth-tube shorter than segments   3. *arenicola*
  3 Leaf-blade obtuse to acute or shortly acuminate at apex; flowers sessile or pedicellate, the pedicel elongating in fruit:
    4 Perianth-tube shorter than segments; leaf-blade oblong, up to 12 cm. broad, narrowed to both ends, the tip usually acute; pedicel obsolete to up to 2 cm. long in fruit
2. *caribaea*
    4 Perianth-tube as long as or longer than segments; leaf-blade strap-shaped, oblanceolate to linear; flowers mostly sessile:
      5 Leaf-blade 4·5–10 cm. broad, acute to obtuse at tip   4. *latifolia*
      5 Leaf-blade not over 4 cm. broad:
        6 Stamen-filaments 2–3 cm. long; leaf-tip obtuse; cultivated; native of subtropical N. Amer.   *H. rotata* (Ker-Gawl.) Herbert
        6 Stamen-filaments 4–6 cm. long; leaf-tipe acute; cultivated widely in tropical countries; native of continental tropical Amer.; Spider Lily
H. littoralis (Jacq.) Salisb.

1. H. fragrans (Salisb.) Salisb. in Trans. Hort. Soc. **1**: 340 (1812).—*Pancratium fragrans* Salisb. (1794).

Perianth-tube 8–10 cm. long; anthers orange; stigma green; seeds obliquely ovoid with shallow longitudinal depressions, 2·9–3·2 cm. long, 2–2·2 cm. broad with the hilum a shallow pit near one end; flowers very sweet-scented.

Rather local (St. Cath., Clar.), on limestone rocks in thickets and open woodlands; SL–2700 feet; fl. Mar, fl. and fr. June–July. *A 7289! H 10225! P 19786!* Barbados.

2. H. caribaea (L.) Herbert in Edw., Bot. Regist. **7** App.: 44 (1821).—*Pancratium caribaeum* L. (1753).

Leaves up to 50 cm. long and 12 cm. broad; scape to 35 cm. long; perianth-tube 5–10 cm. long; perianth-lobes 8–12 cm. long.

Local (Clar., St. Eliz., St. Ann), on limestone rocks in thickets; 2000–2200 feet; fl. July–Mar. *H 9986! Prior!* Mexico, Cuba, Hispaniola, Antigua, Martinique, Barbados. *H. caribaea* and *H. fragrans* may not be distinct. Differences in the shape of the leaf-bases may depend on the size of the bulb. In those plants of *Hymenocallis* in which the pedicel is short, thick and almost indistinguishable from the ovary in the flowering stage, there is a tendency as fruit develops for the pedicel to elongate and become slender so that a definite fruit-stalk is formed which may even allow the ripe fruit to become pendulous.

3. H. arenicola Northrop in Mem. Torr. Bot. Club **12**: 29 (1902).

An obscure species, the Jamaican plants thus identified not being clearly distinct from *H. caribaea*.

In arid rocky coastal woodlands; 25–400 feet; fl. Mar–Sept. *H 10225!* Bahamas.

4. H. latifolia (Mill.) M. J. Roem., Syn. Monogr. **4**: 168 (1847).—*Pancratium latifolium* Mill. (1768). *H. caymanensis* Herbert (1837).

Peduncle up to 60 cm. long; perianth at first green, the lobes becoming white, fragrant.

Locally common along upper beaches and in pastures on coral limestone near the sea; 5–20 (–450) feet; fl. and fr. sporadically. *A 8025! P 16453! Stearn!* Florida, Cuba, Hispaniola, Cayman Is.

7. EUCHARIS Planch. (1853)

1. E. grandiflora Planch. & Linden in Fl. des Serres **9**: 255 (1854).—Eucharist Lily.

Leaves broadly ovate-elliptical, acuminate; scape 40–60 cm. long, terete; flowers stalked below the ovary, fragrant; perianth-limb widely spreading, 7–10 cm. across.

Cultivated and escaped (St. Andr., St. Ann, St. Thom.), along roadsides and in hillside thickets; 500–2000 feet; fl. Nov–Jan. *Skelding UCWI 2799!* Probably native in the northern Andes; Hispaniola, Puerto Rico.

28. AGAVACEAE

Large erect shrubby plants with short massive stems. Leaves usually crowded in a tight spiral, often thick, fleshy and fibrous, linear or lanceolate, entire or with prickly margins. Flowers bisexual, actinomorphic, in racemes or panicles, never in scapose umbels. Perianth-segments 6, more or less uniform, petaloid. Stamens 6, free; anthers 2-locular, opening lengthwise, introrse. Ovary inferior, 3-locular with axile placentas; ovules in 2 rows on each axile placenta, anatropous. Fruit a loculicidal capsule. Seeds with fleshy endosperm; embryo small.

7 genera with about 350 species mostly in tropical America.

1 Stamens exserted, protandrous; style slender; flowers erect   **1. Agave**
1 Stamens included:
  2 Filaments and style thickened below the middle, maturing simultaneously; anthers dorsifixed; flowers pendulous, white or light yellow   **2. Furcraea**
  2 Filaments and style not thickened; anthers basifixed; native of Australia, only in cultivation   *Doryanthes palmeri* W. Hill

1. AGAVE L. (1753); Trelease (1913)

1 Flowers green; leaves straight; pedicels 5–10 mm. long; flowers 40–50 mm. long
1. sisalana

1 Flowers yellow to orange; leaves curved:
  2 Pedicels rarely exceeding 10 mm. long; flowers (35–) 45–60 mm. long:
    3 Ovary 25–30 mm. long, distinctly longer than the perianth
2. harrisii

    3 Ovary 15–20 (–25) mm. long, from slightly shorter to about the same length as the perianth
3. sobolifera

  2 Pedicels 20 mm. long; flowers 60–70 mm. long
4. longipes

1. A. sisalana Perrine, House Rep. Documen. **564**: 8 (1838).—Sisal.

Leaves linear-lanceolate, glaucous grey-green, the margins smooth or armed, spine-tipped, up to about 150 cm. long and 10–12 cm. broad; inflorescence up to about 8 m. high with spreading branches, the flowers 2 or 3 together on the numerous branches of each panicle; stamens epipetalous, inserted at the level of perianth fusion; filaments red spotted, linear, tapered to the connective, 7·5–8 cm. long; anthers versatile, 2·5–2·7 cm. long, withering early; style eventually 7–8 cm. long, the stigma entire until the anthers have withered, then dividing into 3 short lobes and exuding a drop of clear viscous fluid; capsule (rarely formed) oblong, about 6 cm. long; propagation mainly by bulbils formed in the axils of bracteoles below the flowers.

Cultivated, mostly in Clarendon, and escaped, established in many arid areas; SL–700 feet; fl. sporadically. *H 9644! Robertson UWI 3323! Stearn 736!* Indigenous in Yucatan; cultivated in many arid tropical countries.

2. A. harrisii Trelease in Mem. Nat. Acad. Sci. Washington **11**: 34 (1913).

Leaves up to 200 cm. long and 35 cm. broad; inflorescence 8–10 m. high with branches about 60 cm. long; perianth yellow, about 20 mm. long; filaments 30 mm. long.

Occasional, mostly on arid coastal limestone; 10–50 (–2000) feet; fl. Mar–Apr. *H 10934! P 6514! Stearn 652!* Endemic.

3. A. sobolifera Salm-Dyck, Hort. Dyck. **8**: 307 (1834).—*A. morrisii* Bak. (1887). Coratoe, May Pole.

Leaves up to 200 cm. long and 24 cm. broad, 9 cm. thick near the base; inflorescence 5–9 m. high; perianth golden-yellow to light orange.

Locally abundant on well-drained hillsides; 100–2800 feet; fl. Feb–Apr, fr. Apr–May. *H 9643! 10156! P 11873! Stearn 424!* Cayman Brac.

4. A. longipes Trelease in Mem. Nat. Acad. Sci. Washington **11**: 36 (1913).

Plant similar to the preceding but flowers larger; filaments to 5 cm. long.

Local (St. Andr.), on well drained slopes; 3000–3500 feet; fl. Mar. *H 10933!* Endemic.

2. FURCRAEA Vent. (1793); J. R. Drummond (1907)

1. F. hexapetala (Jacq.) Urb., Symb. Ant. **4**: 152 (1903).—*Agave hexapetala* Jacq. (1760). *F. cubensis* Haw. (1812).

Leaves up to 150 cm. long, straight, usually with upwardly hooked marginal prickles; inflorescence up to 7 (–10) m. high with numerous drooping lateral branches to 2 m. long; flowers solitary or clustered on the branches, pendulous; ovary about 2 cm. long; perianth-segments free, elliptical, about 3 cm. long; fruit pendulous, oblong-ovoid, 3-lobed-sulcate, beaked with persistent withered perianth, about 4·5 cm. long and nearly 4 cm. broad, rarely formed in Jamaica; seeds numerous, flat, horizontally disposed in 2 rows in each loculus; glossy ovoid dark green bulbils develop in the inflorescence and fall to grow into new plants.

Common along roadsides and in waste places; 700–2500 (–4100) feet; fl. mostly Sept–Feb, fr. Feb. *J.P. 607A! Stearn 117! Yuncker 17939!* Greater Antilles.

Plants identified as *F. hexapetala* (Jacq.) Urb. (*F. cubensis* Haw.), *F. gigantea* Vent., *F. macrophylla* Bak. and *F. tuberosa* (Mill.) Ait. f. have been reported from Jamaica. Drummond (1907) composed a key to these species, separating them on the basis of minor differences of leaf-shape and the size, shape and spacing of the marginal prickles. The shape of the leaves depends to some extent on the age of the plant and in some variants prickles are entirely absent. Moreover subsequently published diagnoses of these species overlap Drummond's distinctions. The type of *Agave hexapetala* Jacq. consists of a single flower, but it is obviously the common Greater Antillean plant. Pending further field study I am adopting the name *F. hexapetala* for the widespread representative of this genus in Jamaica. *F. tuberosa* is described by Britton and Wilson as having elongated bulbils.

29. HYPOXIDACEAE

Perennial herbs with tuberous rhizomes or corms. Leaves mostly radical, prominently veined, often hairy. Flowers bisexual, actinomorphic. Perianth-tube very short or absent or consolidated into a long beak on top of the ovary; segments 6, spreading, more or less uniform. Stamens 6 or 3, inserted at the bases of the segments; anthers 2-locular, opening lengthwise. Ovary inferior, 3-locular; styles 1 or 3; ovules numerous in each loculus in 2 series on axile placentas. Fruit a capsule or fleshy and indehiscent. Seeds usually small, black; embryo enclosed by copious endosperm.

5 genera with about 140 species mainly in the S. Hemisphere, but our genera with widely distributed species.

Ovary produced into a long beak below the perianth; leaves more or less ribbed or plaited, peduncles shorter than the flowers; fruit fleshy; seeds smooth and shiny, oblong, about 2·3 mm. long and 1·5 mm. broad.   1. **Curculigo**
Ovary not beaked below the perianth; leaves smooth, flat and sedgelike; peduncles longer than the flowers; fruit a capsule.   2. **Hypoxis**

1. CURCULIGO Gaertn. (1788)

1. C. scorzonerifolia (Lam.) Bak. in Journ. Linn. Soc., Bot. **17**: 124 (1878).—*Hypoxis scorzonerifolia* Lam. (1789).

Underground stem fleshy, cylindrical, upright; leaves up to about 1·5 cm. broad; flowers appearing singly from short axillary inflorescences near ground-level; outer perianth-lobes yellow, 1·1–1·2 cm. long, 3 mm. broad with a subterminal adaxial reflexed ascending-barbellate hair; inner perianth-lobes yellow, 1·1–1·2 cm. long, 2–2·5 mm. broad; stamens shorter than perianth with unequal filaments; seeds with a very prominent peglike chalaza remnant.

Very rare (St. Andr.), in grassy places on heavy clay soils; 800–1000 feet; fl. May, Nov. *A 9416!* C. and S. tropical Amer., West Indies.

2. HYPOXIS L. (1759)

Peduncles 2- or more-flowered; leaves more than 2 mm. broad   1. *decumbens*
Peduncles 1-flowered; leaves mostly less than 2 mm. broad   2. *wrightii*

1. H. decumbens L., Syst. Nat. ed. 10, **2**: 986 (1759).—Star of Bethlehem.

Underground tuberous stem upright; leaves up to 40 cm. long and 1 cm. broad; peduncles axillary, up to 15 (–20) cm. long in fruit; outer perianth greenish, hairy; inner perianth tinged greenish abaxially, yellow within, glabrous, the segments 4–9 mm. long; capsule 1–2·5 cm. long, tipped with the subpersistent perianth turned green; seeds subspherical, papillose, about 1 mm. in diameter.

Common on damp grassy pathside and roadside banks, sometimes a weed of pastures on heavy soils; 150–7400 feet; fl. and fr. most of the year. *A 10646! H 8589! P 16750! Stearn 413!* C. and S. tropical Amer., West Indies.

2. H. wrightii (Bak.) Brackett in Rhodora 25: 140, f. 11 (1923).—*H. juncea* Sm. var. *wrightii* Bak. (1878).

Underground tuberous stem cylindrical or ovoid, 5–10 mm. long, up to 7 mm. thick, covered with the fibrous remains of old leaves; leaves 5–18 (–25) cm. long, 1·5–2 (–3) mm. broad, thinly cottony-hairy; peduncles 3–5 cm. long in flower, longer in fruit, very slender; perianth about 8 mm. across, light yellow, the outer members hispid externally, the inner hispid in midline only; ovary pubescent; seeds oblong, minutely muriculate, 1 mm. long.

Very rare (Clar.), among sedges in moist savanna; 2100–2300 feet; fl. Mar, Dec, fr. Dec. *A & D 13078! P 26303!* Florida, Bahamas, Cuba, Hispaniola.

30. HAEMODORACEAE

Herbs with tuberous or rhizomatous stems. Leaves mostly basal and sheathing mostly linear, often hairy. Flowers bisexual, actinomorphic or nearly so, in cymes, racemes or panicles. Perianth 1- or 2-seriate, sometimes the members free but usually tubular, 6-lobed, mostly persistent. Stamens 3 or 6; filaments free; anthers 2-locular, basi- or medifixed, opening lengthwise. Ovary superior to inferior, 3-locular; style usually filiform; ovules 1–many on axile placentas. Fruit a 3-valved loculicidal capsule. Seeds with copious endosperm and a small embryo.

22 genera with about 120 species in the S. Hemisphere and tropical and N. Amer.

1. XIPHIDIUM Aubl. (1775)

Perennial stoloniferous herbs with glabrous strongly distichous leaves with closed sheathing bases; flowers in 2 rows on racemose inflorescence-branches; perianth 2-seriate, segments free and more or less uniform; stamens 3, opposite the inner perianth-segments; ovary superior; ovules numerous in each loculus; 1 species.

1. X. caeruleum Aubl., Hist. Pl. Guiane 1: 33, t. 11 (1775).

Rhizome-branches ascending; shoots to 1 m. in flower, flattened with the leaves distichous; leaves up to 60 cm. long and 6 cm. broad, serrulate near the tip; perianth-segments white, the outer boat-shaped, 6·5 mm. long, about 2·8 mm. broad, greenish at base, the inner nearly flat, 7–7·5 mm. long, about 3·5 mm. broad; fruit a 3-lobed capsule, rarely formed, the plants propagating by slender soboles or deciduous plantlets formed in the axils of bracts at the base of the inflorescence.

Cultivated in gardens and escaped on roadside banks, mostly in damp sheltered places; 400–750 feet; fl. Aug–Jan, May, fr. Apr–May. *A 5574! H 10652! P 7285!* C. and S. tropical Amer., West Indies; introduced and cultivated in Malaysia.

31. IRIDACEAE

Perennial herbs with rhizomes, corms or bulbs. Leaves mostly linear, flat. Flowers bisexual, actinomorphic or zygomorphic, usually in racemes or panicles, rarely solitary. Perianth 2-seriate with 6 petaloid segments. Stamens 3, opposite the outer perianth-segments; anthers 2-locular, opening lengthwise. Ovary inferior, 3-locular with axile placentas or 1-locular with 3 parietal placentas; style slender, 3-lobed in the upper part; ovules usually numerous, anatropous. Capsule loculicidally dehiscent. Seeds with endosperm; embryo small.

70 genera with 1500 species, widely distributed.

1 Perianth with a distinct tube:
  2 Perianth actinomorphic:
    3 Flowers blue, short-lived; perianth-tube short, lobes spreading; stem a rhizome; stamens regular   **1. Aristea**

   3 Flowers usually white or red, never blue, showy and not soon withering; perianth-tube
        longer than the campanulate limb; stem a corm; stamens unilaterally bowed;
        cultivated only; natives of S. Africa   *Watsonia* spp.
   2 Perianth zygomorphic or the limb oblique and the tube more or less curved, with the
        stamens usually more or less together on one side; roots from a corm or bulb:
      4 Spathes emarginate; perianth-tube straight or nearly so; plants stoloniferous
                                                                       2. Crocosmia
      4 Spathes lanceolate; perianth-tube curved; plants without stolons   **3. Gladiolus**
   1 Perianth-segments free or only shortly united:
      5 Stem a rhizome; style-branches divided; filaments free:
        6 Style-branches winged and more or less petaloid; inner perianth white, glabrous;
            inflorescence straight   **4. Dietes**
        6 Style-branches not obviously winged; inner perianth coloured, yellow with brown
            spots basally, outwardly curled and purple-marked distally, pilose within; peduncle
            sharply bent near the base   **5. Neomarica**
      5 Stem a bulb or corm:
        7 Filaments free:
          8 Style-branches divided, not produced beyond the stigmas; capsule exserted on an
              elongated pedicel   **6. Trimezia**
          8 Style-branches undivided, subulate; style short; capsule enclosed by the spath-
              aceous bract   **7. Eleutherine**
        7 Filaments connate; style-branches deeply divided; outer perianth-segments larger,
            usually red distally, spotted proximally; cultivated ornamental; native of Mexico
                                                         Tigridia pavonia (L.f.) Ker-Gawl.

1. ARISTEA Ait. (1789); Weimarck (1940)

Capsule about 6 mm. long, ovoid; seeds angular; perianth-segments 6–9 mm. long,
3–4 mm. broad   **1.** *gerrardii*
Capsule 1·2–2·2 cm. long, sharply angled and transversely sulcate; seeds discoid
                                                                     2. *ecklonii*

1. A. gerrardii Weimarck in Lunds Univ. Årsskr. N.F. Avd. 2, **36** (1): 17 (1940).
—Blue Iris.
   Perennial herb 40–90 cm. high in flower, with flabellately arranged linear leaves;
inflorescence 30–60 (–100) cm. high, branched.
   Locally abundant (St. Andr., Port., St. Thom.), a weed of pathside and roadside
banks and open pastures; 2500–7400 feet; fl. and fr. most of the year. *A 6367!*
H 10505! P 23269! Native of S. Africa.

2. A. ecklonii Bak. in Journ. Linn. Soc., Bot. **16**: 112 (1877).
   Like the last with slightly smaller flowers.
   Very local (St. Andr.), along pathsides; 4500–5000 feet; fl. Jan–Feb, July, fr.
Jan–Apr, July. *A 8776! Martin 14008!* Native of S. Africa and some of the
mountains of tropical Africa.

2. CROCOSMIA Planch. (1851–52)

1. C. × crocosmiflora (Lemoine) N.E. Br. in Trans. Roy. Soc. S. Afr. **20:** 264
(1932).—*Montbretia crocosmiflora* Lemoine (1881). Montbretia.
   Perennial cormous herb with long linear leaves; inflorescence 50–100 cm. high,
branched or simple; perianth bright orange-red; filaments, anthers, style and
stigmas yellow; capsules rarely formed.
   Cultivated and escaped (St. Andr., Port.), gregarious and locally abundant in
cool damp areas; 3400–4000 feet; fl. June–Sept, Jan, Apr. *ACH 370! 421! Robertson
UWI 1776!* This plant is of garden origin being a hybrid between *C. aurea* (Pappe)
Planch., native of Rhodesia, and *C. pottsii* (Bak.) N.E. Br., native of Natal. It has
been widely introduced and has become naturalized in many warm temperate
regions.

3. GLADIOLUS L. (1753)

1. G. cuspidatus Jacq., Ic. Pl. Rar. **2:** 5, t. 257 (1795).
   Erect herb with narrow leaves from a discoid corm, 30–40 cm. high in flower;

perianth greenish-cream sometimes tinged faintly pink, with the tips of the lobes curled; anthers greyish-purple; the plant multiplies by numerous small adventitious buds from the corm.

Rare (St. Andr.), along pathsides and in margins of rough pastures; 4500–5300 feet; fl. July. *A 7409! H 10504!* Native of S. Africa. Several ornamental species and varieties are grown in Jamaican gardens.

4. DIETES Salisb. ex Klatt (1866)

1. D. vegeta (L.) N.E. Br. in Journ. Linn. Soc., Bot. **48**: 36 (1928).—*Moraea vegeta* L. (1762).

Perennial rhizomatous herb with flabellately arranged dark green leaves 30–60 cm. long; inflorescence slender, few-flowered, innovating successively from sub-terminal bracts and sometimes reaching 3 m. in length; outer perianth 3 (–2)-membered, broadly clawed, white spotted yellow with brown spots towards base, densely pubescent on the centre ridge of each segment, about 6 cm. across; inner perianth 3 (–2)-membered, white, glabrous; stamens 3 (–2), free; filaments 8 mm. long, broader below; style-arms 12 mm. long, 6 mm. broad, bifid, mauve; capsule 2–3 (–3·5) cm. long; seeds black, more or less tetrahedral, about 3 mm. long and 4 mm. broad.

Common locally (St. Andr.), gregarious on rocky roadside banks and in low thickets; 3500–5000 feet; fl. and fr. most of the year. *A 6228! Robertson UWI 3054!* Native of S. and S. tropical Africa.

5. NEOMARICA Sprague (1928)

Outer perianth mostly white; plant 50–70 cm. high in flower 1. *northiana*
Outer perianth mostly rich bluish-purple; plant 90–130 cm. high in flower 2. *caerulea*

1. N. northiana (Schneevoogt) Sprague in Kew Bull. **1928**: 280 (1928).—*Moraea northiana* Schneevoogt (1795).

Perennial herb with branched rhizome; leaves flabellate, flaccid; scape flattened and continued above the inflorescence as a leaf-like bract; outer perianth-segments white distally, yellow with brown spots proximally; inner perianth-segments marked blue-purple distally with 2 yellow spots near the mid-line, yellow spotted brown proximally.

Cultivated and escaped locally (St. Andr., Port.), on sheltered rocky banks and limestone cliffs; 1250–5000 feet; fl. Mar–May. *A 10914!* From Nicaragua to Peru and Brazil.

2. N. caerulea (Ker-Gawl.) Sprague in Kew Bull. **1928**: 280 (1928).—*Marica caerulea* Ker-Gawl (1823).

Perennial herb like the last but larger in all its parts; rhizome-branches short with approximated shoots; leaves erect; flowers showy, fragrant, very short-lived; inner perianth-segments rich bluish-purple on distal margins, white with purple lines between, yellow on proximal margins with transverse brown marks across the base.

Cultivated and escaped locally (St. Andr., Clar., Han.), along pathside and roadside banks and in woodland near habitations; (600–) 1000–2000 feet; fl. Sept–Nov, fr. Oct–Dec. *A 7975! H 11233!* Also in Brazil and Paraguay and introduced into S. Tomé and elsewhere in tropical Africa.

6. TRIMEZIA Salisb. ex Herbert (1844)

1. T. martinicensis (Jacq.) Herbert in Edw., Bot. Regist. **30** Misc.: 88 (1844).—*Iris martinicensis* Jacq. (1760). *Cipura martinicensis* (Jacq.) Kunth (1816). Wild Scallion.

Erect herb from a small corm, 20–50 cm. high in flower; scape slender; flowers subglobular 2–2·5 cm. in diameter; perianth-segments yellow, often tinged greyish-blue, brown at base; seeds black, irregularly angled, about 1·5 mm. long and broad.

Common in damp grassy places mostly on heavy soils; 150–5000 feet; fl. and fr. all the year. *A 6371! H 12018! Stearn 153!* Mexico to Brazil, Puerto Rico, Lesser Antilles, Trinidad, Java.

7. ELEUTHERINE Herbert (1843) nom. cons.

1. **E. bulbosa** (Mill.) Urb. in Fedde, Repert. Sp. Nov. **15**: 305 (1918).—*Sisyrinchium bulbosum* Mill. (1768). *E. plicata* (Sw.) Herbert ex Klatt (1862). Alum Root.

Erect herb; leaves plaited with distinct lateral longitudinal veins; inflorescence-branches unequal; bracts striate, 1–1·5 cm. long; perianth white.

Originally cultivated now naturalized (St. Andr., West., Trel.), in grassy waste places; 50–1700 feet; fl. Apr–June, Sept, Dec. *Harris! R 1734!* Tropics of Asia and C. & S. Amer., West Indies.

32. BURMANNIACEAE

Small delicate saprophytic herbs. Leaves mostly reduced to scales. Flowers bisexual, solitary, cymose or racemose. Perianth tubular, usually 6-lobed; tube sometimes winged or angled. Stamens 3, sessile or subsessile within the perianth-tube; anthers 2-locular, the loculi sometimes diverging and independently stipitate, dehiscing transversely. Ovary inferior; style included in the perianth, shortly 3-lobed; ovules numerous, minute. Fruit a capsule, usually crowned by the withered perianth, often 3-winged, usually dehiscing longitudinally. Seeds small with scanty endosperm.

11 genera with about 100 species in the tropics and subtropics.

1 Ovary 3-locular with axile placentas; perianth-segments unequal, persistent; capsule 3-angled; flowers (in Jamaican species) in a tight subcapitate raceme   1. **Burmannia**
1 Ovary 1-locular with 3 parietal placentas:
   2 Perianth wholly persistent, tube not winged or angled; inflorescence more or less elongated, laxly racemiform or reduced to 1 flower; flowers and stems bluish-purple
                   2. **Apteria**
   2 Perianth-limb deciduous, circumscissile, the lower part forming a beak or cupule on the capsule; whole plant white or pale yellow:
     3 Capsule compressed and oblique, more or less winged-keeled on one side; flowers solitary or subracemose   3. **Cymbocarpa**
     3 Capsule terete; inflorescence cymose   4. **Gymnosiphon**

1. BURMANNIA L. (1753)

1. **B. capitata** (J. F. Gmel.) Mart., Nov. Gen. **1**: 12 (1824).—*Vogelia capitata* J. F. Gmel. (1791).

Annual herb 5–25 cm. high with a few subulate leaves mostly towards the base of the simple filiform stem; flower-heads about 1 cm. long and broad; flowers white or light yellow; perianth 3–5 mm. long.

Rare and local (Clar., St. Eliz.), among tufts of small sedge and grass in swampy places; 50–2400 feet; fl. Nov–Dec. *H 12231! 12850!* From Texas and N. Carolina to Paraguay; Greater Antilles.

2. APTERIA Nutt. (1834)

1. **A. aphylla** (Nutt.) Barnh. in Small, Fl. Southeast U.S.: 309, 1329 (1903).—*Lobelia aphylla* Nutt. (1822).

Stem slender, 10–25 cm. high, simple or branched at or above ground-level, bluish-purple above and white below ground-level; flowers 10–13 mm. long; outer perianth light bluish-purple with white transverse bands; inner perianth dark bluish-purple; fruit including base of perianth white, obovoid, flat on top with persistent style.

1a. Var. aphylla
Rare (Port.), in deep leaf litter on clay soil over limestone, in mature forest; 1400–3000 feet; fl. and fr. Aug–Oct. *A 11570! H 9493!* Generally distributed from S. United States to Brazil, West Indies.

1b. A. aphylla var. **hymenanthera** (Miq.) Jonker in Pulle, Fl. Surin. **1** (1): 186 (1938).—*Apteria hymenanthera* Miq. (1850).
Like the last but generally smaller in all its parts; stem 3–16 cm. long, simple or branched at or below ground-level, purple or mauve; leaves 2–3 mm. long; flowers nodding, 8–10 mm. long; perianth pale mauve to purple.
Local (Clar., Port.), in habitats similar to the last or in fern tussocks in open savanna swamp; 1000–2400 feet; fl. and fr. Dec–Aug. *A 5948! H 10704! P 15887!* Within the range of the last but tending to be restricted to the more northerly parts.

3. CYMBOCARPA Miers (1840)

1. C. refracta Miers in Proc. Linn. Soc. **1**: 62 (1840).
Slender totally milky-white herb 6–20 cm. high; capsule opening along the adaxial margin.
Rare and local (St. Andr., Port.), growing among exposed fibrous tree-roots, decaying leaves and rotten wood in forest; 3000–4500 feet; fl. Aug–Dec, fr. Nov. *H 9152! P 9335!* Brazil, Cuba, Hispaniola.

4. GYMNOSIPHON Blume (1827)

1 Stigmas without long appendages, entire; perianth 2–3 mm. long; tube-remnant 1–1·5 mm. long; capsule globose   1. *sphaerocarpus*
1 Stigmas with long filiform appendages; perianth 3·5–5 mm. long; tube-remnant 2–2·5 mm. long; capsule obovoid:
   2 Stigmas sessile or nearly so   2. *fawcettii*
   2 Stigmas on short style-branches   3. *germainii*

1. G. sphaerocarpus Urb., Symb. Ant. **3**: 442 (1903).
Plant 5–12 cm. high, the basal part of the stem trailing, the flowering part more or less erect.
Rare (Port.), among decaying leaves in wet forest on limestone; 900–1600 feet; fl. Mar, Aug, fr. Mar. *A 7901! H 10705!* Puerto Rico, Guadeloupe, Dominica.

2. G. fawcettii Urb., Symb. Ant. **5**: 294 (1907).
Like the last; capsule obovate, 2·3 mm. long.
Local (Trel., Port.), among decaying leaves in wet forest on limestone; 850–2500 feet; fl. Aug–Oct, fr. Oct. *A 7894! H 9494, 12596!* Endemic.

3. G. germainii Urb., Symb. Ant. **3**: 444 (1903).—*G. jamaicensis* Urb. (1907).
Plant 8–18 cm. high; capsule 1·5 mm. in diameter.
Local (Trel., Port.), in deep humus in forest shade; 1250–2500 feet; fl. and fr. Jan, Mar, Aug, fr. Nov. *A 9264! H 10659! P 5264!* Hispaniola, Puerto Rico, Guadeloupe, Dominica.

The Jamaican representatives of *Gymnosiphon* are difficult to distinguish and further study of living plants is needed.

33. ORCHIDACEAE

Herbs, minute to shrubby, mostly perennial with fibrous, fleshy or tuberous roots. Leaves alternate, spiral or distichous, very rarely opposite, entire-margined, often scalelike on corms or rhizomes, entire, emarginate or toothed at the apex, persistent or deciduous, without stipules. Flowers mostly bisexual, zygomorphic, in bracteate spikes racemes or panicles, sometimes solitary. Perianth 2-seriate of 6 segments; the 3 outer usually similar (sepals), valvate or imbricate, mostly free or the lateral connate, or the median variously fused to the lateral petals; the 2 lateral inner

alike (petals), often resembling the sepals; the median inner (lip or labellum)
unlike any other segment and usually strongly modified in association with ento-
mophily, primitively adaxial but frequently becoming abaxial (resupinated) before
anthesis by rotation of the whole perianth through twisting of the ovary or pedicel;
median sepal or lateral sepals together or lip sometimes forming a chin, sac or
nectariferous spur. Stamens 2 or 1 (in Jamaican species), united with the style
into a column bearing the anther(s) terminally and the stigmas subterminally, the
anther-sac (clinandrium) usually separated from the viscid stigmatic surface by a
beaked outgrowth, the rostellum; anthers 2-locular; pollen granular or in 2, 4, 6
or 8 mealy, waxy or bony masses (pollinia), attached by sterile stalks (caudicles) to
viscid glands (viscidia). Ovary inferior, 1-locular with 3 parietal placentas or rarely
3-locular with axile placentas; ovules very numerous, minute, anatropous. Fruit
usually a capsule opening lengthwise. Seeds very numerous, minute, often fusiform,
without endosperm.

Perhaps the largest family of flowering plants with over 600 genera and 20,000
species, predominantly terrestrial in drier or cooler regions and epiphytic in the
tropics.

1 Plants leafy at flowering time:
  2 Leaves plicate with several keels or strongly prominent longitudinal nerves:
    3 Pseudobulbs or corms present; leaves mostly basal or tufted:
      4 Inflorescence terminal; plants terrestrial or on rocks:
        5 Flowers purple; pseudobulbs elongate-ovoid   **24. Liparis**
        5 Flowers white; pseudobulbs globose   **17. Govenia**
      4 Inflorescence lateral:
        6 Corms of several internodes; plants terrestrial:
          7 Pollinia 4; lip chinned between the lateral sepals; perianth green with lip purple
            18. Eulophia
          7 Pollinia 8:
            8 Lip continuous with the base of the column, spurred, 2-keeled but not crested;
              sepals and petals white outside, brown within (discolorous)   **19. Phaius**
            8 Lip jointed to the base of the column, not spurred, 5- or more-crested; sepals and
              petals concolorous, pink or purple, rarely white   **20. Bletia**
        6 Pseudobulbs of 1 internode; plants epiphytic or on rocks:
          9 Flowers in racemes, cream or light yellow:
            10 Leaves elliptic-lanceolate; pollinia 4   **28. Xylobium**
            10 Leaves linear-lanceolate; pollinia 8   **22. Coelia**
          9 Flowers solitary on each peduncle, green   **27. Lycaste**
    3 Pseudobulbs or corms absent; leaves spread along the stems:
      11 Flowers paniculate; bracts small; pollinia 2, pollen granular; terrestrial plants with
        fibrous roots:
        12 Lip linear from the base; sepals cohering into a tube; column long
            2. Corymborkis
        12 Lip saccate at the base; lateral sepals connate into a short chin; column short
            3. Tropidia
      11 Flowers in racemes, spikes or heads; bracts conspicuous; pollinia 8, waxy; roots
        fleshy:
        13 Plants terrestrial; flowers in racemes   **21. Calanthe**
        13 Plants epiphytic; flowers in tight spikes or heads   **26. Elleanthus**
  2 Leaves more or less flat, with a single midrib or keel, or subterete:
    14 Growth monopodial; roots arising at intervals from a creeping, climbing or trailing,
      often branched stem; inflorescence short, axillary; pseudobulbs absent:
      15 Climbing fleshy-stemmed vines usually with ovate or large oblong leaves; fruit
        fleshy   **59. Vanilla**
      15 Trailing herbs with short or narrow leaves; stems not fleshy; fruit typically a
        capsule:
        16 Inflorescence a short often 2-ranked spike; lip shortly spurred; pollinia 2
          62. Campylocentrum
        16 Inflorescence not spicate; flowers stalked, not spurred; pollinia 4:
          17 Flowers solitary in the axils of short 2-ranked leaves; stems compressed
            50. Dichaea
          17 Flowers clustered in the axils of linear-lanceolate leaves; stem more or less terete
            31. Neourbania
    14 Growth sympodial, each shoot of limited growth followed by other shoots from
      axillary, often basal, buds, or the shoots simple and unbranched:
      18 Plants without pseudobulbs:
        19 Terrestrial herbs, growing in leaf-litter or soil, rarely on rocks; leaves membranous,
          rarely fleshy; roots often fleshy and hairy:

20 Anther-loculi distant; one root usually tuberous; leaves basal or on the flower-scape; flowers white or green; lip spurred; pollinia granular   1. **Habenaria**
20 Anther-loculi approximate or united:
  21 Pollinia granular, breaking up on removal, 2 or 4:
    22 Leaves spread along the stem; stem a rhizome or stoloniferous; plants mostly of soggy leafmould in shady woodlands:
      23 Flowers in terminal spikes or racemes:
        24 Lip spurred; perianth white; capsule ellipsoid   4. **Erythrodes**
        24 Lip not spurred; perianth greenish, tinged yellow and purple; capsule erect; leaves rigid, purple beneath   14. **Psilochilus**
      23 Flowers usually solitary in the upper axils; perianth pink; capsule ellipsoid; plant stoloniferous with tuberoid roots; leaves soft   15. **Triphora**
    22 Leaves mostly basal:
      25 Lip abaxial, lowermost in the open flower; median sepal joined to the petals:
        26 Lateral sepals spreading, fused below to form a pointed spur curving free from the ovary at the tip and enclosing the elongated base of the lip   5. **Pelexia**
        26 Lateral sepals directed forwards, free or fused only to form an adnate spur or a blunt chin not or only shortly free from the ovary and not enclosing the elongated base of the lip   6. **Spiranthes**
      25 Lip adaxial, uppermost in the open flower; petals free at least at the tip:
        27 Lateral sepals forming a long spur   7. **Pseudocentrum**
        27 Lateral sepals not forming a spur:
          28 Sepals connate into a cup or tube; lip and petals adnate to the sepal cup; lip deeply concave and hooded, auricled; flowers tightly spicate   8. **Prescottia**
          28 Sepals free:
            29 Lip inserted at the base of the column but embracing it; petals free from the column   9. **Cranichis**
            29 Lip and petals raised on the column   10. **Ponthieva**
  21 Pollinia waxy, not breaking up on removal, 4; small corms resulting from previous growth usually present:
    30 Column long, curved; anther terminal, deciduous   24. **Liparis**
    30 Column very short; anther enclosed, erect, persistent   25. **Malaxis**
19 Plants growing on trees (epiphytic) or on rocks (epipetric):
  31 Leaves flat or sharply keeled beneath:
    32 Inflorescence basal or flowers axillary:
      33 Flowers solitary; column with a foot:
        34 Leaves membranous rather flat:
          35 Perianth up to 1 cm. broad; leaves many in one plane on each shoot   50. **Dichaea**
          35 Perianth more than 2 cm. broad; leaves few, oblanceolate   29. **Cochleanthes**
        34 Leaves leathery, triquetrous   30. **Maxillaria**
      33 Flowers in racemes or panicles; column without a foot:
        36 Inflorescence racemose or paniculate; pollinia 2; lateral lobes of lip, if present, flat; column with petaloid wings   57. **Oncidium**
        36 Inflorescence racemose; pollinia 4; lateral lobes of lip curved upwards; column not winged   58. **Cryptarrhena**
    32 Inflorescence terminal:
      37 Each shoot with 2 or more leaves:
        38 Lip adnate to the column   41. **Epidendrum**
        38 Lip not adnate to the column:
          39 Flowers capitate; lip adaxial; bracteoles conspicuous, ciliate; inflorescence pendulous leaves silvery, shiny   13. **Eurystyles**
          39 Flowers in spikes, racemes or panicles:
            40 Leaves triquetrous with sharp margins and keel; lateral sepals connate; column with petaloid wings   57. **Oncidium**
            40 Leaves not triquetrous:
              41 Leaves linear, less than 6 mm. broad:
                42 Flowers in a 1-sided raceme; perianth pink or purple; pollinia 4   43. **Isochilus**
                42 Flowers in a 2-ranked spike; perianth white or light yellow; pollinia 8   26. **Elleanthus**
              41 Leaves 8 mm. or more broad; inflorescence a raceme or panicle:
                43 Leaves longitudinally ribbed; flowers in a diffuse panicle; perianth streaked mauve; sepals spurred   51. **Ionopsis**
                43 Leaves not ribbed:
                  44 Lip uppermost; flowers rather tight in unilateral raceme(s); perianth cream or yellow   33. **Polystachya**

44 Lip lowermost; flowers in open racemes or rarely the raceme branched; leaves sometimes serrulate towards the tip; perianth white

44. Octadesmia

37 Each shoot with 1 leaf:
  45 Leaf linear:
    46 Inflorescence shorter than the leaf, 1- or more-flowered:
      47 Leaf more than 40 cm. long; flowers numerous; perianth purple; pollinia 8

23. Arpophyllum

      47 Leaf less than 30 cm. long:
        48 Leaf-margin serrulate; pollinia 4   **41.** 12 *Epidendrum serrulatum*
        48 Leaf-margin smooth; pollinia 2   **37. Pleurothallis**
    46 Inflorescence longer than the leaf, usually 1-flowered; perianth orange; pollinia 8

40. Neocogniauxia

  45 Leaf broadly ovate to narrowly oblanceolate, usually rather firm and small, often with 3 small apical teeth or emarginate; lateral sepals more or less connate:
    49 Sepals and petals with long-tailed tips; pollinia 6   34. **Brachionidium**
    49 Sepals and petals without or not all with long-tailed tips; pollinia 2:
      50 Flowers opening by chinks at the side or not opening:
        51 Leaf obovate, up to 9 cm. long and 3 cm. broad, more or less rounded at the tip; perianth crimson   35. **Cryptophoranthus**
        51 Leaf narrow linear-lanceolate or oblanceolate, much smaller than above; perianth brick-red or yellow-green   37. **Pleurothallis**
      50 Flowers opening normally:
        52 Lip and petals similar, free   36. **Stelis**
        52 Lip and petals dissimilar; lip joined to column:
          53 Petals free from column; lip developed medially:
            54 Stigmatic surfaces confluent; flowers usually resupinate

37. Pleurothallis

            54 Stigmatic surfaces widely separated; flowers not resupinate

38. Lepanthopsis

          53 Petal-claws adnate to column; lip developed laterally; flowers usually not resupinate   **39. Lepanthes**
31 Leaves subterete, fleshy, at least the lower surface smoothly rounded:
  55 Leaves linear, spread along the shoot   42. **Jacquiniella**
  55 Leaves tufted and basal or solitary:
    56 Lip deeply 3-lobed, flat, pink   45. **Tetramicra**
    56 Lip not deeply 3-lobed, the proximal margins curled inwards:
      57 Tip of lip extended into a point; flowers not spurred; each shoot with 1 leaf:
        58 Inflorescence terminal; flowers showy   46. **Brassavola**
        58 Inflorescence basal; flowers small   55. **Trichopilia**
      57 Tip of lip rounded or emarginate; sepals forming a spur; leaves usually several   51. **Ionopsis**
18 Plants with pseudobulbs:
  59 Inflorescence terminal:
    60 Peduncles 1-flowered; plant very small, rhizomatous, each pseudobulb with a single fleshy leaf; perianth-members all acuminate-pointed

47. Homalopetalum

    60 Peduncles several-flowered or if short and 1–few-flowered then pseudobulbs with 2 leaves:
      61 Scape stout, very long and canelike, sheathed, with the flowers aggregated at the tip and subtended by large coloured reflexed bracteoles; perianth-segments more or less crisped-undulate   48. **Schomburgkia**
      61 Scape slender or if stout shorter than the flower-bearing part:
        62 Leaves short, thick and fleshy, rounded on the lower surface   45. **Tetramicra**
        62 Leaves more or less flat:
          63 Leaves longitudinally ribbed; pollinia 2   51. **Ionopsis**
          63 Leaves smooth on the surface; pollinia 4:
            64 Flowers in simple or compound 1-sided racemes; lip uppermost in flower

33. Polystachya

            64 Flowers not in 1-sided racemes:
              65 Lip produced basally into a spur adnate to the ovary; petals broader than the sepals   49. **Broughtonia**
              65 Lip not spurred; petals mostly narrower than the sepals

41. Epidendrum

  59 Inflorescence basal or axillary, arising below the pseudobulb:
    66 Peduncles 1-flowered, arising singly or several together or in succession

30. Maxillaria

    66 Peduncles several-flowered:
      67 Flowers distinctly spurred   52. **Comparettia**

67 Flowers not spurred:
  68 Pseudobulbs flat and thin with both margins sharp, never wrinkled; flowers
    showy   53. **Brassia**
  68 Pseudobulbs terete, globose or angular, sometimes wrinkled; flowers small:
    69 Lip with large lateral lobes embracing the column; pseudobulbs terete
                                                  54. **Macradenia**
    69 Lip not lobed:
      70 Scape shorter than the leaves   55. **Trichopilia**
      70 Scape much longer than the leaves:
        71 Lip flat or concave, rounded at the tip; inflorescence racemose; pollinia 2
                                   56. **Leochilus**
        71 Lip folded or angled, more or less pointed; inflorescence subspicate;
          pollinia 4   32. **Bulbophyllum**
1 Plants leafless at least at flowering time:
  72 Terrestrial herbs:
    73 Lip uppermost in the open flower:
      74 Saprophytic herb with numerous long roots; flowers very small; lateral sepals
        forming a short spur   11. **Wullschlaegelia**
      74 Autophytic herb with fleshy roots; perianth 6–7 mm. long, not spurred, dull green
                                        12. **Pterichis**
    73 Lip lowermost in the open flower; roots fleshy or tuberous:
      75 Scape purplish; saprophytic herb; petals free; purple   15. **Triphora**
      75 Scape green; autophytic herbs:
        76 Petals cohering with the median sepal   6. **Spiranthes**
        76 Petals free; lip shortly and broadly spurred, crested   16. **Galeandra**
  72 Plants climbing or tufted epiphytes:
    77 Fleshy vine with usually numerous short roots or rarely a few very long roots;
      fruit fleshy   59. **Vanilla**
    77 Epiphytes with very short obsolete stems bearing long thick greyish-green roots;
      fruit a capsule:
      78 Flowers solitary or few with the lip produced into a long slender spur
                                       60. **Dendrophylax**
      78 Flowers several to many in spikes or racemes with short often swollen spurs:
        79 Capsule dehiscing at the apex; flowers in racemes   61. **Harrisella**
        79 Capsule dehiscing in the middle; flowers in spikes or pedicels very short
                                   62. **Campylocentrum**

1. HABENARIA Willd. (1805)

1 Lip 3-lobed; petals 2-lobed:
  2 Leaves basal as well as spread along the flowering stem:
    3 Spur at least 3 times as long as the ovary; lateral lobes of lip much longer than the
      mid-lobe and strongly curled upwards; flowers white   1. *quinqueseta*
    3 Spur less than twice as long as the ovary; lateral lobes of lip about the same length as
      the mid-lobe, ascending or deflexed, not curled:
      4 Stem erect with short roots of which one develops into an ovoid tuber; leaf-sheaths
        with broad whitish bands mottled brown; flowers white   2. *monorrhiza*
      4 Stem creeping with long fibrous roots; leaf-sheaths green; flowers greenish
                                         3. *repens*
  2 Leaves mostly basal; lobes of lip subequal; flowers greenish:
    5 Lateral lobe of petals as long as or a little longer than the proximal lobe, filiform,
      deflexed at the tip   4. *distans*
    5 Lateral lobe of petals shorter than the proximal lobe, oblong, acute, ascending;
      leaves usually all basal   5. *jamaicensis*
1 Lip and petals entire or nearly so; flowers green:
  6 Spur much longer than the ovary; sepals obtuse   6. *purdiei*
  6 Spur not much longer than the ovary:
    7 Median sepal acute, lateral sepals shortly acuminate; ovary strongly winged
                                              7. *alata*
    7 Median sepal obtuse:
      8 Spur thickened towards the tip; ovary wingless   8. *eustachya*
      8 Spur slender, not thickened towards the tip   9. *socialis*

1. H. quinqueseta (Michx.) Sw., Adnot. Bot.: 46 (1829).—*Orchis quinqueseta*
Michx. (1803). *H. macroceratitis* Willd. (1805).
    Plant in flower 20–90 cm. high; lowest leaves suborbicular, upper elliptical to
ovate, acuminate above, up to 12 cm. long and 3·5 cm. broad; raceme 3–20-
flowered; flowers fragrant; mid-lobe of lip up to about 2 cm. long; spur to 12 cm.
or more long.

Locally abundant in most of the central parishes, in pastures on heavy clay soils and thicket margins on limestone; SL–2800 feet; fl. Oct–Jan, fr. Jan–Feb. *A 8334! H 12428! Powell 634!* S.E. United States to northern S. Amer., Cuba, Hispaniola. A variant with the tips of the petal- and lip-lobes fimbriate has been seen in St. Ann and St. Mary.

2. H. monorrhiza (Sw.) Reichb. f. in Ber. Deutsch. Bot. Gesell. **3**: 274 (1885).

Plant in flower 30–100 cm. high; leaves lanceolate to oblong-ovate up to about 8 cm. long and 3 cm. broad with scarious margins; flowers numerous in compact cylindrical racemes, apparently not fragrant; mid-lobe of lip 7–8 mm. long; spur 1–2 cm. long; petals variable, sometimes undivided; ovary winged.

Widespread and locally common in pastures, moist savannas and on rocky hillsides and roadside banks; 450–5000 feet; fl. Oct–Apr, fr. Dec–June. *A 6355! H 10500! P 4100!* Guatemala to Peru and Brazil, West Indies.

3. H. repens Nutt., Gen. N. Amer. Pl. **2**: 190 (1818).

Plant in flower 15–75 cm. high; leaves mostly rather long and narrow; raceme dense, many-flowered; mid-lobe of lip 4–5 mm. long; spur 9–10 mm. long.

Rare (St. Cath., Trel., St. Ann), in bogs; 900–1200 feet; fl. Jan, June. *P 6173!* S. United States to Brazil, Greater Antilles, Trinidad.

4. H. distans Griseb., Cat. Pl. Cub.: 270 (1866).

Plant in flower 20–40 cm. high; leaves oblong or lanceolate, up to about 15 cm. long and 5 cm. broad, acute; raceme few-flowered; mid-lobe of lip 8–9 mm. long; spur about 1·5 cm. long.

Very rare (Clar., Port.), in rocky thickets and woodland; 2000–3000 feet; fl. Sept–Dec, fr. Oct–Jan. *A 8059! H 7768!* Florida, Greater Antilles, Guatemala.

5. H. jamaicensis F. & R. in Journ. Bot. **47**: 126 (1909).

Plant in flower 15–30 cm. high; leaves suborbicular to oblong-lanceolate, up to about 9 cm. long and 4·5 cm. broad, acute or acuminate; raceme few-flowered; mid-lobe of lip 7–10 mm. long; spur about 1·5 cm. long.

Rather local (St. Andr., St. Thom.), gregarious on stony banks and in rocky thickets; 2800–4500 feet; fl. Dec–Apr, fr. Mar–Apr. *A 6666! H 10499! Arnold von der Porten!* Guatemala, Hispaniola.

6. H. purdiei F. & R. in Journ. Bot. **47**: 263 (1909).

Plant in flower 30–50 cm. high; leaves lanceolate, up to 8 cm. long and 2 cm. broad; raceme rather lax with many flowers; lip about 1·5 cm. long; spur up to about 3 cm. long.

Very rare (Clar.), in moist open savanna; 2300–2400 feet; fl. Nov–Dec. *P 15815!* Endemic.

7. H. alata Hook., Exot. Fl.: t. 169 (1825–26).

Plant in flower 40–100 cm. high; leaves lanceolate, up to 16 cm. long and 2·5 cm. broad; raceme many-flowered; lip 6–8 mm. long; spur 1·2–1·4 cm. long; petals folding inwards and the lip upwards as the flower withers; capsule curved outwards during ripening.

Occasional on stony or clay banks and in thickets and woodland margins on limestone; 1700–4000 feet; fl. Oct–Apr, fr. Nov–Apr. *A 11765! H 12253! P 11035!* Mexico to Bolivia, West Indies.

8. H. eustachya Reichb. f. in Ber. Deutsch. Bot. Gesell. **3**: 274 (1885).—*H. troyana* F. & R. (1909).

Plant in flower up to 60 cm. high; leaves lanceolate, up to 15 cm. long and 5 cm. broad; raceme densely many-flowered; lip about 8 mm. long; spur about 9 mm. long.

Very rare (Trel.), in damp shady forest; 2500 feet; fl. and fr. Nov. *H 10432!* Florida, Guatemala, Greater Antilles.

9. H. socialis F. & R. in Journ. Bot. **47**: 263 (1909).

Plant in flower about 30 cm. high; leaves lanceolate, up to about 8·5 cm. long and 2 cm. broad; raceme laxly many-flowered; lip about 5·5 mm. long; spur about 1 cm. long.

Very rare (Manch.), in marshy soil. Known only from the type, *Purdie*. Endemic.

2. CORYMBORKIS Thouars (1822)

1. C. flava (Sw.) Kuntze, Revis. Gen. Pl. **2**: 658 (1891).

Plant glabrous, up to about 2 m. high; leaves oblong, subsessile, strongly 7–9-nerved, 15–30 cm. long, 3–6 cm. broad, the tip long-acuminate; panicles axillary, 4–10 cm. long; flowers subsessile, yellow; sepals 1·5–2 cm. long; petals and lip a little shorter than the sepals.

Very rare (Clar., Manch.), in damp woodland; about 3000 feet; fl. Oct–Apr. *H 10398!* C. Amer., Greater Antilles, Guadeloupe.

3. TROPIDIA Lindl. (1831)

1. T. polystachya (Sw.) Ames, Orch. **2**: 262 (1908).

Plant 20–60 cm. high, the erect shoots from a short woody rhizome, often branched at the lower nodes; roots numerous, hard, twisted; leaves, 7-nerved, oblong-lanceolate, acuminate, up to about 20 cm. long and 5 cm. broad; panicles 8–15 cm. long; flowers pedicelled, short-lived; sepals and petals 6–7 mm. long, light green; lip obscurely lobed with 2 converging yellow crests on the white terminal lobe.

Rare (Clar., West., Han.), in deep leaf litter among limestone rocks in shady forest; 1300–2250 feet; fl. Aug–Jan, fr. Nov–Mar. *A 8060! Harris!* Florida, Mexico to Costa Rica, Cuba, Hispaniola.

4. ERYTHRODES Blume (1825)

1 Sepals 6 mm. long; stem and spike hairy; leaves 3 cm. or more broad   1. *plantaginea*
1 Sepals 3–3·5 mm. long; leaves mostly less than 2 cm. broad:
   2 Stem and spike sparsely to rather densely pilose; tip of lip transversely lobed
                                                       2. *hirtella*
   2 Stem and spike glabrous; tip of lip 3-lobed, the lateral lobes broadly rounded, the middle subulate   3. *jamaicensis*

1. E. plantaginea (L.) F. & R., Fl. Jam. **1**: 28 (1910).

Stem trailing and rooting from the nodes, up to 2 m. long; flowering branches suberect 30–50 cm. high; leaf-blades elliptical to ovate, narrowed to the sheathing bases, 6–12 cm. long, 3–6 cm. broad, shortly acuminate; spike 5–15 cm. long with numerous flowers.

Occasional and gregarious in leafy soil in shady woodlands; 1500–5500 feet; fl. and fr. Nov–Apr. *A 12689! H 7875! P 9874!* West Indies.

2. E. hirtella (Sw.) F. & R., Fl. Jam. **1**: 29 (1910).

Lower part of stem shortly trailing and rooting; flowering shoot ascending, 15–30 cm. high; leaf-blades ovate to narrowly lanceolate, narrowed to the base, up to 6 cm. long and 2·5 cm. broad; spike many-flowered, elongating in fruit.

Occasional in leaf-litter and among shaded rocks in woodland and woodland margins; 1000–3700 feet; fl. and fr. Feb–Apr. *A 9097! H 9010! P 4096!* West Indies.

3. E. jamaicensis (F. & R.) F. & R., Fl. Jam. **1**: 29 (1910).

Like the last; leaf-blade lanceolate, acute, up to 5 cm. long and 1·6 cm. broad; spike about 4 cm. long.

Very rare (Manch.), in shady woods; 3000 feet; fl. Oct. Known only from the type, *H 10472!* Endemic.

5. PELEXIA Poit. ex Lindl. (1826) nom. cons.

Lateral sepals abruptly acute; terminal lobe of lip broadly rounded, entire, sharply deflexed; spike, bracts, ovary and sepals pubescent; tip of spur only free   1. *adnata*

Lateral sepals long caudate-acuminate; terminal lobe of lip long-acute the margins eroded
  at the broadest part, curved-deflexed; spike etc. glabrescent; spur free for nearly half its
  length   2. *setacea*

1. P. adnata (Sw.) Spreng. in L., Syst. Veg. ed. 16, **3**: 704 (1826).

Plant in flower 30–45 cm. high; scape and petioles pinkish-red; basal leaves 2–5,
blades oval, acute, often spotted white, about 10 cm. long and 5 cm. broad; spike
about 20 (–28)-flowered; sepals about 6 mm. long, light green; lip white.

Uncommon in leafy soil among rocks in woodland on limestone; 1100–2250
feet; fl. Mar–May, fr. Apr–May. *A 9257! H 11988! P 8640!* British Honduras to
Venezuela, Bahamas, West Indies.

2. P. setacea Lindl., Gen. & Sp. Orch.: 482 (1840).

Plant in flower 40–50 cm. high; scape and petioles pinkish-red; basal leaves 1–3,
blades oval or elliptical, acute, up to 15 cm. long and 6 cm. broad; spike 8–10-
flowered; sepals 2·5–3 cm. long, light green; lip green at the base, white distally.

Occasional in leafy soil among limestone rocks in thickets and woodland;
700–2900 feet; fl. Dec–Feb, fr. Feb–Mar. *A 12243! H 7863! H & P 15027!*
Northern S. Amer., Bahamas, Greater Antilles; Cayman Brac. *P. calcarata* (Sw.)
Cogn. of Hispaniola, with entire lip and glabrous inflorescence, is closely related.

6. SPIRANTHES L. C. Rich. (1817) nom. cons.

1 Sepals 4–8 mm. long, not forming a chin; flowers mostly greenish-white or tinged
    brownish-pink:
  2 Leaves linear to linear-lanceolate, usually absent at flowering time; flowers contiguous
      in a regular single spiral; sepals and petals white; lip broad-tipped, fimbriate, green
      within   1. *tortilis*
  2 Leaves broadly ovate to elliptical, always present at flowering time, petioled; flowers
      lax, not obviously in a tight spiral:
    3 Plant usually less than 15 cm. high, tender, with swollen roots less than 2 cm. long;
        leaves basal and on the scape, the blade not exceeding 3·5 cm. long and 2 cm. broad;
        inflorescence usually up to about 5-flowered   2. *fawcettii*
    3 Plant usually more than 20 cm. high, firm, with longer roots; leaves all basal, broader;
        inflorescence several- to many-flowered:
      4 Lip broadly rounded at the tip, shallowly emarginate, crenulate; leaves green with
          petioles 4–7 cm. long   3. *elata*
      4 Lip 3-lobed at the tip, the lateral lobes shorter than the rectangular-truncate mid-
          lobe; leaves purplish with petioles 1–2 cm. long   4. *cranichoides*
1 Sepals 12–30 mm. long, forming a chin; flowers yellow to dull brownish-orange to
    bright red:
  5 Leaves present at flowering time, elliptical, petioled; scape, ovary and sepals glabrous;
      chin very short or obsolete   5. *speciosa*
  5 Leaves usually absent at flowering time, broadly lanceolate to oblanceolate, tapered to
      the base; scape, ovary and sepals densely pubescent with stalked glandular hairs;
      chin prominent:
    6 Perianth from tip of chin to tip of median sepal 12–18 mm. long; spur about half the
        length of the ovary   6. *lanceolata*
    6 Perianth 22–30 mm. long; spur about two-thirds to three-quarters the length of the
        ovary   7. *squamulosa*

1. S. tortilis (Sw.) L. C. Rich., Orch. Eur. Adnot.: 37 (1817).

Plant in flower 15–45 cm. high; leaves up to 30 cm. long, smooth, striate;
sepals 5–6 mm. long.

Widely distributed but not common, in clay soil on banks and in open savannas;
500–5000 feet; fl. and fr. Mar–June. *A 10947! H 9553! P 16246!* S. United States
to C. Amer., Bermuda, Bahamas, West Indies.

2. S. fawcettii Cogn. in Fedde, Repert. Sp. Nov. **7**: 123 (1909).

Plant in flower 5–20 cm. high; lateral sepals linear-oblanceolate, acute, 7–8 mm.
long, white; free tips of lateral petals with 3 brownish lines; lip white with 3 short
green stripes within.

Uncommon on mossy boulders or among leaf-litter in shady woodlands on
limestone, sometimes epiphytic; 1250–5000 feet; fl. Feb–Apr, fr. Mar–Apr.
A 12408! H 10493! P 6250! Greater Antilles.

3. S. elata (Sw.) L. C. Rich., Orch. Eur. Adnot.: 37 (1817).

Plant in flower (15–) 25–60 cm. high; leaf-blades elliptical, up to 15 cm. long and 6 cm. broad; inflorescence variable in pubescence, size of flowers and colour of bracts and flowers; smaller variants with perianth 4·5–5 mm. long usually have bracts, ovary and lip pinkish with the scape and sepals buff or brownish; larger variants with perianth 6–7 mm. long usually have the lip white with green lines in the lower part and the rest of the inflorescence green, the median sepal often darker; petals not lined.

Common on rocky banks and in partial shade in woodlands on limestone; 1000–6000 feet; fl. and fr. Feb–Apr. *A 6769! H 7856! P 6295!* Florida to Argentina, Bahamas, West Indies.

4. S. cranichoides (Griseb.) Cogn. in Urb., Symb. Ant. **6**: 338 (1909).—*Pelexia cranichoides* Griseb. (1866).

Plant in flower 15–35 cm. high; leaf-blades ovate, up to 7 cm. long and 4 cm. broad, shortly acuminate; bracts light green speckled white and mauve; lateral sepals olive-green, about 4 mm. long, median sepal brownish speckled purple; petals dull pink the free tips with 3 brownish-purple lines; lip white.

Rare (Trel., Port.), in deep leaf-litter in woodland on rocky limestone; 1000–2250 feet; fl. Feb–Apr, fr. Mar–Apr. *A 9266! 12440!* Florida, British Honduras, Guatemala, Cuba, Puerto Rico.

5. S. speciosa (Jacq.) A. Rich. in Sagra, Hist. Cuba, Parte 2, **11**: 252 (1850).—*Neottia speciosa* Jacq. (1790). *Stenorrhynchos speciosus* (Jacq.) Lindl. (1840).

Plant in flower 25–60 cm. high; leaf-blades narrowly elliptical, 10–25 cm. long including the petiole, acute or acuminate, sharply keeled beneath the midrib; bracts pink; flowers secund in a compact spike, bright pinkish-red; lateral sepals, 15–20 mm. long; lip acute, somewhat deflexed, white.

Rather uncommon in rocky thickets and low woodlands; 1200–5000 feet; fl. and fr. Oct–Mar. *A 8492! H 9898! P 5145!* Mexico to Venezuela, Cuba, Puerto Rico.

6. S. lanceolata (Aubl.) Fawcett, Prov. List Jam.: 40 (1893).—*Stenorrhynchos lanceolatus* (Aubl.) L. C. Rich. ex Spreng. (1826).

Plant in flower 20–70 cm. high; leaves up to 30 cm. long and 3·5 cm. or more broad; bracts often green or approaching the colour of the flowers; flowers not secund, light dull yellow, buff-orange or rarely bright red; lip acute, white.

Widely distributed but nowhere common, in heavy soil on banks and in pastures and woodland margins; 900–3000 feet; fl. Feb–May, fr. Mar–May. *A 6635! H 7876! P 16240!* Florida, Mexico to Paraguay, Bahamas, West Indies.

7. S. squamulosa (Kunth) León in Contrib. Ocas. Mus. Hist. Nat. Col. de la Salle, Habana **8**: 357 (1946).—*Stenorrhynchos squamulosus* (Kunth) Spreng. (1826).

Like the last with larger flowers, perhaps not really distinct.

Occasional on open clay banks; 500–5000 feet; fl. and fr. Apr–June, Dec. *A 6812! H 10496! P 4552!* Mexico, Colombia, Cuba, Hispaniola, Antigua; Grand Cayman.

7. PSEUDOCENTRUM Lindl. (1859)

1. P. minus Benth. in Hook., Ic. Pl. **14**: 63, t. 1382 (1882).

Plant in flower 30–60 cm. high; stem, leaves and bracts glabrous; leaf-blades ovate-lanceolate, acute or shortly acuminate, 10–14 cm. long, 3–6 cm. broad, smaller on the scape; spike dense, up to about 20 cm. long; ovary and sepals pilose with large pluricellular hairs; median sepal 5–6 mm. long; spur 15 mm. long.

Rare and local on mossy rocks and banks in shade; 5000–6500 feet; fl. Sept–Oct, fr. Jan–Feb. *H 7756! 10088! P 11456!* Endemic.

8. PRESCOTTIA Lindl. (1825)

1 Sepals 1–1·5 mm. long, spreading, white with reddish tips; lip rounded   **1.** *oligantha*

1 Sepals 3–4 mm. long, strongly revolute; lip obliquely ovoid, the lateral margins subcontiguous around the column leaving an elliptical opening:
2 Petioles up to 20 cm. long; lip about 3·5 mm. long, at first light green becoming buff
   then often reddish; sepals pink or white; petioles, bracts and young fruits usually
   dusky pink   2. *stachyodes*
2 Petioles 1·5–3 cm. long; lip about 4·5 mm. long, dark rich green turning light buff;
   sepals greenish-buff; petioles, bracts and fruits mostly green   3. *pellucida*

1. P. oligantha (Sw.) Lindl., Gen. & Sp. Orch.: 454 (1840).
   Plant in flower 20–40 cm. high; petioles furrowed; leaf-blades ovate to elliptical,
2·5–7·5 cm. long, 1·5–3 cm. broad; usually dark greyish-green with paler midrib;
scape usually tinged mauve or pink; flowers very small but variable in size, up to
50 or more in a dense slender spike; perianth white tinged pink or brownish;
petals shorter than the sepals, spreading.
   Rather common in rocky thickets and open woodlands and on rocky banks;
900–5000 feet; fl. Jan–Apr, fr. Feb–Apr. *A 8919! H 7856A! P 6322!* Florida,
C. Amer., Bahamas, West Indies.

2. P. stachyodes (Sw.) Lindl. in Edw., Bot. Regist. **22**: under t. 1916 (1836).
   Plant in flower 30–75 cm. high; bleaf-lades erect up to 20 cm. long and 7 cm.
broad, often with finely denticulate margins, dark green; flowers numerous in a
dense spike; petals narrow, as long as the sepals, reflexed.
   Rather common in leafy soil in woodlands, mostly in the central and eastern
parishes; 2000–5800 feet; fl. and fr. mostly Sept–May. *A 9186! H 10480! P 5820!*
Guatemala to Brazil, West Indies.

3. P. pellucida Lindl. in Ann. & Mag. Nat. Hist. ser. 3, **1**: 335 (1858).
   Like the last and perhaps not more than an ecological variant.
   Very rare (St. Cath., Port., St. Thom.), on mossy rocks or tree-bases in low
thickets on limestone or serpentine; 2000–3000 feet; fl. Jan–Apr, fr. Mar–Apr.
A 9147! 12211! P 16267! Cuba.

9. CRANICHIS Sw. (1788)

1 Leaves several; sheaths on scape open and leafy; petals white; plant glabrous
                                                                    1. *muscosa*
1 Leaves 1–2 (–3); sheaths not leafy; spike glandular-pubescent:
2 Petals oblanceolate, white, glabrous; leaf-blade cordate or narrowed to the petiole
                                                                    2. *diphylla*
2 Petals linear, pink or light brownish-red with white marginal hairs; leaf-blade subcordate   3. *wageneri*

1. C. muscosa Sw., Nov. Gen. & Sp. Pl.: 120 (1788).
   Plant in flower 15–30 cm. high; leaves distinctly petioled, blade ovate, acute or
obtuse, up to 7 cm. long and 3·5 cm. broad; scape purple or reddish; sheaths green;
spike dense-flowered; sepals spreading, white 2·5–3 mm. long; lip white spotted
green within.
   Common on sheltered banks and in open woodlands; 850–4500 feet; fl. Nov–
Mar, fr. Jan–Mar. *A 8600! H 10660! P 4972!* Florida to Venezuela, West Indies.

2. C. diphylla Sw., Nov. Gen. & Sp. Pl.: 120 (1788).
   Plant in flower 10–40 cm. high; leaf-blades ovate, acute or acuminate, up to 6·5
cm. long and 3 cm. broad; sheaths linear-lanceolate, acuminate; spike lax-flowered;
sepals about 2·5 mm. long; lip greenish-white, spotted green within.
   Local (St. Andr., Port., St. Thom.), on mossy banks in forest, rarely epiphytic;
(2000–) 4000–6500 feet; fl. Dec–Apr, fr. Jan–Apr. *A 6244! H 10080! P 4201!*
Mexico, Guatemala, Hispaniola, Puerto Rico.

3. C. wageneri Reichb. f. in Linnaea **41**: 19 (1876).—*C. pilosa* F. & R. (1909).
   Plant in flower 10–25 cm. high; leaf-blades rounded-ovate, acute or acuminate,
up to 10 cm. long and 7·5 cm. broad; sheaths elliptical, shortly acuminate; spike
compact; sepals about 3·5 mm. long; lip greenish-white or tinged pink or brown.
   Rare (St. Andr., Port.), in leafy soil in damp shady forest; 3000–5800 feet; fl.
Nov–Apr. *H 10503! P 5973!* Mexico to Venezuela.

10. PONTHIEVA R. Br. (1813)

1 Margins of petals ciliolate   1. *racemosa*
1 Margins of petals glabrous:
  2 Median sepal glabrous; lip with the lateral lobes poorly developed, 1 mm. broad, attached 2 mm. above the base of the column   2. *harrisii*
  2 Median sepal glandular-pubescent; lip with lateral lobes well developed, 2·5 mm. or more broad, attached nearer the base of the column:
    3 Lateral petals entire, triangular; lip clawed, deeply concave but not saccate at the base, lateral lobes broadly rounded anteriorly with 2 green spots within, median lobe narrow   3. *pauciflora*
    3 Lateral petals lobed; lip hardly clawed, saccate at the base with the proximal margins meeting adaxially, green- or yellow-lined within, 3-lobed at the tip   4. *ventricosa*

1. P. racemosa (Walt.) Mohr in Contrib. U.S. Nat. Herb. **6**: 460 (1901).—
Arethusa racemosa Walt. (1788). *P. glandulosa* (Sims) R. Br. (1813).

Plant in flower 30–60 cm. high; leaves broadly elliptical to oblanceolate, 5–25 cm. long, 2–5 cm. broad; scape and sheaths glandular-pubescent above; pedicels distinct, 3–4 mm. long in flower, longer in fruit; sepals 5–7 mm. long, light green; petals white, lined green; lip green at base, white at tip.

Common on shaded or open banks and in rocky woodlands and thickets; 900–5000 feet; fl. Nov–Mar, fr. Dec–Apr. *A 12699! H 7266! P 15768!* S. United States to Venezuela and Ecuador, West Indies to Grenada.

2. P. harrisii Cogn. in Fedde, Repert. Sp. Nov. **6**: 304 (1909).

Plant in flower 35–50 cm. high, to 60 cm. high in fruit; leaves ovate-elliptical, 6–11 cm. long, 3–5 cm. broad, acutely acuminate; scape, sheaths and bracts glandular-pubescent; pedicels 4–5 mm. long; median sepal 4–5 mm. long, lateral 5·5–7 mm. long, 3–4 mm. broad, olive-green; petals light olive-green; lip white with green blotch; flowers musky fragrant, fading through yellowish to pinkish-brown.

Rare and local (St. Andr., Port.), in deep leafy soil in forest; 3000–4250 feet; fl. Apr–June, fr. June–Aug. *A 7055! H 7618! P 4415!* Hispaniola.

3. P. pauciflora (Sw.) F. & R., Fl. Jam. **1**: 38 (1910).

Plant in flower 15–40 cm. high; leaves ovate to elliptical, obtuse, up to 6 cm. long and 2·5 cm. broad, shortly petiolate; scape glandular-pubescent above; raceme few-flowered; sepals greenish, 4·5–5·5 mm. long; petals clawed, 4–5 mm. long, white.

Rare and local (St. Andr., Port., St. Mary), on shaded banks; 2500–5700 feet; fl. and fr. Nov–Feb. *A 5829! H 7532! P 3771!* Cuba, Hispaniola.

4. P. ventricosa (Griseb.) F. & R., Fl. Jam. **1**: 39 (1910).

Plant in flower 15–35 cm. high; leaves elliptical to obovate, tapered to short or long petioles, 2–7 cm. long, 1–2·5 cm. broad; scape glandular-pubescent above; sepals greenish 5–6 mm. long; petals shortly clawed, about 5 mm. long, white.

Common on open clay banks, rocky cuttings and cliffs, mostly in limestone areas; (300–) 700–2900 feet; fl. Nov–Mar, fr. Dec–Mar. *A 8419! H 12257! HPS 14630!* Greater Antilles.

11. WULLSCHLAEGELIA Reichb. f. (1863)

1. W. aphylla (Sw.) Reichb. f. in Bot. Zeit. **21**: 131 (1863).

Plant in flower 15–35 cm. high; stem slender, erect with clustered fibrous roots arising close together at the base, leafless, white or yellowish, puberulous, scaly; flowers white or cream; median sepal and petals 1·7–1·8 mm. long; lateral sepals and lip about 3 mm. long.

Very rare (Clar., St. Ann), in damp shady woods; fl. Sept–Oct. Not recently found. Guatemala to Brazil and Paraguay, Cuba, Dominica.

12. PTERICHIS Lindl. (1840)

1. P. proctorii Garay in Occ. Pap. Mus. Inst. Jam. **10**: 1–4 (1954).

Leaves not known; scape and spike 40 cm. high; median sepal ovate-lanceolate; lateral sepals shortly connate at the base 6–7 mm. long, 3 mm. broad, glandular-

pubescent outside; petals oblong, obtuse, glabrous, 6 mm. long, 2 mm. broad; lip obscurely 3-lobed, 6 mm. long and broad.
Very rare (St. Thom.), on mossy banks; 6000–7000 feet; fl. Feb. *P 5427!* Endemic.

13. EURYSTYLES Wawra (1863)

1. E. ananassocomos (Reichb. f.) Schltr. in Fedde, Repert. Sp. Nov., Beih. **35**: 39 (1925).—*Stenoptera ananassocomos* Reichb. f. (1863).
Plant in flower 2·5–4 cm. long, the inflorescence pendulous from a tufted rosette of leaves; roots short, thick, hairy; leaves waxy, shiny light grey-green, the margins pectinate-ciliate with short hairs, 1·5–2·5 cm. long, 7–12 mm. broad; bracts denticulate with thick-based hairs; flowers white at the tips (5–) 10–15 (–20) in each head; lateral sepals 5 mm. long, the median shorter; capsule about 8 mm. long.
Rare (St. Andr., St. Ann, St. Thom.), on the trunks and branches of trees; (1000–) 1800–2900 feet; fl. (Dec–) Jan–Mar. *A 12690! J.P. 2283!* Cuba, Hispaniola.

14. PSILOCHILUS J. B. Rodrigues (1882)

1. P. macrophyllus (Lindl.) Ames, Orch. **7**: 45 (1922).—*Pogonia macrophylla* Lindl. (1858).
Plant in flower 20–30 cm. high; roots short, thick; stem, underside of leaves, sheaths and pedicels deep rich purple; leaf-blade ovate, strongly 3-nerved, 3–7 cm. long, 2–3 cm. broad; flowers erect, slightly curved, at first green, latterly yellowish tinged purple; sepals minutely spotted purple, the median about 19 mm. long; lateral petals slightly shorter than the sepals, light green with whitish margins; lip green, with purple submargins and white margins.
Very local and rather rare (St. Andr., Port.), in deep leafy soil in forest; 3500–5000 feet; fl. and fr. Oct–June. *A 6210! H 6252! J.P. 2090!* Guatemala to Colombia and Venezuela, West Indies.

15. TRIPHORA Nutt. (1818)

Stem leafy, decumbent at the base with short fleshy roots; lip trough-shaped
                                                                  1. *surinamensis*
Stem leafless or blades very short, erect from a tuber; lip flat   2. *gentianoides*

1. T. surinamensis (Lindl.) Britton in Britton & Wilson, Sci. Surv. Porto Rico & Virg. Is. **5**: 184 (1924).—*Pogonia surinamensis* Lindl. (1843).
Plant in flower 8–25 cm. high; stem purple, leafy above, scaly below; leaf-blades ovate, 1·5–3 cm. long, acute or acuminate, purplish beneath; flowers solitary in the upper axils; pedicels up to 2 cm. long; sepals pink, 10–12 mm. long, linear; petals white, a little shorter than the sepals, narrowly spathulate; lip mostly white, 3-lobed, the lateral lobes raised, the mid-lobe longer, eroded at the tip, crimson; capsule oblong, up to about 15 mm. long.
Very rare (Clar., Port.), in soggy leafy soil in wet sheltered thickets; 850–2300 feet; fl. and fr. Aug. *A 7893! P 26660!* Puerto Rico, Guadeloupe, Dominica, Trinidad and the Guianas to Brazil.

2. T. gentianoides (Sw.) Ames & Schltr. in Ames, Orch. **7**: 5 (1922).—*Pogonia gentianoides* (Sw.) Spreng. (1826).
Plant in flower 12–40 cm. high; scape purple; bracts green or purplish; pedicels crowded, reddish, 2·5–7·5 cm. long; sepals dark purple, 9 mm. long, narrowly oblong; petals white, a little shorter than the sepals, lanceolate, obtuse; lip narrow at the base, 3-lobed, the mid-lobe oblong, obtuse, entire.
Very rare (St. Thom.), in shady places; fl. June–July. *J.P. 482! Sangster!* Florida, Mexico to Venezuela, Hispaniola; Grand Cayman.

16. GALEANDRA Lindl. (1830)

1. G. beyrichii Reichb. f. in Linnaea **22**: 854 (1850).
Plant in flower 80–120 cm. high; stem robust, rather thick, purplish; roots thick, short, brittle, white, glabrous; raceme 8–12-flowered, the flowers nodding; sepals,

petals and ovary light green; sepals strongly ribbed, 2·3 cm. long; lip broader than long when opened out, with veins green or purplish within, margins creamy-white; spur striate; older flowers completely creamy-yellow.

Very rare (St. Ann, Manch., West.), in deep leaf litter amongst limestone rocks in woodland; 1300–2900 feet; fl. and fr. Sept–Oct. *A 11769! H 9780!* Florida, Costa Rica to Brazil, Cuba, Hispaniola.

17. GOVENIA Lindl. (1831)

1. G. utriculata (Sw.) Lindl. in Edw., Bot. Regist. **25** Misc.: 47 (1839).

Plant in flower 35–60 cm. high from a rhizome with corms; basal two sheaths inflated, purplish; scape streaked purple; leaf-blades elliptical, 15–25 cm. long, 5–9 cm. broad; flowers erect; sepals white, the lateral about 12 mm. long; petals white with faint pink streaks within, 14 mm. long; lip mostly white but light yellow at the base and brown-spotted near the tip; column purplish with brown spot; whole plant blackening on drying.

Uncommon in deep leafy soil among limestone rocks in woodland; 1000–4250 feet; fl. Sept–Feb, fr. Feb–June. *A 9003! H 10483! H & P 15080!* Florida, Mexico to Argentina, Bahamas, Greater Antilles.

18. EULOPHIA R. Br. ex Lindl. (1823) nom. cons.

1. E. alta (L.) F. & R., Fl. Jam. **1**: 112 (1910).

Plant in flower 60–120 cm. high; leaves very long, strongly plicate, mostly not developed at flowering time; raceme laxly many-flowered, the pedicels (ovaries) spreading horizontally; lateral sepals directed upwards, about 2 cm. long, dull yellow-green turning yellow; petals 1·4 cm. long, light-purple; lip developed basally into a rounded greenish saccate spur, lateral lobes short, ascending, rounded, mid-lobe usually purplish or purple-streaked; flowers rarely white.

Widely distributed but not common, in grassy swamps, low pastures and wet thickets; 50–2500 feet; fl. and fr. May–Aug, Oct–Feb. *A 7155! H 7638! P 16486!* Florida to Brazil and Peru, West Indies, W. Africa.

19. PHAIUS Lour. (1790)

1. P. tancarvilleae (Banks ex L'Hérit.) Blume, Mus. Bot. Lugd.-Bat. **2**: 177 (1856).—Nun Orchid.

Plant in flower 60–100 cm. high; stem cormous; leaves long, plicate, up to 10 cm. broad; scape stout; bracts conspicuous, whitish at first, caducous; sepals 4–5·5 cm. long, like the petals white outside, light brown inside; lip reddish-purple distally, whitish or yellowish outside towards the base; flowers fragrant in the morning.

Widely distributed and rather common in grassy savannas and in sheltered woodlands, also cultivated as an ornamental; (600–) 1300–5000 feet; fl. mostly Dec–June, fr. Apr–July. *A 7137! H 9183! P 16125!* Native in the W. Pacific and E. Asia; Cuba.

20. BLETIA Ruiz & Pav. (1794)

Lip with 5 deep white or yellowish crests extending along the mid-lobe nearly to the tip; column 3·5–3·8 mm. broad across the narrowest part   1. *florida*
Lip with 7–9 crests at the highest part with 5 only extending nearly to the tip; column 1·8–2 mm. broad at the narrowest part   2. *purpurea*

1. B. florida (Salisb.) R. Br. in Ait., Hort. Kew. ed. 2, **5**: 206 (1813).

Plant in flower 60–100 cm. high; leaves long, up to 10 cm. broad; inflorescence a lax raceme, usually unbranched; perianth rich reddish-purple, rarely light pink; sepals 22–24 mm. long; base of column not forming a distinct foot so that the flower-buds are not chinned.

Rather local, mostly in the eastern parishes, on clay banks and shale cliffs and cuttings; (600–) 1900–4000 feet; fl. Oct–July. *A 10432! H 7622!* Cuba. Many authors, writing for territories outside our area, include *B. florida* in the synonymy of *B. purpurea*.

2. B. purpurea (Lam.) DC. in Mém. Soc. Phys. & Hist. Nat. Genève **9**: 97, 100 (1841).

Plant in flower 50–120 cm. high; leaves mostly not more than 6 cm. broad; inflorescence racemose, sometimes branched; perianth usually purple, very rarely white; sepals 20–22 mm. long; base of column forming a distinct foot 2 mm. long so that the flower-buds appear chinned.

Common in open undisturbed places; SL–3500 feet; fl. and fr. all the year. *A 7022! P 10315!* Florida, C. Amer. to Venezuela, Bahamas, Cuba, Hispaniola, Barbados. Plants with light purple flowers usually lack purple pigments in the scape and fruits; these are less common than plants with dark purple flowers and purple-pigmented scapes and fruits. A variant lacking purple pigment altogether is var. **alba** Ariza-Julia & Jiménez (1960) reported from St. Ann and originally described from the Dominican Republic.

21. CALANTHE R. Br. (1821) nom. cons.

1. C. mexicana Reichb. f. in Linnaea **18**: 406 (1845).

Plant in flower 30–40 cm. high; leaves oblong to elliptical, up to 30 cm. long and 10 cm. broad; flowers greenish-white at first, becoming white; sepals 11–12 mm. long, 4–5 mm. broad; petals 8 mm. long, 3 mm. broad; plant drying black.

Rare and local (St. Andr., Port.), in rocky thickets; 3500–5000 feet; fl. June– Sept, Dec. *A 7679! H 10437! A. M. Barry!* Mexico to Panama, Cuba, Hispaniola.

22. COELIA Lindl. (1830)

1. C. triptera (Sm.) G. Don ex Steud., Nomencl. Bot. ed. 2, **1**: 394 (1840).

Pseudobulbs narrowly ovate, 3–5 cm. long; leaves several from the top of the pseudobulb, 20–40 cm. long, including the slender basal sheaths, 1–2 cm. broad; scape and raceme 8–15 cm. long; flowers white to light yellow; sepals ovate, obtuse, about 7·5 mm. long; capsule ellipsoid, about 1 cm. long, with 3 broad wings.

Uncommon on exposed boulders and craggy limestone rocks in woodland; 2500– 4000 feet; fl. Jan–Mar, fr. Mar. *H 7614! J.P. 486! P 5551!* Mexico, Guatemala, Cuba.

23. ARPOPHYLLUM Llave & Lex. (1825)

1. A. jamaicense Schltr. in Fedde, Repert. Sp. Nov., Beih. **19**: 32 (1923).—*A. giganteum* of F. & R. (1910), not Hartweg ex Lindl. (1840).

Rhizome creeping and branching; upright shoots 10–25 cm. long of 3–4 inter-nodes; leaf-blade 40–50 cm. long, 2–2·5 cm. broad, leathery; spike with many light purple flowers; sepals 5·5–6 mm. long.

Rare (St. Andr., Port., St. Thom.), on rocks or fallen logs in thickets or epiphytic in moist forest; 1750–4500 feet; fl. Feb–May. *A 9368! H 7652! P 5840!* Endemic.

24. LIPARIS L. C. Rich. (1817) nom. cons.

1 Leaf one subtending each scape:
  2 Leaf oblong, rather obtuse; lip light yellow with reddish-brown veins, 7–8 mm. long
                                                               1. *vexillifera*
  2 Leaf suborbicular, shortly acuminate; lip light green with purple veins, 3·5 mm. long
                                                               2. *neuroglossa*
1 Leaves 2–several on each flowering shoot:
  3 Leaves usually two, more or less flat; lip broadly ovate to oblong, rounded at the tip:
    4 Lip green, veined with purple or brown, 13 mm. long; leaf-tip acute   3. *harrisii*
    4 Lip purple, 5–6 mm. long; leaf-tip obtuse   4. *saundersiana*
  3 Leaves three or more, plicate, acuminate; lip spathulate, emarginate, dark reddish-purple, about 4 mm. long   5. *elata*

1. L. vexillifera (Llave & Lex.) Cogn. in Mart., Fl. Bras. **3** (4): 289 (1895).
Leaf-blade 7–12 cm. long, 1·5–3·5 cm. broad; scape and raceme 10–30 cm. long; flowers light yellow; sepals 6·5–7·5 mm. long.
Rare (St. Andr., St. Ann), on damp clay banks; 2800–5000 feet; fl. and fr. Nov–Feb. *H 7842! J.P. 237!* Mexico to Bolivia and Argentina, Cuba, Puerto Rico, Trinidad.

2. L. neuroglossa Reichb. f., Xen. Orch. **3**: 26 (1900).
Leaf-blade 3–4 cm. long, 2–4 cm. broad, greyish-green; scape and raceme 5–10 cm. long; flowers light green with purple lines; sepals about 4 mm. long.
Rare (St. Andr., Port., St. Thom.), in mossy leaf-litter on rocks in thickets and epiphytic; 4700–6000 feet; fl. and fr. Nov–Feb. *A 8789! H 7733! P 9535!* Bolivia.

3. L. harrisii F. & R. in Journ. Bot. **47**: 7 (1909).
Leaf-blades elliptical, up to 12 cm. long and 7 cm. broad; scape and raceme 8–15 cm. long; sepals linear, greenish margined with purple, 9 mm. long, 2·5 mm. broad.
Rare and local (St. Andr., Port.), in leafy soil in shade; 4000–6000 feet; fl. Aug–Dec, fr. Nov–Apr. *H 9786! P 9537! Skelding UCWI 4456!* Endemic.

4. L. saundersiana Reichb. f. in Gard. Chron. **1872**: 1003 (1872).
Leaf-blades rounded-ovate, cordate, obtuse, up to 6 cm. long and 4 cm. broad; scape and raceme up to 12 cm. long; sepals lanceolate, translucent-greenish, 4–5 mm. long; petals linear, purple, 6 mm. long.
Rare and local (St. Andr.), on sheltered mossy banks; 2600–5000 feet; fl. Oct–Dec. *A 5990! H 9789! P. B. Caws!* Endemic.

5. L. elata Lindl. in Edw., Bot. Regist. **14**: t. 1175 (1828).
Leaf-blades broadly elliptical to lanceolate, 10–30 cm. long, 4–10 cm. broad; pseudobulbs several-noded, 1·5–2 cm. long, formed annually; scape and raceme 15–35 cm. long; scape and flowers mostly reddish-purple; sepals and petals 5–5·5 mm. long; greenish streaked with purple; capsule deep purple, 1·5 cm. long, including the slender twisted pedicel.
Rather common on limestone rocks in thickets and open woodlands; 1200–3000 feet; fl. July–Jan, fr. Aug–Mar. *A 7529! H 7878! P 10619!* Florida to Brazil, West Indies.

25. MALAXIS Solander ex Sw. (1788)

1 Leaves usually two at flowering time, sometimes one withered or withering:
  2 Raceme subumbellately corymbose, not exceeding 1 cm. long, but elongating in fruit;
    lip 3-lobed   1. *umbelliflora*
  2 Raceme laxly subspicate, 3–8 cm. long:
    3 Lip with a distinct apical lobe, 3·5–4 mm. long   2. *spicata*
    3 Lip not lobed, 2·5 mm. long   3. *integra*
1 Leaf one at flowering time; raceme oblong, 3–4 cm. long; lip 3-lobed   4. *unifolia*

1. M. umbelliflora Sw., Nov. Gen. & Sp. Pl.: 119 (1788).
Plant in flower 10–25 cm. high; leaf-blades ovate, shortly acuminate, up to 10 cm. long and 6 cm. broad, petiole sheathing; flowers numerous, green, on pedicels to 12 mm. long; sepals about 3 mm. long; lip 4 mm. long.
Rather local (St. Andr., Port., St. Thom.), on mossy banks; 2000–5000 feet; fl. Sept–Feb, fr. Dec–Feb. *A 8706! H 9798! J.P. 2080!* Mexico, Cuba, Hispaniola, Lesser Antilles.

2. M. spicata Sw., Nov. Gen. & Sp. Pl.: 119 (1788).
Plant in flower 20–30 cm. high; pseudobulbs ovoid, slightly flattened, about 2·5 cm. long, 1·5 cm. broad; leaf-blades as above; flowers numerous, light green, except the lip darker in 2 median lines; pedicels 6–9 mm. long; sepals about 3 mm. long; lip 3·5–4 mm. long.
Occasional, in leafy soil among limestone rocks in woodlands and thickets; 600–4500 feet; fl. Sept–Jan, fr. Oct–Jan. *A 11764! H 10473! H & P 15043!* Greater Antilles, Martinique, Dominica.

3. M. integra (F. & R.) F. & R., Fl. Jam. **1**: 43 (1910).
Like the last, except somewhat smaller in all its parts; perhaps not really distinct.
Rare and local (St. Andr.), on rocky shaded banks; 2500–4000 feet; fl. and fr. Nov. *H 7735!* Endemic.

4. M. unifolia Michx., Fl. Bor. Amer. **2**: 157 (1803).—*M. grisebachiana* (F. & R.) F. & R. (1910).
Plant in flower 10–25 cm. high; leaf-blade elliptical, 2·5–5 cm. long, 1·5–3 cm. broad, sheathing below; flowers yellowish-green on ascending persistent pedicels; sepals about 2 mm. long; lip about 2·3 mm. long; capsule oblong, deeply 6-grooved, 6–7 mm. long.
Rather local (St. Andr., St. Ann, Port.), on shady banks in clay soil; 1600–5000 feet; fl. Oct–Jan, fr. Nov–Mar. *A 8714! H 9790! P 3767!* E. United States to Mexico and Guatemala, Cuba, Hispaniola.

26. ELLEANTHUS C. Presl (1827)

1 Flowers in a compact head; perianth pink; leaves ovate-lanceolate, plicate   1. *capitatus*
1 Flowers in spikes; perianth yellowish:
  2 Leaf-blades elliptic-lanceolate, up to about 2 cm. broad   2. *longibracteatus*
  2 Leaf-blades linear, up to 3 mm. broad   3. *linifolius*

1. E. capitatus (R. Br.) Reichb. f. in Walp., Ann. Bot. **6**: 475 (1862).
Plant tufted with canelike leafy shoots up to 150 cm. long; leaves 10–20 cm. long, 2–5 cm. broad; outer bracts tightly overlapping, acuminate, 2·5–4 cm. long; perianth 12–13 mm. long, embedded in mucilage, the central flowers opening first.
Locally common (St. Andr., Port.), on the branches of trees in humid woodland; 3000–4900 feet; fl. Apr–Oct. *A 8063! H 7717! P 9336!* Mexico to Peru and S. Brazil, Cuba, Hispaniola, Lesser Antilles.

2. E. longibracteatus (Lindl. ex Griseb.) Fawcett, Prov. List Jam.: 38 (1893).
Shoots tufted, slender and canelike, 50–100 cm. long in flower; leaves 8–18 cm. long, 1–2 cm. broad; bracteoles spathaceous 1·5–2·5 cm. long, exceeding the flowers, acuminate; sepals 7·5–8·5 mm. long; lip shorter, saccate at the base, curled, eroded at the tip.
Widespread but nowhere common on limestone or serpentine rocks in thickets and open woodlands or epiphytic on isolated trees; 1400–5000 feet; fl. Mar–July, fr. June–July. *A 7311! H 8556! H & P 14204!* Colombia, Ecuador, Dominica.

3. E. linifolius C. Presl, Rel. Haenk. **1**: 97 (1827).
Shoots clustered, slender, 12–35 cm. long; leaves up to 12 cm. long, 2–3 mm. broad; flowers few in a distichous spike; bracteoles about 10 mm. long, obtuse; sepals ovate-lanceolate, about 3 mm. long; lip longer, obovate-lanceolate, ciliate-denticulate.
Rare, epiphytic in woodland on limestone; 2000–2500 feet; fl. and fr. Mar–May. *A 12848! H 11022! P 11741!* Mexico to Brazil and Peru, Greater Antilles.

27. LYCASTE Lindl. (1843)

1. L. barringtoniae (Sm.) Lindl. in Edw., Bot. Regist. **30** Misc.: 43 (1844).
Pseudobulbs tightly approximated, bluntly 4-angled, 8–9 cm. long, about 5 cm. broad; leaf-blades tapering to both ends, 30–50 cm. long, 5–12 cm. broad; scapes bracteate, 6–12 cm. long; posterior bract hooded, 2·8 cm. long; ovary nearly 2 cm. long in flower, twisted; perianth dull green; chin 1 cm. long, enclosing the foot of the column; median sepal 4 cm. long, 1·6 cm. broad, subacute, lateral slightly shorter, fleshy especially towards the tip; petals similar; lip whitish with the fimbriate margin mauve, 3-lobed, 3 cm. long; capsule with 6 broad rounded ridges, 5 cm. long.
Rather uncommon in rocky woodlands and thickets, a low epiphyte or on cliffs or mossy boulders; 2500–3800 feet; fl. mostly Apr–July. *A 7471! Harris! J.P. 2328!* Cuba, Hispaniola.

28. XYLOBIUM Lindl. (1825)

Pseudobulbs bearing one leaf at the apex; median sepal 18 mm. long, 5 mm. broad
1. *palmifolium*
Pseudobulbs bearing two leaves at the apex; median sepal 14 mm. long, 4 mm. broad
2. *stachyobiorum*

1. X. palmifolium (Sw.) Fawcett, Prov. List Jam.: 39 (1893).
Pseudobulbs tightly approximated, terete, lightly ribbed or smooth, mostly 5–6 cm. long and up to 2 cm. broad; leaf-blades lanceolate, up to about 40 cm. long and 6 (–8) cm. broad; scape simple, bracteate, about 10 cm. long, with 6–9 fragrant flowers; perianth yellowish with minute brown spots; sepals light yellow, reflexed lengthwise; petals lighter yellow attached at side of column; lip white, entire, tip of spur bright yellow; capsule 3–3·5 cm. long, 3-winged, foot and column persistent.
Rather local (St. Andr., Port., Trel.), epiphytic on trees or on rocks in thickets; 2500–5000 feet; fl. July–Dec, fr. Mar. *A 7676! Miss Barrett! Syme!* Greater Antilles, Trinidad.

2. X. stachyobiorum (Reichb. f.) Hemsl., Biol. Centr. Amer. Bot. **3**: 252 (1883).
Like the last with longer narrower leaves and longer more numerous-flowered scapes; flowers smaller; lip 3-lobed distally.
Very rare (St. Andr., St. Mary), on peaty banks; about 3500 feet; fl. Jan–Mar. Not recently collected. *Harris!* Nicaragua, Panama.

29. COCHLEANTHES Raf. (1836)

1. C. flabelliformis (Sw.) R. E. Schultes & Garay in Bot. Mus. Leafl. Harvard **18**: 324 (1959).—*Zygopetalum flabelliforme* (Sw.) Reichb. f. (1863).
Leaf-blades distichous, oblanceolate, up to 25 cm. long and 5 cm. broad, articulated near the base, shortly acuminate; scape simple, solitary from the axil of a sheath or lower leaf, 5–7 cm. long, with 2 overlapping bracts at the base of the pedicel; sepals and petals similar, about 3 cm. long and 1·3 cm. broad, light green, ovate-lanceolate, acute; lip 3 cm. long and 2·5 cm. broad, white with numerous purple lines and crests; capsule 3·5–7·5 cm. long, with 6 rounded ribs.
Locally common in humus on limestone ledges and rocky thickets or epiphytic in shade; 900–5000 feet; fl. Aug–Mar, fr. Nov–Mar. *A 6551! H 7576! P 9326!* Greater Antilles, Trinidad, tropical S. Amer.

30. MAXILLARIA Ruiz & Pav. (1794)

1 Pseudobulbs poorly developed; leaves fleshy, more or less triquetrous, flabellate; flowers appearing singly from the lower axils   1. *crassifolia*
1 Pseudobulbs well developed; leaves thin, more or less flat, not flabellately disposed:
  2 Pseudobulbs approximated; leaf terminal on the pseudobulb; peduncles basal, not subtended by leaves of foliaceous sheaths   2. *rufescens*
  2 Pseudobulbs more or less distant, separated by lengths of creeping scale- or leaf-covered rhizome; peduncles arising in the axils of leaves or foliaceous sheaths:
    3 Fully developed leaves usually solitary on the pseudobulbs only; leaf-blades 10 cm. or more long:
      4 Peduncles clustered in the sheath-axils; perianth about 5 mm. long, mostly yellow
3. *purpurea*
      4 Peduncles solitary in the sheath-axils; perianth 1·5–2 cm. long, white   4. *alba*
    3 Fully developed leaves usually paired on the pseudobulbs and also along the creeping stem; leaf-blades less than 3 cm. long; perianth deep purple   5. *swartziana*

1. M. crassifolia (Lindl.) Reichb. f. in Bonplandia **2**: 16 (1854).—*M. sessilis* (Sw.) F. & R. (1910), not Lindl. (1845).
Stem very short, the leafy branches developing close together; leaf-blades 15–30 cm. long, 15–25 mm. broad; peduncles several together, developing singly, very short; flowers upright, light yellow, sometimes cleistogamous; sepals leathery, about 1·5 cm. long; petals shorter, acute; capsule with 6 equal flattened ribs.

Widespread and locally common on rocks or tree trunks; 500–2500 feet; fl. Oct–Mar, fr. Dec–July. *A 6512! H 9006! P 7493!* Cuba, Hispaniola.

2. M. rufescens Lindl. in Edw., Bot. Regist. **21**: under t. 1802 (1835); **22**: t. 1848 (1836).

2a. Var. rufescens.
Pseudobulbs 3–3·5 cm. long, 1·5 cm. broad; leaf-blades 14–18 cm. long, 2 cm. broad; peduncles 2–3 cm. long; perianth creamy-yellow; sepals oblong, about 2 cm. long, the lateral spreading.
Rare (Clar., Trel.), in limestone crevices and epiphytic on small trees; 1750–2500 feet; fl. and fr. Apr–May, Nov. *A 12424! P 11385!* Honduras to Brazil, Cuba, Hispaniola, Trinidad.

2b. M. rufescens var. **minor** F. & R. in Journ. Bot. **48**: 108 (1910).
Like the last with smaller flowers; perianth green tinged crimson; sepals 1–1·3 cm. long; young fruit crimson, tapered at the base, topped by the thickened persistent column.
Rare and local (St. Andr., Port.), epiphytic on small trees in thickets; 2250–3600 feet; fl. Apr, Sept, fr. Sept. *H 7615! P 8565!* Tropical Amer., Cuba, Hispaniola.

3. M. purpurea (Spreng.) Ames & Correll in Bot. Mus. Leafl. Harvard **11**: 16 (1943).—*Camaridium purpureum* Spreng. (1826). *Ornithidium vestitum* (Sw.) Reichb. f. (1863), not *M. vestita* Schltr. (1924).
Plant trailing and pendulous, branched, to 1 m. long; pseudobulbs more distant on older plants, somewhat compressed, 2–3 cm. long; leaf-blade lanceolate, midrib sharp beneath, 10–20 cm. long, 1·5–2 cm. broad; peduncles bracteate, hardly exserted; sepals broadly elliptical, acute, about 5 mm. long and 3 mm. broad; capsule often purplish, smooth, 7–10 mm. long.
Rather common as an epiphyte on large trees and also on boulders and cliffs in sheltered places; 400–5000 feet; fl. mostly Aug–Apr, fr. Dec–July. *A 9988! H 7839! H & P 15142!* Honduras to Brazil and Peru, Cuba, Hispaniola.

4. M. alba (Hook.) Lindl., Gen. & Sp. Orch.: 143 (1832).
Stem trailing, branched 30–45 cm. long; pseudobulbs ellipsoid-compressed, 4–5 cm. long; leaf-blade linear-lanceolate, 20–30 cm. long, 1·5–2 cm. broad; peduncles bracteate, shortly exserted; sepals oblong, acuminate, about 2 cm. long, and 5 mm. broad; capsule cylindrical, about 3 cm. long.
Rather rare although widely scattered, mostly on rock-ledges but also epiphytic; 2500 feet; fl. July–Mar. *A & S 3222! Harris! J.P. 465! P 8210!* Guatemala to Brazil, Cuba, Trinidad.

5. M. swartziana Adams in Amer. Orch. Soc. Bull. **35** (12): 998 (1966).—*Epidendrum proliferum* Sw. (1788). *Ornithidium proliferum* (Sw.) F. & R. (1910), not *M. prolifera* Ruiz & Pav. (1798).
Stem creeping and branching, 10–25 cm. long; pseudobulbs obovoid-compressed, 1·5–2 cm. long; leaf-blades oblong or oblanceolate, 1·5–2·5 cm. long, 4–7 mm. broad; peduncles 7 mm. long with three bracts at the base; sepals obtuse, 8 mm. long, 3 mm. broad; capsule narrowly spindle-shaped, 1·5 cm. long.
Rare and local (St. Andr., Port.), on rocks in damp shady places and on mossy tree trunks; 2000–4000 feet; fl. Nov–Dec, fr. Dec. *H 7566! P 18376!* Endemic.

31. NEOURBANIA F. & R. (1909)

1. N. adendrobium (Reichb. f.) F. & R. in Journ. Bot. **47**: 125 (1909).
Stem trailing, scrambling and sparingly branched, up to 1 m. or more long; leaves linear-lanceolate, 12–15 cm. long, up to 2 cm. broad; peduncles bracteate, short, several together; perianth about 5 mm. long, white; lip emarginate, lobed near the base; foot 2 mm. long.
Rare and local (St. Ann, Port., St. Thom.), on mossy banks, shrubs and trees; 1800–2500 feet; fl. Sept–Apr, fr. Nov–Apr. *H 7650! P 5698!* Cuba.

32. BULBOPHYLLUM Thouars (1822) nom. cons.

Scape and rachis slender; petals ciliate   1. *jamaicense*
Rachis fleshy and swollen; petals glabrous   2. *pachyrachis*

1. B. jamaicense Cogn. in Fedde, Repert. Sp. Nov. **7**: 122 (1909).

Plant in flower 5–10 cm. high; pseudobulbs about 1 cm. in diameter, wrinkled-globose, tinged brownish, with 1 leaf; leaf-blade acute, green, 1–4 cm. long; scape purple; flowers (2–) 3–6; sepals green, striped or tinged purple, acute, 4·5–5 mm. long; petals shorter, purple; lip bright red, ligulate, ciliate towards the base, mobile.

Rare and local (St. Andr., St. Thom.), epiphytic on small trees in open rocky woodland; 2500–3000 feet; fl. Jan–Mar, fr. Jan–July. *A 12136! H 7998!* Endemic.

2. B. pachyrachis (A. Rich.) Griseb., Fl. Br. W.I.: 613 (1864).

Plant in flower 10–45 cm. high; pseudobulbs about 2 cm. long, 4-winged, each with 2 leaves; leaf-blades leathery, obtuse, 7–20 cm. long, 8–25 mm. broad; flowers numerous, 4–5 mm. long; petals about 2 mm. long; lip thick, triangular in section.

Rare (Han., St. Eliz.), on trees; 500 feet; fl. and fr. Jan. Not recently collected. *H 7634! J.P. 525!* Mexico to Panama, Greater Antilles, Trinidad.

33. POLYSTACHYA Hook. (1824) nom. cons.

Flowers prominently chinned; perianth more than 3 mm. long; capsule 10–12 mm. long
   with a prominent crescent-shaped remnant of column and foot   1. *extinctoria*
Flowers obscurely chinned; perianth 2–2·5 mm. long; capsule 6–9 mm. long   2. *cerea*

1. P. extinctoria Reichb. f. in Walp., Ann. Bot. **6**: 638 (1863).—*P. minuta* (Aubl.)
     Britton (1903), not A. Rich. & Gal. (1845).

Plant in flower 20–35 cm. high; pseudobulbs about 3 cm. long; leaves 10–20 cm. long, 12–25 mm. broad; median sepal ovate, acute, lateral 4 mm. broad, light greenish-cream; petals linear-spathulate, light greenish-cream; lip light green with white margins, fading yellow.

Uncommon on trees and boulders in thickets and woodland margins; 1700–3500 feet; fl. July–Apr, fr. Sept–Apr. *A 8062! H 7765! P 5282!* Florida, Mexico to Brazil and Peru, Bahamas, West Indies.

2. P. cerea Lindl. in Edw., Bot. Regist. **26** Misc.: 86 (1840).—*P. minor* F. & R.
     (1910).

Plant in flower 12–30 cm. high; pseudobulbs up to 2 cm. long; leaves as above; median sepal oblong, lateral about 2 mm. broad, green, ageing deep yellow; petals linear-spathulate, light green; lip yellow-green, fading yellow.

Common on open cuttings, in rocky thickets and epiphytic; 500–3600 feet; fl. Oct–Mar, fr. Nov–Apr. *A 8922! Fawcett 10434! H & P 15177!* Mexico to Guyana, Bahamas, West Indies.

34. BRACHIONIDIUM Lindl. (1859)

1. B. sherringii Rolfe in Kew Bull. **1893**: 4 (1893).

Plant in flower 2–7 cm. high; leaf-blade oblong-lanceolate, 1·5–3 cm. long, 4–12 mm. broad; peduncle longer than the leaf; bracts persistent; perianth about 2·5 cm. long, mostly dull crimson tinged tawny-yellow on the sepals; capsule 9 mm. long, oblique, with 6 cartilaginous ribs.

Rather rare (St. Andr., Port.), epiphytic on small trees or on rotten logs or mossy clay banks; 2500–3500 feet; fl. Oct–June, fr. Dec–July. *A 6672! H 10474! P 9857!* Cuba, Hispaniola, Lesser Antilles to Grenada.

35. CRYPTOPHORANTHUS J. B. Rodrigues (1882)

1. C. atropurpureus (Lindl.) Rolfe in Gard. Chron. ser. 3, **2**: 693 (1887).

Plant rhizomatous, 7–15 cm. high; shoot below the leaf-blade covered with dilated sheaths increasing in size upwards; flowers barely exserted from the top-most sheath; perianth (sepals) 14–15 mm. long; petals 3-toothed, 4 mm. long; lip 4·5 mm. long, 3-lobed; capsule about 1 cm. long, ridged with 6 pairs of wings.

Uncommon as a low-level epiphyte and on banks and fallen logs; 2000–4000 feet; fl. July–Dec, fr. July–Apr. *ACH 660! H 10466! H & P 15140!* Cuba, Hispaniola.

36. STELIS Sw. (1800) nom. cons.

Closed flowers triangular in outline, sharply 3-cornered; sepals broadly deltate, equal
1. *ophioglossoides*
Closed flowers rounded in outline; sepals ovate, unequal, the lateral subrotundate
2. *micrantha*

1. S. ophioglossoides (Jacq.) Sw. in Schrad., Journ. Bot. **1799** (2): 239 (1800).

Rhizome shortly creeping, the shoots approximate; leaf-blade hard, thick, fleshy, oblong-lanceolate, narrowed to the base, 5–12 cm. long, 1–2 cm. broad; racemes 1–several, slender about 10–20 cm. long; bracteoles shortly acuminate, 2–3 mm. long, usually about twice their own length apart; pedicels pink; sepals light greenish-cream or margined crimson, about 2 mm. long; lip and petals purple or deep crimson.

Rather common as an epiphyte in shady woodland, sometimes on logs or boulders; 400–7400 feet; fl. and fr. most of the year. *A 11474! H 7764A! P 5139!* Cuba, Hispaniola, Dominica, Trinidad.

2. S. micrantha (Sw.) Sw. in Schrad., Journ. Bot. **1799** (2): 240 (1800).

Like the last; the shoots mostly more robust; leaf-blade 7–16 cm. long; racemes solitary or paired, 12–15 cm. long, up to about 50-flowered; bracteoles acute, 1·5–2 mm. long, mostly almost contiguous; sepals greenish-cream to white, 2·5–3·5 mm. long; petals and lip purple.

Uncommon as an epiphyte or on fallen logs in woodlands; 3000–7350 feet; fl. and fr. July–Aug, Dec–Apr. *A 10618! A & S 3379! H 7835! P 4187!* Endemic.

A small epiphyte with wholly purple flowers collected in Trelawny (*A 12833!*) is probably undescribed.

37. PLEUROTHALLIS R. Br. (1813)

1 Spathe subtending racemes conspicuous, leathery, 1–3 (–4) cm. long:
  2 Raceme exceeding the leaf, solitary:
    3 Perianth purple; sepals half-connate; lip longer than the petals; leaf-blade narrowed
      to a slender petiole   1. *oblongifolia*
    3 Perianth yellow; sepals connate to the apex; lip shorter than the petals; leaf-blade
      tapered to the base but hardly petiolate   2. *racemiflora*
  2 Racemes shorter than the leaf; flowers yellow; leaf-blade narrowed to a distinct
    petiole; sepals about half-connate; lip shorter than the petals:
    4 Lip entire; sepals about 7 mm. long; leaf including the petiole 10–25 cm. long,
      3–8 cm. broad, blade oblong-elliptical   3. *gelida*
    4 Lip 3-lobed; sepals about 4 mm. long; leaf including the petiole 10–12 cm. long,
      2–2·5 cm. broad, blade oblanceolate   4. *velaticaulis*
1 Spathe inconspicuous, scarious:
  5 Sheaths of the stem straight, more or less appressed:
    6 Leaf-bearing stems evident, mostly as long as or longer than the leaf:
      7 Inflorescences racemose, several-flowered, as long as or slightly longer than the leaf;
        sepals 8 mm. long:
        8 Sepals greenish-yellow; petals 2·5–3 mm. long; lip purple   5. *alpestris*
        8 Sepals purple or purple-veined; petals 4 mm. long; lip purple-spotted   6. *laxa*
      7 Inflorescences 1–few-flowered, shorter than the leaf:
        9 Leaf-blade 10 cm. or more long; peduncles 1-flowered, several to many together:
          10 Sepals 7 mm. long, greenish; petals 3·5 mm. long; flowers numerous, appearing
            together   7. *ruscifolia*
          10 Sepals 15–17 mm. long, purplish; petals 8 mm. long; flowers few, mostly
            appearing singly   8. *uncinata*
        9 Leaf-blade less than 8 cm. long:
          11 Lower sheaths and rhizome covered with reddish-brown setulose hairs; racemes
            4–5 flowered   9. *hirsutula*

11 Lower sheaths glabrous:
  12 Ovary warty; inflorescence 2-flowered   10. *monophylla*
  12 Ovary smooth:
    13 Inflorescence racemose with (2–) 3–6 flowers; plants tufted:
      14 Raceme solitary; sepals light green; flowers cleistogamous; leaf narrowly
        oblong-lanceolate, keeled, thick   11. *pruinosa*
      14 Racemes 2–4; sepals purple; flowers opening normally; leaf narrowly ovate
        to elliptical   12. *odontotepala*
    13 Inflorescence usually 1-flowered, 1–3 together but only 1 flowering at a time;
      plants shortly rhizomatous   13. *wilsonii*
6 Leaf-bearing stems very short, mostly shorter than the leaf:
  15 Leaf-bearing branches widely separated along a creeping rhizome:
    16 Leaf-blade elliptical to suborbicular; flower solitary, purple:
      17 Leaf-blade 15 mm. or more long; sheaths glabrous   14. *testifolia*
      17 Leaf-blade less than 5 mm. long; sheaths setulose   15. *nummularia*
    16 Leaf-blade oblanceolate; flower(s) 1 or 2, filiform-peduncled, yellow
                            16. *sertularioides*
  15 Leaf-bearing branches tufted or rhizome very short:
    18 Inflorescence shorter than the leaf-blade:
      19 Capsule softly prickly; sepals brick-red, usually remaining joined at the tip
                           17. *tribuloides*
      19 Capsule smooth; sepals purple   18. *jamaicensis*
    18 Inflorescence exceeding the leaf-blade:
      20 Leaf-blade obovate, nearly as broad as long   19. *rotundifolia*
      20 Leaf-blade linear-lanceolate to oblanceolate or elliptical:
        21 Inflorescence 1-flowered; lateral sepals connate nearly to the apex; leaf-blade
          narrowly elliptical   20. *corniculata*
        21 Inflorescence 2–several-flowered; lateral sepals not more than one-third
          connate:
          22 Sepals tailed-acuminate; leaf-blade elliptic-oblanceolate   21. *helenae*
          22 Sepals obtuse:
            23 Peduncle usually longer than the leaf-blade; leaf-blade oblanceolate
                           22. *delicatula*
            23 Peduncle about as long as the leaf-blade; leaf-blade linear-lanceolate
                           23. *lanceola*
5 Sheaths of the stem dilated, margins ciliate; peduncle exceeding the leaf-blade; stems
  more or less clustered:
  24 Lateral sepals about half-connate, acuminate; petals obtuse; lip 3-lobed
                            24. *trilobata*
  24 Lateral sepals connate nearly to the apex; petals acute or acuminate; lip entire
                            25. *foliata*

1. P. oblongifolia Lindl. in Comp. Bot. Mag. **2**: 355 (1837).—*P. multirostris*
Reichb. f. (1877). *P. racemiflora* (Sw.) Lindl. (1825), not Lodd. (1824).

Plant 12–30 cm. high in flower; shoots clustered with 2 internodes below the leaf;
leaf-blade oblong, 5–9 cm. long, 1·5–3 cm. broad, rounded at the tip; inflorescence
8–25 cm. long with 10–14 flowers laxly in two rows; sepals 7–9 mm. long; petals
4 mm. long; capsule 7–9 mm. long.

Locally common (St. Andr., Port., St. Thom.), as an epiphyte on mossy tree
trunks and branches; (2750–) 3600–5500 feet; fl. Aug–Jan, fr. Dec–May. *A 7461!*
H 7818! P 9516! Cuba, Hispaniola.

2. P. racemiflora Lodd., Bot. Cab. **10**: t. 949 (1824).—*P. longissima* Lindl. (1859).

Rhizome shortly creeping; plant 30–45 cm. high in flower; shoot 5–13 cm. long
with 2 internodes below the leaf; leaf-blade oblanceolate, 8–16 cm. long, up to about
2·5 cm. broad, obtuse, slightly emarginate, leathery; peduncle 12–15 cm. long;
raceme about 20 cm. long with 20–35 flowers in two rows; sepals 8–9 mm. long;
petals 7–8 mm. long; lip about 4 mm. long, obtusely spathulate; capsule about
1·5 cm. long.

Rather rare in the eastern parishes, on rocks, rotten logs or on trees; 2000–
3600 feet; fl. Mar–July, fr. May–July. *A 11220. ACH 642! H 10424! H & P 14306!*
Mexico to Panama, Puerto Rico.

3. P. gelida Lindl. in Edw., Bot. Regist. **27** Misc.: 91 (1841).

Like the last; racemes 1–3, usually 2, 10–16 cm. long, each with 16–20 secund
very shortly pedicelled flowers; petals about 3 mm. long; capsule about 8 mm. long.

Rare (St. Andr., Manch.), on rocky banks and trees in shady woodland; about 3500 feet; fl. Dec–Feb, fr. Feb. *H 7836!* Florida, Greater Antilles.

4. P. velaticaulis Reichb. f. in Linnaea **22**: 824 (1850).—*P. crassipes* Lindl. (1859).
Like the last; racemes 1–5, 6–10 cm. long, each with up to about 20 flowers; petals about 2 mm. long; capsule oblong, about 8 mm. long.
Very rare and local (St. Andr.), on mossy rocks and trees; 5000 feet; fl. and fr. Nov. *H 7745! A. von der Porten!* C. Amer. to Venezuela and Peru, Greater Antilles.

5. P. alpestris (Sw.) Lindl., Gen. & Sp. Orch.: 7 (1830).
Plant 7–12 cm. high; leaf-blade 3–4·5 cm. long, 7–15 mm. broad, narrowly ovate to lanceolate, rather hard and thick-textured, sharply emarginate; lateral sepals connate nearly to the apex.
Uncommon, in the eastern parishes, on tree trunks and branches; 2400–4000 feet; fl. Nov–Apr, fr. Apr. *A 12691! H 8908! P 5976!* Endemic.

6. P. laxa (Sw.) Lindl., Gen. & Sp. Orch.: 7 (1830).
Like the last; leaf-blade 3–4 cm. long, 10–13 mm. broad, oblong-ovate; racemes longer than the leaf, up to 6-flowered; lateral sepals half-connate.
Rare and local (St. Andr., Port., St. Thom.), on tree trunks; 4000–5500 feet; fl. Aug–Dec. *A 11613! H 7536! J.P. 170!* Endemic.

7. P. ruscifolia (Jacq.) R. Br. in Ait., Hort. Kew. ed. 2, **5**: 211 (1813).
Plant 15–30 cm. high; stems clustered on a creeping rhizome, 4–20 cm. long; leaf–blade (6–) 10–15 cm. long, 1·5–3·5 cm. broad, the apex more or less acuminate; flowers fascicled at the base of the leaf; capsule 10–12 mm. long.
Rather rare (Manch., St. Ann and the eastern parishes), on rocks and trees in shady woodlands; 2000–3100 feet; fl. June–Aug, Dec. *A 8489! ACH 668! Maxon 9461! Yuncker 18249!* Guatemala to Panama, Bolivia and Peru, West Indies.

8. P. uncinata Fawcett in Journ. Bot. **33**: 12 (1895).
Rhizome rather stout, creeping; leaf-blade 15–25 cm. long, 2·5–3 cm. broad, oblong-lanceolate with numerous prominent nerves; peduncles 3–4 cm. long, roughly tomentose; floral bracts sheathing the ovary, 1·3 cm. long; sepals covered with clusters of rough hairs; lip indistinctly 3-lobed at the tip, proximal lobes hookshaped; capsule rough, about 2 cm. long.
Local (St. Andr., Port.), an epiphyte low on tree bases and fallen trunks; 3900–4500 feet; fl. mostly May–July. *A 11618! ACH 296! H 10092! P 7574!* Venezuela.

9. P. hirsutula F. & R. in Journ. Bot. **47**: 3 (1909).
Plant 10–18 cm. high; leaf-blade 5–7·5 cm. long, 10–13 mm. broad, lanceolate; raceme 1·2–2·5 cm. long; sepals about 6 mm. long, deep reddish-purple; petals about 3 mm. long; lip slightly shorter than the petals with an undulate margin.
Rather rare, in the central parishes, on tree trunks and limestone ledges; 2000–3000 feet; fl. May–Aug, fr. Jan, Aug. *ACH 746! H 9890! 10984! P 7987!* Endemic.

10. P. monophylla (Hook.) F. & R., Fl. Jam. **1**: 60 (1910).
Plant about 5 cm. high; leaf-blade 3–4 cm. long, 12 mm. broad, elliptic-lanceolate, obtuse; sepals 7–8 mm. long, green; petals 2–3 mm. long, pinkish in the mid-line with greenish margins; lip as long as the petals, pink or purple.
Very rare; only once collected. *Wiles.* Endemic.

11. P. pruinosa Lindl. in Edw., Bot. Regist. **28** Misc.: 75 (1842).
Plant 5–8 cm. high; leaf-blade 2–4 cm. long, 6–7 mm. broad, thick and sub-triquetrous; racemes slender, up to 6-flowered; sepals 2–3 mm. long, ovate; petals a little shorter than the sepals; lip entire, about 1·5 mm. long; flowers cleistogamous and always forming fruit; capsule oblong, about 6 mm. long.
Uncommon, epiphytic mostly on the trunks of small trees; 2500–3500 feet; fl. Apr, Sept, fr. Jan, Apr. *H 7997! 10477! P 8572!* Costa Rica to Venezuela and the Guianas, West Indies.

12. P. odontotepala Reichb. f. in Flora **48**: 275 (1865).—*P. brachypetala* Griseb. (1866).

Plant 5–8 cm. high; leaf-blade 3–4 cm. long, 10–14 mm. broad; racemes about 1 cm. long, 2–5-flowered; sepals 4 mm. long; petals about 2 mm. long; lip lobed below the middle.

Rare (St. Andr., Port.), on mossy tree trunks; 2500–4500 feet; fl. Nov–Apr. *H 7852! Morris!* Cuba, ? Puerto Rico.

13. P. wilsonii Lindl. in Ann. & Mag. Nat. Hist. ser. 3, **1**: 326 (1858).—*P. confusa* F. & R. (1909). *P. morrisii* F. & R. (1909).

Rhizome shortly creeping; leafy and flowering shoots 5–8 cm. high; leaf-blade 2–5 cm. long, 5–8 mm. broad; peduncles 3–5 mm. long; sepals about 4 mm. long; petals 2–2·5 mm. long; lip about as long as the petals.

Rare (Manch., Port.), epiphytic on tree trunks; 2300–2500 feet; fl. Aug–Nov, fr. Dec–May. *Morris 28! E. C. Tomlinson! Wilson!* Greater Antilles, Guadeloupe.

14. P. testifolia (Sw.) Lindl. in Ann. & Mag. Nat. Hist. ser. 3, **1**: 328 (1858).

Rhizome creeping and branched, 12–20 cm. long; leaf-blade 1·5–3 cm. long, 1–2 cm. broad; sepals 5·5–6 mm. long, dark purple, covered with long white hairs; petals 2·5–3 mm. long, light purple with eroded tips; lip about 2·5 mm. long with white hairs on the margin; ovary hairy.

Rather rare (Manch., St. Andr., Port.), epiphytic on trees; 2200–4000 feet; fl. June, Sept–Apr, fr. Mar–Apr. *ACH 256! H 10484! J.P. 2122!* Costa Rica, Colombia, Venezuela, Cuba, Hispaniola, Martinique.

15. P. nummularia Reichb. f. in Flora **48**: 276 (1865).

Rhizome creeping, crisped-pubescent, branched, appressed to support; leaves on very short branches; leaf-blade 3–4 mm. long, 2–3·5 mm. broad; pedicel sheathed; sheath reddish, setulose; perianth purple, hairy, about 4·5 mm. long.

Rare (Trel.), epiphytic on trees; 2200–2400 feet; fl. July–Aug. *Hespenheide!* Cuba.

16. P. sertularioides (Sw.) Spreng. in L., Syst. Veg. ed. 16, **3**: 731 (1826).

Rhizome slender, creeping and branched extensively; leaf-blade 1·5–2·5 cm. long; sepals about 4 mm. long, the lateral shortly fused, light green to cream with deep yellow tips; petals 3 mm. long, very light green with deep yellow tips; lip about 2·5 mm. long.

Locally common on the branches of low trees in wet areas; 500–2500 feet; fl. July–Nov, fr. Dec–Jan. *A 8244! H 10487! P 16707!* Mexico, Guatemala, Honduras, Cuba, Trinidad.

17. P. tribuloides (Sw.) Lindl., Gen. & Sp. Orch.: 6 (1830).

Plant 3–6 cm. high; leaf-blade narrowly elliptical to oblanceolate, 3–5 cm. long, 8–10 mm. broad, leathery; peduncles very short 1-flowered; sepals papillose, 6 mm. long, the lateral half-connate; petals and lip about 2 mm. long; capsule broadly ellipsoid to subglobose, the prickles whitish.

Common on tree trunks and low branches in thickets, occasionally in limestone crevices; 400–2900 feet; fl. and fr. sporadically throughout the year, most frequently in July. *A 7335! H 7648! P 15611!* Mexico to Costa Rica, Cuba.

18. P. jamaicensis Rolfe in F. & R. in Journ. Bot. **47**: 122 (1909).

Plant 2–4 cm. high; leaf-blade elliptical, obtuse, leathery 1·5–2·5 cm. long; racemes 1–1·5 cm. long, 1–few-flowered; sepals 7–8 mm. long; petals about 6 mm. long, linear-lanceolate, acuminate; lip 2 mm. long.

Very rare; known only from one collection. *Morris!* Endemic.

19. P. rotundifolia Rolfe in Kew Bull. **1895**: 191 (1895).

Plant 3–5 cm. high; leaf-blade fleshy, 6–8 mm. long, 5–7 mm. broad; scape 4–5 cm. long with up to 6 flowers; sepals narrowly ovate, acute 4 mm. long, reddish-purple; petals 1·4 mm. long, yellow-buff with a reddish-purple nerve; lip 1·7 mm. long.

Very rare; known only from one collection. *Morris!* Endemic.

20. P. corniculata (Sw.) Lindl. in Edw., Bot. Regist. **28** Misc.: 83 (1842).

Plant 3–5 cm. high; leaf-blade 1–2 cm. long, 5–7 mm. broad, shortly stalked; peduncle 2·5–3·5 cm. long, filiform, bracteate above the middle; sepals lanceolate, acuminate, 5 mm. long, the lateral connate almost to the apex, light yellow; petals 2·7 mm. long; lip 2·5 mm. long, yellow; capsule oblong, whitish, 6 mm. long.

Rather common on tree trunks at the margins of wet woodland in limestone areas; 400–2400 feet; fl. and fr. most of the year. *A 8085! H 9781! P 7986!* Mexico to Costa Rica, Cuba.

21. P. helenae F. & R. in Journ. Bot. **47**: 4 (1909).

Plant 3–5 cm. high; leaf-blade 10–15 mm. long, 2–3 mm. broad; racemes branched, 1·5–5 cm. long; sepals and petals light green with yellow tips, sepals 3·2 mm. long, petals 2·6 mm. long; lip deep purple, 1·5 mm. long, 3-lobed, the lateral lobes rounded and fimbriate-margined; capsule 3–4 mm. long.

Rare and local (St. Andr., St. Thom.), on the mossy trunks and branches of small trees; 3500–5250 feet; fl. and fr. May–Aug. *A 11355! Harris! P 6674!* Hispaniola.

22. P. delicatula Lindl., Fol. Orch. Pleuroth.: 38 (1859).

Plant up to 4 cm. high in flower; leaf-blade 1–2 cm. long; racemes flexuous, several-flowered; sepals about 3·5 mm. long, green streaked purple or all purple; petals about 2 mm. long, acute; lip about 3 mm. long, lateral lobes serrulate; capsule 6 mm. long.

Rather uncommon on mossy tree trunks in wet woodlands; 1000–4000 feet; fl. and fr. sporadically throughout the year. *ACH 748! 945! H 7826! P 7413!* Endemic.

23. P. lanceola (Sw.) Spreng. in L., Syst. Veg. ed. 16, **3**: 731 (1826).

Plant 4–6 cm. high; leaf-blade 3–4 cm. long, 3–3·5 mm. broad, shortly apiculate; racemes 2–4-flowered; sepals 5–6·5 mm. long, lanceolate, orange or scarlet; petals and lip about 2·5 mm. long; capsule 5 mm. long.

Rare and local (Port.), on trees and rocks; 2500–4000 feet; fl. and fr. Oct–Feb. *H 7742! 10479! J.P. 231!* Endemic.

24. P. trilobata F. & R. in Journ. Bot. **47**: 4 (1909).

Plant 1–2 cm. high; leaf-blade about 7 mm. long and 3 mm. broad; scape usually solitary, 1 or 2-flowered, about 2 cm. long; sepals about 4 mm. long, the lateral half-connate; petals 1·6 mm. long, obtuse; lip 1·5–1·7 mm. long, the lateral lobes spreading.

Rare and local (St. Andr., Port.) on tree trunks; 4000–6000 feet; fl. Feb, Sept. *Britton 267! H 10084!* Endemic.

25. P. foliata Griseb., Fl. Br. W.I.: 610 (1864).—*P. broadwayi* Ames (1908).

Plant in flower 5–8 cm. high; leaves sometimes more than 1 on each shoot, blade 8–12 mm. long, about 5 mm. broad; racemes 15–25 mm. long, 4–8-flowered; sepals 2–3 mm. long, yellow; petals 1–1·5 mm. long; lip linear, obtuse, a little longer than the petals, dark purple-tipped; capsule 3–4 mm. long.

Very rare (St. Andr., Trel.), epiphytic on trees; about 1500 feet; fl. Mar, Dec. *J. G. Hawkes 2276! P 4133! Wilson!* Mexico to Brazil, Cuba, Puerto Rico, Lesser Antilles. *P. broadwayi* may be a distinct species, the fruit being described as 2–3 mm. long, but the affinity with *P. foliata* is very close.

38. LEPANTHOPSIS Ames (1933); Garay (1953)

Racemes longer than the leaf with 5–8 laterally distichous flowers; sepals 1·5–2·5 mm. long; petals rounded   1. *melanantha*
Racemes shorter than the leaf, 2–5-flowered; sepals scarcely 1 mm. long; petals spathulate   2. *microlepanthes*

1. L. melanantha (Reichb. f.) Ames in Bot. Mus. Leafl. Harvard **1** (9): 19 (1933).
    Pleurothallis— melanantha Reichb. f. (1865). *Lepanthes brevipetala* F. & R.
    (1909). *Lepanthes harrisii* F. & R. (1909).

Plant 3–8 cm. high in flower; sheaths with spreading bristles; leaf-blade elliptical to narrowly elliptical, 1·5–4 cm. long, 7–10 mm. broad; flowers dark crimson; lip semicircular, 1–1·3 mm. long.

Rare (St. Ann, Port.), epiphytic on trees; 2600–4000 feet; fl. and fr. Dec–Feb. *H 7539!* Cuba, Hispaniola.

2. L. microlepanthes (Griseb.) Ames in Bot. Mus. Leafl. Harvard **1** (9): 24 (1933).—*Pleurothallis microlepanthes* Griseb. (1864).

Plant 1–2 cm. high; sheath-margins ciliate; leaf-blade elliptical-roundish, 6–8 mm. long, 4–6 mm. broad; flowers mostly yellow; lip oblong, a little shorter than the sepals.

Very rare (Clar., Port.), epiphytic on mossy tree trunks; 2300–3000 feet; fl. Oct–Nov. *P 16721!* Cuba.

39. LEPANTHES Sw. (1799); Hespenheide (1968)

By H. A. Hespenheide

1 Lip essentially simple, at most apically 3-lobulate, broadly emarginate or incised:
  2 Lip long-clawed, subquadrate at apex and there expanded perpendicular to plane of claw; lip conforming closely to relief of anterior petal-lobes   1. *unguicularis*
  2 Lip not clawed, sessile on column; petals planar:
    3 Lip resembling petals, broader than long, lateral lobes quadrate, apically 3-lobulate
             2. *simplex*
    3 Lip very unlike petals:
      4 Lip much broader than long, broadly 3-lobulate; petals deltate, scarcely developed laterally, emarginate at apex; flowers usually all yellow; leaves broadly elliptical to orbicular, rarely obovate   3. *rotundata*
      4 Lip as long as or longer than broad, incised at apex; petals strongly transversely 2-lobed, posterior longer than anterior; flowers all or partly purple; leaves elliptical to oblanceolate:
        5 Base of lip with linear posterior projections; sepals more or less ciliate, acuminate, all strongly cucullate and purple   4. *quadrata*
        5 Base of lip at most slightly auriculate; sepals glabrous, acute, median cucullate and purple, lateral flat and yellow   5. *loddigesiana*
1 Lip complex, 3-partite, usually with 2 large lateral peltate lobes embracing column and a' much smaller median apiculus of varying shape and position:
  6 Lateral sepals connate nearly to apex and forming a tube; racemes longer than leaves; lip and petals less than 1 mm. in largest dimension; pollinia more than 1 mm. long
             6. *tubuliflora*
  6 Lateral sepals not connate nearly to apex, not forming a tube:
    7 Lateral sepals free over most of their length, their tips divaricate:
      8 Sepals broadly deltate and acuminate, connate to about one-third or nearly half of their length, crimson or purple; sheaths of secondary stems strongly dilated, setose; leaves broadly elliptical to orbicular; flowers open near base of leaf:
        9 Petals up to about 1 mm. broad, deltate, lobes occasionally minutely notched
             7. *elliptica*
        9 Petals more than 4 mm. broad, lobes deeply notched with exterior lobules of notches elongate and linear   8. *cochlearifolia*
      8 Sepals narrower, lanceolate, more or less acuminate, free nearly to base; sheaths slightly dilated, ciliolate or muricate; leaves rotundate, ovate, elliptical or lanceolate:
        10 Racemes distinctly longer than leaves; sepals reflexed in flower, purple; leaves rotundate to elliptical   9. *proctorii*
        10 Racemes nearly as long as to barely exceeding leaves; sepals not reflexed, yellow or yellow with crimson nerves or bands:
          11 Sepals ciliate; plant less than 4 cm. high; leaves rotundate   10. *pulchella*
          11 Sepals glabrous; plant more than 5 cm. high:
            12 Leaves ovate-elliptical; petals and lateral sepals typically 2- and 1-veined respectively   11. *divaricata*
            12 Leaves elliptic-lanceolate; petals and lateral sepals 3-veined   12. *brownii*
    7 Lateral sepals connate half or more of their length, coplanar, the tips sometimes falcate:
      13 Lateral lobes of lip not forming axe-shaped expanded surfaces (although broadening otherwise in *L. vinacea*); petals ciliate:
        14 Petals less than 2 mm. broad, deltate, equally lobed; lateral lobes of lip linear, closely appressed to sides of column; flowers yellow with some red
             13. *tridentata*

14 Petals more than 2 mm. broad, the lobes linear, unequal; lateral lobes of lip expanding at middle to broad coplanar surfaces enclosing column; flowers purplish-red　14. *vinacea*

13 Lateral lobes of lip axe-shaped, developing surfaces perpendicular to base of lobe; petals glabrous:

15 Raceme much exceeding leaves; flowers numerous, congested, yellow; sepals narrow, three times as long as broad; petals broad, the posterior lobe elongated and narrow　15. *multiflora*

15 Raceme shorter than or about equal to leaves; sepals less than twice their breadth:

16 Petal-lobes subequal, similar in shape:

17 Flowers entirely red:

18 Sepals ciliate; plants usually less than 5 cm. high　16. *sanguinea*

18 Sepals glabrous; plants 6 cm. or more high　17. *obtusa*

17 Flowers predominantly yellow but more or less red centrally; petals deltate:

19 Plant usually more than 4·5 cm. high; leaves ovate to elliptical, 18 mm. or more long; surfaces of lateral lip-lobes broad; flowers usually more than 3 mm. long or broad　18. *wullschlaegelii*

19 Plant up to 4·5 cm. high; leaves broadly elliptical, up to about 12 mm. long; surfaces of lateral lip-lobes narrow; flowers about 3 mm. in greatest dimension　19. *woodiana*

16 Petal-lobes unequal and/or dissimilar in shape; flowers yellow with more or less orange or red:

20 Posterior petal-lobe conspicuously larger than anterior:

21 Posterior petal-lobe more than twice as long as anterior; anterior petal-lobe short and rounded; leaves elliptical　20. *interiorubra*

21 Posterior petal-lobe conspicuously broader than anterior, expanding in width to tip; anterior petal-lobe deltate, obtuse; leaves lanceolate　21. *lanceolata*

20 Petal-lobes more nearly equal in size, the posterior rarely more than twice the anterior; leaves elliptical, broadly elliptical, ovate or sometimes cordate:

22 Petal-lobes very dissimilar in shape:

23 Anterior petal-lobe deltate, abruptly wider than posterior; posterior petal-lobe falcate; lateral sepals convex; leaves elliptical　22. *convexa*

23 Anterior petal-lobe narrowly acuminate; posterior petal-lobe broadly rectangular and obliquely truncate at apex; sepals coplanar; leaves subcordate　23. *adamsii*

22 Petal-lobes more similar in shape, both quadrate, anterior narrower, posterior rounded at apex; leaves broadly elliptical to ovate or subcordate:

24 Sepals acuminate; anterior petal-lobes as broad as long, not incurved　24. *obtusipetala*

24 Sepals acute; anterior petal-lobes about twice as long as broad:

25 Leaves broadly ovate, subcordate, acute; petal-lobes hardly incurved; plants up to 15 cm. high　25. *intermedia*

25 Leaves ovate to broadly elliptical, acuminate; petal-lobes strongly incurved; plants 15–35 cm. tall　26. *ovalis*

1. L. unguicularis Hespenheide in Proc. Acad. Nat. Sci. Philad. **120** (1): 7, f. 2 (1968).

Plant 2–8 cm. high; sheaths glabrous on ridges, mouths muriculate, shortly aristate; leaf-blade elliptical, 9–27 mm. long, 2·5–6 mm. broad; flowers light violet or partly or all yellow; sepals ovate, acuminate, 3–3·5 mm. long, divaricate petals ear-shaped, both lobes subquadrate the anterior twice the size of the posterior.

Rare and local (St. James), epiphytic low on tree trunks; 2000–2400 feet; fl. Jan, Aug. *HCR 1672! P 23179!* Endemic.

2. L. simplex Hespenheide in Proc. Acad. Nat. Sci. Philad. **120** (1): 6, f. 1 (1968).

Plant 2–6 cm. high; sheaths minutely muriculate on ridges, mouths muriculate, shortly aristate; leaf-blade narrowly elliptical, 9–20 mm. long, 3–6 mm. broad; flowers light violet to yellow; sepals ovate, acuminate, 2–2·5 mm. long, divaricate; petals 0·3 mm. long, 0·6 mm. broad, lobes subquadrate.

Rare and local (Trel.), epiphytic on tree trunks on tops of limestone hills; 2000–2400 feet; fl. Aug. *HCR 1260!* Endemic.

3. L. rotundata Griseb., Fl. Br. W.I.: 610 (1864).—*L. concolor* F. & R. (1904).

Plant 1·5–3·5 (–5) cm. high; sheaths ciliolate on ridges and mouth, subulate-tipped; leaf-blade rather thick, 10–16 mm. long, 5–8 mm. broad; inflorescence usually shorter than the leaf; flowers often resupinate; sepals ovate, obtuse to shortly acuminate, 2–3 mm. long, median sometimes with 3 red veins, lateral connate nearly to apex; petals 1–1·2 mm. long, lobes very short and obtuse; lip broadly expanded around column.

Locally common (St. Andr., Port., St. Thom.), on mossy tree trunks and branches; 2500–7000 feet; fl. June–Feb. *A 6261! ACH 850! H 10465! P 7242!* Endemic.

4. L. quadrata F. & R. in Trans. Linn. Soc., Bot. ser. 2, **7**: 7, t. 1, f. 12–14 (1904).

Plant 2–6 cm. high; sheaths narrow, muriculate on ridges, ciliolate on mouth, aristate; leaf-blade narrowly to broadly elliptical, 10–25 mm. long, 3·5–9 mm. broad; inflorescence shorter than the leaf; sepals ovate-lanceolate, acuminate, 3–3·5 mm. long; anterior lobe of petals obliquely apiculate, posterior lobe oblanceolate, crimson.

Occasional in the central and eastern parishes, common only in south-eastern St. Ann, epiphytic on tree trunks; 1500–4000 feet; fl. all the year. *A 11703! H 7827! P 18291!* Endemic.

5. L. loddigesiana Reichb. f., Xen. Orch. **1**: 145 (1856).—*L. bilabiata* F. & R. (1904).

Plant 2–7 cm. high; sheaths muriculate on ridges, ciliolate on mouth; leaf-blade elliptical, 10–17 mm. long, 4–7 mm. broad; inflorescence shorter or longer than the leaf; sepals ovate, half to two-thirds connate, acute, 2–3 mm. long, yellow tinged purple; petals like the last, about 3 mm. broad, anterior lobe yellow, posterior lobe violet.

Rather local (St. Andr., Port., St. Thom.), epiphytic on tree trunks; 1600–4300 feet; fl. all the year. *H 7786! P 8093!* Endemic.

6. L. tubuliflora Hespenheide in Proc. Acad. Nat. Sci. Philad. **120** (1): 10, f. 4 (1968).

Plant in flower 8–13·5 cm. high; sheaths ciliolate, shortly aristate; leaf-blade narrowly elliptical, 3–4 cm. long, 6–9 mm. broad; racemes 3·5–6 cm. long; sepals lanceolate, 2–2·5 mm. long, probably light violet; petals and lip 0·3–0·4 mm. long; column greatly elongated; pollinia very large, 1·1 mm. long.

Very rare (? St. Cath.), known only from the type; 2800 feet; fl. May. *Shreve!* Endemic.

7. L. elliptica F. & R. in Journ. Bot. **47**: 5 (1909).—*L. arcuata* F. & R. (1909).

Plant 3–4 cm. high; sheaths ciliate on ridges, setose on obliquely ascending mouth; leaf-blade elliptical, 10–20 mm. long, 5–11 mm. broad; sepals ovate, 2·5–3·5 mm. long.

Local (St. Cath., St. Ann, St. Thom.), on tree trunks and limestone rocks; 2000–2700 feet; fl. Jan–Apr, Aug–Sept. *A 11709! H 9894! H & P 15148!* Endemic. Plants intermediate between *L. elliptica* and *L. cochlearifolia* in size and in the degree of lobing of the petals occur in Clarendon (*A 8435!*) and St. Ann.

8. L. cochlearifolia (Sw.) Sw. in Nov. Act. Soc. Sci. Upsal. **6**: 86, t. 5, f. 6 a–b (1799).

Plant 6–8 cm. high; sheaths setose on ridges and transversely dilated mouth; leaf-blade broadly elliptical to orbicular, 10–20 mm. long, 9–18 mm. broad, sometimes purplish beneath; sepals ovate, 5–6 mm. long.

Local (Port.), on mossy tree trunks; 1500–4000 feet; fl. July–Aug, Nov–Apr. *A 9117! H 7788! P 10480!* Endemic. Plants like typical *L. cochlearifolia* except for the petals which resemble those of *L. elliptica* occur in St. Thomas (*A & C 1087!*).

9. L. proctorii Garay & Hespenheide in Proc. Acad. Nat. Sci. Philad. **120** (1): 9, f. 3 (1968).

Plant 4–8 cm. high; sheaths narrow, finely setulose on the ridges and small mouth, aristate; leaf-blade rotundate to elliptical, 8–11 mm. long, 4–6 mm. broad, often purplish beneath; peduncles 1–3, about 3 (–5) cm. long, very slender; sepals ovate, 3·5–4 mm. long.

Rare and local (Port.), on mossy tree trunks and branches in wet primary woodlands and thickets; 1500–2500 feet; fl. Mar–Aug. *A 9150! ACH 939! P 9244!* Endemic.

10. L. pulchella (Sw.) Sw. in Nov. Act. Soc. Sci. Upsal. **6**: 86, t. 5, f. 6 c–e (1799).

Plant 2–3 cm. high; sheaths glabrous on ridges, ciliolate on mouth; leaf-blade elliptical, 10–14 mm. long, 5–6 mm. broad; racemes 1·5–2 cm. long, often pendulous; sepals mostly yellow but often crimson-nerved, caudate-acuminate, 6 mm. or more long; petals subequally lobed, more or less deltate, 2–4 mm. broad, crimson with orange margins.

Locally common (St. Andr., Port., St. Thom.), epiphytic on bare tree trunks; 3000–7400 feet; fl. all the year. *A 12532! H 7761! P 9536!* Endemic.

11. L. divaricata F. & R. in Trans. Linn. Soc., Bot. ser. 2, **7**: 11, t. 2, f. 27–29 (1904).

Plant 7–20 cm. high; sheaths minutely ciliolate; leaf-blade elliptical, acuminate, 20–30 (–40) mm. long, 8–15 mm. broad, sometimes purple beneath; sepals ovate, acuminate, 4–6 mm. long, tips of the lateral pair divergent; petals narrowly and equally lobed, 2·5–3·5 mm. broad, crimson or lobes tipped yellow or orange.

Locally common (St. Andr., Port., St. Thom.), on mossy tree trunks and branches, sometimes on rotten logs or rocks; 4000–7400 feet; fl. and fr. most of the year. *A 10601! H 7833! P 9514!* Endemic.

12. L. brownii Hespenheide in Notulae Naturae Philad. **426**: 1–3 (1969).

Plant 5–8·5 cm. high; sheaths shortly ciliolate; leaf-blade acute, 25–33 mm. long, 9·5–12 mm. broad; sepals lanceolate, acute or acuminate, 5·5–7 mm. long, tips of lateral pair slightly divergent; petals simple, subquadrate, broadly rounded distally, about 2·5 mm. long and 2 mm. broad, greenish tinged pink.

Rare (St. James); 2200–2400 feet; fl. Aug. *H. E. Brown.* Endemic.

13. L. tridentata (Sw.) Sw. in Nov. Act. Soc. Sci. Upsal. **6**: 86 (1799).

Plant 3–8 cm. high with very slender shoots; sheaths muriculate on ridges, ciliolate on mouth, aristate; leaf-blade elliptical, tapered to a narrow base, 15–25 mm. long, 5–7 mm. broad; inflorescences about as long as leaf; sepals ovate, acuminate, about 2·5 mm. long.

Rather rare (St. Andr., Port., St. Thom.), epiphytic on tree trunks; 4000–7000 feet; fl. Jan–Oct. *ACH 636! 852! P 10173!* Endemic.

14. L. vinacea Hespenheide in Proc. Acad. Nat. Sci. Philad. **120** (1): 12, f. 5 (1968).

Plant 1·7–4·5 cm. high; sheaths muriculate on ridges, shortly ciliolate on mouth, shortly aristate; leaf-blade oval, 7–15 mm. long, 4–9 mm. broad; inflorescence not exceeding leaf; sepals narrowly ovate, broadly acute, 2–3·5 mm. long, wine-red; petals with posterior lobe longer, ciliolate.

Rare and local (St. Andr., St. Thom.), epiphytic on trunks of small trees; 2700–5750 feet; fl. June, Nov. *Maxon 9595! P 9534!* Endemic.

15. L. multiflora Adams & Hespenheide in Proc. Acad. Nat. Sci. Philad. **120** (1): 16, f. 8 (1968).

Plant 2·5–6 cm. high in flower; sheaths broadening upwards, minutely ciliolate on ridges and mouth; leaf-blade elliptical, shortly acuminate, 9–12 mm. long, 3–7 mm. broad; peduncles filiform, 1·5–3 cm. long; racemes up to about 50-flowered; sepals lanceolate, median about 2 mm. long, lateral 2·3–2·5 mm. long.

Local (St. James, Trel.), epiphytic on trees in woodland on limestone; 2000–2400 feet; fl. and fr. May–July, Dec. *A 12837! ACH 743! HCR 1183! 1259! P 22988!* Endemic.

16. L. sanguinea Hook. in Curt., Bot. Mag. **70**: t. 4112 (1844).

Plant densely tufted, mostly 1·5–2·5 cm. high; sheaths muriculate on ridges, ciliolate on mouth, acute; leaf-blade ovate, 12–18 mm. long, 5–9 mm. broad; sepals ovate, about 4 mm. long; petals dark red, lobes oblong, ciliolate.

Thinly scattered through central and eastern parishes, a low epiphyte in exposed or open woodlands; 2500–4500 feet; fl. June–Feb. *A 11928! H 7825! H & P 15145!* Puerto Rico.

17. L. obtusa F. & R. in Trans. Linn. Soc., Bot. ser. 2, **7**: 11, t. 2, f. 26 (1904).
Plant 6–15 cm. high; mouth of sheath open, ciliolate; leaf-blade narrowly ovate to elliptical, acuminate, usually bright purple beneath, 3–5·5 cm. long, 1·2–2·5 cm. broad; sepals ovate, 5–6 mm. long; petals deep crimson, about 3·5 mm. broad, the subequal lobes usually blunt, ciliate.
Local in the east-central and eastern parishes, mostly a low epiphyte on tree trunks and logs; 1500–6500 feet; fl. and fr. Feb–Sept. *A 9134! 11736! ACH 943! H 10081! P 16264!* Endemic.

18. L. wullschlaegelii F. & R. in Journ. Bot. **47**: 126 (1909).
Plant 4·5–8·5 cm. high; sheaths muriculate on ridges, ciliolate on mouth; leaf-blade oval, shortly acuminate, 18–28 mm. long, 8–11 mm. broad; sepals ovate, 1·5–2·6 mm. long; petals 1·3–1·4 mm. broad, lobes subequal.
Widespread in central and western parishes, epiphytic especially at bases of trees on limestone hills; 1600–3000 feet; fl. Aug–Feb. *A 8434! HCR 1057! 1668! P 23027!* Endemic.

19. L. woodiana F. & R. in Journ. Bot. **47**: 6 (1909).
Plant 1·5–3·5 cm. high; sheaths muriculate on ridges, acute; leaf-blade oval, 7–14 mm. long, 4–6·5 mm. broad; sepals broadly ovate, obtuse, 1·4–2 mm. long; petals up to about 1 mm. broad.
Locally rather common (St. Cath., St. Ann), epiphytic on tree trunks; 2200–2900 feet; fl. Feb, Aug. *H 9895! HCR 865! 1417!* Endemic.

20. L. interiorubra Hespenheide in Proc. Acad. Nat. Sci. Philad. **120** (1): 14, f. 6 (1968).
Plant 3–6·5 cm. high; sheaths minutely muriculate on ridges, muriculate on mouth, acuminate; leaf-blade 12–20 mm. long, 3–5 mm. broad; sepals ovate, obtuse, apiculate, 1–1·5 mm. long, the lateral red proximally, otherwise yellow; petals 1·1–1·3 mm. broad, posterior lobe 2–3 times as long as anterior.
Occasional (St. Cath., Manch., St. Ann), epiphytic on tree trunks; 2300–3000 feet; fl. Jan, Aug, Oct. *HCR 907! P 11078! 26818!* Endemic.
Species 18–20 are closely related, with putative hybrids in the St. Catherine/St. Ann area where they are sympatric. Variation within the group, possibly also including *L. convexa*, is expressed mainly in the shape of petals and lip.

21. L. lanceolata Hespenheide in Proc. Acad. Nat. Sci. Philad. **120** (1): 18, f. 9 (1968).
Plant 6·5–12 cm. high; sheaths muriculate on ridges, shortly ciliolate on mouth, shortly aristate; leaf-blade lanceolate, 2–4·5 cm. long, 5–9 mm. broad; sepals broadly ovate, abruptly shortly acuminate, 2·5–3 mm. long; petals about 2 mm. broad.
Rare and local (St. James), epiphytic on bases of second-growth trees; 2200–2400 feet; fl. Aug. *HCR 1673!* Endemic.

22. L. convexa Hespenheide in Proc. Acad. Nat. Sci. Philad. **120** (1): 15, f. 7 (1968).
Plant 3·5–9 cm. high; sheaths smooth or granular on ridges, minutely muriculate on mouth, acuminate; leaf-blade elliptical, 15–30 mm. long, 5–9 mm. broad; flowers borne at about middle of leaf-blade; sepals broadly ovate, 2·2–3·1 mm. long; petals 2·6–2·7 mm. broad, lobes subequal.
Local (Trel.), epiphytic on tree trunks; 2000–2400 feet; fl. Jan, May–Sept, fr. May. *A 12836! ACH 744! HCR 1182! P 21356!* Endemic.

23. L. adamsii Hespenheide in Proc. Acad. Nat. Sci. Philad. **120** (1): 19, f. 10 (1968).
Plant 11·5–16·5 cm. high; sheaths ciliolate, acute; leaf-blade broadly ovate to ovate, subcordate, shortly acuminate, 4–6 cm. long, 2–3 cm. broad; sepals broadly ovate, acute, 2·5–3·5 mm. long; petals about 2·5 mm. broad, lobes subequal.

Rare and local (Port.), epiphytic on tree overhanging stream; 750 feet; fl. Aug. *A 11486!* Endemic.

24. L. obtusipetala (F. & R.) F. & R., Fl. Jam. **1**: 72 (1910).
Plant 6–20 cm. high; sheaths ciliolate, acute; leaf-blade ovate to elliptical, acuminate, 2·5–5 cm. long, 1–2·5 cm. broad; sepals deltate, acuminate, 4·5–5 mm. long, yellow turning pink; petals about 2·5 mm. broad, orange or crimson.
Occasional and rather local (St. Andr., Port., St. Thom.), epiphytic on tree trunks, often in very damp places; 1500–6600 feet; fl. Feb–Oct. *A 9116! 9149! ACH 853! P 4331!* Endemic.

25. L. intermedia Hespenheide in Proc. Acad. Nat. Sci. Philad. **120** (1): 20, f. 12 (1968).
Plant 4–15·5 cm. high; sheaths ciliolate, acute; leaf-blade broadly ovate, sub-cordate, 2–4·5 cm. long, 1–3 cm. broad; sepals ovate, acute, 2–3 mm. long; petals 2 mm. broad, lobes subequal.
Occasional in northern parts of the central parishes, epiphytic on trees; 1750–3000 feet; fl. Apr–July. *A 12428! H & P 14348! J.P. 2404!* Endemic.

26. L. ovalis (Sw.) F. & R., Fl. Jam. **1**: 71 (1910).—*L. crassifolia* Reichb. f. (1877).
Plant 12–30 cm. or more high; sheaths ciliolate, acute; leaf-blade ovate to elliptical, acuminate, thick and leathery, sometimes purplish beneath, 3–6 (–7) cm. long, 1·5–3 cm. broad; sepals broadly ovate, shortly acuminate, 2·5–4 mm. long, yellow; petals 2·5–3 mm. broad, orange, anterior (upper) lobe oblong and subacute, posterior lobe obtuse to broadly rounded and crimson at inner margin; capsule 3–winged, long-stalked.
Locally common in the central and eastern parishes as an epiphyte on small trees, low on trunks and branches of large trees and occasionally on rocks; 2000–6600 feet; fl. and fr. all the year. *A 9873! ACH 294! H 9892! J.P. 2268!* Endemic.

40. NEOCOGNIAUXIA Schltr. (1913)

1. N. monophylla (Griseb.) Schltr. in Urb., Symb. Ant. **7**: 496 (1913).—*Laelia monophylla* (Griseb.) N.E. Br. (1882). *Epidendrum brachyglossum* Cogn. (1909).
Shoots approximate from a short rhizome, sheathed with 3 tubular speckled sheaths, bearing a single linear oblong leaf 5–10 cm. long, 7–10 mm. broad; peduncle slender, 5–25 cm. long, with 4–6 bracts; sepals ovate-elliptical, 1·5–2 cm. long, up to 7·5 mm. broad; petals obovate-elliptical, slightly shorter and broader than the sepals; lip obovate, shallowly 3-lobed, papillose on the disk, 7–9 mm. long; anther-cap reddish-purple.
Very local (St. Andr., Port.), epiphytic on trees in mature woodlands; (2500–) 4000–4250 feet; fl. rarely Mar–Apr, commonly Aug–Nov. *A 11595! H 7783!* Endemic.

41. EPIDENDRUM L. (1763) nom. cons.[1]

1 Pseudobulbs present:
  2 Flowers solitary or in a small terminal cluster; rhizome widely creeping and branching; pseudobulbs remote, terete; leaves 2:
    3 Lip entire, the limb flat, free from the column; leaves emarginate; capsule terete
                                        1. *polybulbon*
    3 Lip 3-lobed, the lateral lobes ascending, adnate to the column; leaves apiculate; capsule 3-winged   2. *pygmaeum*
  2 Flowers racemose or paniculate; rhizome mostly short and sparingly branched, the pseudobulbs more or less clustered:
    4 Lip entire or obscurely 2-lobed:
      5 Lip similar to petals; leaf linear, solitary, 6–9 mm. broad; anthers 3; capsule 3-keeled; flowers resupinate   3. *ottonis*

---

[1] The conservation of *Epidendrum* L. (1763) was accompanied by the designation of the type species *E. nocturnum* Jacq. (1760) at which date *E. difforme* Jacq., *E. globosum* Jacq. (*Jacquiniella globosa* (Jacq.) Schltr.), *E. ramosum* Jacq. and *E. rigidum* Jacq. were also published.

5 Lip and petals dissimilar; anther 1:
  6 Leaves less than 1 cm. broad; flowers paniculate, resupinate; lip broadly ovate, clawed; pseudobulbs ovoid; capsule terete   **4.** *subaquilum*
  6 Leaves 1·5 cm. or more broad; flowers in racemes, not resupinate; lip not obviously clawed; pseudobulbs ellipsoid to fusiform; capsule 3-winged:
    7 Sepals and petals directed downwards; lip entire, mostly dusky purple, strongly concave; pseudobulbs slightly flattened, lanceolate, usually with 2 leaves   **5.** *cochleatum*
    7 Sepals and petals spreading; pseudobulbs fusiform with usually 1 leaf:
      8 Lip entire, greenish-white, with about 12 purple lines or rarely purple lines lacking   **6.** *fragrans*
      8 Lip obscurely 2-lobed, reddish-purple with yellowish margins   **7.** *spondiadum*
4 Lip 3-lobed; flowers paniculate; pseudobulbs ovoid, often wrinkled:
  9 Lip adnate to column for about one-third of its length; leaves 2–3 mm. broad   **8.** *angustifolium*
  9 Lip free from column; leaves more than 5 mm. broad:
    10 Lip broader than long with the lateral lobes broadly rounded and larger than the mid-lobe   **9.** *parvilobum*
    10 Lip longer than broad, the lateral lobes narrow and smaller than the mid-lobe:
      11 Ovary and capsule smooth; lateral lobes of lip triangular, acute; leaves 1·5–2 cm. broad   **10.** *belvederense*
      11 Ovary and capsule rugose; lateral lobes of lip oblong, obtuse; leaves about 1 cm. broad   **11.** *sintenisii*
1 Pseudobulbs absent:
  12 Leaf-margins serrulate; shoots with one leaf   **12.** *serrulatum*
  12 Leaf-margins smooth; shoots with several leaves:
    13 Floral bracts small and inconspicuous:
      14 Flower with a bladderlike spur adnate to the ovary; leafy shoots branched; lip entire   **13.** *jamaicense*
      14 Flower not spurred; leafy shoots unbranched, more or less erect; lip variously lobed:
        15 Lateral lobes of lip fimbriate-fringed; basal bracts foliaceous   **14.** *rivulare*
        15 Lateral lobes of lip not fringed:
          16 Flowers clustered on very short peduncles; bracts inconspicuous:
            17 Flowers umbellate, several opening together; perianth including lip greenish-yellow to yellow; lip bifid-emarginate with 2 short distal lobules **15.** *difforme*
            17 Flowers appearing singly; lip 3-fid, the mid-lobe linear, longer than the lateral lobes, white or cream-coloured:
              18 Lateral lobes of lip semi-ovate   **16.** *nocturnum*
              18 Lateral lobes of lip linear   **17.** *angustilobum*
          16 Inflorescence distinctly pedunculate, racemose or paniculate:
            19 Basal bracts spathaceous:
              20 Raceme subcapitate, much shorter than the peduncle; peduncle often re-generating successive racemes; leaves usually tinged mauve, thin, soft; lateral lobes of lip rounded-retuse   **18.** *anceps*
              20 Raceme or panicle lax, nodding, longer than the peduncle, not regenerating; leaves always green; lateral lobes of lip deflexed, lobulate:
                21 Sepals greenish-white, thin, acuminate; lip white; leaves herbaceous, usually acute   **19.** *nutans*
                21 Sepals dark green, leathery, shiny, acute; lip green tinged purple; leaves leathery, obtuse   **20.** *tomlinsonianum*
            19 Basal bracts inconspicuous; flowers in open diffuse panicles:
              22 Lip entire, ovate-cordate, shortly acute or acuminate; leaves oblong, few   **21.** *diffusum*
              22 Lip 3-lobed, the terminal lobe 2-fid; leafy shoot canelike with numerous lanceolate or linear-lanceolate leaves:
                23 Leaf-sheaths covered with minute dark red warts   **22a.** *verrucosum* var. *verrucosum*
                23 Leaf-sheaths smooth   **22b.** *verrucosum* var. *hansenii*
    13 Floral bracts large and spathelike, keeled, covering at least the ovary; leafy shoots often branched and scrambling or trailing; flowers mostly not resupinate, in spikes:
      24 Lip 3-lobed; leaves auriculate-clasping   **23.** *bifarium*
      24 Lip entire:
        25 Floral bracts 2–2·5 cm. long; perianth white; leaves 9–18 cm. long, 1–2 cm. broad   **24.** *paranaense*
        25 Floral bracts up to 1·5 cm. long; perianth green or yellow; leaves shorter and mostly narrower:
          26 Leaves up to three times as long as broad, up to 2 cm. broad, very hard; shoots and spikes strongly compressed with sharp edges; spike longer than the leaves; capsule triquetrous   **25.** *rigidum*

26 Leaves linear, less than 1 cm. broad, softer; shoots not sharp-edged; inflorescence
     shorter than the leaves:
27 Flowers solitary; leaves 1·5–2 (–3) cm. long; bracts 5–7 mm. long   26. *repens*
27 Spikes few-flowered:
28 Leaves 1·5–2 cm. long, linear-lanceolate; bracts 5–6 mm. long
                                                                        27. *strobiliferum*
28 Leaves (2–) 3–9 cm. long, linear; bracts 8–11 mm. long   28. *ramosum*

1. E. polybulbon Sw., Nov. Gen. & Sp. Pl.: 124 (1788).—*Encyclia polybulba*
     (Sw.) Dressler (1961).
Pseudobulbs ellipsoidal, 1·5–2 cm. long; leaf-blades linear-oblong, 2–5 cm. long,
7–15 mm. broad; sepals and petals 15–17 mm. long, dark yellow; lip clawed, as
long as the sepals, 10–12 mm. broad, white with yellow patch at the base; column
purple with 2 white terminal appendages; fragrant.
Occasional, often gregarious and forming large colonies on trees and rocks in
open woodlands; (1000–) 1500–4500 feet; fl. Sept–Mar, fr. Dec–July. *A 7487!*
H 9796! P 8211! Mexico, Guatemala, Honduras, Cuba.

2. E. pygmaeum Hook. in Curt., Bot. Mag. **60**: t. 3233 (1833).—*Hormidium
     tripterum* (Brongn.) Cogn. (1898). *Encyclia pygmaea* (Hook.) Dressler (1961).
Pseudobulbs ovoid-oblong, 2–4 cm. long; leaf-blades elliptical, 3–5 cm. long,
10–15 mm. broad; sepals oblong, long-acuminate, 5–6 mm. long, buff; petals
linear, 4–4·5 mm. long, yellow; lip about 5 mm. broad, white, the mid-lobe shortly
acuminate and purple-tipped; capsule about 1·3 cm. long.
Locally common in the eastern parishes, gregarious and forming colonies in
rocky thickets and around tree-bases; 1500–4300 feet; fl. Oct–Apr, July, fr.
Dec–July. *A 6577! H 7103! H & P 14207!* Florida, Mexico to Brazil, Greater
Antilles, Trinidad.

3. E. ottonis Reichb. f. in Hamburg. Gartenz. **14**: 213 (1858).—*Nidema ottonis*
     (Reichb. f.) Britton & Millsp. (1920).
Pseudobulbs ellipsoidal, 2–3·5 cm. long; leaf-blade 6–15 cm. long; raceme
3–6-flowered, shorter than the leaf; sepals lanceolate, 8–9 mm. long, cream-
coloured; petals and lip similar but shorter; capsule 10–15 mm. long, shortly
beaked.
Rare (St. Eliz., Han.), on trees; 500–1000 feet; fl. Sept–Jan, fr. Jan. *H 7540!*
Panama to Venezuela, Greater Antilles, Trinidad.

4. E. subaquilum Lindl. in Edw., Bot. Regist. **32**: under t. 64 (1846).
Pseudobulbs 2–2·5 cm. long, about 1 cm. broad; leaf-blade solitary, linear, 8–20
cm. long, 3–6 mm. broad; panicle mostly longer than the leaf; sepals linear-
lanceolate, 8–9 mm. long, greenish-yellow striped reddish-brown; petals linear,
7–8 mm. long; capsule ellipsoidal, 10–11 mm. long.
Very rare and local (St. Andr.), on trees; 1000–1500 feet; fl. Jan–Feb, fr. Mar.
H 5583! 7645! Mexico or Guatemala.

5. E. cochleatum L., Sp. Pl. ed. 2, **2**: 1351 (1763).—*Anacheilium cochleatum* (L.)
     Hoffmgg. (1842). *Encyclia cochleata* (L.) Dressler (1961).
Pseudobulbs 9–20 cm. long; leaf-blade oblong-lanceolate, acute, 20–30 cm. long,
(1·5–) 2–3·5 cm. broad; raceme 20–50 cm. long; flowers distinctly pedicelled;
sepals linear-lanceolate, 2·5–4 cm. long, light green turning yellow; petals similar
to sepals but slightly smaller; lip broadly rounded, cordate, about 2 cm. long,
lined alternately purple and yellow inside or rarely all yellow; capsule on a pedicel
1·5 cm. long, 3·5–4·5 cm. long and about 2·5 cm. broad, the adaxial (lip-radius)
wing shorter.
Common as an epiphyte and frequently on rocks in the open or in thickets and
woodlands; 700–4000 (–5000) feet; fl. and fr. all the year. *A 8407! H 10461!*
P 4175! Florida to Venezuela, Bahamas, Greater Antilles, Dominica.

6. E. fragrans Sw., Nov. Gen. & Sp. Pl.: 123 (1788).—*Encyclia fragrans* (Sw.)
     Dressler (1961).
Pseudobulbs 6–15 cm. long; leaf-blade oblong, obtuse, 20–35 cm. long, 2–4 cm.
broad; raceme 6–12 cm. long; sepals lanceolate, long-acuminate, 2·5–3·5 cm. long,

greenish-cream at first, fading yellow; petals shorter and broader; lip broadly ovate, acuminate; capsule 2·5–3 cm. long; flowers fragrant during daylight.

Common as an epiphyte, often forming large colonies on the boughs of large trees, less frequent on rocks; 1500–4000 feet; fl. and fr. all the year. *A 8741! H 10441! H & P 14031!* Mexico to Venezuela, Cuba, Hispaniola, Dominica, Grenada, Trinidad and Tobago.

7. E. spondiadum Reichb. f. in Bot. Zeit. **10**: 731 (1852).

Pseudobulbs 10–17 cm. long; leaf-blade obtuse, 20–30 cm. long, 3–3·5 cm. broad; raceme 8–11 cm. long; sepals oblong, acute, 13–15 mm. long, 3·5–4 mm. broad, greenish-yellow tinged purple; petals similar, broader; lip 11 mm. broad; capsule 3·5–4·5 cm. long.

Rare (St. Andr., St. Thom.), on trees and rocks; 3500–4500 feet; fl. and fr. Oct–Feb. *H 7866!* Costa Rica.

8. E. angustifolium Sw., Nov. Gen. & Sp. Pl.: 123 (1788).

Pseudobulbs about 1·5 cm. long with one linear leaf; leaf-blade 10–22 cm. long; panicle usually longer than the leaf, to 30 cm. long, with lax filiform branches; sepals and petals about 5 mm. long, brownish-orange; lobes of lip subequal.

Rare but widely dispersed, on trees and rocks; 1600–2500 feet; fl. May. *H 10440! P 21420!* Endemic.

9. E. parvilobum F. & R. in Journ. Bot. **47**: 123 (1909).

Pseudobulbs 3–5 cm. long with one linear leaf; leaf-blade up to 22 cm. long, 1·5–2 cm. broad; panicle longer than the leaf, lax; sepals and petals 8–10 mm. long, yellow with light red streaks; lip 8 mm. long, 10 mm. broad.

Very rare and local (Han.), on trees and limestone ledges; 1600–1750 feet; fl. Apr–June. *H 10439! P 10048!* Endemic.

10. E. belvederense F. & R. in Journ. Bot. **47**: 123 (1909).

Pseudobulbs about 4 cm. long with one linear leaf; leaf-blade 15–20 cm. long, 15–18 mm. broad; panicle longer than the leaf, lax; sepals and petals 1·3 cm. long, 3–3·5 mm. broad; lip 1·2 cm. long, 9 mm. broad.

Rare (St. Eliz., Han., Port.), on trees; 500–2500 feet; fr. Dec–Jan. *H 7541! 7620!* Cuba, Puerto Rico.

11. E. sintenisii Reichb. f. in Ber. Deutsch. Bot. Ges. **3**: 277 (1885).—*E. monticola* F. & R. (1909).

Pseudobulbs 1·5–2 cm. long with one or two linear leaves; leaf-blades 10–25 cm. long, up to 1 cm. broad; panicle 30 cm. or more long, lax; sepals and petals 9–10 mm. long; lip 9 mm. long; flowers probably always cleistogamous in Jamaica, at first green, fading yellow; ripe capsules yellow, about 1·5–2 cm. long.

Uncommon and rather local (St. Ann, Manch.), on tree trunks in woodland on limestone; 2500–3000 feet; fl. Nov–Feb, fr. Dec–July. *A 8493! H 10467! P 18367!* Cuba.

12. E. serrulatum Sw., Nov. Gen. & Sp. Pl.: 121 (1788).

Plant up to 10 cm. high, each shoot with one linear leaf folded and sheathed at the base; leaf-blade 2–7 cm. long, 3–5 mm. broad, turning blackish on drying; sepals fleshy, lanceolate, obtuse, 5 mm. long, green tinged dusky purple; petals 4 mm. long, yellowish-green; lip indistinctly 3-lobed, 4 mm. long, 3 mm. broad, yellowish-green; capsule ellipsoidal, 3-winged, pedicelled, 10–11 mm. long.

Local (St. Andr., Port., St. Thom.), epiphytic on mossy tree trunks; 5000–7400 feet; fl. sporadically throughout the year. *A 10674! ACH 924! H 10452! P 4320!* Cuba, Puerto Rico.

13. E. jamaicense Lindl., Fol. Orch. Epid.: 82 (1853).

Plant up to 30 cm. high, lax; leaf-blades oblong-lanceolate, acute, 4–11 cm. long, 1–2·5 cm. broad; flowers few in a short raceme, greenish-yellow turning yellow and deflexed at maturity; sepals oblanceolate, 12–16 mm. long; petals narrower and slightly shorter; lip 12 mm. broad.

Uncommon and local, usually as a low epiphyte on small trees in woodland on limestone; 1400–2700 feet; fl. May–Sept. *A 11682! H 11991! J.P. 481!* Hispaniola, Dominica.

14. E. rivulare Lindl. in Ann. & Mag. Nat. Hist. ser. 3, **1**: 330 (1858).

Plant up to about 80 cm. high, the stems slender, leafy and canelike, sometimes pendulous; leaf-blades linear-lanceolate, acute, 6–12 cm. long, 8–12 mm. broad; peduncle sheathed at the base; raceme 2–3 cm. long, few-flowered; sepals oblong-lanceolate, acute, about 1·5 cm. long, light green with purple spots; petals as long as sepals and similarly coloured; lip about 1 cm. long, white with purple spots.

Uncommon and local, on trees, limestone ledges and craggy boulders in woodland on limestone; 1600–3000 feet; fl. Aug–Mar. *H 10429! J.P. 485! P 9954!* Cuba, Hispaniola.

15. E. difforme Jacq., Enum. Syst. Pl. Carib.: 29 (1760).—*Amphiglottis difformis* (Jacq.) Britton (1924).

Plant 5–25 cm. high, the shoots tufted, rather thick, often pendulous, more or less terete; leaf-blades leathery, oblong-elliptical, obtuse and mostly emarginate, 2–7·5 cm. long, 1–2·5 cm. broad; flowers lustrous and pellucid, faintly scented; sepals and petals about 1·5 cm. long; lip a little shorter.

Frequent as an epiphyte in woodland, often of a secondary nature, and on planted or isolated trees, particularly old *Citrus*; 400–3000 (–4500) feet; fl. all the year. *A 7805! H 7629! P 15669!* Florida, Mexico to Venezuela, West Indies.

16. E. nocturnum Jacq., Enum. Syst. Pl. Carib.: 29 (1760).—*Amphiglottis nocturna* (Jacq.) Britton (1924).

Plant 30–50 (–100) cm. high, the shoots tufted and erect, usually compressed; leaf-blades leathery, oblong, rounded and emarginate at the tip, up to 12 cm. long and 3 cm. broad, the lower early deciduous; sepals linear-lanceolate, up to 5·7 cm. long, yellow to light orange, often tinged reddish outside; petals linear, a little shorter than the sepals but much narrower, 1–2 mm. broad; lateral lobes of lip up to 2·7 cm. long and 1 cm. broad, white, greenish-white or rarely blotched mauve, mid-lobe usually much longer, linear-tapered, mostly undulate; capsule fusiform, about 5 cm. long, long-pedicelled.

Frequent in low woodlands, either epiphytic, on stumps and rocks or rarely terrestrial in rocky soil; 1000–3000 feet; fl. June–Feb, fr. all the year. *A 7497! H 7776! P 16720!* Florida, Bahamas, Mexico to Colombia, West Indies.

17. E. angustilobum F. & R. in Journ. Bot. **47**: 124 (1909).

Like the last with broader leaves, much less conspicuous lip and longer ovary. Very rare (St. Andr.); 4000 feet; fl. Oct. Known only from the type. *H 10485!* Endemic.

18. E. anceps Jacq., Select. Stirp. Amer. Hist.: 224, t. 138 (1763).—*Amphiglottis anceps* (Jacq.) Britton (1924).

Plants up to 1 m. or over in flower; shoots compressed, more or less tufted, usually spreading, sometimes drooping; leaf-blades 5–12 (–18) cm. long, 2–5 (–6) cm. broad; peduncle up to 50 cm. or more long, with imbricate keeled bracts; raceme 2–5 cm. long, terminal or, secondarily on older peduncles, lateral; sepals obovate 5–7 mm. long, fleshy, light brownish-purple at first outside fading through olive-yellow to rose; petals linear-spathulate, 5–6 mm. long; limb of lip about 4 mm. long, 7 mm. broad; capsule about 2 cm. long, with 6 rounded smooth ridges, oblique at base, long-pedicelled.

Common in the central parishes, rare in Portland, epiphytic or on limestone boulders in woodland margins; 1500–3000 feet; fl. all the year. *A 7540! H 10428! P 9961!* Florida, Mexico to Colombia, West Indies.

19. E. nutans Sw., Nov. Gen. & Sp. Pl.: 121 (1788).—*E. nutans* var. *tridentatum* F. & R. (1910).

Plant up to 60 cm. high; shoots tufted, thick, more or less erect and compressed; leaf-blades oblong-lanceolate, acute, 12–20 cm. long, up to about 3 cm. broad; panicle with usually 1 basal branch, up to about 30 cm. long, many-flowered;

sepals lanceolate-spathulate 1·5–1·8 cm. long, 4–5 mm. broad; petals linear-spathulate a little shorter; lip about as long as the sepals.

Occasional as a low epiphyte or on cliffs or rocks in woodland; 2000–4000 feet; fl. Aug–Mar. *A 8484! H 7864! P 23260!* Endemic.

20. E. tomlinsonianum Adams in Amer. Orch. Soc. Bull. **35** (12): 997 (1966).—*E. nutans* var. *obtusifolium* F. & R. (1910), not *E. obtusifolium* Willd. (1805).

Plant 40–80 cm. high; shoots tufted, thick, ascending, more or less terete; leaf-blades leathery; oblong-elliptical, obtuse, up to 24 cm. long and 5·5 cm. broad; inflorescence a simple raceme or rarely shortly branched near the base; few-flowered; sepals ovate, about 1·3 cm. long, 5–6 mm. broad; petals linear; lip longer than the sepals.

Local in the central parishes, on limestone rocks in woodland; 2000–3000 feet; fl. June, Oct–Feb. *H 10420!* Endemic.

21. E. diffusum Sw., Nov. Gen. & Sp. Pl.: 121 (1788).—*Seraphyta diffusa* (Sw.) Pfitz. (1889).

Plant 20–70 cm. high in flower; shoots compressed, spreading, green or some-times whole plant tinged reddish; leaf-blades about 4 on each shoot, mostly oblong, up to 7 cm. long and about 3 cm. broad, mostly obtuse; panicle slender-branched; flowers yellowish-green or light yellow tinged dull rose; sepals about 7·5 mm. long, petals a little shorter; lip as long as lateral sepals; capsule broadly ellipsoidal, about 2·5 cm. long including the pedicel.

Rather common, usually as an epiphyte, more rarely on stumps or boulders, often on isolated planted or relict trees, especially *Citrus* and *Crescentia*; 400–3500 feet; fl. July–Dec. *A 8117! H 7770! P 10623!* Mexico to Brazil, Cuba.

22. E. verrucosum Sw. in Nov. Act. Soc. Sci. Upsal. **6**: 68 (1799).

22a. Var. verrucosum.

Plant in flower 0·5–2 m. high; shoots tufted, more or less erect, terete and cane-like; leaf-blades 8–18 cm. long, 1·3–3 cm. broad, acute; panicle diffusely branched, many-flowered; sepals 9 mm. long, 4 mm. broad, tips green at first, maturing creamy-white, fading light yellow; petals narrower; lip about 12 mm. long; capsule fusiform, beaked.

Locally common at middle elevations in the eastern parishes, rare and local in St. Cath., St. Ann and Manch., mostly on rocks and stony banks in thicket margins, also a low epiphyte; 2500–5000 feet; fl. (Oct–) Nov–Apr, fr. Mar–Sept. *A 8461! H 7819! P 9643!* Mexico to Colombia, Cuba.

22b. E. verrucosum var. **hansenii** Adams in Amer. Orch. Soc. Bull. **35** (12): 997 (1966).—*E. patens* of F. & R. (1910), not Sw.

Like the last, but generally smaller; plant in flower 60–120 cm. high; leaf-blades 6–15 cm. long, 7–17 mm. broad.

Rare (St. Andr., Manch.), epiphytic on the bases of small trees and on rocks; 2600–2900 feet; fl. Nov, fr. Dec–Feb. *A 8364! 12656! J.P. 2008!* Endemic.

23. E. bifarium Sw. in Nov. Act. Soc. Sci. Upsal. **6**: 68 (1799).

Plant 15–30 cm. high in flower; leaf-blades linear-lanceolate, somewhat falcate, 3–6 cm. long, 4–8 mm. broad; spikes 10–25 cm. long, compressed; sepals oblong-elliptical, acute, 6–7 mm. long, green spotted purple; petals 5 mm. long; lip 5–6 mm. long, greenish-yellow; capsule fusiform, black when old.

Local (St. Andr., Port., St. Thom.), mostly epiphytic; (2000–) 4000–5000 feet; fl. Nov–Mar, fr. Jan–July. *A 6456! H 7753! P 9601!* Colombia, Venezuela.

24. E. paranaense J. B. Rodrigues, Gen. & Sp. Orch. Nov. **2**: 139 (1888).—*E. imbricatum* Lindl. (1831), not Lam. (1783).

Plant 40–100 cm. high; shoots lax, branched, somewhat trailing; leaf-blades oblong, obtuse, often twisted at the base; spike subsessile, short, few-flowered; sepals 12–14 mm. long; petals a little shorter and narrower; lip 8–10 mm. long, 6 mm. broad; capsule 2·5 cm. long.

Rare and local (St. Andr., Port.), on trees; 4000–5000 feet; fl. Nov–Apr. *H 7867! J.P. 244!* Costa Rica, Brazil, Hispaniola.

25. E. rigidum Jacq., Enum. Syst. Pl. Carib.: 29 (1760).—*Spathiger rigidus* (Jacq.) Small (1913).

Plant 10–25 cm. high in flower; shoots several from a short much-branched rhizome; leaf-blades 2–6 cm. long, obtuse, shortly emarginate, sharp-edged and keeled; spikes 3–11 cm. long; sepals and petals about 6 mm. long, green or yellowish-green; lip about 7 mm. long, green; capsule 1·7–2 cm. long, triquetrous.

Rather common, on trees, particularly isolated or relict cultivated trees such as *Citrus*, *Crescentia* and *Mangifera*; 400–5000 feet; fl. and fr. most of the year. *A 6820! H 7754! P 15668!* Florida, Bahamas, Mexico to S. Amer., West Indies.

26. E. repens Cogn. in Fedde, Repert. Sp. Nov. **7**: 122 (1909).

Plant trailing with shoots 15–30 (–45) cm. long; leaf-blades 1·5–2 cm. long, 4–6 mm. broad; sepals 7–8 mm. long, 2 mm. broad, green; petals much narrower, a little shorter, green; lip ovate, cordate, acute.

Rare (St. Andr., Port., St. Ann), on mossy tree trunks in shady woodland; 2300–4000 feet; fl. Sept–Feb, fr. Mar. *A 11591! H 7561! P 23296!* Mexico to Venezuela, Cuba, Hispaniola. The St. Ann plant (*ACH 734!*, fl. July) is an altogether larger plant with the flowers tinged purplish-brown.

27. E. strobiliferum Reichb. f. in Nederl. Kruidk. Arch. **4**: 333 (1859).

Like the last; spikes 3–6-flowered; sepals about 4 mm. long, green; petals shorter and narrower; lip subacuminate.

Uncommon on trees; 500–5500 feet; fl. Aug–Jan, fr. Jan–Mar. *H 7519! J.P. 2014! P15763!* Florida, Mexico to Brazil, Cuba, Hispaniola, Lesser Antilles, Trinidad.

28. E. ramosum Jacq., Enum. Syst. Pl. Carib.: 29 (1760).—*Spathiger ramosus* (Jacq.) Britton (1924).

Like the last but generally larger, often much-branched and rather bushy, often pendulous; flowers green, later turning yellow; sepals 8 mm. long, about 2·5 mm. broad; petals linear, 7–8 mm. long, 1 mm. broad; lip acute, 3·5 mm. broad.

Locally common in the eastern parishes, mostly on trees, sometimes on rocky banks; 2000–6300 feet; fl. Aug–Mar, fr. Jan–July. *A 10524! H 7535! P 5706!* Mexico to Brazil, West Indies. A large variant (*A 11600!*) exists in the Hardwar Gap area sympatric with a small population of *E. paranaense* which it tends to resemble.

42. JACQUINIELLA Schltr. (1920)

Flower solitary; leaves 2–4·5 cm. long; perianth about 12 mm. long   1. *teretifolia*
Flowers usually several together in a terminal cluster, rarely solitary; leaves 1·5–2 cm. long; perianth about 3 mm. long   2. *globosa*

1. J. teretifolia (Sw.) Britton & Wilson, Sci. Surv. Porto Rico & Virg. Is. **6** (3): 340 (1926).—*Epidendrum teretifolium* Sw. (1788).

Plant 15–30 cm. high; shoots tufted, erect, the internodes yellowish; peduncle short, terminal, about 7 mm. long; sepals 9–12 mm. long, tapered, yellow; petals 5–7 mm. long, greenish; lip about 8 mm. long; capsule ellipsoid-fusiform, about 1·5 mm. long.

Locally frequent (St. Andr., Port., St. Thom.), on the branches of trees or on rocks; 2900–5300 feet; fl. Oct–Dec, fr. Oct–Apr. *A 5755! J.P. 2054! Watt!* Mexico to Venezuela, Greater Antilles, Guadeloupe.

2. J. globosa (Jacq.) Schltr. in Fedde, Repert. Sp. Nov., Beih. **7**: 124 (1920).— *Epidendrum globosum* Jacq. (1760).

Plant up to 15 cm. high; shoots tufted, rather bushy; peduncles 1–2 mm. long; flowers tinged red or purple; sepals elliptical, about 3 mm. long; petals ovate-rhomboid, a little shorter than the sepals; lip rhomboid, obtuse, 2·5 mm. long; capsule globose-ellipsoid, 4–6 mm. long.

Widely distributed but not common, epiphytic on tree trunks and branches in wet woodlands; 1100–2800 feet; fl. and fr. sporadically throughout the year. *A 9335! H 10445! H & P 14533!* Guatemala to Venezuela, West Indies.

43. ISOCHILUS R. Br. (1813)

1. I. linearis (Jacq.) R. Br. in Ait., Hort. Kew. ed. 2, **5**: 209 (1813).

Plant 30–70 cm. high; shoots slender, tufted, with numerous linear leaves; leaf-blades 3–6·5 cm. long, typically up to 3·5 mm. broad, obtuse or emarginate; raceme 1·5–3·5 cm. long; lateral sepals slightly gibbous, decurrent on the ovary, about 10 mm. long, bright magenta fading mauve; petals a little shorter and narrower than the median sepal; lip uppermost in the open flower, 8–10 mm. long; capsule ellipsoid, 6-ribbed, 7–10 mm. long.

Occasional in the central and eastern parishes, epiphytic and on mossy boulders; 2000–5500 feet; fl. and fr. sporadically throughout the year. *A 7697! H 7752! P 6817!* Mexico to Argentina, West Indies. Larger plants with leaves 4–6 mm. broad occur in Jamaica (*J.P. 32!*) at higher altitudes, and have been named *I. major* Cham. & Schlecht.; originally described from Mexico.

44. OCTADESMIA Benth. (1881)

Plant up to 1 m. high; leaves oblong-lanceolate, often serrulate; perianth 14–18 mm. long
                1. *montana*
Plant up to 2 m. high; leaves broadly lanceolate, entire; perianth 10–15 mm. long 2. *elata*

1. O. montana (Sw.) Benth. in Journ. Linn. Soc., Bot. **18**: 311 (1881).

Shoots canelike, leafy above; leaf-blades 4–11 cm. long, 7–20 mm. broad; flowers creamy-white, in a raceme or sparingly branched panicle up to 20 cm. long, fragrant; sepals obtuse; petals shorter; lip with lateral lobes oblong, ascending, each with a purple spot near the anterior margin, mid-lobe crenulate, yellow in the mid-line with 3 often purple-spotted crests; capsule ellipsoid, about 2·5 cm. long.

Locally common (St. Andr., Port., St. Thom.), epiphytic or on rocky or clay banks; 3500–7000 feet; fl. Oct–Mar, fr. Dec–Mar, July. *A 8183! H 10128! M.L.Farr!* Greater Antilles.

2. O. elata Benth. in Benth. & Hook. f., Gen. Pl. **3**: 525 (1883).

Like the last but larger in all its vegetative parts; leaf-blades 2–3 cm. broad; inflorescence paniculate; flowers fragrant; sepals acute; lip with 2 large proximal lamellae, terminal lobe apiculate, not crested.

Very rare (St. Andr., Clar., Trel.), epiphytic or in leafy forest soil on limestone rocks; 2000–4000 feet; fl. Mar–May. *A 12838! H 11034! J.P. 2382!* Cuba.

45. TETRAMICRA Lindl. (1831)

Plant without pseudobulbs; lip divided nearly to the base, the lateral lobes clawed
                1. *parviflora*
Lower internodes swollen into an ovoid pseudobulb; lip divided about halfway to the base
                2. *bulbosa*

1. T. parviflora (Reichb. f.) Lindl. ex Griseb., Fl. Br. W.I.: 622 (1864).

Shoots tufted or more usually separated by long stolons; leaves 2 or 3 on each shoot, blades boat-shaped, thick, acute, narrowed to the base with sharp margins, 1–4 cm. long, 3–7 mm. broad; scape and raceme up to about 40 cm. long with about 12 flowers, sometimes branched on secondary flowering; sepals and petals mostly green, 4–5 mm. long; lip bright pink with 5 darker lines on the mid-lobe, 5–6 mm. long, mid-lobe obovate, lateral lobes broadly ovate; capsule oblong, about 1·5 cm. long.

Local and rather rare (St. Andr.), mostly on limestone rocks; 300–3750 feet; fl. Feb–May, fr. May–June. *H 11934! Stearn 693!* Hispaniola.

2. T. bulbosa Mansf. in Arkiv Bot. **20A** (15): 18 (1926).

Like the last but the shoots tufted without stolons; leaves 1 or 2 on each shoot, slightly longer and thicker, often tinged purplish; scape and raceme up to about 30 cm. long, 3– or more-flowered; sepals and petals brownish-red; lip bright pink with 5 darker lines on the mid-lobe.

Very rare (Trel.), on limestone rocks; about 2000 feet; fl. Jan–Feb. *Proctor!* Hispaniola.

46. BRASSAVOLA R. Br. (1813)

Blade of lip about 2 cm. long; leaf-blade up to 35 cm. long and usually about 1 cm. broad, dark green or tinged purplish　　　　　　　　　　　　　　　1. *cordata*
Blade of lip 3–5 cm. long; leaf-blade up to about 20 cm. long and 2 cm. broad, light green
　　　　　　　　　　　　　　　　　　　　　　　　　　　　2. *nodosa*

1. B. cordata Lindl. in Edw., Bot. Regist. **22**: t. 1914 (err. 1913) (1836).—*B. sloanei* Griseb. (1864).

Stem shortly rhizomatous, much branched and often forming large clumps; branches 5–8 cm. long below the leaf; leaf-blade thick and fleshy, broadly channelled, subulate-tipped; flowers (1–) 5–9 (–16) in a short erect raceme, fragrant at night; sepals and petals linear-lanceolate to linear, 3·5–4·5 (–5) cm. long, light greenish-yellow, sometimes blotched purplish outside; claw of lip and centre of blade light green or greenish-yellow, otherwise white, tip acuminate, folded basally; capsule beaked, with strong narrow ridges, about 3 cm. long.

Common as an epiphyte on large, often isolated trees and in woodlands in all except the wettest parts of the island; SL–2700 feet; fl. June–Feb, fr. Jan–Apr. *A 6850! H 10626! H & P 14545!* H. G. Jones (Act. Soc. Bot. Polon. **37** (2): 255–259 (1968)) has described as a distinct species *B. harrisii* H. G. Jones, a variant having 14–16 large flowers in the raceme. Flower-number, size and colour do not readily provide taxonomic criteria in Jamaican populations as these characters seem to vary widely with the age of the plant and the ecological situation. Further study is required to confirm the causes of variation in native *Brassavola* in Jamaica. Of the species which Jones attributes to Jamaica only *B. cordata* is truly native. The original plant of *B. cordata* reputedly came from Brazil but Fawcett & Rendle did not confirm this and its existence in any other part of the American tropics has not been proved. *B. subulifolia* Lindl., Gen. & Sp. Orch.: 115 (1831), is an obscure species from Nevis; if this should prove to be the same as the Jamaican plant then that name would have priority over *B. cordata*. *B. nodosa*, common in cultivation, and the closely allied *B. grandiflora* Lindl., are native to the continental American mainland.

2. B. nodosa (L.) Lindl., Gen. & Sp. Pl. Orch.: 114 (1831).
Like the last; flowers few in a short drooping raceme.
It is very doubtful if this handsome species is extant in the wild state in Jamaica, and the paucity of records from other islands suggests that it may not be indigenous in the West Indies; fl. Apr–Sept. *H 10413!* (cult.). Mexico to Peru.

47. HOMALOPETALUM Rolfe (1896)

1. H. vomeriforme (Sw.) F. & R., Fl. Jam. **1**: 107 (1910).
Rhizome slender; leafy-shoots up to 3 cm. high, including the oblong to suborbicular pseudobulb, 2–6 mm. long, and the solitary fleshy ovate to ovate-lanceolate keeled leaf, 1–2 cm. long and 4–7 mm. broad; peduncle 1–6·5 cm. long; flower about 2 cm. long, light dull green; sepals and petals linear-lanceolate; lip as long as the petals; capsule flask-shaped, 1·5 cm. long.

Local (St. Andr., St. Thom.), epiphytic mostly on smooth-barked trees in open woodland margins; 2800–5000 feet; fl. Oct–Jan, fr. Jan, May, July. *A 12137! H 9783!* Cuba.

48. SCHOMBURGKIA Lindl. (1838)

Pseudobulbs elongate-fusiform, smooth, solid, up to 35 cm. long; leaves up to 35 cm. long; Jamaica　　　　　　　　　　　　　　　　　　1. *lyonsii*
Pseudobulbs oblong-conical, sulcate, hollow, shorter; leaves shorter, spreading; Cayman Is.　　　　　　　　　　　　　　　　　*S. brysiana* Lemaire

1. S. lyonsii Lindl. in Gard. Chron. **1853**: 615 (1853).—*Laelia lyonsii* (Lindl.) L. O. Williams (1941).
Shoots tufted from a short branching rhizome, often forming large clumps; pseudobulbs firm, slightly compressed, about 3·5 cm. broad; leaf-blades 2, 1

occasionally or on young plants, suberect, leathery, up to 6 cm. broad; scape 1–2 m. or more long; flowers about 8–12 in a short terminal bracteolate raceme, very showy, short-lived; pedicels 5–6 cm. long; sepals and petals 2–2·5 cm. long, white speckled brownish-purple or rarely all white; lip acute, the lateral lobes margined yellow.

Locally common in some central parishes, otherwise occasional, epiphytic on isolated, mostly large, trees, very rarely on rocks; 1400–3000 feet; fl. June–Nov, fr. Dec–Apr. *H 7748! P 10376!* Cuba.

Schomburgkia brysiana Lemaire has the mid-lobe of the lip emarginate anteriorly; the rest of the perianth is uniformly yellow. Two varieties of this species are reported from the Cayman Islands.

49. BROUGHTONIA R. Br. (1813)

Lip strongly curled at the base, the proximal margins meeting over the column and forming a tube, hairy within; petals acute **1.** *negrilensis*
Lip not strongly curled, the proximal margins sheathing but not meeting over the column, glabrous or very inconspicuously puberulous within; petals obtuse **2.** *sanguinea*

1. B. negrilensis Fowlie in Orch. Digest **25**: 418 (1961).—*B. domingensis* of F. & R. (1910), not (Lindl.) Rolfe.

Pseudobulbs tightly clustered, compressed, of 2 or 3 unequal internodes, up to about 5 cm. long and 2·5 cm. broad, with (1–) 2 leaves; leaves oblong, leathery, mucronate, glaucous greyish-green, often tinged purplish towards the base, up to 20 cm. long and 3 cm. broad; peduncle up to 50 cm. or more long, slender with 7–20 flowers developing in succession towards the apex, rarely branched; sepals linear-lanceolate 2–3 cm. long, about 6 mm. broad, generally pinkish-mauve veined reddish-purple; petals like the sepals, about 1 cm. broad; lip up to about 3 cm. long, broadly rounded distally with a crenulate margin, light mauve with strong purple venation, yellow within with darker purple lines.

Locally common (West., rare in St. Eliz.), on the trunks and branches of trees in woodland on limestone; SL–100 (–500) feet; fl. Oct–Mar, fr. Dec–Mar. *A 8543! Tomlinson!* Endemic.

2. B. sanguinea (Sw.) R. Br. in Ait., Hort. Kew. ed. 2, **5**: 217 (1813).

Like the last; plants usually slightly smaller; leaves 2 (–4), light green, hardly glaucous; peduncle 20–40 cm. or more long; flowers up to about 12, usually bright crimson with darker crimson lines along the base of the lip, but sometimes these lines absent or except for the yellow base to the lip the whole flower white, light yellow, or cream lined pink; lip up to about 2·5 cm. long, almost flat distally; capsule fusiform, about 3 cm. long.

Common in the eastern and central parishes, rare in the west, epiphytic or on exposed limestone rocks; SL–2500 feet; fl. and fr. most of the year, but more abundantly in May–June and during rainy seasons. *A 8853! H 9209!* Endemic. The Cuban plant reported as *B. sanguinea* has 8 pollinia and is probably *B. ortgiesiana* (Reichb. f.) Dressler. The two Jamaican species are sympatric in a small area in western Jamaica where they form putative hybrids. These plants mostly have the lip-shape of *B. negrilensis* combined with the crimson colour of typical *B. sanguinea*.

50. DICHAEA Lindl. (1833)

1 Leafy shoots more or less tufted, suberect, unbranched; leaves glaucous, deciduous; lip obtuse; capsule smooth **1.** *glauca*
1 Leafy shoots generally lax and trailing, often branched; leaves not glaucous:
  2 Leaves deciduous, jointed to the sheath; outer perianth smooth:
    3 Capsule smooth; leaves linear to narrowly oblanceolate; lip acute **2.** *graminoides*
    3 Capsule bristly; leaves oblong **3.** *morrisii*
  2 Leaves persistent, not jointed to the sheath; capsule bristly; outer perianth prickly-papillose:
    4 Leaves oblong, 5 mm. or more broad:
      5 Leaves acuminate-aristate, directed towards the shoot-apex **4.** *pendula*
      5 Leaves rounded at the tip, shortly pointed, directed slightly away from the shoot-apex **5.** *muricata*
    4 Leaves linear, up to 2·5 mm. broad **6.** *trichocarpa*

1. D. glauca (Sw.) Lindl., Gen. & Sp. Orch.: 209 (1833).

Plant 30–50 cm. high; rhizome short, branched, with thick roots; leaf-blades elliptic-oblanceolate, keeled, 3–5·5 cm. long, 9–13 mm. broad; pedicels up to 1·3 cm. long; sepals and petals 7–8 mm. long, mostly white but often spotted or tinged reddish; lip about 6 mm. long and nearly as broad, the short lateral lobes ascending, with a crimson spot at base; capsule jointed to pedicel, elliptical, with 3 narrow ridges, about 1 cm. long.

Local (St. Andr., Port., St. Thom.), an epiphyte usually low on trunks or branches of trees and shrubs in woodland; 3600–5600 feet; fl. June–Aug, fr. most of the year. *A 11357! H 7361! P 6804!* Mexico, Guatemala, Honduras, Cuba, Hispaniola.

2. D. graminoides (Sw.) Lindl., Gen. & Sp. Orch.: 209 (1833).

Plant up to 30 cm. high; shoots sometimes tufted but usually lax and branched from the base, with slender roots; leaf-blades spreading, distant, 3–4 cm. long, 3–6 mm. broad; pedicels slender, up to 2 cm. long; sepals and petals 5–6 mm. long, mostly white; lip about 4·5 mm. long, the terminal lobe broadly cordiform; capsule obovoid, tapered to base, about 2 cm. long.

Locally frequent (St. Andr., Port., St. Thom.), on tree trunks, rotten logs or banks in shady woodland; 3000–5000 feet; fl. Sept–Dec, fr. Dec–July. *A 11592! H 7524! P 9853!* Mexico to tropical S. Amer., Cuba, Hispaniola, Tobago.

3. D. morrisii F. & R. in Journ. Bot. **48**: 107 (1910).

Shoots up to 45 cm. long, usually lax and pendulous; leaf-blades 3·5–5 (–6) cm. long, 10–13 (–15) mm. broad, thin, more or less rounded at the tip; pedicels about 8 mm. long; sepals 11–13 mm. long; petals 10 mm. long; lip white barred with brownish-red, about 9 mm. long; capsule about 1·5 cm. long and 1 cm. broad.

Very rare (St. Andr., Port.), epiphytic on mossy tree trunks; 3500–3600 feet; fl. July, fr. Dec. *A 10143! Morris!* Venezuela, Cuba, Hispaniola.

4. D. pendula (Aubl.) Cogn. in Urb., Symb. Ant. **4**: 182 (1903).—*Limodorum pendulum* Aubl. (1775). *D. echinocarpa* (Sw.) Lindl. (1833).

Shoots up to 50 cm. or more long, branched and trailing; leaf-blades 12–25 mm. long, 5–8 mm. broad; pedicels about 1·5 cm. long; sepals 8–9 mm. long, fleshy, light yellow to pinkish-buff; petals about 8 mm. long, pinkish-buff with or without blue-purple marks near the tips; lip 6 mm. long, white or light mauve with blue-purple blotches and bands; capsule ellipsoid, 12–15 mm. long.

Local (St. Andr., Port.), epiphytic on trees and tree ferns or on mossy banks; 2750–5000 feet; fl. May–Dec, fr. most of the year. *A 11356! H 7527! P 6845!* Venezuela to the Guianas, Greater Antilles, Guadeloupe, Martinique.

5. D. muricata (Sw.) Lindl., Gen. & Sp. Orch.: 209 (1833).

Like the last with slightly smaller leaves; capsule narrower.

Rare (St. Andr., Port., St. Thom.), epiphytic in wet forest or on rocks near streams; 2600–4000 feet; fl. Dec–Jan, fr. Jan–Mar. *H 7649! P 8319!* Mexico to Brazil, West Indies.

6. D. trichocarpa (Sw.) Lindl., Gen. & Sp. Orch.: 209 (1833).

Shoots up to 60 cm. long, slender, much branched and trailing; leaf-blades 9–15 mm. long, 1–2·5 mm. broad; pedicels about 8 mm. long; sepals yellow with green tips, acute, 7 mm. long; petals obtuse, 5 mm. long; lip broad near the tip, 4·5 mm. long; capsule about 8 mm. long.

Rare and local (St. Andr., St. Thom.), on tree trunks and mossy banks; 4000–5000 feet; fl. July–Aug, fr. Dec–May. *A 7433! H 7578! P 6889!* Mexico to Costa Rica, Cuba, Hispaniola.

51. IONOPSIS Kunth (1816)

Leaves flat, thick and shallowly ribbed; lip deeply 2-lobed, much longer than the sepals, mauve with purple streaks   1. *utricularioides*
Leaves terete; lip obtuse or emarginate, about as long as the sepals, white   2. *satyrioides*

1. I. utricularioides (Sw.) Lindl., Coll. Bot.: t. 39A (1825).—Dancing Ladies.

Plant (20–) 30–60 cm. high in flower; very small pseudobulbs sometimes present; leaves leathery up to 12 cm. long and 15 mm. broad; panicle diffuse; sepals about 6 mm. long, acute, the lateral forming a short obtuse spur; petals 4·5 mm. long, obtuse; lip broadened distally, 10–13 mm. long; capsule up to 3 cm. long, including pedicel and beak, smooth, ellipsoid.

Rather common as an epiphyte on trees, especially old *Citrus*, rarely on rocky banks; 500–2600 feet; fl. and fr. most of the year. *A 7017! H 7635! P 10369!* Florida, Mexico to Venezuela, West Indies; Grand Cayman.

2. I. satyrioides (Sw.) Reichb. f. in Walp., Ann. Bot. **6**: 683 (1863).

Plant 5–30 cm. high in flower; pseudobulbs obscure, up to 2 cm. long, terminated by a leaf; shoots tufted; leaves subulate, 3–15 cm. long, 2–4 mm. broad, shallowly grooved adaxially except at the tip; inflorescence axillary, racemose or paniculate; pedicels 1 cm. long; sepals 6–7 mm. long, obtuse, the lateral forming a short sac; petals slightly longer than the sepals, obtuse, white with faint purple lines; lip about 9 mm. long, with 2 yellow calli within; capsule ellipsoid, tapering to base, 13–16 mm. long.

Occasional on trees; 400–1750 feet; fl. and fr. most of the year. *A 7187! H 8666! P 15612!* Venezuela, Greater Antilles, Trinidad.

52. COMPARETTIA Poepp. & Endl. (1836)

1. C. falcata Poepp. & Endl., Nov. Gen. **1**: 42, t. 72, 73 (1836).

Plant 15–40 cm. high in flower; shoots approximate, with very long slender roots; leaf-blades softly leathery, oblong-lanceolate, 5–18 cm. long, 1·5–4 cm. broad; inflorescence at first simple, sometimes branched on secondary flowering; sepals about 10 mm. long, bright pink, the lateral connate and produced at the base into a tapered spur up to 1·5 cm. long; petals 8–10 mm. long, bright pink; lip clawed, the limb up to 15 mm. broad, white at the base and produced into 2 linear spurs enclosed by the sepal spur; capsule stalked, trigonous, 1·5–3 cm. long, 10–13 mm. broad, beak 1 cm. long.

Formerly rather common, but now rare and apparently confined to a few central and eastern parishes, epiphytic on shrubs and small trees in sheltered woodlands; 2300–3500 (–4500) feet; fl. Oct–Mar, fr. Feb–Aug. *H 9009! P 5971!* Mexico to Peru, Greater Antilles.

53. BRASSIA R. Br. (1813)

Lip about 3·5 cm. long; lateral sepals linear, gradually acute, up to 7·5 cm. long; leaves usually 1 on each pseudobulb   1. *maculata*
Lip about 2·5 cm. long; lateral sepals caudate-acuminate, up to 11 (–13) cm. long; leaves usually 2 on each pseudobulb   2. *caudata*

1. B. maculata R. Br. in Ait., Hort. Kew. ed. 2, **5**: 215 (1813).

Plant up to about 60 cm. high in flower; pseudobulbs up to 10 cm. long and 4 cm. broad; leaf-blade oblong-oblanceolate, acute, 10–40 cm. long, 3–5 cm. broad; scape and raceme up to 60 cm. long, 7–14-flowered, usually nodding; flowers musky fragrant; median sepal up to about 6·5 cm. long, like the lateral yellowish-green spotted brownish-purple towards the base; petals falcate, 3–4 cm. long, coloured like the sepals but paler; lip broadly clawed, the limb with broad lateral lobes, the tip acute and more or less deflexed, whitish with brownish-purple spots towards the mid-line; capsule about 7·5 cm. long, up to 1·8 cm. thick, beaked.

Local, often gregarious, usually on sheltered rocky slopes, or epiphytic; (1000–) 1600–3000 feet; fl. Apr-Oct, fr. Aug–Dec. *A 7498! H 10442! P 4594B!* Guatemala, Honduras, Cuba.

2. B. caudata (L.) Lindl. in Edw., Bot. Regist. **10**: t. 832 (1824).

Like the last; pseudobulbs up to 13 cm. long and 3 cm. broad; leaf-blade up to 30 cm. long and 4·5 cm. broad; flowers acrid-musky fragrant; sepals and petals light green at first, later yellow, spotted dark brownish-purple; lip shortly clawed, the tip more or less straight, involute, marked with large reddish-brown blotches towards the base.

Occasional, usually low on tree trunks, sometimes on limestone or shale rocks in woodland; 1500–3000 feet; fl. and fr. mostly May–Sept. *A 8430! P 8005! Watt!* Cuba, Hispaniola, Trinidad.

54. MACRADENIA R. Br. (1822)

1. **M. lutescens** R. Br. in Edw., Bot. Regist. **8**: t. 612 (1822).

Plant 12–20 cm. high; pseudobulbs flask-shaped to cylindrical, 2–5 cm. long with a single terminal leaf; leaf-blade oblong-lanceolate, leathery, (6·5–) 10–15 cm. long, (1–) 1·5–3 cm. broad; inflorescence shorter than the leaf, usually pendulous, (2–) 4–9-flowered; sepals about 1 cm. long, the median scoop-shaped, at first light green then yellow tinged reddish-brown; petals 9 mm. long, light green with a reddish blotch or yellow; lip 5 mm. long and broader than long except the narrow reflexed terminal lobe, white or with light purple lines; capsule ellipsoid, beaked.

Widespread but nowhere common, epiphytic on trees in sheltered places; 500–1750 feet; fl. Oct–May, fr. Jan–Sept. *A 9979! H 7679!* Florida, Colombia to the Guianas, Cuba, Trinidad.

55. TRICHOPILIA Lindl. (1836)

1. **T. subulata** (Sw.) Reichb. f. in Flora **48**: 278 (1865).—*Epidendrum subulatum* Sw. (1788). *T. jamaicensis* F. & R. (1910).

Plant 15–18 cm. high; leaf-blade linear, thick, 11–17 cm. long, 4 mm. broad; inflorescence about 5 cm. long, few-flowered; flowers whitish with violet spots; sepals and petals 16–19 mm. long, 2·5–3 mm. broad; lip acute, saccate at the base, 14 mm. long; capsule 1·5–2 cm. long.

Very rare (St. Thom.), epiphytic; fl. and fr. Sept. Only once collected. *H 7697!* Panama to Peru, Cuba, Trinidad.

56. LEOCHILUS Knowles & Westc. (1838)

1. **L. labiatus** (Sw.) Kuntze, Revis. Gen. Pl. **2**: 656 (1891).

Plant 5–15 (–20) cm. high in flower; roots slender; pseudobulbs clustered, compressed-subglobose, 7–15 mm. long and broad, each with a solitary terminal leaf; leaf-blade oblong-elliptical, hard and leathery, often reddish, 2–7 cm. long, 5–15 mm. broad; inflorescence simple or branched up to about 15 cm. long, nodding, few-flowered; sepals 3·5–4·5 mm. long, the lateral a little longer than the median and united halfway, greenish-yellow speckled reddish-brown; petals like the sepals, smaller; lip oblong-elliptical, rounded at the tip, about 5 mm. long, yellow streaked reddish-brown at the base; capsule subterete, beaked, with 3 shallow ridges, about 3 cm. long including pedicel and beak.

Occasional as an epiphyte, especially on old isolated trees in sheltered areas; 700–2500 feet; fl. and fr. Mar–Nov, mostly Apr–May. *A 7194! Moore 10462! P 10368!* Honduras to Panama, Venezuela, West Indies.

57. ONCIDIUM Sw. (1800)

1 Leaves broadly oblong-ligulate, hard and leathery, 6 cm. or more broad; lateral sepals free; inflorescence usually long, much branched with many showy yellow or brownish flowers   1. *luridum*

1 Leaves linear, triquetrous, rarely more than 1 cm. broad; lateral sepals connate; inflorescence racemose or sparingly branched:

   2 Callus on base of lip at least 3-lobed; distal lobe of lip more or less clawed; flowers about 2·5 cm. from top to bottom, generally white or mauve:

      3 Proximal lobes of lip small, about 3 mm. long and up to 2 mm. broad; distal lobe deeply curved on the proximal margins, shallowly emarginate, usually white

                                                2. *tetrapetalum*

      3 Proximal lobes of lip larger; distal lobe shallowly curved on the proximal margins, deeply bifid, usually mauve   3. *pulchellum*

2 Callus on base of lip absent; base of column confluent with a smooth rounded yellow outgrowth; flowers smaller:
  4 Lip equally 4-lobed or distal lobe emarginate; callus replaced by a raised median ridge; raceme 1 (–3)-flowered   4. *gauntlettii*
  4 Lip obscurely 3-lobed, distal lobe deltate; raceme (6–) 8–12 (–15)-flowered
                                                                                      5. *triquetrum*

Several pseudobulbous species of *Oncidium*, such as *O. sphacelatum* Lindl., natives of the tropical American mainland, have been introduced into Jamaica and are common in cultivation. The records of *O. leucochilum* Bateman ex Lindl. reported by Fawcett & Rendle were almost certainly from cultivated plants.

1. O. luridum Lindl. in Edw., Bot. Regist. **9**: t. 727 (1823).—Brown Gal.
  Plant tufted, often forming large clumps with numerous slender roots some of which are often short and seasonally apogeotropous; leaves gutter-shaped, up to 50 cm. or more long and 10 (–15) cm. broad; panicle 1 m. or more long with 50 or more fragrant flowers: sepals and petals clawed, obovate, about 1·5 cm. or a little over long, undulate-margined, dull yellow with confluent reddish-brown blotches; lateral lobes of lip poorly developed, distal lobe broad, crisped, lighter yellow with faint brownish blotches; callus complicated, bright yellow with white tips; column white; capsule massive, pendulous, beaked, about 6 cm. long.
  Common on trees and steep rocky banks; 50–3000 feet; fl. (Mar–) Apr–Oct, fr. May–Jan. *A 6809! H 8646! P 6940!* Florida, Mexico to Peru, Cuba, Lesser Antilles, Trinidad.

2. O. tetrapetalum (Jacq.) Willd. in L., Sp. Pl. ed. 4, **4**: 112 (1805).—*Epidendrum tetrapetalum* Jacq. (1763). *O. guttatum* (L.) F. & R. (1910), not Reichb. f. (1863). Pimento Orchid.
  Shoots tufted and compact with long slender roots; leaves fleshy (2–) 5–10 (–20) cm. long, 3–7 mm. broad; scape and raceme 30–60 cm. or more long with (5–) 8–12 (–15) flowers; median sepal spathulate, up to 7–8 mm. long, mostly mottled brownish-red, lateral sepals together like the median, fused nearly to the apex; petals like the sepals but broader and clawed, undulate-margined; lip with the distal lobe distinctly clawed, more or less flat, white or light mauve, with usually a reddish spot under the claw, 10–15 mm. long, sometimes smaller; capsule about 2 cm. long, oblong-ellipsoid.
  Locally abundant as an epiphyte on small trees in low open woodland; SL–2750 feet; fl. and fr. most of the year. *A 8061! H 9017! P 7878!* Endemic.

3. O. pulchellum Hook. in Curt., Bot. Mag. **54**: t. 2773 (1827).
  Like the last but with the lobes of the lip larger and more showy; sepals and petals pink mottled olive; lip mostly pinkish-mauve darkening with age, without a reddish spot beneath the broader less distinct claw.
  Widespread and locally common as an epiphyte on small trees in woodland or on limestone ledges; 400–3000 feet; fl. Feb–July, fr. July–Aug. *A 7188! H 8656! P 10220!* Endemic.
  Although extreme variants of the last two species are very different in detail, extensive field observations have shown an almost unbroken series of intermediates between them. Some of these intermediates, including *O. berenyce* Reichb. f., have been named, but the complexity of the whole range of these putative natural hybrids is such as to render the application of any further formal names meaningless at the present time; see Withner & Stevenson (1968).

4. O. gauntlettii Withner & Jesup in Amer. Orch. Soc. Bull. **33**: 463 (1964).
  Plant tufted; leaves falcate, up to 4 cm. long, 3–5 mm. broad; scape to 10 cm. long, few-flowered; flower 13 mm. from top to bottom; median sepal oblanceolate acute; lateral sepals connate nearly to the tip; petals obovate, acute, broader than the sepals; sepals and petals white tinged purple in the mid-line; lip yellow overlaid reddish-brown at the base, white or tinged rosy-purple in the lobes, about 6 mm. broad.
  Very rare (West., Han.), epiphytic; about 1500 feet; fl. Sept–Nov. *Gauntlett.* Endemic.

5. O. triquetrum (Sw.) R. Br. in Ait., Hort. Kew. ed. 2, **5**: 216 (1813).
Plant tufted; leaves softly fleshy with very thin margins and keel, up to 20 cm. long and 10 mm. broad; scape and raceme up to 20 cm. long, regenerating from the upper bracts; flowers more or less clustered; sepals and petals 9–10 mm. long, the petals broader, usually white variously mottled crimson to brownish-red, sometimes light or dark yellow; lip about 10 mm. long and broad, similarly coloured.
Occasional in the western and central parishes, an epiphyte on trees; 400–1250 feet; fl. and fr. almost all the year but mostly June–Sept. *H 7647! J.P. 2286! P 10403!* Endemic; but visual record from Cuba.

58. CRYPTARRHENA R. Br. (1816)

1. C. lunata R. Br. in Edw., Bot. Regist. **2**: t. 153 (1816).
Plant 10–25 cm. high in flower, tufted; leaf-blades linear-oblanceolate, jointed at the base to overlapping sheaths, 6–15 cm. long, 7–13 mm. broad; peduncle simple 10–20 cm. long; flowers olive-green; sepals elliptical, 3·2–3·5 mm. long; petals obovate, obtuse, about 3 mm. long; lip clawed, 4 mm. long, lateral lobes narrow, spreading or curved upwards, mid-lobe shortly divided; capsule cylindrical, 6–8 mm. long.
Rare (Trel., St. Ann, Port.), epiphytic; 2200–3000 feet; fl. Mar, Aug–Oct, fr. Oct–Feb. *H 10430! P 4144!* Mexico to Peru and Brazil, Trinidad.

59. VANILLA Mill (1754)

1 Stem soon leafless; leaves on young shoots lanceolate-subulate; raceme 8–12-flowered
                                                                         1. *claviculata*
1 Stem leafy; raceme usually fewer-flowered:
  2 Bracts conspicuous, leafy 1·5–2·5 (–8) cm. long; leaves elliptical, acuminate, longer
    than the internodes   2. *inodora*
  2 Bracts small, not leafy; leaves not longer than the internodes:
    3 Leaves ovate, shortly acuminate, shorter than the internodes   3. *wrightii*
    3 Leaves oblong or elliptical, acute, about as long as the internodes   4. *phaeantha*

V. planifolia Andr., at one time cultivated as the source of commercial vanilla, has slender capsules 15–25 cm. long, large leaves and small bracts. It is still grown occasionally as an ornamental.

1. V. claviculata (W. Wright) Sw. in Nov. Act. Soc. Sci. Upsal. **6**: 66, t. 5, f. 1 (1799).—*Epidendrum claviculatum* W. Wright (1787). Greenwithe.
Stems 1 cm. or more thick, fleshy, with long branches trailing over small trees and shrubs and on the ground to 20 m. or more; leaves 2–8 cm. long, up to 1 cm. or more broad; curved and more or less involute, sessile; flowers fragrant; perianth about 4·5 cm. long; sepals elliptical, obtuse, glaucous green; petals keeled; lip rounded crenate and crisped, white and purple; capsule fleshy, up to about 10 cm. long.
Locally common in low woodlands on limestone; 650–2250 feet; fl. Apr–June, fr. June–Dec. *H 10411! P 7879!* Greater Antilles; Grand Cayman.

2. V. inodora Schiede in Linnaea **4**: 574 (1829).
Stem about 8 mm. thick with internodes 3·5–10 cm. long; leaves 10–20 cm. long, 5–10 cm. broad, shortly petioled; perianth 4–6 cm. long; sepals and petals lanceolate, light green; lip 3-lobed, white with yellow crests; capsule slender, 12–18 cm. long, not fragrant.
Uncommon on trees in rocky woodland; 50–2600 feet; fl. Dec, fr. Mar–May. *H 8555! 8829!* Mexico to the Guianas, West Indies.

3. V. wrightii Reichb. f. in Flora **48**: 273 (1865).
Stem slender with internodes 5–10 cm. long; leaves distinctly petioled, blade 4–7 cm. long, 2–3·5 cm. broad; bracts 5–7 mm. long, triangular; perianth about 5 cm. long; sepals elliptic-lanceolate, reddish-brown; petals keeled; lip roundish-obovate, acute, margin crenulate, white; capsule linear, 12–14 cm. long.

Rather rare (St. Andr., Clar., Port.), in shady or humid thickets and low woodlands on limestone; 2200–2500 feet; fl. Mar, July, fr. Aug. *H 7885! P 16495!*
Guyana, Suriname, Cuba, Hispaniola, Grenada, Trinidad.

4. V. phaeantha Reichb. f. in Flora **48**: 274 (1865).

Stem up to about 1 cm. thick with internodes 6–14 cm. long; leaves shortly
petioled, blade 10–20 cm. long, (2–) 3–7 cm. broad; bracts ovate, acute, 6–14 mm.
long; perianth 8–9 cm. long; sepals oblanceolate, green; petals broadly keeled; lip
with limb broadly ovate, mucronate below the tip, margin curled and crenulate,
white with greenish-yellow crests; capsule about 8 cm. long.

Very rare (St. Eliz., Han.), in woodland on limestone; 800 feet; fl. May, fr.
May–July. *H 10489!* Cuba, St. Vincent, Trinidad and Tobago.

60. DENDROPHYLAX Reichb. f. (1864)

1 Lip entire, ovate, acuminate, about as long as the lateral sepals; spur curved, 2–2·5 cm.
   long; capsule ellipsoid, 1·5 cm. long   1. *barrettia.*
1 Lip bilobed, white, longer than the sepals; spur more or less straight, 4·5 cm. or more
   long; capsule linear-cylindrical:
   2 Spur about 5 cm. long; lip when flattened out 2–2·5 cm. broad; capsule about 7 cm.
     long   2. *funalis*
   2 Spur up to 15 cm. or more long; lip 3–4 cm. broad; capsule up to about 10 cm. long;
     epiphyte; fl. May, fr. July; Grand Cayman   *D. fawcettii* Rolfe

1. D. barrettiae F. & R. in Journ. Bot. **47**: 266 (1909).

Roots slender, following the fissures of bark; scapes simple, up to 2·5 cm. long;
perianth light green fading to yellow; sepals spreading, linear-lanceolate, acute,
5–6 mm. long; petals similar to sepals, about 5 mm. long.

Rare and local (Clar., St. Ann), on soft-barked shrubs in thickets and open
woodland on limestone; 2250–2500 feet; fl. Aug–Nov, fr. Dec–Apr. *A 11063!
ACH 736! P 11411!* Endemic.

2. D. funalis (Sw.) Benth. ex Rolfe in Gard. Chron. ser. 3, **4**: 533 (1888).

Roots rather thick, diffuse and often loose and pendulous, sometimes proliferating new plants; stem very short, with small dark brown scales; scape up to
10 cm. long, usually 1-flowered; sepals and petals elliptical, light green, about
2 cm. long; lip deeply trough-shaped at the junction with the light yellow-green
spur, the broad lateral lobes curved upwards.

Occasional in the central and western parishes on rough limestone rocks in
thickets and open woodlands or on isolated trees; 800–1600 feet; fl. and fr. Sept–
Apr. *H 7763! P 8618!* Endemic.

61. HARRISELLA F. & R. (1909)

1. H. porrecta (Reichb. f.) F. & R. in Journ. Bot. **47**: 266 (1909).—*Campylocentrum
porrectum* (Reichb. f.) Rolfe (1903).

Roots slender; scapes several, bracteate, zig-zag, (1·5–) 3–5 cm. long; racemes
few-flowered; perianth 2–2·5 mm. long, light golden-yellow; sepals and petals
oblong-elliptical, recurved; lip a little longer than the sepals, hooded over the
column, apiculate; spur globose; capsule 6 mm. long.

Rare and local (St. Andr., Manch., Trel.), perhaps easily overlooked, on shrubs
and small trees in thickets; 300–800 feet; fl. Oct–Nov, fr. Nov. *ACH 553! H 7762!*
Florida, Mexico, El Salvador, Cuba, Hispaniola.

62. CAMPYLOCENTRUM Benth. (1881)

1 Leaves present, in two ranks on an elongated and more or less trailing and rooting stem;
   spikes lateral; lip 3-lobed:
   2 Leaf-tip deeply 2-lobed; spikes 1–2 (–3) cm. long; capsule smooth   1. *micranthum*
   2 Leaf-tip entire or nearly so; spikes 2·5–4 cm. long; capsule 6-ribbed   2. *minus*

1 Leaves absent; racemes terminal or subterminal:
  3 Racemes with up to 40 flowers in 2 ranks, up to 6 cm. long, several appearing in rapid
      succession; bracts less than 2 mm. long; lip 3-lobed   3. *fasciola*
  3 Spikes fewer-flowered, 1–4 cm. long; bracts more than 2 mm. long; lip entire:
      4 Roots less than 2 mm. broad, subterete; sepals obtuse, about 2 mm. long; spur
          subglobose   4. *filiforme*
      4 Roots up to 5 mm. broad, distinctly flattened; bracts spathaceous, 3–4 mm. long;
          sepals acute, about 4·5 mm. long; spur ellipsoid   5. *pachyrrhizum*

1. C. micranthum (Lindl.) Rolfe in Orch. Rev. **9**: 136 (1901).—*Angraecum
micranthum* Lindl. (1835). *C. jamaicense* (Reichb. f. & Wullschl.) Fawcett
(1893). *C. barrettiae* F. & R. (1909).
  Stems up to 30 cm. long and sparingly branched, rooting at intervals or shorter
and unbranched; leaves leathery, concave, sessile on a striate sheath, deciduous,
2–7 cm. long, up to 1·5 (–2) cm. broad; spikes 6–10 (–12)-flowered; sepals and
petals 3–5 mm. long, spreading at the tips, generally pinkish-yellow, paler within;
lip a little shorter than the sepals, spur cylindrical, about 2 mm. long; capsule
11–15 mm. long, ripening orange, often opening laterally along 1 suture.
  Rather common as an epiphyte on old isolated trees and in woodland on limestone; 900–3000 feet; fl. Oct–May, fr. Jan–July. *A 6821! Moore 7801! P 6662!*
Mexico to Venezuela, Greater Antilles, Trinidad and Tobago.

2. C. minus F. & R. in Journ. Bot. **47**: 127 (1909).
  Like the last in habit and dimensions except the flowers slightly smaller and
spur obliquely ellipsoidal and constricted at the base.
  Very rare (Port.), on small trees in damp sheltered area; about 2000 feet; fl.
Dec, fr. Jan. *Harris!* Until this species was rediscovered in 1961 near the type
locality, it was known only from the type.

3. C. fasciola (Lindl.) Cogn. in Mart., Fl. Bras. **3** (6): 520, t. 106, f. 1 (1906).—
Angraecum fasciola Lindl. (1840). *C. sullivanii* F. & R. (1909).
  Stem very short with numerous long thick terete roots; racemes several together,
flower-bearing from near the base; pedicels less than 1 mm. long; perianth about
1·5 mm. long, greenish-white at first, turning light yellow and then orange; sepals
and petals ovate; spur narrowly ellipsoid; capsule oblong-ellipsoid, about 4 mm.
long.
  Occasional as an epiphyte on the branches of small trees, especially *Citrus* and
Guava; (SL–) 500–1750 feet; fl. Jan–May, fr. Apr–Aug. *A 6818! Moore 10464!
P 21330!* Guatemala and Honduras to Venezuela and the Guianas, Puerto Rico,
Trinidad.

4. C. filiforme (Sw.) Cogn. ex Kuntze, Revis. Gen. Pl. **3** (2): 298 (1898).—*Epi-
dendrum filiforme* Sw. (1788).
  Like the last but with more slender roots and shorter inflorescences; flowers
sessile; perianth light green.
  Very rare (St. Ann), known only from one locality on the twigs of cultivated
Guava trees; about 1900 feet; fl. Apr–June, fr. May–June. *P 10367!* Hispaniola.

5. C. pachyrrhizum (Reichb. f.) Rolfe in Orch. Rev. **11**: 246 (1903).
  Like the last but flowers sessile, subtended by conspicuous bracts; perianth and
spur longer, slightly curved, light to brownish orange; capsule 8 mm. long.
  Rare (Han., St. Mary, Port.), on branches of small trees; 500–1500 feet; fl.
Oct–Jan, fr. Jan. *H 7570!* Florida, Guyana, Cuba, Hispaniola, Trinidad and
Tobago.

34. JUNCACEAE

Grass-like herbs with mostly short erect or horizontal rhizomes. Leaves basal,
tufted, linear or filiform, cylindrical or flat, sheathing at the base. Flowers solitary
or in panicles, corymbs or heads, usually small, bisexual or unisexual and dioecious,
actinomorphic. Perianth-segments 6 in 2 series, or rarely 3 segments only, mostly

glumaceous. Stamens 6 or 3, free, opposite the perianth-segments; anthers 2-locular, basifixed, opening lengthwise; pollen in tetrads. Ovary superior, 1-locular or divided by 3 septa or 3-locular; styles 1 or 3; stigmas 3; ovules ascending or parietal. Fruit a loculicidally dehiscent capsule. Seeds sometimes tailed with a small straight central embryo.

8 genera with about 300 species, generally distributed but more numerous in temperate and cold regions.

1. JUNCUS L. (1753)

Tufted herbs with entire glabrous leaves; leaf-sheaths usually open; flowers bisexual; seeds numerous.

1 Flowers subcapitate in tight clusters in a compound inflorescence; plants perennial:
  2 Rhizome horizontal, bulbous at the culm-bases; capsule as long as the perianth; leaves
    flat   1. *aristulatus*
  2 Rhizome shortly branched, without bulbs; capsule shorter than the perianth; leaves
    filiform, canaliculate   2. *tenuis*
1 Flowers cymose along the upper sides of spreading branches; culms densely tufted:
  3 Plant perennial; hyaline tips of leaf-sheath abruptly produced into a short ligular
    appendage; roots wiry   3. *dichotomus*
  3 Plant annual; hyaline tips of leaf-sheath tapered, not usually free; roots cottony
                                               4. *bufonius*

1. J. aristulatus Michx., Fl. Bor. Amer. **1**: 192 (1803).
Plants with compressed erect leafy culms; panicle compound with (2–) 3 (–5)-flowered glomerules; stamens 3; seeds 0·5–0·6 mm. long.
Very rare (St. Eliz.), in silted roadside ditch; 50 feet; fr. Dec. *A 12064!* Mexico, E. United States to Florida, Cuba.

2. J. tenuis Willd. in L., Sp. Pl. ed. 4, **2** (1): 214 (1799).—*J. macer* S. F. Gray (1821).
Low densely tufted and tightly branched perennial forming a sward; sheath auriculate.
Local (Port.); 5200 feet; fl. and fr. Dec. *P 9667!* Native probably in N. Amer., now widely distributed. The Jamaican example of this widespread and tolerant species is much smaller and more compact than typical *J. tenuis*.

3. J. dichotomus Ell., Sketch Bot. S. Carol. & Georgia **1**: 406 (1817).
Tufted wiry perennial with culms (10–) 40–60 (–100) cm. high in flower.
Local (St. Andr., Port., St. Thom.), along paths and roadsides and in damp woodland clearings; 3500–7400 feet; fl. and fr. Nov–July. *A 5785! H 12497! P 7050!* E. United States.

4. J. bufonius L., Sp. Pl. **1**: 328 (1753).
Culms slender, tufted, 15–30 cm. high in flower.
Locally abundant (St. Andr., Port.), in shallow ditches and damp waste places; 2600–4500 feet; fl. and fr. most of the year. *A 7082! H & B 10532! P 6824!* Bermuda, Mexico, Costa Rica and cosmopolitan in temperate regions.

35. CYPERACEAE

Perennial or annual herbs, rarely shrubby, often tufted or rhizomatous, with slender fibrous adventitious roots. Stems usually solid, often triquetrous. Leaves mostly basal and crowded, usually with closed sheaths; blade narrow, usually keeled, sometimes broader and plaited, or absent; ligule rare. Inflorescence often scapose, the branches subtended by leafy bracts. Flowers always small (florets), bisexual or unisexual and monoecious or very rarely dioecious, often protogynous, each subtended by a scalelike bracteole (glume) and distichously or spirally arranged in small spikes (spikelets). Spikelets capitate, umbellate, spicate or paniculate. Perianth represented by hypogynous scales, bristles or hairs or absent. Stamens hypogynous, (1–) 2–3 (–6); filaments free; anthers 2-locular, basifixed, opening lengthwise; connective often produced. Ovary superior, 1-locular with a solitary anatropous ovule erect from the base; style simple below, usually 2–3-branched above, sometimes indurated at the base and persistent on the fruit as a beak. Fruit an indehiscent nutlet (achene), bilateral and lenticular or trigonous or subglobose, sometimes enclosed in a modified glume (utricle).

About 70 genera with over 3500 species, widespread but particularly numerous in high latitudes and mostly growing in wet places.

1 Florets bisexual, rarely the upper in each spikelet staminate only:
  2 Spikelets few-flowered (usually 2, but up to 6); 2 or more of the lower glumes empty:
    3 Style 2-lobed or undivided; stamens 2 (–3); hypogynous bristles present or absent; achene usually beaked; glumes usually broad and thin, mostly 1-nerved, the nerve often excurrent, often reddish-brown or white, spirally arranged
                                                     1. Rhynchospora
    3 Style 3-lobed; achene not beaked:
      4 Glumes distichous; leaves all basal, smooth, wiry, subterete; inflorescence sub-capitate; bristles usually present, 1–6, half as long as achene   **2. Schoenus**
      4 Glumes not distichous; leaves basal and on the flowering stem, denticulate on the margins and keel; inflorescence much-branched; stem hollow; bristles absent
                                                     3. Cladium
  2 Spikelets 1–many-flowered; 1 or rarely 2 of the lower glumes empty:
    5 Glumes distichous; stamens (1–) 2–3; hypogynous bristles absent; achenes not beaked:
      6 Spikelets several–many, flattened or terete; inflorescences usually subtended by conspicuous bracts, often radiate; style-base not enlarged; stigmas 2–3
                                                     4. Cyperus
      6 Spikelet solitary, rarely 2, flattened, subtended by 1 or 2 very small inconspicuous bracts; style-base enlarged; stigmas 3   **5. Abildgaardia**
    5 Glumes spirally arranged:
      7 Plants leafless; culms sheathed at the base; hypogynous bristles usually present:
        8 Spikelet solitary, terminal; culms smooth or septate   **6. Eleocharis**
        8 Spikelets numerous in a compound umbel; culms smooth   **7. Scirpus**
      7 Plants leafy; spikelets rarely solitary:
        9 Hypogynous bristles or scales representing perianth present:
          10 Perianth of bristles or setae only   **7. Scirpus**
          10 Perianth of scales:
            11 Scales 3, sometimes with intervening setae; culm variably hairy, usually 5-angled   **8. Fuirena**
            11 Scales 2; small annual, usually with solitary or few clustered lateral spikelets
                                            9. Hemicarpha
        9 Hypogynous perianth-bristles or scales absent   **10. Fimbristylis**
1 Florets unisexual:
  12 Pistillate florets enclosed by a modified glume (utricle):
    13 Rachilla exserted and hooked at the apex   **11. Uncinia**
    13 Rachilla not exserted or hooked   **12. Carex**
  12 Pistillate florets not enclosed by a utricle:
    14 Spikelets in spikes or panicles   **13. Scleria**
    14 Spikelets in globose axillary clusters   **14. Diplacrum**

1. RHYNCHOSPORA Vahl (1805) nom. cons.; Gale (1944); Kükenthal (1949–51)

1 Style undivided or shortly bifid at apex; plants perennial:
  2 Spikelets in compact globose heads 6–12 mm. in diameter; heads corymbose-paniculate
                                                     1. cyperoides
  2 Spikelets not in globose heads:
    3 Bristles evident, mostly 6, longer than achene but usually not exceeding whole fruit including beak; culms thick and spongy or leafy at base:
      4 Beak of achene long-tapered, longer than spikelet, exserted; spikelets 12–14 mm. long, not clustered; plant stoloniferous   *2. inundata*
      4 Beak of achene more or less conical, shorter than spikelet, not exserted; plant not stoloniferous:
        5 Spikelets (4–) 8–15 (–30) together in compact clusters, about 5 mm. long; beak thick, sulcate, scabrid   *3. gigantea*
        5 Spikelets in small clusters of up to 5 or solitary, 6–8 mm. long:
          6 Spikelets elliptic-fusiform, 2 mm. or more broad, subsolitary or shortly pedi-celled in small clusters; beak of achene thick, deeply sulcate, scabrid
                                                       4. corymbosa

6 Spikelets lanceolate, 1–1·5 mm. broad, not clustered; beak of achene slender, not sulcate, glabrous   5. *eggersiana*
3 Bristles very short and few or absent; culms wiry, the basal leaves short or withered at flowering:
  7 Spikelets in compact clusters, about 4 mm. long; beak of achene black at maturity, covered at first with whitish scurf   6. *uniflora*
  7 Spikelets in lax racemes or panicles:
    8 Panicles elongated, subracemose on long-exserted peduncles; spikelets about 6 mm. long   7. *racemosa*
    8 Panicles ovoid to pyramidal in outline, usually with peduncles rather shortly exserted from the sheaths; spikelets 4–5 mm. long:
      9 Beak of achene short, broader than long; leaves usually antrorsely scabrid adaxially; spikelets light greyish-buff   8. *polyphylla*
      9 Beak of achene longer than broad; leaves mostly smooth; spikelets light buff turning brown   9. *jamaicensis*
1 Style deeply divided:
  10 Hypogynous bristles present; plants perennial:
    11 Bristles plumose, much shorter than achene, rarely 1 or 2 produced into an antrorsely hispid awn   10a. *oligantha* var. *breviseta*
    11 Bristles scabrid from base to apex or smooth, very rarely slightly plumose at base:
      12 Bristles retrorsely scabrid; spikelets in globose clusters
                           11a. *cephalantha* var. *microcephala*
      12 Bristles antrorsely scabrid or smooth; spikelets not in globose clusters:
        13 Surface of achene smooth, biconvex but not turgid, yellow-brown with a prominent paler patch in the centre of each side; bristles mostly shorter than achene   12. *fascicularis*
        13 Surface of achene not smooth:
          14 Surface of achene cancellate (latticed); achene turgid, biconvex; spikelets lanceolate with up to 20 whitish to light brown shiny glumes; stamens 3:
            15 Leaves more or less glabrous; bracts densely ciliate towards base and on keel; spikelets about 6 mm. long, densely packed in an ovoid panicle; glumes mucronate   13. *cephalotes*
            15 Leaves and bracts pilose; spikelets 7–8 mm. long in a more or less open pyramidal panicle; glumes aristate   14. *comata*
          14 Surface of achene transversely rugose:
            16 Transverse rugosities coarse, forming distinct ridges:
              17 Rhizome evident, sometimes stoloniferous; culms distinctly separated:
                18 Partial inflorescences 6–9, corymbose, markedly storied; spikelets 3–4 mm. long; bristles equal to or shorter than achene   15. *miliacea*
                18 Partial inflorescences 3–5, longer than broad with ascending branches, nodding; spikelets 6–9 mm. long; bristles longer than achene   16. *odorata*
              17 Rhizome short without stolons; culms tufted or approximate; bristles shorter than achenes:
                19 Spikelets 2–2·5 mm. long, sessile or subsessile in clusters   17. *perplexa*
                19 Spikelets 3–4 mm. long, filiform-pedicelled, few in open corymbs
                                    20. *rariflora*
            16 Transverse rugosities finely undulate or obscure:
              20 Leaves filiform; culms densely tufted; spikelets 2–3 mm. long, few in branched clusters, blackish-brown; beak subulate, nearly 1 mm. long 19. *lindeniana*
              20 Leaves flat, 1 mm. or more broad:
                21 Spikelets 5–7 mm. long in loose elongated corymbs; beak long-attenuated, 1·5 mm. long, as long as body of achene, overtopped by the bristles
                                    20. *marisculus*
                21 Spikelets up to 5 mm. long in compact clusters; beak deltoid:
                  22 Rhizome knotted or bulbous; spikelets 4·5–5 mm. long; bristles as long as or longer than whole fruit (including beak), spreading at maturity 21. *rugosa*
                  22 Rhizome not conspicuously knotted or bulbous; spikelets up to 4 mm. long; bristles shorter than achene:
                    23 Spikelets 2·5–3 mm. long, subglobose, 2-flowered; surface of achene reticulated   22a. *globularis* var. *pinetorum*
                    23 Spikelets 3–4 mm. long, ovoid, about 3-flowered; surface of achene coarsely cancellate to striate   22b. *globularis* var. *recognita*
  10 Hypogynous bristles absent:
    24 Inflorescence more or less laxly corymbose:
      25 Culms stout, erect; leaves flat; glumes densely and tightly imbricate:
        26 Annual with clustered fibrous roots; culms tufted; spikelets subsolitary, 4–6 mm. long   23a. *nitens* var. *hispaniolica*
        26 Perennial; spikelets clustered:
          27 Spikelets 2–2·5 mm. long; culms tufted, without stolons; leaves 1–2 mm. broad
                                    17. *perplexa*

27 Spikelets 8–10 mm. long; rhizome woody, covered with scales, stoloniferous; leaves 5–8 mm. broad   24. *robusta*
25 Culms and leaves slender; glumes loosely imbricate with achenes often exposed at maturity:
    28 Annual; leaves flat, 1–1·5 (–2·5) mm. broad; spikelets up to about 1·5 mm. long; inflorescence with numerous slender spreading branches; achene coarsely rugose   25. *minutiflora*
    28 Perennial; leaves filiform, canaliculate, less than 1 mm. broad; corymbs rather compact:
        29 Achenes with depressions on either side of beak thereby the fruit conspicuously 3-toothed at apex, coarsely rugose; plant stoloniferous or tufted; spikelets 4–5 mm. long; lower glumes aristate   26. *setacea*
        29 Achenes not as above, finely rugulose; plant not stoloniferous; spikelets 2·5–3·5 mm. long; lower glumes at most mucronulate   27. *intermixta*
24 Inflorescence capitate; achenes marginate; glumes mostly whitish (in life):
    30 Plant annual, tufted, with fibrous roots; bracts green, leafy, sparingly ciliate towards base only; glumes acute   28. *radicans*
    30 Plants rhizomatous or stoloniferous, perennial:
        31 Bracts setaceous, inconspicuous or obsolete, the lowest rarely twice as long as the head; stolons numerous above ground forming mats or sometimes cushions; leaves rarely more than 1 mm. broad   29. *berteroi*
        31 Bracts leafy, many times longer than the leaf, white at least in the lower half:
            32 Bracts ciliate-margined at least towards base; culms stout, approximate, from a woody rhizome, clothed at base with fibrous sheath remnants; glumes acute or obtuse, minutely apiculate   30. *nervosa*
            32 Bracts not ciliate; culms slender from a creeping rhizome or stolon, mostly without fibrous sheath-remants; glumes obtuse, truncate, emarginate or eroded at tip   31. *colorata*

1. R. cyperoides (Sw.) Mart. in Denkschr. Akad. München **6**: 149 (1816–17).— *Schoenus cyperoides* Sw. (1788).

Culm-base and leaves rather fleshy; plants 45–100 cm. high in flower; heads yellowish-brown; stamens 3.

Locally common in the central and western parishes, in ditches and damp open savannas; 50–3000 feet; fl. and fr. all the year. *A 7122! H 11164! P 15804!* Mexico to Paraguay, West Indies, tropical Africa, Madagascar.

2. R. inundata (Oakes) Fernald in Rhodora **20**: 139 (1918).—*Ceratoschoenus macrostachya* (Torr.) Torr. var. *inundatus* Oakes (1841).

Culms spongy, up to 110 cm. high in flower; leaves long, up to 1 cm. broad; spikelets in loose panicles; achene 2·5 (–4) mm. long, beak 16 mm. long.

Very rare and local (St. Eliz.), in swamps; 25 feet; fl. and fr. Sept. *P 15780!* E. United States, British Honduras.

3. R. gigantea Link in Jahrb. Gewächsk. **3**: 76 (1820).

Culms tufted, spongy at base, up to 2 m. or more high in flower; leaves up to 2 cm. broad, sharply keeled, sheath closed by a translucent membrane; corymbs storied from broad inflated sheaths; male florets alongside female, stamens 3; achene 2 mm. long, wrinkled, beak a little longer; bristles short and long.

Very local (Clar., St. Eliz., Port.), gregarious in standing water in swamps and pond margins; SL–2000 feet; fl. and fr. Dec–Apr. *A 12196! Proctor & Mullings 22066!* Guianas, Brazil, Cuba, Puerto Rico, Guadeloupe, Trinidad.

4. R. corymbosa (L.) Britton in Trans. Acad. Sci. New York **11**: 84 (1892).— *Scirpus corymbosus* L. (1756).

Culms 60–120 cm. high in flower; leaves 8–15 mm. broad; achene 2–3 mm. long, beak a little longer.

Rare and local (Clar., St. Eliz., Port.), in swamps and ditches; SL–2200 feet; fl. and fr. May–Aug, Dec. *A 7191! H 12259! P 26599! Robertson UCWI 9212!* Widespread in the tropics.

5. R. eggersiana Boeck., Cyp. Nov. **2**: 26 (1890).

Culms 60–120 cm. high in flower; leaves smooth, flaccid; corymbs several, rather distant, densely flowered; bracteoles persistent, setose, hispid; spikelets dark brown; achene 1·5–1·8 mm. long, dark brown.

Locally common (St. Andr., Port., St. Thom.), on pathside banks and in clearings in montane woodland; 4000–7400 feet; fl. and fr. most of the year. *A 8175! H 12395! J.P. 767!* This species is part of a complex of taxonomically closely related plants extending throughout the mountainous parts of the American tropics. Kükenthal (1949) placed Jamaican specimens in varieties of *R. longiflora* Presl (1832), *R. aristata* Boeck. (1857) and *R. macrochaeta* Steud. ex Boeck. (1873), thus indicating the desirability of further study over a wide geographical range. *R. eggersiana* was originally described from Jamaica.

6. R. uniflora Boeck. in Flora **63**: 439 (1880).—*R. elongata* Boeck. (1888).
Culms slender, numerous, 25–50 cm. long; leaves 1–2 (–3) mm. broad, smooth; partial inflorescences few, small, distant; achene smooth; bristles absent.
Locally common (St. Andr., Port., St. Thom.), on boulders and pathside banks in montane woodland; (2000–) 4000–6000 feet; fl. and fr. Mar–Sept. *A 7447! H 11586! P 16585!* Greater Antilles.

7. R. racemosa C. Wright in Sauv. in Anal. Acad. Cienc. Habana **8**: 86 (1871).
Culms loosely tufted, 80–120 cm. high in flower; leaves 1–2·5 mm. broad; spikelets subsolitary, light brownish-buff.
Rare (St. Andr., St. Thom.), on shaded banks; 3000–4000 feet; fl. and fr. May–July. *A 12578! P 23577!* Cuba, Hispaniola.

8. R. polyphylla (Vahl) Vahl, Enum. Pl. **2**: 230 (1805).—*Schoenus polyphyllus* Vahl (1798).
Culms tufted from a short rhizome, usually rather straggling, 60–120 (–200) cm. long; leaves very short on base of culm, 2–3 (–5) mm. broad above; beak of achene usually green; bristles up to 3 and very short or absent.
Common on roadside banks, in rough pastures and woodland margins; 100–4200 feet; fl. and fr. all the year. *A 5766! 8747! H 12240! P 16247!* Venezuela, Cuba, Lesser Antilles.

9. R. jamaicensis Britton in Bull. Torr. Bot. Club **41**: 1 (1914).
Like the last; culms often scrambling to 2·4 m. long, puberulous; achene with distinct shoulders below the beak.
Occasional in woodland, on shaded banks and pathsides, mostly in the eastern parishes; (2000–) 2700–5000 feet; fl. and fr. June–Aug, Nov–Mar. *A 7827! 8704! H 12332! P 7350!* Puerto Rico.

10. R. oligantha A. Gray in Ann. Lyc. New York **3**: 212, t. 6, f. 22 (1835).

10a. Var. **breviseta** Gale in Rhodora **46**: 129 (1944).
Culms densely tufted, 15–35 cm. high; leaves filiform; spikelets solitary or few, light reddish-brown, appearing disproportionately large (4–) 5–8 mm. long, the lower deflexed; achene transversely rugulose, 2–2·6 mm. long, 1·6–2 mm. broad.
Very rare and local (Clar.), in open grassy savanna; 2300 feet; fl. and fr. Nov–Dec. *A & D 13077! H 12249! J.P. 1518! P 16126!* Florida, Cuba, Hispaniola.

11. R. cephalantha A. Gray in Ann. Lyc. New York **3**: 218, t. 6, f. 30 (1835).

11a. Var. **microcephala** (Britton) Kük. in Engl., Bot. Jahrb. **75** (1): 101 (1950).—*R. axillaris* var. *microcephala* Britton (1892).
Culms tufted, 30–80 cm. high; leaves 1–3 mm. broad; heads 5–10 mm in diameter, 1 terminal and 2–5 lateral; spikelets 3–4 mm. long; achene 1·4–1·6 mm. long.
Rare and local (Clar.), in moist open savanna and bog; 2300 feet; fl. and fr. sporadically. *A 12926! A & S 3213! P 15805!* E. and S.E. United States, Cuba.

12. R. fascicularis (Michx.) Vahl, Enum. Pl. **2**: 234 (1805).—*Schoenus fascicularis* Michx. (1803).
Culms tufted, ascending, 40–120 cm. high; leaves 1–3 mm. broad; spikelets ovate-lanceolate, 3–4·5 mm. long, clustered at the ends of slender branches; achene 1·3–1·5 mm. long, 1·1–1·5 mm. broad.

Rare (Clar., St. Eliz., St. Ann), in swamps and moist open savannas; 50–2500 feet; fl. and fr. sporadically. *A 7118! P 15950! Purdie!* S.E. United States, Bermuda, Greater Antilles, Nevis.

13. R. cephalotes (L.) Vahl, Enum. Pl. **2**: 237 (1805).—*Scirpus cephalotes* L. (1762).
Culms 20–100 cm. high; leaves 5–10 mm. broad; panicle 3–4 cm. long; achene 2 mm. long.
Very rare; record based on a single unlocalized collection. *Masson!* British Honduras to Brazil, Trinidad.

14. R. comata (Link) Schult., Mant. **2**: 50 (1824).—*Schoenus comatus* Link (1821). *R. cephalotoides* Griseb. (1866).
Rhizome creeping with thick roots; culms robust forming loose clumps, 60–100 cm. high; achene 2·5 mm. long.
Rare and local (Clar.), in shady thickets on clay soil; 2300 feet; fl. Mar, Aug. *P 16230! 23365! Weck 7264–4!* Colombia to Brazil, Cuba, Trinidad.

15. R. miliacea (Lam.) A. Gray in Ann. Lyc. New York **3**: 198, f. 4 (1835).—*Schoenus miliaceus* Lam. (1791).
Culms gregarious, robust, 60–140 cm. high; leaves 4–7·5 mm. broad; achene 1–1·3 mm. long, about 1 mm. broad.
Rare and local (Clar., St. Ann), in wet places; 2200–2300 feet; fl. and fr. Mar, July–Sept. *A 12918! P 16483! 23375!* Virginia to Florida, Greater Antilles.

16. R. odorata C. Wright ex Griseb., Cat. Pl. Cub.: 242 (1866).
Culms erect, 70–180 cm. high; leaves 3·5–6 mm. broad; achene 1·4–1·7 mm. long, 1·4–1·6 mm. broad.
Local (St. Eliz., Han., West.), at sandy lagoon and swamp margins; SL–40 feet; fl. and fr. June, Sept–Dec. *A 10194! Britton 1356! P 23624!* N. Carolina to Florida, Bermuda, Bahamas, Greater Antilles.

17. R. perplexa Britton in Small, Fl. Southeast U.S.: 197, 1328 (1903).
Culms wiry, terete, 50–110 cm. high; leaves flat, 1–2 mm. broad; glumes dark brown or blackish; bristles short, rudimentary or absent; achene 1–1·3 mm. long, 0·9–1·2 mm. broad.
Very rare (St. Eliz.), among grass tussocks in swampy pond; 25 feet; fl. and fr. Jan. *P 21924!* Virginia to Florida, Bahamas, Cuba.

18. R. rariflora (Michx.) Ell., Sketch Bot. S. Carol. & Georgia **1**: 58 (1816).—*Schoenus rariflorus* Michx. (1803).
Rhizome knotted; culms 15–50 cm. high; leaves filiform, up to 1 mm. broad; cymes 1–3; achene 1·3–1·4 mm. long, 1·1–1·4 mm. broad.
Very rare (Clar., St. Ann), in grassy savanna; 2300–3000 feet; fl. and fr. Sept–Jan. *A 13179! H 11223! Lewis & duQuesnay 1514!* Virginia to Texas and Florida, Cuba, Hispaniola.

19. R. lindeniana Griseb., Cat. Pl. Cub.: 244 (1866).
Culms numerous, densely tufted, 15–30 (–70) cm. long; leaves capillary to up to 1 mm. broad; cymes 2–5, remote; achene light brown, 1·1–1·4 mm. long, 0·7–0·9 mm. broad; bristles longer than achene, shortly plumose towards base.
Very local (St. Thom.), on rocky serpentine bluffs in open thicket and woodland, co-dominant with terrestrial lichens; 2800–3000 feet; fl. and fr. Jan–Feb, July. *A 12142! 12217!* Cuba, Hispaniola.

20. R. marisculus Lindl. & Nees in Mart., Fl. Bras. **2** (1): 142 (1842); Gale in Rhodora **46**: 273–274 (1944).—*R. borinquensis* Britton (1915).
Culms 60–150 cm. high; leaves 4–7 mm. broad; corymbs 2–3, rather remote; achene 1·4–1·6 mm. long, 1–1·2 mm. broad.
Very rare (Clar., St. Ann, St. Mary), in moist ground and savanna thickets; about 2300 feet; fl. and fr. Aug–Sept. *McNab, Weck 12225–200!* Continental tropical Amer., Greater Antilles, Dominica.

21. R. rugosa (Vahl) Gale in Rhodora **46**: 275 (1944).—*Schoenus rugosus* Vahl (1798). *R. glauca* Vahl (1805).

Culms ascending from the base, 50–130 cm. high; leaves rather glaucous, 1–4 mm. broad; cymes 1–3; achene stipitate, white at first, turning light brown, 1·5–1·7 mm. long, 1·2–1·3 mm. broad.

Rather local, on clay banks and rough open swampy ground; 2000–3000 feet; fl. and fr. Mar–May, Aug–Dec. *A 12495! H 11169! Webster & Proctor 5348!* Colombia and Venezuela to Guianas, Greater Antilles, Dominica, Trinidad, tropical Africa.

22. R. globularis (Chapm.) Small, Man. Southeast. Fl.: 184 (1933).—*R. cymosa* var. *globularis* Chapm. (1860).

22a. Var. pinetorum (Small) Gale in Rhodora **46**: 248 (1944).—*R. pinetorum* Small (1933).

Culms erect from the base, 25–70 cm. high; leaves 1·5–2 mm. broad; cymes 1–4; achene 1·3–1·4 mm. long, 1·4 mm. broad.

Very rare (St. Ann); only once collected in 1850. *Prior!* Florida, Louisiana, Cuba.

22b. R. globularis var. **recognita** Gale in Rhodora **46**: 245 (1944).—*R. cymosa* of authors, not Ell. (1816).

Culms erect, tufted, 15–90 cm. high; leaves 1–3 mm. broad; achene 1·3–1·6 mm. long, 1·2–1·5 mm. broad.

Rather local (St. Andr., Clar.), on moist open clay banks and savanna; 2300–3250 feet; fl. and fr. sporadically. *A 7109! H 11103! P 8256!* Ontario to California and Florida, C. Amer., Greater Antilles, Dominica, Martinique.

23. R. nitens (Vahl) A. Gray, Man. ed. 5: 568 (1867).—*Scirpus nitens* Vahl (1805).

23a. Var. hispaniolica Kük. in Fedde, Repert. Sp. Nov. **32**: 76 (1933).

Culms softly fleshy, 35–65 cm. high; leaves rather fleshy; panicle with spreading branches; achene orbicular-biconvex, 1 mm. long.

Very rare (Han.), at edge of drying brackish swamp; SL–10 feet; fl. and fr. Dec. *A 8585!* Greater Antilles.

24. R. robusta (Kunth) Boeck. in Linnaea **37**: 616 (1873).—*Dichromena robusta* Kunth (1837).

Culms 70–135 cm. high; corymbs 2–3, up to 15 cm. long; spikelets in clusters of 2–3, 3–4 mm. broad, many-flowered.

Very rare and local (Clar., St. Ann), in open grassy savanna; 2000–2300 feet; fl. and fr. June–Sept. *P 10358! 10359! Weck 6235–9!* Mexico to Argentina, Cuba.

25. R. minutiflora (L. C. Rich. ex Spreng.) Adams in Phytologia **21** (2): 70 (1971).—*Scleria minutiflora* L. C. Rich. ex Spreng. (1826). *R. micrantha* Vahl (1805), partly.

Culms (4–) 8–25 (–40) cm. high; achene 0·5–0·7 mm. long.

Frequent in ditches and in open clay or sandy soil in wet areas, especially in the western parishes; 50–2000 feet; fl. and fr. June, Sept–Jan. *A 7984! H 12563! P 16665!* Tropical Amer., W. tropical Africa.

26. R. setacea (Berg.) Boeck. in Vid. Medd. Nat. For. Kjøbenh. **1869**: 159 (1870).—*Schoenus setaceus* Berg. (1772).

Culms usually densely tufted, forming low tussocks, 5–35 cm. high; corymbs 2–4, the terminal with usually 5–6 spikelets, the lower with fewer; achene 1 mm. long.

Occasional in Clarendon and the western parishes, in moist open savannas and pastures on clay soil; 50–2400 feet; fl. and fr. Sept–Dec. *A 8638! H 12232! P 18448!* Honduras to Bolivia and Brazil, West Indies.

27. R. intermixta C. Wright in Sauv. in Anal. Acad. Cienc. Habana **8**: 88 (1871).

Culms tufted, 8–30 cm. high; achene stipitate, white, 0·8–0·9 mm. long; beak very short, bony-white, apiculate.

Rare and local (St. Andr., Clar.), on damp clay banks and in open boggy ground; 2300–3500 feet; fl. and fr. Dec–Mar. *A 6673! H 12261! P 15923!* Florida, Louisiana, Cuba, Hispaniola.

28. R. radicans (Cham. & Schlecht.) H. Pfeiff. in Fedde, Repert. Sp. Nov. **38**: 93 (1935).—*Dichromena radicans* Cham. & Schlecht. (1831).

Culms softly fleshy, 15–40 (–50) cm. high; leaves soft 1·5–4 mm. broad, shorter than the culm; spikelets whitish, drying russet-brown; achene obovoid, brown, 1–1·5 mm. long.

Common on moist banks and in sandy or gravelly waste places, a weed of pastures on clay soil; 400–4000 feet; fl. and fr. all the year. *A 5769! H 12294! P 8527!* Mexico to Paraguay, West Indies.

29. R. berteroi (Spreng.) C. B. Cl. in Urb., Symb. Ant. **2**: 119 (1900).—*Hypolytrum berteroi* Spreng. (1820). *R. pusilla* (Sw.) Griseb. (1857), not Chapm. ex M. A. Curtis (1849).

Culms slender, curved and flexuous, leafy, (2–) 4–10 (–12) cm. long; spikelets lanceolate, 2–6 together in heads about 5 mm. long; achene suborbicular, about 1 mm. long, at first green, turning black.

Very common on shaded banks and exposed limestone in damp places, also in pastures and lawns; 200–3000 feet; fl. and fr. June–Mar. *A 10858! H 12215! P 23898!* Greater Antilles, Guadeloupe, Marie Galante.

30. R. nervosa (Vahl) Boeck. in Vid. Medd. Nat. For. Kjøbenh. **1869**: 143 (1870). —*Dichromena ciliata* Vahl (1805). *D. nervosa* Vahl (1805). Star Grass.

Rhizome short; culms firm, erect, 15–40 (–70) cm. high; leaves 2–4 mm. broad, mostly shorter than the culm; spikelets white, 6–7 mm. long, up to 15 in the head, sometimes viviparous; achene suborbicular, 1–1·5 mm. long.

Abundant in damp pastures and ditches; SL–2900 feet; fl. and fr. all the year. *A 5401! H 11876! P 7262!* Mexico to Chile, West Indies.

31. R. colorata (L.) H. Pfeiff. in Fedde, Repert. Sp. Nov. **38**: 89 (1935).— *Schoenus coloratus* L. (1753). *Dichromena colorata* (L.) Hitchc. (1893). *R. stellata* (Lam.) Griseb. (1857).

Stolons slender, scaly; culms 30–60 cm. high; leaves about 2 mm. broad, shorter than the culm, glabrous; bracts glabrous, reflexed at maturity; head subglobose, 1–2 cm. in diameter; glumes white, smooth; achene obovate, brown.

Occasional, gregarious in undisturbed pastures and low swampy ground, especially near the sea; SL–50 feet; fl. and fr. most of the year. *A 6876! H 12173! J.P. 847! P 8350!* E. and S. United States, Mexico to Brazil, Bermuda, Bahamas, Greater Antilles, Guadeloupe, Marie Galante, Martinique, Grenada; Grand Cayman.

2. SCHOENUS L. (1753)

1. S. nigricans L., Sp. Pl. **1**: 43 (1753).

Densely tufted perennial; culms terete, 15–75 cm. high; lower sheaths dark brown or blackish, shiny; spikelets 3–10 in an ovoid head, blackish; glumes 3–9, keeled, with 1–4 of the upper bearing flowers in their axils; stamens 3; achene 3-gonous.

Very rare (Port.), in crevices of limestone crags; about 2500 feet; fl. and fr. Apr. *P 5726!* Widely distributed in many regions, especially near the sea; Honduras, Cuba, Hispaniola, Bahamas.

3. CLADIUM P. Browne (1756)

1. C. jamaicense Crantz, Inst. Rei Herb. **1**: 362 (1766).—*C. mariscus* (L.) Pohl ssp. *jamaicense* (Crantz) Kük. (1938). *Mariscus jamaicensis* (Crantz) Britton (1913). Saw Grass.

Robust tufted perennial; culms obtusely 3-angled, 1·5–3·5 m. high; leaves up to over 1 m. long, up to 2 cm. broad; inflorescences several forming a large panicle; spikelets in small clusters, 4–5 mm. long; lower glumes empty, the upper hermaphrodite or staminate; stamens 2; achene ovoid, sharply pointed, wrinkled, brown, shiny.

Gregarious and forming large stands at the margins of brackish swamps, very rare inland (Port.), forming large tussocks among limestone crags in wet areas; SL–30 (–2000) feet; fl. and fr. Mar–July. *A 7344! H 8632! J.P. 822! P 5738!* S. United States to Venezuela, Bermuda, Bahamas, Greater Antilles, Lesser Antilles to Grenada; Grand Cayman; also in the islands of the Indian Ocean.

4. CYPERUS L. (1753); McLaughlin (1944)

1 Culms leafy all along their length, tufted at intervals from branches of long creeping rhizomes; scape obsolete, shorter than the simple or tightly compound bracteate head; leaves and bracts uniform, mostly 2–5 cm. long, often folded and subfalcate; spikelets 1-flowered   1. *pedunculatus*
1 Culms leafy at the base or leafless; scape evident; leaves mostly flat or filiform:
  2 Styles bifid; achenes lenticular; spikelets strongly compressed; culms not septate:
    3 Spikelets maturing 1 achene, deciduous complete from the spike-axis; glumes 3–4, unequal, the lower 1 or 2 empty, the uppermost empty or staminate; spikes sessile, short and compact, 1–4 in a dense terminal head:
      4 Leaves reduced to bladeless sheaths; rhizome creeping; culms usually 2 mm. or more thick; head globose:
        5 Bracts obsolete, usually shorter than the head   2a. *peruvianus* var. *peruvianus*
        5 Bracts evident, longer than the head   2b. *peruvianus* var. *foliatus*
      4 Leaves with blades; culms usually slender; heads ovoid, oblong or lobed:
        6 Spikelets 1·3–1·5 mm. broad, ovate; keel of glume usually smooth; culms tufted
                                                3. *sesquiflorus*
        6 Spikelets up to about 1 mm. broad, lanceolate; keel of glume usually spinulose:
          7 Plant annual; culms densely tufted; spikelets narrowly lanceolate; achenes nearly oblong   4. *tenuifolius*
          7 Plant perennial; culms more or less distant on a rhizome; spikelets broadly lanceolate; achenes obovate-oblong   5. *brevifolius*
    3 Spikelets maturing several achenes; glumes several to many, uniform, deciduous from the persistent rachilla, the keel smooth:
      8 Achenes dorsally compressed, i.e. with a face towards the rachilla; spikelets few, sessile, thick, subcapitate in a solitary apparently lateral inflorescence; leaves short or reduced to sheaths   6. *laevigatus*
      8 Achenes laterally compressed, i.e. with an edge towards the rachilla; spikelets very thin and flat:
        9 Spikelets 3–5 mm. broad; glumes 2 mm. broad, 3·5–4 mm. long; rachilla wingless; achenes orbicular, blackish   7. *unioloides*
        9 Spikelets up to 2 mm. broad; glumes 1·2–1·6 mm. broad, 1·7–2·5 mm. long:
          10 Rachilla wingless; achenes elliptical; spikelets 2 mm. broad; rays well developed
                                  8a. *lanceolatus* var. *compositus*
          10 Rachilla winged forming a collar at the base of the achene; achenes truncate at apex; spikelets 1–1·5 mm. broad:
            11 Spikelets fasciculate, forming penicillate clusters at the ends of short rays
                              9a. *polystachyos* var. *polystachyos*
            11 Spikelets divaricate, not penicillate   9b. *polystachyos* var. *texensis*
  2 Styles trifid; achenes trigonous; spikelets not strongly compressed:
    12 Rachilla with conspicuously thickened wings clasping the achene in 1-fruited readily disarticulating joints; culms bulbous at base; annual weedy herbs (if perennial and rhizomatous without bulbs, see 32. *harrisii* and 33. *filiformis*):
      13 Inflorescence a solitary cluster of spikelets 6–9 mm. long; glumes yellowish, 3 mm. long, strongly 7–11-nerved; culms 2–6 cm. high   44. *globulosus*
      13 Inflorescence compound, the spikelets in spikes or racemes; culms taller:
        14 Spikelets radiate, the lower reflexed; glumes red to reddish-brown, 3–9-nerved
                                    10. *flexuosus*
        14 Spikelets not radiate; glumes straw-coloured to reddish-brown; 5–11-nerved:
          15 Spikelets in spikes, mostly 1–1·5 cm. long, the lower not usually reflexed, often yellowish   11a. *odoratus* var. *odoratus*
          15 Spikelets in loosely subglobose compound heads, (1–) 1·5–2 (–3) cm. long, spreading in all directions, usually dark reddish-brown
                                11b. *odoratus* var. *acicularis*
    12 Rachilla thin, not disarticulating:
      16 Culms septate-nodose, leafless; bracts shorter than the inflorescence; rhizome elongated, woody   12. *articulatus*
      16 Culms not septate-nodose, or if slightly so bracts long:

17 Bracts more or less equal in length, numerous; leaves reduced to basal sheaths; rhizome thick; rachilla wingless; inflorescence decompound   13. *alternifolius*
17 Bracts unequal in length:
  18 Rachilla wingless or wings very narrow not exceeding 0·2 mm. broad:
    19 Glumes 7–13-nerved; stamens 3, rarely 2:
      20 Glumes mucronulate; achenes ellipsoid or obovoid, obtuse; spikelets 2–3 mm. broad in globose clusters of 4 or more at the ends of rays; culms 60–100 cm. high; leaves viscous; plant of montane woodland   14. *constanzae*
      20 Glumes mucronate, the keel produced into a mucro 0·4–0·8 mm. long:
        21 Achenes 2–2·5 mm. long, narrowly ellipsoid, the style-base forming a beak 0·5 mm. long, the spongy base of the achene conspicuous; leaves and bracts septate-nodulose, more or less viscous and aromatic   15. *oxylepis*
        21 Achenes 1·4–1·8 mm. long, ovoid to obovoid, beakless, not spongy-torulose at base:
          22 Leaves viscous, septate-nodulose, thick, coriaceous, involute, greyish-green; achenes cuneate and stipitate at base; bracts very long; whole plant aromatic   16. *elegans*
          22 Leaves neither viscous nor septate-nodulose, membranous, flat, bright green; achenes 1·4 mm. long, not stipitate and scarcely cuneate at base:
            23 Rays few, 3–6, unbranched; glumes yellowish-brown with mucros 0·6–0·8 mm. long   17. *confertus*
            23 Rays numerous, branched with 3–5 raylets each; glumes green with mucros 0·4–0·5 mm. long   18a. *diffusus* var. *tolucensis*
    19 Glumes 3–5-nerved or with lateral nerves obscure:
      24 Spikelets in dense spikes, not in heads or clusters; stamens 3; glumes with a strong excurrent mid-nerve with the lateral nerves very close to it   19. *imbricatus*
      24 Spikelets in heads or clusters:
        25 Leaves, if developed, shorter than the culm; achenes up to 0·6 mm. long; stamens 1–3:
          26 Glumes oblong, 0·8–1 mm. long, often reddish; spikelets in clusters of 3–6; secondary rays frequently developed   20. *haspan*
          26 Glumes obovate, 0·5–0·6 mm. long; spikelets very numerous in globose heads terminating short or long primary rays; secondary rays rarely developed   21. *difformis*
        25 Leaves well developed; achenes larger; stamen 1; glumes often appearing chinned when spikelets viewed from the side:
          27 Spikelets contracted into a solitary terminal head; culms smooth, densely tufted; achenes 0·8 mm. long   22. *humilis*
          27 Spikelets in several to numerous heads or clusters; culms loosely tufted:
            28 Culms smooth; spikelets radiating forming more or less globular heads:
              29 Glumes separating at maturity; achenes 1·3–1·5 mm. long, ellipsoid, conspicuously tapering at both ends, iridescent   23. *ochraceus*
              29 Glumes more or less overlapping at maturity; achenes 1 mm. long, linear-oblong, acuminate, scarcely stipitate, not iridescent:
                30 Rays less than 3 cm. long; heads densely congested; rhizome short; glumes opaque with inconspicuous lateral nerves; achenes 0·2–0·3 mm. broad   24a. *luzulae* var. *luzulae*
                30 Rays 3–10 cm. long; heads not densely congested; rhizome long; glumes translucent with lateral nerves evident; achenes 0·3–0·4 mm. broad   24b. *luzulae* var. *entrerianus*
            28 Culms scabrid; spikelets ascending forming obconic heads; rays up to 6 cm. long:
              31 Culms retrorsely scabrid, not septate-nodulose; leaves and bracts up to 3 mm. broad; spikelets 2 mm. broad; achenes 0·6–0·7 mm. long   25. *surinamensis*
              31 Culms antrorsely scabrid, often septate-nodulose; leaves and bracts 6–10 mm. broad, conspicuously septate-nodulose; spikelets 2·5–3 mm. broad; achenes 1·1–1·3 mm. long   26. *virens*
  18 Rachilla winged, the wings 0·3–0·6 mm. broad:
    32 Wings of rachilla conspicuous, readily deciduous, often lustrous, yellowish-brown; plants often 1 m. or more high:
      33 Rays and raylets equal; leaves reduced to bladeless sheaths   27. *giganteus*
      33 Rays and raylets both very unequal, some of them always sessile; leaves 50–100 cm. long   28. *digitatus*
    32 Wings of rachilla persistent, colourless, hyaline:
      34 Spikelets less than 1 mm. broad; glumes 3–5-nerved; spikelets filiform, 1–3 cm. long, distinctly separated in a radiate usually compound panicle; leaves 5 mm. or more broad   29. *distans*
      34 Spikelets (0·6–) 1–4 mm. broad, linear to ovate; glumes 7–11 or more-nerved:

35 Culms conspicuously papillose; leaves glaucous, rigidly spinulose on the margins, thick, coriaceous; spikelets purplish-brown   30. *ligularis*
35 Culms not or hardly papillose:
 36 Lowest bract appearing like a continuation of the culm, usually erect but sometimes deflexed, narrow; inflorescence capitate, never spicate, appearing lateral; rays wanting; perennial:
  37 Spikelets 5–12 (–30)-flowered, few, 5–20 mm. long:
   38 Spikelets 1 or 2 on each culm, 8–30-flowered; rhizome long-creeping; culms, leaves and bracts very slender   31. *trichodes*
   38 Spikelets (2–) 3 or more, 5–12-flowered:
    39 Bracts up to 25–30 cm. long, 4–5 times as long as the culm; rhizome rather long; glumes light brown with a mucro 0·4 mm. long
                         32. *harrisii*
    39 Bracts 2–10 cm. long, less than half as long as the culm; rhizome short; glumes yellowish   33. *filiformis*
  37 Spikelets 1–4 (–12)-flowered, numerous, 1–8 mm. long:
   40 Spikelets 1-flowered, 1–2 mm. long, rhomboidal; lower glumes 1·5 mm. long, 11–15-nerved; rhizome very short or rarely trailing   34. *swartzii*
   40 Spikelets 2–12-flowered, 2–8 mm. long, not rhomboidal; glumes up to 2·3 mm. long; culms densely tufted:
    41 Spikelets 2–2·5 mm. long, 0·6 mm. broad, 2–5-flowered; achenes oblong, 0·3 mm. broad, about as long as the glumes   35a. *nanus* var. *nanus*
    41 Spikelets 3–8 mm. long, 1 mm. broad, 3–12-flowered; achenes ellipsoid, 0·6 mm. broad, shorter than the glumes   35b. *nanus* var. *subtenuis*
 36 Lowest bract ascending or spreading, not appearing like a continuation of the culm, leaflike; inflorescence with 1 or more rays, appearing terminal:
  42 Glumes conspicuously acuminate or long-mucronate; spikelets 12–24-flowered, linear, 2–3 mm. broad; glumes readily deciduous from the persistent rachilla, 2·5–3 mm. broad, 3–3·5 mm. long   36. *compressus*
  42 Glumes obtuse to mucronulate:
   43 Mature spikelets up to 40-flowered, long-persistent on the rachis; glumes 1·5–2·5 mm. broad, more or less readily deciduous from the rachilla:
    44 Glumes light green usually with a dark red blotch, soon deciduous; culms densely tufted; annual plant without stolons or tubers
                         37. *sphacelatus*
    44 Glumes reddish-purple to straw-coloured without blotches, persistent; culms distant; stolons and tubers present; perennial:
     45 Glumes (3–) 5–7-nerved, red to purplish-brown; rhizome and stolons rather thick, few; spikelets 10–20 mm. long, nearly 2 mm. broad, (5–) 12–24-flowered   38. *rotundus*
     45 Glumes with 7–9 widely spaced prominent nerves, usually straw-coloured; stolons very slender, numerous:
      46 Spikelets 5–12 mm. long, about 2 mm. broad, 8–16-flowered
                         39. *esculentus*
      46 Spikelets 15–40 mm. long, 2–3 mm. broad, up to 40-flowered
                         40. *lutescens*
   43 Mature spikelets up to 14-flowered, soon deciduous from the rachis; glumes 0·8–2 mm. broad, persistent on deciduous rachillas; plants without tubers or stolons:
    47 Plants of sandy or rocky places near the sea; spikelets 6–15 mm. long, more than 1 mm. broad, 8–14-flowered, green or reddish-brown; mature achenes broadly ellipsoid, black; inflorescence mostly decompound   41. *planifolius*
    47 Plants of inland waste places, banks and pastures; spikelets up to 10 mm. long and about 1 mm. broad, up to 8-flowered; inflorescence mostly umbellate:
     48 Rays very short with the spikes very often sessile or subsessile or peduncles up to 2 cm. long, but usually shorter than the spikes:
      49 Spikelets 3 mm. long, 1–2-fruited, numerous in dense cylindrical sessile or subsessile spikes 1–2·5 cm. long, golden-yellow; perennial with a short creeping rhizome   42. *flavus*
      49 Spikelets 7–10 mm. long, 4–8-fruited, several in lax oblong shortly pedunculate spikes up to about 1 cm. long, green; annual
                         43. *tenuis*
     48 Rays longer, if some spikes sessile others with peduncles at least 2 cm. long:
      50 Spikes subglobose to shortly oblong, 5–8 mm. in diameter; spikelets 3–6-flowered, 4–6 mm. long, usually yellowish   44. *globulosus*
      50 Spikes cylindrical (1–) 2–5 cm. or more long; perennial:

51 Rays up to 20 cm. long; spikelets 3–6 mm. long, 1–3 (–4)-flowered,
      coppery-brown; spikes dense, 5–11 mm. broad   45. *mutisii*
51 Rays up to 10 cm. long; spikelets 6–10 mm. long, 3–7 flowered,
      greenish-brown; spikes lax, 10–15 mm. broad 46. *hermaphroditus*

1. **C. pedunculatus** (R. Br.) Kern in Act. Bot. Neerl. **7**: 798 (1958).—*Remirea maritima* Aubl. (1775), not *C. maritimus* Poir. (1806). *R. pedunculata* R. Br. (1810).
   Perennial from an underground rhizome; culms (2–) 5–30 cm. high; heads 1–2 cm. long; spikelets 3–5 mm. long; glumes distichous, strongly nerved; stamens 3; style 3-fid, long-exserted.
   In sand on upper sea-beaches, reported from Cayman Islands but not from Jamaica; fl. and fr. June. *Kings LC 81! P 15216!* Tropical shores generally.

2. **C. peruvianus** (Lam.) F. N. Williams in Bull. Herb. Boiss. sér. 2, **7**: 90 (1907).—*Kyllinga peruviana* Lam. (1789).

2a. Var. peruvianus.
   Perennial from a branched scaly rhizome; culms approximate, 15–40 cm. high; heads 6–10 mm. in diameter; spikelets 3–3·5 mm. long; stamens 3.
   Locally common, at sandy lagoon margins and in grassy places near the sea; SL–5 feet; fl. and fr. sporadically. *A 6123! H 9841! J.P. 843!* Honduras to Colombia, Hispaniola, Puerto Rico, Antigua, Trinidad, W. Africa; Grand Cayman.

2b. C. peruvianus var. foliatus (Kük.) Kük. in Engl., Pflanzenr. **4** (20), Heft 101: 587 (1936).
   Like the last but culms more remote on the rhizome and bracts conspicuous, 5–15 (–60) mm. long.
   Rare (St. Ann, St. Mary), in clay soil at roadsides and field margins; 800–1250 feet; fl. and fr. Nov. *A 9916! P 8336!* Endemic.

3. **C. sesquiflorus** (Torr.) Mattf. & Kük. in Engl., Pflanzenr. **4** (20), Heft 101: 39 (1935), 591 (1936).—*Kyllinga sesquiflora* Torr. (1836). *K. odorata* Vahl (1805), not *C. odoratus* L. (1753).
   Aromatic perennial; culms 5–30 cm. high; leaves shorter than the culms, 2–3 mm. broad; bracts 3–4; spikes 6–12 mm. long, 5–6 mm. broad; spikelets 2·5–3·5 mm. long; stamens 2.
   Occasional (St. Andr., St. Eliz., St. Ann), in pastures; 50–2200 feet; fl. and fr. most of the year. *A 8253! H 12144! Powell 836!* S.E. United States, Mexico to Uruguay, Greater Antilles, Martinique, Trinidad.

4. **C. tenuifolius** (Steud.) Dandy in Exell, Cat. Vasc. Pl. S. Tomé: 363 (1944).—*Kyllinga tenuifolia* Steud. (1855). *K. pumila* Michx. (1803), not *C. pumilus* L. (1756).
   Annual; culms 4–30 (–40) cm. high; leaves shorter than the culms, 2–3 (–5) mm. broad; bracts 3–4; spikes 4–6 (–9) mm. long; spikelets 1·5–2·5 mm. long in light green heads; keel of glume spinulose.
   Frequent along damp roadsides, streams and ditches and in moist open savannas; 500–4800 feet; fl. and fr. all the year. *A 7614! H 12238! J.P. 762!* E. and S. United States, Mexico to Argentina, West Indies, tropical Africa.

5. **C. brevifolius** (Rottb.) Endl. ex Hassk., Cat. Hort. Bogor: 24 (1884).—*Kyllinga brevifolia* Rottb. (1773).
   Plant perennial, rhizomatous but usually forming tufts as well; culms 6–25 (–50) cm. high; leaves 1·5–3 mm. broad; bracts 3–4; spikes 4–8 mm. long, usually whitish; spikelets 2–3 mm. long; keel of glume spinulose or smooth.
   Very common in pastures, low-lying undisturbed ground and on open banks; 50–5000 feet; fl. and fr. most of the year. *A 5738! H 9498! P 22809!* General in the subtropics and tropics; Grand Cayman.

6. **C. laevigatus** L., Mant. Pl. Alt.: 179 (1771).
   Rhizomatous perennial; culms tufted at intervals along a horizontal stock or remote, 10–50 cm. high; bracts 2, the lower erect; spikelets 4–10 mm. long, 2 mm.

broad, 12–24-flowered; glumes obtuse, many-nerved; stamens 3; achene ovoid or obovoid.

Rather local along the south coast on stable sea-beaches, banks of brackish streams and salina margins; SL–10 feet; fl. and fr. most of the year. *A 6048! H 12177! J.P. 801!* Pantropical, extending into warm temperate regions.

7. C. unioloides R. Br., Prodr. Fl. Nov. Holl.: 216 (1810).

Culms tufted, 60–90 cm. high; leaves 2–4 mm. broad; bracts 2–4; rays 3–6, up to 5 cm. long; spikelets in terminal clusters of 4–12, 8–18 mm. long, 12–24-flowered, oblong-lanceolate, golden-buff; glumes 3-nerved; stamens 3.

Rare (Clar., St. Eliz.), in wet open savannas and swamps; 10–2000 feet; fl. and fr. May–Aug. *P 10360! 19695!* Honduras, Venezuela, Cuba, Hispaniola and general in the Old World tropics.

8. C. lanceolatus Poir. in Lam., Encycl. Méth. Bot. **7**: 245 (1806).

8a. Var. compositus J. & C. Presl, Rel. Haenk. **1**: 167 (1828).

Culms tufted from slender rhizomes, 20–40 cm. high; leaves 1–2 mm. broad; rays 3–5, up to 2·5 cm. long; spikelets 5–10 in each ray, 4–7 mm. long, 6–16-flowered, light buff; glumes deciduous from below upwards; stamens 2.

Widespread but not common, in clay and silt soils in open low-lying grassy places; SL–1200 feet; fl. and fr. sporadically. *A 7854! H 12422! Yuncker 17878!* S. United States, Mexico to Paraguay and Chile, Cuba, Hispaniola, Philippine Is.

9. C. polystachyos Rottb., Descr. Pl. Progr.: 21 (1772).—*Pycreus odoratus* Urb. (1900), partly, not *C. odoratus* L. (1753).

9a. Var. polystachyos.

Usually perennial; culms tufted, 20–50 cm. high; leaves 2–5 mm. broad; bracts 3–6; rays several, up to 5 cm. long, usually contracted with crowded spikelets; spikelets 8–20 mm. long, 20–40-flowered; glumes 1·5–2 mm. long, 3-nerved; stamens (1–) 2; achene obovoid.

Rather rare and local (Port.), in damp pastures and in sandy grassland near the sea; SL–600 feet; fl. and fr. Aug–Sept. *A 7856! P 22789!* Tropics and subtropics generally; Grand Cayman.

9b. C. polystachyos var. **texensis** (Torr.) Fernald in Rhodora **41**: 530 (1939).— *C. microdontus* var. *texensis* Torr. (1836).

Like the last with a laxer inflorescence, the lateral rays elongated; spikelets 5–15 mm. long, light buff to reddish-brown.

Occasional in the central and western parishes, in ditches and along streambanks and pond margins; SL–1600 feet; fl. and fr. Mar, Sept–Dec. *A 8576! H 12183! P 21453!* Virginia to Florida, Mexico to Ecuador and Venezuela, Cuba, Barbados, Philippine Is.

10. C. flexuosus Vahl, Enum. Pl. **2**: 359 (1805).—*C. vahlii* (Schrad. ex Nees) Steud. (1855).

Culms 30–80 cm. high; leaves and bracts longitudinally plicate, up to 1 cm. or more broad; spikelets 8–15 mm. long in dense spikes; glumes faintly nerved with broad lustrous margins, 2·2–2·5 mm. long.

Uncommon (St. Andr., St. Ann), in damp ditches and marshy ground; 500–2000 feet; fl. and fr. Nov–Dec. *A 9975! J.P. 844!* Mexico to Brazil, Greater Antilles, Virgin Is., Guadeloupe.

11. C. odoratus L., Sp. Pl. **1**: 46 (1753).—*C. ferax* L. C. Rich. (1792).

11a. Var. odoratus.

Culms usually robust, often purplish-red at base, 60–120 cm. high in flower; leaves and bracts longitudinally plicate, usually about 1 cm. broad; spikelets usually in loose spikes; glumes mostly 7–9-nerved, 2–3 mm. long.

Common in ditches and low pastures, also at pond margins and sometimes a weed in cultivated ground; SL–2500 feet; fl. and fr. all the year. *A 7348! H 12218! P 22831!* Widely distributed in tropical and warm temperate regions.

11b. C. odoratus var. **acicularis** (Schrad. ex Nees) O'Neill in Léon, Fl. Cub. **1**: 196 (1946).—*Diclidium aciculare* Schrad. ex Nees (1842).

Culms erect, 20–45 (–90) cm. high in flower; leaves and bracts 5–10 (–15) mm. broad.

Local (St. Cath., St. Thom.), in open low-lying often brackish places; SL–40 feet; fl. and fr. Nov–Mar. *A 10727! 11890! Yuncker 17883!* Texas, Mexico to Brazil, West Indies.

C. flexuosus and *C. odoratus* have extensive synonymy expressing the existence of a complex range of intergrading variants which includes both species.

12 C. articulatus L., Sp. Pl. **1**: 44 (1753).

Culms erect, up to 10 cm. or more apart from the creeping rhizome, up to 10 mm. in diameter, with 1 or 2 broad scale-leaves at base, 1–2 m. high in flower; bracts 2–3, scalelike; inflorescence 4–12-rayed, the rays up to 10 cm. long; raylets to 2 cm. long; spikelets buff tinged reddish, 1–3 cm. long, about 1 mm. broad.

Widely distributed but not common, in ditches and damp open pastures; SL–400 (–3600) feet; fl. and fr. Apr–Sept, Dec–Jan. *A 11657! H 6917! P 23614!* General in the subtropics and tropics.

13. C. alternifolius L., Mant. Pl.: 28 (1767).

Perennial; culms sheathed at base, 40–100 (–150) cm. high in flower; bracts leafy, distinctly spiral in arrangement, 6–10 mm. broad; rays numerous, 4–8 cm. long; spikelets in clusters of 3–7, linear-oblong, 6–10 mm. long, 1·5–2 mm. broad; glumes 3–5-nerved.

Gregarious and locally abundant in gravel along streams and rivers and at margins of ponds; 400–2500 feet; fl. and fr. Nov–Apr. *A 6948! H 11879! P 23043!* Native of Madagascar, now widespread in the subtropics and tropics, often as an escape from cultivation.

14. C. constanzae Urb., Symb. Ant. **7**: 168 (1912).—*C. ignotus* Britton (1916).

Rhizome shortly creeping bearing approximate culms; leaves 4–5 mm. broad; rays up to about 4 cm. long; spikelets 4–6 mm. long; glumes ovate, dark brownish-red with green keel.

Rare and local (Port.); 3500 feet; fl. and fr. May, Sept. *H 12350!* Cuba, Hispaniola.

15. C. oxylepis Nees ex Steud., Synops. Pl. Glum. **2**: 25 (1854).

Culms erect, 15–60 cm. high; leaves rounded abaxially, not keeled; rays up to 16 cm. long; spikelets 10–20 mm. long, 3 mm. broad in heads 1·5–2 cm. in diameter; glumes yellowish with spreading mucros.

Rather local (St. Cath., St. Ann, West.), in open boggy ground and lagoon margins; 15–40 feet; fl. and fr. Sept–Mar. *A 10730! P 9546! Skelding UCWI 3539!* Ecuador to Argentina, Anguilla, Bonaire, Curaçao.

16. C. elegans L., Sp. Pl. **1**: 45 (1753).—*C. viscosus* Sw. (1788).

Culms tufted, spreading or erect, 20–60 (–80) cm. high; leaves very long, rounded abaxially; rays up to 10 cm. long; spikelets 5–8 mm. long, 3 mm. broad in heads about 1 cm. in diameter; glumes greenish sometimes tinged blackish-purple.

Occasional in salina margins and low-lying seasonally inundated pastures, mostly near the sea; SL–40 feet; fl. and fr. June, Sept–Feb. *A 6067! H 12298! P 17439!* S.E. United States, continental tropical Amer., West Indies southwards to St. Lucia; Grand Cayman.

17. C. confertus Sw., Nov. Gen. & Sp. Pl.: 20 (1788).

Annual; culms tufted, 10–30 cm. high; leaves 2–4 mm. broad; rays up to 6 cm. long; terminated by spikes 8–12 mm. in diameter; spikelets 4–6 mm. long, 2 mm. broad; glumes yellowish, the nerve green and extended into a spreading mucro, deciduous; scars on rachilla after achene shedding round and peglike.

Frequent locally as a weed of pastures and open waste ground, especially in sand or gravel; SL–1900 feet; fl. and fr. Oct–June. *A 6583! H 12446! P 23040!* Venezuela, Colombia, Greater Antilles, Virgin Is., Martinique, Trinidad, Galapagos Is.

18. C. diffusus Vahl, Enum. Pl. **2**: 321 (1805).

18a. Var. tolucensis (Kunth) Kük. in Engl., Pflanzenr. **4** (20), Heft 101: 211 (1936).—*C. tolucensis* Kunth (1816).

Culms 30–60 cm. high; leaves prominently 3-veined, 8–16 mm. broad; bracts 5–10; rays to 20 cm. long; spikelets 4–6 mm. long, 1·5–2 mm. broad, in clusters of 2–3.

Common on moist banks and in shaded pasture margins; 10–4000 feet; fl. and fr. all the year. *A 5807! H 12176! J.P. 808! P 22800!* Mexico to Bolivia, Cuba, Hispaniola; var. *diffusus* in S.E. Asia.

19. C. imbricatus Retz., Obs. Bot. **5**: 12 (1788).—*C. radiatus* Vahl (1805).

Culms 70–100 (–120) cm. high; leaves 4–8 mm. broad; lower bracts very long; spikes several, up to 3 cm. long; spikelets 5–9 mm. long; glumes 1·3 mm. long.

Known only from a single unlocalized gathering without collector's name; if still extant, likely to be found near rivers at low elevations. Venezuela, Greater Antilles, Trinidad, a common weed in the Old World tropics.

20. C. haspan L., Sp. Pl. **1**: 45 (1753) (incl. ssp. *juncoides* (Lam.) Kük. (1926)).

Culms (10–) 20–40 (–60) cm. high, rather remote from a short creeping rhizome; leaves, if developed, 2–3 mm. broad; rays many, up to 12 cm. long; spikelets 5–10 mm. long, 1 mm. broad.

Very rare (St. Ann, St. Eliz.), in swampy ponds among grass tussocks; 25–1500 feet; fl. and fr. Jan. *Prior! P 21923!* General in the subtropics and tropics.

21. C. difformis L., Cent. Pl. **2**: 6 (1756).

Annual; culms tufted, erect, (5–) 20–50 cm. high in flower; scape sharply triquetrous; spikelets 4–8 mm. long, numerous in heads 6–12 mm. in diameter.

Very local (St. Cath.), in ditches and open places where water has been standing; 30–75 feet; fl. and fr. Nov–Feb. *A 11893! P 18343!* Venezuela, rare in the New World, common in the subtropics and tropics of the Old World.

22. C. humilis Kunth, Enum. Pl. **2**: 23 (1837).

Annual; culms densely tufted, (2–) 4–12 (–28) cm. high; leaves 1–3 mm. broad; bracts 2–3; capitulum (3–) 8–15 (–18) mm. in diameter; spikelets 3–6 mm. long.

Frequent in shallow pools and alluvial mud or sand; 30–2300 feet; fl. and fr. most of the year. *A 6874! H 12420! P 18407!* Mexico, Honduras, Peru, Cuba, Hispaniola, Martinique.

23. C. ochraceus Vahl, Enum. Pl. **2**: 325 (1805).

Culms 20–50 cm. high in flower; leaves 2–4 mm. broad; bracts 5–7; rays up to 8 cm. long; spikelets dark green, 6–10 mm. long, 2 mm. broad, in globose spikes 1–2 cm. in diameter.

Rather common in swamps and damp pastures and in open sandy places; SL–800 feet; fl. and fr. all the year. *A 6879! H 12323! J.P. 857! P 16459!* Mexico to Argentina, Greater Antilles, Antigua, Guadeloupe.

24. C. luzulae (L.) Retz., Obs. Bot. **4**: 11 (1786).—*Scirpus luzulae* L. (1762).

24a. Var. luzulae.

Culms 20–50 cm. high; leaves 3–5 mm. broad; spikelets 3–4 mm. long in heads 6–10 mm. in diameter; glumes prominently keeled, rather obtuse.

Not clearly confirmed for Jamaica, but some authors do not distinguish the variety *entrerianus*; Kükenthal (1935) cited *H 12440* under *C. luzulae*. McLaughlin (1944) cites specimens from the West Indies, excluding Jamaica, from Cuba to Trinidad.

24b. C. luzulae var. entrerianus (Boeck.) M. Barros in Anal. Mus. Argent. Cienc. Nat. **39**: 309 (1938).—*C. entrerianus* Boeck. (1878); Kük. in Engl., Pflanzenr. **4** (20), Heft 101: 169 (1936).

Like the last with longer rays and leaves 6–10 mm. broad; glumes acute.

Rare and local (St. Cath.), in damp places; 40–150 feet; fl. and fr. June–Feb. *Britton 3074! H 12064! Powell 992!* Mexico to Argentina, West Indies.

25. C. surinamensis Rottb., Descr. Pl. Progr.: 20 (1772).

Like the last; spikelets 3·5–6 mm. long in heads 6–10 mm. in diameter.

Local (St. Mary, St. Thom.), in swampy ground; SL–50 feet; fl. and fr. May, Nov. *A 7228! P 23541! Skelding UCWI 3407!* S. United States, Mexico to Argentina, West Indies.

26. C. virens Michx., Fl. Bor. Amer. **1**: 28 (1803).

Like the last; spikelets 6–8 mm. long in heads 8–12 mm. in diameter.

Very rare (Trel.); 2000 feet; fl. and fr. Oct. *H 12580!* S.E. United States, California, Mexico to Argentina, Greater Antilles.

Species 24–26 are difficult to distinguish and may prove, when better known, to be no more than varietally distinct.

27. C. giganteus Vahl, Enum. Pl. **2**: 364 (1805).

Rhizome thick; culms smooth, terete, with broad ovate or lanceolate scales at the base, 1·2–3 m. high; bracts numerous, 6–20 mm. broad, shortly cuspidate; rays up to 20 cm. long; raylets up to 8 cm. long; spikes 2–5 cm. long; spikelets 4–10 mm. long.

Rather local, in swamps and ditches, gregarious; SL–75 feet; fl. and fr. June, Sept–Dec. *A 11659! H 12452! P 21463!* Mexico, Colombia to Argentina, Greater Antilles, Trinidad.

28. C. digitatus Roxb., Fl. Ind. **1**: 209 (1820).

Culms smooth, 50–150 cm. high; leaves 4–15 mm. broad; bracts 5–7, over-topping the rays; rays up to 18 cm. long; raylets up to 5 cm. long; spikes cylindrical 3–6 cm. long; spikelets about 10 mm. long.

Uncommon (St. Cath., St. Eliz., Han.), in swamps; 50–500 feet; fl. and fr. Sept–Dec. *Britton 1482! H 11647!* General in the subtropics and tropics.

29. C. distans L. f., Suppl.: 103 (1781).

Probably annual; culms tufted, 30–90 cm. high in flower; leaves plaited, whitish-striate beneath, 4–8 mm. broad; rays up to 18 cm. long; raylets up to 8 cm. long; spikelets 1–2 cm. long, 0·7–0·8 mm. broad, usually dark purplish-brown.

Rather common, a weed of damp pastures, grassy waste places and open culti-vated ground; 600–4000 feet; fl. and fr. July–Apr. *A 6909! H 12427! J.P. 784! P 23833!* General in the tropics.

30. C. ligularis L., Syst. Nat. ed. 10, **2**: 867 (1759).—*Mariscus rufus* Kunth (1816).

Perennial with tufted culms 30–120 cm. high in flower; leaves folded, sharply denticulate on the margins, 6–12 mm. broad; rays up to 10 cm. long; spikes 3–7, some sessile, 1·5–2 cm. long; spikelets 4–6 mm. long, dark coppery-brown.

Common in coastal marshes and sandy places near the sea, rare inland at sandy roadsides and along riverbanks; SL–50 (–250) feet; fl. and fr. most of the year. *A 7229! H 12295! J.P. 799!* Florida, tropical Amer. and Africa.

31. C. trichodes Griseb., Fl. Br. W.I.: 564 (1864).

Culms slender, 10–20 cm. high, tufted from a rhizome; bracts 2; spikelets 6–12 mm. long, 1–1·5 mm. broad.

Rare (Manch., St. Eliz.), in pastures; 1000–2500 feet; fl. and fr. May–Sept. *A 12614! Britton 1646! H 12882! P 15426!* Endemic.

32. C. harrisii Kük. in Fedde, Repert. Sp. Nov. **23**: 191 (1926).

Culms filiform, 2–8 (–10) cm. high from an extensively branched rhizome; bracts 3; spikelets 5–7 mm. long, usually breaking up at maturity; glumes 1·5 mm. long, about 11-nerved.

Rare and local (St. Andr.), in shaded pastures; 400–1000 feet; fl. and fr. Sept–Jan. *H 12418! Wynter UCWI 25150!* Endemic.

33. C. filiformis Sw., Nov. Gen. & Sp. Pl.: 20 (1788).

Culms more or less tufted, 10–30 cm. long in flower, elongating in fruit; leaves 0·5–1·5 mm. broad, shorter than the culms; bracts 2–3, short; spikelets 2–7 together, 6–14 mm. long, yellowish.

Widely distributed but not common, in pastures on heavy soils, among limestone rocks and in crevices of old stone walls; SL–2600 feet; fl. and fr. most of the year. *A 7650! H 12636! P 23619!* Venezuela, Bahamas, Greater Antilles, Virgin Is.; Grand Cayman.

34. C. swartzii (A. Dietr.) Boeck. & Kük. in Fedde, Repert. Sp. Nov. **23**: 186 (1926).—*Mariscus swartzii* A. Dietr. (1833). *Kyllinga filiformis* Sw. (1788), not *C. filiformis* Sw. (1788).

Culms spreading and arching from a dense tuft, up to 40 cm. long in fruit; spikelets yellowish.

Occasional in open pastures and in undisturbed waste ground over limestone and abandoned cultivations; SL–2600 feet; fl. and fr. May–Dec. *A 11645! H 12180! P 19697!* Greater Antilles, Guadeloupe; Grand Cayman.

35. C. nanus Willd. in L., Sp. Pl. ed. 4, **1**: 272 (1797).—*Mariscus capillaris* Vahl (1805).

35a. Var. nanus.

Culms capillary, 5–15 cm. long; leaves setaceous, canaliculate, shorter than culm; bracts 2–3, spreading or reflexed; heads globose or ovoid, about 3 mm. long.

Very rare (St. Andr.), in cracks in pavement; about 30 feet; fl. and fr. Oct. *P 8132! Swartz.* Cuba, Hispaniola, St. Croix.

35b. C. nanus var. **subtenuis** Kük. in Engl., Pflanzenr. **4** (20), Heft 101: 536 (1936).—*C. tenuis* of Griseb. (1864), not Sw. (1788).

Like the last but slightly larger in all parts; culms up to 30 cm. high; leaves 1–1·5 mm. broad; heads 6–8 mm. in diameter.

Rare (St. Eliz.), on shaded limestone ledges; 50–1000 feet; fl. and fr. June, Sept. *Britton 1391! H 12359! P 15427!* Cuba.

36. C. compressus L., Sp. Pl. **1**: 46 (1753).

Annual; culms tufted, spreading, 5–35 cm. high in flower; leaves 1·5–3 mm. broad; bracts 3–5; rays up to 12 cm. long; spikelets 3–10 together, 1–2·5 cm. long, 3–4 mm. broad, green at first turning yellow.

Occasional as a weed in moist sandy or gravelly places; SL–1250 feet; fl. and fr. Sept–Dec. *A 9886! H 12686! P 18439!* General in the subtropics and tropics.

37. C. sphacelatus Rottb., Descr. Pl. Progr.: 21 (1772).

Culms tufted, erect, 10–60 cm. high; leaves 2–4 mm. broad; bracts 3–5; rays up to 10 cm. long; spikelets loosely spicate, 6–20 (–40) mm. long, 1·5–2 mm. broad.

Common as a weed of waste ground, roadside banks and pastures; 150–3500 feet; fl. and fr. most of the year. *A 7830! H 12216! J.P. 760! P 18435!* Tropical Amer. and Africa.

38. C. rotundus L., Sp. Pl. **1**: 45 (1753).—Nut Grass.

Culms 15–30 (–50) cm. high in flower; leaves 2–6 mm. broad; bracts 2–4, leafy; rays 3–8, up to 6 cm. long; spikelets loosely spicate.

Common as a weed of disturbed and frequented ground and of cultivations, very persistent and difficult to eradicate; SL–700 feet; fl. and fr. all the year, flowering mostly in wet weather. *A 11547! H 12174! J.P. 776!* In all warm countries.

39. C. esculentus L., Sp. Pl. **1**: 45 (1753).— Tiger Nut.

Culms 10–40 (–70) cm. high; leaves 3–6 mm. broad, often longer than the culms; rays 5–10, up to 4 cm. long, often compound; spikelets numerous, rather distichous in loose spikes.

Rare (St. Cath., Clar.), in waste ground; 300–650 feet; fl. and fr. May, Oct–Nov. *H 11623! 12184! Soutar!* General in the subtropics and tropics, also in parts of S. Europe; frequently cultivated.

40. C. lutescens Torr. & Hook. in Ann. Lyc. New York **3**: 433 (1836).—*C. esculentus* L. var. *macrostachyus* Boeck. (1870).

Like the last; leaves up to about 15 cm. long, mostly shorter than the culms.

Very rare (West., Han.), in sand of upper beach; fl. and fr. Aug–Sept. *A 11661!*
S. United States, Argentina, Uruguay, Cuba, Hispaniola, Galapagos Is., Canary Is.

41. C. planifolius L. C. Rich. in Act. Soc. Hist. Nat. Paris **1**: 106 (1792) (incl. var.
brunneus (Sw.) Kük. (1926)).

Culms tufted, 25–90 cm. high in flower; leaves up to 10 mm. broad; rays several,
up to 9 cm. or more long, often compound; glumes 7-nerved; plants in exposed
places shorter with more compact purplish inflorescences.

Common on sandy and rocky sea-coasts and in thickets near the sea, also on the
cays; SL–50 feet; fl. and fr. most of the year. *A 10976! P 11495! Yuncker 18621!*
Bahamas, Florida, C. Amer. to Venezuela, West Indies southwards to the Grenadines and Barbados, Cayman Is.

42. C. flavus (Vahl) Nees in Linnaea **19**: 698 (1847).—*Mariscus flavus* Vahl (1805).
C. cayennensis (Lam.) Britton (1907).

Culms erect, 10–60 cm. high; leaves 3–6 mm. broad; glumes ovate, 9–13-nerved,
the lowest linear-subulate.

Occasional on grassy banks or moist open savanna; 50–2500 (–4500) feet; fl. and
fr. May–Aug, Nov–Jan. *A 10437! H 12279! P 7435!* E. United States, Mexico to
Paraguay, West Indies.

43. C. tenuis Sw., Nov. Gen. & Sp. Pl.: 20 (1788).—*C. platystachyus* Griseb.
(1864). *Mariscus flabelliformis* Kunth (1816).

Annual or short-lived perennial; culms tufted, 15–30 cm. high; leaves (1·5–)
2–3 mm. broad; spikes shortly cylindrical, overtopped by the bracts; glumes 7–11-
nerved.

Common as a weed of damp waste ground, sandy riverbanks and roadsides;
150–3400 feet; fl. and fr. all the year. *A 7652! H 12292! J.P. 759! P 23837!* Mexico
to Ecuador and Venezuela, Greater Antilles. This plant is allied to *C. globulosus*
and a group of widespread and common tropical weeds with the affinity of *C.
umbellatus* (Rottb.) C.B. Cl. (*Mariscus umbellatus* (Rottb.) Vahl).

44. C. globulosus Aubl., Hist. Pl. Guiane **1**: 47 (1775).—*C. cyclostachyus* Griseb.
(1864).

Culms 10–30 cm. high; leaves up to 4 mm. broad; spikes distinctly pedunculate,
cylindrical; glumes 9–13-nerved.

Occasional as a weed of pastures and waste places; SL–1700 feet; fl. and fr.
Apr–Nov. *A 6910! H 12185!* S.E. United States, Mexico to the Guianas, Bermuda,
Cuba, Puerto Rico, Dominica.

45. C. mutisii (Kunth) Griseb., Fl. Br. W.I.: 567 (1864).—*Mariscus mutisii*
Kunth (1816).

Culms 60–100 cm. high in flower; leaves up to 10 mm. broad; rays 7–10, sometimes sessile, but at least some pedunculate; spikes 2–7, 2–4·5 cm. long; glumes
7–9-nerved, the basal ovate.

Locally common in pastures and on banks; (1200–) 3000–5200 feet; fl. and fr.
Nov–July. *A 5725! H 9500! J.P. 765! P 23586!* Mexico to Brazil and Peru,
Hispaniola, Puerto Rico, Grenada.

46. C. hermaphroditus (Jacq.) Standl. in Contrib. U.S. Nat. Herb. **18**: 88 (1916).
—*Carex hermaphrodita* Jacq. (1791).

Rhizome horizontal; culms densely tufted, erect, 15–60 cm. high; leaves 3–8
mm. broad; rays 6–12; spikes 1–3, broadly cylindrical, 1–3 cm. long; glumes
many-nerved, bulged over the achenes.

Very local (Manch.), in woodland and thicket margins; 2000–2600 feet; fl. and
fr. Aug–Sept. *A 11640! H 12685! Henry!* Mexico to Argentina, Cuba, Hispaniola,
Martinique, St. Vincent, Grenada.

5. ABILDGAARDIA Vahl (1805)

1. A. monostachya (L.) Vahl, Enum. Pl. **2**: 296 (1805).—*Cyperus monostachyos* L.
(1771). *Fimbristylis ovata* (Burm.f.) Kern (1967).

Densely tufted perennial; culms filiform, 6–20 (–40) cm. high; leaves setaceous;

spikelet ovate to ovate-lanceolate, 5–15 mm. long, up to 5 mm. broad; glumes light greenish-brown with white margins; achenes 2–2·5 mm. long, yellowish, tuberculate.

Locally common, in shallow soils over limestone and in grazed pastures especially near the sea; SL–2500 feet; fl. and fr. most of the year. *A 7968! H 12442! H & P 13924!* General in the tropics.

6. ELEOCHARIS R. Br. (1810)

1 Spikelets at least 5 mm. long, lanceolate to linear, up to 5 cm. long in large plants; plants usually over 30 cm. high, perennial:
  2 Scape smooth, not obviously septate-jointed even when dry; sheath-tip acute; glumes pale, broad; plants stoloniferous:
    3 Scape sharply triquetrous; glumes obtuse, about half as broad as spikelet   1. *mutata*
    3 Scape more or less terete; glumes rounded at tip, almost as broad as spikelet:
      4 Spikelet obtuse at tip; glumes broadly oblong to suborbicular, 4–5 mm. long; bristles smooth or nearly so, about as long as achene or a little longer   2. *cellulosa*
      4 Spikelet acute at tip; glumes oblong, about 3·5 mm. long; bristles retrorsely barbellate, nearly three times as long as achene   3. *elongata*
  2 Scape septate-jointed, subterete:
    5 Glumes about half as broad as spikelet, pale; sheath-tip oblique, acute or shortly acuminate; culms tufted, stoloniferous   4. *interstincta*
    5 Glumes about one-fifth as broad as spikelet or less, reddish-brown; sheath-tip transversely truncate or nearly so:
      6 Scape slender, up to 2 mm. thick, septa not always obvious in the living plant; culms numerous and compact from a horizontal rhizome; spikelets up to about 1·5 cm. long, acute, not much broader than scape   5. *nodulosa*
      6 Scape 4–10 mm. thick; culms few, more or less tufted from a short rhizome; spikelets 1–3 cm. long, obtuse, broader than scape   6. *elegans*
1 Spikelets less than 5 mm. long, more or less ovoid; scape smooth; plants (1·5–) 5–15 (–30) cm. high:
  7 Glumes oblong to obovate, rounded at tip and on the back; achenes obovoid, complanate, dark reddish-brown to black; bristles conspicuous:
    8 Scape capillary, 1·5–5 (–10) cm. long; plant annual; culms tufted; achene 0·5–0·6 mm. long, shiny black; bristles as long as achene; in damp hollows among grasses; Grand Cayman   *E. atropurpurea* (Retz.) Kunth
    8 Scape up to 1 mm. thick, mostly more than 10 cm. long; achene black or reddish-brown:
      9 Rhizome creeping; culms spreading, rarely densely tufted; beak of achene pointed; bristles as long as achene or shorter   7. *flavescens*
      9 Culms densely tufted, erect; beak of achene blunt; bristles longer than achene   8. *geniculata*
  7 Glumes lanceolate, keeled; spikelets more or less compressed; achene ovoid, trigonous, often whitish; bristles much shorter than achene; scape filiform:
    10 Achene striate and minutely rugulose; beak triangular-pyramidal, sharply pointed   9. *retroflexa*
    10 Achene smooth; beak short; glumes reddish-brown   10. *microcarpa*

1. E. mutata (L.) Roem. & Schult. in L., Syst. Veg. ed. nov. **2**: 155 (1817).— *Scirpus mutatus* L. (1759). Scallion Grass.

Culms 30–80 (–120) cm. high; spikelet 20–50 mm. long; glumes scarious-margined; achene obovoid, minutely pitted; bristles stiff, retrorsely barbellate, about as long as achene.

Frequent in boggy pastures and swamps, locally dominant in some open freshwater lagoons, marshy riverbanks and ponds; SL–2300 feet; fl. and fr. sporadically throughout the year. *A 7343! H 12310! P 24042!* E. United States, continental tropical Amer. to the Guianas, West Indies; Grand Cayman.

2. E. cellulosa Torr. in Ann. Lyc. New York **3**: 298 (1836).

Culms 40–60 (–70) cm. high, 1–4 mm. thick; spikelet 15–30 mm. long, 2·5–4 mm. broad; glumes with a thin reddish-brown submarginal line and hyaline margin; achene obovoid, nearly 2 mm. long, striate, pitted.

Rather local (Clar., St. Eliz., St. Mary), in wet pastures, muddy swamps and pond margins in water to 2 feet deep; SL–50 feet; fl. and fr. Apr, July, Sept, Dec. *A 11974! H 12558! J.P. 804* partly! S. United States, Mexico to Venezuela, Bahamas, Greater Antilles, St. Lucia.

3. E. elongata Chapm., Fl. South. U.S.: 515 (1860).

Culms numerous, (30–) 50–70 (–100) cm. high, mostly 1–1·5 mm. thick; spikelet 10–20 mm. long, about 2·5 mm. broad; glumes with a thin brown margin; achene obovoid, about 1 mm. long, striate, pitted; bristles 6, stout at the base.

Very rare (West.), in mat of floating vegetation on pond; 50–100 feet; fl. and fr. Nov. *P 27706!* Florida to Louisiana and Texas.

4. E. interstincta (Vahl) Roem. & Schult. in L., Syst. Veg. ed. nov. **2**: 149 (1817). —*Scirpus interstinctus* Vahl (1805).

Culms 40–100 cm. high; sheath dark brownish-red, turning black; spikelet up to 40 mm. long; bristles retrorsely barbellate.

Widely distributed but not common, in swamps, ponds, boggy pastures and wet sandy ground; 20–2300 feet; fl. and fr. most of the year. *A 10377! J.P. 846! P 8693!* S. United States, C. & S. America, West Indies; Grand Cayman.

5. E. nodulosa (Roth) Schult., Mant. **2**: 87 (1824).—*Scirpus nodulosus* Roth (1821).

Rhizome short, apparently with stolons; culms approximate, up to 60 cm. high; spikelet 1–1·5 (–2·5) cm. long, less than 5 mm. broad.

Rare and local (St. Cath., Clar., St. Eliz.), at pond margins and in wet depressions in open savannas; 100–2300 feet; fl. and fr. Oct–Mar. *A 10366! H 12179! P 18411!* Venezuela, Greater Antilles, Antigua, Nevis. Related to *E. montana* (Kunth) Roem. & Schult. with which it has sometimes been combined.

6. E. elegans (Kunth) Roem. & Schult. in L., Syst. Veg. ed. nov. **2**: 150 (1817).— *Scirpus elegans* Kunth (1816). *E. geniculata* of many authors, not (L.) Roem. & Schult.

Culms 30–100 cm. high, often shorter and more slender at higher altitudes; sheath reddish; spikelet purplish-brown, ovoid, 4–9 mm. broad; bristles reddish, retrorsely barbed.

Common in damp open places; SL–4000 feet; fl. and fr. all the year. *A 6568! H 11701! J.P. 675!* Mexico to Paraguay, West Indies.

7. E. flavescens (Poir.) Urb., Symb. Ant. **4**: 116 (1903).—*Scirpus flavescens* Poir. (1804).

Plant mostly perennial; culms slender (2–) 8–15 cm. high; spikelet 2–4 mm. long.

Occasional at swamp and pond margins and in bogs in savannas; SL–2500 feet; fl. and fr. most of the year. *A 9465! H 12265! Proctor & Skelding 16132!* Venezuela, West Indies.

8. E. geniculata (L.) Roem. & Schult. in L., Syst. Veg. ed. nov. **2**: 150 (1817).— *Scirpus geniculatus* L. (1753). *E. capitata* R. Br. (1810), not *S. capitatus* L. (1753). *E. caribaea* (Rottb.) Blake (1918).

Plants mostly annual; culms slender, numerous, 5–30 cm. high; spikelet 3–6 mm. long.

Common on alluvial sand, gravel, wet rock ledges and in damp swamp margins, also a weed of damp roadsides and ditchbanks; SL–1700 feet; fl. and fr. all the year. *A 7212! H 12283! J.P. 805! P 22811!* Widespread in warm countries; Grand Cayman.

9. E. retroflexa (Poir.) Urb., Symb. Ant. **2**: 165 (1900).—*Scirpus retroflexus* Poir. (1804).

Plants annual, gregarious; culms densely tufted, 2–20 cm. long, often nodding or bent down and rooting from below the spikelet; spikelet oblong, 2–4 mm. long.

Occasional in muddy pond margins and in clay soil near streams; 50–2500 feet; fl. and fr. Sept–May. *A 10356! H 12266! P 9739!* General in the tropics.

10. E. microcarpa Torr. in Ann. Lyc. New York **3**: 312 (1836).—*E. minima* Kunth (1837).

Plants annual; culms 5–20 (–30) cm. high; spikelet ovoid to oblong, 2–3 (–5) mm. long; glumes pale.

Rare (Han.), without specific location, but probably coastal. *W. Wright! Britton & Hollick 2142.* United States to Venezuela, Cuba, Puerto Rico.

7. SCIRPUS L. (1753)

1 Bracts conspicuous, long and narrow, spreading; rhizome elongated, almost terete; culms often floating, 3-gonous; heads several, subglobose, stalked; spikelets small
                                                    1. *cubensis*
1 Bracts few, the lowest up to 7 cm. long, usually erect and appearing as a continuation of the culm; culms erect; spikelets up to 10 mm. long:
  2 Culms terete; inflorescence several-branched with 1–few sessile or shortly stalked spikelets terminal on peduncles up to 6 cm. long   2. *validus*
  2 Culms sharply 3-quetrous; spikelets sessile, 5 or more in a single head   3. *olneyi*

1. S. cubensis Poepp. & Kunth in Kunth, Enum. Pl. **2**: 172 (1837).

Culms 1–2 m. long, trailing; leaves few; heads of spikelets about 1 cm. in diameter; spikelets numerous 4–8 mm. long; glumes ovate, acuminate, keeled at the tip with a solitary green thick nerve; connective produced into a slender hair; style-arms 2 (–3); achene ovoid, long-acuminate, smooth.

Rare (St. Cath., St. Eliz., West.), in boggy ponds and swamps; 25–1200 feet; fl. and fr. Jan–Mar. *A 10376! P 11709!* S. America, Cuba, Hispaniola, Trinidad.

2. S. validus Vahl, Enum. Pl. **2**: 268 (1805).

Rhizome perennial; culms light grey-green, 1·5–3 m. high, leafless or with very reduced blades; lowest blade 1–7 cm. long; spikelets ovoid, 5–10 mm. long; glumes broadly ovate, 2·5–3 mm. long, emarginate, apiculate; achene about 2·5 mm. long; bristles usually 6, as long as or a little longer than the achene.

Occasional in brackish or low-lying swamps, mostly near the sea; SL–10 feet; fl. and fr. Nov–June. *A 10972! H 12301! J.P. 824! P 15355!* N. America, Bermuda, Greater Antilles; Grand Cayman.

3. S. olneyi A. Gray in Journ. Nat. Hist. Boston **5**: 238 (1845).

Rhizome short or long; culms more or less tufted, up to 2 m. high, mostly with bladeless sheaths; lowest bract 1–5 cm. long; spikelets ovoid, 4–10 mm. long; glumes entire, mucronulate, scarious-margined, brown; achene obovoid, blackish, 2–2·5 mm. long; bristles as long as or a little shorter than the achene.

Very rare and local (St. Andr./St. Cath., St. Eliz.), in brackish swamps; SL and a little over; fl. Mar–Apr. *Purdie! P 24770!* United States, Bermuda, Greater Antilles, Nevis.

8. FUIRENA Rottb. (1773)

1. F. umbellata Rottb., Descr. Ic. Pl.: 70, t. 19, f. 3 (1773).

Culms tufted (15–) 30–90 (–150) cm. high in flower, glabrous or pilose; leaf-blades flat, up to 2 cm. broad, the lower sheaths glabrous or pubescent; panicles terminal or also axillary; spikelets oblong, clustered, 6–10 mm. long; glumes greenish-brown, pubescent, aristate.

Occasional in swamps and boggy ponds; SL–2400 feet; fl. and fr. all the year. *A 8523! H 11823! P 6920!* General in the tropics.

9. HEMICARPHA Nees (1834)

1. H. micrantha (Vahl) Pax in Engl. & Prantl, Nat. Pflanzenf. **2** (2): 105 (1887).—
Scirpus micranthus Vahl (1805).

Tufted annual, 3–10 cm. high; culms filiform; spikelets broadly ovoid to subglobose, dark brown, 2–3 mm. long; stamen 1.

Very local (St. Cath.), in sandy lagoon beds and salina margins; near SL; fl. and fr. Nov–Dec. *A 10008! Robertson UCWI 3563!* Thinly distributed through the American subtropics and tropics in suitable habitats.

10. FIMBRISTYLIS Vahl (1805) nom. cons.

1 Plant annual; scape tetragonous; spikelets subglobose, about 2 mm. long, in open panicles; glumes keeled; stamens 2; styles 3-fid   1. *miliacea*

1 Plants perennial; spikelets ovoid to linear-lanceolate:
  2 Glumes broadly ovate with obscure midribs, rounded-convex on the back, not at all keeled; spikelets 7–20 mm. or more long; stamens 3; style more or less flattened and winged, ciliate or fimbriate, 2-fid:
    3 Spikelets light reddish-brown; glumes whitish-pubescent near tip with ciliolate margins; leaf-bases light brown   2. *ferruginea*
    3 Spikelets dark brown; glumes glabrous; leaf-bases black, shiny   3. *spadicea*
  2 Glumes ovate to lanceolate with midrib evident and prominent at least in distal half; spikelets mostly less than 10 mm. long:
    4 Glumes weakly keeled, more or less convex, glabrous; stamens 3; style flat with fimbriate margin, 2-fid; inflorescence open   4. *dichotoma*
    4 Glumes strongly keeled; style subterete, not winged or fimbriate:
      5 Leaves poorly developed, the basal sheaths subulate-tipped and rarely more than 4 cm. long in all; spikelets (1–) 2–4 in a compact head; glumes ciliate; stamens 3; style 3-fid   5. *harrisii*
      5 Leaves well developed with normal blades; spikelets numerous; glumes essentially glabrous:
        6 Inflorescence openly paniculate with spreading flexuous branches; stamens 3; style 3-fid   6. *complanata*
        6 Inflorescence compactly corymbose with inner branches ascending and outer often catadromic; stamens 2:
          7 Glumes obtuse or rounded, yellowish-brown, with broad hyaline margins; style 2-fid   7. *cymosa*
          7 Glumes acute, reddish-brown, with narrow hyaline margins; style 3-fid
                                                                              8. *papillosa*

1. **F. miliacea** (L.) Vahl, Enum. Pl. **2**: 287 (1805).—*Scirpus miliaceus* L. (1759).
   Culms weakly divergent, flat at base, angled above, up to 70 cm. high in flower; leaves soft, long-tapered, up to 4 mm. broad; bracts similar but shorter and narrower; panicle decompound; glumes ovate, acute, glabrous; style-arms pilose; achene 3-gonous, obovoid, verrucose.
   Rare and local (St. Cath.), in ditches; SL–40 feet; fl. and fr. Oct–Nov. *A 11891! H 12182!* General in the tropics.

2. **F. ferruginea** (L.) Vahl, Enum. Pl. **2**: 291 (1805).—*Scirpus ferrugineus* L. (1753).
   Culms densely tufted, erect, (30–) 60–100 cm. high; leaves rather short; spikelets few, rarely solitary, ovoid to lanceolate, 8–15 (–20) mm. long, stalked in an umbelliform inflorescence; glumes more or less rounded at tip; achene biconvex, obovate to suborbicular, pale, smooth.
   Common along marshy riverbanks and swamp margins; SL–40 feet; fl. and fr. all the year. *A 6062! 12018! J.P. 802! P 8359!* General in the tropics; Cayman Is.

3. **F. spadicea** (L.) Vahl, Enum. Pl. **2**: 294 (1805).— *Scirpus spadiceus* L. (1753).
   Culms in large dense tufts, erect or spreading, slender, 60–120 cm. high; basal leaves long, involute; inflorescence a simple or compound umbel; spikelets 8–20 mm. or more long; glumes smooth, shiny, apiculate; achene biconvex, obovate, pitted.
   Locally common on dunes, salina margins and along marshy riverbanks near the sea; SL–15 feet; fl. and fr. most of the year. *A 6459! H 11624! J.P. 761!* Continental tropical Amer., West Indies; Grand Cayman.
   A plant intermediate between *F. ferruginea* and *F. spadicea* with light brown leaf-bases, glabrous glumes and long style-arms has been named *F. castanea* (Michx.) Vahl. It is reported from St. Elizabeth (*H & P 14499!*) and Grand Cayman (*Kings GC 193!*) and a similar plant occurs in Bahamas and Cuba. Svenson (N. Amer. Fl. **18** (9): 551 (1957)) puts *F. castanea* in the synonymy of *F. spadicea*.

4. **F. dichotoma** (L.) Vahl, Enum. Pl. **2**: 287 (1805).—*Scirpus dichotomus* L. (1753). *F. diphylla* (Retz.) Vahl (1805). *F. annua* (All.) Roem. & Schult. (1817).
   Culms few, tufted, slender, 10–60 cm. high; leaves flat, 1–3 mm. broad; inflorescence a simple or compound umbel; spikelets ovoid to oblong, 5–10 mm. long; glumes broadly ovate, minutely apiculate; achene biconvex, obovate, usually pale, ribbed and more or less pitted.

Very common in damp grassy places generally; SL–5200 feet; fl. and fr. most of the year. *A 5837! H 12855! J.P. 859! P 22810!* Widespread in warm temperate and tropical regions.

5. F. harrisii (Britton) Adams in Phytologia **21** (2): 66 (1971).—*Stenophyllus harrisii* Britton (1920). *Bulbostylis subefimbriata* Kük. (1929).

Culms densely tufted, 30–80 cm. high in flower, striate, glabrous, with dark brown basal sheaths; inflorescence 1–1·5 cm. long, often viviparous; rays up to 5 mm. long; spikelets 6–8 mm. long, light brown.

Rare and local (Port., St. Thom.), on exposed rocky ground and on rocks near waterfalls; 3000–3500 feet; fl. Sept–Mar. *A 12224! H 12098!* Hispaniola.

6. F. complanata (Retz.) Link, Hort. Bot. Berol. **1**: 292 (1827).—*Scirpus complanatus* Retz. (1789). *S. autumnalis* of Griseb. (1864).

Culms tufted, slender, 20–60 cm. high; leaves shorter than culm; spikelets linear-lanceolate, 5–10 mm. long; glumes brown with single green or brown nerve, apiculate; achene 3-gonous, pale, minutely papillose, 0·5–0·6 mm. long.

Occasional in open ditches, sandy wastes and swamp margins; SL–3000 feet; fl. and fr. May–Jan. *A 7324! H 12237! P 18410!* General in the tropics.

7. F. cymosa R. Br., Prodr. Fl. Nov. Holl.: 228 (1810).—*F. spathacea* Roth (1821). *F. glomerata* Urb. (1900), not (Schrad.) Nees (1834).

Culms slender, stiff, erect, 10–40 cm. high; leaves forming rosettes, stiff, spreading, obtuse, mucronate, 1·5–3 mm. broad; bracts short; glumes shortly emarginate, apiculate, glabrous or minutely pulverulent; style skirted at base; achene biconvex, obovate, minutely verrucose.

Common at lagoon and swamp margins, gregarious and often forming swards, mostly in coastal areas; SL–50 feet; fl. and fr. all the year. *A 5668! H 9381! P 8358!* General in the tropics; Grand Cayman.

8. F. papillosa (Kük.) Alain in Bull. Torr. Bot. Club **92** (4): 290 (1965).— *Bulbostylis papillosa* Kük. (1926).

Culms densely tufted, 10–60 cm. high in flower, glabrous; leaves with broad or narrow blades formed seasonally, the sheaths with long white hairs at mouth; inflorescence up to about 3 cm. long; achenes 3-gonous, obovoid, cream-coloured, 0·8 mm. long.

Rare and local (St. Andr., Clar.), on well or poorly drained heavy soil in grassland which is sometimes burnt-over; 1000–2300 feet; fl. and fr. Sept–May. *H 12219! 12345! P 8461! Weck T 77!* Greater Antilles.

11. UNCINIA Pers. (1807)

1. U. hamata (Sw.) Urb., Symb. Ant. **2**: 169 (1900).—*Carex hamata* Sw. (1788). *U. jamaicensis* Pers. (1807). Bird-catching Sedge.

Perennial; culms tufted, 30–60 (–90) cm. high in flower; leaves flaccid, dark green; peduncles slender; inflorescence spicate, 6–10 cm. long.

Locally common (St. Andr., Port., St. Thom.), by pathsides in montane woodland; 3650–5250 feet; fl. and fr. most of the year. *A 5865! H 11422! J.P. 2010!* C. and S. Amer., Cuba, Hispaniola.

12. CAREX L. (1753)

1 Leaves up to 5 mm. broad; spikelets androgynous, rarely as much as 1·5 cm. long, numerous in panicles:
  2 Utricle (bottle-shaped organ enclosing the true fruit) smooth; leaf-blades 2–4·5 mm. broad   1. *polystachya*
  2 Utricle antrorsely scabrid; leaf-blades 0·5–2·2 mm. broad   2. *scabrella*
1 Leaves 8 mm. or more broad; spikelets 3–6 (–8), the upper all or mostly staminate, the lower all pistillate, up to 6 cm. long; utricle many-nerved, smooth, shiny, spreading widely from the rachilla at maturity   3a. *hystricina* var. *underwoodii*

1. C. polystachya Sw. ex Wahl. in Vet. Acad. Handl. Stockh. **24**: 149 (1803).—
C. *cladostachya* Wahl. (1803).

Perennial; culms tufted, lax, usually exceeding the leaves in flower, up to about
50 cm. high; inflorescence congested or open.

Locally common (St. Andr., Port., St. Thom.), on pathside banks and in rocky
thickets and woodlands; 2300–5300 feet; fl. and fr. Feb–Sept. *A 6356! J.P. 768!
P 10322!* C. and S. Amer., Greater Antilles. A variable species in Jamaica; plants
from Hanover (1200–1600 feet; fl. and fr. Apr, Sept. *A 7999! P 23464!*) with more
compact spikelets and slightly larger minutely scabridulous fruits may be distinct;
plants from eastern Portland with slender more open inflorescences approach
C. *cubensis* Kük. from Cuba and Hispaniola. Both these variants occur on limestone rocks at generally lower altitudes than the typical plants.

2. C. scabrella Wahl. in Vet. Acad. Handl. Stockh. **24**: 149 (1803).

Perennial; culms loosely tufted from a shortly branched rhizome, usually about
20–30 cm. high; inflorescence usually shorter than the leaves.

Rather local (St. Andr., St. Thom., St. Ann), on damp shaded banks and on
rocks in woodland; 250–3800 feet; fl. and fr. most of the year. *A 12581! H 12331!
P 23576!* Cuba, Hispaniola.

3. C. hystricina Muhl. ex Willd. in L., Sp. Pl. ed. 4, **4**: 282 (1805).

3a. Var. **underwoodii** (Britton) Kük. in Engl., Pflanzenr. **4** (20), Heft 38: 700
(1909).—C. *underwoodii* Britton (1905).

Perennial; culms forming large tufts from stoloniferous rhizome, up to 1 m. or
more high; leaf-blades up to 20 mm. broad; tips of glumes antrorsely scabridulous
on the margins.

Very local (St. Andr.), at margin of bog; 3500 feet; fl. and fr. Jan, May–June.
H 11116! 12526! P 8352! Brazil, Hispaniola.

13. SCLERIA Berg. (1765)

1 Culms scrambling; inflorescence paniculate; bracts and bracteoles hispid; achenes not
  beaked; leaf-blades up to 5 mm. broad   1. *secans*
1 Culms erect or at most nodding:
  2 Achenes distinctly beaked at maturity, the beak tardily deciduous:
    3 Inflorescence unbranched, with sessile clusters of spikelets; leaf-blades less than 3 mm.
      broad; glumes much longer than achenes; bracteoles and glumes hispid-setulose
                                                          2. *hirtella*
    3 Inflorescence paniculate; leaf-blades 1·5–3 cm. broad; glumes short, the achenes
      exposed; bracteoles and glumes at most puberulous; plants up to 2 m. or more high:
      4 Achene ovoid; margin of hypogynium at most puberulous   3. *cubensis*
      4 Achene globose; margin of hypogynium ciliate with golden-brown hairs
                                                     4. *eggersiana*
  2 Achenes not beaked but sometimes minutely apiculate:
    5 Inflorescence much branched; leaves up to 1·5 cm. broad:
      6 Achene ellipsoid, 1–1·5 mm. long, white; hypogynium ciliate   5. *microcarpa*
      6 Achene globose, about 2 mm. in diameter, white or blackish-purple; hypogynium
      3-lobed, indurated with a smooth-margined flange   6. *melaleuca*
    5 Inflorescence not or very little branched; leaf-blades less than 4 mm. broad:
      7 Achene warted; rhizome creeping with culms 5–10 mm. apart; leaves and bracts
      hispid; glumes pubescent   7a. *ciliata* var. *elliottii*
      7 Achene smooth; rhizome shortly branched with approximate culms; leaves and
      bracts puberulous or glabrous:
        8 Culms numerous, densely tufted, erect, up to 1 mm. thick; clusters of spikelets
        solitary, pseudo-lateral near the tips of the culms   8. *georgiana*
        8 Culms several, loosely tufted, laxly ascending, more than 1 mm. thick; clusters of
        spikelets 3–5 on each culm, lateral and terminal   9. *lithosperma*

1. S. secans (L.) Urb., Symb. Ant. **2**: 169 (1900).—*Schoenus secans* L. (1759).
  Razor Grass.

Perennial; culms lax, forming tangles, 2–3 m. or more high; sheaths and leaf-
blades very sharply serrulate and capable of causing serious cuts if handled in-
autiously; spikelets blackish.

Occasional in moist savanna thickets and secondary woodlands; 850–2500 feet; fl. and fr. sporadically throughout the year. *A 7895! H 12264! P 16482!* British Honduras, Venezuela to Brazil, West Indies.

2. S. hirtella Sw., Nov. Gen. & Sp. Pl.: 19 (1788).
Perennial; culms pilose or glabrous, remote from a creeping rhizome, up to 40 (–60) cm. high.
Locally common on open banks and grassy savannas on heavy moist clay soils; 1400–3000 feet; fl. and fr. most of the year. *A 12493! H 11101! J.P. 2017! Stearn 984!* C. and S. Amer., Greater Antilles, Trinidad, tropical Africa.

3. S. cubensis Boeck., Cyp. Nov. **2**: 42 (1890).
Culms tufted from a short rhizome, 1·5–2 m. or more high; spikelets brown; achenes at first green maturing greyish-mauve or white.
Frequent, gregarious in glades and at the margins of woodlands usually on heavy soils in limestone areas; 750–2700 feet; fl. and fr. Mar–Oct. *A 6751! J.P. 674! P 18260!* Greater Antilles.

4. S. eggersiana Boeck., Cyp. Nov. **2**: 41 (1890).—*S. grisebachii* C. B. Cl. (1900).
Like the last; culms up to 2 m. or more high.
Occasional at swamp margins and in low-lying poorly drained wet areas; 10–50 feet; fl. and fr. July, Dec–Jan. *A 12013! P 23154! Skelding UCWI 3649!* Greater Antilles, Antigua.

5. S. microcarpa Nees ex Kunth, Enum. Pl. **2**: 341 (1837).
Culms up to about 1·5 m. high; achenes with a very short persistent style-base.
Very rare and local (St. Eliz.), at margins of muddy ponds; about 75 feet; fl. and fr. Jan. *P 23158!* C. and S. Amer., Greater Antilles.

6. S. melaleuca Cham. & Schlecht. in Linnaea **6**: 29 (1831).
Perennial; culms tufted from a short rhizome, up to 90 (–120) cm. high in flower.
Common in ditches, along roadsides and in swamps and damp pastures; 30–3250 feet; fl. and fr. May–Jan. *A 5687! J.P. 810! P 15945!* General throughout the American tropics.

7. S. ciliata Michx., Fl. Bor. Amer. **2**: 167 (1803).

7a. Var. elliottii (Chapm.) Fernald in Rhodora **39**: 392 (1937).—*S. elliottii* Chapm. (1860).
Perennial; culms up to 30 (–50) cm. high; hypogynium with 3 bilobed tubercles.
Local and uncommon (St. Andr., Clar.), in open savannas and exposed grassy hillsides on sandy or clay soils; 2300–3250 feet; fl. and fr. Nov–Mar. *A 6674! P 18409!* Virginia, Cuba, Hispaniola.

8. S. georgiana Core in Brittonia **1**: 243 (1934).
Perennial; culms slender, approximate, up to 30 cm. high.
Rare and local (Clar.), in open wet savanna on sandy or clay soil; 2200–2500 feet; fl. and fr. June, Nov–Dec. *A 5930! H 12262! P 15949!* British Honduras, Cuba, Hispaniola.

9. S. lithosperma (L.) Sw., Nov. Gen. & Sp. Pl.: 18 (1788).—*Scirpus lithospermus* L. (1753). *Scleria filiformis* Sw. (1788).
Perennial; culms up to about 50 (–60) cm. high in flower; spikelets brown; achenes globose, white.
Very common in open or shaded thickets on rocky limestone, often in arid areas; 20–2600 feet; fl. and fr. all the year. *A 5526! H 12355! P 23705! Stearn 828!* General in the American tropics and also in the Old World; Cayman Is.

14. DIPLACRUM R. Br. (1810)

1. D. longifolium (Griseb.) C. B. Cl. in Urb., Symb. Ant. **2**: 153 (1900).— *Pteroscleria longifolia* Griseb. (1864).
Perennial; culms up to about 80 cm. high, rather straggling; leaves long and

narrow; spikelets in globose heads 1–1·5 cm. in diameter; male and female florets each with 2 glumes; style trifid.

Very rare (Clar.), in moist open savanna; 2300 feet; fl. and fr. Aug–Sept. *Skelding UCWI 5013!* Venezuela, Trinidad, West Africa.

36. GRAMINEAE (POACEAE)

Herbs or rarely shrubs or trees. Leafy shoots (culms) usually terete, mostly hollow in the internodes, solid and often slightly swollen at the nodes. Leaves solitary at the nodes, usually 2-ranked (distichous) comprising ideally a sheath, ligule and blade; sheath-margins usually overlapping to the base, rarely the margins connate; ligule at junction of sheath and blade adaxial, membranous or a fringe of hairs, rarely absent; blade usually flat, linear, parallel-veined, rarely narrowed to a petiole, rarely articulated to the sheath and then deciduous. Inflorescence usually terminal consisting of stalked or sessile spikelets in mostly bractless spikes, racemes or panicles. Spikelets of 1–many florets sessile in 2 rows on either side of an axis (rachilla) with 2 empty subopposite bracts (glumes) at the base. Florets when perfect comprising a pistil, stamens and 2 or 3 lodicules subtended by 2 bracteoles, the inner and upper usually hyaline (palea), the outer or lower often green (lemma). Lodicules are small scales or glands representing the perianth. Glumes usually green and more or less conspicuously veined and keeled, the inner and upper usually larger, the outer or both rarely reduced or absent. Florets bisexual or unisexual, monoecious in the same spikelet or in separate spikelets or inflorescences on the same plant, or dioecious. Stamens hypogynous (1–) 3 (–6); filaments slender; anthers 2-locular, basifixed but so deeply sagittate as to appear versatile, usually opening lengthwise. Ovary superior, 1-locular with 1 anatropous ovule often adnate to the carpel; styles (1–) 2 (–3), connate or free; stigmas generally plumose. Fruit usually a caryopsis (the seed adhering to the pericarp) or rarely a nut or berry, often comprising at least the retained and often modified lemma and palea. Seed with a large starchy endosperm and a small unilateral embryo.

600 genera, more or less, with up to 10,000 species; cosmopolitan. The family is usually divided into two subfamilies distinguished as follows:

Spikelets 1–many flowered; reduced florets, if any, above the perfect florets (except in *Phalarideae*); spikelets breaking up at maturity above the more or less persistent glumes or if falling entire 1-flowered; spikelets uniform, more or less laterally compressed
　　　　　　　　　　　　　　　　　　　Subfamily 1. POOIDEAE (*Festucoideae*)
Spikelets 2-flowered (except in monoecious genera or in entirely male or neuter spikelets); upper floret usually fertile; lower floret male or barren and often much reduced; spikelets falling entire at maturity, singly or in groups or together with joints of the rachis; spikelets sometimes unalike, more or less dorsally compressed; one glume, rarely both glumes sometimes wanting　　　　　　　　Subfamily 2. PANICOIDEAE (p. 160★)

Key to the genera of *Pooideae*:

1 Plants woody; culms perennial; leaf-blade articulated to the sheath (Tribe *Bambuseae*):
　2 Culms erect; spikelets 1–many-flowered in branched panicles or panicled spikes; stamens 6　　　　　　　　　　　　　　　　　　　　　　　**1. Bambusa**
　2 Culms climbing or scrambling; spikelets 1-flowered in small panicles; stamens 3
　　　　　　　　　　　　　　　　　　　　　　　　　　　　　2. Chusquea
1 Plants herbaceous; culms usually annual:
　3 Spikelets with (1–) 2 unlike lemmas of sterile florets below that of the perfect terminal floret; panicle compact (Tribe *Phalarideae*):
　　4 Spikelets not strongly compressed; sterile florets exceeding the fertile floret, awned
　　　　　　　　　　　　　　　　　　　　　　　　　3. Anthoxanthum
　　4 Spikelets strongly compressed; glumes winged from the keel; lemmas not awned; a very rare casual; Canary Grass　　　　　　　*Phalaris canariensis* L.
　3 Spikelets without sterile lemmas below the fertile floret or these if rarely present like the fertile ones:
　　5 Lemmas or rachilla-joints bearing long silky hairs which at least in fertile florets envelop the lemma; tall cane or reed grasses with plumelike panicles (Tribe *Arundeae*):

6 Spikelets unisexual; plants dioecious; staminate spikelets glabrous; leaf-blades armed, tardily deciduous leaving closely imbricated leaf-sheaths   4. **Gynerium**
6 Spikelets perfect; leaves persistent, not armed:
  7 Lemmas hairy; rachilla glabrous; leaf-blade bases strongly auricled   5. **Arundo**
  7 Lemmas glabrous; rachilla hairy; leaf-blade bases not strongly auricled
           6. **Phragmites**
5 Lemma and rachilla glabrous or if hairy the hairs not enveloping the lemma:
  8 Spikelets unisexual, monoecious, 1-flowered, the pistillate falling entire:
    9 Pistillate and staminate spikelets in separate panicles; pistillate spikelet sub-globose; glumes wanting; leaf-blades narrowly linear (Tribe *Zizanieae*)
           7. **Luziola**
    9 Pistillate and longer-pedicelled staminate spikelets in pairs in the same panicle; pistillate spikelet terete with the lemma more or less clothed with short hooked hairs; glumes present; leaf-blades broad (Tribe *Phareae*)   8. **Pharus**
  8 Spikelets with at least one perfect floret:
    10 Spikelets articulated below the glumes, falling entire or in small clusters; 1-flowered:
      11 Glumes small or wanting; spikelets pedicelled, paniculate, strongly laterally flattened; lemma and palea about equal, both keeled (Tribe *Oryzeae*):
        12 Glumes present; lemma usually awned   9. **Oryza**
        12 Glumes wanting; lemma awnless   10. **Leersia**
      11 Glumes well-developed; lemma awnless:
        13 Spikelets pedicelled to subsessile in non-unilateral racemes; leaves mostly rather short (Tribe *Zoysieae*):
          14 Plants perennial, tufted and rhizomatous or stoloniferous; leaves narrow; glumes rounded on the back; pedicels broadening upwards:
            15 Spikelets subulate, remote, divergent and later reflexed; lower glume more or less free from the rest of the spikelet, dorsally flattened and ciliate towards the tip; upper glume laterally flattened towards the tip; sandy seashores   11. **Leptothrium**
            15 Spikelets lanceolate, imbricate, erect, about 2·5 mm. long, in racemes 1–1·5 (–2) cm. long; leaves filiform; plants mostly stoloniferous; cultivated for lawns; native of the western Pacific   *Zoysia tenuifolia* Willd.
          14 Plants annual, tufted with ascending geniculate culms; leaves 2–4 mm. broad; spikelets in clusters of 2–5, not connate but falling together as small burs; two lower spikelets with numerous hooked prickles on their upper glumes
              12. **Tragus**
      13 Spikelets sessile, flattened and imbricated in 1-sided spikes; rhizome stout; leaves with the adaxial surface of the blade deeply grooved and mostly enclosed by inrolling; confined to sandy seashores (Tribe *Chlorideae* in part)
           13. **Spartina**
    10 Spikelets usually articulated above the glumes, 1–several-flowered; at least 1 glume well-developed:
      16 Spikelets sessile or subsessile:
        17 Spikelets on opposite sides of a solitary terminal spike; lower glume suppressed except in the terminal spikelet; at some time introduced and cultivated, not recently collected   *Lolium* spp.
        17 Spikelets in two rows on one side of a continuous rachis forming one-sided spikes or spikelike racemes; spikes digitate or racemose (Tribe *Chlorideae*):
          18 Spikes racemose:
            19 Spikelets with 1 perfect floret and an upper floret reduced to awns; spikes short   14. **Botelua**
            19 Spikelets with 2 or more perfect florets; spikes long and slender
              15. **Leptochloa**
           18 Spikes digitate or mostly so:
            20 Spikelets with 1 perfect floret, with or without sterile florets:
              21 Sterile floret wanting; lemma obtuse   16. **Cynodon**
              21 Sterile floret(s) present; lemma awned or mucronate   17. **Chloris**
            20 Spikelets with 2 or more perfect florets:
              22 Rachis extending beyond the spikelets   18. **Dactyloctenium**
              22 Rachis not extending beyond the spikelets   19. **Eleusine**
      16 Spikelets pedicelled in open or contracted, sometimes spikelike, panicles or rarely racemes:
        23 Spikelets 1-flowered (Tribe *Agrostideae*):
          24 Lemmas awned; awn trifid   20. **Aristida**
          24 Lemmas awnless:
            25 Inflorescence an open or contracted panicle; glumes not abruptly mucronate:
              26 Glumes usually much shorter than the floret   21. **Sporobolus**
              26 Glumes longer than the floret   22. **Agrostis**

25 Inflorescence a cylindric spikelike panicle; glumes abruptly mucronate; at one time established on the summit of Blue Mt. Peak, not recently collected; native of Europe　　　　　　　　　　　　　　　　　　　　　　　　*Phleum pratense* L.
23 Spikelets 2–many-flowered:
  27 Glumes as long as or longer than the lowest floret (Tribe *Aveneae*):
    28 Articulation below the glumes; lemma of second floret with a hooked awn; casual introduction; native of Europe　　　　　　　　　　　　　*Holcus lanatus* L.
    28 Articulations above the glumes and between the florets:
      29 Plants annual; rachilla bearded; introduced but not established; Oat　　*Avena* spp.
      29 Plants perennial; rachilla villous; awn from between the teeth of the lemma
　　　　　　　　　　　　　　　　　　　　　　　　　　　　　　23. Danthonia
  27 Glumes shorter than the lowest floret:
    30 Leaf-blades ovate to elliptical, with transverse veinlets between the main veins, petiolate; spikelets 3–5-flowered　　　　　　　　　　**24. Zeugites**
    30 Leaf-blades linear without apparent transverse veinlets, sessile on the sheaths:
      31 Spikelets in 1-sided dense clusters　　　　　　　　　　**25. Dactylis**
      31 Spikelets not in 1-sided clusters:
        32 Lemmas 3-nerved　　　　　　　　　　　　　　　　**26. Eragrostis**
        32 Lemmas 5 or more nerved (Tribe *Festuceae*):
          33 Lowest 1–4 lemmas empty; spikelets firm, strongly compressed; plant perennial:
            34 Plants dioecious, up to 30 cm. high; rhizomatous; brackish coastal swamps and meadows; Grand Cayman　　　　*Distichlis spicata* (L.) Greene
            34 Plants hermaphrodite, usually more than 50 cm. high　　**27. Leptochloopsis**
          33 Lowest lemmas fertile; plants mostly annual and smaller:
            35 Florets spreading horizontally; lemmas cordate at base; spikelets broadly ovate　　　　　　　　　　　　　　　　　　　　**28. Briza**
            35 Florets ascending; lemmas not cordate:
              36 Lemmas awnless; spikelets small　　　　　　　　**29. Poa**
              36 Lemmas awned or mucronate:
                37 Awn from the tip of the lemma　　　　　　　**30. Festuca**
                37 Awn from the bifid apex of the lemma or if awn wanting then spikelets large
　　　　　　　　　　　　　　　　　　　　　　　　　　　31. Bromus

*

Key to the genera of *Panicoideae*:

1 Glumes membranous (leathery in *Scutachne*, the 4 lower indurate and united in *Anthephora*); sterile lemma like the glumes in texture:
  2 Fertile lemma and palea both thin and resembling the glumes in texture (Tribe *Melinideae*):
    3 Awn of lemma bent or twisted　　　　　　　　　　**32. Arundinella**
    3 Awns straight:
      4 Glumes awned; lemma awnless:
        5 Base of spikelet narrowed to a long minutely barbed callus　　**33. Achlaena**
        5 Base of spikelet without a callus　　　　　　　　**34. Reynaudia**
      4 Glumes awnless; sterile lemma awned　　　　　　　**35. Melinis**
  2 Fertile lemma and palea indurate or at least firmer in texture than the glumes (Tribe *Paniceae*):
    6 Spikelets unisexual; plants monoecious; leaves broad:
      7 Panicles large, terminal; pistillate spikelets above and staminate below in the same panicle; large somewhat scrambling undershrub　　　**36. Olyra**
      7 Panicles small, axillary with a single pistillate spikelet above 1–several staminate ones or if terminal staminate only; low tufted perennial　**37. Lithachne**
    6 Spikelets perfect:
      8 Axis thickened and corky, the spikelets sunken in cavities in its joints
　　　　　　　　　　　　　　　　　　　　　　　　　38. Stenotaphrum
      8 Axis not thickened, the spikelets not sunken in it:
        9 Spikelets subtended by 1–many bristles, these distinct or connate to form a false involucre:
          10 Bristles persistent on the rachis, not falling with the spikelets　**39. Setaria**
          10 Bristles falling with the spikelets:
            11 Bristles free or united at the base only, not forming an involucre or bur, often plumose　　　　　　　　　　　　　**40. Pennisetum**
            11 Bristles more or less barbellate, united at the base into an irregularly cleft involucre forming a bur　　　　　　　**41. Cenchrus**
        9 Spikelets not subtended nor surrounded by bristles:
          12 Spikelets in clusters of 4 with the indurate lower glumes united at the base to form a pitcher-shaped involucre, sessile on a flexuous axis　**42. Anthephora**

12 Spikelets free or however clustered glumes not united:
  13 Fruit cartilaginous or papery, not rigid; margins of lemma hyaline, not inrolled; spikelets mostly lanceolate in 2 rows along one side of the rachis; first glume minute or absent; fertile lemma adaxial   43. **Digitaria**
  13 Fruit indurate, rigid; lemma if thin not hyaline-margined:
    14 Spikelets orientated with the fertile lemma abaxial to the rachis, subsessile:
      15 Inflorescence paniculate; rachilla-joint and lower glume forming a swollen ringlike callus at the base of the spikelet   44. **Eriochloa**
      15 Inflorescence racemose; lower glume wanting   45. **Axonopus**
    14 Spikelets orientated with the fertile lemma adaxial to the rachis or distinctly pedicellate in panicles:
      16 Lower glume typically wanting; spikelets plano-convex, subsessile in spike-like racemes   46. **Paspalum**
      16 Lower glume present; spikelets terete or biconvex usually in panicles, rarely partial inflorescences spikelike or racemose:
        17 Glumes or lemmas or both awned or if awn inconspicuous the apex of the palea not enclosed by the inrolled margins of the fertile lemma:
          18 Inflorescence of short 1-sided racemes along a common axis; spikelets awned from between the 2 lobes at the apices of the glumes   47. **Oplismenus**
          18 Inflorescence paniculate:
            19 Spikelets long-silky; lower glume minute; awn short   48. **Rhynchelytrum**
            19 Spikelets scabrid or hispid; lower glume well developed; awn short or long   49. **Echinochloa**
        17 Glumes and lemmas awnless:
          20 Lower floret of spikelet perfect; spikelets small, turgid, obtuse 50. **Isachne**
          20 Lower floret staminate or neuter:
            21 Upper glume and sterile lemma leathery-indurate; fruit mucronate   51. **Scutachne**
            21 Upper glume and sterile lemma membranous; fruit not mucronate:
              22 Inflorescence a compact cylindric or spikelike panicle; aquatic or sub-aquatic:
                23 Upper glume inflated-saccate; fruit stipitate   52. **Sacciolepis**
                23 Upper glume not inflated; fruit not stipitate, scarcely indurate, not closed by the palea above   53. **Hymenachne**
              22 Inflorescence an open or compact panicle, if spikelike the fruit not stipitate nor the upper glume saccate; fertile lemma indurate:
                24 Culms bamboolike; spikelets subglobose, rather large and often dark-coloured   54. **Lasiacis**
                24 Culms not bamboolike:
                  25 Fertile lemma with either lateral appendages or excavations at the base, the margins usually not inrolled   55. **Ichnanthus**
                  25 Fertile lemma without lateral appendages or excavations at the base, the inrolled margins clasping the palea   56. **Panicum**
 1 Glumes indurate; fertile lemma and palea hyaline or membranous, the sterile lemma like the fertile in texture:
  26 Spikelets unisexual, monoecious, the pistillate below, the staminate above in the same or in separate inflorescences (Tribe *Maydeae*):
    27 Staminate and pistillate spikelets conjoint; perennials:
      28 Pistillate spikelets sunken in recesses in the thickened joints of the rachis; inflorescence of solitary or digitate racemes   57. **Tripsacum**
      28 Pistillate spikelets enclosed in a bony beadlike involucre   58. **Coix**
    27 Staminate inflorescence terminal; pistillate inflorescences axillary; annual   59. **Zea**
  26 Spikelets in pairs, one sessile and perfect, the other pedicelled and usually staminate or neuter, sometimes the pedicelled one obsolete or both pedicelled (Tribe *Andropogoneae*):
    29 Spikelets all similar, all perfect:
      30 Inflorescence of 2 to several digitate racemes; leaves narrow; culms branched; sessile spikelets 3·5 mm. long; awn of upper lemma 3–4·5 mm. long; glumes not transversely rugose; casual introduction; native of tropical Asia (*Ischaemum angustifolium* (Trin.) Hack.)   *Eulaliopsis binata* (Retz.) C. E. Hubbard
      30 Inflorescence a densely flowered hairy branched or spikelike raceme:
        31 Spikelets awned   60. **Erianthus**
        31 Spikelets awnless:
          32 Rachis continuous   61. **Imperata**
          32 Rachis disjointing   62. **Saccharum**
    29 Spikelets dissimilar, not all perfect, the sessile usually perfect, the pedicelled usually staminate or rudimentary:

F

33 Lower glume of fertile spikelets not transversely rugose; inflorescence digitate
                                                      63. **Ischaemum**
33 Lower glume of fertile spikelets not transversely rugose:
  34 Fertile spikelet with a hairy pointed callus formed from a rachis disjunction; awns long:
    35 Racemes solitary, not subtended by leaflike spathes; perfect spikelets several to many in each raceme   64. **Heteropogon**
    35 Racemes several in a flabellate cluster, subtended by leaflike spathes; perfect spikelet 1 in each raceme   65. **Themeda**
  34 Fertile spikelet without a callus, the rachis disarticulating immediately below the spikelet; awns usually short or absent:
    36 Pedicel of the sterile spikelet thickened and appressed or adnate to the adjacent thickened rachis-joint; spikelets awnless, appressed to the joint:
      37 Rachis-joint and pedicel adnate; annuals:
        38 Perfect spikelet globose; sterile spikelet conspicuous   66. **Hackelochloa**
        38 Perfect spikelet oblong; sterile spikelet minute   67. **Rottboellia**
      37 Rachis-joint and pedicel distinct, the sessile spikelet appressed to them, its lower glume lanceolate   68. **Hemarthria**
    36 Pedicel of the sterile spikelet distinct, this and the rachis-joint usually slender:
      39 Spikelets in reduced racemes of 1–5 (–7) joints, these peduncled in open panicles; awns, if present, commonly deciduous:
        40 Pedicelled spikelets staminate   69. **Sorghum**
        40 Pedicelled spikelets wanting, the pedicels only present   70. **Sorghastrum**
      39 Spikelets in racemes of several to many joints:
        41 Leaf-blades cordate; pedicelled spikelets obsolete or present only in the lower part of the delicate subdigitate racemes   71. **Arthraxon**
        41 Leaf-blades linear, not cordate; pedicelled spikelets present at least as rudiments:
          42 Inflorescence a large panicle of whorled long-peduncled slender glabrous racemes; spikelets muricate, awnless   72. **Vetiveria**
          42 Inflorescence-branches not whorled; racemes solitary, paired or digitate, sometimes supported by spathes; spikelets not muricate; racemes commonly conspicuously woolly   73. **Andropogon**

1. BAMBUSA Schreb. (1789) nom. cons.

1. B. vulgaris Schrad. ex Wendl., Coll. Pl. 2: 26, t. 47 (1810).—Common Bamboo.
Arborescent to 10 m. or more high, branched and arching above; mature culms 10–12 cm. in diameter.
Cultivated and naturalized, forming extensive groves; SL–3500 feet; fl. infrequently, Nov–Apr. *H 11709! J.P. 731!* Native of the Old World tropics, now very widespread; Grand Cayman.
Several ornamental bamboos of smaller size including *B. multiplex* (Lour.) Räusch. and *Melocanna bambusoides* Trin. have been introduced from Asia and are grown in gardens.

2. CHUSQUEA Kunth (1822)

1. C. abietifolia Griseb., Fl. Br. W.I.: 529 (1864).— Climbing Bamboo.
Extensively branched and forming dense tangles 3–7 m. high; branchlets whorled; leaf-blades rigid, 2–3 cm. long.
Locally abundant (St. Andr., Port., St. Thom.), in thickets and open woodlands; (2750–) 3500–7400 feet; fl. rare and gregarious, Nov–Apr (Seifriz, 1920). *H 12454! J.P. 928! P 4372!* Greater Antilles.

3. ANTHOXANTHUM L. (1753)

1. A. odoratum L., Sp. Pl. 1: 28 (1753).—Sweet Vernal-Grass.
Culms erect, tufted, 30–60 cm. high; panicles spikelike oblong or lanceolate (2–) 4–7 cm. long; spikelets 6–9 mm. long; plant fragrant of coumarin on drying.
Very local (St. Andr.), on well drained path and pathside banks; 4800–5000 feet; fl. Oct–Mar, June–July. *A 6446! H 9499! J.P. 742!* Native of temperate Europe and Asia; Haiti.

4. GYNERIUM Willd. ex Beauv. (1812)

1. G. sagittatum (Aubl.) Beauv., Ess. Nouv. Agrost.: 138, t. 26, f. 6 (1812).—Wild Cane.

Rhizomatous perennial; canelike culms solid growing 4–7 m. or more high in flower; leaf-blades sharply serrulate, up to 2 m. long and 6 cm. broad, flabellately disposed at the top of the culm; panicle 1 m. or more long with drooping branches.

Gregarious and locally abundant at swamp margins, riverbanks and in low-lying ground; 5–2500 feet; fl. July–Sept, Dec. *A 7721! H 11497! J.P. 787!* From Mexico through C. and S. tropical Amer.; West Indies.

5. ARUNDO L. (1753)

1. A. donax L., Sp. Pl. **1**: 81 (1753).—Giant Reed.

Culms stout from a branched robust tenacious rhizome, 5–7 m. high in flower, sparingly branched, often lax and toppling; leaves numerous, evenly spaced along the culm, conspicuously glaucous and distichous, blades up to 60 cm. long and 7 cm. broad, auriculate; panicle large, showy, up to 60 cm. long.

Locally abundant, gregarious along sheltered or open streambanks and river-banks; SL–3700 feet; fl. July–Nov. *A 7838! H 8279!* Native of the Old World, widely cultivated and naturalized.

6. PHRAGMITES Trin. (1820)

1. P. australis (Cav.) Trin. ex Steud., Nomencl. Bot. ed. 2, **2**: 324 (1841).— *Arundo australis* Cav. (1799). *P. communis* Trin. (1820). Reed.

Culms erect from stout creeping rhizomes, 2–4 m. tall; sometimes stoloniferous; leaves to 2 cm. or more wide, long-tapered, deciduous; panicle large, nodding, purplish.

Locally abundant, forming large colonies in slightly brackish swamps, rare away from the south coast; SL–100 feet; fl. Jan–May. *A 6322! H 11576! P 8351!* Widely distributed in temperate and tropical regions.

7. LUZIOLA Juss. (1789)

Leaf-blades 7–10 (–20) mm. broad; inflorescence many-flowered   1. *spruceana*
Leaf-blades 1–4 mm. broad; inflorescence few-flowered   2. *bahiensis*

1. L. spruceana Benth. ex Döll in Mart., Fl. Bras. 2 (2): 18 (1871).

Culms thick, soft and spongy, branched and widely spreading on the surface of water; leaf-blades flat, up to 50 cm. long; spikelets about 5 mm. long.

Very rare (St. Eliz.), at the edge of a small pond; 100 feet; fl. Oct. *P 21458!* Suriname to Brazil, Cuba, Hispaniola, Trinidad.

2. L. bahiensis (Steud.) Hitchc. in Contrib. U.S. Nat. Herb. **12**: 234 (1909).

Culms slender, trailing, with short leaves; erect flowering shoots with longer linear leaves; spikelets about 4 mm. long.

Very rare (St. Eliz.), in grassy swamp; 30 feet; fl. July. *A 11377!* Alabama to Brazil, Cuba, Hispaniola.

8. PHARUS P. Browne (1756)

1 Culms decumbent and rooting   1. *parvifolius*
1 Culms more or less erect:
   2 Lemma of pistillate spikelet pubescent all over   2. *glaber*
   2 Lemma of pistillate spikelet pubescent only at the tip   3. *latifolius*

1. P. parvifolius Nash in Bull. Torr. Bot. Club **35**: 301 (1908).

Leaf-blade narrowly elliptic-lanceolate to oblanceolate, usually about 3 cm. broad or less, rarely more; culms to 40 cm. in flower.

Very rare and local (St. Andr.), in shady woodland on limestone; 80 feet; fl. July–Aug, Nov. *A 8336!* Brazil, Greater Antilles, Trinidad.

2. P. glaber Kunth, Nov. Gen. **1**: 196 (1816).
Leaf-blade oblanceolate, acuminate, mostly 15–25 cm. long, 3–5 cm. broad; culms 50–75 cm. high in flower.
Common on well drained slopes in shady woodlands; 1500–3800 feet; fl. most of the year. *A 7265! H 11397! H & P 14369!* Mexico to Brazil, West Indies.

3. P. latifolius L., Syst. Nat. ed. 10, **2**: 1269 (1759).
Like the last with usually broader leaf-blades and longer fruits.
Occasional in wet shady woodlands on limestone and near streams; 400–1700 feet; fl. most of the year. *A 6101! H 12365! Patrick 292!* C. Amer. to Brazil, West Indies.

9. ORYZA L. (1753)

1. O. sativa L., Sp. Pl. **1**: 333 (1753).—Rice.
Erect annual; leaf-blades flat, up to 1·5 cm. broad; plant in flower up to 1 m. or more high with many-flowered compact panicles, the branches drooping in fruit; spikelets oblong, 7–10 mm. long.
Cultivated in seasonally standing water and casually escaping into swamps and ditches; 40–2000 feet; fl. Oct–Jan. *A 11892! H 12609! P 11104!* Wild variants occur naturally in S.E. Asia; cultivated varieties are grown in all warm countries where an adequate water regime exists.

10. LEERSIA Sw. (1788) nom. cons.

Panicle narrow, the branches ascending, spikelet-bearing from near the base; spikelets hispid, 3–3·5 mm. long   **1. *hexandra***
Panicle open, the branches slender, spreading, naked below; spikelets glabrous, 1·5–2 mm. long   **2. *monandra***

1. L. hexandra Sw., Nov. Gen. & Sp. Pl.: 21 (1788).
Rhizome firm, subterranean; culms up to 1 m. or more high; panicles many-flowered, purplish.
Locally common in rice fields, muddy ditches and roadsides; 30–900 feet; fl. Nov–Dec. *A 10164! H 12450! P 24604!* S. United States to Argentina, West Indies and sporadically through the tropics of the Old World.

2. L. monandra Sw., Nov. Gen. & Sp. Pl.: 21 (1788).
Densely tufted perennial with culms up to about 1 m. high; spikelets imbricate, pale; lemma boat-shaped.
Rare (St. Andr., St. Cath., Clar.), in rocky woodlands on limestone hills; 180–3500 feet; fl. Aug–Dec. *H 11326! J.P. 1525! Weaver 1242!* Florida to Brazil, Greater Antilles.

11. LEPTOTHRIUM Kunth (1829)

1. L. rigidum Kunth, Révis. Gram.: 156 (1829).
Perennial with shortly creeping rhizome; culms tufted, 25–40 cm. high; leaf-blades 1–3 cm. long; spikes 5–10 cm. long; spikelets about 8 mm. long.
Very local (St. Andr., St. Cath.), restricted to loose open seashore sand and sandy thickets, forming colonies; 2–20 feet; fl. Oct–Apr, July. *A 8967! H 11357! J.P. 775!* Venezuela, Colombia.

12. TRAGUS Haller (1768) nom. cons.

1. T. berteronianus Schult., Mant. **2**: 205 (1824).
Annual, branched at the base, with spreading ascending culms 6–30 cm. high; leaf-blades up to 5 cm. long, 2–4 mm. broad with cartilaginous margins; racemes 4–10 cm. long; burs 2–3 mm. long.

Local (St. Andr., St. Cath., St. Thom.), on shingle dunes and open shallow calcareous soil in arid areas; SL–600 feet; fl. Aug–Apr. *A 6945! H 11477! P 9288!* S.W. United States to Brazil, West Indies.

13. SPARTINA Schreb. (1789)

1. S. patens (Ait.) Muhl., Descr. Gram.: 55 (1817).
Rhizomatous perennial; leaf-blades involute, 30 cm. or more long, tapering to a fine point.
Local along the west, south and east coasts forming colonies in the upper parts of unstabilized beaches; fl. very rare. *Robbins 2955!* S.E. United States, Bermuda, Bahamas, Greater Antilles to Antigua and Guadeloupe; Little Cayman.

14. BOTELUA Lag. (1805)

1. B. americana (L.) Scribn. in Proc. Acad. Nat. Sci. Philad. **1891**: 306 (1891).
Perennial with mostly prostrate-spreading branched culms, 10–30 cm. or more long; leaf-blades flat or involute, up to 7 cm. long, 3 mm. broad; spikes several, 2–2·5 cm. long; lemma 3-awned.
A weed of poor lawns and arid waste places, locally common; SL–3800 feet; fl. Oct–Mar. *A 6581! H 11369! P 16011!* C. Amer. to Brazil, West Indies.

15. LEPTOCHLOA Beauv. (1812)

1 Plants annual:
  2 Sheaths, at least the upper, papillose-hispid; glumes nearly as long as the 1–2 mm. long
    spikelets; lemmas awnless   1. *filiformis*
  2 Sheaths glabrous or at most scabrous:
    3 Lemmas acuminate, awned, the awn sometimes short; spikelets 7–10 mm. long
                 2. *fascicularis*
    3 Lemmas truncate, awnless; spikelets 5–7 mm. long   3. *uninervia*
1 Plants perennial:
  4 Sheaths and blades glabrous, often glaucous; lemmas mostly shortly awned   4. *virgata*
  4 Sheaths sparsely papillose-hispid; leaf-blades sparsely villous on the upper surface
    near the base; lemmas mostly long-awned   5. *domingensis*

1. L. filiformis (Lam.) Beauv., Ess. Nouv. Agrost.: 166 (1812).
Culms tufted, erect or branched and geniculate below, 20–40 (–70) cm. high in flower; leaf-blades flat, thin; panicle of numerous slender racemes 5–15 cm. long, spread along an axis 20–30 cm. long, pyramidal when fully open.
Locally common (St. Andr., Clar., St. Thom.), a weed of disturbed ground, plantations and pasture margins; SL–700 feet; fl. Oct–June. *A 6916! H 11238! Jordine 1!* Virginia to California to S. Amer., West Indies.

2. L. fascicularis (Lam.) A. Gray, Man.: 588 (1848).—*Diplachne fascicularis* (Lam.) Beauv. (1812).
Annual, perhaps occasionally perennial, with culms up to 1 m. high in flower; leaf-blades flat; panicle rather compact, up to 20 cm. long, often much smaller; lemmas 4–5 mm. long with the awn short or as long.
Local and rather uncommon in low-lying sometimes brackish situations; SL–50 feet; fl. Dec–Jan, May–July. *A 12058! H 12498! P 15359! 27577!* United States, Mexico, Bahamas, Greater Antilles.

3. L. uninervia (Presl) Hitchc. & Chase in Contrib. U.S. Nat. Herb. **18**: 383 (1917).
Culms tufted, purplish, more or less erect to 60 cm. high in flower; panicle rather dense; spikelets greyish-purple.
Local (St. Cath., Clar.), in salina margins and on ditchbanks; SL–40 feet; fl. Nov–July. *A 6457! H 12309!* S.W. United States to Argentina and Chile, Cu ba.

4. L. virgata (L.) Beauv., Ess. Nouv. Agrost.: 166, t. 15, f. 1 (1812).

Culms tufted, variable in height when in flower, up to 1·5 m.; panicle lax, the lower racemes often distant; spikelets usually dark purple; lemmas 1·5–2 mm. long.

Rather common as a weed of roadsides, waste places and pastures; SL–1000 feet; fl. most of the year. *A 6915! H 12536! A. M. Barry!* Mexico to S. Amer., West Indies.

5. L. domingensis (Jacq.) Trin., Fund. Agrost.: 133 (1820).

Like the last; culms tufted; panicles more elongated with shorter more numerous racemes.

Rather common in old pastures and as a weed of cultivated ground; SL–700 (–3600) feet; fl. Aug–Jan, Apr–May. *A 5891! H 11261! H & P 13766!* Florida, Mexico, West Indies.

16. CYNODON L. C. Rich. (1805) nom. cons.

1. C. dactylon (L.) Pers., Synops. Pl. **1**: 85 (1805).—Bermuda Grass, Bahama Grass.

Low perennial, stoloniferous or rhizomatous according to the condition of the substrate; flowering shoots 10–30 cm. high, rarely more; leaf-blade 2–3 (–5) cm. long, 1·5–2·5 mm. broad; spikes 3–5, 2·5–5 cm. long; spikelets 2–2·5 mm. long.

Commonly cultivated or encouraged as a lawn grass particularly in the drier areas, also a weed of roadsides, pastures and waste places; SL–4000 feet; fl. during wet weather, Apr–Nov. *A 7352! H 11248! P 10300!* Widespread near habitations in warm countries.

17. CHLORIS Sw. (1788)

1 Lemmas awnless; fruiting florets dark brown to black   1. *petraea*
1 Lemmas awned:
  2 Sterile floret narrow, the apex acute or obtuse but not truncate:
    3 Plants 1–1·5 m. high, perennial with stout creeping stolons; cultivated; native of tropical Africa; Rhodes Grass   *C. gayana* Kunth
    3 Plants not so tall; stolons not or poorly developed:
      4 Spikes usually 10 or more, erect or ascending   2. *radiata*
      4 Spikes usually 6 or fewer:
        5 Culms leafy throughout; spikes erect or ascending; short-lived therophyte 3. *mollis*
        5 Culms leafy most towards the base; spikes often reflexed at maturity; mostly perennial:
          6 Spikelets divergent, remote; small delicate plants   4. *cruciata*
          6 Spikelets appressed, imbricate:
            7 Culms slender, wiry; leaf-blades filiform; awn of fertile lemma 7–8 (–10) mm. long; callus pubescent   5. *ekmanii*
            7 Culms stouter; leaf-blades flat, 1–2 mm. broad; awns 1–2 cm. long:
              8 Lemma pilose-ciliate on upper part of margins; callus pilose; rudimentary floret prominent, about 1·5 mm. long   6. *cubensis*
              8 Lemma minutely pilose or nearly glabrous on the margins; callus puberulous; rudimentary floret inconspicuous   7. *sagraeana*
  2 Sterile floret(s) broad, truncate, broadest at the top:
    9 Plants perennial; culms up to 1·2 m. high; spikes flexuous, 8–10 cm. long   8. *dandyana*
    9 Plants annual, usually less than 75 cm. high; spikes less than 8 cm. long; sterile florets 2 (–3):
      10 Awn of fertile lemma not longer than the body; lemma long-ciliate on keel and margins   9. *ciliata*
      10 Awn of fertile lemma at least twice as long as the body; lemma with a few hairs on either side of the keel, margin ciliate, black at maturity   10. *barbata*

1. C. petraea Sw., Nov. Gen. & Sp. Pl.: 25 (1788).

Glabrous glaucous sparingly stoloniferous perennial; culms flattened up to 60 cm. high in flower; sheaths keeled; leaf-blades obtuse; spikes usually 5–6; spikelets 2 mm. long.

Common, especially near the sea, in open limestone pastures and on exposed coral; SL–2300 feet; fl. all the year. *A 6134! H 12464! P 7529!* S. United States, Mexico, Panama, Bermuda, West Indies; Cayman Is.

2. C. radiata (L.) Sw., Nov. Gen. & Sp. Pl.: 26 (1788).

Annual; culms decumbent and rooting; sheaths broad, compressed; leaf-blades light blue-green, obtuse.

A common weed of roadsides, ditches and waste ground; SL–3000 feet; fl. all the year. *A 5579! H 11492! Patrick 244!* Mexico to Paraguay, West Indies, introduced elsewhere.

3. C. mollis (Nees) Swallen in N. Amer. Fl. **17**: 596 (1939).—*Gymnopogon mollis* Nees (1829). *C. rupestris* (Ridley) Hitchc. (1936).

Annual; culms 30–100 cm. high in flower; leaf-blades flat, acuminate; spikes 5–6 (–8); lemma about 5 mm. long the margins shortly ciliate towards the tip.

Rather local, mostly on limestone gravel and broken coral in the more arid parts of the south coast and on some of the cays; SL–1000 feet; fl. Oct–Feb. *A 10003! H 12675!* Ecuador to Brazil, Greater Antilles except Puerto Rico, Curaçao.

4. C. cruciata (L.) Sw., Nov. Gen. & Sp. Pl.: 25 (1788).

Perennial except in the drier habitats where it behaves as an annual; often very small, the culms 6–15 (–30) cm. high in flower; leaf-blades at most 5 cm. long; racemes usually 2–4, delicate, up to 3 cm. long; spikelets distinctly pedicelled; awns 5–10 mm. long.

Locally common on cleared stony mostly shale roadside banks; 300–4500 feet; fl. Oct–Apr, July–Aug. *A 5604! H 12647! J.P. 841!* Cuba, Hispaniola.

(*Purdie* (K!) cited by Griseb. (1864) and *H 12588* (UCWI!) cited by Hitchc. (1936) belong to the next species).

5. C. ekmanii Hitchc., Man. Grasses W.I.: 129 (1936).

Perennial; culms tufted, slender, geniculate, occasionally rooting from the lower nodes, 10–30 cm. high in flower; spikelets loosely overlapping in spikes 2·5–10 cm. long.

Occasional in red bauxitic soils and amongst limestone rocks; 300–2600 feet; fl. Aug–Apr. *A 8932! H 12727! Patrick 266!* Cuba, Hispaniola.

6. C. cubensis Hitchc. & Ekman in Hitchc., Man. Grasses W.I.: 131 (1936).

Perennial, sometimes stoloniferous, some of the culms decumbent and rooting; culms distinctly flattened, up to 50 cm. or more high in flower; racemes 5–7 cm. long; spikelets closely imbricate.

Very local (Clar., Han.), on coral limestone near the sea; SL–20 feet; fl. Sept–Dec. *A 7964! H 12462!* Cuba, Tortuga Is. (Haiti), Antigua.

7. C. sagraeana A. Rich. in Sagra, Hist. Cuba, Parte 2, **11**: 315 (1850).

Perennial; culms tufted, slender, geniculate, 20–60 cm. high in flower; spikes mostly 3–5 cm. long; awn flexuous, about 1 cm. long.

Very rare (Clar., West.), in open arid areas; SL–100 feet; fl. Dec. *H 12738!* Bahamas, Cuba, Hispaniola, Antigua, Guadeloupe, Barbados.

8. C. dandyana Adams in Phytologia **21** (6): 408 (1971).—*C. polydactyla* Sw. (1788), illegitimate name.

Culms rather robust; leaf-blades up to about 1 cm. broad; spikes 5–10; spikelets closely imbricate, silky.

Rather local, more frequent in the western parishes on open slopes, swamp margins and roadsides; SL–1400 feet; fl. Aug–Feb. *A 8052! H 12743! J.P. 929!* Florida to Brazil, West Indies; Grand Cayman.

9. C. ciliata Sw., Nov. Gen. & Sp. Pl.: 25 (1788).

Culms erect or spreading; leaf-blades flat, up to 5 mm. broad; spikes 4–6, flexuous; florets conspicuously silky-ciliate; sterile lemmas strongly green-nerved, purplish distally.

Occasional in low-lying open places and swamp margins; SL–3000 feet; fl. June, Sept–Dec. *A 7258! H 12689!* Texas to Argentina, West Indies.

10. C. barbata Sw., Fl. Ind. Occ. **1**: 200 (1797).—*C. inflata* Link (1821).

Culms tufted; leaf-blades rather long and lax; spikes about 10, 4–6 cm. long, usually flexuous and often purplish; rachilla with a tuft of hairs above the lower glume (below fertile lemma).

Very common as a weed along roadsides and in waste places; SL–500 (–1000) feet; fl. all the year. *A 6267! H 9049! P 11326!* Pantropical; Grand Cayman.

18. DACTYLOCTENIUM Willd. (1809)

1. D. aegyptium (L.) Beauv., Ess. Nouv. Agrost., Expl. Planch.: 10, t. 15, f. 2 (1812).

Culms spreading and rooting at the nodes, repeatedly branching, the flowering tips prostrate or erect to 20 cm. high; leaf-blades flat, ciliate; spikes (1–) 3–4 (–several), 1–3 (–5) cm. long; upper glume mucronate or awned below the tip.

Occasional in open sandy or gravelly waste places and shingle near the sea, a weed; SL–600 (–1250) feet; fl. Oct–June. *A 6043! H 11510! J.P. 870!* Widespread in the warmer parts of the world; Grand Cayman.

19. ELEUSINE Gaertn. (1788)

1. E. indica (L.) Gaertn., Fruct. & Sem. Pl. **1**: 8 (1788).—Yard Grass.

Annual or perennial; culms prostrate or erect, very smooth and tough, dark green, up to 30 cm. or more high in flower, often smaller; leaves keeled; spikes (1–) 2–5 (–6), sometimes with 1 subterminal, 4–17 cm. long; spikelets 2–6-flowered; lemmas about 4 mm. long.

A very common weed especially around habitations and in lawns; SL–3600 feet; fl. all the year. *A 8682! J.P. 753! P 7854!* In all warm countries; Grand Cayman.

20. ARISTIDA L. (1753)

1 Plants annual; awns straight at the base, 10–15 mm. long	1. *adscensionis*
1 Plants perennial:	
2 Awns straight at the base, 15–25 mm. long	2. *swartziana*
2 Awns spirally contorted at the base, about 1 cm. long	3. *refracta*

1. A. adscensionis L., Sp. Pl. **1**: 82 (1753).

Culms usually densely tufted, somewhat geniculate at the base, very variable in number and height; panicles compact, erect or flexuous in larger plants.

Very local (St. Andr., St. Cath.), in open sandy or gravelly soils in arid areas; SL–3000 feet; fl. Nov–Jan. *A 9813! H 11500! P 11434!* Widely distributed in the drier parts of the tropics.

2. A. swartziana Steud., Synops. Pl. Glum. **1**: 137 (1854).— *A. purpurascens* of Griseb. (1864), not Poir.

Culms tufted, 40–70 cm. high; leaf-blades involute; panicles narrow, loose and interrupted, up to 15 cm. long.

Rather rare (St. Andr., St. Cath., Manch.), in dry open places; SL–800 feet; fl. Nov–Jan. *H 12440! J.P. 864! P 21865!* Hispaniola, Antigua especially Barbuda.

3. A. refracta Griseb., Cat. Pl. Cub.: 228 (1866).

Culms usually densely tufted with numerous innovations, erect, slender, 20–60 cm. high; leaf-blades filiform-involute, mostly less than 10 cm. long; panicles narrow, rather lax, 5–15 cm. long.

Local (Clar., Manch.), in clearings in arid woodlands and thickets; 100–800 feet; fl. June, Nov–Dec. *H 12734! P 23609!* Greater Antilles.

21. SPOROBOLUS R. Br. (1810); Clayton (1965)

1 Plants annual; glumes unequal:
  2 Panicles open with very fine hairlike filiform branches; spikelets about 1 mm. long; culms erect, few   1. *tenuissimus*
  2 Panicles compact or pyramidal; spikelets 1·5 mm. long; culms numerous, tufted, suberect or spreading   5. *pyramidatus*

1 Plants perennial; panicles mostly compact or opening diurnally:
  3 Rhizome creeping; panicles spikelike, pale; leaves conspicuouly distichous
                                                              2. *virginicus*
  3 Rhizome very short or absent; leaves not conspicuouly distichous:
    4 Glumes both shorter than the spikelet:
      5 Upper glume obtuse; basal sheaths scarcely compressed; panicle usually rather lax
                                          3. *jacquemontii*
      5 Upper glume acute or cuspidate, sometimes ragged; basal sheaths compressed; panicle usually dense
                                             4. *indicus*
    4 Glumes unequal, the upper as long as the spikelet:
      6 Spikelets 1·5 mm. long; panicle pyramidal   5. *pyramidatus*
      6 Spikelets 2–3 mm. long; panicle elongate-oblong:
        7 Panicle light green or whitish; spikelets about 2 mm. long   6. *domingensis*
        7 Panicle purple; spikelets 3 mm. long   7. *purpurascens*

1. S. tenuissimus (Schrank) Kuntze, Revis. Gen. Pl. **3** (2): 369 (1898).
    Culms tufted, slender, 20–35 (–50) cm. high in flower; spikelets glabrous, about 1 mm. long; lower glume about one quarter and upper glume about half the length of the spikelet.
    Local (St. Andr.), a weed of trampled waste places and lawns on light soil; 600–2000 feet; fl. most of the year. *A 9432! Skelding UCWI 5465!* Hispaniola, Puerto Rico, Lesser Antilles to Brazil, Old World tropics.

2. S. virginicus (L.) Kunth, Révis. Gram.: 67 (1829).
    Flowering culms up to 40 cm. or more, usually shorter; rhizomes hard and scaly; leaf-blades firm, often involuted, sharply pointed, very variable in length and breadth up to 15 cm. long and 5 mm. broad, usually much smaller; spikelets about 2 mm. long; lower glume a little more than half as long as the spikelet, the upper as long as the spikelet.
    Abundant and gregarious, sometimes forming continuous sward along sandy shores and mangrove margins, extending sometimes into neighbouring pastures; SL–20 feet; fl. sparsely and sporadically throughout the year. *A 7358! H 11388! J.P. 774!* All tropical shores; Cayman Is.
    The largest variant of this grass has been distinguished as *S. littoralis* (Lam.) Kunth, but most modern authors regard this merely as an extreme in an intergrading series. Plants identified as *S. littoralis* occur along the Portland coast (*Asprey UCWI 2964! A 7589!*).

3. S. jacquemontii Kunth, Révis. Gram.: 427, t. 127 (1831); W. D. Clayton in Kew Bull. **19** (2): 287–293 (1965).— *S. indicus* of Hitchc. (1936) and other authors, not (L.) R. Br.
    Culms tufted, erect, 50–100 cm. high in flower; spikelets 1·5–1·8 mm. long; lower glume about one-quarter to one-fifth as long as the spikelet, obtuse; upper glume about 0·6 mm. long; grain obovoid-ellipsoid.
    Very common in pastures, waste places and along roadsides; SL–4800 feet; fl. all the year. *A 6912! H 12521! Patrick 122!* S.E. United States to Brazil, West Indies, tropical Africa; Grand Cayman.

4. S. indicus (L.) R. Br., Prodr. Fl. Nov. Holl.: 170 (1810).—*Agrostis indica* L. (1753). *S. poiretii* (Roem. & Schult.) Hitchc. (1932).
    Culms tufted, 40–70 cm. high in flower; spikelets 1·7–2 mm. long; lower glume truncate, about one-third as long as the spikelet; upper glume about 0·9 mm. long; grain oblong.
    Locally common on open banks and along roadsides; 1250–5000 feet; fl. Nov–Mar, June–Aug. *A 5709! H 12486! P 5986!* S.E. United States to Paraguay, West Indies.

5. S. pyramidatus (Lam.) Hitchc., Man. Grasses W.I.: 84 (1936).
    Culms tufted, numerous, often spreading, sometimes prostrate, 10–15 (–25) cm. high; leaf-blades rather thick, about 5 cm. long; ligule conspicuous; spikelets 1·5 mm. long; lower glume about half as long as the spikelet, upper as long as the spikelet.
    Locally common, in the drier southern coastal areas in salina margins and sandy waste places near the sea; SL–240 feet; fl. Sept–Feb, May–July. *A 6068! H 12463! Patrick 168!* S. United States to Argentina, West Indies. A variant with the glumes

unequal and both much shorter than the spikelet, the upper glume being about two-thirds as long as the spikelet, occurs in waste ground near the sea (St. Andr., St. Thom.); SL–20 feet; fl. Nov–Jan, June. *A 6183! 7355!*

6. S. domingensis (Trin.) Kunth, Révis. Gram.: 427 (1831).
Like the last but generally larger; culms 40–80 cm. high in flower; spikelets up to 2 mm. long.
Local around the coasts of the central and western parishes in salinas and on coral limestone; SL–20 feet; fl. Aug–Feb. *A 7965! Patrick 304!* Florida, Bahamas, Greater Antilles.

7. S. purpurascens (Sw.) Ham., Prodr. Pl. Ind. Occ.: 5 (1825).
Culms 25–50 cm. high in flower; leaves mostly basal with blades up to 8 mm. broad, ciliate near the base; spikelets usually brownish-purple, sometimes green, 3 mm. long.
Rather local (St. Andr., St. Thom., Clar.), in open grassy savannas and on pathside banks; 2000–5500 feet; fl. Nov–Jan, June–Aug. *A 7329! H 11434! J.P. 735!* S. United States to Brazil, Cuba, Martinique.

22. AGROSTIS L. (1753)

1 Palea present:
  2 Panicle contracted, the branches flower-bearing nearly to the base; glumes scabrous
    on the back and keels   1. *semiverticillata*
  2 Panicle open, the branches naked at the base; glumes scabrous only on the keels
                                                   2. *stolonifera*
1 Palea wanting; panicle spreading at least in flower; glumes smooth   3. *canina*

1. A. semiverticillata (Forsk.) C. Chr. in Dansk. Bot. Arkiv **4** (3): 12 (1922).
—*Phalaris semiverticillata* Forsk. (1775). *A. verticillata* Vill. (1779). *Polypogon subverticillatus* (Forsk.) Hyl. (1945).
Culms decumbent or spreading, 20–100 cm. long; spikelets about 2 mm. long, falling entire.
Very rare (St. Andr.), in ditches and moist places; 4000 feet; fl. Sept. *H 12895!* Native of the Mediterranean region and subtropical Asia; widely introduced from S. United States to Argentina.

2. A. stolonifera L., Sp. Pl. **1**: 62 (1753).—*A. alba* of Hitchc. (1936), not L.
Culms erect or ascending from a decumbent or even rhizomatous creeping base, 30–60 cm. or more high; spikelets 2–2·5 mm. long, purplish.
Rare and local (St. Andr., St. Thom.); 5000–7400 feet; fl. Sept–Jan, June. *A 11274! H 12664 J.P. 927!* Introduced from Europe; Hispaniola.

3. A. canina L., Sp. Pl. **1**: 62 (1753).
Culms decumbent, up to about 60 cm. high in flower; spikelets 2 mm. long; lemma usually awned from about the middle, awn about 3 mm. long.
Very rare (St. Andr., St. Thom.); 4900–7000 feet; fl. Oct. *H 12667! J.P. 764!* Introduced from Europe; Hispaniola.

23. DANTHONIA DC. (1805) nom. cons.

1. D. domingensis Hack. & Pilg. in Urb., Symb. Ant. **6**: 1 (1909).
Perennial forming large dense tussocks; culms erect to 60 cm. high; leaf-blades long and often overtopping the panicles; spikelets 12–15 mm. long.
Very local (St. Andr.), on cleared slopes; 6000 feet; fl. Jan–June. *H 10914! 11629!* Hispaniola.

24. ZEUGITES P. Browne (1756) nom. cons.

1. Z. americana Willd. in L., Sp. Pl. ed 4, **4**: 204 (1805).
Loosely tufted perennial; culms slender, glossy brown or black, somewhat straggling to 1 m. high; leaf-blades ovate, 2·5–4 cm. long; panicles loosely flowered on capillary pedicels.

Locally common (St. Andr., Port., St. Thom.), on pathside banks and in glades in mountain woodland; 2750–7400 feet; fl. most of the year. *A 10485! H 11418! Patrick 161!* Cuba, Hisapniola.

25. DACTYLIS L. (1753)

1. D. glomerata L., Sp. Pl. **1**: 71 (1753).

Coarse perennial with culms to about 1 m. high in flower; spikelets 5–7 mm. long, secund in tight clusters; lemma lanceolate, ciliate on the keel, shortly awned.

Very rare (St. Andr.); 5000 feet; fl. June. *H 11779! J.P. 744!* Native of Europe, introduced into the United States.

26. ERAGROSTIS Wolf (1776)

1 Palea conspicuously ciliate on the reflexed margins:
  2 Plants perennial, more or less viscid   1. *glutinosa*
  2 Plants annual, not viscid:
    3 Pedicels as long as the spikelets or longer; panicle open, oblong; spikelets 4–8-flowered   2. *tenella*
    3 Pedicels very short; panicle more or less interrupted; spikelets 7–9-flowered   3. *ciliaris*
1 Palea scaberulous, not conspicuously ciliate:
  4 Plants annual; upper margin of sheath pilose:
    5 Plants stoloniferous, forming mats   4. *hypnoides*
    5 Plants not stoloniferous:
      3 Spikelets 2 mm. broad, 6–12-flowered; lower axils of panicle pilose   7. *tephrosanthos*
      6 Spikelets 1–1·5 mm. broad:
        7 At least lower axils of panicle-branches long-pilose; leaf-blade and sheath except upper margin glabrous; spikelets 3–9-flowered   5. *pilosa*
        7 Axils of panicle glabrous; leaf-blade and sheath thinly pilose; spikelets (9–) 12–17-flowered; glands present below culm-nodes   6. *barrelieri*
  4 Plants perennial:
    8 Culms rather short up to 20 (–30) cm. high, stiff and wiry; panicle narrow; spikelets 10–20-flowered; lemmas about 1·8 mm. long   8. *cubensis*
    8 Culms taller, 50 cm. or more high:
      9 Panicle-branches diffuse, capillary; spikelets 8–15-flowered; lemmas about 2 mm. long   9. *elliottii*
      9 Panicle-branches ascending or appressed; spikelets 10–25-flowered; lemmas 1·5–1·8 mm. long   10. *domingensis*

1. E. glutinosa (Sw.) Trin. in Mém. Acad. Sci. St. Pétersb. sér. 6, **1**: 397 (1830).

Culms wiry, branching, 10–30 cm. high; leaf-blades short, slender, involute; panicles open, the branches stiffly ascending; spikelets 4–6 mm. long, mostly 7–11-flowered.

Only once collected. *Swartz.* Cuba.

2. E. tenella (L.) Beauv. ex Roem. & Schult. in L., Syst. Veg. ed. nov. **2**: 576 (1817).—*E. amabilis* (L.) Wight & Arn. (1838).

Culms slender, branching, 10–20 cm. high; leaf-blades 1–5 mm. broad; panicles 4–8 cm. long; spikelets about 2 mm. long, often purplish.

Locally common (St. Andr.), as a weed of lawns, thin open pasture and waste ground; 150–700 feet; fl. Sept–Mar. *A 6587! H 11243! P 16122!* Pantropical; Cayman Brac.

3. E. ciliaris (L.) R. Br. in Tuckey, Narr. Exped. Riv. Zaire: 478 (1818).

Culms tufted, 15–30 cm. high; panicle mostly compact, sometimes loose; spikelets subsessile, crowded, about 2 mm. long.

Common as a weed of waste places and roadsides; SL–5000 feet; fl. most of the year. *A 5980! H 11335! Patrick 183!* Pantropical; Grand Cayman.

4. E. hypnoides (Lam.) Britton, Sterns & Poggenburg, Prel. Cat. New York: 69 (1888).

Culms mostly 5–10 cm. high, spreading; leaf-blades 1–2 cm. long, spreading, puberulous adaxially; panicles compact, 1–2 cm. long; spikelets 2·5–7(–10) mm. long.

Very rare (St. Eliz.), in mud beside stream; 5–10 feet; fl. Apr. *P 26324!* United States to Brazil, West Indies.

5. E. pilosa (L.) Beauv., Ess. Nouv. Agrost.: 162 (1812).
Culms tufted, slender, 15–45 cm. high; panicle diffuse with delicate branches; spikelets 3–5 mm. long, greyish.
Occasional in roadside ditches and damp open ground; SL–3800 feet; fl. most of the year. *A 5982! H 11374! P 18341!* Widely distributed in warm countries.

6. E. barrelieri Daveau in Morot, Journ. Bot. Paris **8**: 289 (1894).
Culms decumbent or erect to 20 cm. high; leaf-sheaths variably hairy; spikelets 7–13 mm. long.
Rather rare (St. Andr.), a weed of disturbed open ground; 20–550 feet; fl. July–Aug, Nov. *A 8302! P 9355! Sternberg!* S. Europe, N. Africa, S. United States, Hispaniola.

7. E. tephrosanthos Schult., Mant. **2**: 316 (1824).
Culms erect or spreading, 5–20 cm. or more high; leaf-blades very narrow; spikelets 4–7 mm. long.
Occasional along sandy roadsides; 100–2000 feet; fl. Oct–Apr. *A 6725! H 11403! P 10131!* West Indies to Brazil.

8. E. cubensis Hitchc. in Contrib. U.S. Nat. Herb. **12**: 243 (1909).
Culms tufted from a hard knotty base; leaf-blades narrow or filiform; spikelets linear, mostly 7–12 mm. long, shortly pedicelled.
Very local (Manch.), in grassy savanna; 300–900 feet; fl. July, Oct–Nov. *H 11664! 12678! P 27532!* Cuba.

9. E. elliottii S. Watson in Proc. Amer. Acad. **25**: 140 (1890).
Culms closely tufted, numerous, erect (30–) 50 cm. high in flower; leaves stiff, erect; spikelets 5–12 mm. long on long stiff slender pedicels.
Local (Manch., St. Eliz.), in open low-lying pastures and ditches; 40–800 feet; fl. Oct–Feb. *A 11999! H 11661! 12435! P 24579!* S.E. United States, Bahamas, Greater Antilles.

10. E. domingensis (Pers.) Steud., Synops. Pl. Glum. **1**: 278 (1854.)
Culms stout, erect, 0·5–1·5 m. high; spikelets shortly pedicelled, 5–10 mm. long.
Local (St. Cath., Clar.), in low-lying probably saline boggy pastures near the sea; SL–20 feet; fl. May–June, Oct–Dec. *A 9896! H 12518!* British Honduras, Cuba, Hispaniola, Guadeloupe, Martinique. This species is probably not really distinct from *E. prolifera* (Sw.) Steud.

27. LEPTOCHLOOPSIS Yates (1966)

1. L. virgata (Poir.) Yates in South-Western Nat. **11** (3): 384 (1966).—*Uniola virgata* (Poir.) Griseb. (1864).
Culms numerous forming large clumps, 1–2 m. high in flower; leaf-blades up to 1 m. long, tapered to a very fine tip, flexuous and curled, coarsely setulose, light green; panicle-branches ascending, up to 4 cm. long; spikelets 3–5 mm. long.
Rather uncommon in southern parishes, in thickets on rocks and rubble in arid limestone areas; SL–400 feet; fl. Oct–Jan, May–Aug. *A 12866! H 11312! P 7496!* Bahamas, Greater Antilles, Virgin Is.

28. BRIZA L. (1753)

Spikelets rather few, 1 cm. broad   1. *maxima*
Spikelets numerous, 5 mm. broad   2. *minor*

1. B. maxima L., Sp. Pl. **1**: 70 (1753).
Culms slender, 30–90 cm. in flower; spikelets maturing light green-yellow, nodding.

Very local (St. Andr.), a weed of pathsides; 3500–5000 feet; fl. May–July, Oct. *A 7399! H 11274! P 23758!* Native of temperate parts of the Old World, Bermuda, an occasional introduction in Amer.

2. B. minor L., Sp. Pl. **1**: 70 (1753).
Culms weak, 20–50 cm. high in flower; panicle-branches spreading; spikelets nodding.
Locally common (St. Andr., Port., St. Thom.), along roadsides and in sheltered moist waste places; 3300–7400 feet; fl. sporadically throughout the year. *A 7400! H 11414! P 23746!* Cosmopolitan, but not frequent in Amer.

29. POA L. (1753)

1 Plants annual; culms tufted	1. *annua*
1 Plants perennial with creeping rhizomes:	
2 Culms distinctly flattened	2. *compressa*
2 Culms more or less terete	3. *pratensis*

1. P. annua L., Sp. Pl. **1**: 68 (1753).
Culms soft, smooth, decumbent at the base, up to 30 cm. high in flower; leaf-blades often transversely wrinkled; panicle open, somewhat triangular in outline; spikelets lanceolate, 3–5 mm. long.
Locally common (St. Andr., Port., St. Thom.), a weed of paths and roadsides; 3600–6500 feet; fl. Oct–Feb, July. *A 10668! H 10920! P 10176!* Cosmopolitan.

2. P. compressa L., Sp. Pl. **1**: 69 (1753).
Culms slender, stiff and wiry, 20–40 cm. high; panicle erect, narrow and compact, rather densely flowered; spikelets ovate-oblong, 3–7 mm. long.
Very rare (St. Thom.); 7400 feet; fl. Oct. *H 12663!* United States, introduced from Europe, Haiti.

3. P. pratensis L., Sp. Pl. **1**: 67 (1753).
Culms 30–50 cm. or more high; panicle open; spikelets ovate, 4–6 mm. long.
Very rare (St. Thom.); 7000–7400 feet. *Orcutt 5311.* Native of Europe, introduced into United States (Kentucky Bluegrass); Haiti.

30. FESTUCA L. (1753)

1 Plants annual; lemmas long-awned:	
2 Spikelets 4–6-flowered; lower glume about 4 mm. long, upper 6–7 mm. long	
	1. *bromoides*
2 Spikelets 2–3-flowered; lower glume 1–2 mm. long, upper 4–4·5 mm. long	
	2. *myuros*
1 Plants perennial; introduced from Europe and cultivated but not recently found:	
3 Lemmas awned or mucronate; Sheep's Fescue	*F. ovina* L.
3 Lemmas awnless; Meadow Fescue	*F. pratensis* Huds.

1. F. bromoides L., Sp. Pl. **1**: 75 (1753).—*F. dertonensis* (All.) Aschers. & Graebn. (1900).
Culms slender, tufted, (10–) 30–50 cm. high in flower; panicles narrow, nodding; awns 10–13 mm. long.
Rather local (St. Andr., Port., St. Thom.), on stony well drained banks; 3000–5000 feet; fl. Oct–Mar, July. *A 6680! H 9504! J.P. 750!* W. United States, introduced from Europe.

2. F. myuros L., Sp. Pl. **1**: 74 (1753).
Like the last with weaker culms and narrower panicles; awns 7–10 mm. long.
Local (St. Andr., St. Thom.), on stony banks; 3000–5000 feet; fl. Nov–Mar, July. *A 10497! H 11564!* United States, Hispaniola, introduced from Europe.

31. BROMUS L. (1753)

Awn of lemma short or none; leaf-sheaths glabrous or pubescent	1. *unioloides*
Awn of lemma (2–) 3–4 (–5) cm. long; leaf-sheaths pilose	2. *diandrus*

1. B. unioloides Kunth, Nov. Gen. **1**: 151 (1816).—*B. catharticus* of Hitchc. (1936), not Vahl (1791). Rescue Grass.

Culms up to 1 m. high in flower; panicle open, up to 20 cm. long; lemmas 9–13 mm. long, 7–9-nerved, uniformly scabrous, green or purple-tipped.

Local (St. Andr.), a weed of roadsides and waste places; (700–) 3000–5000 feet; fl. Oct.–Jan, July. *A 7655! H 10919! J.P. 741!* Native of Europe, S. and W. United States, throughout temperate S. Amer., Bermuda, Hispaniola, Rhodesia.

2. B. diandrus Roth, Bot. Abh. Beobacht.: 44 (1787).—*B. rigidus* of Hitchc. (1936), not Roth (1790).

Culms 40–100 cm. high in flower; panicle open, nodding, 7–20 cm. long lemmas 2–3 cm. long.

Local (St. Andr., St. Thom.), a weed of stony waste places; 3500–5000 feet; fl. May–July, Oct–Dec. *A 7698! H 10916! J.P. 738!* Native of Europe; Hispaniola.

32. ARUNDINELLA Raddi (1823)

1. A. confinis (Schult.) Hitchc. & Chase in Contrib. U.S. Nat. Herb. **18**: 290 (1917).

Culms strong, slender, geniculate, up to 2·5 m. high in flower; leaf-blades scabrous on the upper surface; panicles densely flowered and rather compact, 20–40 cm. long; fertile lemma with a long awn tightly twisted towards the base.

Widely scattered but nowhere very common, on open often north-facing banks; (450–) 1750–3500 feet; fl. Oct–Mar. *A 8420! H 11519! P 15899!* S. Mexico to Paraguay, West Indies.

33. ACHLAENA Griseb. (1866)

1. A. piptostachya Griseb., Cat. Pl. Cub.: 229 (1866).

Densely tufted perennial with stiff erect culms 50–100 cm. high in flower; leaf-blades sharply pointed; panicle long-exserted with spreading or reflexed branches; lower glume 1–2 cm. long including the awn; upper glume about 4 mm. long with an awn about 3 cm. long; fertile lemma awnless.

Rare and local (Han.), in grassy glades and pasture-margins on heavy clay slopes; 1200–1300 feet; fl. Sept. *A 7982!* Cuba.

34. REYNAUDIA Kunth (1830)

1. R. filiformis Kunth, Révis. Gram.: 195 (1830).

Tufted perennial with slender erect or ascending culms 15–40 cm. high in flower, the nodes bearded; leaf-blades involute, slender; panicles erect, contracted, rather densely flowered, 3–6 cm. long; spikelets 3–4 mm. long, purplish; awns 2–3 times as long as the glumes bearing them.

Very rare, only once reported (Manch.). *Wullschlaegel 1370.* Cuba, Hispaniola.

35. MELINIS Beauv. (1812)

1. M. minutiflora Beauv., Ess. Nouv. Agrost.: 54, t. 11, f. 4 (1812).—Molasses Grass, Wynne Grass.

Perennial with numerous lax spreading and sometimes rooting culms 1 m. or more long, pubescent and glandular throughout; leaf-blades up to 15 cm. long and 10 mm. broad; panicles densely flowered, purplish, 15–20 cm. or more long; spikelets about 2 mm. long; awn of sterile lemma straight, about 1 cm. long; the whole plant emits a strong sweet odour.

Locally abundant at middle elevations in St. Andrew and St. Thomas; occasional in Manchester and St. Ann, on steep hillsides and stony banks; 1500–4500 feet; fl. Nov–Mar (–Apr). *A 5708! Asprey UCWI 2314! Powell 4!* This grass was not collected in Jamaica by Harris not any earlier collector; it was not reported by Hitchcock (1936) except for a few cultivated specimens from Hispaniola, Puerto Rico, where it has now also become established, and Dominica. It was introduced into Jamaica about 1925 and is now dominant in some areas, an establishment and spread predicted by Hitchcock. Although native in tropical Africa, it was originally described from Brazil and has been introduced in many other tropical areas.

36. OLYRA L. (1759)

1. O. latifolia L., Syst. Nat. ed. 10, **2**: 1261 (1759).

Culms canelike, scrambling to 2–3 (–5) m. from a woody stock, hollow, glabrous, green or mottled purplish brown, often branched above; leaf-blades oblong-lanceolate, unequal at the base, up to 20 cm. long and 5 cm. broad, glabrous; sheath mostly with short marginal hairs; panicle with stiff ascending or spreading branches; fruit bony.

Fairly common on banks along woodland margins and in thickets in sheltered areas; 50–2000 feet; fl. all the year. *A 5694! H 11346! P 6754!* Tropical Amer. and Africa.

37. LITHACHNE Beauv. (1812)

1. L. pauciflora (Sw.) Poir. in Dict. Sci. Nat. **27**: 60 (1823).

Culms tufted, slender, hard, geniculate and naked below, from a short perennial stock, 25–40 (–50) cm. high; leaf-blades lanceolate, subtruncate at the base, acuminate, spreading, 5–8 cm. long and about 2 cm. broad; fruit 4–5 mm. long, slightly compressed, shiny and bony, usually white, rarely black.

Common on shady banks and in thickets and woodlands on limestone; 50–2000 (–6000) feet; fl. all the year. *A 5550! H 11279! P 10130!* Mexico to Argentina, West Indies.

38. STENOTAPHRUM Trin. (1820)

1. S. secundatum (Walt.) Kuntze, Revis. Gen. Pl. **2**: 794 (1891).—Crab Grass[1], Pimento Grass.

Perennial with extensive creeping and rooting flattened stolons bearing short leafy more or less erect branches; sheaths strongly distichous and contracted at the junction with the blade; leaf-blades often obtuse; rachis fleshy, flattened, un-branched; spikelets 4–6 mm. long.

Common in pastures on heavy poorly drained soils or on sand or coral limestone near the sea; SL–5000 feet; fl. sporadically, mostly May–Dec. *A 7967! H 12618! P 7907!* General throughout the tropics and subtropics; Cayman Is.

39. SETARIA Beauv. (1812) nom. cons.; Rominger (1962)

1 Leaf-blades elliptical, plaited; bristles below only some of the spikelets; inflorescence-branches spreading:
  2 Plant annual; leaf-blades usually less than 2 cm. broad, exceptionally much larger; lower glume subrotundate, about one-third length of spikelet   1. *barbata*
  2 Plant perennial; leaf-blades up to 6 cm. broad, shortly petiolate; lower glume broadly ovate, acute or obtuse, about half the length of spikelet   2. *palmifolia*
1 Leaf-blades linear-lanceolate to linear, flat; bristles below all the spikelets:
  3 Inflorescence spikelike with the primary branches shorter than the bristles:
    4 Bristles below each spikelet more than 4; axis of inflorescence with short hairs only; margins of sheath glabrous:
      5 Plants annual; spikelets 3 mm. long; bristles becoming yellowish to reddish-brown   3. *lutescens*
      5 Plants perennial; spikelets 2–2·5 mm. long; bristles pale yellow or purple   4. *geniculata*
    4 Bristles below each spikelet 1–3; margins of upper sheaths ciliate; plants annual:
      6 Leaf-blades up to 1 cm. broad, more or less pilose on both surfaces; axis of inflorescence with long and short hairs; bristles 1–2 times as long as the spikelets   5. *scandens*

---

[1] This common name is applied elsewhere to species of *Digitaria* for which there seem to be no accurate vernacular equivalents in Jamaica. It is often used incorrectly for the well known lawn grass, popular in wet or sheltered areas, *Axonopus compressus*, q.v.

6 Leaf-blades up to 3·5 cm. broad, not pilose; axis of inflorescence with long hairs, minor axes with short hairs; bristles 5–10 times as long as the spikelets
6. magna

3 Inflorescence shortly branched, paniculate; bristles mostly solitary below each spikelet; plants perennial:

7 Bristles 1–2 cm. long, both retrorsely and antrorsely barbellate; panicle rough to touch; spikelets swollen, subspherical   7. <i>tenax</i>

7 Bristles usually less than 1 cm. long, antrorsely barbellate only; spikelets not obviously swollen
8. setosa

1. S. barbata (Lam.) Kunth, Révis. Gram.: 47 (1829).—Corn Grass.

Culms decumbent, often rooting, spreading and laxly ascending, up to 1 m. or more long, usually smaller; leaf-blades mostly plicate, sometimes flat in small plants, up to 15 (–36) cm. long and 2·5 (–5) cm. broad; bristles 5–10 mm. long; spikelets about 2·5 mm. long green or tinged purplish, mucronate; glumes conspicuously nerved.

Common as a weed of waste ground and thin pastures usually in rather shady places; 50–2250 (–3800) feet; fl. Oct–July. *A 6149! H 11782! P 24074!* Pantropical but apparently absent from Cuba.

2. S. palmifolia (Koenig) Stapf in Journ. Linn. Soc., Bot. **42**: 186 (1914).— *Panicum palmifolium* Koenig (1788).

Culms more or less erect or spreading up to 1·5 m. high, with usually pungent hairs on the leaf-sheaths; leaf-blades strongly plicate, up to 50 cm. long and 6 cm. broad; panicles up to 40 cm. long, the branches nodding; bristles up to about 6 mm. long, inconspicuous; spikelets about 3 mm. long, green, acuminate; glumes obscurely nerved.

Locally common in the eastern parishes on shaded or open sheltered banks and streamsides; (500–) 1750–5000 feet; fl. all the year. *A 6629! H 10911! Patrick 170!* Native in S.E. Asia, introduced.

3. S. lutescens (Weigel) F. T. Hubbard in Rhodora **18**: 232 (1916).

Culms geniculate or spreading, 25–50 cm. high in flower; leaves and bristles often tinged reddish; inflorescence 2–8 cm. long, about 1 cm. thick including the bristles.

Local and uncommon (St. Andr.), a weed of rough ground; 3500–5000 feet; fl. Mar–Apr, July–Nov. *A 7753! H 11272!* Native of Europe, E. United States, Canada, C. Amer., Hispaniola, Puerto Rico, Lesser Antilles, Pacific Is.

4. S. geniculata (Lam.) Beauv., Ess. Nouv. Agrost.: 178 (1812).

Like the last and sometimes annual, usually rhizomatous; culms slender, straggling, erect to nearly prostrate, exceptionally to 1 m. high, the basal internodes often more slender than those above; inflorescence up to 10 cm. or more long.

Very common in pastures, waste places and along roadsides; SL–4000 feet; fl. all the year. *A 5600! H 11382! P 23730!* United States to Argentina, West Indies; Grand Cayman, introduced elsewhere.

5. S. scandens Schrad. in Schult., Mant. **2**: 279 (1824).

Culms numerous, branched below, erect or spreading, 30–50 (–80) cm. long; panicles slender, up to 8 cm. long; spikelets about 1·5 mm. long, green or purplish; bristles retrorsely barbellate at tips, causing the spikes to catch in clothes, etc.

Occasional as a weed of stony waste places and roadsides; 1500–4300 feet; fl. Mar, July–Nov. *A 7704! H 12668! P 16759!* Mexico to Paraguay, Cuba, Hispaniola.

6. S. magna Griseb., Fl. Br.W.I.: 554 (1864).

Culms robust, erect, few, entirely green, up to 2 (–4) m. high in flower; leaf-blades up to 50 cm. long and 3·5 cm. broad; panicles densely flowered, nodding, tapering at each end, to 40 cm. long and 3 cm. broad, much smaller on the branches; bristles 1–2 cm. long; spikelets about 2 mm. long.

Rare in coastal swamps; 10–15 feet; fl. July, Dec. *A 8582!* E. United States to Costa Rica, Cuba, Puerto Rico, Guadeloupe, Martinique.

7. S. tenax (L.C. Rich.) Desv., Opusc.: 78 (1831).

Culms often geniculate at the base, 1–1·5 m. high; panicles narrowed to the summit but not attenuate, 15–30 cm. long, 2–3 cm. broad; spikelets about 2 mm. long.

Uncommon on rocky banks and in thickets and open woodlands along the south coast in the central parishes; 180–1600 feet; fl. June, Sept–Oct. *H 12065! P 23606!* Mexico to Brazil, West Indies.

8. S. setosa (Sw.) Beauv., Ess. Nouv. Agrost.: 178 (1812).

Culms erect, wiry, branching, sometimes decumbent or prostrate, up to about 1 m. long; panicle narrow and attenuate above, 10–20 cm. long; spikelets about 2 mm. long.

Locally common on limestone rocks, in lagoon margin thickets and open waste ground; SL–1500 feet; fl. Oct–Jan, May–June. *A 9858! H 11301! A.M. Barry!* West Indies, Brazil.

40. PENNISETUM L. C. Rich. (1805)

1 Bristles about 4 cm. long; panicles oval in outline; spikelets 10–12 mm. long; garden escape
                                                        1. *villosum*
1 Bristles rarely as much as 2 cm. long; panicles more or less cylindrical; spikelets up to 7 mm. long:
  2 Culms robust, 2–5 m. high; leaves up to 3 cm. or more broad; axis of inflorescence villous
                                            2. *purpureum*
  2 Culms rather slender to 2 m. high; leaves not more than 2 cm. broad:
    3 Spikelets about 3 mm. long, solitary in the sessile involucre; specimen not seen, probably erroneously reported, but widespread in the Caribbean region
                                 P. setosum (Sw.) L. C. Rich.
    3 Spikelets about 5 mm. long, 2 or more in the peduncled involucre
                                 3a. *orientale* var. *triflorum*

1. P. villosum R. Br. ex Fres. in Mus. Senckenb. Abh. **2**: 134 (1837).

Perennial with the culms densely tufted from a knotted crown, 15–50 cm. high in flower; panicles very dense, up to 15 cm. long and 5 cm. broad; spikelets up to 4 in each cluster.

Rare (St. Andr.), either cultivated or a garden escape; 3600 feet; fl. Aug. *H 12402!* Native of E. Africa, introduced into Amer. as an ornamental.

2. P. purpureum Schumach., Beskr. Guin. Pl.: 44 (1827).—Elephant Grass, Napier Grass.

Tufted perennial; leaf-blades up to 70 cm. or more long; inflorescence spikelike, a panicle of numerous sessile clusters of spikelets, about 30 cm. long; spikelets 2 or 3 together, about 7 mm. long, subtended by long bristles.

Locally common and often cultivated; escaped along riverbanks, field margin depressions and ditches; SL–3500 feet; fl. Sept–Apr. *A 11762! P 11247! Skelding UCWI 3010!* Native of tropical Africa, widely introduced elsewhere; Grand Cayman.

3. P. orientale L.C. Rich. in Pers., Synops. Pl. **1**: 72 (1805).

3a. Var. triflorum (Nees) Stapf in Hook. f., Fl. Br. Ind. **7**: 86 (1896).—Himalaya Grass.

Perennial with a tough upright or ascending rhizome, the culms often decumbent from it and rooting, then ascending and branching, 1·5–2 m. high in flower; panicles purplish, rather loose towards the base, 12–20 cm. long; outer bristles short, the inner longer, the innermost strongest and up to 2·5 cm. long.

Locally established (St. Andr., Clar., Manch.), an escape from cultivation; 1500–5600 feet; fl. Oct–Feb, July. *A 7391! H 11300! J.P. 798!* Native of India, introduced in Cuba and Trinidad.

41. CENCHRUS L. (1753)

1 Plants perennial; leaf-blades rarely more than 2 mm. broad; burs 3·5–5 mm. broad excluding the bristles; outer bristles slender, terete, inner and the lobes of the involucre flattened at the base
                                            1. *gracillimus*

1 Plants annual; leaf-blades mostly more than 3 mm. broad:
2 Outer bristles slender, terete, forming a distinct ring around the base of the involucre:
3 Burs not more than 4 mm. broad, numerous, crowded in a long spike; lobes of
involucre acute, not spinelike   2. *brownii*
3 Burs about 5·5 mm. broad, not densely crowded; lobes of involucre acuminately
tapered and spinelike   3. *echinatus*
2 All bristles and lobes of the involucre more or less intergrading in form and flattened
towards the base; burs not densely crowded:
4 Burs glabrous to finely pubescent, 4–6 mm. broad   4. *pauciflorus*
4 Burs usually densely pubescent to woolly, 5–7 mm. broad   5. *tribuloides*

1. C. gracillimus Nash in Bull. Torr. Bot. Club **22**: 299 (1895).

Culms numerous, slender, wiry, erect or ascending, 20–50 (–80) cm. high,
forming dense clumps; spikes 2–6 cm. long; burs tapered to the base, glabrous
except the inner lobes of the involucre ciliate with long white hairs.

Locally common in the western parishes in open pastures on calcareous sand or
in bauxitic soils over limestone, especially near the sea; SL–2300 feet; fl. Nov–
Mar, July–Aug. *A 8511! H 12690! P 11508!* Florida, Cuba, Hispaniola; Grand
Cayman.

2. C. brownii Roem. & Schult. in L., Syst. Veg. ed. nov. **2**: 258 (1817).

Culms erect, sparingly branched, 30–50 (–100) cm. high in flower; spike usually
shortly exserted, 4–10 cm. long, compact; burs villous with spikelets usually 3 in
each, mostly green, sometimes tinged red.

Common as a weed of grassy places; SL–2200 feet; fl. June–Feb. *A 6307!*
H 12617! Patrick 311! S. Florida to Brazil; Grand Cayman; introduced into the
Pacific Is.

3. C. echinatus L., Sp. Pl. **2**: 1050 (1753).

Culms spreading and ascending, branching and sometimes rooting from the
basal nodes, 25–60 (–100) cm. long; spikes finally long-exserted, 3–7 (–10) cm. long;
burs pubescent with usually 4 spikelets in each.

Common as a weed of open fields and stony waste places; SL–3600 feet; fl. all
the year. *A 6155! H 11311! P 7813!* Throughout subtropical and tropical Amer.;
Grand Cayman; introduced into the Pacific Is.

4. C. pauciflorus Benth., Bot. Voy. Sulph.: 56 (1844).—Sandbur.

Culms tufted rarely erect and diffusely branched or trailing and rooting, up to
1 m. long; spikelets usually 2.

Rather rare (St. Eliz., West., Han.), in open sandy places; SL–10 feet; fl. July–
Sept, Dec. *A 8572! 8573!* United States to Argentina, West Indies.

5. C. tribuloides L., Sp. Pl. **2**: 1050 (1753).—Dune Sandbur.

Like the last and perhaps not really distinct; culms usually more robust, trailing
and rooting.

Very local (St. Andr., St. Cath.), on beaches, more frequent on cays; SL–5 feet;
fl. Feb, June, Oct. *A 12264! C.B. Lewis!* E. United States to Brazil, West Indies;
Cayman Is.

42. ANTHEPHORA Schreb. (1810)

1. A. hermaphrodita (L.) Kuntze, Revis. Gen Pl. **2**: 759 (1891).

Culms ascending branched, leafy, 20–50 cm. high; spikes 5–10 cm. long;
glumes 5–7 mm. long.

Common as a weed of roadsides, pastures and waste places; SL–2750 feet; fl.
all the year. *A 6584! H 12679! Patrick 166!* Throughout tropical Amer.; Grand
Cayman.

43. DIGITARIA Heist. ex Fabr. (1759); Henrard (1950)

1 Spikelets long-silky; racemes paniculate; plants perennial   1. *insularis*
1 Spikelets pubescent or glabrous; racemes digitate or occasionally more or less scattered
subterminally:

2 Rachis wing-margined, the green margins as wide as the lighter midrib or wider:
  3 First glume small but evident:
    4 Plant annual; a common weed; leaf-blades more or less pilose but variable
                                                     2. *ciliaris*
    4 Plants perennial; leaf-blade glabrous; cultivated or escaped; native in South Africa:
      5 Spikelets 2·7–3 mm. long; upper glume scantily hairy; plant commonly stoloni-
        ferous; Pangola   *D. decumbens* Stent
      5 Spikelets 3·5 mm. long; upper glume long-ciliate; plant robust, not stoloniferous
                                                    D. smutsii Stent
  3 First glume obsolete or wanting; plants annual:
    6 Rachis with sparsely scattered long spreading hairs; spikelets about 2·5 mm. long
                                                 3. *horizontalis*
    6 Rachis without long hairs:
      7 Pedicels with a ring of short stiff hairs at the summit; spikelets with stripes of dense
        capitellate hairs; spikelets 2 mm. long   4. *argyrostachya*
      7 Pedicels not hairy at the summit; spikelets without capitellate hairs; fertile lemma
        dark brown; spikelets 1·5 mm. long   5. *violascens*
2 Rachis not winged; spikelets pubescent, 1·5 mm. long   6. *panicea*

1. D. insularis (L.) Mez ex Ekman in Arkiv Bot. **11** (4): 17 (1912).—*Trichachne insularis* (L.) Nees (1829).

Culms tufted, coarse, erect, 50–150 cm. high in flower; sheaths sparsely hairy; racemes 10–15 cm. long; spikelets about 4 mm. long.

Locally common as a weed of roadsides and rough pastures in well drained places; SL–3000 feet; fl. all the year. *A 6306! H 11409! Patrick 189!* Throughout the tropics and subtropics of Amer.; Grand Cayman; introduced into the Pacific Is.

2. D. ciliaris (Retz.) Koeler, Descr. Gram.: 27 (1802).—*Panicum ciliare* Retz (1786). *P. adscendens* Kunth (1816). *D. adscendens* (Kunth) Henrard (1934). *D. sanguinalis* of Hitchc. (1936), not (L.) Scop.

Culms often spreading and sometimes rooting at the lower nodes, 60 cm. or more long, sometimes much smaller; sheaths usually bristly; internodes often purplish; leaf-blades somewhat glaucous; racemes spreading or ascending in 1 or 2 whorls, 5–10 cm. long; spikelets about 3 mm. long.

Very common in stony waste places, pastures and roadsides; SL–3500 (–5000) feet; fl. all the year. *A 5803! H 11273! Patrick 281!* S. United States, C. and S. Amer. and generally throughout the tropics. The distinction of *D. ciliaris* and *D. sanguinalis* (L.) Scop. has been discussed by J. E. Ebinger, Brittonia **14**: 248–253 (1962), F. W. Gould, *op. cit.* **15**: 241–244 (1963), and S. T. Blake, Proc. Roy. Soc. Queensland **81**: 10–12 (1969).

3. D. horizontalis Willd., Enum. Hort. Berol.: 92 (1809).

Like the last; culms usually spreading or stoloniferous; leaf-blades often softly pubescent; racemes slender; spikelets narrow.

Apparently not very common in Jamaica, a weed; 650–2000 (–4800) feet; fl. May–Dec. *A 7401! H 12613! P 10327!* Pantropical; Grand Cayman.

4. D. argyrostachya (Steud.) Fernald in Rhodora **22**: 103 (1920).

Culms erect or spreading, 10–60 cm. long; sheaths glabrous; racemes slender, 4–10 cm. long.

Very rare (St. Andr.), a weed of cultivated ground; 5000 feet; fl. Oct–Nov. *H 11413! 11441!* Native in tropical Asia.

5. D. violascens Link, Hort. Bot. Berol. **1**: 229 (1827).

Culms decumbent or erect, 30–60 cm. high; sheaths glabrous; leaf-blades glabrous, mostly 3–6 mm. broad; racemes slender, often curved, with rachis up to 1 mm. broad.

Occasional along pathsides and in open ground especially in shallow soil over rocks; 500–5200 feet; fl. July–Jan. *A 7442! H 12714! J.P. 756!* Native in tropical Asia and Australia, Cuba, Puerto Rico, Trinidad.

6. D. panicea (Sw.) Urb., Symb. Ant. **8**: 23 (1920).

Leaf-blades very narrow, 1 mm. broad; racemes few, erect, 5–20 cm. long, loosely flowered.

Very rare (St. Andr.); about 1000 feet. *Hitchcock 9574!* Florida, C. Amer., Greater Antilles.

44. ERIOCHLOA Kunth (1816)

Lower glume present; plant stoloniferous or culms long-decumbent   1. *polystachya*
Lower glume obsolete; culms erect or decumbent at the base   2. *punctata*

1. E. polystachya Kunth, Nov. Gen. **1**: 95, t. 31 (1816).

Perennial; culms up to 1–2 m. high in flower, with bearded nodes; lower glume a broad loose membrane; upper glume thinly silky with the nerves prominent towards the tip.

Rare (St. Andr., West.), in low-lying swampy ground and along canal banks and pond margins; 15–700 feet; fl. May, Sept, Dec. *A 10154! Johnston 2779!* Ecuador to Brazil, West Indies.

2. E. punctata (L.) Desv. in Ham., Prodr. Pl. Ind. Occ.: 5 (1825).

Perennial; culms ascending from a decumbent base, glabrous, 1 m. or more high in flower; spikelets lanceolate, acute, appressed-pubescent in the lower half, about 4 mm. long; fertile lemma with an awn about 1 mm. long.

Rare (St. Eliz., St. James), in swamps; ?SL–1000 feet. *J.P. 918!* S. United States to Argentina, West Indies.

45. AXONOPUS Beauv. (1812)

1. A. compressus (Sw.) Beauv., Ess. Nouv. Agrost.: 154 (1812).—Carpet Grass.

Perennial with long leafy stolons with short broad leaf-blades; flowering culms more or less glabrous; leaf-blades longer, rather obtuse, often with corrugated margins, bright green, not glaucous; racemes few, mostly erect; spikelets 2–3 mm. long; a plant very variable in habit.

A very common weed, especially in rather damp shaded places; SL–3600 feet; fl. all the year. *A 6925! H 11386! J.P. 833!* In most warm countries.

46. PASPALUM L. (1759)

1 Rachis membranous and more or less winged, 1·5–3 mm. broad; aquatic perennials:
  2 Racemes persistent on the axis; rachis about 3 mm. broad with a spikelet at the apex; spikelets more than 3 mm. long   1. *serratum*
  2 Racemes deciduous; rachis about 1·5 mm. broad, extending beyond the uppermost spikelet; spikelets 1·5–2 mm. long   2. *repens*
1 Rachis not membranous, hardly winged:
  3 Racemes 2, conjugate or nearly so at the apex of the culm, rarely a third or fourth below; plants perennial with short or long creeping rhizomes or stolons:
    4 Spikelets elliptical to narrowly ovate:
      5 Upper glume pubescent; spikelets turgid   3. *paspalodes*
      5 Upper glume glabrous; spikelets flattened:
        6 Leaf-blades spreading, tapered from base to apex, the margins involute
                           4. *distichum*
        6 Leaf-blades erect or ascending, involute-setaceous   5. *distachyon*
    4 Spikelets broadly ovate to obovate or suborbicular:
      7 Spikelets concavo-convex, 1·4–1·8 (–2·2) mm. long, maturing yellow; racemes slender, often bowed, 8–12 cm. long   6. *conjugatum*
      7 Spikelets plano-convex, 2·3–3·5 mm. long, green or tinged purplish; racemes up to 7 cm. long, thick, straight or nearly so:
        8 Spikelets 2·8–3·5 mm. long   7. *notatum*
        8 Spikelets 2–2·5 mm. long   8. *minus*
  3 Racemes 1–many, racemose or fascicled on the axis, not conjugate:
    9 Lower glume developed; racemes usually solitary, bowed, 1–3·5 cm. long
                           9. *decumbens*
    9 Lower glume normally wanting:
      10 Peduncles terminal and axillary:
        11 Leaf-blades ciliate, up to 15 mm. broad; spikelets (1·6–) 1·8–2 mm. long, glabrous or nearly so   10. *propinquum*
        11 Leaf-blades not or hardly ciliate, narrower; spikelets 1·2–1·5 mm. long
                               see couplet 28–28
      10 Peduncles terminal only:
        12 Plants annual; spikelets with broad notched wings to the upper glume and sterile lemma   11. *fimbriatum*

12 Plants perennial:
13 Culms creeping or stoloniferous:
14 Racemes usually solitary, about 10 mm. long; spikelets glabrous, 1–1·4 mm. long; plant terrestrial with flowering branches up to 15 cm. high
12. *breve*
14 Racemes 2–4, (6–)10–40 mm. long; plants of wet places, with flowering branches mostly 10 cm. or more high:
15 Culms 10–60 cm. long; racemes 3–4, 6–23 mm. long; spikelets glabrous, suborbicular, 1–1·2 mm. long   13. *orbiculatum*
15 Culms 30–100 cm. long; racemes 2–3, 10–40 mm. long; spikelets pubescent, obovate-elliptical, 1·5–1·7 mm. long   14. *reptatum*
13 Plants tufted, not stoloniferous:
16 Racemes usually solitary, rarely 2 (–4):
17 Leaf-blades less than 3 mm. broad, often very long:
18 Glume and sterile lemma distinctly crumpled, glabrous:
19 Spikelets 2·5–3 mm. long; glume and sterile lemma pointed beyond the fruit   15. *lindenianum*
19 Spikelets 1·5–2 mm. long; glume and sterile lemma not pointed beyond the fruit   16. *distortum*
18 Glume and sterile lemma not or only very slightly crumpled:
20 Spikelets 1·6–1·8 mm. long, glabrous   17. *filiforme*
20 Spikelets 3 mm. long, pubescent; glume and sterile lemma pointed beyond the fruit   18. *alterniflorum*
17 At least some of the leaf-blades 3 mm. or more broad; spikelets 1·3–1·6 mm. long, appressed-pubescent   19. *saugetii*
16 Racemes 3–many:
21 Spikelets conspicuously silky-ciliate around the margin, the hairs as long as the spikelet or longer, 2·8–3·8 mm. long   20. *dilatatum*
21 Spikelets not conspicuously ciliate:
22 Fruit dark brown and shining; sterile lemma with transverse submarginal wrinkles; spikelets 2·5–2·8 mm. long, glabrous   21. *plicatulum*
22 Fruit usually pale, not shining; sterile lemma smooth:
23 Plants robust; culms commonly more than 1 m. high; leaf-blades firm with sharp cutting edges; spikelets (1·8–) 2–2·5 mm. long:
24 Spikelets pubescent at least towards the apex; leaf-blades flat
22. *virgatum*
24 Spikelets glabrous; leaf-blades V-shaped in section:
25 Spikelets suborbicular; rachis ciliate:
26 Racemes rarely as many as 50; glume and sterile lemma firm in texture   23. *millegrana*
26 Racemes commonly more than 70; glume and sterile lemma fragile
24. *densum*
25 Spikelets obovate-elliptical; rachis not ciliate:
27 Racemes slender; rachis about 0·7 mm. broad   25. *secans*
27 Racemes thick; rachis 1 mm. broad   26. *arundinaceum*
23 Plants not robust; culms, if as much as 1 m. high, slender; leaf-margins not cutting; spikelets 1·3–1·8 (–2) mm. long, more or less pubescent:
28 Spikelets subhemispheric, 1·3 mm. long; leaf-sheaths pungent-hispid; racemes usually numerous   27. *paniculatum*
28 Spikelets elliptical or oval; racemes 3–5 (–8):
29 Spikelets 1·3 mm. long, often brownish; nodes appressed-pubescent, glabrescent   28. *blodgettii*
29 Spikelets 1·5–1·8 (–2) mm. long:
30 Nodes glabrous; spikelets 1·7–2 mm. long, often brownish
29. *laxum*
30 Nodes appressed-pubescent; spikelets about 1·5 mm. long, green or purplish   30. *caespitosum*

1. P. serratum Hitchc. & Chase in Contrib. U.S. Nat. Herb. **18**: 306 (1917).

Plant glabrous and rather glaucous grey-green; culms trailing, up to 2 m. long; leaf-blades 4–9 cm. long, 3–7 mm. broad; racemes usually 2, 12–15 mm. apart, 3–5 cm. long.

Rather uncommon in shallow ponds; 50–2000 feet; fl. July–Jan. *A 10357! H 12582! P 10544!* Cuba.

2. P. repens Berg. in Act. Helvet. **7**: 129, t. 7 (1772).

Culms trailing and rooting; leaf-sheaths inflated, blades 10–20 cm. long, 12–15 mm. broad; racemes numerous, solitary or few in a cluster, 3–5 cm. long, sessile.

Local and uncommon, mostly in the western parishes, in ponds and ditches; 30–1000 feet; fl. Sept–Mar. *A 10403! H 12557! P 18329!* S.E. United States to C. Amer. and Paraguay, Cuba, Trinidad.

3. P. paspalodes (Michx.) Scribn. in Mem. Torr. Bot. Club **5**: 29 (1894).— *Digitaria paspalodes* Michx. (1803). *P. distichum* of Hitchc. (1936) and other authors, not L. (1759).

Flowering culms up to 50 cm. high, usually shorter; leaf-blades flat; racemes spreading at maturity, up to 7 cm. long; spikelets 2·5–3·5 mm. long; lower glume sometimes developed.

Occasional along riverbanks, around ponds and in wet gravel and silted streams; 20–4500 feet; fl. June–Nov. *A 7714! H 11850! J.P. 865!* S. United States to Chile, West Indies.

4. P. distichum L., Syst. Nat. ed. 10, **2**: 855 (1759).—*P. vaginatum* Sw. (1788).

Flowering culms up to 60 cm. high, often much smaller and sometimes diminutive in exposed habitats; leaf-blades rather stiff, usually folded, glabrous; racemes 1–7·5 cm. long; spikelets 3–4·5 mm. long.

Common along all the seashores and on the cays, in brackish sand and swampy grassland and on coral limestone; SL–20 feet; fl. sporadically throughout the year. *A 7359! H 12661! P 15399!* On all warm coasts; Cayman Is.

5. P. distachyon Poit. ex Trin. in Mém. Acad. Sci. St. Pétersb. sér. 6, **3** (2): 142 (1834).

Culms erect, approximate, often forming a sward, 15–40 cm. high in flower, slender from hard yellow rhizomes; leaf-blades 6–15 cm. long, 1–2 (–2·5) mm. broad; racemes up to 5 cm. long; spikelets about 3 mm. long.

Locally common in damp pastures and on ditchbanks; SL–250 feet; fl. July–Mar. *A 11882! H 12548! P 11322!* Cuba, Hispaniola.

6. P. conjugatum Berg. in Act. Helvet. **7**: 129, t. 8 (1772).—Jamaican Sour Grass.

Culms usually creeping and rooting, as much as 2 m. long; flowering branches up to 60 cm. high; leaf-blades flat, 8–12 cm. long, up to 15 mm. broad.

Widespread but only locally abundant in damp pastures and on banks, often in partly shaded places; SL–5000 feet; fl. all the year. *A 7292! H 11308! P 24095!* In all warm countries.

A variant with larger more copiously hairy spikelets, var. *pubescens* Döll (1877), is very rare in St. Thomas, *Orcutt 1985!*

7. P. notatum Flügge, Gram. Monogr. Pasp. **1**: 106 (1810).—Bahia Grass, Bammy Grass.

Rhizomes short, stout, much branched; culms 15–50 cm. high in flower with mostly basal leaves.

Rather common in savannas, pastures and on banks in heavy soils, often cultivated; 100–5000 feet; fl. June–Jan. *A 7527! H 12443! Patrick 268!* Mexico to Argentina, West Indies.

8. P. minus Fourn., Mex. Pl. **2**: 6 (1886).

Very like the last but smaller in all parts, perhaps not really distinct.

Rare in pastures on heavy soils; 100–2900 feet; fl. June–Sept. *A 11255!* Alabama, Louisiana, Mississippi, Texas, Mexico to Paraguay and Bolivia, Greater Antilles.

9. P. decumbens Sw., Nov. Gen. & Sp. Pl.: 22 (1788).

Culms spreading and trailing to 70 cm. long; leaf-blades flat, mostly 5–10 cm. long, 6–12 mm. broad, retrorsely ciliate at the base, antrorsely ciliate towards the tip; spikelets crowded, obovate, 1·7 mm. long; upper glume about half as long as the spikelet.

Very local on shaded banks and trailsides in moist forest; 500–4000 feet; fl. Dec–Apr, July–Aug. *A 7609! H 12255! J.P. 733!* Guatemala to Brazil and Bolivia, Greater Antilles, Trinidad.

10. P. propinquum Nash in Bull. New York Bot. Gard. **1**: 291 (1899).
Culms slender, mostly erect, to 60 cm. high in flower; racemes arching, 7–10 cm. long; spikelets glabrous or minutely pubescent, rather obtuse apically.
Occasional in open sandy waste places, pastures and cultivated ground as a weed; SL–3000 feet; fl. most of the year. *A 7580! P 10296!* Florida, Mexico to Panama, West Indies.
Jamaican specimens of this grass tend to have the small spikelet size of typical *P. propinquum* but the spikelet shape of *P. ciliatifolium* Michx. (1803).

11. P. fimbriatum Kunth, Nov. Gen. **1**: 93, t. 28 (1816).
Culms erect, 25–100 cm. high in flower, or rarely spreading; leaf-blades flat, 10–20 cm. long, 5–12 mm. broad; racemes 3–8, up to 8 cm. long; spikelets including the wings about 3 mm. long and broad.
Very common as a weed of pastures, roadsides and waste places; SL–5000 feet; fl. all the year. *A 5474! H 8596! P 10295!* Panama, northern S. Amer., West Indies; Grand Cayman; introduced into the Pacific Is.

12. P. breve Chase in Urb., Symb. Ant. **7**: 166 (1912).
Culms slender; leaves usually 5–10 cm. long, 1·5–3 mm. broad; spikelets broadly oval.
Rare (Han., St. James), in coral sand and at swamp margin; 20 feet; fl. Sept–Dec. *A 8676! 13240!* Cuba, Hispaniola, Barbados.

13. P. orbiculatum Poir. in Lam., Encycl. Méth. Bot. **5**: 32 (1804).
Stolons leafy, often forming dense prostrate mats; leaf-blades flat, 1–6 cm. long, 1·5–7 mm. broad.
Very rare (St. Thom.), along stream margins; 1500 feet; fl. May. *P 28666!* Mexico to Paraguay, Greater Antilles, Martinique, St. Lucia, Trinidad.

14. P. reptatum Hitchc. & Chase in Contrib. U.S. Nat. Herb. **18**: 318 (1917).
Culms decumbent and rooting at the nodes; leaf-blades flat, 3–10 cm. long, 2–5 mm. broad.
Very rare (Clar.), in wet open savanna; 300 feet; fl. Dec. *H 12717!* Cuba.

15. P. lindenianum A. Rich. in Sagra, Hist. Cuba, Parte 2, **11**: 299 (1850).
Culms numerous forming dense tussocks; leaf-blades 20–50 cm. long, plano-convex in section; racemes erect at first, 6–10 cm. long, the peduncle elongating in fruit; spikelets irregularly ovate-rhomboid.
Occasional in pastures on clay soils and at swamp margins; SL–1500 feet; fl. Apr–Dec. *A 7336! H 12551! P 10072!* Cuba, Hispaniola; Grand Cayman.

16. P. distortum Chase in Contrib. U.S. Nat. Herb. **28**: 142, f. 85 (1929).
Like the last; leaf-blades commonly tortuous; flowering culms to 45 cm. high, often tinged purplish; racemes 2·5–6 cm. long; spikelets ovate to somewhat rhomboid.
Occasional in open pastures on heavy soils; SL–2600 feet; fl. June–Nov. *A 7327! H 12564! P 10335!* Cuba, Hispaniola; Grand Cayman.

17. P. filiforme Sw., Nov. Gen. & Sp. Pl.: 22 (1788).—Wire Grass.
Like the last but culms, leaves and racemes more slender; leaf-blades flexuous, plano-convex in section, up to 40 cm. long, usually short; flowering culms mostly shorter than the leaves at first, elongating in fruit; racemes 3–8 cm. long; spikelets ovate-elliptical.
Occasional in damp hollows in pastures and on shaded banks; 100–3000 feet; fl. June–Nov. *A 8131! P 10273!* Cuba, Hispaniola.
Grisebach's (1864) concept of this species included *P. distortum* (*Prior! Purdie!*) and *P. lindenianum* (*March!*). The close resemblance of these three species is further emphasized by the coincidence of their habitat, geographical distribution and unusual flowering time. Although not difficult to distinguish, these species, and perhaps the next also, constitute, together with some additional Cuban species, a group affording a clear example of local and possibly recent common ancestry; that

they exist in vegetation which is predominantly non-woody may have significant historical implications.

18. P. alterniflorum A. Rich. in Sagra, Hist. Cuba, Parte 2, **11**: 299 (1850).

Culms numerous, 30–100 cm. high; leaf-blades elongated, 1·5–3 mm. broad; racemes 7–12 cm. long; spikelets ovate-oblong.

Rare and local (St. Thom.), in low-lying boggy ground; 25 feet; fl. Aug–Nov. *A 9895! A.M. Barry!* Cuba, Hispaniola.

19. P. saugetii Chase in Contrib. U.S. Nat. Herb. **28**: 147, f. 90 (1929).

Culms slender, 15–40 cm. high; leaf-blades more or less tortuous, 3–15 cm. long, 3–7 mm. broad; racemes 2–4 cm. long; spikelets mostly solitary by abortion.

Occasional and locally common in shallow soil on exposed limestone banks, cliffs and ledges; 1200–2600 feet; fl. Aug–Feb. *A 5587! H 9674! Patrick 256!* Bahamas, Greater Antilles.

20. P. dilatatum Poir. in Lam., Encycl. Méth. Bot. **5**: 35 (1804).

Culms ascending or erect, 40–175 cm. high; leaf-blades flat, commonly 10–25 cm. long, 3–12 mm. broad; racemes 3–5, ascending or drooping, 6–8 cm. long.

Introduced and cultivated (St. Andr.); fl. Nov. *H 12708!* S.E. United States; native in S. Amer. from Brazil to Argentina, sparingly introduced into the West Indies as a pasture grass.

21. P. plicatulum Michx., Fl. Bor. Amer. **1**: 45 (1803).

Culms erect or ascending, 50–120 cm. high in flower, often purplish; leaf-blades mostly channelled and folded towards the base, hispid on the upper surface; racemes usually several, 2–10 cm. long; spikelets oval to obovate.

Very common in open waste ground, pasture margins and along roadsides mostly on heavy soil; 100–5000 feet; fl. all the year. *A 5596! H 11289! P 23607!* S.E. United States to Argentina, West Indies.

22. P. virgatum L., Syst. Nat. ed. 10, **2**: 855 (1759).

Leaf-blades 1–2·5 cm. broad; racemes mostly 10–20, the lower 5–15 cm. long; spikelets obovate, about 2·5 mm. long.

Common in boggy pastures, swamps and ditches; 30–3500 feet; fl. May–Jan. *A 7321! H 12897! P 7961!* Mexico to Brazil, West Indies.

23. P. millegrana Schrad. in Schult., Mant. **2**: 175 (1824).

Leaf-blades long, 7–15 mm. broad; racemes mostly 10–25, rather thick, approximate, 5–15 cm. long; spikelets obovate to suborbicular, 2–2·4 mm. long.

Occasional in swamps and boggy pastures; SL–2000 feet; fl. Apr–Dec. *H 11620! H & P 14536A!* C. Amer. to Brazil, West Indies.

24. P. densum Poir. in Lam., Encycl. Méth. Bot. **5**: 32 (1804).

Culms erect, broadly flabellate and purplish at base, up to about 1·5 m. high; leaf-blades long, 1–2 cm. broad; racemes 50–100, rather thick, the lower 5–9 cm. long, shorter above; spikelets suborbicular, 1·8–2 mm. long.

Local (St. Cath., Clar., St. Eliz.), in open wet savannas; 50–2300 feet; fl. June–Dec. *A 5925! H 11149! P 15947!* Panama to Brazil, Greater Antilles, Guadeloupe, Trinidad.

25. P. secans Hitchc. & Chase in Contrib. U.S. Nat. Herb. **18**: 319 (1917).

Leaf-blades long, 5–10 mm. broad; racemes mostly 7–12, slender, 6–15 cm. long; spikelets obovate-elliptical, about 2·5 mm. long.

Rare and local (St. Eliz., St. James); near SL. *Hitchcock 9649, 9670.* Bahamas, Greater Antilles to Guadeloupe. This grass is easily confused with the next and may not be really distinct.

26. P. arundinaceum Poir. in Lam., Encycl. Méth. Bot. Suppl. **4**: 310 (1816).

Leaf-blades long, 5–10 mm. broad; racemes usually 12–18, rather thick, 8–20 cm. long, approximate or in clusters; spikelets obovate-elliptical, about 2·5 mm. long.

Occasional in ditches along roadsides; SL–50 feet; fl. May–Dec. *A 7231!*
H 12544! P 10456! British Honduras, Guatemala, French Guiana, Greater
Antilles, Antigua.

27. P. paniculatum L., Syst. Nat. ed. 10, **2**: 855 (1759).
Culms erect or decumbent at the base, mostly 50–100 cm. high in flower; leaf-
blades flat, mostly 10–25 cm. long, 10–20 mm. broad, more or less hispid; racemes
4–12 cm. long with numerous crowded spikelets.
Very common in rough pastures and waste places and along damp shady roadsides; SL–5000 feet; fl. all the year. *A 5730! H 11616! P 10297!* Mexico to Argentina, West Indies.

28. P. blodgettii Chapm., Fl. South. U.S.: 571 (1860).
Culms spreading or erect, 30–60 (–100) cm. high in flower; leaf-blades flat, 5–25
cm. long, 5–10 mm. broad, sometimes ciliate along the proximal margins; racemes
2–8 cm. long, in mostly terminal but sometimes axillary inflorescences; spikelets
rather obtuse or rounded apically, often purplish, maturing brown.
Common as a weed of pastures, open sandy wastes and roadsides in limestone
areas; SL–2700 feet; fl. most of the year. *A 7174! H 12620! P 24096!* Florida,
Mexico, Honduras, Bahamas, Greater Antilles; Cayman Is.

29. P. laxum Lam. in Tabl. Encycl. & Méth., Bot. **1**: 176 (1791).
Culms mostly erect, 50–75 cm. high in flower; leaf-blades up to 30 cm. or more
long, 3–8 mm. broad, narrowed to the base; racemes 3–10 cm. long; spikelets
elliptic-obovate, the glume prominently bulged.
Very rare (St. James). *Hitchcock 9674.* S. Florida, Bahamas, West Indies. Some
particularly robust specimens of *P. blodgettii* with spikelets as much as 1·5 mm. long
have been misidentified as this species.

30. P. caespitosum Flügge, Gram. Monogr. Pasp. **1**: 161 (1810).
Culms erect, slender, 30–60 cm. high in flower; leaf-blades flat or involute,
5–20 cm. long, 4–10 mm. broad, narrowed to the base; racemes 1·5–4 cm. long;
spikelets elliptical.
Occasional on rocky banks and in thickets in limestone areas; SL–2700 feet; fl.
July–Jan. *H 12758! J.P. 863! P 10426!* S. Florida, C. Amer., Bahamas, Greater
Antilles. In Jamaican material this species is difficult to distinguish from *P.
blodgettii.*

47. OPLISMENUS Beauv. (1807)

Rachis of racemes mostly 2–3 (–5) mm. long, bearing usually not more than 5 spikelets;
    leaf-blades 1–3 cm. long, 4–10 mm. broad   1. *setarius*
Rachis of lower racemes more than 1 cm. long, bearing usually more than 8 spikelets;
    leaf-blades mostly more than 4 cm. long, 1–2 cm. broad   2. *hirtellus*

1. O. setarius (Lam.) Roem. & Schult. in L., Syst. Veg. ed. nov. **2**: 481 (1817).
Perennial with culms trailing, branching and rooting from the nodes, sometimes
forming loose mats, the ascending flowering shoots 20–30 cm. high; awns 4–8 mm.
long.
Common in moist woodland clearings and shaded roadside banks; 100–5000
feet; fl. July–Feb. *A 8709! H 12577! P 16622!* S. United States to Paraguay,
West Indies.

2. O. hirtellus (L.) Beauv., Ess. Nouv. Agrost.: 54 (1812).
Like the last; leaf-sheaths glabrous to densely papillose-hispid, blades 5–10 cm.
long; awns up to 5–10 mm. long, often purplish.
Rather common in woodland margins and on limestone rocks, often in partial
shade; 80–3500 feet; fl. sporadically throughout the year. *A 8340! H 11253!
P 24181!* Mexico to Argentina, West Indies, warm parts of Africa. The variant
with the sheaths glabrous except along the margins (e.g. *A 8335!*) seems at first
examination to be quite distinct and has been referred to the Asiatic *O. loliaceus*
(Lam.) Beauv. (1812). It is usually found growing with the commoner hispid
plant and there are no other differences.

48. RHYNCHELYTRUM Nees (1836)

1. R. repens (Willd.) C. E. Hubbard in Kew Bull. **1934**: 110 (1934).—*Tricholaena repens* (Willd.) Hitchc. (1936). Natal Grass.

Culms tufted, slender, often decumbent at the base, up to 1 m. high in flower; leaf-sheaths hirsute, blades flat, 5–15 cm. long, 2–7 mm. broad; panicles profusely branched, silky, often tinged purplish to light rose; spikelets delicately pedicelled, about 5 mm. long with longer hairs.

Locally common (St. Andr., St. Cath., St. James), a weed of roadsides and open waste places; SL–2500 feet; fl. most of the year. *A 5471! P 10558! Wynter UCWI 3811!* Native of tropical Africa, now widely distributed in Asia and Amer.; Grand Cayman.

49. ECHINOCHLOA Beauv. (1812)

1 Ligule a dense line of stiff yellowish hairs; plants perennial; fruit pointed, about 4 mm. long; awns 2–10 mm. long   1. *polystachya*
1 Ligule wanting; plants annual:
  2 Racemes simple, rarely exceeding 2 cm. long; spikelets crowded in about 4 rows; awn of sterile lemma reduced to a short point; leaf-blades 3–6 (–10) mm. broad
      2. *colonum*
  2 Racemes more or less branched, usually more than 2 cm. long; spikelets irregularly arranged, usually not in rows; leaf-blades broader:
    3 Sheaths, at least the lower, papillose-hispid; panicle dense; awns up to 2 cm. or more long   3. *walteri*
    3 Sheaths smooth; panicle more open; awns rarely exceeding 1 cm. long:
      4 Panicles lax and nodding; spikelets inconspicuously hispid, 2·5–3 mm. long
        4. *crus-pavonis*
      4 Panicles stiff and erect; spikelets conspicuously hispid, 3–4 mm. long
        5. *crus-galli*

1. E. polystachya (Kunth) Hitchc. in Contrib. U.S. Nat. Herb. **22**: 135 (1920).

Culms coarse to 2 m. high in flower, from a long creeping and rooting base, glabrous, except the nodes hispid with dense yellowish hairs; leaf-blades up to 2·5 cm. broad, scabrid; racemes ascending, 3–6 cm. long, densely hispid at the base; spikelets nearly sessile, plano-convex, about 5 mm. long.

Very rare (West.), by streambanks in low-lying areas. *Hitchcock 9868, Macfadyen!* Mexico to Argentina, West Indies.

2. E. colonum (L.) Link, Hort. Bot. Berol. **2**: 209 (1833).

Culms spreading and ascending, usually much branched at the base, erect in flower, 20–40 cm. high, glaucous, glabrous; leaf-blades rather lax 5–10 cm. or more long, often purple-banded; spikelets plano-convex, about 3 mm. long, setulose and accompanied by long hairs, often purple.

Widely distributed and locally common in ditches, low-lying open waste ground and pond margins; SL–3500 feet; fl. most of the year. *A 8363! H 11241! J.P. 825!* In most warm countries; Grand Cayman.

3. E. walteri (Pursh) Heller, Cat. N. Amer. Pl. ed. **2**: 21 (1900).

Culms erect, often succulent and thick at the base, 1–2 m. high in flower; leaf-blade 1–3 cm. broad, scabrid; racemes appressed or ascending, up to 10 cm. long; spikelets about 3 mm. long, mostly long-awned, often purple.

Very rare (St. Eliz.), in wet places. *Hitchcock 9650.* E. United States, Cuba, Hispaniola; Grand Cayman.

4. E. crus-pavonis (Kunth) Schult., Mant. **2**: 269 (1824).

Culms decumbent at the base or erect, glabrous, 1–1·5 m. high in flower; leaf-blades 5–15 mm. broad, usually scabrid; spikelets hispid on the nerves, hispidulous between; awn of sterile lemma 1–10 mm. long.

Local (St. Cath., St. Eliz., West.), in marshes, rice fields, ditches and ponds; SL–40 feet; fl. July–Dec, Apr. *A 11646! H 11751! P 24012!* Mexico to Argentina, West Indies. Some authors do not distinguish this species from the next.

5. E. crus-galli (L.) Beauv., Ess. Nouv. Agrost.: 161 t. 11, f. 2 (1812).

Like the last; spikelets a little larger with the nerves of the upper glume and sterile lemma more strongly hispid.

Rather local (St. Eliz., West., Han.), in low-lying damp waste ground and rice fields; SL–20 feet; fl. Mar–Apr, Dec. *A 12012! P 24737!* Widely distributed in warm parts of the Old and New Worlds.

50. ISACHNE R. Br. (1810)

1 Panicle contracted, up to 3 cm. long, the branches appressed or ascending; plants low
   and spreading; glumes glabrous   1. *pygmaea*
1 Panicle open, the branches spreading or ascending:
  2 Culms trailing or spreading-suberect; leaf-blades rarely more than 5 cm. long; glumes
     hispidulous   2. *rigens*
  2 Culms clambering; leaf-blades up to 20 cm. long; glumes glabrous or with a few short
     stiff hairs at the apex   3. *arundinacea*

1. I. pygmaea Griseb., Fl. Br. W.I.: 553 (1864).

Culms low, spreading, wiry, 4–15 (–45) cm. long; leaf-blades 0·5–2 (–3) cm. long, up to about 2 mm. broad; spikelets about 1·3 mm. long, nearly sessile.

Local (St. Andr., Port., St. Thom.), on trailside banks in moist areas; 3300–4900 feet; fl. sporadically. *A 5744! H 11314! J.P. 677!* Endemic.

2. I. rigens (Sw.) Trin., Gram. Panic.: 252 (1826).

Culms tufted, slender, wiry, glabrous, short and erect or trailing to 1 or 2 m.; leaf-blades 2–5 cm. long, 2–5 mm. broad, spreading; panicles ovoid or oblong in outline, 2–5 cm. long; pedicels 1–2 mm. long; spikelets 1·8–2 mm. long; lemmas often blackish.

Local (St. Andr., Port., St. Thom.), on damp shady banks; 3500–5800 feet; fl. July–Feb. *A 7417! H 11585! J.P. 788!* Ecuador to Venezuela.

3. I. arundinacea (Sw.) Griseb., Fl. Br. W.I.: 553 (1864).

Culms climbing to 6 m., up to 5 mm. thick at base, with strong canes; leaf-blades narrowly lanceolate, up to 2 cm. broad, long-acuminate; panicles up to 12 cm. long and 10 cm. broad, much branched; spikelets somewhat clustered on the branches, about 1·5 mm. long; pedicels 0·5–2 mm. long.

Locally common (St. Andr., Port., St. Thom.), on steep banks and in thickets; 2000–4500 feet; fl. Nov–Mar, July. *A 5718! H 12487! P 5985!* Mexico to Bolivia, St. Kitts, Nevis, Guadeloupe, Trinidad.

51. SCUTACHNE Hitchc. & Chase (1911)

1. S. dura (Griseb.) Hitchc. & Chase in Proc. Biol. Soc. Washington **24**: 149 (1911).

Tufted wiry perennial; culms erect, 40–70 cm. high in flower; leaf-blades stiff, elongated, sharply acuminate, 4–6 (–10) mm. broad; spikelets elliptical, about 5 mm. long; lower glume glabrous; upper glume and sterile lemma pubescent, brownish.

Rare and local (St. Thom.), in thickets on arid limestone rocks; 100 feet; fl. Oct. *H 12163!* Cuba, Hispaniola.

52. SACCIOLEPIS Nash (1901)

1. S. striata (L.) Nash in Bull. Torr. Bot. Club **30**: 383 (1903).

Culms geniculate and decumbent, rooting at the lower nodes, up to 2 m. high in flower, usually shorter; leaf-blades flat, linear-lanceolate, up to 15 cm. long and 10 mm. broad; panicles elongated, compact, the branches up to 5 cm. long; spikelets about 3·5 mm. long, glabrous, on slender flexuous pedicels.

Local and uncommon (St. Eliz., Han., Trel.), in pond margins and swamps; 15–400 feet; fl. July–Dec. *A 8584! H 11749! J.P. 871!* S.E. United States, Greater Antilles.

53. HYMENACHNE Beauv. (1812)

1. H. amplexicaulis (Rudge) Nees, Agrost. Bras.: 276 (1829).
Culms usually trailing, 1 m. or more long; leaf-blades 20–35 cm. long, 2–3 cm. broad, cordate-clasping; panicles 20–50 cm. long, about 8 mm. broad; spikelets 3–4 mm. long, acuminate.
Locally common in shallow ponds and streams, often floating; SL–900 feet; fl. Sept–Mar. *A 10372! H 12470! P 19748!* West Indies.

54. LASIACIS (Griseb.) Hitchc. (1910)

1 Ligule conspicuous, brownish, 2–5 mm. long; plant shrubby, not usually climbing
                                         1. *oaxacensis*
1 Ligule inconspicuous, hidden within the mouth of the sheath, rarely as much as 1 mm.
  long; plants usually climbing and developing strong canes:
  2 Leaf-blades glabrous on both surfaces, often rather scabrid:
    3 Leaf-blades up to about 5 mm. broad   2. *harrisii*
    3 Leaf-blades more than 5 mm. broad:
      4 Panicles loosely flowered, the branches strongly zigzag and spreading or reflexed;
        spikelets about 4 mm. long   3. *divaricata*
      4 Panicles compactly flowered, the branches not zigzag; pedicels short, stiff, appressed;
        spikelets 4·5–5 mm. long   4. *sloanei*
  2 Leaf-blades usually pubescent on at least one surface:
    5 Leaf-blades narrowly lanceolate, about 8–10 times as long as broad; panicle large and
      open; spikelets (3·5–) 4–5 mm. long   5. *maculata*
    5 Leaf-blades ovate-lanceolate to elliptical, often more or less cordate-clasping; panicle
      rather compact; spikelets 3–4 mm. long   6. *ruscifolia*

1. L. oaxacensis (Steud.) Hitchc. in Proc. Biol. Soc. Washington **24**: 145 (1911).
Culms straggling and branching but not canelike and climbing, often decumbent and rooting below, flowering branches to 2 m. high, glabrous; leaf-sheaths glabrous or pubescent, with villous margins; leaf-blades narrowly lanceolate, up to 25 cm. long and 2 cm. broad, scabrid on both surfaces; panicles rather large and open; spikelets 4 mm. long, often purplish.
Occasional, in the central and western parishes, in clearings in woodland on limestone; 300–2000 feet; fl. Dec–Jan. *A 8632! H 12828! Hunnewell 15225!* Mexico to Ecuador, Hispaniola.

2. L. harrisii Nash in Torreya **13**: 274 (1913).
Culms climbing to a height of 5 m. or more, the branches pendulous, light green or tinged purplish; leaf-blades linear to linear-lanceolate, 5–10 cm. long, glabrous; panicles up to 5 cm. long with branches up to about 1 cm. long; spikelets elliptic-ovoid, about 4 mm. long; fruit black.
Locally common (St. Andr., Port., St. Thom.), in thickets and woodlands in steep rocky places; 3000–4500 feet; fl. July–Mar. *A 9878! H 11552! P 23834!* Colombia, Hispaniola, Puerto Rico, Virgin Is.

3. L. divaricata (L.) Hitchc. in Contrib. U.S. Nat. Herb. **15**: 16 (1910).
Culms straggling or climbing to 4 m. or more long, the sterile shoots shorter and zigzag; leaf-blades narrowly lanceolate, 5–12 cm. long, 5–15 mm. broad, usually glabrous; spikelets ovoid, the glumes and lemma glabrous except the tips minutely puberulous; fruit purplish-black.
Very common in secondary thickets and margins of woodland, mostly on limestone; SL–3000 feet; fl. all the year. *A 5501! H 12740! P 7315!* S. Florida, C. and S. Amer., West Indies; Cayman Is.

4. L. sloanei (Griseb.) Hitchc. in Bot. Gaz. **51**: 302 (1911).
Culms climbing to 3 or 4 m.; leaf-blades firm, commonly 12–15 cm. long, 2–3 cm. broad.
Frequent on roadside banks and wooded hillsides; SL–3000 feet; fl. July–Feb. *A 7735! H 12614! Patrick 187!* C. and S. Amer., Greater Antilles, Antigua.

5. L. maculata (Aubl.) Urb., Symb. Ant. **8**: 751 (1921).—*Panicum maculatum* Aubl. (1775). *L. sorghoidea* (Desv.) Hitchc. & Chase (1917).

Culms much branched and climbing to 3–7 m.; leaf-blades up to 20 cm. long and 3 cm. broad, smaller on the low sterile branches; fruit purplish-black.

Rather common in rocky thickets and on wooded hillsides; 600–3800 feet; fl. Oct–Mar. *A 6399! H 11410! P 9626!* Mexico to Argentina and Bolivia, West Indies.

6. L. ruscifolia (Kunth) Hitchc. in Proc. Biol. Soc. Washington **24**: 145 (1911).

Robust climber; leaf-blades on primary shoot 10–15 cm. long, 3–6 cm. broad, the tip shortly acuminate; spikelets nearly globose at maturity.

Only once recorded, *Swartz*. Mexico to Peru, Cuba.

55. ICHNANTHUS Beauv. (1812)

1 Lower glume equal to or exceeding the lemma, long-attenuate tipped; leaf-blades sparsely pilose with thick-based hairs, 2–8 cm. long, 1–2 cm. broad, long-acuminate; spikelets 3–3·5 mm. long, glabrous or scabrid on the nerves   1. *nemorosus*
1 Lower glume shorter than the lemma, finely acuminate:
  2 Leaf-blades lanceolate, often somewhat falcate, tapered from about the middle, scarcely acuminate, glabrous or nearly so, scabridulous, 3–11 cm. long, 8–20 (–25) mm. broad; spikelets 3–3·5 mm. long, glabrous   2. *pallens*
  2 Leaf-blades ovate or ovate-lanceolate, cordate-clasping, often very scabrid adaxially, pubescent beneath, 1·5–3·5 cm. broad; spikelets often sparsely pilose
                                3. *axillaris*

1. I. nemorosus (Sw.) Döll in Mart., Fl. Bras. **2** (2): 289 (1877).

Culms trailing and rooting, sometimes forming loose mats.

Frequent on sheltered banks and along shaded pathsides in woodland; 400–4000 feet; fl. sporadically throughout the year. *A 6517! H 11476! Patrick 276!* Mexico, C. Amer. to Suriname, West Indies.

2. I. pallens (Sw.) Munro in Benth., Fl. Hongk.: 414 (1861).

Culms trailing and rooting, the flowering branches ascending to 30–50 (–80) cm. high; panicles terminal and axillary, 5–10 cm. long.

Common on banks and among rocks in wooded limestone areas; 500–3000 feet; fl. Nov–Apr. *A 5688! H 12757! P 23022!* Throughout tropical Amer. A variant with supernumerary glumes occurs occasionally.

3. I. axillaris (Nees) Hitchc. & Chase in Contrib. U.S. Nat. Herb. **18**: 334 (1917).

Leaf-blades much broader and panicles larger than the last.

Very rare (St. Thom.); 2000 feet; *A.M. Barry!* C. Amer. to Brazil, Hispaniola, Puerto Rico, Trinidad.

56. PANICUM L. (1753)

1 Plants annual, mostly weeds of pastures, waste places and cultivated land:
  2 Spikelets laterally compressed, about 1·7 mm. long, pilose; upper glume keeled and bulged at the middle; panicles few-flowered   1. *trigonum*
  2 Spikelets terete; upper glume not keeled:
    3 Panicle consisting of spikelike racemes, the spikelets more or less secund:
      4 Spikelets strongly reticulate-veined, often tinged yellow or brown, 2–2·5 mm. long
                                2. *fasciculatum*
      4 Spikelets not or only faintly reticulate-veined, usually green:
        5 Spikelets about 2 mm. long, glabrous; culms usually trailing and prostrate
                                3. *reptans*
        5 Spikelets 3–4 mm. long, pubescent or pilose:
          6 Rachis pilose with bristly hairs; leaf-blades and sheaths softly pubescent   4. *molle*
          6 Rachis scabrid but not pilose:
            7 Spikelets with long ascending tubercle-based hairs and short hairs; culms decumbent   5. *echinulatum*
            7 Spikelets with short hairs only mostly towards the tip; culms ascending
                                6. *adspersum*
    3 Panicles diffuse; lower glume acute, about half as long as the spikelet:
      8 Leaf-blades ovate-lanceolate, cordate; spikelets 1·3 mm. long, sparsely hairy
                                7. *trichoides*

8 Leaf-blades linear, not cordate; spikelets about 2 mm. long, glabrous:
  9 Spikelets obovoid, obtuse; inflorescence elongated and composed of several approximated panicles   8. *cayennense*
  9 Spikelets ovoid-elliptical, acutely acuminate; inflorescence a solitary diffusely branched panicle   9. *capillare*
1 Plants perennial, rarely weedy:
  10 Culms of two kinds, the perennating branches short and forming rosettes, the flowering branches longer, mostly erect and often again branched from the nodes:
    11 Leaf-blades markedly ciliate, often light green, flat, up to 6 cm. long; spikelets glabrous or usually so, about 1·5 mm. long:
      12 Leaf-blades glabrous on the surface   10. *polycaulon*
      12 Leaf-blades thinly pilose on the surface   11. *strigosum*
    11 Leaf-blades not distinctly ciliate or if ciliate also generally hairy, often dark green and longer:
      13 Nodes bearded; spikelets pubescent, elliptical, obtuse, about 2 mm. long:
        14 Leaf-blades glabrous   12. *nitidum*
        14 Leaf-blades villous; flowering culms often bushily branched:
          15 Leaf-blades, at least the basal, lanceolate, up to 15 mm. broad, broad at base
                                                                                                            13. *acuminatum*
          15 Leaf-blades linear, tapered to a sharply pointed tip, up to 5 mm. broad, narrow at base   14. *chrysopsidifolium*
      13 Nodes not bearded:
        16 Spikelets elliptical, about 2 mm. long:
          17 Spikelets pubescent; flowering culms bushily branched; lower glume broadly ovate, cuspidate   15. *aciculare*
          17 Spikelets glabrous; flowering culms usually sparingly branched; lower glume ovate   16. *roanokense*
        16 Spikelets tapered to both ends, 3–3·5 mm. long, pubescent   17. *fusiforme*
  10 Culms of one kind, all more or less elongated:
    18 Spikelets shortly pedicelled along one side of racemose panicle-branches, glabrous; leaf-blades linear to linear-lanceolate:
      19 Fruit transversely rugose:
        20 Nodes bearded; spikelets about 3 mm. long; elliptical; lower glume acute
                                                                                                            18. *muticum*
        20 Nodes glabrous; spikelets about 2 mm. long, ovate, shortly acuminate; lower glume truncate   19. *geminatum*
      19 Fruit not transversely rugose; leaf-blade more or less cordate at the base; spikelets 1–1·5 mm. long acute:
        21 Spikelets ovoid, expanded at maturity by the enlarged sterile palea:
          22 Rachis of racemes pilose   20. *pilosum*
          22 Rachis of racemes not pilose   21. *laxum*
        21 Spikelets lanceolate, not expanded at maturity by the enlarged sterile palea; rachis of racemes sparsely pilose   22. *polygonatum*
    18 Spikelets usually long-pedicelled in open of contracted panicles, not secund, not racemose:
      23 Fruit transversely rugose   23. *maximum*
      23 Fruit not transversely rugose:
        24 Lower glume not more than one-fourth the length of the spikelet; culms trailing or straggling:
          25 Spikelets 4–5 mm. long, lanceolate, acuminate, glabrous; culms thick and spongy
                                                                                                            24. *elephantipes*
          25 Spikelets about 1·5 mm. long, elliptical, acute; culms slender   25. *trichanthum*
        24 Lower glume more than one-fourth the length of the spikelet:
          26 Lower glume not more than one-third the length of the spikelet, blunt; spikelets oblong-lanceolate, acute, 2–2·5 mm. long   26. *bartowense*
          26 Lower glume usually more than one-third the length of the spikelet:
            27 Fruit crested at the apex; spikelets 5·5–6 mm. long   27. *zizanioides*
            27 Fruit not crested; spikelets up to 3·5 mm. long:
              28 Panicles narrow and contracted; plants tufted:
                29 Leaf-blades 1–2 mm. broad, involute, up to 4 cm. long; panicles few-flowered   28. *stenodes*
                29 Leaf-blades 5–8 mm. broad, flat or folded towards the base, up to 25 cm. long; panicles many-flowered   29. *condensum*
              28 Panicles open:
                30 Spikelets viscid and adherent; lower glume nearly as long as the spikelet
                                                                                                            30. *glutinosum*
                30 Spikelets not viscid; lower glume about half as long as the spikelet:
                  31 Culms creeping or straggling; spikelets glabrous or nearly so:
                    32 Leaf-blades 1–3 (–5) cm. long, 2–6 mm. broad; panicles few-flowered; culms very slender; fertile lemma obtuse   31. *parvifolium*

32 Leaf-blades 4–10 cm. long, 10–15 mm. broad, cordate; panicles many-
   flowered; fertile lemma apiculate   32. *sciurotis*
31 Culms erect or geniculate-ascending:
   33 Spikelets hirsute, 3·5 mm. long   33. *rudgei*
   33 Spikelets glabrous:
      34 Culms stout, usually more than 1 m. high; sheaths hirsute; spikelets
         about 2 mm. long   34. *hirsutum*
      34 Culms slender, usually less than 1 m. high:
         35 Leaf-blades 1–3 mm. broad; culms spreading and ascending; spikelets
            2·3 mm. long   35. *diffusum*
         35 Leaf-blades mostly more than 5 mm. broad; culms erect; spikelets
            3 mm. long   36. *ghiesbreghtii*

1. P. trigonum Retz., Obs. Bot. **3**: 9 (1783).—*Cyrtococcum trigonum* (Retz.) A.
Camus (1921).
   Culms slender, branching; leaf-blades lanceolate, flat, 1–3 (–6) cm. long, 3–5
mm. broad, sparsely pilose; panicles few-flowered, 1–3 cm. long.
   Rare and local (Port.), in shaded rough pastures on limestone; 25–600 feet; fl.
Nov–Jan. *A 5565!* Native of tropical Asia, also in W. Africa.

2. P. fasciculatum Sw., Nov. Gen. & Sp. Pl.: 22 (1788).—Browntop Millet.
   Culms spreading and ascending or erect, (20–) 30–100 cm. high in flower; leaf-
blade up to 2 cm. broad; racemes often fascicled, usually ascending; spikelets
obovoid, pointed, sometimes accompanied by long guttating hairs on the pedicels,
particularly in southern Caribbean populations.
   A common weed; SL–1500 (–3500) feet; fl. almost all the year. *A 7226! H 12752!
P 7203!* S.E. United States to S. Amer., West Indies.

3. P. reptans L., Syst. Nat. ed. 10, **2**: 870 (1759).—*Brachiaria reptans* (L.) Gardn.
& C. E. Hubbard (1938).
   Culms freely branching, 10–30 cm. high in flower; leaf-blades lanceolate to ovate-
lanceolate, 1·5–6 cm. long, 4–12 mm. broad, cordate; panicle long-exserted, 2–6 cm.
long; pedicels usually with long hairs; lower glume truncate, very short.
   Occasional on roadside banks and in low pastures and thicket margins; SL–700
feet; fl. May–Jan. *A 7366! H 11539! J.P. 838!* Widely distributed in the tropics.

4. P. molle Sw., Nov. Gen. & Sp. Pl.: 22 (1788).
   Culms spreading and ascending; spikelets 3·5 mm. long, pilose.
   Uncommon and casual (St. Andr., St. Cath.), a weed of cultivated ground and
waste places; 200–3000 feet; fl. Aug–Dec. *A 7943! H 11504! Hitchcock 9350!*
Mexico to Argentina, Cuba, Hispaniola.

5. P. echinulatum Mez in Notizbl. Bot. Gart. Berlin **7**: 62 (1917).
   Habit like the next; not always clearly distinguished from it.
   A rare casual (St. James, St. Ann); SL–50 feet; fl. Aug. *J.P. 785A! Patrick 310!
P 10437!* Bolivia, Paraguay, West Indies.

6. P. adspersum Trin., Gram. Panic.: 146 (1826).
   Culms spreading and ascending, sometimes rooting at the lower nodes, up to
30–100 cm. high in flower; leaf-blades 5–15 cm. long, 8–20 mm. broad; racemes
spikelike, 3–10 cm. long; spikelets fusiform, abruptly acuminate, 3·2–4 mm.
long.
   Common as a weed of old open pastures and waste ground, especially in the
southern parishes; SL–3000 feet; fl. Aug–Apr. *A 5531! H 11343! J.P. 834!*
Florida, Bahamas, Greater Antilles to Antigua and Guadeloupe.

7. P. trichoides Sw., Nov. Gen. & Sp. Pl.: 24 (1788).
   Culms branched, more or less erect, 20–40 (–60) cm. high in flower; leaf-blades
thin, 2–6 cm. long, 1–2 cm. broad; panicle much branched, the branches and pedi-
cels capillary.
   Locally common in shade in areas of high rainfall; SL–2000 feet; fl. July–Feb.
A 10328! H 11328! J.P. 573! Throughout tropical Amer. and closely allied to
P. brevifolium L. of the Old World tropics.

8. P. cayennense Lam. in Tabl. Encycl. & Méth., Bot. **1**: 173 (1791).

Culms erect, often branched at the base, 20–50 cm. high; leaf-sheaths and blades pilose; panicles terminal and axillary, open; spikelets turgid, strongly nerved.

Only once collected (Clar.); 2400 feet; fl. Nov. *H 12226!* Honduras to Brazil, Cuba, Hispaniola.

9. P. capillare L., Sp. Pl. **1**: 58 (1753).

Culms erect, 20–80 cm. high in flower; leaf-sheaths densely hispid, blades 10–25 cm. long, 5–15 mm. broad, hispid on both surfaces; very near to *P. hirsutum* in which the leaf-blades are usually glabrous.

Only once collected, *W. Wright!* E. United States, Bermuda, Virgin Is., Trinidad.

10. P. polycaulon Nash in Bull. Torr. Bot. Club **24**: 200 (1897).

Culms short and tufted with approximate leaves, forming small mats or dense rosettes; flowering culms 10–20 cm. long; leaf-blades flat, 3–6 cm. long, 3–6 mm. broad; panicles ovoid, few-flowered.

Very local (St. Andr., Clar.), on clay banks and in open turf in savannas; 2000–3500 feet; fl. sporadically. *A 6691! H 12852! P 9297!* Florida to British Honduras, Greater Antilles.

11. P. strigosum Muhl. ex Ell., Sketch Bot. S. Carol. & Georgia **1**: 126 (1816).

Very like the last; flowering culms 5–30 cm. high, the nodes pubescent.

Rare and local (Clar.), on clay soil in moist savanna; 2000–2400 feet; fl. Nov– Mar. *A & D 13089! H 12225A! P 8463!* E. and S. United States, Mexico to Panama, Cuba, Hispaniola.

12. P. nitidum Lam. in Tabl. Encycl. & Méth., Bot. **1**: 172 (1791).

Erect culms 30–60 cm. high in flower; sheaths glabrous or pubescent, blades 5–12 cm. long, 5–10 mm. broad; panicles ovoid, 5–8 cm. long, with ascending branches.

Only once collected (Clar.); 3000 feet. *Hitchcock 9532.* S.E. United States, Bahamas, Cuba, Hispaniola.

13. P. acuminatum Sw., Nov. Gen. & Sp. Pl.: 23 (1788).

Erect culms 15–40 (–70) cm. high in flower; leaf-blades variable, up to 8 cm. long, sharply acuminate; panicles up to 10 cm. long with flexuous branches.

Locally common on sand or clay roadside banks and in damp hill savannas; 2000–5600 feet; fl. all the year. *A 5775! H 11315! P 23719!* Colombia, Greater Antilles.

14. P. chrysopsidifolium Nash in Small, Fl. Southeast. U.S.: 100, 1327 (1903).

Erect culms 20–30 cm. high in flower; leaf-sheaths villous, blades 5–10 cm. long, villous on both surfaces; panicles 4–6 cm. long.

Local (Clar., St. Ann), in swampy open savanna and thickets; 2000–3000 feet; fl. Sept–Dec. *A 5949! H 12228! J.P. 1524!* S. United States, Greater Antilles.

15. P. aciculare Desv. ex Poir. in Lam., Encycl. Méth. Bot. Suppl. **4**: 274 (1816).

Very like the last; leaf-sheaths variably pubescent.

Rare and local (Clar.), in clay soil in open wet savanna; 2000–2300 feet; fl. Sept– Jan. *A 12931! H 12812! P 9731! Weck T 56!* S.E. United States, Honduras, Greater Antilles.

16. P. roanokense Ashe in Journ. Elisha Mitchell Sci. Soc. **15**: 44 (1898).

Flowering culms erect or ascending, up to 50 cm. or more high, glabrous; leaf-blades flat, 6–9 cm. long, 3–8 mm. broad; panicles 4–8 cm. long with spreading branches.

Rare (Clar., St. Eliz.), in open wet savanna over clay; 50–3000 feet; fl. Mar– Sept. *A 7136! Hitchcock 9530. P 24772.* S.E. United States.

17. P. fusiforme Hitchc. in Contrib. U.S. Nat. Herb. **12**: 222 (1909).

Flowering culms erect, 30–70 cm. high in flower; leaf-blades flat, stiff, (3–) 5–10 cm. long, 4–8 mm. broad, the lowermost softly pubescent beneath; panicles 4–10 cm. long.

Rare and local (Clar.), in hill savannas; 2400–3000 feet; fl. Nov. *H 12234! Hitchcock 9552*. S.E. United States, Cuba, Puerto Rico.

Species 10–17 above belong to the subgenus *Dichanthelium*. They occur mostly in small areas of Clarendon in central Jamaica where open savannas on heavy clay soils provide suitable conditions. *P. polycaulon* extends to the Richmond Straker soils in St. Andrew and *P. acuminatum* to higher ground in the three eastern parishes. They represent the southernmost limits of a flora which is better developed in the south-eastern United States and Cuba. That the Jamaican populations are marginal to a group which has undergone its major evolutionary radiation in subtropical America is borne out by the restriction to cooler elevations, the seasonal dimorphism combined with a tendency to winter flowering and the lack of local endemism.

18. P. muticum Forsk., Fl. Aegypt.-arab.: lx, 20 (1775).—*Brachiaria mutica* (Forsk.) Stapf (1919). *P. purpurascens* Raddi (1823). Para Grass.

Culms decumbent, rooting at the lower nodes and stoloniferous, up to 6 m. or more long, when ascending 1–3 m. high in flower; sheaths villous; leaf-blades 10–15 mm. broad, glabrous; rachis of racemes puberulous and hispid.

Common in ditches, swamps and along riverbanks, sometimes forming extensive colonies, mostly at low elevations; SL–1500 (–3000) feet; fl. Sept–Mar. *A 8343! H 10930! P 23153!* Throughout the tropics.

19. P. geminatum Forsk., Fl. Aegypt.-arab.: lx, 18 (1775).—*Paspalidium geminatum* (Forsk.) Stapf (1920).

Culms spreading, trailing and rooting in water, glabrous, ascending in flower to 25–80 cm. high; leaf-blades 10–20 cm. long, 3–6 mm. broad, flat; racemes erect, alternately superposed, rather distant.

Rather uncommon in fresh or slightly brackish ponds and streams, mostly along the south coast; SL–750 feet; fl. Oct–Feb, June. *A 12053! H 11783 J.P. 806!* Throughout the tropics; Grand Cayman.

20. P. pilosum Sw., Nov. Gen. & Sp. Pl.: 22 (1788).

Culms decumbent, creeping and rooting at the nodes; leaf-blades 4–20 cm. long, 7–15 mm. broad; racemes numerous, 1–3 cm. long, densely flowered.

Frequent in woodland clearings, pastures and savannas, mostly in areas of higher rainfall; 400–3000 feet; fl. June–Mar. *A 7519! H 12566! P 16757!* Mexico to Paraguay, West Indies.

21. P. laxum Sw., Nov. Gen. & Sp. Pl.: 23 (1788).

Like the last; racemes often again shortly branched, the branchlets bearing 2 or 3 spikelets.

Frequent in swamps, pond margins and low lying pastures; SL–3500 feet; fl. May–Dec. *A 7217! H 11644! Patrick 277!* Mexico to Paraguay, West Indies.

22. P. polygonatum Schrad. in Schult., Mant. **2**: 256 (1824).

Culms spreading and branching and rooting at the nodes, up to 1 m. long; nodes densely pubescent; leaf-blades 3–13 cm. long, 8–15 mm. broad.

Rare and local (Port.), along shaded pathsides and woodland margins in wet areas; 1000–1500 feet; fl. Mar–Aug. *A 9337! duQuesnay 337! Robertson UCWI 4606!* Mexico to Paraguay, Trinidad.

23. P. maximum Jacq., Collect. **1**: 76 (1787).—Guinea Grass.

Culms densely tufted, erect, 1–2·5 m. high in flower; leaf-blades flat, elongated, 1–3·5 cm. broad, more or less glabrous except near the junction with the sheath; nodes densely hirsute; panicle large and open, the lower branches whorled; spikelets elongate-elliptical, usually glabrous, about 3 mm. long, the lower glume about one-third the length of the spikelet.

Very common in rough pastures, ditches and sheltered thickets, often originally planted; 250–1250 (–4225) feet; fl. June–Jan. *A 9825! H 11249! Patrick 151!* Native of Africa, now widespread in the tropics; Grand Cayman. There are small and large variants of this species in Jamaica. The smaller (*A 12727!*) normally stands 45–240 cm. high in flower and has leaf-blades up to about 2·5 cm. broad; the larger (*A 12728! Patrick 151!*) stands 180–500 cm. high in flower and the leaves are up to 4 cm. broad. The fact that these two variants may be found growing together suggests that they may be genetically distinct.

24. P. elephantipes Nees, Agrost. Bras.: 165 (1829).

Culms usually creeping widely and rooting at the nodes, branches erect or ascending, up to 2 cm. thick and 1 m. or more high in flower; leaf-blades flat, up to 2 cm. broad; panicle-branches appressed-ascending.

Local and gregarious along low riverbanks and on level swampy ground in the western parishes; SL–10 feet; fl. Nov–Dec. *A 12021! H 11813! P 11106!* C. Amer. to Argentina, Greater Antilles.

25. P. trichanthum Nees, Agrost. Bras.: 210 (1829).

Culms branching and straggling, 1–2 m. long in flower; leaf-blades oblong-lanceolate, 10–15 cm. long, 10–15 mm. broad, cordate; panicles 10–30 cm. long, open.

Very rare (Port., West.), in wet open places; SL–1000 feet; fl. Jan. *Hitchcock. Purdie!* C. Amer. to Paraguay, Greater Antilles, Trinidad.

26. P. bartowense Scribn. & Merr., U.S. Dept. Agr. Div. Agrost. Circ. **35**: 3 (1901).

Culms erect or decumbent at the base, 20–100 cm. or up to 2 m. high in flower; leaf-sheaths hispid, blades pilose on the upper surface; spikelets glabrous.

Only once collected (St. Eliz.); near SL. *Hitchcock 9645!* Florida, Bahamas, Cuba.

27. P. zizanioides Kunth, Nov. Gen. **1**: 100 (1816).—*Acroceras zizanioides* (Kunth) Dandy (1931).

Culms trailing and rooting at the base, ascending to 50–100 cm. high in flower, glabrous; leaf-blades flat, cordate-clasping, 4–15 cm. long 1–3 cm. broad; panicles 10–25 cm. long, few-branched; spikelets shortly pointed; plant coumarin-scented when dry.

Very common on shaded roadside banks and in boggy pastures; 50–2500 feet; fl. all the year. *A 7199! H 11385! J.P. 726!* Widespread in the tropics.

28. P. stenodes Griseb., Fl. Br. W.I.: 547 (1864).

Culms densely tufted, slender and wiry, mostly erect, 25–50 cm. high, often forming dense colonies; panicles 1–2 cm. long; spikelets about 1·5 mm. long, glabrous.

Rare and local (Clar., Manch.), in moist open savanna in sand or clay soil; 2000–2400 feet; fl. June–Jan. *H 12227! P 15957! Weck T 61!* C. Amer. to Brazil, Greater Antilles, Trinidad.

29. P. condensum Nash in Small, Fl. Southeast. U.S.: 93, 1327 (1903).

Culms erect, 1–2 m. high in flower; lower sheaths compressed and keeled; panicles 10–25 cm. long; spikelets 2–2·5 mm. long, lanceolate, glabrous.

Rare and local (St. Eliz., West., Han.), in swamps; 5–50 feet; fl. June–Dec. *A 8586! H 11750! P 23623!* S.E. United States, Bahamas, Greater Antilles, Guadeloupe; Grand Cayman

30. P. glutinosum Sw., Nov. Gen. & Sp. Pl.: 24 (1788).—Ginger Grass.

Culms decumbent, rooting at the base, laxly ascending, 1–2 m. high in flower; leaf-blades elongate-lanceolate, acuminate, 15–50 cm. long, 15–25 mm. broad; panicle open, 15–30 cm. long, the lower branches verticillate, stiffly ascending; spikelets obovoid, obtuse, 3 mm. long.

Common in woodland margins and thicket glades, mostly in wet upland areas; (500–) 2000–5750 feet; fl. all the year. *A 7662! H 12431! P 9882!* Mexico to Paraguay, Greater Antilles.

31. P. parvifolium Lam. in Tabl. Encycl. & Méth., Bot. **1**: 173 (1791).

Culms branching and long-creeping, rooting at the nodes, glaucous, ascending 20–80 cm. in flower; leaf-blades mostly oblong-lanceolate, rounded or subcordate at the base, glabrous; panicles 2–4 cm. long; spikelets turgid, obtuse, 1·5 mm. long.

Rare and local (Clar.), in pools and boggy savanna; 2000–2400 feet; fl. most of the year. *A 12923! H 12221! P 15813!* British Honduras to Paraguay, Greater Antilles, Trinidad.

32. P. sciurotis Trin., Gram. Panic.: 228 (1826).

Culms tufted, trailing and sparingly rooting, 1–2 m. long; leaf-sheaths hispid, blades linear-lanceolate, thinly pubescent; panicle 6–11 cm. long, the branches spreading and again much branched near the main rachis, not whorled; spikelets oblong-elliptical, 1·7–2 mm. long, glabrous or with a few thinly scattered hairs.

Very rare (Manch.), in rough rocky pasture and woodland margin on limestone; 2900 feet; fl. June, Dec. *A 8457! 10098!* Brazil.

33. P. rudgei Roem. & Schult. in L., Syst. Veg. ed. nov. **2**: 444 (1817).

Culms robust, tufted, 30–100 (–200) cm. high in flower; leaf-sheaths and blades pungent-hirsute, blades linear, thick, 15–40 cm. long, up to 10 mm. broad; panicle stiff-branched; spikelets turgid; glumes and sterile lemma acuminate.

Rare and local (Clar.), in open wet savanna; 2000–2400 feet; fl. Sept–Apr. *A 5915! H 12235! P 15958!* C. Amer. to Brazil, Trinidad.

34. P. hirsutum Sw., Fl. Ind. Occ. **1**: 173 (1797).

Culms erect; leaf-blades flat, mostly glabrous, 2–3·5 cm. broad; panicles 20–35 cm. long; spikelets acute, glabrous.

Known only from the type in Jamaica, *Swartz*. Mexico to Brazil, West Indies.

35. P. diffusum Sw., Nov. Gen. & Sp. Pl.: 23 (1788).

Culms densely tufted, ascending from a decumbent base, 25–50 cm. high in flower; leaf-blades 5–20 cm. long, sparsely pilose on the upper surface; panicles 5–10 cm. long, the branches spreading stiffly; spikelets acuminate.

Very rare (St. Cath., Clar.); 180–1000 feet; fl. Oct. *H 12164! Hitchcock 9463!* Bahamas, West Indies to Martinique.

36. P. ghiesbreghtii Fourn., Mex. Pl. **2**: 29 (1886)

Like the last but mostly larger in all parts; culms hispid, 60–80 cm. high in flower; leaf-blades flat, hirsute; spikelets ovoid-lanceolate, narrowly acuminate.

Very rare (Clar.), in savannas; 180–1000 feet; fl. Nov–Jan. *H 12161! Orcutt 7826!* Mexico to Brazil, West Indies from Bahamas to Guadeloupe.

57. TRIPSACUM L. (1759)

1. T. laxum Nash in N. Amer. Fl. **17**: 81 (1909).—Guatemala Grass.

Culms very robust, tufted, decumbent at the base, arising from short thick rhizomes, up to 4 m. high in flower; leaf-blades glabrous, up to 80 cm. long and 9 cm. broad, narrowed at the base to the keeled sheath; spikes fascicled; staminate spikelets 2-flowered; pistillate spikelets with 1 perfect floret and a sterile lemma, sunken into opposite sides of the rachis; style-arms 2, 2·3 cm. long.

Sparingly cultivated and escaped (St. Andr., St. Thom.); 650–1800 feet; fl. Jan. *A 10257!* Native in C. Amer., introduced into some of the West Indian Is.

58. COIX L. (1753)

1. C. lacryma-jobi L., Sp. Pl. **2**: 972 (1753).—Adlay, Job's Tears.

Tufted annual; culms freely branching, lax, 50–150 cm. high in flower, mostly glabrous and waxy-glaucous; leaf-blades up to 30 cm. long, 2–3 cm. broad, cordate-clasping and undulate-crisped at the base; staminate spikelets 2 or 3 in a short raceme on a slender rachis exserted from the involucre; pistillate spikelets 2 or 3 together enclosed by the bony involucre; involucre 8–10 mm. long, ovoid, hard, shiny, bluish-white.

Locally common and gregarious, a weed of ditches and streambanks; SL–2000 feet; fl. Oct–May. *A 7200! H 11294! P 6565!* Native of Asia, now widespread in the tropics.

59. ZEA L. (1753)

1. Z. mays L., Sp. Pl. **2**: 971 (1753).—Indian Corn, Maize.
Erect annual, 1–2 m. high with broad leaves; staminate spikelets in a large terminal panicle, the tassel, 2-flowered; pistillate spikelets in axillary sheathed cobs, arranged in longitudinal rows, the styles long-exserted from the sheaths. Cultivated. *H 11556!* Apparently of American origin; many cultivars have been bred suitable for cultivation in most warm countries.

60. ERIANTHUS Michx. (1803)

1. E. giganteus (Walt.) Muhl., Cat. Pl.: 4 (1813).
Tufted perennial; culms erect, robust, flattened at the base, 150–200 cm. high in flower; leaf-blades long, harshly pubescent; panicles up to 40 cm. long, densely hairy, purplish; spikelets narrowly lanceolate, about 5 mm. long; glumes equal, copiously clothed with long silvery hairs; lemma with a slender antrorsely barbellate awn about 2 cm. long.
Very local (St. Eliz.), in swamps; 50 feet; fl. July, Dec. *A 12061! H & P 14517!* Coastal E. and S.E. United States, Cuba, Hispaniola.

61. IMPERATA Cyrillo (1792)

Spikelets 4 mm. long; panicle rarely more than 15 cm. long   1. *brasiliensis*
Spikelets 3 mm. long; panicle up to 40 cm. long   2. *contracta*

1. I. brasiliensis Trin. in Mém. Acad. Sci. St. Pétersb. sér. 6, **2**: 331 (1832).
Perennial with rhizomes creeping and branching underground; culms erect, tufted, slender, simple, 50–100 cm. high in flower; leaves clustered, erect, 5–13 mm. broad; spikelets pedicelled on a slender rachis, surrounded by long silky hairs, in pairs, awnless.
Local (St. Eliz., Manch., Trel.), in old pastures on heavy soils and in periodically burnt-over areas; 300–2400 feet; fl. sporadically. *A 12448! H 11660! 12880!* S. Mexico to Brazil, Bahamas, Cuba, Hispaniola, Trinidad.

2. I. contracta (Kunth) Hitchc. in Rep. Miss. Bot. Gard. **4**: 146 (1893).
Like the last but taller with leafy culms.
Occasional in undisturbed pastures on heavy soils; 800–4000 feet; fl. June, Sept–Jan. *A 9917! H 11532!* S. Mexico to Brazil, West Indies.

62. SACCHARUM L. (1753)

1. S. officinarum L., Sp. Pl. **1**: 54 (1753).—Sugar Cane.
Tufted perennial with robust ascending solid culms up to 5 m. high in flower; older leaves falling with the sheath; panicle large, feathery; spikelets rarely forming fruit, paired with one sessile the other pedicelled, 2–2·5 mm. long, the basal hairs longer.
Abundantly cultivated, mostly at low elevations on level ground in deep soils; fl. Nov–Feb. *H 11516!* Widespread in the subtropics and tropics.

63. ISCHAEMUM L. (1753)

1. I. rugosum Salisb., Icon. Stirp. Rar. **1**: t. 1 (1791).
Tufted annual; culms geniculate, branched, ascending, 50–100 high in flower; nodes bearded; leaf-blades flat, 8–12 mm. broad, sparsely pilose; racemes paired, conjugate, 5–10 cm. long, erect, closely appressed; rachis disjointing; spikelets 3–4 mm. long, obtuse, the awn about 1·5 cm. long.
Uncommon (St. Andr., St. Cath., West.), a weed of muddy ditches and rice fields; 40–700 feet; fl. May, Oct–Jan. *A 9417! Skelding UCWI 3624!* Throughout the tropics; Cuba, Trinidad.

64. HETEROPOGON Pers. (1807)

1. H. contortus (L.) Beauv. ex Roem. & Schult. in L., Syst. Veg. ed. nov. **2**: 836 (1817).

Tufted perennial; culms compressed, 40–100 cm. high in flower; leaf-blades scabrid, 2–4 mm. broad; racemes solitary, 3–8 cm. long, excluding the awns; spikelets imbricate, the lower awnless, the upper with brown bent awns 5–8 cm. long.

Rather uncommon, in heavy soil on steep well drained banks; 50–1200 (–3500) feet; fl. almost all the year. *A 7738! H 11259! P 22785!* Widespread in the tropics.

65. THEMEDA Forsk. (1775)

Perfect spikelets villous; glumes of sessile neuter spikelets not strongly papillose; awns 5–7 cm. long   **1. arguens**

Perfect spikelets not villous; glumes of sessile neuter spikelets strongly papillose, the papillae bearing long stiff hairs; awns mostly less than 5 cm. long; cultivated; native in India; Kangaroo Grass   *T. quadrivalvis* (L.) Kuntze

1. T. arguens (L.) Hack. in A. & C. DC., Monogr. Phan. **6**: 657 (1889).—Christmas Grass, Piano Grass.

Annual; culms spreading and branching, compressed, up to 120 cm. high in flower; leaf-blades flat, scabrid; panicles flabellate with each raceme subtended by a leaf-like spathe; fertile spikelets awned.

Common along roadside banks and verges; SL–2000 feet; fl. June–Mar. *A 7333! H 12702! P 8259!* Native of Asia, introduced into Barbados and Guadeloupe.

66. HACKELOCHLOA Kuntze (1891)

1. H. granularis (L.) Kuntze, Revis. Gen. Pl. **2**: 776 (1891).

Annual; culms few, more or less erect, 30–100 cm. high in flower; leaf-blades up to 10 cm. long, 5–10 mm. broad, flat; racemes simple, terminal and axillary, 1–2·5 cm. long; first glume of the sessile spikelet globose, alveolate, clasping the rachis-joint and pedicel of the stalked spikelet, purplish-black at maturity.

Rare (St. Andr., Clar., Han.), in eroded hillside pastures on clay or coarse sandy soils; 500–2500 feet; fl. May, Sept–Nov. *A 8252! H 11334! P 8261!* A common pantropical weed.

67. ROTTBOELLIA L.f. (1779) nom. cons.

1. R. exaltata (L.) L. f., Nov. Gram. Gen.: 37 (1779).—*Aegilops exaltata* L. (1771).

Tufted annual; culms firm, ascending-branched, to 2 m. high in flower; leaf-sheaths pungent-hispid, blades elongated, flat, scabrid; racemes several, axillary, terete, tapered towards the tip with abortive spikelets; rachis-joints hollow above, disarticulating readily in fruit.

Rather common on mostly shaded or sheltered, often rocky, banks; SL–2000 feet; fl. Oct–Feb. *A 5973! H 11387! P 22928!* Widely distributed in the tropics.

68. HEMARTHRIA R. Br. (1810)

1. H. altissima (Poir.) Stapf & C. E. Hubbard in Kew Bull. **1934**: 109 (1934).— *Manisuris altissima* (Poir.) Hitchc. (1934).

Perennial with erect or spreading slender glabrous culms, 50–100 cm. high in flower; leaf-blades flat or folded, 2–5 mm. broad; racemes 5–10 cm. long, falcate, from the upper sheaths, compressed; spikelets awnless, the sessile appressed and sometimes partly adnate.

Very rare (St. Eliz.), in ditches and shallow water; near SL; fl. Sept–Jan. *H 12550!* Native in the Old World tropics; Mexico to Argentina.

69. SORGHUM Moench (1794) nom. cons.

1 Plants perennial with scaly stolons; leaf-blades up to 2 cm. broad; sessile spikelets
   4–5 mm. long and 2 mm. broad, pubescent; awns numerous, few or absent
                                                                     1. halepense
1 Plants annual or short-lived, without stolons or rhizomes:
  2 Leaf-blades up to about 3 cm. broad:
    3 Sessile spikelets villous, 6–7 mm. long, about 2·5 mm. broad   2. *verticilliflorum*
    3 Sessile spikelets thinly pubescent with a few short ascending hairs, about 5 mm. long,
      2 mm. broad; cultivated, a casual introduction; native of the Old World tropics;
      Sudan Grass   *S. sudanense* (Piper) Stapf
  2 Leaf-blades up to about 5 cm. broad; culms erect, up to about 2 m. high; inflorescence
    compact; spikelets turgid, persistent; sparingly cultivated; of Old World origin;
    Guinea Corn, Sorghum; many cultivars   *S. saccharatum* (L.) Moench

1. S. halepense (L.) Pers., Synops. Pl. **1**: 101 (1805).—Johnson Grass.

Culms more or less erect or rarely trailing and rooting, brittle at the nodes, 1–1·5
m. high in flower; leaf-blades usually less than 1·5 cm. broad with a prominent
midrib; panicle 15–25 cm. long, open in flower; awns when present 7–10 mm. long,
deciduous.

Locally common, gregarious and forming colonies, a persistent weed of some
pastures and stony waste ground; 600–3600 feet; fl. Apr–June, Oct–Jan. *A 10211!*
H 11582! A.M. Barry! P 24811! Native of the Old World, introduced in the West
Indies and the Pacific Is.

2. S. verticilliflorum (Steud.) Stapf in Prain, Fl. Trop. Afr. **9**: 116 (1917).—
Andropogon verticilliflorus Steud. (1854).

Culms ascending from the base, stout, supported by numerous strong prop
roots from the lower nodes, 1·2–2·5 m. high in flower; leaf-blades light grey-green,
glaucous; panicle large with spreading and ascending branches; spikelets awned.

Occasional as a weed of cane, field-margins and ditches; 50–2900 feet; fl. May–
Aug, Nov– Feb. *A 9950! P 8302!* Native of E. Africa; Antigua.

70. SORGHASTRUM Nash (1901)

1. S. setosum (Griseb.) Hitchc. in Contrib. U.S. Nat. Herb. **12**: 195 (1909).

Tufted perennial; culms erect, glabrous, up to 2 m. high in flower; leaf-blades
long and narrow, flat or involute; panicle-branches slender; spikelets about 4 mm.
long, light golden-brown tinged green; lower glume and sterile pedicel clothed with
white hairs; awn of the fertile lemma straight or slightly bent, not strongly twisted
at the base.

Local and uncommon (St. Andr., St. Ann), on open grassy hillsides and pathside
banks; 1000–4000 feet; fl. June–Aug, Nov. *A 7496! H 11437! J.P. 812!* Mexico to
Argentina, Greater Antilles, Trinidad.

71. ARTHRAXON Beauv. (1812)

1. A. quartinianus (A. Rich.) Nash in N. Amer. Fl. **17**: 99 (1912).

Culms slender, trailing and much branched, flowering branches ascending
weakly to 20–50 cm. high; leaf-blades ovate, long-acuminate, flat, hispid, ciliate,
3–7 cm. long, 1–2 cm. broad; racemes slender, terminal in flabellate fascicles, 2·5–
3·5 cm. long; perfect spikelet 3 mm. long, awned.

Locally common (St. Andr., Port., St. Thom.), on sheltered banks, pathsides and
ditches; 2600–5000 feet; fl. Oct–July. *A 6233! H 10910! P 23215!* Native of
Ethiopia; Guadeloupe.

72. VETIVERIA Lem.-Lisanc. (1822)

1. V. zizanioides (L.) Nash in Small, Fl. Southeast. U.S.: 67, 1326 (1903).—
Khus Khus.

Tufted perennial forming large compact tussocks; culms erect, branched, 1–2
m. high in flower; leaf-sheaths keeled, blades glabrous, firm, 30–100 cm. long,

4–10 mm. broad, scabrid on the margin; panicle elongate-pyramidal, 20–30 cm. long, the branches whorled; racemes up to 5 cm. long; spikelets about 4 mm. long, covered with short persistent spines, often purplish.

Mostly cultivated and rarely found far from roads or habitations; SL–3750 feet; fl. June, Sept–Jan. *A 8072! H 11557! A.M. Barry!* Native of Asia, now widely cultivated in the tropics.

73. ANDROPOGON L. (1753)

1 Plants aromatic; racemes paired, subtended by spathes, reflexed; lowermost pair of spikelets of one or both racemes alike (homogamous), staminate or neuter and thus similar to the pedicelled spikelets above:
  2 Lower glume of sessile (fertile) spikelets shallowly concave on the back, 5–6 mm. long, 0·7 mm. broad   1. *citratus*
  2 Lower glume of sessile spikelets flat in the upper half on the back, 4–4·5 mm. long, 1–1·1 mm. broad   2. *nardus*
1 Plants not aromatic; racemes solitary, paired or several:
  3 Lowest 1–3 pairs of spikelets homogamous, staminate or neuter; racemes several together; lower glume of sessile spikelet not pitted:
    4 Peduncles slender, flexuous; racemes nodding; joints of rachis and pedicels with a transluscent longitudinal furrow; primary peduncle glabrous   3. *condylotrichus*
    4 Peduncles stiff; racemes erect; joints of rachis and pedicels without a translucent longitudinal furrow; primary peduncle densely pubescent distally   4. *aristatus*
  3 All pairs of spikelets heterogamous, the pedicelled staminate, neuter or suppressed:
    5 Racemes solitary:
      6 Pedicelled spikelets much larger than the sessile; upper glume awned
                                5. *fastigiatus*
      6 Pedicelled spikelets not conspicuously larger; upper glume awnless:
        7 Plants annual; culms weak and trailing; leaf-blades obtuse, mostly 1–3 cm. long; peduncles capillary; awns 4–8 mm. long   6. *brevifolius*
        7 Plants perennial; culms erect:
          8 Rachis slender, densely white-villous with long hairs; leaf-blades involute; spikelets spreading; awns 1–2 cm. long   7. *gracilis*
          8 Rachis more or less thickened; spikelets appressed or ascending; awns about 1 cm. long:
            9 Lower glume of sessile spikelet villous; leaf-blades flat, 2–4 mm. broad
                                  8. *hirtiflorus*
            9 Lower glume of sessile spikelet glabrous; leaf-blades somewhat involute, 1–3 mm. broad   9. *tener*
    5 Racemes paired to several; perennial:
      10 Joints of rachis and pedicels with a translucent midline; racemes usually more than 2:
        11 Pedicelled spikelets about the same size as the sessile, 4·5–5 mm. long:
          12 Lower glume of sessile spikelets pitted   10. *pertusus*
          12 Lower glume of sessile spikelets not pitted   11. *ischaemum*
        11 Pedicelled spikelets much smaller than the sessile; lower glume of sessile spikelet not pitted   12. *saccharoides*
      10 Joints of rachis and pedicels without a translucent midline; racemes usually paired:
        13 Spikelets awnless:
          14 Plants robust with culms to 2 m. high; spathes aggregate in a corymbose, usually dense inflorescence; 1, sometimes 2, of the uppermost pedicellate spikelets larger than the fertile ones, the other pedicellate spikelets rudimentary
                                  13. *bicornis*
          14 Plants slender, up to 1 m. high; spathes not aggregate; sessile spikelets about 3 mm. long   14. *leucostachyus*
        13 Spikelets awned:
          15 Spathes aggregate in a compound flabellate or club-shaped inflorescence
                                  15. *glomeratus*
          15 Spathes not aggregate; peduncles of racemes less than 1 cm. long   16. *virginicus*

1. A. citratus DC., Cat. Hort. Monsp.: 78 (1813).—*Cymbopogon citratus* (DC.) Stapf (1906). Fever Grass, Lemon Grass.

Robust perennial forming dense tussocks; leaf-blades tapered to both ends, up to 1 m. long, 5–15 mm. broad with scabrid margins; panicles, rarely formed in Jamaica, 30–60 cm. long; sessile spikelets linear to linear-lanceolate.

Common in cultivation in gardens and along pathsides; 650–1500 feet; fl. Nov–Jan. *H 11706!* Probably of Indian origin, now widely cultivated in the tropics.

2. **A. nardus** L., Sp. Pl. **2**: 1046 (1753).—*Cymbopogon nardus* (L.) Rendle (1899). Citronella Grass.
Like the last; sessile spikelets lanceolate.
Rare (St. Andr.), in cultivation; 700 feet; fl. Dec. *Harris!* Scattered through the tropics.

3. **A. condylotrichus** Hochst. ex Steud., Synops. Pl. Glum. **1**: 377 (1854).— *Euclasta condylotricha* (Hochst. ex Steud.) Stapf (1917).
Annual with weak branching culms 1–2 m. high; nodes villous; leaf-blades 15–20 cm. long, 4–8 mm. broad, setaceous-pointed; racemes 2–5 cm. long, loosely pilose; pedicelled spikelets about 5 mm. long.
Rare (St. Andr.), perhaps less common than formerly; 900–3500 feet; fl. Nov– Jan. *H 12460! Swartz!* Tropical Amer. and Africa, casual in Asia; Greater Antilles.

4. **A. aristatus** Poir. in Lam., Encycl. Méth. Bot. Suppl. **1**: 585 (1811).—*Dichanthium aristatum* (Poir.) C. E. Hubbard (1939). *A. annulatus* of Hitchc. (1936), not Forsk. (1775).
Perennial; culms decumbent, terete, branched from the base, up to about 80 cm. long, the nodes pubescent otherwise glabrous except immediately below the spikes; leaf-blades up to 20 cm. long, 3–5 mm. broad, tapered to a fine point; racemes 1–3 (–4), 3–7 cm. long; lower glumes overlapping, villous, about 3 mm. long.
Very rare (St. Thom.), in riverside gravel; near SL; fl. Sept. *J. Morley 186!* Native of Asia, widely introduced in the tropics.

5. **A. fastigiatus** Sw., Nov. Gen. & Sp. Pl.: 26 (1788).—*Diectomis fastigiata* (Sw.) Kunth (1816).
Erect sparingly branched annual with slender culms 50–150 cm. high in flower; leaf-blades up to 4 mm. broad; racemes 2–6 cm. long; joint of rachis and pedicel winged towards the upper part, villous with long white hairs; pedicellate spikelets 8 mm. long; awn of fertile lemma geniculate, 4–5 cm. long.
Rather rare and local (St. Andr., Manch.), on open rocky limestone; 800–4000 feet; *A 9953! H 11288! 11507!* Widely distributed in the tropics, Greater Antilles, Grenada.

6. **A. brevifolius** Sw., Nov. Gen. & Sp. Pl.: 26 (1788).—*Schizachyrium brevifolium* (Sw.) Nees ex Buse (1854).
Culms slender, 15–100 cm. long with the internodes often coloured bright red-dish-purple; leaf-blades flat; racemes delicate, 1–2 cm. long; spikelets about 3 mm. long.
Occasional on clay or sand roadside banks and open savannas, sometimes a weed; 300–4500 feet; fl. Nov–Jan. *A 5912! H 12711! P 15956!* Widespread in the subtropics and tropics.

7. **A. gracilis** Spreng. in L., Syst. Veg. ed. 16, **1**: 284 (1824).—*Schizachyrium gracile* (Spreng.) Nash (1903).
Culms wiry, densely tufted, 20–60 cm. high; leaf-blades filiform; racemes long-exserted, 2–5 cm. long; sessile spikelet about 5 mm. long; pedicellate spikelet reduced to an awned or awnless glume, the pedicel very villous.
Frequent on well drained stony or rocky banks; 50–5000 feet; fl. sporadically throughout the year. *A 11452! H 11265! P 6795!* S. Florida, Bahamas, Greater Antilles, Guadeloupe.

8. **A. hirtiflorus** (Nees) Kunth, Révis. Gram.: 569 (1832).—*Schizachyrium hirtiflorum* Nees (1829). *S. domingense* (Spreng. ex Schult.) Nash (1912).
Culms compressed, tufted, 50–150 cm. high; racemes 4–8 cm. long; rachis, joints and pedicels villous; sessile spikelets 5–6 mm. long.
Very rare (Manch.), in savannas, 300–900 feet, fl. Nov. *H 12704!* Florida to Paraguay, Greater Antilles, Trinidad.

9. **A. tener** (Nees) Kunth, Révis. Gram.: 565 (1832).—*Schizachyrium tenerum* Nees (1829).
Culms slender, reclining, 30–100 cm. long, from dense tufts; racemes slender,

2–5 cm. long, nearly glabrous; lower half of lower glume of sessile spikelets lighter coloured giving the raceme a zonate appearance.

Rare (St. Andr., Port.); 1000–4000 feet; fl. Nov–Jan. *H 11439! 12865!* S. United States to Argentina, Cuba, Hispaniola.

10. A. pertusus (L.) Willd. in L., Sp. Pl. ed. 4, **4**: 922 (1806).—*Bothriochloa pertusa* (L.) A. Camus (1931). Seymour Grass.

Stoloniferous with laxly erect culms 20–100 cm. high in flower; leaf-blades 10–20 cm. long, 1–4 mm. broad; racemes villous, 2–6 cm. long; awns about 15 mm. long, twice geniculate, the distal half black.

Abundant in the drier more disturbed areas; SL–2600 feet; fl. Mar–June, Sept–Dec. *A 8014! H 11262! Patrick 148!* Throughout the tropics; Grand Cayman.

11. A. ischaemum L., Sp. Pl. **2**: 1047 (1753).—*Bothriochloa ischaemum* (L.) Keng (1936).
Like the last.

Very rare (St. Andr.); 3500 feet; fl. Nov. *H 11471!* Old World tropics, West Indies.

12. A. saccharoides Sw., Nov. Gen. & Sp. Pl.: 26 (1788).—*Bothriochloa saccharoides* (Sw.) Rydberg (1931).

Culms erect, simple, brittle, 60–150 cm. high; nodes white-hispid; leaf-blades flat, glabrous or nearly so, 2–8 mm. broad; panicles exserted, 5–15 cm. long, pale, silky with numerous racemes; awns 1·5–2 cm. long.

Uncommon; 100–3500 feet; fl. Oct–Mar. *H 12213! J.P. 570!* S.W. United States to S. Amer., Greater Antilles, St. Kitts, Antigua, Martinique.

13. A. bicornis L., Sp. Pl. **2**: 1046 (1753).

Culms densely tufted, more or less terete, weak at the base, 1·2–2 m. high; leaf-blades 2–5 mm. broad, scabrid on the margin; racemes 2–3 cm. long, feathery with hairs as much as 5 mm. long; one, sometimes two, of the uppermost pedicelled spikelets larger than the fertile ones, the other pedicelled spikelets rudimentary.

Common on roadside banks, savannas and rough pastures; 500–5000 feet; fl. most of the year. *A 7820! H 12605! P 6938!* Mexico to Brazil, West Indies.

14. A. leucostachyus Kunth, Nov. Gen. **1**: 187 (1816).

Culms tufted, slender, erect, 30–50 cm. high; leaf-blades acuminate, 1–3 mm. broad; racemes 2 or 3, whitish, 1·5–3 cm. long on slender exserted peduncles; spikelets obscured by copious long silky hairs.

Rather uncommon on open banks; 1200–3600 feet; fl. Nov–Jan, Apr, July. *A 10087! H 12458! P 9729!* S. Mexico to Brazil, Greater Antilles, Trinidad.

15. A. glomeratus (Walt.) Britton, Sterns & Poggenburg, Prel. Cat. New York: 67 (1888).

Culms tufted, robust, flattened and covered with crowded keeled sheaths, 1–1·5 m. high in flower; leaf-blades 3–5 mm. broad; inflorescence dense, club-shaped, the spathes at first green turning pink; racemes 1·5–3 cm. long, villous with whitish or tawny hairs; sessile spikelets 3–4 mm. long; awns about 15 mm. long.

Very common in ditches, swamp margins, pastures and waste places; SL–5000 feet; fl. June–Feb. *A 7500! H 12545! Patrick 251!* S.E. United States and Mexico to northern S. Amer., Greater Antilles, St. Kitts to Martinique; Grand Cayman; Hawaii.

16. A. virginicus L., Sp. Pl. **2**: 1046 (1753).

Culms densely tufted, covered with compressed keeled sheaths, 60–150 cm. high; leaf-blades 2–4 mm. broad; racemes 1·5–3 cm. long, more or less included in the spathes; rachis and pedicels villous with white hairs 3–5 mm. long; awns delicate, straight, 10–15 mm. long.

Rather common on rocky slopes and well drained roadside banks; 100–5000 feet; fl. July–Dec. *A 8588! H 11362! J.P. 734!* Bermuda, Bahamas, E. United States to Panama, Greater Antilles, Hawaii.

37. CASUARINACEAE

Evergreen trees and shrubs with numerous slender jointed whorled and striate branches. Leaves reduced to toothed sheaths. Flowers unisexual, monoecious or dioecious, without perianth. Male flowers in whorls within successive sheaths towards the branch-tips, each comprising 4 serrulate bracteoles and a central solitary stamen; filament lengthening during flowering; anther basifixed, 4-locular, opening lengthwise. Female flowers capitate, each subtended by one bract and a pair of bracteoles; ovary superior, 1-locular; style short, terminal, with 2 elongating linear stigmatic branches; ovules 2, collateral on a single parietal placenta. Fruits crowded into a woody cone of bracts in which the paired bracteoles have become indurated and open like a capsule to expose a samaroid indehiscent 1-seeded nut. Seed without endosperm; embryo straight.

1 genus indigenous in the southern hemisphere; 40–50 species.

1. CASUARINA Adans. (1763)

1. **C. equisetifolia** J. R. & G. Forst., Charact. Gen. Pl.: 104 (1776).—Casuarina, Whistling Pine, Willow.
  Tree 8–25 (–35) m. high with a slender trunk and ascending branches; first season's branches 10–30 cm. long with internodes 4–8 mm. long, (6–) 7–8-grooved, slender, drooping; flowers monoecious; male flowers in terminal spikes mostly about 20 (2–30) mm. long; multiple fruit 12–22 mm. in diameter.
  Common, mostly in sandy coastal areas and often planted; SL– about 1000 feet; male fl. Feb–Apr, fr. most of the year. *Yuncker 17813!* Native of tropical Asia and Australasia, naturalized in West Indies and elsewhere; Grand Cayman.

38. PIPERACEAE

Erect or climbing herbs or shrubs or small trees; stems with distinct and often scattered vascular bundles. Leaves alternate, opposite or verticillate, usually entire, often gland-dotted, sometimes succulent, frequently aromatic when dried; stipules when present adnate to petiole. Flowers mostly bisexual, very small, subtended by sessile or stipulate usually peltate bracts in dense pedunculate spikes. Perianth absent. Stamens (1–) 2–6 (–10), usually free, hypogynous; anthers of 2 distinct or confluent loculi, opening lengthwise. Ovary superior, 1-locular; ovule solitary, basal, orthotropous; stigmas 1–5. Fruit very small, drupaceous, with fleshy or dry pericarp. Seed small, with endosperm; embryo minute.

10–12 genera with 1400 mostly tropical species.

1 Stigma 1; plants herbaceous, erect or creeping, epiphytic or terrestrial; leaves alternate, opposite or whorled; floral bracts round-peltate, usually glabrous; stamens 2
                                                          1. **Peperomia**
1 Stigmas 2–5; undershrubs, shrubs or trees, often with thickened nodes; leaves alternate; floral bracts usually fringed; stamens 2–6:
2 Spikes solitary, leaf-opposed   2. **Piper**
2 Spikes several on a common axillary peduncle   3. **Pothomorphe**

1. PEPEROMIA Ruiz & Pav. (1794); Yuncker (1960)

1 Leaves opposite or verticillate:
2 Leaves mostly more than 2 cm. long:
  3 Stem glabrous; stigma apical   14. *penicillata*
  3 Stem more or less hairy; stigma lateral:
    4 Stem puberulous   15. *stellata*
    4 Stem hirsute or pilose:
      4 Leaves obovate to rounded-subrhomboid, obtuse   16. *polystachya*
      5 Leaves ovate-elliptical, acute   17. *blanda*

2 Leaves mostly less than 2 cm. long:
  6 Leaves usually less than 3 mm. broad and commonly 3 or more times longer than broad:
    7 Stems slender, prostrate; leaves loosely villous with scattered hairs; spikes terminal, mostly solitary   *5. swartziana*
    7 Stems stouter, erect or suberect, moderately to densely hirtellous; spikes terminal and axillary, often several together   *6. galioides*
  6 Leaves broader:
    8 Leaf-apex rounded, emarginate, the blade obovate   *7. quadrifolia*
    8 Leaf-apex not emarginate or only slightly so:
      9 Stem slender, to about 1 mm. thick when dry, pubescent; leaves obovate   *10. lewisii*
      9 Stem thicker:
        10 Stem hirsute with at least some hairs up to 1 mm. long:
          11 Pubescence on stem a mixture of long hairs and an understory of shorter hairs   *8. verticillata*
          11 Hairs all more or less the same length without an understory of shorter hairs   *9a. barbata* var. *barbata*
        10 Stem minutely hirtellous, puberulous or glabrescent:
          12 Leaves very little longer than broad, 3–4 at a node; rachis hirtellous   *11. reflexa*
          12 Leaves mostly up to twice as long as broad or longer; rachis not hirtellous:
            13 Leaves narrowed to a more or less pointed apex:
              14 Leaves mostly 3–5 at a node, somewhat rhomboid, the apex acuminate and bluntly pointed, drying firm and more or less leathery   *12. rhombea*
              14 Leaves mostly 2–3 at a node, elliptical, the apex subacute, drying membranous   *13. discolor*
            13 Leaves rounded at apex, elliptic-obovate:
              15 Leaves mostly 10–15 mm. long, commonly opposite   *9b. barbata* var. *puberula*
              15 Leaves mostly 3–8 mm. long, commonly 3–4 at a node   *8. verticillata*
1 Leaves alternate (occasional upper leaves may appear to be opposite):
  16 Leaves mostly less than 2 cm. long:
    17 Leaves hardly 5 mm. long, the apex emarginate   *2. emarginella*
    17 Leaves 5 mm. or more long, or if smaller, not emarginate:
      18 Leaf-base cordate:
        19 Blade mostly more than 15 mm. broad   *1. pellucida*
        19 Blade mostly less than 10 mm. broad   *3. hispidula*
      18 Leaf-base not cordate:
        20 Leaves elliptic- or round-obovate, the apex retuse   *18. cordifolia*
        20 Leaves not obovate or, if so, the apex not retuse:
          21 Leaves black- or brown-dotted; petiole ciliate; stem with subnodal hairy lines   *19. glabella*
          21 Leaves and stem without the above characters combined:
            22 Leaves rounded or hardly longer than broad:
              23 Leaves mostly 1–2 cm. broad   *27. serpens*
              23 Leaves mostly less than 1 cm. broad:
                24 Blade palmately 3-nerved; stem long, slender, creeping   *20. rotundifolia*
                24 Blade palmately 5-nerved; stems not usually long and creeping   *3. hispidula*
            22 Leaves mostly 1·5–2 times longer than broad:
              25 Leaves usually broadest at or below the middle:
                26 Petioles mostly 2–3 mm. long; leaves 8 mm. or less broad   *4. tenella*
                26 Petioles 1–2 (–3) cm. long; leaves mostly 1–2 cm. broad   *27. serpens*
              25 Leaves usually broadest at or above the middle:
                27 Leaves both opposite and alternate, often on the same plant; spikes mostly 2–3 cm. long   *10. lewisii*
                27 Leaves always alternate; spikes mostly 1–1·5 cm. long   *21. fawcettii*
  16 Leaves mostly more than 2 cm. long:
    28 Stem and/or leaves more or less hairy (obscurely so in *P. glabella* and *P. cubensis*):
      29 Leaves mostly more than 10 cm. long, narrowly peltate   *28. maculosa*
      29 Leaves less than 10 cm. long:
        30 Leaves peltate about 1 cm. from base   *29. hernandiifolia*
        30 Leaves not or only slightly peltate:
          31 Petioles ciliate; stem with subnodal hairy lines, otherwise glabrous; blades black- or brown-dotted beneath   *19. glabella*
          31 Not as above:
            32 Leaf-blade more or less pubescent beneath, at least near the base, glabrous adaxially; spikes 3–4 cm. long   *30. distachya*

32 Leaf-blade glabrous beneath, commonly puberulous adaxially, at least along the
    nerves:
    33 Leaf-base cordate; petiole 3–8 cm. long; spike 1·5–2 cm. long   31. *cubensis*
    33 Leaf-base cuneate; petiole 1–3 cm. long:
        34 Spikes mostly 3–6 cm. long; blades mostly less than 6 cm. long, broadest at
            about the middle   37. *alpina*
        34 Spikes mostly 7–16 cm. long; blades mostly more than 6 cm. long, broadest
            above the middle   36. *crassicaulis*
28 Plants glabrous throughout, except sometimes leaf-tips ciliolate and peduncles of
    P. obtusifolia microscopically hirtellous:
    35 Leaf-blade cordate; stems much branched and tenderly herbaceous
                                                                          1. *pellucida*

    35 Leaf-blade not cordate:
        36 Leaf-blades mostly 2–5 (–6) cm. long:
            37 Peduncles mostly 1–1·5 cm. long; leaves gland-dotted:
                38 Leaves palmately nerved; petiole 2–3 mm. long   22. *proctorii*
                38 Leaves pinnately nerved; petiole 5 mm. long   38. *jamaicana*
            37 Peduncles hardly 1 cm. long:
                39 Stems pendent, up to 1 m. long; internodes 3–5 cm. long; leaves apparently
                    not gland-dotted   23. *harrisii*
                39 Stems not pendent, shorter:
                    40 Leaves mostly broadest at or below the middle, not gland-dotted; upper
                        internodes commonly 1–2 cm. long, winged, more or less zig-zag
                                                                              24. *alata*
                    40 Leaves mostly broadest at or above the middle, gland-dotted:
                        41 Leaf-blades mostly 2–3 times longer than broad, pale pellucid-dotted;
                            internodes 5–10 mm. long   25. *simplex*
                        41 Leaf-blades hardly twice as long as broad, black-dotted; internodes mostly
                            2–4 cm. long   26. *guadaloupensis*
        36 Leaf-blades mostly more than 5 cm. long:
            42 Leaf-apex pointed, acute to acuminate:
                43 Leaves palmately nerved or subplinerved; internodes winged   24. *alata*
                43 Leaves pinnately nerved:
                    44 Leaf-apex sharply and somewhat obliquely acuminate; spikes solitary, 3–5 mm.
                        thick   32. *acuminata*
                    44 Leaf-apex acute to subacuminate, the tip rather blunt, not curved; spikes
                        mostly paired:
                        45 Leaves petiolate, the blade-base decurrent on the petiole; spikes 2–3 mm.
                            thick   33. *talinifolia*
                        45 Leaves sessile, the blade oblanceolate and narrowed to an auriculate base;
                            fruit-beak very short   34. *amplexicaulis*
            42 Leaf-apex obtuse to broadly rounded:
                46 Fruit-beak less than half as long as the ovary, not abruptly hooked:
                    47 Petiole very short; blade-base narrowly auriculate; plants intermediate
                        between *P. amplexicaulis* and *P. clusiifolia*
                    47 Petiole distinct with the blade-base decurrent on it   35. *clusiifolia*
                46 Fruit-beak mostly half to as long as the ovary:
                    48 Leaves strongly gland-dotted beneath; floral bracts 3–4 visible at the same
                        level; fruit-beak strongly subulate, not hooked; peduncle glabrous
                                                                              36. *crassicaulis*
                    48 Leaves not usually strongly gland-dotted; floral bracts 4–5 (–6) visible at the
                        same level:
                        49 Fruit-beak slender, usually abruptly hooked at apex; peduncle commonly
                            microscopically hirtellous   39. *obtusifolia*
                        49 Fruit-beak subulate, more or less curved but not abruptly hooked; peduncle
                            glabrous   40. *magnoliifolia*

1. P. pellucida (L.) Kunth, Nov. Gen. **1**: 64 (1816).—Man-to-Man, Pepper
    Elder, Rabbit Ear, Rat Ear, Ratta-Temper.
    Soft succulent herb up to 30 cm. or more high; leaves broadly ovate, acute,
palmately 5-nerved; spikes slender, mostly 2–5 cm. long; fruit ellipsoidal, longitudinally striate.
    Common as a weed of damp shady places; SL–1250 feet; fl. and fr. all the year.
A 6117! H 6951! P 24613! Yuncker 17829! General in the tropics.

2. P. emarginella (Sw.) C. DC. in DC., Prodr. **16** (1): 437 (1869).
    Creeping herb; stem filiform, pinkish, glabrous; leaves rounded, palmately 3-
nerved, 3–5 mm. broad, glabrous or sparsely hairy; spikes about 1 cm. long; fruit
ellipsoidal, borne on a slender stalk.

Local (Port., St. Thom.), on mossy tree trunks and rotten logs in forest; 700–2000 feet; fl. and fr. Mar–Apr, July–Sept. *A 7886! P 10472! Yuncker 18569!* C. and northern S. Amer., West Indies.

3. P. hispidula (Sw.) A. Dietr. in L., Sp. Pl. ed. 6, **1**: 165 (1831).

Small delicate herb with sparingly hispid or glabrescent stems; leaves ovate-orbicular, rounded or slightly emarginate at tip, palmately 5-nerved, thinly hispid adaxially, glabrous beneath; spikes slender, up to about 1 cm. long; fruit pedicellate, globose, about 1 mm. long, covered with white hairs.

Uncommon (St. Andr., Port., St. Thom.), on damp mossy ground; 4800–7400 feet; fl. and fr. Apr, Aug, Dec. *H 8320! P 23492!* In the strict sense also in Hispaniola and Martinique; a very similar but more robust plant occurs in C. and S. Amer.

4. P. tenella (Sw.) A. Dietr. in L., Sp. Pl. ed. 6, **1**: 153 (1831).

Stem shortly trailing, slender, red-spotted, thinly hirsute, up to 8 cm. long; leaves ovate to ovate-lanceolate, obtusely pointed, palmately 3-nerved, light green or often blotched reddish, 7–15 mm. long, 3–8 mm. broad; spikes solitary, terminal, mostly 2–3 cm. long; fruit pedicellate, obpyriform, about 2 mm. long, brown.

Locally common (St. Andr., Port., St. Thom.), epiphytic on mossy trees or on boulders in wet montane forest; (2000–) 3400–6200 feet; fl. and fr. Dec–Aug. *A 7039! H 8323! P 6680!* Northern S. Amer., West Indies.

5. P. swartziana Miq., Syst. Pip.: 155 (1843).—*P. filiformis* (Sw.) A. Dietr. (1831), not Ruiz & Pav. (1798).

Stoloniferous herb with erect slender shoots up to 6 cm. high; leaves 2–4 at a node, linear-subspathulate to obovate, 5–10 (–15) mm. long, 1·5–3 mm. broad; spikes 1–4 cm. long, the peduncle 5–10 mm. long; fruit globose-ovoid.

Occasional (St. Andr., Port., St. Thom.), on mossy rocks, logs or trees in wet shady places; 3500–5000 feet; fl. and fr. Jan–Sept. *A 7745! 12110! H 7781! P 6835!* Northern S. Amer., West Indies.

6. P. galioides Kunth, Nov. Gen. **1**: 71, t. 17 (1816).

Erect branched herb 20–50 (–100) cm. high; leaves 3–9 at a node, elliptical to oblanceolate, gland-dotted and glabrous beneath, ciliolate near apex, 5–20 (–30) mm. long, 2–5 mm. broad; spikes slender, 3–7 (–15) cm. long; fruit globose-ovoid, 0·8–1 mm. long.

Rare (St. Andr., Port., St. Thom.), in montane woodlands; 5400–5750 feet; fl. and fr. Apr, Nov. *H 5498! Loveless UCWI 1623!* Throughout tropical Amer.

7. P. quadrifolia (L.) Kunth, Nov. Gen. **1**: 69 (1816).

Ascending-branched glabrous succulent herb up to 10 cm. high; leaves (3–) 4 (–6) at a node, elliptical to obovate, palmately 3-nerved, 7–15 mm. long, 5–9 mm. broad; spikes terminal, mostly 2–3 cm. long; fruit ellipsoidal with a short pseudo-cupule.

Frequent (St. Andr., Port., St. Thom.), on rocks and tree trunks in forest; 3600–5800 feet; fl. and fr. most of the year. *A 6363! H 8162! P 24554! Yuncker 18727!* Honduras, northern S. Amer., West Indies.

8. P. verticillata (L.) A. Dietr. in L., Sp. Pl. ed. 6, **1**: 179 (1831).

Branched erect or ascending-stemmed herb 12–20 (–30) cm. high; stem pubescent and with or without longer hairs, often red; leaves 3–6 at a node, elliptical to obovate or rounded, puberulous on both surfaces, thick, often reddish beneath, 5–20 mm. long, 3–8 (–10) mm. broad; spikes up to 7 cm. long; fruit globose-ovoid.

Occasional (St. Andr., Port., St. Thom.), mostly on mossy rocks, sometimes on tree trunks; 1000–5000 feet; fl. and fr. Aug–Feb. *A 9877! H 8331! J.P. 703! P 24052!* Cuba, Hispaniola.

9. P. barbata C. DC. in Urb., Symb. Ant. **5**: 297 (1907).

9a. Var. barbata.

Branched ascending or erect herb up to 15 cm. high; stem red; leaves elliptical or rounded, rounded at apex, acute at base, dark green adaxially, red beneath, villous on both surfaces, 10–15 (–20) mm. long, 6–13 mm. broad; spikes terminal and axillary, 3–7 cm. long; fruit globose, about 0·8 mm. long.

Local in the central parishes, in crevices of limestone rocks; 1650–2500 feet; fl. and fr. Mar–Sept. *A 7553a! H 8531! P 22751!* Endemic.

9b. P. barbata var. **puberula** Yuncker in Bull. Inst. Jam., Sci. ser. **11**: 36 (1960).
Like the last but whole plant puberulous only.

Local (Clar., Manch., St. Ann), in limestone crevices in forest; 2000–3000 feet; fl. and fr. Jan–July, Oct. *A 7553! P 8419! Robertson UCWI 2002!* Endemic.

10. P. lewisii Proctor in Rhodora **61**: 218 (1959).
Stem simple or branched, decumbent and ascending to 5 cm. or more high or pendent, hirsute; leaves obovate to subspathulate, round at apex, cuneate at base, 8–15 (–20) mm. long, 6–10 mm. broad, palmately 3-nerved, hirsute on both sides; fruit globose-ovoid, about 0·8 mm. long.

Rare and local (Port.), in crevices of limestone cliffs in forest; 1500–2250 feet; fl. and fr. Mar–Apr, July–Aug. *A 9123! ACH 966! P 10473!* Endemic.

11. P. reflexa (L. f.) A. Dietr. in L., Sp. Pl. ed. 6, **1**: 180 (1831).
Stems branching, spreading or tufted, the fruiting stems to about 5 cm. long, succulent, minutely hirtellous or glabrescent; leaves rhomboid-elliptical to ovate or suborbicular, fleshy, palmately 3-nerved, 8–15 mm. long, 6–12 mm. broad, fragrant when dry; spikes terminal, 1–3 cm. long, stout; fruit subcylindrical, with basal pseudocupule.

Locally common (St. Andr., Port., St. Thom.), gregarious on mossy tree trunks, logs and boulders; 3800–6250 feet; fl. and fr. Dec–July. *A 7701! H 6515! P 6596! Yuncker 18356!* General in the tropics.

12. P. rhombea Ruiz & Pav., Fl. Peruv. & Chil. **1**: 31, t. 46, f. b (1798).—
P. myrtillus Miq. (1843). *P. rhomboides* Dahlst. (1900).
Stems creeping and branching from the base, ascending to 10–20 cm. high; leaves palmately 3-nerved, glabrous or sparsely puberulous beneath, 1–2 (–3) cm. long, 5–10 mm. broad; spikes terminal, fleshy, up to 6 cm. long on peduncles up to 4 cm. long; fruit subglobose-ovoid, about 1 mm. long.

Occasional (St. Andr., St. Thom.), on the mossy trunks and bases of trees; 4900–5000 feet; fl. and fr. Apr, June, Aug. *H 8326! P 6806!* Northern S. Amer., Cuba, Puerto Rico.

13. P. discolor C.DC. in DC., Prodr. **16** (1): 463 (1869).
Stems slender, branched, rooting below, ascending to 9 cm. high, minutely velvety hirtellous; leaves palmately 3-nerved, minutely puberulous, 12–18 mm. long, 5–8 mm. broad; spikes slender, 3 cm. long, the peduncle 4 mm. long.

Very rare (St. Thom.), on ledges of shaded cliffs; 1000–1250 feet; fl. Feb. *P 7669!* Endemic.

14. P. penicillata C.DC. in Urb., Symb. Ant. **5**: 297 (1907).—*P. septemnervis* of Griseb. (1860), not Ruiz & Pav. (1798).
Glabrous branched herb up to about 30 cm. high; leaves (1–) 2–3 (–4) at a node, lanceolate to oblanceolate, acuminate at tip, cuneate at base, palmately 5–7 nerved, 3–8·5 cm. long, 1–2 cm. broad; spikes terminal or axillary, solitary or paired, up to 14 cm. long on peduncles 2–4 cm. long; fruit ellipsoidal, about 0·8 mm. long, the pseudocupule extending to about the middle.

Rare and local (St. Andr., St. Thom.), on rocky banks or epiphytic; 3600–6000 feet; fl. and fr. Apr–May. *Bengry! H 8327!* Cuba, Hispaniola.

15. P. stellata (Sw.) A. Dietr. in L., Sp. Pl. ed. 6, **1**: 175 (1831).
Branched more or less erect herb 30–50 (–70) cm. high; leaves (2–) 3–4 (–5) at a node, lanceolate to oblong-elliptical, subacuminate to obtuse at tip, acute or obtuse at base, glabrous adaxially, light green, paler beneath and finely gland-dotted, (1·5–) 2–4 (–6) cm. long, 5–15 mm. broad; spikes slender, readily caducous, 3–7 (–8) cm. long; fruit globose-ovoid, about 0·8 mm. long.

Rather common (St. Andr., Port., St. Thom.), terrestrial and on mossy rocks in submontane forest; 3650–5000 feet; fl. and fr. Aug–May. *A 10584! H 8317! P 21950!* Endemic.

16. P. polystachya (Ait.) Hook., Exot. Fl. **1**: t. 23 (1823).

More or less erect herb; leaves (2–) 3 (–4) at a node, cuneate at base, sparsely pubescent on both surfaces, ciliolate, palmately 3-nerved, 3–4 cm. long, about 2 cm. broad; spikes terminal and axillary, solitary or paired, 6–8 cm. long; fruit ovoid.

An obscure species possibly not distinct from the next and not confirmed for Jamaica.

17. P. blanda (Jacq.) Kunth, Nov. Gen. **1**: 67 (1816).

Stems simple or branching, ascending to 30 cm. or more high; leaves (1–) 2–3 at a node, acute at tip, acute or obtuse at base, palmately 3–5-nerved, villous on both surfaces, ciliate, 2·5–5 cm. long, 1–2·5 cm. broad; spikes usually several, terminal and axillary, slender, 3–10 (–14) cm. long; fruit globose-ovoid, about 1 mm. long.

Not confirmed for Jamaica and perhaps erroneously reported. Mexico to Brazil, West Indies.

18. P. cordifolia (Sw.) A. Dietr. in L., Sp. Pl. ed. 6, **1**: 154 (1831).

Scandent trailing slender-stemmed much branched herb, rooting from nodes; leaves fleshy, drying very thin, light green or reddish beneath, often thinly puberulous adaxially, 1–2 cm. long, 5–15 mm. broad; spikes terminal, very slender, readily caducous, 1–3 cm. long; fruit suborbicular.

Locally common, an epiphyte on mossy tree trunks and hanging from branches in wet woodlands; (200–) 1200–4000 feet; fl. and fr. Mar–Aug. *A 6510! H 9011! P 23877! Yuncker 18083!* Dominica.

19. P. glabella (Sw.) A. Dietr. in L., Sp. Pl. ed. 6, **1**: 156 (1831).

Stem branching and trailing extensively, pinkish to reddish-brown, up to 1 m. long; leaves ovate to elliptic-lanceolate, acute to acuminate at tip, acute to obtuse at base, 1·5–4 (–5·5) cm. long, 8–20 (–28) mm. broad; petiole 5–10 mm. long with crisped-ciliate adaxial margins; spikes terminal, and from upper axils, 5–8 (–12) cm. long; fruit globose-ovoid.

Common, on tree trunks, rocks and limestone banks in damp shady places; SL–4000 feet; fl. and fr. Aug–May. *A 6088! 8404! H 8336! P 17474! Yuncker 18213!* Throughout tropical Amer.; Grand Cayman.

20. P. rotundifolia (L.) Kunth, Nov. Gen. **1**: 65 (1816).

Stems branching, filiform, crisped-puberulous or glabrescent; leaves round or rounded-elliptical, more or less pubescent, plano-convex, drying thin, up to 12 mm. long and 10 mm. broad; spikes solitary at the ends of short erect branches, mostly about 2 cm. long; fruit globose-ovoid.

Frequent locally in St. Ann and the eastern parishes, on tree trunks in damp shady forests; 500–2800 feet; fl. and fr. Feb–Aug. *A 7643! H 8363! P 23800! Yuncker 18566!* Throughout tropical Amer., also tropical Africa and Madagascar.

21. P. fawcettii C. DC. in Urb., Symb. Ant. **5**: 295 (1907).

Stems branching, filiform, creeping, hirtellous to pubescent, with erect branches up to 5 cm. high; leaves elliptical to rounded-obovate or suborbicular, acute at base, pubescent on both surfaces, ciliolate, palmately 3-nerved, 5–10 (–15) mm. long, 3–7 (–10) mm. broad, pale gland-dotted beneath; fruit globose-ovoid.

Occasional (St. Andr., Port., St. Thom.), on mossy tree trunks and on logs in montane forests; (1500–) 3900–5500 feet; fl. and fr. Jan, June–July. *Fawcett 8362! P 22527! Yuncker 18580!* Endemic.

22. P. proctorii Yuncker in Bull. Inst. Jam., Sci. ser. **11**: 43 (1960).

Glabrous succulent herb with simple or branched stems, swollen at the nodes, up to 30 cm. or more high; leaves oblong-elliptical to ovate-lanceolate, bluntly pointed, rounded to broadly acute at base, up to 5·5 cm. long and 3·3 cm. broad, thick, 5–7-nerved, black-glandular-dotted; spikes terminal and axillary, slender, 2·5–12 cm. long; fruit globose-ovoid, nearly 1 mm. long.

Occasional in the central parishes, on rock ledges in woodland on limestone; 700–3000 feet; fl. and fr. Aug–Jan. *P 11038! R 1854!* Endemic.

23. P. harrisii C. DC. in Urb., Symb. Ant. **3**: 243 (1902).

Glabrous herb with slender branching stems up to 1 m. long; leaves narrowly oblong-elliptical to lanceolate, acuminate, narrowly acute at base, ciliolate, drying thin, 3–6 cm. long, 1–2 cm. broad; spikes solitary, terminal or axillary, 10–12 cm. long.

Very rare (St. Andr., Han.), epiphytic or trailing and pendent on peaty rocks; 500–5000 feet; fl. Mar–Apr. *H 8322! 8481!* Endemic.

24. P. alata Ruiz & Pav., Fl. Peruv. & Chil. **1**: 31, t. 48, f. b (1798).—*P. dendrophila* of Griseb. (1860), not Schlecht.

Glabrous herb with often simple ascending stems to 30 cm. high, rooting at the base; leaves elliptic-lanceolate to broadly elliptical, acuminate, acute at base, 5–7-nerved, 3–8 cm. or more long, 1–3 (–4) cm. broad, rather fleshy, paler beneath; spikes terminal and from the upper axils, 5–10 (–15) cm. long; fruit globose-ovoid, about 1 mm. long.

Rather frequent, on rocks and epiphytic on tree trunks in shady woodlands; 2000–5000 feet; fl. and fr. Jan–July. *A 5862! H 8330! J.P. 1265! P 21922!* Mexico to Peru, West Indies.

25. P. simplex Ham., Prodr. Pl. Ind. Occ.: 2 (1825).—*P. hamiltoniana* Miq. (1845).

Glabrous herb with trailing or ascending stems up to 20 cm. or more high, rooting at the base; leaves elliptical to elliptical-lanceolate, acute at apex, acute at base, sparsely ciliolate at apex, 2·5–4 (–5·5) cm. long, 1–2 (–3) cm. broad, rather fleshy, paler beneath; spikes solitary or paired, terminal or axillary, up to 8 cm. long; fruit globose, about 0·8 mm. long.

Occasional, mostly on rocks in woodland on limestone; near SL–4400 feet; fl. and fr. Nov–July. *A 6387! P 6332! Robertson UCWI 5480!* Endemic.

26. P. guadaloupensis C. DC. in Journ. Bot. **4**: 139 (1866).

Glabrous like the last but more robust; leaves up to 6 cm. long and 3·5 cm. broad, thick and leathery; spikes up to 14 cm. long.

Rare (Clar., Manch.), on walls, rocks and boulders in shaded places; 2350–2600 feet; fl. and fr. Feb, Sept. *A 12293! H 8334!* Cuba, Guadeloupe, St. Lucia.

27. P. serpens (Sw.) Loudon, Hort. Brit.: 13 (1830).

Stems branched, trailing or scandent, pubescent, speckled red, up to about 30 cm. long; leaves broadly rounded-ovate, obtuse or acute, mostly truncate at base, more or less crisped-puberulous on both surfaces, fleshy, paler beneath, 1–2 cm. long and broad; spikes terminal or axillary, 1–3 cm. long; fruit oblong-ellipsoidal, puberulous, brownish, 1 mm. long including beak.

Rather common, on mossy tree trunks and roots, rocks and walls in wet limestone areas; 50–2200 feet; fl. and fr. all the year. *A 7604! H 9954! P 6251! Yuncker 18406!* Guatemala to south continental Amer., West Indies.

28. P. maculosa (L.) Hook., Exot. Fl. **2**: t. 92 (1825).

Stems thick and fleshy, light green mottled with brownish-crimson blotches, more or less pubescent, ascending to 30 cm. from a rooting base; leaves ovate-elliptical, shortly acuminate, rounded to truncate-subcordate at base, dark green and shiny adaxially, greenish-white beneath, 10–18 cm. long, 6–11 cm. broad; petiole up to 15 cm. long; spikes terminal, solitary or paired, up to 25 cm. long; fruit ellipsoidal, about 1·5 mm. long.

Local (St. Andr., St. Thom.), on rocks, cliffs and mossy tree trunks in woodland; 2400–4200 feet; fl. and fr. Dec–June. *H 8312! Palmer UCWI 9469! P 23097!* Panama to northern S. Amer., Greater Antilles.

29. P. hernandiifolia (Vahl) A. Dietr. in L., Sp. Pl. ed. 6, **1**: 157 (1831).

Stem creeping and rooting, minutely velvety puberulous, reddish-purple; leaves rounded-ovate, sharply acuminate, rounded at base, ciliate, minutely puberulous on both surfaces, paler beneath, 5–9 cm. long, 3–6 cm. broad; petiole up to 8 cm.

long; spikes axillary, 2–3 (–4) cm. long on a longer peduncle; fruit ellipsoidal, 1–1·3 mm. long.

Rather local (St. Andr., Port., St. Thom.), terrestrial or as a low epiphyte in moist forest; 1500–4500 feet; fl. and fr. Oct–July. *A 5853! ACH 385! H 10119! Barkley & Proctor 22J255!* C. and S. Amer., West Indies.

30. P. distachya (L.) A. Dietr. in L., Sp. Pl. ed. 6, **1**: 156 (1831).

Stem creeping and rooting or scandent, sparsely pubescent or glabrescent, speckled red; leaves ovate to ovate-elliptical, acuminate, acute to obtuse or rounded at base, thinly pubescent and paler beneath, 5–9 (–10) cm. long, 3–5 (–6·5) cm. broad; spikes terminal or axillary, mostly paired, sometimes 3, 3–4 cm. long on peduncles up to 2 cm. long; fruit ellipsoidal, about 1 mm. long.

Locally common (St. Ann and the eastern parishes), epiphytic or on rocks, especially near streams; 550–5000 feet; fl. and fr. Nov–Aug. *A 9022! H 8346! P 6439! Yuncker 18565!* Northern S. Amer., Hispaniola, Puerto Rico.

31. P. cubensis C. DC. in Journ. Bot. **4**: 142 (1866).

Stem creeping or scandent, rooting at the nodes, glabrous; leaves broadly ovate, acuminate, 5–7 cm. long, 3·5–5 cm. broad, paler beneath with 5–7 prominent nerves; spikes paired on axillary branches; peduncles 5–15 mm. long; fruit sub-cylindrical, about 1 mm. long.

Rare and local (St. Ann), on limestone boulder in sheltered wooded gully; 900 feet. *P 15545!* Cuba.

32. P. acuminata Ruiz & Pav., Fl. Peruv. & Chil. **1**: 32, t. 51, f.a (1798).—*P. basellifolia* Kunth (1816).

Glabrous robust herb with simple or sparingly branched stem 20–40 (–60) cm. high, rooting at base; leaves elliptical to oblanceolate, decurrent-cuneate at base, minutely dark-dotted and otherwise much paler beneath, 6–13 (–14) cm. long, 2–4 (–5) cm. broad; spikes terminal, solitary, 8–13 (–20) cm. long; fruit obovoid-ellipsoidal, about 1 mm. long.

Local (St. Andr., Port., St. Thom.), epiphytic, on rocks or terrestrial in mossy forest; 4200–7300 feet; fl. and fr. Apr, July, Dec. *A 11350! H 8318! P 6092! Yuncker 18734!* Northern S. Amer., Cuba, Hispaniola.

33. P. talinifolia Kunth, Nov. Gen. **1**: 62, t. 8 (1816).

Simple or branched glabrous herb, rooting at base; leaves oblanceolate to elliptical, glandular-dotted, pinnately nerved with 4–5 nerves on either side of midrib, 6–11 cm. long, 2·5–4 cm. broad; spikes terminal and axillary, solitary or paired, up to 13 cm. long; fruit globose-ellipsoidal.

Very rare (Port., St. Thom.), epiphytic in wet forest; 700–4000 feet; fl. Aug. *A 7883!* Colombia.

34. P. amplexicaulis (Sw.) A. Dietr. in L., Sp. Pl. ed. 6, **1**: 144 (1831).—Jackie's Saddle.

Glabrous rather robust herb, the lower stem trailing, rooting and branched, ascending to 30 cm. high; leaves thick and leathery, 9–22 cm. long, 1·5–4 (–5·5) cm. broad; spikes terminal and axillary, solitary or commonly paired, 8–18 cm. long, peduncles mostly 2–6 cm. long; floral bracts usually (5–) 6–7 visible at the same level; fruit ellipsoidal, about 1 mm. long.

Very common on limestone roadside banks and rocks in woodlands; SL–3000 feet; fl. and fr. all the year. *A 12809! H 8689! P 22895!* Endemic.

35. P. clusiifolia (Jacq.) Hook. in Curt., Bot. Mag. **56**: t. 2943 (1829).

Habit and aspect of the last; leaves obovate to oblong-obovate, 5–15 cm. long, 3–5 (–8) cm. broad; spikes terminal and axillary, 1–3 (–4) together, sympodially stalked, 10–20 cm. long, peduncles up to 7 cm. or more long, often tinged yellow and red; floral bracts 4–5 (–6) visible at the same level; fruit ellipsoidal, nearly 1·5 mm. long.

Common on coral limestone in thickets and open woodlands near the sea, and in exposed rocky places at higher elevations; 15–3000 (–3500) feet; fl. and fr. Dec–Sept. *A 10785! 12148! H 8359! Webster & Wilson 4853!* Probably endemic. Many plants combine characters of *P. clusiifolia* and *P. amplexicaulis* suggesting that hybrids occur (cf. Yuncker, T.G., Bull. Inst. Jam., Sci. ser. **11**: 50 (1960)).

36. P. crassicaulis F. & R. in Journ. Bot. **50**: 177 (1912).

Robust simple or sparingly branched herb rooting at the lower nodes, ascending to 40 cm. or more high; leaves obovate or elliptic-obovate, cuneate-decurrent on petiole, 5–8 (–12) cm. long, 2–5 cm. broad, glabrous on both surfaces or puberulous on the nerves adaxially, drying thick and leathery; spikes terminal, usually 2 or 3, 7–16 cm. long; fruit ellipsoidal, about 1·5 mm. long including the beak.

Local (St. Andr., Port., St. Thom.), terrestrial or a low epiphyte or on rotting logs and boulders, often gregarious in montane forest; 4000–6800 feet, fl. and fr. Dec–Aug. *A 7460! H 8104! J.P. 1481! P 6809!* Endemic.

37. P. alpina (Sw.) A. Dietr. in L., Sp. Pl. ed. 6, **1**: 185 (1831).

Simple or branched-stemmed herb rooting at the lower nodes, with ascending branches to 30 cm. high, often blackish towards the base; leaves broadly elliptical to obovate, puberulous adaxially, paler, gland-dotted and glabrous beneath, 3–5 cm. long, 1·5–3·5 cm. broad, drying firm; spikes terminal or from the upper axils, solitary or paired, 3–6 (–14) cm. long, peduncle up to 3·5 cm. long; floral bracts 2 or 3 visible at the same level; fruit ellipsoidal, the beak about as long as the ovary, altogether about 1·5 mm. long.

Local (St. Andr., Port., St. Thom.), on rocks or in peaty soil in montane forest; 4200–5050 feet; fl. and fr. sporadically throughout the year. *A 11620! 12540! H 8316! P 6903!* ? Mexico, ? Hispaniola. A plant from the same general area and with similar ecology but with glabrous leaves gland-dotted adaxially and other minor differences is represented by *P 22552!* and may be distinct.

38 P. jamaicana Yuncker in Bull. Inst. Jam., Sci. ser. **11**: 51 (1960).

Small herb with stem rooting at the lower nodes, ascending to about 10 cm.; leaves elliptical or elliptic-oblanceolate, bluntly pointed at tip, decurrently cuneate at base, strongly gland-dotted, 2·5–4 cm. long, 8–17 mm. broad; spikes terminal 5-10 mm. long, peduncle 1·5–2 cm. long; ovary ellipsoidal.

Very rare and local (Port.), epiphytic in mossy forest; about 2000 feet. *P 5725!* Endemic.

39. P. obtusifolia (L.) A. Dietr. in L., Sp. Pl. ed. 6, **1**: 154 (1831).

Rather short-stemmed robust herb, trailing and rooting and ascending to 20 cm. or more; leaves elliptic-obovate, cuneate-decurrent at base, thick and leathery, 5–15 cm. long, 3·5–6·5 cm. broad; spikes usually terminal, solitary or paired, 5–15 cm. long, peduncles 2–4 cm. long; fruit ellipsoidal, about 1·3 mm. long including beak.

Rather local and uncommon (St. Andr., Port., St. Thom.), on rocks and tree trunks in damp sheltered places; 1700–3000 feet; fl. and fr. Nov–Jan. *A 9986! A.M. Barry! Robertson UCWI 3492!* Florida, C. Amer., West Indies.

40. P. magnoliifolia (Jacq.) A. Dietr. in L., Sp. Pl. ed. 6, **1**: 153 (1831).

Glabrous herb with a stout stem rooting at the lower nodes, ascending to 25 cm. or more high; leaves obovate-elliptical to spathulate, often emarginate, cuneate-decurrent at base or abruptly contracted, 6–15 cm. long, 3–8 cm. broad; spikes up to 15–18 cm. long, peduncles 1·5–3 cm. long; fruit ellipsoidal.

Very rare (St. Thom.), on rocky banks; 1000–1250 feet. *P 7446!* Panama to Venezuela, West Indies.

2. PIPER L. (1753); Yuncker (1960)

1 Climbing plants introduced from the Old World tropics and occasionally cultivated; flowers usually unisexual:
  2 Spikes up to 2 cm. long with the fruits sunken in the glabrous fleshy rachis; leaves
    rounded-cordate at base   *P. sarmentosum* Roxb.
  2 Spikes longer; leaf-base unequal-sided:
    3 Fruit somewhat embedded in the fleshy rachis; spikes about as long as the leaves
      (10–17 cm. long); leaf-base obtuse to rounded or cordate   *P. betle* L.

3 Fruit rounded, free from the rachis; spikes nearly as long as the leaves (10–14 cm. long); leaf-base acute; Black Pepper   *P. nigrum* L.
1 Erect shrubs or small trees; indigenous species; flowers bisexual; Jointers:
  4 Principal lateral nerves arising at or very near base of blade:
    5 Nerves all arising at the base; margins of blade equal or subequal at base and usually minutely auriculate   1. *amalago*
    5 Nerves of the innermost lateral pair coalescent with the midrib 5–10 mm. above the base; one side of blade about 2 mm. shorter than the other at the base, acute and hardly auriculate   2. *discolor*
  4 Principal lateral nerves arising from the midrib:
    6 Principal lateral nerves arising within the proximal half to two-thirds of the blade:
      7 Leaf-blade deeply unequally cordate-auriculate, the basal lobes 3–5 cm. or more long; petiole strongly winged to the apex   3. *auritum*
      7 Leaves not as above:
        8 Spikes regularly curved, the peduncle longer than the petioles; blades scabrid adaxially   4. *aduncum*
        8 Spikes usually straight or nearly so:
          9 Leaves broadly rhombic-elliptic-subovate; blade smooth or scabridulous adaxially   5. *dilatatum*
          9 Leaves ovate to elliptic-lanceolate:
            10 Stem glabrous or minutely puberulous; adaxial surface of blade smooth, glabrous or minutely scabridulous   6. *murrayanum*
            10 Stem pubescent to densely villous, the hairs clearly visible to the unaided eye; adaxial surface of blade scabrid:
              11 Adaxial surface of leaves scabrid only; stem pubescent to hispid with rather short hairs   7. *hispidum*
              11 Adaxial surface of leaves scabrid and thinly villous; stem densely villous with long hairs   8. *fadyenii*
    6 Midrib giving off lateral nerves more or less throughout its whole length; petiole winged to the blade; blade obviously shorter on one side at the base:
      12 Leaf-tip very shortly and abruptly acute or obtuse; leaf-blade mostly less than 12 cm. long, more or less puberulous   9. *tuberculatum*
      12 Leaf-tip gradually acute to acuminate; leaf-blade mostly longer, glabrous or glabrescent:
        13 Stem smooth or nearly so   10. *arboreum*
        13 Stem heavily warted; leaves glaucous   11. *verrucosum*

1. P. amalago L., Sp. Pl. **1**: 29 (1753).—*P. wullschlaegelii* C. DC. (1869).

1a. Var. amalago.
Usually glabrous shrub 1·2–5 m. or tree to 6 m. high; leaves narrowly lanceolate to broadly ovate, elliptical or oblong, acuminate at tip, acute to rounded at base, 4–14 cm. long, 2–7 cm. broad; spikes shorter or longer than the leaves; fruit ovoid-pointed, light green.
Very common, on gully banks, roadsides and in thickets and woodlands on limestone and on coral sand near the sea; SL–4100 feet; fl. and fr. all the year. *A 6036! H 8353! P 26608!* When construed in a broad sense to include the slightly more hairy *P. medium* Jacq., this species extends throughout the West Indies and tropical continental Amer.; also in Grand Cayman.
Yuncker recognized two additional named variants in Jamaica:

1b. P. amalago var. **nigrinodum** (C. DC.) Yuncker in Bull. Inst. Jam., Sci. ser. **11**: 9 (1960).—*P. nigrinodum* C. DC. (1907). Black Jointer.
Leaves ovate, 4·5–7 cm. long, 2·5–4·5 cm. broad; spikes up to twice as long as leaves.
Occasional mostly in the central parishes, in woodland and pasture margins on limestone; 400–3000 feet; fl. and fr. most of the year. *A. 5601! H 9075! P 23084!* Endemic.

1c. P. amalago var. **variifolium** (Griseb.) F. & R., Fl. Jam. **3**: 21 (1914).
Leaves 3-lobed, the lateral lobes very short.
Very rare (Manch.), and known only from the type, *Wullschlaegel*. Endemic.

2. P. discolor Sw., Nov. Gen. & Sp. Pl.: 15 (1788).
Shrub 1·5–3 m. high with slender glabrescent twigs; leaves elliptical, bluntly acuminate, glabrous, finely granular-dotted and paler beneath, 5–13 cm. long, 3–7 cm. broad; spikes shorter than the leaves, white; fruits distant, ovoid-ellipsoidal, pointed, white with green tips.

Uncommon, in wet shady woodlands on limestone, shale or clay soils; 1000–5500 feet; fl. and fr. Mar–Oct. *A & S 3320! H 7447! P 20680!* Endemic.

3. P. auritum Kunth, Nov. Gen. **1**: 54 (1816).

Soft-wooded aromatic shrub 1–2 m. or tree to 6 m. high; leaves ovate to ovate-elliptical, acute or subacuminate, 20–30 (–40) cm. long, 12–20 (–25) cm. broad, the margin densely white-ciliate; spikes whitish to light yellow, pendulous, 10–25 cm. or more long, 3–5 mm. thick; fruit 3-gonous.

Rare (St. Cath., Clar.), on open shale roadside banks and in thickets on clay soil; 600–2300 feet; fl. and fr. Mar, July, Oct. *P 6972! Robertson UCWI 6761! Yuncker 18360!* Native Mexico to Colombia, cultivated and escaped Cuba.

4. P. aduncum L., Sp. Pl. **1**: 29 (1753).

Shrub or tree 1·5–6 m. high; leaves elliptical to lanceolate, long-acuminate, unequally rounded or shortly cordate with one side 3–4 mm. shorter at base, scabrid and pubescent adaxially, pellucid-dotted, 12–20 (–24) cm. long, 4–7 (–10) cm. broad; spikes 8–15 cm. long; fruit obovoid, somewhat angled.

Common in thickets and secondary vegetation at woodland margins; 200–4000 feet; fl. and fr. most of the year. *H 8355! H & P 14284! Yuncker 18550!* General in the New World tropics.

5. P. dilatatum L. C. Rich. in Act. Soc. Hist. Nat. Paris **1**: 105 (1792).

Shrub 2–3 m. high; leaves acuminate, unequally rounded and shorter on one side at base, sparsely pubescent, 14–20 cm. long, 7–11 cm. broad; spikes 7–8 cm. long; fruit trigonous, usually puberulous at apex.

Probably erroneously reported from Jamaica, the *Caley* specimen referred to by Fawcett & Rendle, Fl. Jam. **3**: 26 (1914) being almost certainly from St. Vincent. This species occurs from Mexico through continental tropical Amer., Lesser Antilles and Trinidad.

6. P. murrayanum C. DC. in DC., Prodr. **16** (1): 280 (1869).—*P. jamaicense* C. DC. (1902). *Artanthe jamaicensis* Griseb. (1860), illegitimate name, partly.

Shrub or tree 1–4 m. high; leaves ovate to elliptic-lanceolate, acuminate, unequally broadly or narrowly rounded at base with one side shorter than the other, finely pellucid-dotted, 6–17 cm. long, 2·5–6 (–9) cm. broad; spikes 5–13 cm. long; fruit oblong.

Common in pastures, thickets and woodland margins; 60–4000 feet; fl. and fr. all the year. *A 7787! Bengry! H 11131!* Endemic, at least in the typical form. Yuncker (1960) distinguished three forms, differing in leaf-size, within his concept of this species. Plants almost identical with the larger-leaved form occur in Dominica.

7. P. hispidum Sw., Nov. Gen. & Sp. Pl.: 15 (1788).—*P. scabrum* Sw. (1797), not Lam. (1791).

Shrub or tree 2–4 m. high; leaves elliptical to ovate-elliptical, acuminate, obliquely rounded and unequal at base, commonly gland-dotted, 11–19 (–25) cm. long, 4–10 (–12) cm. broad; spikes 8–11 cm. long; fruit oblong-trigonous.

Frequent in thickets and rough pastures; 50–4000 feet; fl. and fr. Nov, Feb–Aug. *A 7867! Chalmers UCWI 4642! P 8028!* Hispaniola, Puerto Rico, St. Kitts, Trinidad & Tobago, Suriname. Plants having characters intermediate between this and the last and this and the next are rather common.

8. P. fadyenii C. DC. in DC., Prodr. **16** (1): 284 (1869).—*P. hispidum* of F. & R. (1914), not Sw. *P. otophyllum* C. DC. (1902).

Shrub 1–4 m. high; leaves ovate to oblong-elliptical, acuminate, unequally rounded at base, often bullate, gland-dotted, 8–15 cm. long, 4–8·5 cm. broad; spikes 7–10 cm. long; fruit oblong.

Locally common along pathsides and in gullies in moist montane woodlands; 3300–5000 feet; fl. and fr. all the year. *A 11934! Fawcett 8159! H 8344! P 23741!* Endemic.

9. P. tuberculatum Jacq., Ic. Pl. Rar. **2**: 2, t. 211 (1795).

Shrub or tree 2–4 m. high, the branches often warted, puberulous when young;

leaves elliptical, rounded and more or less cordulate and markedly unequal at base, 7–14 cm. long, 3–6 (–10) cm. broad; spikes 5–12 cm. long; fruit more or less tetragonous.

Very rare; only twice reported and not confirmed by recent collections. Mexico to Suriname, Hispaniola, Trinidad.

10. P. arboreum Aubl., Hist. Pl. Guiane **1**: 23 (1775).

10a. Var. **arboreum**.

Shrub or tree 2–6 m. high; leaves ovate to elliptic-lanceolate, unequally rounded at base, mostly 15–25 cm. long and 6–10 cm. broad; spikes 10–15 cm. long, usually rather thick; fruit oblong.

Widely distributed in rocky thickets and woodlands on limestone and in montane woodlands; 400–6000 feet; fl. and fr. all the year. *A 7742! 7864! H 9242! 11148! P 7614! 8384!* Panama to Suriname.

10b. P. arboreum var. **stamineum** (Miq.) Yuncker in Bull. Inst. Jam., Sci. ser. **11**: 18 (1960).

Shrub or tree 2–10 m. high with slender trunk; leaves elliptical to lanceolate 10–12 (–20) cm. long, 2·5–5 cm. broad, dark green, shiny; spikes light yellow.

Rather common in woodlands on limestone; 900–3500 feet; fl. and fr. Dec–Sept. *A 8597! H 12011! HPS 14657! P 20681!* Cuba. The young leaves of this variety are described as minutely hirtellous in which case there is a possibility of its being confused with *P. tuberculatum*.

11. P. verrucosum Sw., Nov. Gen. & Sp. Pl.: 15 (1788).

Tree 3–8 m. high; leaves elliptical to elliptic-lanceolate, acute, rounded and cordate at the unequal base, leathery, bluish- or greyish-green, often folded lengthwise and usually pendulous, up to 40 cm. long and 16 cm. broad; spikes 12–15 cm. long, 3–4 mm. thick; flowers fragrant.

Occasional, in the central parishes, in open woodlands on craggy limestone; 1750–2500 feet; fl. and fr. sporadically Nov–Aug. *A 7314! Cornman 621! H 8734! P 22570!* Endemic. This species was included by Yuncker (1960) in his broader concept of *P. arboreum*; that author drew attention to the difficulty of constructing a logical classification of the numerous variants close to *P. arboreum*, but *P. verrucosum* seems to differ in a sufficient number of respects to enable its consistent recognition.

3. POTHOMORPHE Miq. (1839); Yuncker (1960)

Leaves peltate, glabrous except near the margin; young internodes commonly glabrous
1. *peltata*
Leaves scarcely peltate, deeply cordate, the nerves puberulous; young internodes commonly hairy in lines
2. *umbellata*

1. P. peltata (L.) Miq., Comm. Phyt.: 37 (1840).—*Piper peltatum* L. (1753).

Soft-wooded shrubby herb 1–2 m. high; leaves rounded-cordate, acute, gland-dotted, 15–35 cm. or more broad; spikes several on solitary or paired axillary peduncles, 5–10 cm. long; fruit trigonous.

Locally common in moist glades and gullies; 350–1750 feet; fl. and fr. July–Apr. *A 9304! H 8441! P 21489! Yuncker 17777!* General in tropical Amer.

2. P. umbellata (L.) Miq., Comm. Phyt.: 36 (1840).—*Piper umbellatum* L. (1753).

Cow Foot.

Shrubby herb like the last; inflorescences commonly paired.

Rather common and widespread in secondary communities in damp sheltered places; 10–2500 feet; fl. and fr. most of the year. *A 7504! H 11142! P 7805! Yuncker 18555!* General in tropical Amer.; W. Africa.

39. CHLORANTHACEAE

Herbs, shrubs or trees, mostly aromatic. Leaves opposite, simple, serrate; stipules small and often fused with the connate petiole-bases into a sheath. Flowers bisexual or unisexual, monoecious or rarely dioecious, small, actinomorphic, in bracteate spikes, heads or panicles; perianth reduced or absent, a 3-toothed calyx in pistillate flowers. Stamens 1 or 3, connate; anthers 1–2-locular, opening lengthwise. Ovary inferior, 1-locular with a solitary pendulous orthotropous ovule; style short or absent; stigma 1. Fruit a small ovoid or globose drupe. Seed with a large oily endosperm and small embryo.

5 genera with about 70 species in the American and east Asian tropics, Polynesia, New Zealand and Madagascar.

1. HEDYOSMUM Sw. (1788)

Flowers unisexual, the male in spikes, the female in panicles or heads; ovary crowned by a minute 3-toothed calyx.

Plants monoecious; leaves lanceolate, long-acuminate; female flowers distant along branches of a cymose panicle
1. *nutans*
Plants dioecious; leaves oblong-elliptical, rather abruptly acuminate; female flowers in clusters of 2 or 3
2. *arborescens*

1. H. nutans Sw., Nov. Gen. & Sp. Pl.: 84 (1788).

Shrub 1·2–5 m. or tree to 6 m. high; leaves fragrant, light green, rather succulent, 6–12 cm. long, 1–2 cm. broad; peduncles deep mauve; male and female inflorescences together in the leaf-axils or female terminal; male spikes 1–2 (–3) cm. long with peduncles up to 3 cm. long; anther-connective produced beyond the loculi; female bracts ovate; drupe 2–3 mm. long, ripening red.

Widely distributed but common only in the eastern parishes, in thickets and on exposed rocky slopes; 1250–5000 feet; fl. and fr. Jan–Mar, July–Aug. *A 6645! H 10204! P 9900!* Cuba, Hispaniola.

2. H. arborescens Sw., Nov. Gen. & Sp. Pl.: 84 (1788).—Cigar Bush, Cold Bush, Headache Bush.[1]

Tree 4–9 m. or shrub 2·5 m. or more high with light grey, friable and finely but not deeply fissured bark, strongly aromatic; branches brittle; leaves 3–12 (–18) cm. long, 1·5–4 (–5·5) cm. broad, usually dark green, but sometimes variegated light green and white; male spikes solitary or paired on a common peduncle, up to 4 (–5) cm. long at maturity; female perianth fleshy, whitish; drupe 3 mm. or more long, red when ripe.

Locally common (St. Andr., Port., St. Thom.), in humid forest; 1500–6000 feet; fl. and fr. most of the year. *A 5875* (♂)! *7445* (♀)! *H 12413* (♀)! *P 5325* (♂)! Puerto Rico, Lesser Antilles.

40. MYRICACEAE

Trees and shrubs, generally aromatic. Leaves alternate, simple, entire to pinnatifid, usually without stipules, glandular-punctate. Flowers mostly unisexual, monoecious or dioecious, in short axillary spikes, usually bracteate, without perianth. Male flower subtended by a single bract, with (2–) 4–8 (–20) stamens; filaments distinct or connate; anthers basifixed, 2-locular, opening lengthwise. Female flower subtended by a bract with or without additional bracteoles; ovary superior, 1-locular; ovule solitary, basal, orthotropous; style short, 2-branched. Fruit a small drupe often warted with waxy papillae; endocarp hard. Seed without endosperm.

2 genera, *Myrica* with a cosmopolitan distribution and 55 of the 56 species.

---

[1] These common names may be applied to either species.

1. MYRICA L. (1753)

Leaves entire or toothed; stipules absent.

1. M. cerifera L., Sp. Pl. **2**: 1024 (1753).—Wax Berry, Wax Wood.

Dioecious shrubs 2–4 m. or trees to 8 m. high, usually with conspicuous whitish lenticels on the twigs, thinly to densely pubescent; leaves elongated-elliptical and deeply serrate-dentate on juvenile shoots and up to 10 cm. long, on older shoots oblanceolate to obovate, crenate, sinuate or entire, 2–6 cm. long, 1–3 cm. broad, more or less glandular on both surfaces but very variable in this respect, the glands in deep or shallow pits; male spikes up to 2 cm. long with bracts about 1 mm. long and broad; stamens 2–6; fruit subglobose, 2·5–3 mm. long.

Common in the eastern mountains (St. Andr., Port., St. Thom.), in woodlands and thickets, often in rather exposed places, rare in the higher parts of Clarendon and Trelawny; (2000–) 2500–5700 feet; fl. and fr. most of the year. *A 12208! 12210! H 9138! P 16583! 23309!* South-eastern United States, Bermuda, British Honduras, Guatemala, Greater Antilles, Bahamas, Guadeloupe.

The variability of *Myrica* in Jamaica and in the Antilles as a whole requires further study. There are great differences between juvenile, mature and flower-shoot foliage on the same plant, in respect to shape, size, margin, vesture and the density of glands. The yellow glands on the leaf-surface are borne in shallow or deep pits; on old specimens the glands dry up and often disappear leaving only the pits. Thus the character used by Fawcett & Rendle to separate the groups of species is meaningless. The lengths of male spikes probably only indicate different stages of development. In these circumstances it is concluded that the names *M. microcarpa* Benth. (1839), *M. microstachya* Krug & Urb. (1892) and *M. jamaicensis* Howard & Proctor (1958) should remain in abeyance pending more detailed study.

41. JUGLANDACEAE

Trees, rarely shrubs, often resinous and aromatic. Leaves alternate, rarely opposite, pinnately compound with an odd terminal leaflet, without stipules. Flowers unisexual, monoecious. Male flowers in erect or pendulous solitary, clustered or paniculate catkins, each flower subtended by a bract and 2 bracteoles and with a perianth of up to 4 (–6) tepals, or bracteoles and perianth absent; receptacle adnate to bracts, bearing 3 to many stamens in 1 or more series; filaments short; anthers erect, 2-locular, opening lengthwise. Female flowers in solitary erect spikes, usually bracteolate, each floral envelope epigynous with 4 teeth or lobes; ovary inferior, 1-locular; ovule solitary, basal, orthotropous; style short with 2 branches. Fruit a drupe or nut, sometimes winged, often with a succulent exocarp derived from the enlarged floral envelope; endocarp hard. Seeds large, lobed, without endosperm.

6 genera with about 60 species, from north temperate to tropical America and Asia.

1. JUGLANS L. (1753)

Leaves alternate; male catkins pendulous; male perianth 3–6-lobed; fruit a large fleshy indehiscent drupe without wings; endocarp rugose.

1. J. jamaicensis C. DC. in DC., Prodr. **16** (2): 138 (1864).

A large tree; twigs, petioles and leaf-rachis rufous-pubescent; leaves 30 cm. or more long with 6–9 pairs of lateral leaflets; leaflets generally alternate, lanceolate, acuminate, serrate, unequal at base, pubescent on the veins beneath, 6–10 cm. long, 3–4 cm. broad; drupes broadly ovoid 4·5–5 cm. long, 4 cm. broad; endocarp corrugated 3–3·5 cm. long, acute at apex, concave at base.

Reported once only from Jamaica, probably erroneously. Hispaniola, Puerto Rico. A distinct but related species, *J. insularis* Griseb., is endemic in Cuba; the fruits of this species are sometimes cast upon Jamaican beaches.

42. PICRODENDRACEAE

Deciduous trees. Leaves 3-foliolate; stipules present, setiform, very small, hardly persistent. Flowers unisexual, dioecious. Male flowers in clusters on peduncled solitary or grouped simple or sparingly branched spikes, each lacking perianth but subtended by (1–) 3 (–7) abaxial imbricate bracts; stamens 3–numerous on an hemispherical receptacle; filaments short, free; anthers 2-locular, basifixed, opening lengthwise; pollen muriculate. Female flowers solitary, axillary on slender pedicels; sepals 4 (–5), unequal, free, valvate; ovary superior, 2-locular; style terminal, deeply 2-lobed, the stigmas with revolute margins; ovules 2 in each loculus, pendulous from top of central axis, anatropous. Fruit a globose drupe with thin fleshy exocarp and brittle endocarp, containing usually 1 seed. Seed with folded cotyledons, lacking endosperm.

1 genus with 3 species; West Indies.

1. PICRODENDRON Planch. (1846)

1. P. baccatum (L.) Krug & Urb. in Engl., Bot. Jahrb. **15**: 308 (1892).—*Juglans baccata* L. (1759). Jamaican Walnut.

Tree 5–13 m. high; new leaves appearing with the flowers; blades of leaflets elliptical to lanceolate, 4–11 cm. long, 1·5–4·5 cm. broad, venation finely reticulate, glabrous adaxially, pubescent especially on midrib abaxially; flowers greenish; bracts of male flowers 1·2–1·5 mm. long; sepals of female flowers linear-lanceolate, 3–5 (–8) mm. long; drupe orange-yellow to light brown, about 2 cm. in diameter.

Rather local in arid rocky limestone thickets and at salina margins in the south-central parishes; SL–650 feet; fl. May, Sept, fr. most of the year. *A 12801! H 9046! 12516! P 22129!* Probably endemic, but *P. macrocarpum* (A. Rich.) Britton, occurring in Bahamas, Cuba, Hispaniola and Grand Cayman, is probably not really distinct.

43. SALICACEAE

Deciduous trees and shrubs. Leaves alternate, simple; stipules free. Flowers unisexual, dioecious, densely aggregated in erect or pendulous catkins, each subtended by a membranous fugacious or persistent bract; perianth absent or possibly represented by a cupular disk or glands. Male flowers with 2 or more stamens; filaments slender, free or more or less united; anthers 2-locular, opening lengthwise. Female flowers with a superior sessile or shortly stipitate unilocular ovary with 2–4 parietal or basal placentas; ovules numerous, ascending, anatropous; style 2–4-fid. Fruit a 2–4-valved capsule. Seeds small, covered with numerous fine hairs arising from the funicle; endosperm lacking.

2 or 3 genera; the 300 species of *Salix* widely distributed.

1. SALIX L. (1753); Schneider (1918)

Male catkins erect with entire bracts; nectaries 1 or 2; style 1 with 2 usually bifid stigmas or stigmas sessile; capsule 2-valved.

1. S. humboldtiana Willd., in L., Spl. Pl. ed. 4, **4** (2): 657 (1806).—*S. chilensis* of Morong & Britton (1892), not Molina (1782). Humboldt's Willow.

Tree to 18 m. high; leaves linear, acuminate, serrulate, glabrous, up to 13 cm. long and 8 mm. broad; catkins 5–7 cm. long; stamens 4–7; capsule about 4 mm. long, stalked.

Planted locally (St. Andr., West., Trel.), in sheltered or wet places; 800–3800 feet. *H 8748! Morley 162! P 23990!* Native probably of Chile, now widespread as an introduction from Mexico to Argentina, Florida, West Indies.

44. FAGACEAE

Trees and shrubs. Leaves alternate, simple with usually early-falling stipules. Flowers usually unisexual, monoecious, apetalous. Male flowers solitary or more often in pendulous heads or catkinlike racemes; calyx with 4–6 (–7) imbricate lobes; stamens 4–many; filaments free, filiform; anthers basifixed, 2-locular, opening lengthwise; pistillode usually present. Female flowers solitary or in small clusters in spikes or at the base of the male inflorescence, each often within an involucre of imbricate bracteoles; calyx 4–6-lobed, adnate to the ovary. Ovary inferior, 3–6-locular; styles as many as the loculi or more; ovules 2 in each loculus, pendulous, anatropous. Fruit a nut, 1–3 being subtended or enveloped by an often spiny or tuberculate involucre of hardened bracteoles. Seed usually solitary with a large embryo, without endosperm.

6 genera with about 600 species, mostly in the temperate and subtropical regions of the northern hemisphere.

1. CASTANEA Mill. (1754)

Male flowers in erect slender spikes; female flowers solitary or few in spikes; male perianth 6-lobed, stamens 8–20; ovary 6-locular; capsule prickly.

1. C. sativa Mill., Gard. Dict. ed. 8 (1768).—Sweet Chestnut.
Tree to 10 m. or more high with lenticellate twigs; stipules leaving small ciliate scars; leaves lanceolate, unequal and more or less rounded at base, margin serrate-dentate to almost entire but always with subulate mucronulae, acuminate, up to about 25 cm. long; male flowers fragrant, in clusters in spikes 10–15 cm. long; fruit up to about 3·5 cm. long, deep brown, shiny.
Rare in cultivation (St. Andr.); 3500–3750 feet; fl. Mar, July, fr. June. *A 6708!* *A. von der Porten!* Native of S. Europe and W. Asia; Hispaniola.

45. ULMACEAE

Trees or shrubs. Leaves alternate, simple, often serrate and unequal at the base; stipules free, usually small, deciduous. Flowers bisexual to unisexual by abortion and monoecious, actinomorphic, small, in cymes, fascicles or solitary. Perianth sepaloid, 4–8-lobed, the lobes imbricate or valvate, persistent or deciduous. Stamens erect in bud, usually the same number as and opposite to the perianth-lobes; filaments free; anthers 2-locular, opening lengthwise. Ovary superior, 1 (–2)-locular; styles 2, divergent, stigmatose on their inner faces; ovule 1, apical, pendulous, anatropous. Fruit a broadly winged samara or a drupe with hard endocarp. Seeds mostly without endosperm; embryo straight or curved.

16 genera and about 300 species widely distributed.

Jamaican representatives share the following characters: flowers unisexual or polygamous; fruit drupaceous; embryo curved.

Segments of male perianth imbricate; fruiting perianth deciduous; leaf-margin entire towards the base at least, adaxial surface usually smooth   1. **Celtis**
Segments of male perianth induplicate-valvate; fruiting perianth persistent; leaf-margin serrate or dentate to base, adaxial surface usually scabrid   2. **Trema**

1. CELTIS L. (1753)

1 Leaf-base subequal; straggling-branched shrub or climber, usually with recurved spines at least on lower part of stem; staminate and pistillate flowers on short pedicels in the same inflorescence; stigmas bifid   1. *iguanaea*
1 Leaf-base markedly unequal; unarmed trees; stigmas entire:
   2 Hermaphrodite or pistillate flowers (or fruits) solitary on pedicels 2–3 times as long as petioles; staminate flowers 3–5 in fascicles   2. *jamaicensis*
   2 Staminate and pistillate or hermaphrodite flowers in loose axillary cymes from the same peduncle, rarely pistillate flowers solitary; pedicels shorter than petioles
          3. *trinervia*

1. C. iguanaea (Jacq.) Sarg., Silva North Amer. **7**: 64 (1895).—*Rhamnus iguanaeus* Jacq. (1760). *C. aculeata* Sw. (1788). *Momisia iguanaea* (Jacq.) Rose & Standl. (1912).

Scrambling shrub to 2·5 m. or a high climber usually with recurved spines on the stem; leaf-blade oblong-elliptical, shortly acuminate, shallowly cordate, thin textured, 4–12 cm. long, 2–5 cm. broad; flowers 5-merous, greenish-yellow in short paniculate axillary cymes; drupe about 14 mm. long.

Rather uncommon (Manch., St. Eliz., St. Ann), in thickets and open woodands on steep well drained or arid rocky limestone; 40–1900 feet; fl. Feb–May, fr.May–Sept. *A 11152! H & B 10619!* General in the American tropics.

2. C. jamaicensis Planch. in Ann. Sci. Nat. sér. 3, Bot. **10**: 290 (1848).—*C. trinervia* of F. & R. (1914), not Lam.

Tree 6–13 m. high with a slender trunk; leaf-blade broadly ovate, acuminate, subcordate or rounded unequally at base, serrate-margined at least distally, 6–13 (–20) cm. long, 3·5–6 (–10) cm. broad, sparsely pubescent; drupe globose-ovoid, bluish-black, about 10 mm. long.

Occasional in the central and eastern parishes, in woodland on steep rocky mostly limestone hillsides; 2000–3800 feet; fr. Dec–July. *A 11036! H 10857! Hunnewell 19770!* Endemic.

3. C. trinervia Lam., Encycl. Méth. Bot. **4**: 140 (1797).—*C. swartzii* Planch. (1848). Bastard Fustic.

Tree 8–15 m. high with smooth bark; leaf-blade broadly lanceolate to ovate-elliptical, long-acuminate, with one side at the base rounded, the other acute, pellucid-dotted, glabrous or nearly so, 3–9 (–12) cm. long, 2–3·5 (–5) cm. broad; drupe ellipsoid, purple, glaucous, about 8 mm. long.

Frequent in woodland on limestone; 50–1700 feet; fl. June–Oct, fr. July–Dec. *A 11958! 12617! H 9695! P 18267!* Guatemala, Greater Antilles, Virgin Is.; Grand Cayman. A variant of this species with hairy twigs and leaves was described as *C. berteroana* Urb. from Jamaica and is also reported from Cuba and Hispaniola.

2. TREMA Lour. (1790)

1 Leaves ovate, broadly and shallowly cordate, more or less pubescent on the veins and lamina beneath   1. *floridanum*
1 Leaves lanceolate, rounded at base:
  2 Leaf-blade 7–15 cm. long, long-tapered acuminate, minutely cordulate, glabrous on the lamina beneath, the minor veinlets not prominent   2. *micranthum*
  2 Leaf-blade up to 6 cm. long, shortly pointed, scabridulous on the prominent veinlets beneath   3. *lamarckianum*

1. T. floridanum Britton in Small, Fl. Southeast. U.S.: 366, 1329 (1903).— *T. micranthum* of F. & R. (1914), partly.

Shrub or tree to 6 m. or tree to 10 m. high; leaves long-acuminate, 6–13 cm. long, 2·5–5·5 cm. broad; inflorescences compact; calyx 5-merous, greenish-white to yellow; drupe subglobose, yellow or orange, 2·5–3·5 mm. in diameter.

Local (St. Andr., Port., St. Thom.), along roadsides and in secondary thickets; 3200–5000 feet; fl. May–Aug, fr. July–Aug. *A 11328! J.P. 1382! 1439! A. von der Porten!* Florida, Mexico to C. Amer., Cuba.

2. T. micranthum (L.) Blume, Mus. Bot. Lugd.-Bat. **2**: 58 (1856).—*Rhamnus micranthus* L. (1759). Jamaican Nettle Tree.

Shrub 2 m. or tree 6–13 (–20) m. high; leaves rigid, paler beneath, 7–12 (–15) cm. long, 2·5–4 cm. broad; flowers greenish-white to greenish-yellow; drupe ovoid, about 3 mm. long.

Occasional on roadside banks and in secondary thickets; 300–2000 feet; fl. Jan, June–Oct. *A 7513! H 9627! 9885! P 19795!* Florida, Mexico to Brazil and Argentina, West Indies.

3. T. lamarckianum (Roem. & Schult.) Blume in Mus. Bot. Lugd.-Bat. **2**: 58 (1856).—*Celtis lamarckiana* Roem. & Schult. (1820). *T. lima* of F. & R. (1914), not (Sw.) Blume.

Tree with spreading branches 4–6 m. high; leaves rigid, 2–6 cm. long, up to about 2 cm. broad; drupe ovoid, about 3 mm. long.

Local (St. Andr., St. James, St. Thom.), in arid thickets on limestone or gravel; 200–2000 feet; fl. Dec–Aug, fr. Aug. *A 7707! J.P. 1274! Webster 4990!* Florida, Bermuda, Bahamas, West Indies south to St. Vincent; Grand Cayman.

46. CANNABACEAE

Erect annual herb; leaves alternate or the lower opposite; embryo curved.

palmately lobed, or digitately compound; stipules free. Flowers small, unisexual, dioecious or occasionally monoecious. Male flowers in panicles; calyx 5-partite with imbricate lobes; petals absent; stamens 5; anthers erect in bud, 2-locular, opening lengthwise. Female flowers clustered and axillary or in bracteate spikes; calyx enveloping the ovary, entire; ovary superior, 1-locular; style bifid; ovule solitary, apical, pendulous, anatropous. Fruit achenial, covered by the persistent perianth. Seed with fleshy endosperm; embryo curved.

2 genera, *Humulus* (Hops) with 3 north temperate species, *Cannabis* with 1 species native in central Asia.

1. CANNABIS L. (1753)

Erect annual herb; leaves alternate or the lower opposite; embryo curved.

1. C. sativa L., Sp. Pl. **2**: 1027 (1753).—Ganja, Indian Hemp, Marijuana.

Erect annual herb 1–2 (–5) m. high; leaves opposite below, alternate above, digitately compound with (3–) 5 (–9) linear-lanceolate serrate leaflets up to about 15 (–18) cm. long and 1·5 cm. broad, scabridulous above, gland-dotted and puberulous on the veins beneath, acrid-odoriferous; male flowers greenish-yellow; female flowers green; calyx with numerous superficial glands; fruit broadly ovoid-biconvex, light brown, shiny, about 3 mm. long.

Cultivated; 600–3000 feet; fl. July–Oct, Jan, fr. July–Aug. *Harris! P 19793! Wynter UCWI 1240!* Widely distributed, mostly in subtropical regions.

47. MORACEAE

Trees or shrubs with usually[1] milky sap, rarely herbs. Leaves alternate, entire, toothed or lobed, stipulate. Flowers unisexual in heads, catkins, expanded disks or hollow receptacles, rarely solitary. Perianth calyxlike or absent, usually 4-merous, the segments imbricate or valvate, free or connate. Male flowers with 1, 2 or 4 stamens opposite perianth-segments; filaments free; anthers 2-locular, versatile, opening lengthwise. Female flowers with superior to inferior usually 1-locular ovary; ovule 1, usually anatropous and pendulous, rarely basal[1] and erect; styles and stigmas mostly 2, filiform. Fruit a small achene or drupe, these often aggregated or united with the perianth or floral axis or embedded in or surrounded by the receptacle.

About 75 genera and over 1500 species, mostly tropical.

1 Herbs with rhizomatous stems; male and female flowers together on the upper surface
    of a peltate stalked discoid lobed receptacle; stipules persistent   **1. Dorstenia**
1 Shrubs or trees; stipules mostly deciduous:
  2 Stipular ring scars on twigs incomplete or absent; stipules small; plants dioecious;
    male flowers in spikes or catkins; leaves distichous:

---

[1] See note under *Poikilospermum* (p. 221).

3 Inflorescence solitary in leaf-axils or at older nodes; female flowers in rounded heads; stigma lateral, simple; leaves blackening on drying; sap creamy, viscid
                                                              2. **Chlorophora**
3 Inflorescence 2 or more in leaf-axils; female flowers in short spikes; stigma terminal, 2-armed; leaves not blackening; sap milky or very thin   3. **Trophis**
2 Stipular ring scars complete or almost so and usually conspicuous; stipules sometimes large:
  4 Leaves peltate, deeply palmately lobed, the lobes independently veined, white-tomentose abaxially; plants dioecious; flowers in digitately clustered spikes; branches hollow   4. **Cecropia**
  4 Leaves not peltate and otherwise not as above:
    5 Shrub with trailing branches, without milky sap; leaves entire; flowers in globose heads in cymes; stigma simple   5. **Poikilospermum**
    5 Erect shrubs or trees (except *Ficus pumila*, a root-climber), with milky sap; stipules usually conspicuous:
      6 Leaves spirally arranged; plants monoecious:
        7 Male and female flowers enclosed within an ostiolate fleshy or fibrous globose to pear-shaped receptacle   6. **Ficus**
        7 Male and female flowers in separate inflorescences, the former spicate, the latter capitate; fruits achenial, if formed embedded in the enlarged fleshy oblong or globose receptacle   7. **Artocarpus**
      6 Leaves distichous:
        8 Plants typically monoecious; each inflorescence with numerous male flowers accompanied by peltate bracts on the surface of a globose receptacle with 1 or 2 female flowers in the centre   8. **Brosimum**
        8 Plants typically dioecious; inflorescence without peltate bracts:
          9 Female flowers solitary, surrounded by an involucre of closely imbricated bracts; male heads discoid, involucrate; leaves glabrous or nearly so
                                         9. **Pseudolmedia**
          9 Female flowers coherent in discoid involucrate heads; male heads reniform; leaves up to 30 (–50) cm. long and 15 (–25) cm. broad, more or less cordate, densely hirsute with golden hairs; stipules large; introduced and planted in several localities; Central American Rubber   *Castilla elastica* Cerv.

1. DORSTENIA L. (1753)

Leaves peltate   1. *jamaicensis*
Leaves not peltate   2. *fawcettii*

1. D. jamaicensis Britton in Bull. Torr. Bot. Club **35**: 567 (1908).

Rhizome short and suberect; leaf-blade ovate to rounded-ovate, scabridulous adaxially, shortly tomentose abaxially, 3–5 cm. long, 1·5–4·5 cm. broad; petiole 3–12 cm. long, pubescent; receptacle 8–10 mm. in diameter, reddish-purple, the margin toothed; peduncle about 2 cm. long.

Rare and local (Manch.), in crevices of limestone cliffs; 1500–2300 feet; fl. and fr. Sept, Jan–Feb. *H & B 10607! Proctor & Stearn 11592! Robertson UCWI 4666!* Endemic.

2. D. fawcettii Urb. in Arkiv Bot. **22A** (10): 2 (1929).—*D. cordifolia* of Sw. (1788), partly, not Lam. (1786).

Rhizome usually elongated, trailing and rooting, with conspicuous stipules; leaf-blade ovate to elliptical, cordate at base, obtuse, the margin entire or sinuate, dark green and rough adaxially, pubescent on veins abaxially, 3–7 cm. long, 2–5·5 cm. broad; petiole 4·5–12·5 cm. long; receptacle 8–10 mm. in diameter, light green or cream-coloured above, sometimes purplish beneath; peduncle 2–7 (–10) cm. long.

Rather common, on shaded banks and in rock crevices in limestone cliffs in forest; 500–2400 feet; fl. and fr. Mar–Dec. *A 6767! P 22161! Yuncker 18562!* Endemic.

2. CHLOROPHORA Gaudich. (1830)

1. C. tinctoria (L.) Gaudich. ex Benth. in Benth. & Hook. f., Gen. Pl. **3**: 363 (1880).—*Morus tinctoria* L. (1753). Fustic Tree.

Tree, often with spreading and drooping branches, sometimes deciduous, up to 20 m. high; leaves ovate to oblong, unequally cordate at base, acuminate, entire or serrate, 3–12 cm. long, 2–6 cm. broad, dark green, glabrous or minutely puberulous; stipules subulate, 2–10 mm. long; male spikes 3–10 cm. long; stamens 4; pistillode about 1 mm. long; female heads 8–10 mm. in diameter in flower, 12–20 mm. in diameter in fruit.

Common in secondary formations, in thickets on river gravel and rocky lime-stone; SL–1500 feet; fl. May–Dec, fr. July–Nov. *A 7356! 8231! Chalmers UCWI 4824! Harris! P 9574! 10216!* Mexico to Argentina, West Indies, introduced into Sierra Leone.

3. TROPHIS P. Browne (1756) nom. cons.

1. T. racemosa (L.) Urb., Symb. Ant. **4**: 195 (1905).—*Bucephalon racemosum* L. (1753). Ramoon.

Tree with slender flexuous twigs, 5–20 m. high; leaves oblong-elliptical, equally rounded at base, acuminate, entire or serrate towards tip, 6–20 cm. long, 2–9 cm. broad, glabrous; stipules subulate, 5–6 mm. long; male spikes 3–5 (–10) cm. long, the flowers interspersed with small peltate bracteoles; stamens 4; pistillode present; female spikes 8–20 mm. long, fruits drupaceous, 8–12 mm. long, red or yellow.

Frequent in woodlands on limestone; 50–2500 feet; fl. Nov–Apr, fr. Jan–May. *H 10211! 10858! P 20588!* Mexico to Brazil, Greater Antilles.

4. CECROPIA Loefl. (1758) nom. cons.

1. C. peltata L., Syst. Nat. ed. 10, **2**: 1286 (1759).—Snake Wood, Trumpet Tree.

Tree up to 20 m. high; sap watery becoming glutinous; leaf-blades with up to 11 lobes, divided to about halfway, up to about 40 cm. broad, strongly veined, scabridulous above, whitish beneath; petioles 30–50 cm. long; stipules 6–9 cm. long; peduncles 7–10 cm. long; male spikes 3–5 cm. long in clusters of 12–30; female spikes 4–6 cm. long in clusters of 2–6, enlarging in fruit; spathes 4–6 cm. long, oblong-conical before anthesis.

Common, especially on recently cleared forested land; 50–2900 feet; fl. and fr. sporadically throughout the year. *A.M. Barry! Stearn 977!* Mexico to Colombia and Venezuela, West Indies.

5. POIKILOSPERMUM Zipp. ex Miq. (1864); Merrill (1934)[1]

1. P. suaveolens (Blume) Merr. in Contrib. Arn. Arb. **8**: 47 (1934).—*Conoce-phalus suaveolens* Blume (1825).

Trailing shrub or climber; leaves broadly ovate to oblong, cuneate to subcordate at base, acute or obtuse, with conspicuous irregular cystoliths in adaxial surface and linear cystoliths along veins abaxially, 10–25 cm. long, 8–18 cm. broad, softly leathery; inflorescences with paired caducous bracts on peduncles; male heads numerous, 5–6 mm. in diameter; female heads 3–5 together, 20–35 mm. in diameter; flowers very fragrant.

Uncommon (St. Andr., St. Ann, St. Mary), in thickets on steep banks, an introduced ornamental escaping from gardens and becoming naturalized; 500–1500 feet; fl. Mar–May. *J.P. 953! P 23495!* Native of E. Indies.

6. FICUS L. (1753)

1 Plant scandent, attached to support by short roots, more or less pilose; stipules sub-persistent; leaves variable from very small to about 8 cm. long and 4 cm. broad; figs about 5 cm. long; cultivated on walls and arbours; native of China and Japan; Climbing Fig   *F. pumila* L.
1 Erect shrubs or trees, sometimes epiphytic but not scandent; leaves mostly glabrous or glabrescent; stipules caducous; figs smaller:
  2 Mouth of fig prominent (slightly prominent in *F. citrifolia*); ripe figs red, usually borne in pairs:

---

[1] Some authors have placed *Poikilospermum* in *Urticaceae* on account of the based ovule; the clear sap and cystolith type support that position.

3 Figs usually sessile or on peduncles up to 5 mm. long, 6–10 mm. broad; stipules
  10–15 mm. long   1. *aurea*
3 Figs on peduncles 2–10 mm. long, 8–15 (–30) mm. broad; stipules 10–25 mm. long
                                                      2. *trigonata*
2 Mouth of fig not or only slightly prominent:
  4 Stipules pilose; figs sessile, 15–20 mm. broad when ripe   3. *membranacea*
  4 Stipules puberulous or glabrous:
    5 Figs sessile, red then finally black; leaves oblong-obovate, rounded to cuneate at
      base, cuspidate, 4–6 (–11) cm. long, 2–4 (–6) cm. broad; cultivated shade and
      ornamental tree; native of S.E. Asia; Chinese Banyan   *F. benjamina* L.
    5 Figs stalked:
      6 Mouth of fig immersed in an umbo; figs paired or rarely solitary on peduncles (2–)
        7–14 mm. long, (5–) 12–20 mm. in diameter, glabrous or very minutely puberu-
        lous   4. *pertusa*
      6 Mouth of fig not immersed in an umbo:
        7 Figs (3–) 7–10 mm. broad, paired, on peduncles (1–) 4–6 mm. long; bracts connate
          at base   5. *perforata*
        7 Figs mostly broader:
          8 Figs solitary, (10–) 15–22 (–28) mm. in diameter; leaves rounded to cuneate at
            base; bracts not connate; male flowers with 2 stamens   6. *maxima*
          8 Figs (1–) 2 together, 8–18 mm. in diameter; leaves rounded, truncate or sub-
            cordate at base; bracts small, shortly connate; male flowers with 1 stamen
                                            7. *citrifolia*

1. F. aurea Nutt., North Amer. Sylva **2**: 4 (1846).

Tree up to 20 m. high; leaves ovate to obovate, rounded to cuneate at base, obtuse or shortly and bluntly acuminate, 6–12 (–15) cm. long, 3·5–6 (–8·5) cm. broad, rather leathery; bracts 3–5 mm. long; mouth of fig closed by prominent scales.

Frequent and generally distributed in woodland on limestone and sea-coast thickets; SL–3200 feet; fl. and fr. most of the year. *A 12602! H 10218! P 19823!* Florida, Bahamas, Cuba, Hispaniola, Cayman Is.

2. F. trigonata L., Pl. Surinam.: 17 (1775).—*F. berteroi* Warb. (1903). *F. mamilli-fera* Warb. (1903).

Tree up to 16 (–25) m. high; leaves elliptical to obovate, mostly more or less rounded at both ends but base sometimes cuneate and tip acute, 3–15 (–25) cm. long, 3·5–10 (–14) cm. broad, the larger leaves on young plants leathery; bracts 2–5 mm. long, glabrous or ciliolate.

Very common, in exposed situations on limestone in coastal and inland areas; SL–5000 feet; fl. and fr. most of the year. *A 6314! H & B 10581! P 20723!* Mexico to Colombia, Greater Antilles, Virgin Is.

3. F. membranacea C. Wright in Sauv. in Anal. Acad. Cienc. Habana **7**: 514 (1871).—*F. harrisii* Warb. (1903).

Tree up to 20 m. high; leaves broadly ovate to oblong-elliptical, broadly rounded and shortly cordate at base, acute to shortly acuminate at tip, 12–30 cm. long, (7–) 9–20 cm. broad, membranous; bracts connate, 5 mm. long, pubescent.

Occasional as an epiphyte or terrestrial on limestone cliffs or in open thickets; 250–2500 feet; fl. and fr. most of the year. *A 7777! H 5221! P 8206! Skelding UCWI 5368!* Cuba.

4. F. pertusa L.f., Suppl.: 442 (1781).—*F. ochroleuca* Griseb. (1860). *F. grabhamii* Britton ex F. & R. (1914). *F. halliana* Britton ex F. & R. (1914). ? *F. morant-ensis* Britton ex F. & R. (1914).

Shrub or tree up to about 12 (–30) m. high; leaves ovate to lanceolate, rounded to cuneate at base, acute to distinctly acuminate at tip, 2·5–15 (–25) cm. long, 1–6 (–8) cm. broad; stipules narrowly deltate, 5–10 mm. long, glabrous or puberulous; bracts 1–4 mm. long, semicircular, glabrous or puberulous; figs yellow or finally red when ripe, soft and sometimes wrinkled and mottled; umbo 1–2 mm. broad, the pit 1–1·5 mm. deep.

Very common as an epiphyte or in rocky woodland margins; SL–3500 feet; fl. and fr. most of the year. *A 8994! 9234! H 7704! 9067! P 23401!* Mexico to Paraguay.

5. F. perforata L., Pl. Surinam.: 17 (1775).—*F. wilsonii* Warb. (1903). Jamaican Cherry Fig.

Shrub or tree up to 16 (–30) m. high; leaves elliptical to oblanceolate, rounded to cuneate at base, shortly acuminate, 2–12 cm. long, 1–5 cm. broad, dark green adaxially, the midrib whitish; stipules 4–20 mm. long, glabrous or ciliolate; bracts 1–2 mm. long, rounded-ovate, glabrous or puberulous; figs green with red spots or red or pinkish-red when ripe; mouth closed by 3 purplish scales, 1–2 mm. broad.

Common as an epiphyte or in woodland margins on limestone; 50–3000 feet; fl. and fr. Dec–Sept. *A 8802! 8986! Britton 2638! H 9452! H & P 13676!* Guatemala to Colombia, Bahamas, Greater Antilles.

6. F. maxima Mill., Gard. Dict. ed. 8 (1768).—*F. suffocans* Griseb. (1860).

Tree 8–30 m. high, the trunk often buttressed; leaves elliptical to oblanceolate, obtuse or cuneate at base, acute to long-acuminate, 6–20 (–24) cm. long, 2·5–8 (–12) cm. broad; stipules 8–15 (–25) mm. long, glabrous or puberulous; bracts 1–2 mm. long, deltate, glabrous or pubescent; peduncles thick, up to 25 mm. long; figs smooth; ostiole 1–2 mm. broad.

Frequent in woodlands on limestone and also in low-lying swampy places; 50–3500 feet; fl. Mar, Aug–Nov, fr. most of the year. *A 10953! H 9445! J.P. 1455! P 24665!* Mexico to Brazil, Cuba.

7. F. citrifolia Mill., Gard. Dict. ed. 8 (1768).—*F. populnea* Willd. (1806).

Shrub or tree up to 13 (–16) m. high, with spreading and drooping branches; leaves ovate to elliptical or obovate, cordate, truncate or broadly cuneate at base, acute or acuminate at tip, 2·5–14 (–20) cm. long, 1·5–8 (–14) cm. broad; stipules 5–15 (–30) mm. long, glabrous; bracts 2–3 mm. long, deltate or broadly rounded, glabrous or puberulous; peduncles up to about 15 mm. long; figs yellow or red when ripe, the ostiole purplish and 2–3 mm. across.

Locally common, particularly in coastal formations, rarer inland; SL–500 (–2500) feet; fl. and fr. most of the year. *A 6864! 8961! H 10382! H & P 15105! Webster & Proctor 5409!* Florida, Mexico to Paraguay, Bahamas, West Indies; Grand Cayman.

Besides the foregoing the following are or have been grown in Jamaica; *F. benghalensis* L., Banyan; *F. carica* L., Edible Fig; *F. elastica* Roxb. ex Hornem., Rubber Plant and *F. religiosa* L., the Peepul Tree of India.

7. ARTOCARPUS J. R. & G. Forst. (1776) nom. cons.; Jarrett (1959)

Leaves pinnatifid with up to 7 lobes along each margin; syncarp globose; stipules (10–) 16–20 (–25) cm. long; inflorescences from young branches   1. *altilis*
Leaves usually entire; syncarp ellipsoid or somewhat irregular; stipules 1·5–5 cm. long; inflorescences often from trunk or large branches   2. *heterophyllus*

1. A. altilis (S. Parkinson) Fosberg in Journ. Acad. Sci. Washington **31**: 95 (1941). —*Sitodium altile* S. Parkinson (1773). *A. communis* J. R. & G. Forst. (1776). *A. incisus* (Thunb.) L.f. (1781). Breadfruit.

Tree up to 15 m. or more high; leaves elliptical in outline, obtuse to cuneate at base, lobes acuminate, up to 100 cm. long and 60 cm. broad, with midrib and veins antrorsely hispid, otherwise shiny adaxially, leathery; stipules pilose; syncarp pedunculate, 20–30 cm. in diameter, with superficial pyramidal prickles, most cultivated variants lacking seeds.

Common in cultivation, but mostly at lower elevations in sufficiently rainy areas; SL–700 (–1500) feet; fl. and fr. all the year. Native to the islands of the S. Pacific and introduced into Jamaica in 1793. The cultivar from Timor has less deeply incised leaves and is similar to seeded variants in this respect. Seeded varieties have been referred to as Breadnut.

2. A. heterophyllus Lam., Encycl. Méth. Bot. **3**: 209 (1789).—*A. integrifolius* of various authors, not L.f. Jack, Jakfruit.

Tree up to 15 m. high; twigs and midrib of young leaves minutely bristly when young, glabrescent; leaves broadly elliptical to obovate, very rarely lobed, cuneate at base, more or less cuspidate at tip, mostly 10–20 cm. long and 5–10 cm. broad, hard and leathery; syncarp sometimes very large, covered with papillae; peduncle 5–10 cm. long, flanged at base of head; endocarps horny, about 30 mm. long and 15 mm. broad.

Frequently planted but sometimes found in remote areas; 400–2300 feet; fl. July, Nov–Feb, fr. Feb–Mar. *A 8935! P 20624!* Native probably of India, now widespread in the subtropics and tropics.

8. BROSIMUM Sw. (1788) nom. cons.

1. **B. alicastrum** Sw., Nov. Gen. & Sp. Pl.: 12 (1788).—Breadnut.

Tree 10–30 m. high, the trunk narrowly buttressed; leaves ovate to oblong-ovate, obtuse to rounded at base, acuminate at tip, or on sapling-leaves caudate-acuminate and remotely toothed, 7–18 (–25) cm. long, 3–6 (–9) cm. broad, glabrous; stipules 4–6 mm. long, caducous leaving nearly complete ring scars; inflorescences 3–8 mm. in diameter; male flowers with 1 stamen; fruit drupaceous, yellow, 1·5–2·5 cm. in diameter.

Locally common, especially in the western parishes, in woodlands on limestone; 10–1750 feet; fl. Mar, July, fr. sporadically; sparingly planted. *A 12773! H 10506! P 9937!* Mexico to Ecuador, Cuba. The name Breadnut is also used improperly for seeded varieties of *Artocarpus altilis*.

9. PSEUDOLMEDIA Trécul (1847)

1. **P. spuria** (Sw.) Griseb., Fl. Br. W.I.: 152 (1860).—*Brosimum spurium* Sw. (1788). Bastard Breadnut.

Tree 8–16 (–20) m. high; leaves ovate to oblong-elliptical, obtuse to rounded at base, cuspidate or abruptly acuminate, 9–15 cm. long, 3–5 cm. broad; stipules narrowly lanceolate, long-acuminate, 10–20 mm. long; heads of male flowers about 8–10 mm. in diameter; female flowers ovoid, 2–2·5 mm. long; fruit ellipsoid, about 1 cm. long.

Rare and not recently collected. British Honduras to northern S. Amer., Greater Antilles.

48. URTICACEAE

Herbs, shrubs or small trees. Leaves opposite or alternate, sometimes with stinging hairs, often with cystoliths present in epidermal cells, mostly stipulate. Flowers unisexual, monoecious or dioecious, very rarely bisexual, in axillary bracteate cymes, panicles or heads or solitary; perianth small, regular, sepaloid or absent. Male flowers (2–) 4 (–5)-merous; perianth-members free or connate, imbricate or valvate; stamens as many as perianth-lobes and opposite to them; filaments free, inflexed in bud, springing outwards elastically at anthesis; anthers 2-locular, opening lengthwise. Female flowers 2–4-merous; ovary free (superior) or adnate to the perianth (inferior), 1-locular; style undivided; ovule solitary, erect. Fruit an achene or drupe, often enclosed in the perianth. Seeds mostly with endosperm; embryo straight.

49 genera with about 2000 species, widespread but mostly tropical.

1 Leaves opposite; plants monoecious or dioecious:
  2 Stipules connate, appearing as an entire intrapetiolar membrane for each leaf, usually obtuse or rounded; inflorescences paniculate or capitate; stigma penicillate; cystoliths linear or punctate; mostly herbs   **1. Pilea**
  2 Stipules free, appearing as two on each side of the node, deciduous, lanceolate, long-acuminate, with a mid-line of ascending white hairs; flowers in clusters in spikes or in the leaf-axils; stigma filiform, persistent; cystoliths punctate; shrubs or small trees   **2. Boehmeria**
1 Leaves alternate:
  3 Stipules absent; herb; female perianth tubular, distal to the male flowers in short axillary inflorescences subtended by green bracteoles; cystoliths punctate
                      3. Parietaria

3 Stipules present:
  4 Herbs or undershrubs; plants mostly monoecious:
    5 Annual herb; stigma short, becoming hooked in fruit; flowers in axillary paniculate cymes; stipules connate, intrapetiolar; bracteoles small; cystoliths poorly developed, punctate   **4. Laportea**
    5 Perennial plants; stigma filiform; flowers in axillary clusters; stipules free; cystoliths punctate:
      6 Female perianth 2–4-toothed; stigma deciduous; female bracteoles foliaceous; creeping herb; leaves entire   **5. Rousselia**
      6 Female perianth lacking; stigma persistent; bracts conspicuous, rusty-brown; erect bushy herbs or undershrubs; leaves serrate-dentate   **6. Phenax**
  4 Shrubs or trees:
    7 Stigma filiform, persistent; plants monoecious or dioecious   **2. Boehmeria**
    7 Stigma penicillate-capitate; plants dioecious:
      8 Female perianth 2-lobed; male perianth 4-lobed; stamens 4; flowers sessile in clusters; plants not armed; cystoliths linear   **7. Gyrotaenia**
      8 Female perianth 4-lobed, fleshy in fruit; male perianth 4–5-lobed; stamens 4–5; flowers stalked or in small clusters in panicles; plants often armed; cystoliths punctate or elongated   **8. Urera**

1. PILEA Lindl. (1821) nom. cons.

1 Leaves all entire or subentire:
  2 Leaves usually less than 1 cm. long; inflorescences very shortly stalked, clustered at nodes:
    3 Leaf-blades mostly more than 5 mm. broad, 3-nerved, heart-shaped, acuminate, rounded and minutely cordate at base; plant glabrous   1. *suta*
    3 Leaf-blades mostly less than 5 mm. broad, 1-nerved, rarely subtriplinerved, or nerveless, obtuse or rounded at tip:
      4 Stems erect or tufted, sometimes becoming shrubby; leaf-blades narrowed to base; plants glabrous:
        5 Leaves obovate:
          6 Plant herbaceous   2a. *microphylla* var. *microphylla* A
          6 Plant woody   2a. *microphylla* var. *microphylla* B
        5 Leaves broadly rounded, fleshy:
          7 Leaves greyish-green or tinged reddish   2b. *microphylla* var. *succulenta*
          7 Leaves light grass-green; cultivated   2c. *microphylla* var. *trianthemoides*
      4 Stems creeping and rooting, at least on non-flowering branches:
        8 Leaves oblanceolate to subspathulate, glabrous, very unequal in each pair, the smaller sessile, strongly margined on both surfaces by large contiguous cystoliths   3. *ordinata*
        8 Leaves suborbicular, more or less equal in each pair, petiolate; stems filiform:
          9 Flowering branches ascending; inflorescence shortly pedunculate; leaves glabrous or with a few hairs   4. *herniarioides*
          9 Flowering and vegetative branches prostrate and forming mosslike carpets; inflorescence subsessile; leaves with long hairs on adaxial surface and margins   5. *brittoniae*
1 Leaves larger, 3-nerved; plants often subshrubby:
  10 Leaves linear-lanceolate:
    11 Cystoliths on adaxial leaf-surface linear, conspicuous; shrubby herb; stipules less than 1 mm. long   6. *nudicaulis*
    11 Cystoliths on adaxial leaf-surface more or less punctate, all very small and inconspicuous; delicate rhizomatous herb; stipules 2 mm. long   7. *lanceolata*
  10 Leaves ovate to lanceolate or oblanceolate:
    12 Leaves oblanceolate, rather thick; petioles short and thick; stipules 1–2 mm. long   26. *flavicaulis*
    12 Leaves ovate to elliptic-lanceolate, more or less acute or acuminate, mostly thin; petioles slender; stipules 1 mm. or less long:
      13 Cystoliths on abaxial leaf-surface large and conspicuous; petioles and both leaf-surfaces usually hispid   8. *parietaria*
      13 Cystoliths on abaxial leaf-surface small and inconspicuous:
        14 Inflorescences distinctly pedunculate much-branched panicles; leaf-margin ciliate or glabrous; nodes often pubescent; leaves markedly unequal in each pair:
          15 Larger leaves (3–) 4–6 (–8) cm. long   9. *weddellii*
          15 Larger leaves 1–2·5 cm. long   10. *alpestris*
        14 Inflorescence sessile or very shortly pedunculate dense clusters of flowers, at least in the male involucrate by subpersistent bracts; plants glabrous; leaves subequal in each pair, acutely acuminate:

H

16 Leaves oblong-lanceolate to narrowly elliptical, mostly about 3 times longer than broad; internodes of flowering shoots subequal, the leaves spread more or less evenly; petioles rarely exceeding half the length of blade   11. *virgata*

16 Leaves broadly ovate, mostly less than twice as long as broad; internodes markedly shorter above, the leaves appearing clustered towards the ends of flowering shoots; petioles mostly more than half the length of blade   12. *laurae*

1 Leaves not all entire:

   17 Leaves of a pair consistently unequal in size or differing in form:

     18 Smaller leaf of pair mostly less than one-third the length of larger, suborbicular to obovate or elliptical, often entire; larger leaf oblanceolate to elliptical; see also 19. *P. ciliata*:

       19 Stem pubescent with brownish hairs; larger leaves mostly elliptical, 10–18 mm. long, serrate-crenate in the distal half   13. *proctorii*

       19 Stem glabrous; larger leaves mostly oblanceolate; cymes simple, capitulate:

         20 Plant much-branched with reddish-brown wiry stems; larger leaves entire to subpinnatifid, 5–12 mm. long, with distal margins minutely white-ciliate   14. *lucida*

         20 Plant little-branched with green or reddish-tinged stems; larger leaves shallowly serrate in distal one-third, long-tapered to base, 10–25 mm. long, glabrous or rarely with a few cilia   15. *crenulata*

     18 Smaller leaf of pair mostly exceeding one-third the length of larger:

       21 Branches hirsute:

         22 Leaves more or less clustered at ends of branches, broadly ovate   43. *inaequalis*

         22 Leaves spread evenly along the branches:

           23 Leaves glabrous, cuneate, with 5–7 deep crenations near the tip; cymes capitulate   16. *wilsonii*

           23 Leaves hirsute, elliptical to suborbicular:

             24 Cymes paniculate; adaxial leaf-surface with long brownish hairs:

               25 Stipules 2–5 mm. long   17a. *rufa* var. *rufa*

               25 Stipules 1 mm. long   17b. *rufa* var. *microstipula*

             24 Cymes umbellate or corymbose; adaxial leaf-surface glabrous   18. *rufescens*

       21 Branches glabrous:

         26 Creeping and rooting herbs; leaves ciliate-margined with a few hairs adaxially; inflorescence subcapitate:

           27 Leaf-blades rhomboid-elliptical, 20–35 mm. long, attenuated at base, serrate   19. *ciliata*

           27 Leaf-blades orbicular to oblong, 3–7 mm. long, up to 6 mm. broad, rounded at base, entire or 3–5-lobed   20. *portlandiana*

         26 More or less erect bushy subshrubs:

           28 Leaves crenate, obovate   21. *saxicola*

           28 Leaves serrate, elliptical:

             29 Leaves serrate in distal two-thirds, some of the larger with petioles 10 mm. or more long; cystoliths on each surface of very different sizes; stipules 3–4 mm. long, membranous   22. *wullschlaegelii*

             29 Leaves serrate in distal one-third; petioles not exceeding 2 mm. long; cystoliths on each surface more or less uniform; stipules short, inconspicuous   23. *clandestina*

   17 Leaves of a pair more or less equal, if unequal the smaller not usually less than half the length of larger:

     30 Stems and leaves glabrous; stems sometimes glandular; rarely distal margin of leaf-blade ciliolate or a very few large pellucid hairs on blade adaxially or petiole thinly hispid:

       31 Leaves up to 2 (–2·5) cm. long:

         32 Leaves 1-nerved, long-tapered to base; plants more or less tufted:

           33 Lateral veins of leaf obvious, pinnate; leaf-margin crenate-serrate in distal half   24. *serrulata*

           33 Lateral veins of leaf inconspicuous; lateral margins of leaf entire, tip 5 (–6)-lobed   25. *yunckeri*

         32 Leaves 3-nerved, crenate-serrate; plant more or less trailing or scandent and rooting from nodes:

           34 Leaves obovate to elliptical, narrowed to base, triplinerved   34. *radicans*

           34 Leaves ovate, broadly cuneate to rounded at base, 3-nerved from the base with distinct ascending secondary veins   32. *rotundata*

       31 Leaves mostly larger, 3 (–5)-nerved at or above the base:

         35 Leaf-margin mostly subentire, serrate only towards the tip; blade oblanceolate to elliptical:

           36 Petiole 3–9 mm. long; lamina more or less decurrent on the petiole   27. *oblanceolata*

           36 Petiole 10 mm. or more long; lamina abruptly rounded at junction with petiole   28. *reticulata*

35 Leaf-margin serrate or dentate at least in distal half:
37 Leaves lanceolate or narrowly elliptical:
    38 Lamina abruptly rounded and often cordulate at base, 3-nerved; cymes stalked; stipules 5–10 (–15) mm. long   29. *crassifolia*
    38 Lamina decurrent on the petiole, triplinerved; cymes subsessile; stipules very small   30. *sessiliflora*
37 Leaves ovate or ovate-lanceolate:
    39 Cymes rounded on peduncles up to about as long as the petioles; leaves often glandular-punctate beneath:
        40 Male perianth glabrous   31a. *impressa* var. *impressa*
        40 Male perianth with long hairs   31b. *impressa* var. *barbata*
    39 Cymes branched or of several distant clusters on peduncles usually longer than the petioles:
        41 Lamina generally exceeding 5 cm. long:
            42 Stipules oblong or elliptical, 7–15 mm. long, early caducous
                                                                        38. *elizabethae*
            42 Stipules broadly ovate to suborbicular, 10–20 mm. long, subpersistent
                                                                        40. *grandifolia*
        41 Lamina generally less than 5 cm. long:
            43 Stipules 3–6 mm. long   47. *silvicola*
            43 Stipules 1–1·5 (–2) mm. long; young stems very shortly puberulous
                                                                        45. *brevistipula*
30 Stems or leaves puberulous, pubescent or pilose:
    44 Peduncles usually shorter than the petioles; whole inflorescences shorter than the leaves; stems creeping and trailing:
        45 Leaf-blade elliptical, about twice as long as broad, serrate   19. *ciliata*
        45 Leaf-blade suborbicular, about as long as broad:
            46 Lamina rounded or obtuse at base, crenate nearly all round, with long hairs adaxially, ciliate   35. *nummulariifolia*
            46 Lamina cuneate at base, crenulate distally only, puberulous adaxially, minutely ciliolate   36. *depressa*
    44 Peduncles usually longer than the petioles:
        47 Leaf-blades mostly 6 cm. or more long, ovate to ovate-lanceolate, coarsely crenate or serrate-dentate, usually broadly or narrowly rounded at base and often shortly cordate; hispid at least on the veins abaxially; stipules (5–) 10–15 (–20) mm. long; shrubby herbs:
            48 Inflorescences at least as to male flowers of compact clusters along a simple rachis; veins abaxially with spreading hairs; stem glabrous   37. *appendicilata*
            48 Inflorescences paniculate, diffusely and divaricately branched:
                49 Leaves ovate to elliptic-lanceolate, rarely as much as 12 cm. long; stipules oblong to elliptical, up to 12 mm. long; stems glabrous to pubescent; petioles and leaf-blades variously hairy   39
                49 Leaves broadly ovate, mostly over 10 cm. long; stipules broadly ovate to suborbicular, 10–20 mm. long:
                    50 Stem and petioles glabrous; veins abaxially with a few short hairs; stipules glabrous, up to 20 mm. long   40. *grandifolia*
                    50 Stem and petioles pilose; lamina pubescent on both surfaces; stipules ciliate, up to 15 mm. long   41. *andersonii*
        47 Leaf-blades mostly less than 7 cm. long; stipules usually less than 6 mm. long, often persistent; stems slender, often flexuous or creeping, sometimes tufted from a rhizome or tuber:
            51 Cymes more or less simple, small, dense-flowered; leaves broadly ovate to orbicular, glabrous or with long pellucid hairs adaxially:
                52 Stems and petioles glabrous; leaf-blade up to 1·5 cm. broad, very shallowly crenate   32. *rotundata*
                52 Stems and petioles pubescent, glabrescent; leaf-blade up to 2·5 cm. broad, boldly crenate   33. *repens*
            51 Cymes compound or branched, laxly flowered; stems and petioles puberulous to pubescent; leaves mostly ovate:
                53 Leaves more or less clustered towards the ends of the shoots:
                    54 Larger leaves not exceeding 2 (–2·5) cm. long; main stem creeping; flowering shoots erect   42. *harrisii*
                    54 Larger leaves 3–4 cm. long; main stems ascending   43. *inaequalis*
                53 Leaves spread evenly along the stem:
                    55 Indumentum of dried plant rufous; stipules usually cuspidate or acuminate
                                                                        44. *nigrescens*
                    55 Indumentum of dried plant greyish or dull brown; stipules rounded or obtuse:
                        56 Stem in younger parts very shortly greyish-puberulous only; stipules 0·5–1·5 mm. long   45. *brevistipula*
                        56 Stem at least in younger parts pubescent to pilose:

57 Stipules 1–2·5 mm. long; leaf-blades rounded to subtruncate at base, mostly
2–3 (–4) cm. long; stems more or less tufted; plant often tuber-bearing:
58 Cystoliths adaxially linear and crescent-shaped only
46a. *lamiifolia* var. *lamiifolia*
58 Cystoliths adaxially linear, crescent-shaped and also dot-like especially in
the midrib region   46b. *lamiifolia* var. *puberula*
57 Stipules 3–6 mm. long; leaf-blades often cordulate in a broadly rounded or
truncate base, mostly (3–) 4–6 (–7) cm. long; main stems trailing:
59 Leaf-margin crenations coarse (4–6 per 2 cm. on the larger leaves), acutely
pointed, the terminal lobule linear-oblong and distinct; stipules glabrous
or ciliate   48. *obtusata*
59 Leaf-margin crenations finer (6–9 per 2 cm.), obliquely obtuse to rounded,
the terminal lobule like the lateral; stipules ciliate   49. *maxonii*

1. **P. suta** Adams in Mitt. Bot. Staatss. München **8**: 107 (1970).
Woody herb 10–25 cm. high; leaves at nodes 3–6 mm. apart, rather thick, 3–12
mm. long, 2·5–7 mm. broad, concave beneath with large white cystoliths on the
veins and within the margin; petioles 1–3 mm. long; stipules very small, persistent;
cymes unisexual.
Very rare (Manch.), among limestone boulders; 3000 feet; fl. and fr. May.
P 8013! Endemic.

2. **P. microphylla** (L.) Liebm. in Dansk Vid. Selsk. Skr., Raekke 5, **2**: 296 (1851).

2a. **Var. microphylla.**—Baby Puzzle, Lace Plant.
Branched prostrate or erect herb; leaves 1·5–7 mm. long; two variants occur:
A. Plant to about 15 cm. high; stem and flowers green or tinged reddish.
Common as a weed of damp ground, paths, pot-plants and walls; 50–4500 feet;
fl. and fr. all the year. *A 6604! H 7257! P 24616!* Throughout the tropics and in
hot-houses in temperate countries.
B. Plant much more robust, profusely branched to 40 cm. high; stem, leaves and
flowers glaucous green.
Rather common locally on walls and pathside rocks in damp shaded places;
2000–5500 feet; fl. and fr. most of the year. *A 11941! H 7407! P 22686!* ?Endemic.

2b. **P. microphylla** var. **succulenta** Griseb., Fl. Br. W.I.: 155 (1860).
Tough-stemmed succulent herb with spreading branches; leaves lustrous
greyish-green, suborbicular, 3–5 mm. broad; flowers pink.
Locally common (St. Andr., Port., St. Thom.), on limestone rocks in semi-arid
areas; 50–1000 feet; fl. and fr. May–July. *A 7093! P 11833! Yuncker 18432!*
Greater Antilles.

2c. **P. microphylla** var. **trianthemoides** (Sw.) Griseb., Fl. Br. W.I.: 155 (1860).
—Artillery Plant.
Low robust shrubby herb with arching branches and readily deciduous leaves.
Commonly cultivated as an edging for flower-beds and on rock walls.

3. **P. ordinata** Adams in Mitt. Bot. Staatss. München **8**: 104 (1970).
Stems glabrous, wiry, more or less tufted, 4–10 cm. high; leaves very regularly
arranged, the distal margin setulose, the larger 5–10 mm. long, 2–4 mm. broad,
darker above, often pinkish beneath; peduncles about 10 mm. long; achenes
glandular.
Very rare (Han.), on moist shaded limestone boulders and ledges; 250–1400
feet; fl. and fr. Aug, Oct. *A 13214! P 26588! 26670!* Endemic.

4. **P. herniarioides** (Sw.) Wedd. in Ann. Sci. Nat. sér. 3, Bot. **18**: 207 (1852).
Stems shortly creeping with flowering branches ascending to about 5 cm. high;
leaves up to about 3 mm. in diameter.
Rare (St. Cath., St. Ann), in hollows in honeycombed limestone near the sea;
SL–20 feet; fl. and fr. Mar, Nov. *A 11900! H 10378!* Costa Rica, Bahamas,
Florida, Greater Antilles, dry islands of the Lesser Antilles, Curaçao; Grand Cay-
man.

5. P. brittoniae Urb., Symb. Ant. **5**: 528 (1908).—*P. herniarioides* of F. & R. (1914), partly.

Like the last but diffusely and prostrately branched to form mats and cushions; leaves about 2 mm. in diameter; perianth purplish.

Locally common (St. Andr., Port., St. Thom.), with mosses on damp sheltered banks and on wet rocks; 3500–6500 feet; fl. and fr. sporadically throughout the year. *A 10654! H 7396! P 8203!* Hispaniola.

6. P. nudicaulis (Sw.) Wedd. in Ann. Sci. Nat. sér. 3, Bot. **18**: 208 (1852).

Glabrous shrubby herb up to 1 m. high; leaves acute, 4–10 cm. long, up to 14 mm. broad, dark green adaxially; peduncles and flowers often tinged reddish.

Locally common (St. Andr., St. Thom.), in thickets and on steep rocky banks; 700–5000 feet; fl. and fr. Mar–Sept. *A 7277! H 8390! J.P. 1210! P 11777!* Cuba.

7. P. lanceolata (Lam.) Wedd. in Ann. Sci. Nat. sér. 4, Bot. **1**: 208 (1854).

Herb like the last, with upright shoots from a creeping rhizome, but altogether smaller and more delicate.

Not recently confirmed for Jamaica, *W. Wright!* Hispaniola, Dominica, St. Vincent.

8. P. parietaria (L.) Blume, Mus. Bot. Lugd.-Bat. **2**: 48 (1856).

Erect or trailing herb with stiff reddish stems 25–60 cm. high; leaves mostly narrowed to base, often obliquely acuminate at tip, 2–8 cm. long, 1–3·5 cm. broad; flowers light green.

Locally common (St. Andr., Port., St. Thom.), rare (St. Cath.), on pathside and streambank rocks; 1250–5000 feet; fl. and fr. Dec–Aug. *A 6336! H 5173! J.P. 1256! P 16612!* When construed in a broad sense this species extends through the West Indies to Grenada and Barbados, however *P. ciliaris* Wedd. with rounder-based leaves has been distinguished in the Lesser Antilles and a glabrous variant from Hispaniola is intermediate.

9. P. weddellii F. & R. in Journ. Bot. **50**: 177 (1912).

Shrubby herb with stiff or brittle often reddish stems, 60–120 cm. high; leaves 1·5–8 cm. long, 1–2·5 cm. broad, paler and very shiny beneath; flowers light green.

Locally frequent (St. Andr., Port., St. Thom.), on damp rocky banks in sheltered places, sometimes epiphytic; 2500–5500 feet; fl. and fr. July–Mar. *A 7436! H 7351! HPS 14795!* Endemic.

10. P. alpestris (Urb.) F. & R., Fl. Jam. **3**: 66 (1914).

Low much-branched shrub with reddish stems thick and woody at the base, up to 60 cm. high; leaves obtuse at base, usually thinly hispid adaxially or glabrous, 5–10 mm. broad.

Locally common (St. Andr., Port., St. Thom.), gregarious forming patches of dense undergrowth in montane woodland; 5200–7200 feet; fl. and fr. Nov–Apr. *A 10626! H 5273! P 9600! Stearn 98!* Endemic.

11. P. virgata Wedd. in DC., Prodr. **16** (1): 112 (1869).

Erect-branched shrubby herb 30–90 cm. high, with reddish stems and petioles; leaves rounded to cuneate at base, dark green adaxially, often tinged reddish especially on the veins adaxially, 3–6 cm. long, up to 2 cm. broad.

Locally common (St. Cath., St. Ann), on rocks in woodland and on limestone roadside banks; 2000–2750 feet; fl. and fr. Jan–May. *A 10809! H 12002! H & P 15152!* Endemic.

12. P. laurae Adams in Mitt. Bot. Staatss. München **8**: 103 (1970), (err. *laurea*).

Shrubby herb with loosely tufted stems 25–40 cm. high; leaves light green, margin slightly irregular, tip abruptly cuspidate-acuminate, 1·3–4·3 cm. long, 1–2·6 cm. broad.

Very rare (Trel.), in humus on shaded limestone ledges; 2000–2200 feet; fl. and fr. Mar–May. *P 24844!* Endemic.

13. P. proctorii Adams in Mitt. Bot. Staatss. München **8**: 106 (1970).

Shoots tufted, pendulous, little-branched, rooted mostly at base, 8–15 cm. long; larger leaves 5–22 mm. long, 4–10 mm. broad, increasing in size towards the branch-tip; smaller leaves suborbicular, 2–4 mm. in diameter, glabrous or with a few large pellucid hairs adaxially, rufous-pubescent on midrib beneath.

Very rare (West.), on moist shaded limestone cliffs; about 1000 feet; fl. not seen. *P 21499!* Endemic.

14. P. lucida (Sw.) Blume, Mus. Bot. Lugd.-Bat. **2**: 48 (1856).

Bushy shrublet 4–20 cm. high; larger pinnatifid leaves with 1–3 lobes on each side, up to 4 mm. broad; peduncles about as long as the larger leaves; flowers greenish.

Occasional on limestone rocks in damp shaded places; 400–2700 feet; fl. and fr. most of the year. *A 11432! H 9082! P 22292!* Endemic.

15. P. crenulata (Sw.) Urb., Symb. Ant. **5**: 308 (1907).

Rhizomatous shrublet 3–15 (–30) cm. high, branched above; leaves mostly obtuse or rounded at tip, the larger 4–10 mm. broad, all green or reddish beneath, the larger cystoliths submarginally parallel; peduncles 10–20 mm. long.

Rare and local (West., Han., Trel.), on damp honeycombed limestone rocks in forest; 1000–1500 feet; fl. and fr. Apr–May, Oct–Dec. *A 8629! H 9238! P 23468!* Endemic.

16. P. wilsonii Urb., Symb. Ant. **5**: 309 (1907).

Stems 4–7 cm. high; larger leaves spathulate, rounded at tip, tapered to base, 6–14 mm. long, 4–7 mm. broad; peduncles 6–7 mm. long.

Very rare (Port.), in wet forest on limestone; about 2000 feet; fl. and fr. ? Mar. *Robertson UCWI 3318!* Endemic.

17. P. rufa (Sw.) Wedd. in Ann. Sci. Nat. sér. 3, Bot. **18**: 220 (1852).

17a. Var. rufa.

Bushy herb or shrub 20–100 (–150) cm. high; larger leaves ovate-elliptical, 1·5–4·5 cm. long, 8–18 mm. broad, paler beneath; peduncles longer than the petioles; flowers white.

Occasional (St. Eliz., Trel., Port.), on cliffs and in crevices in shaded places on limestone; 1100–2000 feet; fl. and fr. Jan–Apr, July–Aug. *A 9122! 12434! HPS 14796! Stearn 967!* Endemic.

17b. P. rufa var. microstipula Adams in Mitt. Bot. Staatss. München **8**: 107 (1970).

A smaller rather more laxly delicate plant than the last with paler hairs.

Very rare (Han.), on shaded limestone ledges; 1000–1400 feet; fl. and fr. Aug, Oct. *P 26669!* Endemic.

18. P. rufescens F. & R. in Journ. Bot. **50**: 178 (1912).

Shrubby herb with tufted much-branched shoots 20–30 cm. high; larger leaves elliptical, 1–2·2 cm. long, 6–10 mm. broad; stipules rounded, 3–4 mm. long; peduncles longer than the petioles; flowers white.

Rare and local (Trel.), on limestone rocks in forest; 2000–2250 feet; fl. and fr. Mar–May. *A 6766! H 8533! Robertson UCWI 958!* Endemic.

19. P. ciliata (Sw.) Blume, Mus. Bot. Lugd.-Bat. **2**: 46 (1856).

Rhizomatous herb with ascending branches up to 20 cm. long; leaves 10–14 (–20) mm. broad, shiny, glabrous, except the ciliate margin, or with a few pellucid hairs adaxially and also sometimes pilose on the nerves beneath; stipules small, triangular; peduncles 1–8 mm. long, reddish; flowers light green.

Occasional, trailing on limestone boulders or low on tree trunks in wet shady woodlands; 600–1750 feet; fl. and fr. Mar, July–Nov. *A 7645! 11572! P 21479!* Endemic.

20. P. portlandiana Adams in Mitt. Bot. Staatss. München **8**: 105 (1970).

Stems filiform, more or less tufted but also rooting, 3–8 cm. long; cystoliths very much larger on the upper leaf-surface than on the lower; smallest leaves with blades above 2 mm. in diameter; peduncles filiform, 8–18 mm. long.

Rare and local (Port.), on limestone rocks in wet forest; 1500–2250 feet; fl. and fr. July–Sept. *A 9133! ACH 938! P 22709!* Endemic.

21. P. saxicola Urb., Symb. Ant. **5**: 311 (1907).
Glabrous branched shrubby herb 25–30 cm. high; larger leaves petiolate, obtuse at apex, cuneate at base, 8–18 mm. long, 6–8 mm. broad; stipules small, persistent; cyme small on a peduncle 5–10 mm. long.
Rare and local (Trel.), on rocks in shady woodland; 1800–2200 feet; fl. and fr. May–June. *H 8540! 8700!* Endemic.

22. P. wullschlaegelii Urb., Symb. Ant. **5**: 310 (1907).
Shrub rooting at the base; leaves ovate-elliptical, obtuse or rounded at base and apex, glabrous or sparingly pilose on the veins beneath, 7–30 mm. long, 7–20 mm. broad; peduncles 5–15 mm. long; cymes small.
Very rare and local (Manch.), on rocks in woodland; about 2000 feet. *H & B 10604! 10611!* Endemic.

23. P. clandestina Wedd. in DC., Prodr. **16** (1): 120 (1869).
Shrublet with woody stems, rooting near the base, 20–30 cm. high, glabrous; larger leaves subsessile, 15–20 mm. long, 5–8 mm. broad, with 3–5 teeth on each side; cymes small, pedunculate.
Very rare (West.), on rocks; 1400 feet; fl. and fr. Sept. *H 9907!* Endemic.

24. P. serrulata (Sw.) Wedd. in Ann. Sci. Nat. sér. 3, Bot. **18**: 213 (1852).
Shrublet 20–30 cm. high; leaves oblong-lanceolate to oblanceolate, obtuse at tip, 10–20 mm. long, 3–4 mm. broad; cymes few-flowered, pedunculate, shorter than the leaves.
Very rare; known only from the type, *Swartz!* A variety occurs in Cuba.

25. P. yunckeri Adams in Mitt. Bot. Staatss. München **8**: 108 (1970).
Stems tufted, branched near the base and above, 6–15 cm. high; leaves obcuneate. with larger more conspicuous cystoliths adaxially, 4–11 mm. long, up to 4 mm. broad; stipules broadly rounded, persistent; cymes subcapitate on peduncles 3–14 mm. long.
Very rare (Port.), on moist rocks in forest; 1000–2000 feet; fl. and fr. June. *Yuncker 18831!* Endemic.

26. P. flavicaulis Urb. & Britton, Symb. Ant. **7**: 194 (1912).
Shrubby herb up to 60 cm. high, with straggling branches forming loose clumps from a short-rhizome; leaves lanceolate to oblanceolate, shortly tapered and irregular-margined at tip, shortly rounded at base, rather succulent, 2–8 cm. long, 1–2 (–2·8) cm. broad; stipules 1–2 mm. long; cymes pedunculate, paniculate; flowers pink.
Local and uncommon (St. Cath., Clar., Trel.); on cliffs, ledges and in crevices of craggy limestone; 1650–2500 feet; fl. and fr. Mar–May, Aug. *A 12832! H 10881! P 24753!* Endemic.

27. P. oblanceolata F. & R. in Journ. Bot. **50**: 179 (1912).
Shrubby herb 10–30 cm. high, with trailing and rooting stems; leaves oblanceolate, acute or shortly acuminate, 3–6 cm. long, 1–1·8 cm. broad; stipules 1–1·5 mm. long; cymes paniculate; peduncles 3–4 cm. long; flowers reddish.
Rare and local (Clar.), in crevices of limestone rocks in shade; 2000–2500 feet; fl. and fr. May–Aug, Dec. *A 12606! H 10946!* Endemic.

28. P. reticulata (Sw.) Wedd. in Ann. Sci. Nat. sér. 3, Bot. **18**: 215 (1852).
Shrubby herb with trailing ascending branches 20–100 cm. high; leaves oblong-elliptical, acute to acuminate at tip, paler beneath, 6–13 cm. long, 2–4·5 cm. broad; stipules up to 4 mm. long; cymes paniculate; flowers greenish-white to crimson.
Rather local (West., Han., Trel.), on limestone rocks in forest; 1200–1750 feet; fl. and fr. Mar–May. *A 12430! 12817! H 9270! Stearn 1051!* Endemic.

29. P. crassifolia (Willd.) Blume, Mus. Bot. Lugd.-Bat. **2**: 52 (1856).
Shrubby herb 30–150 cm. high; leaves elliptic-lanceolate, long-acuminate,

glossy, often bullate between the veinlets, bright green with the main veins often pinkish beneath, 3–13 cm. long, 1–4 cm. broad; stipules oblong, early caducous; cymes paniculate; flowers bright pink to crimson.

Common on limestone walls and on rocks in woodlands; 10–3000 feet; fl. and fr. all the year. *A 12395! H 8963! P 11731! Stearn 441!* Endemic.

30. P. sessiliflora (Sw.) Wedd., Monogr.: 242 (1856).

Shortly rhizomatous herb 20–30 cm. high; leaves narrowly elliptical, long-acute to acuminate, 3–8 cm. long, 0·5–5 cm. broad; petioles up to 3 cm. long; peduncles very short, winged.

Rare (Manch.), in moist woodlands; about 3000 feet; fl. and fr. Feb, Oct. *H 8266!* Endemic.

31. P. impressa Urb., Symb. Ant. **5**: 314 (1907).

31a. Var. impressa.—*P. impressa* var. *troyana* Urb. (1907).

Herb with trailing and rooting stems; flowering branches ascending to 20–30 cm. high; leaves acute to acuminate, broadly rounded and often cordulate at base, fleshy, dark green adaxially, the larger 5–9 cm. long, up to 4 cm. broad; petioles 1·5–3 cm. long, glabrous or thinly hispid; cymes 1–1·5 cm. in diameter.

Rare (Han., Trel.), on rocks in damp forest; 1400–2250 feet; fl. and fr. May. *A 12416! H 8532! 9271!* Endemic.

31b. P. impressa var. **barbata** Adams in Mitt. Bot. Staatss. München **8**: 103 (1970).

Like the last; leaves 1–6 cm. long, up to 2·8 cm. broad, margin crenate in distal half to two-thirds; male perianth with long hairs; female perianth with a few hairs.

Rare and local (Clar.), trailing on jagged limestone rocks in woodland; 2300 feet; fl. and fr. Aug. *A 12605! A & C 1257!* Endemic.

32. P. rotundata Griseb., Fl. Br. W.I.: 158 (1860).—? *P. dauciodora* Wedd. var. *parvifolia* Wedd. (1869).

Diffusely branched herb with often rather thick fleshy stems, rooting from the nodes; flowering shoots 10–20 cm. high; leaves 1–2 (–2·3) cm. long, up to 17 mm. broad; stipules very small; peduncles slender, the male longer than the leaf, or cymes possibly also androgynous.

Rare (Manch., West., Port.), gregarious on moist limestone boulders; 1000–2300 feet; fl. and fr. Mar, Sept–Oct. *H & B 10603! P 7298! Wilson 648!* Endemic.

33. P. repens (Sw.) Wedd. in Ann. Sci. Nat. sér. 3, Bot. **18**: 220 (1852).

Stems creeping and rooting; leaf-blades suborbicular, broadly truncate to rounded at base, 1–3 cm. long; stipules broadly ovate, 1–2 mm. long; peduncles slender, about 2 cm. long; cymes androgynous.

Very rare (St. Ann). *Prior!* Cuba, Hispaniola.

34. P. radicans (Sw.) Wedd. in Ann. Sci. Nat. sér. 3, Bot. **18**: 223 (1852).

Stems creeping and rooting at and between the nodes, branched, up to 50 cm. long; leaves light green, glabrous with 3 or 4 large crenations distally, 10–20 mm. long, 5–12 mm. broad; stipules minute; rarely found in flower.

Rare and local (Trel., Port., St. Thom.), climbing on the stems of shrubs and trees and also on rocks in damp shady woodlands; 1200–3000 feet. *A 12965! H 10574! P 27782!* Endemic.

35. P. nummulariifolia (Sw.) Wedd. in Ann. Sci. Nat. sér. 3, Bot. **18**: 225 (1852).

Stems long-creeping with short erect flowering shoots; leaves up to about 15 mm. long and broad; stipules rounded, 1–2·5 mm. long; peduncles 5–12 mm. long; cymes small, compact; flowers greenish-white.

Locally common (St. Andr., St. Ann, St. Thom.), on shaded roadside banks and limestone walls; 500–2000 feet; fl. and fr. Apr–Aug. *A 6995! H 11608! Webster & Wilson 5226!* Probably originally escaped from cultivation; Bermuda, Greater Antilles, Virgin Is., Antigua, Barbados.

36. P. depressa (Sw.) Blume, Mus. Bot. Lugd.-Bat. **2**: 46 (1856).

Stems long-creeping and rooting, often reddish, up to about 40 cm. long; leaves dark green adaxially, often reddish beneath or glaucous, fleshy, 5–12 (–15) mm. long, 4–9 (–12) mm. broad; petioles 2–7 mm. long, puberulous; stipules rounded, up to 2 mm. long; peduncles usually shorter than the petioles; male flowers pink.

Local (St. Andr., St. Thom.), on damp shaded rocky slopes and cliffs; 400–1500 feet; fl. and fr. July–Oct. *A 9772! H 8433! P 19794!* Reported from Cuba and Hispaniola, but typical plants seem to be restricted to Jamaica.

37. P. appendicilata F. & R. in Journ. Bot. **50**: 179 (1912).

Erect shrubby herb to 1 m. high; leaves ovate to broadly ovate, acuminate, 7–12 cm. long, 5–7 cm. broad, glabrous adaxially; stipules ovate, about 10 mm. long; peduncles glabrous or hispidulous.

Rare (Clar., Manch.), in shaded gullies; 700–2200 feet; fl. and fr. Feb–Apr. *H 10843! P 6520!* Endemic.

38. P. elizabethae F. & R. in Journ. Bot. **50**: 179 (1912).

Glabrous shrubby herb 1–2 (–2·5) m. high; leaves ovate, acuminate, mostly 5–12 cm. long, 2–6 cm. broad; peduncles and cymes usually shorter than the leaves; cymes paniculate, many-flowered.

Occasional, on banks and in thickets on limestone; 100–1500 feet; fl. and fr. Dec–Apr, July–Sept. *A 10925! H 11132! P 6539!* Endemic.

39. A group of populations having the general affinity of *P. elizabethae* but all more or less hairy. Besides variation in the distribution of hairs there are variable patterns of leaf-shape, leaf-margin and leaf-base, ranging between typical *P. crassifolia* and *P. grandifolia*. The combinations of these patterns result in a situation quite beyond the capacity of this author to resolve into a meaningful classification with the information at present available. The following more or less natural division is based on indumentum characters only:

39a. Sub-group A. Stem and adaxial leaf-surface glabrous; petiole and abaxial leaf-surface hairy; stipules glabrous; (Manch,. St. Eliz.); 900–2900 feet; fl. and fr. Feb–Apr. *P 11613!* A further variant with long hairs only on the veinlets abaxially is *A 10202!* (Manch.; 1500 feet; Dec).

39b. Sub-group B. Stem glabrous; petiole and both leaf-surfaces thinly hairy; stipules glabrous or pubescent; (St. Cath., Manch., Trel., Port.); 400–2250 feet; fl. and fr. Jan, Apr, May, Aug, Dec. *P 15606! Wood!* In the Trelawny specimens the stipules are pubescent, in the Portland glabrous and in Manchester both types occur.

39c. Sub-group C. Stem thinly hispid with long hairs on youngest parts; petiole and both leaf-surfaces distinctly hairy; stipules pubescent; (Manch., Trel.); 1500–3000 feet; fl. and fr. Apr–Aug, Oct. *Cornman 584! P 23454! 23455!*

39d. Sub-group D. Stem, petioles and both leaf-surfaces pubescent with whitish hairs; stipules pubescent; (St. Cath., St. Thom.); 800–1000 feet; fl. and fr. Oct. *A 10299! P 19810!*

40. P. grandifolia (L.) Blume, Mus. Bot. Lugd.-Bat. **2**: 52 (1856).—Maroon Bush.

Shrubby herb with thick brittle stems up to 2 m. high; leaves narrowly to broadly ovate, acuminate, often bullate and tinged reddish, shiny adaxially, the veins prominent beneath, up to 22 cm. long and 16 cm. broad; peduncles usually longer than petioles; cymes dioecious, laxly paniculate; flowers often reddish or pink.

Common and locally very abundant in moist shady or sheltered places; 500–5200 feet; fl. and fr. Jan–Sept. *A 7435! H 11129! P 22278!* Endemic.

41. P. andersonii Adams in Mitt. Bot. Staatss. München **8**: 103 (1970).

Shrubby herb like the last 1–2 m. high; leaves long-acuminate, coarsely serrate-dentate, up to 15 cm. long and 10 cm. broad; plants apparently dioecious.

Rare and local (St. Thom.), forming colonies in semi-open areas; about 6500 feet; fl. and fr. July. *ACH 911! Powell 315! P 9444!* Endemic.

42. P. harrisii Urb., Symb. Ant. **1**: 299 (1899).— *P. hollickii* F. & R. (1912).

Shortly rhizomatous herb with erect flowering branches up to 20 cm. high; leaves equal or unequal, ovate, obtuse at both ends, adaxially with a few long hairs or glabrous; peduncles glabrous, up to about 2 cm. long.

Rare (St. Andr., Clar., St. Ann), on honeycombed limestone rocks in woodland; 1000–2500 feet; fl. and fr. Apr, Aug–Nov. *A 7793! Britton & Hollick 2754! H 6881! 11220!* Endemic.

43. P. inaequalis (Juss. ex Poir.) Wedd. in Ann. Sci. Nat. sér. 3, Bot. **18**: 229 (1852).—*P. troyensis* F. & R. (1912). *P. obtusata* of F. & R. (1914), partly. *Urtica inaequalis* Juss. ex Poir. (1816).

Herb with stems ascending to 20 cm. high; leaves more or less unequal in each pair, ovate, with or without large pellucid hairs adaxially, usually ciliate-margined; stipules glabrous or ciliate.

Rare (Clar., Manch., Trel.), in crevices in moist limestone rocks; 1200–2300 feet; fl. and fr. Jan, Apr–Sept. *A 12814! H & B 10608! P 23656!* Cuba, Puerto Rico, Virgin Is., St. Lucia, Trinidad and Tobago.

44. P. nigrescens Urb., Symb. Ant. **1**: 299 (1899).

Shrubby herb with tufted more or less erect flowering shoots (15–) 30–60 cm. high; stem pinkish; leaves ovate, obtuse or shortly acuminate at tip, glabrous adaxially except at base of midrib, bullate, shiny, 2–6 cm. long, 1–3 cm. broad; stipules 3–5 mm. long; peduncles 1–7 cm. long.

Locally common (St. Andr., Port., St. Thom.), gregarious by pathsides and on mossy boulders in montane forest; (2000–) 4800–7000 feet; fl. and fr. July–Dec. *A 11351! H 7350! Maxon 9373!* Endemic.

45. P. brevistipula Urb., Symb. Ant. **6**: 6 (1909).

Shrubby herb with ascending flowering shoots 10–45 cm. high, somewhat rhizomatous at base; leaves rather unequal, ovate to elliptical, obtuse at apex, rounded to obtuse at base, glabrous or with a very few long hairs adaxially, the larger 3–4 cm. long, 1·5–3 cm. broad; cymes long-pedunculate.

Uncommon in the western parishes, on rocks and mossy tree trunks in damp shady places; 700–1700 feet; fl. and fr. Mar–May, Aug–Sept. *H 9905! P 15764! 20855!* Endemic.

46. P. lamiifolia F. & R. in Journ. Bot. **50**: 180 (1912).

46a. Var. lamiifolia.

Stems more or less tufted and laxly ascending, up to 20 cm. high; leaves ovate to ovate-elliptical, sparsely pilose with long hairs adaxially, often coloured mauve beneath, 1–3·5 cm. long, 1–2 (–2·5) cm. broad; stipules ciliate; peduncles glabrous; male flowers often reddish.

Occasional in the central parishes, on rocks in damp shady places; 500–2500 feet; fl. and fr. Aug–Sept, Dec. *A & C 1179! H 11220! H & P 13989! Robertson UCWI 5360!* Endemic.

46b. P. lamiifolia var. puberula F. & R., Fl. Jam. **3**: 79 (1914).—*P. harrisii* of F. & R. (1914), partly.

Stems sometimes tuberous at base, often reddish.

Locally common (Port., St. Thom.), on moist limestone rocks and logs in wet shady woodlands; 100–1750 feet; fl. and fr. Mar, July–Aug, Nov. *A 7647! H 10777! H & B 10683!* Endemic.

47. P. silvicola F. & R. in Journ. Bot. **50**: 181 (1912).

Shrubby herb with ascending branches 20–30 cm. high; leaves ovate-elliptical, shortly acuminate, 3–5 cm. long, 1·5–3 cm. broad, glabrous; peduncles 1–3 cm. long; cymes paniculate.

Very rare (Clar.), in crevices of rocks; 2500 feet; fl. July. *H 10948!* Endemic.

48. P. obtusata Liebm. in Dansk Vid. Selsk. Skr., Raekke 5, **2**: 300 (1851); F. & R. (1914), partly.

Shrubby herb, branches erect or ascending or the older trailing; leaves unequal on sterile shoots, broadly ovate, acuminate, 3–6 cm. long, 2–3·5 cm. broad, with a few large pellucid hairs adaxially, sometimes tinged violet; peduncles 2–7 cm. long; cymes diffusely branched.

Occasional on cliffs and on limestone rocks in woodlands; 100–2300 feet; fl. and fr. Mar–Sept. *A 7570! ACH 710! H 11982! P 15626!* Puerto Rico, Lesser Antilles.

49. P. maxonii Britton in Bull. Torr. Bot. Club **48**: 340 (1922).

Stems trailing to 120 cm. or more and rooting at the nodes, with ascending branches to 25 cm. high; leaves ovate, acuminate or the smaller obtuse, bullate, 1–6 cm. long, 1–4·5 cm. broad; peduncles and cymes up to about 6 cm. long.

Local and uncommon (St. James, Trel., Port.), on rocky limestone in shade; 100–2200 feet; fl. and fr. Apr. *H 8701! Maxon & Killip 1555! P 24036! Stearn 953!* Endemic.

2. BOEHMERIA Jacq. (1760)

1 Leaves alternate; flower-clusters axillary; stipules up to 9 mm. long:
  2 Female perianth glabrous; style 1–1·5 mm. long   1. *jamaicensis*
  2 Female perianth shortly pilose; style 3–5 mm. long   2. *ramiflora*
1 Leaves, at least on main branches, opposite; flower-clusters in spikes:
  3 Spikes pendulous, without leaves at apex, dioecious; stipules 1·5 cm. long, puberulous
    laterally   3. *caudata*
  3 Spikes erect, usually with a few small leaves at apex, monoecious; stipules 8 mm. long,
    almost glabrous laterally   4. *cylindrica*

1. B. jamaicensis Urb., Symb. Ant. **5**: 329, 330, f. B (1907).—Doctor Johnson.

Monoecious or dioecious shrub or tree 2–5 m. high, much branched above; larger leaves ovate-lanceolate to lanceolate, obtuse and unequal at base, smaller lanceolate, rounded and often cordulate at base, all acuminate, 3·5–15 (–30) cm. long, 1–4 (–10) cm. broad, margins serrate-crenate; flower-clusters 2–5 mm. in diameter; stamens 3.

Common in secondary thickets, along pathsides and in gullies, mostly on limestone in moderately wet areas; 50–3000 feet; fl. and fr. June–Feb. *A 8602! H 9560! HPS 14629!* Endemic.

2. B. ramiflora Jacq., Enum. Syst. Pl. Carib.: 31 (1760).

Monoecious or dioecious shrub 2–3 m. high; larger leaves rhomboid-ovate to lanceolate, caudate-acuminate, oblique at base, with petioles up to 4·5 cm. long, smaller ovate-lanceolate with semi-cordate base, 2–20 cm. long, 1·5–6 cm. broad, sparsely hairy adaxially, softly pubescent beneath, serrate; flower-clusters 4–6 mm. in diameter; stamens 3.

Very rare (Manch.), on well drained roadside bank; about 2000 feet; fl. and fr. Apr. *A 12387!* Mexico to Colombia and Venezuela, Lesser Antilles, Trinidad and Tobago.

3. B. caudata Sw., Nov. Gen. & Sp. Pl.: 34 (1788).—Nettle Tree.

Shrub 2–5 m. or tree to 9 m. high; leaves broadly ovate to elliptical, acute or acuminate, rounded or obtuse at base, serrate, unequally petiolate, hispid with broad-based hairs adaxially, pubescent beneath, light green, up to 25 cm. or more long, up to 15 cm. broad; spikes up to 40 cm. long, bracteate at first; fruits obovate, hooked by persistent style-arms.

Locally common (St. Andr., Port., St. Thom.), rare in Manchester, along streamsides and in moist shady woodland margins; 1750–5000 (–7000) feet; fl. and fr. Feb–Sept. *A 6895! 7381! H 9557! P 6668!* Mexico to Argentina.

4. B. cylindrica (L.) Sw., Nov. Gen. & Sp. Pl.: 34 (1788).

Strong-stemmed trailing undershrub with branches ascending, 30–90 cm. high; stem angular-ribbed; leaves ovate-lanceolate to lanceolate, acuminate, rounded to subcordate at base, coarsely serrate, puberulous on veins beneath, 5–18 cm. long, 2·5–7 cm. broad; spikes interrupted, axillary, up to about 6 cm. long; stamens 4.

Occasional, in alluvial sandy or gravelly soil near ponds and rivers; 50–1100 feet; fl. and fr. Jan–Sept. *A 9447! H 9889! P 24651!* Bermuda, widespread in continental Amer. both within and beyond the tropics, Greater Antilles, Trinidad.

B. nivea (L.) Gaudich. var. *candicans* Wedd., China Grass or Ramie, recognized by the silvery-white underleaf, has been cultivated sparingly in Jamaica.

3. PARIETARIA L. (1753)

1. P. judaica L., Fl. Palaest.: 32 (1756).—*P. diffusa* Mert. & Koch (1823).

Branched herb with lax reddish pubescent stems up to 40 cm. long; leaves lanceolate, acuminate with blunt acumen, triplinerved, cystoliths punctate, thinly hispid adaxially, up to 4·5 cm. long and 1·8 cm. broad; bracts shortly connate; female perianth tubular, 4-toothed, tinged pink, 2·5–3 mm. long; male flowers 4-partite.

Very local (St. Andr.), naturalized on stone walls; 3600 feet; fl. and fr. Dec. *A 5787!* Native of S. Europe extending to Asia, also in Bermuda and Brazil.

4. LAPORTEA Gaudich. (1830) nom. cons.

1. L. aestuans (L.) Chew in Gard. Bull. Singapore **21** (2): 200 (1965).—*Fleurya aestuans* (L.) Miq. (1853). *Urtica aestuans* L. (1763).

Erect branched herb 15–120 cm. high; stem and inflorescence-branches often pinkish; sap turbid, slightly viscous; leaves ovate, shortly acuminate, coarsely serrate, bullate, 8–15 (–30) cm. long, up to 12 cm. or more broad, with bristly pungent hairs, cystoliths small, punctate; panicles divaricately branched; flowers 4 (–5)-merous; fruit about 1 mm. long.

Common as a weed of moist or arid waste places; 50–2300 feet; fl. and fr. most of the year. *A 8361! H 11871! P 24162!* General in the tropics.

5. ROUSSELIA Gaudich. (1830)

1. R. humilis (Sw.) Urb., Symb. Ant. **4**: 205 (1905).

Perennial herb with prostrate rooting branches up to 40 cm. long; stems sometimes dull purplish; leaves broadly ovate, acute or obtuse, 3–10 (–30) mm. long; male perianth red, hirtellous; achene compressed, acute with accrescent perianth and persistent bracteole attached.

Widespread and locally common, on stone walls and limestone banks and along roadsides mostly in damp shady places; 100–2350 feet; fl. and fr. most of the year. *A 9933! H 8501! P 21382!* Mexico, Bahamas, Greater Antilles, Saba, Dominica.

6. PHENAX Wedd. (1854)

Annual herb; bracteoles broadly rounded, ciliate; styles up to 2 mm. long; leaves 1–6 cm. long, 0·5–3·5 cm. broad   1. *sonneratii*

Straggling shrubby herb; bracteoles ovate, acute or cuspidate to acuminate, minutely ciliate; styles about 4 mm. long; leaves 4–15 cm. long, 2–6·5 cm. broad   2. *hirtus*

1. P. sonneratii (Poir.) Wedd. in DC., Prodr. **16** (1): 235.37 (1869).—*Parietaria sonneratii* Poir. (1804). *Phenax vulgaris* Wedd. (1854).

Bushy and diffusely branched herb 60–100 cm. high; stems and petioles reddish; lower leaves broadly ovate to lanceolate, upper leaves narrowly elliptic-lanceolate, cuneate to truncate at base; achene ovoid, acute, papillose, olive-brown, 1 mm. long.

Rather common in the wetter parts of the eastern parishes, in pastures and along roads and pathsides, a weed; near SL–2000 feet; fl. and fr. Jan–Mar, July–Aug. *A 7597! P 22094!* Native of tropical Asia, now scattered through tropical Amer.; a comparatively recent adventive in Jamaica.

2. P. hirtus (Sw.) Wedd. in DC., Prodr. **16** (1): 235.38 (1869).

Shrubby herb 0·3–3·0 m. high; stems slender, thinly pubescent; leaves ovate, acuminate, sometimes subcordate at base, somewhat bullate and sparsely hispid adaxially; achene angled, rugulose, reddish-brown, less than 1 mm. long.

Locally common (St. Andr., Port., St. Thom.), on banks in damp sheltered places and by streams; 3500–5300 feet; fl. and fr. Dec–Aug. *A 7079! H 9558! J.P. 1306! P 22517!* Mexico to Bolivia, Hispaniola.

7. GYROTAENIA Griseb. (1861)

Inflorescence spicate   1. *spicata*
Inflorescence paniculate   2. *microcarpa*

1. G. spicata (Wedd.) Wedd. in DC., Prodr. **16** (1): 99 (1869).

Shrub 2·5–4 m. or brittle-branched tree to 10 m. high; leaves oblong-elliptical, shortly acuminate, glabrous, very shiny on both surfaces, softly fleshy, 5–10 cm. long, 2·5–3·5 cm. broad; male spikes 1 cm. long; female spikes fleshy, 1–1·5 cm. long.

Frequent in the central and western parishes, in woodlands on limestone; 250–2700 feet; fl. and fr. June–Mar. *A 8229! 8235! H 8823! Webster & Proctor 5303!* Endemic.

2. G. microcarpa (Wedd.) F. & R., Fl. Jam. **3**: 56 (1914).—*Urera microcarpa* Wedd. (1856).

Shrub 3–5 m. or tree, with diffuse spreading and drooping branches, up to 6 m. high; leaves ovate to elliptic-lanceolate, more or less acuminate, puberulous on the midrib adaxially and on the veins beneath, glabrescent, dull dark green adaxially, paler and shiny beneath, 5–10 cm. long, 3–4·5 cm. broad; cymes 2–5 cm. long.

Occasional in the eastern parishes and St. Ann, in wet woodlands; 600–3500 feet; fl. and fr. July–Apr. *A 9166! 11124! H 6487! 11839! H & P 15156!* Endemic.

8. URERA Gaudich. (1830)

Leaf-blades broadly ovate, up to 30 cm. long and 25 cm. broad, cordate or rounded at base:
2 Cymes dichotomous; leaves crenate-dentate with teeth 5–8 mm. broad; fruiting
    perianth about 1·5 mm. long, vermilion   **1.** *caracasana*
2 Cymes trichotomous or compound-monochasial; leaves shallowly sinuate-dentate with
    teeth 1–2 cm. broad; fruiting perianth 3–5 mm. long, white or pink   **2.** *baccifera*
1 Leaf-blades oblong-ovate to unequally elliptical, smaller, rounded or shortly cordate to
    broadly cuneate at base, shallowly serrate-crenate to subentire; cymes dichotomous:
3 Young branches and leaves usually with stinging hairs; glabrescent   **3.** *elata*
3 Young branches and leaves usually glabrous, but occasionally a few stinging hairs
    present on the midrib beneath; ripe fruiting perianth orange   **4.** *expansa*

1. U. caracasana Griseb., Fl. Br. W.I.: 154 (1860).

Shrub or tree 3–6 m. high; leaves scabrid-hispid adaxially, more or less pubescent beneath, with a few stinging hairs on the veins.

Very rare and not collected since the original gathering in Jamaica, *W. Wright.* Mexico to Costa Rica, Puerto Rico, Lesser Antilles.

2. U. baccifera (L.) Wedd. in Ann. Sci. Nat. sér. 3, Bot. **18**: 199 (1852).

Shrub or soft-wooded tree up to 6 m. high, often with pungent prickles on the trunk and branches as well as the leaves.

Rather local (St. Andr., St. Thom.), in shaded and sheltered waste places on limestone; 150–1500 feet; fl. and fr. Jan–May. *A 10420! Fawcett 7177! P 11776!* Mexico to tropical S. Amer., Greater Antilles, Virgin Is., St. Vincent, Trinidad and Tobago.

3. U. elata (Sw.) Griseb., Fl. Br. W.I.: 154 (1860).

Tree 3–6 m. high, diffusely branched, with stinging hairs present at least on vegetative shoots; leaves caudate-acuminate, 6–15 (–25) cm. long, 3–10 (–15) cm. broad, with stiff hairs adaxially around the thick-bases of which cystoliths radiate; fruits 3 mm. long.

Occasional, mostly in the eastern parishes, in damp shady places and near streams; (750–) 2300–4900 feet; fl. and fr. Apr–Aug. *A 11496! 12390! H 5284! P 23739!* Endemic.

4. U. expansa (Sw.) Griseb., Fl. Br. W.I.: 155 (1860).—*U. ovatifolia* Urb. (1918).
    Shrub 3–5 m. or tree to 12 m. high; leaves acuminate, 8–17 cm. long, 3–8·5 cm. broad; ripe fruiting perianth orange.
    Occasional, mostly in the central and western parishes, in woodland on limestone; 200–1750 feet; fl. and fr. Feb–Sept. *A 11676! H & B 10612! P 10305!* Endemic.

49. PROTEACEAE

Trees or shrubs. Leaves alternate, rarely verticillate or opposite, simple or variously divided, without stipules. Flowers bisexual or sometimes unisexual and dioecious, in racemes, spikes or heads, often conspicuously bracteate and showy. Perianth uniseriate, a petaloid 4-merous valvate-lobed calyx, usually tubular in bud and variously splitting at anthesis. Stamens 4, opposite the calyx-lobes; filaments adnate to the tube or lobes or rarely free; anthers 2-locular, opening lengthwise. Basal scales or disk present or absent. Ovary superior, usually stalked, 1-locular; style simple; stigma often swollen; ovules 1–many, pendulous or parietal. Fruit a follicle, achene, samara or drupe. Seeds without endosperm, sometimes winged.
    62 genera and 1400 species mainly in the drier regions of the southern hemisphere.

1. GREVILLEA R. Br. (1809) nom. cons.

Leaves pinnate; leaflets toothed or divided; flowers clustered in racemes; disk-glands present; stigma bulbous, at first enclosed within the tip of the curved perianth, later withdrawn as the style straightens; ovules 2.

1. G. robusta A. Cunn. in R. Br., Prot. Nov.: 24 (1830).—Silky Oak, Silver Oak.
    Tree to 16 m. or more high; leaves 30 cm. or more long with acute linear segments; flowers yellow or orange in panicled racemes; fruit a 2-valved apiculate woody capsule about 1·8 cm. long and 1·3–1·5 cm. broad, directed downwards.
    Locally common at middle elevations in the more exposed parts of the eastern mountains (St. Andr., Manch., St. Thom.), naturalized and planted; 900–4200 feet; fl. Apr–July. *A.M. Barry!* Native of Australia, now commonly planted in subtropical regions and montane tropical areas.
    Macadamia ternifolia F. von Muell., Queensland Nut, is sparingly cultivated in Jamaica.

50. OLACACEAE

Trees or erect or climbing shrubs. Leaves alternate, simple, mostly entire, without stipules. Flowers usually bisexual, rather small in axillary cymose inflorescences. Perianth biseriate; sepals (3–) 4–6, imbricate or open in bud; petals (3–) 4–6, free or united, valvate. Stamens as many as and opposite the petals or more numerous, free or adnate to the corolla; some occasionally without anthers; anthers 2-locular, opening lengthwise or by porelike slits. Disk present or absent. Ovary superior, sometimes adnate to the disk, 1–3 (–4)-locular, septation sometimes incomplete, placentation typically axile or apical when 1-locular; ovules solitary in each loculus, pendulous, anatropous; style 1, with 2–5-lobed stigma. Fruit a berry or drupe, often surrounded by the persistent enlarged calyx. Seeds with copious endosperm and minute embryo.
    27 genera with about 250 species in the tropics.

Stamens free, about twice as many as the petals; petals free, recurving at anthesis, densely hairy within; disk absent   **1. Ximenia**
Stamens epipetalous, as many as the corolla-lobes, adnate to the base of the campanulate glabrous corolla-tube; disk present and partly enclosing the fruit   **2. Schoepfia**

1. XIMENIA L. (1753)

1. X. americana L., Sp. Pl. **2**: 1193 (1753).—Iguana Berry, Tallow Plum.

Shrub 3 m. or tree 5–13 m. high with short thick spines on the branches; leaves narrowly elliptical, emarginate, 3–7 cm. long, up to 3 (–4) cm. broad, cuneate at base, glabrous; flowers tetramerous in short racemose cymes, fragrant; petals creamy-white, 6–9 mm. long; drupe ellipsoid, yellow, about 3 cm. long.

Occasional in woodland and thickets on calcareous or sandy soils; SL–2600 feet; fl. Oct–May, fr. Apr–July. *Asprey UCWI 2106! H 12098! P 22828!* Widespread in the tropics; Grand Cayman.

2. SCHOEPFIA Schreb. (1789)

1 Flowers sessile, the calyces coherent; leaves elliptical   1. *chrysophylloides*
1 Flowers pedicellate, the calyces more or less separate:
   2 Peduncles wanting or very short, the flowers numerous in clusters at the nodes; leaves ovate, acutely acuminate   2. *multiflora*
   2 Peduncles evident, 1 or 2 at the nodes, 1 cm. or more long; leaves elliptical to lanceolate, tapered to a small rounded tip   3. *harrisii*

1. S. chrysophylloides (A. Rich.) Planch. in Ann. Sci. Nat. sér. 4, Bot. **2**: 261 (1854).—*Diplocalyx chrysophylloides* A. Rich. (1850). *S. angustata* Urb. (1909).

Large shrub or tree 6–8 m. high; bark deeply fissured; leaves obtuse or bluntly acuminate, cuneate at base, light green, 4–7 cm. long, 1·5–3 cm. broad; flowers heterostylous, 1–3 together on peduncles 1–4 mm. long; corolla yellow or with a crimson eye, 3–3·5 mm. long; drupe obovoid, 6–7·5 mm. long, 4–5 mm. broad.

Rare and local (St. Cath., Clar.), in arid coastal thickets on sand or rocky limestone; SL–50 (–100) feet; fl. Nov–Apr, fr. Nov–Jan. *A 9396! H 10186! P 22128!* Florida, Bahamas, Cuba, Hispaniola.

2. S. multiflora Urb., Symb. Ant. **5**: 184 (1907).

Tree 6–8 m. high with fissured bark; leaves broadly cuneate to rounded at base, 6–13 cm. long, 3–7 cm. broad; peduncles 1–2 mm. long; pedicels very short, the calyces sometimes confluent; corolla yellow or greenish yellow, 4–5 mm. long; disk orange; drupe ovoid, 9–10 mm. long, 7·5 mm. broad.

Occasional in the central parishes, in woodland on rocky limestone; 1700–2750 (–3500) feet; fl. Sept–Jan, fr. Dec. *H 9952! 12801! H & P 15153!* Endemic.

3. S. harrisii Urb., Symb. Ant. **5**: 185 (1907).

Slender-stemmed shrub or tree 2–5 (–10) m. high, with corky bark; leaves broadly cuneate at base, 2–10 cm. long, 1·5–3·5 (–4) cm. broad; peduncles 1–1·5 cm. long, 1–3-flowered; pedicels 1–2 mm. long; corolla greenish-yellow; drupe bright red, obovoid, 6 mm. long, 3 mm. broad.

Rare (Clar., Trel.), in woodland on limestone rocks; 500–2500 feet; fl. Aug–Dec, fr. Aug–Jan. *A & C 1164! H 8799! P 9757!* Endemic.

51. LORANTHACEAE

Shrubs parasitic on shrubs or trees or rarely erect trees or shrubs. Leaves usually opposite or whorled, rarely alternate, simple, mostly entire, often leathery, sometimes reduced to scales, without stipules. Flowers bisexual or unisexual and dioecious, actinomorphic, solitary or in cymes, racemes, spikes or panicles. Perianth biseriate, the whorls similar, 2–3-merous, or uniseriate; tepals free or united; with or without an additional cupular calyculus below the perianth (rim of cup-shaped receptacle). Stamens the same number as tepals and inserted on them or at their bases; anthers 1- or 2-locular, opening lengthwise or by terminal pores or transverse slits. Ovary inferior, the loculus and ovules not distinctly differentiated; sporogenous tissue basal; style simple or absent; stigma 1, often sessile. Fruit a berry or drupe, 1-seeded, often viscid within. Seed without testa, often with 2 or 3 united embryos, or embryo with 2–6 cotyledons; endosperm scanty or abundant.

30–40 genera with over 1000 species, mainly in the tropics.

1 Tepals 6, free; calyculus present; leaves opposite; flowers normally bisexual:
2 Perianth petaloid; inflorescence cymose, terminal; flowers pedicellate
　　　　　　　　　　　　　　　　　　　　　　　　　　　　　1. **Psittacanthus**
2 Perianth sepaloid; inflorescences spicate or racemose, axillary:
　3 Flowers sunk in pits in thickened spikes　　　　　　　　　2. **Oryctanthus**
　3 Flowers or flower-clusters stalked:
　　4 Flowers in clusters of 3; bract and bracteoles connate under each cluster
　　　　　　　　　　　　　　　　　　　　　　　　　　　　　3. **Phthirusa**
　　4 Flowers solitary in racemes; bracteoles connate under each flower
　　　　　　　　　　　　　　　　　　　　　　　　　　　　　4. **Dendropemon**
1 Tepals 3, at least in male flowers united at base; calyculus absent; flowers unisexual:
5 Leaves present, opposite; flowers more or less sunk in pits in articulated spikes
　　　　　　　　　　　　　　　　　　　　　　　　　　　　　5. **Phoradendron**
5 Leaves replaced by scales or altogether wanting:
　6 Flowers more or less sunk in pits in articulated spikes; foliar scales opposite
　　　　　　　　　　　　　　　　　　　　　　　　　　　　　6. **Dendrophthora**
　6 Flowers not sunk in pits; internodes of branches continuous; foliar scales spiral, peltate　　　　　　　　　　　　　　　　　　　　　　　　　7. **Eubrachion**

1. PSITTACANTHUS Mart. (1830)

1. P. claviceps (Griseb.) Eichl. in Mart., Fl. Bras. **5** (2): 26 (1868).—*Loranthus claviceps* Griseb. (1860).

Robust branched shrub to about 1 m. high; leaves obliquely ovate-elliptical, rounded at tip, leathery, up to 12 cm. long and nearly as broad; bracteoles forming a cupule below each flower; calyculus cylindrical; perianth 6–7·5 cm. long, at first greenish-yellow later tinged orange, segments deciduous; stamens epitepalous, filaments at first yellow later light red; style red, slightly longer than stamens; fruit black, about 1·5 cm. long; seed green with 4 (–5) cotyledons, about 1 cm. long.

Common on trees,[1] mostly in limestone areas; SL–2900 (–5000 ?) feet; fl. most of the year, fr. Jan–Mar. *A 8376! H 8177! H & P 13706!* Endemic.

2. ORYCTANTHUS Eichl. (1868)

1. O. occidentalis (L.) Eichl. in Mart., Fl. Bras. **5** (2): 89 (1868).—*Loranthus occidentalis* L. (1759). Godbush, Mistletoe, Scorn-the-Earth.[2]

Shrub with rusty-scurfy woody terete stems 30–60 cm. high; roots often super-ficial on the host and inserting haustoria; leaves mostly broadly ovate to rhomboid-elliptical, rounded at the tip, usually yellowish-green, up to 9 cm. long and 7 cm. broad; peduncle 0·5–2 cm. long; spikes 1–3 in the axils, up to 3 cm. long; fruit blackish- or brownish-purple, 4–5 mm. long.

Very common on shrubs and trees; 100–4000 feet; fl. and fr. all the year. *A 8124! H 6812! P 7812!* Endemic.

3. PHTHIRUSA Mart. (1830)

Inflorescence covered with rusty-brown overlapping scales　　　1. *lepidobotrys*
Inflorescence not scaly　　　　　　　　　　　　　　　　　　2. *jamaicensis*

1. P. lepidobotrys (Griseb.) Eichl. in Mart., Fl. Bras. **5** (2): 333 (1868).—*Loranthus lepidobotrys* Griseb. (1860).

Shrub with climbing stem rooting at intervals; roots spreading on external surface of host branches; free shoots up to 1 m. long, the young stems tetragonal

---

[1] The host trees for the more common species of *Loranthaceae* in Jamaica have been recorded by many collectors. These records are sufficient to show that the tolerance of particular species of parasite for different hosts is very high, but they are insufficient to indicate to what extent this is a matter of preference or opportunity. It has been decided therefore to omit detailed lists of host plants.

[2] These common names are applied indiscriminately to all species of *Loranthaceae* in Jamaica.

with corky angles; leaves broadly ovate, cuspidate or acuminate, the midrib sharply prominent beneath when dried, 4–10 cm. long, 4–9 cm. broad; flowers in small panicles, yellow; calyculus a membranous rim persistent on the tip of the fruit; ripe fruit green tipped with rose, 6–7 mm. long.

Occasional in St. Andrew and the central parishes, on shrubs and trees; 2000–5200 feet; fl. and fr. all the year. *A 11089! H 6201! P 26570!* Endemic.

2. P. jamaicensis Krug & Urb. in Engl., Bot. Jahrb. **24**: 15 (1897).

Like the last; leaves ovate to elliptical, somewhat acute, 4–7 cm. long; flowers greenish-white.

Very rare; known only from the type, *Purdie.* Endemic.

4. DENDROPEMON (Blume) Reichb. (1841)

Leaves elliptical, 1–2 cm. long; petioles 1–4 mm. long; ripe fruit about 5 mm. long
1. *parvifolius*
Leaves obovate to rounded-elliptical, 2·5–6 cm. long; petioles 3–7 mm. long; ripe fruit about 7 mm. long
2. *pauciflorus*

1. D. parvifolius (Sw.) Van Tiegh. in Bull. Soc. Bot. France **42**: 170 (1895).— *Loranthus parvifolius* Sw. (1788). *Phthirusa parvifolia* (Sw.) Eichl. (1868).

Much-branched shrublet up to about 50 cm. high; young branches rusty-scurfy; leaves acute to obtuse or rounded, up to 1 (–1·5) cm. broad, the midrib prominent beneath; flowers in bud reddish; perianth white, constricted below the middle, 2–2·5 mm. long; style green.

Locally common (St. Andr., Port., St. Thom.), on shrubs and trees in montane thickets and woodlands; 4900–7000 feet; fl. and fr. June–Nov. *ACH 885! Fawcett 6212! H 9198! Proctor & Cooley 7064!* Hispaniola.

2. D. pauciflorus (Sw.) Van Tiegh. in Bull. Soc. Bot. France **41**: 69 (1894).— *Loranthus pauciflorus* Sw. (1788). *Phthirusa harrisii* (Urb.) F. & R. (1914). *P. pauciflora* (Sw.) Eichl. (1868).

Shrublet with slender flexuous often pendulous branches up to about 60 cm. high; young branches greyish-scurfy, very brittle; roots often spreading on surface of host; leaves mostly rounded at tip, sometimes obtusely pointed, up to 4·5 cm. broad, the midrib prominent beneath; perianth pointed, green outside, tinged red within, about 3 mm. long; ripe fruit green with yellow to red tip; calyculus a persistent subhyaline rim.

Rather common on shrubs and trees; (SL–) 1000–4100 (–5000) feet; fl. and fr. all the year. *A 12284! H 10371! P 7661!* Endemic.

5. PHORADENDRON Nutt. (1847); Trelease (1916)

1 Stem-internodes all with a pair of more or less connate triangular scales above the base (scale-sheath); branchlets more or less terete:
  2 Leaves pinnately veined, the lateral veins obscure, drying blackish; flowers in 4 rows
        1. *piperoides*
  2 Leaves with 5 (–7) prominent main veins radiating from near the base, drying yellowish; flowers in 6 rows
        2. *flavens*
1 Stem-internodes at the base of lateral branches only with one or two scale-sheaths:
  3 Flowers normally in 6 rows in each joint of the spike, each cluster of female flowers of two triads, the middle flower of each triad raised above the other two; leaves triplinerved and pinnately veined
        3. *grisebachianum*
  3 Flowers in 2 or 4 rows in each joint, not in triads except sometimes at the distal end of each joint:
    4 Leaves pinnately veined:
      5 Young branches compressed; spikes 2–4 cm. long; flowers in 4 rows   4. *anceps*
      5 Young branches 4-angled; spikes in fruit 1 cm. long; female flowers in 2 rows, 2 in each joint   5. *campbellii*
    4 Leaves with 3–5 veins radiating within the proximal third of blade, sometimes very obscure:
      6 Young branches more or less distinctly 4-angled or winged:
        7 Branches acutely angled and narrowly winged; spikes 4–9 cm. long, 4–6-jointed, monoecious and androgynous   6. *tetrapterum*

7 Branches 4-angled but not obviously winged; spikes up to 3 (–4) cm. long, mostly
    monoecious:
    8 Leaf-blades elliptical to linear-oblanceolate; spikes 3–7-jointed   7. *quadrangulare*
    8 Leaf-blades obovate; spikes 3–4-jointed   8. *trinervium*
6 Young branches terete or compressed, not 4-angled:
    9 Leaf-margin crenulate; spikes 1·5–2·5 cm. long, 4–6-jointed, with flowers in 4 rows;
        monoecious   9. *crenulatum*
    9 Leaf-margin entire; spikes 1–2 cm. long, 3–4-jointed, probably all dioecious:
        10 Female flowers in 2 rows, 2 in each joint; male flowers in 4 rows; younger
            stem-internodes terete or compressed and broadened below the nodes; fruit
            ovoid   10. *wattii*
        10 Female flowers in 4 rows, 6–13 in each joint; younger stem-internodes compressed
            but not broadened below the nodes; fruit globose   11. *albovaginatum*

1. **P. piperoides** (Kunth) Trelease, Gen. Phorad.: 145 (1916).—*Loranthus
piperoides* Kunth (1820). *P. latifolium* Griseb. (1860). *Viscum latifolium* Sw.
(1797), not Lam. (1789). Mistletoe.
  Branches woody, pendulous; leaves oblong-elliptical, pointed at both ends,
often asymmetrical, the margins undulate especially when dry, yellow-green drying
blackish, 5–10 cm. long, 2–5 cm. broad; spikes solitary or several together in the
axils, 3–7 cm. long, 4–8-jointed, androgynous, orange-coloured; fruits ovoid,
orange or orange-red with yellow tip, 3–4·5 mm. long.
  Occasional on trees and shrubs in the central parishes; 400–2500 feet; fl. and fr.
most of the year. *A 12393! H 10339! P 15679!* Continental tropical Amer., West
Indies.

2. **P. flavens** (Sw.) Griseb., Fl. Br. W.I.: 313 (1860).—*Viscum flavens* Sw. (1788).
  Branches woody, pendulous to 1·2 m. long, the whole plant light green drying
yellowish; leaves ovate-elliptical, shortly acuminate, narrowed to a short petiole,
thick textured, 6–18 cm. long, 2–8 cm. broad; spikes solitary or several together in
the axils, 3–6·5 cm. long, 4–6-jointed, androgynous; fruits ovoid, yellow, about 4
mm. long.
  Rather local (St. Andr., Port., St. Thom.), on trees in submontane woodland;
2200–5700 feet; fl. and fr. Apr–Nov. *A 7826! P 9521! Watt 6219!* Hispaniola; a
variety is known from the Lesser Antilles.

3. **P. grisebachianum** Eichl. in Mart., Fl. Bras. 5 (2): 134m (1868).—*P. fici* Urb.
(1907).
  Branched shrub to 1·5 m. high, erect or pendulous, drying blackish; leaves ovate-
elliptical to oblong-lanceolate, obtuse or bluntly acuminate, often asymmetrical,
5–11 cm. long, 2–5 cm. broad; spikes 1–3 in the axils, dioecious, the female flower-
clusters rather remote with a small scale-sheath between each, up to 4·5 cm. long
in fruit, greenish-yellow to dull orange-red; male clusters compact; fruit subglobose, orange-red tipped with yellow, 3–4 mm. long.
  Widespread but not common, in the central and western parishes, on trees; 50–
4000 feet; fl. Dec (♂), fl. and fr. Jan–Apr, Aug–Oct. *A 8842! Cornman 878!
H 10861! P 26610!* Cuba.

4. **P. anceps** (Spreng.) Krug & Urb. in Engl., Bot. Jahrb. 24: 40 (1897).—*Viscum
anceps* Spreng. (1825).
  Shrub with pendulous branches to 1 m. long, the whole plant orange-brown;
proximal internodes of lateral branches with 2 scale-sheaths; leaves ovate-
lanceolate, the blade more or less rounded at base, tapered to an obtuse tip,
with prominent midrib and veins, up to 9 cm. long and 3·5 cm. broad; spikes
1–3 in the axils, dioecious, 3–5-jointed, up to 4 cm. long, with ciliolate scale-sheaths
2·5–3 mm. broad between the joints; unripe fruit green with orange tip.
  Rare and local (St. Thom.), on trees; 3000 feet; fl. Feb, July, fr. July. *A 12230!
ACH 466!* Hispaniola.

5. **P. campbellii** Krug & Urb. in Engl., Bot. Jahrb. 24: 44 (1897).
  Shrub with long branches; leaves oblong-lanceolate to lanceolate, narrowed or
acuminate at tip, 5–7 cm. long, 1·5–3 cm. broad; spikes 1–2 in the axils, dioecious,
3-jointed; unripe fruit ovoid.

Apparently rare and local (St. Andr.), on trees; about 500 feet; fr. Oct. *Campbell 6398* (partly), *6604* (partly). This obscure species has not been seen since the original collections.

6. P. tetrapterum Krug & Urb. in Engl., Bot. Jahrb. **24**: 35 (1897).
Shrublet 30–60 cm. high; leaves mostly oblong-elliptical, often asymmetric, acute or obtuse, long-tapered to a sessile base, 4–10 cm. long, 1·5–4 cm. broad; spikes 1–3 in the axils, slender, the terminal joints attenuated; fruits subglobose, white, 3–3·5 mm. long.
Local (St. Andr., Trel.), on trees and other Loranthaceae; 400–800 feet; fl. and fr. Aug–Jan. *H 6926! P 15680! Robertson UCWI 24936!* Hispaniola, Puerto Rico, Martinique.

7. P. quadrangulare (Kunth) Krug & Urb. in Engl., Bot. Jahrb. **24**: 35 (1897).— *Loranthus quadrangularis* Kunth (1820). *P. rubrum* Griseb. (1860) and of F. & R. (1914), partly, not *Viscum rubrum* L. (1753). *P. quadrangulare* var. *gracile* Krug & Urb. (1897).
Shrub with pendulous branches up to 1·3 m. long, the whole plant dark green; leaves elliptical to narrowly oblanceolate, obtuse or acute, long-tapered to base, the venation often obscure, 3–5 (–9) cm. long, 0·5–1 (–1·5) cm. broad; spikes 1–2 in the axils, 1–3 cm. long; ripe fruit subglobose, white, yellow or orange, 3–4 mm. long.
Locally common, particularly in the south-eastern parishes, on shrubs and trees; SL–2900 feet; fl. and fr. most of the year. *A 12319! H 6392! P 6230!* Hispaniola, Antigua; Grand Cayman.

8. P. trinervium (Lam.) Griseb., Fl. Br. W.I.: 314 (1860).—*Viscum trinervium* Lam. (1789). *P. verticillatum* F. & R. (1914). *P. trinervium* var. *domingense* (Desv.) Krug & Urb. (1897). *Loranthus domingensis* Desv. (1825).
Shrublet 30–60 cm. high, the plant usually yellowish-green; leaves obovate to oblanceolate, rounded or emarginate at tip, long-tapered to base, rather distinctly 5-veined, 2–6 cm. long, 1–3·5 cm. broad; spikes 1–3 in the axils, up to 2·5 cm. long; ripe fruit ellipsoid, orange-yellow, 6–7 mm. long.
Rather common, in the southern parishes, on shrubs and trees; 50–1500 feet; fl. and fr. all the year. *A 8854! H 9568! P 23661!* Bahamas, West Indies except Cuba.

9. P. crenulatum Urb., Symb. Ant. **5**: 332 (1907).
Plant to 1·2 m. high; leaves obovate-elliptical, rounded at tip, 5–7 cm. long, 1·5–2·5 cm. broad; spikes to 4·5 cm. long, solitary or rarely paired in the axils; fruit rhomboid-ellipsoid, 4 mm. long.
Local (Clar.), on trees; 200–2900 feet; fl. Apr, Dec. *H 6659! P 6570!* Endemic.

10. P. wattii Krug & Urb. in Engl., Bot. Jahrb. **24**: 43 (1897).
Shrub with pendulous branches up to 2·5 m. long; leaves oblong-elliptical to oblong-lanceolate, slightly asymmetrical, obtuse at tip, narrowed abruptly at base or tapered, 4–9 (–12) cm. long, 1–3 cm. broad, venation sometimes obscure; spikes 1–3 in the axils; male flowers in 4 rows; ripe fruits light yellow, 5–7 mm. long.
Common and widespread, on trees; SL–3800 feet; fl. and fr. all the year. *A 10757! H 9696! P 10179!* Endemic.

11. P. albovaginatum Urb., Symb. Ant. **7**: 504 (1913).
Shrub to 75 cm. high; leaves ovate-elliptical to oblong-lanceolate, obtuse or rounded at tip, lateral veins obscure, yellowish-green, 3·5–5 cm. long, 1·5–2·5 cm. broad; female spikes about 2 cm. long.
Very rare (St. Cath.), on trees; about 950 feet; fl. and fr. Nov. Known only from the type, *H 6703!*

6. DENDROPHTHORA Eichl. (1868); Kuijt (1961)

At least the distal internodes compressed; plants dioecious; scale-leaves at nodes very obscure, much smaller than scale-sheaths   1. *opuntioides*
All internodes terete; plant monoecious; scale-leaves at nodes similar to scale-sheaths, spreading, ciliolate   2. *cupressoides*

1. D. opuntioides (L.) Eichl. in Mart., Fl. Bras. **5** (2): 102 (1868).—*Viscum opuntioides* L. (1753). *D. monstrosa* F. & R. (1914).

Stems branched, erect or pendulous, articulated at the nodes; internodes at first succulent, dark green to greenish-yellow, linear-oblong with rounded shoulders or at least broadening slightly towards the apex, 1–5 cm. long, 3–20 mm. broad; spikes mostly solitary at the nodes, up to about 2 cm. long; ripe fruit ellipsoid, yellow or red, 4–5 mm. long.

Rather common in cool upland areas, on shrubs and trees; (1000–) 1300–7300 feet; fl. and fr. all the year. *A 5724! J.P. 1165! P 5969!* Endemic.

2. D. cupressoides (Griseb.) Eichl. in Mart., Fl. Bras. **5** (2): 103 (1868).— *Arceuthobium cupressoides* Griseb. (1860).

Bushy much-branched shrublet 15–30 cm. high, the whole plant dark green; internodes 5–10 mm. long; spikes 6–15 mm. long, 3–5-jointed; ripe fruit about 3 mm. long.

Local (St. Andr., St. Thom.), on trees and shrubs in montane thickets; 5000–7000 feet; fl. and fr. most of the year. *A 10582! Fawcett 6214! P 9610!* Cuba, Hispaniola.

7. EUBRACHION Hook. f. (1846)

1. E. ambiguum (Hook. & Arn.) Engl. in Engl. & Prantl, Nat. Pflanzenf. **3** (1): 192 (1889).—*Viscum ambiguum* Hook. & Arn. (1833).

1a. Var. jamaicense Krug & Urb. in Engl., Bot. Jahrb. **24**: 31 (1897).

Much-branched shrublet 20–40 cm. high; foliar scales persistent, about 2 mm. long; spikes crowded in a racemose arrangement at ends of branches, 3–7 mm. long.

Very local (St. Andr., St. Thom.), on myrtaceous shrubs in montane thickets; 5000–5400 feet; fl. and fr. Aug–Dec. *H 9377! J.P. 663!* Hispaniola; the typical variety occurs in tropical S. Amer.

52. BALANOPHORACEAE

Fleshy herbs, parasitic on the roots of trees and shrubs, without chlorophyll or stomata. Flowers unisexual, very rarely bisexual, densely crowded in monoecious or dioecious spadices on thick scaly peduncles. Male flowers naked or with a valvate 3–8-lobed perianth. Stamens 1–2 in achlamydeous flowers, in those with perianth usually equal in number to and opposite the lobes, with free or connate filaments; anthers 2–4-many-locular, free or connate, opening by pores, slits or rupturing irregularly; pollen globose-trigonal. Female flowers with perianth epigynous, 2-lipped, tubular or absent. Ovary 1–3-locular; styles 1 or 2, terminal, or stigma sessile; ovules solitary in each loculus, pendulous, orthotropous or anatropous. Fruit small, nutlike. Seeds with endosperm and a very small embryo.

18 genera and about 100 species mainly in the subtropics and tropics.

1. SCYBALIUM Schott & Endl. (1832)

Rhizome tuberous, perennial; developing spadices covered with deciduous subpeltate bracts; axis of spadix covered with a compact layer of clavate scales; male flowers 3-merous; female flowers lacking perianth; ovary 2-locular with 2 styles.

1. S. jamaicense (Sw.) Schott & Endl., Melet. Bot.: 12 (1832).—John Crow Nose.

Plants pinkish-red, soon blackening, forming clumps up to 60 cm. or more in diameter in the surface layers of woodland soil; spadices up to 20 cm. long, cylindrical or club-shaped, pedunculate; scales (on peduncle) and bracts broadly triangular to ovate, up to 25 mm. long; male perianth purplish, 5 mm. long; anthers and pollen white; female flowers about 2·5 mm. long; styles yellow.

Locally common in moist shady forest; 500–4500 feet; fl. and fr. all the year. *A 6358! P 21465! Yuncker 17553!* Greater Antilles.

53. ARISTOLOCHIACEAE

Perennial herbs or shrubs, the latter usually twining climbers. Leaves alternate, simple, entire or lobed, without stipules but stipuliform bracts often present. Flowers bisexual, actinomorphic or zygomorphic, solitary, in axillary clusters or racemose. Perianth a petaloid gamosepalous calyx basically of 5 segments forming a regular 3-lobed or bilaterally symmetrical entire or 2-lipped tube, rarely an inner whorl of 3 minute teeth present. Stamens (5–) 6 or more, free or adnate to the style; anthers 2-locular, opening lengthwise. Ovary inferior, rarely half-inferior, 4–6-locular; style 1, short and stout with 3 or more stigma-lobes; ovules several to many in each loculus, parietal, anatropous. Fruit capsular or indehiscent, when dehiscent sometimes pendulous and opening basally. Seeds usually angled or flattened with small embryo and copious endosperm.

10 genera and about 400 species, primarily tropical but extending to most temperate regions.

1. ARISTOLOCHIA L. (1753); Pfeifer (1966)

Vines; flowers irregular, usually solitary and axillary; perianth 1-seriate, the tube usually inflated at the base, bent above with the limb variously shaped and often luridly marked within; capsule septicidally 6-locular, usually parachutelike.

1 Stamens 5; capsule winged, about 2 cm. long; perianth about 3·5 cm. long   1. *pentandra*
1 Stamens 6 or more; capsule ribbed but not winged, 4 cm. or more long; perianth at least 5 cm. long:
   2 Leaves mostly 3-lobed; perianth-limb 2-lipped, one lip long-tailed, the tail 2–3 mm. broad   2. *trilobata*
   2 Leaves not 3-lobed:
     3 Perianth-limb unequally 2-lipped, lower lip straight or slightly curved, channelled, not tailed, up to 16 (–20) cm. long   3. *ringens*
     3 Perianth-limb not 2-lipped, openly cup- or salver-shaped:
       4 Perianth-limb very large, broadly salver-shaped, up to about 20 cm. long with a longer tail   4. *grandiflora*
       4 Perianth-limb smaller, not tailed:
         5 Perianth-limb mucronate at apex, deeply cordate, 6–10 (–13) cm. long; base of tube broadly inflated; capsule obliquely beaked, 6–10 cm. long   5. *odoratissima*
         5 Perianth-limb obtuse, openly cup-shaped, 6–7 (–10) cm. long; base of tube not much inflated; capsule 4–6 cm. long   6. *littoralis*

1. A. pentandra Jacq., Enum. Syst. Pl. Carib.: 30 (1760).

Twining stems from a stout tuberous stock; leaves deeply cordate, shallowly lobed, acuminate, pubescent at least on the veins beneath when young, up to 4 cm. long and 2·5 cm. broad across the hastate base; perianth ochre distally, greenish at base; fruit oblong, about 1·5 cm. broad.

Very local (St. Ann), in upper beach thickets and on limestone sea-cliffs; SL–10 feet; fl. Jan, July, Dec, fr. July, Dec. *P 9561!* Florida, Bahamas, Cuba.

2. A. trilobata L., Sp. Pl. 2: 960 (1753).—Tref.

Twining by petiole-tendrils to 7 m. or more or trailing on the ground and rooting from the nodes; leaves thinly leathery, lobes obtuse, dark green above, paler beneath, up to 15 cm. broad; bracts orbicular, sporadically developed; upper lip of perianth cordate, about 2·5 cm. broad, tail up to 35 cm. long; lower lip emarginate; capsule 7–8 cm. long; seeds ovate-triangular, 7–10 mm. long and broad, broadly thin-margined.

Occasional in coastal woodlands and thickets, rare in woodlands inland; SL–25 (–2100) feet; fl. most of the year. *A 12349! A. M. Barry! Key! P 24656.* Bermuda, British Honduras to Panama, West Indies.

3. A. ringens Vahl, Symb. Bot. 3: 99 (1794).— Dutchman's Pipe.

Robust vine; leaves reniform, 8–12 cm. broad; lower lip of perianth-limb lanceolate, about twice as long as the upper, upper spathulate, green mottled purple;

capsule about 8 cm. long and 3 cm. broad; seeds narrowly cordate, very thin, 12 mm. long, 7 mm. broad.

Sparingly cultivated as an ornamental and locally naturalized (St. Cath., St. Eliz.); 1200–2600 feet; fl. May, Oct–Nov, fr. Nov. *Burns 1! Burrowes 13030!* Native in tropical S. Amer., now widely distributed in the tropics.

4. A. grandiflora Sw., Nov. Gen. & Sp. Pl.: 126 (1788).—Poisoned Hog Meat.

Stems twining to 3 m. or more long; leaves broadly ovate, cordate, acutely acuminate, dark green, 8–12 cm. broad; perianth light purplish-buff outside, limb purple-spotted and veined on cream within, very unpleasantly scented; capsule oblong, 6-ribbed, about 6–10 cm. long and 3–4 cm. broad; seeds triangular, 12 mm. long, 10 mm. broad.

Widely distributed but not common in sheltered thickets near streams and in gullies; 50–1500 feet; fl. and fr. most of the year. *A 9208! H 8438! Lewis & Powell 860! Yuncker 17944!* Mexico to Panama, probably introduced elsewhere.

5. A. odoratissima L., Sp. Pl. ed. 2, **2**: 1362 (1763).

Stems slender, up to 2 m. or more long; leaves broadly ovate to oblong and more or less hastate, often pandurate, usually acuminate at the tip, openly cordate at the base, 5–14 cm. long, 4–9 cm. broad, glabrous above, puberulous beneath; limb of perianth with a brown-spotted purple or pinkish lip, throat yellow within; seeds cordiform with a strong marginal flange, 3–3·5 mm. long, 2–3 mm. broad.

Rare (St. Andr., St. Eliz.), in thickets and on walls and fences; 10–1500 feet; fl. Oct–Jan, fr. Oct–Apr. *H 8450! P 15871! Robertson UCWI 5677!* Mexico to Brazil, West Indies.

6. A. littoralis Parodi in Anal. Soc. Cient. Argent. **5**: 155 (1878).—*A. elegans* Mast. (1885).

Slender-stemmed twiner, scrambling and climbing to 2 m. or more; leaves broadly ovate, cordate, obtuse up to 9 cm. broad; perianth light green at base outside, limb with reddish-purple and dark red blotches on white and a red or pale yellow-green throat; seeds 6 mm. long, 4 mm. broad.

Cultivated and naturalized on fences and stone walls and in thickets; 50–2400 feet; fl. and fr. most of the year; *A 6958! H 10655! Powell 226*: Native probably of Brazil, introduced into the tropics generally.

54. POLYGONACEAE

Herbs, twining or erect shrubs or trees, often with swollen nodes. Leaves usually alternate, simple; stipules usually present and when tubular and sheathing (ochreae) often adnate to the petiole. Flowers bisexual or unisexual and monoecious or dioecious, actinomorphic, in racemes, spikes, panicles or heads, often in bracteate clusters or cymules. Perianth biseriate or apparently uniseriate by reduction or spiral by fusion, of 3–6 free or fused undifferentiated tepals, the tepals often persistent and enlarged in fruit. Stamens usually 6–9, free or united at the base; anthers 2-locular, opening lengthwise. Ovary superior, subtended by an annular glandular disk, often 3-angled, 1-locular; style 1; stigmas 2–4, often free; ovule solitary, basal. Fruit achenial, enclosed by the often accrescent perianth. Seed with copious mealy endosperm; embryo often excentric.

About 40 genera with some 800 species, mostly of north temperate distribution.

1 Plant trailing and climbing by supra-axillary inflorescence-tendrils; flowers bisexual; stipules obsolete; perianth showy, bright pink, 5-membered; stamens 8 (–9)
1. **Antigonon**
1 Plants erect or if trailing not climbing by inflorescence-tendrils; stipules evident; perianth not showy or if conspicuous not bright pink:
  2 Herbs or undershrubs:
   3 Perianth-segments 5, spirally arranged, petaloid, not usually obviously accrescent; flowers bisexual   2. **Polygonum**

3 Perianth-segments 6 in 2 whorls, sepaloid, the inner accrescent; flowers unisexual or
   bisexual   **3. Rumex**
2 Trees or shrubs; flowers unisexual:
   4 Perianth 3-merous; fruit 3-winged by enlargement of inner tepals; stipules entirely
     deciduous, up to over 20 cm. long; cultivated slender-trunked tree to 20 m. or
     more high; Ant Tree, Long John   *Triplaris americana* L.
   4 Perianth (4–) 5-merous; fruit falsely drupaceous by enclosure of the achene
     with accrescent perianth; stipules ochreaceous, much shorter, deciduous, partly
     deciduous or persistent   **4. Coccoloba**

1. ANTIGONON Endl. (1837)

1. A. leptopus Hook. & Arn., Bot. Beech. Voy.: 308, t. 69 (1838).—Coralilla,
Coralita.

Stems angled, puberulous, glabrescent, up to 10 m. or more long; leaves broadly
ovate, cordate, shortly acuminate, shallowly crenate, veins prominent beneath, up
to about 13 cm. long and broad; stipule reduced to a setulose ridge; inflorescences
supra-axillary and terminal; tendrils hooked or coiling, the terminal 3-fid; flowers
jointed-pedicellate, clustered in racemes; perianth-segments cordate, obtuse, the
2 outer larger, about 11 mm. long and 8 mm. broad; filaments fused halfway, the
free portions glandular; stigmas 3, bowed.

Common in cultivation and escaping on to fences and hedges at low elevations;
fl. mainly June–Dec (–Mar). *P 9349! Robertson UCWI 866!* Native of Mexico,
now widespread in the tropics; Grand Cayman.

2. POLYGONUM L. (1753)

1 Flower-clusters cymose-paniculate; leaves not more than three times longer than
   broad:
   2 Leaves cordate-sagittate, acuminate; stipules up to about 1 cm. long; fruit triquetrous,
     dry; casual introduced weed; Buckwheat   *Fagopyrum esculentum* Moench
   2 Leaves ovate, truncate-cuneate at base, cuspidate; stipules up to 5 cm. long; fruit a
     false drupe   1. *chinense*
1 Flower-clusters in solitary or panicled spikes or racemes; leaves at least four times
   longer than broad:
   3 Stipules long-setose ciliate:
     4 Tepals glandular-punctate; leaf-blade glabrous or nearly so adaxially, margin ciliolate;
       peduncle glabrous   2. *punctatum*
     4 Tepals not glandular-punctate; leaf-blade appressed-pubescent on both surfaces;
       peduncle pubescent   3. *acuminatum*
   3 Stipules not ciliate:
     5 Peduncles with short-stalked glands; ochreae thinly pubescent   4. *segetum*
     5 Peduncles and ochreae glabrous   5. *glabrum*

1. P. chinense L., Sp. Pl. **1**: 363 (1753).

Scrambling branched shrubby herb 75–200 cm. high or climbing to about 8 m.;
leaves setose on midrib beneath, otherwise glabrous, 3–12 cm. long, 1·5–6 cm.
broad; petioles and ochreae often crimson; stipules without cilia; flowers white,
capitate; perianth accrescent in fruit, black, fleshy.

Locally naturalized and common (St. Andr., Port., St. Thom.), on open or
shaded sheltered banks; (1700–) 3400–5000 feet; fl. and fr. sporadically throughout
the year. *A 6426! H 11699! C. B. Lewis!* Native from E. Indies to Japan.

2. P. punctatum Ell., Sketch Bot. S. Carol. & Georgia **1**: 455 (1817).

Trailing herb with ascending flowering shoots; stem tough, slender; leaf-blades
glandular-punctate, lanceolate, long-acute, tapered at base, up to 15 cm. long and
2·5 (–3) cm. broad; racemes 1–3, mostly terminal, slender, interrupted below,
5–12 cm. long; perianth-segments white, greenish at base, with reddish glands;
stamens 8; ripe fruit black, about 3 mm. long.

Common in swamps and ditches and at pond margins; SL–3500 feet; fl. and fr.
all the year. *A 7198! H 10531! P 19750!* Canada to Paraguay, West Indies; Grand
Cayman.

3. P. acuminatum Kunth, Nov. Gen. **2**: 178 (1817).

Robust herb with stems trailing to 150 cm. long; flowering branches ascending;
leaf-blades ascending, lanceolate, long-acute, tapered at base, up to about 25 cm.

long and 4 cm. broad; racemes 2–4, branched from a terminal peduncle, stout, 4–10 cm. long; perianth-segments white; stamens 5–8; nutlet biconvex, black, shiny, apiculate, about 2·5 mm. long.

Local in the central and western parishes, in marshes and wet rough pastures; SL–900 feet; fl. and fr. Aug–Apr. *A 8520! H 9044! P 10535!* Continental tropical Amer., West Indies; Grand Cayman.

P. hirsutum Walt., with cordate or subcordate leaf-bases, was reported by Britton (*Britton 394*) in Bull. Torr. Bot. Club **44**: 36 (1917), but has not been confirmed by further collections.

4. P. segetum Kunth, Nov. Gen. **2**: 177 (1817).—*P. mexicanum* Small (1892).

Trailing herb with branches to 1 m. long; flowering shoots ascending; leaf-blades narrowly lanceolate, tapered to both ends, puberulous, glabrescent, up to about 15 cm. long and 1·5 cm. broad; racemes 2–5 from a branched terminal peduncle, rather compact, up to about 6 cm. long; perianth-segments white; stamens 6–8; nutlet biconvex, black, about 3 mm. long.

Rare and local (Clar., Han.), in muddy pond margins; 150–300 feet; fl. and fr. May, Sept, Dec. *Fawcett! H 12721! Robertson UCWI 2014!* S. United States, Mexico to Colombia, Greater Antilles.

5. P. glabrum Willd. in L., Sp. Pl. ed. 4, **2**: 447 (1799).

Robust trailing herb, glabrous; stems up to 120 cm. long, the flowering branches ascending; leaf-blades broadly lanceolate, long-tapered to the tip, up to 20 cm. long and 5 cm. broad; racemes 4–8, 6–10 cm. long, compact; perianth-segments pink or white; stamens 6–8; nutlet biconvex, black.

Rare and local in the central parishes, in marshes and along muddy streams and riverbanks; 50–1500 feet; fl. and fr. sporadically throughout the year. *A 9472! H & B 10530! P 24289!* S. United States, Bahamas, continental tropical Amer., West Indies; Grand Cayman.

Plants intermediate between *P. segetum* and *P. glabrum*, in being glabrous or nearly so but with peduncle-glands and intermediate sized leaves, have been gathered in Trelawny (*P 15707! 15708!*).

3. RUMEX L. (1753); Rechinger (1937)

1 Basal and lower cauline leaves hastate; flowers unisexual, dioecious   1. *acetosella*
1 Leaves not lobed at base; flowers bisexual, markedly protandrous:
   2 Inner perianth-segments in fruit broadly ovate-cordate, entire or minutely denticulate, tubercled, 3·5–6 mm. long; lower leaves cuneate at base, glabrous   2. *crispus*
   2 Inner perianth-segments in fruit triangular, with 3–5 subulate teeth on each margin, one only usually tubercled, 5–6 mm. long; lower leaves mostly cordate at base, shortly pubescent on midrib and veins beneath   3. *obtusifolius*

1. R. acetosella L., Sp. Pl. **1**: 338 (1753).

Perennial herb propagating by short stolons; basal leaves tufted with petioles 2–7 cm. long and obovate blades up to 3 cm. long and 1·5 cm. broad the basal lobes blunt and 4–7 mm. long; cauline leaves elliptical to oblanceolate, the uppermost entire, smaller and narrower; flowering stem profusely branched above, 15–40 cm. high; flowers small, green or blood red, all pistillate in Jamaican plants; fruit not seen.

Rather local (St. Andr., St. Thom.), on open pathside banks and in turf; 4000–7400 feet; fl. Dec–Feb, June–July. *A 10676! 11271! H 10114!* Also in Hispaniola. The species aggregate *R. acetosella* has a very wide temperate distribution, occurring mostly on rather poor non-calcareous soils. Jamaican plants have been identified within the complex as *Acetosella vulgaris* Fourr. and as they are all apparently pistillate and sterile it is presumed that the present populations are clonal derivatives from introduced seed.

2. R. crispus L., Sp. Pl. **1**: 335 (1753).—Curled Dock.

Erect perennial herb up to 1 m. high from a deep strong taproot; basal leaves long-petioled, blade oblong-lanceolate, obtuse, up to about 18 cm. long and 6 cm. broad, margins undulate; flowers pedicelled in numerous whorls, green; tubercles red; fruit 2·5–3 mm. long.

Uncommon (St. Andr., Port., St. Thom.), in damp gravelly waste places; 1700–4100 feet; fl. and fr. most of the year. *A 11212! H 10113! P 23532!* Native in Europe and temperate Africa, naturalized elsewhere; Canada, United States, Bermuda, Mexico, Greater Antilles, Guadeloupe, Martinique.

3. R. obtusifolius L., Sp. Pl. **1**: 335 (1753).—Broadleaved Dock.
   Like the last; basal leaves with petioles up to 25 cm. long, blade ovate-oblong, obtuse, up to 25 cm. long and 12 cm. broad, margins undulate; upper leaves smaller, lanceolate, acute; fruit 3 mm. long.
   Very local (St. Andr., Port.), a roadside weed; 3800–4500 feet; fl. and fr. Feb, Apr–May, Sept. *H 7660! A. von der Porten! Robertson UCWI 1798!* Native in Europe, introduced and naturalized elsewhere; Canada, United States, Mexico, Cuba, Hispaniola.

4. COCCOLOBA P. Browne (1759) nom. cons.;
Howard (1957). Wild Grape

1 Inflorescence paniculate:
   2 Pedicels 0·5 mm. long, shorter than the ochreolae in flower, up to 1·5 mm. long and slightly exceeding the ochreolae in fruit; leaves rounded to cordate at base, deciduous   **1.** *plumieri*
   2 Pedicels 3–6 mm. long, exceeding the ochreolae in flower and in fruit; leaves broadly cuneate at base, the tree apparently never devoid of leaves   **2.** *proctorii*
1 Inflorescence racemose, the racemes solitary or rarely with 1 or 2 short basal branches:
   3 Leaves thick, mostly broader than long; plants mainly coastal   **3.** *uvifera*
   3 Leaves thin, longer than broad:
      4 Pedicels shorter than or only slightly exceeding the ochreolae:
         5 Ochreolae several, persistent, conspicuous; rachis stout; pedicels very short; leaves usually darkening on drying   **4.** *swartzii*
         5 Ochreolae solitary, deciduous; leaves buff or tan on drying:
            6 Inflorescence shorter than the leaves; leaf-blades ovate, acuminate, cordate, usually exceeding 7 cm. in length; inflorescence-rachis conspicuously angled; pedicels ascending; fruit ovoid, about 9 mm. long and 8 mm. broad   **5.** *troyana*
            6 Inflorescence exceeding the leaves; leaf-blades ovate-oblong, acute, obtuse or rounded at base, usually less than 5 cm. long; inflorescence-rachis not angular; pedicels diverging at right-angles; fruit strongly 3-lobed at base, 5 mm. long and 3 mm. broad   **6.** *krugii*
      4 Pedicels conspicuously exceeding the ochreolae in flower and fruit:
         7 Leaves borne above the base of the ochreae, the base of the ochreae at leaf attachment generally conspicuously swollen:
            8 Leaves appearing to be clustered on short lateral shoots, membranaceous, ovate-elliptical; inflorescence generally curved or drooping; pedicels slender
                                                                      7. *tenuifolia*
            8 Leaves coriaceous, oblong-elliptical, generally spaced on the branches which are often wandlike or scrambling; inflorescence stout, generally straight or erect; pedicels stout   **8.** *longifolia*
         7 Leaves borne at the base of the ochreae, the base of the ochreae not conspicuously swollen;
            9 Leaf-blades usually oblong-elliptical, rounded at apex, generally darkening on drying but uniform in colour on both surfaces   **9.** *diversifolia*
            9 Leaf-blades broadly elliptical to rounded-elliptical, acute or acuminate at apex, turning black above on drying but usually paler beneath   **10.** *zebra*

1. C. plumieri Griseb., Fl. Br. W.I.: 162 (1860).—*C. polystachya* Wedd. var. *jamaicensis* F. & R. (1913). Mountain Grape.
   Tree to 15 (–20) m. high, glabrous; petioles attached above the base of the ochreae; leaf-blades mostly broadly ovate, obtuse to broadly rounded at tip, sometimes bluntly cuspidate, in juveniles often very large and leathery, up to 30 (–50) cm. long and 28 cm. broad; panicle branches 2–6 (–11), up to 15 cm. long; flowers solitary; ripe fruit ovoid, 12–18 mm. long, dark purple or black, astringent, edible.
   Occasional in moist savannas, thickets and woodlands on limestone in the central and western parishes; 250–2800 feet; fl. Dec–Jan, May–July, fr. Dec–Apr, July, Aug. *A 10929! 12025! H 12023! P 10593!* Endemic.

2. C. proctorii Howard in Journ. Arn. Arb. **38**: 86 (1957).
   A large tree to over 20 m. high; largest leaves 40 cm. long, 25 cm. broad; flowers solitary or a few together; ripe fruit about 1·5 cm. long.

Rather local (St. Eliz., St. James), on limestone hills; 700–2000 feet; fl. ? Dec, fr. Dec–Jan. *HPS 14718! 14719!* Endemic.

3. **C. uvifera** (L.) L., Syst. Nat. ed. 10, **2**: 1007 (1759).—*Polygonum uvifera* L. (1753). Seaside Grape.
Low shrub or tree 1–5 (–15) m. high; leaves on flowering branches up to about 15 cm. long and 18 cm. broad, shortly cordate, tip and lateral margins rounded; petiole up to 1 cm. long; racemes 8–20 cm. long; rachis tomentellous; pedicels clustered, up to 3 mm. long in fruit; ripe fruit fleshy, purplish, up to about 2 cm. long, edible.
Common and locally dominant along the seacoast on strand, sand dunes and in thickets, rare inland; SL–20 feet; fl. Jan–Aug, fr. Mar–Oct. *A 6136! H 5978! P 11844!* Florida, Bahamas, Atlantic coast from Mexico to the Guianas, West Indies; Grand Cayman; introduced in the Pacific.
C. pubescens L. reported by Fawcett & Rendle (Fl. Jam. **3**: 118) has not been confirmed for Jamaica. This species resembles *C. uvifera* but has the midrib and veins pubescent beneath.

3a. **C.** × **jamaicensis** Lindau in Engl., Bot. Jahrb. **13**: 206 (1890).—*C. litoralis* Urb. (1909).
Much-branched tree 5–12 m. high, resembling the last in habit but with thinner more pointed leaves; racemes up to 30 cm. long; rachis puberulous; pedicels slender.
Occasional in arid woodlands along the south coast; SL–500 feet; fl. Mar, July, fr. July. *H 10228! H & P 14509! Yuncker 17895!* This plant has characters intermediate between *C. uvifera* and *C. tenuifolia* and is presumed to be a hybrid.

4. **C. swartzii** Meisn. in DC., Prodr. **14**: 159 (1856).—*C. diversifolia* of F. & R. (1914), not Jacq. (1760). *C. neglecta* F. & R. (1913).
Shrub or tree 3–12 m. high; petioles attached at or just above the base of the ochreae, up to about 2 cm. long, rather slender; leaf-blades ovate, truncate or shortly cuneate at base, mostly tapered to an obtuse tip, glabrous, up to 12 cm. long and 8 cm. broad, mostly smaller; racemes solitary, 5–15 cm. long; rachis glabrous; flowers solitary or in pairs; ripe fruit purplish-black, about 8 mm. long.
Common in savannas, thickets and woodlands; 300–3500 feet; fl. Mar, July–Dec, fr. Aug–Mar. *A 12628! H 5090! 6739! P 15727!* Bahamas, Cuba, Puerto Rico, West Indies south to Barbados.

5. **C. troyana** Urb., Symb. Ant. **6**: 8 (1909).
Tree 6–15 m. high; petioles attached at the base of the ochreae; leaf-blades shortly and bluntly acuminate, glabrous, 7–10 cm. long, 4–7 cm. broad; racemes 4–9 cm. long; rachis glabrous; flowers solitary; ripe fruit red.
Rather local in two areas (Trel., St. James and Port., St. Thom.) in open woodlands on limestone or serpentine; 1500–2900 feet; fl. Mar, Aug, fr. Aug–Mar. *A 12146! H 9439! H & B 10764! P 22598!* Endemic.

6. **C. krugii** Lindau in Engl., Bot. Jahrb. **13**: 145, t. 5, f. 14 (1890).
Shrub or tree 5–7 m. often developing root suckers when coppiced, glabrous or rarely shortly puberulous; petioles attached at the base of the ochreae; leaf-blades firm, obtuse at tip, up to 8 cm. long and 6 cm. broad; racemes 5–10 cm. long; rachis glabrous or puberulous; flowers usually solitary; ripe fruit black.
Common in thickets on arid limestone particularly along the south coast and in Trelawny; 100–1600 feet; fl. July–Jan. fr. most of the year. *A 6197! 10292! H 10008! H & P 14395!* Bahamas, Hispaniola, Puerto Rico, northern Leeward Is.

7. **C. tenuifolia** L., Syst. Nat. ed. 10, **2**: 1007 (1759).—*C. excoriata* L. (1759).
Shrub 2–5 m. or tree to 8 m. high, deciduous; leaf-blades elliptical to broadly ovate or obovate, glabrous, up to 14 cm. long and 8·5 cm. broad; racemes slender, 6–14 (–17) cm. long; rachis glabrous or puberulous; ripe fruit red, 5–6 mm. in diameter, the nutlet often exposed.
Rather common in woodlands on limestone; SL–3000 feet; fl. Apr–Sept, fr. most of the year. *A 11359! 12043! H 8862! 9809! H & P 13932!* Bahamas, Cuba.

8. C. longifolia Fischer ex Lindau in Engl., Bot. Jahrb. **13**: 161 (1890).
Erect or diffusely branched shrub 2–3 m. or tree to 15 m. high; leaf-blades rounded to cordate at base, usually acuminate, up to 36 (–50) cm. long and 18 (–25) cm. broad, usually glabrous; racemes 7–35 cm. long; rachis puberulous; pedicels solitary to several together; flowers light green, sweet-scented; ripe fruit about 1 cm. long.
Common in woodland on limestone in wet sheltered areas or in swamps; SL–3000 feet; fl. and fr. all the year. *A 9261! 9316! H 8772! 9489! P 10607!* Endemic.

9. C. diversifolia Jacq., Enum. Syst. Pl. Carib.: 19 (1760).—*C. laurifolia* of F. & R. (1914), not Jacq. (1798).
Tree 5–20 m. high, glabrous; leaf-blades rather leathery, prominently veined, up to 14 cm. long and 8 cm. broad; racemes 6–12 cm. long; rachis glabrous; pedicels solitary or paired; flowers greenish-yellow; ripe fruit about 14 mm. long.
Occasional in open woodlands on limestone, mainly in the eastern and central parishes; 150–2700 feet; fl. Mar–July, fr. most of the year. *A 12378! H 5093! H & P 13689!* Florida, Bahamas, West Indies, Venezuela.

10. C. zebra Griseb., Fl. Br. W.I.: 162 (1860).—*C. harrisii* Lindau (1899).
C. priorii F. & R. (1913). Zebra Wood.
Shrub 1·2–1·5 m. or tree to 10 m. high, sometimes much larger; leaf-blades 5–12 (–20) cm. long, 3–8 (–12) cm. broad, rounded at base, glabrous; racemes 6–10 cm. long; rachis tomentellous; pedicels solitary; ripe fruit about 16 mm. long.
Rather rare (Manch., St. Ann, Port.), in woodland on limestone and wet mossy forest; (1500–) 2500–4000 feet; fl. Feb, July, Nov. *H 5481! 7667! H.P.S. 14754! P 11371!* Endemic.

55. CHENOPODIACEAE

Shrubs or annual or perennial herbs; stems sometimes jointed. Leaves alternate, rarely opposite, simple or reduced to scales, without stipules. Flowers bisexual or unisexual, polygamous or dioecious, small, mostly actinomorphic in small dense axillary clusters, cymes or terminal spikes. Perianth uniseriate of (2–) 5 more or less connate imbricate sepals, usually persistent in fruit, or absent. Stamens as many as the calyx-lobes and opposite to them, hypogynous or on a disk or adnate to the calyx; filaments usually free; anthers 2-locular, incurved in bud, opening lengthwise. Ovary superior or inferior, 1-locular; ovule solitary, erect or suspended from a basal funicle, campylotropous; styles 1–3. Fruit a nutlet, usually indehiscent, free or aggregated by union of fleshy calyces. Seed with curved embryo surrounding the endosperm or endosperm lacking.
Over 100 genera and about 1500 species, widely distributed but characteristic of open arid or saline areas.

1 Leaves reduced to opposite scales joined in pairs; stem jointed; inflorescence terminal, spiciform of cymules of usually 3 flowers in cavities at the nodes; stamens 2
                                                  1. Salicornia
1 Leaves evident, alternate, more or less toothed or lobed; flowers in clusters or in simple or branched spikelike cymes; stem not jointed:
  2 Flowers hermaphrodite or polygamous; fruit more or less enclosed by usually 5 perianth-segments; stamens 5   **2. Chenopodium**
  2 Flowers all unisexual; fruit enclosed by 2 appressed bracteoles, the perianth wanting in female flowers; stamens 4   **3. Atriplex**

1. SALICORNIA L. (1753)

1. S. perennis Mill., Gard. Dict. ed. 8 (1768).—*S. ambigua* Michx. (1803).
Glasswort.
Erect-branched suffrutescent herb to 30 cm. high, the main stem often trailing and woody, perennial; stem fleshy, often reddish, usually terete; spikes 3–5 cm. long; flowers green; stigmas 2; seed with hooked hairs.

Local (St. Cath., Clar.), in estuarine swamps and salina and mangrove margins; near SL; fl. and fr. Jan–June. *A 8825! H 10178! C. B. Lewis!* Coasts of Europe, Bermuda, Atlantic and Gulf coasts of N. Amer., Bahamas, Greater Antilles, Virgin Is., Antigua.

So far the above is the only species reported from Jamaica; *S. bigelowii* Torr. is recorded from Little Cayman and *S. peruviana* Kunth from Grand Cayman.

2. CHENOPODIUM L. (1753)

Leaf-blade oblong-lanceolate, shortly petioled, glandular beneath   1. *ambrosioides*
Leaf-blade deltate-ovate, long-petioled, not glandular beneath   2. *murale*

1. C. ambrosioides L., Sp. Pl. **1**: 219 (1753).—Bitter Weed, Hedge Mustard, Mexican Tea, Semicontract, Wormseed.

Bushy taprooted herb up to 120 cm. high, aromatic of garlic and the roots strongly pungent; lower leaves sinuate-toothed, 5–8 cm. long, 1–2 cm. broad, the upper much smaller, entire; inflorescence-branches often attenuated; flowers clustered, greenish; perianth 1 mm. long; stamens yellow or white; stigmas 3; fruit subglobose.

Common as a weed of pathsides and gravelly waste places in inhabited areas; SL–4500 feet; fl. and fr. Dec–July. *A 6039! H 11937! P 23642!* Widespread in warm countries.

2. C. murale L., Sp. Pl. **1**: 219 (1753).

Herb to 60 cm. high; leaves irregularly acutely toothed, 5–8 cm. long, 4–4·5 cm. broad; petioles up to 4 cm. long; perianth-segments 1·3 mm. long, each with prominent midrib; stigmas 2.

Rare (St. Cath., Trel.), in sandy waste places and cultivated fields; SL–75 feet; fl. and fr. Mar–July. *Campbell 6503! Cornman! Thornton!* Widespread in warm countries.

3. ATRIPLEX L. (1753)

1. A. pentandra (Jacq.) Standl. in N. Amer. Fl. **21**: 54 (1916).—*Axyris pentandra* Jacq. (1763). *Spinacia littoralis* Jacq. (1760), not *Atriplex littoralis* L. (1753). *Atriplex cristata* Humb. & Bonpl. ex Willd. (1806).

Annual bushy herb up to 1 m. high, with numerous lax branches; leaves oblong or elliptical, acute, narrowed to base, shallowly repand-dentate; silvery-mealy, 1–2 cm. long, 6–8 mm. broad; flowers densely clustered; bracts of female flowers marginally toothed, crested.

Uncommon and local, on sandy beaches and on some of the cays; SL–10 feet; fl. and fr. June, Sept–Oct. *A 13036! H 9932! P 23663!* Bermuda, Bahamas, Florida, Greater Antilles, Virgin Is., Curaçao, northern S. Amer.

Beta vulgaris L., Beetroot, is cultivated in Jamaica.

56. BATACEAE

Low straggling glabrous shrubs with laxly erect or prostrate stems. Leaves opposite, sessile, simple, entire, fleshy, linear or linear-oblong, subterete, without stipules. Flowers dioecious in conelike bracteate spikes solitary in the leaf-axils. Male flowers: spikes sessile; bracts free, tightly imbricated in 4 rows, persistent with 1 flower in each axil; perianth (calyx) campanulate, shallowly 2-lipped, membranous; stamens 4 (–5), alternating with 4 (–5) clawed petaloid staminodes; filaments free; anthers dorsifixed, 2-locular, introrse, opening vertically. Female flowers: spikes stalked, 4–12-flowered; bracts small, roundish, deciduous; perianth absent; ovary superior, 4-locular, each loculus with a solitary basal anatropous ovule; stigma sessile, capitate. Fruit fleshy, composed of the cohering berrylike pistils. Seeds with endosperm; embryo straight.

One genus *Batis* with 2 species occurring in littoral and sublittoral habitats in subtropical and tropical America and the islands of S.E. Asia and the Pacific.

1. BATIS P. Browne (1756)

1. B. maritima L., Syst. Nat. ed. 10, **2** : 1289 (1759).—Jamaican Samphire.
Stems trailing or ascending to 1 m. high, branched, sometimes looping and rooting, silvery-greyish when old; young stems and leaves light yellow-green; leaves
up to 2·5 cm. long; spikes up to about 1 cm. long in flower, to 2 cm. long in fruit.
Locally abundant and often forming large communities at the margins of salinas
and estuarine flats, mainly along the south coast and on the cays; SL–10 feet; fl.
and fr. most of the year. *A 6063! H 8187! P 11517!* Bahamas, S.E. United States,
continental tropical Amer., Greater Antilles to Antigua; Grand Cayman and
Hawaii.

57. AMARANTHACEAE

Herbs, shrubs or rarely small trees. Leaves opposite or alternate, usually entire,
without stipules. Inflorescence of spikes, fascicles, heads or solitary flowers in
racemose or corymbose arrangements. Flowers small, bisexual or unisexual and
monoecious, polygamous or dioecious, mostly 5-merous, usually bracteate and
2-bracteolate. Perianth uniseriate (calyx) of (2–) 5 free overlapping or shortly
connate often scarious members. Stamens (2–) 5; filaments free or united below
into a tube, often with filamentous pseudostaminodes between the filaments; anthers
2- or 4-locular. Ovary superior, 1-locular; styles 1–3 (–8). Fruit a 1-seeded utricle
or a 1–several-seeded capsule, often circumscissile. Seeds usually cochleate;
endosperm mealy, abundant.
50–60 genera with 500–800 species mostly in the tropics of Africa and America.

1 Leaves alternate; stigmas 2–3 (–4); anthers 4-locular:
  2 Fruit a 1-seeded utricle; style shorter than the stigmas or absent:
    3 Flowers bisexual; scrambling shrub or vine   **1. Chamissoa**
    3 Flowers unisexual; erect or prostrate-branched herbs:
      4 Plants monoecious   **2. Amaranthus**
      4 Plants dioecious; perianth absent in female flowers   **3. Acnida**
  2 Fruit a several-seeded capsule; flowers bisexual; style simple; stigmas very short
        4. Celosia
1 Leaves opposite; ovule 1; perianth-segments 5:
  5 Inflorescence an elongated spike or raceme with the flowers or flower-clusters deflexed
    in fruit; stigma 1; anthers 4-locular:
    6 Flowers solitary along the spike, all perfect; spines formed by bracts and bracteoles
      straight or slightly curved, not hooked   **5. Achyranthes**
    6 Flowers in clusters of (1–) 2–3, accompanied by sterile flowers whose perianth-
      members bear hooked awns   **6. Cyathula**
  5 Inflorescence or partial inflorescence of short spikes or heads; flowers not deflexed in
    fruit; anthers 2-locular:
    7 Flowers in short spikes or heads, not panicled, bisexual:
      8 Perianth forming a hard woolly crested tube in fruit; stigma 1, capitate
        7. Froelichia
      8 Perianth not changing in fruit:
        9 Leaf-bases amplexicaul with a tuft of hairs around the node; leaf-blades linear,
          glabrous or nearly so; stigmas 2:
          10 Stamens 5; staminodes absent; flower-heads mostly pedunculate; leaves fleshy,
            spread along the stem   **8. Philoxerus**
          10 Stamens 2; staminodes 3; flower-heads mostly sessile; leaves linear-oblanceolate,
            the longest clustered in basal rosettes; small perennial tufted herb of rocky
            pastures and sand near the sea; Cayman Is.   *Lithophila muscoides* Sw.
        9 Leaf-bases not or hardly amplexicaul; leaves not linear, not fleshy; filament-tube
          long; staminodes developed:
          11 Stigma 1, capitate; stamens with obvious filaments; bracteoles shorter than sepals
            9. Alternanthera
          11 Stigmas 2, filiform; anthers sessile on the filament-tube; bracteoles equalling or
            exceeding sepals; staminodes 2-lobed; flower-heads usually subtended by
            small leaves   **10. Gomphrena**
    7 Flowers in slender panicled spikes or large terminal panicles, bisexual or unisexual;
      stigmas 2 (–3)   **11. Iresine**

1. CHAMISSOA Kunth (1818) nom. cons.

1. C. altissima (Jacq.) Kunth, Nov. Gen. **2**: 197, t. 125 (1818).—Basket Withe.

A low weedy shrub or large vine with stem to 25 m. long and 3 cm. or more thick, the older stems often strongly ribbed; leaves elliptical to lanceolate, up to 14 cm. long and 8 cm. broad; flowers greenish-yellow in panicles; perianth scarious, 3 mm. long; stamens 5, united below; utricle circumscissile; seed black, discoid, enveloped by a blue translucent aril.

Common in thickets and woodland margins, mostly in rather damp sheltered places; SL–4500 feet; fl. and fr. all the year, but mostly Dec–June. *A 6303! H 8506! P 19675!* Tropical continental Amer., West Indies.

2. AMARANTHUS L. (1753)

1 Plant with spines near the leaf-bases   1. *spinosus*
1 Plant without spines:
  2 Flowers in usually elongated terminal spikes as well as axillary spikes or clusters:
    3 Female flowers with 5 perianth-segments; male flowers with (4–) 5 stamens; utricle smooth, circumscissile   2. *dubius*
    3 Female flowers with 3 perianth-segments; male flowers with (2–) 3 stamens; utricle wrinkled, indehiscent   3. *viridis*
  2 Flowers in axillary clusters or very short spikes; perianth-segments 5:
    4 Female flowers with short thick peduncles deciduous with the fruit; plant entirely green   4. *crassipes*
    4 Female flowers sessile; perianth urceolate, white; stem usually pinkish, glaucous   5. *polygonoides*

1. A. spinosus L., Sp. Pl. **2**: 991 (1753).—Prickly Calalu.

Perennial herb, sometimes forming a massive branched stock, shoots usually erect, up to 60 (–120) cm. high; stem often red; leaves ovate to lanceolate, up to about 6 cm. long; flowers clustered in long terminal spikes and globose axillary clusters; perianth-segments 5, 1·5 mm. long; stamens 5; utricle opening irregularly.

Common as a weed of pastures, lawns and waste places; 50–1500 feet; fl. and fr. all the year. *A 5437! H & P 13940!* General in the tropics.

2. A. dubius Mart. ex Thell. in Mém. Soc. Sci. Nat. Cherbourg **38**: 203 (1912).—*A. tristis* of F. & R. (1914), not L. (1753). Spanish Calalu.

Annual or short-lived perennial herb up to 150 cm. high, usually much smaller; stem green or more usually pinkish, more or less curled-pubescent; leaves ovate, obtuse and often minutely emarginate, mucronate, up to 5 cm. long, usually light green; flowers in long terminal and shorter axillary spikes; perianth-segments 2 mm. long; seed 1 mm. long, dark reddish-black.

Locally common as a weed of cultivation, rough pastures and gravelly waste places; SL–2800 feet; fl. and fr. all the year. *A 5439! H 6802! P 24106!* General in the tropics; Cayman Is.

3. A. viridis L., Sp. Pl. ed. 2, **2**: 1405 (1763).—Garden Calalu.

Annual herb, 1 m. or more high; leaves ovate-rhomboid, rounded and emarginate at the apex, long-petioled, up to 5 cm. or more long; inflorescences as the last; perianth-segments about 1 mm. long; seed about 1 mm. long, black.

Common as a weed of grassy places and open ground; SL–600 feet; fl. and fr. most of the year, especially during wet periods. *A 6920! H & P 15106!* General in the tropics; Grand Cayman.

4. A. crassipes Schlecht. in Linnaea **6**: 757 (1831).

Annual taprooted herb, the branches woody at the base, usually prostrate, rarely erect to 60 cm. high; leaves spathulate, up to 2·5 cm. long, long-petioled; perianth-segments 1·2–1·5 mm. long; utricle fleshy, indehiscent, tubercled.

Occasional as a weed of gravelly waste places; SL–700 (–3500) feet; fl. and fr. all the year. *A 5428! 6028! H 6847! P 23937!* Tropical Amer.

5. A. polygonoides L., Pl. Jam. Pugill.: 27 (1759).

Annual taprooted herb with spreading prostrate branches pubescent on one side, 10–30 cm. long, rarely ascending; leaves obovate, up to 2·5 cm. long, long-petioled;

female perianth-segments white, green in the mid-line, 2 mm. long, enlarging in fruit; utricle indehiscent.

Uncommon (St. Andr., St. Thom.), a weed of waste places on limestone gravel; 20–700 feet; fl. and fr. sporadically. *A 8303! H 11797! P 23930!* Florida, Texas, Mexico to northern S. Amer., Bahamas, West Indies sporadically.

3. ACNIDA L. (1753)

1. A. cuspidata Bert. ex Spreng. in L., Syst. Veg. ed. 16, **3**: 903 (1826).—Water Hemp.

Massive herb with swollen spongy or hollow stems, 1–2·5 (–4) m. high; leaves lanceolate, long-acute, up to 20 cm. or more long; flowers green or pinkish in terminal panicles; perianth-segments in male flowers 5, 2·5 mm. long; utricle 2 mm. long.

Local (St. Cath., Clar., St. Eliz., Han.), in marshes and riverside swamps; SL–100 feet; fl. and fr. most of the year. *A 12075! H 9760! P 23718!* Florida, Cuba, Hispaniola, Trinidad.

4. CELOSIA L. (1753)

Plant herbaceous; stem erect; leaves linear to lanceolate; spikes compact   1. *argentea*
Plant shrubby, scrambling; leaves ovate to lanceolate, broad at the base; spikes lax
     2. *nitida*

1. C. argentea L., Sp. Pl. **1**: 205 (1753).

Annual herb 30–120 cm. high; stem somewhat angled, with scattered vascular bundles; leaves 2–10 cm. long; spikes ovoid to cylindrical, up to about 12 cm. long; perianth-segments 5, scarious, translucent, about 7 mm. long, crimson or silvery-white; stamens 5, united below; staminodes absent; capsule 3–8-seeded, about 4 mm. long; seeds black, shiny.

Rare (St. Andr., Port., St. Thom.), except in cultivation, but sometimes escaping into gravelly waste places, especially near rivers; SL–500 feet; fl. and fr. most of the year. *H 12130! Powell 284!* General in the tropics. *C. argentea* var. *cristata* (L.) Kuntze, Cockscomb, is a variant sometimes grown in gardens or as an ornamental pot-plant.

2. C. nitida Vahl, Symb. Bot. **2**: 44 (1791).

Perennial branched undershrub 0·5–2 m. high; leaves up to about 5 cm. long; spikes lax, often branched and paniculate; perianth-segments 5, rather rigid, striate, often tinged purple, 5 mm. long; capsule 4–5 mm. long, up to 20-seeded.

Common in thickets and woodlands in arid sandy or rocky places; SL–300 feet; fl. and fr. all the year. *A 5660! H 6754! H & P 13795!* Mexico to northern S. Amer., West Indies.

5. ACHYRANTHES L. (1753)

Leaves obovate-rounded, cuspidate; perianth about 4 mm. long   1. *indica*
Leaves elliptical, long-acuminate; perianth about 6 mm. long   2. *aspera*

1. A. indica (L.) Mill., Gard. Dict. ed. 8 (1768).—*Centrostachys indica* (L.) Standl. (1915). Devil's Horse-Whip.

Annual pubescent herb; stems more or less erect to 1 m. high in flower; leaves 3–7 cm. long and broad; spikes up to 30 cm. or more long; bracts and bracteoles pink with hyaline margins; perianth-segments green with hyaline margins, shortly spine-tipped; stamens 5, united at the base; staminodes fringed; ripe utricle indehiscent.

Common as a weed of cultivation and disturbed waste places; 50–2000 feet; fl. and fr. most of the year. *H 6869! H & P 13328!* General in the subtropics and tropics; Grand Cayman.

2. A. aspera L., Sp. Pl. **1**: 204 (1753).—*Centrostachys aspera* (L.) Standl. (1915).

Laxly erect annual herb up to 120 cm. high in flower; lower leaves up to 23 cm. long and 9 cm. broad, upper smaller, long-attenuated to the base; spikes 40 cm. or more long; bracts and perianth-segments reddish.

Rare (St. Andr.), in shady thickets and along riverbanks; 50–1250 feet; fl. Jan. *A 6164! A. von der Porten!* General in the subtropics and tropics.

6. CYATHULA Blume (1825) nom. cons.

Hooked awns nearly as long as the fruiting perianth; leaves shortly acuminate or obtuse at the tip   1. *prostrata*
Hooked awns nearly twice as long as the fruiting perianth; leaves long-acuminate
                                                                                            2. *achyranthoides*

1. C. prostrata (L.) Blume, Bijdr.: 549 (1825).

Lax herb with spreading and geniculate-ascending often reddish branches up to about 30 cm. high; leaves obovate to elliptical, gradually narrowed to the base, usually red-margined, up to 6 cm. long and 3·5 cm. broad; racemes 5–15 cm. long; flower-clusters stalked, usually green; perianth 2·5 mm. long; utricle indehiscent.

Occasional weed of rough pastures and roadsides mostly in wet areas; 150–3900 feet; fl. and fr. June–July, Nov–Mar. *A 10133! H 8638! P 11376!* Widespread in the tropics.

2. C. achyranthoides (Kunth) Moq. in DC., Prodr. **13** (2): 326 (1849).

Herb with decumbent rooting stems, erect flowering branches 30–100 cm. high; leaves ovate-elliptical, long-attenuated at the base, up to 11 cm. long and 4·5 cm. broad; racemes 5–18 cm. long; perianth 2·5–3 mm. long, green; utricle indehiscent.

Uncommon weed of roadside banks and shady waste places; 20–1500 feet; fl. and fr. Jan–Mar, Sept. *A 12317! H 9931!* Costa Rica to Peru and Brazil, Cuba, Hispaniola, Dominica.

7. FROELICHIA Moench (1794)

1. F. interrupta (L.) Moq. in DC., Prodr. **13** (2): 421 (1849).

Perennial herb with spreading or erect branches 25–50 cm. high; radical leaves elliptical, shaggy-woolly beneath, 5 cm. long, 3 cm. broad, cauline obovate to oblanceolate, smaller; spikes terminal, sometimes branched; perianth in fruit 5 mm. long with 2 narrow crests; utricle indehiscent.

Very rare (St. Andr.), a roadside weed, not recently collected; about 100 feet; fl. and fr. Oct–Dec. *H 11234!* Mexico to Paraguay, Cuba, Hispaniola.

8. PHILOXERUS R. Br. (1810)

1. P. vermicularis (L.) Beauv., Fl. Oware & Benin **2**: 65, t. 98, f. 1 (1818).

Perennial taprooted prostrate-branched or bushy herb to 35 cm. high, rarely scrambling with longer branches, sometimes rooting from the lower nodes; stems reddish or pink; leaves 2–4 cm. long; flower-heads globose or ovoid to cylindrical, 8–30 mm. long; perianth 2–3 mm. long, segments unequal, woolly basally.

Locally common in low-lying sandy places near the sea, at brackish swamp-margins and on exposed littoral coral rocks and on the cays; SL–20 feet; fl. and fr. all the year. *A 9815! H 9813! P 10444!* Tropical continental Amer., West Indies, W. Africa; Cayman Is.

9. ALTERNANTHERA Forsk. (1775)

1 Flower-heads mostly long-peduncled; plant more or less erect   1. *ramosissima*
1 Flower-heads sessile; plants mostly prostrate or scrambling:
    2 Leaves of a pair unequal; outer perianth-segments larger than the inner, spine-tipped, more or less villous with glochidiate hairs:
        3 Leaf-blades obovate-rhomboid, unequal-sided; perianth-segments aristate
                                                                                            2. *pungens*

3 Leaf-blades elliptical. attenuated basally; perianth-segments mucronate   3. *peploides*
2 Leaves of a pair more or less equal:
  4 Outer perianth-segments mucronate, rather firm but not sharply spiny:
    5 Young stems and leaves with thin scurfy pubescence; perianth-segments spreading, whitish, transparent; leaf-tips mostly acute   4. *ficoidea*
    5 Young stems and leaves with dense greyish-yellow more or less persistent pubescence; perianth-segments ascending, densely yellowish-hirtellous, opaque; leaf-tips mostly obtuse   5. *halimifolia*
  4 Outer perianth-segments acute or acuminate:
    6 Outer perianth-segments about 4 mm. long, 3-nerved, hairy at base; utricle included in the perianth   6. *paronychioides*
    6 Outer perianth-segments about 2 mm. long, 1-nerved, glabrous; utricle a little longer than the perianth   7. *sessilis*

1. A. ramosissima (Mart.) Chod. in Bull. Herb. Boiss. sér. 2, **3**: 355 (1903).—
Mogiphanes ramosissima Mart. (1826).
Diffusely branched herb 60–120 cm. high; stem and leaves beneath sometimes tinged reddish-purple; leaf-blades rhomboid-elliptical, up to 12 cm. long and 4·5 cm. broad, appressed-pubescent; flower-heads white, subglobose, 1–1·5 cm. in diameter, on peduncles up to 8 cm. long, rarely sessile; bracteoles a little longer than the bracts, half the length of the perianth, aristate, pubescent.
Occasional and often gregarious on roadside banks, probably originally escaped from gardens; 200–2500 feet; fl. and fr. Nov–Mar. *A 10323! Powell 817! Wynter UCW1 1140!* Native of Brazil and Guyana; Florida, Mexico, Hispaniola, Puerto Rico, Lesser Antilles.

2. A. pungens Kunth, Nov. Gen. **2**: 206 (1818).—*A. repens* (L.) Link (1821), not J. F. Gmel. (1791).
Prostrate-branched herb with fleshy roots and pubescent stems; larger leaves up to 2·5 cm. long and 1·5 cm. broad, usually light yellow-green, mucronate; flower-heads 7–12 mm. long; outer perianth-segments 4 mm. long, straw-coloured with glochidiate hairs at base, glabrous distally, inner 2·5–3 mm. long; utricle 1·5 mm. long.
Local (St. Andr., St. Cath., St. Thom.), a weed of arid pathsides, roads and open waste ground; 50–600 feet; fl. and fr. sporadically. *A 9424! H 10517! P 23931!* Widespread in the tropics.

3. A. peploides (Humb. & Bonpl. ex Willd.) Urb. in Fedde, Repert. Sp. Nov. **15**: 168 (1918).—*Illecebrum peploides* Humb. & Bonpl. ex Willd. (1806). *A. parvifolia* (Moq.) F. & R. (1914).
Like the last; larger leaves up to 2 cm. long; flower-heads 7–9 mm. long; outer perianth-segments 3·5 mm. long, softer and with more numerous glochidiate hairs than the last.
Only twice reported (St. Andr.), a weed of waste ground; 5000 feet; fl. and fr. Sept. *Fawcett 6007!* Tropical and subtropical Amer., Atlantic Islands, Spain.

4. A. ficoidea (L.) Roem. & Schult. in L., Syst. Veg. ed. nov. **5**: 555 (1819).—
Crab Withe.
Low bushy or spreading taprooted herb, the branches often trailing to 120 cm. or more and rooting sparingly from the nodes; leaves elliptical, 3–7 cm. long, 1–2 cm. broad; flower-heads 5–10 mm. long; outer perianth-segments about 3·5 mm. long, inner shorter.
Common as a weed of roadsides, rough pastures and waste places, often in low damp localities or in shade; SL–1000 (–5000) feet; fl. and fr. all the year. *A 8122! H 9856! P 23504!* Greater Antilles, Virgin Is., St. Vincent, Trinidad and Tobago.

5. A. halimifolia (Lam.) Standl. in Pittier, Pl. Us. Venez.: 145 (1926).—*Achyranthes halimifolia* Lam. (1785). *Alternanthera ficoidea* var. *flavogrisea* (Urb.) F. & R. (1914).
Robust rather succulent trailing or subscandent herb with branches to 2·5 m. long, rarely shortly erect or bushy; leaves mostly shorter and broader than the last.

I

Common in open sandy places near the sea and on the cays, rare inland; SL–50 (–3600) feet; fl. and fr. most of the year. *A 8968! H 11761! P 21864!* Mexico to Chile, Cuba, Grenada, Curaçao.

6. A. paronychioides St.-Hil., Voy. Brés. **2**: 439 (1833).
Prostrate-branched or spreading herb; stems with long crisped hairs on the younger parts; leaves oblanceolate to elliptical, 1–3 cm. long, 5–10 mm. broad; flower-heads 5–10 mm. long; perianth white.
Uncommon in low-lying damp pastures and in mud at pond margins; SL–1500 feet; fl. and fr. Mar–June, Sept–Dec. *A 10155! H 11942! P 15361!* Tropical Amer., West Indies.
A. bettzickiana (Regel) Nicholson is a closely related species with usually red stems and leaves and low densely bushy habit, grown occasionally as a border for paths and flower-beds. Native of Brazil.

7. A. sessilis (L.) R. Br. ex DC., Cat. Pl. Hort. Bot. Monspel.: 4, 77 (1813).
Prostrate-branched, spreading and ascending taprooted herb, also rooting from the lower nodes; stem often purplish, pubescent along two lines; leaves 1–3 cm. long, up to 1·5 cm. broad, midrib sometimes purplish; flower-heads subglobose at first, about 3 mm. in diameter, often cylindrical later, the older flowers caducous from the base; utricle obovate.
Common in damp pastures, pond margins and low-lying waste places; 20–2100 feet; fl. and fr. most of the year. *A 9456! H 11831! P 10326!* Widespread in warm countries.

10. GOMPHRENA L. (1753)

1 Flower-heads 2–3 cm. long, 2 cm. or more broad, white, mauve or magenta; erect annual herbs   1. *globosa*
1 Flower-heads not over 2 cm. broad; weedy plants of pastures and waste places, with prostrate or ascending branches:
  2 Flower-heads pinkish, 5–10 (–20) mm. long; bracts up to about 2 mm. long; bracteoles 5–6 mm. long, crested for over half their length and broadly so towards apex with several teeth   2. *decumbens*
  2 Flower-heads white or greenish-white, 15–20 (–40) mm. long; bracts about 3·5 mm. long; bracteoles 6–7 mm. long, very narrowly crested only at the tip with 1 or 2 shallow teeth   3. *celosioides*

1. G. globosa L., Sp. Pl. **1**: 224 (1753).—Bachelor's Buttons.
Annual bushy herb 50–80 cm. high in flower; leaves oblong-elliptical, 2–10 cm. long, up to 4 cm. broad; flower-heads magenta, sometimes mauve or white, usually subtended by small leaves.
Cultivated and occasionally escaped; fl. Oct–Apr. *Wood!* Native of tropical Amer., introduced into the tropics generally.

2. G. decumbens Jacq., Pl. Hort. Schoenbr. **4**: 41, t. 482 (1804).—? *G. serrata* L. (1753). *G. dispersa* Standl. (1916).
Branches spreading and ascending from a perennial taprooted stock, up to 60 cm. high, usually smaller, sometimes rooting from the lower nodes; hairs on stem mostly ascending-appressed; flower-heads rosy-pink at first, fading white; bracteoles about as long as the perianth, three times longer than the bract, crested; perianth-segments green at base, hyaline at tip, hairy from the base.
Locally common (St. Andr., Clar., St. Mary), a weed of sandy lawns and pastures and open waste ground; SL–600 (–4500) feet; fl. and fr. most of the year. *A 5534! H 11542! P 9372!* Florida, C. Amer., Cuba, Hispaniola.

3. G. celosioides Mart. in Nov. Act. Nat. Cur. **13**: 301 (1826).
Like the last but the plant more bushy and slightly larger in all vegetative parts; hairs on stem long and spreading.
Rare and local (St. Andr.), at lawn margins and in recently disturbed ground; about 600 feet; fl. and fr. Sept–Dec. *Skelding UCWI 27349!* Native of Brazil, now widespread in the Old World tropics.

11. IRESINE P. Browne (1756) nom. cons.

Flowers bisexual in loose panicles, the spikelets usually pedunculate; sepals 1·5 mm. long, densely villous with long hairs; seed orbicular, 0·7 mm. broad, dark reddish-brown, shiny; perennial   1. *angustifolia*
Flowers dioecious; spikelets sessile or pedunculate; sepals of pistillate flowers 1–1·5 mm. long, whitish-woolly at the base; seed broadly obovoid to suborbicular, 0·5–0·6 mm. broad, dark red, shiny; short-lived perennial or annual   2. *diffusa*

1. I. angustifolia Euphrasen, Beskr. Svensk. Vestind.: 165 (1795).—*I. elatior* L. C. Rich. ex Willd. (1806).

Usually erect herb, branched, 50–200 cm. high; leaves ovate-lanceolate to linear-lanceolate, acuminate or caudate at tip, long-attenuate at base, 5–10 cm. long, 1–4 cm. broad; flowers brownish; utricle orbicular.

Rare and local (Clar.), in thickets in arid rocky places near the sea; fl. and fr. Mar. *H 10187!* Mexico to Ecuador and Brazil, West Indies.

2. I. diffusa Humb. & Bonpl. ex Willd. in L., Sp. Pl. ed. 4, **4**: 765 (1806).— *I. paniculata* (L.) Kuntze (1891), not Poir. (1813). *I. celosia* L. (1759), illegitimate name. Jubba Bush.

Plant usually with trailing or scrambling stems, 1·5 m. or more long; leaves broadly ovate to lanceolate, acute to acuminate, broadly cuneate to truncate at base, 3–14 cm. long, 1·5–7 cm. broad, minutely pellucid-dotted; spikes greenish-white to golden-yellow, fragrant; male perianth glabrous; female flowers develop long hairs on the perianth and on the pedicels.

Common on roadside banks and in thickets; SL–5200 feet; fl. and fr. most of the year. *A 5984! 10460! J.P. 1142! Proctor & Mullings 22028!* Tropical continental Amer., West Indies.

58. NYCTAGINACEAE

Herbs, shrubs or trees. Leaves alternate or opposite, simple, entire, without stipules. Flowers bisexual or unisexual, actinomorphic, in bracteate clusters, corymbs or cymose panicles. Perianth uniseriate of a petaloid typically 5-merous gamosepalous calyx folded or contorted in bud, persistent in fruit. Stamens 1–many, hypogynous, free or basally connate, generally unequal. Ovary superior, unilocular; ovule solitary, basal, anatropous or campylotropous; style slender; stigma simple or divided. Fruit achenial, or utricular by combination of the often modified calyx (anthocarp). Seed with straight or curved embryo; endosperm copious or scanty.

About 30 genera with some 300 species mostly in the American subtropics and tropics.

1 Herbs; distal portion of perianth deciduous in fruit; flowers hermaphrodite; leaves opposite:
  2 Bracteoles formed into a 5-lobed calyxlike cupule; perianth-tube elongated, showy; flowers in corymbose cymes; fruit globose, ribbed, not glandular   1. **Mirabilis**
  2 Bracteoles free, small and inconspicuous; perianth short; flowers small in capitula, umbels or umbelliform cymes; fruits clavate, oblanceolate or obovoid, ribbed, glandular or smooth:
    3 Mature fruits shallowly 10-ribbed, declinate, about 10 mm. long, with large stalked adhesive glands subverticillate distally; distal part of perianth shortly tubular at base, greenish   2. **Commicarpus**
    3 Mature normal[1] fruits strongly 5-ribbed, erect, 3–4 mm. long, with adhesive glandular hairs or patches on or between the ribs or glabrous; distal part of perianth abruptly joined to proximal part, pink or crimson   3. **Boerhavia**
1 Trees or woody climbers:
  4 Bracteoles in threes, conspicuous, coloured, the pedicels adnate to them; woody climbers with supra-axillary spines; leaves alternate; flowers hermaphrodite
                                  4. **Bougainvillea**

[1] See note under *Boerhavia diffusa*.

4 Bracteoles inconspicuous, green; leaves mostly opposite; flowers unisexual, dioecious:
  5 Stamens and style included in perianth; fruit crowned by persistent perianth; unarmed tree   **5. Neea**
  5 Stamens and style exserted; ripe fruit not crowned by persistent perianth:
    6 Fruit leathery, oblong-clavate, 5- or 10-angled with numerous stalked glands on the angles; armed climbing shrub or unarmed tree   **6. Pisonia**
    6 Fruit softly fleshy, drupaceous, angled only when dry, without glands; unarmed erect shrubs or trees   **7. Guapira**

1. MIRABILIS L. (1753)

1. **M. jalapa** L., Sp. Pl. **1**: 177 (1753).—False Jalap, Four O'clock, Marvel of Peru.
Straggly-branched herb to 80 cm. high; leaves ovate, long-acuminate, truncate to subcordate at base, up to 12 cm. long and 6·5 cm. broad; perianth-tube to 5·5 cm. long, usually rose-magenta but sometimes orange, yellow, pink, salmon or white; ripe fruit hard, black, about 13 mm. long.
Common in gardens and as an escape on shady roadside banks; 400–2500 feet; fl. and fr. all the year. *Harris! P 22900!* Native of S. Amer., now widespread in the subtropics and tropics.

2. COMMICARPUS Standl. (1909)

1. **C. scandens** (L.) Standl. in Contrib. U.S. Nat. Herb. **12**: 373 (1909).—
Boerhavia scandens L. (1753). Easy-to-Break, Rat Ears.
Slender weak-stemmed scrambling herb up to 4 m. high; leaves ovate to ovate-lanceolate, cordate at base, obtuse, acute or acuminate, thinly puberulous, up to 5 cm. long and 3·5 cm. broad; partial inflorescences umbellate, the pedicels elongating to 5–10 mm. long in fruit; perianth 5–7 mm. long.
Common, at least in the southern parishes, in hedgerows, waste places, sand dunes and coastal thickets and on the cays; SL–750 feet; fl. and fr. most of the year, *A 6272! H 8426! Webster & Wilson 4932!* Texas and Arizona to Peru, Bahamas, Greater Antilles, Virgin Is. to Antigua, Curaçao.

3. BOERHAVIA L. (1753)

1 Fruits glandular on the ridges and sometimes between them; leaves not glandular-punctate beneath; inflorescence-branches spreading; flowers not all stalked; distal part of perianth magenta or crimson:
  2 Capitula 4–12 (–20)-flowered; inflorescence sparingly branched, often with leafy bracts; stems and inflorescence-branches puberulous; leaf-margins with short curved hairs and a few long hairs or glabrous   **1. coccinea**
  2 Capitula 2–5-flowered; inflorescence much-branched, glabrous, without leafy bracts; stems minutely puberulous and usually thinly pilose or with patches of stalked glands here and there; leaves especially when young ciliate with long pluricellular hairs   **2a. diffusa var. diffusa**
1 Fruits glabrous; stems puberulous; inflorescence-branches glabrous:
  3 Stem spreading and ascending; inflorescence-branches divaricate; leaves not or only very sparingly glandular-punctate beneath; internodes not glutinous in patches; fruits glutinous on the ridges but apparently without separate glandular hairs; distal perianth crimson   **2b. diffusa var. leiocarpa**
  3 Stem erect; inflorescence-branches ascending; leaves glandular-punctate beneath; internodes often glutinous in patches; fruits not at all adhesive; distal perianth pink   **3. erecta**

1. **B. coccinea** Mill., Gard. Dict. ed. 8 (1768).—*B. hirsuta* Jacq. (1770). *B. caribaea* Jacq. (1771). Hogweed.[1]
Branches prostrate at first from a large taproot, trailing to 120 cm. or ascending to 60 cm. high; leaves ovate-rhomboid, mostly obtuse, broadly cuneate to rounded or truncate at base, up to about 4 cm. long and 3·5 cm. broad, paler beneath; perianth about 2·5 mm. long; stamens 2; fruit about 3 mm. long.
Common, as a weed of rough disturbed pastures, waste places and sand dunes; SL–900 (–5000) feet; fl. and fr. all the year. *A 5536! H 11760! Powell 254!* Tropics of America and Africa, sparingly in Asia; Cayman Is.

---

[1] This common name is applied to all species of *Boerhavia*.

58. NYCTAGINACEAE

2. B. diffusa L., Sp. Pl. **1**: 3 (1753).

2a. Var. diffusa.—*B. paniculata* L. C. Rich. (1792).
Branches spreading and ascending from a taprooted stock, up to about 1 m. long;
leaves broadly ovate to oblong-ovate, mostly obtuse, truncate to cordate at base,
up to 7·5 cm. long and 5 cm. broad; perianth about 2 mm. long; stamens (1–) 2
(–3); fruit 3·5–4 mm. long.
Locally common (St. Andr., St. Eliz., St. Mary), in waste sandy or gravelly
places; SL–1000 feet; fl. and fr. Mar–Nov. *A 5467! H 6805! H & P 13869!
Yuncker 17042!* A pantropical weed; Little Cayman. Fruits of this species may
become infected by a Gall Midge, *Asphondylia* sp. These galls are up to 7 (–9) mm.
long, inflated, more or less terete, retain the distal part of the perianth and develop
glands irregularly on the surface. They will be found to contain cottony tissue and
often an insect larva; older anthocarps may be perforated. It is the author's opinion
that this artefact has been contributory to the continued taxonomic recognition of
B. paniculata. Fruits of other species of *Boerhavia* occasionally become similarly
infected and glabrous anthocarps may become hairy but without glands as a result.

2b. B. diffusa var. **leiocarpa** (Heimerl) Adams in Mitt. Bot. Staatss. München **8**:
115 (1970).—*B. paniculata* forma *leiocarpa* Heimerl (1906). *B. coccinea* var.
leiocarpa (Heimerl) Standl. (1931).
Like the last.
Rare and local (St. Andr., St. Cath.), in gravelly waste places; 50–300 feet; fl.
and fr. May–Aug. *A 7097! Prior 340! Weaver 1079! 1286!* Mexico to Argentina,
Hispaniola.

3. B. erecta L., Sp. Pl. **1**: 3 (1753).
Annual taprooted herb to 120 cm. high; leaves ovate to lanceolate, broadly
cuneate to rounded at base, paler and puberulous beneath, red-margined, not
ciliate, up to 6 cm. long and 4 cm. broad; perianth about 2·5 mm. long; stamens 2;
fruit 4 mm. long.
Rather common, a weed of disturbed ground, roadside banks in open areas and
river gravel; SL–1000 (–2000) feet; fl. and fr. all the year. *A 7354! H 6853!
P 11203! Yuncker 17283!* Bermuda, S. United States, Mexico to C. and S. Amer.,
Bahamas, Greater Antilles, Virgin Is., Antigua, Barbados, Grand Cayman, W.
Africa, Malaya.

4. BOUGAINVILLEA Commers. ex Juss. (1789) nom. cons.;
Holttum (1955)

Most of the cultivated ornamental plants belonging to this genus are of garden
origin and have been derived by hybridization or selection from the South American
species to which Holttum has constructed the following key:

1 Perianth-tube very slender (2 mm. in diameter), glabrous
        B. peruviana Humb. & Bonpl.
1 Perianth-tube broader and more or less hairy:
2 Perianth-tube bearing very short hairs curved towards the top   *B. glabra* Choisy
2 Perianth-tube bearing copious spreading hairs up to 1 mm. long *B. spectabilis* Willd.

Numerous varieties with characters intermediate between these species exist and
some of them occur occasionally as relict plants in secondary communities, but
none is naturalized in Jamaica.

5. NEEA Ruiz & Pav. (1794)

1. N. nigricans (Sw.) F. & R., Fl. Jam. **3**: 153 (1914).—*Pisonia nigricans* Sw. (1788).
Saltwood.
Straggling-branched shrub 2–5 m. or gnarled tree to 16 m. high; indumentum
of young twigs reddish; leaves narrowly elliptical to oblong, more or less acuminate,
soft, thinly fleshy, 4–15 cm. long, 2·5–6·5 cm. broad; inflorescence a lax axillary

panicle, whitish-tomentose; male perianth elongate-campanulate, 6–8 mm. long, yellowish-green; stamens 8; pistillode rudimentary; female perianth tubular, 4–4·5 mm. long, yellowish-green; style thick with shortly lobed stigma at mouth of tube; staminodes present; ripe fruits ellipsoidal, 12 mm. long, red.

Frequent in woodlands on rocky limestone; (400–) 1200–3300 feet; fl. Feb–Oct, fr. Apr–Nov. *A 11234! 11257! H 6652! 9180! P 21477! 22601!* Endemic.

6. PISONIA L. (1753)

Glabrescent tree with leathery leaves, without spines; fruit cylindrical, shallowly 10-ribbed with prominent adhesive glands distally on 5 of the ribs   1. *subcordata*
Glabrous or pubescent scrambling shrub, generally armed with curved spines at the nodes; leaves thinly leathery; fruit ellipsoid-clavate, 5-angled with small adhesive glands all along the ribs   2. *aculeata*

1. P. subcordata Sw., Nov. Gen. & Sp. Pl.: 60 (1788).—*Neea rotundifolia* Heimerl (1912).

Tree up to 8 m. or more high; leaves and twigs rusty-pubescent when young; leaves ovate-elliptical to suborbicular, broadly cuneate to subcordate at base, rounded to broadly acute at tip, up to 27 cm. long and 20 cm. broad; inflorescences 3–10 cm. broad, the branches puberulous; flowers greenish-yellow; male perianth campanulate, 3–4·5 mm. long, the 8 stamens longer; female perianth 2–3 mm. long; fruit puberulous or glabrous (7–) 10–14 mm. long, with stalked glands occupying the distal 4–5 mm.

Rare and local (St. Cath., Clar., Manch., Trel.), in thickets and woodlands on rocky limestone; 1200–2500 feet; fl. July–Aug, fr. July. *A 12610! P 27530! 28854!* Puerto Rico to Martinique, Virgin Is.

2. P. aculeata L., Sp. Pl. 2: 1026 (1753).—*P. helleri* Standl. (1918). Cockspur, Fingrigo, Wait-a-bit.

Straggling shrub climbing to 6 m. or more, generally armed with curved spines; leaves elliptical to lanceolate, thinly leathery, variable in shape of tip and pubescence, 3–11 cm. long, 2–6 cm. broad; inflorescence axillary, corymbose-paniculate, puberulous; flowers green or greenish-yellow, fragrant; male perianth campanulate, 3·5–4 mm. long; stamens usually 6; female perianth tubular, 2–2·5 mm. long; fruit about 1·5 cm. long, 3–5 mm. broad.

Common in secondary thickets and woodland margins mostly on limestone; 50–1300 (–2000) feet; fl. Dec–May, fr. Feb–May. *A 6112! 9422! H 10042! 10714! A. von der Porten!* Widespread in the tropics.

7. GUAPIRA Aubl. (1775)

1 Leaves leathery, rigid when dry, rounded or emarginate at tip; branch-tips and inflorescence rusty-puberulous   1. *obtusata*
1 Leaves soft and brittle, thin and more or less papery when dry, with obscure venation:
  2 Leaf-tip obtuse to shortly and sharply acuminate; inflorescence rusty-puberulous, terminal, solitary, erect in flower, drooping in fruit, with a thick peduncle; ripe fruits bluish-black or red   2. *fragrans*
  2 Leaf-tip rounded to obtuse; inflorescences terminal and axillary, often paired; peduncles slender; ripe fruit red   3. *discolor*

1. G. obtusata (Jacq.) Little in Phytologia **17**: 368 (1968).—*Pisonia obtusata* Jacq. (1798). *Torrubia obtusata* (Jacq.) Britton (1904).

Shrub or tree 2·5–8 m. high; leaves suborbicular to oblong-elliptical, cuneate to rounded at base, 3·5–10 cm. long, 2·5–6 cm. broad; panicles slender-branched, 4–6 cm. long including peduncle; male perianth funnel-shaped, 4–5 mm. long; stamens 6–8; female perianth 2·5–3·5 mm. long; ripe fruit red; anthocarp when dry about 4 mm. long.

Occasional in the south-central parishes, in woodlands and thickets in mainly coastal formations on arid limestone and sand; SL–1500 feet; fl. May–July, Nov, fr. July. *H 11722! 11741! P 11335!* Bahamas, Cuba.

2. **G. fragrans** (Dum.-Cours.) Little in Phytologia **17**: 368 (1968).—*Pisonia fragrans* Dum.-Cours. (1814). *Torrubia fragrans* (Dum.-Cours.) Standl. (1916). Beef Wood, Herring Wood.

Shrub to 4 m. or tree to 16 m. high; leaves elliptical to obovate, broad or narrow at base, up to 11 cm. long and 6·5 cm. broad; inflorescences 3–6 cm. long in flower, up to 10 cm. long in fruit; male perianth about 4 mm. long, greenish-yellow; stamens 6–8; female perianth about 3 mm. long, green; fruit about 8 mm. long, sharply ribbed when dry.

Common in woodlands and thickets mostly on limestone; 15–3000 feet; fl. Feb–June, Sept–Nov, fr. June–July, Sept–Jan. *A 11171! 12524! H 9857! 9917! P 8017! Stearn 269!* Tropical S. Amer., West Indies.

3. **G. discolor** (Spreng.) Little in Phytologia **17**: 368 (1968).—*Pisonia discolor* Spreng. (1825). *Torrubia discolor* (Spreng.) Britton (1904).

Erect or straggling shrub 1–4 m. or tree 5–6 m. high; leaves narrowly oblong to elliptical, or rarely suborbicular, very variable even on the same plant, cuneate to rounded at base, paler beneath, up to 11 cm. long and 4 cm. broad; inflorescences 3–7 cm. long in flower, a little longer in fruit; male perianth widely trumpet-shaped, 3–4 mm. long, light green to yellow; stamens 6–8; female perianth about 3 mm. long, green; ripe fruit dull scarlet to bright red, 8–9 mm. long, 6–7 mm. broad, ribbed when dry.

Frequent locally in thickets on arid rocky limestone and on river gravel or old dunes, mostly coastal; SL–750 feet; fl. Apr–Oct, fr. May–Oct. *A 9429! 11261! H 8923! P 23923! 23962!* Bahamas, Greater Antilles; Grand Cayman.

59. PHYTOLACCACEAE

Herbs, climbing or erect shrubs, rarely trees. Leaves alternate, simple, entire; stipules small or absent. Flowers bisexual or unisexual and monoecious, usually actinomorphic, usually bracteate and bracteolate in racemes or cymes. Perianth biseriate or more usually uniseriate; sepals 4–5, more or less connate, imbricate, persistent; petals 5, free, imbricate or absent. Stamens the same number as perianth-segments or more numerous; filaments free or connate at base; anthers 2-locular, opening lengthwise. Ovary superior of 2 or more distinct or connate carpels, 1–many-locular, when 2–more locular each loculus with 1 axile ovule, when 1-locular the ovule(s) 1 or 3–5 and basal; ovules campylotropous or amphitropous; style short or absent; stigmas as many as carpels. Fruit a berry, drupe, achene or capsule. Seeds often arillate; embryo enveloping a large endosperm.

17 genera with over 120 species, widespread but mostly in the American sub-tropics and tropics.

1 Perianth biseriate; stamens 10, unequal, connate at base; ovary 1-locular; stigmas and ovules 3–5; fruit a capsule; seeds arillate (*Stegnospermataceae*)   1. **Stegnosperma**
1 Perianth uniseriate; fruit indehiscent; seeds exarillate:
  2 Ovary 8–12 (–16)-locular with as many styles; ovules 1 in each loculus, axile; stamens 8–20 (*Phytolaccaceae*)   2. **Phytolacca**
  2 Ovary 1-locular; style simple or obsolete; ovule solitary, basal (*Petiveriaceae*):
    3 Fruit dry:
      4 Fruit oblong with 4 hooks at apex; stigma sessile, unilateral, penicillate; stamens usually 8; flowers spicate   3. **Petiveria**
      4 Fruit globose, tuberculate, without hooks; style terminal; stigmas 2; stamens usually 5; flowers racemose   4. **Microtea**
    3 Fruit fleshy, globose:
      5 Stamens 8–16, usually 10; stigma sessile, penicillate; climbing shrub   5. **Trichostigma**
      5 Stamens 4; style short; stigma capitate; shrubby annual herb   6. **Rivina**

1. STEGNOSPERMA Benth. (1844); D. J. Rogers (1949)

1. **S. cubense** A. Rich. in Sagra, Hist. Cuba, Parte 2, **10**: 309 (1845).—*S. halimi-folium* of F. & R. (1914), not Benth. (1844).

Straggling-branched shrub or climber to 8 m. high; leaves elliptical, obtuse, soft, light greyish-green, 2–5 cm. long, 1–3·5 cm. broad; petioles and inflorescences red; racemes 1–10 cm. long, 4–25-flowered; sepals 5, 3 mm. long, reddish in fruit; petals 5, linear to spathulate, white; capsule fleshy, usually 3-valved, dark red, 5–6 mm. long; seeds 1–2 (–3), black, striate; fleshy aril white when young, magenta in ripe fruit.

Rare and local (St. Cath., Manch., St. Thom.), on exposed arid limestone cliffs and in thickets near the sea; 10–330 feet; fl. May, fr. May, Sept, Nov, Feb. *A 9851! 11156! H & B 10522! P 15788!* Mexico to Nicaragua, Greater Antilles.

2. PHYTOLACCA L. (1753)

Inflorescence glabrous; pedicels slender, up to 10 mm. long in fruit; perianth-segments
    about 2 mm. long   1. *rivinoides*
Inflorescence puberulous with short scale-like hairs; pedicels stout, 4–6 (–7) mm. long;
    perianth-segments about 4 mm. long   2. *icosandra*

1. P. rivinoides Kunth & Bouché in Ind. Sem. Hort. Berol. **1848**: 15 (1848).— Jocato.

Tough-stemmed bushy herb 60–120 cm. high; leaves elliptical to ovate-lanceolate, acuminate, 10–20 cm. long, 4–8 cm. broad, covered with minute scattered cystoliths, the tip often calcarate; racemes up to 60 cm. long in fruit, lax-flowered; bracteoles at about the middle of the pedicels; perianth-segments 5, pink; ripe fruit black, depressed, 5–7 mm. in diameter; seeds subreniform, 2–2·2 mm. long.

Occasional, in the central and eastern parishes, in disturbed and waste ground; 500–5000 feet; fl. and fr. most of the year. *A 6659! H 6417! P 16733!* C. and S. Amer., West Indies.

2. P. icosandra L., Syst. Nat. ed. 10, **2**: 1040 (1759).

Shrubby herb like the last; leaves ovate-elliptical, acute or cuspidate; racemes up to 30 cm. long; bracteoles at or above the middle of the pedicels; perianth white at first turning green and then pink in fruit.

Occasional in the central parishes on banks and in open waste places; 50–5000 feet; fl. Dec–Aug, fr. Jan–Aug. *A 6677! J.P. 1266! P 23685!* C. and S. Amer., Bahamas, Greater Antilles, Trinidad; introduced into the Old World tropics.

3. PETIVERIA L. (1753)

1. P. alliacea L., Sp. Pl. **1**: 342 (1753).—Guinea Hen Weed, Strong Man's Weed.

Deeply rooted perennial undershrub with tough stems 60–150 cm. high, the whole plant emitting a heavy garlic odour when broken; leaves elliptical, acuminate, up to 20 cm. long and 7 cm. broad; stipules very small, subulate, caducous; spike elongated, nodding during development, stiff in fruit; bract, bracteoles and perianth-members accrescent, the latter 4, white at first becoming green, the anterior and lateral reflexed, about 4 mm. long; fruit about 1 cm. long.

Locally very common as a weed of semi-shaded roadsides and rough well drained undisturbed ground; 20–1200 feet; fl. and fr. all the year. *A 5480! H 6872! P 25604!* Florida, Mexico to Argentina, Bahamas, West Indies, and now established in some parts of tropical Asia and Africa.

4. MICROTEA Sw. (1788)

1. M. debilis Sw., Nov. Gen. & Sp. Pl.: 53 (1788).

Annual herb 15–45 cm. high; leaves obovate to oblanceolate, 3–7 cm. long, 1·5–3·5 cm. broad; racemes lax; perianth-segments 5, white, 1 mm. long; fruit 1·5 mm. long.

Very rare (West.), a weed near the seashore; fl. and fr. Mar. *H 10214!* Guatemala to Brazil, West Indies, but not recently collected in Cuba.

5. TRICHOSTIGMA A. Rich. (1845)

1. T. octandrum (L.) H. Walt. in Engl., Pflanzenr. **4** (83): 109, f. 31 (1909).—
Rivina octandra L. (1756). Basket Withe, Hoop Withe.

Stems terete with grey bark, trailing or climbing to 7 m. or more, very flexible
with high tensility, the base of petiole persistent and peglike on old stems; leaves
elliptical to lanceolate, acuminate, 5–11 cm. long, 1·5–4·5 cm. broad, reddish when
young; racemes up to 10 cm. long; flowers fragrant; perianth-segments 4, 4–5 mm.
long, at first greenish-white turning red in fruit; ripe fruit blackish-purple, about
5 mm. long.

Common in thickets and woodland margins and on the cays; SL–2800 feet; fl.
all the year, fr. mostly Feb–Aug. *A 6324! 10749! H 11966! P 7795!* Florida,
Mexico to Brazil and Peru, West Indies; Cayman Is.

6. RIVINA L. (1753)

1. R. humilis L., Sp. Pl. **1**: 121 (1753).—Bloodberry, Dogberry.

Taprooted shrubby herb up to 1 m. high; plant variably hairy from nearly
glabrous to softly pubescent; leaves ovate, acuminate, up to 13 cm. long and 6 cm.
broad; perianth-segments 4, white tipped reddish and about 2 mm. long in flower,
green and enlarging to 3 mm. long in fruit; ripe fruit scarlet, about 3 mm. long.

Common as a weed of light open woodlands and shaded waste places, also on
some of the cays; SL–2450 feet; fl. and fr. all the year. *A 5488! H 6854! P 24303!*
Florida and Texas to Argentina, Bahamas, West Indies; Cayman Is; naturalized in
tropical Asia.

60. BASELLACEAE

Perennial herbaceous or subwoody climbers. Leaves alternate, simple, entire,
usually fleshy, without stipules. Flowers bisexual or unisexual, actinomorphic,
small, in axillary or terminal spikes, racemes or panicles, each subtended by a
small bract and 1 or 2 pairs of bracteoles. Perianth uniseriate, the 5 sepals imbricate,
free or basally connate, persistent and often accrescent in fruit. Stamens 5, opposite
the sepals and adnate to their bases; filaments free or joined in a basal ring; anthers
2-locular, versatile, opening lengthwise or by apical pores or slits. Ovary superior,
1-locular; style terminal, usually deeply divided into 3 stigmas; ovule solitary,
basal, campylotropous. Fruit a drupe or utricle surrounded by the often fleshy calyx
or winged bracteoles. Seeds with semi-annular or spiral embryo and usually copious
endosperm.

5 genera with about 20 species in tropical America or Asia (*Basella*).

Filaments straight in bud; flowers sessile in axillary spikes; fruit enclosed by fleshy
   perianth; embryo spirally twisted; endosperm absent or very thin   **1. Basella**
Filaments curved in bud; flowers pedicelled in simple or branched racemes; perianth not
   fleshy in fruit; embryo semi-annular or horseshoe-shaped, curved around the mealy
   endosperm
                                                       2. Anredera

1. BASELLA L. (1753)

1. B. alba L., Sp. Pl. **1**: 272 (1753).—*B. rubra* L. (1753). Country Spinach.

Succulent vine with stems 2 m. or more long; leaves ovate, entire, petioled, 5–7
cm. long; flowers in short or long axillary spikes, white or purplish; fruit green or
purple.

Sparingly cultivated at low elevations. *Robertson UCWI 2278!* Native of tropical
Asia, introduced in the New World.

2. ANREDERA Juss. (1789)

1 Lower bracteoles fused forming a cup at top of pedicel; upper bracteoles flattened in
   flower, orbicular to broadly elliptical, not keeled; flowers white, fragrant; occasionally
   cultivated; native of Brazil; Mignonette Vine   ***A. cordifolia*** (Tenore) Steenis

1 Lower bracteoles free:
  2 Upper bracteoles boat-shaped, longer than perianth and broadly winged in fruit
                                                  1. *vesicaria*
  2 Upper bracteoles oblong, as long as or shorter than perianth, keeled but not winged
                                              2. *leptostachys*

1. A. vesicaria (Lam.) Gaertn. f., Fruct. & Sem. Pl. **3** (2): 176, t. 213 (1807).—
  Basella vesicaria Lam. (1785). *A. scandens* of F. & R. (1914), not *Polygonum
  scandens* L. (1753).

Vine from a tuber-bearing stock; leaves ovate, short- or long-acuminate, up to
9 cm. long and 6 cm. broad; racemes many-flowered, up to 15 cm. long; pedicels
1–2 mm. long; bracteoles about 2 mm. long in flower, the wing developing to about
4 mm. long in fruit; sepals 2 mm. long, white in flower.

Very rare (St. Cath.), climbing on low trees, probably a garden escape; about
100 feet; fl. Oct. *H 12194!* Texas, Mexico to Peru, cultivated in Cuba and introduced into Spain and the subtropics and tropics of the Old World.

2. A. leptostachys (Moq.) Steenis, Fl. Males. **5** (3): 302 (1957).—*Boussingaultia
leptostachys* Moq. (1849).

Glabrous vine with slender pinkish stems and branches up to 8 m. high; roots
tuberous; leaves rather fleshy, ovate to elliptical, acute to acuminate, 2–6 (–8) cm.
long, 1·5–4 (–5) cm. broad, narrowed at the base to a short petiole; racemes longer
than the leaves; pedicels about 1 mm. long; lower bracteoles 0·5–0·6 mm. long;
upper bracteoles 1-nerved, nearly 2 mm. long; flowers white, honey-scented;
sepals 2–2·5 mm. long; fruit not seen.

Rare (St. Andr.), climbing on shrubs and trees, probably a garden escape; 250–
800 feet; fl. Oct–Apr. *A 12730! Campbell 5703! H 11666!* Florida, continental
tropical Amer., West Indies.

61. PORTULACACEAE

Annual or perennial herbs, sometimes undershrubs. Leaves alternate or opposite,
simple, often fleshy; stipules scarious, setose, hairy or absent. Flowers bisexual,
actinomorphic, solitary, cymose, racemose or capitate. Perianth biseriate; sepals
usually 2, herbaceous, free or basally connate, imbricate; petals (2–) 4–6, free or
basally connate. Stamens as many as petals and opposite to them or more numerous,
free, sometimes epipetalous or perigynous; anthers 2-locular, opening lengthwise.
Ovary superior or half-inferior, 1-locular; ovules 1–many, campylotropous, on a
basal or central placenta; style simple below, divided above; stigmas 2–5 (–7).
Fruit a circumscissile or loculicidal capsule, rarely indehiscent. Seed with a curved
embryo surrounding copious endosperm.

19 genera with more than 500 species widely distributed in the subtropics and
tropics, with high concentrations in the Pacific and S. American regions.

Flowers solitary or cymose-capitate; ovary half-inferior; capsule circumscissile; stipules
  scarious or hairy   **1. Portulaca**
Flowers in racemes or panicles; ovary superior; capsule 3-valved; stipules absent
                                                                 2. Talinum

1. PORTULACA L. (1753)

1 Leaves obovate to spathulate, more than 5 mm. broad; leaf-axils with very short inconspicuous hairs   1. *oleracea*
1 Leaves elliptical to linear-lanceolate or subulate, at most 4 mm. broad; leaf-axils with
  long conspicuous persistent hairs:
  2 Branches spreading, numerous, creeping and rooting:
    3 Leaves lanceolate to elliptical or narrowly obovate, mostly less than 5 mm. long and
      2 mm. broad; stems very slender; petals yellow   2. *quadrifida*
    3 Leaves linear-lanceolate, 5–15 mm. long; stems thick; petals purple or crimson
                                                        3. *pilosa*
  2 Branches ascending or erect, not rooting; leaves more or less caducous:
    4 Stems and taproot slender, not very succulent; leaves linear-subulate, 5–8 mm. long;
      petals 2·5–3 mm. long; capsule-cap hemispherical to onion-shaped; capsule-base
      about 1·5 mm. in diameter; seeds black   4. *halimoides*

4 Stems and roots thick and succulent, the latter tuberous; leaves linear-lanceolate, 5–15 (–25) mm. long; petals 5–7 mm. long; capsule-cap a low flattish dome; capsule-base about 3 mm. in diameter; seeds brown   5. *phaeosperma*

1. P. oleracea L., Sp. Pl. **1**: 445 (1753).—Pussley.

Annual diffusely branched herb; leaves softly fleshy, fragile, up to 22 m. broad; flowers in terminal involucrate heads, variable in size up to 14 mm. across when open; petals (4–) 5 (–6), deeply notched, light or dark yellow; stamens (7–) 9–15; stigmas up to 7; capsule-base about 3 mm. in diameter; seeds black.

Very common, a weed of cultivated ground and waste places; SL–1000 feet; fl. and fr. all the year. *H 9993! A. von der Porten!* Native of the Old World tropics, now in all warm countries; Grand Cayman.

2. P. quadrifida L., Mant. Pl.: 73 (1767).

Perennial and variably persistent, usually creeping, rarely erect; leaves thin and red during drought, thicker and green in wet weather, opposite; petals 4, usually yellow, sometimes tinged red; flowers open in the sun from midday; stamens 6–8; stigmas 3–4.

Local (St. Andr., St. Cath., Clar.), a weed of lawns and on bare sandy clay at lagoon and salina margins; SL–700 feet; fl. and fr. most of the year. *H 6944! P 7610!* Native of the Old World tropics, now widespread.

3. P. pilosa L., Sp. Pl. **1**: 445 (1753).—Crimson-flowered Purslane, Kiss-me-quick.

Low somewhat shrubby annual; leaves fleshy, bitter to taste; flowers in terminal heads surrounded by hairs; petals 4–6 mm. long; stamens 15 or more; capsule-base about 2·5 mm. in diameter.

Occasional on alluvial gravel and rocky roadsides; 150–2000 feet; fl. and fr. Jan–June. *A 7251! H 6092! A. von der Porten!* S.E. United States, Mexico to S. Amer., West Indies; Grand Cayman.

4. P. halimoides L., Sp. Pl. ed. 2, **1**: 639 (1762).

Annual; branches usually numerous forming a small bush up to 20 cm. high or rarely spreading; stem often pink; sepals dull pink; petals 4–5, yellow or very rarely white.

Locally common on sand dunes and ridges or limestone gravel in southern coastal areas; SL–100 (–500) feet; fl. and fr. July–Mar. *A 6327! 10007! H 9942! P 23942!* Mexico to Venezuela, West Indies southwards to Guadeloupe, Bonaire, Curaçao; Grand Cayman.

5. P. phaeosperma Urb., Symb. Ant. **4**: 233 (1905).

Perennial or long-lived annual herb with rather few fleshy branches, mostly 5–15 cm. high; petals yellow.

Local (St. Cath.), in crevices of honeycomb limestone rock out-cropping on salinas; SL–5 (–100) feet; fl. and fr. Mar, Sept–Nov. *A 6470! H 9104! P 11670!* Florida, Bahamas, Greater Antilles, Virgin Is., Curaçao; Grand Cayman. A larger variant with reddish petals has been collected (*A 9782! 12886! Yuncker 17294!*) on beaches around Kingston.

P. umbraticola Kunth (*P. lanceolata* Engelm.) was reported from a coastal area in St. Elizabeth by N. L. Britton in 1907 but no specimen has been seen.

P. grandiflora Hook., a usually double magenta-flowered species, is sparingly cultivated and is sporadically established along sandy roadsides.

2. TALINUM Adans. (1763)

Inflorescence loosely paniculate; sepals deciduous; petals 4–5 mm. long, usually pink, rarely yellow   1. *paniculatum*
Inflorescence in a simple or branched raceme; sepals persistent; petals 8–9 mm. long, usually yellow, rarely pink   2. *triangulare*

1. T. paniculatum (Jacq.) Gaertn., Fruct. & Sem. Pl. **2**: 219, t. 128, f. 13 (1791).—
    Portulaca paniculata Jacq. (1760).

Diffusely branched herb up to 120 cm. high; leaves elliptical, fleshy, 4–8 cm.

long, up to 4·5 cm. broad; panicle-branches cymose; pedicels terete; sepals 3–4 mm. long; stamens 15–20; capsule 3–5 mm. in diameter; seeds black.

Occasional on well drained rocky banks; 10–1500 feet; fl. and fr. all the year. *A 6959! Fawcett! P 15879!* S. United States, continental tropical America, West Indies. A variant with variegated leaves is in cultivation in Jamaica.

2. T. triangulare (Jacq.) Willd. in L., Sp. Pl. ed. 4, **2**: 862 (1799).—*Portulaca triangularis* Jacq. (1760).

Annual taprooted herb to 50 cm. or more high; leaves obovate to oblanceolate, fleshy, 2–6 cm. long, up to 2·5 cm. broad; inflorescence few- or many-flowered; pedicels trigonous; sepals 5 mm. long; stamens numerous; style-arms 2–3; capsule about 5 mm. long; seeds black.

On limestone rocks and sandy waste places in arid areas mostly along the south coast; SL–300 (–1000) feet; fl. and fr. all the year. *A 8685! H 9601! P 23698!* S. Amer., Greater Antilles, Lesser Antilles on the drier islands, W. Africa. Local populations seem to be consistent in their flower colours, being in Jamaica very light pink in bud, opening yellow and fading white; in some of the islands of the Lesser Antilles and in W. Africa the flowers are pink throughout.

62. AIZOACEAE

Herbs or low shrubs, often fleshy. Leaves simple, alternate, opposite or whorled, with or without stipules. Flowers bisexual, actinomorphic, in contracted cymes or umbels, or paniculate or solitary in leaf-axils. Perianth 1–several-seriate; segments imbricate (5 in Jamaican genera). Stamens few to many, hypogynous or perigynous; anthers 2-locular, opening lengthwise. Ovary superior to inferior, 1–several-locular; stigmas and style-arms as many as the loculi; style simple or obsolete; placentation usually axile; ovules anatropous or campylotropous. Fruit a capsule (sometimes a berry or nutlike in non-Jamaican genera), enclosed by the persistent calyx; seeds bilateral, mostly cochleate with curved embryo; endosperm mealy.

Over 100 genera with some 600 species; widespread but concentrated in S. Africa.

1 Capsule loculicidal; stamens 3–5; tepals 5, free; leaves alternate or apparently whorled:
  2 Seeds strophiolate with a long filiform appendix curved round the seed, resembling a funicle but not attached to the placenta; flowers clustered in axillary cymes; leaves mostly not very narrow, often stellate-pubescent   **1. Glinus**
  2 Seeds without strophioles; flowers distinctly stalked in terminal cymes, pseudo-racemes or umbels; radical leaves spathulate, cauline narrow, glabrous   **2. Mollugo**
1 Capsule circumscissile; stamens (5–) 6 or more; perianth 3–8-lobed, gamophyllous, corolline or calycine; flowers axillary; leaves opposite or whorled:
  3 Styles 3–5; capsule 3–5-locular with numerous seeds; leaves linear or oblong, equal in each pair   **3. Sesuvium**
  3 Style 1 (–2); capsule 1-locular with 6–8 seeds; leaves obovate to elliptical, unequal   **4. Trianthema**

1. GLINUS L. (1753)

1. G. radiatus (Ruiz & Pav.) Rohrb. in Mart., Fl. Bras. **14** (2): 238, t. 55, f. 1 (1872).—*Mollugo radiata* Ruiz & Pav. (1798).

Taprooted annual herb covered with stellate hairs; branches numerous, spreading from the stock, 10–20 cm. long; leaves elliptical to spathulate, mostly 4–10 mm. long, 3–7 mm. broad; flowers 3–6 in a cluster; perianth-segments hooded, mucronate, 4 mm. long; seeds reniform, reddish-brown, shiny.

Only once collected (Clar.), a weed round edges of pond; 240 feet; fl. and fr. Sept. *H & B 10634!* Brazil, Chile, Cuba, Hispaniola.

2. MOLLUGO L. (1753)

Leaves spread along the stem in whorls, oblanceolate to linear; flowers in small mostly axillary umbels; seeds longitudinally ribbed   *1. verticillata*
Leaves all radical, oblanceolate to obovate; flowers in terminal cymose panicles; seeds muricate   *2. nudicaulis*

1. M. verticillata L., Sp. Pl. **1**: 89 (1753).

Taprooted annual herb; branches numerous and again diffusely branched, spreading from the stock, up to 40 cm. or more long; smaller leaves subulate, larger up to 3 cm. long and 5 mm. broad, glabrous, 4–7 (–10) in a whorl; pedicels slender, with short glandular hairs, up to 14 mm. long; perianth-segments about 2 mm. long, scarious margined; capsule 3-valved, whitish; seeds shiny red or reddish-brown, 0·7 mm. broad.

Locally common (St. Andr., St. Cath., St. Eliz.), in open sandy pastures and waste places; SL–50 feet; fl. and fr. most of the year. *A 8689! P 19763! Skelding UCWI 3527!* Continental tropical Amer., West Indies, W. Africa.

2. M. nudicaulis Lam., Encycl. Méth. Bot. **4**: 234 (1797).

Taprooted annual herb 7–35 cm. high; leaves forming a basal rosette, up to 4 (–6) cm. long and 1·5 (–2) cm. broad, glabrous; scapes erect; pedicels slender, glabrous; perianth-segments about 2 mm. long, greenish outside, white within; seeds dark reddish-brown or black, 0·7 mm. broad.

Not reported in Jamaica since 1781, (St. Cath.), *Masson*. Cuba, Hispaniola, Puerto Rico and thinly scattered through the New World tropics; a common weed of sandy waste places in Asia and Africa.

3. SESUVIUM L. (1759)

Leaves usually more than 2 cm. long; stamens numerous   1. *portulacastrum*
Leaves usually much less than 2 cm. long; stamens 5; annual prostrate herb; flowers 3–4 mm. long; Grand Cayman   *S. maritimum* (Walt.) Britton, Sterns & Poggenburg

1. S. portulacastrum (L.) L., Syst. Nat. ed. 10, **2**: 1058 (1759).—*Portulaca portulacastrum* L. (1753). Seaside Purslane.

Perennial; branches trailing and rooting at nodes; stem and leaves succulent, often reddish; leaves obtuse, 2·5–5 cm. long, up to 1 cm. broad; flowering branches shortly ascending; flowers solitary, pedicellate; perianth deeply 5-lobed, up to about 1 cm. long, green outside, pink to deep rose within; ovary superior, usually 3-locular; seeds black, nearly smooth.

Common on salinas, at mangrove margins and on sandy or rocky brackish wastes; SL–50 feet; fl. and fr. most of the year. *ACH 810! H 8183! P 11497!* General on subtropical and tropical shores.

4. TRIANTHEMA L. (1753)

1. T. portulacastrum L., Sp. Pl. **1**: 223 (1753).—Horse Purslane.

Perennial; branches prostrate, spreading, up to 1 m. long; stem and leaves rather fleshy, sparsely pubescent or glabrous; larger leaves up to 3·5 (–5) cm. long, 1·5–3 cm. broad; flowers usually solitary, partly enclosed by the sheathing leaf-base; perianth pink; capsule 4–5 mm. long, crested, the lid caducous in one piece, 3–4 mm. in diameter; seeds reniform.

Rather rare and local (St. Andr., St. Cath.), on salinas and open waste places near the sea; SL–20 feet; fl. and fr. most of the year. *A 6469! 11907! H 9538!* A common pantropical weed.

63. CARYOPHYLLACEAE

Annual or perennial herbs, very rarely undershrubs. Leaves opposite, verticillate or rarely alternate, simple, entire, often united at the base; stipules scarious, fugacious, or absent. Flowers mostly bisexual, actinomorphic, solitary or in cymose, rarely racemose, inflorescences. Perianth usually biseriate; sepals free or united,

4–5, imbricate, often with membranous margins, persistent; petals as many as the sepals, free, often clawed, entire or bilobed, or absent. Stamens up to 10 in 1 or 2 whorls, free or filaments shortly united; anthers 2-locular, opening lengthwise. Ovary superior, 1-locular with free-central placentation, or placentation basal; ovules 1–many, campylotropous or rarely anatropous; styles (1–) 2–5, free or united below. Fruit a capsule opening by valves or teeth or a utricle or achene. Seeds with a usually curved embryo, with endosperm.

About 80 genera with over 2000 species mostly in north temperate regions.

1 Sepals united into a 5-toothed tube; petals mostly pink, each with a bifid corona-scale; stipules absent; stamens 10; styles 3 (–4), free; capsule dehiscent at apex by 6 teeth
    1. Silene
1 Sepals free or nearly so; petals white, without corona-scales:
  2 Styles fused below, 3; stipules present, small, scarious; leaves rotundate, subcordate at base; stamens 5 or fewer, shortly connate at base; capsule 3-valved **2. Drymaria**
  2 Styles free; stamens 10 or fewer:
    3 Stipules present, small, scarious; leaves linear, some clustered in the axils of other leaves; inner whorl of stamens, if present, opposite the entire petals and shortly adnate to them at the base; styles 5; capsule 5-valved **3. Spergula**
    3 Stipules absent; leaves not clustered:
      4 Petals entire; styles 3; capsule 6-toothed **4. Arenaria**
      4 Petals bilobed:
        5 Styles 5; capsule cylindrical, slightly curved, 10-toothed, about twice as long as calyx, many-seeded; plant glandular-pubescent **5. Cerastium**
        5 Styles 3; capsule subglobose, 6-valved, a little longer than the calyx; plant not glandular, except the calyx; stem with a line of hairs along each internode
    6. Stellaria

1. SILENE L. (1753)

Plant villous; inflorescence racemose   1. *gallica*
Plant glabrous; inflorescence corymbose   2. *armeria*

1. S. gallica L., Sp. Pl. **1**: 417 (1753).

Herb 30–50 cm. high; leaves sessile, oblanceolate, about 2 cm. long; calyx 8 mm. long, enlarging slightly in fruit; petals mostly pink, edged with white; capsule 8 mm. long.

Local (St. Andr., St. Thom.), by pathsides; 5000–5500 feet; fl. and fr. Apr–July. *ACH 860! Bengry! J.P. 1403! Robertson UCWI 1343!* Native of C. & S. Europe, now widespread, Hispaniola.

2. S. armeria L., Sp. Pl. **1**: 420 (1753).

Annual taprooted herb (10–) 30–75 cm. high; internodes viscous in the upper half; leaves sessile, broadly lanceolate, subcordate, 3–5 cm. long; calyx 16–18 mm. long, usually dull pink; petals bright rose-pink; capsule 15–16 mm. long; seeds suborbicular, radially striated, dark reddish-brown.

Local (St. Andr., St. Thom.), a weed of open earth banks and cultivations; 3500–5200 feet; fl. and fr. most of the year. *A 7390! H 9598! P 6014!* Native of C. & S. Europe, now widespread.

2. DRYMARIA Willd. (1819)

1. D. cordata (L.) Willd. in Roem. & Schult. in L., Syst. Veg. ed. nov. **5**: 406 (1819).—West Indian Chickweed.

Slender-stemmed much branched trailing herb with fragile nodes; stem glabrous or if puberulous also with a few gland-tipped hairs; leaves very small or up to 2 cm. broad; flowers in terminal cymes; pedicels usually densely glandular at least in the lower two-thirds; sepals glabrous or with a patch of short stiff hairs in the middle and also glandular, 2–3·5 mm. long; petals bifid; stamens often 3; capsule shorter than the calyx, shortly stalked; seeds 2–8, usually about 4.

Very common as a weed of undisturbed ground, often in shady or damp places; 200–4000 feet; fl. and fr. all the year. *A 5519! H 8444! P 24003!* Tropics generally.

3. SPERGULA L. (1753)

1. S. arvensis L., Sp. Pl. **1**: 440 (1753).

Annual herb 8–40 cm. high; stems erect, branched, with a line of pubescence along one side; leaves channelled beneath, fleshy, obtuse, 1–3·5 cm. long, up to 1 mm. broad, more or less glandular-sticky; stipules deciduous; pedicels 1–2·5 cm. long; sepals 2·5–3 mm. long; petals a little longer than the sepals; capsule ovoid, longer than the sepals; seeds black, tuberculate or papillose, about 1 mm. long.

Very rare (St. Thom.), a weed of cultivated ground, reported to be calcifuge; 4000–5500 feet; fl. and fr. Nov. Not recently found, *Harris!* General in temperate regions, tropical mountains.

4. ARENARIA L. (1753)

1. A. lanuginosa (Michx.) Rohrb. in Mart., Fl. Bras. **14** (2): 274, t. 63 (1872).

Slender-branched straggling herb, rarely more or less erect, the stems puberulous, tufted from a perennial stock, up to about 60 cm. long; leaves linear to lanceolate, the basal often spathulate, whitish papillose-punctate on both surfaces, up to 3 cm. long; pedicels 10–28 mm. long, mostly solitary in the axils of leaves; sepals keeled, about 4 mm. long; petals about as long as the sepals; seeds lenticular, shiny black or reddish-brown, obscurely marginate, less than 1 mm. broad.

Locally common (St. Andr., Port., St. Thom.), a weed of sheltered banks and pathsides; 1900–4300 (–5000) feet; fl. and fr. June–Sept, Dec–Mar. *A 5702! P 19757! Skelding UCWI 6704!* Virginia to temperate S. Amer., Bermuda, Cuba, Hispaniola.

5. CERASTIUM L. (1753)

1. C. glomeratum Thuill., Fl. Paris, ed. 2: 226 (1799).—*C. viscosum* of F. & R. (1914), not L.

Annual, viscous-glandular, with more or less tufted spreading or ascending shoots 5–25 (–40) cm. long; leaves sessile, broadly ovate to oblanceolate, with long white hairs, 1–2 cm. long; flowers shortly pedicelled; sepals hairy, 4–5 mm. long; petals about as long as the sepals; capsule 9 mm. or more long; seeds compressed, reddish-brown, tuberculate, finely marginate.

Locally common (St. Andr., Port., St. Thom.), a weed of open banks, pathsides and mountain pastures; 3000–7400 feet; fl. and fr. Apr–Sept, Dec–Feb. *A 5741! J.P. 2028! P 24610!* General in temperate regions, tropical mountains, ? Hispaniola.

6. STELLARIA L. (1753)

1. S. media (L.) Vill., Hist. Pl. Dauph. **3**: 615 (1789).

Annual straggling-branched herb; stems soft, up to about 40 cm. long; leaves ovate to broadly elliptical, acute, petiolate or the upper (bracts) sessile, 1–2 (–3) cm. long, glabrous or thinly ciliate, papillose-punctate; pedicels recurved during fruit maturation, finally erect; sepals obtuse, membranous-margined, glandular with long pluricellular hairs, 3–4 mm. long; petals slightly shorter than sepals; stamens 3 or more; seeds lenticular, verrucose, shallowly notched, reddish-brown, about 1 mm. broad.

Local (St. Andr., Port., St. Thom.), a weed of damp shady banks and waste places; 2750–4700 feet; fl. and fr. Aug–Sept, Dec–Mar. *A 6000! H 10544! P 24637!* General in N. temperate regions, tropical mountains, Cuba, Hispaniola.

64. CACTACEAE

Succulent herbs, shrubs or trees, the stems often cylindrical and fluted or with flattened joints. Leaves alternate, flat and fleshy or awl-shaped or scalelike or much more usually altogether absent, mostly with clusters of radiating spines or bristles in the axils (areoles) or at the nodes. Flowers borne on the areoles, mostly bisexual, actinomorphic, usually solitary. Perianth indistinctly differentiated into sepals and petals borne on a short or long receptacle-tube. Stamens numerous, arranged spirally or in groups on the inner face of the hypanthium, free or some adnate to bases of petals; anthers 2-locular, opening lengthwise. Ovary inferior, sometimes sunken in the stem, 1-locular; ovules anatropous, numerous, on 3 or more parietal placentas; style simple; stigmas as many as carpels, radiating. Fruit a juicy berry and often spiny or bristly or rarely dry and dehiscent. Seeds small, hard, with straight or curved embryo, often arillate or strophiolate; endosperm present or absent.

Genera mostly ill-defined and of uncertain number; species about 1800, almost all native in subtropical and tropical America.

1 Leaves broad and flat, more or less persistent; flowers pedicellate   1. **Pereskia**
1 Leaves small and terete or vestigial and early deciduous or absent; flowers sessile:
  2 Glochids[1] present; stems segmented, the units more or less flattened, areolate but not ribbed; branched terrestrial shrubs or trees   2. **Opuntia**
  2 Glochids absent; stems not uniformly segmented, spirally tuberculate, radially fluted and ribbed or if terete flat or smooth then plants epiphytic or climbing:
    3 Epiphytic or climbing plants often producing adventitious and aerial roots:
      4 Plants evidently spiny, scrambling and more or less monopodial; stem 3- or more-angled; flowers large, nocturnal and strongly fragrant; fruit 5 cm. or more long:
        5 Receptacle-tube bearing broad leafy scales without spines, hairs or bristles in their axils; stems mostly 3-gonous   3. **Hylocereus**
        5 Receptacle-tube with non-leafy scales usually with spines, hairs or bristles in their axils; stems 8–13-ribbed   4. **Selenicereus**
      4 Plants not evidently spiny but perhaps with tufts of minute pungent hairs on young parts or juveniles; habit more or less sympodial; fruits up to about 1 cm. long, white, green or pinkish:
        6 Stem terete; receptacle-tube scarcely produced and not narrowed beyond the ovary; flowers small; fruit not scaly   5. **Rhipsalis**
        6 Stem at least in younger parts flattened; receptacle-tube produced and narrowed beyond the ovary:
          7 Flowers small, the receptacle-tube shorter than the limb; stigma-lobes 3–5; fruits scaly; wild epiphytic plants   6. **Disocactus**
          7 Flowers large, nocturnal, fragrant, the receptacle-tube elongated; stigma-lobes numerous; fruit not formed in Jamaica; cultivated ornamental; native of Mexico *Epiphyllum oxypetalum* (DC.) Haw.
    3 Terrestrial spiny plants normally rooting from the base only:
      8 Stems globular or shortly cylindrical, not branched except occasionally as basal propagation; flowers diurnal:
        9 Stem ridged, not tuberculate; flowers arising in a dense terminal mass of woolly hairs and reddish bristles (cephalium) capping the plant   7. **Melocactus**
        9 Stem tuberculate, each tubercle crowned by a tuft of spines accompanied by radiating bristles; flowers arising laterally below the apex in axils or tubercles, a cephalium not developing   8. **Mammillaria**
      8 Stems more or less elongated, upright and columnar, sometimes branched, ridged; flowers nocturnal (in Jamaica):
        10 Receptacle-tube in the proximal part and ovary naked or with a few scales which occasionally bear hairs in their axils, thus fruit smooth even at first:
          11 Flowering and non-flowering areoles similar; perianth 20 cm. or more long, the inner segments white; sparingly cultivated ornamental; native of S. Caribbean and northern S. Amer.   *Cereus hexagonus* (L.) Mill.
          11 Flowering and non-flowering areoles dissimilar, the flowering areoles long-woolly; perianth 5–6 cm. long, the inner segments pinkish to greenish-yellow; stems 10–12-ribbed; spines yellow   9. **Cephalocereus**
        10 Receptacle-tube with conspicuous scales and it and the ovary with felted bristly areoles, thus fruit not at first smooth; stems mostly 8–12-ribbed; spines grey:
          12 Perianth up to about 8 cm. long, the inner segments pinkish or green; branches mostly 10–15 cm. thick; longer spines 25–40 mm. long; ripe fruit ellipsoid, red, with clusters of spines at first   10. **Stenocereus**
          12 Perianth up to 20 cm. long, the inner segments white; branches mostly 4–5 cm. thick; longer spines 20–25 mm. long; ripe fruit globose, orange, at first tuberculate and with fleshy bracts   11. **Harrisia**

---

[1] Short *barbed* bristles accompanying spines or present even when spines are lacking.

1. PERESKIA Mill. (1754)

Petals usually white; scrambling or climbing shrub; leaves mostly 3–7 cm. long
1. *aculeata*
Petals usually pink; erect shrub or small tree; leaves mostly 8–15 cm. long   2. *grandifolia*

1. P. aculeata Mill., Gard. Dict. ed. 8 (1768).—*Cactus pereskia* L. (1753). West Indian Gooseberry.

Branches up to 3 (–10) m. long with long spines; younger leafy parts with short spines; leaves ovate or elliptical, fleshy, short-petioled; flowers numerous in paniculate cymes; petals numerous, spreading, about 2 cm. long; ripe fruit yellow, globose, bearing leafy perianth-members towards the apex, 1–2 cm. in diameter.

Occasional in cultivation at low elevations; fl. June, Oct–Nov, fr. Mar, Oct. *Harris!* Native of tropical Amer., now widespread in the tropics.

2. P. grandifolia Haw., Suppl. Pl. Succ.: 85 (1819).

Tree up to 5 m. high with a trunk to 10 cm. thick, covered with long spines; leaves oblong, acute or obtuse, short-petioled; flowers clustered; petals about 2 cm. long; fruit pear-shaped; seeds black.

Rare as an introduced ornamental at low elevations; fl. June. *P 15400!* Native of Brazil, widely cultivated.

2. OPUNTIA Mill. (1754)

1 Areoles without large spines or rarely solitary or small clusters of spines developed:
  2 Petals erect, crimson; stamens longer than petals   1. *cochenillifera*
  2 Petals spreading, yellow fading pinkish; stamens shorter than petals; arborescent shrub with stem-joints up to 50 cm. or more long; fruit 5–9 cm. long, usually red, edible; sparingly cultivated; original locality unknown; Indian Fig.
O. ficus-indica (L.) Mill.
1 Areoles consistently with long spines; stamens shorter than petals:
  3 Plant developing an erect unjointed trunk; petals orange-yellow turning red, about 1 cm. long; joints of branches narrowly oblong or falcate, up to 50 cm. or more long; plants up to 4 (–5) m. high, the trunks usually solitary but sometimes several together
2. *spinosissima*
  3 Plants not developing simple unjointed trunks; petals light yellow sometimes fading pinkish, 2 cm. or more long; joints of stem obovate to oblong; plants often forming low thickets:
    4 Spines thick, the larger often flattened and curved, yellow more or less mottled brown; joints grey-green, glaucous; fruit pear-shaped to subglobose, purplish, smooth
3. *dillenii*
    4 Spines slender, needlelike, terete, white, light grey or light yellow, straight; fruit red, prickly:
      5 Spines yellowish at least at fruit, becoming grey, often reflexed, mostly 3–5 at each areole; joints oblong to elliptical, 8–18 cm. long   4. *tuna*
      5 Spines whitish, spreading, mostly 2 at each areole; joints obovate, 7–13 cm. long
5. *jamaicensis*

1. O. cochenillifera (L.) Mill., Gard. Dict. ed. 8 (1768).—*Cactus cochenillifer* L. (1753). *Nopalea cochenillifera* (L.) Salm-Dyck (1850). Cochineal Cactus, Roast Pork, Smooth Pear.

Shrubby much-branched plants up to 4 m. high; small leaves soon falling; joints up to about 45 cm. or more long; areoles distant, small, roundish; flowers developing singly along the upper margins of joints, 5·5–7 cm. long overall; fruit rarely formed, about 5 cm. long, red.

Common in gardens at low elevations; fl. sporadically. *H 5614! P 9832! Robertson UCWI 2057!* Native probably of Mexico, now widespread in the tropics.

2. O. spinosissima Mill., Gard. Dict. ed. 8 (1768).—*Consolea spinosissima* (Mill.) Lemaire (1862). Prickly Pear Tree.

Shrub and then eventually a tree with a blackish scarred and spiny trunk up to about 30 cm. thick; branches again branched with somewhat pendulous joints; areoles numerous, close; spines short or up to 5 cm. or more long, slender, greyish;

perianth 2·5–3 cm. in diameter; ovary and receptacle fusiform, 3–8 cm. long; fruit oval, pendulous, 6–8 cm. long.

Locally common in coastal parts of the southern parishes, on arid exposed limestone or gravel; SL–200 feet; fl. Jan–Sept, fr. Sept–Feb. *H & P 13935! Robertson UCWI 2371!* Cayman Brac.

3. O. dillenii (Ker-Gawl.) Haw., Suppl. Pl. Succ.: 79 (1819).—Prickly Pear, Seaside Tuna.

Low shrubby branched plant mostly 1–2 m. high; joints 7–40 cm. long; areoles rather few, slightly raised, with 1–4 (–10) spines up to 7 cm. long; petals 4–5 cm. long, light yellow sometimes tinged salmon-pink towards base; ripe fruit 5–7·5 cm. long.

Frequent on sandy upper beaches, salina margins, in arid scrub thickets and on some of the cays; SL–25 (–400) feet; fl. and fr. most of the year. *A 12277! Asprey UCWI 2066! P 23688!* S.E. United States, Mexico to northern S. Amer., West Indies; Grand Cayman; also in Canary Is., S. India and Australia where it has become known as Pest Pear.

4. O. tuna (L.) Mill., Gard. Dict. ed. 8 (1768).—Tuna.

Shrubby with spreading branches, up to about 1 m. high; areoles numerous; spines up to 3·5 (–5) cm. long; flowers light yellow, about 5 cm. across; ripe fruit obovoid, deep crimson, about 3 cm. long; seeds 3–4 mm. in diameter.

Uncommon (St. Andr., St. Thom.), in rocky or sandy waste places, gullies and scrubby thickets; SL–400 feet; fl. Mar–Sept, fr. May–July. *H 6947! P 23508! Robertson UCWI 5252! 9297!* Cayman Is.; cultivated elsewhere and naturalized in S. Europe.

5. O. jamaicensis Britton & Harris in Torreya **11**: 130 (1911).

More or less erect branched shrub up to 1 m. high; areoles about 2·5 cm. apart; spines up to about 2·5 cm. long; flowers light yellow, 4–5 cm. across, with usually 14–16 petals; ripe fruit pear-shaped, red, 3·5–4 cm. long; seeds about 4 mm. in diameter.

Very locally abundant (St. Cath., Manch.), forming pure stands near mangrove and logwood thickets and salinas in heavy silt alluvium; SL–25 feet; fl. and fr. sporadically. *H & B 10887!* Endemic.

3. HYLOCEREUS (A. Berger) Britton & Rose (1909)

1. H. triangularis (L.) Britton & Rose in Contrib. U.S. Nat. Herb. **12**: 429 (1909).—*Cereus triangularis* (L.) Haw. (1812). *Cactus triangularis* L. (1753). God Okra, Prickle Withe.

Much-branched climber or creeper with 3 (–5)-angled stems 3–4 cm. broad; roots adventitious, woody; areoles about 2 cm. apart with 6–8 spines; flowers 20 cm. or more long, green outside, white within; scales on receptacle-tube 2–5 cm. long; ripe fruit ovoid, crimson, about 10 cm. long.

Locally common, in thickets, on rocks and on large old trees; SL–4000 feet; fl. June–Dec, fr. Sept–Oct. *A 7280! C. B. Lewis! Robertson UCWI 6781!* Endemic.

4. SELENICEREUS (A. Berger) Britton & Rose (1909)

1. S. grandiflorus (L.) Britton & Rose in Contrib. U.S. Nat. Herb. **12**: 430 (1909). —*Cactus grandiflorus* L. (1753). *Cereus grandiflorus* (L.) Mill. (1768). Queen-of-the-Night.

Habit like the last; stem with (4–) 7–8 or more ribs, 1·5–2·5 cm. broad; areoles approximated, with 5–12 spines; flowers about 20 cm. long, pinkish-buff outside, the outer perianth members yellow within, the inner white; filaments and style declinate; ripe fruit ovoid, yellow or orange, hairy at first, 5–8 cm. long.

Common, climbing on trees and on rocks; SL–2500 feet; fl. and fr. most of the year. *H & P 14984! Yuncker 17963!* Cuba.

5. RHIPSALIS Gaertn. (1788) nom. cons.

1. R. baccifera (J. S. Mill.) Stearn in Cact. Journ. **7**: 107 (1939).—*Cassytha baccifera* J. S. Mill. (1771–77). *R. cassutha* Gaertn. (1788). Currant Cactus, Mistletoe.

Pendulous epiphyte up to 2 m. long, with forking or whorled branches up to about 4 mm. thick; joints terete, 4–8 cm. long; flowers white; perianth-members 6–10, the inner petaloid and up to 4 mm. long; stamens 9–12; fruit globose, whitish-translucent, about 6 mm. long; seeds oblong, curved, black, shiny, 1·1–1·2 mm. long.

Common, mostly in rather wet areas, on the trunks and branches of trees or shrubs and occasionally on rocks; 500–4900 feet; fl. Oct, Jan–Mar, fr. most of the year. *A 8918! Barkley & Proctor 22J339! Robertson UCWI 2496!* Widespread in the tropics.

6. DISOCACTUS Lindl. (1845); Kimnach (1961)

Branches mostly 3–6 cm. broad; flowers about 15 mm. long, the receptacle-tube constricted above ovary for 4–5 mm.; tepals 10; berry ovoid, about 1 cm. long, yellowish-green turning pink   1. *alatus*
Branches mostly 5–40 mm. broad; flowers 6–9 (–12) mm. long, the recptacle-tube constricted above ovary for 1–2 mm.; tepals 6–9; berry subglobose, 4–8 mm. long, white or pink   2. *ramulosus*

1. D. alatus (Sw.) Kimnach in Cact. & Succ. Journ. Amer. **33**: 14 (1961).—*Cactus alatus* Sw. (1788). *Rhipsalis alata* (Sw.) K. Schum. (1890).

Pendulous epiphyte up to 5 m. long; joints compressed, 20–40 cm. long, irregularly notched or crenate; flowers white or yellowish, borne on the margins of joints; petaloid tepals lanceolate, acute; seeds ovoid to pyriform, black, not shiny, about 1·3 mm. long, punctulate.

Occasional in the central and western parishes, on trees and limestone ledges; 400–2350 feet; fl. Oct, fr. sporadically. *H 7619! P 9953!* Endemic.

2. D. ramulosus (Salm-Dyck) Kimnach in Cact. & Succ. Journ. Amer. **33**: 14 (1961).—*Cereus ramulosus* Salm-Dyck (1834). *Rhipsalis jamaicensis* Britton & Harris (1909).

Pendulous epiphyte like the last but smaller in all its parts; juvenile plants have tufted linear-oblanceolate shoots with tufts of long hairs at the nodes; margins indistinctly notched; flowers greenish-white; petaloid tepals oblong, obtuse; seeds as above, minutely verrucose or smooth.

Frequent mainly in wetter areas, on trees, cliffs and ledges; 400–3300 feet: fl. Feb–Mar, fr. Jan–Aug. *A 8981! H & B 8562! H & P 14200!* Mexico to Brazil and Peru, Hispaniola.

7. MELOCACTUS Link & Otto (1827) nom. cons.

1. M. communis Link & Otto in Verh. Beförd. Gartenb. **3**: 417, t. 11 (1827).—*Cactus melocactus* L. (1753). Turk's Cap, Turk's Head.

Stem subglobose to eventually barrel-shaped, up to about 40 cm. or rarely more high and about 30 cm. in diameter; ribs (10–) 12–13 (–14), 2–3 cm. deep; spines in areoles 10–12, 2–5 cm. long, the innermost straight, the marginal curved, thicker at base; cephalium up to about 10 cm. broad; perianth bright rose-pink; fruit clavate, about 5 cm. long, smooth, shiny, brilliant light magenta; seeds black shiny, about 1 mm. long.

Local in the southern parishes, in sand or gravel or on rocks in arid places; SL–400 feet; fl. Mar–Sept, fr. most of the year. *P 8674! Robertson UCWI 5459!* Endemic, but doubtfully distinct from Cuban and Haitian plants.

8. MAMMILLARIA Haw. (1812) nom. cons.

1. M. aff. columbiana Salm-Dyck, Cact. Hort. Dyck. **1849**: 99 (1850).

Stem cylindrical, 12–15 cm. high, about 6 cm. in diameter; tubercles with (2–) 3–4 central brownish-yellow spines 6–8 mm. long and about 20 white radial spines 4–6 mm. long; axils of tubercles with whitish fuzz; flowers pink, 4–5 mm. in diameter; fruit clavate, orange-red, 7–8 mm. long; seeds brown, pyriform, finely punctate, 1·1–1·2 mm. long.

Very rare (Trel.), on vertical limestone cliff; 1500 feet; fl. Dec–Mar, fr. Apr–July. *Proctor!* Very close allies range from Mexico to Venezuela.

9. CEPHALOCEREUS Pfeiff. (1838)

1. C. swartzii (Griseb.) Britton & Rose in Contrib. U.S. Nat. Herb. **12**: 420 (1909).—*Cereus swartzii* Griseb. (1860). *Pilosocereus swartzii* (Griseb.) Byl. & Rowl. (1957).

Columnar cactus with few erect branches, up to 4 (–5) m. high; flowers protandrous; ripe fruit brilliant magenta both outside and within, subglobose, 3–3·5 cm. long and broad; seeds black, shiny, obliquely pyriform, faintly punctate, 1·8–2 mm. long, sinking in water when fresh.

Locally common in coastal areas on sand, gravel or rocks; SL–300 feet; fl. and fr. Aug–Apr. *H & P 13816! Robertson UCWI 1084!* Cayman Brac.

10. STENOCEREUS (A. Berger) Riccobono (1909)

1. S. hystrix (Haw.) Buxb. in Bot. Studien **12**: 100 (1961).—*Cereus hystrix* Haw. (1819). *Lemaireocereus hystrix* (Haw.) Britton & Rose (1909). *Ritterocereus hystrix* (Haw.) Backeberg (1944). *Cereus peruvianus* of F. & R. (1926), partly. Dildo, Dildo Pear.

Columnar cactus like the last, to 4 (–5) m. high; branches 10–15 cm. in diameter; spines short and long, 14–16 in each areole, straight; ripe fruit bright scarlet, about 4 cm. long and 3·5 cm. broad, covered at first with clusters of about 8 small spines but these falling; seeds similar to the above, 1·6–1·8 mm. long, sinking in water when fresh.

Locally abundant in southern coastal areas, in sand, gravel, along salina margins and on arid limestone hillsides; SL–100 feet; fl. Oct–May, fr. Oct–June. *P 9426! Robertson UCWI 1085!* Greater Antilles.

11. HARRISIA Britton (1908)

1. H. gracilis (Mill.) Britton in Bull. Torr. Bot. Club **35**: 563 (1908).—*Cereus gracilis* Mill. (1768). Torchwood Dildo.

Shrubby cactus 2–6 m. high with mostly divergent-ascending branches 4–5 cm. in diameter; spines dark or pale with dark tips, unequal, 8–11 (–16) in each areole, straight; ripe fruit yellow with green fleshy bracteoles, later orange, 5–6·5 cm. long and 4·5–5 cm. broad; pulp white; seeds black, shiny, obliquely pyriform, minutely punctate at the narrow hilum end, coarsely papillose at the broader end, 2–2·4 mm. long, floating in water when fresh.

Rather common with other cacti in thickets mainly in arid areas of the south coast; SL–700 feet; fl. June–Sept; fr. July–Nov. *H 6946! H & P 13936! Robertson UCWI 2333!* Cayman Is.

65. MAGNOLIACEAE

Trees or shrubs. Leaves alternate, simple, mostly entire; stipules large, deciduous, leaving scars. Flowers mostly large and usually bisexual, actinomorphic, solitary, terminal or axillary. Perianth of 6–many free imbricate tepals cyclically arranged in 2 or more series; outer tepals sometimes sepallike. Stamens numerous, free, hypogynous and spirally disposed on the proximal part of the floral axis; anthers 2-locular, opening lengthwise. Carpels few to many, spirally arranged on the often elongated distal part of the floral axis; ovaries 1-locular; ovules 2–several on ventral placentas; style and stigma solitary. Fruiting carpels follicular, samaroid or baccate, rarely united. Seeds with very small embryo and large oily endosperm, in dehiscent fruits, suspended by a threadlike funicle.

12 genera with over 200 species in the temperate northern hemisphere and in the tropics of America and E. Asia; none native in Jamaica.

1. MICHELIA L. (1753)

Stipules adnate to petioles of young leaves; flowers axillary; anthers opening laterally; gynoecium stalked; fruiting carpels follicular.

1. M. champaca L., Sp. Pl. **1**: 536 (1753).—Champac.
    Tree to 10 m. or more high; twigs olive-grey, pubescent, with ring scars and conspicuous lenticels; leaves ovate-lanceolate, long-acuminate, up to about 25 cm. long and 10 cm. broad; flowers light yellow, fragrant; tepals 12 or more, linear, the longer about 4 cm. long; follicles 2–2·5 cm. long; seeds rounded-angular, pink, 6–8 mm. long.
    Occasionally planted as an ornamental; 600–3700 feet; fl. June–Dec, fr. Aug–Feb. *Chalmers UCWI 4612! Deans!* Native of tropical Asia.

66. ANNONACEAE

Trees, shrubs or woody climbers; wood and leaves aromatic. Leaves alternate, simple, entire, without stipules. Flowers bisexual, very rarely unisexual, actinomorphic, solitary or in few-flowered clusters, often fragrant. Perianth typically 3-seriate, hypogynous; sepals 3, free or connate, valvate; petals 6 in 2 series, valvate or imbricate in each series, the inner series sometimes small or wanting, the outer often fleshy. Stamens numerous; filaments short and thick, inserted spirally on an often enlarged axis below the carpels; anthers 4-locular at anthesis, opening lengthwise, the connective variously produced and enlarged. Carpels few or numerous, free or united; styles separate or stigmas sessile; ovules 1–many, basal or parietal, anatropous. Fruit of free and often stalked or variously connate carpels forming a berrylike aggregate. Seeds large with hard shiny testa, often arillate with large endosperm and very small embryo.

100 or more genera with about 2000 mostly tropical species.

1 Carpels united; fruit large, unitary; petals distinct   **1. Annona**
1 Carpels free; fruiting carpels distinct, more or less stalked, berrylike or drupaceous:
  2 Connective dilated and truncate beyond the anther; ovules several in each carpel
                                                  2. Xylopia
  2 Connective produced beyond the anther into a long tapering appendage:
    3 Ovule 1 in each carpel; flowers less than 1 cm. long   **3. Oxandra**
    3 Ovules several in each carpel; tree to 15 m. high with drooping branches; leaves thin, in two rows, up to 20 cm. long and 8 cm. broad; flowers greenish-yellow, showy, very fragrant with 6 petals 7–12 cm. long; ripe carpels distinctly stalked, black; cultivated; native of Malaya; Ylang Ylang
                            Cananga odorata (Lam.) Hook. f. & Thoms.

1. ANNONA L. (1753)

1 Petals 6, the 3 inner conspicuous, ovate; leaves glabrous:
  2 Fruit smooth   **1.** *glabra*
  2 Fruit spiny:
    3 Spines of fruit large, curved   **2.** *muricata*
    3 Spines of fruit short, straight   **3.** *montana*
1 Petals 3, the inner reduced to minute scales or wanting; leaves mostly at least puberulous beneath:
  4 Fruit with hooked tubercles; leaves downy beneath:
    5 Petals oblong; leaves elliptical   **4.** *praetermissa*
    5 Petals ovate; leaves oblong-elliptical   **5.** *jamaicensis*

4 Fruit without hooked tubercles; petals oblong, keeled on the inner side:
  6 Fruit marked out with rounded bulging segments; leaves pubescent on nerves beneath
    or glabrous   6. *squamosa*
  6 Fruit smooth:
    7 Leaves elliptical, velvety beneath   7. *cherimola*
    7 Leaves elliptic-oblong, glabrate beneath   8. *reticulata*

1. A. glabra L., Sp. Pl. **1**: 537 (1753).—*A. palustris* L. (1762). Pond Apple.

Tree (2–) 3–8 (–12) m. high, the trunk narrowly buttressed at the base; leaves oblong-elliptical, acute or shortly acuminate, 7–15 cm. long, up to 6 cm. broad; pedicel curved, expanded distally; sepals 4·5 mm. long, 9 mm. broad, apiculate; outer petals valvate, ovate-cordate, cream-coloured with a crimson spot at base within, 2·5–3 cm. long, 2–2·5 cm. broad; inner petals subimbricate, shortly clawed, 2–2·5 cm. long, 1·5–1·7 cm. broad, whitish outside, dark crimson within; stigmas sticky, deciduous; fruit up to 12 cm. long, 8 cm. broad, yellow outside when ripe, pulp pinkish-orange, rather dry, pungent-aromatic; seeds light brown, 1·5 cm. long, 1 cm. broad.

Occasional at the margins of swamps near the sea, local in marsh forest inland; SL–50 feet; fl. Mar–Apr, fr. Apr–Aug. *A 7371! Powell 303!* Coasts of tropical Amer. and W. Africa.

2. A. muricata L., Sp. Pl. **1**: 536 (1753).—Sour Sop.

Tree 5–8 m. high; leaves obovate-oblong, acuminate, 8–25 cm. long, 3·5–8 cm. broad; pedicel curved, the flower pendent; outer petals greenish-yellow, 2 mm. thick, shortly acuminate, 3–3·5 cm. long, 2–3 cm. broad, inner nearly as long as the outer; fruit obliquely ovoid, 15–20 cm. long, up to 10 cm. broad, green when ripe, pulp white, acid; seeds like the last, black or brown.

Commonly cultivated; 500–2000 feet; fl. June–Nov, fr. Sept–Jan. *A 8233! Chalmers UCWI 2700! P 22871!* Tropics, mostly Amer., introduced into W. Africa.

3. A. montana Macf., Fl. Jam. **1**: 7 (1837).—Mountain Sour Sop.

Tree about 5 m. high; leaves as above, 7–14 (–22) cm. long, 3–6 (–7·5) cm. broad, with domatia; outer petals olive-green to yellow, as above; fruit oblate-subglobose, 5–12·5 cm. in diameter, dry; seeds orange-brown, up to 2 cm. long, 10–12 mm. broad.

Uncommon on wooded hillsides and in pasture margins; 500–1700 feet; fl. and fr. May–Oct. *H 9446! J.P. 1457! P 19812!* Venezuela and the Guianas to Brazil and Peru, West Indies.

4. A. praetermissa F. & R. in Journ. Bot. **52**: 74 (1914).—Wild Sour Sop.

Tree up to 10 m. high; leaves elliptical or ovate-elliptical, 12–18 cm. long, 4–8·5 cm. broad; petals 2–2·5 cm. long, 5 mm. broad; fruit globose, about 6 cm. in diameter, dull brownish-green with soft yellow pulp; seeds 1·7 cm. long, 1 cm. broad.

Rare (St. Andr., St. Thom.), on wooded hillsides; 1000–2500 feet; fl. May–June, fr. June–Oct. *H 11648! P 19809!* Endemic.

5. A. jamaicensis Sprague in Bull. Herb. Boiss. sér. 2, **5**: 701 (1905).

Tree 6–12 m. high; leaves oblong-elliptical to broadly ovate, acuminate, 10–24 cm. long, 5–11 cm. broad; petals ovate, densely rusty-tomentose, 11–12 mm. long, 8 mm. broad, yellowish-green; fruit globose, up to about 6 cm. in diameter, drying and dehiscing irregularly into 3 or 4 segments; seeds 1·4–1·5 cm. long, 6–7 mm. broad, brown, shiny.

Occasional in pasture margins and woodlands on limestone in the central parishes; 1000–2500 feet; fl. Jan, Apr–May, fr. May–Oct. *A 12593! H 12006! P 19681!* Endemic.

6. A. squamosa L., Sp. Pl. **1**: 537 (1753).—*Xylopia glabra* L. (1759). Sweet Sop, Sugar Apple.

Shrub or tree 5–8 m. high; leaves oblong to narrowly elliptical, glaucous, 7–15 cm. long, 3–4 (–5) cm. broad; flowers usually clustered; petals narrow, oblong, 1·5–2 cm. long, light green outside, yellow or white within, reddish purple at the base; fruit globose, 6–10 cm. in diameter, glaucous, green when ripe, pulp white; seeds dark brown or black.

Commonly cultivated, escaping near habitations and along roadsides and pasture margins; 80–800 feet; fl. and fr. Apr–July and sporadically. *A 6953! J.P. 592!* Tropics, introduced into the Old World; Grand Cayman.

7. **A. cherimola** Mill., Gard. Dict. ed. 8 (1768).—Cherimoya.
Shrub about 1 m. or tree up to 6 m. high; leaves shortly acuminate to subacute, 7–14 cm. long, 4–9 cm. broad; petals 2–2·5 cm. long, rusty pubescent; fruit up to 20 cm. long and 10 cm. broad with white pulp; seeds black at first, drying yellowish-brown, 14–15 mm. long, 8–9 mm. broad.

Uncommon, sparingly cultivated and escaped into hillside thickets; 3500–5000 feet; fl. June–Sept, fr. Sept. *A 12713! H 9373! P 16609!* Native of the Andes, now widely introduced.

8. **A. reticulata** L., Sp. Pl. 1: 537 (1753).—Custard Apple.
Tree 5–10 m. high; leaves 10–15 cm. long, 2–5 cm. broad; flowers fragrant; petals greenish, concave, 2–3 cm. long; fruit variable in shape, often subglobose, up to 10 (–12) cm. in diameter with granular yellowish pulp; seeds dark brown or black, 1·5 cm. long.

Cultivated and escaped into waste places near habitations; about 750 feet; fl. July, fr. Feb. *A. M. Barry! P 27547!* S. Florida, Mexico to Peru and Brazil, West Indies, introduced into the Old World tropics.

2. XYLOPIA L. (1759) nom. cons.

Leaves elliptical, obtuse or shortly and bluntly acuminate, 3–7 cm. long, 1·5–3 cm. broad, strigose-pubescent beneath, more or less glabrescent; outer petals 8–9 mm. long
1. *hastarum*
Leaves lanceolate, caudate-acuminate, the acumen narrow and sharply pointed, 6–9 cm. long, 1·3–2·5 (–3) cm. broad, appressed-pilose on the midrib beneath; outer petals (1·0–) 1·4–1·5 mm. long
2. *muricata*

1. **X. hastarum** M. L. Green in Kew Bull. **1926**: 255 (1926).—*X. glabra* of L. (1763), partly, not L. (1759). White Lancewood.
Slender straight-trunked tree 5–12 m. high, or smaller and shrubby when coppiced; leaves gland-dotted, rather leathery with short transverse ridges on the midrib beneath; flowers in clusters of 2–4, white; fruiting carpels oblong, torulose, up to about 2 cm. long, with usually three seeds; seeds about 8 mm. long.

Widley distributed but much less common than formerly in woodlands on limestone; 1500–2500 feet; fl. Dec, Apr–July, fr. Dec–Jan, May–Sept. *A 7315! H 11067! HPS 14686!* Endemic.

2. **X. muricata** L., Syst. Nat. ed. 10, 2: 1250 (1759).—Lancewood.
Shrub or straight-trunked tree 5–10 m. high with smooth bark; petiole 2–4 mm. long; leaves gland-dotted; flowers often paired; calyx yellow, 2·5 mm. long; petals white; fruiting carpels ellipsoidal, 2·5–3 cm. long, with usually 3 seeds; seeds about 8 mm. long.

Generally distributed in woodlands on limestone in the central parishes; 400–3000 feet; fl. Jan, Aug–Sept, fr. Jan, May, Oct–Dec. *A 13014! H 11064! P 10518! 11043!* Endemic.

X. grandiflora St.-Hil., native of S. tropical Amer., with rusty-pubescent flowers over 3 cm. long, has not been confirmed for Jamaica. The plant previously given this name in Cuba and Trinidad is *X. aromatica* (Lam.) Mart.

3. OXANDRA A. Rich. (1845)

Leaves oblong-elliptical, 8–16 cm. long, 3–5 cm. broad; petioles 3 mm. long; petals 7 mm. long
1. *laurifolia*
Leaves elliptical, 4–10 cm. long, 2–4 cm. broad; petioles 1–2 mm. long; petals 4–5 mm. long
2. *lanceolata*

1. O. laurifolia (Sw.) A. Rich. in Sagra, Hist. Cuba, Parte 2, **10**: 20, t. 8 (1845).

Straight-trunked tree to 16 m. or more high; leaves mostly narrowed to both ends, villous beneath when young; flowers solitary or in small clusters, axillary; outer petals oblong, obtuse, white; fruiting carpels drupaceous, 11–16 mm. long, 8–10 mm. broad.

Rare and local (Port., St. Thom.), in forest on limestone in wet areas; about 2000 feet; fr. Jan. *A. M. Barry! Deans!* Greater Antilles, Guadeloupe.

2. O. lanceolata (Sw.) Baill. in Adansonia **8**: 168 (1868).—Black Lancewood.

Like the last, 3–10 m. high; leaves gland-dotted, short- or long-acuminate, the acumen obtuse; flowers usually solitary; outer petals broadly obovate, white; fruiting carpels drupaceous, 11–12 mm. long, 7–9 mm. broad, black when ripe.

Occasional in woodland on limestone mostly in the central and western parishes; 25–1400 feet; fl. July–Nov, fr. Sept–Feb, May. *Asprey UCWI 2207! H 6863! P 9482!* Greater Antilles.

67. MYRISTICACEAE

Trees, or rarely shrubs, with aromatic wood. Leaves alternate, simple and entire, often with pellucid dots, without stipules. Flowers small, unisexual and usually dioecious, actinomorphic, in racemes, corymbs, fascicles or heads. Perianth an open saucer-shaped to subglobose (2–) 3 (–5)-lobed calyx. Male flowers with 2–30 stamens; filaments united into a column; anthers 2-locular, extrorse, free or connate, opening lengthwise. Female flowers with a solitary 1-locular superior ovary; ovule solitary, basal or parietal, anatropous; stigma subsessile or sessile. Fruit a dehiscent drupe. Seed arillate, embryo small; endosperm large, usually ruminate.

15 genera with over 250 species in the tropics.

1. MYRISTICA Gronov. (1755) nom. cons.

Flowers dioecious; bracteoles enfolding base of perianth; stamens 12–30; anthers connate; aril laciniate; testa hard.

1. M. fragrans Houtt., Nat. Hist. 2 (3): 333 (1774).—Nutmeg.

Dense-crowned tree up to 8 m. or more high; leaves oblong-elliptical, acutely acuminate, pinnately veined, rather glaucous beneath, up to 13 cm. long and 6·5 cm. broad; flowers small, creamy-yellow; fruit pyriform to subglobose, 4–5 cm. in diameter, yellow or reddish, fleshy, 2-valved; aril yellow or red.

Sparingly cultivated on sheltered hillsides in wet areas at low and middle elevations; fl. and fr. sporadically throughout the year. *H & P 14289! P 22912A!* Native of E. Indies, introduced into some W. Indian islands and commercially cultivated especially in Grenada.

68. LAURACEAE

Trees or shrubs (or leafless parasitic twiner—*Cassytha*), often aromatic. Leaves mostly alternate, simple, usually entire, often glandular-punctate, without stipules. Flowers mostly bisexual, rarely unisexual and polygamous or dioecious, actino-morphic, in cymes, panicles, heads, umbels, spikes or racemes, rarely solitary, sometimes when umbellate or capitate subtended by an involucre of bracts. Peri-anth (calyx) small, shortly tubular at base with usually 6 segments, imbricate in 2 series or valvate in 1 series, usually sepaloid, the base at least persistent and often accrescent in fruit as a cup. Stamens ideally 4 whorls of 3 each, adnate to the perianth-tube, the inner (fourth) series usually staminodal, the third series always present and usually with a pair of glands on the filaments; filaments free or rarely united; anthers 2- or 4- or rarely 1-locular, opening by valves moving upwards, those of the outer series usually introrse and of the third series often extrorse.

Ovary usually superior, 1-locular; ovule solitary, anatropous, pendulous, parietal; style terminal; stigma entire or 2–3-lobed. Fruit usually a 1-seeded berry, rarely a drupe, often partly covered by the calyx cupule. Seed with a large embryo, without endosperm.

45 genera with about 2000 species mostly in the subtropics and tropics of Asia and America.

1 Leafless filiform twiner; flowers spicate or racemose; calyx-segments 6 in 2 series, the outer much smaller; perfect stamens 9 in 3 rows; anthers 2-locular, of the outer 2 rows introrse, of the third row extrorse; fruit enclosed by the succulent calyx
                         1. Cassytha
1 Trees with well developed leaves; calyx-segments 6, subequal or rarely the outer smaller:
 2 Anthers 2-locular:
  3 Perfect stamens 9 (3 outer series); anthers introrse; staminodes of the fourth series present; calyx deciduous in fruit   **2. Beilschmiedia**
  3 Perfect stamens 3 (third series); anthers extrorse; staminodes present (2 outer series), fourth inner series absent; calyx cupular in fruit, double-margined **3. Licaria**
 2 Anthers mostly 4-locular; perfect stamens 9:
  4 Leaves prominently 3-nerved from base or subtriplinerved **7. Cinnamomum**
  4 Leaves pinnately nerved:
   5 Staminodes of the fourth series small, stalked or completely aborted; anthers all 4-locular; limb of calyx deciduous in fruit, the tube becoming cupular:
    6 Anther-loculi superposed in pairs; flowers usually unisexual, the plants polygamo-dioecious   **4. Ocotea**
    6 Anther-loculi superposed in an arc; flowers usually bisexual **5. Nectandra**
   5 Staminodes of the fourth series usually large, sagittiform (up to 1 mm. or more long); anthers of third series sometimes 2-locular, extrorse; calyx more or less persistent in fruit:
    7 4-locular anthers with the lower loculi touching the upper loculi laterally, introrse; perianth generally with the outer lobes smaller than the inner; base of calyx unchanged in fruit, the limb-segments caducous   **6. Persea**
    7 4-locular anthers with the bases of the upper loculi touching the apices of the lower ones obliquely, extrorse; perianth-lobes subequal; 6-lobed calyx hardening in fruit   **7. Cinnamomum**

1. CASSYTHA L. (1753)

1. C. filiformis L., Sp. Pl. **1**: 35 (1753).

Slender twining parasite with yellow or somewhat greenish stems, attached to host plants by haustorial outgrowths; leaves represented by minute scales; flowers white, about 2 mm. long; fruit globose, light green, about 6 mm. in diameter.

Locally common, especially in the central and western parishes, on a variety of woody and herbaceous hosts in arid and semi-arid open thickets, on strand behind beaches and on some of the cays; SL–1500 feet; fl. all the year, fr. July–Aug, Nov–Mar. *A 9862! H 9525! P 23244!* General in the tropics; Grand Cayman.

2. BEILSCHMIEDIA Nees (1831)

1. B. pendula (Sw.) Hemsl., Biol. Centr. Amer. Bot. **3**: 70 (1882).—*Hufelandia pendula* (Sw.) Nees (1833). Slogwood, Slugwood.

Tree 8–20 (–30) m. high; young branches light buff-tomentose; leaves elliptical to obovate or oblanceolate, cuneate at base, shortly acuminate at tip, pinnately veined, the veins prominent beneath, shiny adaxially, 5–16 cm. long, 2–5·5 cm. broad; panicles lax, few-flowered; flowers greenish-yellow, 2–2·5 mm. long; fruit ellipsoid, black, 2·5–3·3 cm. long, 13–15 mm. broad, without a cupule at base.

Occasional in the wetter woodlands and at the margins of pastures; 2100–4350 feet; fl. and fr. sporadically throughout the year. *H 5575! 9447! P 23126!* West Indies from Cuba to St. Vincent.

3. LICARIA Aubl. (1775)

1. L. triandra (Sw.) Kostermans in Rec. Trav. Bot. Néerl. **34**: 588 (1937).—*Misanteca triandra* (Sw.) Mez (1889). Pepperleaf Sweetwood.

Tree 7–16 m. high; young branches sparsely yellowish-tomentellous, glabrescent;

leaves broadly ovate-elliptical, cuneate at base, broadly acuminate at tip, glabrescent, 6–12 (–15) cm. long, 2·5–4·5 (–5·5) cm. broad; flowers numerous in lax
panicles 5–8 (–12) cm. long, more or less clustered on peduncles 1–2 cm. long,
about 2·5 mm. long, cream-coloured; fruit purple, ellipsoid, up to 20 mm. long and
12 mm. broad; cupule red, 9–12 mm. high, 12–18 mm. in diameter.

Frequent in woodlands on limestone or shale; 400–3000 feet; fl. Sept–Nov, fr.
Jan–Sept. *A 7163! 11807! H 5374! 5630! P 22687!* Florida, West Indies southwards to Martinique.

4. OCOTEA Aubl. (1775)

1 Leaves thick and leathery:
  2 Leaf-surfaces both quite smooth, the lateral veins and veinlets obscure; leaf-blades up
    to 7 cm. long and 3 cm. broad; fruit ellipsoid   1. sp. *A*
  2 Leaf-surfaces with lateral veins at least prominulous abaxially and veinlet-reticula
    visible adaxially; leaf-blades mostly more than 7 cm. long and 2·5 cm. broad; outer
    anthers sessile:
    3 Flowers hermaphrodite, 9 mm. in diameter; leaves usually flat   2. *staminea*
    3 Flowers unisexual, monoecious, the male about 4 mm. in diameter; leaves often
      folded and recurved   3. *robertsoniae*
1 Leaves thinner, papery when dry; outer anthers stalked:
  4 Leaves with patches of hairs (domatia) in vein-axils abaxially; flowers hermaphrodite;
    filaments hairy   4. *martinicensis*
  4 Leaves without domatia; flowers dioecious or polygamous:
    5 Anther-filaments slightly hairy; fruit ellipsoid with cupule 4·5 mm. in diameter;
      veins and veinlets of leaves rather prominent on both surfaces   5. *jamaicensis*
    5 Anther-filaments glabrous; fruit globose or subglobose with cupule 5–8 mm. in
      diameter; flowers dioecious:
      6 Cupule simple-margined, 5–7 mm. broad; leaf-blade broadly cuneate to rounded at
        base; outer perianth-segments 1·5–2 mm. long   6. *leucoxylon*
      6 Cupule double-margined, 6–8 mm. broad; leaf-blade narrowly cuneate at base;
        outer perianth-segments 2–3 (–4) mm. long   7. *floribunda*

1. O. sp. A.

Tree 12 m. high; leaves elliptical, cuneate at base, obtuse at tip, the margin
slightly reflexed, glabrous, (3–) 4–7 cm. long, 1·5–3 cm. broad; petiole 3–5 mm.
long; inflorescences racemose, 5–8 cm. long from the upper axils; cupules saucer-
shaped, about 6 mm. broad; fruit ellipsoid, 13–15 mm. long.

Rare and local (Clar.), in woodland on limestone; 2500–2800 feet; fr. Mar.
H 10872! Endemic.

2. O. staminea (Griseb.) Mez in Jahrb. Bot. Gart. Berlin 5: 240 (1889).—Lignum
Dorum, Spicewood.

Tree 5–15 (–20) m. high; leaves elliptic-lanceolate to oblong, narrowed to both
ends, acute or acuminate, 7–24 cm. long, up to 9 cm. broad; flowers greenish-
yellow to white, fragrant; fruit oblong-ellipsoid, up to about 25 mm. long and
10 mm. broad, with 2-margined cupule and accrescent persistent perianth-segments.

Frequent in central and western parishes, in woodland on limestone; 750–3000
feet; fl. most of the year, fr. Jan–Mar. *A 8483! H 7169! 9184! H & P 14340!*
Endemic.

3. O. robertsoniae Proctor in Bull. Inst. Jam., Sci. ser. 16: 7, t. 1 (1967).

Shrub or tree up to 8 m. high; leaves oblong-elliptical, rounded to subcordate
at base, obtuse at tip, glabrous, very hard and brittle, 9–15 cm. long, up to 7 cm.
broad; female inflorescences up to 10 cm. long, the male shorter; flowers white,
fading pinkish then brown; mature fruits not known; cupule about 7 mm. broad.

Rather local (Clar., St. Eliz., St. Ann), in exposed aspects of woodland on limestone; 1300–2500 feet; fl. Apr–May, fr. May–July. *A 12481! Fosberg 42899!
P 19833! Robertson UCWI 2036!* Endemic.

4. O. martinicensis Mez in Jahrb. Bot. Gart. Berlin 5: 270 (1889).

Large tree; leaves broadly elliptical to obovate-elliptical, cuneate at base, acute
or shortly acuminate at tip, 10–20 cm. long, up to 12 cm. broad; flowers 3 mm.
long; fruit about 25 mm. long and 14 mm. broad in a simple-margined cupule,
the perianth-segments deciduous.

Not confirmed by recent collections. A specimen from the parish of Westmoreland (*H 9834!*), lacking flowers or fruits, has very well developed domatia in the vein-axils abaxially but the leaf-blades are smaller and relatively broader at the base than in typical specimens from Guadeloupe, Dominica and Martinique.

5. O. jamaicensis Mez in Bull. Herb. Boiss. sér. 2, **5**: 241 (1905).

Shrub or tree; leaves narrowly elliptical, cuneate at base, acuminate at tip, 6–16 cm. long, 2·5–5·5 cm. broad; flowers scarcely 2 mm. long; perianth-tube conical, very short; stamens shorter than perianth-lobes; filaments of series 3 2-glandular at base; staminodes capitellate; fruit 14 mm. long and 8–9 mm. broad.

Subsequent to the syntype collections (*H 5114! 5267!*), plants from the ridge of the Port Royal Mts., above Newcastle dividing the parishes of St. Andrew and Portland, have been identified with *Nectandra patens* (Sw.) Griseb. Further study is needed to determine whether there are one or two taxa of this affinity and their correct genera.

6. O. leucoxylon (Sw.) Gómez Maza, Dicc. Bot. Nom. Vulg. **1**: 12 (1888).— Loblolly Sweetwood.

Tree up to 25 m. high; leaves oblong-elliptical to oblanceolate, broadly cuneate to rounded at base, shortly bluntly acuminate at tip 10–20 (–30) cm. long, up to 9 cm. broad; flowers light yellow to white; perianth-segments of male flowers about 2 mm. long, of female deciduous in fruit; fruit about 10 mm. long and broad.

Rather common, in woodland in wet areas and thickets along streambanks; 800–4000 feet; fl. Apr–Nov, fr. most of the year. *H 9839! 10854! Loveless UCWI 2447! P 16488!* West Indies.

7. O. floribunda (Sw.) Mez in Jahrb. Bot. Gart. Berlin **5**: 325 (1889).—Black Candlewood, Black Sweetwood.

Tree to 20 m. high; leaves lanceolate to oblong-elliptical, cuneate at base, acuminate at tip, sparsely reddish-pubescent when young, 6–18 cm. long, 2–6 cm. broad; flowers greenish-white; perianth-segments of male flowers about 2·5 mm. long, of female more or less persistent in fruit; fruit about 13 mm. long and 10 mm. broad.

Occasional in woodland on limestone in wet areas; 1200–3000 feet; fl. Oct–Jan, fr. Dec–Mar, July. *H & B 10706! H & P 13613! 15018!* C. and northern S. Amer., West Indies.

5. NECTANDRA Roland. ex Rottb. (1778) nom. cons.

1 Fruit globose; leaves with lateral veins impressed or flat adaxially, prominent abaxially:
  2 Flowers 8–11 mm. in diameter; anther-filaments wanting   1. *antillana*
  2 Flowers 4–5 mm. in diameter; anther filaments glabrous   2. *membranacea*
1 Fruit ellipsoid or rounded-oblong; leaves with veins and veinlets prominent on both surfaces; filaments hairy:
  3 Flowers 4–5 mm. in diameter; filaments of outer series about half as long as anthers; fruit 10–25 mm. long; cupule 8 mm. broad   3. *patens*
  3 Flowers (5–) 6–8 (–10) mm. in diameter; filaments of outer series about as long as anthers; fruit 8–15 mm. long; cupule 4·5–6 mm. broad   4. *coriacea*

1. N. antillana Meisn. in DC., Prodr. **15** (1): 153 (1864).—Long-leaved Sweetwood, Shinglewood, Yellow Sweetwood.

Tree to 15 m. or more high; leaves lanceolate to oblong-elliptical, cuneate to rounded or rarely subcordate at base, acuminate at tip, glabrous or puberulous on midrib, 6–25 cm. long, 2–8 cm. broad; panicle with spreading branches, many-flowered; outer perianth spreading, inner more or less erect, white or tinged pink; flowers fragrant; fruit purplish-black, 8–10 mm. in diameter; cupule red, 7–9 mm. broad.

Very common in pastures and woodland margins and along roadsides on limestone; 50–2800 feet; fl. and fr. most of the year. *A 8442! H 5346! J.P. 723! P 24520!* In the strict sense confined to the Greater Antilles and St. Thomas, perhaps also extending to Guadeloupe. *N. globosa* (Aubl.) Mez replaces this species in continental America.

2. N. membranacea (Sw.) Griseb., Fl. Br. W.I.: 282 (1860).

Tree up to 20 m. high; leaves elliptical to lanceolate, cuneate at base, acute or acuminate at tip, glabrous or puberulous, drying a deep olive colour, 8–15 (–24) cm. long, 2–8 cm. broad; flowers white in axillary panicles mostly shorter than the leaves; perianth-segments ovate or rounded, 1·5–2 mm. long; fruit black, 10 mm. in diameter; cupule thick, tapered from pedicel to a diameter of 6–7 mm.

Rare and local (Manch., Port.), in forest; about 3000 feet; fl. July–Oct. *H 5344! 8299!* West Indies.

3. N. patens (Sw.) Griseb., Fl. Br. W.I.: 281 (1860).—Capberry Sweetwood.

Shrub or tree 2·5–20 m. high; leaves ovate to narrowly elliptical, cuneate at base, more or less acuminate at tip, shiny and dark green adaxially, puberulous abaxially, gland-dotted, conspicuously reticulate-veined, 7–20 cm. long, 3–9 cm. broad, drying greyish; inflorescences probably more or less unisexual and monoecious or functionally so, the male much branched, the female few-branched or certainly becoming so, sericeous-puberulous; perianth-segments about 2 mm. long; fruit not adequately known; cupule with pedicel tapered to thick rim.

Typical specimens, i.e. more closely matching the type (*Swartz!*), come from the woodlands of the Port Royal and Blue Mts., and have the smallest flowers and fruits; these are probably not different from *Ocotea jamaicensis* with which they are sympatric; other specimens with larger flowers tend to resemble typical *N. coriacea* which is a species characteristically of lower drier habitats. The large-fruited plants (*H 5682!*) may represent a further distinct taxon. In the circumstances of the uncertain limits of the application of the name *N. patens* it is considered inadvisable to provide ecological and distributional notes and cite specimens. Further investigations should include careful study of floral details, sexuality and fruits.

4. N. coriacea (Sw.) Griseb., Fl. Br. W.I.: 281 (1860).—*N. sanguinea* of F. & R. (1914), not Roland ex Rottb. (1778). *N. martinicensis* of F. & R. (1914), not (Jacq.) Mez (1889). Small-leaved Sweetwood, Timber Sweetwood.

Tree to 12 m. high; leaves ovate or lanceolate to elliptical, cuneate at base, acute or bluntly acuminate, shiny adaxially, leathery with prominent venation, light green beneath, drying brown, 6–12 cm. long, 2·5–4·5 cm. broad; inflorescences shorter than leaves; flowers hermaphrodite, white, fragrant; perianth-segments mostly about 3 mm. long; ripe fruit black; cupule red.

Common in woodland and woodland margins on limestone, less so in the western parishes; SL–3000 feet; fl. and fr. sporadically throughout the year. *A 11138! H 5674! Webster & Proctor 5318! Yuncker 17323!* Florida, C. Amer., West Indies especially the drier islands.

6. PERSEA Mill. (1754) nom. cons.; Kopp (1966)

1 Cultivated or naturalized plants with fruits at least 5 cm. long; perianth-segments pubescent on both surfaces, those of the outer and inner whorls subequal, reflexed at anthesis and finally deciduous in fruit; gynoecium pubescent   1. *americana*
1 Wild plants with fruits not over 2 cm. long; perianth-segments of the outer whorl glabrous adaxially, those of the inner whorl longer, patent at anthesis:
  2 Perianth-segments reflexed in fruit, the tips of the inner deciduous; gynoecium glabrous
                                                                                      2. *alpigena*
  2 Perianth-segments patent in fruit and persistent; gynoecium pubescent   3. *urbaniana*

1. P. americana Mill., Gard. Dict. ed. 8 (1768).—Alligator Pear, Avocado Pear, Pear.

Tree up to 14 (–40) m. high; leaves narrowly to broadly elliptical or obovate, cuneate to rounded at base, usually acuminate at tip, glabrous adaxially, glaucous abaxially, 6–25 (–30) cm. long, 3·5–15 (–19) cm. broad; inflorescence several-branched and many-flowered, tomentellous, shorter than leaves; flowers greenish-yellow; outer perianth-segments 4–6 mm. long; inner a little longer; fruit usually pear-shaped, shiny, green or purplish, up to about 15 cm. long; seed fleshy, loose in the pericarp at maturity, about 6 cm. long.

Common in cultivation up to about 3500 feet; fl. mostly Jan–Apr, fr. Aug–Dec, Apr. *A 8790! H 5603! Yuncker 18281!* Native of Mexico, now widespread in cultivation in the subtropics and tropics.

2. **P. alpigena** (Sw.) Spreng. in L., Syst. Veg. ed. 16, **2**: 268 (1825).—Wild Pear.

2a. Var. **alpigena.**
    Tree to 13 (–22) m. high; leaves elliptical to obovate-elliptical, cuneate at base, acute to rounded at tip, densely tomentellous abaxially and there becoming glaucous, glabrescent adaxially, 7–18 cm. long, 3–9 cm. broad; inflorescences shorter, equal to or longer than leaves; flowers about 5 mm. long; outer perianth-segments 1·3–1·7 mm. long, inner 3·5–4 mm. long; fruit globose, shiny, becoming glaucous, 8–10 mm. in diameter, green at first, finally black.
    Rather local (St. Andr., Port., St. Thom.), in submontane and montane thickets and woodlands, mostly on shale, rare in Trelawny on limestone; (1200–) 2000–5000 feet; fl. Feb, June–Sept, fr. mostly Aug–Jan. *A 8730! H 5492! 5971! P 11423!* Endemic.

2b. **P. alpigena** var. **harrisii** (Mez) Kopp in Mem. New York Bot. Gard. **14** (1): 40 (1966).—*P. harrisii* Mez (1897).
    Like the last but leaf-blades 4–9 (–12) cm. long, 2–5 (–7) cm. broad, minutely strigulose abaxially.
    Locally common (St. Andr., Port.), in montane thickets on shale; 2500–5050 feet; fl. mostly July–Sept, fr. sporadically throughout the year. *H 5382! P 21947!* Endemic. Taken together these two varieties show a range of leaf-pubescence within which the type of *P. harrisii* and several other specimens identified as this species by Fawcett & Rendle fall intermediate between extremes.

3. **P. urbaniana** Mez in Jahrb. Bot. Gart. Berlin **5**: 143 (1889).
    Tree up to 7 m. high; leaves elliptical to oblanceolate, obtuse or rounded at base, acute to acuminate at tip, puberulous abaxially, 7–20 cm. long, 2·5–10 cm. broad; inflorescences usually shorter than leaves; flowers 3–4 mm. long; outer perianth-segments 1·3–1·9 mm. long, inner about 4 mm. long; fruit globose or nearly so, 1·5–2 cm. long.
    Known only from the type (*W. Wright*) in Jamaica and otherwise from Puerto Rico and the Lesser Antilles from Montserrat to St. Lucia.

7. CINNAMOMUM Schaeffer (1760); Kostermans (1957), (1961)

1 Leaves opposite and subopposite, strongly 3-nerved from base to near apex, shiny adaxially, somewhat glaucous abaxially, drying yellowish-brown, glabrous, ovate to elliptic-lanceolate, rounded at base, up to about 15 cm. long; cultivated; native of S. Asia; Indian Cinnamon Tree   *C. zeylanicum* Nees
1 Leaves spirally arranged, weakly 3-plinerved proximally and pinnate medially, with hair-filled pits (domatia) at axils of basal veins, hardly shiny adaxially, cuneate at base, acuminate, up to about 9 cm. long:
    2 Leaves drying reddish, glaucous beneath, strongly aromatic of camphor when crushed; occasional in cultivation; native of Japan, E. China and Formosa, introduced elsewhere; Japanese Camphor Tree   *C. camphora* (L.) Sieb.
    2 Leaves drying greyish-brown, puberulous beneath, odoriferous but not pleasantly of camphor; wild species   1. *montanum*

1. **C. montanum** (Sw.) Bercht. & Presl, Priroz Rostlin **2**: 36 (1825).—*Phoebe montana* (Sw.) Griseb. (1860).
    Tree (4–) 6–16 m. high; leaves ovate-lanceolate to elliptical, veins impressed adaxially, nerves prominent abaxially, 4–10 cm. long, 2–5 cm. broad; inflorescence paniculate, pubescent, many-flowered, the branches ascending; flowers cream or greenish; perianth-segments elliptical, about 2 mm. long, remaining distinct in fruit; fruit ellipsoid, 8–11 mm. long.
    Occasional, in woodlands on limestone; 1250–4000 feet; fl. Nov–Mar, fr. Feb–July. *A 10048! H 5099! 5540! P 22007! 22200!* Greater Antilles.

69. CANELLACEAE

Trees, rarely shrubs, with aromatic bark. Leaves alternate, simple, entire, leathery, with pellucid glands, without stipules. Flowers bisexual, solitary or in cymose or racemose inflorescences. Perianth bi-triseriate, actinomorphic; sepals 3, broadly imbricate; petals 4–10 (–12) in one or more series, usually free, imbricate. Stamens 5–12, the filaments connate; anthers 2-locular, extrorse, opening lengthwise. Ovary superior, 1-locular, of 2–6 carpels; placentas parietal each with 2 or more half-anatropous ovules; style short, persistent; stigma-lobes as many as carpels. Fruit a berry. Seeds hard, shiny, with oily endosperm.

6 genera with about 20 species in tropical America and Africa.

Petals 5; stigma obscurely 2-lobed; placentas not evident in fruit: inflorescence terminal
1. **Canella**
Petals 8–10, spiral or in two series; stigma 4–5-lobed; placentas pulpy in fruit; inflorescence lateral
2. **Cinnamodendron**

1. CANELLA P. Browne (1756) nom. cons.

1. **C. winterana** (L.) Gaertn., Fruct. & Sem. Pl. 1: 373, t. 77, f. 2 (1788).—*Laurus winterana* L. (1753). Canella, Wild Cinnamon.

Shrub 2–3 m. or tree to 10 m. high; bark grey, deeply fissured into lozenge-shaped patches; leaves obovate to oblanceolate, rounded at tip, cuneate and decurrent on petiole at base, 2·5–7 (–10) cm. long, 1·5–3 (–4) cm. broad, glossy or dull adaxially, paler abaxially; inflorescence a cymose panicle; sepals 2–3 mm. long, glaucous; petals 4–5 mm. long, crimson marked yellowish at base within; anthers bright red; berry about 1 cm. long, turning red, 1–4-seeded; seeds black, curved at one end, rounded on one side, 5 mm. long, 4–5 mm. broad.

Common in thickets and woodlands in arid areas; SL–1200 feet; fl. Apr–July, fr. all the year. *A 10171! H 11757! P 9552! 23693!* Florida, Bahamas, West Indies on the drier islands to Barbados; Cayman Is.

2. CINNAMODENDRON Endl. (1840)

1. **C. corticosum** Miers in Ann. & Mag. Nat. Hist. ser. 3, 1: 351 (1858).— Mountain Cinnamon, Red Canella.

Tree 3–10 m. high; bark brownish-grey; leaves obovate-elliptical, shortly acuminate, unequally broadly cuneate at base, 5–12 cm. long; cymes few-flowered, axillary; sepals about 2 mm. long; petals about 5 mm. long, scarlet; berry about 1 cm. in diameter.

Rare and local (Port., St. Thom.), in forest in wet areas; 1000–3000 feet; fl. June–Aug, fr. Aug. *H 5963! 6458!* Endemic.

70. HERNANDIACEAE

Trees or shrubs. Leaves alternate, simple or digitately compound, without stipules. Flowers bisexual or unisexual and monoecious or polygamous, actinomorphic, in axillary corymbose or paniculate cymes. Perianth biseriate or uniseriate, with 3–5 valvate calyxlike segments in each whorl or 4–8 in one whorl, free or united at base. Stamens 3–6, opposite the outer sepals; anthers 2-locular, opening longitudinally introrsely or laterally by 2 valves; staminodes glandlike in one or two whorls outside the stamens or absent. Ovary inferior, 1-locular; ovule solitary, pendulous. Fruit dry, achenial, partly covered by the expanded receptacle. Seed solitary, without endosperm; embryo straight.

4 genera with about 65 species; pantropical.

1. HERNANDIA L. (1753)

Leaves simple, entire, often peltate but not so in Jamaican species; cymes corymbose, pedunculate, bracteate; flowers monoecious; female flowers involucrate; fruit 8-ribbed, not winged, partly enclosed in a bladderlike cupule.

Leaves 5-nerved at base, acuminate; fruit smooth with 8 longitudinal lines, depressed-globose, about 2 cm. long and 2·5 cm. broad   **1.** *catalpifolia*
Leaves 3-nerved at base, obtuse; fruit 8-ribbed, broadly ovoid, about 2·2 cm. long and 2 cm. broad   **2.** *jamaicensis*

1. H. catalpifolia Britton & Harris in Torreya **11**: 174 (1911).—Water Mahoe, Water Wood.

Tree 10–20 m. high with rather smooth bark and white wood; leaves broadly elliptical to broadly ovate, subtruncate at base, up to 37 cm. long and 26 cm. broad, nerves prominent and minutely puberulous beneath, softly leathery; petiole up to 30 cm. long, channelled near the blade; inflorescence 9–18 cm. long; bracts elliptical, 6 mm. long; male perianth-segments 6; stamens 3; female perianth-segments 8, white, 4 mm. long.

Locally common (Port., St. Thom.), by streams and in damp ravines in sub-montane woodlands; 1500–2100 feet; fl. Feb, Sept, fr. Feb–Mar. *A 12955! H & B 10566! 10685! P 22697!* Endemic. This species has been placed in the synonymy of *H. sonora* L. by recent authors (Gooding, Loveless & Proctor, Fl. Barb.: 161 (1965)) but differs in never having peltate leaves, the flowers being greenish-white and the fruit being flatter. *H. sonora* is widespread throughout most of continental tropical America, the West Indies and Pacific, but is known in Jamaica only in cultivation.

2. H. jamaicensis Britton & Harris in Bull. Torr. Bot. Club **35**: 338 (1908).—Pumpkin Wood, Suck Axe.

Tree 8–26 m. high with straight bole and shallowly striate almost smooth bark; leaves broadly elliptical to oblong-elliptical or obovate, rounded at base, up to 22 cm. long and 9 cm. broad, nerves prominent beneath, glabrous, thinly leathery; petiole up to 12 cm. long; inflorescence about 15 cm. long; bracts oblong, 4–5 mm. long, minutely tomentellous; involucel in fruit about 3 cm. broad and long, veiny, light green; fruit black, hard.

Local in the western and north-western parishes, in woodlands on limestone; 1200–2200 feet; fl. July–Sept, fr. Dec–Mar. *A 12795! H 9835! 10312! P 22604!* Endemic.

71. RANUNCULACEAE

Annual or perennial herbs, climbing or erect shrubs, rarely trees. Leaves alternate, rarely opposite, usually compound or if simple palmately divided, usually without stipules. Flowers bisexual or sometimes unisexual and monoecious or dioecious, actinomorphic or zygomorphic, solitary or in cymose, racemose or paniculate inflorescences. Perianth typically biseriate but corolla often reduced and calyx then petaloid; sepals (4–) 5 or more, free, imbricate or valvate, rarely spurred at base, deciduous; petals few to many, often 5, free, often with a nectary at base. Stamens usually numerous, spiral, hypogynous, free; anthers 2-locular, opening lengthwise. Carpels 1–many, free or rarely partly connate, the ovaries superior and 1-locular; ovules 1–many, anatropous, basal and erect or pendulous, or adaxial. Fruit achenial, follicular, baccate or capsular. Seed with small straight embryo and copious endosperm.

Some 50 genera with about 2000 species, cosmopolitan but more numerous in north temperate regions.

Herbs; leaves radical and alternate, simple and palmately divided or 3-foliolate; sepals 5, imbricate; petals (2–) 5, more or less equal; achenes shortly beaked   **1. Ranunculus**
Climbing shrub; leaves opposite, 3-foliolate or pinnately compound, the leaflets opposite, the petiolules tendriliform; sepals 4, valvate; petals absent; achenes each with a long feathery beak   **2. Clematis**

1. RANUNCULUS L. (1753)

1 Stems creeping; plant perennial, stoloniferous; sepals spreading; petals 1–1·5 cm. long;
   achenes smooth   1. *repens*
1 Stems erect or diffuse; plant tufted, not stoloniferous; sepals reflexed; petals 3–10 mm.
   long:
   2 Plant annual; achenes with minutely hooked tubercles; receptacle glabrous
                                                                                 2. *parviflorus*
   2 Plant perennial; achenes punctulate or smooth; receptacle hispid   3. *recurvatus*

1. R. repens L., Sp. Pl. **1**: 554 (1753).
   Stems villous or pubescent, ascending to 30 cm. high in flower; leaves 3-foliolate;
leaflets 2–4·5 cm. long, lobed and toothed below, entire above; sepals 6–7 mm. long;
petals 5, yellow; achenes 2·5 mm. long, excluding curved beak; receptacle slightly
hairy.
   Local (St. Andr., Port., St. Thom.), along damp sheltered paths and roadsides;
3900–6000 feet; fl. Dec–Aug, fr. July–Aug, Dec. *A 7675! 10698! Bengry! J.P. 1414!*
Europe, Asia, N. Africa, N. and S. Amer.

2. R. parviflorus L., Syst. Nat. ed. 10, **2**: 1087 (1759).
   Stems villous, diffuse, up to about 30 cm. high in flower; leaves 3-partite, the
segments coarsely toothed or lobed, or upper entire; sepals about 3 mm. long;
petals 2–5, yellow, about 3 mm. long; achenes 2 mm. long, with a short triangular
curved beak.
   Rare (St. Andr.), in damp shady places; 4000–5000 feet; fl. and fr. May–July.
H 8585! 12389! Europe, W. Asia, N. Africa, Madeira, Canary Is., S. United
States, Bermuda, Hispaniola, Australia and New Zealand.

3. R. recurvatus Poir. in Lam., Encycl. Méth. Bot. **6**: 125 (1804).

3a. Var. **recurvatus.**
   Stems villous, erect to 35 cm. high in flower; leaves mostly radical with long
hairy petioles, the blade rounded in outline and cut to beyond the middle or less,
the upper 3–5-partite to entire; sepals 3–5 mm. long; petals 5, yellow, narrowly
obovate-elliptical, up to 10 mm. long; achenes 2–3 mm. long, excluding the long
slender circinate beak.
   Occasional (St. Andr., Port., St. Thom.), on shaded pathside banks; 4800–5250
feet; fl. Apr–Aug, fr. May–Aug. *A 11338! H 6304! P 23786!* Nova Scotia to
Florida, Hispaniola.

3b. R. recurvatus var. **tropicus** (Griseb.) F. & R., Fl. Jam. **3**: 188 (1914).—*R.
   repens* var. *tropicus* Griseb. (1859). *R. cubensis* Griseb. (1866).
   Like the last but the leaves more deeply divided or 3-foliolate.
   Locally common (St. Andr., St. Thom.), along damp shaded pathsides and on
banks; 5400–7400 feet; fl. and fr. Feb–Aug. *A 10607! ACH 901! H 5891! Webster
& Wilson 5448!* Cuba, Hispaniola.

2. CLEMATIS L. (1753)

1. C. dioica L., Syst. Nat. ed. 10, **2**: 1084 (1759).—Wild Clematis.
   Woody climber to 15 m. high; stem ribbed, the older outer layers stripping
easily; leaflets ovate to elliptical, acuminate, subcordate to obtuse at base, rather
thick and fleshy, glabrous, up to 9 cm. long and 7 cm. broad; flowers numerous in
panicles, dioecious; sepals 6–8 mm. long, greenish-white to cream-coloured;
carpels in female flowers surrounded by a ring of staminodes; achenes about 4 mm.
long, excluding beak.
   Common in thickets and woodland margins in limestone areas; 200–2800 feet;
fl. Oct–Jan, fr. Nov–Feb. *A 5504! 8381! H 8810! 12318! P 24201! 24202!* Mexico
to Brazil, Greater Antilles, Guadeloupe, Dominica, Martinique, Tobago.

72. MENISPERMACEAE

Trees or usually climbing shrubs. Leaves alternate, simple and usually entire or rarely trifoliolate, often palmately veined, without stipules. Flowers unisexual, dioecious, small, generally actinomorphic and greenish or yellow. Perianth usually several-seriate, 2-merous or 3-merous in each whorl, the outer smaller. Male flowers with 6, 3, 2 or many stamens, opposite the petals when of the same number, free or variously united; anthers 4-locular, short, opening lengthwise. Female flowers with (6) or without staminodes; carpels 3, 6 or rarely one or several, free, sessile or stalked; style terminal, short or stigma sessile; ovules 2 aborting to 1, anatropous, marginal. Fruiting carpels drupaceous or achenial. Seed often curved, with or without endosperm.

About 70 genera and 400 species, mostly tropical.

Sepals 6 in two series; stamens 6, free; carpels 3; fruiting carpels laterally compressed and obliquely stalked; leaves glabrous, leathery, not obviously cordate; trees
1. **Hyperbaena**
Sepals 4 in male flowers, 1 in female flowers; stamens 4, fused; carpel 1; drupe subglobose; leaves variably hairy, rounded, cordate; climber
2. **Cissampelos**

1. HYPERBAENA Miers ex Benth (1861) nom. cons.

Leaves oblong to broadly ovate-elliptical, 8–18 cm. long, 4–10 cm. broad; inflorescences cauliflorous; fruiting carpels 1·5–2·5 cm. long
1. *valida*
Leaves ovate to lanceolate, 5–10 cm. long, 2–5 cm. broad; inflorescences axillary; fruiting carpels 7–10 mm. long
2. *prioriana*

1. **H. valida** Miers in Ann. & Mag. Nat. Hist. ser. 3, **19**: 95 (1867).—*H. domingensis* of F. & R. (1914), not *Cocculus domingensis* DC. (1818). *H. laurifolia* of F. & R. (1914), not *Cissampelos laurifolia* Poir. (1804).

Shrub 1 m. or slender-trunked tree to 10 m. high; twigs finely striate; leaves mostly obtuse; inflorescences clustered, rusty-puberulous, the male 6–12 cm. long, the female up to 5 cm. long; flowers yellowish-green; ripe carpels solitary or 2 or 3 together, at first scarlet, finally black.

Rare, in coastal thickets and woodlands on limestone; 10–250 feet; fl. June, Sept, fr. Nov–Mar. *A 10182! Cornman 627! H 10235! H & P 13793! P 11189!* Probably endemic.

2. **H. prioriana** Miers, Contrib. Bot. **3**: 301 (1871).—Beef Bone.

Shrub 4 m. or tree to 15 m. high; twigs striate; leaves ovate to lanceolate, obtuse or rarely acute, rounded to very shortly cordate at base, 5–10 cm. long, 2–5 cm. broad; male inflorescences 6–8 cm. long, the female shorter; flowers light green.

Occasional in the central parishes, in rocky limestone woodland; 400–2300 feet; fl. July–Sept, fr. Sept–Mar. *A 12612! H & P 14456! P 23012! Robertson UCWI 2491!* Endemic.

2. CISSAMPELOS L. (1753)

1. **C. pareira** L., Sp. Pl. **2**: 1031 (1753).—Pareira Brava, Velvet Leaf.

Twiner; leaves 2–12 cm. long and broad; male flowers in corymbose panicles, numerous, with corolla 1–1·5 mm. in diameter, white tinged maroon; female flowers in simple cymes clustered in the axils of rounded bracts; ripe drupe red, 4–5 mm. long.

Very common in thickets, woodland margins and on fences; 50–5000 feet; fl. all the year, fr. May–Dec. *A 8217! H 8737! H & P 14484! P 6485!* General in the tropics. In different populations the density and distribution of hairs varies distinctly.

73. NYMPHAEACEAE

73. NYMPHAEACEAE

Aquatic herbs. Leaves usually alternate, submerged, floating or emersed, simple or finely dissected. Flowers bisexual, actinomorphic, solitary, long-peduncled, often showy. Perianth biseriate or segments spirally arranged; sepals 3–5 or many, free, green or petaloid; petals 3–indefinite, imbricate. Stamens 3–6 or indefinite, the filament often produced beyond the anthers; anthers 2-locular, opening lengthwise. Ovary apocarpous and superior or syncarpous and superior or inferior; carpels 2–many; ovules 1–many, anatropous, mostly on parietal placentas. Fruit indehiscent or a tardily dehiscent berry or of nutlets. Seeds usually with endosperm; embryo straight.

8 genera with about 80 species; cosmopolitan.

1 Carpels sunken in and united by adnation to a cuplike receptacle; ovules numerous in each loculus; stamens numerous, introrse; fruit an irregularly dehiscent berry; perianth-segments spreading; sepals 4; petals numerous; seeds arillate; leaves with a basal sinus   1. **Nymphaea**
1 Carpels free; ovules 1–3 in each carpel; stamens few to many, if many extrorse; fruits indehiscent; leaves peltate:
  2 Perianth of numerous segments, the flowers large and showy; receptacle large, top-shaped, with many uniovulate carpels sunken in separate cavities, the stigmas only protruding; carpels maturing into nuts; stamens very numerous, extrorse, hypogynous; all leaves floating or emergent, centrally peltate   2. **Nelumbo**
  2 Perianth of 6–8 segments; carpels 4–18, superior; fruit 1–3 seeded; plants often mucilaginous:
    3 Plants with opposite dissected submersed leaves and peltate floating leaves; stamens 3–6   3. **Cabomba**
    3 Plants with alternate undivided floating leaves only; stamens 18–36   4. **Brasenia**

1. NYMPHAEA L. (1753) nom. cons.

1. N. ampla (Salisb.) DC., Regn. Veg. Syst. **2**: 54 (1821).—*Castalia ampla* Salisb. (1805). White Water Lily.

Stem short, thick, submersed, forming persistent barky corms; petiole terete, flexible, variable in length; leaf-blade floating, rounded, the margin subentire to toothed, the veins prominent beneath, up to about 40 cm. broad; peduncle terete; flowers fragrant, generally white; perianth-segments up to 7–9 cm. long, 1·5–2·3 cm. broad; stamens very numerous, yellow with white sterile tips; carpels 14–23.

Common in freshwater lagoons and ponds; SL–900 feet; fl. most of the year, fr. Mar, Sept. *A 10740! H 11002! H & P 15186! Robertson UCWI 3337!* Texas, continental tropical Amer., West Indies; Grand Cayman. This is a variable species showing considerable differences between young and old plants and probably also responding to habitat conditions; it is here construed to include the varieties mentioned in the Fawcett & Rendle Flora. The other species listed by Fawcett & Rendle, *N. amazonum* Mart. & Zucc. and *N. rudgeana* G.F.W. Meyer, have not been confirmed by recent collections and may have been cultivated plants.

2. NELUMBO Adans. (1763)

1. N. lutea (Willd.) Pers., Synops. Pl. **2**: 92 (1806).—*Nelumbium luteum* Willd. (1799). *Nelumbo jamaicensis* (DC.) F. & R. (1914). Pancake Rose, Water Bean.

Rhizome horizontal; petioles erect, terete; leaf-blade circular in outline, flat when floating, concave when emersed, 60 cm. or more in diameter; scape to 130 cm. high; petals rounded at tip, light yellow, 7–9 cm. long, 3·5–4·5 cm. broad; stamens yellow.

Rare and local (St. Eliz.), in marshy ponds; near SL–75 feet; fl. Sept. *Britton 1491! P 24013!* United States, Cuba, Hispaniola.

3. CABOMBA Aubl. (1775)

1. C. piauhiensis Gardn. in Hook., Ic. Pl. **7**: t. 641 (1844).
Stems slender, mainly submersed, up to 50 cm. or more long; submersed leaves petiolate, 2–4 cm. long; floating leaves linear; flowers on long peduncles from the upper axils, white, mauve or pink; sepals and petals 6–7 mm. long; carpels more or less echinate.

Rare and local (St. Eliz., Han., Trel.), in swampy ponds, slow streams and lakes; 25–400 feet; fl. and fr. Nov–Feb. *P 11168! 21929!* Mexico to Brazil, Greater Antilles, Trinidad.

4. BRASENIA Schreb. (1789)

1. B. schreberi J.F. Gmel. in L., Syst. Nat. ed. 13, **2**: 853 (1791).
Rhizome slender; petioles 10–20 cm. long; leaf-blade elliptical, entire, 4–10 cm. long, 2–5 cm. broad, purple and coated with jelly beneath, with veins radiating from the centre; flowers solitary on axillary peduncles up to 10 cm. long; perianth 14–15 mm. long, deep dull purple; carpels 6–8 mm. long.

Rare (Clar., St. Eliz.), in ponds; 25–2300 feet; fl. Feb. *Proctor & Mullings 21965!* Canada and United States to Mexico and C. Amer., Cuba, Hispaniola.

74. CERATOPHYLLACEAE

Perennial submerged aquatic rootless herbs. Leaves whorled, sessile, dichotomously divided into linear or filiform segments with serrulate margins, without stipules. Flowers unisexual, monoecious, actinomorphic, minute, solitary at the nodes, sessile. Perianth of 6–15 similar sepaloid segments joined at the base. Male flowers with 10–20 stamens arranged spirally on a flat receptacle; filaments short; anthers erect, 2-locular, opening lengthwise; connective produced into a thick coloured toothed appendage. Female flowers: ovary solitary, superior, sessile, ovoid, 1-locular; ovule 1, pendulous from a parietal placenta, anatropous; style slender. Fruit an indehiscent nut terminated by the persistent style, often spiny. Seeds with a straight embryo, without endosperm.

1 genus with 4 or more species, cosmopolitan.

1. CERATOPHYLLUM L. (1753)

1. C. demersum L., Sp. Pl. **2**: 992 (1753).—Morass Weed.
Stems trailing, branched, up to about 1 m. long; leaves stiff, once or twice forked, 1–2 cm. long, dark green; flowers pinkish; fruit with 2 spines at the base, tuberculate, about 4 mm. long, tipped by the longer persistent style.

Widely distributed but not common, in sluggish streams and ponds, sometimes in brackish swamps in coastal areas; SL–25 (–250) feet; fl. Jan–May, but seldom recorded. *A 10854! H 9983! P 19749!* Cosmopolitan.

75. DILLENIACEAE

Trees, erect or climbing shrubs or rarely herbs. Leaves alternate, simple, entire or toothed, rarely lobed, usually with prominent parallel primary veins; stipules caducous or adnate to the petiole as wings or absent. Flowers bisexual, or unisexual and dioecious or less commonly monoecious, actinomorphic, solitary or in clusters or panicles. Perianth biseriate; sepals (4–) 5 (–6), strongly imbricate, persistent; petals 5 or fewer, imbricate, often crumpled in bud, deciduous. Stamens usually numerous, hypogynous, free or variously united into bundles; anthers 2-locular, opening lengthwise or by apical pores. Carpels several, usually free, rarely 1, each with 1 or more erect basal anatropous ovules; styles as many as carpels. Fruiting carpels follicular or berrylike. Seeds with minute embryo and copious fleshy endosperm, usually arillate.

11 genera with over 300 species, mostly subtropical and tropical.

1 Erect shrubs or trees; flowers showy, more than 5 cm. in diameter; anther-loculi parallel, the connective not thickened:
2 Leaves broadly elliptical, rounded at tip, up to about 25 cm. long; flowers 7–11 cm. in diameter; petals light golden-yellow; fruiting carpels dehiscent; seeds with scarlet aril; shrub to 10 m. high, cultivated and escaped locally (St. Mary); native of Malaya and Indonesia *Dillenia suffruticosa* (Griff.) Martelli
2 Leaves oblanceolate-elliptical, shortly acuminate, up to 35 cm. long, much larger on saplings; flowers 15–20 cm. in diameter; petals white; fruit indehiscent, 8–10 cm. in diameter, enclosed by fleshy sepals; seeds exarillate, black; tree to 30 m. high; cultivated; native of S.E. Asia *Dillenia indica* L.
1 Climbing shrubs; flowers small in panicles; anther-loculi divergent or separated by connective, the filaments thickened towards the apex; seeds arillate:
3 Sepals 5, very unequal, the 2 inner concave and enlarging to enclose the fruit; carpels 1–3 with 2 ovules in each, when ripe indehiscent or bursting irregularly
1. **Davilla**
3 Sepals 4–5, slightly unequal, spreading, enlarging but not enclosing the fruit; carpels 3–5 with several ovules in each, when ripe opening adaxially or by 2 valves
2. **Tetracera**

1. DAVILLA Vandelli (1788)

1. D. rugosa Poir. in Lam., Encycl. Méth. Bot. Suppl. **2**: 457 (1812).—Red Withe.
Twining woody climber, rarely erect; leaf-blade oblong-elliptical, shortly acuminate to obtuse or rounded, decurrent on the short petiole, lateral veins prominent beneath, rough or smooth adaxially, pilose at least on the midrib and veins beneath, up to 13 cm. long and 7 cm. broad; petals mostly 4, oblanceolate, emarginate, soon falling, yellow, 6–8 mm. long; ripe fruit reddish-brown, about 7 mm. in diameter; seed shiny black with a thin whitish aril.
Occasional in the central and western parishes; on fences and on trees in thickets; 75–1500 feet; fl. Mar, Dec, fr. sporadically. *A 11982! H 8826! P 20673!* Honduras, Cuba.

2. TETRACERA L. (1753)

1. T. volubilis L., Sp. Pl. **1**: 533 (1753).—*T. jamaicensis* DC. (1817). Briar Withe.
Woody climber; leaves oblong-elliptical, mostly rounded at tip, base of blade decurrent on petiole, rather rough with scattered stellate hairs, coarse-textured, up to 20 cm. long and 9 cm. broad; panicle diffuse or compact, 7–12 cm. long; sepals 5, unequal, 2–5 mm. long in flower; petals 5, yellow; ripe carpels glabrous, shiny, beaked, 8–15 mm. long.
Rare (Trel., Port., St. Thom.), on fences and on trees in thickets; 200–1000 feet; fl. and fr. Apr, Sept. *Asprey UCWI 1101! H & B 10577! P 19781!* Mexico to Colombia and northern Brazil, Cuba.

76. OCHNACEAE

Trees, shrubs or herbs. Leaves alternate, simple or rarely pinnately compound; stipules present. Flowers bisexual, actinomorphic, in racemes, cymes or panicles. Perianth biseriate; sepals 4–5 (or 10), free or basally connate, imbricate or rarely contorted; petals 4–5 (or 10), free, imbricate or contorted. Stamens 5, 10 or many, free, hypogynous; 1–3 series of staminodes sometimes present, sometimes connate into a tube; anthers 2-locular, basifixed, opening lengthwise or by a terminal pore. Ovary superior, entire or deeply lobed, 1–10 (–15)-locular, borne on an enlarged torus or gynophore; ovules 1–many, erect or pendulous on axile placentas, or if ovary 1-locular ovules on intrusive parietal placentas; style simple; stigmas 1–5. Fruit a capsule or fruiting carpels separating into distinct fleshy drupelets. Seeds with or without endosperm; embryo usually straight.
About 25 genera with 400 species in the subtropics and tropics.

1 Herbs; petals white; stamens 5; anthers opening lengthwise; staminodes present; stipules laciniate-ciliate, persistent; ovary 1-locular; fruit a septicidal 3-valved capsule
1. **Sauvagesia**

1 Trees or shrubs; petals yellow; anthers opening by pores or short slits; staminodes absent; stipules entire, striate, caducous; ovary lobed; fruit of several drupelets on an expanded receptacle:
2 Stamens 10; leaf-margin entire or at most shallowly crenate, rarely serrulate
        2. Ouratea
2 Stamens about 30; leaf-margin serrulate; cultivated ornamental shrub; native of E. Africa   *Ochna mossambicensis* Klotzsch

1. SAUVAGESIA L. (1753)

1. S. brownei Planch. in Linden & Planch., Trois. Voy. Linden **1**: 64 (1863).— Iron Shrub.

Bushy or straggling-branched taprooted herb, rather woody at base, with branches 5–20 cm. long; leaves lanceolate to elliptical, pointed at both ends, serrulate, subsessile, up to 16 mm. long and 6 mm. broad; flowers usually solitary in the axils; sepals 4·5–5 mm. long, with a hairlike tip; petals obovate, 5 mm. long; seeds pitted.

Locally common in damp savannas and on moist shale or clay banks; 200–2400 feet; fl. and fr. all the year. *A 8257! H 12256! P 7289!* Cuba.

2. OURATEA Aubl. (1775) nom. cons.

1 Panicle compact, many-flowered, subcorymbose and rounded in outline; sepals 3–4 (–5) mm. long; drupelets ellipsoid, 9–11 mm. long, 5–6 mm. broad   1. *laurifolia*
1 Panicle lax, more or less elongate-pyramidal; sepals 7 mm. or more long:
2 Pedicels 1·5–2 cm. long; leaf-blade narrowly elliptical, gradually long-acuminate; sepals 8–11 mm. long   2. *elegans*
2 Pedicels up to 1·2 cm. long; leaf-blade elliptical to oblong-elliptical, acute or cuspidate:
3 Sepals 7–8 mm. long; drupelets subglobose, 7 mm. long, 6 mm. broad   3. *nitida*

3 Sepals 8–10 mm. long; drupelets obovoid, 10–13 mm. long, 7–8 mm. broad
        4. *jamaicensis*

1. O. laurifolia (Sw.) Engl. in Mart., Fl. Bras. **12** (2): 350 (1876).—*Gomphia laurifolia* Sw. (1798).

Shrub 3 m. or tree 4–8 m. high; leaves lanceolate to elliptic-lanceolate, acute to a usually small blunt tip, shiny above, pendulous latterly, 6–12 cm. long, 2–4·5 cm. broad; pedicels 1–2 cm. long; receptacle red and swollen in fruit, the base of pedicels often persistent; drupelets black.

Frequent in the central and eastern parishes in woodlands and thickets on rough limestone and shale; 600–3500 feet; fl. Dec–Mar, fr. Mar–May. *A 9273! 11161! H 6978! 10875! P 17410!* Cuba.

2. O. elegans Urb., Symb. Ant. **5**: 428 (1908).

Tree 10 m. high; leaf-blades 9–13 cm. long, 2–3·5 cm. broad; petals broadly obovate, 13–15 mm. long; fruit not known.

Very rare (West.); fl. Sept. *H 9912!* Endemic.

3. O. nitida (Sw.) Engl. in Mart., Fl. Bras. **12** (2): 310 (1876).—*Ochna nitida* Sw. (1788).

Tree; leaf-blades 5–12 cm. long, cuspidate, with minutely serrulate margins; pedicels 5–7 mm. long; petals slightly longer than sepals.

Known only from the original collection in Jamaica, *Shakespear*. Cuba.

4. O. jamaicensis (Planch.) Urb., Symb. Ant. **1**: 362 (1899).—*Gomphia jamaicensis* Planch. (1847).

Tree 5–10 m. high; leaf-blades mostly acute, rarely cuspidate, entire or with a few small teeth distally, 8–13 cm. long, 2·5–6 cm. broad; pedicels 5–12 mm. long; petals 10–11 mm. long.

Occasional in central and western parishes in woodland and pasture margins on limestone; 1200–2600 feet; fl. June–Nov, fr. Nov. *A 11746! H 9071! P 26482!* Endemic.

77. THEACEAE

Trees and shrubs. Leaves alternate, simple, without stipules. Flowers mostly bisexual, actinomorphic, solitary in the leaf-axils or in clusters, more rarely in racemes or panicles, often with paired bracteoles below the calyx. Perianth more or less biseriate, the sepals and petals spirally arranged and not always clearly distinguished, usually 5-merous; sepals and petals free or shortly connate, strongly imbricate. Stamens usually numerous, rarely 10–15, hypogynous in several series, free or basally connate or in fascicles often adnate to the petals; anthers 2-locular, mostly opening lengthwise, very rarely by pores. Ovary superior, 2–5 (–10)-locular; ovules (1–) 2 – numerous in each loculus, anatropous, on axile placentas; styles as many as carpels, free or connate. Fruit usually a loculicidal capsule with a central column persistent after dehiscence, rarely a drupe or berry, or dry and indehiscent. Seeds sometimes winged; embryo straight or curved; endosperm usually scanty.

About 35 genera with 600 species in the subtropics and tropics of both hemispheres.

1 Fruit a loculicidal capsule; flowers showy with white petals usually more than 1·5 cm.
　　long; anthers short, versatile:
　　2 Ovary 3-locular; seeds 1 or 2 in each loculus of the fruit, not winged; shrub formerly
　　　　cultivated and occasionally naturalized in the mountains; native of W. China; Tea
　　　　　　　　　　　　　　　　　　　　　　　　　　　　　Camellia sinensis (L.) Kuntze
　　2 Ovary 5 (–6)-locular; seeds 4–6 in each loculus of the fruit, winged; native shrubs or
　　　　trees   1. Laplacea
1 Fruit indehiscent; petals mostly less than 1·5 cm. long; anthers more or less elongated,
　　basifixed; seeds curved, not winged:
　　3 Style very short or wanting; ovary 3 (–5)-locular, stigmas 3; anthers glabrous, opening
　　　　at the base; flowers mostly unisexual; petals 5–6 mm. long, white; seeds numerous;
　　　　plants hairy   2. Freziera
　　3 Style evident; flowers bisexual; petals mostly 7–13 mm. long; connective produced;
　　　　seeds usually few:
　　　　4 Style simple, stigma minute, entire or 2-lobed; anthers glabrous, longer than fila-
　　　　　　ments; ovary 2-locular; petals white; plants glabrous   3. Ternstroemia
　　　　4 Style 3–4-fid; anthers setose, much shorter than the filaments; ovary 3-locular;
　　　　　　petals usually yellow; plants puberulous at least on youngest parts   4. Cleyera

1. LAPLACEA Kunth (1822) nom. cons.; Kobuski (1949)

Leaves obovate, villous abaxially, subsessile   1. villosa
Leaves elliptical, with inconspicuous hairs mostly on the midrib abaxially, petiolate
　　　　　　　　　　　　　　　　　　　　　　　　　　　　　　　　　　　　　2. haematoxylon

1. L. villosa (Macf.) Griseb., Fl. Br. W.I.: 104 (1860).—*Gordonia villosa* Macf. (1837).

Shrub 1·5–2·5 m. high; leaves rounded or obtuse at apex, crenate to subentire, leathery, 5–10 cm. long, 3–5·5 cm. broad; peduncles 3–6 mm. long, solitary in the leaf-axils; sepals 12–15 mm. long; petals 5–7, obovate, showy, white turning pink, silky-hairy outside; capsule 2·5 cm. long.

Very rare (St. Andr.), in montane woodland; 4700–5000 feet. *A 12979! J.P. 987!* Endemic.

2. L. haematoxylon (Sw.) G. Don, Gen. Syst. 1: 569 (1831).—*Gordonia haematoxylon* Sw. (1800). Bloodwood.

Tree 5–16 m. high, with hard deep red wood; leaves obtuse or shortly acuminate, crenulate or serrate-crenate, thin, with scattered pellucid glands; peduncles 2–5 mm. long; sepals 9–11 mm. long; petals 5–6 obovate, emarginate, white, sub-glabrous; capsule 1·5–2 cm. long.

Locally common, in montane woodlands in the eastern parishes, rare at lower elevations in the central parishes; 1700–5000 feet; fl. Oct–Mar, fr. Nov–Jan. *A 6235! H 10852! P 24550!* Endemic.

2. FREZIERA Willd. (1799) nom. cons.; Kobuski (1941)[a]

1. F. grisebachii Krug & Urb. in Engl., Bot. Jahrb. **21**: 542 (1896).

Tree to 13 m. high, the youngest parts with reddish tomentose hairs drying golden; leaves ovate-lanceolate to elliptical, acute or acuminate; peduncles 2–3 in the leaf-axils, 2–4 mm. long; sepals 4–5 mm. long; petals 5–6 mm. long; female flowers with about 25 staminodes 1·6 mm. long in a single series; ovary and style 4 mm. long; male or hermaphrodite flowers and fruit not known in Jamaica.

Very rare and local (St. Andr.), in submontane woodland; 3150 feet; fl. Mar, Nov. *H 5658!* Honduras, British Honduras, Cuba.

The Lesser Antillean species *Freziera undulata* (Sw.) Willd. was reported for Jamaica by Fawcett & Rendle on the basis of a collection by W. Wright. This record has not been confirmed by Kobuski (1941)[a] nor by any recent collections and Beard (1949) regarded the species as endemic from St. Kitts to Grenada (see Introduction p. 23). It is distinguished by the much shorter calyx, 1·5–2 mm. long.

3. TERNSTROEMIA Mutis ex L. f. (1781) nom. cons.; Kobuski (1943)

1 Calyx-lobes more than 1 cm. long:
  2 Calyx-lobes glandular-denticulate, 13–17 mm. long; bracteoles 7–9 mm. long, ovate, eglandular; pedicels 1·5–2 cm. long   1. *calycina*
  2 Calyx-lobes eglandular, 10–12 (–14) mm. long; bracteoles less than 3 mm. long, acuminate; pedicels up to 8 cm. long   2. *rostrata*
1 Calyx-lobes less than 1 cm. long, glandular-denticulate:
  3 Lateral veins of leaf conspicuous abaxially, 12–13 pairs, with both surfaces densely granular-punctate; pedicels 3·5–4·5 cm. long; sepals 8–9 mm. long; petals 12–13 mm. long   3. *granulata*
  3 Lateral veins of leaf obscure and both surfaces smooth:
    4 Sepals 6–7·5 mm. long; petals about 8 mm. long and hardly exserted from the calyx; pedicels 1·5–5 cm. long; lateral veins of leaf 5–8 pairs   4. *hartii*
    4 Sepals 4–5 mm. long; petals about 7 mm. long and clearly exserted from the calyx; pedicels about 1·5 cm. long; lateral veins of leaf 8–10 pairs   5. *howardiana*

1. T. calycina F. & R. in Journ. Bot. **60**: 363 (1922).

Tree 3–10 m. high; leaves obovate-elliptical, obtuse, rounded to cuneate at base, margin slightly recurved, leathery, 5–10 cm. long, up to 5 cm. broad, midrib prominent abaxially; sepals crimson.

Rare and local (Clar., Trel.), in woodland on limestone; 2000–2500 feet; fl. Mar–May, fr. July. *H 10979! 11035! P 20757!* Endemic.

2. T. rostrata Krug & Urb. in Engl., Bot. Jahrb. **21**: 533 (1896).

Tree 10–13 m. high; leaves obovate to obovate-elliptical, obtuse and occasionally emarginate at apex, cuneate at base, 7–12 cm. long, 3–5·5 cm. broad; pedicels mostly 6–8 cm. long, thickened at apex; sepals green; flowers fragrant; fruit 1·5–2 cm. in diameter, acuminately beaked.

Uncommon (Han., Port.), in rocky woodlands; 250–5600 feet; fl. July, fr. Mar. *H 10259! J.P. 964! P 11284!* Endemic.

3. T. granulata Krug & Urb. in Engl., Bot. Jahrb. **21**: 534 (1896).

Tree; leaves obovate-elliptical, obtuse at apex, abruptly cuneate at base, midrib evident the entire leaf-length on both surfaces, 7–9 cm. long, 3–4 cm. broad; bracteoles suborbicular, 4–5 mm. long, glandular-denticulate; outer sepals glandular-denticulate.

Very rare and known only from the type, *Purdie!* Endemic.

4. T. hartii Krug & Urb. in Engl., Bot. Jahrb. **21**: 532 (1896).

Tree 6–15 m. high; leaves obovate to obovate-elliptical, obtuse or rounded at apex, abruptly or narrowly cuneate at base, margin slightly revolute or flat, the midrib adaxially becoming obscure below the tip, 6–11 cm. long, 3–5·5 cm. broad; bracteoles ovate, 3–5 mm. long, usually not glandular-denticulate; outer sepals glandular-denticulate, about 1 cm. long in fruit; filaments of stamens uniformly slender, fruit 15–19 mm. long and broad, 2–3-seeded.

Generally distributed but nowhere common, in woodland on limestone hills; 400–2500 (–5500) feet; fl. June–Sept, fr. Sept–Dec. *A 7806! H 5767! 12369! Stiven!* Endemic.

5. T. howardiana Kobuski in Rhodora **59**: 36 (1957).

Tree 4–5 m. high with verticillate branches; leaves obovate, broadly obtuse to rounded at tip, cuneate at base, leathery, 4–7 cm. long, 2–4 cm. broad, margins subrevolute when dry; bracteoles 1·3 mm. long; flowers solitary, fragrant; filaments of stamens broader in basal two-thirds.

Rare and local (Port.), in mossy woodland on limestone; 1500–2500 feet; fl. Aug–Sept. *H & P 14841! P 10465!* Endemic.

Ternstroemia in Jamaica is still insufficiently known to enable species to be fully described and defined. *T. hartii*, as construed by Kobuski, seems to be the commonest and most widespread species, but the fruit is not adequately known. For the rest, specimens from or near type localities are easy enough to match but others are more difficult to place. Thus the reservations held by Fawcett & Rendle (1926) still hold and peculiarities of distribution and altitudinal range remain to be explained.

4. CLEYERA Thunb. (1783) nom. cons.; Kobuski (1941)[b]

1. C. theaeoides (Sw.) Choisy in Mém. Soc. Phys. & Hist. Nat. Genève **14**: 110 (1855).—*Eroteum theaeoides* Sw. (1788). Wild Damson.

Shrub 2–3 m. or tree to 10 m. high; leaves obovate to elliptical, 3–8 (–10) cm. long, 2–4·5 cm. broad, crenate-margined, leathery, very dark green adaxially; pedicels solitary (–2), 1–2 cm. long; sepals unequal, up to 6 mm. long; petals 8–9 mm. long, white or yellow; mature fruit globose, purple, 8–10 (–12) mm. long; seeds 6–8.

Frequent in the upland parts of the eastern parishes, occasional in the central parishes, in thickets and montane woodland margins on limestone or shale; 2300–7300 feet; fl. May–Feb, fr. July–Feb. *A 10570! 10689! H 9514! P 16474! 26399!* Kobuski (1941)[b] regarded this species as endemic to Jamaica, but subsequent authors (e.g. L. O. Williams in Fieldiana, Bot. **29**: 353 (1961) and Robyns in Ann. Miss. Bot. Gard. **54**: 51 (1967)) have included *C. panamensis* (Standl.) Kobuski and other species in the synonymy, thus adding an extended range from Mexico to Panama.

78. MARCGRAVIACEAE

Climbing and epiphytic shrubs, rarely arborescent. Leaves alternate, simple, entire, without stipules. Flowers bisexual, actinomorphic, in short umbelliform or elongated terminal racemes, the bracts more or less adnate to pedicels and some of them modified into sacs, pitcherlike glands or spurs. Perianth biseriate; sepals (4–) 5 (–7), free or connate, broadly imbricate, persistent; petals 5, free or connate or completely united into a calyptra. Stamens (3–) 5–numerous in a single series, hypogynous, free or shortly united, free from or attached to base of petals; anthers 2-locular, opening lengthwise. Ovary superior, 1 or more-locular or septa indefinite; ovules numerous on thick axile placentas; stigmas sessile, radiate. Fruit a globose fleshy or leathery berry, indehiscent or later opening loculicidally from the base. Seeds numerous, small, without endosperm; embryo slightly curved.

5 genera with about 120 species, restricted to tropical America.

Leaves all of one kind; inflorescence elongate-racemose; flowers and bracts red; cultivated and an occasional escape (St. Andr., St. Mary); native of tropical S. Amer.
Norantea guianensis Aubl.

Leaves polymorphic; inflorescence umbellate; flowers greenish-cream   1. **Marcgravia**

1. MARCGRAVIA L. (1753)

Pedicels 4–5 cm. long; sepals 4–6 mm. long; stamens 25–30; stalk (pedicel of aborted flower) of nectary obsolete; leaves of pendulous branches sessile or subsessile
1. *brownei*

Pedicels 2–3 cm. long; sepals about 1 mm. long; stamens 8–16; stalk of nectary 10–15 mm. long; leaves of pendulous branches distinctly petiolate   2. *brachysepala*

1. M. brownei (Triana & Planch.) Krug & Urb. in Urb., Symb. Ant. **1**: 367 (1899).—*M. rectiflora* Triana & Planch. var. *brownei* Triana & Planch. (1862).

Juvenile branches low-climbing or more or less terrestrial, with short adventitious roots, angled and winged with subsessile light green or reddish obtuse distichous leaves; mature stems up to 16 m. long and 8 cm. or more thick; leaves on pendulous branches oblong, unequal at base, cuspidate-acuminate at tip, 7–13 cm. long, 3–5 cm. broad, dark green, shiny; inflorescence with up to about 20 radiating pedicels; flowers erect at anthesis; corolla green; about 1 cm. long; fruit globose, about 11–12 mm. in diameter, tardily splitting; seeds red.

Locally abundant (St. Andr., Port.), on trees in montane forest; 3500–4700 feet; fl. Aug–Mar, fr. May–Nov. *A 5771! 11594! H 6022!* Endemic.

2. M. brachysepala Urb., Symb. Ant. **6**: 17 (1909).

Glabrous scandent shrub like the last; stem up to 10 m. long; leaves of juvenile shoots up to 4·5 cm. long and 2·5 cm. broad, crenulate, of flowering branches 5–11 cm. long and 2–4 cm. broad; inflorescence 15–35-flowered; corolla 8–9 mm. long; ripe fruit red, about 1 cm. in diameter.

Occasional in sheltered woodlands on limestone, mostly in rather wet areas, on rocks and trees; 900–2900 feet; fl. most of the year, fr. Mar, June–July. *A 8460! H & B 10771! P 16725!* Endemic.

79. QUIINACEAE

Trees or erect or climbing shrubs. Leaves opposite or whorled, entire or pinnately divided, with interpetiolar stipules. Flowers bisexual or polygamous, actino-morphic, in axillary or terminal racemes or panicles. Perianth biseriate; sepals 4–5, free, imbricate; petals 4–5 (–8), free, imbricate. Stamens numerous, hypogynous or perigynous, free or united at base and with the petals; anthers 2-locular, opening lengthwise. Ovary 2–3 or 7–14-locular, superior or fused to calyx; ovules anatro-pous, paired in each loculus, ascending from the base; styles 2–3 or wanting; stigmas peltate. Fruit a 1–4 (or many) -seeded berry. Seeds tomentose; embryo straight; endosperm present or absent.

3 genera with 37 species in tropical America.

1. QUIINA Aubl. (1775)

Stamens 15–30, hypogynous; ovary 2–3-locular, superior; fruit usually 1-locular with 1–4 seeds; endosperm wanting.

1. Q. jamaicensis Griseb., Fl. Br. W.I.: 105 (1860).—Mountain Bay, Velvet Seeds.

Shrub 2 m. or tree 5–12 m. high; leaves opposite, oblong-elliptical, shortly acuminate, cuneate at base, glabrous except on midrib beneath when very young, up to 15 cm. long and 5·5 cm. broad; flowers in axillary racemes 2–3 cm. long, opposite or subopposite on the rachis, cream-coloured at first, turning brown; pedicels about 4 mm. long; sepals ciliate, 1·5–2 mm. long; petals obcordate, 2·5–3 mm. long; fruit about 1·5 cm. long; seeds about 1 cm. long, covered with reddish-brown velvety hairs.

Local in north-western parishes, in woodland on limestone; 1400–2500 feet; fl. Mar–June, fr. June. *H 8721! 11985! P 24716!* Endemic.

80. GUTTIFERAE (CLUSIACEAE)

Trees, shrubs or herbs, usually with resinous sap. Leaves opposite or whorled, rarely alternate, sometimes with pellucid glands, without stipules. Flowers unisexual or bisexual, actinomorphic, solitary, clustered or in cymose panicles. Perianth 2–several-seriate; sepals 2–6 or more, free or basally connate, decussate or imbricate; petals 2–6, rarely indefinite or absent, free, imbricate or contorted, rarely valvate. Stamens usually numerous, hypogynous, free or variously united, often in fascicles; anthers 2-locular, opening lengthwise; often reduced to staminodes in female flowers. Disk present or absent. Ovary superior, 1–many-locular; ovules anatropous, 1–many in each loculus on axile, basal or rarely parietal placentas; styles free, connate or obsolete; stigmas as many as ovary-loculi. Fruit a capsule, berry or drupe. Seed with straight or curved embryo, with or without an aril; endosperm lacking.

About 45 genera and nearly 1000 species mostly in the subtropics and tropics; *Hypericum* extends to most temperate regions.

1 Wiry-stemmed undershrub with clear sap; leaves up to 2 cm. long, pellucid-dotted; flowers 4-merous, hermaphrodite; sepals unequal; petals 4, yellow; ovary 1-locular; styles 2, free; fruit a septicidal capsule; seeds exarillate—(characters of the Jamaican species only)   1. **Hypericum**
1 Large shrubs or trees with creamy, yellow or turbid sap; leaves more than 3 cm. long; style solitary, compound or obsolete:
  2 Stamens about 15, united into a 5-fid tube around the ovary, subtended by a cupular disk above the petals; flowers hermaphrodite; sepals 5, imbricate; petals 5, contorted, scarlet; ovary 5-locular; fruit a berry   2. **Symphonia**
  2 Stamens numerous at least in male flowers, free or united only at base; flowers not generally 5-merous, not all hermaphrodite, mostly polygamous:
    3 Fruit a leathery thick-walled capsule; stigmas more or less distinct, usually 5 or more, sessile or subsessile; bracteoles sepaloid in one or more pairs below the calyx; ovary 4–10-locular; leaves often thick, leathery and smooth, mostly obovate   3. **Clusia**
    3 Fruit indehiscent, a berry or drupe; stigma unitary or 2-lobed; leaves not thickly leathery, mostly elliptical:
      4 Ovary 1-locular with 1 ovule; leaves with numerous parallel lateral veins without veinlets; stigma shield-shaped on evident style; sepals 2–4; petals 0–4   4. **Calophyllum**
      4 Ovary 2–4 (–8)-locular; leaves reticulate-veined; style short or wanting; sepals 2; petals 4–6:
        5 Petiole with a margined pit adaxially at base; ovary 2– or more-locular with a solitary ovule in each loculus; disk thick and fleshy; fruit ellipsoid, subglobose or oblate, up to about 7 cm. long or broad, smooth; seeds with thin smooth testa   5. **Garcinia**
        5 Petiole without a margined pit at base; ovary 2-locular with (1–) 2 ovules in each loculus; disk wanting; fruit ovoid to globose, apiculate, with a thick brownish skin, 10–15 cm. in diameter; seeds rough with reddish fibrous coat   6. **Mammea**

1. HYPERICUM L. (1753)

1. **H. hypericoides** (L.) Crantz, Inst. Rei Herb. **2**: 520 (1766).—*Ascyrum hypericoides* L. (1753).

Slender-stemmed much branched shrub up to 120 cm. high; leaves oblong-oblanceolate, 5–20 mm. long, 1·5–7 mm. broad; flowers terminal; bracteoles linear, about 4 mm. long; outer sepals about 10 mm. long; petals about the same length; seeds oblong, black, about 1 mm. long.

Locally common (St. Andr., St. Thom.), on steep rocky banks in montane thickets and clearings; 2000–7400 feet; fl. and fr. most of the year. *A 5777! H 8584! P 20804!* S. United States, Cuba, Hispaniola.

2. SYMPHONIA L. f. (1781)

1. **S. globulifera** L. f., Suppl.: 302 (1781).—Boar Wood, Hog Doctor, Hog Gum Tree.

Slender straight-trunked tree up to 25 m. or more high; prop roots descend from about 1 m. above ground; branching rectangular, ultimately pendulous; all parts exude a yellow gum which hardens on drying; leaves lanceolate to oblong-elliptical, acuminate, 5–12 cm. long, 1·5–3·5 cm. broad; flowers scarlet in umbelliform cymes; petals about 12 mm. long; fruit ovoid or globose, 1·5–2·5 cm. long.

Locally common in the east (Port., St. Thom.), in wet submontane woods and in the west (St. Eliz., West.), in marsh forest; near SL–75 feet and 2100–3500 feet; fl. July–Jan, fr. Sept–Dec. *A 12787! H 9854! P 18452!* Tropical continental Amer., Hispaniola, Guadeloupe, Dominica, St. Lucia, Trinidad, tropical Africa, Madagascar.

3. CLUSIA L. (1753)

1 Fruit globose or subglobose, 6- or more-locular, 2 cm. or more in diameter; stigmas 6 or more:

  2 Stigmas 10–14 (–16); bracteoles 6–14 (–30); sepals 9–11 mm. long   1. *flava*
  2 Stigmas 6–9; bracteoles 2–4; sepals about 20 mm. long   2. *rosea*

1 Fruit oblong to ellipsoid, (4–) 5-locular, not exceeding 15 mm. in diameter; stigmas (4–) 5; sepals less than 9 mm. long:

  3 Sepals 8 mm. long; fruit 2–2·5 cm. long; leaves 5–11 cm. long, 4–8 cm. broad; female inflorescence about 3-flowered; bracteoles 2   3. *clarendonensis*
  3 Sepals 4–6 mm. long; fruit less than 2 cm. long; female inflorescence (1–) 3–15-flowered; bracteoles 2 or 4:

    4 Leaves (4–) 7–12 cm. long and (2·5–) 4–9·5 cm. broad; fruit 12–18 mm. long   4. *havetioides*
    4 Leaves 13–18 cm. long and 7·5–12 cm. broad; fruit 10–15 mm. long   5. *portlandiana*

1. C. flava Jacq., Enum. Syst. Pl. Carib.: 34 (1760).—Card Gum, Tar Pot.

Tree 3–10 m. high, often growing on exposed rocks or other trees, without red coloration in twigs and leaves; leaves 6–15 cm. long, 3–10 cm. broad; inflorescences few-flowered, the female flower usually solitary; petals 4, 2–2·5 cm. long, greenish-white turning yellow or tinged pink; fruit 2–3·5 cm. in diameter; aril orange-red; seeds yellow.

Common as an epiphyte or also terrestrial in woodland on limestone and in swamp forest; SL–2950 feet; fl. Dec–June, fr. Feb, June–Sept. *A 12209! 12616! H 9976! 10213! P 11689! 22088!* British Honduras, Grand Cayman.

2. C. rosea Jacq., Enum. Syst. Pl. Carib.: 34 (1760).—Balsam Fig.

Shrub or tree up to 16 m. high; leaves 6–18 cm. long, 6–10 cm. broad; inflorescences with up to 3 flowers; petals 6–8, 3–4 cm. long, creamy-white or pink; fruit 5–8 cm. in diameter.

Common, especially in western parishes, epiphytic or on rocks; SL–3000 feet; fl. July–Jan, fr. July, Nov–Apr. *H 10197! P 11705!* Florida, Bahamas, Greater Antilles, Virgin Is.; Cayman Brac.

3. C. clarendonensis Britton in Bull. Torr. Bot. Club **39**: 7 (1912).

Tree about 5 m. high; leaves retuse or rounded at tip; male flowers not known; seeds 3–4 mm. long.

Very rare and local (Clar., St. James, Trel.), in woodland on limestone; 2000–2500 feet; fr. July–Dec. *H 10992! 12793! P 21366!* Endemic.

4. C. havetioides (Griseb.) Planch. & Triana in Ann. Sci. Nat. sér. 4, Bot. **13**: 368 (1860).—*C. stenocarpa* Urb. (1908).

Tree 4–8 (–13) m. high; leaves usually with somewhat recurved margin, paler beneath; petioles and branches of inflorescence often tinged red; petals 5 (–6), 6–8 mm. long, cream-coloured or yellowish; seeds about 3 mm. long.

Common in open woodlands; 1500–5600 feet; fl. and fr. most of the year. *A 8717! 12615! H 5356! P 9886!* Endemic.

5. C. portlandiana Howard & Proctor in Journ. Arn. Arb. **39** (2): 103 (1958).

Shrub or tree 4–7 m. high; leaves flat, sessile; inflorescence reddish; petals 5, 5–6 mm. long, yellow.

Rare and local (Port.), epiphytic in mossy forest over limestone; 1500–2500 feet; fl. Aug–Sept, fr. Mar. *P 9797! 22701! Swabey UCWI 13018.* Endemic.

4. CALOPHYLLUM L. (1753)

1. C. calaba L., Sp. Pl. **1**: 514 (1753); Howard in Journ. Arn. Arb. **43**: 398 (1962).—*C. jacquinii* F. & R. (1926). *C. longifolium* Willd. (1811). Galba, Santa Maria

Straight-trunked tree up to 25 (–40) m. high; leaves elliptical to oblong-elliptical, rounded or rounded-emarginate at tip 4–11 (–30) cm. long, 2·5–5 (–7) cm. broad; racemes axillary; sepals 5–8 mm. long; petals white; stamens in male flowers up to 50, in female flowers fewer in a single series; fruit a globose drupe, 2–2·5 cm. in diameter; seed about 13 mm. in diameter.

Common in woodlands on limestone in areas of higher rainfall; SL–2500 feet; fl. Nov–Jan, fr. Mar–Apr, Aug, Nov. *A 11875! P 19663! Robertson UCWI 25531!* West Indies.

5. GARCINIA L. (1753)

1 Fruit purple, subglobose or somewhat flattened; leaves elliptical, acuminate, up to 25 cm. long and 10 cm. broad; flowers yellowish, 5–6 cm. in diameter; sparingly cultivated in areas of moderately high rainfall; native of Malaysia; Mangosteen   *G. mangostana* L.
1 Fruit yellow, ellipsoid, markedly acuminate-beaked; flowers creamy-yellow or white, not over 2·5 cm. in diameter; wild plants:
    2 Leaves obtuse, acute or cuspidate, the blade narrowly to broadly elliptical, (8–) 10–18 (–28) cm. long, (3–) 4·5–11 (–16) cm. broad; petals 6–10 mm. long; fruit 3–6 cm. long   1. *humilis*
    2 Leaves acutely acuminate, ovate to elliptic-lanceolate, 4–10 (–12) cm. long, 1–4 (–5) cm. broad; petals about 3 mm. long; fruit 4–8 cm. long   2. *decussata*

1. G. humilis (Vahl) Adams in Phytologia **20** (5): 312 (1970).—*Mammea humilis* Vahl (1798). *Rheedia lateriflora* L. (1753), not *G. lateriflora* Blume (1825). *R. sessiliflora* Planch. ex Vesque (1889). Wild Mammee.

Shrub to 2·5 m. or tree up to 10 m. high; leaves broadly cuneate to shortly cordate at base; inflorescences axillary or lateral, the flowers clustered; pedicels mostly 8 mm. or more long, but shorter in female flower and longer in fruit; seeds 1–3, about 2·5 cm. long.

Rather uncommon, in the central parishes, in sheltered aspects of woodland on limestone; 500–3000 feet; fl. Apr, July–Dec, fr. Aug–Sept, Jan. *A 11689! H 8668! 9471! P 11023! 18351!* West Indies.

2. G. decussata Adams in Phytologia **20** (5): 312 (1970).—*Rheedia pendula* Urb. (1899), not *G. pendula* Engl. (1908). Hat-Stand Tree.

Shrub 2–2·5 m. or tree to 10 m. high, with markedly regular rectangular lateral branching; leaves cuneate to rounded at base; flowers mostly clustered at the nodes of leafless branches; pedicels 3–8 mm. long; seeds 1–3, about 2·5 cm. long.

Uncommon and local (Clar., Manch., Port.), in damp shady woodlands on limestone; 1500–3900 feet; fl. Sept–Oct, Feb–Apr, fr. Jan, Apr, Aug. *A 9118! 11583! H 7888! 11208! P 9991!* Endemic.

6. MAMMEA L. (1753)

1. M. americana L., Sp. Pl. **1**: 512 (1753).—Mammee.

Tree mostly 8–18 m. high; leaves dark green, pellucid-dotted, up to 25 cm. long and 10 cm. broad; inflorescences axillary, solitary or clustered, 1-flowered; flowers fragrant; petals white, 17–20 mm. long; seeds 1–4, large, embedded in yellow pulp.

Cultivated and wild, generally distributed in limestone areas; 400–2500 feet; fl. June–Sept, fr. June–Aug. *A 12561! H 8971! H & P 14414! Powell 307!* Mexico to Brazil, West Indies, introduced into the Old World tropics.

81. DROSERACEAE

Annual or perennial insectivorous herbs, rarely subshrubs, mostly plants of bogs. Leaves alternate, usually in basal rosettes, generally furnished with alluring, trapping and digestive sessile or stalked glands. Flowers bisexual, actinomorphic, in determinate simple or branched racemes or panicles. Perianth biseriate, (4–) 5 (–8)-merous; sepals mostly shortly connate, imbricate, persistent; petals free imbricate. Stamens up to 20, usually in whorls of 5, mostly free, hypogynous; anthers 2-locular, opening lengthwise, the connective broad; pollen in tetrads. Ovary superior or partially inferior, unilocular with 3–5 parietal placentas or placentation basal, or rarely 3–5-locular with axile placentation ; ovules anatropous, few to many on each placenta; styles 3–5, free, often branched. Fruit a loculicidal capsule. Seeds numerous; embryo straight; endosperm granular.

4 genera, 3 of them monotypic; *Drosera* with about 100 species widely distributed, very numerous in Australia.

1. DROSERA L. (1753)

Leaves fringed with stalked glands; stipules usually present; inflorescence circinate, the flowers unilateral; stamens as many as petals; seeds winged or not winged.

1. **D. capillaris** Poir. in Lam., Encycl. Méth. Bot. **6**: 299 (1804).—Sundew.
    Small annual or short-lived perennial rosette herb; leaves spathulate, the blade suborbicular to obovate, reddish, up to 4 cm. long including the petiole; inflorescence simple or forked, scapose, up to 15 cm. high; pedicels up to 2 mm. long; calyx 3–4 mm. long, glabrous; petals a little longer than the calyx, white; seeds obovoid, minutely papillose, black or dark brown, nearly 0·5 mm. long.
    Very rare (Clar.), known only from one locality in open boggy ground where it may be locally abundant; 2200–2300 feet; fl. and fr. most of the year. *A 5927! 7161! P 15797! Skelding & Loveless UCWI 2982!* Canada to southern Brazil, Cuba, Trinidad.

82. PAPAVERACEAE

Annual or perennial herbs, shrubs or small trees, often with milky or coloured sap. Leaves alternate, entire or lobed, without stipules. Flowers bisexual, actinomorphic, solitary or in panicles. Perianth biseriate; calyx of 2–3 free or connate sepals, usually caducous; petals 4–12, in 1, 2 or 3 whorls, free, imbricate or lacking, often crumpled in bud. Stamens usually numerous in several whorls, free, often petaloid; anthers 2-locular, opening lengthwise. Ovary superior, usually 1-locular of 2–many carpels; ovules numerous on each parietal placenta, rarely solitary and basal, anatropous or campylotropous; style simple or obsolete; stigmas as many as carpels. Fruit a capsule opening by pores or valves, rarely follicular or indehiscent. Seeds with small or large embryo and copious oily or mealy endosperm.

Perhaps 30 genera with some 300 species mostly in temperate and subtropical regions of the northern hemisphere.

Annual or short-lived prickly herbs; petals present, yellow; ovules numerous; flowers solitary with 2–3 leafy bracteoles; ovary sessile   1. **Argemone**
Shrub, not prickly; petals absent; ovule solitary, basal; flowers in panicles with small bracts at the nodes; ovary stalked   2. **Bocconia**

1. ARGEMONE L. (1753)

1. **A. mexicana** L., Sp. Pl. **1**: 508 (1753).—Mexican Poppy, Yellow Thistle.
    Glaucous herb 20–90 cm. high with yellow sap; leaves pinnatifid with spine-tipped lobes, up to 20 cm. long and 7 cm. broad; sepals 3, spine-tipped, caducous,

about 2 cm. long; petals (5–) 6 (–8), about 2·5 cm. long, yellow or rarely cream-coloured; capsule prickly, 4–6-valved, 2·5–3·5 cm. long; seeds globose, about 2 mm. in diameter.

A rather common weed of roadsides and waste places; SL–2300 feet; fl. and fr. most of the year. *H 6979! 8456! P 23888!* Bermuda, S. United States, Mexico to Argentina, Bahamas, West Indies; Grand Cayman and widely naturalized in the Old World tropics.

2. BOCCONIA L. (1753); Hutchinson (1920)

1. B. frutescens L., Sp. Pl. **1**: 505 (1753).—Celandine, John Crow Bush.

Shrub 2–3 m. or tree to 5 m. with dark yellow or orange sap; leaves oblong-elliptical in outline, pinnatifid, glabrous adaxially, tomentose beneath when young, up to 50 cm. long and 25 cm. broad, light greyish-green; panicles up to 50 cm. long with numerous small protogynous flowers; sepals 2, 6–8 mm. long, caducous at anthesis of stamens; capsule 2-valved from the base leaving a persistent replum; seed covered at base with a pulpy aril.

Frequent in secondary thickets and low woodlands or as a weed of clearings and roadsides in damp sheltered places; 1200–7000 feet; fl. and fr. sporadically through-out the year. *A 5798! H 12671! P 23287!* Mexico to Peru, West Indies southwards to St. Vincent, introduced into Hawaii.

83. CAPPARACEAE

Trees, erect or climbing shrubs, or herbs. Leaves alternate, simple or digitately compound; stipules small, glandular or spinose, or absent. Flowers bisexual, rarely unisexual and monoecious, in elongated or corymbose often bracteate racemes. Perianth usually biseriate, actinomorphic or zygomorphic; sepals 4 (–8), free or connate; petals (0–) 4 (–many), clawed or sessile. Stamens 4, 6, 8 or many, free on the receptacle or raised on the gynophore (androgynophore); anthers 2–4-locular, opening lengthwise. Ovary superior, often stalked (gynophore), 1–several-locular; ovules campylotropous, few–many on parietal placentas; style short or filiform; stigma 2-lobed or capitate. Fruit a 2-valved capsule, or a berry or nut. Seeds usually reniform; embryo curved; endosperm absent.

46 genera with 700–800 species, mostly in the drier parts of the subtropics and tropics.

1 Fruit an elongated capsule dehiscing by 2 valves separating from a persistent replum; annual or short-lived herbs or undershrubs with mostly digitate leaves   1. **Cleome**
1 Fruit an elongated, ovoid or globose berry or drupe, sometimes hard and rounded or soft and torulose, dehiscent or indehiscent without replum; perennial shrubs or trees:
   2 Leaves simple; flowers bisexual; petals present   2. **Capparis**
   2 Leaves 3-foliolate; flowers often or all unisexual:
     3 Petals present; ovary and fruit with conspicuous gynophore; fruit several-seeded
                                                3. **Crateva**
     3 Petals absent; ovary and fruit subsessile; fruit 1-seeded   4. **Forchhammeria**

1. CLEOME L. (1753); Iltis (1960), (1967)

1 Leaves simple, glabrous, minutely serrulate; petals yellow; stamens 6; gynophore evident   1. *procumbens*
1 Leaves digitately compound:
   2 Stamens 8–30; leaflets (3–) 5 (–6), pubescent with simple and glandular hairs, strongly aromatic; plants unarmed; petals yellow with reddish-purple spot at base; gyno-phore obsolete   2. *viscosa*
   2 Stamens 6:
     3 Plants armed with paired stipular prickles at petiole-bases and sometimes below the bracts:

4 Leaflets (1–) 3; gynophore hardly evident in fruit; flowers solitary in axils of some or
   all uppermost simple shortly petiolate leaves, not obviously racemose; pedicels
   slender, up to about 2 cm. long; petals white; seeds wrinkled   3. *aculeata*
4 Leaflets (3–) 5–7; gynophore evident in fruit; flowers in elongated racemes with
   simple cordate sessile or subsessile leafy bracts; pedicels stout, 2·5–3·5 cm. long;
   seeds scaly, otherwise rather smooth:
   5 Gynophore in fruit 5–7 mm. long; petals shortly clawed, rose-pink, about 15 mm.
      long, as long as or longer than stamens   4. *houstonii*
   5 Gynophore in fruit 2–7 cm. long; petals long-clawed, at least 20 mm. long, but
      shorter than stamens:
      6 Bracts obtuse or rounded at tip; petals white; ovary puberulous   5. *spinosa*
      6 Bracts acute; petals pink; ovary glabrous   6. *hasslerana*
3 Plants without paired stipular prickles:
   7 Fruit-stalk once-articulated, androgynophore very short or wanting; leaflets mostly
      3:
      8 Stems, petioles and leaf-blades thinly hispid; leaf-margins ciliate; gynophore in
         fruit 4–8 mm. long; seeds rough   7. *rutidosperma*
      8 Stems, petioles and leaf-blades glabrous; leaf-margins serrulate; gynophore very
         short; seeds rather smooth   8. *serrata*
   7 Fruit-stalk twice-articulated, androgynophore developed; leaflets mostly 5; seeds
      wrinkled:
      9 At least the lower bracts 3-foliolate; petals 11–16 mm. long including the slender
         claw, usually white; androgynophore (middle section of fruit-stalk) 12–24 mm.
         long, as long as or longer than gynophore (distal section of fruit-stalk)
                                                                          9. *gynandra*
      9 Bracts simple; petals 25–30 mm. long, usually pink; androgynophore 3–10 mm.
         long, much shorter than gynophore   10. *speciosa*

1. C. procumbens Jacq., Enum. Syst. Pl. Carib.: 26 (1760).
Slender-branched taprooted herb, 6–25 cm. high; leaves narrowly elliptical,
acute, 9–20 mm. long; flowers in short terminal bracteate racemes; petals 3–5 mm.
long; capsule 14–25 mm. long, shortly stipitate, beaked; seeds muricate, reddish-
brown, about 1 mm. broad.
Rather uncommon, a weed of low-lying sandy ground in the southern central
parishes; SL–600 feet; fl. and fr. sporadically throughout the year. *H 12051!*
P 24344! Robertson UCWI 20422! Cuba, Hispaniola.

2. C. viscosa L., Sp. Pl. 2: 672 (1753).—*Polanisia viscosa* (L.) DC. (1824).
C. icosandra L. (1753). Wild Caia.
Bushy herb 30–120 cm. high, all vegetative parts and young fruit aromatic-
glandular; leaflets rhomboid-obovate to elliptical, mostly pointed at both ends, up
to 8 cm. long and 3 cm. broad; flowers solitary in the axils of leaflike bracts; petals
obovate, about 1 cm. long; pedicels in fruit 2–3·5 cm. long; capsule 5–10 cm. long,
beaked; seeds dark reddish-brown, radiately striate, minutely verrucose, 1·4–1·6
mm. broad.
Common as a weed of disturbed ground and gravelly waste places; 50–800 feet;
fl. and fr. all the year. *A 9993! H 12121! Barnett!* Native of the Old World, now
widespread in the tropics of both hemispheres.

3. C. aculeata L., Syst. Nat. ed. 12, **3**: 232 (1768).
Straggling-branched annual herb up to about 60 cm. high; stems and leaves
puberulous; leaflets ovate or the terminal rhomboidal, acute or obtuse, 2–6 (–8)
cm. long, 1·5–3 (–4) cm. broad; petals about 5 mm. long, easily falling; ovary
glabrous; capsule 2·5–6 cm. long; seeds yellowish, about 2·5 mm. broad.
Rare and local (St. Andr.), a weed of recently disturbed ground; about 550 feet;
fl. and fr. Nov–Jan. *A 12880!* Native from Mexico to Argentina and in the West
Indies, introduced into India, Java and W. tropical Africa, apparently absent from
Cuba.

4. C. houstonii Ait. f. in Ait., Hort. Kew. ed. 2, **4**: 131 (1812).
Erect bushy glandular-pubescent herb up to 1·5 m. high; usually prickly on
petioles and at bases of bracts; leaflets mostly 5, lanceolate, 3–9 cm. long; raceme
10–25 cm. long; capsule 5–9 cm. long; seeds about 1·5 mm. broad.
Very rare (St. Cath.), known in Jamaica only from a single collection made
nearly 300 years ago. *Sloane!* C. Amer., Cuba.

5. C. spinosa Jacq., Enum. Syst. Pl. Carib.: 26 (1760).

Tough-stemmed herb 1–2 m. high, usually glandular and without prickles on petioles and inflorescence; leaflets elliptic-lanceolate, narrowed to base, long-acuminate, up to 9 cm. long and 3 cm. broad; racemes 10–40 cm. long; stamen-filaments purple; gynophore in fruit up to about 4 cm. long; capsule 6–10 (–13) cm. long; seeds yellowish- or greyish-brown, scaly at first, 1·5–1·8 mm. broad.

Locally common weed of open waste ground, dry river beds, gully banks and roadsides; SL–1500 feet; fl. and fr. all the year. *A 6047! H 11775!* Mexico to northern S. Amer., West Indies.

6. C. hasslerana Chodat in Bull. Herb. Boiss. **6**, App. **1**: 12 (1898).—*C. spinosa* var. *horrida* F. & R. (1914), partly, not *C. horrida* Mart. ex Schult. (1829).

Erect bushy pubescent herb 1–2 m. high, usually prickly on petioles and at bases of bracts; leaflets 5–7, lanceolate, 3–10 cm. long; raceme up to 40 cm. long; bracts 10–18 mm. long, 8–12 mm. broad; capsule up to 12 cm. long and 5 mm. broad; seeds yellowish.

Very rare (Trel.), a roadside weed, probably a garden escape; 1500 feet; fl. and fr. May. *H 8547!* Native of Brazil, Argentina and Paraguay, elsewhere as a culti-vated ornamental.

7. C. rutidosperma DC., Prodr. **1**: 241 (1824).—*C. ciliata* Schumach. (1827).

Annual taprooted herb with erect or trailing branches to 1 m. long; leaflets ovate-rhomboid to elliptical, acuminate, up to 3 cm. long and 1·5 cm. broad; flowers solitary in the axils of leaflike bracts; petals usually 4, oblanceolate, acute, 6–9 mm. long, white or light mauve; capsule (2·5–) 3·5–5 (–6) cm. long; seeds reddish-brown, radiately striate-verrucose, with adhesive strophiole, 1·1–1·3 mm. broad.

Common as a weed of disturbed ground, rough pastures and sandy waste places; SL–1500 feet; fl. and fr. all the year. *A 5621! H 6551! P 24365!* Native of West tropical Africa, introduced into several parts of the West Indies and probably elsewhere in the tropics.

8. C. serrata Jacq., Enum. Syst. Pl. Carib.: 26 (1760).

Stiff-stemmed undershrub 30–100 cm. high, probably perennial; leaflets ovate to lanceolate, acuminate, thin, 3–12 cm. long, (1–) 1·5–3·5 cm. broad; racemes few-flowered, elongating greatly in fruit; bracts subulate, less than 1 mm. long; petals shortly clawed, about 1 cm. long, white tinged pink or crimson; capsule (3–) 6–8 cm. long, shortly beaked; seeds dark grey, almost smooth with short scattered scales, 1·5–1·7 mm. broad.

Rather common in ditches, damp waste places, thickets and pastures; 50–900 feet; fl. and fr. most of the year. *A 9926! H 6623! P 11194!* Tropical continental Amer., Cuba, Hispaniola, Trinidad.

9. C. gynandra L., Sp. Pl. **2**: 671 (1753).—*Gynandropsis gynandra* (L.) Briq. (1914). *G. pentaphylla* DC. (1824).

Annual herb to 1 m. high, pubescent and somewhat glandular; leaves 3–5 (–7)-foliolate; leaflets elliptical to obovate, thin, 2–6 cm. long, 1–3·5 cm. broad; in-florescence terminal, up to 40 cm. long at least in fruit; bract-leaflets elliptic-oblanceolate, up to about 1 cm. long; pedicels, androgynophore, gynophore and filaments purple; capsule 4–9 cm. long; seeds coarsely scaly-rugose, about 1·5 mm. broad.

Rare and local (St. Andr.), a weed of waste ground; 50–600 feet; fl. and fr. May–Jan. *A 12726! P 6725! Wynter UCWI 25110!* Native of the Old World, now general in the subtropics and tropics.

10. C. speciosa Kunth, Nov. Gen. **5**: 84, t. 436 (1821).—*Gynandropsis speciosa* (Kunth) DC. (1824).

Like the last but larger in all its parts, sometimes glabrous; bracts broadly ovate, cordate; capsule 6–10 cm. long; seeds muricate, about 3 mm. broad.

An escape from gardens; mostly grown at middle elevations; fl. Oct–Nov, fr. Nov. Native of continental tropical Amer., now widely distributed.

2. CAPPARIS L. (1753)

1 Leaves glabrous or puberulous beneath:
  2 Stamens as long as the petals; leaves up to 30 cm. long, acute, clustered; fruit up to
    7·5 cm. long   1. *baducca*
  2 Stamens much longer than petals; leaves smaller, up to 10 cm. long (on mature plants),
    obtuse, rounded or emarginate at tip, uniformly distributed; stem with prominent
    gland above the axil of each leaf; fruit 10–30 cm. long   2. *flexuosa*
1 Leaves scaly or hairy beneath:
  3 Leaf-blade abaxially with a dense covering of fine stellate hairs; fruit ovoid to globose,
    not exceeding 2 cm. in length, not torulose   3. *ferruginea*
  3 Leaf-blade abaxially scaly; fruit elongated, torulose:
    4 Calyx valvate in bud, angled by the 4 appressed margins, 7–11 mm. long; leaves
      shiny above with obscure venation, the veinlets beneath hidden by scales.
                                          4. *cynophallophora*
    4 Calyx open in bud, the sepals subulate or lanceolate, 2–3 mm. long; leaves dull
      above with prominent venation, the veinlets beneath exposed between the scales
                                          5. *indica*

The sepals, petals and ovary of these species of *Capparis* characteristically bear the same kind of indumentum as the abaxial surfaces of the leaves. Most species of *Capparis* have juvenile leaves much longer and narrower than those of mature plants; these can usually be equated on indumentum and texture characters.

1. C. baducca L., Sp. Pl. **1**: 504 (1753).
    Shrub or tree to 8 m. high; leaves mostly oblong-elliptical, acutely acuminate, base cordulate, variable in size in the same flush, 5–30 cm. long, 2·5–10 cm. broad; stipules lateral, minute, triangular; flowers in terminal corymbs; calyx-lobes suborbicular, 1·5–2 mm. long; petals obovate, greenish-white or purplish, 8–11 mm. long; fruit irregularly torose, 3–7·5 cm. long; seeds light brown, about 6 mm. broad.
    Uncommon and local (St. Andr., St. Cath., St. Ann), in thickets and woodland margins on limestone; 50–1000 feet; fl. Feb–May, fr. sporadically. *Powell 269!* *Thornton UCWI 987!* Continental tropical Amer., West Indies.

2. C. flexuosa (L.) L., Sp. Pl. ed. 2, **1**: 722 (1762).—*Morisonia flexuosa* L. (1759). Bottle-cod Root.
    Shrub 2–5 m., sometimes scrambling, or tree to 9 m. high with drooping branches, branches sometimes trailing in undergrowth; bark silvery-grey, smooth with transverse ridges; twigs, petioles and midrib beneath puberulent, glabrescent; leaves of mature plant mostly oblong, of juvenile linear, rather leathery, 4–12 cm. long, 1–5 cm. broad; corymbs few-flowered; flowers fragrant, opening in the evening; sepals imbricate, unequal, 4–7 (–10) mm. long; petals usually white, sometimes rose-pink, 1·5–2 cm. long; filaments numerous, white; gynophore red; fruit follicular, yellow or reddish outside, red within; seeds ellipsoid, white, covered with oily white pulp, 8–11 mm. long.
    Common in thickets, mainly in arid parts of the south coast and on the cays, occasional inland; SL–1500 feet; fl. Mar–Oct, fr. June–Nov. *A 7086! 7772!* *P 22266!* Florida, Honduras to northern S. Amer., West Indies; Cayman Is.

3. C. ferruginea L., Syst. Nat. ed. 10, **2**: 1071 (1759).—Mustard Shrub.
    Shrub 1–3 m. or tree to 8 m. high; bark pinkish-buff; leaves elliptical to oblanceolate, acute or acuminate, apiculate, narrowly rounded and cordulate at base, 3–10 (–12) cm. long, 1–4 cm. broad; inflorescences small, corymbose in the upper axils; sepals 2–2·5 mm. long, reflexed; petals about 5 mm. long, white or cream; stamens usually 8, pilose at base; seeds compressed, brown, smooth, 6–7 mm. broad.
    Common in thickets and woodlands on arid limestone in coastal areas and on the cays, rare inland; SL–800 (–1000) feet; fl. and fr. most of the year. *A 6184!* *H 9624! P 23717!* Cuba, Hispaniola, Grand Cayman.

4. C. cynophallophora L., Sp. Pl. **1**: 504 (1753).—*C. longifolia* Sw. (1788). Black Willow.
    Shrub 2–5 m. or tree to 10 (–15) m. high; leaves (from plants in open coastal

or lowland areas) oblong-elliptical, obtusely rounded or emarginate at tip, or (from plants in woodlands or upland areas) broadly elliptical and acute or acuminate at tip, 4–15 (–20) cm. long, 2–6·5 cm. broad; juvenile leaves can be 15 cm. or more long although not more than 6 mm. broad; inflorescences few-flowered, subterminal and in the upper axils; flowers fragrant; sepals reflexed; petals scaly outside, tomentose within, 10–13 mm. long, white turning purple; stamens 15–30, filaments cream turning reddish-purple, pilose at base; fruit 5–30 cm. or more long, eventually dehiscent, scarlet within; seeds compressed, 5–6 mm. broad.

Common in coastal woodlands and thickets on limestone and around salina margins, rarer in interior woodlands; SL–500 (–2600) feet; fl. Jan–Sept, fr. May–Sept. *A 10750! H 9816! Powell 978!* Florida, Mexico to northern S. Amer., Bahamas, West Indies; Grand Cayman.

5. C. indica (L.) F. & R. in Journ. Bot. **52**: 144 (1914).—*Breynia indica* L. (1753). White Willow.

Like the last; petals 10–12 mm. long, white, scaly tomentose on the outer surface, tomentose within; fruit 5–25 cm. long, scarlet within.

Very rare (St. Cath.); low elevations; fl. Apr–May, fr. July–Aug. Not recently collected. C. Amer. to Venezuela, Puerto Rico to Barbados.

3. CRATEVA L. (1753)

1. C. tapia L., Sp. Pl. **1**: 444 (1753).—*C. gynandra* L. (1762). Garlic Pear Tree.

Tree 5–15 m. high, glabrous; trunk grey, smooth; leaflets elliptical, acuminate, the lateral unequal at base, 5–15 cm. long, 2·5–6 (–8) cm. broad; inflorescence a corymbose raceme terminating a short lateral leafy branch; flowers fragrant; sepals 3–6 mm. long; petals of male flowers up to 25 mm. long, of hermaphrodite flowers smaller, white with light crimson lines; stamens about 16–20, up to 6 cm. long, purple; ripe fruit yellow, subspherical, up to 3·5 (–6) cm. in diameter, the woody gynophore 3–5·5 (–7) cm. long; seeds about 8 mm. in diameter.

Rather uncommon, mostly in damp low-lying areas, occasionally on the sea-shore; SL–700 feet; fl. Mar–July, fr. May–Aug. *A 10786! H 9569! Powell 1098! P 19700!* Continental tropical Amer., West Indies.

4. FORCHHAMMERIA Liebm. (1853)

1. F. trifoliata Radlk. ex Millsp. in Field Mus. Bot. **1**: 399 (1898).

Glabrous shrub or tree from about 3 (–12) m. high; leaflets pulvinate, oblanc-eolate-elliptical, cuneate at base, shortly acutely acuminate, with numerous spreading more or less parallel lateral veins and reticulate veinlets, hard-textured, drying light greyish, 8–16 cm. long, 3·5–5·5 cm. broad; inflorescences racemose-paniculate, on the older branches, 5–18 cm. long; sepals up to about 8, under 1 mm. long; fruit subglobose, 7–10 mm. in diameter when dry.

Very rare (St. James), in thickets on rocky limestone; 15–350 feet; fr. Mar. *P 11926!* Mexico, Cuba.

84. CRUCIFERAE (BRASSICACEAE)

Annual, biennial or perennial herbs, rarely undershrubs. Leaves usually alternate, simple or pinnate, without stipules. Flowers bisexual, usually actinomorphic, typically in ebracteate racemes. Perianth triseriate; sepals 4, free, imbricate in decussate pairs; petals 4, free, clawed, imbricate or contorted, sometimes absent. Stamens typically 6 in 2 whorls, tetradynamous, the 2 outer shorter and opposite the lateral sepals, the 4 inner longer and opposite the petals, usually free, but filaments of each inner pair sometimes connate; anthers (1–) 2-locular, opening lengthwise. Ovary superior, usually 2-locular by a complete false septum; ovules anatropous or campylotropous, few to many on parietal placentas; style simple or obsolete;

stigmas mostly 2. Fruit usually valvately dehiscent, a long (siliqua) or short (silicula) capsule, or an indehiscent 1–few-seeded nut, or dividing transversely into 1-seeded portions (lomentum). Seed with large embryo; endosperm scant or absent; cotyledons variously orientated to the radicle.

350 genera with up to 3000 species, mostly in the cooler parts of the northern hemisphere.

1 Fruit transversely 2-jointed (lomentum), indehiscent; leaves fleshy; plant of sandy sea-shores   1. **Cakile**
1 Fruit a siliqua or silicule; leaves thin; inland plants, mostly weeds:
  2 Capsule a dehiscent siliqua at least 3 times as long as wide:
    3 Petals absent; leaves toothed, the lower lobed   2. **Rorippa**
    3 Petals present:
      4 Petals white; leaves pinnate:
        5 Seeds in 2 rows in each loculus; stems trailing and rooting in water; stamens 6; valves of fruit not curling, convex   3. **Nasturtium**
        5 Seeds in 1 row in each loculus; rosette herb with slender erect flowering shoot; stamens usually 4; valves of fruit curling upwards elastically, flat   4. **Cardamine**
      4 Petals yellow; leaves deeply pinnatisect to subentire; seeds in 1 row in each loculus:
        6 Siliqua appressed to the inflorescence-axis, subulate, pointed, 1–2 cm. long   5. **Sisymbrium**
        6 Siliqua spreading, narrowed at both ends, subtorulose, 3–3·5 cm. long   6. **Brassica**
  2 Capsule a silicula; petals white, rarely absent; at least the basal leaves pinnatifid:
    7 Silicula indehiscent, the valves subglobose, 1–seeded; stamens usually 2; branches spreading and ascending   7. **Coronopus**
    7 Silicula dehiscent, the valves compressed; erect herbs:
      8 Silicula triangular-obcordate; seeds several in each loculus; stamens 6   8. **Capsella**
      8 Silicula orbicular; seed 1 in each loculus; stamens 2 or 6   9. **Lepidium**

1. CAKILE Mill. (1754)

1. **C. lanceolata** (Willd.) O. E. Schulz in Urb., Symb. Ant. **3**: 504 (1903).—*Raphanus lanceolatus* Willd. (1800).

Annual; stem weak and straggling from a woody base, 30–100 (–150) cm. long; leaves mostly oblanceolate, entire or coarsely serrate-dentate, up to 8 cm. long and 1·5 (–2·5) cm. broad; racemes terminal, elongating in fruit to 35 cm. or more long; sepals hyaline-margined; petals white, 7 mm. long; fruit fleshy, 2–3 cm. long, articulated below the middle.

Locally common on sandy beaches and on the cays; SL–10 feet; fl. and fr. all the year. *A 7588! H 10353! C. B. Lewis!* S. United States to northern S. Amer., West Indies; Grand Cayman.

2. RORIPPA Scop. (1760)

1. **R. heterophylla** (Blume) R. O. Williams in Fl. Trin. & Tob. **1**: 24 (1929).—*Nasturtium heterophyllum* Blume (1825).

Annual branched taprooted herb up to 25 cm. high; leaves all stalked, the lower with a large terminal lobe otherwise serrate-dentate to subentire, up to 10 cm. long, including petiole, and 3 cm. broad; racemes terminal, up to 12 cm. long in fruit; sepals about 1·5 mm. long; fruit spreading, slender, straight, about 3 cm. long.

Rather rare (St. Andr., St. Mary, Port.), a weed of cultivated land and roadsides in damp sheltered places; 490–3900 feet; fl. and fr. July, Nov–Mar. *A 6658! H 11703! P 6908!* General in tropical Asia, continental tropical Amer., Trinidad.

3. NASTURTIUM R. Br. (1812) nom. cons.

1. **N. officinale** R. Br. in Ait., Hort. Kew. ed. 2, **4**: 111 (1812).—*Sisymbrium nasturtium-aquaticum* L. (1753). *N. fontanum* Aschers. (1864). Water Cress.

Trailing branched herb with soft stems and leaves, perennial; lower leaves 5–10 cm. long with up to 4 (–5) pairs of leaflets, the terminal leaflet the largest; sepals 2 mm. long; petals about 4 mm. long; fruit 15–18 mm. long.

Cultivated in permanent streams here and there and sometimes escaping, not common; 50–3600 feet; fl. Apr–May. *Johnson 28!* Native of Europe and W. Asia, now cultivated and naturalized almost throughout the world.

4. CARDAMINE L. (1753)

1. C. hirsuta L., Sp. Pl. **2**: 655 (1753).—Hairy Bitter Cress, Lady's Smock.

Annual herb with erect or straggling flowering shoots 10–30 (–45) cm. high; leaves mostly basal with 3–7 pairs of orbicular or oblanceolate entire or toothed leaflets, terminal leaflet largest up to 1·8 cm. broad; inflorescence at first corymbose, elongating in fruit; sepals about 1·5 mm. long; petals 2·5–3 mm. long; fruit more or less erect, 15–20 (–25) mm. long.

Locally common (St. Andr., Port., St. Thom.), in sheltered cultivations and on roadside banks and rocks in damp places; 1900–5000 feet; fl. and fr. Nov–July. *A 5983! H 9204! P 20629!* Widespread in the northern hemisphere and on many tropical mountains; Hispaniola.

5. SISYMBRIUM L. (1753)

1. S. officinale (L.) Scop., Fl. Carniol. ed. 2, **2**: 26 (1772).—*Erysimum officinale* L. (1753). Hedge Mustard.

Annual taprooted herb 30–90 cm. high, rectangular-branched above, often with bristly retrorse hairs; lower leaves deeply pinnatifid, up to 12 cm. long, upper much smaller; inflorescence at first corymbose, elongating greatly in fruit; sepals 2 mm. long, hispid; petals 3 mm. long; fruit 10–15 mm. long, broadest at base.

Rather rare and local (St. Andr., Manch., St. Thom.), a weed of well-drained banks, roadsides and waste places; 2300–4500 feet; fl. and fr. Jan–May, Aug. *A 10701! Hart! P 21935!* Native of Europe, N. Africa and W. Asia, now widespread; Hispaniola.

6. BRASSICA L. (1753)

1. B. willdenovii Boiss in Ann. Sci. Nat. sér. 2, **17**: 88 (1842).—*Sinapis integrifolia* Willd. (1804), not Vahl (1793). *B. integrifolia* Rupr. (1860). Wild Mustard.

Annual taprooted herb to about 1 m. high, glaucous; lower leaves broadly elliptical to obovate in outline, pinnatifid and dentate, up to 25 cm. long and 8 cm. broad, the upper much smaller, narrower and more entire; racemes elongating in fruit; pedicels about 1 cm. long; sepals 5–6 mm. long; petals 9–10 mm. long; anthers sagittate.

Local and uncommon, a weed of pathsides and waste places; about 1000–3500 feet; fl. and fr. Feb–Apr, Aug. *A 7754! Bengry! P 16174!* Native of S. Asia, naturalized elsewhere, Bermuda, Bahamas, Greater Antilles, Antigua, Virgin Is., Guadelope to Trinidad.

7. CORONOPUS Zinn (1757) nom. cons.

1. C. didymus (L.) Sm., Fl. Brit. **2**: 691 (1800).—*Lepidium didymum* L. (1767).

Annual taprooted herb with spreading or ascending branches up to 30 cm. long, foetid; lower leaves stalked, deeply pinnatisect with 3–5 pairs of oblanceolate toothed segments, upper leaves sessile with narrower entire segments; racemes terminal and axillary, up to 3·5 cm. long in fruit; sepals 0·5 mm. long; petals much shorter than sepals or absent; fruit emarginate, about 2 mm. broad, valves reticulate-pitted.

Rare and local (St. Andr.), a weed of cultivated and waste ground; 3500–5000 feet; fl. and fr. Feb, May. *H 8579! P 24642!* Native probably in S. Amer., now widespread.

8. CAPSELLA Medic. (1792) nom. cons.

1. C. bursa-pastoris (L.) Medic., Pflanzengatt.: 85 (1792).—*Thlaspi bursa-pastoris* L. (1753). Shepherd's Purse.

Annual or short-lived perennial herb with flowering shoots up to 40 cm. high;

lower leaves in a basal rosette, pinnatifid to subentire, oblanceolate in outline, up to 12 cm. long, upper leaves clasping the stem with acute basal auricles; racemes terminal and in the upper axils; sepals about 1 mm. long; petals nearly 2 mm. long; fruits 6–9 mm. long on spreading pedicels; seeds up to 12 in each valve.

Very rare, in the mountains, not reported since Macfadyen (1837). A cosmopolitan weed of temperate regions and many tropical mountains..

9. LEPIDIUM L. (1753)

Valves of fruit 3 mm. long; pedicels spreading; stamens usually 2; upper stem-leaves mostly toothed   1. *virginicum*
Valves of fruit 5–6 mm. long; pedicels erect or ascending; stamens 6; upper stem-leaves sessile, entire   2. *sativum*

1. L. virginicum L., Sp. Pl. 2: 645 (1753).—Wild Peppergrass.

Annual or short-lived perennial taprooted herb, developing at first a basal rosette of leaves; flowering shoots branched, 15–30 (–45) cm. high, puberulous; lower leaves deeply to shallowly pinnatifid, up to about 10 cm. long, upper leaves subentire; racemes terminal on main and lateral branches, up to about 8 cm. long; sepals 0·8–0·9 mm. long; petals as long as to nearly twice as long as sepals; fruiting pedicels 3–5 mm. long.

Common as a weed of pathsides, cultivations and waste gravelly ground; SL–4500 feet; fl. and fr. all the year. *A 5789! 8360! H 6914! H & P 13533!* Native of N. Amer., now widely distributed.

2. L. sativum L., Sp. Pl. 2: 644 (1753).—Peppergrass.

Annual taprooted herb with a single erect flowering shoot to 40 cm. high; basal leaves petiolate, lyrate with toothed obovate lobes, uppermost leaves linear; raceme terminal; petals twice as long as sepals; pedicels in fruit 3–4 mm. long.

Very rare, a casual relict from cultivation, reported by Macfadyen. Native of Egypt and W. Asia, cultivated almost all over the world.

85. TOVARIACEAE

Shrubby herbs. Leaves alternate, 3-foliolate, without stipules. Flowers bisexual, actinomorphic, in terminal bracteolate racemes. Perianth biseriate; sepals (7–) 8, free; petals (7–) 8, free, inserted on the disk. Stamens (7–) 8; filaments free, flat, pilose; anthers 2-locular, opening lengthwise. Disk platelike with glands between the filaments. Ovary superior, shortly stipitate, subglobose, 6–8-locular; ovules numerous, campylotropous, on spongy axile placentas; stigma subsessile, 5–8-lobed. Fruit a globose berry. Seeds small; embryo curved and enclosed in endosperm.

One genus with 2 species in the mountains of tropical America.

1. TOVARIA Ruiz & Pav. (1794) nom. cons.

1. T. diffusa (Macf.) F. & R., Fl. Jam. 3: 246 (1914).—*Bancroftia diffusa* Macf. (1837).

Plant about 2 m. high; stem lax, branched; petioles 2–6 cm. long; leaflets lanceolate, long-acuminate, cuneate at base, unequally so on lateral leaflets, 6–15 cm. long, up to 2·5 cm. broad, glabrous; racemes 15–20 cm. long; flowers greenish-white, 12–15 mm. in diameter; berry 10–11 mm. in diameter, crowned by persistent stigma-lobes.

Rare and local (Port.), in wet montane woodland; 4500 feet; fl. fr. Oct. *H 7449!* Endemic.

86. MORINGACEAE

Deciduous trees; bark with gum. Leaves alternate, bi- to tripinnate, the pinnae opposite; stipules and stipels reduced to glands or lacking. Flowers bisexual, zygomorphic, in axillary cymose panicles. Perianth biseriate; sepals 5, unequal, imbricate, reflexed; petals 5, free, perigynous on the hypanthium, the anterior petal larger. Disk present, lining the hypanthium with a short free margin. Stamens 5, free, declinate, alternating with 3–5 staminodes; anthers 1-locular, dorsifixed, opening lengthwise, surrounding the style at anthesis. Ovary superior, stipitate, 1-locular; ovules numerous, in 2 rows on each of 3 parietal placentas, anatropous, pendulous; style terminal, slender, tubular. Fruit an elongated 3-valved capsule. Seeds with or without wings, without endosperm.

One genus with about 10 species native from N. Africa to the E. Indies.

1. MORINGA Adans. (1763)

1. M. oleifera Lam., Encycl. Méth. Bot. **1**: 398 (1785).—Ben Nut Tree, Horse Radish Tree.

Tree 5–6 m. high, shrubby when coppiced; roots thick and soft; leaflets obovate, entire, 4–24 mm. long, 3–12 mm. broad; flowers fragrant; sepals and petals white distally, greenish at base, sometimes tinged crimson; sepals 10–13 mm. long; fruit 24–30 cm. long; seeds with 3 broad papery wings, the seed-body globose and 6–7 mm. in diameter.

Rather common near habitations in the drier southern parishes; SL–700 feet; fl. and fr. Nov–Aug. *Harris! Molina & Barkley 22ʃ434! Yuncker 17880!* Native of India and the E. Indies now naturalized throughout the tropics and subtropics.

87. CRASSULACEAE

Annual or perennial succulent herbs or shrubs. Leaves alternate, opposite or whorled, usually simple and entire, without stipules. Flowers bisexual, actinomorphic, in bracteate cymose inflorescences. Perianth biseriate; sepals free or united; petals the same number as the sepals, usually 4 or 5, free or rarely connate, hypogynous. Stamens as many as the petals and alternate with them or twice as many, hypogynous, free or shortly connate, or epipetalous when corolla gamopetalous; anthers 2-locular, opening lengthwise. Ovary superior of 3 or more free or connate unilocular carpels, each often subtended by a scalelike gland; ovules anatropous, usually numerous on adaxial placentas; style and stigma simple. Fruit a follicle, opening adaxially. Seeds small with straight embryo; endosperm usually present, fleshy.

An indefinite number of genera (? about 30) with some 1400 species, widely distributed in the northern hemisphere and numerous in S. Africa, mostly in dry areas.

1. BRYOPHYLLUM Salisb. (1805)

Leaves opposite, simple or imparipinnate; inflorescence cymose-paniculate; calyx 4-toothed; corolla 4-lobed; stamens 8, epipetalous; carpels 4.

1. B. pinnatum (Lam.) Oken in Allgem. Naturgesch. **3** (3): 1966 (1841).—
    Cotyledon pinnata Lam. (1786). *Kalanchoe pinnata* (Lam.) Pers. (1805).
    Leaf-of-Life.

Glabrous more or less erect fleshy herb up to 1·2 m. high; lower leaves simple, upper pinnate, the blades up to 18 cm. long and 12·5 cm. broad, when compound the terminal much larger than the lateral leaflets, crenate, producing at times plantlets in the marginal notches; flowers nodding or pendulous; calyx light green turning yellowish, inflated, the tube 2·5–3·5 cm. long; corolla dull brownish-red, the tube constricted below the middle, lobes acute, 2·5–4·5 cm. long; carpels 12–15 mm. long.

Common on pathside and roadside banks; 550–4000 feet; fl. Dec–May, fr. Jan–May. *P 23282! Wynter UCWI 1072!* Native of Madagascar, now widespread in the tropics.

Several species of *Kalanchoe* and other species of *Bryophyllum* are grown in gardens in Jamaica; perhaps the commonest of these is *K. tubiflora* (Harv.) Hamet with subterete speckled leaves and dusky pink flowers.

88. PITTOSPORACEAE

Trees, shrubs or woody climbers, sometimes spiny, often with secretory ducts. Leaves alternate or whorled, simple, without stipules. Flowers mostly bisexual and actinomorphic, in cymes or panicles, or solitary. Perianth biseriate; sepals 5, free or basally connate, imbricate; petals 5, sometimes basally connivent, imbricate. Stamens 5, alternate with petals, usually free, hypogynous; anthers 2-locular, introrse, opening by pores or lengthwise. Ovary superior, 2–5-locular with axile placentation or 1-locular with parietal placentation; ovules numerous, anatropous; style simple, short; stigmas as many as carpels or coalesced. Fruit a loculicidal capsule or a berry. Seeds immersed in viscid pulp; embryo small; endosperm large, fleshy.

9 genera with over 200 species in warmer parts of the Old World.

1. PITTOSPORUM Banks ex Gaertn. (1788) nom. cons.

Erect trees or shrubs; fruit a leathery capsule with wingless seeds.

1. **P. undulatum** Vent., Jard. Cels: t. 76 (1802).—Mock Orange.

Tree up to 10 m. high; buds covered with numerous imbricate scales; leaves oblong-elliptical to lanceolate, acute or acuminate, entire, usually strongly undulate, leathery and very shiny, crowded, up to about 10 cm. long and 3–4 cm. broad; flowers white, fragrant, about 12 mm. long, in terminal cymose panicles; sepals and petals at length reflexed; sepals mostly connate; petals free; capsule 2-valved, 10–14 mm. broad, with persistent style, orange; seeds red, covered with glutinous adhesive.

Locally common (St. Andr., Port., St. Thom.), naturalized along pathsides and in montane thickets; 2000–5700 feet; fl. Feb–July, fr. Mar–Nov. *A 6442! Loveless UCWI 2976! P 23911! 24694!* Native of eastern Australia.

Other ornamental species of *Pittosporum* have been introduced and are grown in gardens in the hills.

89. BRUNELLIACEAE

Trees, sometimes spiny, often tomentose. Leaves opposite or whorled, simple, trifoliolate or imparipinnate with small deciduous stipules. Flowers functionally unisexual, dioecious, in panicles. Perianth uniseriate; calyx 4–5-partite, the segments valvate; disk adnate to the calyx, 8–10-lobed. Stamens 8–10, inserted at the base of the disk; rudimentary ovary present in male flowers. Carpels 4–5, free, sessile, 1-locular; ovules 2 in each carpel, collateral; styles subulate, recurved with simple stigmas; rudimentary stamens present in female flowers. Fruiting carpels capsular, 2-valved, 1–2-seeded. Seeds with fleshy endosperm and flat cotyledons.

One genus with 35 species in tropical America.

1. BRUNELLIA Ruiz & Pav. (1794)

1. B. comocladiifolia Humb. & Bonpl., Pl. Aequin. **1**: 211, t. 59 (1808).—West Indian Sumach.

Tree with spreading crown 8–15 m. high, young twigs and leaves rusty-pubescent; leaves up to 40 cm. long with 5–11 pairs of leaflets; leaflets elliptical to oblong, rounded unequally at the base, usually acuminate, serrate, tomentose beneath, up to 15 cm. long and 6 cm. broad; panicles 6–8 cm. long with rusty-pubescent branches; calyx olive-green, the lobes about 2·5 mm. long; filaments yellow; fruiting carpels pungent-setose, 3–4 mm. long; seeds reddish-brown, shiny, oval in outline, narrower at one side with a short oblique beak.

Locally common (St. Andr., Port., St. Thom.), in the margins of montane woodlands; 3500–5600 feet; fl. Apr–Aug, fr. June–Oct. *A 7380! J.P. 1271! P 23912!* Mexico to Peru, Greater Antilles, Guadeloupe.

90. CUNONIACEAE

Trees and shrubs. Leaves opposite or whorled, simple, trifoliolate or pinnate; stipules present. Flowers small, bisexual or unisexual and dioecious, in terminal racemes, cymes or heads, rarely solitary. Perianth biseriate; sepals (3–) 4–5 (–6), imbricate or valvate, free or basally connate; petals the same number as the sepals or lacking, sometimes lobed, mostly free. Stamens mostly numerous, sometimes as many as the petals and then alternate with them, inserted on an annular glandular disk; filaments free; anthers 2-locular, short, opening lengthwise. Carpels 1–5, superior, free or united into a 2–5-locular pistil; placentation axile when ovary compound; ovules few to numerous, mostly anatropous, in 2 rows on each placenta; styles and stigmas distinct, as many as the carpels, straight or coiled in bud. Fruit a 2-valved capsule or a nut. Seeds with endosperm and small straight embryo.

25 genera with about 350 species, mostly in the southern hemisphere.

1. WEINMANNIA L. (1759) nom. cons.

Leaves opposite, mostly imparipinnate; stipules interpetiolar, deciduous; flowers usually in racemes; petals mostly present; ovary compound, 2-locular; fruit a capsule.

Leaves with (2–) 3–4 (–12) pairs of leaflets, the terminal leaflet up to 3 cm. long and 1·5 cm. broad   **1.** *pinnata*
Leaves with (2–) 3 leaflets, the terminal leaflet up to 6 cm. long and 2·5 cm. broad   **2.** *portlandiana*

1. W. pinnata L., Syst. Nat. ed. 10, **2**: 1005 (1759).—Bastard Braziletto.

Shrub 2–4 m. or tree to 10 m. high; leaves with winged petiole and rachis, lateral leaflets oblong-elliptical, mostly about 2 cm. long and 1 cm. broad or a little larger, smaller than the terminal leaflet, pubescent on the veins beneath or glabrous, serrate-margined distally; stipules oblong-ovate, about 6 mm. long and 5 mm. broad; racemes usually longer than the leaves; pedicels filiform; flowers white; petals 1·5–1·8 mm. long; stamens exserted; capsule 3·5–4·5 mm. long.

Locally frequent (St. Andr., Port., St. Thom.), in montane thickets and open woodlands; 3600–6500 feet; fl. Sept–Dec, fr. Jan. *A 12709! H 9513! J.P. 653! P 11468!* Mexico to Brazil, West Indies.

2. W. portlandiana Howard & Proctor in Journ. Arn. Arb. **39** (2): 106 (1958).

Shrub 2 m. or tree to 5 m. high; leaves with winged petiole 1·2–1·4 cm. long, lateral leaflets 2·5–2·7 cm. long, light green beneath, glabrous or glabrescent; inflorescence terminal, pseudoracemose, 8 cm. long, with flowers in clusters of 2–5; bracteoles 1 mm. long; pedicels 4·5 mm. long, sparsely pubescent; calyx 4-merous; petals 2–3 mm. long; capsule reddish, about 5 mm. long.

Rare and local (Port.), in wet mossy thickets on limestone; 2500 feet; fl. Dec. fr. Dec–Jan. *A 9154! ACH 952! P 11351!* Endemic.

Ceratopetalum gummiferum Sm., a shrub with white flowers, is grown in some hill gardens; introduced from Australia.

91. ROSACEAE

Trees, shrubs or herbs. Leaves alternate, rarely opposite, simple or compound; stipules usually present and paired, sometimes adnate to petiole. Flowers usually bisexual, actinomorphic or rarely zygomorphic in various types of inflorescence. Perianth biseriate or one or both series absent; calyx often perigynous, typically of 5 basally connate imbricate sepals; petals the same number as calyx-lobes, free, imbricate or rarely contorted, or absent. Stamens (1–3) 5 (–8) –many, in one or several whorls, free or rarely connate; anthers small, 2-locular, usually opening lengthwise. Ovary superior or inferior, of 1–several–many simple carpels or compound with 2–5 loculi each with 2–few basal or axile ovules; styles (or style-branches) and stigmas as many as carpels. Fruits achenial, drupaceous, follicular, pomaceous or aggregate. Seed with small embryo and usually without endosperm.

About 100 genera with some 3000 species, cosmopolitan but more numerous in temperate regions.

1 Trees or shrubs without prickles; leaves simple:
  2 Fruit inferior, pomiform, enclosed by hypanthium and crowned by persistent calyx, 2–5-carpelled; ovary-loculi 2-ovuled; leaves strongly veined and toothed, rusty-tomentose beneath; inflorescence rusty-tomentose; cultivated; native of China and Japan; Loquat   *Eriobotrya japonica* (Thunb.) Lindl.
  2 Fruit superior, drupaceous, 1-carpelled, 1-seeded; ovules 2; leaves entire:
    3 Style terminal; stamens 15 or more, regular   **1. Prunus**
    3 Style arising from one side of the ovary near the base:
      4 Stamens 10 or more, regular in a ring   **2. Chrysobalanus**
      4 Stamens 3–10 or more, perfect on one side of the oblique receptacle, staminodal on the other   **3. Hirtella**
1 Herbs or prickly shrubs; carpels numerous; leaves compound:
  5 Shrubs with prickles; epicalyx absent:
    6 Receptacle conical bearing numerous drupelets in fruit   **4. Rubus**
    6 Receptacle urceolate enclosing the achenes in fruit   **5. Rosa**
  5 Herbs with stolons or decumbent rooting branches, without prickles; epicalyx present:
    7 Leaves 3-foliolate; achenes numerous, dry, partly embedded in the enlarged ovoid to subglobose fleshy receptacle   **6. Fragaria**
    7 Leaves mostly 5-foliolate; achenes inserted in the hemispheric dry receptacle   **7. Potentilla**

1. PRUNUS L. (1753)

Leaves oblong-elliptical to elliptical, 10–20 cm. long, 4·5–7·5 cm. broad; flowers 8–9 mm. across; drupe ellipsoid, about 20 mm. long   1. *occidentalis*
Leaves ovate to oblong-ovate, 5–12 cm. long, 2·5–4·5 cm broad; flowers 5–7 mm. across; drupe subglobose, 8–12 mm. in diameter   2. *myrtifolia*

1. P. occidentalis Sw., Nov. Gen. & Sp. Pl.: 80 (1788).—Prune or Pruan Tree.
Tree with spreading branches, 8–10 (–13) m. high; leaves subacuminate, glabrous; racemes 1 or 2 together, sometimes branched, 2–8 cm. long; pedicels 5–7 mm. long; flowers fragrant; petals white or cream, rounded-obovate, about 3 mm. long.
Occasional, in woodlands; 1700–3500 feet; fl. Jan–Apr, fr. Jan–May. *H 5591! 10145! P 21889!* Guatemala, Panama, Greater Antilles, Lesser Antilles to St. Vincent.

2. P. myrtifolia (L.) Urb., Symb. Ant. **5**: 93 (1904).—*Celastrus myrtifolius* L. (1753). Ant's Wood, Wild Cassada.
Slender-trunked tree (2–) 5–20 m. high; wood hard, heavy, red; leaves acuminate, mostly rounded at base, glabrous, usually with a few flat glandular areas in the lamina; stipules intrapetiolar, paired; racemes 1 or rarely 2 together in the axils,

3–5 cm. long; pedicels 2–4 mm. long; petals white or cream, rounded, shortly clawed, about 2·5 mm. long.

Common in open situations and woodland margins on limestone; 1200–3150 feet; fl. Dec–Apr, fr. Dec–May, Aug–Sept. *A & S 3017! H 12814! P 8493!* Mexico to Brazil, Bahamas, Florida, Greater Antilles, Montserrat, Trinidad and Tobago.

Species of temperate fruit trees which have been introduced into the cooler parts of Jamaica include *Prunus dulcis* (Mill.) D. A. Webb, Almond, *P. persica* (L.) Batsch, Peach and varieties of *Malus domestica* Borkh. (*Pyrus malus* L.), Apple.

2. CHRYSOBALANUS L. (1753)

1. C. icaco L., Sp. Pl. **1**: 513 (1753).—Coco Plum.

Shrub 1–3 m. or bushy tree 2–6 (–10) m. high; leaves elliptical to obovate, rounded and often emarginate at tip, 3–10 cm. long, 2·5–7 cm. broad; cymes axillary and terminal, pubescent, 3–6 cm. long; sepals (4–) 5; petals (4–) 5, white, about 5 mm. long, perigynous; carpel solitary, protogynous; drupe green turning brownish-purple and then black, subglobose to obovoid, 1·5–3 cm. in diameter, angled when dry.

Rather common in sublittoral thickets and woodlands, local in damp open savannas on heavy soils in the interior; SL–3000 feet; fl. Nov–June, fr. Nov–Sept. *A 5899! 12005! H 8515! P 9722! 15903!* Bahamas, Florida, Mexico to S. Amer., West Indies; Grand Cayman; a very closely related species occurs in tropical Africa.

3. HIRTELLA L. (1753)

1 Inflorescence a raceme 10–30 cm. long; stamens 5–7	1. *racemosa*
1 Inflorescence paniculate, much shorter than above; stamens generally 3:	
2 Fruit 2–2·5 cm. long, densely hairy; leaves shortly acuminate	2. *jamaicensis*
2 Fruit about 1·4 cm. long, sparingly puberulous; leaves usually long-acuminate	3. *multiflora*

1. H. racemosa Lam., Encycl. Méth. Bot. **3**: 133 (1789).

Shrub or small tree to 3 m. high; leaves oblong to oblong-lanceolate, acute to long-acuminate, glabrous or pubescent beneath, shiny above; pedicels 6–7 mm. long; petals about 3 mm. long; fruit obovoid-oblong, glabrous, 8–12 mm. long.

Very rare and known only from a single collection, *McNab!* C. and S. tropical Amer., St. Vincent, Trinidad.

2. H. jamaicensis Urb., Symb. Ant. **5**: 355 (1908).

Shrub 2·5 m. or tree to 6 m. high; leaves oblong to oblong-elliptical, with short appressed hairs beneath, 4–12 cm. long, 2–4 cm. broad; petals about 3 mm. long, white; stamens exserted, purple; fruit oblong, truncate apically, narrow at base, obscurely ribbed.

Rather rare (Clar., St. Ann, St. Thom.), in thickets; 2300–4000 feet; fl. Sept–Mar, fr. Mar–May. *H 5604! P 19791!* Endemic.

3. H. multiflora Urb., Symb. Ant. **5**: 356 (1908).

Tree 4–9 m. high; leaves ovate-lanceolate to oblong, sparingly hairy on both surfaces, 5–12 cm. long, 2–3·5 cm. broad; petals 3–3·5 mm. long, white; stamens exserted, white; fruit angled, ribbed.

Uncommon and local (Port., St. Thom.), in moist woodland on limestone; 1500–3000 feet; fl. July–Sept, Mar. *H 5317! P 9231!* Endemic.

4. RUBUS L. (1753)

1 Calyx usually and petiole, rachis and young stem always bristly; leaves whitish-tomentellous beneath:
  2 Leaflets rounded or broadly obtuse at tip; bristles of calyx rarely lacking, not gland-tipped; petals white; ripe fruit yellow   1. *ellipticus*
  2 Leaflets acute or shortly acuminate; calyx and stem with prickles and gland-tipped bristles; petals purple; ripe fruit purplish-black   2. *racemosus*

1 Calyx hairy but not prickly or bristly; leaflets acute or acuminate:
  3 Leaves imparipinnate; leaflets 5–7; plant with numerous minute sessile spherical
    glands; petals shorter than sepals   3. *rosifolius*
  3 Leaves digitate; leaflets 3–5; plant not glandular; petals as long as or longer than sepals:
    4 Leaflets glabrous or nearly so between the veins beneath, 3 (–4); calyx thinly tomen-
      tose   4. *alpinus*
    4 Leaflets beneath and calyx villous-tomentellous; leaflets 3 or 5   5. *jamaicensis*

1. R. ellipticus Sm. in Rees, Cyclop. 30 Rubus: no. 16 (1813).—Cheeseberry.

Shrub 2·5–5 m. high; stem armed with curved thorns and reddish spreading tapered bristles and with an understory of white thinly shaggy hairs; leaflets 3, the terminal larger on a rachis half to as long as petiole, up to 7 (–10) cm. long and 5 cm. broad; flowers in axillary and terminal panicles; petals about 1 cm. long.
Locally common (St. Andr., St. Thom.), forming thickets in mountain pastures, woodland margins and along roadside banks; 3500–5000 feet; fl. most of the year, fr. Mar–May, Aug. *A 5757! 8722! H 9131! Patrick 88!* Introduced from the mountains of subtropical Asia, now naturalized; also in Puerto Rico.

2. R. racemosus Roxb., Fl. Ind. **2**: 519 (1832).

Shrub, low to 2 m. high with glaucous stems; leaves pinnate with (3–) 5 (–7) leaflets, the lateral leaflets opposite and sessile, the terminal larger, broadly ovate, lobed, denticulate, up to 6 cm. long and broad; flowers in few-flowered axillary corymbose racemes; petals about 8 mm. long.
Local (St. Andr., St. Thom.), once cultivated and now naturalized in clearings; 4100–7400 feet; fl. Dec–July, fr. Mar–July. *A 10643! P 6088! Weaver 1147!* Native in the mountains of S. India.

3. R. rosifolius Sm., Pl. Ic. Ined. **3**: t. 60 (1791).

Scrambling shrub, sparsely prickly; leaflets ovate-lanceolate, doubly serrate, the lateral leaflets opposite, the terminal largest, up to 7 cm. long and 3·5 cm. broad; flowers subsolitary; sepals attenuate-aristate, about 15 mm. long; petals white, about 10 mm. long; ripe fruit oblong-ellipsoid, about 15 mm. long, red.
Rare (Manch., Trel.), in waste places and clearings; 1500–3000 feet; fl. Apr, Nov, fr. Nov. *P 18346!* A fairly recent introduction; native of S.E. Asia; Greater Antilles, Lesser Antilles on the wetter islands south to St. Vincent, Trinidad, Brazil, Hawaii.

4. R. alpinus Macf. ex Griseb., Fl. Br. W.I.: 232 (1860).

Scrambling shrub up to 1 m. or more high, with purplish stems; leaflets ellip-tical, acutely serrate, up to 8 cm. long and 3·5 cm. broad; sepals 3–4 mm. long; petals white tinged purplish, 5–6 mm. long; fruit broadly ovoid, 5 mm. in diameter, dark purple.
Uncommon (St. Andr., St. Thom.), in montane woodland clearings; 4200–7400 feet; fl. Mar, Nov–Dec, fr. Dec. *A 12969! J.P. 1376! P 6089! Yuncker 17541!* Endemic.

5. R. jamaicensis L., Mant. Pl.: 75 (1767).—Bramble.

Sprawling shrub with arching prickly stems 1–5 m. long; leaflets ovate to obovate-elliptical, serrate-dentate, 5–17 cm. long, up to 9 cm. broad; flowers numerous in terminal and axillary panicles; petals white usually tinged pink or purplish, 7–8 mm. long; fruit depressed-globose, 8–10 mm. in diameter, at first red, finally black.
Common in the central and eastern parishes, in thickets and woodland margins and on shady banks; 1000–5000 feet; fl. and fr. all the year. *A 6405! 8480! H 9141! P 6309!* Hispaniola.

5. ROSA L. (1753)

1 Petals deep rose-pink, about 2·5 cm. long; leaflets up to 7 cm. long and 3 cm. broad
  glabrous; peduncles glandular   1. *indica*
1 Petals white, at least 3·5 cm. long:
  2 Calyx bristly, the lobes about 3 cm. long; leaves mostly 3-foliolate, the leaflets ovate-
    lanceolate, acuminate, glabrous, up to 7 cm. long and 2·5 cm. broad   2. *laevigata*

2 Calyx not bristly, densely tomentose outside, the lobes about 2·5 cm. long, long-acuminate; leaflets 3–7, elliptical to obovate, pubescent beneath, 1·5–3·5 cm. long, 0·8–2 cm. broad   3. *bracteata*

1. R. indica L., Sp. Pl. **1**: 492 (1753).—Blush Rose, China Rose.

Shrub to 2 m. high, sparingly prickly; leaflets 3–5, ovate-lanceolate, acuminate; flowers fragrant, in corymbs; sepals lanceolate, glandular on margins; ripe fruit orange-yellow.

Rather uncommon (St. Andr.), an escape from cultivation; 500–3600 feet; fl. June, Dec. *H 6912!* Native of China, also in Mexico, Hispaniola, Puerto Rico and possibly Cuba.

2. R. laevigata Michx., Fl. Bor. Amer. **1**: 295 (1803).—Cherokee Rose.

Shrub scrambling to 6 m., glabrous, with stout curved prickles; leaflets serrulate; flowers solitary, sometimes fragrant; petals 3·5–4 cm. long and broad; ripe fruit pyriform, orange-red, bristly, 2–2·5 cm. long.

Occasional, on roadside banks and in thickets along pasture margins; 1000–4700 feet; fl. Jan–May, Aug, fr. Jan. *A 6653! P 23529! Yuncker 18302!* Native of China and Japan, also in S. United States and Cuba.

3. R. bracteata Wendl., Bot. Beobacht.: 50 (1798).—Macartney Rose.

Trailing and climbing shrub or low bush, 60 cm. to 2 m. or more high, the young branches densely woolly, with short-stalked glands and prickles, the thorns on lower branches larger; leaflets obtuse, minutely crenate-denticulate; flowers sub-solitary, faintly fragrant, subtended by deciduous laciniate bracts; petals obcordate, up to 4 cm. long and 4·5 cm. broad; ripe fruit spherical, orange-red, tomentose.

Locally common (St. Andr., Port.), on dry banks forming tangled thickets; 2500–4000 feet; fl. sporadically. *A 5826! A. M. Barry! H 5711* Native of China, also in S. United States

6. FRAGARIA L. (1753)

Petals white; petioles with spreading hairs; epicalyx-segments not toothed; ripe fruit 8–15 mm. in diameter, bright red, edible   1. *vesca*
Petals yellow; petioles with loose fine hairs; epicalyx-segments 3-toothed; ripe fruit about 1 cm. in diameter, bright red, tasteless   2. *indica*

1. F. vesca L., Sp. Pl. **1**: 494 (1753).—Wild Strawberry.

Perennial herb with a stout rhizome covered with more or less persistent leaf-bases and stipules; plants stoloniferous; leaflets paler beneath, up to 4 cm. long and 2·5 cm. broad; flowers 10–15 mm. across.

Locally rather common (St. Andr., Port., St. Thom.), on banks and in pastures; 3500–7400 feet; fl. and fr. most of the year. *A 5767! H 9216! P 23782!* Native in north temperate regions; Cuba, Hispaniola.

2. F. indica Andr., Bot. Repos. **7**: t. 479 (1807).—*Duchesnea indica* (Andr.) Focke (1888).

Stoloniferous herb like the last, altogether a little more robust; leaflets up to 3 cm. long and 2·5 cm. broad; flowers 15–20 mm. across.

Rare and local (St. Andr., West., Port.), on grassy slopes and in moist glades in montane woodland; (2000–) 4400–4900 feet; fl. and fr. July–Aug. *P 23783!* Native of subtropical and tropical Asia, now widely dispersed.

7. POTENTILLA L. (1753)

1. P. anglica Laicharding, Veg. Eur. **1**: 475 (1790).—*P. procumbens* Sibth. (1794).

Tufted herb with long trailing branches rooting at the nodes; petioles variable up to about 15 cm. long; leaflets obovate-elliptical, narrowed to base, toothed, up to 4·5 cm. long and 2 cm. broad; flowers solitary on long peduncles; petals obovate, 6–12 mm. long, golden-yellow.

Very local (St. Andr., Port.), on stone walls, roadside banks and waste places; 3600–4000 feet; fl. Dec–Aug. *A 5791! A. M. Barry! H 11938!* Native of W. and C. Europe.

92. CONNARACEAE

Trees or erect or climbing shrubs. Leaves alternate, compound, 1–3-foliolate or imparipinnate; leaflets entire; without stipules. Flowers bisexual, rarely unisexual and dioecious, small, actinomorphic or slightly zygomorphic. Perianth biseriate; (4–) 5-merous; sepals free or basally connate, segments imbricate or valvate, often persistent in fruit; petals free or shortly connate, imbricate or rarely valvate. Stamens (4–) 5 or (8–) 10, the filaments alternately short and long (heterostyle), free or united at the base; anthers 2-locular, opening lengthwise. Carpels (1–) 5, free, 1-locular with 2 collateral ovules ascending from the inner angle of each; ovules orthotropous or anatropous; style simple capitate. Fruit 1–5-carpellary, each follicle stalked or sessile, dehiscent, usually 1-seeded. Seeds often arillate, with or without endosperm.
24 genera with over 300 species, pantropical.

1. ROUREA Aubl. (1775) nom. cons.

Panicles terminal and axillary; sepals persistent but not much enlarged in fruit; stamens unequal, 10; carpels 5, but 4 usually imperfectly developed; follicle sessile; seed without endosperm.

1. R. glabra Kunth, Nov. Gen. **7**: 41 (1824).—*R. paucifoliolata* Planch. (1850). *R. surinamensis* Miq. (1854).
Woody vine climbing to 30 m. or more, lateral branches becoming tendrils or hooks; leaves glabrous, long-petioled; leaflets (1–) 3–5 (–9), oblong-elliptical, acuminate, 3–11 cm. long, 2–5 cm. broad; panicles 5–10 cm. long; pedicels 2–8 mm. long; calyx-lobes ovate, ciliolate, about 3 mm. long; petals lanceolate, acute, about 5 mm. long, white or cream-coloured; follicle 1–1·5 cm. long; seed brown, 1 cm. long; aril 4 mm. long.
Occasional in wet sheltered woodlands on limestone, mostly in the central parishes; 400–2200 feet; fl. Apr–Sept, fr. Sept–Apr. *A 8936! H8753! P 8643! 23087!* Honduras to the Guianas, Greater Antilles, St. Lucia, Grenada, Trinidad.
R. paucifoliolata has been distinguished from *R. glabra* primarily on the relative lengths of the carpels and stamens. In view of the heterostyle conditions described elsewhere in this family, a taxonomic distinction based on such characters may not be valid.

93. CAESALPINIACEAE

Trees, shrubs or herbs. Leaves alternate, pinnate or bipinnate, rarely 2- or 1-foliolate or simple, usually stipulate; stipels mostly absent. Flowers bisexual, zygomorphic or subregular in spikes, racemes or panicles, rarely cymose or solitary. Perianth usually biseriate; sepals 5, free and imbricate or valvate or variously connate; petals up to 5 or absent, the adaxial (upper) petal inside in bud, the others imbricate, free. Stamens ideally 10 in 2 whorls of 5, rarely more numerous, free or connate, when fewer than 10 the others often replaced by infertile or rudimentary stamens (staminodes). Ovary superior, 1-locular, sometimes stipitate on a concave or convex receptacle; ovules anatropous in 2 adaxial marginal rows; style simple. Fruit typically a legume but sometimes indehiscent. Seeds with a large embryo, with or without endosperm.
About 150 genera with some 2200 species mostly in the tropics and subtropics.

1 Leaves apparently simple:
  2 Petal 1, with or without 2 rudimentary petals; calyx closed in bud, rupturing irregularly; stamens numerous, of two kinds; leaves 1-foliolate   **1. Swartzia**
  2 Petals 5; calyx of 5 more or less valvately fused sepals, spathaceous at anthesis; stamens 1–10; leaves composed of 2 united leaflets   **2. Bauhinia**
1 Leaves obviously compound with 2 or more free leaflets:

3 Leaves bipinnate; flowers 5-merous; stamens 10:
   4 Sepals subvalvate:
      5 Pinnae in 1 or 2 pairs, long, flattened, winged, with minute leaflets, the primary rachis developed into a spine; fruit slender, constricted between the seeds
                                                          3. Parkinsonia
      5 Pinnae in several to many pairs and otherwise not as above; fruit massive, woody, not constricted between the seeds
                                                          4. Delonix
   4 Sepals strongly imbricate, the anterior scoop-shaped:
      6 Abaxial sepal distinctly larger than the others, glandular-pectinate; fruit flat, indehiscent, thin at the margins; racemes terminal; plant not prickly
                                                      5. Peltophorum
      6 Abaxial sepal only slightly larger than the others, not glandular-pectinate:
         7 Fruit thin, papery, splitting in the middle of the valves; racemes axillary; plant unarmed or spines formed by reduced axillary branches   **6. Haematoxylum**
         7 Fruits woody, fleshy or leathery, not papery, if dehiscent opening at the margins; racemes axillary or in terminal panicles; plant with curved epidermal thorns or unarmed
                                                      7. Caesalpinia
3 Leaves pinnate; sepals imbricate:
   8 Calyx 5-merous; leaflets opposite:
      9 Petals absent; flowers small in panicles of spikes; stamens 10, equal; fruit woody, indehiscent, 1-seeded; leaflets in 1 or 2 pairs   **12. Prioria**
      9 Petals present, 5:
         10 Fruits splitting at one or both margins or indehiscent; stamens often unequal; anthers basifixed, sometimes opening by pores or short slits   **8. Cassia**
         10 Fruit splitting in the middle of the valves, not at the margins; stamens 10, equal; anthers versatile
                                                       6. Haematoxylum
   8 Calyx 4-merous:
      11 Leaflets 2; petals 5; stamens 10; fruit large, woody, indehiscent, with few seeds
                                                      9. Hymenaea
      11 Leaflets several to many:
         12 Petals 3 with in addition 2 rudimentary petals; stamens 3, united into a sheath; leaflets small, opposite in 10–20 pairs; fruit elongated, sometimes constricted between the 2–8 seeds, indehiscent   **10. Tamarindus**
         12 Petals absent; stamens (8–) 10, free; leaflets alternate, 7–13; fruit oblong with 1 or 2 seeds, tardily dehiscent   **11. Crudia**

1. SWARTZIA Schreb. (1791) nom. cons.; Cowan (1968)

1. S. simplex (Sw.) Spreng. in L., Syst. Veg. ed. 16, **2**: 567 (1825).—*Possira simplex* Sw. (1788).

Glabrous or glabrescent shrub or tree to 6 m. high; leaflet lanceolate to elliptical, long-acuminate in the Jamaican variant, 8–14 cm. long, up to about 4·5 cm. broad; stipules linear, about 5 mm. long, caducous; racemes axillary and terminal, 2–4-flowered; petal roundish, yellow, up to about 4 cm. broad; fruit inflated, 1–2 (–3)-seeded, beaked with the persistent style.

Apparently now known only in cultivation in Jamaica, mostly in sheltered humid areas; 200–500 feet; fl. June–July, Dec–Feb, fr. Feb, July. *Collins UCWI 6733! H 9279! J.P. 1011!* Mexico to Brazil, Dominica to Trinidad.

2. BAUHINIA L. (1753)

1 Fertile stamen 1:
   2 Leaves mostly less than 8 cm. broad; petals less than 3 cm. long; wild species
                                                      1. *divaricata*
   2 Leaves mostly 12 cm. or more broad; petals more than 3 cm. long; cultivated species; natives of tropical Asia:
      3 Petals white; shrub   *B. acuminata* L.
      3 Petals pink dotted red, with one petal mostly red; tree   *B. monandra* Kurz
1 Fertile stamens 3–5; cultivated plants:
   4 Leaves up to 7 cm. broad; petals with limb about as long as broad, abruptly clawed, clear light scarlet, about 4 cm. long; stamens usually 3; shrub; native of S. Africa
                                                B. galpinii N.E. Br.
   4 Leaves rarely less than 9 cm. broad; petal-limb mostly about twice as long as broad, not scarlet; trees; native of tropical Asia; Poor Man's Orchid
      5 Fertile stamens 3 (–4); petals oblanceolate and gradually clawed, mostly pinkish with a darker midline, about 5·5 cm. long and less than 2 cm. broad   *B. purpurea* L.

5 Fertile stamens 5; petals obovate, the lateral petals hardly clawed, the median more distinctly so, light violet, the mid-petal also with light crimson centre, or rarely petals all white, about 6 cm. long and more than 2 cm. broad   *B. variegata* L.

1. B. divaricata L., Sp. Pl. **1**: 374 (1753).—Bull Hoof, Moco John.

Shrub 2–4 m. or tree to 6 m. high; leaves 2-lobed, 3–8 (–11) cm. long, up to 8 cm. broad; racemes mostly terminal, elongating during flowering; petals 1·2–2·5 cm. long, white turning pink; filament of solitary anterior fertile stamen 3–4 cm. long, white; anther black; gynoecium sometimes rudimentary; fruit flat, dehiscent, 6–12 cm. long, 1–1·5 cm. broad; seeds flat, 6–8 mm. long.

Common in thickets and open woodlands on limestone, mostly in rather dry or well-drained areas; SL–2500 feet; fl. and fr. all the year. *A 5642! H 8980! P 22847!* British Honduras, Greater Antilles, St. Kitts; Grand Cayman.

3. PARKINSONIA L. (1753)

1. P. aculeata L., Sp. Pl. **1**: 375 (1753).—Jerusalem Thorn.

Tree up to 6 (–10) m. high; young stems green, drooping; stipules forming lateral spines on either side of the leaf-rachis spine; pinnae up to 30 cm. long; leaflets elliptical to oblong or obovate, obtuse, 2–6 mm. long; flowers in small axillary racemes; petals yellow, the adaxial with red spots, about 12 mm. long; fruit pointed, striate, 5–10 (–15) cm. long with 1–3 (–5) seeds; seeds compressed-ellipsoid, 10–12 mm. long, about 5 mm. broad.

Established locally (St. Andr., St. Cath., St. Eliz.), in arid south-coast areas, often planted elsewhere; SL–20 (–700) feet; fl. and fr. sporadically throughout the year. *A 8909! P 15332! Stearn 850!* Native of tropical Amer., extending from southern United States to Argentina, also Greater Antilles and drier islands of the Lesser Antilles; introduced into the Old World tropics.

4. DELONIX Raf. (1837)

1. D. regia (Boj. ex Hook.) Raf., Fl. Tellur. **2**: 92 (1837).—*Poinciana regia* Boj. ex Hook. (1829). Flamboyant, Poinciana.

Tree, usually with wide-spreading branches, 5–10 (–15) m. high; leaves up to 45 (–60) cm. long with 11–18 pairs of pinnae 5–10 cm. long; leaflets 12–25 pairs oblong, about 1 cm. long and 2·5–4 mm. broad; stipules negligible to bipinnate; flowers in lax racemes; petals clawed, light or dark scarlet, more rarely orange or yellow, 3–5 (–6) cm. long, up to 5 cm. broad across the limb; fruit up to 60 cm. long, 5–6 cm. broad; seeds up to 2·5 cm. long and 6 mm. in diameter.

Commonly cultivated and occasionally naturalized; SL–1750 feet; fl. mostly June–Sept, fr. July–Nov. *Harris!* Native of Madagascar, now widely introduced in the tropics.

5. PELTOPHORUM (T. Vogel) Benth. (1840) nom. cons.

1. P. linnaei Benth. in Hook., Journ. Bot. **2**: 75 (1840).—*P. brasiliense* of F. & R. (1920). *Caesalpinia brasiliensis* L. (1753), partly. Braziletto.

Tree with flaky bark, 5–15 (–25) m. high; pinnae 2–4 pairs, up to about 15 cm. long; leaflets 6–8 pairs, elliptical, puberulous, 2–4 (–5) cm. long, 1–2 (–2·5) cm. broad; racemes 8–20 cm. long; fringed sepal orange; petals yellow, 6–7 mm. long; fruit 6–10 cm. long, about 3 cm. broad, brown, thinly puberulous.

Locally common, in coastal areas of the central and western parishes, in thickets and open woodlands on arid limestone; SL–1500 feet; fl. Apr–July, Nov, fr. July–Feb. *A 8852! Campbell 6481! P 8670! 26392!* Yucatan, Cuba, cultivated Barbados and Puerto Rico. *Peltophorum pterocarpum* (DC.) Back. ex K. Heyne of S.E. Asia and N. Australia has been introduced as an ornamental; it can be distinguished by the rusty pubescence of at least the young branches, the smaller oblong more numerous leaflets and the larger flowers. The inflorescence of this species is a large panicle and the exterior sepal is not obviously pectinate.

6. HAEMATOXYLUM L. (1753)

1. H. campechianum L., Sp. Pl. **1**: 384 (1753).—Logwood.

Gnarled tree with the trunk and larger branches fluted, up to 10 m. high; bark light grey, rather smooth; leaves pinnate or bipinnate; leaflets in 2–4 pairs on each rachis, obovate to obcordate, 1–3 (–4·5) cm. long, 7–20 (–32) mm. broad, closely veined; racemes 5–12 cm. long; flowers fragrant; sepals purplish-red, 4·5 mm. long; petals yellow, 6–7 mm. long; fruit pointed at both ends, 3–5 cm. long, up to 3-seeded.

Common on exposed limestone hillsides in dry secondary thickets and planted in fences; SL–1500 feet; fl. Dec–May, fr. Dec–Mar. *A 6277! H & P 13467! Wynter UCWI 1091!* Native of C. Amer., introduced and widely naturalized in the Caribbean and some parts of the Old World tropics.

7. CAESALPINIA L. (1753)

1 Leaves imparipinnate; plant a small tree without prickles; leaflets linear-oblong, very
   numerous, 5–9 mm. long, 1–2 mm. broad   **1.** *coriaria*
1 Leaves paripinnate; leaflets larger and fewer; stems thorny or prickly:
  2 Fruits prickly, with 1 or 2 globose to ellipsoid seeds; scrambling shrubs:
    3 Stipules leafy; caducous; leaflets shortly acuminate, obtuse, rounded or emarginate;
      bracts spreading; seeds grey   **2.** *bonduc*
    3 Stipules wanting; leaflets acuminate; bracts erect; seeds yellow   **3.** *major*
  2 Fruits smooth, with usually 3 or more flattish seeds:
    4 Pinnae 2–3 pairs; leaflets 2 pairs; small tree   **4.** *vesicaria*
    4 Pinnae 5–10 pairs; leaflets 8–13 pairs:
      5 Erect shrub; stamens more than twice as long as petals; fruit glabrous
                                             5. *pulcherrima*
      5 Scrambling or climbing very prickly shrub; stamens less than twice as long as petals;
        fruit thinly puberulous   **6.** *decapetala*

1. C. coriaria (Jacq.) Willd. in L., Sp. Pl. ed. 4, **2**: 532 (1799).—*Poinciana coriaria* Jacq. (1763). Divi-divi, Libi-dibi.

Tree 5–10 m. high with crooked branches; leaves 6–8 cm. long with 4–8 pairs of opposite pinnae; leaflets 18–28 pairs; panicles terminal and axillary with numerous fragrant white flowers; fruit curled, indehiscent, 3·5–5 cm. long, about 2 cm. broad.

Rather local, (St. Andr., Manch.), in arid open areas; 250–700 feet; fl. May, Aug–Jan, fr. Oct–Apr. *A. M. Barry! Wynter UCWI 1258!* Venezuela, Colombia, Greater Antilles and to Trinidad on the drier islands.

2. C. bonduc (L.) Roxb., Fl. Ind. **2**: 362 (1832).—*Guilandina bonduc* L. (1753). *C. bonducella* (L.) Flem. (1810); F. & R., Fl. Jam. **4**: 93 (1920). Grey Nickal or Nicker.

Shrub trailing or scrambling to 5 m. on shrubs and trees, rarely erect; stem puberulous with straight prickles; leaves up to 60 cm. long, the rachis with curved prickles; pinnae 5–9 pairs, up to about 15 cm. long; leaflets 5–7 pairs, puberulous, up to 4 (–5) cm. long and 2·5 cm. broad; flowers yellow, in axillary racemes; petals about 12 mm. long; stamens shorter than petals; fruit 5–7 cm. long, 3–5 cm. broad, with 1 or 2 seeds; seeds subglobose, 1·5–2 cm. in diameter.

Common in thickets near the sea; SL–20 feet; fl. and fr. Aug–Apr. *A 8316! H 9824! P 11143!* General in the subtropics and tropics; Grand Cayman.

3. C. major (Medic.) Dandy & Exell in Journ. Bot. **76**: 180 (1938).—*Bonduc majus* Medic. (1786). *C. bonduc* of F. & R. (1920), not (L.) Roxb. Yellow Nickal or Nicker.

Scrambling shrub like the last; leaflets up to 7 cm. long and 4 cm. broad; seeds oblong-globose, up to about 23 mm. long and 18 mm. broad.

Uncommon in thickets and woodlands on limestone, mostly in the central parishes; 1150–2500 feet; fl. May–Sept, fr. Sept, Mar. *A 12468! H 12391! Powell 825! Webster & Proctor 5514!*

When correcting the name of the common widespread Grey Nicker to *C. bonduc*

(L.) Roxb., Dandy and Exell (1938) adopted the next available legitimate epithet for the closely related Yellow Nicker. They accepted, following Urban (Symb. Ant. **2**: 272 (1900)) and Fawcett & Rendle (1920), that this was also a widespread species. *Bonduc majus* Medic. was however based on an East Indian plant with oblong long-beaked pods with up to 4 small round grey seeds and therefore the use of this name for Jamaican plants is doubtfully correct. When inflorescences and flowers are better known, Jamaican plants will undoubtedly be shown to be distinct, and possibly endemic, with their closest relatives amongst the yellow-seeded species of Cuba.

4. C. vesicaria L., Sp. Pl. **1**: 381 (1753).—Indian Savin Tree.
Strongly aromatic tree, often with a thick trunk with corky spines, 5–8 m. high; branchlets prickly; leaves 5–10 cm. long; leaflets broadly obliquely obovate; emarginate, 1·5–3·5 cm. long, 1–2·5 cm. broad; flowers in simple or compound terminal racemes; anterior sepal 8 mm. long, greenish, the other sepals yellow; petals yellow, the posterior with red marks; fruit somewhat swollen, 3·5–5 cm. long, reddish, 1–5-seeded; seeds 7–8 mm. in diameter, rather flat.
Rather local in arid thickets on limestone near the sea, mostly in the southern parishes; SL–600 feet; fl. Nov–Apr, fr. Jan–Apr, Aug. *A 6061! H 11858! Loveless UCWI 3800! P 16086!* Mexico, Cuba, Hispaniola, Curaçao.

5. C. pulcherrima (L.) Sw., Obs. Bot.: 166 (1791).—*Poinciana pulcherrima* L. (1753). Pride of Barbados.
Shrub or small tree with scattered prickles on the branches, 2–5 m. high; young twigs glabrous and somewhat glaucous; leaves mostly 30 cm. or more long; leaflets oblong, 9–18 mm. long, up to 8 mm. broad; racemes terminal; pedicels up to about 5 cm. long; larger petals up to 2·5 cm. long, yellow, yellow streaked red, or rarely crimson, adaxial petal smaller with a tubular claw; fruit flat, 8–11 cm. long, 1·5–2 cm. broad; seeds flat.
Common in cultivation at lower elevations; fl. and fr. most of the year. *A. M Barry! H 8231!* Origin uncertain but probably native in tropical Asia, now cultivated as an ornamental throughout the tropics.

6. C. decapetala (Roth) Alston in Trimen, Handb. Fl. Ceylon **6**, Suppl.: 89 (1931).
—*Reichardia decapetala* Roth (1821). *C. sepiaria* Roxb. (1832). Wait-a-bit.
Densely prickly shrub, erect to 2 m. or usually climbing, tomentose; leaves 20–30 cm. long; leaflets oblong, 8–20 (–25) mm. long, up to 10 mm. broad; racemes terminal and axillary; pedicels up to about 2 cm. long; larger petals about 12 mm. long, light yellow; fruit slightly curved and thickened, 7–10 cm. long, up to about 3 cm. broad, long-beaked; seeds olive, about 10 mm. long.
Frequent on open banks along roads and streams, rarer in the western parishes; 300–3500 feet; fl. Nov–June, fr. Dec–Apr. *A 8768! H & P 13591! Robertson UCWI 1052!* Native of India and S.E. Asia, naturalized in many parts of the American tropics, introduced in W. Africa.

8. CASSIA L. (1753)

1 Trees:
  2 Leaf-rachis with a cylindrical gland between at least the proximal pair of leaflets; small
    cultivated ornamental trees:
    3 Leaflets in 1–2 pairs, obliquely ovate, 2·5–12 cm. long; fruit cylindrical, opening
      along one suture; native of tropical S. Amer.; (*C. bacillaris* L. f.)   *C. fruticosa* Mill.
    3 Leaflets in 4–6 pairs, elliptical, 3–7 (–10) cm. long; fruit flat, dehiscent along both
      margins; native of E. Indies and Australasia; (*C. glauca* Lam.)   *C. planisiliqua* L.
  2 Leaf-rachis glandless:
    4 Fruit more or less flat, dehiscent; petals yellow; leaflets rounded at tip:
      5 Leaflets (2–) 3 (–5) pairs   1. *emarginata*
      5 Leaflets 6–14 pairs, oblong, 3–7 cm. long; racemes corymbose; fruit 20–25 cm. long,
        12–15 mm. broad; commonly planted; native of Indo-Malaysia; Siamese Cassia
                                        C. siamea Lam.
    4 Fruit cylindrical or quadrangular; introduced ornamental and occasionally natural-
      ized trees:

6 Petals yellow:
7 Fruit indehiscent, cylindrical, up to 60 cm. long and 2 cm. in diameter; leaflets 4–8 pairs, ovate, minutely emarginate, 6–15 (–22) cm. long; inflorescence pendulous; native of tropical Asia; Cassia Stick Tree   *C. fistula* L.
7 Fruit tardily dehiscent along one suture, cylindrical or quadrangular, up to 25 cm. long, about 1 cm. in diameter; leaflets (4–) 12–15 (–18) pairs, lanceolate, shortly acutely acuminate, 3–8 cm. long; inflorescence erect; native of northern S. Amer., Trinidad and Tobago   *C. spectabilis* DC.
6 Petals red, pink or white; fruit indehiscent; leaflets about 10–20 pairs, hairy:
8 Fruit more or less smooth and not obviously constricted between the seeds, up to 50 cm. long and 14–15 mm. in diameter; leaflets about 6 cm. long; native of tropical Asia; Pink Cassia (*C. nodosa* Ham.)   *C. javanica* L.
8 Fruit not as above:
9 Fruit rough with transverse ridges, obtusely ribbed on one side, up to 60 cm. long and about 4 cm. in diameter; leaflets 2·5–6 cm. long; native of continental tropical Amer.; Horse Cassia   *C. grandis* L. f.
9 Fruit somewhat constricted between the seeds, otherwise smooth, 20–30 cm. long and about 2 cm. in diameter; leaflets 2–3·5 cm. long; native of Indo-Malaysia, reported as naturalized in Jamaica; (*C. marginata* Roxb.)   *C. roxburghii* DC.
1 Herbs or erect or climbing shrubs; petals mostly yellow:
10 Fruit terete, turgid or tetragonal, not flat or winged; leaves with glands:
11 Leaflets (1–) 2 pairs, with a cylindrical gland between the proximal pair; fruit dehiscent by one suture:
12 High-climbing vine with ribbed stem; leaflets mostly up to 6 cm. long   2. *viminea*
12 Shrub; leaflets up to 12 cm. long   *C. fruticosa* Mill.
11 Leaflets (2–) 3–5 (–6) or rarely more pairs:
13 Gland at base of petiole adaxially; herbs or undershrubs; leaflets acuminate or acute:
14 Leaflets hairy   14. *hirsuta*
14 Leaflets subglabrous, margins ciliate; see 31–31 ahead:
13 Gland between at least the proximal pair of leaflets:
15 Fruit indehiscent or bursting irregularly, terete; large shrubs:
16 Leaflets acuminate, 3–4 pairs; gland cylindrical   3. *floribunda*
16 Leaflets rounded at tip, 3–5 pairs; gland oblong or clavate   4. *bicapsularis*
15 Fruit dehiscent; leaflets 2–4 (–5) pairs, obtuse or rounded at tip; glands awl-shaped; erect herbs:
17 Fruit linear, straight, turgid with prominent margins, deeply impressed between the seeds, villous; glands long-stalked   5. *uniflora*
17* Fruit very narrow, usually curved, subtetragonal, not impressed between the seeds, thinly pubescent; glands subsessile or short-stalked:
18 Leaves with a gland between the proximal pair of leaflets only; pedicels in fruit 12–22 (–35) mm. long; three anticous anthers narrowed at apex 6. *obtusifolia*
18 Leaves with a gland between the proximal and middle pairs of leaflets; pedicels in fruit 5–11 mm. long; three anticous anthers rounded-truncate at apex
                    7. *tora*
10 Fruit flat, compressed or winged, dehiscent:
19 Fruit 4-winged, straight; short-lived shrub; leaflets 6–12 pairs, broadly oblong; glands wanting   8. *alata*
19 Fruit not 4-winged:
20 Fruit 2-winged with crests over the seeds, oblong, curved; glaucous annual undershrub; leaflets 3–7 pairs, obovate; glands wanting; stamens 10   9. *italica*
20 Fruit not winged or crested:
21 Gland(s) on petiole and/or rachis wanting:
22 Erect annual viscid-glandular herb; leaflets 2 pairs, 2–3 cm. long; flowers racemose; stamens 5–7   10. *absus*
22 Slender-branched, prostrate, spreading or ascending subwoody herbs; leaflets mostly less than 2 cm. long; flowers axillary:
23 Leaflets 1 pair, obovate-elliptical; stipules ovate, cordate   18. *rotundifolia*
23 Leaflets 2–5 pairs, oblong; stipules lanceolate   19. *pilosa*
21 Gland(s) on petiole and/or rachis present:
24 Gland between at least the leaflets of the proximal pair; stipules subulate to linear-lanceolate, mostly deciduous:
25 Leaflets 2–4 (–5) pairs:
26 Gland subglobose-tuberculate; leaflets oblong to oblong-lanceolate or elliptical, 2·5–6 cm. long; inflorescence corymbose-paniculate
                    11. *bahamensis*
26 Gland(s) cylindrical; flowers solitary or paired in the leaf-axils; see 17* above:
25 Leaflets (4–) 7–15 pairs, mostly elliptical:
27 Gland bluntly conical; leaflets 10–15 pairs; terminal leaflets oblong, up to 2 cm. long   12. *clarendonensis*

27 Gland cylindrical, often pointed, sometimes stalked; leaflets oblong to obovate:
28 Leaflets 5–15 pairs, 4–10 mm. long; pedicels 1 or 2 on a common peduncle
without glands at the base; leaf-gland solitary, between the proximal leaflets only, subulate; cultivated ornamental; native of Hispaniola and
Puerto Rico   C. polyphylla Jacq.
28 Leaflets mostly (4–)7–10(–13) pairs, 1·5–3 (–4) cm. long; pedicels 2–3 or if more
corymbose on a common peduncle, or rarely solitary, with glands at the base;
leaf-gland(s) 1 or 2 between the leaflets of the proximal pair(s)   13. biflora
24 Glands mostly on the petiole below the proximal leaflets, otherwise on the rachis
between the pairs:
29 Petals subequal; leaflets (3–) 4–8 (–10) pairs, mostly acuminate and more than
3 cm. long; stipules mostly caducous; inflorescences racemose or paniculate;
sepals rounded on back, not keeled; seeds transverse:
30 Leaflets and fruits conspicuously hairy; gland cylindrical   14. hirsuta
30 Leaflets and fruits glabrous or nearly so, the leaflet-margins often ciliolate:
31 Gland cylindrical-conical; leaflets 4–8 pairs, narrowly lanceolate, mostly
acute; fruit flat   17. ligustrina
31 Gland ovoid-globose to globose; leaflets ovate to lanceolate, mostly acuminate:
32 Leaflets (3–) 4–5 (–6) pairs; fruit more or less compressed even at maturity;
seeds in one row; pedicels pubescent   15. occidentalis
32 Leaflets (3–) 5–6 (–9) pairs; fruit swollen-terete at maturity but the sutures
still prominent; seeds in two rows; pedicels glabrescent   16. sophera
29 Petals unequal; leaflets 2–25 pairs, less than 3 cm. long and 1 cm. broad;
stipules mostly persistent; flowers solitary or in small axillary or supra-
axillary clusters; seeds oblique:
33 Pedicels slender, elongated, mostly 1·5 cm. or more long in fruit; leaflets
ciliate, otherwise usually glabrous:
34 Stems laxly erect, long-pilose; leaflets oblong, 2–5 pairs, up to 20 (–25) mm.
long and 10 mm. broad   19. pilosa
34 Stems prostrate, puberulous or glabrescent; leaflets linear-oblong, 4–9
pairs, 4–8 mm. long, 1·5–2·5 mm. broad   20. serpens
33 Pedicels shorter:
35 Leaflets leathery, oblong to obovate, mostly fewer than 10 pairs; fruit 5–6
mm. broad:
36 Leaflets tomentellous abaxially, glabrous to tomentellous adaxially, 2–7
(–11) pairs, 6–20 mm. long, 2–7 mm. broad; gland solitary, sessile; fruit
2·5–6 cm. long   21. lineata
36 Leaflets nearly glabrous, 4–8 pairs, 13–25 mm. long, 5–9 mm. broad;
glands 1–3 (–5), long-stalked; fruit 3–4 cm. long   22. jamaicensis
35 Leaflets thin, oblong-linear, less than 5 mm. broad, (5–) 8–25 pairs:
37 At least the larger petals more than 10 mm. long, the open flower more than
1·5 cm. across:
38 Gland(s) 1–several, stout, long-stalked; stem pubescent; leaflets 2–4 mm.
broad, glabrous above, puberulous beneath   23. glandulosa
38 Gland solitary, short-stalked; stem glabrescent; leaflets up to about 2 mm.
broad, glabrous   29. fasciata
37 Larger petals not exceeding 10 mm. long, the open flower less than 1·5 cm.
across:
39 Petiolar gland(s) sessile, depressed, orbicular, 1 or 2 together; plant densely
pubescent   24. patellaria
39 Petiolar gland stalked, elevated or obconic:
40 Petiolar gland slender-stalked, abruptly expanded to the limb:
41 Plant shrubby; stem with long yellowish spreading and short curled
hairs; leaflets 10–25 pairs, up to 3 mm. broad, ciliate; sepals 6–8 mm.
long; fruit 4 mm. broad, loosely villous   25. stenocarpa
41 Plant usually herbaceous; long spreading hairs on stem few or absent;
leaflets (5–) 10–15 pairs, about 2 mm. broad, ciliate or subglabrous on
margins; sepals 4–5 mm. long; fruit 3 mm. broad, pubescent or
glabrescent   26. chamaecrista
40 Petiolar gland short-stalked, usually broadened upwards gradually;
leaflet-margin ciliolate or minutely ciliolate:
42 Stem and leaf-rachis long-pilose with straight yellowish hairs and
shorter curved hairs; leaflets 6–12 mm. long, about 1·5 mm. broad;
fruit loosely villous   27. caymanensis
42 Stem and leaf-rachis shortly or thinly pubescent with mostly curved
hairs; leaflets (6–) 10–17 mm. long, up to 2–2·5 mm. broad:
43 Fruit appressed-pubescent to puberulous with curved hairs; 3–3·5 cm.
long, 4 mm. broad   28. smaragdina
43 Fruit hispid with straightish hairs; 3–5 cm. long, 5–6 mm. broad
                                                          29. fasciata

1. C. emarginata L., Sp. Pl. **1**: 376 (1753).—*Isandrina emarginata* (L.) Britton &
Rose (1924). Senna Tree, Yellow Candle Wood.
Shrub to 2 m. or tree to 6 m. high; wood and leaves with pungent odour;
leaflets elliptical to obovate-elliptical, rounded or emarginate at apex, puberulous
adaxially, pubescent abaxially, 1·5–6·5 cm. long, 1–3 cm. broad; stipules bristlelike,
3 mm. long; racemes corymbose, axillary; sepals and petals unequal, the smaller
petals 8–10 mm. long; fruits 12–25 (–35) cm. long, about 1 cm. broad.
Rather common, mostly in coastal areas on arid limestone; SL–800 feet; fl. Mar–
Aug, Nov–Dec, fr. Nov–Aug. *A 6954! Deans! H 6933! Stearn 801!* Mexico to
Venezuela, Greater Antilles, Antigua, Guadeloupe.

2. C. viminea L., Syst. Nat. ed. 10, **2**: 1016 (1759).—*Chamaefistula viminea* (L.)
Britton & Rose (1930).
Trailing and climbing shrub to 13 m. high with broad rounded ribs on the old
stems and swollen nodes; stems up to 4 cm. thick; wood bright yellow; leaflets
obliquely ovate-lanceolate, acute or acuminate, up to 8 cm. long and 3·5 cm. broad;
stipules bristlelike, 5–10 mm. long; racemes forming terminal panicles; petals
12–16 mm. long; fruit 3–7 (–10) cm. long.
Frequent in thickets and woodlands in hilly areas; (300–) 1000–5000 feet; fl.
July–Feb, fr. Oct–Feb. *A 7374! 7996! H & P 13597! J.P. 1112!* Endemic.

3. C. floribunda Cav., Descr.: 132 (1802).—*C. laevigata* Willd. (1809). *Adipera
laevigata* (Willd.) Britton & Rose (1924).
Shrub 1–2 m. high or small tree; leaflets ovate-lanceolate to elliptical, 3–8 cm.
long, up to 3·5 cm. broad; stipules caducous, 5–8 mm. long; racemes axillary;
petals 12–16 mm. long; fruit 6–9 cm. long.
Rather rare (St. Andr., Manch., St. Ann), in sheltered waste places; 2000–5000
feet; fl. May, Nov, fr. Nov. *A 11214! J.P. 601! Wynter UCWI 3042!* Continental
tropical Amer., Cuba, Puerto Rico, Trinidad and widespread in the Old World
tropics.

4. C. bicapsularis L., Sp. Pl. **1**: 376 (1753).—*Adipera bicapsularis* (L.) Britton &
Rose (1924).
Shrub 1·2–3 m. high; leaflets obovate-elliptical, 1·5–3·5 cm. long, 1–2 cm. broad;
stipules subulate, 3 mm. long, early caducous; racemes axillary or terminal and
paniculate; petals 10–18 mm. long; fruit straight or slightly curved, blunt, 6–15
cm. long, 1–1·5 cm. broad.
An occasional weed of roadsides; 10–2500 feet; fl. and fr. Nov–May. *H 10719!
P 11249!* Bermuda, continental tropical Amer., West Indies, introduced and
widespread in the Old World tropics.

5. C. uniflora Mill., Gard. Dict. ed. 8 (1768).—*Sericeocassia uniflora* (Mill.)
Britton (1930).
Annual shrubby herb 15–120 cm. high; leaflets elliptical to obovate, rather
thick, bluish-green adaxially, with yellowish or reddish hairs on both surfaces, 1–5
cm. long, 8–30 mm. broad; stipules bristlelike, rather persistent, 6–11 mm. long;
racemes axillary, very short; petals 5–7 mm. long; fruit 2·5–5 cm. long.
Frequent, usually in rather swampy pastures, roadsides and upper beaches;
SL–600 feet; fl. and fr. all the year. *A 9788! 11821! H 9927! H & P 14459!*
Bahamas, continental tropical Amer., Cuba, Hispaniola, Barbados.

6. C. obtusifolia L., Sp. Pl. **1**: 377 (1753).—*C. tora* of F. & R. (1920), not L.
Annual shrubby herb 15–120 cm. high; leaflets oblong-obovate, glabrous
adaxially, (2–) 3 pairs, 1·5–4 cm. long, 1–2·5 cm. broad; stipules subulate to linear-
lanceolate, 8–13 mm. long; flowers mostly paired on short peduncles in the leaf-
axils; petals light to orange-yellow, about 1 cm. long; fruit 15–20 (–25) cm. long;
areole on side of seed almost linear, 0·3–0·5 mm. broad.
Common as a weed of open waste places; SL–1200 feet; fl. and fr. all the year.
A 8211! H 12632! P 24246! Throughout the Old and New World tropics except
Australasia.

7. C. tora L., Sp. Pl. **1**: 376 (1753).
Annual shrubby herb like the last; leaflets up to 4·5 cm. long; flowers solitary or paired in the axils; areole on side of seed 1·5–2 mm. broad.
Very rare (Port.), a weed of pasture; 100 feet; fl. and fr. Aug. *A 11530!* Tropical Asia, Polynesia, Guadeloupe.

8. C. alata L., Sp. Pl. **1**: 378 (1753).—*Herpetica alata* (L.) Raf. (1838). Candlestick, King-of-the-Forest, Ringworm Shrub.
Glabrous short-lived shrub to 3·5 m. high; leaflets obtuse at tip, unequal at base, the terminal pair obovate, up to 15 (–20) cm. long and 7·5 (–13) cm. broad; stipules unequally broadened at base, falcate, up to 15 mm. long; racemes axillary, the flowers at first covered by broad orange deciduous bracteoles; petals golden-yellow, up to about 2 cm. long; ripe fruit black, 10–15 cm. long; seeds angular, arranged transversely in the pod, about 7 mm. long.
Locally common in swampy places but often cultivated; SL–3000 feet; fl. and fr. mostly Oct–Apr. *H 10045! P 11160!* Native of tropical Amer., now widespread in warm countries.

9. C. italica (Mill.) Lam. ex F. W. Andr., Fl. Pl. Angl.-Egypt. Sudan **2**: 117, f. 49 (1952).—*Senna italica* Mill. (1768). *C. obovata* Collad. (1816). Italian Senna, Port Royal Senna.
Glaucous annual shrubby herb 30–90 cm. high; leaflets broadly rounded and mucronulate at tip, obliquely oblong to obovate, 1·5–3 (–4) cm. long, 7–20 mm. broad; stipules triangular to lanceolate, acuminate, persistent, 4–6 mm. long; racemes axillary; pedicels deflexed; petals oblanceolate, light yellow, about 1 cm. long; ripe fruit purplish-brown, 3–6 cm. long; seeds broadly triangular, acuminate, grey-scaly with a dark oblong smooth area on each side, 5 mm. long, 3 mm. broad.
Rare (St. Andr.), in low-lying sandy thickets, escaped from past cultivation; about 10 feet; fl. and fr. mostly Aug–Mar. *A 12704! Grabham!* Native of the drier parts of the Old World tropics, cultivated and naturalized elsewhere.

10. C. absus L., Sp. Pl. **1**: 376 (1753).—*Grimaldia absus* (L.) Britton & Rose (1930).
Annual much-branched herb 20–60 cm. high; leaflets obliquely elliptical, glabrous adaxially, 2–3 cm. long; petioles slender, 1–3 cm. long; stipules lanceolate, about 3 mm. long; petals yellow, 5–7 mm. long; fruit linear, long-hispid, 2·5–4 cm. long, 3–6 mm. broad; seeds obovate, black, 3·5–4 mm. long.
Very rare (St. Andr.), in sandy waste places. *H 6907!* Generally but sporadically distributed throughout the tropics; Mexico to Paraguay, Grenada, Bonaire.

11. C. bahamensis Mill., Gard. Dict. ed. 8 (1768).—*Peiranisia bahamensis* (Mill.) Britton & Rose (1930).
Shrub 1–3 m. high; leaflets glabrous above, sparingly pubescent at base of mid-rib abaxially; pedicels 8–20 mm. long; sepals about 7 mm. long; petals 10–15 mm. long; fruit flat, 7–10 cm. long, 4–8 mm. broad.
Fawcett & Rendle included this species on the basis of a collection by Hitchcock from Port Morant. This specimen has not been seen by the present author; Britton & Rose (1930) did not confirm the species for Jamaica.

12. C. clarendonensis Britton in Bull. Torr. Bot. Club **42**: 388 (1915).—*Peiranisia clarendonensis* (Britton) Britton & Rose (1930).
Puberulous shrub about 1·2 m. high; leaflets elliptical to oblong, 1–2 cm. long; stipules linear-lanceolate, 3 mm. long; racemes axillary, corymbose, with 2–6 flowers; sepals 7–9 mm. long; petals yellow, 13–14 mm. long; fruit slightly curved, impressed between the seeds, 7–10 cm. long, 6 mm. broad; seeds ellipsoidal, 4 mm. long.
Very rare (Clar.); 200 feet; fl. and fr. Oct–Dec. *H 11693! 12149!* Endemic.

13. C. biflora L., Sp. Pl. **1**: 378 (1753).—*C. oxyadena* DC. (1825). *C. fulgens* Macf. (1837). *Peiranisia jamaicensis* Britton (1930).
Shrub or shrubby tree 2–5 m. high, the young twigs, petioles and inflorescence densely villous to puberulous or glabrescent; leaflets elliptical to oblong-elliptical, the terminal oblong-obovate and larger; stipules bristlelike, 5–8 mm. long, deci-

duous; racemes corymbose-umbellate in the upper axils, (1–) 2–6-flowered; sepals unequal, 5–9 mm. long; petals clear rich yellow, 1·5–2·3 cm. long; fruit slightly curved, impressed between the seeds, 6–15 cm. long, 7–9 mm. broad.

Rare and local (St. Andr., St. Thom.), in thickets on open hillsides; 2000–4800 feet; fl. July–Nov, fr. Oct. *A 7404! 12584! H 9031!* Mexico southwards in continental Amer., Bahamas, Cuba, Hispaniola, Trinidad.

14. C. hirsuta L., Sp. Pl. **1**: 378 (1753).—*Ditremexa hirsuta* (L.) Britton & Rose (1924).

Shrubby herb mostly 1–2 m. high; leaflets ovate to ovate-lanceolate, 3–5 pairs, 4–10 cm. long, up to 4 cm. broad, larger distally, bad-smelling when bruised; stipules linear, sometimes persistent, 7–10 mm. long; inflorescences few-flowered; petals yellow, about 1 cm. long; fruit linear, slightly curved, (4·5–) 10–15 (–20) cm. long, 5–6 mm. broad; seeds olive-brown, 3·5 mm. long.

Locally common in the southern parishes in weedy pastures and rocky waste places; 50–1500 feet; fl. Oct–Jan, fr. Nov–Mar. *A 9806! Powell 864! Yuncker 17301!* Native of continental tropical Amer., now widely scattered in the Old and New World as a weed; apparently of quite recent introduction to Jamaica but spreading rapidly.

15. C. occidentalis L., Sp. Pl. **1**: 377 (1753).—*Ditremexa occidentalis* (L.) Britton & Rose (1924). Dandelion, Piss-a-bed, Stinking Weed, Wild Coffee.

Erect shrubby herb like the last; leaflets ovate to lanceolate, 3–10 cm. long, up to 3·5 cm. broad, often with a narrow red margin; stipules triangular, long-acute, early deciduous, 7–8 mm. long; petals yellow, drying paler or white, 10–14 mm. long; fruit linear, slightly curved, 5–12 cm. long, 6–8 mm. broad; seeds broadly ovoid, minutely punctulate except for a smooth flat areole on each side, dark olive, about 4 mm. long.

Common in waste places, especially in coastal areas; SL–1000 feet; fl. and fr. most of the year. *A 9804! P 9401! Thompson 7993! Yuncker 17087!* General throughout the subtropics and tropics; Cayman Is.

16. C. sophera L., Sp. Pl. **1**: 379 (1753).—*Ditremexa sophera* (L.) Britton & Rose (1924).

Shrubby herb like the last up to about 1·2 m. high; leaflets acute or acuminate, up to about 5 cm. long and 1·5 cm. broad, the terminal pair not always the largest; petals slightly larger and more deeply coloured than the last; pedicels with a few small glands, otherwise almost glabrous; fruit beaked, up to about 9 cm. long and 8 (–10) mm. broad; seeds obliquely D-shaped, 4 mm. long and 3·5 mm. broad.

Rather common in open waste places; SL–1500 feet; fl. and fr. most of the year. *A 6862! H 5999! H & P 13348! J.P. 610! Wedderburn 60! 61!* Native of India and widespread in the Old World tropics; thinly scattered through continental tropical Amer. and the West Indies.

17. C. ligustrina L., Sp. Pl. **1**: 378 (1753).—*Ditremexa ligustrina* (L.) Britton & Rose (1924).

Shrubby herb mostly 1–2 m. high; leaflets thinly pubescent on midrib adaxially, 2–8 cm. long, 8–15 mm. broad; stipules linear, leafy, deciduous, about 7 mm. long; racemes in a terminal corymbose panicle; petals yellow or orange-yellow, 12–15 mm. long; fruit flat or slightly biconvex, glabrous, 7–14 cm. long, 6–8 mm. broad; seeds ovoid-compressed, 4 mm. long, the margin grey, areole black.

Rather common in waste ground, pasture margins and thickets, 20–2500 feet; fl. Feb, Apr, July–Sept, fr. July–Dec. *A 12718! H 9653! Powell 569!* Florida, Bermuda, Greater Antilles; Grand Cayman; cultivated in W. Africa.

18. C. rotundifolia Pers., Synops. Pl. **1**: 456 (1805).—*Chamaecrista rotundifolia* (Pers.) Greene (1899).

Stems prostrate, 30–45 cm. long; leaflets obliquely obovate-elliptical, 6–18 (–24) mm. long, glabrous except margin minutely ciliate; petiole hairy; stipules 5–7 mm. long; pedicels solitary, 1·5–3·5 cm. long, longer than the leaves; petals yellow, 4–5 mm. long; fruit flat, linear-oblong, slightly curved, 1·8–3·5 cm. long, 2·5–4 mm. broad.

Very rare (St. Mary); only once collected in 1838, *McNab!* General from Florida and Mexico to Brazil, Cuba, Puerto Rico, Ghana, Nigeria.

19. C. pilosa L., Syst. Nat. ed. 10, **2**: 1017 (1759).—*Chamaecrista pilosa* (L.) Greene (1899).

Bushy herb with spreading or ascending branches, 15–120 cm. high; stipules ovate-lanceolate; gland on petiole, if present, very small, sessile or shortly stalked; pedicels 1–4, 2–4 cm. long in flower; sepals usually crimson, 4–5 mm. long; petals yellow darkening to orange, 5–7 mm. long; fruit linear, 2–4 cm. long, 3–4 mm. broad, 9–13-seeded.

Occasional as a weed of waste places on alluvial soils, possibly calciphobe; 10–2500 feet; fl. and fr. Sept–Mar. *A 8691! H 12268! P 21526!* Guatemala to El Salvador, Colombia, Venezuela, Cuba, Trinidad.

20. C. serpens L., Syst. Nat. ed. 10, **2**: 1018 (1759).—*Chamaecrista serpens* (L.) Greene (1899).

Taprooted woody-based herb with spreading branches to 45 cm. long; stipules lanceolate; gland on petiole filiform-stalked; pedicels 1–2, 1–2·5 cm. long in flower; sepals yellowish-green, 4–5 mm. long; petals yellow turning orange, 4–6 mm. long, sometimes larger; fruit linear, 1·5–2·5 cm. long, 3–4 mm. broad, 5–10-seeded.

Uncommon (St. Andr., St. Eliz.), a casual weed of disturbed sandy or gravelly ground, possibly calciphobe; 10–600 feet; fl. and fr. July–Jan. *A 10265! H 6911! P 18441!* Mexico?, Guatemala, Colombia, Venezuela, Brazil, Cuba.

21. C. lineata Sw., Nov. Gen. & Sp. Pl.: 66 (1788).—*Chamaecrista lineata* (Sw.) Greene (1909).

Bushy shrub 60–100 cm. high near the sea, up to 2 m. high elsewhere, the twigs puberulous or glabrous; stipules lanceolate; gland on petiole sessile or subsessile, flat or shortly cylindrical; pedicels 1–3, 5–15 (–25) mm. long; sepals 7–10 mm. long; petals yellow, up to about 1·5 cm. long; fruit pubescent or glabrescent, 10–12-seeded.

Locally common, on coastal limestone and in arid rocky thickets; SL–2200 feet; fl. and fr. sporadically throughout the year. *A 8549! H 11725! P 11178! 27535!* Bahamas, Cuba, Hispaniola. Typical *C. lineata* has leaflets glabrous adaxially. Jamaican specimens with leaflets tomentellous on both surfaces have been equated with *C. clarensis* (Britton) Howard, otherwise endemic to Cuba, but both variants occur together in several Jamaican localities.

22. C. jamaicensis (Britton) Adams in Phytologia **20** (5): 311 (1970).—*Chamaecrista jamaicensis* Britton (1915). *Cassia polyadena* of F. & R. (1920), not DC.

Straggling shrub up to 120 cm. high, the twigs appressed-pubescent; stipules subulate; glands on petiole and rachis capitate or cylindrical; pedicels 1–3, 1–2 cm. long; sepals greenish-yellow, tinged red, 8–10 mm. long; petals yellow, 1–1·5 cm. long; fruit thinly ascending-hispid, 5–7 seeded.

Very local (St. Andr.), on arid rocky limestone banks and cliffs; 300–500 feet; fl. June–July, Oct–Feb, fr. Jan, June–July. *H 9615! P 11435!* Endemic. This species has several features, both morphological and ecological, which suggest its origin as a hybrid between *C. lineata* and *C. glandulosa*.

23. C. glandulosa L., Syst. Nat. ed. 10, **2**: 1017 (1759).—*Chamaecrista glandulosa* (L.) Greene (1899). Broom Cassia, Jamaican Broom.

Erect much branched shrub 60 cm. to 2 m. high; leaflets oblong, 8–16 pairs, 8–20 mm. long; stipules lanceolate, 5–8 mm. long; pedicels 1–3, 5–20 mm. long; sepals yellow, often with crimson margins, 8–9 mm. long; petals yellow, sometimes tinged crimson, 10–13 mm. long; fruit 3–4 cm. long, about 4 mm. broad, shortly pubescent.

Locally common on banks and in thickets on shale or serpentine; 750–4200 feet; fl. and fr. most of the year. *A 7832! J.P. 1197! P 20808!* Endemic. *C. glandulosa* var. *swartzii* (Wikstr.) Macbr. extends the range of this species to Puerto Rico and the drier Lesser Antillean islands southwards to Grenada and Barbados. *C. polyadena* DC., type from Guadeloupe, is probably not different from this variety.

24. C. patellaria DC. ex Collad., Hist. Cass.: 125, t. 16 (1816).—*Chamaecrista patellaria* (DC. ex Collad.) Greene (1899).

Annual or short-lived shrubby herb 30–60 (–100) cm. high; leaflets 10–25 pairs, mucronate and very oblique at tip, 8–15 mm. long, 1·5–3 mm. broad; stipules lanceolate, 8–12 mm. long; pedicels 1–4, up to 8 mm. long; sepals lanceolate, 5–6 mm. long; petals bright yellow, as long as or nearly as long as sepals; fruit 1·5–3·5 cm. long, about 4 mm. broad, hispid, 6–10-seeded.

Rather uncommon, in open sandy places and on shale banks, possibly calciphobe; 100–3500 feet; fl. and fr. sporadically throughout the year. *A 11461! Fawcett! P 20602!* Mexico to Brazil, Cuba, Puerto Rico, Trinidad and Tobago.

25. C. stenocarpa T. Vogel, Syn. Cass.: 68 (1837).—*C. broughtonii* F. & R. (1917). *Chamaecrista stenocarpa* (T. Vogel) Standl. (1916).

Shrubby herb to 1 m. or more high; leaflets 8–18 mm. long; stipules narrowly lanceolate, 10–14 mm. long; pedicels 1–3, 4–8 mm. long; petals bright yellow, up to 8 mm. long; fruit 2·5–5 cm. long, villous, up to about 16-seeded.

Rare and local (St. Andr., St. Thom.), in grassy or waste places; 600–1500 feet; fl. Oct–Nov, fr. Nov. *H 6949!* Mexico to Brazil, Cuba, Hispaniola, Bonaire.

26. C. chamaecrista L., Sp. Pl. **1**: 379 (1753).—*C. nictitans* of F. & R. (1920), not L.

Perennial or short-lived bushy herb up to 60 cm. high or branches sometimes prostrate or spreading; leaflets mostly about 10 mm. long, commonly glabrous adaxially and more or less pubescent abaxially, short- or long-ciliate on margins; stipules lanceolate, 7–8 mm. long; pedicels 1–3, 2–5 mm. long; petals yellow, as long as sepals except one up to 7 mm. long; fruit 1·5–3·5 cm. long, pubescent, 6–14-seeded.

Frequent in the central and eastern parishes, along roadsides and in low-lying sandy or grassy waste places; SL–2900 feet; fl. and fr. July–Mar. *A 5542! P 24117! Robertson UCWI 9177!* Mexico, Cuba to Grenada. Typical *C. nictitans* L. has glabrous eciliate leaflets and extratropical distribution. The type of *C. chamaecrista*, and as described by Britton & Rose (1930), is a plant of distinctly smaller dimensions from Curaçao. Modern collections seen from Curaçao resemble the Jamaican plants in size but have long hairs on the adaxial leaflet surface and long cilia.

27. C. caymanensis Adams in Phytologia **20** (5): 310 (1970).—*Chamaecrista riparia* of Britton & Millsp. (1920), not *Cassia riparia* Kunth (1824). *Chamaecrista confusa* Britton (1930), not *Cassia confusa* Phil. (1894).

Annual herb becoming woody at base, 30–50 cm. high; branches erect or prostrate; leaflets 6–25 pairs, linear, aristate, without or with very short cilia; stipules linear-lanceolate, 4–10 mm. long; gland on petiole narrowly obconic; pedicels solitary, up to about 5 mm. long; sepals 4·5–5 mm. long; petals yellow, shorter than calyx except one 6–8 mm. long; fruit 1·5–4 cm. long, 4–4·5 mm. broad.

Uncommon in St. Andrew and the westernmost parishes, on upper beaches and rocky limestone wastes; near SL–1500 feet; fl. and fr. sporadically throughout the year. *A 8563! Webster & Proctor 5346!* Bahamas, Grand Cayman.

28. C. smaragdina Macf., Fl. Jam. **1**: 347 (1837).—*Chamaecrista smaragdina* (Macf.) Britton (1930).

Herb, woody at base, 30–40 cm. high; leaflets 8–17 pairs, linear-oblong, 6–15 mm. long, mucronate, shortly ciliate, bright emerald-green; stipules lanceolate, 6–10 mm. long; gland on petiole cupular, subsessile; pedicels 1–4, 3–8 mm. long; sepals 7–8·5 mm. long; petals yellow, often with a red spot at base, 7–10 mm. long.

Local (St. Andr., Port., St. Thom.), on sandy or gravelly waste ground, possibly calciphobe; 1900–5000 feet; fl. and fr. Dec–Mar, July. *A 11932! J.P. 1258! P 16066!* ?Hispaniola.

29. C. fasciata (Britton) F. & R., Fl. Jam. **4**: 115 (1920).—*Chamaecrista fasciata* Britton (1910).

Erect, branched herb up to 1 m. high; leaflets 10–22 pairs, linear-lanceolate, 10–17 mm. long, aristulate, minutely ciliolate; stipules lanceolate, 8–10 mm. long; gland subsessile; pedicels 1–4, 5–8 mm. long; sepals 8–9 mm. long; petals yellow, often with a red spot at base, 9–13 mm. long.

Rare (Clar., St. Thom.), on waste ground, near SL–2400 feet; fl. and fr. Sept–Dec. *H 10641! 12838! P 19821!* Cuba.

9. HYMENAEA L. (1753)

1. **H. courbaril** L., Sp. Pl. **2**: 1192 (1753).—Stinking Toe, West Indian Locust.
Tree 10–25 (–30) m. high; bark smooth, grey; stipules boat-shaped, 3·3–3·4 cm. long; leaflets oblong to ovate, shortly acuminate, curved inwards, glabrescent but often remaining puberulous on midrib beneath, 6–9 cm. long, 2–4 cm. broad; inflorescence terminal, corymbose-paniculate; receptacle campanulate; sepals about 15 mm. long, softly sericeous; petals white with pellucid glands, about 17 mm. long; stamens inflexed in bud, 2·5–3·5 cm. long; fruit rough, reddish-chestnut to dark brown, up to 10 cm. long and usually not as broad; seeds 2–6, oblong-ovoid, 2·2 cm. long, 1·6 cm. broad.

Locally common, along roadsides, in pastures and on wooded hillsides; 50–2200 feet; fl. June–Sept, fr. Sept–Feb. *A 7816! Campbell 6414! P 19811! 22627!* Mexico to Brazil, West Indies.

10. TAMARINDUS L. (1753)

1. **T. indica** L., Sp. Pl. **1**: 34 (1753).—Tamarind.
Tree, often of great girth, up to 16 m. or more high; bark with large hard brownish-grey flakes; leaves paripinnate, up to about 12 cm. long; leaflets oblong, rounded at tip, unequal at base, 1–2 (–3) cm. long, 3–6 (–10) mm. broad, glabrous; racemes axillary or terminal, 2–4 cm. long; petals yellow marked red, rarely white, about as long as the calyx, the larger 13–15 mm. long; fruit 7–15 cm. long, 2·5–3 cm. broad, light brown and corky outside; seeds reddish-brown, 13–15 mm. long, 11–13 mm. broad.

Cultivated and naturalized; 50–2000 feet; fl. June–Oct, fr. Oct–June. *A. M. Barry! Thornton UCWI 1141!* Probably native to tropical Africa, now cultivated and naturalized throughout the tropics and subtropics of both hemispheres.

11. CRUDIA Schreb. (1789) nom. cons.

1. **C. spicata** (Aubl.) Willd. in L., Sp. Pl. ed. 4, **2**: 539 (1799).—*Apalatoa spicata* Aubl. (1775). *C. antillana* Urb. (1909). Cacoon Tree.
Tree up to 20 m. high with spreading drooping branches; leaflets oblong or oblong-elliptical, acuminate, glabrous, 7–13 cm. long, 2–4 cm. broad; racemes simple, terminal or lateral on older twigs; sepals about 6 mm. long, puberulous; stamens exserted; fruit flat, prominently nerved, rusty tomentellous, 7–10 cm. long, 5–6·5 cm. broad; seeds compressed, 4·5–6 cm. long, 3·5–4 cm. broad.

Very locally common (St. Eliz., West.), in riparian forest beside swamps and streams; 25–50 feet; fl. and fr. Sept–Mar. *H 9915! P 24523!* Guyana.

12. PRIORIA Griseb. (1860)

1. **P. copaifera** Griseb., Fl. Br. W.I.: 215 (1860).
Glabrous tree up to 30 m. high; leaflets elliptical or ovate-elliptical with unequal sides, acuminate, pellucid-dotted, 7–16 cm. long, 2·5–6 cm. broad; spikes about 12 cm. long; flowers small, light yellow, fragrant; fruit obovate-roundish, 6–10 cm. long, convex on one side.

Very local (West., St. Thom.); SL–600 feet; fl. Dec. fr. Sept, Dec. *H 10575! 11814!* Nicaragua, Panama, Colombia.

94. MIMOSACEAE

Trees, shrubs or herbs. Leaves alternate, bipinnate or less usually pinnate, often with large glands on the rachis; stipules present, sometimes spinescent. Flowers mostly bisexual, actinomorphic, in racemes, spikes or heads. Perianth biseriate; sepals (4–) 5 (–6), connate or rarely free, valvate or rarely imbricate; petals (4–) 5 (–6), free or connate, valvate, usually hypogynous, equal. Disk present or absent. Stamens as many as or twice as many as the petals or numerous, free or monadelphous or adnate to base of corolla-tube; anthers small, 2-locular, opening lengthwise, often with a deciduous gland at the apex. Ovary superior, 1-locular; ovules usually numerous, anatropous. Fruit dehiscent or indehiscent, sometimes breaking into 1-seeded segments. Seeds usually compressed; embryo large; endosperm lacking or very thin; aril rarely present.

About 50 genera and 2500 species in the subtropics and tropics.

1 Filaments united into a short or long tube; stamens more than 10; pollen grains usually in 2–6 masses in each anther-loculus; shrubs or trees, not undershrubs or woody herbs:
  2 Leaves pinnate with 3–6 pairs of broad opposite leaflets; leaf-rachis broadly winged; flowers in short spikes   **1. Inga**
  2 Leaves bipinnate, the pinnae sometimes with only one pair of leaflets:
    3 Inflorescences clustered on older leafless branches; leaves with one pair of pinnae, each with up to 5 broad leaflets, the proximal leaflets subopposite; fruit thin   **2. Zygia**
    3 Inflorescences borne on leafy twigs:
      4 Fruits thick, indehiscent, septate between seeds; flowers shortly stalked in compact umbels; large unarmed introduced trees:
        5 Fruit broadly circinate; leaflets 20–30 pairs, oblong, 8–15 mm. long; filaments yellowish; gland near base of petiole and between 2 distal pairs of pinnae; bark smooth   **3. Enterolobium**
        5 Fruit linear; leaflets 3–8 pairs, unequally rhomboid, up to 4 (–5·5) cm. long; filaments pink distally; glands on rachis between pinnae and between all except proximal leaflets; bark rough   **4. Samanea**
      4 Fruits thin or dehiscent:
        6 Fruits curved or coiled, the valves usually twisting after dehiscence   **5. Pithecellobium**
        6 Fruits straight or if slightly curved valves not twisting:
          7 Fruits elastically dehiscent, revolute from apex   **6. Calliandra**
          7 Fruits not elastically dehiscent, thin:
            8 Fruit breaking transversely between the seeds   **7. Pseudalbizzia**
            8 Fruit not breaking transversely, not septate   **8. Albizia**
1 Filaments free or nearly so; leaves bipinnate:
  9 Stamens more than 10; pollen grains usually in 2–6 masses in each anther-loculus; armed or unarmed shrubs and trees   **9. Acacia**
  9 Stamens 10 or fewer; pollen grains separate:
    10 Anthers at least in bud with an apical gland (except *Neptunia oleracea*):
      11 Inflorescence capitate; proximal flowers sterile; fruit dehiscent, septate between the seeds; herbs or undershrubs of wet places   **10. Neptunia**
      11 Inflorescence spicate or racemose; plants not herbaceous:
        12 Robust climbing shrub, unarmed; terminal pinnae modified to tendrils; fruit massive, splitting transversely into 1-seeded joints, the sutures persistent   **11. Entada**
        12 Trees or spreading or erect shrubs, not climbing:
          13 Fruit dehiscent longitudinally by valves; seeds bright red; unarmed tree with oblong-elliptical leaflets   **12. Adenanthera**
          13 Fruit indehiscent, neither splitting longitudinally nor fracturing transversely; usually armed shrub or tree with linear-oblong leaflets   **13. Prosopis**
    10 Anthers lacking apical glands:
      14 Valves of fruit not separating from the sutures; unarmed undershrubs, shrubs or trees; flowers mostly white or cream-coloured, not usually pink:
        15 Fruit narrowly linear; seeds arranged lengthwise or very oblique to sutures   **14. Desmanthus**
        15 Fruit broadly linear; seeds arranged transversely or only slightly oblique to sutures   **15. Leucaena**
      14 Valves of fruit separating from persistent sutures; plants often armed with prickles; flowers pink or white; leaves mostly sensitive to touch:

16 Fruit more or less flat, not beaked, the valves entire or jointed and broader than
  the sutures   16. **Mimosa**
16 Fruit subtetragonal, beaked, the valves entire and narrower than the sutures
                                                                 17. **Schrankia**

1. INGA Mill. (1754)

1. **I. vera** Willd. in L., Sp. Pl. ed. 4, **4** (2): 1010 (1806).—Panchock, River Koko.
  Tree up to 15 m. high; twigs pubescent; leaflets oblong-elliptical to lanceolate,
acuminate or acute, 4–16 cm. long, 2·5–6 cm. broad, with a circular gland on the
rachis between each pair; peduncles 1·5–6 cm. long; flowers few, subsessile in
spikes; calyx 9–15 mm. long; corolla 13–16 mm. long, silky-hairy; stamens, including tube, up to 6 cm. long, white turning light yellow; pod 4-ribbed, tomentose,
10–15 cm. long, 8–15 (–20) mm. broad.
  Locally common, in the eastern parishes, along riverbanks and in sheltered
ravines; 200–2300 feet; fl. Jan–May, Aug–Sept, fr. May–Dec. *A 6575! 8104!*
H 10364! P 16563! 20636! Mexico to northern S. Amer., Greater Antilles.

2. ZYGIA P. Browne (1756)

1. **Z. latifolia** (L.) F. & R., Fl. Jam. **4**: 150 (1920).—Hoop Wood, Horse Wood.
  Tree up to 8 m. high; twigs glabrescent; leaflets elliptical, more or less pointed
at both ends, 3–15 cm. long, 1·5–7 cm. broad, with glands between the pinnae and
the opposite leaflets; flowers-head subsessile or shortly stalked in clusters; calyx
1·5 mm. long, green; corolla 6–7 mm. long, dull yellow; stamens, including tube,
up to about 2 cm. long, the style longer; filaments bright magenta; anthers blackish-
purple; pod 10–30 cm. long, up to 2·5 cm. broad.
  Locally common in the north-eastern parishes and eastern St. Catherine, along
riverbanks and on steep sheltered shale slopes; 200–900 feet; fl. Oct–Mar, fr. Oct–
July. *A 9188! H 5610! P 20599! 20832!* Panama to northern S. Amer., Cuba,
Hispaniola, Martinique, St. Vincent, Trinidad.

3. ENTEROLOBIUM Mart. (1837)

1. **E. cyclocarpum** (Jacq.) Griseb., Fl. Br. W.I.: 226 (1860).—Elephant Ear,
  Monkey Soap.
  Tree to 20 m. high with wide-spreading branches, more or less deciduous;
pinnae 4–15 pairs with gland at base of lowest pair and rarely distally also (on
seedlings); peduncles clustered in axils, 1·5–5 cm. long; pedicels very short; calyx
2–2·5 mm. long and corolla 4–5 mm. long, both light green; stamens creamy-
yellow; flowers acrid scented; ripe fruit chocolate-brown, 8–10 cm. broad; seeds
brown with a light yellow ring on each side.
  Occasional, mostly where planted and hardly naturalized; 10–600 feet; fl. Nov–
May, fr. Jan–May. *Duncan UCWI 5526! P 19702! Thompson 7248!* Native of C.
and tropical S. Amer., introduced and cultivated in West Indies and W. Africa.

4. SAMANEA (Benth.) Merr. (1916)

1. **S. saman** (Jacq.) Merr. in Journ. Acad. Sci. Washington **6**: 47 (1916).—
  Enterolobium saman (Jacq.) Prain (1897). Guango, Rain Tree.
  Tree up to about 16 m. high, with stout trunk and rather low spreading branches,
more or less deciduous; pinnae 2–4 pairs; leaflets obtuse, pubescent; peduncles
one or 2 together, 5–9 cm. long; calyx 6 mm. long, greenish; corolla 10–12 mm. long,
yellow or red; stamens 3–5 cm. long, white proximally, pink distally; ripe fruit
blackish, 10–25 cm. long, 15–18 mm. broad; seeds rounded-truncate at one end,
pointed at the other, dark reddish-brown, with a paler ring on each side, 10–11
mm. long, 5–6 mm. broad.
  Common in inhabited areas and in old pastures where planted, naturalized in
riparian forest and in secondary communities on level ground; 50–700 feet; fl. Mar–
May, July–Nov, fr. Dec–May. *H 9276! P 26347!* Native of C. and tropical S. Amer.,
introduced in other tropical areas and often naturalized.

5. PITHECELLOBIUM Mart. (1837) nom. cons.

1 Leaflets in one pair; pinnae in one pair; seeds with a pulpy aril; stipules spinescent;
  leaflets obovate to elliptical:
  2 Calyx and corolla glabrous; inflorescence-branches glabrescent   1. *unguis-cati*
  2 Calyx, corolla and inflorescence-branches puberulous   2. *dulce*
1 Leaflets in 3 or more pairs; seeds not arillate;
  3 Flowers racemose:
    4 Pinnae 4–9 pairs; leaflets 10–14 pairs, rhomboid   3a. *alexandri* var. *alexandri*
    4 Pinnae 1–3 pairs; leaflets 3–6 pairs, ovate to elliptical or obovate
                                         3b. *alexandri* var. *troyanum*
  3 Flowers capitate; leaflets linear-oblong to linear-lanceolate:
    5 Fruit subterete and constricted between the seeds, red; leaflets (15–) 20–30 pairs,
      mostly 5–10 mm. long; pinnae 8–16 pairs; unarmed tree   4. *arboreum*
    5 Fruit flat, not constricted between the seeds, green turning brown; leaflets 8–30
      pairs, mostly 4–7 mm. long, pinnae mostly 5–10 pairs; spines scattered, slender
                                         5. *mangense*

1. P. unguis-cati (L.) Benth. in Hook., Lond. Journ. Bot. **3**: 200 (1844).—
Mimosa unguis-cati L. (1753). Bread-and-Cheese, Privet.

Shrub or tree up to 6 (–8) m. high; leaflets 1–5 cm. long, 8–28 mm. broad, the
young foliage reddish, with a gland between pinnae and leaflets; flower-heads
globose or shortly racemose, with about 20 fragrant flowers; calyx about 2 mm. long;
corolla 5–6 mm. long; stamens, including tube, 16–17 mm. long, white or purplish;
pod reddish, spirally curled; seeds black; aril white.

Common in thickets on arid limestone and forming open woodland at salina
margins, cultivated as a hedge; SL–1200 feet; fl. Sept–May, fr. Dec–May, Aug.
A 6271! H 10376! Powell 695! Florida to northern S. Amer., W. Indies, introduced
elsewhere.

A plant with often slightly larger leathery leaflets, occasionally in 2 pairs, is
represented by *Asprey UCWI 619!* and *P 16169!* from localities on the north coast
and has been identified with *P. keyense* Britton (1928); the taxonomic distinctness
of this entity requires confirmation.

2. P. dulce (Roxb.) Benth. in Hook., Lond. Journ. Bot. **3**: 199 (1844).—*Mimosa
dulcis* Roxb. (1795).

Tree up to 16 m. high; bark smooth, light grey; spines often wanting; leaflets
1–5 cm. long, up to 2 cm. broad; inflorescence-branches pendulous; flower-heads
with 18–24 greenish fragrant flowers; calyx about 1·3 mm. long; corolla 3–4 mm.
long; stamens, including tube, 7–8 mm. long, white; fruits as above.

Locally naturalized (St. Andr.), along roadsides and in thickets; 20–700 feet;
fl. and fr. Sept–Apr. *A. M. Barry! H 12336!* Native of continental tropical Amer.,
introduced and often naturalized elsewhere.

3. P. alexandri (Urb.) Urb., Symb. Ant. **5**: 358 (1908).

3a. Var. alexandri.—*P. jupunba* Urb. var. *alexandri* Urb. (1900). *Jupunba alexandri* (Urb.) Britton & Rose (1928). Tamarind Shadbark.

Tree up to 14 m. high; young twigs and leaves rusty-tomentellous; leaflets 8–25
mm. long, strongly reticulate-veined; peduncles 3–8 cm. long; racemes 3–6 cm.
long; flowers numerous, fragrant; calyx 3 mm. long; corolla 6 mm. long; fruit
about 8 cm. long; seeds white and blue, (5–) 6–7 (–8) mm. long.

Occasional, at pasture margins and in woodland on limestone in moderately
wet areas; SL–2500 feet; fl. and fr. Feb–Mar, Aug. *H 9182! P 26640!* Endemic.

3b. P. alexandri var. **troyanum** Urb., Symb. Ant. **5**: 359 (1908).—*Jupunba
troyana* (Urb.) Britton & Rose (1928). Shadbark.

Tree up to 15 m. high; young twigs densely rusty-tomentellous, leaflets 3–5
(–7·5) cm. long, up to 3·5 (–5·5) cm. broad, obtuse, reticulate-veined; peduncles
2–6 cm. long; racemes 4–8 cm. long; flowers greenish-white; calyx 2–2·5 mm. long;
corolla 5–6 mm. long; fruits as above.

Locally common, in the western parishes, in woodlands on limestone; 175–2500
feet; fl. Sept–June, fr. Sept–May. *A 10931! H 8800! HPS 14666! P 11211!*
Endemic.

4. P. arboreum (L.) Urb., Symb. Ant. **2**: 259 (1900).—Wild Tamarind.

Tree up to 20 m. high with lightly fissured flaky bark with reddish powder between the flakes; twigs rusty-pubescent; leaflets (3–) 5–10 (–12) mm. long, up to 3·5 mm. broad; flowers sessile, pinkish; calyx 2 mm. long; corolla 6–8 mm. long; stamens about 12 mm. long, white; seeds 8–12 mm. long, black.

Widely distributed, mostly in woodlands on limestone; 100–2700 feet; fl. Apr–Oct, fr. most of the year. *A 8150! H 6791! P 18324! 20836!* Mexico to Costa Rica, Greater Antilles.

5. P. mangense (Jacq.) Macbr. in Contrib. Gray Herb. **59**: 3 (1919).—*Enterolobium mangense* (Jacq.) F. & R. (1920). *Chloroleucon mangense* (Jacq.) Britton & Rose (1928).

Tree up to 5 (–8) m. high; old bark stripping in long flakes, new bark whitish; twigs and leaves crisped-puberulous or glabrous; spines present or absent; peduncles 7–15 mm. long; flowers shortly pedicellate, fragrant, light greenish-yellow; calyx about 3 mm. long; corolla about 5 mm. long; stamens 12–14 mm. long; fruit 6–14 cm. long, 10–12 mm. broad.

Rather local (St. Andr., St. Cath., St. Thom.), in open thickets and arid pastures on limestone; 50–500 feet; fl. Sept, Mar–June, fr. Oct–Jan, May. *A 6990! 12337! H 9231! P 18271!* Panama, Colombia to Brazil.

Pithecellobium lentiscifolium (A. Rich.) C. Wright (1869) (*Chloroleucon lentiscifolium* (A. Rich.) Britton & Rose (1928)), originally described from Cuba, was reported for Jamaica by Britton & Rose (1928) and by León & Alain (1951), but no specimen has been seen.

A specimen (*Maxon & Killip 312*), collected from a coastal area west of Kingston, was identified with *Pithecellobium tortum* Mart. (1837); a fragment only of this specimen has been seen and is inadequate for further determination. Britton & Rose (1928) did not report the species which was originally described from Brazil.

6. CALLIANDRA Benth. (1840) nom. cons.

1 Low shrubs armed with straight stipular spines 2–10 mm. long; pinnae in one pair; leaflets 4–10 (–13) pairs, subsessile, oblong to narrowly obovate, obtuse, 4–7 mm. long; peduncles solitary or rarely 2 in the axils, up to 2 cm. long and stiffly ascending in fruit; flowers about 12–18 in a head   1. *pilosa*
1 Unarmed shrubs or trees; leaflets larger than above:
  2 Heads solitary or few together, mostly axillary; pinnae 2–5 (–6) pairs; leaflets (7–) 10–25 (–30) pairs, subsessile, rectangular and very obliquely attached to costa, linear-oblong, obtuse; flowers about 15 in a head   2. *portoricensis*
  2 Heads in mostly terminal racemes or panicles; pinnae (1–) 2–3 pairs; leaflets (5–) 7–10 pairs:
    3 Leaflets oblong, somewhat falcate, subacute or subacuminate, sessile; panicle corymbose with racemes 1–2·5 cm. long; flowers 4–8 in a head; corolla-buds pointed
                                                        3. *comosa*
    3 Leaflets elliptical to oblong-oblanceolate, rounded at both ends, shortly petiolulate; panicle-axis elongated; racemes up to 8 cm. long; flowers 2–5 in a head; corolla-buds rounded at tip   4. *paniculata*

1. C. pilosa (Bert. ex DC.) Urb. in Arkiv Bot. **24** A (4): 4 (1932).—*Acacia pilosa* Bert. ex DC. (1825). *C. haematomma* of F. & R. (1920), not Benth. (1844). *Annesliapilosa* (Bert. ex DC.) Britton & Rose (1928).

Shrub to 2 m. or tree up to 4·5 m. high; branches spreading and arching, glabrescent; peduncles pilose, thicker in fruit; flowers sessile; calyx about 2 mm. long; corolla 5 mm. long, green; stamens white, greenish-yellow or reddish, nearly 2 cm. long including tube; fruit 3–4 cm. long, pubescent.

Occasional, in the south-eastern parishes, on arid rocky banks and in thickets; 50–1000 feet; fl. Oct–Mar. fr. Dec–Mar, July. *A 6335! H 9318! P 17413! 22074* Endemic.

2. C. portoricensis (Jacq.) Benth. in Hook., Lond. Journ. Bot. **3**: 99 (1844).—*Annesliaportoricensis* (Jacq.) Donnell Sm. (1891).

Shrub 2–3 m. or tree up to 8 m. high; twigs glabrous or pubescent; leaflets 6–16 mm. long, 1–2·5 (–4) mm. broad; peduncles slender, up to 10 cm. long, solitary or few together, sometimes reddish; flowers sessile, fragrant; calyx 2 mm. long; corolla 3–4 mm. long, greenish-white; stamens cream-coloured, 15–20 mm. long including tube; fruit 4–10 cm. long, 5–13-seeded; seeds ovoid-elliptical, black.

Common in secondary thickets and in open woodlands, mostly on limestone; 500–3000 (–5000) feet; fl. and fr. most of the year. *A 8193! H 8904! J.P. 1316! Powell 1099!* Mexico, El Salvador, Hispaniola, Puerto Rico, Virgin Is., introduced in Grenada and W. Africa.

3. C. comosa (Sw.) Benth. in Hook., Lond. Journ. Bot. **5**: 104 (1846).—*Anneslia comosa* (Sw.) Britton & Rose (1928).

Shrub 1–3 m. or tree up to 7 m. high, branching only at the top; twigs and leaves glabrous; leaflets 6–12 mm. long, 2–6 mm. broad; inflorescence shortly pubescent; flowers sessile; calyx about 1·5 mm. long; corolla 6 mm. long, white, 3–5-toothed; stamens white, about 18 mm. long including tube; fruit curved, 5–7 cm. long, puberulous, 3–8-seeded.

Rare and local (Manch., Trel.), on jagged limestone cliffs; 1500–2300 feet; fl. Dec–Feb, May–July, fr. Feb. *A 12843! P 16179!* Endemic.

4. C. paniculata Adams in Phytologia **20** (5): 310 (1970).

Shrub 1–2 m. high; young branches glabrescent; leaflets ciliate, (4–) 7–15 (–18) mm. long, 2·5–5 (–7) mm. broad; inflorescence thinly pubescent; flowers sessile; calyx 1·5 mm. long; corolla 6 mm. long, yellowish-green, shortly deltate-toothed; stamens 20–25 mm. long including tube; fruit slightly curved, tapered at base, about 6 cm. long and 7 mm. broad, canescent.

Rare (Manch.), on arid limestone rocks; 20 feet; fl. and fr. May. *A 11154!* Endemic.

Several ornamental species of *Calliandra* with pink or red flowers, native to Central and South America, are cultivated in gardens.

7. PSEUDALBIZZIA Britton & Rose (1928)

1. P. berteroana (Balb. ex DC.) Britton & Rose in N. Amer. Fl. **23** (1): 48 (1928). —*Acacia berteroana* Balb. ex DC. (1825). *Albizia berteroana* (Balb. ex DC.) Gómez Maza (1889).

Tree 6–11 m. high; petiole with an oblong gland and glands also between the distal pinnae; pinnae 6–15 pairs; leaflets 15–40 pairs, linear to linear-oblong, 4–8 mm. long, 1·5–2 mm. broad, glabrous; flower-heads numerous, racemose-panicu-late; flowers sessile in small heads, greenish-white, fragrant; calyx 1·5 mm. long; corolla about 3 mm. long; stamens 5–6 mm. long including tube; fruit 7–12 cm. long, 10–17 mm. broad.

Rather local (Clar., Manch., St. Eliz.), in open woodlands on limestone; SL–2600 feet; fl. June–Sept, fr. Sept–Mar. *A 13031! H 9654! 9946! H & P 14452!* Cuba, Hispaniola, Antigua.

8. ALBIZIA Durazz. (1772)

Leaflets (20–) 30–50 pairs, linear-oblong, 5–15 mm. long, mucronate; flowers capitate; fruit at length dehiscent, 10–15 cm. long, 2–2·5 cm. broad   1. *julibrissin*
Leaflets 4–9 pairs, obliquely oblong to obovate, 1·5–5 cm. long, obtuse; flowers umbellate; fruit not or tardily dehiscent, 15–30 cm. long, 2–4 cm. broad   2. *lebbeck*

1. A. julibrissin Durazz. in Mag. Tosc. **3** (4): 11 (1772).

Tree up to 12 m. high; pinnae 4–12 pairs; leaflets with midvein close to margin, 2·5–3·5 mm. broad; peduncles clustered or subracemose, 3–5 cm. long; calyx about 3 mm. long; corolla pink; stamens red.

Once reported as naturalized in Jamaica, but not gathered recently and not confirmed by Britton & Rose (1928). *Wilson.* Native of subtropical Asia, planted in S.E. United States and Puerto Rico.

2. A. lebbeck (L.) Benth. in Hook., Lond. Journ. Bot. **3**: 87 (1844).—Woman's Tongue Tree.

Tree 5–12 (–20) m. high; pinnae 2–4 (–6) pairs; gland near base of petiole and at

94. MIMOSACEAE

least between distal pinna-pairs; leaflets with midvein about one-third to two-fifths of width from distal margin, 5–20 mm. broad; peduncles 2–4-together, 3–10 cm. long; flowers greenish-white, strongly fragrant; pedicels pubescent; calyx about 4 mm. long; corolla about 6 mm. long; stamens about 3 cm. long including tube; seeds light brown, the areole depressed.

Locally common, naturalized in open secondary woodlands, mostly on gravelly soils near habitations; 300–1250 feet; fl. Apr–July, fr. most of the year. *A 7372! 8159! Wynter UCWI 5621!* Native of the Old World tropics, now widespread.

9. ACACIA Mill. (1754)

1 Climbing or scrambling shrubs with numerous recurved prickles on the branches; inflorescences paniculate or subpaniculate; flowers white in globose heads:
  2 Calyx about 1 mm. long; corolla 2–3 mm. long; fruit tomentellous, flat   1. *riparia*
  2 Calyx 3 mm. long; corolla 4 mm. long; fruit glabrous, thick and fleshy, constricted between the seeds   2. *rugata*
1 Erect shrubs or trees:
  3 Plants without stipular spines or prickles:
    4 Flowers light yellow in globose heads in panicles; fruit moniliform; large tree with green branches; introduced and naturalized in the mountains   3. *mearnsii*
    4 Flowers white or cream in short peduncled racemes; fruit flat, thin, with ribbed entire margins; indigenous shrub or small tree in lowland thickets   4. *villosa*
  3 Plants with stipular spines or paired prickles below the stipules only:
    5 Flowers white to light yellow in long slender spikes, the spikes sometimes in terminal panicles; prickles paired below the subulate stipules   5. *polyacantha*
    5 Flowers yellow or orange-yellow, in pedunculate globose heads, the peduncles usually axillary and rarely paniculate; stipules spiny:
      6 Peduncle bracteate at or about the middle; fruit flat, moniliform, canescent-velvety; cultivated throughout the West Indies; native of N. Africa and S.W. Asia; Gum Arabic   *A. nilotica* (L.) Willd. ex Delile
      6 Peduncle bracteate below the flower-head; fruit-margin entire or nearly so, sometimes torulose in *A. tortuosa*:
        7 Pinnae 8–60 pairs, 5–20 cm. long; leaflets 15–40 pairs; fruit somewhat compressed to subterete, densely velutinous when young   6. *macracantha*
        7 Pinnae 2–8 pairs, up to about 4 cm. long; leaflets 10–20 (–25) pairs; fruit terete or subterete:
          8 Fruit tomentellous, at least at first, obscurely striate, 8–16 cm. long, 7–8 mm. broad   7. *tortuosa*
          8 Fruit glabrous, conspicuously striate, 4–7 cm. long, up to nearly 2 cm. broad   8. *farnesiana*

1. A. riparia Kunth, Nov. Gen. **6**: 276 (1824).
Climbing shrub; pinnae 4–12 pairs; leaflets 15–40 pairs, oblong-linear, 5–10 mm. long; fruit 6–20 cm. long, 1·5–2·5 cm. broad.
Reported by Fawcett & Rendle but not confirmed by Britton. This species seems to occur throughout the West Indies except Cuba and Jamaica. The *Wullschlaegel* specimen cited by Fawcett & Rendle may have come from Antigua.

2. A. rugata (Lam.) Buch.-Ham. ex F. & R., Fl. Jam. **4**: 141 (1920).—*Mimosa rugata* Lam. (1783).
Climbing shrub with branches up to 10 m. long, pubescent; pinnae 4–8 pairs; leaflets 12–30 pairs, oblong-linear, 7–14 mm. long; fruit 8–13 cm. long, 1·5–2 cm. broad.
Naturalized very locally (St. Andr.), in thickets and waste places; 900–1000 feet; fl. and fr. Feb–Apr. *H 8253! J. P. 1133! A. von der Porten!* Native of tropical Asia.

3. A. mearnsii De Wild., Pl. Bequaert. **3**: 61 (1925).—Black Wattle.
Unarmed tree up to 20 m. or more high; trunk dull rusty-brown; younger branches green; twigs pubescent; pinnae (3–) 7–15 pairs, each pair with a prominent gland on the rachis on the proximal side, 3–4·5 cm. long; leaflets 15–50 pairs, 1·5–4 mm. long; flowers fragrant in numerous small heads; fruit moniliform, 4–8-seeded, velvety-pubescent.
Cultivated and naturalized locally (St. Andr.), an escape from gardens; 3500–4900 feet; fl. Jan–June, fr. Jan, Apr. Aug. *A 12685! A. von der Porten!* Native of Australia, naturalized in E. Africa and Mediterranean region, cultivated elsewhere.

4. A. villosa (Sw.) Willd. in L., Sp. Pl. ed. 4, **4** (2): 1067 (1806).—*Acaciella villosa* (Sw.) Britton & Rose (1928). Yellow Tamarind.

Shrub or tree 3–4 (–9) m. high; twigs pubescent; pinnae 4–7 (–10) pairs, without glands, 3–6 cm. long; leaflets 9–20 (–30) pairs, 4–10 mm. long, thinly pubescent, paler beneath; peduncles several together in axils or verticillate on a subterminal axis; calyx 0·7 mm. long; corolla white, 2–2·5 mm. long; fruit 3·5–5 cm. long, 6–9 mm. broad.

Locally common (St. Andr., St. Thom.), in thickets on gravel; 400–1700 feet; fl. and fr. most of the year. *A 7739! H 12907! J.P. 1350! P 17453!* Endemic. This species is maintained distinct from *A. glauca* (L.) Moench which has broader fruits and less hairy twigs and is known from Guadeloupe, Barbados and Curaçao.

5. A. polyacantha Willd. in L., Sp. Pl. ed. 4, **4** (2): 1079 (1806).—*A. suma* (Roxb.) Buch.-Ham. ex J. O. Voigt (1845).

Tree 5–10 m. high; prickles below stipules small or wanting; leaves up to 20 cm. long with a large elliptical gland on the petiole; pinnae 10–40 pairs, 2–6 (–9) cm. long; leaflets 25–50 pairs, linear, acute, 3–8 mm. long; spikes 6–10 cm. long; calyx 2 mm. long; corolla 3 mm. long; fruit 6–12 cm. long, 10–18 mm. broad, flat, thin, glabrous, reticulate-veined, 3–6-seeded.

Naturalized locally (St. Andr.), in thickets and secondary woodlands; 400–700 feet; fl. July–Sept, fr. Sept–Dec. *A. M. Barry! H 8790!* Native of tropical Asia, with a subspecies in tropical Africa; planted and naturalized in the West Indies.

6. A. macracantha Humb. & Bonpl. ex Willd. in L., Sp. Pl. ed. 4, **4** (2): 1080 (1806).—*A. lutea* (Mill.) Britton (1889), not Leavenw. (1823). Park Nut.

Tree 5–14 m. high; twigs pubescent; stipular spines 0·5–9 cm. long, the larger 2-edged; pinnae up to 60 pairs, with glands sometimes present between the pairs or on the petioles; leaflets 14–40 pairs, oblong-linear, 1·5–4 mm. long; peduncles 1–2 cm. long; calyx about 1·5 mm. long; corolla 2–2·5 mm. long; fruit 5–10 cm. long, 9–12 mm. broad.

Common locally (St. Andr., St. Cath., St. Thom.), in secondary thickets on arid limestone; 150–2000 feet; fl. and fr. sporadically throughout the year. *A 8191! H 11857! P 16171!* Florida, continental tropical and subtropical Amer., Greater Antilles and the drier islands of the Lesser Antilles.

7. A. tortuosa (L.) Willd. in L., Sp. Pl. ed. 4, **4** (2): 1803 (1806).—*Poponax tortuosa* (L.) Raf. (1838). Wild Poponax.

Shrub or tree with spreading branches 3–5 (–8) m. high; stipular spines thick, the larger terete, 1–4 cm. long; leaves with an oblong gland on the petiole and circular glands between distal pinnae; leaflets oblong-linear, (3–) 4–5 (–7) mm. long; peduncles solitary or clustered, 1·5–3·5 cm. long; calyx 1–1·5 mm. long; corolla about 2 mm. long; fruit up to 13-seeded.

Locally very common, along the south coast and on some of the cays, on arid limestone and at salina margins; SL–150 (–750) feet; fl. Jan–Apr, July–Sept, fr. Jan–Mar, July–Nov. *A 6269! H 9386! J.P. 997! P 11528!* Hispaniola, Puerto Rico, Virgin Is. and the drier islands of the Lesser Antilles.

8. A. farnesiana (L.) Willd. in L., Sp. Pl. ed. 4, **4** (2): 1083 (1806).—Cassie Flower.

Shrub to 3 m. or tree to 9 m. high; stipular spines 0·5–2 (–5) cm. long, whitish; leaves with a small gland on petiole; leaflets linear to oblong-linear, 3–5 mm. long; peduncles 2–4 cm. long; calyx 1–1·5 mm. long; corolla 2–3 mm. long.

Naturalized locally (St. Andr., St. Eliz., Han.), in secondary woodlands and thickets on arid limestone; 50–600 feet; fl. and fr. sporadically. *P 11535! Wynter UCWI 2072!* Native of the Old World, now widespread in subtropical and tropical parts of both hemispheres.

10. NEPTUNIA Lour. (1790); Windler (1966)

Stems rather woody, branched, more or less erect or ascending; leaves with a large sessile gland below the proximal pinna-pair; fruit 8–20-seeded; leaflets usually more than 20 pairs   1. *plena*

Stems soft, sparingly branched, usually floating; leaves without glands; fruit 4–8-seeded; leaflets up to 20 pairs
2. oleracea

1. N. plena (L.) Benth. in Hook., Journ. Bot. **4**: 355 (1841).—*Mimosa plena* L. (1753). Water Dead-and-Awake.

Low sprawling undershrub up to about 1 m. high; pinnae 2–4 (–5) pairs; leaflets 9–38 pairs, 4–15 mm. long, 1–3 mm. broad; spikes 10–18 mm. long on peduncles 3–15 cm. long, each with 30–60 flowers; proximal flowers sterile with yellow petaloid stamens; distal flowers fertile; fruit oblong, up to 5·5 cm. long and 1·1 cm. broad; seeds brown, ovoid, about 4 mm. long.

Rather local, in swamps and swamp margins; 25–300 feet; fl. and fr. mostly July–Jan. *A 11443! H 11826! P 9543!* Texas, Mexico to the Guyanas, Bahamas, West Indies, India.

2. N. oleracea Lour., Fl. Cochinch.: 654 (1790).

Glabrous herb with prostrate or floating spongy stems up to 1·5 m. long; pinnae 2–3 (–4) pairs; leaflets 8–20 pairs, 5–18 mm. long, 1·5–3·5 mm. broad; spikes on peduncles 5–20 (–30) cm. long, each with 30–50 flowers similar to the above but fertile flowers greenish with anthers lacking a terminal gland; fruit oblong, 2–3 cm. long, 8–10 mm. broad; seeds brown, ovoid, 4–5 mm. long.

Very rare (St. Eliz.), in muddy pond; 5–10 feet; fl. and fr. Sept. *P 24017! Purdie!* Mexico to Brazil, tropical Asia and Africa.

11. ENTADA Adans. (1763) nom. cons.

1. E. gigas (L.) F. & R., Fl. Jam. **4**: 124, f. 38 (excl. fruit) (1920).—*Mimosa gigas* L. (1759). Cacoon.

Climber to 15 m., the stems often much longer; pinnae 1–2 pairs; leaflets (2–) 4–5 pairs, obliquely oblong-elliptical, 2–8 cm. long, up to about 2·5 cm. broad; spikes supra-axillary, solitary or 2 together, usually longer than the leaves; flowers greenish-yellow, numerous; bracteoles persistent; calyx 1·5 mm. long; corolla of free or nearly free petals 3–4 mm. long; stamens 10, 6–8 mm. long; fruit (0·6–) 1–2 m. long, about 10 cm. broad, 10–12-seeded; seeds compressed, roundish, 3–5 cm. in diameter.

On trees along riverbanks and the margins of tall forest in wet areas; 400–2000 feet; fl. Mar, July–Aug, fr. Jan–Mar. *A 11330! P 15602! 20655!* C. and tropical S. Amer., West Indies, tropical Africa.

12. ADENANTHERA L. (1753)

1. A. pavonina L., Sp. Pl. **1**: 384 (1753).—Circassian Seed, Red Bead Tree.

Tree up to 15 m. high, almost everywhere glabrous; pinnae 2–5 pairs, 10–20 cm. long; leaflets 11–21 pairs, often alternate, oblong-elliptical, obtuse, 1·5–4·5 cm. long, 10–24 mm. broad; racemes simple and axillary or in terminal panicles, 5–15 cm. long; flowers yellow or orange-yellow; pedicels to about 3 mm. long; calyx about 1 mm. long; corolla of nearly free petals 3 mm. long; fruit 15–25 cm. long, 12–16 mm. broad; seeds bright scarlet, 6–10 mm. broad.

Rather common, planted as a shade tree and naturalized; 400–1600 feet; fl. July, fr. sporadically throughout the year. *A 11424! H 11075! A. von der Porten!* Native of tropical Asia, introduced in Africa and America.

13. PROSOPIS L. (1767); Benson (1941)

1. P. juliflora (Sw.) DC., Prodr. **2**: 447 (1825).—Cashaw.

Shrub, sometimes with prostrate branches on exposed dunes, or tree up to 6 (–13) m. high, with spreading branches; spines when developed axillary, 1–3 cm. long; pinnae 1–2 pairs; leaflets 10–21 (–30) pairs, oblong to linear-oblong, 7–16 (–25) mm. long, 2–6 (–7·5) mm. broad, sparsely ciliate; racemes 5–10 cm. long, 3–4 together; calyx a little over 1 mm. long; corolla light green, about 3 mm. long; stamens yellow, about 4 mm. long; fruit curved, yellowish, up to 25 cm. long and 10–16 mm. broad.

Locally common (St. Andr., St. Cath., Clar.), in low pastures in arid areas and on sand and shingle dunes; SL–150 (–700) feet; fl. and fr. most of the year. *A 11894! H 8525! Powell 696! P 9383!* S. United States to northern S. Amer.; naturalized in the West Indies.

14. DESMANTHUS Willd. (1806) nom. cons.

Erect shrub; petiole-gland elliptical; leaflets 10–30 pairs; fruits 3–7 cm. long   1. *virgatus*
Undershrub with prostrate branches; petiole-gland circular; leaflets 8–16 pairs; fruits 2–3·5 (–5) cm. long   2. *depressus*

1. D. virgatus (L.) Willd. in L., Sp. Pl. ed. 4, **4** (2): 1047 (1806).—Ground Tamarind.

Slender-branched almost glabrous shrub 0·6–2·5 m. high; stipules setaceous; pinnae (1–) 2–5 (–7) pairs with a large gland below the proximal pair; leaflets linear or linear-oblong, 3–6·5 (–9) mm. long, about 1 mm. broad, unequally truncate at base, ciliate; peduncles up to 5 cm. long in fruit; flowers white or creamy-yellow; calyx 2 mm. long; petals free, green, 3·5 mm. long; stamens twice as long as petals, 10, free; fruit 3–4 mm. broad.

Common in waste places and thickets; near SL–2400 feet; fl. and fr. all the year. *A 8170! H 11931! P 15385!* Native of subtropical and tropical Amer. now widespread in the Old World.

2. D. depressus Humb. & Bonpl. ex Willd. in L., Sp. Pl. ed. 4, **4** (2): 1046 (1806). —*Acuan depressum* (Humb. & Bonpl. ex Willd.) Kuntze (1891).

Taprooted woody herb with slender prostrate or ascending branches up to 40 cm. long; pinnae 1–4 pairs with a small gland below the proximal pair; otherwise as above but smaller in almost all dimensions.

Thinly scattered in low-lying situations, mostly in rocky pastures near the sea; 20–600 feet; fl. and fr. Feb–Sept. *A 6331! C. B. Lewis!* Florida, Texas to Brazil, West Indies.

15. LEUCAENA Benth. (1842)

Pinnae 3–10 pairs; leaflets 10–20 pairs, 7–15 mm. long, 2–3·5 mm. broad   1. *leucocephala*
Pinnae 10–25 pairs; leaflets 22–55 pairs, 2–4·5 mm. long, 0·6–1 mm. broad
   2. *brachycarpa*

1. L. leucocephala (Lam.) De Wit in Taxon **10** (2): 54 (1961).—*Leucaena glauca* Benth. (1842), partly, not *Mimosa glauca* L. (1753). Lead Tree.

Shrub to 3 m. or tree to 6 m. high, reported as taller elsewhere; petiole with or without a gland; leaflets linear-oblong, acute, paler beneath; peduncles solitary or 2–3 together, 3–5 cm. long, or racemose at ends of branches; calyx 2·5 mm. long; corolla cream or greenish-yellow, nearly 4·5 mm. long; stamens 10, free, filaments white, nearly twice as long as petals; fruits 10–20 cm. long, up to 2 cm. broad, pubescent, 8–25-seeded, often several developing from the same head.

Common along roadsides and in sandy waste places and thickets; SL–1600 feet; fl. and fr. most of the year. *A 8300! H 6713! A. M. Barry!* Florida to Brazil, West Indies, introduced and widely naturalized in the Old World, including Hawaii.

2. L. brachycarpa Urb., Symb. Ant. **2**: 265 (1900).

Tree 5–10 m. high; glands borne on the rachis; leaflets linear-oblong, acute or subacuminate; peduncles solitary or 2–6 together, 1–2 cm. long; calyx about 2 mm. long; corolla white, about 4 mm. long; fruit 6–12 cm. long, 1–2 cm. broad, puberulous, 1–5 developing from each head.

Probably all planted or naturalized in Jamaica (St. Andr., St. Thom.), on steep wooded hillsides; 600–4000 feet; fl. May–July, fr. June–Nov. *H 12342! 12451! P 23721!* Native of Mexico, introduced in Martinique.

16. MIMOSA L. (1753)

1 Low prostrate glabrous herb without prickles; fruit reduced to a single joint; pinnae in 1 pair; stamens 4–5   **1. viva**

1 Woody herbs, shrubs or small trees; fruit several-jointed; plant usually prickly:
2 Pinnae in 1 pair or 2 subdigitate pairs; fruits with bristly margins, otherwise glabrescent and smooth, 2–5-jointed:
3 Woody herb or undershrub; pinnae usually 2 approximate pairs; leaflets 15–25
   pairs, linear, 6–10 mm. long; stamens 4   2. *pudica*
3 Vine; pinnae in 1 pair; leaflets 3–4 pairs, oblong or oblong-lanceolate, 1–4 cm. long;
   stamens 8   3. *casta*
2 Pinnae 4–15 pairs; leaflets small, in many pairs; fruit at most hispid or setose; stamens
   twice as many as petals, 8 or 10; shrubs or small trees:
4 Fruit 10–13 or more-jointed; pinnae 7–15 pairs   4. *pigra*
4 Fruit 3–8-jointed; pinnae 4–9 pairs:
5 Shrub or small tree; flowers creamy-white in small globose heads in panicles
                                                                  5. *bimucronata*
5 Scrambling vine with slender stems with numerous recurved prickles; flowers rosy
   or purplish in solitary or few heads   6. *invisa*

1. M. viva L., Sp. Pl. 1: 517 (1753).

Stems creeping and rooting at nodes, up to about 20 cm. long; pinnae 1–2 cm.
long; leaflets in 4 pairs, oblong, 3–5 mm. long; peduncles solitary, axillary, 1–3 cm.
long; heads few-flowered; fruit long-stalked, 5–6 mm. long, setose or naked.

Rare (St. Ann, St. Thom.), in pastures; near SL–1200 feet; fl. and fr. Feb.
Fawcett 8132! H & P 14256! Cuba.

2. M. pudica L., Sp. Pl. 1: 518 (1753).—Shame Weed.

Stems straggling, 30–100 cm. long, glabrous or hispid, sometimes suberect,
sometimes trailing and rooting; stipules ciliate; pinnae 2–6 (–9) cm. long; leaflets
1·5–2 mm. broad; peduncles solitary or 2–3 together, axillary, 1–2 cm. long;
flower-heads ovoid or globose; corolla about 2 mm. long, red-tipped; filaments
pink; fruit 2–5-jointed, 1–1·5 cm. long.

A common weed of pastures and open stabilized waste places; SL–2000 feet;
fl. and fr. all the year. *A 11499! J.P. 1428! P 23072!* Native of tropical S. Amer.,
now widespread.

3. M. casta L., Sp. Pl. 1: 518 (1753).

Spreading vine up to 4 m. high; leaflets 3–5-veined from base, bristly-margined;
peduncles slender, up to about 2 cm. long; flower-heads globose, up to about 1 cm.
in diameter; fruit 3–5-jointed, 2–4 cm. long, 10–12 mm. broad.

Very rare (St. Thom.), in roadside waste places; about 750 feet; fr. Jan. Known
only from a single recent collection, *Cornman 440!* Panama to Brazil, Guadeloupe
to Trinidad.

4. M. pigra L., Cent. Pl. 1: 13 (1755).

Shrub with straggling branches up to 2 (–3) m. high, with setose and hooked
prickles; rachis often with intercostal paired hooks and spines between the pinnae;
leaflets pubescent, 5–8 mm. long, about 1 mm. broad; peduncles solitary or
paired, 2–5 cm. long; flower-heads light rosy-lilac; fruits hispid, often several
together, 3–9 cm. long, 10–12 mm. broad.

Occasional, in ditches and along riverbanks; 20–100 feet; fl. Mar, Aug, Dec, fr.
June–Dec. *A 10961! 11502! H 11815! P 26598!* General in the tropics.

5. M. bimucronata (DC.) Kuntze, Revis. Gen. Pl. 1: 198 (1891).—Brazil Macca.

Shrub or tree up to 8 m. high, with or without prickles; leaflets oblong-linear,
glabrous, 5–10 mm. long, about 1 mm. broad; panicles up to 30 cm. or more long;
flower-heads 5–6 mm. in diameter; corolla about 2·5 mm. long; fruit 4–5 cm. long,
7–8 mm. broad, glabrous.

Locally common (Clar., Manch., St. Eliz.), in hillside thickets and pastures and
riverside alluvium; (120–) 700–3000 feet; fl. July–Sept, fr. Aug–Dec. *A 12587!
H & B 10625! P 24009! 26661!* Native in the Guianas, Brazil, Paraguay; introduced in S.E. Asia.

6. M. invisa Mart. in Flora 20, Beibl. 2: 121 (1837).

Rampant undershrub with branches up to 2 m. long; leaflets oblong-linear,
ciliate, 3–5 mm. long, about 1 mm. broad; peduncles solitary or 2 together,

axillary, up to about 1 cm. long; flower-heads small; corolla about 2 mm. long; fruit 1–2·5 cm. long, 5–6 mm. broad, pubescent.

Rare (St. Andr., St. Eliz.), recently seen only in cultivation; 450–1500 feet; fl. May, Sept. *H 9863! Skelding UCWI 3202!* Native of tropical Amer., now widespread.

17. SCHRANKIA Willd. (1806) nom. cons.

1. S. leptocarpa DC., Prodr. **2**: 443 (1825).—*Leptoglottis leptocarpa* (DC.) Standl. (1925).

Slender-branched, straggling or climbing undershrub with numerous recurved prickles on the angled stems, 1 m. or more high; pinnae 2–3 pairs; leaflets 10–20 pairs, linear, glabrous, 6–10 mm. long; peduncles axillary, 4–15 mm. long; flowers pink; fruit 7–13 cm. long, armed with short squarrose prickles.

Rare (Clar.), in moist ground near pond; about 150 feet; fr. Feb. *P 11533 partly!* Native of tropical Amer., now widespread in the Old World.

Piptadenia peregrina (L.) Benth. was included by Fawcett & Rendle on the strength of an undoubtedly erroneous report by Macfadyen. Britton & Rose (1928) did not confirm the presence of this species in Jamaica.

95. PAPILIONACEAE (FABACEAE)

Trees, shrubs or herbs, the herbs and shrubs often twining. Leaves alternate, pinnate, digitate, 3-foliolate or 1-foliolate, rarely 2-foliolate; stipules typically present but sometimes very small or obsolete; stipels often present. Flowers mostly bisexual, zygomorphic, in terminal or axillary racemes, heads, clusters, panicles or solitary. Perianth biseriate; sepals usually 5 and connate; petals 5, imbricate, unequal, the adaxial exterior forming the standard (vexillum), the lateral pair the wings (alae), and the abaxial interior pair enclosing the stamens and style often connate by their lower margins to form the keel (carina). Stamens usually 10, usually monadelphous or diadelphous, rarely all free, mostly all fertile; anthers 2-locular, uniform or often alternately small and large, usually opening lengthwise. Ovary superior, 1-locular; ovules anatropous; style simple. Fruit a legume, lomentum or rarely drupaceous. Seeds with a large embryo, with little or no endosperm.

About 400 genera with some 9000 species generally distributed.

1 Leaves on mature plants reduced to spines; stipules wanting; shrub; corolla yellow
                              3. Ulex
1 Leaves evident:
 2 Leaves 1-foliolate or digitately 2-, 4- or 5 (–9)-foliolate, not 3-foliolate or pinnate:
  3 Shrubs or trees; fruits flat, indehiscent, mostly 1-seeded or up to 4-seeded:
   4 Flowers in axillary clusters, orange-yellow; stipules spinescent; leaflets leathery, less than 2 cm. long   **42. Brya**
   4 Flowers in axillary panicles, white; stipules very small and early caducous; leaflets mostly more than 3 cm. long   **11. Dalbergia**
  3 Herbs, undershrubs or shrubs; fruits dehiscent or jointed, if 1-seeded not flat:
   5 Twining or climbing herbs:
    6 Twiners with terete stems; leaves 1-foliolate; corolla small   **25. Galactia**
    6 Climber by leaf-tip tendril; stem winged; leaves 2-foliolate; corolla showy, purple; rare escape from cultivation in the mountains; Tangier Pea
                    Lathyrus tingitanus L.
   5 Erect, spreading or trailing herbs or shrubs:
    7 Flowers in small cymules enclosed by broad inflated distichously arranged bracts; shrub; leaves gland-dotted beneath   **18. Moghania**
    7 Flowers not enclosed by inflated bracts; herbs or undershrubs:
     8 Fruit inflated, not jointed, not septate   **5. Crotalaria**
     8 Fruit not inflated, if thick jointed or septate between the seeds:
      9 Fruit 2-valved, dehiscent, transversely septate:
       10 Leaves digitately 4–9-foliolate; flowers whorled in terminal racemes
                     4. Lupinus

10 Leaves 1-foliolate; flowers in axillary racemes 13. **Indigofera**
9 Fruit jointed, the joints 1-seeded and indehiscent, or dehiscent in *Desmodium motorium* and *D. gyroides*:
11 Leaflets 2 or 4; each flower subtended by 2 appressed basally appendaged bracts
44. **Zornia**
11 Leaflet solitary:
12 Leaflet broader than long; fruit folded at the joints, enclosed in the inflated calyx 46. **Christia**
12 Leaflet not broader than long; fruit straight or curved, not folded:
13 Fruit terete, the joints thick 47. **Alysicarpus**
13 Fruit compressed, the joints more or less constricted 48. **Desmodium**
2 Leaves 3-foliolate or pinnate with 2 or more pairs of leaflets:
14 Leaves 3-foliolate:
15 Trees:
16 Trunks and branches often thorny; deciduous trees with broad usually acuminate leaflets; standard petal much longer than wings or keel, red, vermilion or rarely white; fruit often torulose 26. **Erythrina**
16 Twigs with stipular spines; evergreen trees with small leaflets; standard, wings and keel about equal in length, orange-yellow; fruit flat 42. **Brya**
15 Herbs, shrubs or climbers:
17 Erect, straggling or creeping herbs or shrubs, not normally twining or climbing:
18 Leaflets toothed; fruit usually indehiscent; stipules adnate to petiole:
19 Flowers in rather compact racemose heads; corolla persistent in fruit; low creeping or trailing herbs 40. **Trifolium**
19 Flowers in axillary racemes; corolla caducous in fruit; erect bushy herb
41. **Melilotus**
18 Leaflets entire or at most repand or sinuate-margined:
20 Fruit jointed:
21 Stipules united to petiole; stipels absent; fruit indehiscent, 1–2-jointed; leaflets subdigitate 45. **Stylosanthes**
21 Stipules not united to petiole, striate; stipels present; fruit indehiscent and often with short hooked hairs or rarely dehiscent, 2- or more-jointed; leaves pinnately 3-foliolate 48. **Desmodium**
20 Fruit 2-valved, dehiscent, not jointed:
22 Leaves gland-dotted beneath:
23 Fruit more or less flattened, obliquely depressed between the seeds, about 5-seeded; stipules about 4 mm. long; corolla yellow streaked red, or yellow only 17. **Cajanus**
23 Fruit turgid, 1–2-seeded; stipules 7–9 cm. long; corolla purple
18. **Moghania**
22 Leaves not gland-dotted beneath:
24 Fruit globose or oblong, inflated, smooth within, without septa; stipels absent 5. **Crotalaria**
24 Fruit flat and compressed or linear; stipels present:
25 Keel petals twisted; calyx 5-toothed; fruit linear; petals blood-red
30. **Macroptilium**
25 Keel petals not twisted:
26 Inflorescences terminal and in the upper axils, racemose; fruit flat, compressed, covered with small hooked hairs, smooth within; corolla rose-red 14. **Pseudarthria**
26 Inflorescences axillary racemes; fruit linear, more or less septate between the seeds:
27 Herbaceous annual with ovate leaflets 7–15 cm. long; nodes of raceme not swollen; calyx 5-toothed; petals pink, mauve or whitish; fruit bristly, 2–4-seeded; cultivated; origin unknown; Soya Bean
Glycine max (L.) Merr.
27 Perennial undershrub with elliptical leaflets 2·5–7 cm. long; nodes of raceme swollen; calyx 4-toothed; petals rose-red
25. **Galactia**
17 Herbaceous twiners or climbing shrubs; fruits 2-valved, dehiscent:
28 Leaves with scattered resinous dots beneath; stipels absent or small:
29 Ovules and seeds (1–) 2 19. **Rhynchosia**
29 Ovules and seeds 3 or more 20. **Atylosia**
28 Leaves without resinous dots beneath; stipels usually present:
30 Standard petal much shorter than keel; fruit hispid sometimes with irritant pungent hairs; calyx sericeous and bristly 27. **Mucuna**
30 Standard petal as long as or longer than keel:
31 Standard petal much larger than other petals; bracteoles conspicuous:
32 Bracteoles persistent, shorter than calyx; fruit tetragonal, ribbed near the middle; seeds globose, sticky 34. **Clitoria**

32 Bracteoles deciduous, nearly as long as to longer than calyx and appressed
    to it; fruit compressed, not ribbed   35. **Centrosema**
31 Standard petal not much longer or shorter than other petals; bracteoles
    inconspicuous:
  33 Calyx unequally 2-lipped, the upper with 2 lobes united, the lower with 3
      small teeth; fruit with 1 or 2 pairs of longitudinal ribs; endocarp papery,
      white   23. **Canavalia**
  33 Calyx 4–5-toothed:
    34 Fruit 4-winged, 5–6 (–15)-seeded; corolla mauve or bright blue; stipules
        produced basally; vexillary stamen free   28. **Psophocarpus**
    34 Fruit not winged; corolla not bright blue:
      35 Stamens all united or vexillary stamen fused above and free below:
        36 Woody climber; nodes of raceme swollen; fruit 4 cm. or more broad;
            seeds 2·5 cm. or more long with a long hilum   21. **Dioclea**
        36 Herbaceous or shrubby twiners; fruit less than 1 cm. broad:
          37 Nodes of raceme swollen; all anthers fertile; fruit not with a hooklike
             or oblique beak   22. **Pueraria**
          37 Nodes of raceme not swollen; alternate anthers sterile; fruit with a
             hooklike or oblique beak   36. **Teramnus**
      35 Vexillary stamen free; anthers uniform; fruit septate or more or less
        filled between the seeds:
        38 Style glabrous; stigma terminal:
          39 Standard petal limb narrowed into the claw, not auricled or appen-
             daged; calyx 4-toothed; fruit septate   37. **Shuteria**
          39 Standard appendaged with 2 inflexed auricles:
            40 Calyx 5-lobed; fruit impressed between the seeds
                  24. **Calopogonium**
            40 Calyx 4-lobed; fruit more or less smooth externally:
              41 Leaflets linear to ovate, elliptical or oblong; stipules small or
                 deciduous   25. **Galactia**
              41 Leaflets broadly rhomboid-ovate; stipules ovate   22. **Pueraria**
        38 Style bearded on one side or around the stigma:
          42 Keel-petals spirally twisted; stigma lateral:
            43 Calyx 4-toothed; leaflets ovate to ovate-rhomboid   29. **Phaseolus**
            43 Calyx 5-toothed; leaflets ovate-lanceolate to linear, mostly obtuse
                  30. **Macroptilium**
          42 Keel-petals not spirally twisted but sometimes curved:
            44 Stigma terminal; style flattened laterally; fruit flat, obliquely oblong-
              falcate, warted along the margins   31. **Lablab**
            44 Stigma lateral; fruit turgid:
              45 Stigma oblique, not globose   32. **Vigna**
              45 Stigma globose, lateral to the inferior side; roots tuberous
                  33. **Pachyrhizus**
14 Leaves pinnate with 4 or more leaflets:
  46 Herbaceous or slender woody climbers:
    47 Leaves imparipinnate with 2–3 (–4) pairs of lateral leaflets, with stipels; bracteoles
      persistent; flowers resupinate   34. **Clitoria**
    47 Leaves paripinnate; leaflets without stipels:
      48 Terminal leaflet replaced by a bristle; flowers clustered in racemes; fruits
        partitioned within; seeds usually red and black, not arillate   38. **Abrus**
      48 Terminal leaflet replaced by a tendril; flowers 1 or 2 in the axils; fruits smooth
        within; seeds dull brown, thinly arillate   39. **Vicia**
  46 Herbs, erect or robust climbing shrubs, or trees:
    49 Trees or erect or climbing shrubs:
      50 Filaments free, 10; leaves imparipinnate; leaflets opposite or subopposite, rarely
        alternate:
        51 Tree; leaflets 4–5 (–8) pairs, opposite; fruit 2-valved, 1–2-seeded; seeds scarlet
          with a black mark near the hilum   1. **Ormosia**
        51 Shrub; leaflets 5–12 pairs, opposite to alternate; fruit torulose, indehiscent,
          usually several-seeded; seeds yellow or brown   2. **Sophora**
      50 Filaments connate, monadelphous or diadelphous:
        52 Robust trailing or scrambling shrubs; fruit indehiscent; leaves impari-
        pinnate:
          53 Leaflets opposite, 2–4 pairs   9. **Lonchocarpus**
          53 Leaflets alternate, 3–5 (–7)   11. **Dalbergia**
        52 Trees or large erect shrubs:
          54 Fruit dehiscent; leaflets opposite or mostly so:
            55 Leaves paripinnate; leaflets 10–20 or more pairs, subopposite; fruit septate;
              corolla very large and showy   15. **Sesbania**
            55 Leaves imparipinnate; leaflets 4–12 pairs, opposite:

56 Hairs medifixed; corolla purplish-red, about 1 cm. long; leaves finely
   gland-dotted; fruit septate; shrub   13. **Indigofera**
56 Hairs simple; corolla 2–2·5 cm. long; leaves not gland-dotted; fruit more or
   less continuous within:
   57 Calyx obliquely cupular; corolla mostly pink; leaflets 4–8 pairs, thinly
      pubescent; deciduous tree   6. **Gliricidia**
   57 Calyx toothed; corolla white or yellowish; leaflets 9–12 pairs, velvety-
      pubescent; shrub or small tree   12. **Tephrosia**
54 Fruit indehiscent; trees; leaves imparipinnate:
   58 Leaflets alternate, 5–9; fruit compressed, broad, with thin margins, 1–2
      seeded; corolla yellow   7. **Pterocarpus**
   58 Leaflets opposite; corolla not yellow, except sometimes in *Lonchocarpus
      latifolius*:
      59 Fruit 4-winged, 3–7-seeded; panicles axillary; lateral leaflets 3–5 pairs
                                                         8. **Piscidia**
      59 Fruit not 4-winged:
         60 Fruit flat, 1–5-seeded; inflorescences axillary; lateral leaflets 2–5 pairs
                                                         9. **Lonchocarpus**
         60 Fruit drupaceous, 1-seeded, globose-ellipsoidal; panicles terminal; lateral
            leaflets 4–7 pairs   10. **Andira**
49 Herbs or undershrubs:
   61 Leaflets serrate distally, 9–17, 1–1·5 cm. long; flowers white or purple; fruit
      oblong, 2-valved, 2–3 cm. long; native of Mediterranean region; Chick Pea,
      Gram   *Cicer arietinum* L.
   61 Leaflets entire:
      62 Leaves paripinnate:
         63 Fruit indehiscent, ripening below ground; leaflets 2 pairs; stipules fused to
            petiole; corolla yellow; cultivated plant of uncertain origin; Ground Nut,
            Monkey Nut, Pea Nut   *Arachis hypogaea* L.
         63 Fruit dehiscent, not ripening below ground; leaflets 10–30 or more pairs;
            stipules free, caducous   15. **Sesbania**
      62 Leaves imparipinnate:
         64 Fruit jointed, breaking up at maturity; leaflets numerous
                                                         43. **Aeschynomene**
         64 Fruits 2-valved, dehiscent:
            65 Fruit not septate between the seeds; hairs simple   12. **Tephrosia**
            65 Fruit septate between the seeds:
               66 Hairs medifixed; fruit terete or angled   13. **Indigofera**
               66 Hairs simple; fruit flat, grooved between the seeds 16. **Cracca**

1. ORMOSIA G. Jackson (1811) nom. cons.

1. O. jamaicensis Urb., Symb. Ant. **5**: 366 (1908).—Red Nickel.
   Tree up to 25 m. high; leaflets elliptical to oblong-lanceolate, acuminate, 5–12
(–25) cm. long, 2–4 (–7) cm. broad; panicle terminal; calyx-tube cupular, densely
puberulous, 3·5 mm. long; corolla rose-purple; fruit 4–6 cm. long; seed about 1·5
cm. in diameter and 1 cm. thick.
   Very local (Han.), in woodland on limestone hillsides; 1000–1700 feet; fl. July–
Aug, fr. Mar–July. *H 9241! P 10414! 26577!* Endemic.

2. SOPHORA L. (1753)

Erect coastal shrub; leaflets tomentose; corolla light yellow; fruit 5–9-seeded; seeds
subglobose, about 6 mm. long   **1. tomentosa**
Scrambling inland shrub; leaflets glabrescent beneath; corolla bright yellow; fruit 1–3
(–5)-seeded; seeds flattened-subglobose, about 9 mm. long   **2. saxicola**

1. S. tomentosa L., Sp. Pl. **1**: 373 (1753).
   Shrub 1·2–3 m. high; leaflets rounded-elliptical, obtuse, 2·5–4 cm. long; raceme
up to 40 cm. long; calyx truncate, 7–10 mm. long; standard petal elliptical, folded
lengthwise, 2–3 cm. long; fruit 5–15 cm. long.
   Occasional, in dry coastal thickets and on some of the cays; near SL; fl. Mar,
July, fr. Jan. *Asprey UCWI 2173! Fawcett! P 11491! 11659!* General in the subtropics and tropics.

2. S. saxicola Proctor in Bull. Inst. Jam., Sci. ser. **16**: 15, t. 4 (1967).
Straggling shrub with branches up to 5 m. long; leaflets narrowly ovate to elliptical, 2–5·5 cm. long; raceme 40–50 cm. long; calyx and corolla as above; fruit 5–10 cm. long, markedly torose.
Rare and very local (Trel.), on limestone cliffs and ledges; 1500–1750 feet; fl. July, Oct–Nov, fr. June, Nov. *P 18355! 21431! Weaver 1000!* Endemic.

3. ULEX L. (1753)

1. U. europaeus L., Sp. Pl. **2**: 741 (1753).—Gorse.
Ascending-branched glabrescent shrub 1–2·4 m. high; young plants with trifoliolate leaves; older plants leafless but with copious furrowed spines 1·5–2·5 cm. long; flowers pedicellate, axillary, about 15 mm. long; pedicels and calyx hairy; fruit explosively dehiscent, about 15 mm. long, several-seeded; seed carunculate.
Locally frequent (St. Andr.), in montane thickets; 3600–5050 feet; fl. Apr–June, Nov, Feb, fr. May–June, Nov–Feb. *A 8728! P 7335! Robertson UCWI 3053!* Native of W. Europe now widely naturalized in temperate regions and tropical mountains.

4. LUPINUS L. (1753)

1. L. angustifolius L., Sp. Pl. **2**: 721 (1753).—Lupin.
Annual herb 60–90 cm. high; petiole up to 4 cm. long; leaflets 2–3·5 cm. long, 2–4 mm. broad, rounded at tip; racemes up to 10 cm. long; corolla bright blue; fruit villous, yellowish-brown, 2–2·5 cm. long, 4–5-seeded.
Local (St. Thom.), a weed of pastures and cleared stony ground; 4000–5500 feet; fl. Jan–Feb, Aug, fr. Jan, July. *A 10547! A. M. Barry! Royer 3821! Stearn 116!* Native of the Mediterranean region, widespread in cultivation.

5. CROTALARIA L. (1753)

1 Leaves 3- or 5-foliolate; corolla yellow; stipules minute, deciduous or wanting:
  2 Leaflets 5; pod glabrous   1. *quinquefolia*
  2 Leaflets 3; pod more or less puberulous or pubescent:
    3 Peduncles axillary, 1–few-flowered; pod 2–2·5 cm. long; perennial shrub
                                    2. *lotifolia*
    3 Peduncles terminal or leaf-opposed, several–many-flowered; annual or short-lived sometimes bushy herbs:
      4 Pod with spreading hairs   3. *incana*
      4 Pod with appressed hairs or merely puberulous:
        5 Leaflets up to 3 cm. long; racemes contracted   4. *pumila*
        5 Leaflets 3·5–10 cm. long; racemes elongated:
          6 Leaflets narrowly elliptical, at least three times as long as broad; ripe pods blackish   5. *zanzibarica*
          6 Leaflets broader, not more than twice as long as broad; ripe pods buff-coloured:
            7 Leaflets elliptical, more or less pointed at both ends, mucronate; pod straight
                                  6. *pallida*
            7 Leaflets obovate, rounded or emarginate at tip; pod curved   7. *falcata*
1 Leaves 1-foliolate:
  8 Stipules conspicuous at least on younger parts of plant, decurrent or semi-amplexicaul, subpersistent:
    9 Corolla blue; stipules lunate; leaf-blade ovate to roundish-elliptical; pod thinly appressed-pubescent   8. *verrucosa*
    9 Corolla yellow; pods glabrous:
      10 Leaf-blade oblong-oblanceolate, pubescent beneath; racemes long, many-flowered; stipules semi-sagittate, reflexed, persistent; pod 3–5 cm. long, secund; bracteoles broad, persistent   9. *spectabilis*
      10 Leaf-blades linear to linear-lanceolate or narrowly elliptical; racemes short, few-flowered; stipules decurrent with acute tips; bracteoles narrow:
        11 Stem unequally winged in continuity of acutely tapered decurrent stipules; leaves spreading, sparsely hairy; bracts and bracteoles linear; pod 2–3 cm. long
                                  10. *sagittalis*
        11 Stem long-winged in upper part in continuity of small triangular decurrent stipules; leaves closely appressed to stem, densely hairy; bracts and bracteoles lanceolate; pod 3–4 cm. long   11. *pilosa*

8 Stipules minute or wanting; corolla yellow:
  12 Leaves 1–2 cm. long, 4–6 mm. broad, oblong-linear to oblanceolate; pods about
    7 mm. long; low bushy annual herbs:
    13 Pod appressed-silky, yellowish-brown, about as long as calyx, 1-seeded
                                       12. *ramosissima*
    13 Pod glabrous, black, a little longer than calyx, several-seeded   13. *nana*
  12 Leaves 3 cm. or more long; shrubs or shrubby herbs:
    14 Pod short, 2-seeded, enclosed by the calyx; inflorescence paniculate
                                       14. *berteroana*
    14 Pod at least 3 cm. long, several-seeded, not enclosed by the calyx at maturity;
      inflorescence more or less racemose:
      15 Leaf-blade oblanceolate, apex rounded or emarginate; pod glabrous, black when
        ripe   15. *retusa*
      15 Leaf-blades linear to linear-lanceolate; pods brown, velvety-tomentose:
        16 Standard petal bright yellow speckled blackish-purple outside; leaf-blade
          linear-lanceolate, acuminate, 10–30 cm. long; stem 4-angled   16. *tetragona*
        16 Standard petal yellow, dull crimson outside; leaf-blade linear to elliptic-linear,
          obtuse to rounded at tip, 4–10 cm. long; stem multi-striate   17. *juncea*

1. C. quinquefolia L., Sp. Pl. 2: 716 (1753).

Annual herb, erect, 60–90 cm. high; leaflets linear-lanceolate, 2–8 cm. long; racemes elongated; corolla yellow with fine brownish-red lines on the standard petal; pod about 5 cm. long.

Very local and rare (Trel.), by muddy lake margin; about 400 feet; fl. and fr. Aug–Oct. *P 15704!* Native of S.E. Asia and tropical Australia, scattered through the West Indies in Cuba, Hispaniola, Antigua, Guadeloupe, Martinique and Trinidad.

2. C. lotifolia L., Sp. Pl. 2: 715 (1753) (err. *latifolia*).

Shrub 1·5–3 m. high; leaflets up to about 5 cm. long; corolla sulphur-yellow with a band of brown across and dark purple lines along the standard.

Rather local (St. Andr., St. Cath., Clar.), in thickets and woodlands on arid limestone; SL–300 feet; fl. and fr. June–July, Oct–Mar. *A 11904! H 12520! A. M. Barry!* Also in Mexico, Honduras, Bahamas, Greater Antilles, Grenadines and Barbados.

3. C. incana L., Sp. Pl. 2: 716 (1753).

Annual herb 50–120 cm. high; leaflets elliptical to obovate, 2–3·5 (–5) cm. long; corolla light yellow, standard slightly darker, keel greenish; pod about 3 cm. long.

Common, in rather arid waste places, roadsides and pastures; SL–3000 feet; fl. and fr. almost all the year. *A 7960! Harris! H & P 13310!* Generally distributed in the tropics.

4. C. pumila Ort., Nov. Pl. Hort. Matrit.: 23 (1797).

Diffusely branched perennial herb to about 1 m. high; leaflets variable in shape from cuneate-obovate to elliptical; racemes 1–2 cm. long; peduncle 4–6 cm. long; corolla yellow; pod 1–1·5 cm. long.

Very local (St. Ann), on sandy banks; fl. and fr. Jan–Mar. *H 10368!* Florida, tropical C. and S. Amer., Bahamas, Greater Antilles and through the drier islands of the Lesser Antilles to Barbados.

5. C. zanzibarica Benth. in Hook., Lond. Journ. Bot. 2: 584 (1843).—*C. usaramoensis* Bak. f. (1914).

Annual herb up to 60 (–200) cm. high, becoming rather woody at the base; leaflets obtuse or acute at tip; standard yellow with crimson lines outside, wings striate, keel greenish-yellow; flowers open in the morning; pod about 3 cm. long.

Very local (Manch., Trel., Clar.), on open roadside banks; 1800–2725 feet; fl. and fr. Apr, Sept. *A 12407! 12853! Powell 847!* Scattered throughout the tropics of both hemispheres.

6. C. pallida Ait., Hort. Kew. 3: 20 (1789).—*C. mucronata* Desv. (1814). *C. striata* DC. (1825).

Shrubby little-branched herb, 1–2 m. high; leaflets 4–13 cm. long, (1–) 2–6 cm. broad, standard narrowly ovate; standard and keel strongly purple-striate; otherwise dull yellow; pod minutely puberulous, about 4 cm. long, the adaxial suture more or less prominent and not sharply curved upwards near tip.

Rather common, along roadside banks and in waste places; 400–4000 feet; fl. and fr. most of the year. *A 8246! P 22942!* General in tropical Africa and Asia, recorded in the Caribbean from Puerto Rico, Trinidad and Tobago.

7. C. falcata Vahl ex DC., Prodr. **2**: 132 (1825).—*C. obovata* G. Don (1832).

Like the last, but differing in the leaflets and the pod; leaflets 3–8 cm. long, 1·5–4·5 cm. broad; pod minutely puberulous, 3·5–4 cm. long, the adaxial suture sunken between the valves and sharply curved upwards behind the tip.

Occasional and local (St. Andr.), on sandy roadside banks and in waste places; 550–1250 feet; fl. June–July, Oct–Jan, fr. Oct–Jan. *A 6157! Allwood UCWI 1398! Asprey UCWI 663! A. M. Barry!* Scattered through E. tropical Asia, the Pacific Islands and Africa, recorded in the Caribbean from Hispaniola, Puerto Rico, Dominica, St. Lucia, Grenada, Trinidad and Tobago, usually, as in Jamaica, from lower elevations and often in sandy places near the sea.

8. C. verrucosa L., Sp. Pl. **2**: 715 (1753).—Blue Rattleweed.

Laxly branched herb 50–90 cm. high; leaf-blade 3–7 cm. long, 2–5 cm. broad; standard streaked with blue or mauve, yellowish at tip, wings blue or mauve, greenish at base, keel greenish with blue or mauve lines; pod 3–3·5 cm. long.

Common, along roadsides and in waste places and pasture margins; SL–2000 feet; fl. and fr. all the year. *A 7839! H 8814! P 11137!* Throughout the tropics; Cayman Is.

9. C. spectabilis Roth, Nov. Pl. Sp.: 341 (1821).—*C. sericea* Retz. (1789), not Burm. f. (1768).

Robust herb 1–2·2 m. high; leaf-blade up to 16 cm. long and 7·5 cm. broad, glaucous and with prominent veins beneath; standard clear yellow except the mid-line dark outside and a few short dark lines inside; wings oblong-rotundate, saccate near the base; flowers open in the late morning.

Rare and local (St. Andr., St. Thom.), established in waste places, otherwise cultivated as an ornamental; (50–) 250–3500 feet; fl. and fr. July–Mar. *A 6710! P 8518!* Scattered through the tropics.

10. C. sagittalis L., Sp. Pl. **2**: 714 (1753).

Annual herb up to 50 cm. high; leaf-blade 1·5–7 cm. long; corolla white or light sulphur to dark yellow; standard purple-lined; variable throughout its range, a larger variant (var. *fruticosa* (Mill.) F. & R.) up to 125 cm. high having been reported from Jamaica.

Locally common in the east-central region, along roadsides and in sandy savannas and river gravel; 300–3800 feet; fl. and fr. all the year. *A 5846! H 11855! P 18387!* Throughout tropical Amer.

11. C. pilosa Mill., Gard. Dict. ed. 8 (1768).—*C. pterocaula* Desv. (1814).

Brownish-hairy erect herb up to 60 cm. high; ascending leaf-blades 2–4 (–8) cm. long; corolla canary-yellow; the Jamaican variant in this widely diverse species has been referred to var. *skutchii* Senn (1939).

Very local (Clar., St. Ann), in swampy ground in hill savannas; about 2400 feet; fl. and fr. Dec. *H 12844! C. pilosa* var. *skutchii* is also reported from Costa Rica, Panama and Cuba; the species extends throughout tropical Amer.

12. C. ramosissima Roxb., Fl. Ind. **3**: 268 (1832).

A densely bushy aromatic annual herb up to 60 cm. high; bracts and sepals with marginal glands; standard pink outside, yellow inside; other petals yellow; seeds 4 mm. long, 3 mm. broad, flat, shiny, buff.

Rare and casual (St. Andr.), in sandy waste places; 200–600 feet; fl. and fr. Apr–May. *P 22285! Skelding UCWI 3253!* Native of India and apparently not otherwise recorded in the New World.

13. C. nana Burm. f., Fl. Ind.: 156, t. 48, f. 2 (1768).

Annual much branched herb to about 30 cm. high.

Only twice reported by *Wilson* and *Wullschlaegel*. A casual introduction from the Old World tropics.

14. C. berteroana DC., Prodr. **2**: 127 (1825).—*C. fulva* Roxb. (1832).
Shrub up to 2·5 m. high; stipules about 1 cm. long, setaceous, deciduous, or wanting; leaf-blades oblanceolate to elliptical, densely silky-hairy, up to 9·5 cm. long and 3 cm. broad; calyx 10–13 mm. long; corolla mostly yellow, keel greenish, fragrant.
Local (St. Andr., St. Thom.), in gravelly waste places mostly near rivers; SL–2000 feet; fl. and fr. July–Dec. *A 7709! H 6825! P 24060!* Native in the Old World tropics and naturalized in many oceanic islands; the type was from a plant cultivated in Guadeloupe.

15. C. retusa L., Sp. Pl. **2**: 715 (1753).—Rattleweed.
Glaucous shrubby herb up to about 1 m. high, the plant drying blackish; leaf-blades pellucid-dotted, 3–8 cm. long, 1–3·5 cm. broad; corolla yellow, the standard tinged red outside; flowers open in the afternoon; pod 3–4 cm. long.
Common, along roadsides and in waste places; SL–1250 feet; fl. and fr. all the year. *A 6159! H 6784! H & P 14472!* Throughout the subtropics and tropics.

16. C. tetragona Roxb. ex Andr., Bot. Repos. **9**: t. 593 (1809).
Straggly-branched herb 1–1·6 m. high; corolla yellow; pod dark brown, 3·5–5 cm. long.
Very local (St. Andr., St. Thom.), along roadsides and in gravelly waste places; 2000–4200 feet; fl. and fr. Oct–Mar. *A 6420! H 5597!* Native of S.E. Asia; St. Vincent.

17. C. juncea L., Sp. Pl. **2**: 714 (1753).—Sunn Hemp.
Erect slender-branched shrubby herb 1–3 m. high; corolla mostly yellow, keel light green; pod light reddish-brown, velvety, about 3 cm. long.
Rather local (St. Andr., St. Thom.), sporadic in sandy and gravelly waste places; SL–1250 feet; fl. and fr. Oct–Feb. *A 8505! H 8274! P 8519!* Cultivated and probably native in tropical Asia; also in Australia.

6. GLIRICIDIA Kunth ex Endl. (1840)

1. G. sepium (Jacq.) Kunth ex Griseb. in Abh. Gött. Akad. **7**: 52 (1857).—
Robinia sepium Jacq. (1760). Aaron's Rod, Grow Stick, Quick Stick.
Tree up to 6 m. or rarely more high; stipules intrapetiolar, minute; leaflets ovate to elliptic-lanceolate, 3–6 cm. long, 1·5–3 (–4) cm. broad, glabrescent; racemes 5–7 cm. long; calyx obliquely cupular; standard petal erect, about 2 cm. broad, mostly pink but yellow centrally; wings oblong, 2 cm. long, purplish; keel pink at tip; pod 10–12 cm. long; seeds about 1 cm. long.
Common, mostly planted in fences and hedgerows; SL–1600 feet; fl. Jan–Apr, fr. Jan–May. *H 8453! Proctor & Skelding 16123!* Native of tropical Amer., now widespread and often planted as a fast-growing shade plant.

7. PTEROCARPUS Jacq. (1763) nom. cons.

1. P. officinalis Jacq., Select. Stirp. Amer. Hist.: 283, t. 183, f. 92 (1763).—
Dragon's Blood Tree.
Tree up to 10 m. high with flanged buttresses; leaflets elliptical to oblong-elliptical, shortly acuminate, 5–15 cm. long, 3–6 cm. broad; panicles lax; standard petal long-clawed, 11–13 mm. long; fruit veiny, about 4 cm. in diameter.
Rare and local (Port.), cultivated in St. Mary, in swamps; SL–580 feet; fl. July–Aug, fr. July–Sept. *Harris! Harty! Robbins UCWI 702!* Throughout tropical Amer.

8. PISCIDIA L. (1759) nom. cons.

1. P. piscipula (L.) Sarg. in Gard. & For. **4**: 436 (1891).—*Erythrina piscipula* L. (1753). Dogwood.
Deciduous tree 5–15 (–20) m. high; young twigs and flowers silky-puberulous;

stipules broadly orbicular, very early caducous, about 4 mm. broad; leaflets oblong or elliptical, obtuse or acute, 4–11 cm. long, 2·5–5·5 cm. broad; panicles many-flowered; calyx purplish, 4-toothed, the adaxial tooth emarginate; corolla white or tinged purple or standard greenish, standard 13–15 mm. long; fruit 4–8 cm. long with wings 1–2 cm. broad, light golden-buff.

Common, mostly in rather arid areas on sand or gravel or in woodlands on lime-stone, also on some of the cays; SL–500 (–2500) feet; fl. Jan–Aug, fr. Mar–Oct. *A 6980! Bengry! H 8518! J.P. 1319!* Florida, Mexico, Bahamas, Antilles, Trinidad.

9. LONCHOCARPUS Kunth (1824) nom. cons.

1 Primary branches of inflorescence racemose; standard petal glabrous outside; fruit conspicuously reticulate, thin and broadly winged; lax climbing shrub; cultivated; native of W. Africa; Yoruba Indigo   *L. cyanescens* (Schumach.) Benth.
1 Primary branches of inflorescence (peduncles) 1–2-flowered; standard petal silky-hairy outside:
  2 Woody climber; peduncles of usually paired flowers slender, glabrescent, 7–17 mm. long; bracteoles oblong, spreading, 0·7–0·8 mm. long; fruit with reticulate surface-marking   1. *patens*
  2 Trees; peduncles obsolete or up to 5 mm. long, densely puberulous; fruits smooth:
    3 Flowers less than 1 cm. long; peduncles up to 2 mm. long; calyx 2–2·5 mm. long; bracteoles triangular-subulate, 0·3–0·6 mm. long, erect; fruit thin-textured; leaflets thinly puberulous beneath   2. *latifolius*
    3 Flowers more than 1 cm. long; calyx 3 mm. or more long; bracteoles orbicular, 1·5–3 mm. broad, spreading; fruit woody:
      4 Leaflets densely puberulous and silky beneath; peduncles up to 2 mm. long; calyx 3–4 mm. long; corolla 12–14 mm. long   3. *sericeus*
      4 Leaflets glabrous; peduncles up to 5 mm. long; calyx about 5 mm. long; corolla about 15 mm. long   4. *domingensis*

1. L. patens Urb., Symb. Ant. 5: 367 (1908).

Straggling and trailing shrub to 10 m.; leaflets 2–4 pairs, ovate to oblong, obtuse to long-acuminate at tip, 2·5–11 cm. long, 1·5–5·5 cm. broad; inflorescence glabrescent; corolla reddish-purple usually with the centre of the standard petal white, 12–15 mm. long; fruit thick-margined, 6–9 cm. long, nearly 2 cm. broad.

Local (St. Eliz., Trel.), in thickets and woodlands on limestone; 1000–2000 feet; fl. Feb–Sept, fr. Aug, Nov, Jan. *A 6776! H 8739! H & P 14425!* Endemic.

2. L. latifolius (Willd.) DC., Prodr. 2: 260 (1825).—*Amerimnon latifolium* Willd. (1803).

Tree 5–16 m. high; leaflets 2–4 pairs, lanceolate to oblong, acuminate, up to 20 cm. long, 3·5–9 cm. broad; inflorescence up to 12 cm. long, puberulous; corolla usually pink, sometimes yellow, 6–8 mm. long; fruit papery, 4–9 cm. long and about 2 cm. broad.

Occasional, in all the central parishes, in thickets on limestone and bauxitic soil; SL–2200 feet; fl. May–July, fr. Sept. *A 7308! H 12366! P 19693! 23860!* Mexico to Brazil, West Indies.

3. L. sericeus (Poir.) DC., Prodr. 2: 260 (1825).—*Robinia sericea* Poir. (1804).

Tree up to 20 m. high; leaflets mostly in 4–5 pairs, oblong, acuminate, truncate to rounded at base, 6–18 cm. long, 3·5–8 cm. broad; inflorescence densely puberu-lous; bracteoles 1·5–1·8 (–3) mm. broad; corolla light purple, pinkish or white; fruit up to 10 cm. long and 2 cm. broad, densely rusty-tomentose.

Very local but sometimes gregarious (St. Eliz., West.), along riverbanks and in marsh forest; 25–150 feet; fl. Oct–Jan, fr. Jan–Feb. *H 9153!* (cult. St. Mary), *P 24519! 24673!* Mexico to tropical S. Amer., Grenada, Trinidad, W. Africa.

4. L. domingensis (Turp.) DC., Prodr. 2: 259 (1825).—*Dalbergia domingensis* Turp. (1807).

Tree 6–13 m. high, in most respects very like the last but all parts slightly larger and less hairy; leaflets 1–3 pairs, suborbicular to oblong-elliptical, obtuse to shortly and bluntly acuminate, broadly rounded at base, 5–15 cm. long, 4–8·5 cm. broad, sometimes with a few ascending hairs on midrib beneath; bracteoles mostly 2–2·5 mm. broad; corolla bright rose-purple, the standard white at base; fruit up to 10 cm. long, 2–2·5 cm. broad.

Rather local (Manch., St. Eliz., Trel.), in woodlands on limestone; 50–2000 feet; fl. May–Aug. fr. Sept–Dec. *A 13026! 13034! H 9153! 9491! P 10608! 27551!* Greater Antilles, Guadeloupe, Martinique. This species replaces *L. sericeus* in the Greater Antilles and the northern part of the Lesser Antillean chain, the only overlap in known distribution being in the parish of St. Elizabeth in Jamaica. The slightly hairy-leaved variants of *L. domingensis* occur there but it is noteworthy that the recorded flowering times do not coincide.

10. ANDIRA Juss. (1789) nom. cons.

1. A. inermis (W. Wright) DC., Prodr. **2**: 475 (1825).—*Geoffroea inermis* W. Wright (1787). Angelin, Cabbage Bark Tree.

Tree 6–15 (–25) m. high; stipules subulate, early caducous, about 3 mm. long; leaflets oblong, acuminate, 5–11 cm. long, 2·5–4 cm. broad; panicle many-flowered, tomentose; calyx dark purple or brownish, 3 mm. long; corolla rosy-pink or purple, fragrant; standard petal about 10 mm. long; fruit ellipsoidal, about 3·5 cm. long.

Very common, in woodlands and woodland margins, often secondary; 100–2000 (–2500) feet; fl. and fr. all the year. *A 9214! H 9053! P 19754!* Florida and throughout the American tropics, also in W. Africa.

11. DALBERGIA L. f. (1781) nom. cons.

1 Leaflets on mature plants (2–) 3–5, acuminate; petals subequal; stamens 9, diadelphous; fruit 1-seeded   1. *monetaria*
1 Leaflets at least on mature plants solitary; petals unequal; stamens 10, monadelphous:
  2 Leaflets glabrous adaxially, very thinly puberulous beneath; juvenile plants with compound leaves; older branches developing lateral spines or tendrils; fruit oblong 1–4-seeded, glabrous   2. *brownei*
  2 Leaflets puberulous on both surfaces, always solitary; older branches without spines or tendrils; fruit roundish, 1-seeded, puberulous   3. *ecastaphyllum*

1. D. monetaria L. f., Suppl.: 317 (1781).

Climbing shrub with long trailing and twining branches; leaflets elliptical or rounded-elliptical, 5–13 cm. long, up to about 10 cm. broad; panicles short, corymbose; calyx about 3 mm. long; petals about 6 mm. long, keel slightly shorter than the wings; fruit glabrous, 2·5–3·5 cm. long.

Rare and local (St. Mary), on steep rocky hillside and in thickets near stream; 300–1000 feet; fl. Apr, July–Sept, fr. July. *H 12080! P 24019!* Tropical S. Amer., West Indies.

2. D. brownei (Jacq.) Urb., Symb. Ant. **4**: 295 (1905).—*Amerimnon brownei* Jacq. (1760).

Trailing and climbing shrub up to 5 m. or more high, with young branches rusty-puberulous; juvenile plants with prostrate branches with 2–7-foliolate leaves; older branches suberect, spreading or twining with distichous laterals; stipules very small, oblong, early caducous; leaflets ovate-elliptical, obtuse to bluntly acuminate, very shortly cordate at base, (1·5–) 4–7 (–9) cm. long, 1–4 (–5) cm. broad; flowers fragrant in short corymbose panicles; calyx 4–5 mm. long; standard petal 9–10 mm. long; keel 5·5 mm. long and shorter than the wings; fruit 1·5–5 cm. long, glabrous except on the margins when young.

Very common, forming thickets on banks and in mangrove margins and coastal woodlands; SL–50 feet; fl. Jan–Aug, fr. May–Oct. *A 9247! J.P. 1004! P 10564!* Florida, C. and S. Amer., Cuba, Hispaniola, Grand Cayman.

3. D. ecastaphyllum (L.) Taub. in Engl. & Prantl, Nat. Pflanzenf. **3** (3): 335 (1894).—*Hedysarum ecastaphyllum* L. (1759).

Lax shrub or small tree with twining or straggling branches, 3–6 m. high; stipules obsolete; leaflets elliptical, shortly acuminate, truncate to shortly cordate at base, 4–13 cm. long, 3–8 cm. broad; flowers in short panicles, fragrant; calyx tomentose, 3–3·5 mm. long; standard petal 7 mm. long; wings longer than the standard; keel 6 mm. long; fruit 2–3 cm. long.

Rather common in thickets near the sea; SL–15 feet; fl. and fr. most of the year. *A 8071! 12283! H 5997! Powell 286!* Florida, C. and S. Amer., West Indies; Grand Cayman; W. Africa.

12. TEPHROSIA Pers. (1807) nom. cons.

1 Flowers 2 cm. or more long:
  2 Stipules broadly ovate, persistent; corolla crimson; pod glabrous on the sides, about
    5 cm. long; leaflets 5–7 pairs, hardly exceeding 2 cm. long　　　　1. *grandiflora*
  2 Stipules linear-lanceolate, deciduous; corolla white; pod more or less villous on the
    sides:
    3 Pod 9–12 cm. long, up to 1·5 cm. broad, densely shaggy-villous; leaflets 6–9 pairs,
      4–8 cm. long, rounded at the tip　　　　　　　　　　　　　　2. *vogelii*
    3 Pod 7–9 (–10) cm. long, about 8 mm. broad; thinly villous; leaflets 9–12 pairs, 3–5 cm.
      long, acute at the tip　　　　　　　　　　　　　　　　　　3. *candida*
1 Flowers less than 2 cm. long:
  4 Leaflets 12–20 pairs; corolla white; calyx and pod densely villous　　4. *sinapou*
  4 Leaflets up to 10 pairs:
    5 Calyx densely villous with brownish hairs; corolla white marked with bluish-purple;
      pod thinly villous with spreading brownish hairs　　　　　　　5. *noctiflora*
    5 Calyx thinly villous to puberulous; corolla usually tinged pink, red or purple:
      6 Pod falcate, 3–4 cm. long, bulged at the seeds; plants drying green; many of the
        leaflets oblanceolate and emarginate　　　　　　　　　　6. *purpurea*
      6 Pod straight or nearly so, (3·5–) 4–6 cm. long, not conspicuously bulged at the seeds;
        plants drying silvery especially in the younger parts; hairs on calyx usually rather
        long; most leaflets oblong-elliptical, acute to rounded at tip:
        7 Hairs on pod appressed-ascending, contiguous　　　　　　7. *cinerea*
        7 Hairs on pod spreading　　　　　　　　　　　　　　　8. *littoralis*

1. **T. grandiflora** (Vahl) Pers., Synops. Pl. 2: 329 (1807).—*Galega grandiflora* Vahl (1791).

Erect slender-branched shrub 50–120 cm. high; leaflets obtuse, aristate, up to 6 mm. broad; corolla bright rose-purple to crimson, sometimes marked white, 2–2·5 cm. long; pod acuminately beaked, 8–9 mm. broad, (7–) 10–14-seeded.

Locally common (St. Andr., St. Thom.), on exposed well drained banks; 3000–5000 feet; fl. and fr. Dec–Aug. *A 6669! J.P. 1176! P 23584!* Native of S. Africa.

2. **T. vogelii** Hook. f. in Hook, Niger Fl.: 296 (1849).

Shrub 1–2 m. or arborescent to 3 m. high; plant densely hairy with silky rust-coloured hairs on almost all parts; leaflets oblong-elliptical, rounded at both ends, up to 2·5 cm. broad; flowers clustered towards the end of a stout peduncle, about 2·5 cm. long.

Very local (St. Andr., St. Thom.), on open hillsides and planted with coffee; 3250–4050 feet; fl. Mar–Aug, fr. May–Aug. *A 7763! A. M. Barry! P 23728!* This is an introduced species indigenous to tropical Africa where the populations of W. Africa tend to have purple flowers and small leaflets. The Jamaican plants most resemble those from Rhodesia which usually have white flowers and larger round-tipped leaflets.

3. **T. candida** DC., Prodr. 2: 249 (1825).

Shrub 1–3 m. high with velvety-pubescent branches; leaflets up to about 10 mm. broad; corolla white or rarely yellow, 2–2·5 cm. long.

Very local (St. Andr., St. Mary), along gravelly riverbanks and courses; 480–2000 feet; fl. and fr. Aug–Nov. *H 9040! 11804!* Native of tropical Asia also reported from Hawaii, S. Tomé, St. Lucia and the wetter parts of W. Africa.

4. **T. sinapou** (Buc'hoz) A. Chev. in Compt. Rend. 180: 1522 (1925).—*Galega sinapou* Buc'hoz (1775). *T. toxicaria* (Sw.) Pers. (1807). *G. toxicaria* Sw. (1788). Suriname Poison.

Shrub 60–180 cm. high with velvety branches; leaflets narrowly oblong, silky-pubescent, 2·5–7 cm. long, up to 12 mm. broad; standard petal 1·2–1·5 cm. long; pod 4·5–7 cm. long.

Rare and local (St. Andr.), probably only in cultivation; 1000–2000 feet; fl. and fr. Sept. *Thompson 8045!* Widespread on the mainland of tropical continental Amer. from Mexico to Brazil and Bolivia, Hispaniola.

5. T. noctiflora Boj. ex Bak. in Oliv., Fl. Trop. Afr. **2**: 112 (1871).

Spreading herb or undershrub 70–200 cm. high; branches and leaves brownish-pubescent; leaflets in 6–9 pairs, narrowly oblanceolate, 2·5–4·5 cm. long, 5–8 mm. broad; stipules filiform; standard petal about 10 mm. long and broad, usually marked with purple lines and a central dark purple spot; pod 4–5 cm. long, 5–6 mm. broad.

Local (St. Andr., St. Mary), in waste ground and thicket margins; 150–1000 feet; fl. and fr. sporadically, mostly Aug–Dec. *A 11877! H 6952! P 20773!* Widespread in warm countries.

6. T. purpurea (L.) Pers., Synops. Pl. **2**: 329 (1807).—*Cracca purpurea* L. (1753). *T. wallichii* Graham ex F. & R. (1917).

Therophyte of varied habit, usually bushy with spreading or erect branches up to 60 cm. high, rarely prostrate; leaflets 4–6 (–9) pairs, linear-oblanceolate, usually glabrous adaxially, 10–25 (–32) mm. long, 3–7 (–10) mm. broad; flowers usually about 8·5 mm. long, reddish-purple to white; standard petal 7–8 mm. long and broad, often with a large green spot near base; wings spreading; pod appressed-pubescent, 5–7-seeded.

Rather local (St. Andr., Manch., St. Thom.), but sometimes common in sandy or gravelly waste places and open pastures; less frequent near the sea in St. Andrew than *T. cinerea*; SL–1000 feet; fl. and fr. all the year. *A 8314! H 12205! P 10213!* Widespread in the subtropics and tropics.

7. T. cinerea (L.) Pers., Synops. Pl. **2**: 328 (1807).—*Galega cinerea* L. (1759).

Therophyte of varied habit, erect up to 60 cm. high or prostrate with branches to 120 cm. long; leaflets 4–5 (–6) pairs, oblong-elliptical, often thinly appressed-puberulous adaxially, 15–50 mm. long, 3–8 (–10) mm. broad; flowers up to 14 (–15) mm. long, usually deep pink, the standard petal sometimes dull yellowish outside and the keel bright pink; limb of standard 10–12 mm. long; wings directed forward, longer than keel; pod 6–10-seeded.

Very local (St. Andr.), but sometimes abundant in sandy waste places near the sea, less frequent as a weed inland; SL–800 (–1000) feet; fl. and fr. almost all the year. *A 6266! A. M. Barry! H 6752!* Throughout the American tropics; Cayman Is.

8. T. littoralis (Jacq.) Pers., Synops. Pl. **2**: 329 (1807).—*Vicia littoralis* Jacq. (1760). *T. cinerea* var. *littoralis* (Jacq.) Benth. (1859).

Low herb, sometimes mats with trailing branches; hairs on stem, pedicels and pod spreading; leaflets 3–6 pairs, oblanceolate, mostly rounded at tip, glabrescent adaxially, 15–40 mm. long, 5–10 mm. broad; flowers up to about 10 mm. long, purplish-pink; limb of standard petal about 8 mm. long; wings shorter than keel.

Rare and local (St. Andr., Clar.), a weed of roadsides and cultivated land; SL–600 feet; fl. and fr. June, Nov–Feb. *A 6019! H 6956! A. von der Porten.* Mexico to Venezuela, Hispaniola, Cape Verde Is.

13. INDIGOFERA L. (1753)

1 Leaflets mostly alternate, (4–) 6–10, oblong to obovate-elliptical, up to 25 mm. long and
　　11 mm. broad, glabrous adaxially　　　　　　　　　　　　　1. *spicata*
1 Leaflets opposite or solitary:
　2 Leaflets tapered to a long acute tip, the laterals in about 8 pairs, up to 7 cm. long and
　　　2 cm. broad, thinly puberulous on both surfaces　　　　　2. *zollingerana*
　2 Leaflets obtuse to rounded or subemarginate at tip, rarely subacute, fewer, rarely as
　　　much as 5 cm. long and 1·5 cm. broad:
　　3 Fruits strongly curved on reflexed pedicels, 1–1·5 cm. long; leaflets subacute to
　　　　obtuse, the laterals in 5–6 (–7) pairs, usually glabrous adaxially; indumentum of
　　　　inflorescence brownish　　　　　　　　　　　　　3. *suffruticosa*
　　3 Fruits straight or only slightly curved:
　　　4 Fruits spreading, straight; plant with stalked glands on stem, leaves, pedicels and
　　　　fruits; leaflets mostly less than 1 cm. long　　　　　7. *colutea*

4 Fruiting pedicels reflexed; plants not glandular; leaflets mostly more than 1 cm. long:
    5 Fruit straight, turgid, shortly beaked; leaflets 1 (–2), 3–5 (–8), elliptic-oblanceolate
                                                    5. *lespedezioides*
    5 Fruit slightly curved, acuminate-beaked; lateral leaflets 2–6 pairs:
      6 Lateral leaflets 3–6 pairs, obovate and often submarginate, usually glabrous
        adaxially; racemes shorter than the leaves, with whitish indumentum; fruit
        more or less torulose, curved mostly at the tip   4. *tinctoria*
      6 Lateral leaflets mostly 2 pairs, ovate-elliptical to oblong, appressed-puberulous on
        both surfaces; racemes longer than the leaves; fruit slender, not torulose,
        somewhat tetragonal   6. *subulata*

1. I. spicata Forsk., Fl. Aegypt.-arab.: cxviii, 138 (1775).

Lax herb with trailing branches; stipules scarious; racemes axillary, 10 cm. or more long; flowers about 5 mm. long, the calyx more than half as long; corolla crimson; pod more or less straight.

A rare casual weed (St. Mary), along sandy roadsides; 1700 feet; fl. Nov. *A 9943!* General in the Old World tropics.

2. I. zollingerana Miq., Fl. Ind. Batav. **1** (1): 310 (1855).

Shrub about 2 m. high; leaves about 20 cm. long; leaflets lanceolate, gland-dotted, 3–7 cm. long, 1·3–2 cm. broad; racemes axillary, 10 cm. or more long; flowers about 1 cm. long; corolla purplish-red.

Very rare (St. Ann), probably an introduced cover crop escaped into pasture; 1500 feet; fl. Sept. *A 11752!* Native of mainly coastal regions in the East Indies and Philippines.

3. I. suffruticosa Mill., Gard. Dict. ed. 8 (1768).—Guatemala Indigo, Markham Gungo, Wild Indigo.

Shrub 0·6–2·4 m. high, usually much branched above; leaves 5–12 cm. long; leaflets oblong, (1–) 1·5–3 (–5) cm. long, 4–8 (–15) mm. broad; racemes 2–5 cm. long; flowers 3·5–4·5 mm. long, vermilion to light crimson; pod thickened along the margins, 3–6 (–8)-seeded.

Common in rough pastures and waste places; SL–4000 feet; fl. and fr. all the year. *A 7250! 8210! J.P. 1252! H & P 13938!* Native of tropical Amer., now widespread in the tropics; Grand Cayman.

4. I. tinctoria L., Sp. Pl. **2**: 751 (1753).—Indigo.

Straggling-branched herb or undershrub 60–120 cm. high; leaves 6–10 cm. long; leaflets up to 2 cm. long and 1 cm. broad; racemes 2–7 cm. long; flowers 5–6 mm. long, orange- or rose-pink; pod 2–3·5 cm. long, (6–) 8–12-seeded.

Frequent in rather arid sandy places and in thickets and on waste ground on open limestone; SL–1000 feet; fl. and fr. most of the year. *A 7249! H 6265! P 11515!* Native of the Old World tropics, formerly widely cultivated.

5. I. lespedezioides Kunth, Nov. Gen. **6**: 457 (1824).

Perennial herb or undershrub 15–90 cm. high; leaflets 2–5 cm. long, 4–10 mm. broad, puberulous on both surfaces; flowers about 5 mm. long, pink or reddish; pod 1·5–3 cm. long.

Rare and local (St. Andr., Manch.), in pastures subject to periodic burning; 300–1250 feet; fl. July–Aug, Nov, fr. Nov. *H 11745! 12204! Skelding UCWI 23922!* Honduras to tropical S. Amer., Cuba, Hispaniola.

6. I. subulata Poir. in Lam., Encycl. Méth. Bot. Suppl. **3**: 150 (1813).

6a. Var. scabra (Roth) Meikle in Kew Bull. **5**: 352 (1951).—*I. scabra* Roth (1821).

Herb or undershrub up to about 120 cm. high; leaflets 8–45 mm. long, 4–20 mm. broad, appressed-puberulous on both surfaces; flowers about 5 mm. long, vermilion to pink, drying mauve; pod 2·5–4·5 cm. long, 10–15-seeded.

Frequent as a weed of cultivated and waste places; 500–3500 feet; fl. and fr. most of the year. *A 6638! Powell 751! Thompson 8066!* Native in the Old World subtropics and tropics from W. Africa to India; Cuba, Hispaniola.

7. I. colutea (Burm. f.) Merr. in Philipp. Journ. Sci. **19**: 355 (1921).—*Galega colutea* Burm. f. (1768).

Shrubby much branched annual herb to 60 cm. high; lateral leaflets 3–5 pairs, oblong to obovate-elliptical, rounded at tip, 5–10 mm. long, up to 5 mm. broad, strigose-puberulous on both surfaces with much larger hairs beneath; racemes up to 8 cm. long, longer than the leaves; flowers 3·5 mm. long, deep rose to flame pink; pod 12–22 mm. long, up to about 14-seeded.

Local and sporadic (St. Andr.), in sandy or gravelly waste ground; 500–1250 feet; fl. and fr. May, Sept–Jan. *A 6158! Skelding UCWI 9202!* Scattered through the drier parts of the Old World tropics and Australia; Hispaniola.

14. PSEUDARTHRIA Wight & Arn. (1834)

1. P. viscida (L.) Wight & Arn., Prodr. Fl. Pen. Ind. Or. **1**: 209 (1834).— *Hedysarum viscidum* L. (1753).

Straggling undershrub; leaflets 3, broadly rhomboid-ovate to subrotundate, thinly hairy on both surfaces, up to about 5 cm. long and broad; inflorescence a terminal raceme or panicle; flowers about 3·5 mm. long; pod oblong, papery, 5–18 mm. long, about 6 mm. broad, 2–5-seeded.

Very rare (St. Andr.), on shaded earth banks; 1900–3500 feet; fl. and fr. Nov– Dec. *P 18370! A. von der Porten!* Native of tropical Asia; St. Vincent.

15. SESBANIA Scop. (1777) nom. cons.

1 Flowers large, (5–) 6–8 (–10) cm. long, curved in bud; leaflets 12–22 pairs, oblong, 2–4 cm. long, 7–12 mm. broad; usually a small tree   1. *grandiflora*
1 Flowers smaller, up to 2·5 cm. long, more or less straight in bud; shrubs:
  2 Plant silky-pubescent at least on stems and lower surfaces of leaves, without prickles; leaflets 12–25 pairs; flowers 8–9 mm. long   2. *sericea*
  2 Plants glabrous or thinly pubescent:
    3 Leaflets 10–18 (–25) pairs; branches smooth; flowers 15–18 mm. long   3. *sesban*
    3 Leaflets 20–50 pairs; branches rough or prickly:
      4 Flowers 10–12 mm. long; branches with short prickles   4. *bispinosa*
      4 Flowers 20–24 mm. long; branches rough but not usually prickly   5. *exasperata*

1. S. grandiflora (L.) Pers., Synops. Pl. **2**: 316 (1807).—*Robinia grandiflora* L. (1753). *Agati grandiflora* (L.) Desv. (1813). Baby Boots.

Deciduous tree 3–5 m. high; leaves 15–30 cm. long, thinly pubescent; racemes 2–4-flowered, rarely branched; calyx campanulate, 2–2·5 cm. long; corolla crimson, pink or white; pod linear, up to 30 cm. or more long.

Occasional in cultivation or as an escape; 50–600 feet; fl. Mar–Apr, Nov–Dec, fr. Apr, Dec. *Fawcett! Johnston UCWI 3383! C. B. Lewis! Wynter UCWI 709!* Native of India or Australia, widely cultivated in the tropics as an ornamental.

2. S. sericea (Willd.) Link, Enum. Hort. Berol. **2**: 244 (1822).—*Coronilla sericea* Willd. (1809).

Shrub with slender branches, 2–5 m. high, annual; leaflets oblong-linear, glabrous adaxially, irritable, 15–25 mm. long, 4–7 mm. broad; racemes slender, (2–) 5–8-flowered; calyx 3–4 mm. long; corolla about 8 mm. long, yellow, the standard petal spotted blackish-purple; pod 10–20 cm. long, about 3 mm. broad.

Local (St. Andr., St. Cath., St. Thom.), in brackish swamp and sandy salina margins; 5–100 (–500) feet; fl. Oct–Mar, fr. Nov–Mar. *A 6458! H 9051!* Suriname, Bahamas, Antilles southwards to Martinique, Trinidad.

3. S. sesban (L.) Merr. in Philipp. Journ. Sci. **7**: 235 (1912).—*Aeschynomene sesban* L. (1753).

Shrub or rarely arborescent, 2–3 (–6) m. high; leaves 10–15 cm. long, thinly pubescent especially on rachis and petiolules; leaflets oblong-linear, 20–30 mm. long, about 5 mm. broad; racemes up to 12-flowered; calyx about 5 mm. long; corolla yellow, the standard petal suffused or spotted reddish-purple; pod up to 24 cm. long, slightly torulose.

M

Very rare (St. Andr.), cultivated and in wet sandy places mostly near the sea; SL (–700) feet; fl. and fr. Jan. *H 8617!* Native of the Old World tropics, widely scattered.

4. S. bispinosa (Jacq.) Steud. ex F. &. R., Fl. Jam. **4**: 24 (1920).—*Aeschynomene bispinosa* Jacq. (1792).

Shrubby annual herb 1–2·4 m. high; leaves up to about 30 cm. long, glabrous and glaucous with short prickles on the rachis; leaflets about 40 pairs, oblong-linear, 10–20 mm. long, 1·5–5 mm. broad; racemes mostly 2–4-flowered; calyx about 4 mm. long, hairy within the margin; corolla yellow, the standard petal speckled blackish-purple; pod up to 30 cm. long, 3 mm. broad, often slightly torulose.

Rare and local (St. Andr., St. Thom.), at the margins of brackish swamps; SL–20 (–500) feet; fl. and fr. Oct–Nov. *A 8312! E. Brown UCWI 9251!* Native of the Old World tropics, now widely scattered.

5. S. exasperata Kunth, Nov. Gen. **6**: 534 (1824).

Herb or undershrub 1–2·4 m. high; leaves up to 30 cm. long; leaflets mostly in 30–40 pairs, oblong-linear, irritable, 15–25 mm. long, 3–6 mm. broad; racemes 4–6-flowered; calyx about 7 mm. long; pod 20–25 cm. long, long-beaked.

Not recently confirmed for Jamaica. *Purdie*, the only specimen seen, is without precise locality and may be one of that collector's Colombian gatherings. Trinidad, continental tropical Amer.

16. CRACCA Benth. (1853) nom. cons.

1. C. caribaea (Jacq.) Benth. in Benth. & Oersted in Vid. Medd. Nat. For. Kjøbenh. **1853**: 9 (1853).—*Galega caribaea* Jacq. (1763).[1] *Benthamantha caribaea* (Jacq.) Kuntze (1898).

Perennial straggling undershrub; stipules setose-filiform, persistent; leaflets about 5 lateral pairs, elliptical, mucronate, thinly silky-pubescent on both surfaces, the terminal the largest, up to 30 mm. long and 13 mm. broad; flowers pink; pod linear, up to 7 cm. long and 3 mm. broad, 10–25-seeded.

A rare casual (St. Andr.), cultivated or adventive; about 500 feet; fr. Mar. *A 12369!* Mexico to Paraguay, Hispaniola, Lesser Antilles, Trinidad.

17. CAJANUS DC. (1813) nom. cons.

1. C. cajan (L.) Millsp. in Field Mus. Bot. **2**: 53 (1900).—*Cytisus cajan* L. (1753).
Gungo Pea, Pigeon Pea.

Shrub to about 3·5 m. high; stipules ovate, acuminate, very hairy; leaflets elliptical, acute, densely silvery-tomentellous beneath, up to about 9 cm. long and 3 cm. broad; racemes pedunculate, subcapitate, 6–8 cm. long; corolla 15–18 mm. long; pod up to 7 cm. long, 13–14 mm. broad, finely pubescent and glandular; seeds smooth, green, about 8 mm. broad.

Common in cultivation up to about 3400 feet; fl. and fr. mostly Dec–Feb. *H 6167! A. von der Porten! Wynter UCWI 2054!* Native probably of E. Indies, now widespread in the subtropics and tropics.

18. MOGHANIA J. St.-Hil. (1813)

Leaves 1-foliolate; stipules subulate-lanceolate, about 3 mm. long; flowers in cymules enclosed by large folded persistent bracts   1. *strobilifera*
Leaves (1–) 2–3-foliolate; stipules lanceolate, up to 9 cm. long; flowers in dense subspicate racemes with deciduous bracts   2. *macrophylla*

---

[1] Jacquin described a plant with glabrous leaves and fruit and his plate illustrates leaves with more numerous leaflets with longer mucronate tips than are found on plants consistently named *Cracca caribaea* by many workers.

95. PAPILIONACEAE (FABACEAE)

1. M. strobilifera (L.) J. St.-Hil. ex Kuntze, Revis. Gen. Pl. **1**: 199 (1891).—
Hedysarum strobiliferum L. (1753). *Flemingia strobilifera* (L.) Ait.f. (1812).
Wild Hops.

Shrub 0·6–2 m. high; leaflets ovate to elliptical, acute or obtuse at tip, mostly
rounded at base, 5–20 cm. long, 3–10 cm. broad; inflorescences axillary and terminal; bracts distichous in racemes, cordate, 10–25 mm. long, green becoming
brown; calyx about 5 mm. long, pubescent; corolla 5–6 mm. long, greenish-
yellow or pinkish; pod about 1 cm. long, puberulous and with red glands; seeds
subglobose, mottled, 3 mm. in diameter.

Common at field margins, waste places near streams and in abandoned pastures,
often gregarious; SL–3000 feet; fl. Dec–Aug, fr. Dec–Mar, July–Aug. *A 5841!*
H 11770! H & P 13586! Native of E. Indies extending to the islands of the Indian
Ocean; Hispaniola, St. Vincent, Barbados, Trinidad, Grand Cayman.

2. M. macrophylla (Willd.) Kuntze, Revis. Gen. Pl. **1**: 199 (1891).—*Crotalaria
macrophylla* Willd. (1803). *Flemingia stricta* Roxb. (1812).

Shrub 0·6–2·4 m. high; branches 3-angled; leaflets lanceolate to narrowly
elliptical, acuminate, up to 30 cm. long and 9 cm. broad; racemes axillary, 5–15 cm.
long; bracts linear to lanceolate, 15–20 mm. long; calyx-segments unequal, 9 mm.
long overall; corolla rosy-purple, the standard petal 9 mm. long; pod 13 mm. long.

Cultivated and naturalized locally in Trelawny and St. Ann, along roadsides and
in low-lying places; 400–1200 feet; fl. Jan–Feb, fr. Feb–Mar. *H 7269! Powell 1572!*
Native of India, scattered thinly through tropical Asia and Africa.

19. RHYNCHOSIA Lour. (1790) nom. cons.

1 Calyx up to 2 cm. long, at least the anterior lobe exceeding the greenish-white or cream-
   coloured corolla, densely pubescent with long hairs; stipels present   1. *hirta*
1 Calyx not exceeding 12 mm. long; corolla yellow or yellow with the standard petal
   streaked crimson or purple
   2 Robust woody vine with subterete or ribbonlike stems; pods constricted between the
      seeds; stipels absent; calyx 5 mm. long, the anterior lobe lanceolate, the laterals ovate
                                                                                        2. *phaseoloides*
   2 Slender vines with ribbed and angled stems; pods not constricted between the seeds;
      stipels present; calyx-lobes lanceolate, slender-tipped
      3 Calyx 7–8 (–11) mm. long, about as long as the yellow corolla   3. *reticulata*
      3 Calyx 3·5–4·5 mm. long; corolla about 6 mm. long, yellow with the standard petal
         streaked crimson or purple   4. *minima*

1. R. hirta (Andr.) Meikle & Verdcourt in Taxon **16**: 462 (1967).—*Dolichos hirtus*
Andr. (1807). *Cylista albiflora* Sims (1816). *R. albiflora* (Sims) Alston (1931).

Twining climber; leaves ovate, obtuse to caudate-acuminate, pubescent, up to
11 cm. long and 7 cm. broad; bracts broadly ovate, acuminate, about 18 mm. long,
deciduous; calyx-lobes oblanceolate, acute or obtuse; seeds blue or blue-black,

Very rare (St. Thom.), in waste places; about 1000 feet; fl. Dec. *Cornman 348!*
Tropical Asia and Africa.

2. R. phaseoloides (Sw.) DC., Prodr. **2**: 385 (1825).—*Glycine phaseoloides* Sw.
(1788).

Climber to 6 m. or more high; stipules small and very early deciduous; leaflets
broadly ovate to rhomboid, obtuse to shortly acuminate, subglabrous adaxially,
tomentellous beneath, up to 13 cm. long and 8 cm. broad; bracts linear-lanceolate,
about 3·5 mm. long, early deciduous; standard petal streaked dark crimson, 8–10
mm. long; pod 2–2·5 cm. long; seeds black sometimes also marked red.

Rather common, in thickets and woodland margins; 1000–4000 feet; fl. Dec–
May, fr. Feb–Aug. *A 6887! H 8775! J.P. 872! P 16032!* Continental tropical Amer.,
West Indies.

3. R. reticulata (Sw.) DC., Prodr. **2**: 385 (1825).—*Glycine reticulata* Sw. (1788).

Low perennial scrambling twiner; young stems densely tomentose; stipules ovate,
about 5 mm. long, more or less persistent; leaflets broadly lanceolate, acute,
tomentellous on both surfaces, up to 9 (–10) cm. long and 3·5 (–4·5) cm. broad,

prominently veined beneath; bracts like the stipules, deciduous; standard petal and wings flattened together; pod about 2 cm. long; seeds dark brown with black spots.

Occasional (St. Andr., St. Eliz., St. Thom.), in thickets and on open roadside banks; 560–2600 feet; fl. Oct–July, fr. Dec–June. *A 5805! Morris! P 15331!* Continental tropical Amer., West Indies.

4. R. minima (L.) DC., Prodr. 2: 385 (1825).—*Dolichos minimus* L. (1753).

Slender-stemmed herbaceous twiner; young stems tomentose; stipules lanceolate-subulate, 3 mm. long, persistent; leaflets ovate-rhomboid, acute or obtuse, puberulous, 1–2·5 (–4) cm. long and broad; bracts 2 mm. long, deciduous; pod 13–17 mm. long; seeds black.

Common, in waste places and cultivated land; SL–1500 feet; fl. and fr. all the year. *A 5892! H 11980! H & P 14461!* General in the subtropics and tropics; Grand Cayman.

R. caribaea (Jacq.) DC. was reported doubtfully by Fawcett & Rendle (1920) and has not been confirmed. It resembles *R. minima* but has larger flowers and fruits.

20. ATYLOSIA Wight & Arn. (1834)

1. A. scarabaeoides (L.) Benth. in Miq., Pl. Jungh.: 242 (1852).—*Dolichos scarabaeoides* L. (1753).

Slender-stemmed twiner; stems, leaves, calyx and fruit densely tomentellous; stipules ovate, 1·5 mm. long, deciduous; leaflets obovate to elliptical, rounded at tip, 1–3·5 cm. long, up to 2 cm. broad, with veins prominent beneath; stipels wanting; flowers usually 2 on a short axillary peduncle; calyx about 8 mm. long; corolla yellow, about 10 mm. long, glabrous; pod 2–2·5 cm. long, 3–6-seeded, depressed and septate between the seeds; seed oblong-globose, 3 mm. broad, dark brown with paler mottling, caruncled.

Very local (St. Thom.), on herbs and shrubs in alluvial gravel area; near SL; fl. and fr. June–Sept. *Robertson UCWI 3606! 5616!* Coasts of tropical Asia and Africa.

21. DIOCLEA Kunth (1824)

1. D. reflexa Hook. f. in Hook., Niger Fl.: 306 (1849).

Woody climber with pilose branches; stipules peltate, acute; leaflets ovate-elliptical, acuminate, thinly pilose, 8–16 cm. long, up to 10 cm. broad; racemes stiff and woody, longer than the leaves; bracts narrowly lanceolate, about 1·5 cm. long; calyx about 1·5 cm. long, brownish-silky pubescent; standard petal reddish-violet with a yellow spot at base; pod oblong, 9–13 cm. long, 4–6 cm. broad, 1–3-seeded; seeds rounded, 2·5–3·5 cm. long, with a long hilum.

Not collected in Jamaica for over 100 years. *Wilson.* General in the tropics.

22. PUERARIA DC. (1825)

Calyx and fruit densely hairy with usually appressed yellowish or brownish hairs; ripe pod about 6 mm. broad; cultivated as a cover crop; native of Asia; Kudzu (*P. thunbergiana* (Sieb. & Zucc.) Benth.)   *P. triloba* (L.) Makino
Calyx sparsely setose with a few ascending hairs; ripe pod glabrescent, about 4 mm. broad
                                                                                                1. *phaseoloides*

1. P. phaseoloides (Roxb.) Benth. in Journ. Linn. Soc., Bot. **9**: 125 (1867).— *Dolichos phaseoloides* Roxb. (1832).

Robust but slender-stemmed twiner, with hairs variously spreading or ascending or sometimes reflexed on the stems, petioles, peduncles and leaves; stipules ovate, acuminate, striate, about 5 mm. long; leaflets broadly rhomboid-ovate, acute at tip, with long appressed and very short erect hairs adaxially, whitish-sericeous beneath, 2·5–12 cm. long, 1·5–9 cm. broad; racemes axillary, 10–20 cm. long, with stout peduncles, many-flowered, with swollen nodes; flowers about 1·5 cm. long; standard petal deep mauve-pink, drying bluish; pod linear, curved towards the tip, 6–9 cm. long, 10–15-seeded.

Locally common (Port.), naturalized in damp roadside and riverside thickets; 20–200 feet; fl. May, Aug–Dec, fr. Aug–Dec. *A 11503! 11633!* Native of tropical Asia, introduced into parts of Africa and Amer. as a cover crop.

23. CANAVALIA DC. (1825) nom. cons.; Sauer (1964)

1 Nodes of inflorescence each with 3 or more flowers; 3 teeth of anterior calyx-lip united
   into a common base 2 mm. long; fruit with sutural ribs only; seeds nearly spherical;
   hilum nearly equalling the seed; Grand Cayman, on arid rocky cliffs, and also reported
   from Mexico, Bahamas, Cuba, Hispaniola, Puerto Rico and Virgin Is. (*C. rusiosperma*
   Urb.), but not from Jamaica   *C. nitida* (Cav.) Piper
1 Nodes of inflorescence each with 2 or 3 flowers; 3 teeth of anterior calyx-lip free from
   one another; fruit with an additional rib on each side; seeds more or less compressed:
  2 Calyx leathery, veins not prominent; standard petal without auricles above the claw;
    keel and wings spirally twisted; each valve of fruit with the extra rib about 8 mm. from
    ventral rib; hilum 4–5 mm. long   1. *altipendula*
  2 Calyx papery, with prominent veins; standard petal auricled above the claw; keel and
    wings not twisted; each valve of fruit with the extra rib about 3–5 mm. from ventral
    rib:
    3 Leaflets subrotundate, emarginate, fleshy; hilum about 8 mm. long; plant almost
      confined to sea-beaches   2. *maritima*
    3 Leaflets ovate to elliptical, not emarginate, not fleshy; mostly cultivated plants:
      4 Leaflets obtuse to acute or slightly acuminate; seeds white, about 20 mm. long;
        hilum 8–9 mm. long   3. *ensiformis*
      4 Leaflets acuminate; seeds brownish-red, rarely white, 20–35 mm. long; hilum 15–
        20 mm. long   4. *gladiata*

1. C. altipendula (Piper) Standl. in Field Mus. Bot. **11**: 160 (1936).—*Wende-rothia altipendula* Piper (1925). *C. altissima* Macf. (1837), partly, not *Dolichos altissimus* Jacq. (1760). Wild Overlook Bean.

High climbing vine, often with purplish glabrescent stems; young stems with short reflexed hairs; leaflets elliptical to obovate-elliptical or oblong, obtuse, up to 9·5 (–13) cm. long and 4·5 (–6) cm. broad; petiolules pubescent; racemes up to 18 cm. long with peduncles to 5 cm. long; standard petal about 2 cm. long, purple; keel, wings and stamens white, fading yellowish; pod up to 20 (–24) cm. long and 2·5 (–3) cm. broad, appressed-puberulous; seeds greyish- or reddish-brown with black blotches, 12–13 (–18) mm. long, 9–12 (–15) mm. broad.

Common, on open slopes and climbing in thickets and woodlands on limestone; 50–3000 feet; fl. June–Jan, fr. July–Feb. *A 5529! H 8729! J.P. 988! HPS 14707!* Endemic.

2. C. maritima (Aubl.) Urb. in Fedde, Repert. Sp. Nov. **15**: 400 (1919).—*Dolichos maritimus* Aubl. (1775). *C. rosea* (Sw.) DC. (1825). *C. obtusifolia* (Lam.) DC. (1825). Seaside Bean.

Stem trailing, up to 10 m. long, sometimes climbing, glabrescent; leaflets rounded to obovate, 5–10 cm. long, 4–10 cm. broad, thinly appressed-puberulous on both surfaces; racemes up to 30 cm. long with peduncles to 15 cm. long; standard petal about 3 cm. long, usually purple fading rosy-violet, very rarely white; pod 5–15 cm. long, about 2·5 cm. broad, thinly puberulous; seeds brown, marbled, about 18 mm. long and 12 mm. broad.

Very common, on the strand and sandy wastes near the sea; near SL; fl. and fr. all the year. *A 7842! H 9544! Stearn 723! P 11489!* Tropical and subtropical shores; Cayman Is.

3. C. ensiformis (L.) DC., Prodr. **2**: 404 (1825).—*Dolichos ensiformis* L. (1753). Horse Bean, Jack Bean, Overlook Bean.

Plant climbing or sometimes bushy; corolla like the last, purple; pod (20–) 30–35 cm. long, 2·5–3·5 cm. broad; seeds ivory or white, up to 21 mm. long, 15 mm. broad and 10 mm. thick.

Cultivated sporadically as a cover crop and occasionally for food. *Harris!* Native probably of the American tropics, now widespread by introduction.

4. C. gladiata (Jacq.) DC., Prodr. **2**: 404 (1825).—*Dolichos gladiatus* Jacq. (1788). Sword Bean.

Climber; pod 25–40 cm. long, about 5 cm. broad; seeds red, pink or brown, up to 35 mm. long, 20 mm. broad and 14 mm. thick.

Cultivated and occasionally escaping. Native of the Old World tropics.

24. CALOPOGONIUM Desv. (1826)

1. C. mucunoides Desv. in Ann. Sci. Nat. sér. **1**, **9**: 423 (1826).

Stem twining and trailing, densely pilose with spreading brownish hairs; leaflets ovate to rhomboid-ovate, 4–10 cm. long, 2–5 cm. broad, appressed-pubescent on both surfaces; axillary racemes up to 20 cm. long; peduncles up to 12 cm. long or very short; calyx with subulate teeth, 6–8 mm. long; corolla mauve or blue, a little longer than calyx; pod straight or falcate, densely brownish-hirsute, 2–4 cm. long, 4–5 mm. broad, 5–8-seeded.

Sparingly cultivated (St. Andr., St. Thom.), as a cover crop at low elevations and escaping; fl. and fr. Dec, fr. Jan. *Allwood & Robertson UCWI 1802!* Native of tropical Amer., cultivated and naturalized in Africa and Asia.

25. GALACTIA P. Browne (1756)

1 Low shrub up to 1 m. high, densely pubescent; leaves 3-foliolate with elliptical or oblong leaflets; calyx 6 mm. long; petals pink, the standard 10–12 mm. long; fruit (3–) 5–7 cm. long, about 8 mm. broad; seeds black; once reported but not confirmed, otherwise from tropical C. and S. Amer., Cuba, Hispaniola and Trinidad
            G. jussiaeana Kunth
1 Twining vines:
  2 Flowers 2–3·5 cm. long; woody climbers; leaflets 3:
    3 Leaflets ovate, mostly acute or acuminate, 7–12 cm. long; calyx about 20 mm. long; wing petals shorter than keel, about as long as or a little longer than standard; seeds reddish-brown, 7–8 mm. long   1. *sangsterae*
    3 Leaflets mostly elliptical and obtuse with an often emarginate tip, rarely as much as 8 cm. long; calyx 7–9 mm. long; wing petals a little shorter than keel and standard; seeds brown with blackish spots, 5–6 mm. long   2. *pendula*
  2 Flowers up to about 1·2 cm. long; slender-stemmed twiners:
    4 Petiolule of terminal leaflet (beyond insertion of lateral leaflets) (3–) 7–10 (–12) mm. long; leaflets 3, the larger (terminal) mostly 4–6 (–8) cm. long:
      5 Calyx 3·5–4·5 mm. long, the lobes about as long as the tube; leaflets ovate-oblong to ovate-lanceolate   3. *laxiflora*
      5 Calyx (4·5–) 5–7 (–8) mm. long, the lobes about twice as long as the tube; leaflets ovate to broadly elliptical   4. *striata*
    4 Petiolule of terminal leaflet 1–2 (–4) mm. long; leaflets 1–3, the larger mostly 1·5–3 cm. long:
      6 Leaflets mostly linear to linear-oblong or oblanceolate, glabrous or nearly so adaxially   5. *parvifolia*
      6 Leaflets mostly elliptical and softly pubescent on both surfaces   6. *uniflora*

1. G. sangsterae Proctor in Bull Inst. Jam., Sci. ser. **16**: 10, tt. 2, 3 (1967).

Leaf-rachis and petiolules tomentose; leaf-blades glabrous, up to about 7 cm. broad; inflorescences racemose, sometimes branched, up to 20 cm. or more long, often arising from older stems; petals bright rose, the standard 2–2·6 cm. long, shorter than the rest; pod up to 12 cm. long, 1·3 cm. broad, appressed-puberulous.

Rare and local (St. Eliz.), on a wooded limestone hillside; 750–1000 feet; fl. Apr–June, fr. May–June. *P15420!* Endemic.

2. G. pendula Pers., Synops. Pl. **2**: 302 (1807).

Climber with slender woody stem up to 6 m. long; leaflets variably hairy, usually glabrescent adaxially and thinly puberulous abaxially, 1·5–7 (–8) cm. long, 1–3 (–4) cm. broad; racemes compound, up to 30 cm. long, each short thick lateral branch bearing up to 20 flowers in spiral succession but basally fewer to solitary-flowered; petals light crimson to deep rose, the standard 2–2·5 cm. long; pod 6–9 cm. long, 6–9 mm. broad, thinly puberulous.

Common, on steep banks and limestone rocks in thickets and woodlands in moderately dry areas; SL–2500 feet; fl. July–Feb, fr. most of the year. *A 10167! H 8864! P 18354!* Endemic.

3. G. laxiflora Urb., Symb. Ant. **2**: 315 (1900).
Stem very slender, 1–2 m. long; leaflets acute or narrowly obtuse, glabrescent adaxially, appressed-puberulous abaxially, 2–4 cm. long, 8–20 mm. broad; inflorescence short (in *A 8353*) or up to 20 cm. long (in *Type*); petals pink, the standard 7–8 mm. long, wings 6 mm. long and a little shorter than the keel; pod 5–6 cm. long, 5–5·5 mm. broad, sparingly puberulous.
Very rare (St. Andr.), in thickets; about 300 feet; fl. and fr. Nov. *A 8353!* Hispaniola. A rather obscure species related to and perhaps only an extreme (? shade) variant of the next.

4. G. striata (Jacq.) Urb., Symb. Ant. **2**: 320 (1900).—*Glycine striata* Jacq. (1770).
A very variable climber, like the last but usually more robust in vegetative growth; leaflets rounded to obtuse or acute at tip, thinly pubescent or glabrescent adaxially, usually silky-tomentose abaxially, 2–8 cm. long, up to 4 cm. broad; inflorescence 1–24 cm. long; petals bright pink, fading bluish or mauve, the standard 7–9 (–11) mm. long, the wings and keel about as long; pod 4–7 cm. long, 6–9 mm. broad, narrowly marginate when mature, with longer hairs along the middle of each valve; seeds usually mottled pinkish and brown, kidney-shaped, 3·5–4 mm. long.
Common in the southern parishes, a weed of pastures, roadsides, thickets and arid sandy places; SL–1200 feet; fl. Sept–Feb, fr. Sept–Mar. *A 6220! 8205! 11813! H 10162! 12288! P 19764!* Mexico, Venezuela, Greater Antilles, Antigua, Guadeloupe, Tobago; Grand Cayman.

5. G. parvifolia A. Rich. in Sagra, Hist. Cuba, Parte 2, **10**: 176 (1845).
Branched slender-stemmed twiner; leaflets rounded at tip, glabrous or nearly so and strongly reticulate-veined adaxially, appressed-pubescent abaxially, 1–3 (–4) cm. long, 4–10 mm. broad; inflorescence short with the peduncle less than 1 cm. long, 1–3-flowered; calyx 4–8 mm. long, the lower lobe commonly twice as long as the tube; petals purple, the standard 9–12 mm. long; pod 2–5·5 cm. long, 4–6 mm. broad, appressed-pubescent; seeds subrectangular, mottled brownish-black, about 2·5 mm. long.
Local (St. Cath.), in arid thickets on limestone; 5–300 feet; fl. Nov, fr. Nov–Feb. *A 8901! 12885! Skelding UCWI 3521!* Florida, Cuba, Hispaniola.

6. G. uniflora Urb., Symb. Ant. **2**: 325 (1900).
Trailing and climbing twiner; leaflets rounded at the tip, reticulate-veined on both surfaces, 8–30 mm. long, 5–10 mm. broad; inflorescence very short, 1–3-flowered; calyx 6 mm. long; petals light mauve or bluish, the standard 8–9 mm. long; pod 3–4 cm. long, 4–4·5 mm. broad, densely appressed-pubescent; seeds rounded-compressed, olive-green with dark purplish markings, 2·5–3 mm. long.
Rare and local (St. Cath., Manch.), among grasses and low shrubs; 50–900 feet; fl. and fr. July–Aug, Nov. *duQuesnay 554! H 11727! 12439!* Bahamas, S. Caicos. Plants included here are probably part of a continuous population with the last in Jamaica. Affinity with *G. dubia* DC. of Cuba, Puerto Rico, Virgin Is. and Anguilla to Guadeloupe, is close.

26. ERYTHRINA L. (1753); Krukoff (1939), (1969)

1 Standard petal straight in open flower; wings and keel included in calyx, about 1 cm. long; calyx truncate, about 1 cm. long, puberulous, glabrescent; leaflets broadly ovate-triangular, obtuse or acute at tip   1. *corallodendrum*
1 Standard petal recurved in open flower; wings and keel conspicuous; calyx oblique, beaked abaxially, puberulous to densely velvety-tomentose:
  2 Calyx about 1 cm. long in the open flower, puberulous; standard vermilion:
    3 Calyx split adaxially; keel less than half as long as standard and less than twice as long as wing petals; leaflets obtuse   2. *glauca*
    3 Calyx not split adaxially; keel more than half as long as standard and four times longer than wing petals; leaflets long-acute to caudate-acuminate, rarely obtuse
                                                                                      3. *poeppigiana*

2 Calyx more than 2 cm. long, densely velvety-tomentose with stellate indumentum at
    least when young, split adaxially in the open flower; keel and wing petals subequal:
4 Pedicels up to 1 cm. long; standard about 7 cm. long and 3 cm. broad, deep scarlet or
    white suffused pink; leaflets broadly acute or shortly cuspidate at tip; filaments
    almost straight   4a. *variegata* var. *orientalis*
4 Pedicels up to 2 cm. long; standard up to 6 cm. long and 3 cm. broad, deeply curved,
    vermilion; leaflets rounded-obtuse to emarginate at tip; filaments deeply curved
                                                    5. *velutina*

1. **E. corallodendrum** L., Sp. Pl. 2: 706 (1753).—Cutlass Bush, Duppy Machete,
    Spanish Machete.
Tree up to 10 m. high but usually smaller, rarely taller, with or without prickles;
bark buff, greenish within; leaflets truncate at base, glabrescent, 5–15 cm. long,
4–12 (–13) cm. broad; flowers 2 or 3 together appearing mostly when tree leafless;
standard petal folded lengthwise, 6–7 cm. long, 1·7–2 cm. broad when opened out,
crimson or deep scarlet; filaments shorter than standard, dull purplish, remaining
clustered and mostly included; pod 15–25 cm. long; seeds scarlet, about 1 cm. long.
Rather common, in woodlands and thickets on rough limestone, often planted as
live fence posts; near SL–3000 feet; fl. and fr. Dec–July. *A 8550! H 7309! Proctor
& Mullings 22038!* Hispaniola, Grand Cayman.

2. **E. glauca** Willd. in Ges. Naturf. Freunde, Neue Schr. 3: 428 (1801).—Bucare,
    Swamp Immortelle.
Soft-wooded tree 5–20 m. high with prickly branches; bark golden-buff, prominently flaking in longitudinal strips, green beneath; leaflets broadly ovate to
elliptical, glabrous, glaucous abaxially, 8–18 cm. long, 5–12 cm. broad; flowers 2 or
3 together, opening in successive whorls; pedicels deflexed; calyx fleshy, brown;
standard petal opening incompletely, 5–5·5 cm. long, 4·5–5 cm. broad when
opened out; wing petals white at base, dusky-crimson distally; keel creamy-
brown; stamens exposed, rather short; style crooked; pod up to 30 cm. long;
seeds mottled black, 1·5 cm. long.
Very local (Port., St. Thom.), in coastal freshwater swamps; SL–10 feet; fl.
Jan–Mar. *A 12305! HPS 14810!* Guatemala to Brazil, Greater Antilles, St. Vincent,
Trinidad and Tobago.

3. **E. poeppigiana** (Walp.) O. F. Cook in Bull. U.S. Dept. Agric. 25: 57 (1901).—
    Micropteryx poeppigiana Walp. (1850). Anauca, Mountain Immortelle.
Tree 8–20 m. high; trunk and branches sparingly prickly; leaflets rhomboid-
ovate, glabrous, (5–) 10–25 cm. long, (4–) 8–20 cm. broad; stipels cupular, glandular, up to 5 mm. broad; flowers 2 or 3 together; standard petal opening, elliptical,
5·5 cm. long, 2·5 cm. broad; wing petals ovate, yellow at base, scarlet distally;
keel yellow at base, scarlet in middle, dark orange distally, enclosing stamens and
style except their tips exserted; style straight; pod up to 13 cm. long; seeds brown,
1 cm. long.
Rather local in the east-central parishes, on hillsides and in pastures, mostly in
sheltered fruit growing areas, probably naturalized; 500–2000 feet; fl. Jan–Mar,
fr. Feb–Mar. *A 12310! P 11885! Yuncker 17899!* Native of Peru, now spread
from Panama to Brazil and throughout the West Indies, introduced into W. Africa.

4. **E. variegata** L., Herb. Amboin.: 10 (1754).

4a. **Var. orientalis** (L.) Merr., Interpr. Rumph. Amb.: 276 (1917).—*E. corallo-
    dendrum* L. var. *orientalis* L. (1753). *E. indica* Lam. (1786). Dabdab, Indian
    Coral Tree.
Low-branched tree up to 12 m. high; bark greenish with rather few blackish
prickles; leaflets rhomboid-ovate, glabrous, up to about 15 cm. long and 12 cm.
broad; inflorescence about 30 cm. long; calyx 3 cm. long; wing and keel petals
broadly obliquely obovate, about 15 mm. long and 10 mm. broad, deep scarlet or
salmon-pink; pod up to 25 cm. long; seeds purple, 1·5 cm. long.
Locally common (St. Andr., West.), cultivated for ornament and as a live fence,
probably naturalized; 50–600 feet; fl. Jan–Mar, fr. Apr. *A 12285! H 76! Johnston
3400!* Native of E. Asia and Pacific Is., introduced elsewhere in Asia and Africa
and scattered throughout the West Indies.

5. E. velutina Willd. in Ges. Naturf. Freunde, Neue Schr. **3**: 426 (1801).—Coral Tree, Red Bean Tree.

Tree to 20 m. high with prickles on trunk and branches; leaflets rhomboid-ovate to subrotundate, tomentose beneath, 5–15 (–20) cm. long, up to 10 (–20) cm. broad; inflorescence 12–25 cm. long; calyx 2·5–3 cm. long; wing and keel petals olive-brown with crimson margins; pod (5–) 8–13 cm. long; seeds red, about 1·5 cm. long.

Local, cultivated and probably naturalized along roadsides; 500–1700 feet; fl. Feb–Apr, fr. Apr. *A 12503! Bengry! Fawcett!* Native of tropical S. Amer., introduced elsewhere.

27. MUCUNA Adans. (1763) nom. cons.

1 Leaflets glabrous beneath; peduncle up to about 1 m. long, the flowers in a short terminal raceme; standard petal and wings purplish, keel yellowish; fruit 10–20 cm. long, 5 cm. broad, transversely ribbed, pubescent and with urticating hairs   1. *urens*
1 Leaflets thinly to densely appressed-pubescent beneath:
  2 Flowers yellow, clustered at the end of a peduncle as long as to twice as long as the petiole; calyx nearly 2 cm. long; leaflets silvery-sericeous beneath, obscurely mucronate; hilum almost surrounding the seed, flat:
    3 Flowers subumbellate; standard petal 3–3·5 (–4) cm. long; wings 5–5·5 (–6·5) cm. long; fruit with distinct marginal wings 2–5 mm. broad; seeds mostly about 2·5 cm. in diameter, the hilum 5 mm. broad   2. *sloanei*
    3 Flowers on an expanded receptacle; standard petal 5–5·5 cm. long, wings 7–8 cm. long; marginal wings of fruit obsolete; seeds 3·5–4 cm. in diameter, the hilum about 10 mm. broad   3. *fawcettii*
  2 Flowers usually purplish, about 3·5 cm. long, in racemes or rarely solitary; peduncle often shorter than petiole; calyx about 1 cm. long; leaflets thinly sericeous beneath, conspicuously mucronate; hilum 5–8 mm. long; prominently carunculate:
    4 Fruit brown, pubescent and with caducous urticating bristly hairs; seeds brown mottled black, 10–12 mm. long; keel as long as to a little longer than wings   4. *pruriens*
    4 Fruit greyish, velvety-appressed-tomentose without urticating hairs; seeds greyish-white often mottled dull brown, 15–18 mm. long; keel longer than wings, sharply ascending; sparingly cultivated; native probably of tropical Asia, but first described from Florida; Velvet Bean   *M. deeringiana* (Bort) Merr.

1. M. urens (L.) Medic. in Vorles. Churpf. Ges. **2**: 399 (1787).—*Dolichos urens* L. (1759). Ox-eye Bean.

Climber; leaflets ovate to oblong, abruptly acuminate, 8–16 cm. long, 4–7·5 cm. broad; raceme zigzag, 5–10 cm. long; pedicels solitary or 2–3 together, about 1 cm. long; calyx about 1 cm. long; standard petal 3–4 cm. long, nearly as long as wings; keel slightly longer than the wings, turned up sharply at tip; seeds roundish-compressed, (2–) 3–4 cm. in diameter, yellowish-grey, almost surrounded by the black hilum.

Very rare and not recently collected; reported from St. Mary and Portland; fl. Sept–Oct. C. and S. Amer., Cuba, Hispaniola, Puerto Rico, St. Kitts to Trinidad.

2. M. sloanei F. & R. in Journ. Bot. **55**: 36 (1917).—Horse-eye Bean.

Climber to 6 m. or more high; leaflets ovate to elliptical, shortly acuminate, 7–16 cm. long, 4·5–9 (–14) cm. broad; pedicels 6–12 mm. long; flowers rich lemon-yellow; keel slightly longer than wings; pod 10–18 cm. long, 4–6 cm. broad, obliquely transversely ridged, 2–4-seeded, hispid with urticating hairs.

Occasional, on trees at woodland margins and along streambanks; 500–2500 feet; fl. Nov–Feb., fr. Jan–Apr. *A 9967! H 5590! P 20552!* Throughout tropical Amer. and also W. Africa.

3. M. fawcettii Urb., Symb. Ant. **5**: 371 (1908).

Climber like the last; leaflets rounded-elliptical, shortly acuminate, 13–17 cm. long, 12–15 cm. broad; pedicels about 10 mm. long; flowers yellow; keel longer than wings; pod 12 cm. long, 5·5 cm. broad, 1–2-seeded, hispid with bristly hairs.

Very rare (Trel.); about 2000 feet; fl. Nov–Dec. *H 8818! 9102!* Endemic.

4. M. pruriens (L.) DC., Prodr. 2: 405 (1825).—*Dolichos pruriens* L. (1754). Cowitch.

Twining climber; leaflets rhomboid-ovate, obtuse or acute at tip, 7–15 cm. long, 6–10 cm. broad, with petioles up to 20 (–30) cm. long; inflorescence longer or shorter than petiole; pedicels up to 8 mm. long; corolla deep or reddish-purple, sometimes yellowish at base or rarely all yellow; pod 4–8 cm. long, about 1·5 cm. broad, 2–6-seeded.

Frequent, in cultivations, thickets and woodland margins; 250–2500 feet; fl. Oct–Mar, fr. Nov–Mar, Aug. *A 8350! J.P. 636! Powell 693! P 20704!* General in the tropics; Grand Cayman.

28. PSOPHOCARPUS DC. (1825) nom. cons.

Fruit 4–6 cm. long, with denticulate wings 4–5 mm. broad; inflorescence 15–30 cm. long
 including stout peduncle; flowers about 2 cm. long   **1. palustris**
Fruit 15–20 cm. long, 12–15 seeded, with subentire wings 7–8 mm. broad; inflorescence
 10–12 cm. long including slender peduncle; flowers about 2·5 cm. long, light violet;
 leaflets ovate-triangular, acute, subglabrous; at one time cultivated; native of tropical
 Asia; Manila Bean   ***P. tetragonolobus*** (L.) DC.

1. P. palustris Desv. in Ann. Sci. Nat. sér. 1, **9**: 420 (1826).

Rather robust trailing and spreading climber; leaflets broadly ovate-rhomboid, sometimes shallowly 3-lobed, shortly acuminate, 4–9 (–12) cm. long, 3·5–8 (–9) cm. broad, thinly pubescent on the veins beneath; stipels oblong; paired bracteoles broadly ovate, longer than calyx, deciduous; calyx 6–7 mm. long, campanulate, the lateral lobes short; standard petal and wings bright blue within, standard greenish outside; keel petals joined at tip only, greenish-cream; seeds black, subglobose, puberulous, about 6 mm. long.

Naturalized locally (St. Cath., Clar., Port.), in damp places at streamsides, swamp margins and in roadside thickets; SL–1600 feet; fl. Nov–Mar, fr. Dec–Mar. *A 8441! 11847! Powell 704! Yuncker 17955!* Native of tropical Africa.

29. PHASEOLUS L. (1753)

1 Keel forming more than 1 and up to 2 spirals:
 2 Corolla 2–2·5 cm. long; flowers aggregated or in a short raceme; peduncle stout;
  bracts (at base of pedicel) deciduous; pedicels very short; bracteoles (at apex of
  pedicel) shorter than calyx; lateral calyx-teeth lanceolate-falcate; fruit linear,
  10–15-seeded   **1. adenanthus**
 2 Corolla 1–1·5 cm. long; bracts persistent; lateral calyx-teeth ovate to deltate, not
  falcate:
  3 Flowers few in a cluster; peduncle very short, less than 1 cm. long; bracteoles broad,
   usually longer than calyx; corolla white or purplish; fruit linear, often slightly
   curved, 10–15 cm. long, 4–6-seeded; seeds reniform; cultivated; Red Peas
        P. vulgaris L.
  3 Flowers in elongated racemes; peduncle up to 5 cm. or more long; bracteoles ovate,
   shorter than calyx; flowers 3 or more together at each node of raceme; fruit much
   narrower near base than towards tip, slightly curved adaxially, deeply curved
   abaxially, 2–4-seeded; seeds flat   **2. lunatus**
1 Keel forming 1 or an incomplete spiral; peduncles 5 cm. or more long:
 5 Stipules appendaged at base; flowers clustered at tip of a peduncle (5–) 8–12 cm. long;
  bracteoles longer than calyx; calyx-teeth broadly ovate; petals yellow; standard
  11–13 mm. long; keel forming one complete spiral; fruit 6–7 mm. broad, pilose
        3. trichocarpus
 5 Stipules truncate at base; peduncles (6–) 10–30 cm. long; bracteoles shorter than calyx;
  fruit about 5 mm. broad, thinly appressed-pubescent:
  6 Lateral calyx-teeth deltate, not falcate; flowers more or less racemose; petals white to
   light purple; standard about 15 mm. long; keel sigmoid **4. peduncularis**
  6 Lateral calyx-teeth falcate; flowers in 1 or a few clusters; petals mauve fading yellowish or dull orange; standard about 20 mm. long; keel twisted and bent to one side
        5. antillanus

1. P. adenanthus G. F. W. Meyer, Prim. Fl. Esseq.: 239 (1818).

Perennial twiner; stipules truncate at base; stem and petioles thinly hispid, the petiolules more densely pubescent with shaggy brownish hairs; leaflets ovate to ovate-elliptical, acute, 7–12 cm. long, 4–6 cm. broad; peduncle about 10 cm. long; raceme with several approximated swollen nodes; calyx-tube about 4 mm. long; corolla mauve or pink, sometimes with white or yellow; pod straight or falcate, appressed-pubescent, 7–10 cm. long, about 10 mm. broad; seeds about 2 mm. long.

Rare (St. Andr., West., St. Thom.), in waste places and in thickets; SL–150 feet; fl. Dec, fr. Dec–Mar. *A 5679!* General throughout the tropics.

2. P. lunatus L., Sp. Pl. **2**: 724 (1753).—Broad Bean.

Annual or short-lived perennial twiner; stipules truncate; stem glabrescent; leaflets ovate to rhomboid, acuminate, 5–10 cm. long, 3–6 cm. broad; inflorescence 10–15 cm. long including the long or short peduncle; bracts 1–1·5 mm. long, ovate; calyx 3–4 mm. long, the lateral teeth deltate; petals white or yellowish, sometimes mauve; standard 1 cm. long; pod 4–9 (–12) cm. long, 1·5–2 cm. broad; seeds variable in colour, often whitish, about 15 mm. long.

Cultivated at the lower elevations; fl. and fr. Aug–Jan. *Johnston 45! P 23903!* Native of tropical Amer., now widespread in the tropics.

3. P. trichocarpus C. Wright in Sauv. in Anal. Acad. Cienc. Habana **5**: 337 (1868).

Slender-stemmed twiner with retrorse yellowish hairs; leaflets oblong-lanceolate, acute, 4–10 cm. long, 1–2·5 cm. broad; bracts and bracteoles subulate; calyx about 3 mm. long; pod 3–4 cm. long, black at maturity with yellowish hairs; seeds dark brown with white hilum, shiny, 3–4 mm. long.

Very rare (St. Eliz.), among grasses in swamps; about 25 feet; fl. and fr. Jan. *P 21926!* Guyana, Suriname, Cuba, Puerto Rico.

4. P. peduncularis Kunth, Nov. Gen. **6**: 447 (1824).—*Vigna peduncularis* (Kunth) F. & R. (1920).

Stem herbaceous, twining or trailing and rooting, glabrescent; leaflets broadly ovate, acuminate, 2–7 (–10) cm. long, 1·5–4 (–8) cm. broad; bracts and bracteoles minute; calyx about 4 mm. long; pod linear, 4–8 cm. long; seeds brownish, about 2 mm. long.

Very rare (Port.), in thickets in damp areas; about 500 feet; fl. Sept. *A 8133!* Panama to Suriname, Greater Antilles, Trinidad. *Harris 5979*, the only exsiccata cited by Fawcett & Rendle, is *Macroptilium lathyroides*.

5. P. antillanus Urb., Symb. Ant. **4**: 309 (1905).—*Vigna antillana* (Urb.) F. & R. (1920).

Stem twining to 6 m., sparsely hairy; leaflets oblong-ovate to deltate, acutely acuminate, 3–8 cm. long, 2–4 cm. broad; bracts minute, roundish, 1 mm. long; bracteoles oblong-elliptical, ribbed, about 3 mm. long; calyx about 8 mm. long, puberulous, becoming thick and warted; pod linear, 8–14 cm. long; seeds, compressed-reniform, dark brown, 4·5–5 mm. long.

Rare and local (St. Andr.), on waste ground and in thickets on hillsides; 800–3000 feet; fl. and fr. Apr–June. *A 11216! 12714! Powell 543! 1094!* Great erAntilles and Lesser Antilles south to Barbados.

30. MACROPTILIUM (Benth.) Urb. (1928)

1. M. lathyroides (L.) Urb., Symb. Ant. **9**: 457 (1928).—*Phaseolus lathyroides* L. (1763).

Erect tough-stemmed herb to 1 m. high or later twining, rarely prostrate; stipules lanceolate, subulate-tipped, 8–10 mm. long, not appendaged at base; leaflets variable from linear to ovate-elliptical or lanceolate, mostly obtuse at tip, (2–) 3–7 cm. long, (2·5–) 12–25 mm. broad; stipels subulate, 3–4 mm. long; inflorescence up to 30 cm. or more long; flowers often paired; bracteoles shorter than calyx, subulate, deciduous; calyx 4–6 mm. long; petals bright deep pink to dark crimson, whitish towards base and keel whitish, rarely all white; standard about 15 mm. long; pod linear, straight, 8–10 cm. long, 2–3 mm. broad; seeds oblong-reniform, dark and light brown, about 3·5 mm. long.

Very common, a weed of waste places and cultivations; SL–2300 feet; fl. and fr. all the year. *A 5545! 12915! H 5979! J.P. 1357! P 19660!* Tropical Amer. and Asia; Grand Cayman.

31. LABLAB Adans. (1763)

1. **L. purpureus** (L.) Sweet, Hort. Brit.: 481 (1827).—*Dolichos purpureus* L. (1763). *L. niger* Medic. (1787). *D. lablab* L. (1753). Banner Bean, Bonavist Bean.

Perennial strong-stemmed twiner, glabrescent; stipules deltate to oblong-lanceolate, not appendaged; leaflets broadly rhomboid-ovate, abruptly acutely acuminate, 6–13 cm. long, 5–11 cm. broad; peduncles massive, up to 30 cm. long; racemes 8–16 cm. long, the flowers several at each swollen node and appearing tiered; bracts and bracteoles deciduous; calyx 6–7 mm. long, ciliate; petals white, rosy or violet, the standard about 15 mm. long; pod rough, with warty margins, 4–8 cm. long, 18–20 mm. broad; seeds variously coloured but often black with prominent white hilum, 9–10 mm. long.

Occasional, escaping into waste places after cultivation; SL–1750 feet; fl. and fr. Dec–Mar, July. *A 8763! Harris! J.P. 1254!* Probably native of tropical Africa, now widely cultivated in the subtropics and tropics.

32. VIGNA Savi (1824)

1 Stipules appendaged at base; fruit 10–40 cm. long, usually glabrous; cultivated in many varieties; native of tropical Asia; Black Eye Pea, Cow Pea, etc.
          V. unguiculata (L.) Walp.
1 Stipules not appendaged at base; fruit 4–10 cm. long;
  2 Stipules truncate at base; plant typically glabrescent; flowers in a short raceme at the tip of the smooth peduncle; calyx 4–5 mm. long; petals greenish-yellow or yellow; standard about 15 mm. long   1. *luteola*
  2 Stipules shortly cordate at base; plant thinly bristly-pilose with reflexed hairs; flowers subumbellate at the tip of the peduncle; peduncle rough with minute recurved prickles; calyx about 10 mm. long; petals light purple turning yellow; standard 20–25 mm. long   2. *vexillata*

1. **V. luteola** (Jacq.) Benth. in Mart., Fl. Bras. 15 (1): 194, t. 50, f. 2 (1859).—*Dolichos luteolus* Jacq. (1770). *V. repens* (L.) Kuntze (1891), not Bak. (1876).

Stem trailing and twining; stipules up to 5 mm. long; leaflets ovate to lanceolate, acute or acuminate, 3–9 cm. long, 1–5 cm. broad; inflorescence up to 20 cm. long; pod linear, slightly curved, sparsely pubescent, 4–7 cm. long, about 7 mm. broad, 8–12-seeded; seeds black with white hilum, 4–5 mm. long.

Common, at swamp margins and on banks and in ditches, mostly in sandy soil near the sea; SL–20 (–450) feet; fl. and fr. most of the year. *A 7227! 12703! H 9260! P 11118!* General in the subtropics and tropics; Grand Cayman.

2. **V. vexillata** (L.) A. Rich. in Sagra, Hist. Cuba, Parte 2, 10: 191 (1845).—*Phaseolus vexillatus* L. (1753).

Stem twining, more or less pilose; stipules lanceolate, about 8 mm. long; leaflets lanceolate to ovate, acute or narrowly acuminate, rarely obtuse, 5–9 (–12) cm. long, 1·5–3 (–5) cm. broad; peduncles up to 30 cm. long; flowers 2–3 (–5) together; pod linear, pilose, 7–10 cm. long, about 5 mm. broad; seeds oblong, black or brown with white hilum, 4–5 mm. long.

Rare and local (St. Cath.), in roadside thickets; about 250 feet; fl. and fr. July. *Weaver 915!* General in the tropics.

33. PACHYRHIZUS L. C. Rich. ex DC. (1825) nom. cons.

Terminal leaflet typically entire; fruit up to 30 cm. long; seeds ovate, about 12 mm. long, poisonous; corolla about 2 cm. long, white or mauve; sparingly cultivated; native of tropical Amer., now widespread; Yam Bean   *P. tuberosus* (Lam.) Spreng.
Terminal leaflet typically angular-dentate; fruit mostly 7–15 cm. long; seeds squarish, about 7 mm. long, otherwise like the last; sometimes escaping   *P. erosus* (L.) Urb.

34. CLITORIA L. (1753)

Leaflets 5 (–9), thinly puberulous; fruits compressed or turgid, without lateral ribs; seeds compressed, dry and smooth   1. *ternatea*

Leaflets 3, thinly pubescent beneath; fruits tetragonous by virtue of a midline rib on each valve; seeds globose, viscid   2. *rubiginosa*

1. C. ternatea L., Sp. Pl. **2**: 753 (1753).—Blue Pea.

Perennial twiner from a large taproot; leaflets ovate to elliptical, mostly obtuse and shortly emarginate at tip, 2–5 cm. long, 1–3 cm. broad; flowers solitary on very short axillary pedicels, usually resupinate; bracteoles rounded-oblong, up to 9 mm. long; petals commonly deep blue, paler and tinged yellow at base, or light blue or white; standard 3–5 cm. long; wing petals lightly fused to keel; vexillary stamen free; pod 6–11 cm. long; seeds ellipsoidal-compressed, 5 mm. long.

Common in cultivation as an ornamental, and escaping into waste places, field margins and thickets; SL–600 feet; fl. and fr. all the year. *A 5489! H 10370! Powell 969!* Native of the Old World tropics, now widespread in cultivation; Grand Cayman.

2. C. rubiginosa Juss. in Pers., Synops. Pl. **2**: 303 (1807).—*Martiusia rubiginosa* (Juss.) Britton (1924).

Perennial twiner with seasonal regrowth; young stems pubescent; leaflets ovate to elliptical, obtuse or rather acute, often mucronate, rarely emarginate, glabrous adaxially, 3–9 cm. long, 1·5–5 cm. broad; flowers solitary or paired at the tip of a stiff axillary peduncle (1–) 3–10 cm. long; bracts and bracteoles lanceolate; calyx pilose; petals white streaked with purple or mauve; standard 4–5 cm. long, pilose outside; flowers fragrant; stamens monadelphous; pod 2·5–3·6 (–5) cm. long, 6–8 mm. broad, 4–6-seeded, seeds about 4 mm. in diameter.

Uncommon, in sheltered or low-lying thickets and woodland margins; 900–2000 feet; fl. Sept–Nov, fr. Sept–Jan. *A 10387! Gauntlett! H 9429!* Native of and throughout tropical Amer., introduced into W. Africa.

35. CENTROSEMA (DC.) Benth. (1837) nom. cons.

1 Bracteoles distinctly longer than calyx, glabrous; leaflets broadly ovate to oblong, the terminal usually more than 8 cm. long; peduncles stout; standard petal 4 cm. or more broad; fruit winged about one-third from each valve-margin, 7–9 mm. broad, glabrous:

2 Peduncle thick, 1–3 cm. long, with 2 or more flowers; bracteoles oblong, rounded at tip; keel abruptly turned upwards   1. *plumieri*

2 Peduncle slender, (3–) 5–10 cm. long, solitary-flowered; bracteoles acuminate; keel curved upwards   2. *haitiense*

1 Bracteoles equal to or shorter than calyx, ovate, puberulous; leaflets ovate to linear-lanceolate, the terminal usually less than 7 cm. long; peduncles slender; standard not exceeding 4 cm. broad; fruit ribbed close to each valve-margin or obscurely, 4–6 mm. broad, glabrescent:

3 Leaflets pubescent on both surfaces at least when young, but the margins ciliate, ovate; veinlets inconspicuous; calyx-teeth very unequal, the lateral deltate about as long as tube, the anterior subulate-tipped and much longer; fruit 8–10 (–15) cm. long, about 6 mm. broad   3. *pubescens*

3 Leaflets glabrous or nearly so beneath, minutely puberulous and glabrescent adaxially, ciliate, ovate to linear-lanceolate; calyx-teeth subequal, longer than tube, the lateral and anterior subulate-tipped, the anterior slightly longer; fruit (6–) 8–12 cm. long, about 4 mm. broad   4. *virginianum*

1. C. plumieri (Turp. ex Pers.) Benth., Comm. Leg. Gen.: 54 (1837).—*Clitoria plumieri* Turp. ex Pers. (1807). *Bradburya plumieri* (Turp. ex Pers.) Kuntze (1891). Fee-fee.

Robust herbaceous climber; leaflets acuminate, glabrous, 5–12 cm. long, 3–10 cm. broad; bracteoles up to 20 mm. long and 15 mm. broad; calyx 5 mm. long, the anterior and lateral teeth deltate; standard petal up to 5·5 cm. broad, white fading to cream and tinged crimson or dark purple towards the centre; wings white tinged purple distally; pod 10–15 cm. long, 9 mm. broad with longitudinal wings about 3 mm. from each margin; seeds dark brown or black, about 7 mm. long.

Frequent, in thickets; 10–750 feet; fl. and fr. Oct–Dec. *A 5417! H 8446! P 21494!* Throughout tropical Amer., introduced into tropical Asia and Africa.

2. **C. haitiense** Urb. & Ekman in Arkiv Bot. **20**A (5): 13–14 (1926).
Climber like the last; leaflets broadly subtruncate at base, the lateral often slightly
pandurate, roughish; standard petal deep rose-violet with whitish centre; pod
about 13 cm. long, about 7 mm. broad with longitudinal wings about 2 mm. from
each margin.
Rare and local (Clar.), in roadside thickets; 1500–1700 feet; fl. Nov–Dec, fr.
Feb. *P 18412! 24628!* Hispaniola.

3. **C. pubescens** Benth., Comm. Leg. Gen.: 55 (1837).—*Bradburya pubescens*
(Benth.) Kuntze (1891).
Widely scrambling perennial twiner with slender wiry stems; leaflets mostly
shortly and bluntly acuminate, 3–7 cm. long, 2–5 cm. broad; racemes 2–8-flowered;
bracteoles ovate, acute, pubescent, up to 10 mm. long and 5 mm. broad; petals
generally mauve fading to cream; standard often with a purplish patch or two purple
spots at the centre and about 3·5 cm. long; keel paler; seeds about 4·5 mm. long
and 4 mm. broad.
Common, in waste places and thickets and on banks and fences; SL–1600 feet;
fl. Oct–Mar, fr. Nov–Mar. *A 5419! A. M. Barry! H 8258!* Native of tropical Amer.,
introduced into Asia and Africa.

4. **C. virginianum** (L.) Benth., Comm. Leg. Gen.: 56 (1837).—*Clitoria virginiana*
L. (1753).
Diffuse twiner like the last, the stems more slender and smaller in most parts;
leaflets mostly obtuse, 2–7 cm. long, 1–2·5 cm. broad; racemes 1–4-flowered;
bracteoles ovate, acuminate, pubescent, up to 8 mm. long; petals reddish-purple at
first, opening bluish-mauve, fading paler; standard often with a white patch at
centre, about 2·5 cm. long; seeds oblong, greenish or brown, about 2 mm. long.
Common, in waste places, rough pastures and thickets; SL–2500 feet; fl. and fr.
all the year. *A 6014! H 6939! P 18383!* Generally distributed from Bermuda and
S. United States to Argentina, throughout the West Indies and introduced into
tropical Africa.

36. TERAMNUS P. Browne (1756)

1 Fruit densely long-villous, the beak 3–5 mm. long and hooked; hairs on stems and
  petioles often yellowish; calyx-segments 5, longer than tube   1. *uncinatus*
1 Fruit thinly and shortly pubescent, the beak up to about 2 mm. long; hairs on stems and
  petioles whitish; calyx-segments not longer than tube:
  2 Leaflets ovate to elliptical; calyx-segments 5; fruit often slightly curved, the beak
    porrect or almost so   2. *labialis*
  2 Leaflets lanceolate to oblong-elliptical; calyx-segments 4; fruit usually straight, the
    beak antrorsely oblique   3. *volubilis*

1. **T. uncinatus** (L.) Sw., Nov. Gen. & Sp. Pl.: 105 (1788).—*Dolichos uncinatus*
L. (1763).
Taprooted twiner with densely hairy stems up to about 2 m. long; leaflets
variable from ovate to oblong-lanceolate, acute or obtuse, sericeous on both surfaces but more densely so beneath, up to 9 cm. long and 3·5 cm. broad; flowers
solitary in the axils to racemose; calyx about 6 mm. long; corolla a little longer than
calyx, pink, white or rarely yellowish, fading and drying bluish; pod (2·5–) 4–7 cm.
long, 3–4 mm. broad; seeds oblong, reddish-brown, 3–3·5 mm. long.
Occasional, in thickets and on waste ground; 800–3500 feet; fl. and fr. Nov–
Mar. *A 9173! 9960! H 6612!* Continental tropical Amer., Bahamas, Greater
Antilles; Grand Cayman, introduced into tropical Africa.

2. **T. labialis** (L. f.) Spreng. in L., Syst. Veg. ed. 16, **3**: 235 (1826).—*Glycine
labialis* L. f. (1781).
Taprooted twiner, the trailing branches rooting freely; leaflets mostly subacute,
thinly appressed-pubescent to glabrous adaxially, appressed-pubescent abaxially,
1–5·5 cm. long, up to 3 cm. broad; flowers racemose; calyx about 4 mm. long;
corolla longer than calyx, white or tinged pink especially on the standard or wings;
pod 2·5–5 cm. long, 3–3·5 mm. broad; seeds oblong, brown, 2–3 mm. long.

Common, in thickets and on stony waste ground; SL–1000 feet; fl. and fr. most of the year. *A 5614! H 10040! P 11932!* General in the tropics.

3. T. volubilis Sw., Nov. Gen. & Sp. Pl.: 105 (1788).
Twiner with slender hairy stems; leaflets mostly narrowly rounded at tip, glabrous adaxially, thinly appressed-pubescent abaxially, 2–7 cm. long, up to 2·5 cm. broad; flowers solitary to few in the leaf-axils or racemose; calyx 3·5–4 mm. long; corolla about 5 mm. long, white tinged pink or pink, fading to mauve, drying blue; pod 2·5–3·5 cm. long, about 2 mm. broad; seeds oblong, olive-green, about 2 mm. long.
Uncommon, in thickets and waste places; 25–1000 feet; fl. and fr. May, Oct–Feb. *A 12092! A. M. Barry! H 11424!* Tropical S. Amer., Trinidad.

37. SHUTERIA Wight & Arn. (1834) nom. cons.

1. S. vestita Wight & Arn., Prodr. Fl. Pen. Ind. Or. **1**: 207 (1834).
Trailing and twining vine; stems slender, hairy, often tinged crimson; stipules ovate-lanceolate, adnate to stem at base, hairy, 5–6 mm. long; leaflets elliptical to rhomboid-obovate, obtuse, up to 4 cm. long and 3 cm. broad, thinly villous on both surfaces, darker adaxially but often with a pale midrib region; inflorescence racemose, up to about 5 cm. long; bracts persistent; calyx about 6 mm. long including the teeth; corolla 8 mm. long; standard petal pink at margin, cream or yellow in the centre with pink or reddish lines; wings and keel purplish at tips; pod villous, tinged crimson, 2–3·5 cm. long, 4 mm. broad, 3–7-seeded; seeds reniform, black, 2·5 mm. long.
Locally common (St. Andr., Port., St. Thom.), on roadside and pathside banks; 3300–4800 feet; fl. Jan–Apr, fr. Feb–Apr. *A 6370! 12374! P 23272! Yuncker 18646!* Native at higher elevations in the Himalayas, Ceylon and the East Indies, not otherwise reported from the western hemisphere.

38. ABRUS Adans. (1763)

1. A. precatorius L., Syst. Nat. ed. 12, **2**: 472 (1767).—Crab's Eyes, Red Bead Vine, Wild Liquorice.
Twiner, becoming woody, much branched; leaves 5–13 cm. long; leaflets 10–20 pairs, oblong, rounded at both ends, glabrous adaxially, 1–2 cm. long, 3–6 mm. broad; racemes 3–8 cm. long; calyx 2–3 mm. long; corolla dusky purplish-pink, 8–12 mm. long; pod oblong, beaked, 2–4 cm. long, 1–1·5 cm. broad, pubescent; seeds ovoid, scarlet with a black spot, 5–7 mm. long.
Common, in thickets, hedgerows and on fences, mostly in rather dry areas; SL–800 feet; fl. Oct–Apr, fr. Nov–June. *A 8308! Harris! P 15231!* Throughout the subtropics and tropics; Grand Cayman.

39. VICIA L. (1753)

1. V. sativa L., Sp. Pl. **2**: 736 (1753).—Vetch.
Herb climbing by leaf-tendrils; stems finely ribbed, very thinly hairy; leaflets 4–6 pairs, opposite or subopposite, mostly oblong and truncate at tip, mucronate, (10–) 20–30 mm. long, 5–8 (–10) mm. broad; flowers mostly paired in the leaf-axils, the pedicels and common peduncle short; calyx-tube 6 mm. long, the teeth as long, subequal; corolla light purple, drying blue, about as long as calyx; pod 3–4·5 cm. long, 5 mm. broad; seeds about 3 mm. in diameter.
Rare and local (St. Andr., St. Thom.), naturalized in thickets and waste places; 3500–3600 feet; fl. Feb–Mar, fr. Feb–Apr. *A 6701! H 10802!* Native of western Asia, now naturalized in Europe and elsewhere; Hispaniola.

40. TRIFOLIUM L. (1753)

1 Flowers 2–3 mm. long; corolla yellow; stems more or less hairy, procumbent; flower-heads on short axillary peduncles   1. *dubium*

1 Flowers larger; corolla white, white tinged pink, or purple:
  2 Stems glabrous, creeping and rooting at nodes; flower-heads on long axillary peduncles;
    flowers 7–10 mm. long; corolla white or tinged pink   2. *repens*
  2 Stems more or less pubescent, erect or decumbent, not rooting; flower-heads terminal,
    sessile; flowers up to about 18 mm. long; corolla usually purple; fruit 1-seeded,
    dehiscing transversely; sparingly cultivated on high ground in St. Thomas; native of
    Europe and W. Asia; Red Clover   *T. pratense* L.

1. T. dubium Sibth., Fl. Oxon.: 231 (1794).—Shamrock.
Slender-stemmed annual herb with branches up to 30 cm. or more long; leaflets
obovate, rounded and often very shortly emarginate at tip, cuneate at base, up to
9 mm. long and 6 mm. broad; peduncles 5–20 mm. long, always much longer than
petioles; flower-heads 5–9 mm. across; corolla yellow to light greenish-yellow,
turning brown; pod 2 mm. long, 1 (–2)-seeded.
Frequent locally (St. Andr., Port., St. Thom.), a weed of pathsides and pastures;
2800–7400 feet; fl. Dec–Aug, fr. Feb–Sept. *A 6349! P 22068!* Native of Europe
and W. Asia, naturalized in N. Amer. and elsewhere.

2. T. repens L., Sp. Pl. 2: 767 (1753).—Dutch or White Clover.
Perennial herb with creeping branches up to about 50 cm. long; leaflets roundish-
obovate or elliptical, shortly emarginate at tip, broadly cuneate at base, 8–15 (–30)
mm. long, 6–15 (–25) mm. broad; peduncles up to 30 cm. long; flower-heads 15–20
mm. across; flowers fragrant; pod 4–5 mm. long, 3–6-seeded but fruit apparently
not formed in Jamaica.
Locally common (St. Andr., Port., St. Thom.), a weed of roadsides and damp
pastures; 2800–7400 feet; fl. most of the year. *A 5697! J.P. 1223! P 20818!*
Native of Europe and N. and W. Asia, widely naturalized elsewhere; Cuba,
Hispaniola.

41. MELILOTUS Mill. (1754)

1. M. alba Medic. in Vorles. Churpf. Ges. 2: 382 (1787).
Shrubby glabrous annual herb up to about 2 m. high; leaflets obovate to oblanceolate, up to 22 mm. long and 8 mm. broad; stipules setaceous; racemes slender;
flowers about 4 mm. long, fragrant; corolla white; pod 3–4 mm. long, reticulate.
Very rare (St. Ann), probably only a casual weed of waste ground by roadside;
1400–1500 feet; fl. and fr. June. *A 9466!* Native of Europe and Asia, introduced
elsewhere; Hispaniola, Puerto Rico.

42. BRYA P. Browne (1756)

1. B. ebenus (L.) DC., Prodr. 2: 421 (1825).—*Aspalathus ebenus* L. (1763).
Coccus Wood, West Indian Ebony.
Deciduous hard-wooded shrub 1–2 m. or tree up to 8 m. high; bark fissured;
leaflets more or less clustered, elliptical to obovate-elliptical, mostly 1–1·5 cm. long
and up to 6 (–8) mm. broad; calyx green, about 3 mm. long, the lobes ovate and
about as long as tube; petals deep orange-yellow; standard about 1 cm. long;
staminal sheath white; fruit ripening brownish with short and long whitish hairs,
mostly 1-seeded and about 13 mm. in diameter, the fruiting carpels stipitate.
Locally common, in thickets and pastures on limestone hills or rocky alluvial
plains, mostly in rather arid areas; SL–2000 feet; fl. and fr. sporadically throughout
the year after rain. *A 11131! H 8635! P 21525!* Cuba.

43. AESCHYNOMENE L. (1753)

1 Stipules not peltate, not appendaged below the point of attachment; calyx campanulate
    with 5 subequal lobes; leaflets mostly obovate; fruit 2–3 (–4)-jointed   1. *brasiliana*
1 Stipules peltate, appendaged below the point of attachment; calyx 2-lipped, the adaxial
    lip 2-parted, the abaxial 3-parted; leaflets oblong; fruit (3–) 6–10 (–14)-jointed:
  2 Leaflets 2- or more-costate; fruit with upper margin entire or shallowly crenate and
    lower margin deeply indented:
    3 Fruit glabrous or puberulous   2. *americana*

3 Fruit hispid with thick-based hairs   3. *villosa*
2 Leaflets 1-costate; costa central; lower margin of fruit not deeply indented:
   4 Leaflets usually denticulate or serrulate-ciliate, up to 20 (–30) mm. long and 4 (–8)
      mm. broad; plant generally hispid; fruit 5–7 mm. broad, hispid at maturity, not
      muricate, with both margins entire or nearly so   4. *ciliata*
   4 Leaflets entire or subentire, up to 10 mm. long and 2·5 mm. broad; plants generally
      glabrous or sparsely hispid; fruit with upper margin straight and lower margin
      crenate:
      5 Fruit 2·5–3·5 mm. broad, the joints rugulose   5. *evenia*
      5 Fruit 4–5 mm. broad, the joints sometimes muricate   6. *rudis*

1. **A. brasiliana** (Poir.) DC., Prodr. **2**: 322 (1825).—*Hedysarum brasilianum*
Poir. (1804). *A. biflora* F. & R. (1920).
Diffusely branched shrubby herb, prostrate or up to 1 m. high; stems glandular-
hispid and also crisped-pubescent; leaflets ciliate, 8–12 mm. long, 4–6 mm. broad;
panicle laxly several-flowered; flowers about 5–8 mm. long; corolla yellow; segments of fruit 3–4 mm. in diameter.
Very rare and known only from a single collection dated 1730, *Houston!* Mexico
to Brazil, Cuba, St. Lucia, Trinidad.

2. **A. americana** L., Sp. Pl. **2**: 713 (1753).—Bastard Sensitive Plant.
Low slender-branched or robust bushy annual 30–120 cm. high; stems thinly
hispid to glabrous, often with glandular hairs; stipules persistent; leaflets up to
about 30 pairs, 5–15 mm. long, 1–2 mm. broad; inflorescences few-flowered;
corolla usually white tinged mauve or light violet, sometimes pinkish-orange, the
standard petal 5–7 mm. long; fruit up to about 8-jointed, each segment about 4 mm.
long.
Common in pastures, on waste ground and sandy roadsides and strand; SL–
2800 feet; fl. and fr. Nov–May. *A 8643! 9919! H 6957! P 24316!* Florida to
Argentina, West Indies, Sierra Leone.

3. **A. villosa** Poir. in Lam., Encycl. Méth., Bot. Suppl. **4**: 76 (1816).
Like the last but the plant often smaller and the stems slender; hairs sometimes
glandular; corolla usually yellow or orange; standard petal 5–6 mm. long; fruit
bristly, mostly 4–6-jointed.
Occasional in the central parishes, in pastures and along roadsides on heavy
soils; SL–2900 feet; fl. and fr. Dec–Mar. *A 8414! H & P 13441! Yuncker 17912!*
Described originally from Puerto Rico, this species has been amalgamated so
frequently that its distribution is not known.

4. **A. ciliata** T. Vogel in Linnaea **12**: 84 (1838).
Annual herb up to 120 cm. high; stem spongy and thick at base, with numerous
adventitious roots, clothed with long yellowish glandular hairs above; stipules subpersistent, ciliate; leaflets oblong, 10–20 pairs; inflorescence few-flowered, with
ciliate bracts; standard petal orange lined crimson, about 8 mm. long, ciliate; keel
light yellow; fruit mostly 6–7-jointed.
Rare and local (St. Eliz., West.), in open swamps and rice fields; SL–50 feet;
fl. and fr. June, Dec. *A 10163! Britton 1498! P 24590!* S. Mexico to Ecuador and
Brazil.

5. **A. evenia** C. Wright in Sauv. in Anal. Acad. Cienc. Habana **5**: 334 (1868).
Shrubby herb; leaflets oblong to lanceolate, acute or rounded, 5–20 pairs; flowers
axillary or in short racemes; petals cream or rosy; wings longer than standard;
fruit up to 10-jointed, 2·5–3·5 cm. long.
Very rare (Clar.), in shallow pools in pasture; about 200 feet; fl. and fr. Dec.
P 24345! S. United States to Argentina, Cuba, Hispaniola.

6. **A. rudis** Benth., Pl. Hartw.: 116 (1843).
Shrubby herb up to 120 cm. high, with sparsely setulose zig-zag branches;
stipules ovate-elliptical, very early caducous; leaflets oblong, obtuse, up to about
20 pairs; flowers in short racemes; petals yellow; standard striate, about 9 mm.
long; fruit up to about 9-jointed, up to 4 cm. long, very thinly pubescent with thick-
based hairs.

Very rare (St. Eliz., West.), at pond margins; 50–100 feet; fl. and fr. Jan. *P 26787! Robertson UCWI 5384!* N. Carolina to Texas, C. Amer., Puerto Rico. This species is closely allied to *A. indica* L., a widespread Old World species.

44. ZORNIA J. F. Gmel. (1792)

Flowers solitary, pedicelled, supra-axillary; leaflets 4, retuse; fruit with up to 15 segments, without bristles, densely glandular   1. *myriadena*
Flowers in axillary or terminal spikelike racemes, sessile or subsessile; leaflets 2, acute or acuminate; fruit with 4–7 segments, bristly without glands   2. *reticulata*

1. Z. myriadena Benth. in Mart., Fl. Bras. **15** (1): 85 (1859).—*Z. tetraphylla* (L.) F. & R. (1920), not Michx. (1803). *Ornithopus tetraphyllus* L. (1759).
Perennial, woody and branched near the base with straggling shoots up to 1 m. high, often much shorter; leaflets obovate to oblanceolate, punctate, (5–) 10–15 mm. long, 3–6 mm. broad; pedicels 2–12 mm. long; calyx 3–3·5 mm. long; corolla bright yellow; standard petal 12–15 mm. long; fruit with rectangular segments 1·8–2·0 mm. long, 1·2–1·6 mm. broad without reticulate markings.
Very rare (Clar.), among grasses; about 100 feet; fl. and fr. Dec. *H 12731!* E. Brazil, Cuba.

2. Z. reticulata Sm. in Rees, Cyclop. **39**: 2 (1819).—*Z. diphylla* Pers. (1807), partly, not *Hedysarum diphyllum* L. (1753).
Perennial taprooted herb with erect branches to 30 cm. high or branches trailing to 1 m. or more long; leaflets lanceolate or elliptical, 1·5–4 cm. long, 3–13 mm. broad; inflorescence compact, flattened, with conspicuous bracts attached above the base and mostly about 1 cm. long; calyx 2-lipped, about 3 mm. long; corolla light or dark yellow; standard petal 8–10 mm. long with a reddish-purple spot centrally or red-veined distally; segments of fruit broadly kidney-shaped, 2·5 mm. long, 2 mm. broad.
Frequent, in pastures and open waste land and on roadside banks; 380–2600 feet; fl. and fr. almost all the year. *A 6574! 12854! Campbell 6028! P 8310!* S. United States to Paraguay, West Indies.
Mohlenbrock, Webbia **16**: 1–141 (1961), identifies *Harris 6910* (F) as *Z. gemella* (Willd.) T. Vogel. I am unable to distinguish the duplicate of this exsiccata at the British Museum from *Harris 12070* which that author designates the neotype of *Z. reticulata*.

45. STYLOSANTHES Sw. (1788)

Plant viscid and aromatic with glandular hairs; leaflets mostly elliptical; terminal segment of fruit shortly beaked   1. *viscosa*
Plant not viscid nor aromatic; leaflets mostly lanceolate; terminal segment of fruit with a beak as long as itself   2. *hamata*

1. S. viscosa Sw., Nov. Gen. & Sp. Pl.: 108 (1788).—Poor Man's Friend.
Low shrubby herb up to about 1 m. high; stems ascending, with short curled hairs and longer spreading gland-tipped hairs; leaflets 5–10 mm. long, up to 5 mm. broad, viscid-pubescent on both surfaces; stipules with large gland-tipped hairs; flowers solita ry; standard petal deep yellow or orange-yellow, about 4 mm. long; fruit about 4 mm. long.
Occasional, in sandy, gravelly, clay or limestone waste places and in open pastures; 100–2250 feet; fl. July–Aug, Nov–Mar, fr. July, Dec. *A 6662! H 12406!* Mexico to Brazil, Cuba, Hispaniola.

2. S. hamata (L.) Taub. in Verh. Bot. Brand. **32**: 22 (1889).—*Hedysarum hamatum* L. (1759). Cheesy Toes, Donkey Weed, Pencil Flower.
Herb with slender tough spreading stems up to 60 cm. or more long; internodes pubescent with longer hairs on one side; leaflets 7–27 mm. long, 2–6 mm. broad, glabrous adaxially, prominently veined abaxially; stipules bristle-tipped, pubescent; flowers solitary, sometimes accompanied by a sterile branchlet; standard petal yellow streaked purple, about 5 mm. long; fruit including beak about 9 mm. long.

Common, especially in waste places on limestone and exposed pastures near the sea; SL–600 (–2500) feet; fl. and fr. most of the year. *A 5535! Campbell 6019! P 24166!* Florida, Guatemala to Colombia, Bahamas, West Indies south to Barbados and the Grenadines; Grand Cayman.

46. CHRISTIA Moench (1802)

1. C. vespertilionis (L. f.) Bakh. f. in Reinwardtia **6**: 89 (1961).—*Hedysarum vespertilionis* L. f. (1781). *Lourea vespertilionis* (L. f.) Desv. (1813).

Slender erect sparingly branched subwoody herb up to 60 cm. or more high; stem puberulous with minutely hooked hairs; stipules lanceolate-subulate, early caducous; leaflet up to 1·5 cm. long, 3–8 cm. broad, purplish-tinged adaxially, lighter along the veins, lateral leaflets if present much smaller; flowers single or paired in racemes up to 12 cm. long; pedicels and calyx with long simple and short hooked hairs; bracts deciduous; calyx inflated in fruit up to 11 mm. long; corolla yellow, greenish-cream or white, standard petal 6 mm. long; fruit 4–6-jointed, each segment about 3 mm. long.

Locally common (St. Andr., St. Mary), a weed of pastures and roadsides on sand, shale or gravel; 300–2400 feet; fl. and fr. most of the year. *A 5409! H 12108! P 19672!* Native of E. Indies, naturalized in Hispaniola, St. Kitts, Martinique, St. Vincent and Trinidad.

47. ALYSICARPUS Desv. (1813) nom. cons.

1. A. vaginalis (L.) DC., Prodr. **2**: 353 (1825).—*Hedysarum vaginale* L. (1753).

Herb with tough ascending or more usually trailing glabrous branches 20–100 cm. long; stipules scarious, more or less joined opposite the leaflet and enclosing 2 stipels; leaflet suborbicular to lanceolate, cordate at base, usually rounded but rarely acute at tip, glabrous except the margins minutely ciliolate, 0·3–2·0 (–4·5) cm. long, up to 14 mm. broad; racemes axillary or terminal with 6–12 (–18) flowers; calyx 5 mm. long; corolla deep reddish-purple, fading violet or pink, rarely orange; fruit 1–2 cm. long, 4–7-jointed.

Frequent, in sandy waste places, cultivations and rough pastures; near SL–2500 (–4000) feet; fl. Aug–Mar, fr. Sept–Mar. *A 8693! H 12159! P 17435!* Native of the Old World now widely naturalized in the American subtropics and tropics.

48. DESMODIUM Desv. (1813) nom. cons.

1 Leaves generally 1-foliolate, or upper leaves only 3-foliolate:
  2 Petiole winged; leaflets lanceolate; fruit shallowly notched along both margins, indehiscent; stem more or less triquetrous   1. *triquetrum*
  2 Petiole not winged; stem not triquetrous:
    3 Fruit dehiscent along lower margin; bracts broadly ovate; stipules and bracts early deciduous:
      4 Leaflets oblong-lanceolate; corolla about 6 mm. long, mostly orange-yellow; standard streaked violet   2. *motorium*
      4 Leaflets broadly round-sided-elliptical, rounded at tip, cordate at base; corolla about 10 mm. long, bluish-purple   3. *gyroides*
    3 Fruit indehiscent; bracts subulate to narrowly lanceolate, persistent at least at anthesis; stipules persistent:
      5 Leaflets ovate-elliptical, broadly acute; stipules subulate, free; corolla 3–4 mm. long, light yellow or white tinged rose   4. *gangeticum*
      5 Leaflets mostly linear-oblong; stipules lanceolate, fused at first; corolla about 5 mm. long, pink   12b. *canum* var. *angustifolium*
1 Leaves 3-foliolate, except sometimes basal leaves on juvenile shoots 1 foliolate:
  6 Fruit dehiscent:
    7 Flowers in short corymbose racemes; bracts and sepals long-ciliate; leaflets oblong-elliptical, rounded at both ends   5. *barbatum*
    7 Flowers in elongated racemes:
      8 Corolla rose-red; leaflets broadly rhomboid-ovate
                      14. 1. *Pseudarthria viscida* (p. 353)
      8 Corolla mostly yellow-orange; leaflets oblong-lanceolate   2. *motorium*
  6 Fruit indehiscent, or tardily dehiscent in *D. barbatum* and *D. triflorum*:

 9 Flowers in axillary or leaf-opposed umbels or clusters:
  10 Low herb, creeping and rooting; flowers in leaf-opposed or axillary clusters;
      leaflets less than 1 cm. long, mostly emarginate at tip   6. *triflorum*
  10 Shrubs; flowers in dense axillary umbels or heads; leaflets mostly 5 cm. or more
      long, not emarginate; fruit 1–5-jointed; rare and casual introductions from
      tropical Asia:
     11 Branches terete, downy when young; leaflets elliptical to roundish
                                                          D. umbellatum (L.) DC.
     11 Branches triquetrous, densely pubescent; leaflets elliptical with many parallel
         lateral veins (*D. cephalotes* (Roxb.) Wall. ex Wight & Arn.)
                                                          D. triangulare (Retz.) Santapau
 9 Flowers in terminal or axillary racemes or panicles:
  12 Racemes compact, corymbose, terminal, 1·5–3 cm. long; calyx long-ciliate
                                                                    5. *barbatum*
  12 Racemes or panicles lax, elongated at least in fruit, (5–) 8–30 cm. or more long:
     13 Fruit notched at the joints equally on both margins, the segments elliptical,
         orbicular or rhomboidal in outline, reticulate-veined on the surface:
        14 Segments of fruit 2, the terminal only fertile, emarginate on one side, puberulous,
            6–7 mm. long   7. *glabrum*
        14 Segments of fruit (2–) 4–8, all fertile, puberulous, with hooked hairs; glandular
            hairs present at least in inflorescence:
           15 Segments of fruit narrowly elliptical, more than twice as long as broad, not
               twisted; stipules broadly amplexicaul, acuminate   8. *scorpiurus*
           15 Segments of fruit orbicular or rhomboidal, often twisted:
              16 Stem erect; stipules conspicuous, semicordate, broadly ovate-lanceolate,
                  filiform-tipped; leaflets oblong-ovate, obtuse, uniformly green; bracts
                  lanceolate, early deciduous   9. *tortuosum*
              16 Stem procumbent; stipules small, subulate from an abruptly expanded base;
                  leaflets rhomboid-ovate to ovate-lanceolate, subacute, usually with the midrib
                  area distinctly paler; bracts subulate, subpersistent   10. *procumbens*
     13 Fruit notched at the joints unequally and more deeply on the lower margin, the
         upper margin often almost straight; segments with lower margin always more
         rounded than upper, often D-shaped:
        17 Shrub to about 3 m. high; leaflets lanceolate, up to 9 cm. long; flowers numerous
            in panicles; corolla blue, 8–9 mm. long; fruit with 5–8 segments 4–6 mm. long,
            deeply notched nearly to the upper margin; casual adventive only once reported;
            native of C. and S. Amer.   *D. cajanifolium* (Kunth) DC.
        17 Undershrubs or herbs; corolla not blue except on fading or drying:
           18 Stem triquetrous, rough with hooked hairs on the angles; terminal leaflet up to
               12 cm. long and 7 cm. broad; stipules free, deciduous; bracts broadly ovate to
               ovate-lanceolate, acuminate, ciliate, early deciduous; fruit subequally notched
                                                                    11. *intortum*
           18 Stem terete; terminal leaflet smaller than above; fruit deeply notched on lower
               margin:
              19 Fruit (3–) 5–7 (–8)-segmented; bracts narrowly lanceolate, 2·5–3 mm. long,
                  0·5 mm. broad, not overlapping in the young inflorescence, persistent at
                  least until anthesis, thinly pubescent; branches caespitose, rooting sparingly
                  only from the lowest nodes, not stoloniferous; stipules half-united at first,
                  persistent:
                 20 Leaflets 3, oblong-elliptical or oblong-ovate, up to 4 cm. broad
                                                               12a. *canum* var. *canum*
                 20 Leaflets 3 or 1, mostly oblong-linear and up to 1·5 cm. broad
                                                          12b. *canum* var. *angustifolium*
              19 Fruit (1–) 2–4 (–7)-segmented; bracts ovate to lanceolate, deciduous before
                  anthesis, at least ciliate; branches rooting freely, often trailing or stoloni-
                  ferous:
                 21 Stipules free even if overlapping at first, persistent; fruit usually with 2–4
                     segments:
                    22 Leaflets subrotundate to elliptical, rounded to acute at tip, thin-textured;
                        bracts ovate-lanceolate, long-acuminate, 5–7 mm. long, up to 2·5 mm.
                        broad, overlapping in the young inflorescence; branches mostly ascending
                        and rooting near the base, sometimes prostrate   13. *adscendens*
                    22 Leaflets rhomboid-ovate, mostly obtuse; bracts lanceolate, 2·5–4 mm. long,
                        0·5 mm. broad; main stem trailing and rooting   14. *affine*
                 21 Stipules fused at first; main stem trailing with stoloniferous branches; bracts
                     ovate, acuminate, 3–4 mm. long, about 1·5 mm. broad, pilose; fruit with
                     1–3 segments:
                    23 Stem and petioles puberulous only, without long hairs; leaflets broadly
                        ovate, acutely acuminate; stipules deciduous
                                                          15b. *axillare* var. *stoloniferum*

23 Stem and petioles with long hairs; leaflet-tips not acuminate:
  24 Leaflets broadly rhomboid-ovate, obtuse; stipules deciduous; long hairs on
    stem sparse   15a. *axillare* var. *axillare*
  24 Leaflets ovate-elliptical, acute; stipules more or less persistent; long hairs
    on stem dense   15c. *axillare* var. *acutifolium*

1. D. triquetrum (L.) DC., Prodr. **2**: 326 (1825).—*Hedysarum triquetrum* L.
(1753).
  Undershrub 60–120 cm. high; leaflet linear-lanceolate to ovate-lanceolate, the
blade up to 20 cm. long and 3·5 cm. broad, acute; racemes terminal and axillary,
8–30 cm. long; bracts linear-lanceolate, deciduous; calyx red; corolla pink or white
marked purple, about 5 mm. long; fruit 6–8-jointed, the segments subrectangular,
5–7 mm. broad.
  Uncommon and local (St. Andr.), in clay soils along roadsides and on waste
ground; 1000–2000 (–3000?) feet; fl. and fr. Apr–May, Aug–Dec. *A 10090!*
12500! H 12677! P 8311! 23498! Native of E. Asia.

2. D. motorium (Houtt.) Merr. in Journ. Arn. Arb. **19**: 345 (1938).—*Hedysarum motorium* Houtt. (1779). *D. gyrans* (L. f.) DC. (1825).
  Undershrub or woody taprooted herb 30–120 cm. high; leaflets obtuse, often
paler along the midline, the terminal up to 8 cm. long and 2 cm. broad, the lateral
when present always much smaller and undergoing jerking movements; bracts
about 5 mm. long, glabrous except margin ciliolate; wing petals orange with upper
margin dark blue; fruit 6–10-jointed, often curved, 4–5 mm. broad.
  Locally common (St. Andr., St. Mary), on rocky roadside banks and in thickets;
500–3700 feet; fl. and fr. most of the year. *A 6397! H 12133! P 19670!* Native of
E. Asia.

3. D. gyroides DC., Prodr. **2**: 326 (1825).
  Shrub up to 4 m. high; leaflets up to 10 cm. long and 5 cm. broad; racemes
terminal and axillary, up to 10 cm. long; bracts about 7 mm. long and 4·5 mm.
broad, striate, glabrous except margins ciliate; fruit 2–8-jointed, straight adaxially,
shallowly notched abaxially, pubescent without hooked hairs, 4–5 mm. broad.
  Rare and very local (St. Andr.), in thickets near streams; about 2800 feet; fl.
July–Sept, fr. Aug–Dec. *A 7741! P 19756!* Native of tropical Asia.

4. D. gangeticum (L.) DC., Prodr. **2**: 327 (1825).—*Hedysarum gangeticum* L.
(1753).
  Lax herb up to about 1 m. high; leaflets rounded at base, up to 10 (–15) cm. long
and about 6 cm. broad; inflorescence elongated and more or less nodding in fruit;
fruit 5–10-jointed, curved, the segments 2–2·5 mm. long and about 2 mm. broad.
  Locally common in the eastern parishes, on open rocky waste ground and in
river gravel; 500–1400 feet; fl. and fr. most of the year. *A 5817! H & B 10568!*
Proctor & Powell 20538! Native of the Old World tropics; St. Lucia, Trinidad.

5. D. barbatum (L.) Benth. in Miq., Pl. Jungh.: 224 (1852).—*Hedysarum
barbatum* L. (1759).
  Low bushy or spreading hairy taprooted herb 5–60 (–100) cm. high; leaflets
elliptical, up to 3·5 cm. long and 1·5 (–1·8) cm. broad, mostly much smaller;
stipules persistent, up to 15 mm. long; calyx 4 mm. long; corolla shorter than
calyx, pink fading blue and on drying; fruit 2–4-jointed, the segments 2·3–3 mm.
long with rather long hooked hairs.
  Locally common in the eastern and central parishes, on open sandy roadside
banks and in grassy savannas; 2250–3500 feet; fl. and fr. most of the year. *A 5836!*
H 7260! J.P. 1477! P 18392! Continental tropical Amer., Greater Antilles,
Antigua, Barbados, Trinidad and Tobago; varieties occur in the Old World
tropics.

6. D. triflorum (L.) DC., Prodr. **2**: 334 (1825).—*Hedysarum triflorum* L. (1753).
  Much branched creeping herb with slender more or less pubescent stems;
leaflets broadly obovate to subrotundate, mostly 4–8 mm. long; stipules persistent;
flowers 2–4 together; calyx 2–3 mm. long; corolla reddish-purple; standard usually

bright purple, about 4 mm. long; fruit 3–6-jointed, the segments about 3 mm. long, reported as tardily dehiscent.

Very common, in pastures, lawns and waste ground; SL–2000 feet; fl. and fr. June–Mar. *A 9440! H 9084! Powell 207!* General in the tropics; Grand Cayman.

7. D. glabrum (Mill.) DC., Prodr. 2: 338 (1825).—*Hedysarum glabrum* Mill. (1768). *D. molle* (Vahl) DC. (1825).

Shrubby herb up to 1 m. or more high; stems velvety-pubescent with a few hooked hairs; leaflets ovate to lanceolate, 2–8 cm. long; stipules half-cordate, acuminate; calyx about 2 mm. long; corolla about 3 mm. long, purplish; fertile segment of fruit thin with a broad reticulate margin and prominent seed.

Rare and local (St. Andr., St. Cath.), in sandy waste places and dry secondary thickets; SL–200 feet; fl. Nov, fr. Nov–Jan. *P 24499! 31095!* Continental tropical Amer., Greater Antilles, Virgin Is., Bahamas, Martinique, Curaçao.

8. D. scorpiurus (Sw.) Desv. in Journ. Bot. Appliq. **1**: 122 (1813).—*Hedysarum scorpiurus* Sw. (1788).

Perennial herb with prostrate or spreading angular branches, rough with hooked hairs, up to about 1 m. long; leaflets elliptical to obovate-elliptical, mostly rounded at both ends, 1–4 (–5) cm. long, up to about 2 cm. broad; corolla 4·5 mm. long, pink rapidly fading to light blue or whitish; fruit 4–7-jointed, the segments 4–5 mm. long.

Rather common, a weed of sandy pastures and roadsides and rocky or stony waste ground; SL–1100 feet; fl. and fr. all the year. *A 6594! 12185! H 11854! P 17419!* Mexico to Peru, Greater Antilles, Virgin Is., St. Kitts, Antigua to Trinidad, introduced into W. Africa and E. Indies.

9. D. tortuosum (Sw.) DC., Prodr. 2: 332 (1825).—*Hedysarum tortuosum* Sw. (1788). *D. purpureum* (Mill.) F. & R. (1920), not Hook. & Arn. (1832).

Erect or straggling annual or perennial herb up to 2 m. high; stem, petioles and inflorescence with hooked and simple gland-tipped hairs; leaflets 2–8 (–11) cm. long, up to about 4·5 cm. broad; corolla about 4 mm. long, white or pink fading light blue or mauve; pedicels exceeding 1 cm. and mostly ascending in fruit; fruit 2–6-jointed, the segments 4–5 mm. long, rhomboidal.

Rather common, on roadside banks and in waste places, SL–3500 feet; fl. May–Feb, fr. May–Dec. *A 7736! 11380! P 19671! Thompson 7963!* Florida, continental tropical Amer., West Indies, tropical Africa.

10. D. procumbens (Mill.) Hitchc. in Rep. Miss. Bot. Gard. **4**: 76 (1893).—*Hedysarum procumbens* Mill. (1768).

Herb with spreading or diffusely scrambling branches up to 120 cm. long; leaflets variable in shape, 1–3 (–6) cm. long, up to 1·5 (–2·5) cm. broad; corolla about 3 mm. long, pink fading mauve; pedicels up to about 1 cm. long in fruit, patently spreading; fruit 2–5 (–8)-jointed, the segments 3–4 mm. long, orbicular to rhomboidal.

Occasional (St. Andr., Clar., Han.), on rocky roadside banks and in stony dry pastures; 80–2500 feet; fl. and fr. Apr, Oct–Dec. *A 5472! H 11852! P 18423!* General in the tropics.

11. D. intortum (Mill.) F. & R., Fl. Jam. **4**: 34 (1920).—*Hedysarum intortum* Mill. (1768).

Robust trailing and scrambling undershrub with strong stems up to 1·5 m. or more long; leaflets ovate to ovate-lanceolate, acute or shallowly acuminate at tip, rounded or truncate at base, 5–12 cm. long, 2–7 cm. broad; racemes up to 30 cm. long, covered with hooked hairs; bracts 7–10 mm. long and about 3 mm. broad; corolla about 10 mm. long, pink fading to blue; fruit (3–) 5–10-jointed, the segments 3–3·5 mm. long, unequally elliptical.

Locally common (St. Andr., Port., St. Thom.), on rocky banks mostly in moist sheltered places; 1500–3750 feet; fl. Aug–May, fr. Sept, May. *A 5705! H 6665! P 22963!* Continental tropical Amer.. Hispaniola, Puerto Rico.

12. D. canum (J. F. Gmel.) Schinz & Thell. in Mém. Soc. Neuchâtel Sci. Nat.
5: 371 (1914).—*Hedysarum canum* J. F. Gmel. (1792). *D. supinum* DC. (1825).
Sweetheart.
Perennial herb up to about 70 cm. high; stems and petioles often reddish;
leaflets of firm texture with distinct venation, pubescent beneath with whitish hairs;
corolla about 5 mm. long, reddish-purple or pink fading to light blue; fruit often
slightly curved, the upper margin almost continuous, the segments 3–4 mm. long.

12a. Var. canum.
Leaflets 1·5–6·5 (–8) cm. long, 1–4 cm. broad.
Common in pastures and on banks; SL–3500 feet; fl. and fr. all the year. *A 7127!*
H 11759! H & P 13363! Stearn 980! Florida, Bahamas, continental tropical Amer.,
West Indies, tropical Africa and the Mascarene, Atlantic and Pacific Is.; Cayman Is.

12b. D. canum var. angustifolium (Griseb.) León & Alain in Contrib. Ocas.
Mus. Hist. Nat. Col. de la Salle, Habana 9: 18 (1950).—*D. incanum* var.
angustifolium Griseb. (1860).
Leaves mostly 1-foliolate and narrow, mostly up to 6·5 cm. long and 13 mm.
broad, but occasional broader leaflets occur.
Uncommon and local in the central parishes, in dry grassy savannas and on well
drained slopes; 300–800 feet; fl. Nov–Mar, fr. Jan–Mar. *H 12742! HPS 14738!*
Greater Antilles, Antigua, Barbados.

13. D. adscendens (Sw.) DC., Prodr. 2: 332 (1825).—*Hedysarum adscendens*
Sw. (1788).
Taprooted herb up to about 45 cm. high; leaflets thin with indistinct veins,
1–4·5 cm. long, 7–23 mm. broad, slightly paler beneath; stipules lanceolate;
corolla about 5 mm. long, pink fading through mauve to light blue; fruit usually
straight, the upper margin continuous, the segments mostly 2–4 and about 4 mm.
long.
Common as a weed of pastures and cultivations and along roadsides in moderately wet areas; SL–3700 feet; fl. and fr. most of the year. *A 5909! 11398!*
J.P. 1446! P 22825! Continental tropical Amer., West Indies, tropical Africa.

14. D. affine Schlecht. in Linnaea 12: 312 (1838).
Trailing and straggling herb with erect or suberect branches up to about 40 cm.
high; leaflets 2–6·5 cm. long, 1·5–3·5 cm. broad, thinly pubescent adaxially; stipules
ovate, acuminate, 8–9 mm. long, 4–6 mm. broad; corolla light yellow in bud, opening
white or white marked with pink, fading yellow tinged dull crimson; standard petal
limb about 5 mm. long and 6·5 mm. broad; fruit usually 3–4-jointed, the segments
5–6 mm. long and 3·5 mm. broad.
Uncommon and local, on rocky banks and in pastures and cultivations, mostly in
rather shaded places; 50–1700 feet; fl. Aug–Dec, fr. Oct–Dec. *A 8330! 9829!*
Tai! Mexico, Guatemala, El Salvador, Greater Antilles, Trinidad, tropical Africa.

15. D. axillare (Sw.) DC., Prodr. 2: 333 (1825).—*Hedysarum axillare* Sw. (1788).
Wild Pindar.
Stem trailing and rooting with erect flowering branches up to about 40 cm. high;
leaflets more or less silvery-tomentose beneath; calyx 2 mm. long; corolla 4 mm.
long; pedicels spreading in fruit; upper margin of fruit shallowly notched.

15a. Var. axillare.
Leaflets 3–6·5 (–8) cm. long, 2–4·5 cm. broad; corolla rich crimson fading
through purple to blue; fruit usually 2-jointed, the segments about 8 mm. long
and 5 mm. broad.
Common along roadsides and in pastures in sheltered areas and in woodland
clearings; 150–3800 feet; fl. and fr. all the year. *A 9201! T. J. Harris 8290!*
J.P. 1430! P 21400! Continental tropical Amer., West Indies.

15b. D. axillare var. stoloniferum (L. C. Rich. ex Poir.) B. Schubert in Journ.
Arn. Arb. 44: 289 (1963).—*Hedysarum stoloniferum* L. C. Rich. ex Poir.
(1804).
Leaflets up to 7 cm. long and 4·5 cm. broad, thinly tomentose beneath and paler;
bracts glabrescent; corolla pink or light purple.

Rather rare, along pathsides and in woodland margins in moderately wet areas on limestone; 850–1000 feet; fl. and fr. Aug–Sept. *A 7924!* C. Amer., Greater Antilles.

15c. D. axillare var. **acutifolium** (Kuntze) Urb., Symb. Ant. **4**: 292 (1905).
Leaflets 3–7·5 cm. long, 1·5–3·5 cm. broad, often with whitish marking, velutinous beneath; corolla pink or mauve; fruit 2–3-jointed, the segments about 6·5 mm. long and 4 mm. broad.
Rather rare, in moderately wet areas; 600–2000 feet; fl. and fr. July–Aug. *H & P 13462. Morley 82! Thompson 6750! 8029A!* C. Amer., Puerto Rico.

Assistance on several points of difficulty in the genus *Desmodium* has been given by Dr. Bernice Schubert whose co-operation is gratefully acknowledged.

96. OXALIDACEAE

Herbs, shrubs or trees. Leaves alternate, pinnately or digitately compound, or simple by suppression of leaflets, with or without stipules. Flowers normally bisexual, actinomorphic, solitary, cymose-umbellate, racemose or paniculate. Perianth ideally biseriate; sepals 5, free or connate, imbricate; petals 5, free or united or basally free and connate above, contorted in bud, rarely absent in cleistogamous flowers. Stamens 10, sometimes 5 reduced to staminodes and petaloid, basally connate, in 2 unequal series, the shorter outer series opposite the petals; anthers 2-locular, opening lengthwise. Ovary superior, 5-locular; ovules 1 or more in each loculus, anatropous on axile placentas; styles 5, free, homo- or heterostyle in relation to stamens; stigmas usually capitate. Fruit a loculicidal capsule or a berry. Seeds sometimes arillate; endosperm fleshy; embryo straight.
7 genera with about 950 species in the subtropics and tropics; *Oxalis* has 850 species with a few extending to temperate regions.

Trees; leaves imparipinnate; flowers in small axillary panicles; fruit a 5-ridged berry; cultivated; natives of E. Indies and China[1]   **1. Averrhoa**
Herbs or subshrubs; leaves 3-foliolate; inflorescence cymose-umbellate; fruit a loculicidal capsule or, if not developed, plant forming bulbils   **2. Oxalis**

1. AVERRHOA L. (1753)

Fruit fusiform-ellipsoid, deeply angular-ridged, 7–12 cm. long; leaflets 9–12, the terminal larger, ovate, acuminate, puberulous beneath; Carambola   *A. carambola* L.
Fruit oblong, shallowly ridged, 5–10 cm. long, 1–2·5 cm. broad; leaflets 17–31, the terminal not usually much larger than those in the middle, oblong-lanceolate, acutely acuminate, puberulous on both surfaces; Bilimbi, Long Jimbelin[2]   *A. bilimbi* L.

2. OXALIS L. (1753)

1 Plants with trailing leafy branches; corolla yellow; capsules commonly developed, explosive; leaflets 8–18 mm. broad   1. *corniculata*
1 Plants tufted; petioles long, all basal; corolla pink or purple; plants reproducing by bulbils; leaflets 2 cm. or more broad:
  2 Leaflets obcordate; bulbils basally clustered; petals 12–15 mm. long   2. *corymbosa*
  2 Leaflets widely 2-lobed, the lobes broadly triangular and obtuse; bulbils formed at the end of stolons 3–5 cm. long; petals 10–14 mm. long   3. *intermedia*

1. O. corniculata L., Sp. Pl. **1**: 435 (1753).—Edge Teeth, Yellow Sorrel.
Annual or short-lived perennial herb with upright flowering shoots 5 cm. or more high, more or less appressed-puberulous except adaxial leaflet surfaces glabrous and margins ciliate; stipules 1·5–2 mm. long; flowers few on pedicels up

---

[1] Webster (Journ. Arn. Arb. **38**: 71 (1957)) postulates a New World origin for both these species of *Averrhoa* with post-Columbian introduction to the Old World. Neither species is known in the truly wild state.
[2] I am pleased to acknowledge the assistance of Dr. David DeCamp in establishing legitimate common names for the fruits of *Averrhoa bilimbi* and *Phyllanthus acidus* (p. 407).

to 1·5 cm. long; sepals light green, 4 mm. long; petals 7–8 mm. long; pollen of different sizes from the long and short stamens; capsule oblong, beaked, 9–17 mm. long; seeds flat, circular in outline, brownish-red, covered by a white fleshy aril, about 1·3 mm. in diameter.

Very common, in usually shady waste places; 250–7400 feet; fl. and fr. most of the year. *A 5972! H 12335! W. B. Lewis!* Throughout the tropics, extending to warm temperate regions.

2. **O. corymbosa** DC., Prodr. 1: 696 (1824).—Pink Sorrel, Shamrock.

Perennial herb; petioles, leaflets and inflorescence thinly pilose; petioles up to 22 cm. or more long, expanded at the base; scapes usually longer than petioles; pedicels several, up to 2 cm. or more long; sepals green with 2 small orange glands at tips, hyaline-margined, 5 mm. long; petals pink drying bluish-mauve, about 15 mm. long; style-length irregular; fruit not known.

Locally common, in moist shady places; 150–4000 feet; fl. most of the year. *A 9202! H 12334! Powell 5!* Florida to northern S. Amer., West Indies; introduced as a cultivated ornamental into the Old World tropics.

3. **O. intermedia** A. Rich. in Sagra, Hist. Cuba, Parte 2, **10**: 129 (1845).

Like the last but with 2-lobed obdeltate leaflets up to 5 cm. broad; inflorescence 4–13-flowered; pedicels 1–2 cm. long; sepals 4–5 mm. long; petals pinkish-lilac; fruit not known.

Rather local (St. Andr., Manch., Trel.), a lawn weed and along roadsides and in waste places, an escape from gardens; (700–) 1150–3600 feet; fl. July, Nov–Dec. *A 5793! W. Lumsden! Powell 6!* Florida, Bahamas, Cuba, Puerto Rico, Virgin Is. and southwards to Martinique, also in Trinidad.

97. GERANIACEAE

Herbs or shrubs. Leaves alternate or opposite, usually simple but always lobed or divided, usually hairy, stipulate. Flowers bisexual, actinomorphic or rarely zygomorphic in cymose or umbellate bracteate inflorescences. Perianth biseriate; sepals usually 5, free, imbricate; petals usually 5, rarely otherwise or absent, free, imbricate, alternating with glands. Stamens in 1–3 whorls of 5 each, some of which may be reduced to staminodes, free or connate; anthers 2-locular, opening lengthwise. Ovary superior, 5-locular, rarely 3- or 8-locular; ovules 1–2 in each loculus, pendulous, anatropous, on axile placentas, the axis produced into a beak in fruit; styles 3–5, slender and more or less combined with the ovarian beak. Fruit a septicidal capsule dehiscent by mericarps, the styles often adherent to the beak and recurving elastically. Seeds with curved embryo, endosperm usually absent.

11 genera with about 850 species mostly in temperate and subtropical regions.

1. GERANIUM L. (1753)

Herbs with palmately veined and divided leaves; perianth actinomorphic, 5-merous; stamens 10; ovary 5-locular; stigmas 5; styles separating elastically from the ovary-axis in fruit.

1. **G. carolinianum** L., Sp. Pl. 2: 682 (1753).

Annual or short-lived herb, 25–45 cm. high; stem branched, pubescent, often pinkish; petioles of lower leaves up to 10 cm. long, shorter above; leaf-blade 3–6 cm. broad, divided almost to the base into 5–7 again-divided segments; stipules 3–8 mm. long; peduncles about 1·5 cm. long, mostly 2-flowered; sepals 5–8 mm. long, mucronate; petals white to mauve tinged light rose; beak of fruit 12–18 mm. long; seeds reticulate.

Locally common (St. Andr., St. Thom.), a weed of roadsides and cultivation in shallow rocky well drained soil; (2000–) 3800–5000 feet; fl. and fr. Mar–July. *A 6704! 7694! H 11939! Powell 1074!* S. Canada to Mexico, Bermuda.

Ornamental Geraniums (*Pelargonium* spp.) are commonly grown in Jamaica.

98. TROPAEOLACEAE

Succulent scrambling or twining herbs with acrid sap. Leaves mostly alternate and simple, sometimes pinnately compound, peltate, usually without stipules. Flowers bisexual, zygomorphic, solitary or rarely in axillary umbels. Perianth biseriate; calyx bilabiate, coloured, of 5 free imbricate or valvate sepals, the abaxial one produced into a spur; corolla of 5 free, usually clawed, imbricate unequal petals. Stamens 8 in 2 series, free, perigynous, declinate, unequal; anthers 2-locular, opening lengthwise. Ovary superior, sessile, 3-locular; ovules 1 in each loculus, axile, pendulous, anatropous; style 1; stigmas 3. Fruiting carpels separating from the short axis, 1-seeded, indehiscent; seeds with straight embryo, thick cotyledons, without endosperm.

2 genera with all except one of the 80 species in the genus *Tropaeolum*; native from Mexico to Argentina.

1. TROPAEOLUM L. (1753)

Fruiting carpels not winged.

1. T. majus L., Sp. Pl. **1**: 345, **2**: 1231 (1753).—Nasturtium.
Glaucous trailing annual herb; leaves peltate, the blade suborbicular and up to about 8·5 cm. or more in diameter; flowers showy, yellow, orange or crimson; spur about 25 mm. long; petals obtuse, the 3 anterior narrower and fringed on the claws.
Cultivated in gardens in the hills and occasionally escaping on to roadsides. *P 23199! Wynter!* Native of northern S. Amer., introduced elsewhere.

99. BALSAMINACEAE

Soft-stemmed succulent herbs or rarely shrubs. Leaves alternate, opposite or in whorls of 3, simple, denticulate with marginal glands, without stipules. Flowers bisexual, zygomorphic, solitary or few on axillary peduncles, often resupinate. Sepals 3 (or 5), often coloured, imbricate, unequal, the posterior often enlarged and prolonged into a tubular spur. Petals 5, alternate with the sepals, all free or the lateral united. Stamens 5, hypogynous; filaments flattened sheathing the ovary and often the style; anthers 2-locular, coherent or connate. Ovary superior, 5-locular; ovules pendulous, anatropous, 3–many on each axile placenta; stigmas 1–5, more or less sessile. Fruit a fleshy 5-valved capsule dehiscing explosively by elasticity and twisting of the valves, rarely a berry. Seeds with a straight embryo, without endosperm.

2 genera of which most of the 450 species belong to the genus *Impatiens* and are indigenous in tropical Asia and Africa.

1. IMPATIENS L. (1753).

Lateral petals united in pairs; fruit a dehiscent capsule.

I. balsamina L., Garden Balsam, native of Asia, and *I. wallerana* Hook. f., native of E. tropical Africa are grown in Jamaican gardens, the latter often escaping and becoming temporarily established at low and high elevations.

100. LINACEAE

Herbs or shrubs. Leaves alternate or opposite, rarely whorled, simple, entire, with or without stipules. Flowers bisexual, actinomorphic, in cymes or racemes. Perianth biseriate; sepals (4–) 5, free or basally connate, imbricate; petals usually (4–) 5, free, contorted, fugacious. Stamens 5, rarely more, alternate with petals; filaments connate to form a tube with or without glands, sometimes alternated with toothlike staminodes; anthers 2-locular, opening lengthwise. Ovary superior, (3–) 5-locular or falsely 10-locular by intrusion of carpel walls; ovules anatropous, typically 2 in each loculus, pendulous from axile placentas; styles 3–5, free or united, with capitate stigmas. Fruit a septicidal capsule or a drupe. Seeds with straight embryo; endosperm copious, scanty or absent.

When construed in the broadest sense (Engler, *Syllabus*, 1964) a family of 25 genera and 500 species; cosmopolitan.

Slender annual or short-lived perennial herbs; stipules absent or glandular; flowers in racemes; staminodes present or absent; styles 5; capsule 5-locular with false septa
  1. **Linum**
Small shrub; leaves obovate-elliptical, up to 6 cm. long and about 2·5 cm. broad; stipules small, deciduous; flowers solitary or in corymbs; corolla trumpet-shaped, about 3·5 cm. long, orange-yellow; staminodes present; styles 3; capsule 6-locular; cultivated and escaped; native of S.E. Asia   *Reinwardtia indica* Dumort.

1. LINUM L. (1753); C. M. Rogers (1963)

1. **L. floridanum** (Planch.) Trelease in Trans. Acad. Sci. St. Louis **5**: 13 (1892).—
   L. jamaicense (Small) F. & R. (1920). *L. virginianum* L. var. *floridanum* Planch.
   (1848). *Cathartolinum jamaicense* Small (1913).
Erect short-lived perennial herb 20–60 cm. high; stem slender, glabrous sparingly branched above or simple; leaves mostly alternate, linear-oblanceolate, up to about 13 mm. long and 2 mm. broad, without stipular glands; flowers remote in elongated racemes; sepals about 3 mm. long, lanceolate; petals 4·5–7·5 mm. long, yellow; capsules pyriform, about 2·5 mm. long; seeds compressed-ellipsoid, narrowly winged on one side, about 1·8 mm. long.

Rare and local (Clar.), in grassy open savanna; 2000–2400 feet; fl. and fr. Mar, June–Dec. *A 5934! H 11159! P 15935!* S.E. United States.

Drift Fruit (*Sacoglottis amazonica* Mart., sometimes separated as *Humiriaceae*) is the hard 5-segmented ellipsoid endocarp of this species often to be found washed up on Jamaican beaches. The endocarp is 3–4·5 cm. long and 2–3·5 cm. broad, hollowed into chambers which become pits as the tissues break away. The living plant occurs in Trinidad and tropical S. Amer.

101. ERYTHROXYLACEAE

Shrubs or trees. Leaves alternate, rarely opposite, simple, entire or crenate; intrapetiolar stipules usually present. Flowers bisexual, actinomorphic, small, solitary or in axillary clusters. Perianth biseriate; calyx campanulate, lobes 5 (–6), imbricate, persistent; petals 5, free, imbricate or convolute, with ligulate appendages or callosities at their base, deciduous. Stamens usually 10 in 2 whorls, basally connate; anthers 2-locular, opening lengthwise. Ovary superior, 3-locular, but mostly 2 of these sterile; ovule(s) 1–2 in the fertile loculus, anatropous, pendulous from an axile placenta; styles 3, free or basally connate; stigmas capitate. Fruit drupaceous. Seed with straight embryo; endosperm fleshy or rarely absent.

4 genera of which *Erythroxylum* with nearly 200 species occurs mainly in the New World tropics; other small genera in Africa.

1. ERYTHROXYLUM P. Browne (1756)

Stipules persistent, 2-ridged or keeled, spurred when leaf not developed, persistent; flowers often dimorphic, heterostyled; ovule solitary.

1 Leaf-blades rounded at base, rounded and often emarginate at apex, elliptical, 9·5 cm. long, 2·5–4·5 cm. broad; stipules 2–3 mm. long, not keeled; pedicels thickened upwards, up to 3–4 mm. long   1. *jamaicense*
1 Leaf-blades cuneate at base:
   2 Leaf blades acute or acuminate at apex, elliptical, 8–10 (–14) cm. long, 3·5–4·5 (–7) cm. broad; stipules 1–1·5 mm. long; pedicels thickened upwards, 8–10 mm. long
                               2. *incrassatum*
   2 Leaf-blades rounded or emarginate at apex; stipules more or less obviously keeled:
     3 Leaves beneath usually with a distinct central area bounded by 2 longitudinal lines; stipules broadly triangular-ovate; pedicels slender, slightly thickened upwards:
       4 Leaf-blades oblong-elliptical, 3–14 cm. long, 2–6·5 cm. broad; stipules 2·5–4 mm. long; pedicels in fruit (3–) 4–8 mm. long   3. *areolatum*
       4 Leaf-blades obovate to oblanceolate, 2–5 cm. long, 0·6–2 cm. broad; stipules about 2 mm. long; pedicels about 3 mm. long; occasionally cultivated; native of Peru; Cocaine Tree   *E. coca* Lam.
     3 Leaves mostly without a distinct central area or if this evident not usually bounded by longitudinal lines; pedicels broadening and winged below the calyx, 2–4 mm. long:
       5 Leaf-blades obovate to oblong-elliptical, (1–) 3–6 cm. long, (0·7–) 1·5–3·5 cm. broad, lateral veins making angles of 50–60° with the midrib; stipules ovate to lanceolate, 3–5 mm. long   4. *confusum*
       5 Leaf-blades obovate to oblanceolate, 1–2·5 (–3) cm. long, 5–12 (–15) mm. broad, lateral veins making angles of 30–40° with the midrib; stipules triangular-ovate, 1–2 mm. long   5. *rotundifolium*

1. E. jamaicense F. & R. in Journ. Bot. **55**: 38 (1917).
Tree 4–5 m. high; calyx 1·5–2 mm. long; petals oblong, yellow, nearly 3 mm. long; drupe oblong, curved, 1·3 cm. long, 2–2·5 mm. broad.
Rare (Clar., St. Ann), in woodland on rocky limestone hills; 1750–2500 feet; fl. July–Oct, fr. Sept–Oct. *duQuesnay 127! H 11203! P 19831!* Endemic.

2. E. incrassatum O. E. Schultz in Urb., Symb. Ant. **5**: 210 (1907).
Tree 5–10 m. high; flowers 1–3 in leaf-axils; calyx about 1·5 mm. long; petals obovate-oblong, cream with white appendages, 4–4·5 mm. long; fruit not known.
Rare (Manch., St. Ann), in woodland on limestone; 1500–2700 feet; fl. July. *H & P 14249!*. Endemic.

3. E. areolatum L., Syst. Nat. ed. 10, **2**: 1035 (1759).—Coca Shrub.
Shrub about 3 m. or tree to 8 m. high, deciduous; flowers often appearing before or with the new leaves, fragrant; calyx 1·5–2 mm. long; petals oblong-elliptical, white, about 3 mm. long; drupe oblong, orange-scarlet, 7–9 mm. long.
Common in thickets and open woodlands on limestone or alluvial gravel in semi-arid areas; SL–3500 feet; fl. Feb–May, Oct, fr. May–July, Oct–Nov. *A 5613! H 9333! 12349! P 20889!* Bahamas, C. Amer., Greater Antilles; Grand Cayman.

4. E. confusum Britton in Britton & Millsp., Bah. Fl.: 199 (1920).—*E. obovatum* of Griseb. (1860), not Macf. (1837). Barberry Bullet, Greenheart.
Shrub 2–3 m. or tree to 8 m. high, deciduous; flowers appearing with the new leaves; calyx about 2 mm. long; petals elliptical, greenish-yellow, about 2·7 mm. long; stamens (8–) 10; styles (2–) 3; drupe oblong, scarlet, 8–9 mm. long.
Rather common in the central parishes, in exposed thickets and pastures on limestone; 25–2400 feet; fl. May–Sept, Dec, fr. May–Sept, Dec–Feb. *A 11169! H 9661! P 8720!* Bahamas, Cuba.

5. E. rotundifolium Lunan, Hort. Jam. **2**: 116 (1814).—*E. obovatum* Macf. (1837).
Shrub 1·2–3 m. or tree to 12 m. high; young leaves reddish; flowers 1–4 in leaf-axils; calyx about 2 mm. long; petals elliptical, white, 1–2 mm. long; drupe obliquely ellipsoid, scarlet, 5–6 mm. long.

Rather common in lagoon margin or inland thickets on well drained limestone, gravel or serpentine; SL–2500 feet; fl. Jan, (May–) June–Sept, fr. Jan, July–Nov. *A 11909! H 10942! P 9493!* Bahamas, Greater Antilles, Virgin Is.; Grand Cayman.

102. ZYGOPHYLLACEAE

Herbs, shrubs or trees; branches often jointed at nodes. Leaves opposite or rarely alternate, mostly pinnate, rarely simple or 2–3-foliolate; leaflets entire; stipules present, persistent, paired, often spinescent. Flowers usually bisexual, rarely unisexual and dioecious, actinomorphic or rarely zygomorphic, solitary, paired or in cymes. Perianth biseriate; sepals (4–) 5, free or rarely connate at base, imbricate or rarely valvate; petals (4–) 5, free, imbricate, contorted or rarely valvate, sometimes lacking. Disk usually present. Stamens in 1, 2 or 3 whorls of 5, hypogynous, free, often with basal scales; anthers 2-locular, opening lengthwise. Ovary superior; (2–) 4, 5, 10 (–12)-locular; ovules (1–) 2–many in each loculus on axile placentas, style simple, short, or stigmas sessile. Fruit a loculicidal or septicidal capsule, schizocarpic, or rarely drupaceous. Seeds with straight or curved embryo, usually with endosperm.

30 genera with 250 species, mostly in the drier subtropics.

1 Tree with glabrous leathery leaflets; petals blue or white; capsule fleshy, yellow with red
   seeds, usually 2-valved   **1. Guaiacum**
1 Herbs with thin hairy leaflets; stems weak and trailing; petals yellow; fruit dry, the
   segments eventually separating and not releasing seeds:
  2 Ovary 5-locular; sepals deciduous; fruit spiny; stipules lanceolate   **2. Tribulus**
  2 Ovary 10 (–12)-locular; sepals persistent; fruit tuberculate, beaked; stipules linear,
    acuminate   **3. Kallstroemia**

1. GUAIACUM L. (1753)

1. G. officinale L., Sp. Pl. **1**: 381 (1753).—Lignum Vitae.

Dense-crowned squat gnarled tree with mottled bark, (3–) 4–8 m. high; wood very hard, resinous; leaflets 2 (–3) pairs, sessile, obovate-elliptical, 1·5–3·5 (–6) cm. long, and 2·5 (–3·5) cm. broad; stipules 1 mm. long; peduncles 1-flowered, few or clustered in the distal leaf-axils; sepals tomentose, 4–5 mm. long; petals puberulous, 12 mm. long; stamens 10; capsule 1·5–2 cm. long.

Locally common particularly in the arid limestone hills of the south coast and on alluvial plains, often planted; SL–800 feet; fl. mostly during dry weather, Dec–Apr, June–Sept, fr. May–Feb. *A 10233! H 8639! Powell 1645!* Continental tropical Amer., West Indies, introduced into the Old World tropics.

2. TRIBULUS L. (1753)

1. T. cistoides L., Sp. Pl. **1**: 387 (1753).—Kill Bukra, Kingston Buttercup, Turkey Blossom.

Low spreading herb with jointed branches up to 1 m. or more long; leaves often unequal (3–) 6–9 pairs, oblong, unequal-sided, silvery-silky, 7–17 mm. long, 3–7 mm. boad; stipules 4–9 mm. long; flowers solitary in the axil of the smaller leaf of a pair; petals yellow or rarely cream-coloured, variable up to 2·5 cm. long; fruit up to about 1 cm. long and broad with stiff spines and long white hairs.

Locally abundant along well drained roadsides and in dry waste places; SL–800 (–2000) feet; fl. and fr. all the year. *H 11967! P 26767!* Bahamas, S.E. United States, continental tropical Amer., West Indies; Grand Cayman, S.W. Pacific; introduced into Africa and Asia.

3. KALLSTROEMIA Scop. (1777); Porter (1969)

Fruits glabrous   **1.** *maxima*
Fruits with long simple hairs   **2.** *pubescens*

1. K. maxima (L.) Torr. & A. Gray, Fl. N. Amer. **1**: 213 (1838).—*Tribulus maximus* L. (1753). Police Macca.

Much branched herb with stems spreading from a central taproot, up to 50 (–100) cm. or more long; leaflets 2–4 pairs, unequal-sided, with at least hairy margins, (0·5–) 1–2 cm. long, 3–10 mm. broad; flowers solitary; petals creamy-to orange-yellow, yellow-green at base, 5–8 mm. long; fruit ridged and warted, 8–9 mm. long.

Common, a weed of lawns, field margins and waste places; 25–750 feet; fl. and fr. all the year. *A 6600! H 6638! P 23938!* Bahamas, S.E. United States, continental tropical Amer., West Indies.

2. K. pubescens (G. Don) Dandy in Kew Bull. **10**: 138 (1955).—*Tribulus pubescens* G. Don (1831).

Prostrate-branched herb like the last; petals light yellow.

Very rare (St. Andr.), a weed of waste ground; 25–50 feet; fl. and fr. Aug, Nov. *A. M. Barry! Porter 1034! P 23941!* Continental tropical Amer., West Indies, W. Africa.

103. RUTACEAE

Shrubs or trees, rarely herbs. Leaves alternate or opposite, usually pinnate, sometimes digitately compound or simple, often gland-dotted and aromatic, without stipules. Flowers bisexual or unisexual, mostly actinomorphic, solitary, clustered, cymose, racemose or paniculate. Perianth typically biseriate, often gland-dotted; calyx 3–5-merous, the sepals free or connate, imbricate; corolla (0–) 3–5 (–8)-merous, the petals usually free, imbricate or rarely valvate. Stamens as many as or twice as many as the petals or more numerous, free or united, rarely adnate to petals; anthers 2-locular, opening lengthwise; connective often glandular at tip. Disk present between stamens and ovary. Ovary superior, syncarpous or carpels more or less free or united by the styles and free below, typically 4–5-carpelled, sometimes 1-locular or many-locular; ovules 1–2-several in each loculus, usually on axile placentas; styles united with a single capitate or lobed stigma or free. Fruit a capsule, berry or drupe, sometimes schizocarpic, sometimes the whole fruit winged. Seed with straight or curved embryo, sometimes polyembryonous, cotyledons often fleshy and green; endosperm present or absent.

About 150 genera with over 1500 species; widely distributed in warm regions, most numerous in S. Africa and Australia.

1 Fruit dry, capsular, coccoid or winged:
  2 Petals united at base; 2 outer sepals larger than inner; fruit coccoid; leaves opposite, 1– or 3-foliolate; peduncles axillary   **1. Ravenia**
  2 Petals free; sepals equal or 1 only enlarged or outer smaller than inner; leaves alternate, 3- or more-foliolate:
    3 Ovary and fruit entire, (2–) 3-locular; fruit 3-winged; leaves imparipinnate with numerous leaflets; panicle terminal, large; monocarpic slender-trunked sparingly branched trees   **2. Spathelia**
    3 Ovary and fruit lobed or fruit 1–8-coccoid; polycarpic shrubs or trees:
      4 Flowers dioecious or polygamous; stamens as many as petals; leaves pinnate; carpels becoming follicular and separating in fruit, rounded, less than 1 cm. long and broad   **3. Fagara**
      4 Flowers all perfect; leaves digitate; fruits larger:
        5 Stamens as many as petals; capsule 4–5 cm. broad, splitting septicidally into 5 cocci, each carpel opening loculicidally   **4. Esenbeckia**
        5 Stamens numerous; flowers showy; capsule about 5 cm. broad, of usually 8 beaked cocci   **5. Peltostigma**
1 Fruit fleshy, drupaceous or baccate; ovary entire:
  6 Stames 15 or more; ovary 8–15-locular; leaves 1-foliolate, the petiole often winged; fruit at least 2 cm. in diameter, the internal sections filled with large vesicular cells; plants often spiny, usually cultivated   **6. Citrus**
  6 Stamens 6–10; ovary 1–5-locular; leaves 1–9-foliolate, the petiole and rachis not winged; fruit smaller:
    7 Fruit drupaceous, black, 1-seeded; ovary 1-locular; flowers 4-merous, white, in terminal and axillary panicles; stamens 8; leaves 3- or 5-foliolate; plants in natural thickets on limestone, not spiny   **7. Amyris**

7 Fruit baccate, pink or red, 2- or more-seeded; ovary 2–5-locular; flowers rarely 4-merous; introduced cultivated plants, sometimes naturalized:
  8 Petals usually 3; stamens 6; style evident; flowers solitary or in small axillary clusters; leaves 3-foliolate; plants with solitary or paired axillary spines; cultivated as an ornamental and hedge plant; native of E. Indies; Chinese Lemon, Limeberry *Triphasia trifolia* (Burm. f.) P. Wilson
  8 Petals (4–) 5; stamens 10; flowers paniculate; plants without spines:
    9 Leaves 1–3 (–5)-foliolate; flowers small, petals about 3 mm. long; style obsolete
                                                                        8. Glycosmis
    9 Leaves 3–9-foliolate; flowers rather showy, white, fragrant, petals about 12 mm. long; style evident; shrub with small obtuse shiny leaflets; cultivated ornamental; native of tropical Asia; China Box, Mock Orange *Murraya paniculata* (L.) Jack

1. RAVENIA Vell. (1825)

Leaves 3-foliolate; peduncles 1–3-flowered, 2–6 cm. long   1. *spectabilis*
Leaves simple; peduncles mostly 2-flowered, up to 1 cm. long   2. *swartziana*

1. R. spectabilis (Lindl.) Planch. ex Griseb. in Mem. Amer. Acad. n.s. **8**: 170 (1860).
Shrub 2–3 m. high; leaflets digitate, glabrous, narrowly elliptical to oblong, 3–5 cm. long; petiole 1–2 cm. long; outer sepals ovate, 11–13 mm. long, 7–10 mm. broad, inner shorter, roundish; corolla fleshy, red or pink, the tube 13 mm. long, slightly zygomorphic; mid-stamen 3-lobed; seeds ellipsoid, about 3·5 mm. long.
Probably only ever in cultivation in Jamaica; fl. Feb. *Harris 59!* Native of Cuba, Hispaniola and Guadeloupe, introduced and cultivated in Asia and elsewhere.

2. R. swartziana (Miers) F. & R. in Journ. Bot. **55**: 38 (1917).
Leaves elliptical, shortly acuminate, glabrous, 4–8·5 cm. long; petiole 2–3 mm. long; outer sepals ovate, 6–7 mm. long, 2–3 mm. broad, inner lanceolate and about half as long; corolla fleshy, the tube 7 mm. long.
Known only from the type, *Swartz.* Endemic.

2. SPATHELIA L. (1762) nom. cons.

Leaflets sessile or subsessile; stamen-filaments with hairy winglike appendages
                                                                        1. *sorbifolia*
Leaflets generally distinctly stalked; stamen-filaments without or with only rudimentary basal winglike appendages   2. *glabrescens*

1. S. sorbifolia (L.) F. & R., Fl. Jam. **4**: 193 (1920).—*Spathalea sorbifolia* L. (1760). *Spathelia simplex* L. (1762). Mountain Pride.
Unbranched or sparingly branched slender tree 6–16 m. high; young twigs and leaves rusty-pubescent; leaflets 22–45 pairs, oblong-lanceolate, often truncate at base, crenate or crenate-serrate, 7–12 (–19) cm. long, up to 2·5 (–3·5) cm. broad, more or less velvety beneath with stellate and simple hairs; panicle much-branched; first flowers male, later hermaphrodite and male, pinkish-lilac to bright magenta; sepals 3·5–4 mm. long; petals 5–8 mm. long; fruit 2–3 cm. long.
Rather common in open thickets or woodlands on well drained shale or limestone; SL–1500 feet; fl. June–Nov, Jan, fr. July–Nov. *A 7278! 9854! Campbell 6421! H & P 13827! P 7385!* Endemic.

2. S. glabrescens Planch. in Hook., Lond. Journ. Bot. **5**: 581 (1846).
Tree like the last up to 24 m. high; leaflets up to 25 or more pairs, oblong-lanceolate, mostly rounded at base, crenulate to subentire, up to about 15 cm. long and 3·5 (–5) cm. broad, glabrescent; panicle and fruits more or less as above; flowers usually mauve-pink.
Local and gregarious in hilly parts of the central parishes; 1250–2400 feet; fl. Mar–Aug, fr. Mar–Sept. *H 9659! P 6488! 15652! R 1952! Robbins UCWI 1827!* Endemic.
Specimens of what is almost certainly a distinct species, related to *S. glabrescens* but with fewer smaller blunt shiny leaflets and scarlet flowers, have been gathered recently in Trelawny (*A 12844! R 1910!*)

3. FAGARA L. (1759) nom. cons.

1 Petiole and rachis winged; leaflets opposite, with marginal and also sometimes a few
  superficial pellucid glands; inflorescences mostly lateral; flowers 4-merous; carpels in
  fruit solitary or paired:
  2 Leaflets each with 2 prominent glands abaxially at base, entire or emarginate, oblong-
    linear to elliptical, up to 15 mm. long and 7 mm. broad; flowers usually few in
    axillary clusters; cocci 2·5–3 mm. long; branchlets spiny　　　　　　　　1. *spinifex*
  2 Leaflets with marginal glands not usually prominent, at most the proximal blade-
    margins thickened, larger than above; flowers in simple or compound spikes; branch-
    lets rarely spiny but trunk often so:
    3 Cocci of fruit 3–4 mm. long with slender stipes 1–4 mm. long; leaflets mostly obovate,
      up to about 2·5 cm. long, the margin entire or crenate distally, mostly emarginate
      　　　　　　　　　　　　　　　　　　　　　　　　　　　　　　　　　2. *pterota*
    3 Cocci of fruit 5–7 mm. long with short thick tapered stipes up to 1 mm. long; leaflets
      mostly elliptical, up to 4 (–5) cm. long, the margin usually crenate
      　　　　　　　　　　　　　　　　　　　　　　　　　　　　　3. *culantrillo*
1 Petiole and rachis not winged, at most grooved or narrowly ribbed:
  4 Leaflets with several to numerous marginal and superficial pellucid glands:
    5 Flowers 3-merous; inflorescence small, axillary; leaflets often 3　　　9. *trifoliata*
    5 Flowers 4–5-merous; inflorescences terminal and subterminal, rarely lateral:
      6 Leaf-rachis, leaflets and inflorescences pubescent; leaflets entire to crenulate:
        7 Twigs, petioles and inflorescence puberulous with simple hairs; leaf-rachis often
          with broad-based prickles; leaflets opposite or alternate, with mostly rather few
          pellucid glands　　　　　　　　　　　　　　　　　　　　　4. *martinicensis*
        7 Twigs, petioles and inflorescence with minute stellate hairs; leaf-rachis without
          prickles; leaflets opposite or subopposite, with very numerous pellucid glands
          　　　　　　　　　　　　　　　　　　　　　　　　　　　　　　　5. *flava*
      6 Leaf-rachis and leaflets glabrous or nearly so; branches without spines or prickles:
        8 Leaves imparipinnate; leaflets mostly opposite, crenate, thin, with large con-
          spicuous pellucid glands; inflorescence-branches glabrous, soon becoming corky;
          sepals overlapping deeply with broad rounded hyaline margins; base of trunk
          only spiny　　　　　　　　　　　　　　　　　　　　　　　6. *elephantiasis*
        8 Leaves paripinnate; leaflets opposite or alternate, entire, with small or incon-
          spicuous pellucid glands; inflorescence-branches puberulous, glabrescent but not
          becoming corky; sepals small, deltate; unarmed trees:
          9 Leaflets 1–2 (–3) pairs, coriaceous, obovate to oblanceolate, with recurved margins,
            mostly obtuse; flowers 4-merous　　　　　　　　　　　　　7. *rhodoxylon*
          9 Leaflets 4–6 (–10) pairs, subcoriaceous, elliptical to elliptic-lanceolate, flat-
            margined, acuminate; flowers 5-merous　　　　　　　　　　8. *harrisii*
  4 Leaflets without or with few mostly marginal pellucid glands; flowers 3-merous;
    carpels in fruit barely stipitate, gland-dotted, the endocarp separating easily;
    leaflets opposite, mostly closely and prominently reticulate-veined and shiny
    adaxially; trees usually armed with spines but often variably so, glabrous:
  10 Inflorescence axillary clusters or small racemes or panicles often shorter than the
      petioles; leaflets 2–7 (–13), often 3, superficially gland-dotted　　　9. *trifoliata*
  10 Inflorescence terminal and corymbose; leaflets mostly more than 3, not or only
      sparsely superficially gland-dotted:
    11 Ovary of 1 carpel, the style lateral and adnate; leaves paripinnate; leaflets elliptic-
        lanceolate, distinctly acuminate　　　　　　　　　　　　　　10. *acuminata*
    11 Ovary of 2 or 3 carpels, the style terminal; leaflets ovate, elliptical or obovate,
        mostly obtuse or if acuminate only shortly and broadly so:
      12 Leaves usually without an odd terminal leaflet; leaflet-margins mostly entire, the
          tip more or less emarginate; pellucid glands wanting or very few:
        13 Leaflets 1·5–6·5 cm. long; plants usually spiny　　　　　　　11. *spinosa*
        13 Leaflets 7–13 cm. long; plants unarmed　　　　　　　　　　12. *negrilensis*
      12 Leaves usually with an odd terminal leaflet; leaflet-margins more or less crenate;
          pellucid dots evident at least adaxially on some of the leaves:
        14 Leaflets 3–11 cm. long, shortly acuminate, the acumen emarginate
          　　　　　　　　　　　　　　　　　　　　　　　　　　　13. *jamaicensis*
        14 Leaflets 2–5 cm. long, rounded at tip, not obviously emarginate　　14. *hartii*

1. **F. spinifex** Jacq., Fragm.: 10, t. 6, f. 2 (1801).—*Zanthoxylum spinifex* (Jacq.)
  DC. (1824).
  Shrub 1–5 m. high, the branchlets usually with straight stipular spines; leaflets 5
(–11); flowers 4–5-merous; petals nearly 2 mm. long, exceeded by stamens; carpels 2.
  The Jamaican record of this species is based solely on a specimen in the Sloane
Herbarium. There has been no subsequent report but the plant is known from all

major islands of the Antillean chain from Cuba to Barbados and from Venezuela in coastal thickets and woodlands.

2. F. pterota L., Syst. Nat. ed. 10, **2**: 897 (1759).—*Zanthoxylum fagara* (L.) Sarg. (1890). Bastard Ironwood, Saven Tree.

Shrub or tree 2–10 m. high; spines on trunk with broad conical bases; prickles on branches more or less recurved; leaflets (5–) 7–9 (–11), obovate to elliptical, 1–2·5 (–3·5) cm. long, up to 13 (–17) mm. broad; inflorescences up to 2 cm. long, dioecious; petals greenish, about 3 mm. long; seeds shiny black, 2·5–3 mm. long.

Locally common in southern parishes, in coastal thickets and woodlands on arid limestone; SL–1000 feet; fl. Jan–July, fr. July–Jan. *H 8384! 8841! P 20802! 21524! Yuncker 17994!* Subtropical and tropical continental Amer., Bahamas, Cuba, Hispaniola, Martinique, Trinidad.

3. F. culantrillo Krug & Urb. in Engl., Bot. Jahrb. **21**: 574 (1896).—*Zanthoxylum insulare* Rose (1899).

Aromatic tree 5–12 m. high, often gnarled and low-branched, the base of trunk sparingly armed with needlelike spines on broad bases; leaflets 7–15, oblanceolate to elliptical, up to 6 cm. long and 2·7 cm. broad; inflorescences 4–10 cm. long; flowers 4 (–5)-merous, light yellowish-green; petals of male flowers about 3 mm. long, of female flowers 2 mm. long; seeds subglobose, 3–4 mm. in diameter.

Locally common, in hilly parts of the southern parishes in open woodlands and thickets on limestone; 1000–3000 feet; fl. Jan–Sept, fr. July–Mar. *A 12278! H 9662! P 11631! 24871!* Tropical S. Amer., Cuba, Hispaniola.

4. F. martinicensis Lam. in Tabl. Encycl. & Méth., Bot. **1**: 334 (1792).—*Zanthoxylum martinicense* (Lam.) DC. (1824). Prickly Yellow, Yellow Hercules.

Tree 6–15 (–18) m. high; trunk and branches armed with thick conical spines; young twigs rusty-pubescent; leaflets 4–7 (–15) pairs, oblong-elliptical to oblanceolate, unequal at base, obtuse to shortly acuminate, 3–9 (–13) cm. long and up to about 3·5 cm. broad; inflorescences dioecious, paniculate with numerous subsessile white flowers; male flowers 5–6-merous with 5 carpellodes; petals 1·5–2 mm. long, inflexed; fruit-cocci about 5 mm. long; seeds shiny black, about 4 mm. long.

Common, especially in secondary formations; SL–2500 feet; fl. and fr. most of the year. *A 8051! H 9092! P 19760! 22211! Wynter UCWI 3287!* Northern S. Amer., West Indies.

5. F. flava (Vahl) Krug & Urb. in Engl., Bot. Jahrb. **21**: 571 (1896).—*Zanthoxylum flavum* Vahl (1807). Jamaican Satinwood.

Shrub or tree 3–8 (–13) m. high, without spines or prickles; leaflets (1–) 3–6 pairs, ovate to oblong-lanceolate, unequal at base, acuminate, 5–12 cm. long, up to 4·5 cm. broad; inflorescences dioecious, paniculate with numerous flowers on pedicels 1–4 mm. long; flowers 4–5-merous; petals greenish-white, 3–4 mm. long; fruit-cocci 4–6 mm. long.

Frequent in central and western parishes, in thickets and woodlands on rocky limestone; 50–1500 feet; fl. Mar, fr. Aug–Nov. *Asprey UCWI 1830! H & P 13701!* Florida, Bermuda, Bahamas, Greater Antilles, Lesser Antilles south to St. Lucia; Grand Cayman.

6. F. elephantiasis (Macf.) Krug & Urb. in Engl., Bot. Jahrb. **21**: 564 (1896).—*Zanthoxylum elephantiasis* Macf. (1837). Yellow Sanders.

Tree 5–16 m. high; leaflets (2–) 5–6 (–8) pairs, ovate-elliptical to oblong-lanceolate, unequal at base, crenate, acuminate, (3–) 5–10 cm. long, 1·3–4 cm. broad; inflorescence paniculate with numerous flowers on pedicels 1–2·5 (–5) mm. long; flowers 5-merous; petals light green, 5–6 mm. long; fruit-cocci warted, 6–8 mm. long.

Common, mostly in central and western parishes, in open woodlands; SL–3000 feet; fl. Feb–June, fr. most of the year. *A 12469! H 10018! P 26574!* Mexico, Costa Rica, Cuba, Hispaniola.

7. F. rhodoxylon Urb., Symb. Ant. **5**: 530 (1908).—*Zanthoxylum rhodoxylon* (Urb.) P. Wilson (1910). Caesarwood, Rosewood.

Tree 8–16 m. high; leaflets sessile, obtuse or rounded at tip, 4–7 cm. long, 1·5–

3·5 cm. broad, shiny adaxially; panicles 3–5 cm. long; petals white, 2–2·5 mm. long; fruit-cocci warted, 5–6 mm. long.

Locally common, in central parishes, in woodlands on rocky limestone, 700–2500 feet; fl. May, Oct, fr. Mar–May, Dec. *H 12780! P 11058! 22993!* Endemic.

8. F. harrisii (P. Wilson ex Britton) Adams in Phytologia **20** (5): 311 (1970). —*Zanthoxylum harrisii* P. Wilson ex Britton (1922).

Tree 15–18 m. high; leaflets stalked, 7–15 (–20) cm. long, 4–6 cm. broad, shiny adaxially; petals 3 mm. long.

Very rare (Port.), in humid woodland; 3500 feet; fl. Mar. *A 10872! H 12878!* Endemic.

9. F. trifoliata Sw., Nov. Gen & Sp. Pl.: 33 (1788).—*Zanthoxylum punctatum* Vahl (1793). *Z. trifoliatum* of W. Wright (1828), not L. (1753). Toothache Tree.

Aromatic shrub or tree 2–6 m. high; leaflets obovate-elliptical to elliptic-lanceolate, variable at tip, margin entire or nearly so, 2·5–10 cm. long, 1–4·5 cm. broad; inflorescence glabrous, 2–6 cm. long; petals creamy-white, about 2 mm. long; fruit-cocci globose, pinkish, warted, 3·5–5 mm. long.

Occasional, on wooded limestone hills; 250–3000 feet; fl. Oct–Nov, fr. July–Oct. *P 11027!* Jamaican plants thus identified have longer inflorescences and larger leaflets than typical plants from the Lesser Antilles and Trinidad. Specimens from Hispaniola and Puerto Rico have harder more sharply pointed leaves and may also be distinct.

10. F. acuminata Sw., Nov. Gen. & Sp. Pl.: 33 (1788).—*Zanthoxylum acuminatum* (Sw.) Sw. (1797). Satinwood.

Tree 5–13 m. high; trunk and sometimes larger branches with spines from compressed conical bases; leaflets 2–4 pairs, the margin entire or finely crenulate, shiny, 5–13 cm. long, (2–) 3–6 cm. broad; panicles 5–10 cm. long; petals about 2·5 mm. long; fruit-coccus globose, glandular, warted, 5 mm. long.

Occasional, localities scattered, in woodlands on limestone or shale; 500–3000 feet; fl. Dec–Apr, fr. June–Aug. *H 6549! 7100! J.P. 2087! P 24796!* Cuba.

F. cubensis (P. Wilson) Urb. is a very shiny much branched tree with imparipinnate leaves, resembling the above in floral structure. It occurs in Grand Cayman and Cuba and was reported from Jamaica by León & Alain, Flora de Cuba **2**: 385 (1951), but has not been confirmed.

11. F. spinosa (L.) Sw., Nov. Gen. & Sp. Pl.: 33 (1788).—*Zanthoxylum spinosum* (L.) Sw. (1797). Licca Tree, Lignum Rorum, Suarra Wood.

Shrub or tree up to 8 m. high; leaflets 2–6 (–10) pairs, distinctly shortly petiolulate, (1–) 1·5–3·5 (–4) cm. broad; panicles with peduncles 3–6 cm. long; pedicels up to 2 mm. long in flower and 4 mm. long in fruit; flowers fragrant; petals white, hooded, 2·5–3 mm. long; fruit-cocci roundish-ellipsoid, 5–6 mm. long.

Common in mostly arid exposed thickets on limestone; near SL–3050 feet; fl. June–Mar, fr. Nov–June. *A 6846! 9852! H 9575! 10375! H & P 15120!* Cuba, Hispaniola, Dominica, Grand Cayman. Urban (1896), maintained the species *F. sapindoides* (DC.) Krug & Urb. and *F. swartzii* Krug & Urb. distinct depending on the presence of a terminal leaflet, the sizes and number of leaflets, the presence of glands on the leaflets and the positions of inflorescences. Similar considerations decide the taxonomic recognition of the three following species also. There is no doubt that species 11–14 belong to a group of closely related plants in which intergrading variants exist.

12. F. negrilensis (F. & R.) Engl. in Engl. & Prantl, Nat. Pflanzenf. ed. 2, **19**A: 223 (1931).—*Zanthoxylum negrilense* F. & R. (1917).

Unarmed tree 3 m. high; leaflets 2–4 pairs, 4–6 cm. broad; panicles about 8 cm. long; fruit-cocci ellipsoid, about 5 mm. long.

Rare and local (West.), in rocky woodland, on limestone; 300 feet; fr. Mar. *H 10242!* Endemic.

13. F. jamaicensis (P. Wilson) Engl. in Engl. & Prantl, Nat. Pflanzenf. ed. 2, **19**A: 223 (1931).—*Zanthoxylum jamaicense* P. Wilson (1909).

Shrub 2–3 m. or tree to 10 m. high; trunk and usually lower branches spiny;

leaflets 1–4 pairs, the proximal pair often distinctly smaller, 1·5–5 cm. broad; panicles about 10 cm. long; petals white, about 2 mm. long; fruit-cocci globose, reddish, 4 mm. long.

Occasional in central and western parishes, in thickets and woodlands on limestone; 450–3000 feet; fl. Aug–Mar, fr. July–Mar. *A 8592! Britton 2318! H & P 14333! Howard, Proctor & Wagenknecht 20511!* Endemic.

14. F. hartii Krug & Urb. in Engl., Bot. Jahrb. **21**: 556 (1896).—*Zanthoxylum hartii* (Krug & Urb.) P. Wilson (1910).

Low-branching tree 2–7 m. high, armed basally; leaflets 1–4 pairs, 10–20 mm. broad; panicle 1·5–2 cm. long; petals white, 2·3 mm. long; fruit-cocci ovoid, blue-black or dark reddish-purple, about 5 mm. long; seeds shiny black.

Very local (St. Andr., Port.), in montane woodland; 5500–5800 feet; fl. Mar, Nov, fr. Aug. *A & S 3462! H 6524! Loveless UCWI 1822! P 9540!* Endemic.

4. ESENBECKIA Kunth (1825)

1. E. pentaphylla (Macf.) Griseb., Fl. Br. W.I.: 135 (1860).—Wild Orange.

Tree 8–25 m. high; leaves digitately 3–5-foliolate; leaflets oblong-elliptical to obovate, obtuse, rounded or broadly acuminate at tip, entire, 6–20 cm. long, up to 7 cm. broad, with numerous pellucid dots; panicle terminal, corymbose; petals oblong, 3–4 mm. long, yellowish; seeds brown, compressed, 10–13 mm. long, smooth and shiny.

Occasional in woodlands, more common in western parishes; 500–2500 feet; fl. June–Dec, fr. Nov–Mar. *H 10285! 12367! H & P 13561!* Endemic.

5. PELTOSTIGMA Walp. (1846)

1. P. pteleoides (Hook.) Walp., Repert. Bot. **5**: 387 (1846).—Candlewood, Cantoo.

Aromatic tree 5–8 m. high; leaves digitately 3–5-foliolate; leaflets elliptical to lanceolate or oblanceolate, acute or shortly acuminate, entire or serrate distally, the terminal leaflet usually larger, 4–20 cm. long, with numerous pellucid dots; flowers large in long-stalked axillary corymbs; petals oval to suborbicular, 14–25 mm. long, creamy-white; seeds brown, oblong.

Very rare (St. Eliz.), in woodland on limestone; about 2600 feet; fl. Sept, Dec, fr. Sept. *H 9800! 12875!* Mexico, Guatemala.

6. CITRUS L. (1753); Swingle (1943)

1 Petioles wingless, not apparently articulated with the leaf-blade; spiny shrub or tree up to 3 m. high; leaf-blade 8–20 cm. long, 3–9 cm. broad; flowers 5-merous, pinkish; stamens 30–40; fruit oblong, 10–20 cm. long; peel very thick, usually bumpy; rare in Jamaica; native of S.W. Asia; Citron   *C. medica* L.
1 Petioles distinctly articulated with the leaf-blade, often but not always winged:
  2 Petioles narrowly margined but not winged; flowers tinged pink or purplish outside, 4–5-merous; stamens 20–40; spiny tree 3–6 m. high; leaf-blade 5–10 cm. long, 3–6 cm. broad; fruit ovoid with terminal nipple, 5–10 cm. long; peel rough or smooth; native of S.E. Asia; Lemon (*C. limonum* Risso)   *C. limon* (L.) Burm. f.
  2 Petioles broadly or narrowly winged; flower-buds white:
    3 Twigs pubescent; petioles broadly winged; flowers mostly 4-merous, large; stamens 25–35; spiny tree 5–15 m. high; leaf-blade 5–20 cm. long, 2–12 cm. broad; fruit globose, pyriform or oblate, 10–30 cm. in diameter and weighing up to 2 Kg. or more, the pulp-vesicles easily separated; native of S.E. Asia; Pummelo, Shaddock
                                        C. grandis (L.) Osbeck
    3 Twigs glabrous or nearly so; flowers 4–5-merous; fruit smaller; pulp-vesicles cohering:
      4 Petiole usually broadly winged; fruit spherical; moderately spiny trees; stamens 19–25:

5 Fruits 8–15 cm. in diameter, rather smooth, maturing light-yellow; tree 4–10 m. high; leaf-blade up to 14 cm. long and 10 cm. broad; originated in cultivation probably in the West Indies; Grapefruit   *C. paradisi* Macf.
5 Fruits 7–9 cm. in diameter, rather rough, maturing bright orange; tree to 10 m. high; leaf-blade up to 20 cm. long and 7 cm. broad; native of S.E. Asia; Seville or Sour Orange (*C. vulgaris* Risso)   *C. aurantium* L.
4 Petiole usually narrowly winged:
  6 Fruit with loose peel, oblate, 5–6 cm. in diameter; embryo green; small tree with spines mainly on inner branches; leaf-blade 4–8 (–9·5) cm. long, 1·5–4 (–5·5) cm. broad; stamens 15–20; native of China; Mandarin, Tangerine
                    C. reticulata Blanco
  6 Fruit with adherent peel; embryo not green; stamens 20–25:
    7 Fruits ovoid or subglobose, light yellow when ripe, the pulp greenish and very acid, mostly 3·5–5 cm. in diameter; shrub or tree 2–5 m. high, with very numerous spines; leaf-blade 4–7 (–8) cm. long, 2–4 (–5) cm. broad; native of tropical Asia; Lime   *C. aurantifolia* (Christm.) Swingle
    7 Fruits subglobose, greenish-yellow to orange when ripe, the pulp yellow or orange, sweet, mostly 5–10 cm. in diameter; tree 4–12 m. high with usually rather few spines; leaf-blade 5–15 cm. long, 2–8 cm. broad; native of China; Sweet Orange   *C. sinensis* (L.) Osbeck

The foregoing key enables the identification of those types of *Citrus* which are regarded as species by most authorities. In addition a number of hybrids or putative hybrids having intermediate characters occur in Jamaica. The more important of these are Tangelos (*C. paradisi* × *C. reticulata*), the commonest local example being known as the Ugli, and Tangors (*C. reticulata* × *C. sinensis*), the commonest in our area being the Ortanique. The Calamondin (*C. reticulata* × *Fortunella* sp.), a small-leaved shrub or tree with very small extremely acid fruits is occasionally grown for ornament.

7. AMYRIS P. Browne (1756)

1 Leaves alternate or subopposite; leaflets usually 3, 3–10 cm. long, 2–6 cm. broad; ovary glabrous; drupe globose, 4–7 mm. long   1. *plumieri*
1 Leaves opposite or subopposite:
  2 Ovary glabrous; drupe globose, 5–8 mm. long; leaflets usually 3, 2–5 (–7) cm. long, 1–3 cm. broad   2. *elemifera*
  2 Ovary and young fruit puberulous; drupe ellipsoid, 9–14 mm. long; leaflets usually 5, 4–13 cm. long, 2–6 cm. broad   3. *balsamifera*

1. A. plumieri DC., Prodr. 2: 81 (1825).—Candlewood.
Shrub 1·2–4 m. or tree to 6 m. high; bark light grey; leaflets broadly ovate, rounded to acuminate at tip, crenate; panicle usually terminal; flowers fragrant; petals elliptical to obovate, 2·5–3 mm. long.
Common in southern parishes, mostly in woodlands and thickets on arid limestone; 10–2600 feet; fl. Mar–Sept, fr. May–Jan. *A 11444! H 5194! 8630! H & P 13692!* Mexico, Costa Rica, Colombia, Venezuela.

2. A. elemifera L., Syst. Nat. ed. 10, 2: 1000 (1759).—Torchwood.
Aromatic shrub 2·5–4 m. or tree to 11 m. high, with slender trunk; leaflets lanceolate to broadly ovate, acute to long-acuminate, entire or crenulate; petals elliptical to obovate, 2–3 mm. long.
Common usually in open thickets on arid limestone; 10–3000 feet; fl. Aug–Apr, fr. Dec–Mar. *A 10223! 10742! H 9192! P 21851! 24465!* Florida, Bahamas, C. Amer., Greater Antilles, Lesser Antilles south to St. Vincent; Cayman Is.

3. A. balsamifera L., Syst. Nat. ed. 10, 2: 1000 (1759).—Amyris Wood, West Indian Sandalwood, White Cantoo.
Aromatic tree 5–13 m. high; leaflets lanceolate to ovate, acute to long-acuminate, entire or crenulate; petals elliptical to obovate, about 3 mm. long.
Common in woodlands on limestone; 800–3000 feet; fl. sporadically throughout the year, fr. Sept–Dec. *A 10228! H 11992! P 22770! 24306!* Florida, Honduras, Ecuador, Colombia, Venezuela, Greater Antilles.

8. GLYCOSMIS Correa (1805)

1. **G. parviflora** (Sims) Little in Phytologia **2**: 463 (1948).—*G. pentaphylla* of F. & R. (1920), not (Retz.) DC. (1824).

Shrub 1·5–3 m. or tree to 6 m. high; leaflets elliptical to lanceolate, usually entire, narrowed to both ends, 7–19 cm. long, 2–6 cm. broad; panicles axillary; flowers white; petals slightly reflexed; ripe fruits light red or white, becoming globose, 7–9 mm. in diameter.

Locally common, in secondary thickets, usually in low-lying areas near ponds and rivers, and in pastures; 200–1200 feet; fl. and fr. sporadically throughout the year. *A 9428! Brinsley UCWI 998! H 9036! P 15671!* Native of China and Indo-China naturalized in the Guianas and in many West Indian Islands.

104. SIMAROUBACEAE

Shrubs or trees often with a bitter principle present in bark and leaves. Leaves alternate, rarely opposite, simple or more usually pinnately compound, without stipules. Flowers bisexual or more usually unisexual and mostly dioecious, often small, actinomorphic, solitary, clustered in the axils of leaves or in spikes, racemes or panicles. Perianth biseriate or uniseriate; calyx 3–8-lobed, the lobes imbricate or valvate; petals 3–5, free, imbricate, or absent. Stamens as many as the petals or twice as many, free, inserted in or at the base of a disk; anthers 2-locular, opening lengthwise. Ovary superior, often raised on the disk, syncarpous and 1–5 (–8)-locular or apocarpous with 2–5 free carpels, with 2–8 free or connate styles; ovules 1 or 2 in each loculus on basal or axile placentas. Fruit drupaceous or a berry, rarely capsular. Seeds with straight or curved embryo, with or without endosperm.

About 30 genera with some 200 mostly tropical species.

1 Leaves simple:
  2 Flowers bisexual; carpels and styles free; stamens 10; petals 5; ovules 2 in each carpel, basal, erect   **1. Suriana**
  2 Flowers unisexual; plants dioecious; apices of carpels and bases of styles fused; stamens 8; petals 4; ovules 1 in each carpel, lateral   **2. Castela**
1 Leaves pinnately compound:
  3 Carpels free but joined at the top by the connate styles; leaflets opposite:
    4 Stigma simple; petals 5, red; flowers bisexual; petiole and rachis broadly winged; cultivated shrub; native of tropical S. Amer.; Bitter Quassia   *Quassia amara* L.
    4 Stigmas 2–3; petals 4–5, light green; flowers mostly unisexual; petiole and rachis not winged   **3. Picrasma**
  3 Carpels fused; styles 2–3 or stigmas spreading; leaflets mostly alternate; flowers mostly unisexual:
    5 Fruit a 3-winged, 1-seeded capsule; stamens 5; leaflets oblong, usually revolute-margined, up to 3·5 cm. long; inflorescence racemose   **4. Alvaradoa**
    5 Fruit drupaceous or berrylike; leaflets usually flat, larger:
      6 Inflorescences axillary and terminal, paniculate; stamens 10; ovary 5-lobed, the carpels separating in fruit; drupelets 1–5, 1-seeded; leaflets mostly obtuse, paler beneath   **5. Simarouba**
      6 Inflorescences leaf-opposed, racemose, pendulous; stamens 3–5; ovary entire, 2–3-locular, the carpels not separating in fruit; berry 1–2-seeded; leaflets acuminate, not paler beneath   **6. Picramnia**

1. SURIANA L. (1753)

1. **S. maritima** L., Sp. Pl. **1**: 284 (1753).—Bay Cedar.

Diffusely branched shrub 1–2·5 m. high or rarely a tree to 6 m.; leaves clustered, linear-spathulate, pubescent with simple and capitellate hairs, 1–3·5 cm. long, up to about 7 mm. broad; flowers solitary or in short axillary racemes; sepals 6–10 mm. long; petals shorter than the sepals, yellow; ripe carpels 2–5, pubescent, 5 mm. long.

Frequent on coastal limestone rocks and in sandy places near the sea, also on the cays; SL–20 feet; fl. and fr. Sept, Dec–June. *A 8030! H 9526! P 23422!* General in the tropics but not in W. Africa; extends in the Caribbean area southwards to the Grenadines, Aruba and Curaçao; Grand Cayman.

2. CASTELA Turp. (1806) nom. cons.

1. C. macrophylla Urb., Symb. Ant. **5**: 377 (1908).

Sparingly spiny shrub 60 cm.–3 m. high, with spreading or trailing branches; leaves mostly elliptical, apiculate, rigid, reticulate-veined, glabrous above, minutely and thinly puberulous beneath, 2–6 cm. long, 1–2·5 cm. broad; flowers clustered; calyx 0·7 mm. long; petals rosy-pink, about 3 mm. long; drupelets bright scarlet, ellipsoidal-compressed, shortly beaked, 1–1·5 cm. long.

Occasional in thickets and woodlands on arid rocky limestone; 25–900 feet; fl. and fr. July, Nov–Apr. *A 10770! M. L. Farr! H 9347! P 9556!* Endemic.

3. PICRASMA Blume (1825)

1. P. excelsa (Sw.) Planch. in Hook., Lond. Journ. Bot. **5**: 574 (1846).—*Quassia excelsa* Sw. (1788). *Picraena excelsa* (Sw.) Lindl. (1838). Bitterwood, Jamaican Quassia.

Tree 6–25 m. high; lateral leaflets in (4–) 5 (–6) pairs, narrowly elliptical, acuminate, unequally cuneate at base, glabrescent above, puberulous on the veins beneath, 5–13 cm. long, up to 4·5 cm. broad; flowers small, fragrant, numerous in axillary corymbose panicles; petals of male flower about 2 mm. long, of hermaphrodite flower about 3 mm. long; ripe drupelets bluish-black, 6–7 mm. long.

Locally common in hill pastures, relict woodlands and along roadsides; 300–2750 feet; fl. Aug–Dec, fr. Feb–Mar, July–Oct. *A 8359! H 7017! 9984! P 21438! 22861!* Greater Antilles.

4. ALVARADOA Liebm. (1854); Cronquist (1944)

1 Leaflets strigose-pubescent beneath; petals filiform; capsule lanceolate to ovate-lanceolate in outline, 8–18 mm. long, with ciliate margins; Grand Cayman
A. amorphoides Liebm.
1 Leaflets glabrous; capsules glabrous or nearly so:
  2 Capsule ovate-lanceolate in outline, broadest at the middle, cuneate at base, 12–18 mm. long   1. *lewisii*
  2 Capsule broadly ovate in outline, broadest at or below the middle, cuneate to sub-truncate-rounded at base, 10–12 mm. long   2. *jamaicensis*

1. A. lewisii Howard & Proctor in Journ. Arn. Arb. **39**: 102 (1958).

Shrub 4 m. high; leaflets elliptical or oblong, rounded or retuse at apex, narrowed or rounded at base, 1·5–2·5 cm. long 0·7–1 cm. broad; racemes pendulous, the rachis yellowish-puberulous; pedicels 9–11 mm. long in fruit; immature capsule reddish, yellow when ripe.

Very rare (Trel.), in limestone ravine; about 1500 feet; fr. Sept. *H & P 14128!* Endemic.

2. A. jamaicensis Benth., Pl. Hartw.: 344 (1857).

Shrub 1–5 m. or tree to 8 m. high; leaflets oblong or oblong-elliptical, rounded and emarginate at apex, abruptly narrowed at base, dark green above, pale beneath, 1–3·5 cm. long, up to 1 cm. broad; male flowers on pedicels up to 2 mm. long with calyx 2 mm. long, shaggy-pubescent, triangular-lobed to halfway; petals absent; stamens 5 mm. long, pubescent at base, pink; ripe fruit scarlet, glabrous except the proximal margins shortly ciliate, on pedicels up to 12 mm. long.

Rare and local, in woodlands and thickets on rocky limestone in the central and north-western parishes; 1200–2650 feet; fl. Mar–June, fr. June–Feb. *A 10795! 11691! H 8982! P 10021! 26685!* Endemic.

5. SIMAROUBA Aubl. (1775) nom. cons.

1. S. glauca DC. in Ann. Mus. Hist. Nat. Paris **17**: 424 (1811).—Bitter Damson.

Tree 6–8 (–16) m. high; leaflets 5–15 (–19), oblong-elliptical to oblong, rounded and sometimes emarginate to obtuse and shortly pointed at tip, narrowed at base, paler and puberulous or glabrescent beneath, 4–10 cm. long and up to 3·5 cm. broad; calyx-lobes ciliate or glabrous; petals 5–6 mm. long, yellow, glabrous; filaments glabrous; ovary hairy; ripe fruits dull blue or purplish, about 1·5 cm. long.

Common in woodlands on limestone; SL–2000 feet; fl. Mar–Apr, fr. May–July. *A 6811! Asprey UCWI 2433! H 9358! H & P 14223!* Bahamas, Florida, Mexico to Panama, Cuba, Hispaniola.

6. PICRAMNIA Sw. (1788) nom. cons.

Flowers 3-merous   1. *antidesma*
Flowers 5-merous   2. *pentandra*

1. P. antidesma Sw., Nov. Gen. & Sp. Pl.: 27 (1788).—Macary Bitter, Majoe Bitter.

Shrub 2–3 m. or tree to 6 m. high; leaflets 5–9 (–11), ovate to lanceolate, acuminate, broadly rounded to unequally cuneate at base, glabrous or puberulous on midrib and lamina beneath, 2·5–14 cm. long, 1·5–6 cm. broad, larger towards tip of leaf; inflorescences up to 26 cm. long; flowers very small, greenish to yellow; in male flowers sepals (2–) 3, acute; stamens (2–) 3 (–4); petals 2–3; fruiting pedicels 5–20 mm. long; berries scarlet then finally black, 12–15 mm. long.

Very common, on well drained slopes and in woodland on limestone; 50–4000 feet; fl. May, Sept–Dec, fr. Sept–June. *A 5894! 8552! H 8855! 9770! P 11068! 11236!* C. Amer., Cuba (?), Hispaniola.

2. P. pentandra Sw., Fl. Ind. Occ. 1: 220 (1797).

Tree 4–6 m. high; leaflets 5–9, elliptical to lanceolate, acuminate, unequal at base, glabrous, 3–12 cm. long; calyx about 2 mm. long, the lobes acute; stamens 5, about 3 mm. long; petals 5, linear to linear-lanceolate; berries reddish-brown, 10–13 mm. long.

Reported for Jamaica but not conclusively confirmed; specimens of recent collections identified as this species are without flowers. Bahamas, Florida, Colombia, Venezuela, West Indies.

105. BURSERACEAE

Trees or shrubs with resinous bark. Leaves alternate, rarely opposite, mostly imparipinnately compound, rarely simple, usually aromatic but without pellucid glands, often deciduous, without stipules. Flowers bisexual or unisexual and then polygamo-dioecious, small, actinomorphic, solitary or in clusters, racemes or panicles. Perianth biseriate, rarely corolla absent, 3–5-merous; sepals more or less connate; petals usually free, valvate or imbricate. Stamens the same number as or twice as many as the petals, inserted on a disk; filaments free; anthers 2-locular, opening lengthwise. Ovary superior, 2–5-locular; ovules (1–) 2, collateral and axile in each loculus; style simple or absent; stigmas usually as many as carpels. Staminate and pistillate flowers with pistillode or staminodes respectively. Fruit drupaceous, the pyrenes adherent or tardily separating. Seeds without endosperm; embryo straight or curved.

About 20 genera with 600 species in the tropics.

Ovary 3-locular; stigma 3-lobed; fruit usually with a solitary 1-seeded pyrene; leaves imparipinnate or simple   **1. Bursera**
Ovary 4–5-locular; stigma 5-lobed; fruit with 1–5 1-seeded pyrenes tardily separating; leaves imparipinnate   **2. Protium**

1. BURSERA Jacq. ex L. (1762) nom. cons.

1 Leaves simple   1. *lunanii*
1 Leaves pinnate:
  2 Petiole, leaf-rachis and inflorescence-branches glabrous   2. *aromatica*

2 Petiole, leaf-rachis and inflorescence-branches pubescent:
  3 Leaflets thin; bark reddish, peeling in papery layers   3. *simaruba*
  3 Leaflets leathery; bark grey, not peeling in papery layers   4. *hollickii*

1. **B. lunanii** (Spreng.) Adams & Dandy, unpublished.—*Amyris lunanii* Spreng.
(1825). *B. simplicifolia* DC. (1825). Black Birch.

Tree 5–13 m. high; bark rough, scaly, resinous, greyish; leaves borne usually
towards the ends of upcurved brittle branches, elliptical to obovate, obtuse or
rounded at tip, mostly cuneate at base, rather leathery, glabrous, 3–7 cm. long,
1·5–4·5 cm. broad; inflorescences terminal and axillary, subracemose few-flowered
panicles up to 7 cm. long; flowers greenish-yellow to white, fragrant, male 4—5-
merous with 8 or 10 stamens, hermaphrodite 3–5-merous with 6 stamens; fruit
8–9 mm. long.

Locally common in thickets and open woodlands on arid limestone near the sea;
SL–900 feet; fl. Jan–Apr, July–Sept, fr. Feb–Sept. *A 6201! H 10166! P 9972!*
Endemic.

2. **B. aromatica** Proctor in Bull. Inst. Jam., Sci. ser. **16**: 16, t. 5 (1967).—Siboney.

Shrub 5 m. or tree 8–13 m. high; bark grey; leaves 3–5-foliolate; leaflets ovate,
acuminate, 6–10 cm. long, 3–5 (–7) cm. broad; inflorescences terminal and axillary,
up to about 8 cm. long; flowers greenish-cream; fruit slightly oblique, shortly
beaked, 9–11 mm. long.

Rare and local (Han., St. James, Trel.), in hillside woodland on limestone;
1500–2200 feet; fl. Jan–May, fr. Mar–Sept. *P 22597! 23457!* Endemic.

3. **B. simaruba** (L.) Sarg., Gard. & For. **3**: 260 (1890).—*Pistacia simaruba* L.
(1753). Red Birch.

Tree 5–15 (–25) m. high; bark smooth or sometimes mealy; underbark green;
leaves (1–) 3–9 (–13)-foliolate; leaflets ovate to elliptic-lanceolate, long- or short-
acuminate, rounded to cuneate (terminal leaflet) to subcordate (proximal leaflets) at
base, 5–10 cm. long, 2–4·5 cm. broad, often pubescent at least on the midrib
abaxially; inflorescences racemose or paniculate; perianth mostly trimerous,
sometimes tetramerous; petals greenish tinged crimson; stamens 6 (–8); fruit
1–1·3 cm. long, reddish-brown, shiny; pyrene trigonal.

Common in woodland on limestone mostly in arid areas near the sea and on the
cays; SL–2300 feet; fl. Mar–Apr, Sept, fr. Jan–Aug, Nov. *A 12363! H 6768!
P 7792!* Florida, Mexico to Venezuela and the Guianas, West Indies; Grand
Cayman.

4. **B. hollickii** (Britton) F. & R., Fl. Jam. **4**: 207 (1920).—*Terebinthus hollickii*
Britton (1908).

Tree 5–6 m. high; bark rather smooth, not peeling, mottled grey and buff;
underbark green; leaves 3–5 (–7)-foliolate; leaflets broadly elliptical to obovate-
elliptical, shortly acuminate, rounded to cuneate at base, 2·5–6 cm. long, 1·5–4·5
cm. broad, shiny adaxially, paler beneath, glabrous; inflorescence racemose, 4–7
cm. long; fruit 8–9 mm. long.

Very local (St. Cath.), in thicket on arid rocky limestone hill; about 400 feet;
fr. Mar. *A 12800! H 10149!* Endemic.

2. **PROTIUM** Burm. f. (1768) nom. cons.

1. **P. attenuatum** (Rose) Urb., Symb. Ant. **7**: 240 (1912).—*Icica attenuata* Rose
(1911).

Tree to 15 (–20) m. high; bark smooth, mealy, grey; leaves (3–) 5–7 (–11)-
foliolate; leaflets ovate to oblong-elliptical, acuminate, the lateral unequal-sided at
base, 7–13 (–18) cm. long, 3–5·5 cm. broad, glabrous; panicle 3–5 cm. long,
glabrous; flowers 4–5-merous; petals about 3 mm. long, green; fruit up to about
2 cm. long.

The Jamaican record is based on only one early collection; *W. Wright* (before
1786). Guadeloupe to St. Lucia, ? St. Vincent.

106. MELIACEAE

Shrubs or trees, often with scented wood. Leaves usually alternate, mostly pinnately compound, without stipules. Flowers mostly bisexual, actinomorphic, usually small in cymose panicles, often fragrant. Perianth biseriate; calyx 4–5 (–6)-lobed, imbricate, valvate or open in bud; petals (3–) 4–5 (–8), free or connate, imbricate or contorted or, if adnate to staminal tube, valvate. Stamens (5–) 8–10 or rarely numerous, often monadelphous with a disk more or less developed as a gynophore between the staminal tube and ovary; anthers 2-locular, opening lengthwise. Ovary superior, 2–5-locular; ovules (1–) 2 (–12), pendulous, anatropous, on axile placentas; style solitary or stigma sessile and capitate or discoid. Fruit a berry, capsule or rarely a drupe. Seeds with or without endosperm, sometimes winged.

50 genera with about 1000 species mostly in the tropics.

1 Leaves mostly bipinnate; leaflets serrate; fruit a drupe; corolla mauve, the staminal tube much darker; petals imbricate   1. **Melia**
1 Leaves pinnate or digitate:
  2 Fruit a drupe; leaves mostly imparipinnate with a small terminal leaflet; leaflets very unequal at base, serrate-dentate, about 8 pairs; flowers creamy-white; sparingly cultivated; native of India; Neem   *Azadirachta indica* A. Juss.
  2 Fruit a capsule; leaflets entire:
    3 Seeds winged; ovules numerous; capsule woody; leaves paripinnate:
      4 Stamen-filaments free above the gynophore to which keels of petals also attached; petals imbricate; leaflets usually with domatia in vein-axils beneath; capsule opening from apex   2. **Cedrela**
      4 Stamen-filaments fused into a 10-toothed staminal tube; petals free, contorted; leaflets without domatia; capsule opening from base   3. **Swietenia**
    3 Seeds not winged, arillate; ovules few; capsule more or less leathery; petals imbricate or valvate:
      5 Leaves imparipinnate or digitate; capsule 2–3 (–4)-valved; inflorescence paniculate   4. **Trichilia**
      5 Terminal leaflet lacking and often replaced by an undeveloped bud; capsule 4-valved; inflorescence subracemose, simple or very sparingly or shortly branched   5. **Guarea**

1. MELIA L. (1753)

1. M. azedarach L., Sp. Pl. **1**: 384 (1753).—China Berry, Persian Lilac, West Indian Lilac.

Tree 2–6 m. high with slender straggling branches; young shoots, inflorescence and calyx with rusty plumose and stellate hairs; leaflets lanceolate, glabrous, 2–7 cm. long; panicles axillary; sepals ovate, acute, 1·5–2 mm. long, open in bud; petals (4–) 5 (–6), oblong-oblanceolate, 11 mm. long, with dark stalked glands outside; staminal tube 8 mm. long, blackish-violet, ascending-hairy within; drupe ellipsoidal, yellow when ripe, about 1·5 cm. long.

Common in cultivation and in waste places; SL–1000 feet; fl. and fr. almost all the year. *A 9805! H 8372! A. von der Porten!* Native of tropical Asia, now widespread in the tropics, subtropics and S. Europe.

2. CEDRELA P. Browne (1756); C. E. Smith (1960)

Leaflets strongly oblique at base, often glabrous; calyx usually glabrous, irregularly dentate; petals thin, uniformly light in colour; column in capsule with wings extending to base of broad apex   1. *odorata*
Leaflets slightly oblique at base, puberulous to pubescent, particularly along veins beneath; calyx puberulous, generally regularly 5-dentate; petals moderately thick, often darker in colour at the apical margin; column in capsule with wings extending to base of narrow apex; rare cultivated tree; native Mexico to Argentina   *C. angustifolia* DC.

1. C. odorata L., Syst. Nat. ed. 10, **2**: 940 (1759).—*C. mexicana* M. J. Roem. (1846). West Indian Cedar.

Tree up to 20 m. high; leaves up to 80 cm. long, with (5–) 6–7 (–14) pairs of leaflets; leaflets ovate to lanceolate, acute to rounded at base, acute, acuminate or obtuse at tip, 8–20 cm. long, 2·5–5·5 (–8) cm. broad, generally glabrous; flowers with a heavy malty odour, 6–9 mm. long; petals greenish-cream in bud, opening white; fruit 2·5–4·5 cm. long, septicidally 5-valved; seeds flat, chestnut-brown, about 25 mm. long and 6–7 mm. broad.

Common in places where probably planted, especially in pastures and along roadsides; 100–3360 (–4000) feet; fl. Dec–Feb, fr. mostly Jan–Apr. *A 10262! H 11705! P 8607! 21867!* Mexico to Brazil and Peru, West Indies, introduced into many parts of the Old World tropics.

3. SWIETENIA Jacq. (1760)

Leaflets 2·5–8 cm. long, 1·5–3·5 cm. broad, in 2–5 pairs; capsule 6–10 cm. long, 4–6 cm. broad   1. *mahagoni*
Leaflets 7·5–15 cm. long, 2·5–7 cm. broad, in 4–6 pairs; capsule 12–15 cm. long, about 7 cm. broad, cultivated tree; native of C. Amer.; Honduras Mahogany
     S. macrophylla G. King

1. S. mahagoni (L.) Jacq., Enum. Syst. Pl. Carib.: 20 (1760).—*Cedrela mahagoni* L. (1759). West Indian Mahogany.

Deciduous tree up to 20 m. high and 5 m. girth, with greyish or brown flaking bark; leaves up to 30 cm. long; leaflets ovate to lanceolate, strongly unequal-sided at base, acuminate at tip, glabrous; panicles up to 15 cm. long; flowers fragrant; petals obovate-elliptical, greenish-yellow to cream-coloured, 3–4 mm. long; capsule ovoid; seeds 5–6 cm. long, the reddish-brown wing about twice as long as body.

Occasional in pastures and along roadsides, rare in natural woodland and then mainly in the central and western parishes; (SL–) 500–1500 (–3000) feet; fl. May–July, fr. Nov–Jan. *H 8636! P 26433!* Florida, C. Amer. to Peru, Bahamas, West Indies, Cayman Is., introduced into W. Africa and elsewhere in the tropics.

4. TRICHILIA P. Browne (1756) nom. cons.

1 Leaves digitately compound, leaflets (2–) 3 (–5), oblong-obovate to oblanceolate, rather leathery, conspicuously reticulate-veined   1. *reticulata*
1 Leaves pinnately compound; leaflets (3–) 5–15 (–21) and otherwise not as above:
   2 Leaflets 9–15 (–21), rarely fewer, opposite or alternate, thinly hairy beneath on midrib and veins; panicles from leaf-axils on the younger branches, long-peduncled
     2. *hirta*
   2 Leaflets up to 9:
    3 Leaflets alternate on rachis, mostly 7 or 9, pubescent on midrib and puberulous on lamina beneath, glabrescent; panicles terminal and axillary   3. *moschata*
    3 Leaflets mostly opposite:
     4 Inflorescence terminal and subterminal; leaflets mostly 5–9, glabrous except for hairs sometimes in axils of veins beneath   4. *glabra*
     4 Inflorescence axillary, the flowers congested, corymbiform; leaflets (2–) 3–7 (–9), glabrous   5. *havanensis*

1. T. reticulata P. Wilson in N. Amer. Fl. **25** (4): 278 (1924).—*T. polyneura* Urb. (1908), not C. DC. (1905). *T. rendlei* Urb. (1924).

Shrub or tree with slender trunk up to 6 m. high; leaflets cuneate at base, mostly rounded or obtuse at tip, glabrous, 3–8 cm. long, 1–3 cm. broad; panicles axillary, racemelike with very short branches, up to 4·5 cm. long; petals 4 or 5, valvate or imbricate, 3–4 mm. long, creamy-white; fruit 13–16 mm. long, velvety-pubescent with brownish hairs.

Widely scattered but nowhere common, in woodland on well drained limestone; near SL–1800 feet; fl. Nov–Feb, July–Aug, fr. Jan, May–July. *A 10205! H 8380! 8860! HPS 14667!* Endemic.

2. T. hirta L., Syst. Nat. ed. 10, **2**: 1020 (1759).—Wild Mahogany.
Tree up to about 12 m. high; leaflets lanceolate, cuneate to rounded at base,
acute or acuminate at apex, 3–9 (–13) cm. long, 1·5–3 (–4) cm. broad; pedicels
jointed below the middle; petals 5, light green to cream-coloured, glabrous, 5–6
mm. long; staminal tube paler; capsule (2–) 3-valved, velutinous, 10–15 mm.
broad; seeds 6–8 mm. long, with light scarlet aril.
Locally common, especially in the southern parishes, in thickets and along
roadsides on alluvial gravel; 25–1600 feet; fl. Apr–July, Oct–Dec, fr. Dec–July.
A 5494! H 7693! J.P. 1303! P 15436! Mexico to Brazil, Greater Antilles, Virgin
Is., Grenada.

3. T. moschata Sw., Nov. Gen. & Sp. Pl.: 67 (1788).—*T. harrisii* Britton (1908).
Muskwood.
Tree up to 12 m. or more high; leaflets elliptical to oblong-elliptical, acute at
base, acuminate at tip, up to 17 cm. long and 7·5 cm. broad; petals 4 or 5, greenish-
yellow, puberulous, about 3 mm. long; staminal tube glabrous; capsule 3–4-
valved, densely velutinous, ovoid to subglobose, 15–20 mm. long; seed 8–12 mm.
long.
Widely distributed in thickets and woodlands on limestone; 20–2500 feet; fl.
Sept–Apr, fr. Mar–July. *A 12465! H 5637! 10230! P 23821! 26848!* Guatemala.

4. T. glabra L., Syst. Nat. ed. 10, **2**: 1020 (1759).
Tree up to about 12 m. high; trunk slender; leaflets ovate to oblong-elliptical,
acute to rounded at base, acuminate at apex, 4–8 (–10) cm. long, 2–3·5 (–5) cm.
broad; petals 4 or 5, greenish-white, minutely tomentose, 7–8 mm. long; staminal
tube tomentose; capsule (2–) 3-valved, tomentose, globose, the valves ovate, 10–18
mm. broad; seeds 7–8 mm. broad, black with scarlet aril.
Frequent in woodlands on limestone hills; mostly between 800 and 2500 feet,
with reports from near 5000 feet unconfirmed; fl. Dec–Apr, fr. June–Aug. *A 12485!
H 9592! 10332! H & P 14457! P 26322!* Cayman Is.

5. T. havanensis Jacq., Enum. Syst. Pl. Carib.: 20 (1760).
Shrub to 5 m. or tree up to 11 (–15) m. high; leaf-rachis often narrowly winged;
leaflets narrowly obovate to oblong-elliptical, broadly or narrowly cuneate at base,
rounded to acute or acuminate at apex, 2–9 (–13) cm. long, 1–4 (–4·5) cm. broad;
panicles up to about 2·5 cm. long; pedicels jointed at base; petals 4 or 5, yellow or
rarely light mauve, glabrous or nearly so, 3·5–5 mm. long; staminal tube villous
within, glabrescent outside; capsule (2–) 3-valved, glabrous, ovoid to subglobose,
9–13 mm. long; seeds 5–6 mm. long, with red aril.
Occasional in woodlands and woodland margins on limestone or shale; 500–
2750 feet; fl. Nov–Mar, July–Aug, fr. Jan–Feb, July–Sept. *A 8803! H 8207!
H & P 14966! J. P. 1134! P 9862!* Mexico to S. Amer., Cuba, Hispaniola.

5. GUAREA Allam. ex L. (1771) nom. cons.

Leaflets ovate to elliptical, 4–17 (–20) cm. long, 3–8 cm. broad, shortly acuminate; inflorescence with occasional 2–5-flowered branches, up to 25 cm. long   1. *swartzii*
Leaflets obovate to elliptical, 4–8 cm. long, 2–3·5 cm. broad, obtuse; inflorescence with
occasional 3-flowered branch, up to 3·5 cm. long   2. *jamaicensis*

1. G. swartzii DC., Prodr. **1**: 624 (1824).—*G. glabra* of C. DC. (1878) and F. & R.
(1920), partly, not Vahl (1807). Alligator Wood.
Tree 5–16 (–20) m. high; leaflets mostly opposite, 2–5 pairs, perhaps more
numerous on saplings, cuneate at base, glabrous above, hirsute beneath in the
vein-axils, with crooked pellucid lines in the lamina; pedicels up to 2 mm. long.
often shorter than stipelike base of calyx; petals 4 or 5, white, puberulous, 5–6 mm.
long; staminal tube glabrous or nearly so; capsule subglobose or obscurely 4-
angled, about 2 cm. broad; seeds (3–) 4 (–5), angled inwards, with broad white hilum,
10–15 mm. long; aril orange-scarlet.
Frequent in mostly rather wet and sheltered woodlands on limestone or shale;
500–5600 feet; fl. Aug–Nov, fr. most of the year. *A 7384! 11565! H 5384! 7040!
P 8444! 24023!* Endemic.

2. G. jamaicensis Proctor in Bull, Inst. Jam., Sci. ser. **16**: 19, t. 6 (1967).

Tree about 8 m. high; leaflets opposite, 1–3 pairs, broadly cuneate at base, glabrous, rather leathery; pedicels about 1·5 mm. long; petals 4, light buff, silky-puberulous outside, about 6 mm. long; staminal tube glabrous; fruit subglobose, woody, about 1·5 cm. broad.

Rare and local (St. Andr., St. James), on limestone hills; 2000–2500 feet; fl. Jan, fr. Feb. *P 9870! 23188!* Endemic.

107. MALPIGHIACEAE

Trees or erect or climbing shrubs. Leaves usually opposite, simple and usually entire, sometimes with pungent hairs especially beneath, often with glands on petiole or lamina; stipules present or absent: Flowers mostly bisexual, actino-morphic or often slightly zygomorphic, in usually bracteate clusters, umbelliform cymes, racemes or panicles. Perianth biseriate; sepals 5, mostly free, imbricate or rarely valvate, often with paired glands; petals 5, free, often convolute, clawed, usually unequal and often toothed or fringed. Stamens 10 in 2 whorls, some some-times reduced to staminodes, hypogynous, often basally connate; anthers 2-locular, opening lengthwise. Ovary superior, (2–) 3 (–5)-locular and -lobed; ovules 1 in each loculus, semianatropous, pendulous from axile placentas; styles as many as carpels, free or rarely connate; stigmas entire or lobed, often oblique. Fruit a samara, schizocarp, capsule, berry or drupe. Seeds with a straight or curved embryo; endosperm lacking.

About 60 genera with some 850 species mostly in the American tropics.

1 Erect shrubs or trees:
  2 Fruit schizocarpic; leaves with a gland on each margin near the base; inflorescence racemose; flowers fragrant; calyx without glands; petals yellow; filaments glabrous; styles free; cultivated ornamental, sparsely naturalized; native of Mexico; Galphimia, Shower of Gold (*Thryallis gracilis* (Bartl.) Kuntze)   *Galphimia gracilis* Bartl.
  2 Fruit drupaceous with 1–3 pyrenes; calyx with glands:
    3 Filaments hairy at base; styles free, subulate-tipped; fruit with a single pyrene; inflorescence racemose; leaves without glands, with intrapetiolar stipules
                                       1. Byrsonima
    3 Filaments glabrous; stigmas capitate; fruit with 1–3 pyrenes, usually red:
      4 Styles free; leaves mostly without glands; inflorescence corymbose or umbelliform; petals pink, purple or white; pyrenes crested   **2. Malpighia**
      4 Styles united; leaves usually with 2 glands on lower surface; inflorescence racemose; petals yellow; pyrenes smooth or reticulate-sculptured   **3. Bunchosia**
1 Woody climbers; carpels of fruit separating, samaroid:
  5 Fruiting carpels with a single dorsal wing or crest:
    6 Flowers in racemes or panicles; wing of samara thickened on lower margin
                                       4. Heteropteris
    6 Flowers in corymbs or subumbellate:
      7 Fruiting carpels with a short thick triangular crest   **5. Brachypterys**
      7 Fruiting carpels with an oblong-obovate wing, 1·5 cm. or more long, the thickened upper margin lobed adaxially at its base   **6. Stigmaphyllon**
  5 Fruiting carpels laterally as well as dorsally winged or crested:
    8 Petals pink or mauve; spreading wing of samara 3-lobed, Y-shaped
                                       7. Triopteris
    8 Petals yellow:
      9 Samara with 4 lateral wings arranged in a X-shape; wings oblong
                                       8. Tetrapteris
      9 Samara with 2 lateral wings; wings semicircular or obversely deltate
                                       9. Mascagnia

1. BYRSONIMA L. C. Rich. ex Juss. (1811)

Petals yellow; leaves usually tomentose beneath but varying to nearly glabrous
                                                1. *coriacea*

Petals white to pink or purplish; leaves more or less glabrous   2. *trinitensis*

1. B. coriacea (Sw.) DC., Prodr. **1**: 580 (1824).—Hogberry, Locust-berry Tree.
Tree 3–16 m. high; trunk often slender; crown spreading; young leaves on both surfaces silvery- or ochraceous-tomentose; leaf-blades elliptical, oblong-elliptical to lanceolate, sometimes obovate, cuneate at base, obtuse, acute, acuminate or cuspidate at tip, 5–17 cm. long, 2–8·5 cm. broad, variably brownish-tomentose beneath, usually becoming dark green and glossy adaxially; racemes up to 14 cm. long; pedicels up to 13 mm. long; sepals ovate, about 4 mm. long; petals including claw 8–9 mm. long, fading reddish or orange-red; fruit subglobose, ripening brownish-yellow, about 1 cm. in diameter.
Common in secondary woodlands; 200–3000 feet; fl. Apr–Oct, fr. July–Feb. *A 7817! 7973! H 8411! 9093! H & P 13922! P 19687!* Northern S. Amer., West Indies.

2. B. trinitensis A. Juss. in Ann. Sci. Nat. sér. 2, **13**: 334 (1840).—*B. glaberrima* Niedenzu (1901). *B. bracteata* F. & R. (1917). *B. craigiana* F. & R. (1917). *B. smallii* F. & R. (1917).
Shrub 2 m. or tree up to 13 m. high; leaf-blades oblong, elliptical or obovate, cuneate at base, rounded, obtuse or subacute at tip, 2–9 cm. long, 1·5–4·2 (–5·5) cm. broad, rusty-puberulous at least at first, becoming glossy adaxially; racemes 3–8 (–11) cm. long, rachis and pedicels brown-tomentose at first; pedicels up to 13 mm. long; sepals ovate, 4–5 mm. long, the glands about half as long; petals including claw 8–9 mm. long, darkening on drying, the mid-petal often acquiring orange tints; fruit globose, up to about 1 cm. in diameter.
Local and uncommon (Clar., Manch., St. James, Trel.), in woodlands on rocky limestone; 1500–2500 feet; fl. and fr. May–Dec. *A 12609! H 10595! P 11397! Webster & Proctor 5417!* Dominica, Martinique, Trinidad. Several minor poorly understood variants are included here.

2. MALPIGHIA L. (1753)

1 Leaf-margin sinuate-dentate with spines; leaf-blade rounded, mostly less than 2 cm. long, glossy adaxially; petals pink or white; probably only in cultivation at the present time; native of West Indies   *M. coccigera* L.
1 Leaf-margin entire, not spiny:
   2 Leaves glabrous or usually so, membranous; petals not or only narrowly keeled, thus unopened corolla smoothly rounded or at most shallowly angled; drupe oblate to subglobose:
     3 Leaves acute or acuminate at tip; inflorescence mostly axillary, pedunculate   1. *glabra*
     3 Leaves obtuse or rounded at tip:
       4 Inflorescences axillary and on older twigs or cauliflorous; petals at least crenate   2. *punicifolia*
       4 Inflorescence terminal; petals subentire   3. *obtusifolia*
   2 Leaves with large pungent medifixed hairs mostly on abaxial surface at least when young; petals keeled, thus unopened corolla winged:
     5 Keel of petals about 1 mm. broad, hooked at free tip; leaves elliptic-lanceolate; drupe subglobose but slightly narrowed to flat tip; inflorescence subsessile   4. *incana*
     5 Keel of petals about 0·5 mm. broad, decurrent at tip:
       6 Leaves oblong to broadly elliptical, firmly papery to subcoriaceous:
         7 Primary branches of inflorescence evident, each raceme slender-peduncled   5. *harrisii*
         7 Primary branches of inflorescence obsolete or very short and massive   6. *fucata*
       6 Leaves narrower, elliptical to elliptic-lanceolate, thin; inflorescence subsessile or primary branches less than 1 cm. long:
         8 Leaf-tip mostly long-acute to acuminate; drupe somewhat pyramidal   7. *biflora*
         8 Leaf-tip acute or obtuse; drupe subglobose   8. *urens*

1. M. glabra L., Sp. Pl. **1**: 425 (1753).—Wild Cherry.
Shrub 0·6–3 m. or tree to 6 m. high with numerous often ascending branches; leaves ovate, elliptical or lanceolate, sometimes rhombic, narrowly cuneate to rounded at base, up to 8 (–9) cm. long and 4 (–5) cm. broad, essentially glabrous but often with patches of erinosity especially beneath, probably as a response to insect infestation; peduncles 2–15 mm. long; glands on sepals about half as long; petals about 8 mm. long; stigmas bilateral, each lobe truncate; fruit about 1 cm. in diameter.

Very common in rough pastures, thickets and on rocky ground; SL–3000 feet; fl. and fr. most of the year. *A 8208! 9903! H 8848! 9646! Powell 757! 1016!* Continental tropical Amer. south to Venezuela, Cuba, Hispaniola, Virgin Is.

2. **M. punicifolia** L., Sp. Pl. ed. 2, **1**: 609 (1762).—Barbados or West Indian Cherry.

Shrub to 5 m. or tree to 12 m. high; leaves thin, up to about 8 cm. long; inflorescences sessile or shortly pedunculate; fruit globose to depressed-globose, up to about 2·5 cm. broad in cultivars.

Uncommon in the wild state, common in cultivation at low elevation; fl. and fr. sporadically, mostly during wet weather. *A 8039! Asprey UCWI 2161! 2169! C. B. Lewis!* The wild variants occurring in thickets on limestone in Jamaica tend towards arborescence and cauliflory; when taxonomic criteria in this genus are better understood these variants may prove to be separable. Cultivated variants are grown throughout the West Indies and have been introduced into tropical Asia and Africa.

3. **M. obtusifolia** Proctor in Bull. Inst. Jam., Sci. ser. **16**: 20, t. 7 (1967).

Dense-crowned shrub or tree up to 4 m. high; leaves subsessile, elliptical to obovate, broadly cuneate to rounded at base, sometimes shortly emarginate at tip, 3–5 cm. long, 1·5–3·5 cm. broad; peduncles 5–15 mm. long; secondary branches up to 10 mm. long; pedicels up to 15 mm. long; sepals about 2 mm. long.

Rare (Clar., St. James, Trel.), on cliffs and in woodlands on limestone; 1600–2500 feet; fl. June–Aug. *P 22571!* Endemic.

4. **M. incana** Mill., Gard. Dict. ed. 8 (1768).

Shrub 1·2–2 m. or tree to 5 m. high; leaves elliptic-lanceolate to lanceolate, cuneate to rounded at base, mostly acute at tip, firmly papery, 3–8 cm. long, 1–2·5 (–3) cm. broad; peduncles mostly less than 5 mm. long; pedicels up to 15 mm. long; fruit up to 12 (–15) mm. broad.

Local (St. Andr., St. Cath., St. Thom.), in thickets on gravel and arid limestone; 25–300 feet; fl. July, Oct–Dec, fr. July–Feb. *A 9777! H 10048! Loveless UCWI 2377! P 23935!* Mexico, Cuba.

5. **M. harrisii** Small in Torreya **13**: 77 (1913).

Slender-branched shrub or tree up to 5 m. high; leaves mostly broadly elliptical and rounded at base and apex, 7–12 cm. long, 4–6·5 cm. broad, the abaxial pungent hairs caducous; racemes clustered, the peduncles several together, up to 15 mm. long; pedicels jointed; sepals about 3 mm. long; petals 12–13 mm. long; fruit not known.

Rare (Clar., St. Ann), in woodland on limestone; about 2500 feet; fl. Aug–Oct. *H 11189! Webster & Proctor 5402!* Endemic.

6. **M. fucata** Ker-Gawl. in Edw., Bot. Regist. **3**: t. 189 (1817).

Shrub 2–3 m. or tree up to 6 m. high; branches glabrescent, sometimes densely hairy when young; leaves as above, up to 13 (–21) cm. long and 7·5 (–11) cm. broad; pedicels up to 2·5 cm. long; sepals 3–4·5 mm. long; petals 6–7 mm. long; fruit subglobose, about 7 mm. long.

Local (Manch., Trel., St. Ann), in woodland on limestone; 1500–3000 feet; fl. Mar, June, Nov, fr. Jan, July. *A 11235! 12744! H & P 14326! P 9908! 23111!* Puerto Rico.

7. **M. biflora** Poir. in Lam., Encycl. Méth. Bot. **4**: 326 (1797).—Cowitch Cherry.

Shrub 0·6–3 m. high, usually erect, sometimes scrambling; leaves narrowly ovate to lanceolate, cuneate to rounded at base, up to 8 (–12) cm. long and 2·5 (–4) cm. broad, the pungent hairs orange-yellow and up to 6 mm. long; peduncles 4–10 (–15) mm. long; pedicels up to 20 mm. long; sepals 2·5 mm. long; larger petal up to 12 mm. long; fruit mostly less than 1 cm. broad.

Common in thickets and woodlands on limestone; SL–3000 (–5000) feet; fl. and fr. all the year. *A 5663! 11250! H 12676! C. B. Lewis!* Cuba, Hispaniola.

8. M. urens L., Sp. Pl. **1**: 426 (1753).

Shrub like the last. This species, as diagnosed by Fawcett & Rendle (1920), has not been matched by recent collections. It is certain that the specimens cited by those authors do not refer to the same taxon. Niedenzu (Pflanzenr. 4 (141): 628 (1928)) has referred specimens cited by Fawcett & Rendle under *M. urens* and *M. biflora* to *M. martinicensis* Jacq. var. *jamaicensis* Urb. & Niedenzu (1899), but these are probably all *M. incana*.

3. BUNCHOSIA L. C. Rich. ex Juss. (1811)

Leaves mostly narrowly elliptical or rhombic-elliptical, cuneate at base, up to 10 cm long and 5 cm. broad; drupe less than 10 mm. long and broad   **1.** *media*
Leaves broadly elliptical, broadly cuneate to rounded at base, up to 18 cm. long and 10 cm. broad; mature drupe more than 10 mm. long and broad   **2.** *jamaicensis*

1. B. media (Ait. f.) DC., Prodr. **1**: 581 (1824).—*Malpighia media* Ait. f. (1811). *B. swartziana* Griseb. (1860).

Shrub or tree 1–6 m. high, sometimes scrambling especially in arid areas; leaves usually acuminate, racemes variable, shorter or longer than leaves; pedicels pubescent at first; sepals 3–4 mm. long; petals 5–9 mm. long.

Very common in thickets and open woodlands on limestone; SL–3000 feet, fl. and fr. all the year. *A 8048! 11236! H 9633! HPS 14729! P 24718!* Cuba; Hispaniola.

2. B. jamaicensis Urb. & Niedenzu in Ind. Lect. Lyc. Brunsb., Bunchosia: 10 (1898).

Tree 5–10 m. high with drooping branches; leaves usually rounded or obtuse at apex, rarely shortly acuminate; racemes shorter than leaves; pedicels almost glabrous; sepals about 3 mm. long; petals about 5 mm. long.

Rare and local (Trel., St. Ann), in woodland on limestone; 1500–3000 feet; fl. July, Dec, fr. Feb, Aug. *H 6463! 8897! H & P 13497! P 15740!* Endemic.

4. HETEROPTERIS Kunth (1822) nom. cons.

1 Petals purple or pink; petiole with 2 glands at about the middle; leaves membranous, less than 6 cm. long; sepals erect-tipped; samara 2–3 cm. long with oblong-ovate wing   **1.** *purpurea*
1 Petals yellow or tinged reddish; petiole without glands; leaves leathery, mostly much more than 8 cm. long; sepals revolute-tipped; samara 2·5–4 cm. long:
   2 Samara-wing elliptical with a slight proximal constriction and curved parallel striations; panicle pyramidal   **2.** *laurifolia*
   2 Samara-wing suborbicular without a constriction, with more or less straight flabellate striations   **3.** *multiflora*

1. H. purpurea (L.) Kunth, Nov. Gen. **5**: 164 (1822).

Slender climbing shrub to 5 m. or more high; leaves ovate to elliptical, usually rounded at base, acute, obtuse or rounded at apex, glabrous, 1–5 cm. long, 1–3 cm. broad; flowers in terminal and axillary sparingly pubescent corymbs; sepals oblong to oblong-ovate, 2·5–3 mm. long; petals up to 5·5 mm. long.

Very rare and not recently collected. Colombia, Venezuela, Hispaniola to Trinidad.

2. H. laurifolia (L.) A. Juss. in Ann. Sci. Nat. sér. 2, Bot. **13**: 276 (1840).—*Banisteria laurifolia* L. (1762). Dragon Withe.

Strong-stemmed scrambling or climbing shrub; leaves ovate-lanceolate to elliptical, cuneate to rounded at base, obtuse to acuminate at tip, pubescent when young, glabrescent, 7–18 cm. long, 2–9 cm. broad; panicle reddish-pubescent; petals 5–7 mm. long.

Rather common, at least in eastern parishes, on shrubs and trees in thickets; 10–2000 feet; fl. and fr. most of the year. *A 6150! 11852! H 8519! 11773! P 11878!* C. Amer., Greater Antilles.

3. H. multiflora (DC.) Hochr. in Bull. New York Bot. Gard. **6**: 277 (1910).—
Byrsonima multiflora DC. (1824). *H. reticulata* (Poir.) Niedenzu (1903), not
Griseb. (1858). *Malpighia reticulata* Poir. (1816).

Climber; leaves oblong-lanceolate to elliptical, obtuse or rounded at base, acute
or acuminate at tip, glabrous or with a few scattered hairs beneath, up to 30 (–40)
cm. long and 13 (–15) cm. broad; panicle rusty-tomentose; petals unequal, 5–6
mm. long.

Very rare and local (St. Eliz.), on trees by river; about 50 feet; fl. and fr. Sept.
H 9753! Northern S. Amer., Trinidad.

5. BRACHYPTERYS A. Juss. (1837)

1. B. ovata (Cav.) Small in N. Amer. Fl. **25**: 138 (1910).

Shrub usually about 1 m. high, erect at first then twining; leaves ovate to
lanceolate, rounded at base, acute at tip, glabrescent, fleshy and rather brittle, up to
9 (–11) cm. long and 3·5 (–5) cm. broad, midrib with 2 glands near to base; inflorescence subumbellate, 2–8-flowered; peduncle up to 7 cm. long; pedicels up to
3 cm. long; sepals 3–4 mm. long; petals yellow fading orange, including claw 9–16
mm. long; fruit 1–2-coccous, with dorsal wing triangular, thick with irregular
prominent ridges, 3–5 mm. high; cotyledons unequal.

Uncommon (St. Mary, Port., St. Thom.), in coastal swamp and rocky mangrove
margins; SL–5 feet; fl. Aug, Nov–Mar, fr. Mar, Aug. *A 11827! Webster & Wilson
5229! Yuncker 18173!* C. and northern S. Amer. from British Honduras to Brazil,
Greater Antilles, Guadeloupe, St. Vincent, Trinidad and Tobago, West Africa.

6. STIGMAPHYLLON A. Juss. (1832)

Leaves broadly cuneate to rounded or subcordate at base, acuminate at tip; anterior style
longer than the two posterior styles and the accompanying stamens also longer and
thicker than posterior; style-tips expanded, foliaceous   1. *puberum*
Leaves cordate, rounded or truncate at base, usually emarginate and apiculate at tip;
anterior style, i.e. that opposite the glandless sepals, shorter and thinner than the two
posterior styles as are also the accompanying stamens; style-tips foot-like
                                                                2. *emarginatum*

1. S. puberum (L. C. Rich.) A. Juss. in Ann. Sci. Nat. sér. 2, Bot. **13**: 289 (1840).

Climbing shrub; leaves ovate to ovate-elliptical, silky-pubescent abaxially,
7–10 (–16) cm. long, 5–6 (–10) cm. broad; inflorescence of corymbose or subumbellate racemes; peduncles 7–20 cm. long; pedicels up to 8 mm. long; sepals
2·5–3·5 mm. long; petals yellow tinged red, up to about 12 mm. long; samara 2·5–
3 cm. long.

Very rare (St. Mary). *McNab!* Costa Rica to Peru, West Indies.

2. S. emarginatum (Cav.) A. Juss. in Ann. Sci. Nat. sér. 2, Bot. **13**: 290 (1840).

Trailing and twining shrub; leaves very variable from linear (mostly in juveniles and on sterile branches) to suborbicular, silky beneath, glabrate, 1–10 (–14) cm.
long, 1–6 (–9) cm. broad; sepals 3–4 mm. long; petals yellow, up to about 13 mm.
long; samaras up to about 2 cm. long.

Common on shrubs and trees in thickets, especially near the sea; SL–2600 feet;
fl. Oct–June, fr. Nov–Apr. *A 5652! H 9232! 10350! P 20597!* Greater Antilles,
Virgin Is. and Lesser Antilles south to St. Lucia.

7. TRIOPTERIS L. (1753)

Leaves oblong-ovate or ovate, obtuse or acute at tip; inflorescence corymblike
                                                                1. *paniculata*
Leaves broadly ovate, distinctly acuminate; inflorescence pyramidal   2. *brittonii*

1. T. paniculata (Mill.) Small in N. Amer. Fl. **25**: 124 (1910).

Woody climber or scrambler; leaves rounded to subcordate at base, 3–10 cm.
long, 2–4 (–7) cm. broad; sepals, 4 with paired glands, 1 without, 3–3·5 mm. long;
petals pink, mauve or light violet, 5–6 mm. long; samara 13–24 mm. broad.

Locally common, especially in secondary growth along the south coast; SL–800 feet; fl. Aug–May, fr. Sept–June. *A 9826! H 7263! P 15344! 23583!* Endemic.

2. T. brittonii Small in N. Amer. Fl. **25**: 124 (1910).
Climbing shrub; leaves cordate at base, reticulate-veined, 4–10 cm. long, up to about 6 cm. broad; sepals about 2 mm. long; petals purple, 4·5–5 mm. long; fruit not known.
Very rare (Manch.), known only from the type, *Britton 1069!* Endemic.

8. TETRAPTERIS Cav. (1790)

1. T. citrifolia (Sw.) Pers., Synops. Pl. **1**: 508 (1805).—*Triopteris citrifolia* Sw. (1788).
Climbing shrub; leaves elliptical to oblong-elliptical, cuneate to rounded at base, acute to shortly acuminate at tip, without glands, 5–13 (–17) cm. long, 2·5–8 cm. broad; inflorescence shortly white-hairy; bracts foliaceous; sepals 3–4 mm. long; petals 4–6 mm. long; upper wings of samara 13–16 mm. long, lower wings 9–10 mm. long.
Uncommon (Trel., St. Mary, St. Thom.), in thickets; 400–1250 feet; fl. Sept, fr. Aug. *McNab! Powell 853! P 15604!* Hispaniola.

9. MASCAGNIA Bert. ex Colla (1824)

1. M. hiraea (Gaertn.) F. & R., Fl. Jam. **4**: 240 (1920).
Robust woody climber to 6 m. high; leaves ovate to ovate-lanceolate, cuneate to rounded at base, acutely acuminate, glossy, 6–15 cm. long, up to 7 cm. broad; inflorescence of paniculate racemes; sepals ovate to roundish, yellowish, inflexed, 3–4 mm. long; petals spreading, one deeply fimbriate, 7–11 mm. long; wings of samara about 2 cm. broad.
Occasional in woodlands on limestone; 25–2250 feet; fl. Apr–Sept, fr. Apr–Nov. *A 6806! 11002! H 9487! P 15678! Robertson UCWI 5743!* Endemic.

Species-limits in the West Indian *Malpighiaceae* are in general very difficult to define. It is noticed in some of the commoner species that there is marked local phenotypic and ontogenetic variability which could suggest that some differences are more apparent than real in other species. For example if *Triopteris ovata* Cav. were not different from *T. paniculata*, the range of the latter would be extended to include Hispaniola and Mexico; if *Tetrapteris citrifolia* were deemed to be not different from *T. inaequalis* Cav., the range would be extended to Tobago. Parallel situations exist in *Malpighia, Byrsonima, Bunchosia* and *Stigmaphyllon* and taxonomic interpretations will continue to differ widely until the causes of variation are understood.

108. POLYGALACEAE

Herbs, erect or climbing shrubs or small trees. Leaves mostly alternate, simple, without stipules or with small stipular glands. Flowers bisexual, zygomorphic, with one bract and two bracteoles, in spikes, racemes or panicles of racemes. Perianth biseriate; sepals 5, usually free, imbricate, the two inner (wings) larger and often petaloid, often persistent; petals ideally 5 but mostly only 3 present, generally more or less adnate to the androecium, the anterior median petal (keel) with or without a fringed crest and often concave and enclosing the androecium. Stamens basically 10 in two whorls of 5, but usually only 8; filaments monadelphous in a split sheath; anthers mostly 1-locular, opening by a pore. Ovary superior; loculi 1–3 or 5; ovules pendulous, anatropous, solitary on each axile placenta; style simple; stigmas or stigma-lobes as many as carpels. Fruit usually a loculicidal capsule, rarely a nut, samara or drupe. Seed often hairy, often arillate; endosperm fleshy or absent; embryo straight.
13 genera with about 800 species, widely distributed.

Fruit a capsule; ovary 2-locular; erect herbs, shrubs or trees; inflorescences racemose; lateral petals 2   **1. Polygala**
Fruit samaroid; ovary 1-locular; scrambling shrubs, rarely erect; inflorescence mostly paniculate; lateral petals 4 (2 rudimentary)   **2. Securidaca**

1. POLYGALA L. (1753)

1 Shrub or tree; inflorescences corymbose; leaves ovate to lanceolate; lateral sepals white turning yellow; keel not crested; capsule 7–9 mm. long, narrowly winged; calyx more or less deciduous; ovary puberulous   1. *jamaicensis*
1 Annual herbs; inflorescence racemose; leaves linear to narrowly lanceolate; lateral sepals pink or purple in flower; calyx persistent; ovary glabrous:
  2 Keel crested; leaves linear-acicular; capsule 2 mm. long; plant glandular
                                                                       2. *paniculata*
  2 Keel not crested; leaves elliptic-lanceolate; capsule 3·5–4 mm. long; plant not glandular
                                                                       3. *angustifolia*

1. P. jamaicensis Chod. in Mém. Soc. Phys. & Hist. Nat. Genève **31** (2, 2): 11, t. 13, f. 14 (1893).—*Badiera diversifolia* DC. (1824), not *P. diversifolia* L. (1753). White Lignum Vitae.

Shrub or tree 2–8 m. high; leaves yellowish-green, 3–8 cm. long, 1–2·5 cm. broad; racemes axillary, short; wing sepals 1·5–2 mm. long; keel petals 4–4·5 mm. long, green, clawed and attached to the staminal tube.

Frequent in woodlands, particularly in the southern parishes; SL–4300 feet; fl. and fr. all the year. *A 7684! H 9651! J.P. 641! P 21912!* British Honduras, Guatemala.

2. P. paniculata L., Syst. Nat. ed. 10, **2**: 1154 (1759).

Bushily branched herb 15–50 cm. high, sometimes woody at the base, with fragrant roots; leaves 7–22 mm. long, up to 2 mm. broad; racemes terminal and axillary, 3–9 cm. long; flowers rosy or purplish, rarely all white, 2–2·5 mm. long.

A common weed of roadsides and waste ground on limestone or shale in moderately cool areas; 500–5000 feet; fl. and fr. all the year. *A 7214! H 9400! P 16751!* Mexico to Brazil, West Indies, Malaysia and Oceania.

3. P. angustifolia Kunth, Nov. Gen. **5**: 405, t. 511 (1823).—*P. brizoides* St.-Hil. (1829).

Sparingly branched herb 10–30 cm. high; branches puberulous but not glandular; leaves 2–5 cm. long, 3–10 mm. broad; racemes axillary, lax-flowered, 2–6 (–10) cm. long; wing sepals 3–4 mm. long; keel petal about 4 mm. long, dull pink with yellow tip, clawed and attached to the staminal tube.

A rare weed (St. Andr., Manch.), in stony waste places and pastures and on earth banks; 300–2800 feet; fl. and fr. Sept–Nov. *A 7947! 11772! H 12705!* Continental tropical Amer., Cuba, St. Thomas, Trinidad, Curaçao. *P. angustifolia* Kunth is antedated by *P. angustifolia* Gilibert, Fl. Lituan. **2**: 113 (1781), but most authors concur that as this work did not consistently employ the binomial nomenclature the new names in it are not validly published.

2. SECURIDACA L. (1759) nom. cons.

Wing of fruit 3–4 cm. long, 1·5–2 cm. broad distally, 9–10 mm. broad at base; racemes 4–8 cm. long, numerous in a leafy panicle; lower leaves up to about 5 cm. diminishing to about 1 cm. long in the inflorescence   1. *brownei*
Wing of fruit 2·5–3 cm. long, 10–12 mm. broad distally, 4–6 mm. broad at base; racemes up to 15 cm. long, numerous at the ends of branches; lower leaves 1·5–2 cm. diminishing to about 5 mm. long in the inflorescence   2. *virgata*

1. S. brownei Griseb., Fl. Br. W.I.: 30 (1859).

Scrambling shrub or woody vine to 8 m. high, rarely low and straggling or erect; leaves elliptical, obtuse to rounded or emarginate at tip; flowers numerous; wings up to nearly 1 cm. broad, bright reddish-purple to dull mauve-pink; larger petals about 6 mm. long; keel about 8 mm. long, white or light mauve with a yellow spot; seed-case of fruit with the anterior margin widening above and produced into a horn.

Common in thickets and woodland margins on limestone; 300–3000 feet; fl. July–Apr, fr. July–May. *A 6999! 8237! H 8992! P 8251!* Endemic.

2. **S. virgata** Sw., Nov. Gen. & Sp. Pl.: 104 (1788).—*S. erecta* of F. & R. (1920), not Jacq. (1760).

Like the last; flowers and fruit smaller, the seed-case with the anterior margin very narrow and not produced into a horn.

Very rare and not collected recently in Jamaica. *Macfadyen!* Greater Antilles.

S. lamarckii Griseb. was reported either erroneously from Jamaica or possibly confused with *S. virgata*. It is a species occurring in Cuba, Martinique and St. Vincent. *S. diversifolia* (L.) Blake (*S. erecta* Jacq.) occurs throughout C. Amer., the Lesser Antilles and Trinidad.

109. EUPHORBIACEAE

Trees, shrubs or herbs, sometimes with viscid or milky sap. Leaves alternate, rarely opposite or whorled, mostly simple, sometimes reduced to scales, rarely absent, with or without stipules. Flowers unisexual, monoecious or dioecious. Perianth biseriate, uniseriate or wanting, usually small and actinomorphic; sepals free or united, valvate or imbricate; petals usually free, rarely united. Male flowers: stamens 1–many, free or united; anthers 2 (–3–4)-locular, erect or inflexed in bud, opening lengthwise or rarely by pores; rudimentary ovary (pistillode) present or absent; disk often present. Female flowers: ovary superior, usually 3-locular; ovules anatropous, 1 or 2 collateral in each loculus on axile placentas, usually pendulous; staminodes sometimes present; disk often present as a ring or lobed or of separate glands. Fruit usually capsular, splitting mostly into 3 2-valved cocci separating from a persistent axis, rarely drupaceous and indehiscent. Seeds often with a conspicuous caruncle, usually with endosperm; embryo straight or curved.

About 250 genera with over 7000 species; widely distributed but mainly in the tropics.

1 Flowers, one female and several male, enclosed in a cup-shaped involucre (cyathium) forming a partial inflorescence; male flowers consisting of 1 pedicelled stamen; female flower of a pedicelled ovary exserted in fruit; ovules 1 in each ovary-loculus; trees, shrubs or herbs, sometimes without leaves, always with milky sap; fruit a 3-seeded capsule:
  2 Involucre regular, calyxlike with 1 or more external glands   32. **Euphorbia**
  2 Involucre irregular, shoe-shaped, the glands enclosed in a lateral appendix
                                                          33. **Pedilanthus**
1 Flowers solitary or in various kinds of inflorescence or in clusters but not enclosed in a gland-bearing involucre:
  3 Leaves digitately 3-foliolate; perianth 2-seriate; seeds large, rounded, mottled; tree to 20 m. high; native of tropical S. Amer.; Para Rubber
                                       Hevea brasiliensis (Kunth) Müll. Arg.
  3 Leaves simple, entire to deeply divided:
    4 Slender twiner with stinging hairs   25. **Tragia**
    4 Erect herbs, shrubs or trees:
      5 Fruit indehiscent, fleshy outside at least at first; mostly trees:
        6 Leaves with 1 or 2 glands at or near apex of petiole and base of blade:
          7 Leaf with 1 gland at junction of petiole and blade adaxially; inflorescence spicate; endocarp woody, 6–9-seeded; perianth single; male flowers with 2 (–3) stamens
                                                   21. **Hippomane**
          7 Leaf with 2 glands at apex of petiole; blade simple or acutely lobed; inflorescence paniculate; fruit 1–3-seeded; perianth double, white; male flowers with 15–20 stamens; native of E. Indies, widely introduced and naturalized; Candlenut, Jamaican Walnut   *Aleurites moluccana* (L.) Willd.
      6 Leaves without glands; perianth 1-seriate:
        8 Ovary 3- or more locular; branchlets pinnatiform; plants monoecious
                                                1. **Phyllanthus**
        8 Ovary 1–2-locular; branchlets simple, uniform; plants dioecious:
          9 Leaves with stellate scales; inflorescences paniculate; fruit 1-seeded
                                                6. **Hyeronima**

9 Leaves without stellate scales; flowers in axillary clusters; fruit 1–2-seeded
7. **Drypetes**
5 Fruit capsular or coccoid, sometimes fleshy externally at first but always breaking up
eventually, or plant not in fruit:
10 Capsule of up to 20 2-valved segments; tree usually with spiny trunk and branches;
styles of female flower united into a fleshy column with a radially lobed stig-
matic limb, 3–4 (–5) cm. long					22. **Hura**
10 Capsule of (2–) 3 (–5–10) 2-valved segments or cocci, or plant not in fruit:
11 Ovules or seeds 2 in each carpel:
12 Shoots of two kinds (branching phyllanthoid— main to penultimate or prepenul-
timate branches with spiral phyllotaxy with leaves modified to cataphylls;
ultimate and sometimes penultimate branchlets deciduous and flower-
bearing with leaves in 2 rows or leafless):
13 Leaves variegated; ornamental shrub; native of Oceania; Snow Bush
Breynia disticha J. R. & G. Forst.
13 Leaves not variegated					1. **Phyllanthus**
12 Shoots all of one kind (branching not phyllanthoid):
14 Flowers in axillary panicles; dioecious tree; leaves stellate-scaly; calyx cam-
panulate; petals absent					6. **Hyeronima**
14 Flowers solitary, in clusters or reduced cymules; leaves without stellate scales;
calyx-lobes imbricate:
15 Pistillode absent or very small; petals absent:
16 Monoecious herbs or shrubs; stamens 3; ovary 3-locular; styles bifid
1. **Phyllanthus**
16 Dioecious trees:
17 Stamens 3–10; ovary 1–2-locular; style(s) entire; disk crenate	7. **Drypetes**
17 Stamens 4; ovary (3–) 4–5-locular; styles divided; disk annular
2. **Margaritaria**
15 Pistillode present; stamens 5; shrubs or trees:
18 Petals absent; twigs spiny					4. **Securinega**
18 Petals present at least in male flowers; plants not spiny:
19 Petals about as large as or larger than calyx-lobes; pistillode peltate;
petiole one third or more of length of leaf-blade		3. **Astrocasia**
19 Petals much smaller than calyx-lobes; pistillode slender, usually lobed;
petioles short; plants dioecious				5. **Savia**
11 Ovules or seeds 1 in each carpel:
20 Perianth absent in male and female flowers (bracteoles calyxlike in *A. integer*)
31. **Ateramnus**
20 Perianth present at least in female flowers:
21 Perianth double (calyx and petals present at least in male flowers):
22 Inflorescence spicate or racemose:
23 Leaves variegated usually with light and dark green and often with yellow
and/or red spots or blotches, various shapes; plants monoecious but
inflorescence unisexual; sepals strongly imbricate; male flowers pedicellate; stamens about 25; female flowers sessile; styles 3, simple, united
at base; ornamental shrubs; native of Oceania; Garden Croton
Codiaeum variegatum (L.) Blume
23 Leaves not variegated; sepals valvate or narrowly imbricate; styles once or
more divided:
24 Capsule spiny; inflorescences axillary, bisexual; stamens 10; herb of
swamps					8. **Caperonia**
24 Capsule often hairy, scaly or warted but not spiny; dry land herbs, shrubs
or trees:
25 Stamens usually 10–20; inflorescences terminal and sometimes also
axillary; plant not drying bluish			8. **Croton**
25 Stamens 4; inflorescences axillary; plant drying bluish; leaves mostly
3-nerved from base				10. **Argythamnia**
22 Inflorescence not spicate or racemose:
26 Flowers in clusters on short lateral branchlets formed as cushions covered by
prickly stipules; leaves borne on similar branchlets	11. **Acidocroton**
26 Flowers in cymose panicles, the female solitary in primary and sometimes
secondary forks:
27 Calyx imbricate; stamens up to 10; mostly shrubs		12. **Jatropha**
27 Calyx valvate; stamens 15–20; trees	*Aleurites moluccana* (L.) Willd.
21 Perianth single (calyx):
28 Leaves deeply palmately divided, usually glaucous beneath; racemes simple
or branched:
29 Lower flowers male, upper female; stipules united over the bud; calyx
valvate; stamens numerous with branched filaments; leaf-lobes serrate-
dentate					13. **Ricinus**

29 Lower flowers female, upper male; stipules small; calyx imbricate; stamens 10, simple; leaf-lobes entire 14. **Manihot**
28 Leaves entire, serrate, crenate or dentate:
  30 Inflorescence neither spicate nor racemose:
    31 Flowers in panicles; bracts subtending cymules on main axis of inflorescence leafy; calyx-segments imbricate; stamens 2–3, the filaments united 15. **Omphalea**
    31 Flowers clustered on reduced shoots or in the axils of small leaves; filaments arising from a central column:
      32 Calyx-segments valvate; capsule smooth; stamens (8–) 15–20 16. **Adelia**
      32 Calyx-segments imbricate; capsule prickly; stamens 6–7 17. **Chaetocarpus**
  30 Inflorescence at least as to male flowers spicate or racemose, (male flowers paniculate in *Alchornea*):
    33 Leaves opposite or verticillate; sepals 3:
      34 Herbs; stamens 8–15; carpels and styles 2 18. **Mercurialis**
      34 Shrubs or trees; stamens 3; carpels and styles 3; bracts with paired glands 19. **Sebastiania**
    33 Leaves alternate or clustered:
      35 Bracts with paired glands:
        36 Male flowers solitary in the axils of bracts; sepals free; stamens 3; filaments free 19. **Sebastiania**
        36 Male flowers (2–) 3 (–15) together in the axils of bracts; sepals united; stamens 2 or 3:
          37 Spikes axillary; filaments free; ovary 2–3-locular 20. **Sapium**
          37 Spikes terminal; filaments united; ovary 6–9-locular 21. **Hippomane**
      35 Bracts without paired glands or bracts inconspicuous:
        38 Spikes terminal or pseudoterminal, with female flower(s) at or near the base; filaments united:
          39 Ovary 5–20-locular; calyx cupular; stigma radiately lobed at tip of 3–4 cm. long columnar style; stamens 8–20; leaf-blade with 2 glands at base 22. **Hura**
          39 Ovary 3-locular; calyx valvately lobed at least in male flowers; styles simple, united below; stamens 3; 2 small glands at junction of petiole and blade 23. **Grimmeodendron**
        38 Spikes axillary or on the older stems:
          40 Male sepals imbricate; stigmas 2, sessile 24. **Tetrorchidium**
          40 Male sepals valvate or calyx closed in bud and splitting valvately:
            41 Styles 3, united at base into a thick column; stamens numerous on a raised receptacle; young leaves and fruits with stinging hairs 26. **Acidoton**
            41 Styles free or very shortly united at base; plants without stinging hairs:
              42 Styles 2 (–3), 12–20 mm. long, simple; male inflorescence paniculate; stamens up to 8 27. **Alchornea**
              42 Styles 3, much shorter than above; male inflorescence or part of inflorescence catkinlike with closely packed flowers; stamens mostly more than 8:
                43 Styles short, thick and entire or longer and divided about halfway; anther-loculi oblong, parallel 28. **Lasiocroton**
                43 Styles much-branched:
                  44 Female bracts entire, small; anther-loculi very short, subglobose 29. **Bernardia**
                  44 Female bracts often leafy, accrescent and usually toothed or lobed; anther-loculi pendulous, oblong or flexuous 30. **Acalypha**

1. PHYLLANTHUS L. (1753); Webster (1956–58)

1 Branching not phyllanthoid (ultimate axes persistent, their subtending leaves not reduced to scales); stamens 3; fruit a dehiscent capsule:
  2 Herbs; stamens free; pedicel of female flower and seeds less than 2 mm. long 1. *caroliniensis*
  2 Shrubs or trees; stamens connate; pedicel of female flower and seeds more than 2 mm. long 10. *nutans*
1 Branching phyllanthoid (ultimate axes deciduous, their subtending leaves usually reduced to scales):
  3 Branchlets of mature plants without normal leaves, at least the lateral axes modified into phylloclades; shrubs or trees; fruit a dehiscent capsule; stamens 3 (–4), united at least at base:

4 Phylloclades each representing a simple branchlet borne spirally on the main stem
14. *eximius*
4 Phylloclades borne distichously on main axis of a compound deciduous branchlet:
  5 Branchlets long-persistent with their main axes greyish- or brownish-barked;
    pedicels of female flowers 0·5–1·5 mm. long   15. *montanus*
  5 Branchlets soon deciduous, their main axes greenish and of the consistency of the
    phylloclades:
    6 Pedicels of female flowers 0·5–1·5 mm. long; disk of female flower lobed; phyllo-
      clades more or less rhombic-lanceolate; apical cone 10–15 mm. broad; cataphylls
      scarcely or not ciliate; ovary smooth   16. *latifolius*
    6 Pedicels of female flowers mostly 2 mm. or more long; disk of female flower entire;
      phylloclades linear to obovate-lanceolate; apical cone up to about 10 mm. broad:
      7 Cataphylls of main axis remaining brown or grey, ciliate if at all only towards the
        base; phylloclades mostly 1–2 cm. broad with 20–40 nodes; ovary smooth
        17. *arbuscula*
      7 Cataphylls of main axis becoming dark- or blackish-brown, copiously ciliate on
        margins at least when young; phylloclades up to about 10 mm. broad with
        7–25 nodes, often subopposite at end of branchlet axis:
        8 Phylloclades lanceolate, (2–) 5–12 mm. broad with (8–) 10–25 nodes; styles not
          dilated, the tips 3–4-lobed; ovary rugulose   18. *angustifolius*
        8 Phylloclades narrowly linear, 1–3·5 mm. broad with 7–10 (–13) nodes; styles
          dilated, the tips crenulate, reflexed over the smooth ovary   19. *proctoris*
3 Branchlets with normal leaves:
  9 Branchlets bipinnatiform; stamens 3, united; male disk of 3 segments; pedicels of
    female flowers 5–12 mm. long, scabridulous; fruit a capsule   9. *acuminatus*
  9 Branchlets pinnatiform:
    10 Shrubs or trees, plant rarely less than 1 m. high:
      11 Fruits indehiscent or tardily dehiscent:
        12 Fruits baccate; calyx-lobes 5–6; ovary with 9–10 carpels; stamens 5, united in 2
          sets   3. *reticulatus*
        12 Fruits drupaceous, 6–8-lobed, often 1-seeded; calyx-lobes 4; ovary with 3 (–4)
          carpels; stamens (3–) 4 with free or almost free filaments   4. *acidus*
      11 Fruits dry, dehiscent; flowers mostly in naked thyrses; stamens 2–3, united;
        ovary of 3 carpels; branchlets often clustered and radiate near the branch-tips:
        13 Stipules lanceolate, reflexed, not fused with branchlet; branchlets subterete;
          leaves oblong-lanceolate; inflorescences cauline; male flower with 5 (–6) calyx-
          lobes and 3 stamens   11. *cladanthus*
        13 Stipules triangular, massive, not reflexed, more or less fused to branchlet; male
          flower with 4 calyx-lobes and 2 stamens:
          14 Branchlets angled, not flattened; leaves oblong-lanceolate; inflorescences
            cauline and axillary; pedicels of female flower 2–3 mm. or more long; ovary
            extended into a long stylar column   12. *cauliflorus*
          14 Branchlets flattened; leaves mostly elliptical, leathery; inflorescences strictly
            axillary to leaves on branchlets; female flowers subsessile; ovary ellipsoid
            13. *axillaris*
    10 Annual herbs, rarely shrubby, plant rarely as much as 75 cm. high; fruit capsular:
      15 Stamens 5, free; calyx-lobes 5; capsules pendulous on slender pedicels 3–7 mm.
        long; seeds papillose   2. *tenellus*
      15 Stamens (2–) 3, free or united; calyx-lobes 5–6; capsules on pedicels up to 2·5
        (–3·5) mm. long; seeds not papillose:
        16 Ovary rugose; seeds transversely ribbed; calyx-lobes 6; stem usually reddish and
          with acute ridges decurrent from the nodes; stamens 3, completely united
          5. *urinaria*
        16 Ovary smooth or nearly so; seeds longitudinally ribbed or striate; calyx-lobes
          (4–) 5 (–6); stem terete:
          17 Filaments free; stem rather woody; capsule and seeds not known
            6. *fadyenii*
          17 Filaments completely united; stem not markedly woody:
            18 Weed of waste and cultivated ground; stem green; seeds with 5–6 longitudinal
              ribs   7. *amarus*
            18 Herb of swamps and pond margins; stem reddish, often spongy; seeds with
              10–12 fine longitudinal striations   8. *stipulatus*

1. P. caroliniensis Walt., Fl. Carol.: 228 (1788).

Herb 5–30 cm. high; leaves elliptical to obovate, 5–18 mm. long, 2–8 mm. broad;
calyx-lobes 6; capsule 1·5–2 mm. in diameter; seeds usually light greyish-brown,
minutely warted.

Two subspecies have been recognized for Jamaica:

1a. Ssp. caroliniensis.
Branches smooth; tertiary veinlets of leaf usually evidently reticulate beneath; calyx-lobes of female flower mostly linear-spathulate, 1–1·2 mm. long.
Rare and local (St. Andr., St. Mary), a weed of shady places; 500–4000 feet; fl. and fr. Jan, Sept, *H 12143! Robertson UCWI 24391!* Illinois to Argentina and Paraguay, Martinique, Trinidad and Tobago.

1b. P. caroliniensis ssp. saxicola (Small) Webster in Contrib. Gray Herb. **176**: 46 (1955).—*P. saxicola* Small (1905).
Branches scabridulous; tertiary veinlets of leaf obscure or invisible beneath; calyx-lobes of female flower broadly oblong or spathulate, usually rounded or obtuse, 0·6–0·9 (–1) mm. long.
Very rare (St. Cath.), in damp hollows in pastures; about 300 feet; fl. and fr. May. *H 12046!* S. Florida, Bahamas, Cuba, Hispaniola.

2. P. tenellus Roxb., Fl. Ind. **3**: 668 (1832).—*P. minor* F. & R. (1919).
Annual or short-lived perennial herb 30–60 cm. high, often rather woody at base; leaves thin, broadly elliptical to obovate, acute or obtuse, paler beneath, mostly 10–20 mm. long, 5–10 mm. broad; calyx-lobes whitish except for narrow green midrib; capsule 1·7–1·9 mm. in diameter, obscurely reticulate-veiny; seeds trigonous, light brown.
Uncommon and local, a weed of pathsides and shady places; 700–4000 feet; fl. and fr. most of the year. *A 8151! H 12208! P 10331!* Native of Mascarene Is., introduced sporadically into the subtropics and tropics of the New World.

3. P. reticulatus Poir. in Lam., Encyl. Méth. Bot. **5**: 298 (1804).
Much branched shrub or small tree to 4 m. high; leaves smooth, elliptical, obtuse or rounded at tip, cuneate at base, mostly 1·5–2·5 cm. long, 1–1·5 cm. broad; pedicels 4–10 mm. long; calyx-lobes green tinged purplish with white margins, in the male 2–2·5 mm. long, in the female smaller and unequal; mature fruit 4–6 mm. in diameter, black with dark purplish pulp; seeds blackish, 1·5–2 mm. long.
Rare and local (St. Mary, St. Thom.), mostly an escape from cultivation near riverbanks; 30–500 feet; fl. and fr. sporadically. *A 12358! H & B 10677! Webster & Wilson 5237!* Native of Old World tropics; St. Vincent.

4. P. acidus (L.) Skeels in U.S. Dept. Agric. Bur. Pl. Indust. Bull. **148**: 17 (1909).—*Averrhoa acida* L. (1753). *P. distichus* (L.) Müll. Arg. (1866).
Cheramina, Jimbling, Otaheite Gooseberry, Short Jimbelin.[1]
Tree with rough grey bark, up to 10 m. high; leaves broadly ovate to ovate-lanceolate, acute at tip, obtuse or rounded at base, (4–) 5–9 cm. long, (2–) 2·5–4·5 cm. broad; flowers in dense cymules on naked branchlets; calyx-lobes 1–1·5 mm. long; female flowers with up to 4 staminodes; fruit yellow or whitish, mostly 1·5–2 cm. in diameter, with a hard bony endocarp.
Almost always cultivated and rather uncommon; fl. mostly Jan–Mar. *Wynter UCWI 3701!* Probably native of Brazil but introduced into tropical Asia at an early date and now widespread (Webster in Journ. Arn. Arb. **38**: 68–72 (1957)).

5. P. urinaria L., Sp. Pl. **2**: 982 (1753).—Chamber Bitter.
Erect or lax-branched annual herb 15–60 cm. high; leaves oblong, obtuse or acute, mucronulate, usually asymmetric at base, mostly 8–20 mm. long and 2·5–6 mm. broad, usually dark green adaxially and hispidulous-margined, sensitive to touch; pedicels up to 0·5 mm. long; calyx-lobes yellowish, less than 1 mm. long; capsule 2·0–2·2 mm. in diameter; seeds light greyish-brown with 12–15 sharp transverse ridges.
Common as a weed of usually damp shady places; 40–4000 feet; fl. and fr. all the year. *A 5416! H 12142! P 7420!* Native of tropical Asia; Hispaniola, Puerto Rico, Lesser Antilles and here and there in many other tropical regions.

---

[1] Refer to footnote 2 on p. 376.

6. P. fadyenii Urb., Symb. Ant. **6**: 13 (1909).

Stem terete with brown or grey bark; leaves broadly oblong or ovate, rounded or obtuse at tip, cordulate at base, 4–5·5 mm. long, 2·5–4 mm. broad; male flowers apparently at separate axils from the female and smaller; female calyx-lobes up to 2 mm. long.

Very rare and known only from the type, *Macfadyen*. Endemic.

7. P. amarus Schumach., Beskr. Guin. Pl.: 421 (1827).—*P. niruri* of F. & R. (1920), not L. (1753). Carry-me-Seed, Egg Woman.

Annual herb 10–50 cm. high; leaves elliptic-oblong, obtuse or rounded at both ends, 5–11 mm. long, 3–6 mm. broad, light green, glabrous, glaucous beneath; most axils with one male and one slightly larger female flower; capsule about 2 mm. in diameter; seeds light brown, about 1 mm. long.

Very common as a weed of waste and disturbed ground; SL–700 feet; fl. and fr. all the year. *A 5433! H 12136! Webster & Wilson 5059!* Native of tropical Amer., although first described from W. Africa, now widespread.

8. P. stipulatus (Raf.) Webster in Contrib. Gray Herb. **176**: 53 (1955).—*Moeroris stipulata* Raf. (1838). *P. aquaticus* C. Wright (1866).

Sparingly branched flexuous slender-stemmed herb to 1 m. high; leaves oblong-elliptical, obtuse or rounded at both ends, 3–13 mm. long, 2–4 (–6) mm. broad, light green and often tinged reddish; male flowers several together in the proximal axils, female solitary in the distal axils; capsule about 2·5 mm. in diameter; seeds sharply trigonous, light brown, 1·1–1·3 mm. long.

Rare and local (St. Cath., Clar., St. Eliz.), in ponds and boggy pastures; SL–2300 feet; fl. and fr. Sept–Jan. *A 10406! H 12555! P 6177!* Widespread in tropical Amer.

9. P. acuminatus Vahl, Symb. Bot. **2**: 95 (1791).

Shrub or tree 2–8 m. high; leaves ovate to broadly elliptical, abruptly cuspidate-acuminate, usually obtuse at base, 2–4·5 cm. long, 1–2·5 cm. broad, more or less scabridulous adaxially; flowers in bisexual cymules; calyx-lobes 6, green with whitish margins; capsule reticulate-veiny, 4·5–5 mm. in diameter; seeds smooth, reddish-brown, sometimes mottled, 2·2–2·8 mm. long.

Very local (Han., St. Mary), possibly an escape from cultivation; SL–500 feet; fl. Apr–Nov, fr. July–Dec. *H 8409! 11147! Shaw!* Mexico to Argentina and Paraguay, Cuba, St. Vincent, Trinidad.

10. P. nutans Sw., Nov. Gen. & Sp. Pl.: 27 (1788).

Shrub 1–3 m. or tree to 10 m. high; leaves variable but often ovate or elliptical, obtuse or rarely acute at tip, cuneate to rounded at base, (1·5–) 3·5–8 (–11) cm. long, (1–) 2–5 (–8) cm. broad; stipules variable, sometimes leafy; inflorescences nodding, racemiform; pedicels 8–30 mm. long, often red; calyx-lobes 6, about 2–3 mm. long; capsule 6-ribbed rugulose, about 10 mm. in diameter; seeds trigonous, smooth, mottled light brown, 4–7 mm. long.

Common and widespread, in rocky woodlands and on well drained slopes in limestone or shale areas, SL–2600 feet; fl. and fr. most of the year. *A 5682! 12482! H 10260! P 15874!* Grand Cayman. A subspecies distinguished by Webster (1958) in having dark persistent stipules and pointed revolute leaves is named *P. nutans* ssp. *grisebachianus* (Müll. Arg.) Webster and is reported from Little Cayman and Cuba.

11. P. cladanthus Müll. Arg. in Linnaea **32**: 46 (1863).

Slender tree 2·5–10 m. high, often unbranched; leaves chartaceous, ovate to oblong-lanceolate, abruptly acuminate, obtuse to rounded or subcordate at base, 9–15 cm. long, 3–7 cm. broad; cymules with 1 female and up to 10 male flowers in naked thyrses 5–20 cm. long; calyx-lobes pinkish, 1–1·5 mm. long; capsule smooth, about 8 mm. in diameter.

Uncommon, in thickets and woodlands on limestone in wet areas; 1000–2500 feet; fl. Mar, June, Aug, fr. Mar, Aug. *H 8722! P 11800!* Endemic.

12. P. cauliflorus (Sw.) Griseb., Fl. Br. W.I.: 33 (1859).—*Omphalea cauliflora* Sw. (1788).

Slender tree 1·5–6 m. high, like the last in habit and foliage; thyrses 3–18 cm. long; calyx-lobes green or red, unequal and 0·9–1·8 mm. long in the male and 1·0–1·2 mm. long in the female flowers; capsule subglobose but flattened basally, beaked, 5 mm. in diameter; seeds trigonous, pointed at one end, dark brown and minutely papillose, about 3·5 mm. long.

Rare and local (West., Han.), on limestone rocks in thickets and woodlands; SL–1700 feet; fl. Mar, fr. Mar–Apr. *A 12765! Cornman 520!* Endemic.

13. P. axillaris (Sw.) Müll. Arg. in DC., Prodr. **15** (2): 412 (1866).

Shrub 0·5–4 m. high, like the last in habit; leaves coriaceous, ovate to elliptic-lanceolate, shortly acuminate, obtuse to rounded at base, 6–11 cm. long, 2·5–5·5 cm. broad; thyrses 1–3 cm. long; calyx-lobes yellowish-green, 1–1·5 mm. long; capsule trigonous, acuminate, about 7 mm. long; seeds about 3·5 mm. long, light brown.

Very local (Trel.), in scrubby woodland on dry rocky hillsides; 1500–1700 feet; fl. Apr–Nov, fr. May–Jan. *Cornman 590! H & P 14131! P 18349!* Endemic.

14. P. eximius Webster & Proctor in Journ. Arn. Arb. **41**: 283 (1960).

Shrub or small tree 2–4 m. high with ascending branches; phylloclades simple, rigid, spathulate to oblanceolate, obtuse, rounded or emarginate at tip, long-tapered to base, with up to about 50 notched nodes, 8–17·5 cm. long, 1–4·5 cm. broad; cymules mostly bisexual; calyx-lobes (5–) 6, 1–1·5 mm. long, yellowish-or greenish-white or tinged reddish; capsule about 3·5 mm. long.

Rare and local (Port.), in mossy thickets and woodland on limestone in wet areas; 1500–2500 feet; fl. May, fl. and fr. June. *A 9155! duQuesnay 322! Proctor & Webster 9228! Skelding UCWI 9997!* Endemic.

15. P. montanus (Sw.) Sw., Fl. Ind. Occ. **2**: 1117 (1800).

Shrub 2–4 m. or tree to 8 m. high; apical cone inconspicuous, about 2 mm. long; phylloclades mostly elliptical to lanceolate, obtuse to long-attenuate at tip, with usually 10–30 nodes, 3–20 cm. long, 0·5–3 cm. broad; flowers usually red; capsule about 4 mm. in diameter; seeds trigonous, reddish, 1·7–1·9 mm. long.

Frequent in the central and western parishes, on limestone cliffs and rocks in thickets and open woodlands; 700–3000 feet; fl. and fr. sporadically throughout the year. *A 8367! H 12797! P 7255!* Endemic.

16. P. latifolius Sw., Fl. Ind. Occ. **2**: 1109 (1800).

Shrub 0·3–1 m. or small tree 4 m. high with erect or straggling branches; phylloclades rigid, rhombic to narrowly lanceolate, blunt to caudate at tip, with usually 20–50 nodes, (3·5–) 5–8 (–10) cm. long, 1–3 cm. broad; flowers pink to deep crimson; calyx-lobes 6; capsule about 4 mm. in diameter; seeds reddish-brown, about 2 mm. long.

Locally common (St. Andr., St. Cath.), on limestone rocks, cliffs and ledges in arid thickets; near SL–1000 feet; fl. Mar–Sept, fr. Mar–July. *A 12320! H 9631! P 19703!* Endemic.

17. P. arbuscula (Sw.) J. F. Gmel. in L., Syst. Nat. ed. 13, 2 (1): 204 (1791).—*Xylophylla arbuscula* Sw. (1788). *P. speciosus* Jacq. (1789). *P. inaequaliflorus* F. & R. (1919). *P. coxianus* F. & R. (1919). *P. swartzii* F. & R. (1919).

Shrub 0·3–4 m. or tree to 7 m. high, sparingly branched; apical cone 5–10 (–15) mm. broad; phylloclades thin to rigid, elliptical to lanceolate, acute or acuminate, with usually 20–40 conspicuously notched nodes, (2·5–) 4–11 cm. long, (0·5–) 1–2 (–2·7) cm. broad; male pedicels mostly 2–5 mm. long, female 3·5–10 mm. long; calyx-lobes yellowish-green, pink or crimson; capsule 4–5 mm. in diameter, rugulose; seeds about 3 mm. long with reddish-brown dots.

Common in middle-elevation and montane woodlands on limestone; (250–) 1250–4700 feet; fl. Jan–Nov, fr. Mar–Dec. *A 11476! 12549! H 8714! 9191! 9368! P 6722!* Endemic. This species is very variable and although distinctive populations can be recognized, Webster (1958) regards the variation as clinal and not at the present time susceptible of formal taxonomic treatment. He proposes a working division into four races grouped more or less around the types of *P. dingleri* Webster (1956), *P. coxianus*, *P. inaequaliflorus* and *P. speciosus*.

18. P. angustifolius (Sw.) Sw., Fl. Ind. Occ. **2**: 1111 (1800).—Seaside Laurel.
Shrub 0·6–3 m. high, rarely a tree 5 m. high; bark often reddish; apical cone 3–4 (–6) mm. broad; phylloclades usually flexible, narrowly elliptical to obovate-lanceolate, obtuse to bluntly acuminate, 3–10 (–13) cm. long; pedicels mostly 2–6 mm. long; calyx-lobes mostly 1–1·5 mm. long, greenish-white, pink or light red; capsule 3–4 mm. in diameter, rugulose; seeds brown, 1·5–2·5 mm. long.
Common on rocks, especially near the sea; SL–2500 (–5000) feet; fl. and fr. all the year. *A 6856! H 10262! P 15335! 23714!* Cayman Is.; cultivated elsewhere.

19. P. proctoris Webster in Journ. Arn. Arb. **39**: 195 (1958).—*P. linearis* of Griseb. (1859) and F. & R. (1920), not (Sw.) Sw. (1800).
Shrub up to 3 m. high; apical cone roundish, 2·5–3 mm. long; phylloclades thin, linear-lanceolate, attenuate to tip, 3–11 cm. long; pedicels 2–3·5 mm. long; calyx-lobes about 1 mm. long, greenish-white.
Uncommon in the western parishes, in rocky places on limestone; 100–1300 feet; fl. Jan–Mar, fr. Mar. *H 12871! P 20697!* Endemic. *P. linearis* (Sw.) Sw. differs from *P. proctoris* in having pedicels about 8 mm. long and bifid styles. Webster suggests that *P. linearis*, which is imperfectly known, may be a hybrid between *P. proctoris* and *P. arbuscula*.
Phyllanthus portoricensis Urb. (1910), reported by Fawcett & Rendle (Fl. Jam. **4**: 254), is *Securinega virosa* (Roxb. ex Willd.) Baill. (1866), a cultivated introduction from the Old World tropics.

2. MARGARITARIA L. f. (1781)

1. M. nobilis L. f., Suppl.: 428 (1781).—*Phyllanthus nobilis* (L. f.) Müll. Arg. (1866). *P. antillanus* (A. Juss.) Müll. Arg. (1863). Bastard Hog Berry.
Tree 6–12 m. high; leaves narrow- to oblong-elliptical, acute to subacuminate at both ends; calyx-lobes 4, of male 1·5 mm. long, of female 2 mm. long; ripe fruit about 1 cm. in diameter, fleshy outside at first, breaking up irregularly; seeds smooth, greenish-brown, 3 mm. long.
Rare and local (Trel.); 1600–2200 feet; fl. June, Sept, fr. Sept. *H 8698! 9450! 9490!* Mexico to Brazil, West Indies.

3. ASTROCASIA Robins. & Millsp. (1905)

1. A. tremula (Griseb.) Webster in Journ. Arn. Arb. **39**: 208 (1958).—*Phyllanthus tremulus* Griseb. (1859). *P. glabellus* F. & R. (1919), not *Croton glabellus* L. (1759).
Shrub 2–3 m. or tree 4–8 m. high; leaves with slender petioles, blade broadly elliptical, glabrous, paler beneath, 3–12 cm. long, 2·5–8 cm. broad; male flowers: pedicels 9–11 mm. long; sepals 1·5 mm. long; petals to 2·5 mm. long; filaments united into a column; female flowers: pedicels at first as those in male but elongating to 3·5–4·5 cm. in fruit; sepals to 2 mm. long; petals to 3·5 mm. long; capsule 8–10 mm. in diameter; seeds buff to shiny dark brown, about 4 mm. long.
Occasional, in arid coastal thickets and woodlands on limestone; 20–1500 feet; fl. Sept, Mar–May, fr. Sept–May. *A 11148! H 9903! P 9508! 15334! Thornton UCWI 790!* Cuba, Grand Cayman.

4. SECURINEGA Commers. ex Juss. (1789) nom. cons.

1. S. acidoton (L.) F. & R. in Journ. Bot. **57**: 68 (1919).—Green Ebony.
Shrub 2–3 m. or tree to 6 m. high, much-branched with zig-zag spinescent twigs; leaves clustered, obovate, prominently veined, 5–15 mm. long, 3–9 mm. broad, with petioles 1–2 mm. long; sepals roundish, greenish-yellow, up to 2 mm. long in female flowers; capsule 4–5 mm. in diameter; seeds light brown, about 2 mm. long.
Uncommon and local (St. Cath., Clar., St. Thom.); in arid thickets on limestone and at salina margins; SL–400 feet; fl. July–Aug, fr. Nov. *A 10761! H 10513! 12191! C. B. Lewis!* Greater Antilles, Virgin Is.

5. SAVIA Willd. (1806)

1 Leaves acuminate, elliptical to ovate-lanceolate; petiole and midrib at base densely
pubescent   1. *sessiliflora*
1 Leaves rounded or obtuse at tip:
 2 Leaves reticulated on both surfaces, dark green; blade obovate-elliptical; ovary and
 young capsule tomentellous   2. *erythroxyloides*
 2 Leaves beneath smooth and light green; blade oblong-obovate, narrowed at base, up to
 6 cm. long, 1–3 cm. broad; Grand Cayman   *S. bahamensis* Britton

1. S. sessiliflora (Sw.) Willd. in L., Sp. Pl. ed. 4, **4** (2): 771 (1806).

Shrub or tree 4–8 m. high, the young twigs pubescent; leaves 4–8 cm. long;
flowers clustered or female solitary; sepals pubescent; ovary glabrous; capsule
glabrous, 8–10 mm. in diameter; seeds broadly ellipsoidal, 5 mm. long.

Known in Jamaica only from the original collection by Swartz. Greater Antilles,
Virgin Is.

2. S. erythroxyloides Griseb. in Mem. Amer. Acad. n. s. **8**: 157 (1860).

Shrub or tree 2–6 m. high; leaves leathery, glabrous, 2–6 cm. long, 1·2–2·5 cm.
broad; sepals pubescent; capsule glabrous at maturity, about 8 mm. in diameter;
seeds trigonous, brown, 4–5 mm. long.

Rare (St. Andr., Trel., St. Thom.), in thickets and woodlands on rough limestone; 200–2200 feet; fl. May, Nov, fr. July, Nov. *duQuesnay 372! H 12196!
12198! H & P 14390!* Cuba, Hispaniola, Cayman Brac.

6. HYERONIMA Allem. (1848)

1. H. jamaicensis Urb., in Fedde, Repert. Sp. Nov. **16**: 137 (1919).—*H. clusioides*
of F. & R. (1920), not (Tul.) Müll. Arg. (1865).

Tree 8–20 m. high; leaves elliptical, acute or obtuse at apex, midrib and primary
veins prominent and hairy beneath, (7–) 9–16 (–20) cm. long, 4–8 cm. or more
broad; stipules about 1 mm. long; panicles scaly; calyx yellow-green, less than
1 mm. long, in fruit expanding to 1·3 mm.; fruit fleshy with hard endocarp,
globose, 2 mm. long.

Rare (St. Andr., Clar., St. Ann, Port.), in woodlands on limestone in wet areas;
2300–3000 feet; fl. Mar., fr. July. *H & B 10733! P 19651!* Endemic.

7. DRYPETES Vahl (1807)

1 Leaves spiny-toothed; ovary 1-locular; stigma solitary; stamens shorter than sepals
                                                                       1. *ilicifolia*
1 Leaves entire, wavy-margined or with small teeth; stamens longer than sepals:
 2 Ovary 1-locular; stigma solitary; sepals about 1 mm. long; fruiting pedicels 3–8 mm.
 long   2. *alba*
 2 Ovary 2-locular; stigmas 2; sepals about 2 mm. long; fruiting pedicels 1–3 mm. long
                                                                       3. *lateriflora*

1. D. ilicifolia Krug & Urb. in Engl., Bot. Jahrb. **15**: 352 (1892).—Rosewood.

Shrub 2–3 m. or tree to 10 m. high; leaves ovate to oblong-elliptical, apex acute
and spine-tipped, unequal at base, leathery, 7–15 cm. long, 2·5–7 cm. broad;
male flowers greenish, 4–10 in a cluster; stamens 7–10 with filaments less than
1 mm. long; fruit ellipsoidal, tomentellous, 1-seeded, 1·5–2 cm. long.

Rare and local in woodlands on limestone in north-western and western coastal
areas; SL–1750 feet; *A 8655! H 7089!* Puerto Rico.

2. D. alba Poit. in Mém. Mus. Hist. Nat. Paris **1**: 157, t. 7 (1815).

Tree 5–20 m. high; leaves ovate to lanceolate, with short obtuse acumen,
cuneate and often unequal at base, 5–12 cm. long, 1·5–3 cm. broad; flowers
clustered, greenish-white, fragrant; stamens 3–6; fruit obliquely ellipsoidal, white,
8–13 mm. long.

Rather local and uncommon on wooded limestone hillsides in the western
parishes; 700–2500 feet; fl. Jan–Feb, fr. Feb–Mar. *H 7098! P 9956! 11652!*
Greater Antilles, Guadeloupe, ? Antigua.

3. D. lateriflora (Sw.) Krug & Urb. in Engl., Bot. Jahrb. **15**: 357 (1892).— Whitewood.

Shrub 1–5 m. or tree to 13 m., with slender trunk and drooping branches; leaves oblong or elliptical, acute to subacuminate, obtuse and often unequal at base, 5–11 cm. long, 2–6 cm. broad; flowers clustered, fragrant; stamens 3–5, with filaments up to 5 mm. long; ripe fruit ovoid, puberulous at least during ripening, orange-scarlet, about 1 cm. long; seed solitary, 5–7 mm. long.

Frequent, mostly in the central and western parishes, on limestone cliffs and in rocky woodlands; SL–2750 feet; fl. Jan–Mar, Sept, fr. Mar–Sept. *A 6847! H 8709! H & P 15051!* Florida, C. Amer., Bahamas, Greater Antilles.

8. CAPERONIA St.-Hil. (1824)

1. C. castaneifolia (L.) St.-Hil., Hist. Pl. Rem. Brés.: 245 (1824).

Weak-stemmed straggling herb 30–60 cm. high with sparse glandular hairs only on youngest parts; leaves linear to oblong-lanceolate or obovate, distantly serrate, 2·5–7·5 cm. long, 1–4 cm. broad; spikes 2·5 cm. long; calyx of male flowers 2·5 mm. long; petals slightly longer, white, unequal; female flowers with smaller petals; styles divided into 6–7 pointed lobes; capsule about 6 mm. broad.

Rather uncommon, in swamps, wet meadows and rice fields; SL–400 feet; fl. June–Jan, fr. July–Jan. *A 10160! H 12096! P 24383!* Mexico to Paraguay, Cuba, Hispaniola, Barbados.

9. CROTON L. (1753)

1 Annual weedy herbs or short-lived undershrubs, mostly less than 60 cm. high; racemes
  up to about 3 cm. long:
  2 Leaves entire, without glands, hispid with stellate hairs beneath, glabrous adaxially;
    stipules glandular   **1. *ovalifolius***
  2 Leaves serrate, with stalked glands at base of blade, usually with flat stellate hairs on
    both surfaces; stipules not glandular:
    3 Glands at base of blade saucer-shaped, the stalks less than 1 mm. long; stipules about
      2 mm. long, caducous   **2. *glandulosus***
    3 Glands at base of blade cup-shaped, the stalks tapered, 1·5–2 mm. long; stipules up to
      5 mm. long   **3. *hirtus***
1 Perennial shrubs mostly more than 60 cm. high, or trees, often aromatic:
  4 Leaves serrate or serrulate:
    5 Glands at apex of petiole wanting; stipules subulate, stellate-hairy, up to 12 mm. long,
      early caducous; adaxial surface of leaf usually with sparse stellate hairs
                                    4. *grisebachianus*
    5 Glands at apex of petiole or base of blade usually present:
      6 Stipules inconspicuous; leaves crowded as if whorled in groups of 3–5, with few
        stellate hairs; glands 1 or 2 at base of midrib beneath or wanting   **5. *wilsonii***
      6 Stipules 5 mm. or more long; leaves not crowded:
        7 Stipules linear-acuminate, 5–6 mm. long, early caducous; young shoots, petioles
          and leaves with numerous stellate hairs; mature capsule about 10 mm. long
                                    6. *corylifolius*
        7 Stipules 5–10 mm. long, cut into linear gland-tipped segments; leaves with few
        stellate hairs on both surfaces; capsule 6–7 mm. long; shrub about 2 m. high
                                  7. *populifolius*
  4 Leaves entire or subentire (denticulate in *C. flavens*):
    8 Leaf-margin ciliate with glands; stipules obsolete or of minute glands:
      9 Leaf-blade stellate-pubescent beneath, otherwise variably hairy
                                    8a. *humilis* var. *humilis*
      9 Leaf-blade glabrous   **8b. *humilis* var. *adenophyllus***
    8 Leaf-margin not ciliate with glands:
      10 Glands at base of leaf-blade and at apex of petiole wanting; blade glandular-
        punctate; stipules less than 1·5 mm. long or wanting:
        11 Leaves with silvery scales beneath, soft to touch; capsule often warted, scaly
                                    9. *eluteria*
        11 Leaves usually with stellate hairs or glabrous; capsule scabridulous with stellate
        hairs, not warted:
        12 Stems and leaves scabrous with scaly stellate hairs; female calyx 4 mm. long with
          lanceolate segments; leaves not usually crowded; leaf-blade broadly elliptical,
          glabrescent   **10. *laurinus***

12 Stems and leaves variably stellate- or simple-hairy to glabrous; female calyx
5–6·5 mm. long with oblong to spathulate segments, accrescent in fruit with
the basal margins becoming reflexed; leaf-blades mostly elliptic-lanceolate
                                                                    11. *lucidus*
10 Glands at base of leaf-blade and/or at apex of petiole present; stipules absent:
  13 Leaves ovate to lanceolate, denticulate, densely pale stellate-pubescent on both
    surfaces; lateral veins ascending; plants monoecious   12. *flavens*
  13 Leaves linear to oblong or narrowly lanceolate, entire, adaxial surface not densely
    stellate-pubescent, abaxial surface densely tomentose and whitish or yellowish
    with stellate hairs; lateral veins mostly at rightangles to midrib:
    14 Plants monoecious; leaves narrowly lanceolate to oblong and gradually tapered to
      apex; adaxial surface with very small stellate hairs; petioles 4–15 mm. long
                                                                    13. *priorianus*
    14 Plants usually dioecious; leaves glabrous or pustulate adaxially; petioles 2–8
      (–15) mm. long:
      15 Leaf-blades linear, mostly more than four times longer than broad, up to 7 cm.
        long and 8 mm. broad, the lateral veins obscure on both surfaces
                                                                    14. *linearis*
      15 Leaf-blades oblong to roundish-elliptical, mostly less than five times longer
        than broad, up to 8 cm. long and 2 cm. broad, the lateral veins impressed
        above and evident beneath   15. *discolor*

1. C. ovalifolius Vahl in West, Bidr. St. Croix: 307 (1793).
Woody herb 15–45 cm. high; leaves ovate to oblong-elliptical, rounded or
obtuse at apex, cuneate to obtuse at base, up to 3·5 cm. long and 1·5 cm. broad;
calyx of male flower 2·5 mm. long; stamens about 10, filaments hairy below; calyx
of female flower 4 mm. long, accrescent to 5·5 mm. long in fruit; capsule globose,
4·5 mm. long.
Rare (St. Andr.), a weed of sandy or gravelly soil and on dry banks; 100–150
feet; fl. and fr. Feb, July, Dec. *H 12119!* Hispaniola, Virgin Is., Guadeloupe,
Aruba, Curaçao, Bonaire and the coasts of Colombia and Venezuela.

2. C. glandulosus L., Syst. Nat. ed. 10, **2**: 1275 (1759).
Annual shrubby herb 15–80 cm. high; leaves ovate to elliptical, obtuse or acute
at tip, obtuse at base, coarsely serrate, 2–3·5 (–7) cm. long, 1–4 cm. broad; male
calyx about 2 mm. long; stamens 8–10, filaments glabrous; female calyx with
unequal segments up to 3 mm. long; capsule ellipsoid-globose, 5–6 mm. long.
Rare (St. Andr., St. Eliz.), in open sandy places; 10–15 feet; fl. and fr. Feb.
Proctor & Mullings 21971! Continental tropical Amer., Greater Antilles, Antigua,
Guadeloupe, Martinique, Trinidad, Curaçao.

3. C. hirtus L'Hérit., Stirp. Nov. **1**: 17, t. 9 (1785).
Annual shrubby herb 30–100 cm. high with unequally rayed stellate hairs on the
twigs; leaves broadly ovate, serrate to crenate-serrate, cuneate to truncate at base,
2·5–7 cm. long, 2–5 cm. broad; male calyx 1·5 mm. long; stamens about 10,
filaments glabrous; female calyx with unequal segments up to 2·5 mm. long,
accrescent to 4 mm. long in fruit; capsule subglobose, about 4 mm. long.
Uncommon and local (St. Andr.), a weed of pastures and waste ground; 150–
600 feet; fl. and fr. Aug–Dec. *A 7948! H 6959!* Mexico to tropical S. Amer.,
Trinidad, W. Africa, Malaya.

4. C. grisebachianus Müll. Arg. in Flora **47**: 484 (1864).—? *C. micans* of F. & R.
(1920), not Sw. (1800).
Shrub 1·5–4 m. high; leaves ovate to broadly lanceolate, usually distinctly
acuminate, shortly cordate at base, doubly serrate to serrulate, 2–10 cm. long, 1–5
cm. broad; racemes axillary and terminal, 2–4 cm. long; male calyx about 2 mm.
long; stamens about 12, filaments glabrous; female calyx 3·5–4 mm. long; capsule
not seen.
Local (St. Andr., St. Cath.), in thickets on arid limestone; 500–1400 feet; fl.
June–Sept, Dec. *A 5636! H 8944! P 19722!* Endemic.

5. C. wilsonii Griseb., Fl. Br. W.I.: 40 (1859).—Wild Camphor.
Straggling or erect shrub 1–3 m. or tree to 5 m. high; leaves oblong-lanceolate,
acuminate, dark green and shiny above, paler beneath, aromatic, 4–10 cm. long,

1–4·5 cm. broad; racemes 3–9 cm. long; male calyx about 2 mm. long; stamens about 30, filaments glabrous; female calyx 2·5–3 mm. long; capsule globose, with pungent stellate hairs, about 5 mm. long.

Frequent, in the central and western parishes, in thickets and woodlands on limestone rocks and cliffs; 50–2700 feet; fl. Dec–Sept, fr. Feb–Sept. *A 12466! H 9167! P 7656!* Endemic.

6. C. corylifolius Lam., Encycl. Méth. Bot. **2**: 205 (1786).

Shrub 4–5 m. or tree to 10 m. high; leaves broadly ovate to oblong, obtuse, acute or shortly acuminate, truncate-subcordate at base, aromatic, often tinged pink or red, 3–14 cm. long, 2–9 cm. broad; racemes 10–20 cm. long; flowers subsessile; male calyx 2·5 mm. long; stamens 16–18, filaments glabrous; female calyx 3 mm. long.

Occasional, but rare in the wettest areas, in hillside pastures and woodland margins, mostly on limestone; 1200–3000 feet; fl. Apr–June, Sept, fr. Feb, May–July. *A 7261! H 11093! H & P 14349!* Venezuela, Cuba, Hispaniola, Guadeloupe to St. Vincent.

7. C. populifolius Mill., Gard. Dict. ed. 8 (1768).

The existence of this species in Jamaica has not been confirmed, but it is otherwise widespread in the southern Caribbean region. It was reported by Swartz (1788) perhaps erroneously or as a misidentification of *C. corylifolius* which was not reported before Geisel (1807).

8. C. humilis L., Syst. Nat. ed. 10, **2**: 1276 (1759).—Pepper Rod.

8a. Var. humilis.

Aromatic shrub 0·3–1·5 m. high; leaves ovate, mostly long-acute, rounded or cordate at base, 1·5–5·5 (–7) cm. long, up to 3 cm. broad; stipules minute, glandular; racemes 3–14 cm. long; male calyx 2–3 mm. long, stamens more than 20; female calyx about 4 mm. long, accrescent in fruit; capsule subglobose, 5–6 mm. long.

Common in rough pastures and rocky thickets; 25–2600 feet; fl. and fr. all the year. *A 6719! H 9676! H & P 13773!* Florida, Bahamas, Mexico, Hispaniola, Puerto Rico, Virgin Is., Grenadines.

8b. C. humilis var. adenophyllus (Spreng.) Adams, unpublished.—*C. adenophyllus* Spreng. (1826). *C. laurinus* var. *adenophyllus* (Spreng.) F. & R. (1920).

Like the last but leaves and flowers glabrous.

Locally common (St. Andr., St. Thom.), in thickets on limestone rocks and on gravel in arid areas; 25–2500 feet; fl. Feb, June–Aug, Nov, fr. June–Aug, Nov. *A 5505! H & B 10791! P 19706!* Endemic.

9. C. eluteria (L.) Sw., Nov. Gen. & Sp. Pl.: 100 (1788).—*Clutia eluteria* L. (1753). *Croton glabellus* of F. & R. (1920), not L. (1759). *Croton nitens* Sw. (1788). Cascarilla Bark.

Shrub 1·5–3 m. or tree to 12 m. high; leaves oblong-elliptical, usually obtuse at apex and base, sometimes rounded and subcordate at base, gland-dotted, 2–15 (–20) cm. long, up to 6·5 cm. or more broad; racemes usually branched, 3–12 cm. long; sepals 2–2·5 mm. long; stamens 10–13; capsule subglobose to obovoid, 7–9 mm. long; seeds 6 mm. long.

Common, in thickets and woodlands on limestone in dry and wet areas; SL–2900 feet; fl. and fr. all the year. *A 6965! H 9750! P 21900!* Bahamas, Mexico to Colombia and Ecuador, Cuba, Hispaniola, Grand Cayman; introduced into Nigeria. *C. nitens* Sw. is sometimes kept distinct and refers to the Jamaican plant.

10. C. laurinus Sw., Nov. Gen. & Sp. Pl.: 100 (1788).

Shrub 1·5–3 m. or slender-branched tree to 8 m. high; sap often turning red or orange on exposure to air; leaves broadly elliptical, obtuse, retuse, shortly apiculate or acutely acuminate at tip, rounded to cuneate at base, 7–28 cm. long, up to 15 cm. broad; racemes to 30 cm. long; male calyx 2 mm. long, stamens 5–10; female calyx 4 mm. long; capsule oblong-ellipsoid, 12–15 mm. long.

Occasional, in open woodlands on craggy limestone; 1500–3150 feet; fl. and fr. most of the year. *A 7533! 12403! H 8693! P 15982!* Endemic.

11. C. lucidus L., Syst. Nat. ed. 10, **2**: 1275 (1759).—Basket Hoop.

Shrub 1–2·5 m. or tree to 12 m. high; leaves ovate to elliptic-lanceolate, shortly acuminate, narrowly rounded to subcordate at base, gland-dotted, aromatic, 4–10 cm. long, up to 4·5 cm. broad; racemes 3–8 cm. long with white or greenish-white flowers; male calyx about 2 mm. long, stamens 11–12; female calyx 5–6 mm. long; capsule oblong-ellipsoid, smooth, about 5 mm. long; seeds 3 mm. long.

Frequent, on rocky banks and in open thickets and woodlands on limestone; 50–3000 feet; fl. and fr. most of the year. *A 8840! 12475! H 10966! P 24018!* Bahamas, Greater Antilles; Grand Cayman.

12. C. flavens L., Syst. Nat. ed. 10, **2**: 1276 (1759).—Yellow Balsam.

Aromatic shrub 1–3 m. high; leaves ovate to lanceolate, acute, mucronate, cordate at base, 2–10 cm. long, 1–4 cm. broad; racemes 10–12 cm. long; male and female calyx about 2 mm. long; stamens 12–16; capsule subglobose, about 5 mm. long; seeds dark grey, 3–3·5 mm. long.

Locally common (St. Andr., St. Cath., St. Thom.), on exposed rocks and in open thickets in arid limestone areas; SL–750 feet; fl. and fr. Nov–July. *A 6979! H 9586! P 16508!* Yucatan, Cuba, Hispaniola, St. Lucia, St. Vincent, Grenadines, Grenada.

13. C. priorianus Urb., Symb. Ant. **3**: 295 (1902).

Shrub 1–2·2 m. high; leaf-apex obtuse or rounded, base rounded or subcordate, blade 2·5–6 cm. long, 7–15 mm. broad; racemes 2–10 cm. long; sepals 1·5–2 mm. long; stamens about 16; capsule globose, about 5 mm. in diameter.

Local and uncommon (Manch., St. Eliz., Han.), in thickets and woodlands on limestone; 750–2300 feet; fl. Jan–Feb, June, fr. June. *H & P 14952! P 15446!* Endemic.

A plant with a strong superficial resemblance to *C. priorianus* and almost identical with *C. plumieri* Urb. of Hispaniola, occurs in a small population with *C. flavens* and *C. linearis* and has been shown to be a hybrid between these two species; St. Catherine, in thickets on arid limestone; 40–200 feet; fl. and fr. Nov–Feb. *A 8886! 12732!*

14. C. linearis Jacq., Enum. Syst. Pl. Carib.: 32 (1760).—Rosemary.

Aromatic shrub 1–3 (–4) m. high; plants dioecious; racemes 2–12 cm. long; sepals about 1·5 mm. long; stamens about 12; capsule 5 mm. long; seeds 2·5 mm. long.

Very common, mainly coastal on limestone rocks but also inland on well drained calcareous or serpentine soils and gravel; SL–3000 feet; fl. and fr. all the year. *A 5591! 7237! H 9580! 9733! H & P 13737!* Florida, Bahamas, Hispaniola, Grand Cayman.

15. C. discolor Willd. in L., Sp. Pl. ed. 4, **4** (1): 532 (1805).

Aromatic shrub like the last, 0·3–3 m. high; plants dioecious but sometimes monoecious.

Rather common, in thickets and on exposed limestone rocks and on sand dunes; SL–25 (–500) feet; fl. and fr. sporadically throughout the year. *A 7370! 11836! H 10261! P 6649! 16252!* Bahamas, Greater Antilles, Virgin Is. The type of *C. discolor* Willd. from St. Croix has not been seen; it was stated by N. L. Britton that the Jamaican plant thus named is not identical with true *C. discolor* but merely a broad-leaved variant of *C. linearis*.

10. ARGYTHAMNIA P. Browne (1756); Ingram (1966)

1. A. candicans Sw., Nov. Gen. & Sp. Pl.: 39 (1788).

Straggling branched shrub 15–140 cm. high; leaves elliptical to lanceolate, acute or obtuse, narrowed to base, serrulate, often greyish or tinged purple, from very small up to 8 cm. long and 3·5 cm. broad; racemes up to 1 cm. long; male flowers 4-merous, white or greenish-white; female flowers 5-merous, styles 3, twice divided; capsule 3–4 mm. long, 5–6 mm. broad; seeds obovoid, 2–2·5 mm. long.

Rather common in arid thickets and woodlands on limestone; 20–2200 feet; fl. Sept–May, fr. Nov–May. *A 5658! H 9055!* Bahamas, Greater Antilles, Virgin Is., Anguilla. *A. proctorii* Ingram, described from Grand Cayman, differs in having the glands of the pistillate flowers squarish and almost as thick as broad.

11. ACIDOCROTON Griseb. (1859)

1. A. verrucosus Urb., Symb. Ant. **7**: 513 (1913).

Shrub 1–3 m. or tree about 4 m. high with arching branches; leaves obovate to elliptical, obtuse at tip, 1·5–2·5 cm. long, 5–9 mm. broad; petioles 3–5 mm. long; male flowers 5-merous, greenish; fruiting calyx foliaceous.

Rare and local (Clar., Trel.), on limestone rocks in woodlands; 1500–2500 feet; fl. May–June, Oct, fr. July–Nov. *H 11018! H & P 14381! P 21350!* Endemic.

12. JATROPHA L. (1753)

1 Petals green, white or yellow; leaves not or only shortly lobed; stamens about 10 (9–15), more or less united; stipules small:
  2 Leaves peltate; inflorescence-branches forking but not diverging; capsule about 1·5 cm. long   *1. hernandiifolia*
  2 Leaves not peltate:
    3 Inflorescence repeatedly 2-forked, the branches diverging; leaf-blade rounded at base; capsule about 2 cm. long   *2. divaricata*
    3 Inflorescence of corymbose cymes; leaf-blade openly cordate at base; capsule 2·5–4 cm. long   *3. curcas*
1 Petals red or crimson:
  4 Leaves peltate, palmately 3–6-lobed to subentire; stipules glandular-fringed, hardening and persistent after leaf-fall; base of stem swollen; petals scarlet, 6–7 mm. long; filaments 6–9, free; cultivated and occasionally naturalized; native of C. Amer.; Coral Plant, Gouty Foot   *J. podagrica* Hook.
  4 Leaves not peltate; stipules not persistent; stems not swollen:
    5 Leaves not palmately lobed, often hastate at base, narrowed at the middle, the terminal lobe acuminate; stipules subulate, 1–2 mm. long; petals rose-crimson, 12 mm. long; filaments 10 in two series, united into a central column; cultivated ornamental shrub; native of Cuba, Hispaniola and Puerto Rico   *J. integerrima* Jacq.
    5 Leaves palmately lobed; stipules divided into threadlike segments:
      6 Lobes of leaf 3–5, obovate-elliptical; blade-margin, petiole and stipules glandular; petals deep purplish-red within, 4 mm. long; filaments 10–12, united   *4. gossypiifolia*
      6 Lobes of leaf up to 11, linear-lanceolate or pinnatifid; blade-margin, petiole and stipules not glandular; petals rose-scarlet, 6 mm. long; filaments 7–8, free; ripe fruits yellow, ridged on the backs of the cocci; cultivated in gardens and occasionally escaping; native of tropical Amer.; French or Spanish Physic Nut   *J. multifida* L.

1. J. hernandiifolia Vent., Jard. Malm. **1**: under t. 52 (1804).—*Curcas hernandiifolius* (Vent.) Britton (1924).

Shrub 2–4 m. high; leaves entire or 3-lobed, 8–17 cm. long, 4–10 cm. broad.

This species, if really distinct from *J. divaricata*, requires confirmation for Jamaica. Some saplings found in a locality where *J. divaricata* is common (Clar.; 2500 feet. *Morley 301!*), had both peltate and non-peltate leaves. Hispaniola, Puerto Rico.

2. J. divaricata Sw., Nov. Gen. & Sp. Pl.: 98 (1788).—Wild Oil Nut.

Shrub 2·5–4 m. or tree to 8 m. high with flexible branches and resinous aromatic latex; leaves ovate to elliptical, softly leathery, acute or acuminate, rounded to cordate at base, 7–16 cm. long, 3–9 cm. broad, glabrous except often with patches of hairs between the main lateral veins abaxially; peduncles 3–10 cm. long; male petals about 5 mm. long; female flowers larger, solitary at the panicle-branches; seeds purplish, 10–12 (–14) mm. long.

Frequent in the central and western parishes, in thickets and woodland margins on limestone; 750–2500 feet; fl. and fr. all the year. *A 11038! H 8715! H & P 15062!* Endemic.

3. J. curcas L., Sp. Pl. **2**: 1006 (1753).—Physic Nut.

Shrub to about 4 m. high or rarely a tree with viscid milky or reddish sap; leaves broadly ovate, entire or shallowly lobed, acute or acuminate, up to 15 cm. long and 13 cm. broad; cymes branched from the base or pedunculate; flowers green; petals 6–7 mm. long; seeds blackish, about 2 cm. long.

Frequent, mostly near habitations; SL–3000 feet; fl. and fr. sporadically. *A. M. Barry! H 7666!* Native of tropical Amer., now widespread; Grand Cayman.

4. J. gossypiifolia L., Sp. Pl. **2**: 1006 (1753).—Belly-ache Bush, Cassada Marble.

Shrub with low tortuous branches 60–120 (–200) cm. high; sap turbid, yellowish; leaves cordate at base, the lobes acute or acuminate, more or less pubescent, 5–14 cm. in diameter; panicles 5–20 cm. long; capsule about 1 cm. long; seeds greyish-brown, mottled, 7–8 mm. long.

Locally common especially in sandy or gravelly waste places near the sea; SL–1000 feet; fl. and fr. all the year. *A 8160! H 6819! P 23508!* General in the tropics and subtropics.

13. RICINUS L. (1753)

1. R. communis L., Sp. Pl. **2**: 1007 (1753).—Castor Oil Plant, Oil Nut.

Short-lived shrub or small tree; leaves peltate, divided into 7 or more serrate lobes, up to 60 cm. broad; male flowers 12–15 mm. across, the calyx 3–5-partite; female calyx about 8 mm. long, deciduous; capsule 12–24 mm. long, usually spiny, sometimes smooth; seeds variously mottled, carunculate, 10–17 mm. long.

Common as a cultivated plant and on waste ground; SL–1500 feet; fl. and fr. most of the year. *H & P 13484! Wynter UCWI 5615!* Native of the Old World tropics, now widespread; Grand Cayman. There are many varieties of this species differing in stature, colour of foliage, size and shape of fruit and size and colour of seeds.

14. MANIHOT Mill. (1754)

Leaves peltate, deeply 3–5-lobed, the lobes elliptical; tree to 12 m. high; capsule not winged, about 2 cm. long and broad; planted locally; native of Brazil; Ceara Rubber

M. glaziovii Müll. Arg.

Leaves not peltate, deeply 3–7-lobed, the lobes linear- to elliptic-lanceolate; shrub or small tree with clustered tuberous roots; capsule 6-winged, about 1·5–2 cm. long

1. *esculenta*

1. M. esculenta Crantz., Inst. Rei Herb. **1**: 167 (1766).—*M. utilissima* Pohl (1826). *Jatropha manihot* L. (1753). Cassava, Tapioca.

Usually a glabrous shrub with thick-noded stems 2–3 m. high; lobes of leaves glabrous or sparingly puberulous, acuminate, 8–15 cm. long; stipules linear, 6–7 mm. long; seeds mottled, about 12 mm. long.

Cultivated locally on the heavier soils; SL–3500 feet; fl. and fr. sporadically when allowed to mature, being mostly propagated by cuttings. Native probably in Brazil, now widespread in all tropical countries; Grand Cayman.

15. OMPHALEA L. (1759) nom. cons.

Anthers 3; young branches glabrous or glabrescent; ovary glabrous; leaves usually narrowed to base; tree

1. *triandra*

Anthers 2; young branches and ovary tomentose-pubescent; leaves cordate at base; climber

2. *diandra*

1. O. triandra L., Syst. Nat. ed. 10, **2**: 1264 (1759).—Pop Nut.

Shrub, when coppiced, or tree up to 15 m. high; sap viscid, clear; leaves with two glands at apex of petiole, blade subentire, obovate-elliptical, fleshy, shiny above, up to 30 cm. long and 12 cm. broad; panicle glabrous or shortly golden-puberulous; bracts up to 15 cm. long, narrowly spathulate; capsule depressed, 3·5–4·5 cm. long, 6–7·5 cm. broad; seeds globose, about 2·5 cm. in diameter.

Frequent, in wet or moderately wet woodlands on limestone, or rare at higher elevations on shale; SL–2000 (– ? 4000) feet; fl. most of the year, fr. Feb–Mar, Aug, Nov. *A 10175! H 9272! H & P 13852!* Endemic; introduced into Haiti for the apparently edible seeds.

2. O. diandra L., Syst. Nat. ed. 10, **2**: 1264 (1759).
Trailing and climbing shrub; leaves broadly ovate-elliptical, shortly cuspidate-acuminate, 9–17 cm. long; bracts 1–7 cm. long; seeds 4·5 cm. long.
Reported from Portland but not recently confirmed. Panama to Brazil and Peru, Cuba, St. Kitts, Guadeloupe, St. Vincent, Trinidad.

16. ADELIA L. (1759) nom. cons.

1. A. ricinella L., Syst. Nat. ed. 10, **2**: 1298 (1759).—*A. haemiolandra* (Griseb.) Pax (1912). Wild Lime.
Deciduous more or less spiny shrub 2–3 m. or tree to 10 m. high; bark very pale; wood lemon-yellow; branches erect or drooping with short shoots spine-tipped; leaves obovate to oblanceolate, with or without domatia, up to 8 (–12) cm. long and 3 (–4·5) cm. broad; flowers dioecious, fragrant; male flowers with pedicels 4–5 mm. long, sepals about 3 mm. long; female flowers with pedicels 1–5 cm. long, sepals green, reflexed in fruit; capsule about 8 mm. in diameter.
Rather common in arid thickets on limestone mostly in coastal areas; SL–800 feet; fl. Feb–May, fr. Mar–Aug. *A 12289! 12331! H 10222! 10234! P 11955!* Greater Antilles, Virgin Is., Grenada, Curaçao, Tobago; Grand Cayman.

17. CHAETOCARPUS Thwaites (1854) nom. cons.

1. C. globosus (Sw.) F. & R. in Journ. Bot. **57**: 312 (1919).
Tree, much branched and densely leafy, 4–10 m. high; twigs puberulous; leaves broadly ovate, obtuse at tip, rounded at base, shiny, 2·5–4 (–5) cm. long, 2–3 (–3·5) cm. broad; flowers dioecious; calyx about 2 mm. long; capsule reddish-brown, 11–12 mm. long.
Locally common (St. Andr., Clar., Port.), in montane and submontane woodland; 2000–5000 feet; fl. May–Aug, fr. July, Oct, Dec. *A 12529! A & S 3123! H 9112! P 16732!* Cuba.

18. MERCURIALIS L. (1753)

1. M. annua L., Sp. Pl. **2**: 1035 (1753).

1a. Var. **ambigua** (L. f.) Duby in DC., Bot. Gall. ed. 2, **1**: 417 (1828).—*M. ambigua* L. f. (1762).
Annual herb with glabrescent branches, 7–20 cm. high; leaves ovate to elliptic-lanceolate, crenate-serrate, 1·5–5 cm. long; flowers monoecious, male clustered, female subsessile in the leaf-axils; fruit hispid, 3–4 mm. broad.
Rare and local (St. Andr.), on earth banks by pathsides; about 3200 feet; fl. and fr. July. *P 23852!* Native of W. Europe and the Azores; the var. *annua* introduced into Hispaniola.

19. SEBASTIANIA Spreng. (1821)

1 Leaves sessile, ovate to oblong-elliptical, alternate below, opposite or in whorls of three
   above; female spike terminal, very short and subsessile   1. *fasciculata*
1 Leaves petiolate, alternate, opposite or subverticillate:
   2 Leaf-blade elliptical, 3·5–8 cm. long; spikes terminal, usually unisexual or sometimes
      with female flowers at base of male spike   2. *spicata*
   2 Leaf-blade ovate to elliptical, 2·5–3·5 cm. long; male spikes terminal and axillary
                                        3. *alpina*

1. S. fasciculata (Millsp.) Pax & K. Hoffm. in Engl., Pflanzenr. 4 (147), VII: 422 (1914).—*Dendrocousinsia fasciculata* Millsp. (1913).

Shrub 2–3 m. or tree to 5 m. high; leaves obtuse at both ends, thinly leathery, glabrous, 3–9 (–12) cm. long, 2–5·5 cm. broad; female flowers sessile or subsessile; sepals 3, broadly ovate, about 2 mm. long.

Rare and local (Han.), in woodland on limestone; 1650–1780 feet; fl. Mar–Apr, Sept–Oct, fr. Apr. *A 8002! H 10266! P 10052!* Endemic.

2. S. spicata (Millsp.) Pax & K. Hoffm. in Engl., Pflanzenr. 4 (147), VII: 422 (1914).—*Dendrocousinsia spicata* Millsp. (1913).

Shrub 3–4 m. or tree to 5 m. high; leaves rounded or emarginate at both ends with distinct revolute margins, thickly leathery, up to 5·5 cm. broad; male spikes to 10 cm. long, the flowers sessile; female spikes 2–3·5 cm. long, the flowers subsessile or with pedicels to 4 mm. long; capsule 7–8 mm. in diameter; seeds grey, 4 mm. long.

Very local (Clar.), on limestone rocks in woodland; 2500 feet; fl. Dec–Jan, May–Sept, fr. Jan, May–Sept. *H 10980! 10981! P 9767! Robertson UCWI 1997!* Endemic.

3. S. alpina (F. & R.) Pax & K. Hoffm. in Engl. & Prantl, Nat. Pflanzenf., ed. 2, 19C: 193 (1931).—*Dendrocousinsia alpina* F. & R. (1919).

Shrub 2–3 m. or tree to 6 m. high; leaves obtuse at both ends, margin with a few small teeth and slightly revolute, up to 2 cm. broad, thinly leathery; male bracts and sepals roundish-ovate.

Rare and local (St. Andr., Port.), in wet montane woodland on limestone; 2500–5750 feet; fl. Nov–Mar, fr. Jan. *H 12906! HPS 14752! P 9542!* Endemic.

Sebastiania in Jamaica requires further collection and study. The described species are not well known and others, as yet insufficient for description, exist.

20. SAPIUM P. Browne (1756); Jablonski (1968)

1 Glands paired at apex of petiole; leaf-blade oblong-elliptical, broadly cuneate to rounded at base, the lateral veins conspicuous; ovary sessile   1. *jamaicense*
1 Glands two or few, very small, on basal margins of blade or obsolete; leaf-blade obovate to oblong-oblanceolate, narrowly cuneate at base, the lateral veins obscure:
2 Ovary on a stalk about 3 mm. long, 2-locular; styles united halfway   2. *harrisii*
2 Ovary sessile or subsessile, 2–3-locular; styles shortly united at base only   3. *cuneatum*

1. S. jamaicense Sw., Adnot. Bot.: 62 (1829).—Beyacca, Gum Tree, Milk Wood.

Tree 10–21 m. high with copious milky sap; leaves rounded or cuspidate or shortly acuminate at tip, 7–28 cm. long, up to 8 cm. broad; spikes monoecious, up to 15 cm. long; capsule globose, 7–8 mm. in diameter; seeds roundish-lenticular, about 4 mm. long.

Frequent, in sheltered valleys; 120–2200 feet; fl. Dec–Mar, Aug–Sept, fr. Jan–June. *A 9213! H 9195! J.P. 943! P 21896!* Mexico, Guatemala, Honduras, Greater Antilles.

2. S. harrisii Urb. ex Pax in Engl., Pflanzenr. 4 (147), V: 236 (1912).

Tree with numerous ascending branches, sap milky, 6–20 m. high; leaves rounded or shortly cuspidate at tip, 10–17 cm. long, up to 6 cm. broad; male spikes 3·5–5 cm. long; female spikes axillary and at leafless nodes, erect or ascending, up to 7 cm. long; capsule ovoid, about 9–10 mm. long.

Rather local (St. Andr., Port.), in submontane woodland and forest; 3500–4300 feet; fl. Oct–Feb, fr. Dec–Jan. *A 12079! 12678! H 10117! P 22072!* Endemic.

3. S. cuneatum Griseb., Fl. Br. W.I.: 49 (1859).—Burn Eye.

Tree with copious milky sap, 8–13 m. high; leaves rounded or cuspidate at tip, 7·5–16 cm. long, 3·5–7·5 cm. broad; male spikes up to 7 cm. long; female spikes 3–4 cm. long; capsule 11–12 mm. in diameter; seeds red, ellipsoidal, about 8 mm. long.

Occasional in the central parishes, in woodland on limestone; 1500–3100 feet; fl. Apr–June, Sept, fr. Jan–Mar, Sept. *A 10999! H 9817! 11214! P 17476!* Endemic.

21. HIPPOMANE L. (1753)

1. H. mancinella L., Sp. Pl. 2: 1191 (1753).—Manchineel.

Tree with copious milky sap, 4–20 m. high, deciduous; bark silvery- to blackish-grey, longitudinally and transversely fissured; leaves ovate to ovate-elliptical, rounded-subcordate at base, apex obtuse to acuminate, serrulate, soft, shiny adaxially, 2·5–10 (–12) cm. long, up to 8 cm. broad; spikes 4–15 cm. long; fruit depressed-globose, exocarp yellow when ripe, endocarp bony and warted, about 3 cm. in diameter, (5–) 6–9-locular.

Rather local and not now common (mostly St. Cath., Clar., St. Thom.), in thickets on limestone and at salina-margins, never far from the sea; SL–40 feet; fl. Apr, fr. July, Nov. Feb–Mar. *A 11905! H 10189! Powell 694! P 23421!* Florida, Bahamas, Mexico to Venezuela, West Indies; Grand Cayman, also Pacific coast of C. Amer. and introduced into W. Africa.

22. HURA L. (1753)

1. H. crepitans L., Sp. Pl. 2: 1008 (1753).—Sandbox Tree.

Tree 10–25 (–30) m. high, often of great girth, the trunk and branches usually spiny, otherwise smooth; leaves rounded-ovate, acuminate, cordate at base, margin often wavy-toothed, up to 20 (–50) cm. long and 15 (–30) cm. broad; male spike 3–5 cm. long on peduncle 5–11 cm. long, solitary to each bract; anther-loculi usually in 2 whorls; female flower dull red; capsule up to 8 cm. in diameter; seeds about 2 cm. in diameter.

Frequent, mostly as a planted tree; 50–1400 feet; fl. May–Oct, Jan, fr. Oct–Jan. *H 8383! Powell 252!* Native Costa Rica to Peru and Brazil and West Indies; introduced Florida, California, Bahamas, and in the Old World tropics.

23. GRIMMEODENDRON Urb. (1908)

1. G. jamaicense Urb., Symb. Ant. 5: 399 (1908).—Burn Eye, Redlight Wood.

Shrub 2·5 m. or tree 6–16 m. high, with milky sap; leaves oblong-oblanceolate, rounded, obtuse or acuminate at tip, cuneate at base, the margin more or less remotely serrulate, 7–20 cm. long, 3–8 cm. broad, paler beneath; spikes 12–17 cm. long, peduncle 1–2·5 cm. long; male flowers several to each bract; capsule ovoid-globose, 10–12 mm. long.

Local and uncommon (St. Eliz., Trel.), in forest on craggy limestone rocks; 1300–2200 feet; fl. Mar–Apr, fr. Jan–Feb, June–July. *A 12437! H 12380! P 9945! 24451!* Endemic.

24. TETRORCHIDIUM Poepp. & Endl. (1841)

1. T. rubrivenium Poepp. & Endl., Nov. Gen. 3: 23, t. 227 (1841).—Gum Wood.

Tree 5–16 m. high, exuding clear gummy sap from leaf-glands and from buds during drying; leaves obovate to elliptic-oblanceolate, obtuse or shortly and bluntly acuminate, cuneate at base with paired glands at junction of petiole and blade, midrib prominent beneath, 8–20 cm. long, up to 5·5 cm. broad; clusters of male flowers greenish, in spikes or panicles up to 20 cm. long; female racemes usually up to 5 cm. long, but sometimes male and female flowers mixed; capsule ovoid, about 8 mm. long.

Rather widespread in the central and eastern parishes but not common, in woodlands and thickets on limestone; 1000–3000 feet; fl. Dec–May, fr. Feb–May. *A 11082! H 8991! H & P 13621! P 9778!* C. and tropical S. Amer., St. Vincent.

25. TRAGIA L. (1753)

1. T. volubilis L., Sp. Pl. 2: 980 (1753).—Twining Cowitch.

Shrub with trailing and twining branches to 3 m. high, rough with stinging hairs; leaves triangular-ovate to oblong, acute, truncate or shallowly cordate at base, strongly serrate-dentate, up to 9 cm. long and 4·5 cm. broad; stipules lanceolate, about 3 mm. long; racemes very slender up to 9 cm. long, with narrow persistent bracts to the male flowers 1·5 mm. long; female flowers basal with pedicels elongating in fruit to 1·5–3 cm. long; fruit deeply divided into 3 sub-globose cocci, about 7 mm. in diameter, pungent-setose.

Common in pastures, thickets and woodland margins; near SL–2400 feet; fl. and fr. most of the year. *A 8869! H 9697! P 24367!* Throughout tropical Amer., also in W. Africa.

26. ACIDOTON Sw. (1788) nom. cons.

1. A. urens Sw., Nov. Gen. & Sp. Pl.: 84 (1788).—Mountain Cowitch.

Shrub 3–5 m. or slender tree to 8 m. high, sometimes branches of coppice-regrowth trailing to 6 m. long; leaves elliptical to elliptic-oblanceolate or sublinear, sharply acute or acuminate, narrowly rounded at base, the margin entire or with a few pointed teeth below the apex, strongly beset and margined on saplings and regrowth with long deciduous pungent hairs, 5–14 cm. long, 1·5–4·5 cm. broad; stipules 4–5 mm. long, persistent, becoming corky; flowers fragrant; perianth green; male flowers in short axillary racemes; female flowers in racemes up to 20 cm. long drooping in fruit; capsule 3-lobed, with stinging hairs, about 1·5 cm. in diameter; seeds globose, mottled, 4–5 mm. in diameter.

Common except in the extreme eastern parishes, in woodlands and thickets on limestone; 400–2600 feet; fl. and fr. most of the year. *A 6781! 7313! H 8733! 9433! H & P 14423! P 20842!* Endemic.

27. ALCHORNEA Sw. (1788)

1. A. latifolia Sw., Nov. Gen. & Sp. Pl.: 98 (1788).—Dovewood, Lablab Tree.

Tree (3–) 6–16 m. high with hard striate bark; leaves broadly ovate, acuminate, mostly rounded to shortly cordate at base with the lamina there glandular; margin with short remote glandular teeth, thinly stellate-puberulous beneath and with more or less well developed domatia in the vein-axils, curling and pendulous in dry weather, (5–) 10–30 (–40) cm. long, (3–) 6–18 (–25) cm. broad; female spikes up to 20 cm. long, male panicles shorter; capsule often purplish, 9–10 mm. in diameter; seeds subglobose, about 5 mm. in diameter.

Common in woodlands and thickets; 250–5600 feet; fl. Dec–Apr, fr. Feb–Aug. *A 8996! 12262! H 10707! HPS 14597! Patrick 212!* Mexico to Panama, Greater Antilles, introduced in S. Florida.

28. LASIOCROTON Griseb. (1859)

1 Primary and secondary veins of leaf very prominent beneath; leaf-blade densely tomentellous beneath; styles short entire or nearly so:
  2 Leaves broadly ovate, more or less cordate, entire or very shallowly sinuate
                                                                 1. *macrophyllus*
  2 Leaves elliptic-lanceolate, shortly rounded at base, denticulate   2. *trelawniensis*
1 Primary slightly and secondary veins not prominent beneath; leaf-blade rounded to
  cuneate at base, entire; styles each divided halfway into several spreading fimbriate
  style-arms:
  3 Leaves beneath glabrous on the lamina, thinly pubescent on the veins; petioles 2·5–
    7 cm. long   3. *fawcettii*
  3 Leaves beneath thinly to densely tomentose; petioles 1–3 cm. long   4. *harrisii*

1. L. macrophyllus (Sw.) Griseb., Fl. Br. W.I.: 46 (1859).

Shrub 2–4 m. or tree 5–8 m. high, the stems simple or sparingly branched; leaves usually bluntly pointed, 5–14 cm. long, 3–10 cm. broad; petioles 1·5–4 cm. long; male racemes axillary, compact, 3–4 cm. long; stamens up to about 15; female racemes pendulous in fruit, up to 15 cm. long; capsule olive-green, about 5 mm. long; seeds mottled, 3 mm. in diameter.

Locally common, in thickets on arid limestone along the south coast; 10–300 feet; fl. Jan–May, Sept, fr. Feb–Mar. *A 8902! H 11868! P 11661!* Endemic.

2. L. trelawniensis Adams in Phytologia **20** (5): 312 (1970).

Shrub up to 4 m. high; leaves acute, 2·5–7 cm. long, 1–2·5 cm. broad; petioles up to 1 cm. long; male racemes unknown; female racemes erect in fruit, 3–6 cm. long, 5–11-flowered; capsule olive-green, 5–6 mm. long, tawny-tomentose.

Very rare (Trel.), on wooded limestone hilltop; 2000–2200 feet; fr. Mar. *P 20746!* Endemic.

3. L. fawcettii Urb., Symb. Ant. **6**: 14 (1909).

Shrub to 4 m. or tree to 10 m. high with slender sparingly branched stems; young stems, petioles, inflorescence, flowers and fruits tomentose; leaves oblong-elliptical, acuminate, up to 25 cm. long and 10 cm. broad; male flowers clustered along a simple or shortly branched rachis up to 26 cm. long; stamens 15–20; female flowers solitary in racemes; capsule about 5 mm. long; seeds olive-grey, 3 mm. in diameter.

Rare and local (West., Han., Trel.), in thickets and woodlands on limestone; 300–1600 feet; fl. Mar–Apr, Oct, fr. Mar–Apr. *A 10955! H 10283! 10306! P 10016!* Endemic.

4. L. harrisii Britton in Bull. Torr. Bot. Club **41**: 16 (1914).

Shrub to 4 m. or tree to 8 m. high; very like the last and perhaps not really distinct, recent collections in and between the typical areas of both species tending to show intermediate characteristics.

Rare and local (Clar., St. Ann), in thickets on limestone hillsides; 2300–2500 feet; fl. Apr–May, fr. Apr–May, Sept. *A 12484! H 11192! P 10228!* Endemic.

29. BERNARDIA Mill. (1754)

1. B. dichotoma (Willd.) Müll. Arg. in Linnaea **34**: 172 (1865).—*Croton dichotomus* Willd. (1805). *B. carpinifolia* Griseb. (1859).

Shrub with slender straggling branches 1–6 m. or tree to 8 m. high; under-bark orange in colour; leaves ovate to elliptic-lanceolate, obtuse, more or less rounded at base, crenate-dentate, tomentose on both surfaces, 3–10 cm. long, 1–4 (–5) cm. broad; spikes axillary with overlapping bracts, 1–3·5 cm. long; male flowers 3–5 together with green perianths; stamens 15–25; capsule globose, 3-lobed, tomentose, 7 mm. long; seeds ovoid-globose, mottled black, 6 mm. long.

Rather common, on sea-cliffs and in thickets and woodlands on arid limestone; 10–2300 feet; fl. Sept–Apr, fr. Mar, May, Sept. *A 8844! H 12192! H & P 13808! 14914!* Bahamas, Greater Antilles, St. Vincent, Grenadines; Grand Cayman.

30. ACALYPHA L. (1753)

1 Herbs, annual or short-lived perennial, taprooted:
  2 Perennial; petioles short, up to 10 mm. long; young branchlets densely pubescent
    with light tawny hairs, without glands; fruiting bracts dentate; spikes bisexual
                                                        1. *chamaedrifolia*
  2 Annual; petioles slender, often as long as the blades; spikes unisexual:
    3 Inflorescence densely cylindrical; fruiting bracts lobed to base with 3–5 filiform
      appendages with long gland-tipped hairs; young branchlets and leaf-margins
      thinly pubescent with whitish curved hairs and long spreading gland-tipped hairs
                                          2. *alopecuroidea*
    3 Inflorescence narrowly spicate; fruiting bracts lobed to about the middle into about
      20 acute lobes; plant minutely pubescent, with few glandular hairs   3. *ostryifolia*
1 Shrubs or rarely small trees:
  4 Leaves ovate:
    5 Leaves variegated; introduced ornamental shrubs:
      6 Leaves mostly red marked with green, crimson or bronze; native of Pacific Is.;
      Copper Leaf   *A. wilkesiana* Müll. Arg.
      6 Leaves mostly green with white or cream-coloured margins; native of New Guinea
                                          A. godseffiana Mast.
    5 Leaves not variegated:
      7 Spikes very showy, pendulous, red, much longer than the leaves; cultivated orna-
      mental; native of tropical Asia; Cat Tail   *A. hispida* Burm. f.

7 Spikes not as above; wild plants:
  8 Spikes unisexual; leaves 3–5-nerved and broadly rounded, truncate or rarely shortly cordate at base, margin serrate-dentate, not glandular   4. *elliptica*
  8 Spikes bisexual; leaves (5–) 7–9-nerved and cordate at base, margin crenate, glandular-punctate adaxially and on the petiole and veins beneath   5. *cuspidata*
4 Leaves elliptical to lanceolate; spikes generally unisexual:
  9 Females bracts few (3–4), subcontiguous, usually pubescent, in spikes less than 1·5 cm. long; male catkins up to 4 cm. long; leaves acute or obtuse, elliptical, up to 8 cm. long and 3 cm. broad, usually much smaller; plants mostly dioecious
                                                    6. *scabrosa*
  9 Female bracts (3–) 5–20, remote at least at base of spike, usually setose-ciliate distally; male catkins mostly more than 4 cm. long; leaves acuminate, often more than 10 cm. long; plants monoecious:
    10 Leaf-margin strongly serrate; twigs, petioles and inflorescence densely pubescent; female bracts about half-divided with 5–9 oblong obtuse teeth, about 5 mm. long
                                                  7. *jamaicensis*
    10 Leaf-margin shallowly crenate to serrulate:
      11 Young twigs, petioles and inflorescence puberulous; female bracts shortly divided into 7–9 acute deltate teeth   8. *virgata*
      11 Young twigs, petioles and inflorescence glabrous or glabrescent; female bracts divided to at least halfway:
        12 Lobes of female bracts lanceolate, tapered to a usually acutish tip   9. *laevigata*
        12 Lobes of female bracts oblong, rounded to obtuse at tip   10. *pruinosa*

1. A. chamaedrifolia (Lam.) Müll. Arg. in DC., Prodr. **15** (2): 879 (1866).

Woody-based herb bushily branched, 15–30 cm. high; leaves lanceolate, obtuse, truncate to broadly rounded at base, crenate-margined, up to 22 mm. long and 11 mm. broad; spikes 1–3 cm. long; capsule 3-lobed, 1·6 mm. long; seeds ellipsoidal, dark brown, 1–1·2 mm. long.

Rare and local (St. Andr., St. Cath., Manch.), in grassland and thickets; 300–900 feet; fl. and fr. July–Aug, Nov. *H 12404! P 27542!* Florida, Bahamas, Greater Antilles, Lesser Antilles south to Guadeloupe.

2. A. alopecuroidea Jacq., Collect. **3**: 196 (1790).

Erect herb branched from near the base, 15–90 cm. high; leaves ovate, acuminate, shortly cordate at base, serrate, thinly hispid with short and long hairs, 3·5–8 cm. long, up to 4·5 cm. broad; petiole 1–4 (–7) cm. long; male spikes axillary, 3–9 mm. long; female spikes terminal or axillary, 5–20 mm. long, with a bristlelike appendage tipped by an aborted flower; capsule 1·3–1·4 mm. long; seeds ellipsoidal, reddish-brown about 1 mm. long.

Rather common as a weed of damp shady roadside ditches, cultivations and waste places on limestone; SL–2500 feet; fl. and fr. June–Mar. *A 6070! H 8210! P 23049!* Bermuda, Bahamas, Mexico to Venezuela, Greater Antilles, Grenada, Trinidad; Grand Cayman.

3. A. ostryifolia Ridd., Syn. Fl. W. States: 33 (1835).—*A. corchorifolia* of A. Rich. (1850), not Willd. (1805). *A. persimilis* Müll. Arg. (1865).

Erect herb 20–80 cm. high; leaves ovate, shortly acuminate, obtuse or cordate at base, closely serrate, puberulous, 3·5–7 (–10) cm. long; petiole nearly as long as blade; male spikes axillary, slender, to 1·2 cm. long; female spike usually terminal, occasionally axillary, 4–5 cm. long; capsule tuberculate, about 2·5 mm. long; seeds globose-ovoid, reddish-brown or grey, warty-wrinkled, about 2 mm. long.

Very rare and local (St. Eliz.), on roadside banks; 25–50 feet; fl. and fr. Apr. *P 22247! Sangster 517!* S.E. United States, Bahamas, Mexico, Greater Antilles.

4. A. elliptica Sw., Nov. Gen. & Sp. Pl.: 99 (1788).

Shrub 2·5–5 m. high, rarely a small tree; monoecious or dioecious; leaves ovate-elliptical, more or less acuminate, light yellowish-green, 6–18 cm. long, 2·5–11 cm. broad; petiole very variable, up to 15 cm. long; spikes 5–20 cm. long; female bracts green, 1-flowered; styles crimson; capsule globose, 3-lobed, setose-hispid, about 2 mm. long; seeds ovoid, acute, brown or black, about 1·5 mm. long.

Widespread but not common, in shaded limestone gullies especially near streams; 500–3000 feet; fl. most of the year, fr. Jan–Apr. *A 7779! 7782! 12385! H 5123! P 20820!* Endemic.

5. A. cuspidata Jacq., Pl. Hort. Schoenbr. **2**: 63, t. 243 (1797).

Shrub 60–120 cm. high; leaves ovate, cuspidate-acuminate, 4–10 cm. long, 2·5–6 cm. broad; petiole 1·5–3 cm. long; stipules lanceolate, 3–6 mm. long; spikes axillary, 2–4 cm. long; female bracts few at base of spike, 2-flowered; capsule globose, 3-lobed, villous, nearly 2 mm. long; seeds ellipsoidal, punctulate, about 1·2 mm. long.

Local and uncommon, in thickets and open woodland on arid limestone along the south coast; 50–1800 feet; fl. Sept, Nov, Mar, fr. sporadically. *A 10779! H 10185! P 24236!* Mexico to S. Amer., Cuba, Hispaniola, St. Vincent, Grenadines, Grenada.

6. A. scabrosa Sw., Nov. Gen. & Sp. Pl.: 99 (1788).

Shrub 30 cm. to 2 m. high; dioecious or rarely monoecious; leaves variably hairy, often pubescent or tomentose, 1–8 cm. long, 0·5–3 cm. broad, serrate or crenate; fruiting bracts acutely toothed, ribbed, enclosing 1 or 2 flowers; capsule globose, 3-lobed, villous above, about 1·7 mm. long; seeds ellipsoidal, brown, about 1·4 mm. long.

Common, in open thickets and woodlands on limestone rocks and cliffs, especially in arid locations; near SL–2000 (–3500) feet; fl. and fr. all the year. *A 6195! 7833! 11460! H 9578! 12104! P 9460! 23708!* Endemic.

7. A. jamaicensis Britton in Bull. Torr. Bot. Club **39**: 7 (1912).

Shrub 2–2·5 m. high; leaves elliptic-lanceolate, subcordate at narrow base, pubescent on both surfaces, with hairs spreading on midrib and nerves beneath, 5–18 cm. long, 2–5·5 cm. broad; spikes up to 10 cm. long; male catkins dark red; female bracts prominently ribbed, hirtellous on margins, enclosing 2 flowers; capsule about 1 mm. long, hirtellous.

Rare (Clar., Trel.); 1750–2500 feet; fl. and fr. Jan–May. *A 12823! H 10856!* Endemic.

8. A. virgata L., Syst. Nat. ed. 10, **2**: 1275 (1759).

8a. Var. virgata.

Shrub up to 2·5 m. high or slender tree; leaves elliptical to narrowly oblong-elliptical, obtuse or subcordate at base, smooth, 5–15 cm. long, 1·5–4·5 (–6) cm. broad; spikes axillary, male 3–14 cm. long, sometimes with 1 or 2 female flowers at base, female 3–7·5 cm. long; female bracts enclosing 1 or 2 flowers; capsule about 2 mm. long, puberulous; seeds ellipsoidal, about 1·8 mm. long.

Locally frequent (St. Andr., Port., St. Thom.), on riverside rocks and on rocky pathsides in montane woodland; 400–6000 feet; fl. Dec–Aug, fr. Feb. *A 11118! 12537! H 5507! J.P. 1159! P 23264!* Endemic.

8b. A. virgata var. **pubescens** F. & R. in Journ. Bot. **57**: 314 (1919).

Like the last with young branches, petioles, midrib and nerves densely pubescent; leaves sparingly pubescent on both surfaces.

Rare and local (Port.), in montane forest; about 4500 feet; fl. Jan, Sept, fr. Sept. *J.P. 1421! P 7569!* Endemic.

9. A. laevigata Sw., Nov. Gen. & Sp. Pl.: 99 (1788).

Erect or straggling shrub 30 cm. to 3 m. high, glabrous except on very young leaves; leaves lanceolate to narrowly elliptical, cuneate at base, margin crenate-serrulate, sometimes reddish beneath, 3–15 cm. long, up to 5 cm. broad; male catkins red or tawny-yellow, often at leafless nodes, 8–17 cm. long; female bracts enclosing 2 flowers, up to 6 mm. long in fruit; capsule about 3 mm. long, puberulous.

Frequent in the central and western parishes, on rough limestone rocks in woodlands and woodland margins; 500–2600 feet; fl. most of the year, fr. Mar–June, Dec. *A 11013! H 12504! P 9938!* Endemic.

10. A. pruinosa Urb., Symb. Ant. **5**: 388 (1908).

Diffusely branched glabrous shrub 1·5–2·5 m. high; leaves like the last, cordulate at base; female spikes on younger twigs, 3–5 cm. long; female bracts enclosing 1 flower, up to 8 mm. long in fruit; capsule about 2·5 mm. long, hirtellous.

Occasional (Manch., St. James, Trel.), on limestone rocks and cliffs in wood-land; 1750–3000 feet; fl. Dec–June, fr. Dec–May. *A 12445! H 8738! P 16117! Robertson UCWI 5477!* Endemic. This taxon is perhaps not more than varietally distinct from the last.

31. ATERAMNUS P. Browne (1756)

1 Leaves obtusely pointed to rounded at tip, not acuminate; sap not usually milky; filaments free:
  2 Leaves ovate to elliptical, broadly cuneate to rounded at base; male bracts 1-flowered; ovary sessile; peduncle shorter than ovary, not increasing in fruit   1. *ellipticus*
  2 Leaves oblanceolate, tapered to a narrowly rounded base; male bracts 3-flowered; ovary stalked above the bracteoles; peduncle and pedicel increasing in fruit
                                                                              2. *lucidus*
1 Leaves acuminate, rounded at base; ovary sessile or subsessile; sap milky; filaments united:
  3 Leaves entire, leathery; female bracteole cupular, divided or toothed; female flower shortly stalked above the bract; male bracts 3-flowered   3. *integer*
  3 Leaves crenate-serrate to subentire thin; female bracteoles 3; female flower sessile above the bract; male bracts 3–7-flowered   4. *glandulosus*

1. A. ellipticus (Sw.) Rothm. in Fedde, Repert. Sp. Nov. **53**: 5 (1944).—*Gymnanthes elliptica* Sw. (1788). Parrot Wood.
  Shrub 1–4 m. or tree to 10 m. high, also reported as a large forest tree; leaves usually entire, rather leathery, shiny adaxially, 2·5–10 cm. long, 1·5–4·5 cm. broad, without glands at base; male spikes 3–9 cm. long, sometimes with a few female flowers at base, bracts glandular, 1-flowered; female racemes 1–2 cm. long; bracts 1-flowered; pedicel 2–4 mm. long; capsule globose, 3-lobed, 7–8 mm. in diameter; seeds dark grey, about 5 mm. long.
  Common in thickets and woodlands, mostly on limestone; 300–3600 feet; fl. Dec–June, fr. May–Dec. *A 5638! 7164! H 8946! 8947! H & P 14411! 15073!* Endemic.

2. A. lucidus (Sw.) Rothm. in Fedde, Repert. Sp. Nov. **53**: 5 (1944).—*Gymnanthes lucida* Sw. (1788). Crab Wood.
  Shrub with straggling branches 1·5–4 m. or tree to 8 (–10) m. high, without milky sap; leaves rather leathery, shiny adaxially, with or without glands near base, 3–13 cm. long, 1·5–3 cm. broad; spikes terminal or axillary, male or bisexual, 1–3 cm. long; bracts without glands; peduncle up to 3 cm. and pedicel up to 2 cm. long in fruit; capsule globose, 3-lobed, about 10 mm. in diameter; seeds globose, dark brown, 4–5 mm. in diameter.
  Common in thickets on limestone rocky or sandy soils especially near the sea; SL–1500 feet; fl. Feb–June, Sept, fr. June–July, Nov. *A 11135! H 8643! J.P. 1106! H & P 13756! P 7150!* Florida, Bahamas, Mexico to Guatemala, Greater Antilles, Lesser Antilles, south to Guadeloupe; Grand Cayman.

3. A. integer (F. & R.) Rothm. in Fedde, Repert. Sp. Nov. **53**: 5 (1944).—*Gymnanthes integra* F. & R. (1920).
  Shrub 1·5–2·5 m. or tree to 10 m. high; leaves elliptical, without glands, slightly revolute, 4–9 cm. long, 1·5–3·7 cm. broad; male spikes axillary; bracts without glands; female spikes terminal, very short with 1 or 2 flowers; capsule about 7 mm. long; seeds globose-ellipsoidal, buff mottled black or brown, 4 mm. long.
  Uncommon in the central and western parishes, in thickets and woodlands on limestone; 1750–3000 feet; fl. Jan–Apr, Aug, fr. Jan–Apr, Aug–Oct. *H 10159! 11225! Henry! P 10051! 15733!* Endemic.

4. A. glandulosus (Sw.) Adams in Phytologia **20** (5): 309 (1970).—*Excoecaria glandulosa* Sw. (1800). *Gymnanthes glandulosa* (Sw.) Müll. Arg. (1863).
  Shrub with slender straggling branches 1·2–3 m. or tree 4–10 m. high; leaves ovate to oblong-lanceolate, very shiny adaxially, rarely with 1 or 2 glands at apex of petiole, 3–8 (–9·5) cm. long, 1·5–3·5 (–4·9) cm. broad; spikes bisexual, terminal, up to 4·5 cm. long; capsule subglobose, 7–8 mm. in diameter; seeds globose, dark brown and finely mottled, 3·5 mm. in diameter.

Occasional in thickets on limestone in usually rather dry areas; SL–2900 feet; fl. May–Sept, fr. June–Oct. *A 11146! 11241! H 10826! 10936! P 19721!* Cuba.

32. EUPHORBIA L. (1753); Dressler (1962), Burch (1966)

1 Leaves wanting or very small and early deciduous; shrubs or sometimes trees; stems terete or angled, glabrous:
  2 Stems narrowly 2-winged, otherwise terete; branches whorled, not armed   1. *alata*
  2 Stems not winged; succulent cultivated plants used mainly for boundaries and hedges:
    3 Stems 3–4-angled, green marbled whitish between the angles, stout, armed at the nodes with dark paired spines; leaves succulent, suborbicular, cuspidate, 4–6 mm. long; native of E. Indies   *E. lactea* Haw.
    3 Stems much-branched, terete, at first 4–6 mm. thick, smooth, lightly striate, unarmed, jointed, leaves linear-lanceolate, acute, about 2 cm. long and 3 mm. broad; native of S. Asia   *E. tirucalli* L.
1 Leaves consistently present or if lacking then only seasonally deciduous:
  4 Leaves alternately or spirally arranged at least on vegetative shoots:
    5 Shrubs or trees:
      6 Plants armed with nodal spines; bracts paired, rounded; cultivated plants:
        7 Stems and leaves succulent; stems erect, mostly 5-angled; leaves oblong-oblance-olate, apiculate, the midrib sharply prominent beneath, up to 15 cm. long and 3·5 cm. broad; peduncle short and thick; bracts inconspicuous, green, soon withering; shrub or tree to 10 m. high; native of E. Indies   *E. neriifolia* L.
        7 Stems woody, diffusely scrambling; leaves obovate to oblanceolate, thin, up to 8 cm. long; peduncle slender; bracts bright red, persistent; ornamental shrub; native of Madagascar; Crown of Thorns   *E. milii* Ch. des Moulins
      6 Plants not armed; leaves subtending inflorescence and bracts partly or entirely bright red:
        8 Leaves all entire, oblong-lanceolate to oblanceolate or linear-elliptical; inflorescence dichasial with equal broad paired bracts; glands of involucre 4–6; trees or shrubs   2. *punicea*
        8 Leaves of vegetative shoots repand-dentate, ovate to lanceolate; ultimate inflorescence-branches monochasial with very unequal bracts, one being an entire coloured leaf, the other reduced to a scale; glands of involucre solitary; ornamental shrubs; native of Mexico; Poinsettia
                  E. pulcherrima Willd. ex Klotzsch
    5 Annual herbs; uppermost leaves or bracts opposite or whorled:
      9 Glands of involucre 2, bilabiate, with a petaloid appendage; stem internodes hollow; leaves entire; seeds ellipsoidal, about 2·5 mm. long, with mammiform warts   3. *oerstediana*
      9 Glands of involucre without petaloid appendages:
        10 Inflorescence compound-umbellate with broad paired green bracts on the branches; involucral glands 4, transversely oblong with a subulate horn at each end; seeds 6-angled, about 1·2 mm. long, pitted, light grey, carunculate   4. *peplus*
        10 Inflorescence cymose-corymbose; involucral glands solitary, not horned; seeds ovoid-ellipsoidal, about 2·5 mm. long, tuberculate:
          11 Gland more or less funnel-shaped with a circular opening; upper leaves without red colour; stems usually pilose   5. *heterophylla*
          11 Gland more or less bilabiate with a narrowly oblong opening; upper leaves usually red at base; stems glabrous or sparsely pilose   6. *cyathophora*
  4 Leaves all opposite or whorled, often at swollen nodes:
    12 Leaves whorled, equal-sided, entire, distinctly petiolate; shrubs or small trees; involucral glands 5 (–6):
      13 Leaves (2–) 4–11 at a node:
        14 Leaf-blades elliptical; bracts leaflike, white; glands of involucre with triangular entire acute white petaloid appendages; cultivated ornamental; native of Guatemala   *E. leucocephala* Lotsy
        14 Leaf-blades broadly ovate to suborbicular; bracts inconspicuous; glands of involucre with transversely oblong appendages; involucres 1 (–3), glabrous; petioles slender   7. *petiolaris*
      13 Leaves (2–) 3 (–4) at a node; petioles slender; bracts inconspicuous; appendages to involucral glands transverse, often eroded:
        15 Leaves tinged coppery-purple, broadly ovate, up to about 7·5 cm. long and 6·5 cm. broad; ornamental shrub; native origin unknown but probably tropical Amer.   *E. cotinifolia* L.
        15 Leaves light green, ovate to elliptical or oblanceolate, puberulous beneath, 3–6·5 cm. long, 1·5–3·5 cm. broad; petiole 5–13 mm. long; involucres 10–16 in a cyme, puberulous; seeds irregularly tuberculate   8. *nudiflora*

12 Leaves opposite, usually unequal-sided, entire or serrulate, often with very short
    petioles or subsessile, mostly small; herbs or small shrubs:
16 Branches usually prostrate or creeping; involucres 1–3 (–5) in clusters or short
    cymes in alternate leaf-axils; leaves mostly less than 1 cm. long; annual weedy
    herbs:
17 Stem puberulous on upper side; seeds tranversely ridged:
18 Capsule deflexed, with short bristly spreading hairs on the keeled margins only
          9. *prostrata*
18 Capsule more or less erect and only partially exserted from the involucre,
    generally covered with ascending hairs   10. *thymifolia*
17 Stem and capsule glabrous:
19 Branches spreading, rarely suberect, but not rooting; seeds brown, obscurely
    transversely ridged; stipules often more than 1 mm. long   11. *blodgettii*
19 Branches creeping and rooting; seeds light pinkish-brown, smooth; stipules
    about 0·5 mm. long   12. *serpens*
16 Branches usually erect, sometimes decumbent:
20 Shrubs (5–) 30–100 cm. high; leaf-margin entire; involucres solitary:
21 Stems and leaves glabrous; leaves cordate-auriculate at base; capsule glabrous;
    plant of sea-coast   13. *mesembrianthemifolia*
21 Stems and leaves beneath puberulous; leaves rounded to broadly cuneate at
    base; capsule puberulous; rare plant of rocky inland situations
          14. *myrtillifolia*
20 Annual or sometimes perennial weedy herbs; leaf-margin usually serrulate at
    least towards the tip; at least the larger leaves more than 1 cm. long; involucres
    in cymose glomerules:
22 Capsule glabrous:
23 Capsule more than 2 mm. long; seeds 1·2–1·8 mm. long, grey; rare plant of
    coastal sand   15. *bahiensis*
23 Capsule less than 2 mm. long; seeds mostly about 1 mm. long, not exceeding
    1·5 mm. long; common weeds of cultivations, pastures and waste places, not
    restricted to coastal sand:
24 Capsule less than 1·4 mm. long; involucres usually in peduncled leafless
    compactly cymose glomerules; seeds wrinkled, brown   16. *glomerifera*
24 Capsule more than 1·6 mm. long; involucres few together at the ends of
    divaricate leafy axillary branches; seeds with 2–4 ridges on each face,
    black or grey   17. *hyssopifolia*
22 Capsule pubescent; leaves often with a central reddish blotch, more or less
    pubescent:
25 Stems rigidly erect, few-branched, greyish-pubescent; gland-appendages
    white or pink, conspicuous; capsule 2 mm. in diameter; seeds brownish-
    grey, prominently wrinkled and pitted in rows   18. *lasiocarpa*
25 Stems lax or bushily branched, with short curved and usually also longer
    spreading hairs; gland-appendages small or obsolete; capsule about 1 mm.
    or less in diameter; seeds pinkish:
26 Stems sparingly branched mostly from near the base, the longer hairs
    numerous, yellowish and about 1 mm. or more long; inflorescence axillary
    and terminal   19. *hirta*
26 Stems divaricately branched above the base, the longer hairs few, not yellow-
    ish and about 0·5 mm. long; inflorescences mostly terminal and paired
          20. *ophthalmica*

1. E. alata Hook., Ic. Pl. **7**: t. 700 (1844).

Erect branched shrub up to 2·5 m. high; stems long-jointed with whorled
branches, 4–5 mm. thick when young; involucres solitary or clustered, terminal or
at the distal nodes; glands fleshy with roundish appendages.

Very rare (Manch., Trel.), on honey-combed limestone rocks in forest; about
2500 feet. *H 9098!* Endemic.

2. E. punicea Sw., Nov. Gen. & Sp. Pl.: 76 (1788).—*E. troyana* Urb. (1908).

Shrub to 3 m. or tree to 10 m. high with rather thick terete branches marked
with leaf-scars; leaves narrowly oblong to obovate, thin at least when dry, obtuse
or acute, dark green adaxially, 4–15 cm. long, 1–3 cm. broad; bracts 2–7, elliptical
to obovate, shorter than the leaves, brilliant scarlet, crimson or pinkish-orange;
cymes corymbose, rarely involucres solitary; involucres 5–7 mm. long, puberulous
or glabrous; glands 4–8; ovary puberulous or glabrous; capsule about 1 cm. in
diameter.

Widespread and rather common in the central and western parishes, on limestone rocks in open woodlands and thickets; 50–2900 feet; fl. and fr. most of the

year. *A 8664! H 10241! P 21429!* Endemic. The pattern of variation within this species requires further study, particularly in respect to the number of cyathia formed in the inflorescence, the position and number of glands on them and the presence of hairs; the grounds on which *E. troyana* is distinguished are such that several other variants representing populations in different parts of the island could also have species rank.

3. **E. oerstediana** (Klotzsch & Garcke) Boiss. in DC., Prodr. **15** (2): 59 (1862).—
Poinsettia oerstediana Klotzsch & Garcke (1860).
Annual herb 20–45 cm. high, erect or geniculate; leaves ovate to oblong-elliptical, paler and puberulous beneath, 3·5–7·5 cm. long, 1·5–4 cm. broad; involucres long-stalked, glabrous or puberulous apically; capsule puberulous.
Occasional on damp shady banks and in low-lying pastures; 50–1700 feet; fl. and fr. most of the year. *A 6114! H 12075! P 22079!* Continental tropical Amer., Lesser Antilles, Trinidad.

4. **E. peplus** L., Sp. Pl. **1**: 456 (1753).
Annual taprooted erect glabrous herb 8–45 cm. high; leaves elliptical to obovate, narrowed to base, thin, light green, 5–20 mm. long, up to 13 mm. broad; involucres solitary, 5-lobed; capsule trigonous, about 2 mm. in diameter.
Locally common (St. Andr., St. Thom.), along roadsides and on stony cultivated ground; 2800–5250 feet; fl. and fr. Dec–Aug. *A 6616! H 8580! P 9581!* Native of Europe and western Asia; Hispaniola.

5. **E. heterophylla** L., Sp. Pl. **1**: 453 (1753).—*E. geniculata* Ort. (1797). *Poinsettia heterophylla* (L.) Klotzsch & Garcke (1860); Dressler in Ann. Miss. Bot. Gard. **48**: 339 (1962).
Annual taprooted herb to 1 m. high; leaves ovate, elliptical to lanceolate, entire or variously repand-dentate, up to 14 cm. long and 5 cm. broad; cymes terminal; involucre glabrous; seed angular, coarsely and bluntly tuberculate.
Occasional in the central and eastern parishes, a weed of roadside banks and open waste places; 10–3500 feet; fl. and fr. most of the year. *A 8279! H 11957! P 27669!* General in the subtropics and tropics.

6. **E. cyathophora** Murr., Comm. Gött. **7**: 81, t. 1 (1786).—*E. graminifolia* Michx. (1803). *Poinsettia cyathophora* (Murr.) Klotzsch & Garcke (1860). *E. heterophylla* var. *cyathophora* (Murr.) Griseb. (1859).
Annual or short-lived perennial herb like the last; leaves variable in shape as above to linear; seed not angular, finely and sharply tuberculate.
Locally common in the central and western parishes, on limestone rocks in exposed places and waste ground; SL–2600 feet; fl. and fr. all the year. *A 6822! Fawcett 8398! H 10079! P 24159!* E. United States to C. and S. Amer., West Indies, Pacific Is.; Grand Cayman; an occasional escape from cultivation in West Africa and other parts or the Old World tropics.

7. **E. petiolaris** Sims in Curt., Bot. Mag. **23**: t. 883 (1805).
Shrub or tree to 6 m. high with slender stems; leaf-blade shallowly emarginate, truncate at base, glabrous, up to 25 mm. long and 22 mm. broad; petioles up to about 2·5 cm. long; capsule glabrous, 4·5 mm. broad; seeds pitted.
Very rare (St. Thom.), on gravel river bank; 500–700 feet; fl. and fr. Feb–Apr. *P 24800! 26847!* Bahamas, Hispaniola, Puerto Rico, Virgin Is., Guadeloupe, Martinique, Margarita.

8. **E. nudiflora** Jacq., Collect. **3**: 180 (1790).
Shrub with forking slender branches, 1–2·5 m. high, weakly erect or scrambling; leaves obtuse or rounded at apex and base; involucral glands 5, greenish with broader white appendages; capsule glabrous, long-stalked, 3·5 mm. long; seed 3 mm. long.
Locally common (St. Andr., St. Cath., St. Thom.), in thickets and open places on arid limestone rocks; 100–2500 feet; fl. and fr. Dec–Mar. *A 5626! Fawcett! P 8362!* Colombia, St. Vincent.

9. E. prostrata Ait., Hort. Kew. **2**: 139 (1789).—*Chamaesyce prostrata* (Ait.) Small (1903). Milkweed.[1]

Branches spreading around a central taproot forming small mats, 5–20 cm. long; leaves elliptical to narrowly obovate-elliptical, obtuse, minutely toothed distally, 3–7 mm. long; capsule 1–1·4 mm. in diameter; seeds pinkish, about 1 mm. long.

Locally common, a weed of sandy waste places and lawns; SL–4000 feet; fl. and fr. sporadically throughout the year. *A 6608! H 11803! P 15771!* General in the tropics; naturalized in S. Europe. This species is often infected with the rust fungus *Uromyces euphorbiae* which in the aecidial stage causes the stems to grow erect; other species in this part of the genus are also known to become similarly infected.

10. E. thymifolia L., Sp. Pl. **1**: 454 (1753).—*Chamaesyce thymifolia* (L.) Millsp. (1916).

Habit like the last, with pinkish branches up to 30 cm. long, sometimes forming dense mats; leaves oblong to oblong-elliptical, obtuse, often tinged purple, puberulous at first, glabrescent, serrulate-crenate, 4–9 mm. long; capsule about 1·5 mm. in diameter; seeds oblong, pink, about 1 mm. long.

Rather local (St. Andr., St. Cath., St. Mary), a field weed and in waste places on sandy soil; SL–600 feet; fl. and fr. July–Mar. *A 8687! 11853! Robertson UCWI 3333!* General in the tropics and subtropics, not reported from Bahamas, Cuba, Australia.

11. E. blodgettii Engelm. ex Hitchc. in Rep. Miss. Bot. Gard. **4**: 126, t. 13 (1893).
—*Chamaesyce blodgettii* (Engelm. ex Hitchc.) Small (1903).

Habit like the last; plant glabrous with branches 8–50 cm. long; leaves elliptical to obovate-elliptical, more or less serrulate and rounded at apex, 3–9 (–13) mm. long, up to 5 (–6) mm. broad, glaucous; stipules pale, triangular, fimbriate; capsule 1·5–2 mm. in diameter; seeds oblong, tetragonal, about 1 mm. long.

Rather common on sandy or gravelly ground, mostly near the sea, also on the cays; SL–200 feet; fl. and fr. July–Mar. *A 9812! 11966! H 9536! J.P. 1007! P 21392!* Bermuda, Bahamas, Florida, Cuba, Puerto Rico, Virgin Is., Grand Cayman.

12. E. serpens Kunth, Nov. Gen. **2**: 52 (1817).—*Chamaesyce serpens* (Kunth) Small (1903).

Stems slender, branched, up to 50 cm. long; plant glabrous; leaves oblong-orbicular, rounded at apex, entire, 3–6 mm. long, 3–5 mm. broad; stipules short, denticulate, whitish; capsule about 1 mm. long, the cocci sharply keeled; seeds ovoid, tetragonal.

Uncommon (St. Andr., St. Cath., Clar.), on sand dunes and salinas; near SL; fl. and fr. June–Mar. *A 6466! 12890!* Throughout tropical Amer., Sierra Leone, S. Europe.

13. E. mesembrianthemifolia Jacq., Enum. Syst. Pl. Carib.: 22 (1760).—
E. buxifolia Lam. (1781). *Chamaesyce buxifolia* (Lam.) Small (1903).

Shrub or perennial shrubby herb usually branched and bushy; leaves ovate to broadly oblong, fleshy, rather acute, up to 12 mm. long and 7 mm. broad; stipules conspicuous, about 1 mm. long, more or less fringed, white; capsule 2 mm. in diameter; seeds whitish, about 1 mm. long, inconspicuously angled and marked.

Very common on limestone rocks and in pastures near the sea, also on the cays; SL–20 feet; fl. and fr. all the year. *A 7587! 8015! H 11627! P 20660!* Bermuda, Bahamas, Florida and throughout tropical Amer.; Grand Cayman.

14. E. myrtillifolia L., Syst. Nat. ed. 10, **2**: 1048 (1759).—*Chamaesyce myrtilli-folia* (L.) Millsp. (1916).

Shrub 30–90 cm. high with long slender branches; leaves obovate-elliptical to suborbicular, retuse at apex, roundish at base, paler beneath, 5–15 mm. long;

---

[1] The common name Milkweed is given to all the smaller-leaved herbaceous species of *Euphorbia* comprising the subgenus *Chamaesyce* (spp. 9–20).

stipules very short, truncate, fringed; capsule 2 mm. in diameter, trigonous, the cocci slightly keeled; seeds tetragonal, reticulate-wrinkled.

Rare and local (St. Andr.), on cliffs and steep rocky banks; 1400–2000 feet; fl. and fr. Dec–Mar. *H 5196! P 9595!* Endemic.

15. E. bahiensis (Klotzsch & Garcke) Boiss. in DC., Prodr. **15** (2): 24 (1862).— *Anisophyllum bahiense* Klotzsch & Garcke (1860).

Taprooted glabrous herb up to 20 cm. or more high with branches spreading and ascending from a central stock; stem and leaves tinged crimson; leaves oblong-elliptical to obovate, 6–15 (–30) mm. long, 3–9 (–12) mm. broad, more or less evidently serrate to the obtuse or rounded tip; gland-appendages white; capsule broadly ovoid, up to about 3 mm. in diameter; seeds ovoid to suborbicular, angled, the faces wrinkled.

Very rare (St. Cath.), on upper beach; near SL; fl. and fr. Nov–Apr. *A 10013! Asprey UCWI 1010!* Costa Rica to Brazil.

16. E. glomerifera (Millsp.) L. C. Wheeler in Contrib. Gray Herb. **127**: 78 (1939).—*Chamaesyce glomerifera* Millsp. (1913). *C. hypericifolia* Millsp. (1909), partly. *E. hypericifolia* of F. & R. (1920), not L. (1753).

Branched glabrous or minutely puberulous herb 20–60 (–80) cm. high; leaves elliptical to oblong, more or less serrulate to the obtuse or rounded tip, often tinged reddish, 1–3·5 cm. long, 5–13 mm. broad; gland-appendages white or pink; capsule 1·2–1·5 mm. in diameter; seeds tetragonal, about 1 mm. long.

Common as a weed of cultivations and in rough pastures and along roadsides; SL–1800 (–2700) feet; fl. and fr. Mar–Dec. *A 12041! H 12646! P 10330!* Throughout the subtropics and tropics of the New World and introduced into the Old; Grand Cayman.

17. E. hyssopifolia L., Syst. Nat. ed. 10, **2**: 1048 (1759).—*E. brasiliensis* Lam. (1786). *Chamaesyce hyssopifolia* (L.) Small (1905). *E. nirurioides* (Millsp.) F. & R. (1920).

Glabrous or thinly pubescent herb 30–120 cm. high, with slender somewhat dichotomous and flexuous branches; leaves oblong to narrowly obovate or sometimes almost linear, entire or sparingly toothed, in the broader-leaved variants up to 3·5 cm. long and 1·4 cm. broad, in the narrow-leaved variants up to 4 cm. long but only 4 mm. broad; gland-appendages white; capsule 1·5–1·9 mm. in diameter; seeds ovoid, tetragonal, shallowly transversely rugose, about 1·3 mm. long.

Common and abundant locally as a weed of waste places; SL–2800 feet; fl. and fr. all the year. *A 5413! 5678! 11755! H 9035! 12112! P 8514!* Bermuda, Florida and throughout tropical Amer., adventive in the Old World. The broad- and narrow-leaved variants are usually quite distinct and from available data seem to differ in their flowering and fruiting times, the former mainly from June to October, the latter from October to February.

18. E. lasiocarpa Klotzsch in Nov. Act. Acad. Caes. Leop-Carol. Nat. Cur. **19**, Suppl. 1: 414 (1843).—*Chamaesyce lasiocarpa* (Klotzsch) Arthur (1911).

Herb with rather thick rigid pubescent stems 15–90 cm. high, rarely prostrate; leaves elliptical to oblong-lanceolate, obtuse at apex, serrulate, 1–4 cm. long, 4–13 mm. broad, often tinged reddish; cymes terminal, densely corymbose; seeds elongated-ovoid, tetragonal, about 1·1 mm. long.

Locally common (St. Andr., St. Eliz., St. Thom.), on roadside banks and open gravelly waste places; SL–2100 (–5000) feet; fl. and fr. June–Mar. *A 5463! H 11136! P 9587!* Mexico to Peru, Hispaniola, Puerto Rico, Antigua to Barbados sporadically.

19. E. hirta L., Sp. Pl. **1**: 454 (1753).—*Chamaesyce hirta* (L.) Millsp. (1909).

Herb with spreading branches up to 60 cm. long, sometimes prostrate especially when trampled; leaves ovate to rhomboid or oblong-lanceolate, acute, unequal-sided at base, serrulate, up to 4 cm. long and 1·6 cm. broad; stipules minute; involucres less than 1 mm. long; gland-appendages minute; seeds oblong, sharply tetragonal, transversely wrinkled, about 0·8 mm. long.

Very common, a weed of roadsides, waste places, lawns, pastures and cultivated ground; 30–3700 feet; fl. and fr. all the year. *A 10984! H 9517! P 10329!* General in the subtropics and tropics.

20. E. ophthalmica Pers., Synops. Pl. **2**: 13 (1806).—*Chamaesyce ophthalmica* (Pers.) Burch (1966).

Suberect or sometimes prostrate-branched herb 5–20 high; leaves obliquely ovate-rhomboid, acute, strongly unequal-sided at base, serrulate, 4–17 mm. long, up to 8 mm. broad; stipules minute; involucres 0·5 mm. long; gland-appendages obsolete; seeds oblong, pointed at one end, obscurely tetragonal, pink, 0·6–0·7 mm. long.

Locally common, a weed of sandy roadsides, lawns and pasture margins; 30–1700 feet; fl. and fr. all the year. *A 5530! T. J. Harris 8291! H & P 14462! 14470!*

Bahamas, southern Florida, Greater Antilles, Antigua, Barbados; Grand Cayman.

33. PEDILANTHUS Poit. (1812) nom. cons.; Dressler (1957)

1 Leaves less than twice as long as broad, acute or acuminate at tip; cultivated plants:
  2 Leaf-base cuneate; blade not markedly glossy, often variegated
                                               1a. *tithymaloides* ssp. *tithymaloides*
  2 Leaf-base obtuse to subcordate; blade glossy   1b. *tithymaloides* ssp. *parasiticus*
1 Leaves more than twice as long as broad, acute to obtuse at tip, not variegated; wild
  plants   1c. *tithymaloides* ssp. *jamaicensis*

1. P. tithymaloides (L.) Poit. in Ann. Mus. Hist. Nat. Paris **19**: 390, t. 19 (1812).
—*Euphorbia tithymaloides* L. (1753).

Shrubs with thick terete stems up to 2 m. or more high; leaves thick and succulent; stipules glandular; inflorescences terminal and axillary; involucres bright red; capsule pedicellate, beaked, more or less 3-lobed and -angled.

1a. Ssp. tithymaloides.—Monkey Fiddle.

Cultivated shrub with stems and leaves usually mottled light green and white and sometimes with pink; involucres glabrous outside but with ciliate margin. Native from Mexico to Suriname, cultivated elsewhere in the tropics.

1b. P. tithymaloides ssp. **parasiticus** (Klotzsch & Garcke) Dressler in Contrib. Gray Herb. **182**: 148 (1957).—*P. parasiticus* Klotzsch & Garcke (1860). *P. latifolius* Millsp. & Britton (1915).

Cultivated shrub like the last with shorter and relatively broader leaves, not variegated; involucres glabrous except the ciliate margin. Native Mexico, British Honduras; also reported from Hispaniola, Puerto Rico, Virgin Is., Grand Cayman.

1c. P. tithymaloides ssp. **jamaicensis** (Millsp. & Britton) Dressler in Contrib. Gray Herb. **182**: 159 (1957).—*P. jamaicensis* Millsp. & Britton (1913). *P. grisebachii* Millsp. & Britton (1913).

Shrubs diffusely branched and more or less erect to 2 m. high or branches trailing to 4 m.; young stems and leaves puberulous; leaves 1–10 cm. long, very narrow to 3 cm. broad, linear-lanceolate to obovate or oblanceolate; involucres pubescent to glabrous.

Rather local in thickets and woodlands in moderately arid localities along the south coast, on limestone rocks; 20–150 feet; fl. and fr. Nov–May. *A 10168! 11133! H 10238! 12117! P 9831! 11152! 11957!* Endemic. *Pedilanthus grisebachii* and *P. jamaicensis* are not identical. Dressler (1957) draws attention to the geographical cline existing between them within the Jamaican subspecies of *P. tithymaloides*. The most easterly populations (St. Thomas–type locality of *P. grisebachii*) have narrow leaves and pubescent involucres; the most westerly populations (Westmoreland–type locality of *P. jamaicensis*) have broad leaves and glabrous involucres; intermediate populations (south Clarendon and Manchester) have narrow leaves and glabrous involucres.

110. CALLITRICHACEAE

Annual terrestrial or aquatic herbs with slender stems. Leaves opposite, entire, without stipules. Flowers minute, unisexual, solitary or rarely male and female paired in the same axil. Perianth absent; each flower subtended by 2 hornlike bracteoles. Male flower with 1 stamen; anther 2-locular, opening vertically and across the top. Female flower with superior 4-lobed, 4-locular ovary with a solitary axile pendulous anatropous ovule in each loculus; styles 2, filiform. Fruit 4-keeled or -winged, indehiscent, schizocarpic. Seed with fleshy endosperm and straight embryo.

One genus *Callitriche* with as many as 44 species and a cosmopolitan distribution including Cuba and Hispaniola; none as yet reported from Jamaica.

111. BUXACEAE

Trees, shrubs or herbs without milky sap. Leaves opposite or alternate, often leathery and evergreen, usually entire, without stipules. Flowers usually unisexual, actinomorphic, mostly monoecious in bracteate spikes or racemes. Perianth uniseriate or usually 4 basally connate imbricate sepals or absent. Male flowers with 4, 6 or numerous stamens, when 4 then opposite the sepals; filaments free; anthers 2-locular, opening lengthwise or by valves; pistillode sometimes present. Female flowers often larger and fewer or solitary; calyx 4-, 6- or more-parted; ovary superior (2-) 3-locular with usually 2 axile collateral pendulous anatropous ovules in each loculus; styles simple as many as the carpels, basally connate or distinct. Fruit a loculicidal capsule or drupaceous. Seeds glossy, black, usually with a caruncle; embryo straight; endosperm fleshy, rarely absent.

6 genera with up to 60 species mostly in temperate and subtropical regions of the Old World, but *Buxus* well represented in the West Indies.

1. BUXUS L. (1753)

Shrubs or trees; leaves opposite; sepals 4 in male flowers; stamens 4; female flowers terminal in the racemes, with 6 sepals in 2 series; styles distinct, spreading, persistent as horns on the capsule.

1 Leaves less than 10 cm. long:
　2 Leaves elliptical, 4–9 cm. long, apex usually acuminate:
　　3 Bracteoles and sepals of male flowers 2–3 mm. long; pedicels of male flowers 2–4 mm. long; capsule about 5 mm. long; leaves thin　　　　　　1. *laevigata*
　　3 Bracteoles and sepals of male flowers 1·5–2 mm. long; pedicels of male flowers less than 1 mm. long; capsule about 7 mm. long; leaves leathery　　　2. *arborea*
　2 Leaves mostly obovate to oblanceolate, 1·5–5 cm. long, apex acute to obtuse, usually mucronate, leathery; capsule about 5 mm. long:
　　4 Leaves oblanceolate, usually acute, with venation rather obscure but faintly evident on the adaxial surface　　　　　　　　　　　　　　　　　3. *bahamensis*
　　4 Leaves obovate to oblanceolate, apex rounded to obtuse, with venation evident on both surfaces　　　　　　　　　　　　　　　　　　　　4. *vahlii*
1 Leaves more than 15 cm. long, elliptic-lanceolate to narrowly elliptical, acute at both ends; pedicels and bracteoles about 2 mm. long; capsule 7–8 mm. long
　　　　　　　　　　　　　　　　　　　　　　　　　　　　5. *macrophylla*

1. B. laevigata (Sw.) Spreng. in L., Syst. Veg. ed. 16, **3**: 847 (1826).
Shrub 1–3 m. or tree to 8 m. high, with yellow wood and deeply furrowed bark; racemes sessile or stalked; flowers green or light yellow; stamen-filaments 2–2·5 mm. long, 1 mm. broad; seeds black, shiny.
Occasional in thickets and woodlands on limestone; 1800–3000 feet; fl. and fr. most of the year. *A 12472! H 9492! P 15906!* Endemic.

2. B. arborea Proctor in Bull. Inst. Jam., Sci. ser. **16**: 23, t. 8 (1967).

Like the last; tree 9–13 m. high; racemes subsessile, glomerate; seeds about 4 mm. long, black, shiny.

Rare and rather local (St. James, Trel.), in woodland on limestone; 2000–2200 feet; fl. and fr. Jan–June. *P 21347! 23189!* Endemic.

3. B. bahamensis Bak. in Hook., Ic. Pl. **19**: t. 1806 (1889).

Shrub to 3 m. or small tree to 5 m. high; leaves rigid, 1·5–3 cm. long; flowers light yellowish-green; filaments narrower than the anthers.

Rare and local (St. Andr.) in thickets and woodlands on steep arid limestone hillsides; 200–500 feet; fl. and fr. July. *H 9607! P 24501!* Bahamas, Turks and Caicos Is., Cuba.

4. B. vahlii Baill., Monogr. Bux.: 67 (1859).

Glabrous shrub 1–3 m. high with slender twigs; leaves 2–5 cm. long; racemes sessile, shorter than the leaves; flowers light green.

Not confirmed for Jamaica and said by Britton & Wilson (1924) to be endemic in Puerto Rico and St. Croix.

5. B. macrophylla (Britton) F. & R., Fl. Jam. **5**: 3 (1926).

Unbranched or sparingly branched shrub, 50–180 cm. high with rough bark; leaves more or less clustered towards the stem apex, 15–30 cm. long, 5–8 cm. broad; racemes supra-axillary and on the stem below; flowers dull cream-coloured; capsule minutely puberulous with short horns.

Rare and local (St. Thom.) in thicket margins on limestone hillsides; 1500–2000 feet; fl. and fr. Mar. *H & B 10770! Proctor & Stearn 11819!* Cuba.

112. ANACARDIACEAE

Trees and shrubs usually with resinous bark, the resin often caustic and blackening on exposure to air. Leaves alternate, very rarely opposite, simple, trifoliolate or pinnate with an odd leaflet, exstipulate or stipules obscure. Flowers typically bisexual or unisexual by reduction, mostly actinomorphic, small, in terminal or axillary panicles. Perianth usually biseriate; calyx of 3–7 free or connate sepals; corolla of 3–7 free or rarely basally connate petals. Disk present, generally ringlike. Stamens equal in number to or twice as many as the petals; filaments free or less often basally connate, inserted on or at the base of the disk; anthers 2-locular, opening lengthwise. Ovary superior, usually 1-locular, less often 2–5-locular; styles 1–5; ovules solitary in the loculi, anatropous. Fruit usually a drupe with resinous mesocarp. Seeds without or with very little endosperm; cotyledons fleshy.

Over 70 genera with some 600 species, mainly subtropical or tropical, but widely distributed.

1 Leaves simple; carpel 1; stamens markedly unequal:
  2 Gynophore in fruit much thickened and enlarged; mesocarp hard   1. **Anacardium**
  2 Gynophore inconspicuous; mesocarp soft   2. **Mangifera**
1 Leaves pinnate, mostly imparipinnate; pistil compound:
  3 Carpels and styles (3–) 5, each loculus developing a cavity in fruit with a single seed
    from its upper end   3. **Spondias**
  3 Carpels and styles 3 with styles fused and 3-cleft or -lobed or free; ovary 1-locular and
    developing a single seed in fruit:
    4 Stamens twice as many as petals; flowers white; pericarp fleshy, red; endocarp of
      drupe bony; tree planted in the mountains of St. Andrew, (550–) 3000–5000 feet;
      native of Brazil   *Schinus terebinthifolia* Raddi
    4 Stamens as many as petals:
      5 Sepals free; endocarp thick and bony   4. **Mosquitoxylum**
      5 Sepals fused below; endocarp thin:
        6 Leaflets usually more than three pairs, often sinuate-margined or toothed, rarely
          entirely glabrous or leathery; calyx usually 3-cleft; leaves crowded near the top of
          the stem   5. **Comocladia**
        6 Leaflets mostly 5-foliolate, subentire, glabrous, rather leathery; calyx (4–) 5 (–6)-
          cleft; leaves not crowded   6. **Metopium**

1. ANACARDIUM L. (1753)

1. A. occidentale L., Sp. Pl. **1**: 383 (1753).—Cashew.

Tree with spreading branches, 4–8 (–12) m. high; leaves obovate-elliptical, rounded or emarginate at tip, mostly narrow at base or rounded and cordulate, glabrous, up to about 18 cm. long and 10 cm. broad; panicles exceeding the leaves; flowers polygamous, crowded towards the ends of panicle branches, tomentose, yellow turning pink then rose-red, fragrant; calyx 4–5 mm. long; petals 5, 7–13 mm. long; stamens 8–10, one larger, up to 11 mm. long; gynophore in fruit 6–10 cm. long, pear-shaped; drupe nutlike, kidney-shaped, 2–3·5 cm. long.

Common, mostly in cultivation, in arid to moderately rainy areas; near SL–1750 feet; fl. Feb–June, fr. June–Aug. *A 10307! H 11899! Proctor & Mullings 21972!* Native in the American tropics, now widely distributed.

2. MANGIFERA L. (1753)

1. M. indica L., Sp. Pl. **1**: 200 (1753).—Mango.

Tree 5–10 (–15) m. high, usually heavy-branched from a stout trunk; leaves lanceolate-elliptical, acuminate, mostly shortly cuneate at base, 10–25 (–50) cm. long, 2·5–8 (–13) cm. broad, glabrous, reddish when young, dark green later; panicles erect or ascending, up to 30 cm. or more long, puberulous; flowers polygamous, yellow, fragrant; calyx about 2·5 mm. long, puberulous; petals 5, longer than calyx, glabrous; stamens 4–5, ripe drupe ellipsoid to obliquely reniform, 5–15 cm. long.

Cultivated and naturalized; SL–4500 feet; fl. Dec–Mar, fr. (Feb–) Apr–Aug. *H 11929! P 23280! Wynter UCWI 2446!* Native of S.E. Asia, now widely distributed; Grand Cayman. Many varieties differing in the shape, size, colour, texture and flavour of fruit occur in Jamaica; the varieties ' Bombay ', ' East Indian ' and ' Julie ' are usually regarded as superior to others; all have been introduced.

3. SPONDIAS L. (1753); Airy Shaw & Forman (1967)

1 Leaflets nearly equal-sided at base, 6–9 (–11) cm. long, shortly and acutely acuminate, margin distantly serrate to crenate; fruit 6–8 (–10) cm. long; flowers creamy-white
1. *dulcis*
1 Leaflets unequal-sided at base; fruit up to 4 cm. long:
2 Leaflets 5–9 (–13) cm. long, glabrous, subentire, acuminate, the acumen obtuse; flowers greenish-white   2. *mombin*
2 Leaflets 2–4 cm. long, pubescent on the midrib, serrulate distally, acute; flowers red
3. *purpurea*

1. S. dulcis Parkinson, Journ. Voy. S. Seas: 39 (1773).—*S. cytherea* Sonn. (1782). Jew Plum.

Tree 6–16 m. high, glabrous; leaves on vegetative shoots up to about 50 cm. long with up to 12 pairs of leaflets; panicle as long as or longer than subtending leaf; petals 2–2·5 mm. long; ripe fruit yellow; stone spiny-fibrous.

Occasional in cultivation; SL–about 2000 feet; fl. Mar–Apr, fr. Sept–Dec. *H & P 13531! Thomas 7772!* Native in the S. Pacific, introduced generally in the tropics.

2. S. mombin L., Sp. Pl. **1**: 371 (1753).—Hog Plum.

Tree 8–18 m. high, glabrous; leaves up to about 25 cm. long; panicle usually longer than the leaf; petals about 3 mm. long; flowers fragrant; ripe fruit yellow, ovoid to oblong, 3–4 cm. long.

Common, mostly along roadsides and field margins; 100–2500 feet; fl. mostly Apr–June, fr. mostly July–Aug. *A 7844! H 9230! C. B. Lewis!* Doubtfully native in the New World tropics, now widely distributed.

3. S. purpurea L., Sp. Pl. ed. 2, **1**: 613 (1762).—*S. mombin* of L. (1759), not L. (1753). *S. lutea* Macf. (1837). Jamaican Plum.

Tree 3–10 m. high; bark grey with vertical lines of lenticels; leaves deciduous usually re-appearing with the flowers, up to about 15 cm. long; panicle short, few-flowered, puberulous; petals 3–4 mm. long; ripe fruit purple or yellow, obovoid, 2·5–3 cm. long.

Occasional in cultivation and in hedgerows and along roadsides; 450–2500 feet; fl. and fr. most of the year. *A 7192! P 8610! Wynter UCWI 9138!* Native from Mexico to Peru and Brazil, introduced elsewhere, Florida, West Indies. The yellow-fruited variant (forma *lutea* F. & R.) flowers and fruits later in the year (fr. Sept–Nov).

4. MOSQUITOXYLUM Krug & Urb. (1895)

1. M. jamaicense Krug. & Urb. in Notizbl. Bot. Gart. Berlin **1**: 79 (1895).— Mosquito Wood.

Tree 10–16 m. high, with a straight smooth trunk, densely puberulent on shoots and inflorescence at least when young; leaves up to 35 cm. long with up to about 8 pairs of leaflets; leaflets glabrous, oblong-elliptical, unequal-sided at base, apex obtuse or shortly acuminate, 5–13 cm. long, 1·5–3·5 cm. broad; flowers dioecious, numerous, sessile in spikes in a panicle longer than the leaves, white; sepals 1·2–1·5 mm. long; ripe fruit 1-celled with fleshy red pericarp, 7–8 mm. long.

Local in woodland on limestone in the western parishes; 1500–2000 feet; fl. Feb, fr. Feb, Aug–Sept. *H 9173! 9185! J.P. 1287!* British Honduras, Panama.

5. COMOCLADIA P. Browne (1756)

1 Leaflets stalked, truncate-subcordate to broadly cuneate at base:
  2 Leaflets entire to remotely denticulate or shallowly sinuate, dentate in juvenile plants,[1]
    subglabrous to pubescent, mostly oblong and acutely acuminate   1. *pinnatifolia*
  2 Leaflets consistently dentate in mature plants; Grand Cayman   *C. dentata* Jacq.
1 Leaflets sessile or nearly so, cordate or more rarely truncate at base, entire:
  3 Leaves and panicle-branches densely rusty-tomentose; leaflets rounded to bluntly
    pointed at tip   2. *velutina*
  3 Leaves and panicle-branches glabrous or at most puberulous; leaflets more or less
    acuminate at tip, rather leathery:
    4 Leaflets 3–5 pairs, up to 5 cm. long and 2·5 cm. broad   3. *parvifoliola*
    4 Leaflets (5–) 6–7 pairs, up to 9 cm. long and 4 cm. broad   4. *cordata*

1. C. pinnatifolia L., Syst. Nat. ed. 10, **2**: 861 (1759).—*C. pilosa* Britton (1910). *C. troyensis* F. & R. (1921). Maiden Plum.

Tree or treelet 1–8 m. high with a slender trunk, sparingly or not branched; leaves up to 70 cm. or more long, longer and probably with more numerous pairs of leaflets when plant not in flower; leaflets 6–11 (–14) pairs, 4–17 cm. long, 2·5–7 cm. broad; panicles usually shorter than the leaves; flowers very small, sessile or subsessile, dark crimson; drupe slightly curved, blackish-red; 7–13 mm. long.

Common in thickets and woodlands on limestone hills; 50–4100 feet; fl. Nov–June, fr. Feb–June. *A 6189! 8915! H 10786! H & P 13529!* Hispaniola. The indumentum of leaves and inflorescence ranges from subglabrous to rather densely tomentose and may comprise short or long spreading hairs or a mixture of both types; it has not been possible as a result of examining a wide range of specimens from many localities to propose any subdivision. Individual populations show some consistency but there is no obvious geographical correlation between them.

2. C. velutina Britton in Torreya **7**: 6 (1907) —Velvet-leaved Maiden Plum.

Tree 4–12 m. high, mostly low-branched; leaves 25–40 cm. long; leaflets (4–) 6–8 pairs, oblong, up to 7·5 cm. long and 4 cm. broad; panicles shorter than the leaves; flowers subsessile, deep crimson; drupe ellipsoid, about 1 cm. long.

Locally common from St. Andrew to St. Elizabeth along the south coast on arid limestone; 50–200 feet; fl. Sept, Dec–Apr, fr. Mar–Apr. *A 8963! H 9208! Powell 697! P 11674!* Endemic.

---

[1] Several species of *Comocladia* with variously dentate leaflets have been described for Jamaica. They were based on inadequate and possibly juvenile or precocious material; these are *C. grandidentata* Britton (1910), *C. hollickii* Britton (1910) and *C. jamaicensis* Britton (1910). The pattern of variation within this genus in Jamaica is extremely complicated and requires much further study before a satisfactory classification can be made.

3. C. parvifoliola Britton in Bull. Torr. Bot. Club **37**: 346 (1910).

Tree to 10 m. high, glabrous; leaves up to 20 cm. long; leaflets ovate to oblong-lanceolate with gland pits in the vein-axils beneath; panicle as long as or longer than the leaves; flowers crimson.

Rare and local (Han.), in woodland on limestone hill; 1750 feet; fl. Mar–May. *H 10267! P 31308!* Endemic.

4. C. cordata Britton in Torreya **7**: 6 (1907).

Tree with slender trunk 4–12 m. high; leaves up to 30 cm. long; leaflets ovate-oblong to oblong, the gland pits not conspicuous; panicles longer than the leaves; flowers crimson.

Local (Trel.), in woodland on limestone; 1550–2200 feet; fl. Mar, June, Sept. *H 9416! P 9907! 21341!* Endemic.

6. METOPIUM P. Browne (1756)

1. M. brownii (Jacq.) Urb., Symb. Ant. **5**: 402 (1908).—*Rhus metopium* L. (1759). *Terebinthus brownii* Jacq. (1760). Burn Wood.

Tree (3–) 5–12 m. with light reddish-brown scaling and scurfy bark and viscid caustic sap turning black on exposure to air; leaves up to about 25 cm. long; leaflets long-petiolulate, broadly ovate, shortly acuminate to retuse, glossy, up to about 9 cm. long and 6 cm. broad; panicles mostly about as long as the leaves; flowers about 4 mm. long, the pistillate slightly smaller than the staminate or hermaphrodite, glabrous, greenish turning yellow; drupe ellipsoid, about 1 cm. long, orange-red to reddish-brown.

Common in thickets and open woodlands on rocky limestone; SL–1750 feet; fl. Feb–Apr, fr. Mar, July–Aug. *A 10922! H 9224! 9317! Powell 976!* Mexico, Honduras, Cuba, Hispaniola.

On account of the caustic sap *Metopium brownei* and species of *Comocladia* may be dangerous to handle, some subjects being apparently more susceptible to the burning effects than others.

113. SAPINDACEAE

Trees or erect or climbing shrubs, rarely herbs. Leaves alternate or very rarely opposite, simple or compound; stipules present in some climbers. Flowers bisexual or functionally unisexual (polygamo-dioecious), actinomorphic or zygomorphic, usually small, in racemose or paniculate cymes. Perianth biseriate or uniseriate; sepals (4–) 5, free or connate, imbricate and often unequal or rarely valvate; petals 3–5, free, equal or unequal, often with scales on the inside, or absent. Glandular disk present. Stamens hypogynous, inserted within the disk, typically 10 in 2 whorls of 5, but often 8, rarely 4, 5 or many, free; anthers 2-locular, opening lengthwise; pistillode present in male flowers. Ovary superior, usually 3-locular, rarely 1-, 2- or 4-locular; ovules anatropous or campylotropous, usually 1 or 2, ascending on usually axile, sometimes parietal, placentas; style usually 1, rarely 2–4, terminal or between the ovary-lobes, simple or divided. Fruit a capsule, drupe, berry or samaroid. Seeds with usually curved embryo, often arillate; endosperm lacking.

140 genera with about 1500 subtropical and tropical species.

1 Shrubs climbing by inflorescence-tendrils; leaves biternate, the blades often pellucid-dotted; flowers zygomorphic; petals 4; seeds usually arillate:
  2 Fruit of 3 samaras, indehiscent; inflorescence thyrsoid-paniculate   **1. Serjania**
  2 Fruit a capsule:
    3 Capsule 3-valved, leathery, opening septifragally, the valves winged or ribbed; inflorescence thyrsoid   **2. Paullinia**
    3 Capsule 3-lobed, membranous, inflated, opening loculicidally; inflorescence corymbose   **3. Cardiospermum**

1 Erect shrubs or trees:
  4 Leaves simple (unifoliolate); capsule 3-winged; flowers regular; petals absent; seeds
    exarillate   **4. Dodonaea**
  4 Leaves compound; petals present:
    5 Flowers 4-merous; fruit usually drupaceous, 1-seeded:
      6 Leaflets 3, petiole not margined, blades often pellucid-dotted; flowers subregular;
        fruit softly fleshy or dry, less than 1 cm. in diameter; aril small, fleshy
                                                                        5. Allophylus
      6 Leaflets 2–3 pairs, the rachis winged on young plants; flowers regular; fruit leathery,
        at least 2 cm. in diameter; aril pulpy   **6. Melicoccus**
    5 Flowers 5-merous, regular:
      7 Fruit indehiscent; seeds exarillate:
        8 Leaves digitately (2-) 3-foliolate; petiole margined; sepals deciduous; fruit a
          1-seeded drupe 6–8 mm. long, without rudiment of a second loculus
                                                                        7. Hypelate
        8 Leaves not mostly 3-foliolate, usually pinnate:
          9 Leaflets (1–) 3–4 (–6), petiole and rachis not margined or winged; sepals persistent;
            fruit a 1-seeded drupe about 1 cm. in diameter, with rudiment of a second
            loculus   **8. Exothea**
          9 Leaflets 3–6 pairs, petiole and rachis with or without a margin or wing; sepals
            deciduous; fruit of up to 3 1-seeded cocci, 2 or 1 not usually developed, the
            cocci 1·5–2 cm. in diameter   **9. Sapindus**
      7 Fruit dehiscent; leaves pinnate, petiole and rachis not winged; seeds arillate:
        10 Capsule 7–8 cm. long, thick-walled, leathery; aril cream-coloured; seeds 3,
           shiny, black, about 2 cm. long   **10. Blighia**
        10 Capsule 10–16 mm. long; seeds less than 1 cm. long:
           11 Petals 5, with 1 or 2 scales joined to the outer margin; leaflets crenate-sinuate to
              serrate, up to about 6 pairs; capsule loculicidally 3-valved   **11. Cupania**
           11 Petals rudimentary; leaflets entire, up to about 9 pairs; capsule 1–3-lobed, the
              lobes globose, 2-valved   **12. Matayba**

1. SERJANIA Mill. (1754)

1 Cymules umbelliform with common peduncles less than 1 mm. long; leaf-rachis distal
  to basal pinnae plainly winged; stem 5-ridged, often prickly; fruit glabrous or nearly
  so, 2·0–2·7 cm. long   1. *mexicana*
1 Cymules unilaterally racemose with common peduncles at least 3 mm. long; leaf-rachis
  hardly or not winged; stem obscurely ridged, not prickly; fruit pubescent:
  2 Fruit 1·3–1·5 cm. long   2. *laevigata*
  2 Fruit 2·3–3·2 cm. long   3. *equestris*

1. S. mexicana (L.) Willd. in L., Sp. Pl. ed. 4, 2 (1): 465 (1799).

High climber; leaves up to 30 cm. long; leaflets ovate to oblong-elliptical,
obtuse, acute or acuminate, entire or with a few blunt teeth, 6–13 cm. long;
inflorescences racemose or paniculate; flowers white, fragrant; sepals unequal,
about 3 mm. long; petals a little shorter.
  Occasional, in thickets and woodlands in rather dry limestone areas; 200–2600
feet; fl. Feb, fr. May. *H 8364! P 23228! Wynter UCWI 3102!* Mexico to Colombia
and Venezuela.

2. S. laevigata Radlk. in Urb., Symb. Ant. **1**: 347 (1899).

Trailing or climbing shrub with stems to 10 m. long; leaves up to about 15 cm.
long; leaflets ovate to elliptical, obtuse or emarginate, subentire, leathery, glabrous
except domatia beneath, 6·5–8 cm. long; panicles 10–15 cm. long; flowers white with
yellow spot; sepals unequal, 3·5–4·5 mm. long; petals obovate, about 5 mm. long.
  Occasional, in thickets and woodlands mostly on arid coastal rocky limestone;
SL–2000 feet; fl. Aug–Jan, fr. Jan, July. *A 11957! H 6679! 9024! H & P 15095!
Weaver 1060!* Endemic.

3. S. equestris Macf., Fl. Jam. **1**: 156 (1837).—Mountain Supple Jack.

Climber to about 6 m. high; stems 3-angled; leaflets ovate, subacuminate,
distantly toothed beyond the middle, papery, glabrous except domatia beneath,
about 5 cm. long; sepals very unequal; petals as long as the larger sepals.
  Uncommon and rather local (St. Andr., St. Thom.), on trees in submontane
woodlands; 1500–2750 feet; fl. June–Aug, Nov–Jan, fr. June, Nov–Jan. *A 6006!
H 5742!* Endemic.

2. PAULLINIA L. (1753)

Capsule 3-winged, abruptly contracted at base into a short stalk, retuse; leaflets entire to
  shallowly crenate, obtuse to rounded at tip, thick, opaque   1. *barbadensis*
Capsule 6-ribbed, tapered at base into a long stalk, more or less apiculate; leaflets shallowly
  serrate, acute to narrowly obtuse, thin, with fine pellucid lines   2. *jamaicensis*

1. P. barbadensis Jacq., Enum. Syst. Pl. Carib.: 36 (1760).—Supple Jack.

Shrub climbing by eventually woody tendrils to 6 m. or more; leaves up to 8 cm.
long; leaflets elliptical to obovate, the proximal usually very small, the terminal
3–5 cm. long; inflorescence a thyrse with many short branches; sepals unequal,
2–3 mm. long; petals elliptical, longer, white or cream-coloured; capsule about 1·5
cm. long and broad, puberulous; seeds shiny black with white aril.

Frequent, in thickets and woodlands on arid limestone, particularly in southern
coastal areas; SL–1500 feet; fl. July–Mar, fr. Dec–Feb. *A 8899! H & B 10795!
P 5999! Skelding UCWI 5035!* Endemic; apparently known only in cultivation in
Barbados.

2. P. jamaicensis Macf., Fl. Jam. **1**: 158 (1837).—Supple Jack.

Climbing shrub like the last; leaves up to 20 cm. long, the terminal leaflets 4–8
cm. long; petals obovate, white; ripe capsule red, puberulous, about 1·5 cm. long
including the stalk.

Frequent, in thickets and on rocky limestone banks; 100–3000 feet; fl. May–Jan,
fr. Aug–Mar. *A 8859! H 8416! H & P 13488! P 15724!* Cuba.

3. CARDIOSPERMUM L. (1753)

1 Flowers 8–10 mm. long; capsule 6–7 cm. long, pointed at both ends; leaves up to 20 cm.
    long   1. *grandiflorum*
1 Flowers 4–6 mm. long; capsule and leaves smaller:
  2 Capsule broader than long, truncate or depressed apically, about 1 cm. long and 2 cm.
      broad; seed about 3 mm. in diameter, with a 2-lobed aril   2. *microcarpum*
  2 Capsule subglobose:
    3 Capsule 3–4 cm. long; seed about 5 mm. in diameter with a heart-shaped aril
                                                                               3. *halicacabum*
    3 Capsule 2–3 cm. long; seed 2·5–4 mm. in diameter with a kidney-shaped or barely
        notched semicircular aril   4. *corindum*

1. C. grandiflorum Sw., Nov. Gen. & Sp. Pl.: 64 (1788).—Heart Pea, Wild Supple
Jack.

Stems up to 8 m. or more long, variably hairy from long-hispid to thinly pubescent; leaflets serrate, the tips long and sharply pointed, the terminal up to 9 cm. long
and 4·5 cm. broad; inflorescence sometimes longer than the leaves; appendage to
anterior petal orange; seed 7 mm. in diameter, black, with a white roundish aril.

Rather common, in secondary thickets; 400–2700 feet; fl. July–Feb, fr. July–Dec.
A 8230! H 9969! H & P 14237! P 22876! Scattered through continental tropical
America and sparsely in the West Indies; throughout tropical Africa.

2. C. microcarpum Kunth, Nov. Gen. **5**: 104 (1821).

Small climber up to 3 m. high; leaves up to 8 cm. long; leaflets ovate to lanceolate up to 3·5 cm. long and 2 cm. broad.

Occasional, in secondary thickets and trailing on the ground in rocky waste
places; SL–300 feet; fl. Dec, fr. Dec–Mar. *H 12754! H & P 13589! Yuncker 18459!*
General in the tropics. This species has been regarded by recent authors as a variety
of *C. halicacabum* or not distinguished from that species (Alain, Fl. Cub. **3**: 190
(1953); Keay in Fl. W. Trop. Afr. **1** (2): 711 (1958)), but the smaller, relatively
broader, truncate fruits are diagnostic.

3. C. halicacabum L., Sp. Pl. **1**: 366 (1753).

Like the last; leaves up to 12 cm. long; leaflets ovate-elliptical, acute-tipped, up
to 5 cm. long and 2·5 cm. broad.

Uncommon (St. Cath., Clar.), a weed of thickets and waste places; 50–200 feet;
fl. May–June, fr. Feb. *H 12060! A. von der Porten!* General in warm countries.

4. C. corindum L., Sp. Pl. ed. 2, **1**: 526 (1762).

Like the last; leaves up to 15 cm. long; stem and leaves pubescent to glabrous; leaflets ovate-elliptical, deeply serrate-incised, up to 5 cm. long and 2 cm. broad.

Occasional, in thickets and waste places; 100–1200 (–3800) feet; fl. May–Feb, fr. June–Mar. *A 5455! 8537! H 8837! P 7520!* General in the tropics; Grand Cayman.

4. DODONAEA Mill. (1754)

1. D. viscosa Jacq., Enum. Syst. Pl. Carib.: 19 (1760).—Switch Sorrel.

1a. Var. **viscosa.**

Erect or straggling-branched shrub 1–3 m. or tree to 5 m. high; stems and leaves viscous-glandular; leaves oblong-obovate, 3–12 cm. long, 1·5–3·5 cm. broad; flowers green or greenish-yellow or tinged reddish; sepals oblong, 2–3 mm. long; capsule variable in size up to 2·5 cm. broad; seeds black, lenticular.

Rather local (St. Cath., Clar., St. Eliz.), in sandy beach thickets; SL–20 feet; fl. and fr. most of the year. *A 12759! Campbell 6052! P 15333! Stearn 848!* Widespread on tropical shores; Cayman Is.

1b. D. viscosa var. **angustifolia** (L. f.) Benth., Fl. Austral. **1**: 476 (1863).

Leaves linear-oblong to narrowly oblanceolate, 4–12 mm. broad; capsule up to about 1·5 cm. broad.

Locally common, mainly on cleared well drained hillsides in the mountains; (900–) 2000–6000 feet; fl. and fr. most of the year. *A 8513! H 8578! P 23306!* Widespread in the mountainous parts of the tropics of both hemispheres.

5. ALLOPHYLUS L. (1753)

1 Leaflets serrate, generally pubescent, usually more or less glabrescent adaxially, rarely so beneath; ripe fruit drupaceous with orange or red fleshy pericarp   1. *cominia*
1 Leaflets entire, glabrous or nearly so but often with conspicuous domatia in the vein-axils beneath; ripe fruit with thin rather dry pericarp:
  2 Leaves 15 cm. or more long; larger leaflets 10–19 cm. long, with distinct petiolules up to 15 mm. long; petioles (4–) 5·5–10 cm. long; panicle with (5–) 8–20 or more primary branches; peduncle about as thick as the subtending petiole
                                                 2. *jamaicensis*
  2 Leaves up to 15 cm. long; larger leaflets 5–11 cm. long, with petiolules 1–7 mm. long; petioles 1·5–6 cm. long; panicle with 2–4 (–5) primary branches; peduncle much thinner than the subtending petiole
                                                 3. *pachyphyllus*

1. A. cominia (L.) Sw., Nov. Gen. & Sp. Pl.: 62 (1788).

Tree 4–13 m. high; leaves light green, 10–20 cm. long; leaflets mostly obovate-elliptical, acuminate, (5–) 8–15 cm. long, up to 6 cm. broad; panicles 4–many-branched, usually shorter than the leaves; flowers numerous, small, fragrant; sepals unequal the inner larger to nearly 1·5 mm. long, cream-coloured; petals white; stamens 8, inserted in the lower side of the flower; fruit when dry sub-spherical, about 5 mm. in diameter.

Common on well drained slopes and in woodlands and thickets on limestone; 30–3000 feet; fl. mostly June–Nov, fr. Aug–Jan. *A 6046! 7740! J.P. 1131! P 7472!* British Honduras, Bahamas, Cuba, Hispaniola.

2. A. jamaicensis Radlk. in Urb., Symb. Ant. **5**: 407 (1908).

Shrub 2 m. or tree 5–11 m. high with spreading branches; leaflets elliptical to elliptic-lanceolate, shortly and obtusely acuminate, up to 8 cm. broad, domatia present or on the larger older leaves absent; calyx yellowish or green; petals white; fruit obovoid, 7–10 mm. long, ribbed.

Occasional in the central and western parishes, in woodland on limestone; 1000–3000 feet; fl. July–Dec, fr. Nov–Jan. *A 12719! H 9440! H & P 13625!* Reported from British Honduras.

3. A. pachyphyllus Radlk. in Urb., Symb. Ant. **5**: 406 (1908).
Bushy tree like the last up to 8 m. high; leaflets elliptical, narrowed to base and decurrent on the petiolule, up to 5 cm. broad; domatia usually present; fruit obovoid-globose, 7–10 mm. long, ribbed, sometimes splitting.
Rare (Manch., St. Ann), in woodland on limestone; 2000–2600 feet; fl. Sept, fr. Jan. *A 12695! Cornman 699! H 12857!* Endemic.
Leenhouts ["A conspectus of the genus *Allophylus* ", Blumea **15** (2): 301–358 (1967)], after a detailed and reasoned discussion, concludes provisionally that phenetically speaking *Allophylus* consists of but one species. Of the 255 names of currently accepted species in the genus Leenhouts accepts *A. cobbe* (L.) Räusch.

6. MELICOCCUS P. Browne (1756)

1. M. bijugatus Jacq., Enum. Syst. Pl. Carib.: 19 (1760).—*Melicocca bijuga* L. (1762). Ginep, Guinep.
Tree, deciduous, 6–18 m. high; trunk often very thick; bark elephant grey, rather smooth; leaflets elliptical to ovate-elliptical, acute, acuminate or obtuse, somewhat unequal-sided at base, (5–) 7–10 (–14) cm. long, up to 6 cm. broad; racemes simple or paniculate; flowers 6–8 mm. in diameter, fragrant; petals white; drupe yellowish-green, 2–3 cm. in diameter; pulp gelatinous around a usually solitary spherical seed.
Common along roadsides and in secondary thickets and woodlands; 50–2000 (–3000) feet; fl. Apr–June, fr. May–Dec. *A 11136! H 9228! Powell 555! P 10144!* Native of tropical Amer., cultivated and naturalized in the West Indies; Grand Cayman. Introduced in tropical Asia and Africa.

7. HYPELATE P. Browne (1756)

1. H. trifoliata Sw., Nov. Gen. & Sp. Pl.: 61 (1788).—Ketto.
Shrub or slender-trunked tree 5–10 (–13) m. high; leaflets sessile, obovate to oblanceolate, more or less rounded at tip, 2–4 (–4·5) cm. long, 1–2 (–2·5) cm. broad; inflorescence-branches slender; pedicels and flower-buds glutinous; petals 5 (–6), white, about 2 mm. long; drupe ellipsoidal black, 6–8 mm. long.
Common in the southern parishes in thickets and woodlands on arid limestone; SL–1250 feet; fl. Feb–Aug, fr. July–Dec. *A 11360! H 9570! P 23709!* Florida, Bahamas, Greater Antilles, St. Martin, Anguilla; Cayman Is.

8. EXOTHEA Macf. (1837)

1. E. paniculata (Juss.) Radlk. in Durand, Ind. Gen. Phan.: 81 (1888).—Wild Ginep or Guinep.
Tree with reddish or brown bark 5–16 m. high; leaflets elliptical to oblong-elliptical, obtuse at tip, 6–11 cm. long, 2–4 cm. broad, glossy; panicles as long as or longer than the leaves, yellowish- or orange-tomentose; sepals about 3 mm. long; petals about as long as sepals, white; drupe purplish-black.
Occasional in the central parishes, in woodland on rocky limestone; (100–) 1200–3000 feet; fl. Dec–Mar, July, fr. Jan–May. *A 8472! H 5121! P 21940! Stearn 343!* Guatemala, Florida, Bahamas, West Indies south to St. Vincent.

9. SAPINDUS L. (1753)

1. S. saponaria L., Sp. Pl. **1**: 367 (1753).—Soap Berry Tree.
Tree (4–) 8–15 m. high; leaflets elliptical, oblong or lanceolate, equal- or unequal-sided, rounded to long-acute at tip, 7–18 cm. long, up to 3·5 cm. or more broad; panicles up to 30 cm. long; flowers small; sepals 1–2 mm. long; petals shorter than sepals, hairy, white; ripe fruit yellow; seed globose, black, 12 mm. in diameter.
Occasional or sometimes planted, mostly in rather dry or well drained areas on limestone; SL–2900 feet; fl. Dec, fr. Feb–Apr, July; *H 5815! 6666! Proctor & Mullings 21966!* Florida to Argentina, West Indies, introduced in tropical Asia and Africa.

10. BLIGHIA Konig (1806)

1. B. sapida Konig in Konig & Sims, Ann. Bot. **2**: 571, tt. 16, 17 (1806).—Ackee.
Tree 8–15 (–22) m. high; leaflets opposite or subopposite, elliptical to oblong-obovate, shortly acuminate, 6–18 cm. long, up to 7 cm. broad; inflorescences of simple or compound pendulous racemes; flowers fragrant; calyx 5 (–6)-parted, light green; petals cream-coloured, about 4·5 mm. long; ripe fruit pear-shaped, red or yellow tinged red.
Commonly cultivated and naturalized; SL–2950 feet; fl. and fr. most of the year. *H 10924! P 23899!* Native of W. Africa, commonly cultivated in Jamaica, but rare or absent in many of the W. Indian islands.

11. CUPANIA L. (1753)

Capsule glabrous, top-shaped-triangular; leaves glabrous or more or less pubescent
  beneath; inflorescence puberulous   1. *glabra*
Capsule velvety-tomentose, sub-globose; leaves pubescent on nerves adaxially, densely
  pubescent beneath; inflorescence tomentose   2. *americana*

1. C. glabra Sw., Nov. Gen. & Sp. Pl.: 61 (1788).—Toadwood, Wild Ackee.
Tree (3–) 4–13 m. high; leaflets oblong to oblong-elliptical, rounded and minutely emarginate at tip, cuneate at base, 5–10 (–15) cm. long, 2–6 cm. broad, glossy adaxially; panicle as long as or longer than the leaves; sepals and petals about 2 mm. long; capsule 10–13 mm. long and broad; (2–) 3 (–4)-locular, apiculate; seeds roundish-ellipsoidal, about 7 mm. long.
Common, in thickets and secondary woodlands on limestone; 50–2750 feet; fl. June, Sept–Jan, fr. July, Oct–Apr. *A 6286! 8218! H 5548! 9772! P 21936!* Florida, C. Amer., Cuba.

2. C. americana L., Sp. Pl. **1**: 200 (1753).—Wild Ackee.
Tree like the last to 12 m. high; leaflets larger up to 20 cm. long, mostly obovate-elliptical; leaves on juvenile plants simple or with fewer leaflets and often very large; petals creamy-white; capsule 1–1·5 cm. long, 1·5–2 cm. broad, dark golden- or olive-brown.
Widely scattered and rather local in thickets and woodlands mostly in low-lying poorly drained areas; 200–1000 feet; fl. Jan–Feb, fr. Feb–Apr. *A 9229! 10474! H 11920! P 20607!* Venezuela, Greater Antilles, Dominica, Martinique, Barbados, Trinidad and Tobago.

12. MATAYBA Aubl. (1775)

1. M. apetala (Macf.) Radlk. in Sitz. K. Bayer. Akad. **9**: 535, 624 (1879).—Coby
  Wood.
Tree 5–20 m. high; leaves about 20 cm. long; leaflets oblong-elliptical, rounded at apex or rarely subacuminate, cuneate and often unequal at base, 6–12 cm. long, 2–3 cm. broad; panicles up to as long as leaves; flowers greenish- or light yellow, fragrant; calyx about 1 mm. long; capsule distinctly stalked, reddish, 10–12 (–16) mm. long; seed ellipsoidal, black, 6–8 mm. long, with light orange aril.
Widely distributed, in rough pastures, secondary woodlands and thickets along streambanks; 700–4300 feet; fl. Feb–Aug, fr. July–Oct. *A 11215! H 9166! P 16610! 19717!* Honduras, Hispaniola.

114. CYRILLACEAE

Shrubs and trees. Leaves alternate, simple, entire, without stipules. Flowers bisexual, actinomorphic, small, in racemes. Perianth biseriate; calyx (4–) 5 (–8)-partite, imbricate or rarely valvate, often enlarged in fruit; petals the same number as the sepals, free or shortly connate, imbricate or contorted. Stamens 10 in 2 whorls or 5, hypogynous; filaments free; anthers 2-locular, opening lengthwise or by apical pores. Disk present. Ovary superior, 2–4-locular; ovules 1–4 in each loculus, collateral, apical, pendulous, anatropous; style 1, simple; stigmas 2–5, usually 3. Fruit dry, indehiscent, often angled or winged. Seeds with fleshy endosperm and small embryo; testa lacking.

3 genera with 13 species in subtropical and tropical America.

1. CYRILLA Garden ex L. (1767); Thomas (1960)

Stamens 5; anthers opening by longitudinal slits; disk glandular; style shortly 2–4-lobed; ovules mostly 3 in each loculus; fruit 2–3-sulcate.

1. C. racemiflora L., Mant. Pl.: 50 (1767).—Beet Wood.

Tree (3–) 8–25 (–30) m. high, often branched near the base or shrubby when coppiced; bark and heartwood red; leaves mostly clustered towards the ends of branches, elliptical to oblanceolate, leathery, 2–12 cm. long, up to 2·5 cm. broad, the margin hyaline and produced as an undulate wing to the petiole; racemes clustered, up to 15 cm. long, many-flowered; pedicels 2–4 mm. long; sepals and petals white; petals free, 2·3–2·5 mm. long, sometimes tinged pink; anthers yellow or purplish; fruit about 2 mm. long, at first green turning pink.

Occasional in the central parishes, common in the eastern parishes in submontane thickets and woodlands; (1500–) 2100–6200 feet; fl. and fr. mainly Apr–Aug. *A 7764! H 8725! J.P. 890! P 15922!* S.E. United States, Mexico to Brazil, West Indies.

Thomas (1960), has investigated variation in this species and concludes that clonal reproduction is correlated with a high incidence of seedless parthenocarpy in populations where self-pollination is likely. Sexual reproduction being minimal, these factors have resulted in a pattern of variation in which morphologically distinct populations occupy different environments.

115. AQUIFOLIACEAE

Trees or shrubs. Leaves alternate, simple, often leathery; stipules minute or absent. Flowers bisexual or unisexual, small, actinomorphic, in axillary or internodal cymes or fascicles or solitary. Perianth biseriate; sepals 3–6 (–7), more or less connate basally, the segments imbricate; petals 4–6 (–7), free or shortly united, rotate, the segments imbricate or valvate. Stamens hypogynous, as many as or rarely more than the petals, alternate with them and sometimes adhering to their bases; filaments free; anthers 2-locular, opening lengthwise. Disk absent. Staminodes and pistillodes usually reciprocally present in unisexual flowers. Ovary superior, 3–many-locular; style terminal or absent; stigma lobed or capitate; ovules 1–2 in each loculus, pendulous on axile placentas, anatropous. Fruit drupaceous with 3 or more 1-seeded pyrenes. Seed with small straight embryo and large endosperm.

3 genera with 450 species widely distributed.

1. ILEX L. (1753)

Petals imbricate, more or less connate basally; ovary up to 8-locular, usually fewer, the stigmas as many as the loculi.

1 Leaf-blade with small but distinct marginal teeth, flat-margined or only narrowly revolute when dry, rounded to cuneate at base:
  2 Leaves mostly less than 3 cm. long and 2 cm. broad:
    3 Lamina broadly obovate, obtuse or rounded and toothed distally, glabrous; inflorescences clustered   **1. vaccinoides**
    3 Lamina narrowly ovate, long-acute, puberulous at least on the midrib adaxially; inflorescences solitary   **2. puberula**
  2 Leaves mostly more than 3 cm. long and 2 cm. broad:
    4 Inflorescences solitary, axillary and internodal; leaves elliptical, acuminate to long-acute, glabrous or puberulous on the midrib adaxially   **3. macfadyenii**

4 Inflorescences clustered; leaves elliptical to obovate-elliptical, shortly acuminate, glabrous   4. *nitida*
1 Leaf-blade entire or very indistinctly crenate-toothed, glabrous:
  5 Leaves leathery, margins more or less revolute at least when dry; flowers 4–5 (–6)-merous:
    6 Lamina rarely as much as 3 cm. long, less than 1·5 cm. broad   5. *obcordata*
    6 Lamina 3–12 (–15) cm. long, 1·5–8 cm. broad:
      7 Leaf-blade roundish-elliptical, rounded to emarginate at tip, rounded or truncate at base; inflorescences clustered, corymbose, many-flowered   6. *florifera*
      7 Leaf-blade elliptical to obovate, obtuse to shortly and bluntly acuminate at tip:
        8 Mature fruits 7–8 mm. long on pedicels 4–7 mm. long; stigma flat; leaf-base broadly cuneate to rounded   4. *nitida*
        8 Mature fruits up to 5 mm. long; stigma prominent; leaf-base cuneate:
          9 Fruits 5 mm. long on pedicels 7–9 mm. long; leaf-tip bluntly acuminate; lamina with veins prominent beneath   7. *jamaicana*
          9 Fruits 3·5–4 mm. long on pedicels 4–8 mm. long; leaf-tip obtuse; lamina with veins obscure beneath   8. *sideroxyloides* var. *occidentalis*
  5 Leaves at most only thinly leathery, margins flat; flowers 4–5-merous:
    10 Leaf-tip acute to shortly acuminate; blade elliptical to obovate with conspicuous veins beneath   9. *harrisii*
    10 Leaf-tip obtuse:
      11 Inflorescences clustered; petioles up to 1 cm. long; leaf-blade elliptical to obovate, thinly leathery, the veins obscure beneath   8. *sideroxyloides* var. *occidentalis*
      11 Inflorescences usually solitary; petioles 1–2 cm. long; leaf-blade elliptical to ovate, thin   10. *subtriflora*

1. **I. vaccinoides** Loes. in Urb., Symb. Ant. **7**: 270 (1912).
Shrub 2 m. or tree to 13 m. high; leaves 1–3 (–3·5) cm. long, 0·8–2 cm. broad with petioles up to 6 mm. long; stipules awl-shaped, up to 1 mm. long; inflorescences 1–few-flowered; flowers 4-merous; petals 2–2·5 mm. long; drupe ellipsoidal, 4 mm. long with (3–) 4 pyrenes.
Rare and very local (St. Andr.), in montane woodland; 5800 feet; fl. Apr, Aug. *H 9217! 9379!* Endemic.

2. **I. puberula** Proctor in Bull. Inst. Jam., Sci. ser. **16**: 26 (1967).
Shrub or tree to about 10 m. high; leaves 1–3 (–3·5) cm. long, 0·5–1·5 cm. broad with petioles up to 5 mm. long, punctate beneath; inflorescences 1–few-flowered; flowers (5–) 6-merous; petals white, about 2 mm. long; drupe subspherical, 3 mm. long, with persistent discoid stigma.
Rather local (St. Andr., Port.), in exposed montane thickets and open woodlands on steep slopes; 4000–5000 feet; fl. Feb–Apr, fr. July. *P 15547! Skelding UCWI 4460!* Endemic.

3. **I. macfadyenii** (Walp.) Rehder in Journ. Arn. Arb. **3**: 215 (1922).—*Prinos macfadyenii* Walp. (1842). *P. montanus* Sw. (1788). *I. montana* (Sw.) Griseb. (1860), not Torr. & A. Gray (1848). Winter Berry.
Shrub 2·5–5 m. or tree to 7 (–10) m. high; young stems purplish; leaves 3–8 (–10) cm. long, 1·5–3·5 (–4·5) cm. broad, punctate beneath; petioles 1–2 cm. long; male inflorescences usually branched; flowers (5–) 6 (–7)-merous; petals free to stamen adnation, about 2 mm. long; drupe subspherical, 5–7 mm. broad, purple or white with persistent discoid stigma.
Locally abundant (St. Andr., Port., St. Thom.), in submontane and montane woodlands and thickets; (1500–) 3000–7400 feet; fl. and fr. almost all the year. *A 10479! H 5632! P 8753!* Cuba, Hispaniola, Nevis, Dominica.

4. **I. nitida** (Vahl) Maxim. in Mém. Acad. Sci. St. Pétersb. sér. 7, **29** (3): 27 (1881).—*Prinos nitidus* Vahl (1798).
Shrub or slender bushy tree 5–10 (–20) m. high; leaves 6–12 (–14) cm. long, 3–5·5 cm. broad, shiny; male inflorescences corymbose, female 1-flowered; flowers 4 (–5) -merous; corolla 3–4 mm. long, white; drupe ovoid, 7–8 mm. long, orange-scarlet, with obscure stigma.
Occasional in the central and eastern parishes in woodlands on limestone or shale; 1500–5000 feet; fl. mainly Dec–Mar, fr. May–Sept. *H 11073! P 23362!* Mexico, Cuba, Puerto Rico, Montserrat, Guadeloupe, Martinique.

5. I. obcordata Sw., Fl. Ind. Occ. **1**: 338 (1797).

Shrub 1–3 m. or slender dense-crowned tree to 10 m. high; twigs ridged below the petioles; leaves (0·7–) 1–3 cm. long, up to 1·3 cm. broad, mucronulate, paler beneath; stipules spinelike; male inflorescences 2–3-flowered, female 1-flowered; flowers 4 (–6)-merous; corolla 3 mm. long, white to cream; drupe subspherical, smooth, red, about 5 mm. long, with prominent stigma.

Locally common (St. Andr., Port., St. Thom.), in margins or montane woodlands and in exposed thickets; (3500–) 4000–7400 feet; fl. Dec–Feb, May–July, fr. most of the year. *A 12528! H 9139! P 24025!* Endemic.

6. I. florifera F. & R. in Journ. Bot. **59**: 18 (1921).

Tree to 13 m. high; leaves 6–10 (–15) cm. long, 4·5–8·5 cm. broad, glossy; flowers 4-merous, rank-scented; petals 2·3–2·5 mm. long, white; drupe subspherical to pear-shaped, about 3 mm. long including prominent stigma.

Rare and local (St. Ann), in woodland on limestone; 1300–2000 feet; fl. Feb–May, fr. Apr–May. *Cornman! H 12012! Stearn 581!* Endemic.

7. I. jamaicana Proctor in Bull. Inst. Jam., Sci. ser. **16**: 24, t. 9 (1967).

Tree to 7 m. high; leaves obovate, 3–7 cm. long, 1·5–3 cm. broad; petioles 4–8 mm. long; inflorescences solitary or clustered, female 1-flowered; flowers 4–5-merous.

Rare and local (Port.), in wet mossy woodland on limestone; 2500 feet; fr. June. *P 19738!* Endemic.

8. I. sideroxyloides (Sw.) Griseb. in Abh. Gött. Akad. **7**: 224 (1857).—*Prinos sideroxyloides* Sw. (1788).

8a. Var. occidentalis Loes. in Urb., Symb. Ant. **1**: 345 (1899).—*I. occidentalis* Macf. (1837), illegitimate name, partly.

Shrub 4–6 m. or tree 5–13 m. high; leaves 2–8 cm. long, 1·5–5 cm. broad; petioles 5–10 mm. long; male inflorescences 3 (–6)-flowered, female 1-flowered; flowers 4 (–5)-merous, heavily fragrant; drupe ovoid with a conical stigma, turning red.

Uncommon in the eastern parishes and local in Clarendon, in hill savannas and high thickets and woodlands; 2300–5400 feet; fl. and fr. July–Jan. *H 6088! H & P 14814! P 16726!* Cuba.

9. I. harrisii Loes. in Urb., Symb. Ant. **1**: 346 (1899).

Shrub or tree 5–8 m. high; leaves (4–) 5–8 (–10) cm. long, mostly 2–3 cm. broad, with some very small hairs adaxially when young; petioles 10–15 mm. long; male inflorescences usually 3-flowered, female usually 1-flowered; drupe ellipsoid-globose, lightly grooved, 4 mm. long, 3–3·5 mm. broad; stigma shallowly convex.

Local and uncommon (St. Andr., Port.), in montane woodlands; 3500–5600 feet; fl. Feb–Mar, fr. July. *H 5673! 7673! P 9887!* Endemic.

10. I. subtriflora Griseb. ex Loes. in Engl., Bot. Jahrb. **15**: 312 (1892).

Tree; leaves 4–11 cm. long; inflorescences 1–3-flowered or racemose or paniculate; corolla about 2 mm. long.

Very rare (St. Ann), an obscure species not recently collected. *March, Prior.* Endemic.

Some species of *Ilex* resemble species of *Maytenus* (*Celastraceae*); they may be distinguished by the midrib adaxially which is characteristically depressed in *Ilex* and prominent in *Maytenus*, and in flower by the absence of disk in *Ilex*.

116. CELASTRACEAE

Trees, shrubs or woody climbers. Leaves opposite or alternate, simple, often leathery; stipules small, deciduous or absent. Flowers unisexual or bisexual, actinomorphic, small, in clusters, cymes or panicles. Perianth biseriate or petals sometimes absent; calyx with 4–5 imbricate lobes, persistent; petals mostly free, 4–5, imbricate or valvate. Disk present, conspicuous. Stamens (2–) 3–5 (–10), free or attached to the margin of the disk; anthers 2 (–4)-locular, opening lengthwise or when very short transversely. Ovary superior, although sometimes sunken in the disk, 2–5-locular; ovules 2 or more in each loculus, axile, anatropous; style short, branched or with a 2–5-lobed stigma. Fruit a capsule, drupe, berry, samara or schizocarp. Seed with or without endosperm or aril.

About 80 genera and over 1000 species, widely distributed.

1 Woody climbers; inflorescence cymose, pedunculate; calyx and corolla 5-merous; ovary
 3-locular; fruit a schizocarp with winged carpels; leaves opposite (*Hippocrateaceae*[1])
                        1. Cuervea
1 Erect shrubs or trees:
  2 Flowers in axillary clusters or solitary; leaves alternate; ovary 2–3-locular:
   3 Fruit a leathery capsule, dehiscing by 2–3 valves; perianth (4–) 5-merous; seed
    arillate; leaves thick with a prominent midrib and often drying a pewter colour;
    flowers polygamous   **2. Maytenus**
   3 Fruit a drupe containing 2 stones; perianth 4-merous; seed not arillate; leaves alter-
    nate or clustered, thin; flowers unisexual, monoecious or dioecious
                      3. Schaefferia
  2 Flowers in pedunculate cymes or panicles; leaves mostly opposite; fruit a drupe; seed
  not arillate; perianth 4-merous (unless otherwise stated):
   4 Flowers bisexual; ovary 4-locular; ovule 1 in each loculus, erect from the base
                    4. Crossopetalum
   4 Flowers unisexual; plants dioecious:
    5 Ovules 2 in each of 2–5 loculi, erect from the base; leaves sometimes alternate;
     perianth 4–5-merous; staminodes petaloid   **5. Cassine**
    5 Ovule 1 in each loculus, pendulous; staminodes absent:
     6 Ovary-loculi 2; drupe usually 1-seeded   **6. Gyminda**
     6 Ovary-loculi 4; drupe 2- or more-seeded   **7. Tetrasiphon**

1. CUERVEA Triana ex Miers (1872)

1. C. kapplerana (Miq.) A. C. Sm. in Brittonia **3**: 399 (1940).—*Hippocratea kapplerana* Miq. (1854). *H. oblongata* (Miers) Solander ex F. & R. (1926).

Robust woody vine; leaves ovate-elliptical, acute, 6–15 cm. long, 2–6 cm. broad, rather prominently reticulate-veined on both surfaces; cymes axillary, up to about 3 cm. long; peduncles 1–1·5 cm. long; flowers light yellow, fragrant; calyx-segments unequal; petals obovate, 4–5 mm. long; carpels in fruit capsular, broadly obovate to suborbicular, 5–9 cm. long, up to 10 cm. broad.

Very rare and local (Han., West., St. Thom.), in woodland on arid limestone; 50–100 feet; fl. Nov. *A 10185! P 11153!* Costa Rica to Brazil and Peru, St. Vincent, Trinidad.

2. MAYTENUS Molina (1782)

1 Leaf-tip rounded, obtuse or shortly acuminate:
  2 Petals 1·5–2·5 mm. long:
   3 Capsule smooth; pedicel slender:
    4 Capsule 12–15 mm. long, distinctly stipitate; pedicels 4–8 mm. long
                    1. jamaicensis
    4 Capsule 7–9 mm. long, shortly stipitate; pedicels 2·5–5 mm. long **2. microcarpa**

---

[1] *Celastraceae* and *Hippocrateaceae* are closely allied and although some modern authors keep them distinct, others unite them. *Cuervea kapplerana* is the only representative of the latter family in the Jamaican flora.

3 Capsule rough with corky patches; pedicel thick:
  5 Leaf-margin revolute; petals 2–2·5 mm. long; capsule 15–19 mm. long
                                                 3. *clarendonensis*
  5 Leaf-margin flat; petals 1·5–1·8 mm. long; capsule 10–17 mm. long   4. *crassipes*
2 Petals about 1 mm. long; capsule slightly rough, 13–20 mm. long; pedicel slender,
  4–5 mm. long   5. *virens*
1 Leaf-tip caudate-acuminate; capsule 15–17 mm. long; pedicel 6–10 mm. long
                                                  6. *harrisii*

1. M. jamaicensis Krug & Urb. in Notizbl. Bot. Gart. Berlin **1**: 78 (1895).

Shrub 3–5 m. or densely leafy slender-trunked tree to 10 m. high; leaves mostly ovate to elliptical but variable from suborbicular to oblong, apex acute to emarginate, leathery or thin, drying pewter grey with the midrib prominent adaxially, up to 17 cm. long and 7 cm. broad, mostly smaller; flowers clustered in the axils, greenish-yellow, fragrant; calyx-lobes broadly rounded, 1·2–1·5 mm. long.

When construed in the broadest sense this species is frequent in thickets and woodlands on limestone or shale in the central and eastern parishes; 300–4500 feet; fl. July, Nov–Mar, fr. Feb–Sept. Hispaniola. Urban, Symb. Ant. **5**: 57–58 (1904) recognizes three varieties, but the pattern of variation requires further study. Plants with small leathery often emarginate leaves occur in thickets on arid limestone and elsewhere at elevations up to about 4000 feet (*A 8850! 10704! H 5460!*); plants with larger thin obtuse or shortly acuminate leaves occur in montane woodlands in the eastern parishes from 3000–4500 feet (*A 11616! H 5505! 7609!*).

2. M. microcarpa F. & R. in Journ. Bot. **59**: 19 (1921).

Shrub about 3 m. high; leaves ovate-elliptical to oblong, shortly and bluntly acuminate, 5–11 cm. long, 2–5 cm. broad, shiny adaxially; flowers clustered; calyx-lobes about 1 mm. long.

Rare and local (Clar., Trel.), on limestone rocks in woodland; 1250–2500 feet; fl. Aug, Dec–Jan, fr. May. *H 12800! 11054! P 9777!* Endemic.

3. M. clarendonensis Britton in Bull. Torr. Bot. Club **39**: 8 (1912).—*Ilex uniflora* F. & R. (1921).

Shrub or small tree 3–4 m. high with straggling branches or erect tree to 15 m.; leaves broadly elliptical, 6–13 cm. long, 4–9 cm. broad, obtuse or rounded at both ends; flowers light green; pedicels 2–7 mm. long; calyx about 1·6 mm. long; seeds about 10 mm. long.

Rather local in woodland on limestone hills in the central parishes; 1500–3000 feet; fl. Nov–Jan, fr. July–Sept. *A 8485! H 10947! HPS 14606!* Endemic.

4. M. crassipes Urb., Symb. Ant. **5**: 404 (1908).

Shrub 2–4 m. or tree 8 m. high; leaves ovate-lanceolate to elliptical, obtuse to subacuminate at tip, rounded to obtuse at base, thinly leathery, 5–15 cm. long, 3–5 cm. broad; flowers light green; pedicels in flower 1–3 mm. long, longer and thickened in fruit; calyx-lobes about 1 mm. long; seeds 8–10 mm. long, black.

Occasional in woodland on rough limestone in the central and western parishes; 1500–3000 feet; fl. June–Aug, Dec, fr. Aug–Oct, Dec–May. *A 11700! H 8966! P 10288!* Endemic.

5. M. virens Urb., Symb. Ant. **5**: 60 (1904).

Shrub 2–3 m. or tree to 6 m. high; stem slender; branches drooping; leaves ovate to oblong, distinctly acuminate, mostly broadly rounded at base, olive-buff when dry, 3–7 (–10) cm. long, 1·5–4 (–5) cm. broad; flowers green; calyx-lobes 0·7–0·8 mm. long; seeds flat, ellipsoid, 6–8 mm. long, almost covered by the lobed aril.

Frequent in woodland on limestone, mostly in the central parishes; 1500–3000 feet; fl. July–Dec, fr. most of the year. *H 10346! 12796! H & P 14397! P 16213!* Endemic.

6. M. harrisii Krug & Urb. in Notizbl. Bot. Gart. Berlin **1**: 78 (1895).

Leaves ovate to broadly elliptical, acute at base, 6–9 cm. long, 3–5 cm. broad; seeds ellipsoid, flat, 10–11 mm. long.

Very rare (St. Andr.), in montane woodland; about 4500 feet; fr. July. Known only from the type, *H 5266* Endemic.

Improvement of the classification of *Maytenus* in Jamaica depends on further careful study of the flowers; these are fragile and shortlived and can be of at least two types. It is noteworthy that, although some of the Jamaican species are quite common, none is known to have been described under any name before 1895.

3. SCHAEFFERIA Jacq. (1760)

1 Leaves mostly acute or acuminate, a few usually obtuse or rarely emarginate, elliptical to lanceolate, 2·5–7 cm. long, 1–2·5 (–3·5) cm. broad, secondary veins conspicuous
    1. *frutescens*
1 Leaves all obtuse or emarginate, obovate, much smaller, less than 1·5 cm. broad, secondary veins inconspicuous:
  2 Leaf-blades 1–2·5 cm. long, mostly rounded at the tip   2. *obovata*
  2 Leaf-blades mostly less than 1 cm. long, emarginate   3. *marchii*

1. S. frutescens Jacq., Enum. Syst. Pl. Carib.: 33 (1760).

Shrub 1–3 m. or tree up to 6 m. high; leaves narrowed to both ends, light green; flowers green or greenish-yellow; petals spreading or reflexed, 3–4 mm. long; stamens 4; ripe drupe reddish-orange or bright red, 4–6 mm. long, spherical to ovoid.

Rather common at least in the southern parishes, in thickets and open woodlands on arid limestone; 50–2400 feet; fl. sporadically Mar–Oct, fr. May–Dec. *A 12323! H 9694! P 23612!* Florida, Bahamas, Mexico, Ecuador, West Indies south to Grenada; Cayman Is.

2. S. obovata Urb., Symb. Ant. **5**: 405 (1908).

Shrub 1·5–2 m. high; leaves thin; flowers greenish-yellow; ripe drupe orange, about 3·5 mm. long, ovoid to ellipsoid.

Rare and local (St. Andr.), in coastal thickets; SL–20 feet; fl. July, fr. July–Sept. *H 9384! 10820!* Endemic.

3. S. marchii Griseb. ex Urb., Symb. Ant. **5**: 86 (1904).

Shrub about 2·5 m. high; leaves rather leathery; ripe drupe 3·5–4 mm. long, broadly ellipsoid.

Rare and local (St. Andr., St. Cath.), in thickets on arid limestone; SL–100 feet; fr. Sept. *H 9383! Holness!* Cuba.

4. CROSSOPETALUM P. Browne (1756)

1. C. rhacoma Crantz, Inst. Rei Herb. **2**: 321 (1766).—*Rhacoma crossopetalum* L. (1759). Poison Cherry.

Shrub 0·6–5 m. high, rarely a small tree; branches slender, straggling or drooping; leaves opposite or ternate, very variable in shape, linear-lanceolate to obovate, rounded or emarginate, the margin usually crenulate, 1–4 (–5) cm. long; flowers green tinged red; petals 1–1·2 mm. long; stamens strongly recurved between petals; anthers 2-locular; ripe drupe red, obliquely obovoid to subglobose, 6–7 mm. long; stone usually solitary, pear-shaped.

Rather common in thickets and open woodlands near the sea on arid limestone or sandy lagoon margins; SL–800 (–1700) feet; fl. and fr. all the year. *A 6265! H 9572! P 21866!* Florida, Bermuda, Bahamas, Colombia, West Indies south to St. Lucia; Grand Cayman.

5. CASSINE L. (1753)

1. C. xylocarpa Vent., Choix Pl.: 23, t. 23 (1803).—*Elaeodendron dioicum* (Macf.) Griseb. (1864). *E. xylocarpum* (Vent.) DC. vars. *dioicum, obovatum, acuminatum* and *dolichocarpum* Urb. (1904).

Glabrous tree 5–12 m. high; trunk usually slender; bark grey, rather smooth with an odour of licorice when broken; leaves mostly elliptical, rounded to shortly acuminate at tip, tapered to an acute or obtuse base, margin serrate-crenate, 7–15 cm. long, 3·5–6·5 cm. broad, thinly leathery; flowers light green; drupe globose to oblong-ellipsoid, 1·7–3 cm. long, 1–1·8 cm. broad, yellow.

Occasional in woodlands; SL–3500 feet; fl. Apr–July, fr. June–Mar. *H 5125! 10108! P 16618! 22287!* St. Vincent, Grenadines.

This species is represented in Bahamas, Cuba, Hispaniola, Puerto Rico and Grand Cayman by the variety *C. xylocarpa* var. *attenuata* (A. Rich.) Kuntze. Size and shape of leaf and fruit are very variable throughout the range of the species. *C. glauca* (Vahl) Kuntze, an Asian plant at some time cultivated in Jamaica, has a lax inflorescence of hermaphrodite flowers and a 2-locular ovary.

6. GYMINDA Sarg. (1891)

1. G. latifolia (Sw.) Urb., Symb. Ant. **5**: 80 (1904).

Shrub 2–4 m. or tree 5–10 m. high; bark grey, flaking in narrow strips; leaves variable, generally obovate, rounded at the tip, narrowed to the base, 1·5–6 cm. long, obscurely pellucid-dotted, drying brownish beneath; flowers white; petals elliptical, 1·6–2·2 mm. long; ripe drupe bluish- or reddish-black, narrowly to roundish-ellipsoidal, 4–8 mm. long.

Rather common in woodland on limestone in arid areas; SL–1100 feet; fl. Apr–May, Aug, Dec, fr. sporadically. *A 11837! H 10174! P 7562! 9498!* Florida, Bahamas, Mexico, West Indies south to St. Vincent; Grand Cayman.

7. TETRASIPHON Urb. (1904)

1. T. jamaicensis Urb., Symb. Ant. **5**: 84 (1904).

Tree 5–10 m. high; trunk slender; bark rough, vertically striate; leaves obovate to elliptical, rounded at the tip, narrowed at the base into the petiole, 2·5–5·5 cm. long; flowers greenish-yellow; petals about 2 mm. long; ripe drupe dark purple, roundish-ellipsoid, about 1 cm. long.

Rare and local (St. Andr., St. Thom.), in thickets on arid limestone; 150–200 feet; fl. Jan–Feb, June, fr. Jan–Feb, June–July. *H 8604! 11865!* Endemic.

117. STAPHYLEACEAE

Trees and shrubs. Leaves usually opposite, pinnately compound; stipules and stipels usually present. Flowers usually bisexual, actinomorphic in terminal or axillary racemes or panicles. Perianth biseriate, 5-merous; sepals imbricate; petals free, imbricate, inserted on or outside a lobed hypogynous disk. Stamens 5, free, alternate with the petals; anthers 2-locular, opening lengthwise. Ovary superior, lobed or entire, 2–3-locular; ovules anatropous, few or many, usually ascending and inserted in 2 rows on each axile placenta; styles 2 or 3, usually free. Fruit a berry or a membranous inflated capsule dehiscing apically. Seeds with straight embryo and fleshy endosperm.

7 genera and about 50 species mostly in the north temperate region.

1. TURPINIA Vent. (1807) nom. cons.

Leaves opposite; calyx persistent; disk large; carpels completely united; fruit leathery, indehiscent.

1. T. occidentalis (Sw.) G. Don, Gen. Syst. **2**: 3 (1832).—*Staphylea occidentalis* Sw. (1788). Mutton Wood.

Tree 4–15 m. high; bark grey, striate; slash light buff-brown, pungent-aromatic; branches spreading and drooping; leaves with (3–) 5–7 (–9) leaflets; stipules inconspicuous; leaflets ovate-elliptical, acuminate with obtuse acumen, margin serrate-crenate to subentire, rounded at base, glabrous, 4–12 cm. long, 2–6·5 cm. broad; panicle-branches opposite; sepals unequal, 3–4 mm. long; petals unequal, white; fruit 1, 2 or 3-lobed, about 1 cm. long, about 1·5 cm. broad when 3-lobed, with short persistent styles.

Rather common in woodlands on limestone or shale; 250–5500 feet; fl. Feb–June, fr. May–Jan. *A 10545! 11722! H 9354! P 22274!* Guatemala, Puerto Rico (?), Lesser Antilles.

118. ICACINACEAE

Trees and shrubs. Leaves mostly alternate, simple, without stipules. Flowers bisexual or unisexual by abortion, actinomorphic, usually small in paniculate cymes. Perianth biseriate; calyx 4–5-lobed; petals 4–5, free or sometimes connate, valvate. Stamens as many as the petals and alternate with them, free; anthers 2-locular, sometimes 4-lobed, opening lengthwise. Disk cup-shaped, lobed or absent. Ovary superior, 1-locular or rarely 3-locular; ovules usually 2, pendulous, anatropous; style 1; stigmas (2–) 3 (–5). Fruit a 1-seeded drupe or rarely samaroid. Seeds usually with endosperm.

45 genera with 400 species; mostly tropical.

1. MAPPIA Jacq. (1797) nom. cons.

Flowers bisexual; disk lobed; ovary 1-locular; fruit a drupe; embryo large with foliaceous cotyledons.

1. M. racemosa Jacq., Pl. Hort. Schoenbr. **1**: 22, t. 47 (1797).

Shrub or tree 2–10 m. high; leaves oblong-lanceolate, gradually acuminate, cuneate at base, with porose domatia in the vein-axils beneath, 6–15 cm. long, 2–3·5 cm. broad; panicles axillary, much-branched, mostly shorter than the leaves; petals 4 mm. long, thickened at apex, light greenish-yellow; drupe 1·6–1·8 cm. long.

Uncommon (Clar., Manch., St. Ann), in woodlands on limestone hills; 2000–3100 feet; fl. sporadically throughout the year, fr. Dec. *H 5387! J.P. 1342! P 17475!* Greater Antilles.

119. RHAMNACEAE

Trees, erect or climbing shrubs or rarely herbs. Leaves usually alternate, sometimes opposite, simple, entire or serrate; small stipules usually present, sometimes prickly. Flowers bisexual, rarely unisexual and monoecious, small, actinomorphic, generally in axillary cymose inflorescences. Perianth biseriate or uniseriate, more or less perigynous or epigynous; calyx shortly tubular, 4–5 (–6)-lobed, lobes valvate, often keeled within; petals mostly 4–5 or absent, arising between and smaller than the calyx-lobes, often concave and clawed. Stamens 4–5 (–6), opposite the petals and often hooded by them, arising outside the margin of a usually conspicuous perigynous disk; anthers 2-locular, opening lengthwise. Ovary superior but often sunken in the disk or inferior, 2–4-locular; ovules 1 or 2 in each loculus on basal placentas, anatropous; style 2–3 (–4)-lobed. Fruit capsular or drupaceous, rarely samaroid; persistent lower part of calyx often evident. Seed with large straight embryo; endosperm fleshy, scanty or absent.

About 50 genera and 900 species, widespread but mostly in the tropics and warmer temperate regions.

1 Shrub climbing by tendrils modified from branch-tips at bases of inflorescences; fruit inferior and crowned by the calyx, 3-winged, the 3 cocci separating from the axis; leaves alternate; petals present   **1. Gouania**
1 Erect or scrambling shrubs or trees without tendrils; fruit superior to half-inferior:
  2 Leaves opposite or subopposite, pinnately veined; petals present; fruit drupaceous:
    3 Ovary 2-locular; endosperm wanting; cotyledons very convex; inflorescence sessile
        2. Rhamnidium

3 Ovary imperfectly 2-locular; endosperm present; cotyledons flat; inflorescence a
　　small panicle　　　　　　　　　　　　　　　　　　　　　　　　　3. **Auerodendron**
2 Leaves consistently alternate; or if opposite then petals absent:
　　4 Fruit capsular, the calyx forming a cupule around its base; flowers in umbelliform
　　　　simple or compound cymules; petals present　　　　　　　　　4. **Colubrina**
　　4 Fruit drupaceous, the calyx evident but free and not cupular around its base:
　　　　5 Leaves distinctly serrate, thin, glabrous or merely puberulous beneath; inflorescence
　　　　　　a pedunculate cyme, the cymules umbelliform; petals present　　5. **Rhamnus**
　　　　5 Leaves entire or if minutely denticulate densely woolly beneath:
　　　　　　6 Leaves 3-nerved; petals present or absent　　　　　　　　6. **Ziziphus**
　　　　　　6 Leaves pinnately veined:
　　　　　　　　7 Petals present; leaves always alternate　　　　　　6. **Ziziphus**
　　　　　　　　7 Petals absent; leaves often opposite　　　　　　7. **Krugiodendron**

As with many families which of themselves are very natural and easy to recognize,
generic distinctions in the *Rhamnaceae*, particularly in the drupaceous-fruited
groups, are tenuous and difficult to define. Further study of the little known genera
Auerodendron, Krugiodendron and *Rhamnidium* may indicate that broader generic
concepts are preferable to those reluctantly employed here.

1. GOUANIA Jacq. (1763)

1. **G. lupuloides** (L.) Urb., Symb. Ant. **4**: 378 (1910).—*Banisteria lupuloides* L.
(1753). Chew Stick.
　　A vine sometimes of large size climbing on trees to 12 m. or more; bark deeply
fissured, underbark wine-red, wood white with very large vessels; leaves ovate,
acuminate, rounded-subcordate at base, shallowly crenate, thinly pilose, up to
10 cm. long and 5 cm. broad; inflorescence a raceme of cymules; flowers light
greenish-yellow; mature capsule straw-coloured, 10–13 mm. broad.
　　Common, except in the western parishes, in thickets and woodlands mostly on
limestone; SL–3300 feet; fl. Oct–Mar, fr. Nov–Mar. *A 5673! 8156! H 6841!
Powell 1024!* Florida, Bahamas, C. Amer., West Indies south to Grenada.

2. RHAMNIDIUM Reissek (1861)

1. **R. dictyophyllum** Urb. in Fedde, Repert. Sp. Nov. **13**: 459 (1914).
　　Shrub or tree to 11 m. high; leaves elliptical, shortly acuminate, rounded-
subcordate at base, 6–10 cm. long, 3–5·5 cm. broad; flowers 5-merous, 3–3·5 mm.
in diameter, yellow-green; petals very small; disk orange; drupe ellipsoid-globose,
about 1·5 cm. long.
　　Very rare (Manch.), on rocky limestone in woodland; 2000–2300 feet; fl. Feb,
fr. Sept. *H & B 10606! P 11601! Stearn 352!* Endemic.

3. AUERODENDRON Urb. (1924)

1. **A. jamaicense** (Urb.) Urb., Symb. Ant. **9**: 228 (1924).—*Rhamnidium jamaicense* Urb. (1908). Turtle Fat.
　　Tree 10–20 (–25) m. high; leaves ovate-elliptical, obtuse or acute at apex, subcordate at base, minutely gland-dotted, with short spreading hairs on either side of
midrib beneath, 2·5–14 cm. long, 1·7–8·5 cm. broad; inflorescence few-flowered,
2–3·5 cm. long; drupe ellipsoid-globose, with persistent style about 1 mm. long,
2-locular, about 1·5 cm. long.
　　Occasional in coastal woodland on limestone and on old sand dunes; SL–500,
feet; fl. Mar, fr. Sept, Dec. *Asprey UCWI 1753! Cornman 80 AF! H 9708! C. B.
Lewis!* Endemic.

4. COLUBRINA L. C. Rich. ex Brongn. (1826) nom. cons.

1 Leaves serrate, 3-nerved at base, acuminate, shiny, thinly pilose on the veins; scrambling
　　shrub with trailing branches; branchlets glabrous or nearly so　　　1. *asiatica*
1 Leaves entire, pinnately veined, rusty-tomentose to glabrescent at least beneath; erect
　　shrubs or trees:
　　2 Leaf-tip obtuse to rounded; leaves and twigs glabrescent; inflorescence rusty-tomen-
　　　　tose　　　　　　　　　　　　　　　　　　　　　　　　　　　2. *obscura*

2 Leaf-tip acute to acuminate:
   3 Leaves beneath and twigs woolly-ferrugineous tomentose at least when young, the
     lamina sometimes glabrescent; calyx-cupule covering nearly half the capsule
                                                     3. *arborescens*
   3 Leaves beneath glabrescent; twigs light brownish-buff pubescent; calyx-cupule
     hardly covering one-third of capsule   4. *elliptica*

1. C. asiatica (L.) Brongn. in Ann. Sci. Nat. sér. 1, **10**: 369 (1827).—*Ceanothus
asiaticus* L. (1753). Hoop Withe.
Diffuse scrambling shrub with branches to 5 m. long; leaves broadly ovate, 4–9
cm. long, up to 5·5 cm. broad; common peduncle nearly glabrous, 1–2 mm. long;
flowers yellowish-green; capsule subspherical, 8 mm. in diameter; seeds with 2 flat
inner faces and the outer face rounded, matt grey, smooth.
Common in coastal thickets and on sandy and rocky shores and cays: SL–40
feet; fl. Apr–Sept, fr. most of the year. *A 8026! H 11952! Powell 970!* Native of the
E. Indies and Pacific Is., also in Florida, Cuba, Martinique, Grand Cayman.

2. C. obscura (Schrank) M. C. Johnston in Brittonia **23** (1): 13 (1971).—*Rhamnus
obscurus* Schrank (1824). *C. ferruginosa* of F. & R. (1926), partly.
Shrub 1·5–4 m. or tree to 8 m. high; leaves elliptical, cuneate to rounded and
slightly unequal at base, with about 6 pairs of inconspicuous lateral veins, glabrous
or minutely puberulous on midrib beneath, 4·5–9 (–12) cm. long and 3–5 (–6) cm.
broad; common peduncle very short; flowers 5–6-merous, yellow or greenish-
yellow; disk orange.
Rather rare (Manch., Trel., St. Ann), in thickets and woodlands on limestone;
1750–3000 feet; fl. Jan–Apr, fr. Mar–June. *A 12463! P 21888! 26878! Swartz!*
Endemic.

3. C. arborescens (Mill.) Sarg., Trees & Shrubs **2**: 167 (1911).—*Ceanothus
arborescens* Mill. (1768). *Colubrina ferruginosa* Brongn. (1827). Black Velvet,
Snake Wood.
Tree 2·5–13 m. high; leaves ovate-lanceolate to oblong-elliptical, rounded at
base, with about 6 pairs of rather conspicuous lateral veins, 3·5–12 cm. long, 1·5–6
cm. broad; common peduncle usually at least 5 mm. long, stout, rusty-tomentose;
flowers 4–5-merous, greenish; capsule subglobose, black, 7–8 mm. long; seeds
black, shiny, about 4 mm. long.
Frequent in some of the central parishes in rocky limestone woodlands; 500–
2100 feet; fl. May–Sept, fr. May–Jan. *A 12715! H 9686! 12026! P 26861!* Florida,
Bahamas, Mexico to Honduras, West Indies south to Barbados on the drier islands;
Grand Cayman; cultivated in W. Africa. This is the Mabie of Antigua.

4. C. elliptica (Sw.) Briz. & Stern in Trop. Woods **109**: 95 (1958).—*Rhamnus
ellipticus* Sw. (1788). *C. reclinata* (L'Hérit.) Brongn. (1827).
Tree 4–6 m. high with slender trunk and branches; leaves elliptical, mostly
broadly cuneate at base, with usually 7 or more pairs of conspicuous lateral veins,
2·5–7 (–9) cm. long, up to 4·5 cm. broad; common peduncle very short up to 8 mm.
long, mostly exceeded by the pedicels, buff-pubescent; flowers 5-merous, greenish;
capsule globose, reddish-brown, 6–7·5 mm. in diameter; seeds blackish, shiny,
about 5 mm. long.
Occasional (St. Andr., St. Cath., St. Eliz.), in thickets and woodlands on arid
limestone; 50–400 (–2000) feet; fl. July–Mar, fr. Sept–Mar. *A8962! 12874! H 9540!
P 7504!* Florida, Bahamas, Mexico to Guatemala, Venezuela, West Indies south to
St. Vincent; Grand Cayman. This is the Mabi of Puerto Rico and Mawbie of the
Lesser Antilles.

5. RHAMNUS L. (1753)

1. R. sphaerospermus Sw., Nov. Gen. & Sp. Pl.: 50 (1788).—Pumpkin Wood,
West Indian Buckthorn.
Tree 5–10 m. high; bark dark grey with paler streaks and patches; wood yellow;
youngest shoots rusty- or buff-pubescent; leaves oblong-elliptical, acutely acumi-

nate, broadly cuneate to subcordate and slightly unequal at base, up to 17 cm. long
and 6·5 cm. broad; flowers 5-merous, green; petals very small, olive green or yellow;
fruit globose, about 7 mm. long.

Frequent, in the central and eastern parishes, in woodlands on limestone or
shale; 1500–7000 feet; fl. Feb–June, Sept, fr. Apr–Aug. *A 10540! H 5219! P 17468!*
Greater Antilles.

6. ZIZIPHUS Mill. (1754); M. C. Johnston (1964)

1 Leaves pinnately veined, leathery, broadly ovate, entire, often emarginate; cymules
    pedunculate, often in panicles; petals present   1. *sarcomphalus*
1 Leaves 3-nerved:
  2 Leaf-blade entire, acute to long-acuminate, glabrous; cymules on short axillary
    branched peduncles; plants unarmed; petals absent   2. *chloroxylon*
  2 Leaf-blade minutely denticulate, obtuse, densely whitish-woolly beneath; cymules
    subcapitate in the leaf-axils; plants with stipular thorns; petals present
                                        3. *mauritiana*

1. Z. sarcomphalus (L.) M. C. Johnston in Amer. Journ. Bot. **50** (10): 1021
(1963).—*Rhamnus sarcomphalus* L. (1759). *Sarcomphalus laurinus* Griseb.
(1860). *S. laurinus* var. *fawcettii* Krug & Urb. (1897). Bastard Lignum Vitae.

Shrub 3 m. or tree 4–15 m. high; bark silvery-grey; wood very hard; juvenile
branches diffusely spreading with very small ovate-subcordate leaves and axillary
needlelike spines; leaves broadly cuneate to rounded-subtruncate at base, shiny,
up to 8 (–11) cm. long and 5 (–7) cm. broad; flowers (4–) 5-merous, greenish-
yellow to orange or tinged crimson; drupe subspherical, smooth, brownish, ripening
at various sizes from 1–2 (–3) cm. in diameter; seeds hemispherical.

Locally common in open rocky thickets along the south coast;
SL–750 feet; fl. and fr. all the year. *A 12338! H 9207! P 11522! 20568!* Endemic.

2. Z. chloroxylon (L.) Oliv. in Kew Bull. **1889**: 127, t. (1889).—*Laurus chloroxylon*
L. (1759). Cogwood, Jamaican Greenheart.

Tree to 20 m. high; leaves ovate-elliptical to lanceolate, up to 20 cm. long and
10 cm. broad; flowers 5-merous, greenish-yellow, the buds ovoid and conical-
pointed, in cymose panicles up to 2 cm. long; fruit subglobose, 16–20 mm. in
diameter.

Generally distributed but much less common than hitherto, in woodland on
limestone; 1000–3000 feet; fl. June–Nov, fr. Sept–Mar. *H 11200! H & B 10605!
H.P.S. 14653! P 10289!* Endemic.

3. Z. mauritiana Lam., Encycl. Méth Bot. **3**: 319 (1789).—*Rhamnus jujuba* L.
(1753). *Z. jujuba* (L.) Lam. (1789), not Mill. (1768). Coolie Plum, Crab Apple,
Jujube.

Tree with slender drooping prickly branches, 4–8 (–15) m. high; leaves broadly
ovate to rounded-elliptical, slightly unequal at base, up to 8 cm. long and 5 cm.
broad; flowers 5-merous, greenish-cream, acrid scented; fruit ellipsoid to subglobose, ripening brownish-orange, fleshy, 2–2·5 cm. long.

Established and fairly common in some waste places, occasionally forming
thickets; 25–1000 feet; fl. June–Nov, fr. Sept–Dec. *A 5426! A. M. Barry! H 12150!*
Native of warm parts of the Old World, now in Florida and throughout the drier
parts of the West Indies and elsewhere in the tropics.

7. KRUGIODENDRON Urb. (1902)

1. K. ferreum (Vahl) Urb., Symb. Ant. **3**: 314 (1902).—*Rhamnus ferreus* Vahl
(1794). Black Ironwood.

Tree 5–12 (–17) m. high; bark hard, rough, whitish or grey, cracking in regular
longitudinal lines; leaves ovate to roundish-elliptical, obtuse and often emarginate,
more or less rounded at base, thinly leathery, shiny, 1·5–6 (–8) cm. long, 1–3
(–4·5) cm. broad; inflorescences short, axillary, few-flowered; flowers (4–) 5 (–6)-
merous, greenish-yellow with a strong odour of almonds; drupe smooth, corky or
warted, black when ripe, broadly ellipsoid, about 7 mm. long.

Widely dispersed but nowhere common, at rocky lagoon margins and in woodlands on limestone; SL–3000 (–5000) feet; fl. Mar–Sept, fr. May–Jan. *A 11906! H 10944! P 9176! 26423!* Florida, Turks and Caicos Is., Mexico to Honduras, West Indies south to Bequia, Curaçao.

A species of *Reynosia* Griseb. has been discovered recently. The genus is distinguished by the drupaceous fruit and seeds with ruminate endosperm.

Reynosia jamaicensis M. C. Johnston in Journ. Arn. Arb. **52** (2): 366 (1971).
Very rare and local (Han.), on wooded limestone hilltop; 1450 feet; fl. May–June, fr. Aug. *P 26686! 31306.*

120. VITACEAE

Shrubs, mostly climbing by leaf-opposed tendrils (modified terminal shoots or inflorescences), rarely erect, or small trees, the nodes often swollen or jointed. Leaves alternate, the lower sometimes opposite, simple or compound; stipules present or absent. Flowers bisexual or unisexual and usually monoecious, actinomorphic, in leaf-opposed spikes, racemes, cymes or panicles. Perianth biseriate; calyx small, entire or 4–5-toothed or -lobed; petals as many as sepals, free or united, valvate, caducous, or obsolete. Stamens as many as petals and opposite to them, inserted at the base of the disk; anthers free or connate, 2-locular, opening lengthwise. Ovary superior, 2 (–6)-locular; ovules anatropous, 1–2 in each loculus on axile placentas; style simple, short; stigma discoid or capitate. Fruit a berry. Seeds with small straight embryo and copious endosperm.

12 genera with 700 species, mostly in the tropics but extending to temperate regions.

1 Erect shrubs without tendrils; leaves bi-ternate to decompound; flowers mostly 5-merous; petals connate; stipules enveloping the stem apex, deciduous from the node and petiole-base; ornamentals, occasionally naturalized; natives of the Old World tropics:
  2 Flowers white   *Leea sambucina* Willd.
  2 Flowers red; Jamaican Holly   *Leea coccinea* Planch.
1 Climbers, usually with tendrils; leaves simple or 3-foliolate; membranous stipules usually present on young shoots:
  3 Flowers 4-merous; leaves simple or 3-foliolate; inflorescence corymbose; petals free or becoming so   **1. Cissus**
  3 Flowers 5-merous; leaves simple:
    4 Inflorescence paniculate with cymose branches; petals cohering at the tips and caducous together at anthesis   **2. Vitis**
    4 Inflorescence spikelike; petals free   **3. Ampelocissus**

1. CISSUS L. (1753)

1 Leaves simple, softly fleshy:
  2 Stem 4-winged   1. *quadrangularis*
  2 Stem terete, often with pendulous roots   2. *sicyoides*
1 Leaves 3-foliolate; plants without pendulous roots:
  3 Leaflets fleshy, obovate-cuneate, toothed distally; cymes longer than the opposing leaf; flowers greenish- or creamy-yellow   3. *trifoliata*
  3 Leaflets thin, elliptical to ovate, serrate; cymes shorter than the opposing leaf; flowers red   4. *microcarpa*

1. C. quadrangularis L., Mant. Pl.: 39 (1767).
Diffusely branched climber forming tangles; branches 2–6 m. long; wings of stem red-margined; leaves only on the younger parts, ovate triangular, truncate or subcordate at base, coarsely toothed, gland-pitted, up to 11 cm. long and broad; flowers few in short cymes, light yellow speckled red; ripe fruits red.
Very local (St. Cath.), scrambling on fences and trees in thorn scrub on arid limestone and at salina margin; SL–50 feet; cultivated St. Andrew, 700 feet; fl. Sept–Nov, fr. Nov–Jan. *A 12735! H 10651! P 24333!* Native in the drier parts of tropical Africa and Asia.

2. C. sicyoides L., Syst. Nat. ed. 10, 2: 897 (1759).—Pudding Withe, Snake Withe.

Glabrous or thinly pubescent climber to 20 m. or more; leaves oblong to ovate, base truncate to cuneate, margin remotely toothed, (2–) 5–15 cm. long, up to 7·5 cm. broad; cymes (1–) 2–3 times branched; flowers light yellow, sometimes tinged purplish at base; ripe fruits bluish-black, 6–9 mm. long.

Very common, on trees, walls, fences and in thickets; SL–3800 feet; fl. Mar–Apr, July–Dec, fr. July–Apr. *A 5561! 8037! H 6767! C. B. Lewis!* Bahamas, Florida, continental tropical Amer., West Indies; Grand Cayman. This species is frequently infected with the smut fungus *Mycosyrinx cissi* which causes a " witches broom ".

3. C. trifoliata (L.), L. Syst. Nat. ed. 10, 2: 897 (1759).— *Sicyos trifoliata* L. (1753). Sorrel Vine.

Glabrous climber to 4 m.; leaflets 1·5–5 cm. long, up to 3 cm. broad; berries ovoid, bluish-black, 6–8 mm. long.

Common, mainly in arid coastal regions on shrubs and trees in thickets on lime-stone; SL–600 feet; fl. Feb, May–Sept, fr. Feb, May–Oct. *A 8894! 12868! H 9300! P 23756!* Bahamas, Florida, northern S. Amer., Greater Antilles, Virgin Is., Guadeloupe, Martinique, Aruba; Grand Cayman.

4. C. microcarpa Vahl, Eclog. Amer. **1**: 16 (1796).

Glabrous climber to 30 m. or more, becoming woody; stems sometimes winged; terminal leaflets rhomboid-elliptical, lateral ovate, unequal-sided, mucronate-serrate, 6–10 cm. long, up to 6 cm. broad; pedicels sparingly hairy, bright crimson; berry ovoid-globose, blackish-purple, about 8 mm. long.

Common, in thickets and on fences, walls and rocks; 500–2800 feet; fl. Apr–Dec, fr. July–Nov. *A 7295! H 8759! P 23635!* Mexico to Brazil, Cuba.

2. VITIS L. (1753)

1. V. tiliifolia Humb. & Bonpl. ex Willd. in Roem. & Schult. in L., Syst. Veg. ed. nov. **5**: 320 (1819).—Water Withe, Wild Grape.

High climbing vine, often reaching the tops of large trees, with stems up to 20 cm. thick; younger stems finely striate, with or without bristles, at first tomentose with white or rusty indumentum; tendrils branched; leaves cordate, usually repand-lobed, mucronate-toothed, 5–15 (–20) cm. broad; tomentose beneath; flowers polygamo-dioecious, greenish, fragrant; ripe berry globose, dull purplish-blue, 8–10 mm. in diameter.

Occasional, on banks, fences and over trees in woodland on limestone; 300–2800 feet; fl. Feb–May, fr. May, Oct–Nov. *A 10958! H 7086! 9367! P 19808! 20869!* Florida to Texas, Mexico to Ecuador, West Indies southwards to Guadeloupe.

Vitis vinifera L., the European or Wine Grape is cultivated.

3. AMPELOCISSUS Planch. (1884) nom. cons.

Young branches and undersurface of leaves with reddish woolly hairs; leaves 3–5-lobed
    1. *robinsonii*
Young branches and undersurface of leaves with whitish hairs; leaves not or sparingly small-lobed on each margin
    2. *alexandri*

1. A. robinsonii Planch. in A. & C. DC., Monogr. Phan. **5** (2): 415 (1887).—*Cissus rugosa* DC. (1824), not *A. rugosa* (Wall.) Planch. (1884).

Vine with slender branches; leaves cordate, mostly 3-lobed, toothed, 4–10 cm. long and broad; calyx subentire; petals 5, oblong; seeds triangular-ovate.

Very rare (Clar.), not collected in Jamaica since the type, *Robinson*. Cuba, Hispaniola.

2. A. alexandri Urb., Symb. Ant. **6**: 15 (1909).

Vine; leaves broadly rounded-ovate, deeply cordate, narrowly long-acuminate, irregularly crenate, 14–20 cm. long, 11–14 cm. broad; calyx 5-lobed; petals 1·5 mm. long; berry globose, black, 1·5 cm. in diameter; seeds depressed-obovoid, 5 mm. long, 4 mm. broad.

Very rare and local (St. Ann), not recently collected. *Prior. Britton & Hollick 2767.* Endemic.

121. TILIACEAE

Trees, shrubs or herbs. Leaves alternate, rarely opposite, simple, often with stellate indumentum; stipules paired or wanting. Flowers bisexual, rarely unisexual and monoecious, actinomorphic, solitary or in usually cymose axillary or terminal inflorescences. Perianth typically biseriate; sepals (3–) 5 (–7), free or united at base, usually valvate; petals usually as many as sepals, sometimes sepaloid, free, usually imbricate, or wanting. Stamens 10 or many, hypogynous, free or shortly united or in fascicles; anthers 2-locular, opening lengthwise or by apical pores. Ovary superior, 2–10-locular; ovules 1–many in each loculus, anatropous on axile placentas; style simple or obsolete; stigmas as many as ovary-loculi. Fruit 1- or more locular, the loculi sometimes again divided by means of longitudinal and transverse partitions, a capsule, drupe or berry, or dry and indehiscent. Seeds with usually straight embryo; endosperm scanty or copious.

About 50 genera woth 600 or more mostly tropical species.

1 Shrubs or herbs; petals when present yellow, imbricate; leaves dentate, serrate or crenate:
  2 Fruits subglobose, indehiscent or tardily splitting, covered with hooked bristles; leaves toothed and often angled or lobed; stamens inserted on the gynophore
                                                         1. Triumfetta
  2 Fruits ellipsoidal to linear-cylindrical, capsular, glabrous, puberulous or woolly, without bristles; leaves serrate but not angled or lobed; stamens inserted on the receptacle
                                                         2. Corchorus
1 Trees; stamens inserted on or around a disk, indefinite; (*Elaeocarpaceae*):
  3 Leaves lanceolate with hairs and glands; stipules longer than petioles; petals white, imbricate, thin; stigma sessile; fruit a berry; leaves serrate
                                            3. Muntingia
  3 Leaves elliptical to obovate, glabrous; stipules small, soon falling; petals sepaloid, valvate; style subulate; fruit a woody bristly loculicidal capsule; leaves entire or shallowly toothed
                                            4. Sloanea

1. TRIUMFETTA L. (1753)

1 Bristles of fruit glabrous; body of fruit tomentose; petals present; leaves with minute stellate hairs adaxially
                                             1. *rhomboidea*
1 Bristles of fruit retrorsely hispid:
  2 Body of fruit glabrous or thinly puberulous; petals present; stem and adaxial surface of leaf with small stellate hairs
                                             2. *semitriloba*
  2 Body of fruit densely tomentose with stellate hairs:
    3 Petals absent; leaves adaxially and stem with large and small stellate hairs; stipules ovate-triangular, attenuate-tipped; carpels 2
                                        3. *lappula*
    3 Petals present:
      4 Leaves with minute stellate hairs adaxially; stipules filiform; carpels 2
                                        4. *sloanei*
      4 Leaves with long simple hairs adaxially; stem with long simple hairs and an underfelt of short curled hairs; stipules lanceolate; carpels 3
                                        5. *bogotensis*

1. T. rhomboidea Jacq., Enum. Syst. Pl. Carib.: 22 (1760).—*T. bartramia* L. (1759), illegitimate name. *Bartramia indica* L. (1753), not *T. indica* Lam. (1791). Paroquet Bur.

Weedy undershrub to about 1 m. high; leaves broadly ovate, sometimes more or less 3-lobed, 2·5–9 cm. long; stipules linear, 3–5 mm. long; sepals narrowly oblong, 6–8 mm. long; petals shorter than sepals; stamens about 15; fruit 2–6-locular, about 3 mm. in diameter.

Apparently rare and not recently collected. General throughout the tropics and reported from most of the West Indian islands.

2. T. semitriloba Jacq., Enum. Syst. Pl. Carib.: 22 (1760).—Bur Weed.

Undershrub 60–150 cm. high; leaves broadly ovate and usually 3-lobed or -angled, the upper narrower, midlobe acuminate, base subcordate or broadly cuneate, margin toothed, (2–) 4–12 (–20) cm. long, up to about 9 cm. broad; stipules lanceolate with a subulate tip, 5–8 mm. long; sepals often crimson, narrowly oblong, 5–8 mm. long; petals shorter than sepals; stamens about 20; fruit 3-locular, 4–5 mm. in diameter.

Very common, a weed of roadsides, rough pastures and thickets; SL–5000 feet; fl. and fr. June–Aug, Nov–Mar. *A 8640! J.P. 939! 1144! P 24363!* Bermuda, Florida and throughout continental tropical Amer. and the West Indies; Grand Cayman; naturalized in Pacific Is.

3. T. lappula L., Sp. Pl. **1**: 444 (1753).—*T. heterophylla* Lam. (1789).

Robust herb or shrub 0·6–3 m. high; leaves rounded-ovate, often with 3–5 shallow lobes or angles, shortly acuminate, lower truncate at base, upper cuneate, 4–12 cm. long and up to about 11 cm. broad; stipules 5–6 mm. long; sepals linear-oblong, 3–4 mm. long; stamens about 10; fruit 2 (–4)-locular, 4–5 mm. in diameter.

Rather common, a weed of waste places and thickets; SL–1750 feet; fl. Dec–Apr, fr. Dec–Mar, Sept. *A 6056! H 8217! P 24368!* Mexico, C. Amer., West Indies; Grand Cayman, Cape Verde Is., Mauritius.

4. T. sloanei F. & R. in Journ. Bot. **59**: 225 (1921).

Undershrub; leaves ovate to lanceolate, often somewhat 3-lobed, base broadly or narrowly cuneate; sepals linear, 5–7·5 mm. long; petals shorter than sepals; stamens about 15; fruit 2-locular, 3–3·5 mm. in diameter.

An obscure species not recently collected; older records from St. Andrew and St. Catherine. Endemic.

5. T. bogotensis DC., Prodr. **1**: 506 (1824).—*T. hispida* A. Rich. (1845).

Shrubs 80–150 cm. high with straggling branches; leaves broadly ovate, sometimes 3-angled, abruptly acuminate, tomentose abaxially with mostly quadrifid stellate hairs, 2·5–9 cm. long, up to 6·5 cm. broad; stipules hispid and ciliate, 6–7 mm. long; sepals linear-oblong, densely hispid, subulate-tipped, 5–8 mm. long; petals shorter than sepals; stamens about 20; fruit 3–6-locular, 3–4 mm. in diameter.

Rather local (St. Andr., St. James, Trel., St. Thom.), a weed of shaded roadsides and woodland margins; 15–1650 feet; fl. Dec–Apr, fr. Jan–Mar. *A 8806, Campbell 6123! P 17420!* Mexico to Colombia and Ecuador, Greater Antilles, Trinidad.

2. CORCHORUS L. (1753)

1 Leaf-blade with the two basal teeth prolonged into hairs; stipules linear-lanceolate, 1 cm. or more long; leaves ovate to lanceolate:
  2 Capsule 5-locular, 3–9 cm. long, abruptly narrowed to apex, with 2 narrow ridges on the back of each valve, the valves distinctly obliquely transversely ridged within; plant almost glabrous except for a few short bristles on the petioles and base of the blade beneath; leaf-margin serrate   *1. olitorius*
  2 Capsule 3-locular, 1·5–3 cm. long, with 3 divergent often bifid beaks and 2 broad wings on the back of each valve, the valves obscurely transversely ridged within; plant thinly hispid with thick-based hairs, rarely glabrous; leaf-margin crenulate
        2. aestuans
1 Leaf-blade without hairlike prolongations at the base; stipules subulate, less than 5 mm. long:
  3 Capsule woolly, oblong-ellipsoidal, about 12 mm. long, 4-locular, with short erect beak; stem rusty-stellate pubescent   *3. hirsutus*
  3 Capsule glabrous or puberulous, linear, 2·5–8 cm. long; stems with lines of short hairs:
    4 Capsule 3-locular, with an erect beak and transverse partitions between the seeds
          4. orinocensis
    4 Capsule 2-locular, with 4 short erect horns, without transverse partitions between the seeds   *5. siliquosus*

1. C. olitorius L., Sp. Pl. **1**: 529 (1753).—Jew's Mallow.

Robust annual herb 0·3–3 m. high, usually woody towards the base; leaves oblong-ovate, acute, truncate at base, (2–) 7–10 (–14) cm. long, up to 6 cm. broad; flowers solitary or paired; sepals as long as or shorter than the petals, 7–8 mm. long; seeds 1·5–2 mm. long, rhomboidal.

Occasional as a weed of waste places, rough pastures and marshy areas; SL–600 feet; fl. Jan–Feb, July, Nov, fr. Feb, July. *A 8282! H 9961! A. von der Porten!* Native of E. Indies, cultivated and naturalized elsewhere.

2. C. aestuans L., Syst. Nat. ed. 10, **2**: 1079 (1759).

Taprooted annual herb with prostrate branches or, in damp areas, branches erect to 60–90 cm. high; leaves roundish, ovate or oblong-lanceolate, acute, rounded at base, 2–8 cm. long, up to 4 cm. broad; flowers 2 or 3 together; sepals often reddish-purple, 3–4 mm. long, as long as the petals; seeds up to about 1 mm. long, discoid.

Occasional as a weed of open sandy or muddy waste places; SL–250 feet; fl. July, Oct–Mar, fr. July, Nov–Mar. *A 9792! 10981! P 18364!* Tropics generally.

3. C. hirsutus L., Sp. Pl. **1**: 530 (1753).

Much-branched shrub 0·6–2 m. high; leaves lanceolate to elliptical, crenate, softly stellate-pubescent, paler beneath, 2–7 cm. long, up to 2·5 cm. broad; flowers 2–8 together on a common axillary peduncle; sepals and petals 6 mm. long; seeds 1·5–2 mm. long, ellipsoidal, black.

Locally common (St. Andr., St. Cath., St. Thom.), on well drained rocky slopes and in thorn scrub on arid limestone; SL–1000 feet; fl. and fr. Sept–June. *A 7255! H 6543! J.P. 1351! P 17449!* Bahamas, Mexico to tropical S. Amer., Greater Antilles, St. Martin, N.E. tropical Africa.

4. C. orinocensis Kunth, Nov. Gen. **5**: 337 (1823).

Bushy herb 60–90 cm. high; stem sometimes reddish; leaves ovate to lanceolate, long-acute, cordulate or rounded at base, glabrous except midrib usually with a few hairs, 3–10 cm. long, up to 2·5 cm. broad; flowers generally 2 together; sepals 7–8 mm. long; petals 4–5 mm. long; capsule 4–6 cm. long; seeds 1·2 mm. long, irregularly cubical, black.

Uncommon, a weed of roadsides and waste places; SL–700 feet; fl. and fr. Nov–Jan. *A 5436! H 6856! Proctor & Powell 20537!* Texas, Mexico to Venezuela and Bolivia, Cuba, Puerto Rico, Antigua southwards to St. Vincent.

5. C. siliquosus L., Sp. Pl. **1**: 529 (1753).—Slippery Bur.

Erect bushy herb, woody at base, taprooted, 50–120 cm. high; leaves ovate to lanceolate, acute or acuminate, glabrous or with a few hairs on veins beneath, 0·5–7 cm. long, 0·4–4 cm. broad; flowers solitary or 2 together; sepals yellow tinged crimson, 6–7 mm. long; petals 5–6 mm. long; capsule 3·5–8 cm. long; seeds about 1 mm. long, trigonous, bluish-black.

Very common, a weed of roadsides, stony waste places, rough pastures and thickets, mostly on limestone; SL–2300 feet; fl. and fr. all the year. *A 5650! H 5955! P 11195!* Bahamas, Florida, Texas, continental tropical Amer., West Indies; Grand Cayman.

3. MUNTINGIA L. (1753)

1. M. calabura L., Sp. Pl. **1**: 509 (1753).—Jamaican Cherry.

Spreading-branched shrub or tree 3–10 (–12) m. high; twigs pubescent and glandular; leaves oblong to oblong-lanceolate, acuminate, unequally subcordate at base, 4–14 cm. long, up to 4 cm. broad, tomentose with stellate hairs beneath; peduncles 1-flowered, 1–3 together; sepals lanceolate, 8–12 mm. long; petals obovate; berry at first yellow, finally red, about 12 mm. in diameter.

Locally common on well drained limestone banks and in thickets; SL–2000 feet; fl. and fr. all the year. *A 6977! J.P. 684! P 8658! Yuncker 17774!* Tropical continental Amer., Cuba, Hispaniola, St. Vincent, Trinidad, introduced elsewhere and into the Old World.

4. SLOANEA L. (1753)

1. S. jamaicensis Hook., Ic. Pl.: **7**: tt. 693–696 (1844).—Break-axe Tree, Lignum Durum.

Tree 6–20 (–30) m. high; heartwood very dark and hard; leaves shortly acuminate or obtuse, entire or wavy-toothed near tip, shiny on both surfaces, 10–20 cm. long, up to 12 cm. broad; flowers solitary in the leaf-axils, yellowish-brown or tinged dull rose; peduncle 1·5–3 cm. long; sepals 4, broadly ovate, 1·2–1·7 cm. long; petals smaller, dull yellow; stamens shorter than petals with linear bright yellow anthers;

capsule 10–11 cm. long, 7–10 cm. in diameter, 4–5-valved; seeds 2 or more in each loculus, covered with a yellow fleshy aril.

Frequent in forest and woodland on limestone; 1200–2500 feet; fl. Mar, Sept, fr. Jan–Feb, July. *A 7534! 9140! H 8877! 10760! P 26294! Howard, Proctor & Wagenknecht 20510!* Endemic. When better known the populations of eastern Portland may prove to belong to a species distinct from *S. jamaicensis*.

122. MALVACEAE

Herbs, shrubs or trees, often with stellate hairs on the young stems, leaves and inflorescences. Leaves alternate, simple, entire or lobed, usually palmately veined; stipules present, free. Flowers bisexual or rarely unisexual and dioecious, actinomorphic, solitary in the leaf-axils or racemose or paniculate; bracteoles when present free or united into an involucel around, or an epicalyx on, the calyx. Perianth biseriate; sepals (3–) 5, usually united, the lobes valvate; petals 5, free distally and contorted or rarely imbricate in bud, usually adnate at base to the staminal tube. Stamens numerous, hypogynous, monadelphous, the filaments ultimately free; anthers 1-locular, opening lengthwise; pollen grains usually spiny and large. Ovary superior, (2–) 5–many-locular; ovules anatropous, 1–many in each loculus on axile placentas; style usually branched at the tip; stigmas as many as or twice as many as carpels, often capitate, peltate or decurrent. Fruit a loculicidal capsule or schizocarp, rarely berrylike or samaroid. Seeds often pubescent or comose; embryo straight or curved; endosperm often oily. Some general features of variation in this family are noted below.[1]

About 80 genera with some 1500 species in all temperate and tropical regions.

1 Staminal column bearing anthers at the apex:
  2 Involucel of 3 bracteoles; carpels separating, 2-valved:
    3 Corolla red; carpels 14–20, transversely septate between 2 seeds; leaves palmately divided   **1. Modiola**
    3 Corolla yellow; carpels 5 or more, 1-seeded; leaves serrate   **2. Malvastrum**
  2 Involucel wanting:
    4 Corolla blue or rarely white; inner walls of fruiting carpels disappearing, the outer carpel walls separating with the single seed   **3. Anoda**
    4 Corolla yellow, white, pink or red:
      5 Carpels constricted with a transverse ridge inside, if not constricted and internally ridged then seeds in each carpel 3; carpels (4–) 5; carpel-beaks short, smooth, spreading; seeds pyriform   **4. Wissadula**
      5 Carpels not constricted and internally ridged; carpels 5–20 (–30):
        6 Ovules (2–) 3 or more in each carpel:
          7 Carpels leathery or parchmentlike, with more or less prominent often spreading or rarely erect beaks   **5. Abutilon**
          7 Carpels membranous, bladderlike, rounded at the top and not beaked
                           6. Herissantia
        6 Ovule 1 in each carpel; carpel-beaks if present curved inwards or erect:
          8 Carpels separating, more or less tardily dehiscent by 2 valves or splitting irregularly   **7. Sida**
          8 Carpels united into a loculicidal capsule   **8. Bastardia**
1 Staminal column bearing anthers on the outside, the apex truncate or toothed:
  9 Style-arms twice as many as carpels, 8–10; carpels usually 5, separating at maturity:
    10 Flowers in heads surrounded by an involucre of leafy bracts   **9. Malachra**

[1] Plants of this family are frequently infected with viruses which cause a patchy yellowing of the leaves and often stunted growth. Herbicides can cause drastic changes in leaf-shape mostly expressed in a narrowing of the blade-base combined with fasciation of the veins, fimbriation of the distal part of the blade and reflexion of the margin. Natural variation is found in the density of the indumentum, particularly affecting the stellate hairs on the undersurface of the leaf. The flowers are mostly diurnal and in many species the corolla changes in colour during the course of the day; such changes normally take place through increasing intensity of pink or red anthocyanin pigments so that white tends to change to pink, light yellow to orange, orange to crimson. Although the flowers are generally strictly actinomorphic, the staminal column may be pendulous or deflexed downwards; in the latter event the free filaments may be declinate.

10 Flowers not in heads, with involucels but without an involucre of bracts:
11 Leaves with 1–3 narrow split glands on median veins abaxially; carpels with
    numerous glochidiate spines on outer surface, 5, indehiscent; involucel 5-cleft
                                              10. **Urena**
11 Leaves without such glands; carpels without glochidia:
12 Carpels dry and hard with barbed spines at apex or crested or winged; bracteoles
    of involucel 4 or more, free or more or less united with each other and the base
    of the calyx; corolla-limb spreading, not bright red; petals not auriculate
                                            11. **Pavonia**
12 Carpels at first fleshy, smooth; bracteoles of involucel 5 or more, free from base of
    calyx; corolla-limb campanulate, hardly opening and exceeded by the exserted
    staminal column, bright red or pink; petals auriculate on claw
                                      12. **Malvaviscus**
9 Style-arms or lobes and carpels 5; fruit a loculicidal capsule or the carpels also rarely
  separating at maturity, or indehiscent:
13 Style thick at apex, shortly 5-lobed or undivided; seeds obovoid or angular:
14 Bracteoles 3–5 (–8), small, entire, deciduous; young shoots with peltate scales;
    fruit indehiscent (in native plants), leathery   13. **Thespesia**
14 Bracteoles 3, conspicuous, cordate and usually toothed distally, persistent; stems
    and leaves often with black glandular dots; capsule dehiscent   14. **Gossypium**
13 Style-arms or stigmas distinct, spreading; seeds reniform:
15 Ovule 1 in each ovary-loculus; bracteoles small or obsolete; corolla white drying
    yellow; fruiting carpels separating, the valves with marginal hooks, strongly
    angled   15. **Kosteletzkya**
15 Ovules 2 or more in each ovary-loculus; bracteoles often attached to calyx-tube;
    corolla rarely white; capsule not usually strongly angled, but winged in *H.*
    vitifolius:
16 Calyx regularly 5-toothed or -lobed, persistent   16. **Hibiscus**
16 Calyx spathaceous, irregularly toothed or lobed, splitting laterally at anthesis,
    caducous in fruit   17. **Abelmoschus**

1. MODIOLA Moench (1794)

1. **M. caroliniana** (L.) G. Don, Gen. Syst. 1: 466 (1831).—*Malva caroliniana* L.
(1753).

Herb with prostrate, rooting, hairy branches; leaf-blade deeply or shallowly
divided, 1–5 cm. in diameter; stipules leafy, persistent, ciliate, 4–5 mm. long;
flowers solitary or 2 together, axillary, the peduncles 2–4 cm. long; bracteoles
about 1 mm. below calyx; calyx 6–7 mm. long; petals 7–8 mm. long, deep crimson;
fruiting carpels cuspidate, black, about 4 mm. long; seeds glabrous.

Locally common (St. Andr., St. Thom.), by pathsides, on banks and in waste
places; 2700–5000 feet; fl. and fr. Feb–July. *A 6343! H 11954! P 24807!* Bermuda,
S.E. United States, C. and S. Amer., Hispaniola, S. Africa, E. Java, W. Europe.

2. MALVASTRUM A. Gray (1849) nom. cons.

1 Flowers in dense, sometimes more or less interrupted, terminal spikes or heads; indumentum stellate-tomentose; carpels beaked at apex, dorsal spines wanting
                                         1. *americanum*
1 Flowers solitary or in few-flowered clusters, mostly axillary; indumentum strigose with
  appressed simple or 2–4-armed hairs:
2 Carpels with 1 apical and 2 dorsal spines   2. *coromandelianum*
2 Carpels without spines   3. *corchorifolium*

1. **M. americanum** (L.) Torr. in Emory, Rep. U.S. & Mex. Bound. Surv. 2 (1):
38 (1859).—*Malva americana* L. (1753). *Malvastrum spicatum* (L.) A. Gray
(1849).

Shrubby herb up to 2·4 m. high; leaves broadly ovate to lanceolate, broadly
cuneate to truncate at base, dentate, 2–9 (–10) cm. long, 1–5 (–6) cm. broad, rough
with stellate hairs; calyx about 5 mm. long; petals 6–8 mm. long, yellowish-orange,
open in the afternoon; carpels about 2 mm. long.

Locally common along roadsides, ditches and in waste places; SL–2000 feet;
fl. and fr. Aug–Mar. *A 5440! H 8168! H & P 13583!* General in the subtropics and
tropics.

2. M. coromandelianum (L.) Garcke in Bonplandia **5**: 295 (1857).—*Malva coromandeliana* L. (1753).

Tough-stemmed herb up to about 1 m. high; leaves rounded-ovate to lanceolate, broadly cuneate to truncate at base, serrate-dentate, 2–6 (–8) cm. long, 1–4 cm. broad; stipules lanceolate, 5–8 mm. long; calyx about 5 mm. long, enlarging in fruit; petals about 8 mm. long; carpels up to 3 mm. broad.

Common weed of cultivated ground, pastures and waste places; 150–2500 feet; fl. and fr. most of the year. *A 5435! H 6851! P 26460!* Native of the New World now general in the tropics and subtropics.

3. M. corchorifolium (Desr.) Britton in Small, Fl. Miami: 119 (1913).—*Malva corchorifolia* Desr. (1792).

Like the last and possibly not really distinct; stipule and inflorescence characters sometimes used to distinguish these two species do not hold when examined in a wider range of specimens and the spines on the carpels vary considerably in length.

Rather common on sandy or gravelly waste ground in shale or limestone areas, especially near the sea; SL–600 (–2000) feet; fl. and fr. July–Mar. *A 8162! Britton 1022! A. von der Porten! P 9470!* Florida, continental tropical Amer., Bahamas, Greater Antilles, Virgin Is., Grenadines, Barbados; Grand Cayman. The preference of this species for coastal situations and the discrepant flowering time in comparison with *M. coromandelianum* require further investigation.

3. ANODA Cav. (1785)

1. A. acerifolia DC., Prodr. **1**: 459 (1824).

Woody herb with slender prostrate or erect branches 15–60 cm. long; leaves ovate, deltate or pentagonal, truncate or shallowly cordate at base, the upper hastate, dentate, 2–7 cm. long, long-petioled; peduncles slender, mostly as long as or longer than the leaves; calyx enlarging in fruit to 15–25 mm. broad; petals bright blue, 8–15 mm. long; carpels 9–16, shortly beaked, about 4 mm. long.

Local as a weed of cultivations and waste places; 1400–4500 feet; fl. and fr. Jan–July, Oct. *A 12903! H 8414! P 22858!* Mexico to Peru, Greater Antilles, Barbados, Trinidad and Tobago, naturalized in E. Indies. This species requires further careful comparison with *A. hastata* Cav. (1785) from which it may not be distinct.

4. WISSADULA Medic. (1787); R. E. Fries (1908)

1 Ripe carpels with 1 seed, 4 mm. long including the rather distinct beak; leaf-blade ovate-triangular, truncate or openly cordate at base, gradually narrowed to a long tip; panicle large, open, with slender branches   1. *fadyenii*
1 Ripe carpels with (2–) 3 seeds, 5–8 mm. long; ovary-loculi with 3 ovules:
2 Flowers in a compact panicle with short erect branches; carpels hardly constricted, not 2-chambered in fruit; seeds of one kind; leaf-blade broadly rounded, cordate at base, abruptly acuminate   2. *contracta*
2 Flowers in open panicles with long slender spreading branches; carpels constricted and incompletely divided into proximal and distal chambers; proximal seed with tufts of long hairs and longer hairs at hilum, distal seed(s) puberulous in patches:
3 Leaf-blade usually with a very narrow sinus or even the two margins overlapping at base, acuminate   3. *amplissima*
3 Leaf-blade usually with a broad open sinus at base, long-acute   4. *periplocifolia*

1. W. fadyenii Planch. ex R. E. Fries in Kungl. Svensk. Vet. Akad. Handl. **43** (4): 30 (1908).

Tough-stemmed herb up to 120 cm. high, with yellowish-brown stellate hairs on younger parts; leaves 5–8 (–10) cm. long, up to 4 cm. broad, with small scattered whitish stellate hairs adaxially and larger more numerous stellate hairs abaxially; calyx 3–3·5 mm. long with both kinds of hairs; petals creamy or tawny-yellow, 4–5 mm. long; seed about 2 mm. long.

Locally common in the drier parts of the southern central parishes, as a weed of pastures and gravelly waste places; SL–800 feet; fl. and fr. July–Jan. *A 10271! H 6830! P 23056!* Colombia, Trinidad, Grand Cayman.

2. W. contracta (Link) R. E. Fries in Kungl. Svensk. Vet. Akad. Handl. 43 (4): 60 (1908).—*Sida contracta* Link (1822).

Erect undershrub up to 2 m. or more high, greyish-tomentellous; leaves over 4–12 (–18) cm. long, 2–10 (–12) cm. broad, glabrescent adaxially, whitish-tomentellous abaxially; panicles 20–30 cm. long; calyx 3–3·5 mm. long, tomentellous outside; petals light yellow or white, 3·5–4 mm. long; seeds 2–2·5 mm. long, glabrescent.

Very rare (St. Andr.); fl. and fr. Oct; only once reported, *Purdie!* Native Guatemala to Brazil and possibly the Lesser Antilles and Trinidad, introduced elsewhere and casual in the Indian Ocean islands and S.E. Asia.

3. W. amplissima (L.) R. E. Fries in Kungl. Svensk. Vet. Akad. Handl. 43 (4): 48 (1908).—*Sida amplissima* L. (1753). *W. hernandioides* (L'Hérit.) Garcke (1890).

Erect robust herb up to 2 m. or more high, the stem tomentose with small pale and large stalked brownish stellate hairs; leaves broadly ovate, acuminate, 4–12 (–15) cm. long, up to 8 (–10) cm. broad, sometimes crenulate, more less tomentose on both surfaces with usually fewer smaller hairs on darker upper surface; calyx 3–4 mm. long; petals light orange-yellow, 4–6 mm. long; seeds about 2 mm. in diameter.

Common, in sandy and gravelly waste places; 50–2500 feet; fl. and fr. Sept–June. *A 8351! H 11949! P 24376!* Mexico to Paraguay, Greater Antilles, Virgin Is., Guadeloupe; the variety *rostrata* (Schumach.) R. E. Fries occurs in tropical Africa and some of the Atlantic Is.

4. W. periplocifolia (L.) C. Presl ex Thwaites, Enum. Pl. Zeyl.: 27 (1858).—*Sida periplocifolia* L. (1753).

Undershrub up to 1·5 m. high, the stem with large and small stellate hairs; leaves ovate to lanceolate and more or less triangular, 3–7 (–15) cm. long, 1–3 (–7) cm. broad, sparsely stellate-pubescent adaxially, whitish-tomentose and with large ferrugineous stellate hairs abaxially; calyx 2–3 mm. long; petals light yellow with darker veins, rarely orange or white, 4–5 mm. long; seeds 2–2·5 mm. in diameter.

Uncommon (St. Andr., St. Cath., Manch.), a weed of roadsides and waste places; 50–700 feet; fl. and fr. Mar, June–Aug. *A 10726! Burrowes 13058! Thornton UCWI 1309!* S. Amer., Greater Antilles, Virgin Is., Grenadines, Trinidad, early introduced into S.E. Asia.

5. ABUTILON Mill. (1754)

1 Flowers pendulous; corolla light yellow turning pink, appearing tubular by rolling of petals; calyx often reddish; filaments deep crimson; cultivated ornamental shrub occasionally escaping from hill gardens (4200–5000 feet)
  A. megapotamicum (Spreng.) St.-Hil. & Naud.
1 Flowers not as above:
  2 Petals rose-pink, up to about 25 mm. long, longer than calyx; carpels 10–15, cuspidate, with 6–13 ovules in each 1. *hulseanum*
  2 Petals mostly yellow or orange, sometimes reddish at base, rarely white; carpels with usually 3 ovules (except in *A. striatum* 7–9):
    3 Carpels usually 5; petals yellow, with or without a dark red spot at base, about 5 mm. long; capsule oblong, truncate, about 8 mm. long and 7 mm. broad
  2. *trisulcatum*
    3 Carpels more than 5; petals 8 mm. or more long; capsule usually rounded on sides or apex:
      4 Petals 25–40 mm. long, deep yellow with red veins; leaves deeply 3–5-lobed, serrate; cultivated ornamental shrub, rarely escaping; native of Brazil
  A. striatum Dicks. ex Lindl.
      4 Petals up to about 20 mm. long; leaves not or only shallowly or rarely lobed; wild or naturalized plants:
        5 Petals becoming reflexed, yellow or orange-yellow, 9–15 mm. long; flowers in a usually ample terminal panicle 3. *elatum*
        5 Petals as far as known not becoming reflexed:
          6 Stipules up to 15 mm. long and 10 mm. broad near base, unequally auriculate; flowers solitary or in loose terminal panicles; petals 10–13 mm. long, deep yellow 4. *auritum*
          6 Stipules smaller, much narrower, not auriculate:

7 Carpels and styles 12–30; introduced Old World species:
  8 Stems, petioles and pedicels with spreading long simple hairs, minute stellate hairs and short glandular hairs; petals 15–20 mm. long, orange-yellow with a basal purple spot   5. *hirtum*
  8 Stems, petioles and pedicels densely covered with stellate hairs, without glandular hairs and with few or very local long simple hairs; petals variable, 15 mm. or more long, yellow, without purple bases   6. *indicum*
7 Carpels and styles (5–) 6–10; petals 8–15 mm. long, yellow or orange-yellow:
  9 Calyx-lobes subcordate and overlapping at base, at least in bud; petals 10–15 mm. long; carpels 7–10, 13–14 mm. long in fruit, including erect acuminate beak; seeds reticulated with lines of minute hairs   7. *abutiloides*
  9 Calyx-lobes not subcordate; seeds not reticulated with minute hairs, minutely tuberculate otherwise glabrous:
    10 Stems and leaves rough-pubescent; leaves concolorous or nearly so; petals usually about 8 mm. long; carpels 5–7, long-aristate with spreading awns in fruit   8. *umbellatum*
    10 Stems and leaves softly pubescent, at least when young; leaves discolorous; petals 12–15 mm. long; carpels 9–10, shortly beaked in fruit   9. *permolle*

1. A. hulseanum (Torr. & A. Gray) Torr. ex Chapm., Fl. Southern U.S.: 56 (1860).—*Sida hulseana* Torr. & A. Gray (1838). *A. pauciflorum* of F. & R. (1926), not St.-Hil. (1825).

Erect robust herb or undershrub up to 2·5 m. or more high; leaves ovate, cordate at base, acuminate, 5–15 cm. long, up to 14 cm. broad, softly pubescent with stellate hairs on both surfaces, paler beneath; flowers solitary, axillary; peduncles stout, jointed 1–2 cm. below calyx in fruit; calyx 12–15 mm. long in flower, longer in fruit; ripe carpels 13–16 mm. long, villous outside, separating and exposing glabrous interfaces; seeds black, deeply reniform, pubescent, about 2 mm. long.

Locally common, a weed of waste ground; 80–1200 feet; fl. and fr. Feb–Nov. *A 6219! H 9864! J.P. 1340 P 7695!* S. United States, Mexico, Honduras, Cuba, Martinique.

2. A. trisulcatum (Jacq.) Urb. in Redde, Repert. Sp. Nov. **16**: 32 (1919).—*Sida trisulcata* Jacq. (1760).

Much branched shrubby herb up to 1·5 m. high; indumentum of very short stellate hairs, cinereous; leaves broadly ovate, openly cordate at base, acuminate, serrate-crenate, up to 12 cm. long and 8 cm. broad; inflorescence open with spreading branches; base of staminal column angled, puberulous; each carpel with a short spreading beak; seeds with an irregular pattern of short stellate hairs.

Local (St. Cath., Clar.), a weed of roadsides and thickets; 30–50 feet; fl. and fr. Oct–Jan. *A 6022! 9787!* Mexico, Bahamas, Cuba, Hispaniola.

3. A. elatum (Macf.) Griseb., Fl. Br. W.I.: 79 (1859).—*Sida elata* Macf. (1837). *A. giganteum* of F. & R. (1926), partly, not *Sida gigantea* Jacq. (1797).

Shrub 2–4 m. high; leaves broadly rounded-ovate, cordate, acuminate, shallowly denticulate, 6–20 cm. long, up to about 15 cm. broad, finely tomentellous with cinereous stellate hairs; stipules lanceolate; inflorescence eventually broadly paniculate; calyx about 10 mm. long; petals bright orange-yellow, longer than the calyx, open in the afternoon; carpels 8–14, about 1 cm. long, shortly beaked; seeds with a peglike glabrous appendage, obliquely reniform, white-pubescent.

Rather local, in thickets and at woodland margins in arid south coast districts; 40–900 feet; fl. and fr. Dec–Mar. *A 10751! Yuncker 18148!* Cuba.

4. A. auritum (Wall. ex Link) Sweet, Hort. Brit.: 53 (1826).—*Sida aurita* Wall. ex Link (1822). *Abutilon elatum* Griseb. (1859), partly.

Shrubby herb 1–2 m. high, generally velvety-tomentose with small stellate hairs and a few longer hairs especially on veins of leaves beneath; leaves broadly rounded-ovate, cordate with rounded sinus, caudate-acuminate at tip, crenate-denticulate, up to 14 (–20) cm. long and nearly as broad; calyx 5–7 mm. long; petals spreading, yellow or light orange with purple veins; carpels 8–12, apiculate, 10–12 mm. long and slightly exceeding calyx in fruit; seeds punctate by minute stellate hairs.

Very rare and known only from three unlocalized early collections. *Distin! J.P. 1054! Wilson 202!* Tropics of S.E. Asia and Australasia, introduced and rare elsewhere.

5. A. hirtum (Lam.) Sweet, Hort. Brit.: 53 (1826).—*Sida hirta* Lam. (1783). *A. indicum* of F. & R. (1926).

Annual undershrub up to 3 m. high, glandular and aromatic; leaves broadly ovate, cordate, shortly acuminate, coarsely or shallowly crenate-dentate, up to 9 cm. long and 8 cm. broad; flowers mostly solitary in the axils; calyx 7–9 mm. long; carpels 20–25, 10–12 mm. long in fruit, shortly acuminate; seeds as above.

Occasional (St. Andr., St. Ann, St. Thom.), in arid gravelly and rocky waste places; 20–150 feet; fl. and fr. Oct–Jan. *A 8287! Bengry & Patrick! Campbell 6160! H 8869!* Native in the drier parts of the Old World tropics, introduced and generally scattered throughout the American subtropics and tropics.

6. A. indicum (L.) Sweet, Hort. Brit.: 54 (1826).—*Sida indica* L. (1756). *A. leiospermum* Griseb. (1859).

Undershrub up to 2 m. high; leaves broadly ovate, cordate, hardly acuminate, coarsely dentate and shallowly repand, up to 7 cm. long and nearly as broad, generally stellate-tomentose but with long simple hairs at apex of petiole; flowers solitary, axillary; pedicels up to 6 cm. long and distinctly jointed about 1 cm. below calyx; calyx 4–5 mm. long; carpels 15–22, about 12 mm. long in fruit, shortly acuminate; seeds stellate-hairy or glabrous.

Very rare and known only from two unlocalized early collections. *Distin! March 129!* A variable subtropical and tropical weedy species with its main abundance and evolutionary radiation in S.E. Asia; least well represented in tropical Amer.

7. A. abutiloides (Jacq.) Garcke in Engl., Bot. Jahrb. **15**: 485 (1893).—*Sida abutiloides* Jacq. (1764). *A. americanum* (L.) Sweet (1826), not Panzer (1797). *Lavatera americana* L. (1759).

Shrubby herb up to 1·5 m. high, generally tomentose with small and large stellate hairs; leaves broadly ovate, cordate, shortly acuminate, crenate or irregularly toothed, 2–13 cm. long and nearly as broad; flowers solitary and axillary, becoming crowded at ends of branches; calyx folded outwards, 12–14 mm. long, the lobes acuminate.

Rare and local (St. Andr., St. Cath.), a weed of arid waste places near the sea; SL–100 feet; fl. and fr. Sept–Feb. *A. M. Barry! H & B 10793! Prior 711!* S. United States, Mexico, Colombia, Bahamas, Greater Antilles, Virgin Is.

8. A. umbellatum (L.) Sweet, Hort. Brit.: 53 (1826).—*Sida umbellata* L. (1759).

Bushy herb up to 2m. high, generally tomentose with stellate hairs; leaves broadly ovate, truncate to openly cordate at base, shortly and abruptly acuminate, crenate or dentate, up to 12 cm. long and 10 cm. broad; flowers corymbose and subumbellate in axillary and terminal clusters, the whole inflorescence a leafy panicle; calyx 5–6 mm. long.

Occasional (St. Andr., St. Cath., St. Thom.), along roadsides and on waste ground on limestone and also in salina thickets; 10–800 feet; fl. and fr. May–Jan. *A 10266! 11913! Campbell 5976! P 8165!* Mexico, Colombia, Greater Antilles, Virgin Is., Antigua, Grenadines, Trinidad.

9. A. permolle (Willd.) Sweet, Hort. Brit.: 53 (1826).—*Sida permollis* Willd. (1809).

Shrub with slender branches, 1–1·5 m. high; leaves stellate-tomentose and soft to touch; stem glabrescent; leaf-blade rounded-ovate, deeply cordate at base, acuminate to obtuse, crenulate, 2–9 cm. long, nearly as broad; flowers solitary and axillary or in subcorymbose terminal clusters; calyx 8–10 mm. long, enlarging slightly in fruit.

Occasional (St. Cath., St. Eliz., St. James), in arid open waste places and thickets; 100–750 feet; fl. and fr. Apr–May, Aug–Sept. *Asprey UCWI 1282! H 6757! H & P 13930!* Florida, Mexico, Bahamas, Cuba, Cayman Is.

6. HERISSANTIA Medic. (1788)

1. H. crispa (L.) Briz. in Journ. Arn. Arb. **49** (2): 278–9 (1968).—*Sida crispa* L. (1753). *Abutilon crispum* (L.) Medic. (1787). *Gayoides crispa* (L.) Small (1903). *Bogenhardia crispa* (L.) Kearney (1954).

Annual shrubby herb with slender trailing or spreading branches 60–120 cm.

long; leaves ovate, cordate, acute to shortly acuminate, crenate-dentate, 2–8 cm.
long and nearly as broad, stellate-tomentose; flowers mostly solitary and axillary;
peduncles 1·5–5 cm. long; calyx 4–8 mm. long; petals as long as to twice as long as
calyx, white distally, yellow, orange or green at base; carpels about 12, 10–15 mm.
long in fruit; seeds 1–3 in each carpel, black, thinly setulose.

Common on pebbly limestone and as a weed of disturbed ground; 20–1200
feet; fl. and fr. July–Feb. *A 5441! H 9529! P 8163!* Native of subtropical and
tropical Amer., now widespread in the tropics; Grand Cayman.

7. SIDA L. (1753); Kearney (1954)[a]; Clement (1957)

1 Peduncle adnate to stalk of subtending leaf or bract; flowers mostly few in terminal
    clusters; petals all red or yellow at tips; carpels 5–8, shortly muricate   1. *ciliaris*
1 Peduncles free:
  2 Leaf-blade entire, linear to lanceolate; petals white with dark red spot at base; carpels
    7–9, bluntly 2-beaked   2. *linifolia*
  2 Leaf-blade variously toothed:
    3 Flowers numerous in open or thyrsoid more or less leafy panicles; peduncles mostly
      long; shrubby plants often more than 1 m. high:
      4 Petals dark purple or red, 3–5 mm. long, often reflexed; leaves ovate to ovate-
        lanceolate, hardly cordate, without glandular hairs; stipules and peduncles fili-
        form; calyx 2·5–3 mm. long; carpels 5, cuspidate, finely reticulate   3. *paniculata*
      4 Petals yellow or orange, 5–8 mm. long, not reflexed; leaves broadly ovate, cordate;
        calyx 3·5–4·5 mm. long; carpels awned:
        5 Leaves narrowly cordate, glandular-pilose with also a few long simple spreading
          hairs; carpels 5, the awns slender and antrorsely pilose   4. *glutinosa*
        5 Leaves openly cordate, stellate-puberulous, with or without a few glandular or long
          simple hairs; carpels 5–8, the awns short with stellate hairs   5. *pyramidata*
    3 Flowers not in many-flowered panicles or, if so, then shortly pedunculate and
      crowded:
      6 Carpels 5; calyx angular:
        7 Stipules broadly lanceolate, acuminate, several-nerved, ciliate; leaves markedly
          distichous, with long appressed simple hairs adaxially; flowers subcapitate,
          axillary; carpels reticulate-veined, pointed   6. *glomerata*
        7 Stipules not as above:
          8 Peduncles and pedicels very slender, usually as long as or longer than petioles:
            9 Plant erect or lax, not creeping; inflorescence eventually few-flowered: paniculate;
              leaves ovate to lanceolate, up to 5 cm. or more long:
              10 Petals light orange; carpels with antrorsely hispid awns   7. *glabra*
              10 Petals reddish-purple; carpels cuspidate   3. *paniculata*
            9 Plant with laxly prostrate or creeping branches; leaves seldom more than 3 cm.
              long; flowers at least at first solitary in the axils:
              11 Stems creeping and often rooting; leaves orbicular to broadly ovate, deeply
                cordate, coarsely crenate; petals yellow; carpels awned, opening irregularly
                below   8. *javensis*
              11 Stems decumbent, not rooting; leaves ovate to lanceolate, shallowly cordate to
                truncate, coarsely crenate-serrate; petals light yellow or whitish; carpels
                shortly beaked, opening at apex   9. *procumbens*
          8 Peduncles and pedicels mostly shorter than petioles (but variable in *S. spinosa*):
            12 Flowers several together in terminal and axillary clusters or racemes; leaves
              mostly cordate and long-petiolate; carpels hardly beaked:
              13 Leaves ovate to lanceolate, acuminate, coarsely dentate or serrate; indumentum
                mostly of simple or few-armed hairs; carpels apiculate; margins of calyx
                darker green   10. *urens*
              13 Leaves broadly ovate, obtuse to subacuminate, finely dentate; indumentum
                mostly a felt of short stellate hairs with also some long hairs on calyx and
                sometimes on petioles; carpels obtuse; margins of calyx not darker
                  11. *aggregata*
            12 Flowers solitary or few together in axillary clusters; beaks of carpels antrorsely
              hispid:
              14 Ripe carpels pale, opening irregularly below by a white membrane, shortly
                beaked, striate on back; petals white or whitish; indumentum minutely
                stellate-pubescent   12. *alba*
              14 Ripe carpels dark brown, opening at apex, more or less aristate, transversely
                rugulose on back:
                15 Stems and leaves minutely stellate-pubescent; stems erect, often with infra-
                  petiolar tubercles; leaves ovate to narrowly lanceolate; petals yellow turning
                  orange; carpels with 2 short spines   13. *spinosa*

15 Stems and leaves softly stellate-tomentose; stems often spreading, without infrapetiolar tubercles; leaves rounded-ovate to subrhomboidal; petals white distally, yellow at base; carpels with 2 short beaks 14. *jamaicensis*
6 Carpels more than 5, mostly 7–14:
16 Leaves (especially on the veins beneath), stems, pedicels and calyx with small scattered stellate hairs, otherwise more or less glabrous; stipules unequal in each pair, ciliate; carpel-beaks shortly hispid; leaves mostly lanceolate 15. *acuta*
16 Leaves, at least beneath and at first, tomentellous to velutinous:
17 Leaves broadly rounded-ovate, velvety on both surfaces; stem and calyx velvety:
18 Carpels with erect aristate retrorsely hispid awns
16a. *cordifolia* var. *cordifolia*
18 Carpels without awns 16b. *cordifolia* var. *althaeifolia*
17 Leaves rhomboid-elliptical to lanceolate or linear:
19 Stem and calyx velutinous; leaf-blades linear to narrowly oblong, pale-concolorous; awns of carpels slender distally, erect, retrorsely hispid
17. *salviifolia*
19 Stem and calyx with minute stellate hairs; leaf-blades mostly lanceolate, discolorous; awns of carpels tapered, spinelike, usually spreading, thinly pubescent or glabrous 18. *rhombifolia*

1. S. ciliaris L., Syst. Nat. ed. 10, **2**: 1145 (1759).
Taprooted herb with prostrate or ascending stellate-hairy branches up to about 20 cm. long; leaves oblong, shallowly cordate or rounded at base, obtuse, serrate-dentate, glabrescent adaxially, up to 25 mm. long and 7 mm. broad; calyx 4–5 mm. long; carpels about 2 mm. long in fruit.
Occasional, as a weed of open waste ground and pastures; SL–700 feet; fl. and fr. most of the year. *A 6956! H 9768! P 26797!* Subtropical and tropical Amer., casual and rare in the Old World.

2. S. linifolia Juss. ex Cav., Monad. Cl. Diss. Dec. **1**: 14, t. 2, f. 1 (1785).
Erect herb up to about 1 m. high; leaves rounded or truncate at base, long-acuminate, 2–8 (–15) cm. long, rarely as much as 10 mm. broad; flowers few in terminal corymbs or in upper axils; calyx about 5 mm. long; carpels subtetrahedral, 2·5 mm. long in fruit.
Very rare (Clar.), a weed in open situations; about 2400 feet; fl. and fr. Dec. *H 12250!* Mexico to Paraguay, Cuba, Hispaniola, Trinidad and Tobago, W. Africa, Fiji.

3. S. paniculata L., Syst. Nat. ed. 10, **2**: 1145 (1759).
Straggling-branched shrubby herb 0·6–2·4 m. high; leaves long-acuminate, serrate-dentate, up to about 8 cm. long and 3 cm. broad, with yellowish stellate hairs; flowers solitary at first, on hairlike peduncles 1–4 cm. long; carpels about 3 mm. long in fruit.
Occasional in open or shaded sandy, gravelly or rocky places; 10–1700 feet; fl. and fr. Jan–Sept. *A 8808! H 12049! HPS 14742!* Texas, Mexico to Paraguay, Cuba, tropical Africa.

4. S. glutinosa Commers. ex Cav., Monad. Cl. Diss. Dec. **1**: 16, t. 2, f. 8 (1785).
Shrubby herb 0·6–1·5 m. high, glandular and aromatic; leaves usually acuminate, irregularly serrate-dentate, 2–6 (–9) cm. long, 1–4 (–6) cm. broad; calyx-lobes acute or cuspidate; carpels about 3 mm. long in fruit, including awns.
Occasional on shaded roadside banks and waste places; 300–2750 feet; fl. and fr. Nov–Feb. *A 8749! H 6619! P 22936!* Mexico to northern S. Amer., Bahamas, Greater Antilles, Virgin Is., Antigua, Guadeloupe, Trinidad, introduced into Mauritius, Madagascar and locally in S.E. Asia.

5. S. pyramidata Desportes ex Cav., Monad. Cl. Diss. Dec. **1**: 11, t. 1, f. 10 (1785).
Shrubby herb 1–2·5 m. high; leaves acuminate, crenate-dentate, 4–12 cm. long, up to 9 cm. broad; calyx-lobes acuminate to a subulate tip; carpels about 3 mm. long in fruit, including 2 very short awns.
Occasional weed of shaded roadside banks, moist thickets and woodland margins on limestone; 20–2500 feet; fl. and fr. Jan–Mar. *A 8931! H 10216! P 24688!* Mexico to northern S. Amer., Greater Antilles, Martinique.

6. S. glomerata Cav., Monad. Cl. Diss. Dec. **1**: 18, t. 2, f. 6 (1785).

Shrubby herb up to about 1 m. high; leaves lanceolate, truncate to subcordate at base, long-acute, coarsely serrate, 1·5–5 (–7) cm. long; peduncles short; calyx 5–7 mm. long, with acuminate lobes; petals yellow, a little longer than calyx; carpels about 2 mm. long in fruit, glabrous.

Very rare (St. Ann); fl. and fr. Jan. *Prior!* S. Mexico to Paraguay, Bahamas, West Indies.

7. S. glabra Mill., Gard. Dict. ed. 8 (1768).

Shrubby herb with slender straggling branches 30–90 cm. long; leaves lanceolate, cordate, acuminate, serrate-dentate, 2–7 cm. long, 1–3 cm. broad, thinly puberulous; peduncles slender, 1–4 cm. long; calyx 5–6 mm. long, with acuminate dark green-margined lobes; carpels, except the awns, about 2·5 mm. long in fruit, glabrous.

Rather common, in rocky thickets and waste places; 20–1400 feet; fl. and fr. Sept–Feb. *A 5602! H 12687! P 25603!* Mexico to Venezuela, West Indies southwards to Martinique.

8. S. javensis Cav., Monad. Cl. Diss. Dec. **1**: 10, t. 1, f. 5 (1785).— ? *S. hederifolia* Cav. (1785), of F. & R. (1926).

Herb with trailing and rooting branches up to about 30 cm. long; stems, petioles and peduncles with minute stellate and often longer simple hairs; leaves up to 25 mm. broad; peduncles up to about 20 mm. long; calyx 3–3·5 mm. long; carpels, except the awns, 3 mm. long in fruit; awns about 1 mm. long, antrorsely hispid.

Very rare (St. Ann), along shaded woodland pathsides; about 2000 feet; fl. and fr. Oct. *P 22846! W. Wright!* Generally but thinly scattered through the subtropics and tropics.

9. S. procumbens Sw., Nov. Gen. & Sp. Pl.: 101 (1788).

Taprooted herb with trailing or rarely erect branches up to 60 cm. long; stems, petioles and peduncles usually with a dense felt of small stellate hairs and also long simple hairs but sometimes the latter very few; leaves up to 20 mm. long and 14 mm. broad; peduncles up to 20 mm. long; calyx 4–5 mm. long; carpels 3–4 mm. long in fruit, including retrorsely or spreading-hispid beaks.

Occasional to locally abundant, in sandy pastures and on strand and in thickets in arid areas; SL–1200 feet; fl. and fr. July–Jan. *A 10004! H 9904! P 24335!* S. United States, Mexico to Venezuela, Bahamas, Greater Antilles, Virgin Is., Anguilla, Guadeloupe, Aruba.

10. S. urens L., Syst. Nat. ed. 10, **2**: 1145 (1759).

Straggling-branched slender-stemmed undershrub up to 1·5 m. high; leaves mostly cordate but sometimes subtruncate, acute or shallowly acuminate, 2–6 (–10) cm. long, up to about 4 (–6) cm. broad, with numerous stellate hairs on both surfaces at least when young; peduncles 0–2 (–6) cm. long; calyx 6–8 mm. long, the lobes long-acuminate; petals yellow or orange, often darker or reddish towards base and becoming pinkish in the afternoon; carpels about 2·5 mm. long in fruit.

Rather common, a weed of abandoned cultivations, roadside banks and shaded or seasonally damp waste places; 400–4200 feet; fl. and fr. Nov–Feb. *A 6408! Allwood UCWI 1134! P 26697!* General in subtropical and tropical Amer. and Africa. The range of variation in this species may be due to phenotypic responses within a broad habitat tolerance.

11. S. aggregata C. Presl, Rel. Haenk. **2**: 106 (1835).

Undershrub 1–1·5 m. high, aromatic; leaves 2·5–8·5 cm. long, up to about 6 cm. broad; calyx 7–8 mm. long, with acute triangular lobes; petals 11–12 mm. long, light orange to pinkish with red lines; carpels about 2 mm. long in fruit.

Local (St. Andr., St. Thom.), a weed on gravel and in arid thickets on limestone; SL–500 feet; fl. and fr. Oct–May. *A 5507! Barnett 17! H 6988!* Mexico to Colombia, Guyana, St. Croix.

12. S. alba L., Sp. Pl. ed. 2, **2**: 960 (1763).

Erect annual herb 30–50 cm. high; leaves broadly ovate to lanceolate, subcordate to rounded at base, serrate, 1·5–4·5 cm. long, up to about 2 cm. broad; peduncles 8–13 mm. long; calyx 5–6 mm. long; carpels about 2 mm. long in fruit.

Very rare (St. Eliz.), a weed; about 2000 feet; fl. and fr. Sept. *T. J. Harris 8288!* Texas, S. Amer., Bahamas, Cuba, Puerto Rico, Montserrat and thinly in the Old World tropics. This species is very similar to the next; further study may confirm that their continued separation cannot be maintained.

13. S. spinosa L., Sp. Pl. 2: 683 (1753).

Erect annual shrubby herb, 30–120 cm. high; leaves rounded at base, obtuse or acute at tip, serrate, 1–4 cm. long, up to about 2 cm. broad; peduncles 2–5 mm. long but accrescent to 10 mm. or more; calyx 5–7 mm. long, the lobes deltate, acute; petals sometimes with pink veins; carpels about 2·5 mm. long in fruit.

Common, as a weed of cultivation and in pastures and sandy waste places; 20–3200 feet; fl. and fr. June–Mar. *A 11886! H 11910! P 24203!* General in the subtropics and tropics.

14. S. jamaicensis L., Syst. Nat. ed. 10, 2: 1145 (1759).

Undershrub with low spreading branches or erect to 1 m. high; stem sometimes reddish; leaves unequally rounded at base, obtuse at tip, serrate-crenate, 1–5 cm. long, 1–3 cm. broad; peduncles very short; calyx 5–7 mm. long, the lobes acuminate; carpels 2–3 mm. long in fruit, including beaks.

Occasional in open sparsely vegetated pastures and on eroded banks; 10–1500 feet; fl. and fr. Oct–Jan. *A 9946! P 18440! Robertson UCWI 1669!* Mexico to Colombia, Cuba, Hispaniola, Virgin Is., Antigua, Guadeloupe and some of the smaller dry islands of the Leeward group.

15. S. acuta Burm. f., Fl. Ind.: 147 (1768).—Broomweed.

Short-lived undershrub up to about 1 (–1·5) m. high; leaves ovate to linear-lanceolate, cuneate to rounded at base, acute at tip, serrate, 1–7 (–15) cm. long, 0·5–3 (–6) cm. broad; stipules up to 11 (–15) mm. long; peduncles 2–5 mm. long; calyx about 7 mm. long, the lobes acuminate; petals light yellow or white and more or less yellow at base; carpels 6–10, 3–4 mm. long in fruit.

Very common in pastures, waste places and cultivations; SL–1500 feet; fl. and fr. all the year. *A 5438! 8863! H 6759! H & P 13311!* General in the tropics. A variable species with an extensive synonymy; Borssum Waalkes (1966) recognizes two subspecies, both occurring in the American tropics and distinguished as follows:

Leaves lanceolate to linear, rarely the lower ovate to oblong, at base mostly acute, rather coarsely and remotely serrate; indumentum without or with few simple hairs; flowers in cluster of 2–3   ssp. *acuta*

Leaves ovate to oblong, at base mostly rounded, finely and densely serrate, the teeth ending in a simple hair; indumentum with many simple hairs; flowers in clusters of up to 8   ssp. *carpinifolia* (K. Schum.) Borssum Waalkes

16. S. cordifolia L., Sp. Pl. 2: 684 (1753).

16a. Var. cordifolia.

Bushy undershrub up to about 1 m. high; leaves cordate, acute or obtuse, serrate to crenate, 1·5–9 cm. long, 1–6·5 cm. broad; flowers solitary or clustered; peduncles short in flower elongating to up to 20 mm. in fruit; calyx 6–8 mm. long; petals yellow turning orange or reddish-orange; carpels 7–12, about 3·5 mm. long in fruit with awns equally as long or longer.

Occasional as a weed of waste places; SL–560 feet; fl. and fr. Aug–Apr. *A 6992! Barnett 16! Campbell 5972!* General in the subtropics and tropics.

16b. S. cordifolia var. althaeifolia (Sw.) Griseb., Fl. Br. W.I.: 76 (1859).— *Sida althaeifolia* Sw. (1788).

Like the last but sometimes more robust, up to 1·5 m. high; leaves up to 7 cm. long and 5 cm. broad.

Locally common as a weed of hillside pastures and sandy or gravelly waste ground; 10–2600 feet; fl. and fr. Oct–Mar. *A 8203! P 18438! Robertson UCWI 1335!* Mexico, Honduras, Greater Antilles, Cape Verde Is.

17. S. salviifolia C. Presl, Rel. Haenk. **2**: 110 (1835).

Shrubby erect annual herb 30–75 cm. high; leaves rounded at base acute or obtuse at tip, serrulate, 1–5 cm. long, up to about 5 mm. broad; flowers solitary in the axils or corymbose distally; peduncles short, but usually longer than petioles; calyx 5–6 mm. long; petals white to light yellow becoming pinkish; carpels 7–10, about 3 mm. long in fruit excluding bristles, the latter about 2 mm. long.

Local (St. Andr., St. Cath.), in sandy and gravelly waste places in areas of low rainfall; SL–300 feet; fl. and fr. Oct–Feb. *A 6270! H 10000! P 8162!* Mexico Colombia, Venezuela, Puerto Rico, Virgin Is.

18. S. rhombifolia L., Sp. Pl. **2**: 684 (1753).—*S. troyana* Urb. (1908).

Erect shrubby herb 30–90 cm. high, usually bushy with several slender branches; leaves lanceolate to rhombic-oblong, narrowed and minutely cordate at base, acute to broadly obtuse at tip, serrate-dentate, 2–8 (–10) cm. long, up to 4 (–5·5) cm. broad, darker and glabrescent adaxially; peduncles up to 40 mm. long; calyx 6–7 mm. long; petals yellow to orange-yellow becoming pinkish; carpels 9–14, 3–4 mm. long in fruit including awns.

Very common in pastures and waste places, particularly at middle elevations; SL–5200 feet; fl. and fr. almost all the year. *A 5552! 11558! H 12604! P 7593!* General in the subtropics and tropics.

8. BASTARDIA Kunth (1822)

1 Valves of capsule beaked; petals 7–10 (–15) mm. long   1. *bivalvis*
1 Valves of capsule without beaks; petals about 5 mm. long:
   2 Leaves mostly more than 3 cm. long, green on both surfaces; stem with many short
      gland-tipped and long simple hairs   2a. *viscosa* var. *viscosa*
   2 Leaves mostly less than 2·5 cm. long, canescent; long hairs on calyx and distal part of
      petiole only; glandular hairs few or absent   2b. *viscosa* var. *parvifolia*

1. B. bivalvis (Cav.) Kunth, Nov. Gen. **5**: 255 (1822).—*Sida bivalvis* Cav. (1785).

Coarse herb or shrub 0·6–2 m. high, aromatic-glandular; indumentum of short stellate and long simple hairs; leaves broadly ovate, cordate, acuminate to a broad tip, crenate, 1–9 (–13) cm. long, up to about 7·5 cm. broad; flowers axillary; peduncles up to 7 mm. long; calyx 7–8 mm. long, stellate-tomentose; flowers fragrant; petals light yellow to yellow-orange; capsule 3·5–4 mm. long, with beaks 1·5–2 mm. long.

Occasional in the south-eastern parishes, at roadsides and in waste places on limestone in arid areas; 20–700 feet; fl. and fr. Nov–Mar. *A 10005! H 8914! P 24497!* Mexico to Peru and Brazil, Greater Antilles.

2. B. viscosa (L.) Kunth, Nov. Gen. **5**: 256 (1822).—*Sida viscosa* L. (1759).

2a. Var. viscosa.

Bushy herb up to 120 cm. high, glandular, glutinous and aromatic; leaves broadly ovate, cordate with a narrow sinus, more or less caudate-acuminate, denticulate, 1·5–12 cm. long, up to 6 (–10) cm. broad; flowers axillary; peduncles up to 25 (–40) mm. long; calyx 3·5–4 mm. long, tomentose; petals yellow to orange-yellow; capsule about 3 mm. long.

Local (St. Andr., St. Cath.), in gravelly or sandy waste places mostly near the sea; SL–150 feet; fl. and fr. July–Nov, Mar. *A 11817! Campbell 6326! 6362! J. P. 628! P 8164!* Mexico to Venezuela and Peru, Bahamas, Hispaniola, Puerto Rico, Virgin Is., Antigua, Barbados, Grand Cayman, Galapagos Is.

2b. B. viscosa var. **parvifolia** (Kunth) Griseb., Fl. Br. W.I.: 80 (1859).—*B. parvifolia* Kunth (1822).

Bushy tough-stemmed herb 30–90 cm. high; leaves paler beneath, gradually acuminate, serrulate-dentate to coarsely repand, 1–4 cm. long, 8–25 mm. broad; peduncles 4–10 (–14) mm. long; petals yellow or light orange-yellow.

Locally common, in the southern parishes in salina margins and open ground near the sea; SL–20 (–150) feet; fl. and fr. Sept–Jan, May. *A 6467! 11818! H 5977! 9721! P 9365!* Ecuador, Colombia, Brazil, Greater Antilles, Trinidad. This variety is distinct in the type locality (Brazil) and in Jamaica; plants from Trinidad lack the long hairs on the calyx; plants from Cuba and Hispaniola resemble more var. *viscosa* in having numerous short glands and large leaves, but they also lack long simple hairs.

9. MALACHRA L. (1767)

1 Flower-heads terminal; large hairs on stem stellate, on leaves abaxially mostly 3-fid, persistent; small hairs restricted to band(s) on stem and petiole; leaves deeply 3–5-lobed, the lobes again lobed   1. *radiata*
1 Flower-heads axillary:
  2 Large hairs on stems and leaves mostly simple and spreading and not usually accompanied by small stellate hairs except in bands on stem and petiole; flower-heads sessile or subsessile; calyx 4–5 mm. long:
    3 Leaves mostly 3–5-lobed to about halfway, margin serrate; carpels reticulate-veined, warted; glabrescent: corolla white turning pink   2. *fasciata*
    3 Leaves mostly not lobed or shortly angular towards base, dentate or denticulate; carpels faintly reticulate beneath a virgate indumentum; corolla yellow or orange, sometimes reddish   3. *urens*
  2 Large hairs on stems and leaves usually stellate, often but sometimes only thinly accompanied by numerous small stellate hairs; calyx 6–8 mm. long; corolla yellow or rarely white:
    4 Large hairs 3- or more-fid, caducous; flower-heads often stalked on peduncles up to 8 cm. long; leaves undivided or often sinuate-lobed, crenate-dentate; carpels glabrous   4. *capitata*
    4 Large hairs 2–3-fid or simple, mostly persistent, pungent; flower-heads sessile or shortly stalked on peduncles rarely more than 2 cm. long; leaves undivided or angular-pointed, serrate-dentate; carpels tomentellous   5. *alceifolia*

1. M. radiata (L.) L., Syst. Nat. ed. 12, **2**: 459 (1767).—*Sida radiata* L. (1763).
Erect herb or undershrub 0·5–1·5 (–2·5) m. high; leaf-blades up to about 12 cm. long and broad, with some simple hairs adaxially; stipules about 1 cm. long; outer bracts up to about 4 cm. long; epicalyx of 9–12 filiform bracteoles 10–12 mm. long; calyx 8–10 mm. long; corolla pink, 11–13 mm. long; carpels glabrous, minutely warted, about 4 mm. long.
Very rare (Trel.), growing in shallow water; about 400 feet; fl. and fr. Aug. *P 15711! W. Wright!* C. and S. Amer., Greater Antilles, tropical Africa.

2. M. fasciata Jacq., Collect. **2**: 352 (1789).
Annual coarse herb, woody at base, up to 1·5 m. or more high; leaf-blades up to about 15 cm. along the median midrib; stipules 1·5–3 cm. long; outer bracts subcordate, long-acuminate at tip and often reflexed; corolla about 10 mm. long; carpels 3–3·5 mm. long.
Occasional, in damp grassy places; 30–600 feet; fl. Oct–Jan, May, fr. Nov–Feb. *A 11883! P 21462! Skelding UCWI 5421!* Mexico to the Guianas and Paraguay, Cuba, Puerto Rico, Lesser Antilles, Trinidad, Pacific Is.

3. M. urens Poit. in Ledeb. & Adlerstam., Diss. Pl. Doming.: 22 (1805).
Undershrub or annual herb becoming woody at base up to about 60 cm. high; leaves ovate to lanceolate, truncate at base, sharply and unevenly dentate, 2–6 (–9) cm. long and up to about 4 cm. broad; stipules 5–10 mm. long; bracts cordate, long-acuminate, up to about 15 mm. long; corolla about 12 mm. long; carpels 3 mm. long.
Uncommon, in the southern and western parishes, in low swampy pastures; 5–900 feet; fl. and fr. Nov–Apr. *A 8518! 8829! P 11115! 26329!* S. Florida, Greater Antilles.

4. M. capitata (L.) L., Syst. Nat. ed. 12, **2**: 458 (1767).—*Sida capitata* L. (1753).
Erect shrubby herb up to about 2 m. high; leaves ovate to orbicular in outline, openly cordate, the angles or lobes obtuse, up to 9 cm. long and 8 cm. broad; stipules filiform, 10–15 mm. long; bracts broadly ovate, acute with slightly recurved tip; up to about 2 cm. long; petals 9–10 mm. long; carpels 3–3·5 mm. long.
Rather rare (St. Eliz., Han.), pondside weed and in open swamp; 2–200 feet; fl. and fr. Mar, June. *H 10252! P 27513!* Texas, Mexico and C. Amer., West Indies southwards to Martinique; Grand Cayman; an introduced weed in tropical Africa and Asia.

5. M. alceifolia Jacq., Collect. **2**: 350 (1789).—Wild Okra.

Robust herb or undershrub up to 3 m. high; leaves ovate to suborbicular, 3–5-angled or -lobed, up to 10 (–12) cm. long and 10 cm. broad; stipules 10–20 mm. long; bracts deeply cordate, acute, up to 2·5 cm. long and broad; petals about 15 mm. long; carpels 3–3·5 mm. long.

Locally common, mainly in south-eastern parishes, a weed of roadsides and low-lying waste places; SL–100 feet; fl. and fr. most of the year. *A 8813! H 9050! P 9545!* Mexico to the Guianas and Peru, Greater Antilles, Virgin Is., Antigua, Grenadines, Barbados, Trinidad and Tobago; sparsely naturalized in the Old World tropics.

The recognition of 5 species of *Malachra* in Jamaica is the result of attempts to group a range of specimens around the types involved. Some of these specimens have intermediate character-combinations and are difficult to place. Because of this authors have been persuaded to combine some of these species but have completely failed to agree on the taxonomic affinity or status of their units. Comparison of two recent publications will suffice to indicate the futility of any further attempts to adjust the classification of species of *Malachra* until at least some of the factors controlling phenotype expression are known and their mode of operation understood. [A. Robyns, " Flora of Panama – Malvaceae ", in Ann. Miss. Bot. Gard. **52** (4): 524–527 (1965); J. van Borssum Waalkes, " Malesian Malvaceae Revised ", in Blumea **14** (1): 145–149 (1966)].

10. URENA L. (1753)

1. U. lobata L., Sp. Pl. **2**: 692 (1753).—Ballard Bush, Bur Mallow.

Erect shrubby herb 30–150 cm. high; leaves very variable, the lower orbicular and often rather small, those on the middle of the stem larger, orbicular, angled or lobed to deeply incised, the upper ovate to lanceolate, variously serrate, crenate or sometimes entire, 1–12 cm. long and as broad, covered with numerous small stellate and fewer large simple hairs; flowers axillary; calyx 5–7 mm. long; petals pink, darker at base, about 15 mm. long; mature carpels 5–6 mm. long.

Common weed of roadsides, cultivations and waste places; 50–5000 feet; fl. and fr. all the year. *A 6433! J. P. 1060! P 18437!* General in the tropics.

Most modern authors concur in recognizing only one species of *Urena* but usually maintain a number of infraspecific taxa. *U. lobata* var. *sinuata* (L.) Hochr. (1901) (*U. sinuata* L. (1753)), is sometimes distinguishable on account of the deeply lobed leaves with rounded sinuses (*H 6616! P 18437!*), but intermediates (*P 18442!*) occur.

11. PAVONIA Cav. (1786) nom. cons.; Kearney (1954)[b]

1 Carpels without spines or with short weak spines on back or merely mucronate or crested; bracteoles 6–12; petals greenish-yellow to orange-yellow:
 2 Inflorescence an open usually leafy panicle; robust annual herb; stems and leaves usually glandular-pilose, the stems with or without long simple hairs also; leaves distinctly serrate-dentate; bracteoles subulate or filiform, much longer than calyx
                     1. *paniculata*
 2 Inflorescence usually of simple terminal racemes or corymbs; shrubs; young stems and leaves stellate-tomentose, glabrescent; leaves entire or very shallowly serrate; bracteoles broader, shorter than calyx:
  3 Bracteoles lanceolate or oblong-lanceolate, 2–3·5 mm. broad at base; carpels not wing-margined, lightly reticulate, with a prominent keel and 3 thick triangular apical crests, the latter sometimes obscure   **2a.** *spicata* var. *spicata*
  3 Bracteoles oblong or elliptic-oblong, up to 5 mm. broad at base; carpels rather broadly wing-margined   **2b.** *spicata* var. *troyana*
1 Carpels with usually 3 retrorsely barbed hispid or pilose awns or cusps; bracteoles 5–10:
 4 Petals yellow, 2–3 (–4) cm. long; leaves cordate to rounded at base, broadly ovate to ovate-lanceolate; bracteoles 5–8, nearly distinct, narrowly oblanceolate, long-ciliate; flowers solitary; awns of carpels variable in length, the lateral often strongly divergent   **3.** *spinifex*
 4 Petals white or white turning pink, less than 2 cm. long; leaves cuneate or at most rounded at base; awns of carpels erect or weakly spreading:
  5 Flowers solitary or 2–4 together in stalked axillary corymbs; stems hirsute or hispid with yellowish or brownish hairs; bracteoles 8–10, linear, shortly united
                     4. *pseudotyphalaea*

5 Flowers several to numerous in mostly terminal inflorescences; indumentum of short
   whitish or yellowish forked or stellate hairs; bracteoles united below:
6 Inflorescence corymbose or paniculate with clustered and solitary flowers; bracteoles
   7–11, narrowly lanceolate to linear, 0·5–2 mm. broad   5. *rosea*
6 Inflorescence capitate and usually stoutly pedunculate; bracteoles 5–7, ovate to
   broadly lanceolate, commonly 2–4 mm. broad   6. *fruticosa*

1. P. paniculata Cav., Monad. Cl. Diss. Dec. 3: 135, t. 46, f. 2 (1787).

Herb or undershrub 1–3 m. high; leaves ovate to broadly ovate, deltate or 3-
lobed, cordate at base, acute or acuminate, crenate or serrate, up to 12 cm. long and
10 cm. broad; stipules lanceolate, up to 10 mm. long; bracteoles 7–10 mm. long;
calyx 5–6 mm. long; petals 10–15 mm. long; ripe carpels 3–4 mm. long.
Very rare, recorded only once, *Swartz*, and not since confirmed. Mexico to
Argentina, Hispaniola, Puerto Rico.

2. P. spicata Cav., Monad. Cl. Diss. Dec. 3: 136, t. 46, f. 1 (1787).—*Malache scabra* B. Vogel (1772). *P. scabra* (B. Vogel) Ciferri (1936), not C. Presl (1835).

2a. Var. spicata.—Smaller Mahoe.

Erect or scrambling shrub or tree 1–3 (–5) m. high; leaves broadly ovate or
rotundate, cordate, acuminate, 6–18 cm. long, 4–10 cm. broad; stipules linear-
lanceolate, 6–13 mm. long; bracteoles 7–14 mm. long; calyx 10–13 mm. long;
petals 15–22 mm. long; ripe carpels 8–11 mm. long.
Occasional, in thickets on rocks near brackish lagoons and in marsh forest;
SL–100 feet; fl. and fr. most of the year. *A 6040! 12059! H 8629! P 24655!*
S. Florida, C. to northern S. Amer., Bahamas, Greater Antilles, Antigua ?,
Guadeloupe, Trinidad and Tobago.

2b. P. spicata var. troyana (Urb.) F. & R., Fl. Jam. 5: 132 (1926).

Shrub or tree 2–6 m. high; wing on carpel broad or narrow.
Local in the central and north-western parishes, in open woodlands on limestone;
900–2800 feet; fl. and fr. most of the year. *A 12435! H 10964! P 21415!* Endemic.

3. P. spinifex (L.) Cav., Monad. Cl. Diss. Dec. 3: 133, t. 45, f. 2 (1787).— *Hibiscus spinifex* L. (1759). Spur Bur.

Shrub with trailing branches 60–120 cm. long, rarely larger in Jamaica; leaves
acuminate, crenate-dentate, 4–12 cm. long, 2–8 cm. broad; stipules subulate, 7–11
mm. long; bracteoles 7–12 mm. long; calyx 9–13 mm. long; ripe carpels 4–6 mm.
long with awns up to 10 mm. long.
Occasional in the central parishes, in shaded thickets and on rocky banks and
waste ground; 30–2300 feet; fl. and fr. July–Feb. *A 9855! H 12445! P 24163!*
S.E. United States, Bermuda, Mexico to Brazil and Peru, Bahamas, Greater
Antilles, Virgin Is., Antigua, Barbados, Trinidad.

4. P. pseudotyphalaea Planch. & Linden in Ann. Sci. Nat. sér. 4, Bot. 17: 160 (1862).

Undershrub; leaves oblong-elliptical or lanceolate, acute, serrate with large
teeth, 6–16 cm. long; stipules linear-subulate, 8–10 mm. long; bracteoles 7–9 mm.
long; calyx 5–6 mm. long; petals 12–15 mm. long; fruit not known.
Only once reported, *Swartz*. Colombia.

5. P. rosea Schlecht. in Linnaea 11: 355 (1837).—Conger Watchman.

Shrubby herb up to about 1·5 m. high; leaves elliptical to obovate, acute or
acuminate, irregularly serrate or crenate, 4–20 cm. long, up to about 8 cm. broad;
stipules linear-subulate, 7–10 mm. long; bracteoles 5–11 mm. long; calyx 3–5 mm.
long; petals 8–13 mm. long; ripe carpels 5–7 mm. long with awns up to 6 mm. long.
Very common in damp areas in clearings and on banks; 50–2500 (–3500) feet;
fl. and fr. all the year. *A 8093! H 11875! P 23856!* Mexico to Colombia and Brazil,
Cuba.

6. P. fruticosa (Mill.) F. & R., Fl. Jam. 5: 130 (1926).—*Sida fruticosa* Mill. (1768). *Typhalea fruticosa* (Mill.) Britton (1924).

Like the last and perhaps not really distinct; calyx 4–7 mm. long; ripe carpels
with awns up to 7·5 mm. long and apically approximated.

Very rare (St. Ann, St. Mary); fl. and fr. Oct–Dec. *Cornman! McNab! Prior!* Panama to Brazil and Paraguay, Hispaniola, Puerto Rico, Dominica, Trinidad and Tobago.

12. MALVAVISCUS Fabr. (1759); Schery (1942)

1. M. arboreus Cav., Monad. Cl. Diss. Dec. **3**: 131, t. 48, f. 1 (1787).—*Hibiscus malvaviscus* L. (1753). *M. sagraeanus* A. Rich. (1845). Mahoe Rose, Sugar Bark.

Diffuse scrambling shrub up to 5 m. high, densely to thinly stellate-pubescent or subglabrous; leaves broadly ovate to lanceolate, entire or 3–5-lobed, cordate at base, unequally crenate or serrate, more or less acuminate; flowers mostly solitary and axillary; bracteoles 7–12, 11–20 mm. long and about equal to calyx; flowers very variable in size, the corolla (2–) 2·5–3 (–5·5) cm. long, usually pinkish-red; fruit 6–10 mm. long, at first red.

Common, in rocky woodlands and thickets mostly on limestone; 50–5650 feet; fl. July–Sept, Nov–Apr, fr. Nov–Apr. *A 10310! H 11835! P 24537!* Texas, Mexico to Peru and Brazil, West Indies. This is a very variable species within which Schery (Ann. Miss. Bot. Gard. **29**: 183–245 (1942)) distinguishes a number of varieties mostly by making arbitrary divisions. A common plant in cultivation in Jamaica, originating from Central America, is *M. arboreus* var. *penduliflorus* (DC.) Schery, Pepper Hibiscus or Sleeping Hibiscus, distinguished by its red or pink flowers being (5–) 6–7 cm. long. A variant with very small flowers known from Cayman Brac and Grand Cayman is *M. arboreus* var. *cubensis* Schlecht. (1837).

13. THESPESIA Solander ex Correa (1807) nom. cons.; Howard (1949)

Leaves with a linear nectary on midrib beneath, usually 3–5-lobed; fruit a dehiscent capsule with numerous seeds; uncommon cultivated ornamental, native of S.E. Asia
                                    T. lampas (Cav.) Dalz. & Gibs.
Leaves without a nectary on midrib beneath, not lobed; fruit indehiscent; seeds few (up to 20); common seashore plant   1. *populnea*

1. T. populnea (L.) Solander ex Correa in Ann. Mus. Hist. Nat. Paris **9**: 290, t. 25, f. 1 (1807).—*Hibiscus populneus* L. (1753). Seaside Mahoe.[1]

Shrub or tree (1–) 3–6 (–20) m. high; leaves ovate, cordate, acute or acuminate, 5–15 (–22) cm. long, 3–11 (–16) cm. broad; stipules subulate, 3–10 mm. long; flowers solitary, axillary, showy; bracteoles lanceolate, 5–15 mm. long; calyx cupular, entire or minutely 5-toothed, 9–14 mm. long; petals 4–7 cm. long, yellow becoming purplish, with a red spot at base of each; fruit depressed-globose, 2–4·5 cm. broad, the mesocarp soft with yellow sap, drying blackish.

Common in littoral situations; SL–20 feet; fl. Apr–Jan, fr. May–Jan. *A 7364! M. L. Farr! Harris! Philipson 524!* General on sandy or gravelly shores and at mangrove margins throughout the tropics.

14. GOSSYPIUM L. (1753); J. B. Hutchinson (1943)

1 Staminal column long, sometimes 3–4 cm. long, filaments compactly arranged and of uniform length:
  2 Capsule 3–5 cm. long, broadest near base; seeds free
                                    1a. *barbadense* var. *barbadense*
  2 Capsule 5–7 cm. long, broadest about the middle; seeds connate
                                    1b. *barbadense* var. *acuminatum*
1 Staminal column short, usually 1–2 cm. long and about half as long as corolla, filaments loosely arranged, the distal longer:
  3 Bracteole-teeth 3–11, usually about 6; anthers numerous on rather uniformly long filaments; perennial shrubs usually strongly branched from lower part of main stem
                                    2c. *hirsutum* var. *marie-galante*

[1] This name is also applied, and more correctly, to the much less common *Hibiscus tiliaceus*.

3 Bracteole-teeth 6–14, usually about 10; anthers sparse and rather irregular:
4 Short-lived sparingly branched undershrubs with large leaves and widely expanding
    corolla   2a. *hirsutum* var. *hirsutum*
4 Perennial much branched shrubs with small leaves; corolla forming a narrow tube
                      2b. *hirsutum* var. *punctatum*

1. **G. barbadense** L., Sp. Pl. 2: 693 (1753).

1a. Var. **barbadense.**—*G. peruvianum* Cav. (1788). Long Staple Cotton, Sea
    Island Cotton.
Shrub up to 5 m. high; leaves broadly ovate, entire or 3–5 (–7)-lobed, the central
segment often longer, up to 16 cm. long; stipules leafy; bracteoles with (3–) 7–9
(–11) teeth, shorter than the petals; calyx densely gland-dotted, shortly 5-toothed;
corolla light yellow, turning pinkish, 3·5–5·5 cm. long; seeds with long white lint,
only partly covered with fuzz.
    Annual forms cultivated; perennial forms naturalized in moist sheltered places.
Native of S. Amer., now widespread in cultivation.

1b. **G. barbadense** var. **acuminatum** (Roxb.) Mast. in Hook. f., Fl. Br. Ind. 1:
    347 (1875).—*G. acuminatum* Roxb. (1832). *G. lapideum* Tussac (1818). *G.
    brasiliense* Macf. (1837). *G. barbadense* var. *brasiliense* (Macf.) J. B. Hutch.,
    Silow and Stephens (1947). Kidney Cotton.
Bracteoles with 9–13 teeth, as long as or longer than petals; lint white.
    An early introduction into Jamaica. *Fawcett! Harris 138!* Eastern S. Amer.,
C. Amer., West Indies, also in India and thinly in other parts of tropical Asia and
Africa, thriving in wet areas.

2. **G. hirsutum** L., Sp. Pl. ed. 2, 2: 975 (1763).—Short Staple Cotton, Upland
    Cotton.

2a. Var. **hirsutum.**
This variety has been introduced into Jamaica but has not become established.
Native of subtropical N. and C. Amer.

2b. **G. hirsutum** var. **punctatum** (Schumach.) J. B. Hutch., Silow and Stephens,
    Evol. Gossyp.: 40 (1947).—*G. punctatum* Schumach. (1827).
    Sparingly cultivated in Jamaica. Native of the coasts and islands of the Gulf of
Mexico, Florida, Bahamas, Hispaniola, Cayman Is.

2c. **G. hirsutum** var. **marie-galante** (Watt) J. B. Hutch., Silow and Stephens,
    Evol. Gossyp.: 43 (1947).—*G. marie-galante* Watt (1927). *G. harrisii* Watt
    (1927). *G. arboreum* of F. & R. (1926), not L. *G. purpurascens* of F. & R.
    (1926), not Poir. *G. hirsutum* of F. & R. (1926), for most part.
Large shrub or small tree up to 5 m. high; leaves 3–5-lobed and cut halfway or
more deeply to base on lower branches, on lateral branches often entire, cordate,
lobes acuminate, up to about 15 cm. long; stipules caducous, up to 10 mm. long;
bracteoles mostly with 4–7 long acuminate teeth; lint copious, with or without fuzz.
    Common, especially in southern coastal areas, in arid thickets and waste ground
near habitations; SL–700 feet; fl. and fr. Nov–May. *A 11862! H 10179! P 23509!
Stephens & Griddi-Papp!* Native along the coasts and islands of northern S. Amer.
and the Caribbean; Cuba, Jamaica, Puerto Rico, Lesser Antilles. S. G. Stephens
considers that Jamaican populations to-day are not uniform and can be distinguished as " dooryard " races into which *G. barbadense* has introgressed considerably, and " wild " races which have much in common with wild plants from
Hispaniola and Puerto Rico thus approaching *G. hirsutum* var. *punctatum*. It is
clear from recent studies that bract-tooth number is subject to environmental
control but branching habit is the result of genetical differences.

15. **KOSTELETZKYA** C. Presl (1835) nom. cons.

1. **K. pentasperma** Griseb., Fl. Br. W.I.: 83 (1859).—*Hibiscus pentaspermus* Bert.
    ex DC. (1824), not Nutt. (1822).
Shrubby herb up to about 2 m. high, the stem with pungent yellow stellate hairs;

leaves lanceolate to ovate, often angular or subhastate, irregularly serrate-dentate, acute to acuminate, up to 11 cm. long and 8 cm. broad; flowers solitary in upper axils; bracteoles 2–3·5 mm. long; calyx 3–5 mm. long; petals 10–13 mm. long; capsule about 4–5 mm. long and 8–10 mm. broad; seeds blackish-brown, about 2·5 mm. long, thinly covered with curved hairs.

Occasional, in the southern parishes in swampy meadows, ditches and along riverbanks; 20–50 feet; fl. and fr. May–Jan. *A 8811! H 9054! P 24332!* Mexico to Venezuela, Ecuador, Cuba, Hispaniola.

16. HIBISCUS L. (1753) nom. cons.

1 Fruiting carpels separating from receptacle at maturity, winged, scarious and strongly veined; stem velvety with short hairs and also with scattered pungent branched hairs; bracteoles 7–9 (–12), filiform, shorter than calyx; corolla showy, lemon-yellow with dark crimson or purple eye 1. *vitifolius*
1 Fruiting carpels not separating from receptacle and not winged:
  2 Trees; bracteoles united into 8 or more-toothed involucel; stipules broad, 1–6 cm. long:
   3 Calyx and bracteoles more or less persistent in fruit; petals usually less than 8 cm. long; seeds minutely tufted with stellate hairs, otherwise glabrous; plant of brackish swamps 2. *tiliaceus*
   3 Calyx and bracteoles caducous in fruit; petals usually more than 9 cm. long; seeds shortly and densely tomentose; plant of inland situations, often cultivated
    3. *elatus*
  2 Shrubs or herbs; bracteoles free or only shortly united at base:
   4 Petals changing colour from white to pink during the day; leaves shortly 5-lobed; indumentum greyish-stellate; bracteoles 8–12, lanceolate; cultivated ornamental; native of China and Japan; Changeable Rose, Rose of Sharon *H. mutabilis* L.
   4 Petals not changing colour as above:
    5 Calyx becoming inflated and mostly or wholly enclosing fruit, thin or membranous, the lobes rather short and 3-nerved excluding the margin; bracteoles linear-subulate, shorter than calyx; capsule obtuse, bristly:
     6 Annual weedy herb; leaves deeply 3–5-lobed; calyx scarious, hispid on main nerves, glabrous on reticulate veins between; capsule about 1 cm. long; seeds reniform, minutely verrucose 4. *trionum*
     6 Robust herbs or shrubs; leaves shallowly or angular-lobed; calyx hairy on and between veins; capsule 2·5 cm. or more long:
      7 Stems at least when young armed with prickles, otherwise thinly pubescent; seeds velvety; plant of wet places 5. *trilobus*
      7 Stems unarmed, shortly velvety; seeds glabrous; plant of well drained places
       6. *clypeatus*
    5 Calyx not becoming inflated or, if accrescent deeply lobed and not enclosing fruit:
     8 Calyx with thickened marginal nerves, those to the sinus forming the margins to adjacent lobes; bracteoles of epicalyx sometimes lobed or bifurcate at or near tip; leaves deeply 3–5-lobed or rarely simple; capsule more or less pointed, with ascending bristly hairs; stems often prickly:
      9 Calyx becoming fleshy in fruit, the hairs soft and not from tuberculate bases; stem mostly without prickles; leaves with a cleft gland at base of midrib beneath, glabrescent; seeds puberulous 7. *sabdariffa*
      9 Calyx not becoming fleshy in fruit; stems mostly prickly:
       10 Calyx-lobes attenuated and greatly exceeding capsule, like the bracteoles beset with pungent tubercle-based hairs; leaves with a swollen usually closed gland near base of midrib beneath; seeds scaly 8. *cannabinus*
       10 Calyx about as long as capsule, the lobes acute or acuminate; leaves mostly with a slotted gland at or near base of midrib beneath:
        11 Peduncles aculeate and pubescent, up to 3 cm. long in fruit; calyx bristly, green; corolla 8–9 cm. long; lobes of leaves acuminate 9. *bifurcatus*
        11 Peduncles glabrescent, up to 1 cm. long in fruit; calyx glabrescent; corolla 3–4 cm. long, yellow tinged wine-red; lobes of leaves obtuse; resembles most examples of *H. sabdariffa* in presence of deep wine-red colour in leaves and calyx; casual introduction (also in Puerto Rico); native of tropical Africa; (*H. eetveldeanus* De Wild. & Dur.) *H. acetosella* Welw. ex Hiern
     8 Calyx essentially herbaceous without thickened margins; leaves not lobed or at most shallowly angular:
      12 Peduncles short and stout, up to about 6 mm. long, articulated at base; calyx-lobes similar to bracteoles, linear-lanceolate; leaves cordate at base with 3 or 5 caudate-acuminate angles, dentate; stem, petioles and pedicels with yellow caducous pungent hairs; valves of capsule beaked; seeds minutely stellate-hairy 10. *lunariifolius*

12 Peduncles longer, slender, articulated at about or beyond the middle; calyx-lobes much broader than bracteoles; leaves not lobed capsule obtuse:
13 Bracteoles of epicalyx dilated at tip into a reniform or suborbicular blade; leaves deeply cordate, crenate, obtuse; calyx-lobes broadly ovate, exceeding capsule, indistinctly veined, stellate-pubescent   11. *sororius*
13 Bracteoles of epicalyx at most oblanceolate or expanded to a spathulate tip; leaves mostly truncate to cuneate at base:
  14 Cultivated ornamental shrubs with showy flowers; leaf-blades glabrescent but petioles always pubescent adaxially; leaf-margins mostly serrate; bracteoles 6–8; fruit formed but rarely so in Jamaica:
    15 Petals entire; bracteoles one-third to one-half as long as calyx; calyx divided about halfway; stipules more or less persistent   12. *rosa-sinensis*
    15 Petals laciniate; bracteoles very small, less than one-tenth as long as calyx; calyx shortly divided; stipules early deciduous   13. *schizopetalus*
  14 Herbs or undershrubs, not or rarely cultivated; leaf-blades stellate-pubescent at maturity; leaf-margins mostly dentate; bracteoles about 10; seeds with long cottony hairs:
    16 Bracteoles linear, mostly longer than calyx; corolla 1·5–2 cm. long, with a few large stellate hairs outside; androecium mostly included   14. *brasiliensis*
    16 Bracteoles mostly oblanceolate or spathulate and shorter than calyx:
      17 Calyx about 1 cm. long; corolla 1·5–2·5 cm. long, glabrous or nearly so outside; androecium clearly exserted   15. *pilosus*
      17 Calyx about 2 cm. long; corolla 2·5–4·5 cm. long, with numerous large stellate hairs outside; androecium included   16. *lavateroides*

1. H. vitifolius L., Sp. Pl. **2**: 696 (1753).—*Fioria vitifolia* (L.) Mattei (1917).

Herb or undershrub with lax branches, 60–150 cm. high; leaves 3–5-angled or -lobed, cordate, coarsely dentate or serrate, up to 13 cm. long and 12 cm. broad, with large 3-fid and smaller simple hairs; corolla 3–6 cm. long; capsule 10–15 mm. broad; seeds minutely warted, glabrous, about 3 mm. long.

Occasional in south-eastern parishes, a weed of waste ground; SL–100 feet; fl. and fr. most of the year. *A 8283! H & B 10794! P 24242!* Native and wide-spread in the Old World subtropics and tropics, naturalized in the drier parts of the West Indies.

2. H. tiliaceus L., Sp. Pl. **2**: 694 (1753).—*Pariti tiliaceum* (L.) Britton (1918). Seaside Mahoe.

Shrub 1–2 m. or tree up to 12 m. high, with weak dipping and ascending branches; leaves rounded-ovate, cordate, acuminate, entire or crenate, the midrib and central lateral nerves often with a split gland near the base beneath, 8–20 (–30) cm. long, up to about 25 cm. broad; calyx 1·5–3 cm. long; corolla 4–7 (–8·5) cm. long, lemon-yellow fading tawny-orange with or without a dark red spot at base; capsule 1·5–3 cm. long; seeds about 4 mm. long.

Rather local (Han., St. Ann, Port., St. Thom.), in brackish swamps and inner margins of mangroves; near SL; fl. Mar, Aug, Nov–Dec, fr. Nov–Dec. *A 11825! P 20698!* General throughout the tropics in seashore situations. A variable species which although quite distinct in E. Asia, approaches *H. elatus* closely in the West Indies.

3. H. elatus Sw., Nov. Gen. & Sp. Pl.: 102 (1788).—Blue Mahoe, Cuba Bark, Mountain Mahoe.

Typically a tall straight tree branching sparingly in the upper part, up to 20 (–25) m. high; leaves as of the last, variably hairy, sometimes glabrescent; stipules oblong, 2–4 cm. long, caducous; calyx mostly 3–5 cm. long; corolla 8–12 cm. long, orange-yellow or orange-red fading to deep crimson; capsule 3–4 cm. long; seeds nearly 5 mm. long.

Common but mostly as a result of planting; near SL–4000 feet; fl. and fr. all the year. *A 9171! 12803! Harris! H & P 14335! P 24662!* Native of Cuba and Jamaica, introduced elsewhere. Authors who reduce this species to infraspecific status in *H. tiliaceus* do so with some justification as there are strong affinities between Seaside Mahoe and Blue Mahoe in the West Indies, sufficient to suggest amal-gamation – at least on the basis of herbarium characters. The main distinguishing feature of deciduous calyx and epicalx does not always hold as the involucel tends to fall in Jamaican plants of both taxa, in overall contrast to typical Old World *H. tiliaceus*. Several plants of intermediate character are known occupying inland habitats and having the large size of *H. elatus* but the flower-colour and branching habit of *H. tiliaceus*. Pollen is not formed by the flowers of those intermediate-character trees which have been studied and it is believed that they are hybrids. It would seem that in our area *H. tiliaceus* is being diluted by introgression from *H. elatus* and if this is so, and there is also limited reproductive potential in their hybrids, the two species should be kept distinct.

4. H. trionum L., Sp. Pl. **2**: 697 (1753).

Herb 30–60 cm. high; lobes of upper leaves narrowly elliptical, crenate-serrate, up to about 3 cm. long; lower leaves entire, rounded; bracteoles 7–12; calyx 1·5–2 cm. long in fruit; corolla yellow with purple centre, 2–4 cm. long; seeds 2·5 mm. long.

Rare and local (St. Andr.), in waste places; 5000 feet; fl. and fr. Feb. *H 12328!* S. United States, a casual introduction from S. Europe and dry parts of the Old World tropics.

5. H. trilobus Aubl., Hist. Pl. Guiane **2**: 708 (1775).

Coarse prickly undershrub or shrub up to 3 m. high; leaves broadly ovate in outline, 3–5-lobed, cordate, the lobes acute or acuminate, 7–11 cm. long; bracteoles 12–14; calyx 3–3·5 cm. long; corolla reddish-purple, darker at base, 7–9 cm. long; seeds about 3·5 cm. long.

Rare (West., St. Eliz.), in swamps; SL–50 feet; fl. May–Aug, Dec, fr. May, Nov–Dec. *T. L. Lewis 1551! P 11107! W. Wright!* Hispaniola, Puerto Rico, Trinidad, French Guiana.

6. H. clypeatus L., Syst. Nat. ed. 10, **2**: 1149 (1759).—Congo Mahoe.

Erect or trailing-branched shrub up to 6 m. high; leaves angularly 3-lobed, cordate, subacuminate, up to about 15 cm. long and broad; bracteoles 9–11; calyx 3·5–4 cm. long; corolla dull tawny-yellow tinged and lined crimson within becoming more red on ageing, velvety, 4·5–6 cm. long; androecium declinate; seeds globose, about 3·5 mm. in diameter.

Frequent in the southern parishes, in open arid woodland and thickets on rocky limestone; 10–1000 feet; fl. June–Feb, fr. Aug–Apr. *A 8837! 10774! H 9060! P 6124! 9487!* Mexico, Greater Antilles, St. Croix; Cayman Is.

7. H. sabdariffa L., Sp. Pl. **2**: 695 (1753).—Jamaican Sorrel, Red Sorrel, Roselle, Sorrel.

Shrubby annual up to about 1·5 m. high, glabrous or glabrescent; leaf-lobes linear to elliptical, acute, serrulate, up to 15 cm. long and 4 cm. broad; flowers shortly peduncled in upper axils; calyx 1·5–2 cm. long in flower, increasing to 3–3·5 cm. long in fruit and becoming bright red; corolla light yellow with pinkish to maroon spot at base, 4–5 cm. long; capsule 2–2·5 cm. long; seeds about 5 mm. long.

Common in cultivation; near SL–2500 feet; fl. Aug–Nov, fr. Nov–Jan. *Henry! P 18443!* Native of the Old World tropics and widely cultivated.

8. H. cannabinus L., Syst. Nat. ed. 10, **2**: 1149 (1759).—Deccan Hemp.

Shrubby annual 1 m. or up to 4 m. high; leaves deeply divided into 3, 5 or 7 narrow serrate lobes; flowers axillary or racemose on short stout pedicels; bracteoles 7–10; calyx 12–25 mm. long, the lobes each with a swollen gland on midrib; corolla light yellow with reddish-purple spot at base, 4–8 cm. long; capsule up to about 2 cm. long; seeds about 5 mm. long.

Uncommon and mostly relict from past cultivation; 100–2300 feet; fl. and fr. Nov–Mar. *H 12326! P 18379!* Native of E. Indies, now general in the subtropics and tropics. Some of the plants in Jamaica lack the glands on calyx and leaves and have thus been identified as *H. radiatus* Cav. which is considered by most modern authors to be a distinct species.

9. H. bifurcatus Cav., Monad. Cl. Diss. Dec. **3**: 146, t. 51, f. 1 (1787).

Shrub 2–5 m. high with rather weak straggling branches; upper leaves bractlike and simple; lower leaves 3–5-lobed, the lobes deltate to lanceolate, serrate; flowers solitary in axils of upper leaves; bracteoles 9–13, unequally forked at tip; calyx

13–15 mm. long in flower, expanding to 2 cm. long in fruit; corolla purple or pink with crimson spots at base, 7–9 cm. long; capsule 2 cm. long; seeds glabrous, angular, 3–4 mm. long.

Rare (Port.), in swamp in forest; near SL; fl. and fr. Oct–Mar. *A 12351!* Mexico to Brazil, Cuba, Puerto Rico, Guadeloupe, Martinique, St. Vincent, Trinidad.

10. H. lunariifolius Willd. in L., Sp. Pl. ed. 4, 3 (1): 811 (1800).

Shrub with erect flowering branches covered with pungent hairs, 1–2 m. high; leaves broadly rounded to ovate in outline, up to 9 cm. long and 8 cm. broad; flowers axillary or in terminal racemes; bracteoles 5–6 rarely more; calyx 10–12 mm. long, expanding slightly in fruit; corolla light yellow tinged pink with a dark purple eye, 5–6 cm. long; capsule 15–18 mm. long; seeds about 3 mm. long.

Local (St. Andr.), on roadside banks and in gravelly waste places; 550–2500 feet; fl. Nov–Mar, fr. Jan–Mar. *A 10065! H 8260! P 24262!* Native of E. Indies and tropical Africa.

11. H. sororius L. f., Suppl.: 311 (1781).

Woody herb up to 1 m. high; stem and petioles with small appressed yellow stellate hairs; leaves suborbicular to broadly ovate, angled, stellate-hairy on both surfaces, 4·5–12 (–16) cm. long, up to about 13 cm. broad; flowers solitary in upper axils on peduncles up to 12 cm. long; bracteoles 8–10 up to 1·5 cm. long; calyx 2–3 cm. long; corolla pink with a darker spot at base, 5–7·5 cm. long; capsule 1·2–2·5 cm. long, black, bristly; seeds glabrous, 1·5–2 mm. long.

Very rare (West.), in floating mat of vegetation; 50–100 feet; fl. and fr. Nov. *P 27705!* Panama to Suriname and Argentina, Cuba, Guadeloupe.

12. H. rosa-sinensis L., Sp. Pl. **2**: 694 (1753).—Hibiscus, Shoe Black.

Shrub to 6 (–8) m. high; leaves ovate, up to about 15 cm. long and 9 cm. broad, often dark green and glossy adaxially, smooth or scabridulous; flowers solitary in the axils on short or long peduncles, showy; bracteoles 5–7, about 1 cm. long; calyx about 2 (–3) cm. long; corolla very variable in size and colour in the many cultivated varieties but commonly red and 6–10 cm. long; capsule oblong, about 3 cm. long; seeds globose, about 3 mm. in diameter.

Very common in cultivation and relict; SL–4000 feet; fl. all the year. Native origin uncertain but possibly tropical Asia.

13. H. schizopetalus (Mast.) Hook. f. in Curt., Bot. Mag. **106**: t. 6524 (1880).—
H. rosa-sinensis var. *schizopetalus* Mast. (1879).

Shrub like the last but rather smaller in most vegetative parts; peduncles usually longer and more slender, up to 14 cm. long, with the flowers pendulous; calyx 1·5–2 cm. long; corolla usually pink or red, the lobes reflexed, up to about 7 cm. long; staminal column greatly exserted and pendulous, bearing anthers distally only; capsule oblong; seeds glabrous.

Common in cultivation. Native origin uncertain but possibly E. tropical Africa. Many commonly cultivated ornamental varieties of *Hibiscus* comprise elements of both *H. rosa-sinensis* and *H. schizopetalus* combined by hybridization.

14. H. brasiliensis L., Sp. Pl. ed. 2, **2**: 977 (1763).—*H. phoeniceus* Jacq. (1776).

Annual herb or shrub up to 1·5 m. high; leaves deltate to ovate, truncate at base, subacuminate, 3–5 (–9) cm. long, up to 4·5 cm. broad; flowers solitary in axils; peduncles 1–5 cm. long; calyx 7–12 mm. long; corolla rose-crimson or rarely white; capsule subglobose, shorter than calyx.

Rare in cultivation but also possibly indigenous; fl. and fr. most of the year. *H 11873!* Mexico to Colombia and Venezuela, Greater Antilles, Virgin Is., Antigua, Guadeloupe, St. Vincent, Curaçao.

15. H. pilosus (Sw.) F. & R., Fl. Jam. **5**: 137 (1926).—*Achania pilosa* Sw. (1788)

Undershrub 45–150 cm. high; leaves triangular-ovate, openly cordate to subtruncate at base, obtuse at tip, 2–8 cm. long, up to 5·5 cm. broad; flowers solitary in axils on long peduncles; calyx 9–15 mm. long; corolla vermilion to dark crimson; capsule longer than calyx.

Occasional in the northern and western parishes, on rocky limestone near the sea; 10–1000 feet; fl. and fr. Aug–Mar. *Cornman 671! H 10379! P 11764! Stearn 463!* Mexico, Cuba.

16. H. lavateroides Moric. in Mém. Soc. Phys. & Hist. Nat. Genève **7**: 263, t. 16 (1836).

Shrubby herb 60–120 cm. high; leaves deltate-ovate, cordate to truncate at base, acute, acuminate or obtuse at tip, irregularly serrate, 3–9 cm. long, up to about 6 cm. broad; flowers solitary in upper axils; calyx 17–22 mm. long; corolla pink with whitish veins; capsule shorter than calyx.

Rare (Manch., St. James, St. Ann), in thickets on limestone and on roadside banks; 300–1750 feet; fl. and fr. Jan–Mar. *A 8737! E. G. Britton 2898! P 24752!* Mexico, Honduras.

17. ABELMOSCHUS Medic. (1787)

Stem with reflexed bristly hairs; peduncles mostly 3 cm. or more long; capsule ovoid, membranous, not markedly ribbed, usually hispid, 5–8 cm. long   1. *moschatus*
Stem glabrous or almost so; peduncles 5–15 mm. long in flower, up to 25 mm. long in fruit; capsule oblong, fleshy becoming fibrous, usually sulcate, at first densely hirsute, glabrescent, (5–) 8–16 (–25) cm. long   2. *esculentus*

1. A. moschatus Medic., Malv.-Fam.: 46 (1787).—*Hibiscus abelmoschus* L. (1753). Musk Okra.

Undershrub up to 1·5 (–3) m. high; leaves very variable, suborbicular in outline to broadly elliptical, up to 15 (–20) cm. long and 18 (–24) cm. broad, usually cordate at base, angular or up to 7-lobed, the lobes deltate to narrowly ovate, acute or acuminate, serrate-dentate, hirsute or pubescent; bracteoles 8–10, 10–17 mm. long; calyx 2–3·5 cm. long; corolla yellow with a dark purple spot at base, 4–8 cm. long; seeds glabrous, striate, musky-scented, 3–5 mm. long.

Uncommon, sparingly cultivated as a fibre plant and naturalized at low altitudes; fl. Sept–Nov. *McNab! Tomlinson!* Native of S.E. Asia, introduced elsewhere.

2. A. esculentus (L.) Moench, Meth. Pl.: 617 (1794).—*Hibiscus esculentus* L. (1753). Okra.

Herbaceous undershrub 1–2 (–4) m. high; leaves mostly suborbicular in outline and up to about 20 cm. long and broad, cuneate to cordate at base, often 5-lobed to the middle, the lobes deltate-ovate to narrowly elliptical, acute, serrate, glabrous or with a few stiff hairs on veins beneath; bracteoles 8–12, 10–15 mm. long; calyx 1·5–2·5 cm. long; corolla yellow with a dark red or purple spot at base, 3–6 cm. long; seeds glabrous or puberulous or pilose, 3–6 mm. in diameter.

Common in cultivation. Native of the Old World tropics and now widespread in all warm countries.

123. BOMBACACEAE

Trees, sometimes with thick trunks. Leaves alternate, simple or digitately compound often lepidote; stipules free, caducous. Flowers bisexual, actinomorphic, large and showy, often bracteate, solitary or clustered. Perianth typically biseriate, sometimes subtended by an epicalyx or involucre; sepals 5, free or basally connate, the lobes valvate, sometimes closed in bud and opening irregularly; petals 5, contorted, or absent. Stamens 5–many, free or monadelphous; anthers 1 (–2)-locular, opening lengthwise; staminodes often present. Ovary superior, 2–5-locular; ovules anatropous, 2 or more in each loculus, erect on axile placentas; style simple, capitate or lobed; stigmas 1–5. Fruit a loculicidal capsule or indehiscent or berrylike. Seeds smooth, sometimes embedded in pith or wool derived from the endocarp; cotyledons flat, plicate or contorted; endosperm scanty or absent.

About 25 genera with some 150 species, mostly in the American tropics.

1 Leaves simple, cordate, subentire to toothed, angular or lobed; anthers spirally twisted
and adnate to the distal part of the staminal tube; stigmas 5, spirally twisted; capsule
elongated, 5-valved with brownish cottony hairs within **1. Ochroma**
1 Leaves digitately compound:
  2 Calyx deeply 5-lobed, 5–8 cm. long, tomentose; petals white; anthers very numerous,
free at the apex of a broad staminal tube; fruit massive, indehiscent, yellowish-
brown tomentose, pendulous on a long peduncle; seeds embedded in a dry pith;
occasional in gardens; native of tropical Africa; Baobab Tree
Adansonia digitata L.
  2 Calyx truncate and cupular to shallowly 5-lobed; fruit usually dehiscent:
    3 Calyx shallowly 5-lobed; flowers fascicled; stamens 5–15 in 5 bundles of 1–3; petals
creamy-white; trunk often spiny **2. Ceiba**
    3 Calyx cupular; flowers solitary; stamens numerous; planted ornamental trees:
      4 Petals erect or spreading, red; trunk usually spiny; branches whorled; native of S.E.
Asia; Red Silk Cotton Tree *Bombax malabaricum* DC.
      4 Petals strongly recurved and coiled, purplish; stamens deep pink; trunk smooth,
green; native of Mexico *Bombax ellipticum* Kunth

1. OCHROMA Sw. (1788)

1. O. pyramidale (Cav.) Urb. in Fedde, Repert. Sp. Nov., Beih. **5**: 123 (1920).—
Bombax pyramidale Cav. (1788). *O. lagopus* Sw. (1788). Balsa, Bombast
Mahoe, Corkwood, Down Tree.

Tree 6–20 m. or more high; wood very light and spongy; leaf-blade up to about
35 cm. or more long and broad; flowers terminal, pedicellate, subtended by a
deciduous involucre of 3 bracteoles 1·5–2 cm. long; calyx 7–9 cm. long; petals
12–14 cm. long, white or yellowish, often tinged reddish; ripe capsule to 30 cm. or
more long; seeds obovoid.

Locally common in disturbed sheltered moderately wet areas; 500–1500 (–3000)
feet; fl. Jan–Mar, fr. Mar. *H 11962! P 20728! A. von der Porten!* Mexico to
Brazil, West Indies.

2. CEIBA Mill. (1754)

1. C. pentandra (L.) Gaertn., Fruct. & Sem. Pl. **2**: 244, t. 133 (1791).—*Bombax
pentandrum* L. (1753). Silk Cotton Tree.

Deciduous or partly deciduous tree 10–40 m. high, with or without prickles on
the trunk and branches; trunk buttressed; leaflets 5–8 (–15), lanceolate to oblanceolate, acutely acuminate, cuneate at sessile base, 10–20 cm. long, 3–4 cm. broad,
glabrous; pedicels about 3 cm. long; calyx 1–1·5 cm. long, glabrous; petals 2–4 cm.
long, silky-villous outside; capsule 10–30 cm. long; seeds subglobose, 4–6 mm. in
diameter.

Occasional, perhaps mostly planted; SL–2500 feet; fl. Jan–Feb, fr. Mar. *H 8448
12333. HPS 14727!* Native of tropical Amer., now distributed throughout the
tropics.

124. STERCULIACEAE

Trees, shrubs or herbs. Leaves alternate, simple or digitately compound, often
with stellate indumentum; stipules present, usually caducous. Flowers generally
bisexual, rarely unisexual and monoecious, usually actinomorphic, solitary or in
various types of mostly axillary, sometimes cauline, inflorescences. Perianth typically biseriate; sepals 3–5, united, the lobes valvate; petals 5, free or adnate to the
base of the androecium, contorted in bud, or reduced or absent. Stamens in 2
whorls, the outer staminodal, free or more usually monadelphous in a tubular
column with or without the sterile filaments alternating with solitary or grouped
anthers; anthers 2-locular, opening lengthwise, sometimes coherent. Ovary superior
with 1–5 more or less united carpels with as many loculi or loculi rarely 10–12;
ovules (1–) 2–several in each loculus, anatropous, on axile placentas; styles as many as
carpels, free or more usually connate or style simple. Fruit a leathery or woody
berry or more usually the carpels separating as dehiscent dry cocci or woody
follicles. Seeds with straight or curved embryo, with or without endosperm.
About 70 genera with 1000 mostly tropical species.

1 Petals absent; flowers unisexual; ovary apocarpous; fruiting carpels follicular; trees:
    2 Anther-loculi in 2 superposed whorls; seeds usually without endosperm, compact and
        angular, pinkish; ripe carpels greenish-brown; leaves elliptical, entire; cultivated;
        native of W. tropical Africa; Bissy, Cola   *Cola acuminata* (Beauv.) Schott & Endl.
    2 Anther-loculi irregularly arranged; seeds usually with endosperm, loose and oblong-
        rounded, blackish; ripe carpels red or brownish-red; cultivated ornamental trees:
        3 Leaves simple, palmately divided; fruiting carpels red; seeds greyish-black; native of
            tropical Amer.; French Peanut   *Sterculia apetala* (Jacq.) Karst.
        3 Leaves digitately compound; fruiting carpels reddish-brown; seeds bluish-black;
            native of tropical Asia; Java Olive   *Sterculia foetida* L.
1 Petals present; flowers bisexual; ovary syncarpous:
    4 Androgynophore present; follicular carpels twisted in fruit; petals flat, soon falling;
        shrub or small tree   1. **Helicteres**
    4 Androgynophore absent; fruit not twisted:
        5 Trees; fruit indehiscent; petals hooded and appendaged:
            6 Petals with a linear 2-cleft appendage; flowers in panicles, mostly on the young
                branches; fruit subglobose, tuberculate   2. **Guazuma**
            6 Petals with a spathulate appendage; flowers mostly in clusters on the trunk and
                older branches; fruit a large ellipsoid leathery berry; native of continental
                tropical Amer.; Chocolate Tree, Cocoa   *Theobroma cacao* L.
        5 Shrubs or herbs, rarely trees; fruit capsular:
            7 Ovary and fruit distinctly stalked (except in *A. magna*), 5-locular; petals hooded,
                caducous; capsule prickly   3. **Ayenia**
            7 Ovary and fruit sessile or only shortly stalked; petals flat, more or less persistent:
                8 Native shrubs or herbs with small flowers; stipules small, up to about 8 (–10) mm.
                    long, often falling early:
                    9 Ovary 5-locular; capsule 5–10-seeded; flowers heterostylous (except in *M.
                        manducata*)   4. **Melochia**
                    9 Ovary 1-locular; capsule 2-valved, 1-seeded; flowers homostylous
                        5. **Waltheria**
                8 Introduced shrubs with showy flowers and mostly rather large coarse leaves:
                    10 Inflorescence extra-axillary, few-flowered; petals red; fruit a large (4–) 5-winged
                        capsule; stipules 5–10 mm. long; naturalized in moist thickets in St. Thomas;
                        native of tropical Asia and Australasia   *Abroma augustum* (L.) L.f.
                    10 Inflorescences axillary, corymbose-umbellate, many-flowered; fruits not formed
                        in Jamaica; stipules 1 cm. or more long; cultivated ornamentals flowering
                        mostly November–January:
                        11 Petals white; peduncles stiff, spreading; bracts and stipules early deciduous;
                            native of E. tropical Africa   *Dombeya mastersii* Hook. f.
                        11 Petals pink; peduncles long, pendulous; bracts and stipules persistent, stipules.
                            2 cm. or more long; hybrid of garden origin   *Dombeya × cayeuxii* André

1. HELICTERES L. (1753)

1. H. jamaicensis Jacq., Enum. Syst. Pl. Carib.: 30 (1760).—Screw Tree.
    Shrub or straggly-branched tree up to 5 m. high; twigs slender, often drooping,
densely tomentose; leaves broadly ovate, cordate, doubly dentate, shortly acuminate, up to 20 cm. long and 15 cm. broad, usually smaller; peduncles few-
flowered; calyx unequally toothed; petals white; stamens 10; staminodes 5; fruit
2–4·5 cm. long on a carpophore up to 10 cm. long, densely tomentose.
    Common in thickets and woodlands mainly in arid coastal areas on limestone;
SL–1800 feet; fl. Apr–Dec, fr. all the year. *A 5631! H 9632! P 7500!* Bahamas,
Greater Antilles, Virgin Is.; Grand Cayman.

2. GUAZUMA Mill. (1754)

1. G. ulmifolia Lam., Encycl. Méth. Bot. **3**: 52 (1789).—Bastard Cedar.
    Tree 5–10 (–16) m. high with spreading branches; leaves lanceolate to ovate,
unequal at base, crenate-dentate, very variably hairy from subglabrous to tomentose, mostly less than 10 cm. long and 5 cm. broad but sometimes much larger
especially on vegetative shoots; panicles axillary, rather small; sepals mostly 3,
reflexed, green turning buff; petals clawed, yellow, the appendages purple; stamens
fused in groups of 2 or 3; fruit black, subglobose, up to about 2·5 cm. in diameter.

Very common along roadsides, in pastures and open secondary woodlands; SL–1250 (–2000) feet; fl. May–Feb, fr. Nov–Apr. *A 5518! 6815! H 11008! P 20608!* Throughout tropical and subtropical continental Amer. and the West Indies, introduced into Asia and W. tropical Africa.

3. AYENIA L. (1756); Cristobal (1960)

1 Leaves crenate-dentate; short-lived taprooted subwoody herbs or undershrubs; twigs, petioles and pedicels distinctly hairy:
  2 Leaves ovate, cordate, caudate-acuminate, mostly up to 8 cm. long and 5 cm. broad; twigs and leaves pubescent; petals greenish; capsule 5–6 mm. long and about 8 mm. broad   1. *magna*
  2 Leaves oblong-lanceolate to suborbicular, up to 3 cm. long and 1·5 cm. broad; twigs, petioles and pedicels curled-puberulous; petals purple; capsule 3 mm. long and about 4 mm. broad   2. *ardua*
1 Leaves ovate-lanceolate, entire; perennial shrubs or trees; petals dark red; capsule 10–12 mm. long; young twigs, petioles and calyx with a few small stellate hairs otherwise plants glabrous:
  3 Leaves mostly obtuse or acute at tip, 2–6 cm. long, 1–2·5 cm. broad   3a. *laevigata* var. *laevigata*
  3 Leaves mostly acuminate at tip, (5–) 7–11 cm. long, (2–) 3–4·5 cm. broad   3b. *laevigata* var. *acuminata*

1. A. magna L., Syst. Nat. ed. 10, **2**: 1247 (1759).
Annual or short-lived perennial shrubby herb 1–2 m. high; flowers numerous in axillary stalked cymes; sepals about 3 mm. long; ovary and fruit subsessile.
Local, in thickets on arid limestone in the southern parishes; 20–800 feet; fl. Oct–Feb, fr. Oct–Mar. *A 12653! 12879! H 6950! P 24195!* Northern S. Amer., Curaçao, Bonaire. Loveless, A. R. & G. F. Asprey in Journ. Ecol **45**: 806 (1957) describe a chamaephytic mode of behaviour for this species correlated with marked seasonality of rainfall.

2. A. ardua Cristobal in Opera Lilloana **4**: 128 (1960).—*A. pusilla* of F. & R. (1926) and other authors, not L. (1759).
Wiry-stemmed taprooted undershrub branching from the base, up to about 40 cm. high; flowers solitary or 2–3 together; sepals 2 mm. long; ovary and fruit on a stalk about 2 mm. long.
Rare and local (St. Andr., St. Cath.), in weedy thickets in areas of low rainfall; 20–60 feet; fl. and fr. Jan, June–Aug. *H 12069! P 25705!* Mexico, Cuba, Hispaniola.

3. A. laevigata Sw., Nov. Gen. & Sp. Pl.: 97 (1788).

3a. Var. laevigata.
Shrub or tree 2–5 m. high; leaves rather leathery; inflorescences mostly 3-flowered; sepals about 4 mm. long; petals with 2 lateral teeth apically, about 4 mm. long; ovary and fruit on a stalk about 1·5 mm. long; seeds unequally fusiform, pointed at one end, 7 mm. long.
Rare and local (St. Andr., St. Thom.), in thickets and woodlands on arid limestone hills; 150–750 feet; fl. Jan–Feb, May–July, fr. Jan, May–July, Oct. *H 8932! 9612!* Endemic.

3b. A. laevigata var. acuminata Adams in Phytologia 20 (5): 309 (1970).
Like the last but more robust and larger in all the vegetative parts.
Rare and local (Trel.), in thickets on rocky limestone hillside; 1500–1770 feet; fl. Apr–May, fr. Apr–July. *Cornman 795! H & P 14384! R 1909!* Endemic.

4. MELOCHIA L. (1753); Goldberg (1967)

1 Fruit pyramidal and 5-winged; petals pink to purple:
  2 Capsule rounded on the outer angles, the wings obscure; branches trailing, tomentose; leaves finely crenate, rounded at apex; inflorescences extra-axillary   1. *crenata*
  2 Capsule prominently winged; branches erect or ascending; leaves coarsely dentate, mostly acute at tip:

3 Inflorescences mostly axillary; leaves densely tomentose and canescent, with stellate hairs; shrub with woody twigs   2. *tomentosa*
3 Inflorescences mostly leaf-opposed; leaves subglabrous; branches usually sparsely pubescent with simple, forked or stellate hairs; weedy undershrub with more or less herbaceous twigs   3. *pyramidata*
1 Fruit subglobose or ovoid, mostly shallowly 5-sulcate, variously dehiscent but cocci eventually separating:
  4 Petals bright yellow or orange-yellow; flowers in pedunculate open or contracted panicles along a rather leafless terminal inflorescence   4. *pilosa*
  4 Petals not at all or only tinged or spotted yellow:
    5 Calyx accrescent in fruit, turning brown; flowers distinctly pedicellate in axillary umbellate clusters; petals white with a yellow spot at base   5. *lupulina*
    5 Calyx not or hardly accrescent:
      6 Stipules, bracts and calyx-lobes mostly ovate or deltate; inflorescences axillary or distally at leafless nodes, tightly compact and rounded; petals usually white with crimson veins   6. *nodiflora*
      6 Stipules, bracts and calyx-lobes mostly lanceolate to linear-lanceolate; petals mauve, pink or purple:
        7 Stems and leaves with numerous long spreading sometimes glandular hairs; inflorescences predominantly terminal and interrupted-spiciform   7. *villosa*
        7 Stems glabrous or with a line of short hairs; leaf-blade subglabrous; inflorescences axillary and terminal   8. *manducata*

1. M. crenata Vahl, Symb. Bot. **3**: 86, t. 68 (1794).

Shrubby herb with trailing prostrate branches up to 2 m. long; stipules ovate to lanceolate, 2·5–3·7 mm. long; leaves ovate, rather thick, the veins prominent beneath, light greyish-green, 1·4–3·5 cm. long, 1–2·5 cm. broad; pedicels 4–7 mm. long; capsule about 10 mm. long, beaked.

Very local (St. Andr.), on sand near the sea; SL–10 feet; fl. and fr. Jan–Feb, July–Aug. *Asprey UCWI 1928! Cornman! J. P. 1002! McDonnell 3025!* Colombia.

2. M. tomentosa L., Syst. Nat. ed. 10, **2**: 1140 (1759).—Raichie, Tea Bush.

Shrub 1–4 m. high; stipules deltate to linear, 2–9 mm. long; leaves ovate to lanceolate, rounded-subcordate at base, acute to rounded at tip, silvery-grey, 2–6 cm. long, 1–4 cm. broad; pedicels 2–6 mm. long; capsule 7–10 mm. long, long-beaked.

Frequent in thickets on arid limestone and well drained waste places; SL–1000 feet; fl. and fr. most of the year. *A 6200! H 7222! Powell 1028!* Florida, Texas, Bahamas, Mexico to Brazil, West Indies; Cayman Is.

3. M. pyramidata L., Sp. Pl. **2**: 674 (1753).

Shrubby herb with straggling branches, 30–120 cm. high; stipules deltate to lanceolate, 1·5–5 mm. long; leaves ovate to lanceolate, mostly rounded at base, acute to rounded at tip, 2–6 (–10) cm. long, 1–3 (–7) cm. broad; pedicels 1–5 mm. long; capsule 5–9 mm. long, shortly beaked.

Common in open grassy places on rocky ground; SL–1500 feet; fl. and fr. most of the year. *A 6716! H 12751! Proctor & Powell 20540!* S. United States, Bahamas, Mexico to Argentina, West Indies; Grand Cayman; casually introduced into the Old World tropics.

4. M. pilosa (Mill.) F. & R., Fl. Jam. **5**: 164 (1926).

Undershrub with few long slender branches, up to 3 m. high; stipules deltate to lanceolate, 3–10 mm. long; leaves ovate to lanceolate, subcordate at base, usually acute at tip, more or less hairy, 2–9 (–11) cm. long, 1–7·5 cm. broad; pedicels mostly 1–3 mm. long; capsule 4–6 mm. long, hardly beaked.

Local and uncommon (St. Andr., Clar.), on open steep shale banks; 1000–1700 feet; fl. Apr–Oct, fr. May–Oct. *A 7180! H 9026! P 19815!* Mexico, Colombia to Argentina.

5. M. lupulina Sw., Nov. Gen. & Sp. Pl.: 97 (1788).

Weak-stemmed shrubby herb 40–90 cm. high or sometimes scrambling to 3 m.; stipules deltate, acuminate, 3·5–8 mm. long; leaves ovate, rounded to cordate at base, acute to subacuminate at tip, 3–12 cm. long, 2–8 cm. broad; pedicels 1–5 mm. long; capsule globose, 2·5–3 mm. long, not beaked.

Frequent in the extreme eastern and western parishes, on steep roadside banks and in waste places; 50–2700 feet; fl. Dec–Apr, fr. Jan–Apr. *A 8517! H 10328! P 23323! Stearn 485!* Mexico to Peru.

6. **M. nodiflora** Sw., Nov. Gen. & Sp. Pl.: 97 (1788).
   Shrub 0·5–2·5 (–4) m. high; stipules deltate to ovate, 1·5–5 mm. long; leaves ovate to ovate-lanceolate, rounded or cordate at base, acute at tip, with a few scattered hairs, 3–8 (–11) cm. long, 1·5–5 (–7) cm. broad, often mottled pink and white; peduncles obsolete; pedicels up to 3 mm. long but often wanting; fruit depressed-globose, 3–3·5 mm. long, not beaked, often reddish.
   Common on limestone roadside and gully banks and in gravelly waste places; SL–1500 feet; fl. and fr. Nov–May. *A 5645! H 6697! P 24370!* Mexico to Brazil, Bahamas, West Indies; Grand Cayman.

7. **M. villosa** (Mill.) F. & R., Fl. Jam. **5**: 165 (1926).
   Shrubby herb 0·3–1·5 (–2) m. high; stipules deltate to lanceolate or linear, mostly 4–8 mm. long; leaves mostly ovate, rounded to subcordate at base, acute at tip, 2–10 cm. long, 1–8 cm. broad; peduncles and pedicels obsolete or very short; capsule depressed-globose, about 3 mm. long, sometimes shortly beaked.
   Local, in the central and western parishes, in open grassy places on wet heavy soils; 10–2400 feet; fl. and fr. most of the year. *A 10412! H 12251! P 9720!* Florida, Mexico to Paraguay, Greater Antilles, Trinidad.

8. **M. manducata** C. Wright in Sauv. in Anal. Acad. Cienc. Habana **5**: 241 (1868).
   Shrubby herb up to 1·5 m. high; stipules lanceolate, 3–7·5 mm. long; leaves lanceolate to broadly ovate, rounded at base, acute at tip, 1–5 (–7) cm. long, 1–6 cm. broad; peduncles and pedicels obsolete or each up to about 15 mm. long; capsule globose, about 3 mm. long, not beaked.
   Very rare (West.), among weeds at edge of temporary pond; 50–100 feet; fr. Nov. *P 27716!* Mexico to Brazil, Cuba, Hispaniola.

5. WALTHERIA L. (1753)

Leaves ovate to lanceolate, mostly 3–9 cm. long; calyx about 5 mm. long   1. *indica*
Leaves obovate to suborbicular, 1–3·5 cm. long; calyx 7–8 mm. long   2. *calcicola*

1. **W. indica** L., Sp. Pl. **2**: 673 (1753).—*W. americana* L. (1753). Raichie.
   Shrubby herb or subarborescent 0·5–2 m. high; leaves rounded or subcordate at base, usually obtuse at tip, serrate, tomentose, paler beneath, (1–) 3–9 (–10) cm. long, 1–4 (–5) cm. broad; flowers crowded in axillary sessile or stalked inflorescences; petals yellow, about 6 mm. long; fruit about 2·5 mm. long.
   Common in open sandy ground and waste places, especially near the sea; SL–2250 feet; fl. and fr. all the year. *A 8201! J.P. 1460! Powell 1029!* General in the subtropics and tropics.

2. **W. calcicola** Urb., Symb. Ant. **1**: 475 (1900).
   Shrub 1–2 m. high; leaves rounded or subtruncate at base, rounded at tip, crenulate, tomentose, 1–3·5 cm. long, 5–25 mm. broad; flowers in small heads at the ends of branches; petals yellow, about 8 mm. long; fruit not known.
   Rare and local (St. Cath.), in evergreen thicket on arid limestone; about 300 feet; fl. Nov–Mar. *H 10155! Skelding & Robertson UCWI 3528!* Hispaniola, Puerto Rico.

125. THYMELAEACEAE

Trees and shrubs, often with strongly fibrous inner bark, or rarely herbs. Leaves alternate or opposite, simple, entire, without stipules. Flowers bisexual or unisexual and mostly dioecious, actinomorphic or rarely slightly zygomorphic, solitary or more usually in terminal or axillary heads, umbels, spikes or racemes. Calyx usually tubular (hypanthium) and often corolloid, with 4–5 imbricate lobes;

petals small and scalelike, 4–12, inserted usually at or near the mouth of the hypanthium tube, or absent. Stamens as many as sepals and alternate with them or twice as many, or 2, inserted on the hypanthium tube; anthers 2-locular, introrse, opening lengthwise. Cupular or annular hypogynous disk or scalelike nectaries often present. Ovary superior, 1 (–2)-locular; ovule 1 in each loculus, anatropous, pendulous; style simple or lacking; stigma capitate or discoid. Fruit a drupe or nut, rarely a capsule or berry. Seeds with straight embryo; endosperm present or absent.

Over 40 genera and 600 species with wide distribution, but concentrated in S. Africa, Australia, the Mediterranean region and central and western Asia.

Jamaican genera have: leaves alternate; flowers 4-merous, without petals; stamens 8 in 2 series; ovary 1-locular:

Flowers bisexual in spikes; fruit dry, enclosed by persistent perianth, red	1. **Lagetta**
Flowers unisexual, dioecious or monoecious, in heads or umbels; fruit drupaceous, not enclosed, white	2. **Daphnopsis**

1. LAGETTA Juss. (1789)

1. L. lagetto (Sw.) Nash in Journ. New York Bot. Gard. **9**: 117 (1908).—*Daphne lagetto* Sw. (1788). *L. lintearia* Lam. (1792). Lace Bark, Lagetto.
Tree 5–12 m. high; leaves ovate to elliptical, shortly acuminate, more or less rounded at base, 4–15 cm. long, up to 10 cm. broad; spikes 4–10 cm. long, with 10–25 flowers; fruit 5 mm. long.
Occasional, mostly in the central parishes, in woodlands on limestone hills; 1500–2600 feet; fl. Apr–Nov, fr. May–Sept. *A 11068! H 8762! H & P 14398!* Cuba, Hispaniola.

2. DAPHNOPSIS Mart. & Zucc. (1824)

Peduncles simple; perianth-lobes acute	1. *occidentalis*
Peduncles dichotomously branched; perianth-lobes obtuse	2a. *americana* ssp. *cumingii*

1. D. occidentalis (Sw.) Krug & Urb. in Engl., Bot. Jahrb. **15**: 349 (1892).— *Daphne occidentalis* Sw. (1788). Burn Nose.
Shrub 1–3 m. or tree 4–12 m. high; leaves lanceolate to oblanceolate, rounded to long-acute or acuminate at tip, long-tapered to base, 5–11 (–15) cm. long, 1·5–3·5 cm. or more broad, light green, paler beneath; peduncle 1–5 cm. long; flowers greenish- to yellowish-white, fragrant; male perianth-tube funnel-shaped, about 8 mm. long, lobes 4 mm. long; anthers orange; female perianth-tube campanulate, 4 mm. long, lobes 2–2·5 mm. long; ripe fruit ellipsoid, 14–18 mm. long, with sticky pulp.
Widespread but not common, in open or shady woodlands and moist thickets on limestone; 1600–3000 feet; fl. Apr–Dec, fr. Apr–Oct. *A 7536! 11776! H 8967! 11218! P 10257! 10585!* Endemic.

2. D. americana (Mill.) J. R. Johnston in Contrib. Gray Herb., n.s. **37**: 242 (1909).—*Laurus americana* Mill. (1768). *D. tinifolia* (Sw.) Meisn. (1857).

2a. Ssp. cumingii (Meisn.) Nevling in Journ. Arn. Arb. **41**: 413 (1960).— *D. tinifolia* var. *cumingii* Meisn. (1857). Burn Nose.
Shrub 3 m. or more usually a tree up to 20 m. high; bark smooth, dull brown; more or less deciduous in dry areas; leaves oblong-lanceolate to elliptical, acute or acuminate at tip, cuneate at base, 6–12 cm. long, up to 6 cm. broad; inflorescences terminal, mostly 3-branched; flowers greenish- to yellowish-white, fragrant; male perianth-tube 9 mm. long, lobes recurved, 2·5 mm. long; anthers orange; female perianth-tube 3·5–5 mm. long, lobes spreading, 1 mm. long, stigma exserted; ripe fruit ovoid, crowned by persistent style-base, 9–13 mm. long, with sweet pulp.
Rather common, in thickets and woodland margins in rocky or sandy or gravelly alluvial places; 450–5000 feet; fl. Apr–Dec, fr. July–Feb. *A 5615! 6173! H 6478! 11084! P 22674! 26622!* Hispaniola.

126. VIOLACEAE

Shrubs or herbs, usually perennial. Leaves alternate or sometimes opposite, simple; stipules leafy or small. Flowers usually bisexual, zygomorphic or actinomorphic, solitary to paniculate. Perianth biseriate; sepals 5, usually free, imbricate; petals 5, hypogynous or perigynous, the lower often spurred, imbricate or contorted. Stamens 5, hypogynous or perigynous; anthers erect, closely connivent around the ovary, one often spurred, 2-locular, opening lengthwise. Ovary superior, 1-locular; carpels 3–5; ovules anatropous, 1–2–many, on parietal placentas; style simple. Fruit a loculicidal capsule or a berry. Seeds with straight embryo and fleshy endosperm, sometimes winged or tomentose, sometimes arillate.
16 genera with 850 species, in all regions.

1. VIOLA L. (1753)

Herbs; leaves alternate; stipules leafy; flowers solitary or paired in leaf-axils, zygomorphic; sepals produced at base; capsule elastically 3-valved; seeds smooth.

Stem short; a perennial rosette herb; stipules entire, linear or linear-lanceolate; sepals
    lanceolate   1. *patrinii*
Stem erect or ascending; annual herb; stipules leafy, pinnately lobed; sepals with large
    auricles   2. *tricolor*

1. V. patrinii DC., Prodr. **1**: 293 (1824).
Taprooted herb with woody stem; petioles up to 15 cm. long; leaf-blade ovate to oblong, hastate and cordate and decurrent on petiole at base, margin crenate or serrate, up to 8 cm. long and 6 cm. across the basal lobes; corolla bluish-purple; spur usually short and broad, pink; capsule 6–12 mm. long; seeds pyriform, light brown.
Locally common (St. Andr., Port., St. Thom.), a weed in open places and on mossy banks; 3750–6600 feet; fl. Dec–Apr, fr. sporadically. *A 10605! A & S 3414! H 10928! P 23268!* Native of C. Asia.

2. V. tricolor L., Sp. Pl. **2**: 935 (1753).—Heartsease.
Basally branched lax-stemmed herb 10–30 cm. or more high; leaves ovate to lanceolate, long-petioled, coarsely crenate-serrate, 1·5–4 cm. long; petals flat, purple and yellow; spur thick, blunt; capsule 5–8 mm. long, open valves broadly elliptical, up to 4 mm. broad; seeds ellipsoid, obtuse at one end, acute at the other, about 1·5 mm. long, whitish or reddish.
Very rare (St. Andr.), a weed in fields; 4900–5300 feet; fl. and fr. Feb, May. *H 11976! Powell 1593!* Native of Eurasia and N. Africa, also in N. Amer. and Hispaniola.

127. FLACOURTIACEAE

Trees or shrubs. Leaves alternate, rarely opposite, simple; stipules small and caducous or absent. Flowers bisexual or unisexual and monoecious, polygamous or dioecious, actinomorphic, often with jointed pedicels, in cymose or rarely racemose terminal or lateral inflorescences. Perianth biseriate or uniseriate; sepals 2–15, free or shortly united, imbricate, rarely valvate; petals as many as or more than the sepals or absent, imbricate, with or without a basal scale. Stamens as many as sepals or more numerous, free or in bundles, sometimes alternating with staminodes; anthers 2-locular, opening lengthwise. Disk often present. Ovary usually superior, sometimes half-inferior or inferior, unilocular; placentas 2–10, parietal; ovules numerous on each placenta; style 1 or as many as the carpels (2–10). Fruit a loculicidal capsule or rarely indehiscent or a berry. Seeds sometimes arillate, with straight embryo and usually abundant endosperm.
84 genera with about 1000 species of subtropical and tropical distribution.

1 Inflorescence racemose or racemose-paniculate; leaves not gland-dotted; stipules very
   small or wanting; flowers bisexual; fruit capsular:
2 Petals present, persistent; stamens in small bundles opposite petals, alternating with
   glands; leaves crenate-serrate to subentire, pinnately nerved; seeds exarillate
          1. Homalium
2 Petals absent; stamens in a ring attached to a lobed cup-shaped disk; leaves entire,
   3-nerved from the base; seeds arillate   **2. Lunania**
1 Inflorescence cymose or flowers solitary or in shortly stalked or sessile clusters; petals
   absent:
3 Flowers dioecious or polygamous; leaves not gland-dotted, without stipules; often
   with slender axillary spines on the twigs, these thicker and branched on the lower
   stem; fruit a unilocular berry; seeds exarillate   **3. Xylosma**
3 Flowers bisexual; leaves gland-dotted, usually serrate or crenate-margined; stipules
   present, small, caducous; fruit a fleshy or dry capsule; seeds arillate:
4 Staminodes present, alternating with the stamens in a ring:
5 Stamens usually 8 or 10 (6–15); style short, stigma capitate or 3-cleft
          4. Casearia
5 Stamens 20 or more; stigma sessile or nearly so, peltate   **5. Zuelania**
4 Staminodes absent:
6 Stamens hypogynous, numerous; calyx reflexed   **6. Laetia**
6 Stamens perigynous, (8–) 10–12 (–13); calyx campanulate   **7. Samyda**

1. HOMALIUM Jacq. (1760)

1. H. racemosum Jacq., Enum. Syst. Pl. Carib.: 24 (1760).—*H. integrifolium*
Britton (1910). White Cogwood.

Shrub to 3 m. or tree to 16 m. high; leaves elliptical, apex acute or acuminate,
often with hairy domatia in the vein-axils beneath, 5–15 cm. long, up to 6 cm.
broad; racemes simple or panicled, pubescent distally, up to 15 (–18) cm. long
including peduncle; flowers pubescent; calyx-segments mostly lanceolate, 2–3 mm.
long; petals 3·5–5 mm. long, cream-coloured; stamens 3 or 4 together; styles 3;
capsule conical with 1 or 2 seeds.

Uncommon in woodlands and marsh forest; 50–3500 feet; fl. Feb–Mar, July–
Sept, fr. Sept. *A 6415! H 9981! P 24662!* Mexico to northern Brazil, West Indies.

2. LUNANIA Hook. (1844) nom. cons.

Racemes solitary, pendulous, 30 cm. or more long; pedicels to 2 mm. long; sepals 2, about
3 mm. long   1. *racemosa*
Racemes several together, erect, subumbellate on a terminal peduncle, up to about 10 cm.
long; pedicels to 1 mm. long; sepals usually 3, about 1·5 mm. long   2. *polydactyla*

1. L. racemosa Hook., Lond. Journ. Bot. **3**: 317, tt. 11, 12 (1844).

Tree 8–16 m. high; leaves ovate, acuminate, broadly rounded at base, glabrous,
6–15 cm. long, up to 7·5 cm. broad; racemes terminal, sometimes forked; flowers
chestnut-brown; fruit obovoid, 3-valved, crowned by the style; seeds numerous.

Rare (St. Ann, St. Mary, St. Thom.), in woodlands; about 2000 feet; fl. and fr.
Mar. *H & B 10738!* Endemic.

2. L. polydactyla Urb., Symb. Ant. **6**: 18 (1909).

Tree 4–10 m. high; young branches zig-zag; leaves ovate to ovate-elliptical,
rather abruptly and obtusely acuminate, broadly rounded to subcordate at base,
glabrous except for a few long hairs at base of midrib beneath, 4–10 cm. long, up to
5·5 cm. broad; calyx pinkish, reflexed; staminal ring becoming dark red, persistent;
fruit subspherical, 3-angled, up to 5 mm. in diameter; seeds numerous.

Rather rare (Clar., Port., St. Thom.), in hillside woodlands; 1600–3000 feet;
fl. Mar–Apr, June, Aug, fr. Mar–Apr. *A 11119! H 10851! H & B 10686!* Endemic.

3. XYLOSMA Forst. f. (1786) nom. cons.

1 Leaves cordate at base, glabrous and smooth adaxially   1. sp. *A*
1 Leaves truncate, rounded or cuneate at base:
2 Leaf-blade scabrid adaxially, lanceolate, truncate at base, the margin coarsely serrate-
   dentate   2. sp. *B*

2 Leaf-blade smooth adaxially:
  3 Leaf-tip acuminate; blade narrowly ovate to elliptical, mostly more than 8 cm. long,
    shallowly serrate   3. *nitida*
  3 Leaf-tips acute to obtuse or rounded; blades mostly less than 6 cm. long; at least the
    trunk and larger branches armed with branched spines:
    4 Leaves mostly 3–6 cm. long, elliptical to ovate or obovate, remotely serrate-crenate
      from the base   4. *fawcettii*
    4 Leaves smaller, obovate to oblanceolate, entire in the proximal half at least:
      5 Leaf-tip rounded smoothly or with a few teeth, veins ascending; flowering twigs not
        usually spiny; berry globose, about 5 mm. long   5. *schaefferioides*
      5 Leaf-tip mostly acute, entire or bluntly toothed; twigs with subulate spines to
        1·5 cm. long; berry obovoid-oblong, about 6 mm. long; Grand Cayman
                                              X. bahamensis (Britton) Standl.

1. X. sp. A.
    Slender-branched shrub to 2·5 m. high; leaves crenate, acuminate, oblong, 8·5–
17 cm. long, 3·5–7·5 cm. broad, shiny above; inflorescences axillary or below the
leaves, very shortly racemose; flowers green; berry subglobose, about 5 mm. in
diameter, 6-seeded.
    Rare and local (Trel.), on wooded limestone hillside; 1700 feet; fl. Aug, fr.
Mar, Aug. *P 15736!* Endemic.

2. X. sp. B.
    Shrub 1–3 m. high; twigs pubescent; axillary spines glabrous, 1 cm. long;
petioles and main veins of leaves pubescent on both surfaces; blade 6–13 cm. long,
2–4·5 cm. broad; flowers and fruit unknown.
    Rare and local (St. Andr.), on rocks in limestone gorge; 400 feet. *A 8975!*

3. X. nitida (Hellenius) A. Gray ex Griseb., Fl. Br. W.I.: 21 (1859).—*Hisingera
nitida* Hellenius (1792).
    Glabrous shrub 2–4 m. or tree to 7 m. high; lower trunk with slender sometimes
branched spines, upper stems not usually spiny; leaves 5–13 cm. long, 1·5–5 cm.
broad, gradually acuminate; flowers in umbelliform clusters at nodes, fragrant;
sepals 4, light green to greenish-yellow, up to nearly 3 mm. long, ciliate; stamens
15–25; berry globose, 5–6 mm. in diameter.
    Occasional in the central parishes, more common at middle elevations in the
eastern parishes, in woodlands; (1250–) 3500–5700 feet; fl. Aug, Nov–Feb, fr.
Dec–June. *A 11944! H 5412! H & P 15151! Powell 229!* Endemic.

4. X. fawcettii Urb., Symb. Ant. **5**: 440 (1908).
    Shrub 2–3 m. or tree to 13 m. high; twigs and petioles puberulous or glabrous;
male flowers fragrant, in clusters, on pedicels up to 8 mm. long; sepals 4 or 5,
1·5–2 mm. long, ciliate; stamens 10–17, filaments 3–4 mm. long.
    Occasional in the central parishes, in thickets and woodlands on limestone;
1200–2900 feet; fl. sporadically, fr. Jan. *H 9776! 10672! HPS 14649!* Endemic.

5. X. schaefferioides A. Gray in Mem. Amer. Acad. n.s. **8**: 155 (1860).
    Tree to 6 m. high; leaves 2–3·5 cm. long, 8–18 mm. broad; pedicels 4–7 mm.
long; sepals ovate or rounded, about 1 mm. long, yellow; stamens 10–16; ripe
berries scarlet.
    Rare (St. Cath., St. Eliz.), in thickets on limestone; 1200–1500 feet; fl. and fr.
Sept. *A 8944! H 9746!* Greater Antilles.

4. CASEARIA Jacq. (1760)

1 Flowers in corymbose cymes; leaves rounded to cordulate at base, glabrous, very shiny;
  stamens 8   1. *nitida*
1 Flowers in shortly stalked or sessile clusters; leaves mostly cuneate at base:
  2 Leaves and twigs conspicuously hairy; leaves oblong-elliptical; stamens 8 (–10)
                                                     2. *hirsuta*
  2 Leaves and twigs more or less glabrous:
    3 Leaves oblong-obovate to ovate-elliptical; stamens usually 8; capsule 6–12 mm. long:
      4 Leaf-blade up to 8 cm. long and 3·5 cm. broad, usually smaller; pedicels jointed at or
        just below the middle; scrambling or scandent shrub or small tree often armed
        with branchlet-spines   3. *aculeata*

4 Leaf-blade (7–) 10–18 cm. long and (2·5–) 4–7 cm. broad; pedicels jointed near the
   base; erect shrub or tree without spines   4. *guianensis*
3 Leaves lanceolate to narrowly ovate; stamens 10; capsule 3–5 mm. long:
   5 Leaf-margin distinctly serrulate, acumen acute; pedicels jointed beyond the middle;
      style simple   5. *arborea*
   5 Leaf-margin obscurely serrulate to subentire, acumen obtuse to emarginate, apicu-
      late; pedicels jointed below the middle; style 3-fid   6. *sylvestris*

1. C. nitida (L.) Jacq., Enum. Syst. Pl. Carib.: 21 (1760).—*Samyda nitida* L.
   (1759). *C. bahamensis* Urb. (1902).
   Shrub or tree 2·5–6 m. high, semi-deciduous; leaves ovate to ovate-elliptical,
usually shortly and bluntly acuminate, 3–10 cm. long, up to about 4·5 cm. broad;
stipules membranous, ovate to subrotundate, about 1 mm. long, falling very early;
inflorescences 1–2·5 cm. long; flowers fragrant; calyx about 4 mm. long, greenish-
white; staminodes light red; anthers dark red; fruit 15–20 mm. in diameter, at first
fleshy drupaceous, later dehiscent by 3 or 4 valves, yellow when ripe; aril orange.
   Locally common in thickets on rocky limestone and on dunes; SL–750 feet;
fl. Apr–May, Sept, fr. Apr–July, Dec. *A 9430! H 9225! P 23507!* Bahamas,
Hispaniola. Plants from the continental mainland hitherto identified with this
species are distinct.

2. C. hirsuta Sw., Fl. Ind. Occ. **2**: 755 (1798).—Cloven Berries, White Wattle,
   Wild Coffee.
   Shrub 2·5–4 m., rarely a tree to 8 m. high with a slender trunk, deciduous and
often flowering with or before the young leaves; leaves obtuse to broadly acuminate,
up to 18 cm. long and 8 cm. broad, shallowly serrate, with golden-yellow hairs
spreading on the midrib and veins; stipules subulate, 2–3 mm. long; flowers white
to greenish-cream, heavily scented; calyx about 4·5 mm. long; staminodes and
filaments white; capsule ovoid, up to 2 cm. long, 3–4 valved; seeds irregularly
ovoid, compressed, with a narrow strophiole; aril waxy, bright red.
   Common in thickets and woodland margins and invading pastures, mostly on
limestone; 250–2750 feet; fl. Jan–June, Sept–Oct, fr. most of the year. *A 6723!
H 8365! P 10183!* Panama to Guyana, Cuba, Hispaniola, Grand Cayman.

3. C. aculeata Jacq., Enum. Syst. Pl. Carib.: 21 (1760).—*C. odorata* Macf.
   (1837). Cockspur.
   Shrub 1·5–4 m. or branches scrambling to 10 m., rarely a tree 4–5 m. high with
spines on the old trunk and vegetative shoots, or spines lacking on young plants;
leaves acute to shortly and obtusely acuminate, remotely serrate to subentire,
hairy on the veins when young; glabrescent; stipules broadly subulate, about 1·5
mm. long; flowers light green, fragrant; calyx 4–5 mm. long; staminodes and
filaments white; capsule subglobose, about 9 mm. in diameter, 3-valved; seed
usually solitary enclosed in the reddish-orange aril.
   Rather common in pasture margins and damp low-lying thickets; SL–2600 feet;
fl. June–Dec, fr. July–Jan. *A 8277! H 9808! J.P. 1463! Powell 867!* Continental
tropical Amer., Greater Antilles.

4. C. guianensis (Aubl.) Urb., Symb. Ant. **3**: 322 (1902).—*Iroucana guianensis*
   Aubl. (1775). Wild Coffee.
   Shrub 2–3 m. or tree with slender trunk to 10 m. high; leaves acutely or obtusely
acuminate, shallowly serrate-crenate, sparsely hairy when young, glabrescent;
stipules subulate, 2–5 mm. long; flowers white to greenish-yellow, fragrant; calyx
4–5 mm. long; staminodes and filaments white; capsule ellipsoidal to subglobose,
6–12 mm. long, (2–) 3 (–4)-valved, light green, usually marked purple; seeds
solitary or few, arils scarlet.
   Rather common, especially in the northern parishes, in open woodlands and
secondary in pastures; SL–2500 feet; fl. Feb–Apr, Sept, fr. Mar–June, Sept–Dec.
A 8139! 11954! H 10369! H & P 14047! Continental tropical Amer., West Indies.

5. C. arborea (L. C. Rich.) Urb., Symb. Ant. **4**: 421 (1910).—*Samyda arborea*
   L. C. Rich. (1792).
   Evergreen shrub 1·5–2 m. or tree 5–7 m. high; leaves oblong to oblong-
lanceolate, long-acuminate, variously tomentose or glabrescent beneath, 3–10 cm.

long, 1–3 cm. broad; stipules filiform, up to 10 mm. long; flower-clusters shortly stalked; flowers light greenish-yellow; calyx about 4 mm. long; capsule ellipsoidal, about 4 mm. long.

Uncommon (Clar. and eastern parishes), in woodlands and wet thickets; 750–3000 feet; fl. sporadically, fr. Mar, Aug. *A & C 1154! H & B 10734! P 20769!* Honduras to Brazil and Peru, Greater Antilles.

6. C. sylvestris Sw., Fl. Ind. Occ. 2: 752 (1798).
Evergreen shrub 1·5–3 m. or tree 5–10 m. or more high; leaves ovate to elliptical or oblong, usually long-acuminate, often unequal at base, glabrous, 3–15 cm. long, up to 3·5 cm. broad; stipules deltate, ciliate, about 0·4 mm. long; flower-clusters sessile; flowers light green or yellowish, strongly fragrant; calyx 2–2·5 mm. long; staminodes and filaments white; anthers orange to light brown; ripe fruit red or orange, subglobose, 4–5 mm. in diameter, 3-valved.

Common on open limestone banks and in hillside thickets and woodland margins; 400–3500 (–5000) feet; fl. sporadically throughout the year, fr. Aug–Oct. *A 7831! H & B 10555! P 19653!* Mexico to Argentina, West Indies, Grand Cayman.

5. ZUELANIA A. Rich. (1845)

1. Z. guidonia (Sw.) Britton & Millsp., Bah. Fl.: 285 (1920).—*Laetia guidonia* Sw. (1788). Cuffey Wood.
Tree 6–20 (–30) m. high, with rather smooth bark, bole long, deciduous, the flowers usually appearing before or with the young leaves; leaves elliptical to oblong-lanceolate, obtuse to acute at apex, rounded to shortly cordate at base, serrate-crenate to subentire, pubescent at least on midrib and veins, 5–20 cm. long; stipules ovate-lanceolate, about 3 mm. long; flowers numerous in compact cymose clusters; pedicels up to 11 mm. long, jointed near the base, rusty-pubescent; sepals suborbicular, about 7 mm. long; fruit subglobose, 3–5 cm. in diameter.

Rather rare in the central and western parishes, in woodland on rocky limestone; 500–2600 feet; fl. Feb–May, fr. *A 12398! H 7088! P 24734!* Bahamas, British Honduras, Panama, Cuba, Hispaniola, Trinidad, Grand Cayman.

6. LAETIA Loefl. ex L. (1759) nom. cons.

1. L. thamnia L., Pl. Jam. Pugill.: 31 (1759).—Scarlet Seed.
Shrub 2–5 m. or tree to 6 m. high; leaves elliptical, shortly acuminate, crenate to entire, glabrous, 5–10 cm. long, up to about 4·5 cm. broad; stipules broadly deltate, up to 1 mm. long; flowers in pedunculate corymbs, fragrant, sepals 4, 5–7 mm. long, white within, mostly pinkish outside; filaments white; fruit globose, 2–4 cm. in diameter, shortly tomentose, 4–5-valved, many-seeded, bright crimson inside; aril red or orange; seeds white.

Common in thickets and woodlands on limestone; 25–3000 (–5000) feet; fl. all the year, fr. Apr–Dec. *A 5629! 6857! H 9619! P 10187!* Cuba, Hispaniola.

7. SAMYDA Jacq. (1760) nom. cons.

1 Leaves serrulate, rounded to obtuse and shortly cuspidate at tip, hairy on both surfaces; flowers pinkish-rose or white; calyx 5-keeled   1. *pubescens*
1 Leaves subentire, broadly acute to shortly acuminate at tip; flowers white or greenish; calyx not keeled:
  2 Indumentum of young parts, petioles and midribs of leaves of golden-brown appressed more of less persistent hairs; leaf-base broadly rounded   2. *villosa*
  2 Indumentum puberulous, sparing on leaves, glabrescent; leaf-base broadly cuneate to rounded   3. *glabrata*

1. S. pubescens L., Sp. Pl. ed. 2, **1**: 557 (1762).
Shrub 1–3 m. high with densely villose branchlets; leaves oblong to obovate, rounded at base, soft to touch, 3–7 (–10) cm. long, up to about 3 (–4·5) cm. broad; flowers 2–4, pedicelled in the leaf-axils; calyx 9–16 mm. long; stamens 10–12; fruit subangular-globose, about 2 cm. long, scarlet when ripe, crimson within; aril orange.

Occasional in cultivation, rarely escaping; SL–1100 feet; fl. and fr. sporadically. *Burrowes 13038! Harty!* Cuba, Hispaniola.

2. S. villosa Sw., Nov. Gen. & Sp. Pl.: 68 (1788).

Shrub 1·5–3 m. or tree to 5 m. high; leaves 4–11 cm. long, typically puberulous on both surfaces, villose on the nerves beneath, but glabrescent in some specimens; flowers solitary, fragrant; calyx green outside, white within, tube 6–8 mm. long; stamens 10; fruit ovoid, acuminate, thinly hispid outside, about 2·5 cm. long, flesh crimson; arils orange-red.

Local and uncommon (Clar., Manch., Trel.), in pastures, thickets on limestone rocks and in woodlands; 1900–3000 feet; fl. Apr, fr. Apr–May. *A 11010! 12520! H & P 14329!* Endemic.

3. S. glabrata Sw., Nov. Gen. & Sp. Pl.: 68 (1788).

Shrub or tree to 6 m. high; young branchlets puberulous; leaves glabrescent; leaves oblong, acuminate, 2–18 em. long, up to about 6 cm. broad; flowers maturing singly from compact axillary bracteate cymose clusters; calyx-tube greenish-white, lobes white, tube 6–7 mm. long; capsule thinly hispid, long-acuminate, about 3·5 cm. long, 3–4-valved; aril orange.

Uncommon and local (Port., St. Thom.), in thickets on limestone in wet areas; 850–3000 feet; fl. June, Oct–Dec, fr. Dec. *A 5689! H 5176!* Endemic.

Modern collections of *S. villosa* and *S. glabrata* are intermediate between the typical plants and tend to diminish the distinction between them.

128. LACISTEMATACEAE

Shrubs and small trees. Leaves alternate, simple, with small paired, lanceolate, imbricate, caducous stipules. Flowers bisexual, small in axillary clustered spikes or racemes; bracts imbricate, concave; bracteoles 2 at the base of each flower. Perianth uniseriate of 4–6 distinct unequal segments, or absent. Stamen 1, inserted on or within a fleshy disk; anther with 2 ovoid loculi widely separated on the forked connective, opening lengthwise. Ovary superior, 1-locular with 2–3 parietal placentas; stigmas 2–3, short; ovules 1–2 on each placenta, pendulous, anatropous. Fruit at first berrylike, later dehiscing by valves. Seeds with endosperm.

2 genera with about 20 species in tropical America.

1. LACISTEMA Sw. (1788)

Bracts imbricate, conspicuous; leaves entire; flowers spicate.

1. L. aggregatum (Berg.) Rusby in Bull. New York Bot. Gard. **4**: 447 (1907).—
Piper aggregatum Berg. (1772).

Shrub 2–3 m. or tree to 6 m. high; leaves entire or subentire distally, elliptical or oblong-elliptical, shortly acuminate, glabrous, 8–11 (–15) cm. long, 3–5 (–6·5) cm. broad; stipules 10–13 mm. long, linear-lanceolate, light green; spikes up to 7 together in the axils or on the older wood, 5–11 mm. long, greenish-yellow; sepals 4; fruit ellipsoidal, often tinged brownish-purple, shiny, 6–8 mm. long.

Occasional, in woodland and damp thickets on limestone; 1200–2900 feet; fl. Dec–June, fr. Mar–July. *A 11248! 12415! H 11059! P 19652!* Mexico to the Guianas, Trinidad.

129. TURNERACEAE

Herbs or shrubs, rarely trees. Leaves alternate, simple, entire, toothed or lobed; stipules small or wanting. Flowers bisexual, actinomorphic, axillary, solitary, few together or racemose. Perianth biseriate; calyx tubular, deciduous, the lobes 5, imbricate; petals 5, free, perigynous, clawed, caducous, contorted. Stamens 5, opposite the sepals, inserted on the calyx-tube or rarely hypogynous; anthers 2-locular, opening lengthwise. Ovary superior, 1-locular with 3 intrusive parietal placentas; ovules numerous, anatropous, ascending; styles 3, linear or flattened; stigmas fimbriate. Fruit a loculicidal capsule opening by 3 valves with a placenta in the middle of each. Seeds arillate, often pitted, with straight embryo and hard or fleshy endosperm.

8 genera with about 120 species in the American and African subtropics and tropics.

Styles simple; corona absent; pedicel sometimes united with petiole of subtending leaf; bracteoles paired; plants perennial, shrubby **1. Turnera**
Styles divided below apex; corona a narrow fringed membrane at base of petals; pedicels free, without bracteoles; annual herb with stellate indumentum **2. Piriqueta**

1. TURNERA L. (1753)

1 Pedicels not united with petioles; flowers axillary; leaves mostly less than 1·5 cm. long, punctate beneath with yellow glands **1. *diffusa***
1 Pedicels and petioles united; leaves mostly 2 cm. or more long, not glandular-punctate beneath:
  2 Annual weedy subwoody herb up to 25 cm. high; flowers subcapitate at ends of branches; leaf-blade usually without a pair of glands at base **2. *pumilea***
  2 Perennial shrubby herbs or undershrubs; flowers arising from petioles of upper leaves; leaf-blade with a pair of glands at base:
    3 Bracteoles with toothed margins; capsule depressed globose; leaves mostly ovate to lanceolate, toothed **3. *ulmifolia***
    3 Bracteoles entire; capsule ovoid; leaves elliptic-oblanceolate, subentire **4. *zeasperma***

1. T. diffusa Willd. in Schult. in L., Syst. Veg. ed. nov. **6**: 679 (1820).
Shrub 75–120 cm. high; leaves obovate, narrowed to base, crenate, pubescent on both surfaces or glabrous adaxially, aromatic, 5 –15 (–30) mm. long, up to 10 mm. broad; bracteoles linear-lanceolate, 2–5 mm. long; corolla bright orange-yellow, 6–7 mm. long; capsule 3–4 mm. long; seeds curved, pyriform, 1·5–2 mm. long.
Uncommon (St. Andr., St. Cath., St. Eliz.), in thickets on arid limestone; 100–1600 feet; fl. July–Sept, fr. July–Oct. *A 12871! H 9608! 9679!* Mexico, Honduras, Brazil, Bahamas, Greater Antilles, St. Croix.

2. T. pumilea L., Syst. Nat. ed. 10, **2**: 965 (1759).
Annual herb becoming woody at base, 10–30 cm. high; leaves elliptical to oblanceolate, coarsely toothed, with long spreading hairs, 1–4 (–6) cm. long, up to about 12 mm. broad; flowers clustered terminally; bracteoles linear to subulate, 4–8 mm. long; calyx 6–7 mm. long; petals about 4 mm. long, yellow sometimes drying white; capsule 3–6 mm. long; seeds obovoid-oblong, curved, about 2 mm. long.
Very rare (St. Cath.), amongst grasses in savanna; about 50 feet; fl. and fr. June. *H 12066! Swartz!* Mexico, Greater Antilles, Curaçao, tropical S. Amer.

3. T. ulmifolia L., Sp. Pl. **1**: 271 (1753).—Ram-Goat Dashalong.
Aromatic shrubby herb or shrub up to 120 cm. high; leaves mostly lanceolate to oblong-lanceolate, rounded to narrowly cuneate at base, usually acute at tip, subentire to deeply and doubly coarsely dentate, very variably hairy from sub-glabrous to densely pubescent, mostly 4–13 cm. long and 2–5 cm. broad; bracteoles leafy, lanceolate, 1–2·3 cm. long; calyx-tube and lobes together about 2 cm. long, greenish-yellow; petals obovate, 2–3·5 cm. long, 1–3 cm. broad, light or dark yellow, sometimes with a brown spot at base; capsule 7–8 mm. long and slightly broader; seeds obovoid-oblong, slightly curved, about 2·5 mm. long.
Common along roadsides and on waste ground and in thickets on limestone or coral sand near the sea, also on the cays; SL–2500 (–3500) feet; fl. and fr. all the year. *A 10153! H 9236! 11029! P 23657!* Bermuda, Bahamas and generally throughout tropical continental Amer. and West Indies; Grand Cayman; introduced into tropical Africa and Asia. This is one of the most morphologically variable weedy indigenous species of the American tropics. To the extent that its breeding systems

also differ in different areas, the species can be said to be undergoing active evolution. Martin (Bull. Torr. Bot. Club **92** (3): 185–192) has described self-incompatibility associated with pin and thrum type distyly in *Turnera ulmifolia* in Puerto Rico. Of all flowers examined on Jamaican plants none had anthers positioned above the tips of the styles although anther-height varied appreciably.

4. T. zeasperma Adams & Been in Phytologia **20** (5): 313 (1970).

Perennial undershrub 30–120 cm. high with slender woody twigs, reddish when young; leaves narrowly elliptical to oblanceolate, cuneate at base, acute at apex, serrate-dentate, subglabrous, mostly 2·5–7·5 cm. long and 5–15 mm. broad; bracteoles linear-lanceolate, about 10 mm. long and 1 mm. broad; calyx 2 cm. long, the lobes 1·5 cm. long; petals obovate, about 30 mm. long and 25 mm. broad, bright yellow; capsule up to about 8 mm. long and 5–6 mm. broad; seeds as above.

Rare and local (St. Andr.), on limestone rocks and streamside calcareous gravel in gorge; 250–500 feet; fl. and fr. Apr, July, Nov. *A 6969! Elliott 17! Weaver 1078! Yuncker 17384!* Endemic. Crosses made between this species and *T. ulmifolia* were in a few instances brought to flowering; self-pollinations and back-crosses produced no viable seed. The F_1 hybrids could be distinguished by the red stem character which they inherited in every case from the *T. zeasperma* parent; in other respects such as hairiness and capsule-shape the hybrids were intermediate.

2. PIRIQUETA Aubl. (1775)

1. P. cistoides (L.) Griseb., Fl. Br. W.I.: 298 (1860).

Annual herb, erect or branched and ascending from the base, 15–60 cm. high; hairiness and leaves variable; stem with stellate hairs with short subequal rays or one ray exaggerated and spreading; leaf-blade linear to oblong-lanceolate, usually more or less acute, stellate-hairy, mostly 3–8 cm. long, (2–) 3–15 mm. or more broad; flowers axillary, usually solitary; pedicels up to about 3 cm. long; calyx 4–9 mm. long; petals obovate, orange-yellow, 6–9 mm. long; capsule ellipsoidal, 5–9 mm. long; seeds obovoid, striate-pitted, 1·4–1·8 mm. long.

Occasional, a weed of pastures, open cultivations and well drained shaly banks; SL–2300 feet; fl. and fr. most of the year after rain. *A 6224! H 8273! P 15931!* Georgia, Florida, Mexico, Honduras, northern S. Amer. to Paraguay, West Indies.

130. PASSIFLORACEAE

Herbs, erect shrubs, tendril climbers or rarely trees Leaves alternate, usually simple, entire or lobed, rarely compound; stipules paired. Flowers usually bisexual, actinomorphic, in axillary cymes, panicles or solitary, bracteate. Perianth biseriate or uniseriate; sepals (4–) 5, free or basally connate, imbricate, often petaloid and horned below the tip; petals (4–) 5, free or shortly connate or wanting. Corona usually present between perianth and androecium of 1–several whorls of filamentous or membranous appendages. Stamens 5 or more, usually opposite petals, inserted on the receptacle or raised on the gynophore, free or basally connate; anthers 2-locular, opening lengthwise; staminodes sometimes present. Ovary superior, usually raised on a gynophore, 1-locular; ovules numerous, anatropous, pendulous from 3 (–5) parietal placentas; styles as many as carpels, free or connate; stigmas 3–5, often capitate. Fruit a berry or loculicidal capsule, rarely indehiscent. Seeds surrounded by pulpy arils; endosperm fleshy; embryo straight.

12 genera with 600 species in the subtropics and tropics.

1. PASSIFLORA L. (1753); Killip (1938)

Slender lianes climbing by tendrils; petioles and/or leaf-blades often glandular; andro-gynophore present; stigmas 3.

1 Bract and bracteoles forming a conspicuous 3-segmented involucre below the flower:
  2 Involucre-segments pinnately divided into glandular filaments; leaves 3-lobed; petioles
    without glands:
    3 Plants villous or velvety   1. *foetida*
    3 Plants glabrous   2. *ciliata*
  2 Involucre-segments entire or leafy; petioles with glands:
    4 Segments of involucre united above the base; leaves entire or 3-lobed:
      5 Stipules broadly lanceolate, up to 3·5 cm. long, persistent; glands on petiole linear,
        up to 1 cm. long   3. *ligularis*
      5 Stipules linear to lanceolate, about 1 cm. long:
        6 Glands paired at apex of petiole with occasionally 1 or 2 other small glands lower
          down; stipules linear, persistent   4. *seemannii*
        6 Glands all on petiole, flat, sessile; stipules lanceolate, caducous   5. *maliformis*
    4 Segments of involucre free to the base:
      7 Leaves entire:
        8 Petioles with 2 or 3 pairs of glands   6. *quadrangularis*
        8 Petioles with 1 pair of glands   7. *laurifolia*
      7 Leaves deeply 3-lobed:
        9 Paired glands at apex of petiole; sepals within and petals white; leaf-margin serrate
                                        8. *edulis*
        9 Paired glands not at apex of petiole; cultivated ornamentals; natives of C. and S.
          Amer.:
          10 Paired or subopposite glands at about middle of petiole; leaf-margin entire;
            petals white to light rose; corona violet banded white; stipules broadly falcate
                                      P. stipulata Aubl.
          10 Paired glands at base of petiole; leaf-margin sinuate-crenate; petals and corona
            scarlet; stipules subulate, curved, 3–4 mm. long   *P. vitifolia* Kunth
1 Bract and bracteoles small, usually at the junction of the peduncle, or wanting:
  11 Petioles with 2 glands; petals wanting:
    12 Calyx yellowish-green; fruit subspherical to ovoid; leaves very variable from entire
      and ovate to 3-lobed, subglabrous to very hairy   9. *suberosa*
    12 Calyx not green:
      13 Leaves entire or with indistinct basal lobes; calyx dull crimson or reddish-purple;
        fruit globose   10. *lancifolia*
      13 Leaves 3-lobed; calyx scarlet; fruit fusiform   11. *macfadyenii*
  11 Petioles without glands; petals present:
    14 Leaves without glands beneath; stems hairy: leaves usually velvety beneath, 2 (–3)-
      lobed, the lateral lobes spreading and pointed with a broad bay between; sepals
      green; petals white, cream-coloured or pinkish:
      15 Peduncles 1-flowered; leaves cordate; flowers 3–4 cm. in diameter   12. *rubra*
      15 Peduncles 3 or more-flowered; leaves subcordate; flowers 1·5–2 cm. in diameter
                                      13. *sexflora*
    14 Leaves with large superficial glands beneath:
      16 Peduncles 3-flowered; leaves glabrescent, 3-lobed with lateral lobes acute, base
        subcordate; stem pubescent   14. *triflora*
      16 Peduncles 1-flowered; leaves glabrous, mostly rather leathery:
        17 Flowers green   15. *penduliflora*
        17 Flowers pink or purple:
          18 Leaves truncate at apex or variously 2- or 3-lobed, the midrib clearly exceeding
          half the length of the main lateral veins; petiole more than 1 cm. long; corona
          filamentous   16. *oblongata*
          18 Leaves 2-lobed with the midrib less than half as long as the main lateral veins;
          petiole less than 1 cm. long:
            19 Corona at throat filamentous; leaf-blade deeply cordate, the basal lobes em-
            bracing the stem; stem usually puberulous   17. *perfoliata*
            19 Corona at throat tubular, entire; leaf-blade obtuse or truncate at base; stem
            glabrous; leaves membranous   18. *murucuja*

1. P. foetida L., Sp. Pl. 2: 959 (1753).

Annual or perennial; stem, petioles and leaves variably hairy; leaves cordate, lobes entire or minutely toothed, up to 12 cm. long; bracts of involucre exceeding the sepals, light green; up to 2·5 cm. long; petals white with purple veins; outer corona filamentous, inner series very short; ripe fruit yellow, about 2 cm. in diameter.

Common in thickets, hedgerows and waste places; SL–2000 feet; fl. and fr. all the year. *A 8762! H 8520! P 15687!* A pantropical weed.

2. P. ciliata Ait., Hort. Kew. 3: 310 (1789).

Like the last and by some authors considered as one of several varieties of that species; bracts less divided; sepals whitish; petals white tinged mauve; ripe fruit oblong-globose, about 2 cm. long and broad, orange-red to crimson.

Local (St. Eliz., West.), on fences and in thickets at swamp margins; SL–100 feet; fl. and fr. sporadically throughout the year. *E. G. Britton 2880! H 11816! P 24599!* Bahamas, Cuba.

3. P. ligularis Juss. in Ann. Mus. Hist. Nat. Paris **6**: 113, t. 40 (1805).—Granaditta.

High-climbing vine; leaves ovate, acutely acuminate, deeply cordate, entire, 15 cm. or more long; petiole with 6 long glands; peduncle about 2·5 cm. long, 1-flowered; sepals green; petals white; fruit about 6 cm. in diameter.

Locally naturalized (St. Andr., Port.), in damp sheltered clearings; 3500–4000 feet; fl. May–June, fr. June. *Powell & Garay 967! P 22272! Robertson UCWI 6731!* Native from Venezuela to Peru.

4. P. seemannii Griseb. in Bonplandia **6**: 7 (1858).

Robust glaucous vine; leaves broadly ovate, cuspidate, deeply cordate, margin shallowly crenate-serrate, up to about 12 cm. long and 11 cm. broad; flowers solitary, fragrant, purple, the corona-filaments violet and white in transverse bars.

Locally naturalized (Clar., Port.), in moist thickets and woodland margins; 300–1650 feet; fl. Mar, June–Aug. *P 20647!* Native of Panama.

5. P. maliformis L., Sp. Pl. **2**: 956 (1753).—Sweet Cup.

Terete-stemmed vine; leaves oblong-elliptical and entire or 3-lobed, cordate to rounded at base, up to 16 cm. long and 8 cm. broad; petiole with 2 (–6) glands; flowers solitary, 5–6 cm. in diameter; sepals green with reddish spots; petals greenish-white with red or purplish spots; fruit subspherical, slightly longer than broad, 4·5–5 cm. long, smooth, hard, the wall finally brittle and about 1 mm. thick; seeds flat, ovate, pitted.

Cultivated and naturalized; 500–1200 feet; fl. and fr. Sept–Dec. *H 33!* Northern S. Amer., West Indies.

6. P. quadrangularis L., Syst. Nat. ed. 10, **2**: 1248 (1759).—Granadilla.

High-climbing robust vine with a 4-angled or -winged stem; leaves broadly ovate, abruptly acuminate, cordate to rounded at base, up to 20 cm. long and 17 cm. broad; flowers 8–10 cm. in diameter; petals white or pinkish; corona-filaments bluish-purple with white speckles; fruit ovoid to ellipsoid, 12–20 cm. long, up to 15 cm. broad, the pericarp thick and fleshy.

Mostly cultivated; 700–1500 feet; fl. July, Nov–Apr, fr. Dec. *Harris! P 11824!* Native of C. Amer., cultivated throughout tropical Amer. and in W. Africa.

7. P. laurifolia L., Sp. Pl. **2**: 956 (1753).—Golden Apple.

High-climbing vine with striate grooved stem; leaves elliptical, shortly acute, rounded at base, glabrous, up to 14 cm. long; peduncles 1-flowered; flower 6–10 cm. in diameter, fragrant; petals cream-coloured with minute red speckles; corona bright purple with white bands towards base; ripe fruit egg-shaped, yellow-orange, exocarp leathery, 7–8 cm. long, about 5·5–6 cm. broad; seeds flat, about 6·5 mm. long and 4 mm. broad.

Cultivated and naturalized. Native of tropical S. Amer.; West Indies and widespread by introduction in the Old World.

8. P. edulis Sims in Curt., Bot. Mag. **45**: t. 1989 (1818).—Mountain Sweet Cup, Passion Fruit.

High-climbing vine; lobes of leaf oblong to elliptical, acuminate, serrate, shiny adaxially; flowers 4 cm. or more in diameter; sepals green outside, white inside; petals white; corona-filaments blue-violet at base, white distally; fruit subglobose, 4–8 cm. long, greenish-yellow or purple-spotted when ripe.

Cultivated and naturalized; 300–5600 feet; fl. Apr, July, Dec. *H 11953! P 23060!* Native of tropical S. Amer., cultivated in many subtropical and tropical areas.

9. P. suberosa L., Sp. Pl. **2**: 958 (1753).

Slender-stemmed twiner; entire leaves lanceolate to broadly elliptical, up to 9 cm. long and 5·5 cm. broad, 3-lobed leaves with the median lobe always longer, linear-lanceolate to oblong, mucronate; petiole with 2 stalked glands nearer the blade than the base; peduncles often paired, slender, up to about 1 cm. long; flower mostly about 1·5 cm. in diameter; corona in 4 series, the two outer filamentous, the third a continuous pleated membrane, the innermost a low ridge on the disk; fruit ellipsoid to suglobose, glaucous blue-black, variable in size but mostly about 12–13 mm. long and 9 mm. broad.

Common in thickets and waste places, especially in semi-arid woodland on limestone; SL–3800 feet; fl. and fr. all the year. *A 5496! 8327! H 12747! H & P 13866!* Florida, Turks and Caicos Is., Mexico to Paraguay, West Indies.

10. P. lancifolia Desv. in Ham., Prodr. Pl. Ind. Occ.: 48 (1825).—*P. regalis* Macf. ex Griseb. (1860).

Slender angled-stemmed vine to 6 m.; branches often pendulous, pilose; leaves mostly ovate-lanceolate, tip usually long-acute, often shallowly lobed at base, up to 10 cm. long and 5·5 cm. broad; peduncles mostly 2 at a node, very slender; sepals about 2·5 cm. long; berry blue-black, 10–12 mm. in diameter.

Locally common (St. Andr., Port., St. Thom.), on trees and shrubs in open woodlands and thickets; 2800–4200 feet; fl. May–Jan, fr. May–Jan. *A 5723! 8152! Burrowes 13017! P 22948!* Endemic.

11. P. macfadyenii Adams in Bull. Inst. Jam., Sci. ser. **16**: 27 (1967).—*P. regalis* of F. & R. (1926), not Macf. ex Griseb. (1860).

Slender vine with an angled pubescent stem 2–4 m. long, the flowering and fruiting branches pendulous; leaves with the lobes usually obtuse, rounded to subcordate at base, 2–5 cm. long, up to 6 cm. broad across the lateral lobes; peduncles 1–2, the proximal section rather short and stiff; sepals 18–23 mm. long; berry purplish-black, shiny, pointed at both ends, 2–2·5 cm. long, 8–10 mm. broad.

Rare and local (St. Andr., St. Thom.), on shrubs in gullies and on banks on yellow shale; 200–1000 feet; fl. July–Feb, fr. Sept–Feb. *A 10232! P 15884!* Endemic.

12. P. rubra L., Sp. Pl. **2**: 956 (1753).—Bat Wing, Dutchman's Laudanum.

Slender pubescent climber; stem, tendrils and petioles often tinged crimson; leaves 2–10 cm. along the midrib, 4–10 cm. across the tips of the lateral lobes; peduncle 1–6 cm. long, longer in fruit, jointed 1–2 mm. below the calyx; calyx usually speckled pink within; corona pink to reddish-purple, darker at base; berry red, soft, hairy, 6-lined, ellipsoidal, about 2·5 cm. long.

Common in thickets and woodland margins and on rocky banks; 250–2600 feet; fl. and fr. all the year. *A 8197! 9771! J.P. 1072! Webster & Miller 8259!* Tropical continental Amer., West Indies south to St. Lucia; Trinidad.

13. P. sexflora Juss. in Ann. Mus. Hist. Nat. Paris **6**: 110, t. 37, f. 1 (1805).—Goat Foot.

Softly puberulous slender vine; leaves 2–8 cm. along the midrib, 5–14 cm. across the tips of the lateral lobes; common peduncle very short, 3–6 (–8)-flowered; pedicels 7–8 mm. long, with small linear caducous bracteoles, not jointed; corona-filaments white distally, purple at base; berry black or bluish-black, puberulous, globose, up to 9 mm. in diameter.

Common in thickets and woodland margins; 1500–5000 feet; fl. and fr. all the year. *A 10032! H 7707! P 15908!* Florida, Mexico, Greater Antilles, St. Kitts.

14. P. triflora Macf. ex Griseb., Fl. Br. W.I.: 293 (1860).

Stem pubescent; lateral lobes of leaf diverging, acute, median lobe indistinct; peduncles paired; sepals longer than the white petals; corona in 3 series, outer threads purple tipped with white.

This species, reported from Portland, is known only from Macfadyen's description.

15. P. penduliflora Bert. ex DC., Prodr. **3**: 326 (1828).—Handsome Gal.

Robust high-climbing vine with long pendulous branches; leaves semi-elliptical, broader distally, shallowly 3-lobed at apex, obtuse or rounded at base, with 2 rows of glands beneath, 2·5–6·5 cm. along the midrib, 2–7 cm. across the tips of the lateral lobes; peduncles 1 or 2, 7–14 cm. long, pendulous; flowers campanulate,

2–3·5 cm. long; corona-glands orange; berry globose-ellipsoidal, purplish-black 2–2·5 cm. long.

Frequent, climbing on trees in rocky woodlands, mostly in limestone areas; 30–5200 feet; fl. and fr. most of the year. *A 6045! 7412! H 9414! P 21446!* Cuba.

16. P. oblongata Sw., Nov. Gen. & Sp. Pl.: 97 (1788).—*P. lyrifolia* Tussac (1808). *P. tacsonioides* Griseb. (1860). Puss Gut.

Strong corky-barked vine to 10 m. or more high; leaves very variable in shape from oblong to oblate-semiorbicular, at the tip truncate and obsoletely 3-lobed to distinctly 3-lobed with the two lateral lobes greatly attenuated, at the base obtuse to rounded or subcordate, 3-nerved, 5–18 cm. long, those on the flowering branches often much smaller; peduncles mostly paired accompanied by leaves or not, in leafless racemes from the older stems or usually leafy on the younger, lower section below the joint 4–23 mm. long, minutely bracteate; calyx about 3·5 cm. long; petals nearly 2 cm. long; gynophore purplish, nearly 3 cm. long in fruit; berry oblong-ellipsoid, green turning yellow, 2–3 cm. long.

Widely distributed but not common, in thickets and woodlands on limestone; 300–4225 feet; fl. Jan–Aug, fr. Mar–July. *A 9311! Morley 306! H & P 14312! P 8562!* Endemic.

17. P. perfoliata L., Sp. Pl. 2: 956 (1753).

Stem slender, striate, puberulous; lobes of leaf mostly oblong, rounded, retuse, mucronate, up to 7 cm. long, 1–3 (–4) cm. broad; peduncles mostly paired, proximal section 1–2 cm. long, rather stiff; flowers dull reddish-purple to bright crimson, 3–4 cm. long; ripe berry globose, dark blue, 1·5–2 cm. in diameter.

Common, in semi-arid woodlands and thickets mostly on rocky limestone; SL–2700 feet; fl. and fr. all the year. *A 6854! H 9516! J.P. 1311! P 8504!* Endemic.

18. P. murucuja L., Sp. Pl. 2: 957 (1753).

Like the last; flowers crimson; calyx 3–3·5 cm. long; berry ellipsoidal, 1·5–2·5 cm. long.

Possibly erroneously reported; no authentically localized specimen has been seen. Hispaniola, Puerto Rico.

131. BIXACEAE

Shrubs or small trees with red or yellow sap. Leaves alternate, simple, entire, palmately veined, with a pair of deciduous stipules. Flowers bisexual in a terminal panicle, showy; pedicel with 5 glands below the calyx. Perianth biseriate; sepals 5, imbricate, deciduous; petals 5, free, imbricate, twisted in bud. Stamens numerous; filaments hypogynous, free; anthers horse-shoe shaped, opening by terminal slits or a pore. Ovary superior, 1-locular with 2 parietal placentas; ovules numerous, anatropous; style terminal, simple; stigma shortly 2-lobed. Fruit a 2-valved capsule, opening between the placentas, usually covered with spines. Seeds obovoid with red fleshy testa; endosperm present.

One monotypic genus native in tropical America, now widely distributed in the tropics.

1. BIXA L. (1753)

1. B. orellana L., Sp. Pl. 1: 512 (1753).—Anatto.

Shrub to 3 m. or tree to 6 m. high; leaves ovate, cordate or truncate at the base, acuminate, long-petioled, often scaly when young, 8–20 cm. long, 4–15 cm. broad; flowers 3–5 cm. in diameter; sepals 1·2–1·4 cm. long, covered with scaly hairs; petals pink, rarely white, about 2·5 cm. long; capsule 3–4 cm. long and broad, bristly; seeds angular, about 5 mm. long, covered with a soft reddish-orange pulp.

Common in cultivation and occasionally naturalized in secondary growth and on waste ground, especially on alluvial soils; 250–1500 feet; fl. June–Nov, fr. most of the year. *P 8133! Skelding UCWI 3013! Yuncker 17152!* Mexico to Brazil, mostly cultivated and naturalized elsewhere.

132. CARICACEAE

Soft-wooded rarely branched relatively short-lived trees with milky sap. Leaves alternate, large, mostly long-petioled, in a terminal crown, without stipules. Flowers unisexual or bisexual, of several forms[1], actinomorphic, in axillary racemes, corymbs or solitary. Perianth biseriate; calyx small, gamosepalous, 5-lobed; corolla of 5 petals, usually fused, but in fertile flowers usually with a much shorter tube and becoming free and falling, aestivation contorted or valvate. Stamens epipetalous in two whorls of 5 each, or one whorl lacking, the filaments free or connate at the base; anthers 2-locular, opening lengthwise; rudimentary ovary present or absent in male flowers. Ovary superior, 1-locular; ovules numerous, on 5 intrusive parietal placentas sometimes fusing to form a falsely 5-locular ovary; style short or wanting; stigmas 3–5, simple or fimbriate; staminodes absent in female flowers. Fruit a large berry with numerous seeds. Seeds with fleshy endosperm and straight embryo.

4 genera with 45 species in the tropics of Africa and America; *C. papaya* pantropical in cultivation.

1. CARICA L. (1753)

Stems smooth, without prickles; trees; sepals and petals alternating; filaments not connate.

Fruits subglobose, but shortly pointed or acuminate apically, up to about 6 cm. long,
    borne either solitary or more usually 2–6 together in a short corymb   1. *jamaicensis*
Fruits variously shaped, globose and lightly ribbed, oblong-ellipsoid or obovoid, always
    much larger than above and usually solitary   2. *papaya*

1. C. jamaicensis Urb., Symb. Ant. **6**: 20 (1909).—Wild Papaw.

Unbranched tree 2–5 m. high; leaf-blade up to 60 cm. long, deeply 7-lobed, the lobes acuminate and usually again pinnatifid; petiole long and thick; older leaves deciduous leaving conspicuous scars on the stem; plants dioecious; male flowers cymose-clustered in pendent panicles, the corolla-tube 13–15 mm. long; female flowers in short corymbs, about 4 cm. long, creamy-white to yellow; ripe fruit yellow-orange.

Local on open rocky banks or in thickets and rough pastures on coral limestone or sandy soil near the sea; 5–2000 feet; fl. Nov–Apr, fr. Nov–May. *A 11830! H 10673! P 24360!* Endemic.

2. C. papaya L., Sp. Pl. **2**: 1036 (1753).—Papaw, Papaya.

Tree like the last, usually larger; dioecious or polygamous; male and polygamous flowers on long pendent racemes or panicles; male flowers 2–3·5 cm. long, fragrant, yellowish; female flowers up to 3 in a short cyme, the corolla 4–5 cm. long, the segments almost free, greenish-yellow; ripe fruit light green or yellow, rarely marked with red or purple; seeds with a pulpy coat.

Common in cultivation, hardly naturalized; up to about 2500 feet; fl. and fr. sporadically. Native probably in the American tropics, now widespread and in some tropical countries fully naturalized.

---

[1] W. B. Storey (1938) has described the flowers of *Carica papaya* as follows: type 1) the bisexual flower which is sympetalous and has 10 stamens and 5 carpels; type 2) the bisexual flower which is obscurely sympetalous and has 5 stamens and 5 carpels; type 3) the staminate flower which is sympetalous and has 10 stamens and no carpels; and type 4) the pistillate flower which appears to be choripetalous and has no stamens and 5 carpels. The Papaya produces many kinds of teratological flowers, most of which result from carpellody of stamens, i.e. stamens assuming the structure and function of carpels.

133. LOASACEAE

Herbs or shrubs, sometimes climbing, usually with forked or barbed scabrid or stinging multicellular hairs. Leaves opposite or alternate, entire to pinnatifid, without stipules. Flowers bisexual, actinomorphic, solitary or in cymes, often leaf-opposed. Perianth biseriate; calyx-lobes (4) 5 (–7), imbricate or open, persistent; petals equal in number to sepals, free or connate, induplicate-valvate, imbricate or contorted. Stamens usually numerous, free or connate in bundles opposite the petals or basally connate, the outer sometimes sterile and petaloid and alternating with the petals; anthers usually 2-locular, opening lengthwise. Ovary inferior or nearly so, 1–3-locular; ovules anatropous, 1–many on usually parietal or sometimes axile placentas; style solitary, simple of shortly divided; stigma entire. Fruit a loculicidal capsule, the valves often spirally twisted. Seeds with endosperm; embryo small.

15 genera with 250 species mostly in western S. Amer., one genus in Africa and Arabia.

1. MENTZELIA L. (1753)

Herbs, rough with barbed hairs; leaves alternate; flowers solitary, sessile at nodes; perianth 5-merous; petals imbricate, united at base with the stamens in a ring; ovary 1-locular with 3 intrusive placentas; capsule valved at apex; seeds not winged.

1. **M. aspera** L., Sp. Pl. **1**: 516 (1753).
Straggly-branched herb 15–120 cm. high, annual; leaves broadly ovate to lanceolate, serrate-dentate, more or less lobed, covered with glochidiate and spine-like hairs on both surfaces, up to 15 cm. long, smaller above; calyx hairs, 7 mm. long, lobes triangular, 6 mm. long; petals obovate, yellow-orange, 8 mm. long, 6 mm. broad, glabrous except for a minute apical tuft of hairs; filaments in two series, outer series 20 or more, inner series fewer, broader; capsule 15–18 mm. long, obconic, rough with hairs, the calyx tardily deciduous in one piece.

Locally common in the south-eastern parishes, on limestone banks in the open or in arid thickets; 20–900 feet; fl. and fr. Oct–Feb. *A 6050! 8344! Aspery UCWI 1673! P 8274!* Continental tropical Amer., Greater Antilles, St. Lucia, Aruba, Bonaire, Curaçao.

134. BEGONIACEAE

Undershrubs or herbs with jointed often soft stems, sometimes tuberous. Leaves alternate, simple, sometimes lobed and often unequal-sided; stipules paired, usually deciduous. Flowers unisexual, monoecious in bracteate cymes, zygomorphic or actinomorphic. Male flowers: sepals 2 opposite (or 5), valvate, usually coloured; petals 2 (or 5) or wanting, imbricate; stamens numerous, free or connate basally; anthers 2- locular, basifixed, opening lengthwise or rarely by pores. Female flowers: tepals undifferentiated, (2–) 5 (–many), imbricate; ovary inferior or half-inferior, (1–) 3 (–5)-locular, mostly angled or winged; placentation usually axile with very numerous anatropous ovules; styles 2–5, free or connate, the stigmas often twisted. Fruit a loculicidal often unequally winged capsule, rarely a berry. Seeds minute with reticulate testa, without endosperm.

5 genera with over 800 species, mostly of *Begonia*; widespread in the subtropics and tropics.

1. BEGONIA L. (1753)

Perianth-members free; ovary inferior, (2–) 3-locular; fruit a loculicidal capsule.

1 Leaf-blade equal-sided or almost so; inflorescences large much branched panicles with many flowers:
  2 Plant climbing by rooting from nodes; stipules more or less persistent, up to 2·5 cm.
    long   1. *glabra*
  2 Plant laxly scambling, upper nodes at least without roots; stipules deciduous about
    2 cm. long   2. *convolvulacea*

1 Leaf-blades unequal-sided; inflorescences smaller with few flowers:
   3 Stem obsolete or shortly rhizomatous; stipules about 1 cm. long, persistent
                                                                  3. *purdieana*
   3 Stem evident:
     4 Plants herbaceous, annual or short-lived; stipules and bracts fimbriate-ciliate,
       persistent:
       5 Leaves reniform, the larger lobe rounded; plant succulent, growing in very wet open
         places   4. *patula*
       5 Leaves with the larger lobe obtusely or acutely pointed; plants not succulent,
         growing on well drained shaded banks:
         6 Stem villous   5. *hirtella*
         6 Stem glabrous   6. *humilis*
     4 Plants shrubby, perennial; stipules entire, mucronate or aristate:
       7 Stipules more or less persistent; leaf-margin sharply dentate and denticulate; leaf-
         blade thinly hispid adaxially   7. *acutifolia*
       7 Stipules deciduous; leaf-margin at most shallowly repand-denticulate; leaf-blade
         glabrous adaxially:
         8 Leaves ovate to lanceolate, short to long caudate-acuminate, pubescent on the
           veins beneath; stamens 8–10   8. *purpurea*
         8 Leaves broadly ovate, acute to shortly acuminate, glabrous, fleshy; stamens 30–50
                                                                     9. *minor*

1. B. glabra Aubl., Hist. Pl. Guiane **2**: 916, t. 349 (1775).

Stem 3–10 m. or more long; leaf-blades broadly ovate, rounded at base, acuminate, denticulate, up to 14 cm. long and 12 cm. broad, shiny on both surfaces;
inflorescence a large pendulous cymose panicle; inner perianth white, segments
3–4 mm. long; larger wing of capsule up to 14 mm. broad, other wings obsolete or
sharply triangular.

Locally common in damp shady woodland, on trees, rocks and stone walls;
1000–4000 feet; fl. and fr. Aug–Mar. *A 5876! H 12860! P 21873!* C. and tropical
S. Amer., Cuba, Dominica, Trinidad.

2. B. convolvulacea (Klotzsch) A. DC. in DC., Prodr. **15** (1): 365 (1864).—
  Wageneria convolvulacea Klotzsch (1855).

Thick-stemmed undershrub; leaf-blades oblate, openly cordate, shortly and
subacuminately 3-lobed, dentate and denticulate, up to 15 cm. long and 25 cm.
broad, thinly pubescent, gland-dotted beneath; peduncles stout, subterminal and
in the upper axils; perianth-segments 6–7 mm. long, white.

Very local (Manch.), sheltered amongst rocks; 2100–2500 feet; fl. and fr. Nov–
Dec. *P 16024! Robertson UCWI 22618!* Native of Brazil, probably an escape from
cultivation.

3. B. purdieana A. DC. in Ann. Sci. Nat. sér. 4, Bot. **11**: 124 (1859).

Stem very short, covered with stipules and roots; petioles 2–6 cm. long; leaf-
blades obliquely ovate, acute or shortly acuminate, irregularly crenate, with
scattered hairs on both surfaces, 8·5–9·5 cm. long, 5 cm. broad; scapes 11–18 cm.
long; flowers white; male perianth-segments 6 mm. long; female perianth-segments
3–5 mm. long; larger wing of capsule up to 15 mm. broad.

Known only from the type (Han.), fl. and fr. Jan. *Purdie*. Endemic.

4. B. patula Haw., Suppl. Pl. Succ.: 100 (1819).—*B. tovarensis* Klotzsch (1855).

Erect herb up to about 50 cm. high; stem pink or red, trailing and rooting at
base, glabrous or nearly so; leaves up to about 6 cm. long and broad, sparsely
ciliate around the shallowly crenate margin and with a few hairs on veins beneath;
stipules lanceolate, 9–12 mm. long; perianth-segments 5–7 mm. long, white;
larger wing of capsule up to 15 mm. broad.

Very rare (West.), in floating mat of vegetation, 50–100 feet; fl. and fr. Nov.
P 27701! Mexico to Bolivia and Peru, Cuba.

5. B. hirtella Link, Enum. Hort. Berol. **2**: 396 (1822).

Stem soft and succulent, ascending from the base, 10–50 cm. high; leaves
obliquely ovate, acute, serrulate-ciliate, thinly hispid, glabrescent, up to 8 cm. long
and 5 cm. broad, thin; inflorescences mostly axillary, few-flowered; inner perianth-
segments of male flowers about 3 mm. long, white or light pink; largest wing of
capsule 9–14 mm. broad.

Local (St. Andr., St. Cath., St. Mary), probably introduced and naturalized on shaded rock cuttings and roadside banks; 250–500 feet; fl. and fr. Jan, Apr. *A 6116! 9024! P 22144! Robertson UCWI 5496!* Native to tropical S. Amer., also Guadeloupe, Martinique.

6. B. humilis Ait., Hort. Kew. **3**: 353 (1789).
Like the last and perhaps not more than varietally distinct.
Reported only once from Jamaica, possibly erroneously, *McNab.* Dominica, Trinidad, Tobago, cultivated elsewhere.

7. B. acutifolia Jacq., Collect. **1**: 128 (1787).—*B. obliqua* L. (1753), partly.
Diffusely branched shrub with red stem, swollen at the nodes, up to 1 m. high; leaves oblong-lanceolate, acuminate, 3·5–10 cm. long, 1–3·5 cm. broad; stipules up to about 1 cm. long; inflorescences terminal and axillary, few-flowered; inner perianth-segments of male flowers 10–17 mm. long, oblanceolate, white, outer pink; largest wing of capsule 14–15 mm. broad.
Locally common (St. Andr., Port., St. Thom.), on sheltered rocky banks and walls; (100–) 2800–6500 feet; fl. and fr. most of the year. *A 8731! H 9148! P 21953!* Cuba.

8. B. purpurea Sw., Nov. Gen. & Sp. Pl.: 86 (1788).
Like the last, up to 120 cm. high; leaves ovate-reniform to oblong-lanceolate, up to 11 cm. long and 4 cm. broad; stipules 8–9 mm. long; inflorescences mostly terminal, several-flowered; inner perianth-segments of male flowers 7–10 mm. long, elliptic-oblanceolate, white, outer pink; largest wing of capsule 15–25 mm. broad.
Common throughout the central and western parishes on rough limestone banks and walls; 1200–3000 feet; fl. and fr. all the year. *A 6522! H 8308! P 21884!* Endemic.

9. B. minor Jacq., Collect. **1**: 126 (1787).
Sparingly branched shrub 60–120 (–200) cm. high with thick stems swollen at the nodes; leaves subentire, 6–15 (–20) cm. long, up to about 12 cm. broad; stipules 1·5–2·5 cm. long; inflorescences terminal and subterminal, many-flowered; inner perianth-segments of male flowers oblong, up to about 10 mm. long, white or pink, rarely red, much smaller than the outer; largest wing of capsule up to about 17 mm. broad.
Locally common in the eastern parishes, rarer in the central parishes; 400–4100 feet; fl. and fr. most of the year. *A 5811! J.P. 896! P 20642!* Endemic.

Several other species of *Begonia* are grown in gardens. *B. heracleifolia* Cham. & Schlecht., native of Mexico, has escaped on to roadside banks in some areas; it has a short thick rhizome, thick petioles with numerous bristly red-based hairs, a ring of larger hairs below the blade and a blade divided deeply into about 8 lobes with dark green margins; the inflorescence is scapose with numerous light pink flowers.

135. CUCURBITACEAE

Climbing or trailing annual or rarely perennial monoecious or dioecious herbs, rarely shrubs or small trees; stems often 5-angled; simple or branched tendrils usually present inserted laterally above the petiole-base. Leaves alternate, simple or rarely compound, usually lobed or divided, without stipules. Flowers unisexual, or very rarely bisexual, actinomorphic, solitary, racemose or paniculate. Perianth biseriate; calyx tubular with (3–) 5 (–6) imbricate or open lobes; corolla with (3–) 5 (–6) free or united petals, imbricate or induplicate-valvate. Male flowers: stamens basically 5 with unilocular anthers opening lengthwise or variously fused into a) 2 pairs plus a single stamen, the pairs often appearing as simple stamens with 2-locular anthers, b) a single central filament-column with free anthers or anthers fused and twisted or variously horizontally or vertically disposed, c) anthers fused but filaments free; rudimentary ovary often present. Female flowers: staminodes sometimes present; ovary inferior; carpels basically 5, mostly reduced to (4–) 3;

unilocular with parietal placentas or 3-locular with apparently axile placentas; ovules anatropous, (1–) few to numerous, style(s) 1 with as many style-arms or stigma-lobes as carpels or rarely 3. Fruit normally berrylike with a soft or hard pericarp, dehiscent or indehiscent. Seeds with straight embryo; endosperm absent. 100 genera with 850 species, mainly tropical and subtropical.

1 Leaves mostly 3-foliolate, fleshy; tendrils branched, the tips holding by appression not often coiling; inflorescence paniculate; male flowers small, 5-merous, light yellow, fragrant; stamens 5; sparingly cultivated in gardens; native of Burma, Thailand and Philippines   *Neoalsomitra sarcophylla* (Wall.) Hutch.
1 Leaves simple, entire or variously lobed; tendrils coiling:
  2 Flowers in panicles, small:
    3 Plants dioecious; stamens 5, anthers free; fruit large, woody; seeds 10–12, discoid, 5–6 cm. in diameter; tendrils 2-branched distally   1. **Fevillea**
    3 Plants monoecious; stamens 3, anthers coherent; fruit 1–1·5 cm. long, fleshy, leathery-skinned; seeds 1–3, compressed-ellipsoid, about 8 mm. long; tendrils unbranched   2. **Cayaponia**
  2 Flowers solitary or in clusters or racemes:
    4 Ovule and seed solitary; monoecious; male flowers racemose or clustered on a long peduncle; stamens 3, filaments connate; fruit 10–12 cm. long, light green, fleshy, with soft scattered prickles; tendrils 3-branched   3. **Sechium**
    4 Ovules and seeds several to many:
      5 Fruit not very large, rarely as much as 15 cm. long; plants wild or completely naturalized, not generally cultivated:
        6 Fruit spherical or subspherical, smooth, about 7 cm. in diameter; flowers solitary, rarely clustered, rather large, monoecious:
          7 Tendrils simple; corolla about 4 cm. long; filaments free; anthers connate, the loculi folded vertically into an erect column   4. **Cionosicys**
          7 Tendrils 3–5-branched; corolla 7–8 cm. long; filaments united; anthers twisted together   5. **Sicana**
        6 Fruit otherwise; tendrils simple; perianth not usually over 2 cm. long; corolla yellow:
          8 Fruit smooth, a soft berry not more than 1·5 cm. long; leaves shallowly lobed to subentire; perianth 4–5 mm. long:
            9 Female flowers solitary on slender axillary peduncles; monoecious; berry ellipsoid to subglobose, finally purplish-black; seeds compressed   6. **Melothria**
            9 Female flowers in short racemes or clustered; dioecious; berry oblong; shortly beaked, ripening orange-red; seeds pyriform   7. **Doyerea**
          8 Fruit tuberculate or prickly, at least 3 cm. long; leaves deeply lobed; perianth 1–2 (–2·5) cm. long; plants monoecious:
            10 Male flowers solitary or racemose, peduncle usually with a bract; ripe fruit tuberculate, orange or red, dehiscent   8. **Momordica**
            10 Male flowers solitary or clustered, peduncles without bracts; ripe fruit spiny, yellow, indehiscent   9. **Cucumis**
      5 Fruit large, at least 15 cm. long, mostly edible or having some definite use; monoecious cultivated annuals, occasionally established near habitations but not indigenous:
        11 Petiole with a pair of hornlike glands at apex; tendrils branched; flowers white; fruits woody, of various shapes and colours; seeds ridged; native of tropical Asia; Bottle Gourd (*Lagenaria vulgaris* Ser.)   *Lagenaria siceraria* (Molina) Standl.
        11 Petiole without glands; flowers yellow:
          12 Mature fruits dry, mostly 15–40 cm. long, up to 8 (?–10) cm. broad, not prickly; pericarp membranous, brittle, opening apically by a lid; endocarp fibrous; tendrils branched; leaves mostly 5–7-lobed; seeds flat, not winged or with only a narrow hyaline margin; petals large, free or nearly so; anthers free; natives of tropical Asia:
            13 Fruit cylindrical, nearly terete with 10 shallow darker grooves; sepals not keeled; petals rounded at tip; stamens usually 5; Loofah, Strainer Vine (*Luffa cylindrica* M. J. Roem.)   *Luffa aegyptiaca* Mill.
            13 Fruit clavate, with 10 sharp longitudinal ridges; sepals keeled; petals emarginate; stamens 3; East Indian Okra   *Luffa acutangula* (L.) Roxb.
          12 Mature fruits fleshy; pericarp leathery, indehiscent:
            14 Leaves deeply pinnatisect; stem and leaves softly woolly; tendrils usually 3- or more-branched; flowers rather small; anthers free; fruit globose, smooth, dark green, the flesh bright pink; seeds shiny; native of tropical Africa; Water Melon (*Citrullus vulgaris* Schrad.)
                Citrullus lanatus (Thunb.) Matsum. & Nakai
            14 Leaves lobed to entire, not pinnatisect; flowers large:
              15 Tendrils simple; corolla deeply lobed, rotate; stamens free:

16 Ovary more or less bristly; fruit tuberculate or prickly, sometimes obscurely
    so, usually cylindrical, flesh greenish-white; leaf-lobes pointed or angular;
    native of S. Asia; Cucumber   *Cucumis sativus* L.
16 Ovary densely pubescent; fruit smooth or rough, glabrescent, often ellip-
    soid, flesh usually yellow to light orange; leaf-lobes mostly rounded;
    natives of subtropical and tropical Asia; Cantaloupe, Musk Melon
                                                          Cucumis melo L.
15 Tendrils branched; fruits not prickly but sometimes warted:
17 Fruit globose to oblong, smooth, glabrescent, white-waxy on the surface;
    corolla deeply lobed; anthers free; seeds white, smooth; native of tropical
    Asia; Chinese Preserving Melon, White Gourd
                                      Benincasa hispida (Thunb.) Cogn.
17 Fruits variously depressed-globose or oblong to pyriform, often with rounded
    ridges, not usually waxy; corolla campanulate, lobed about halfway;
    anthers united; seeds not shiny, bordered; probably originally natives of
    tropical and subtropical America:
18 Fruit-stalk soft and spongy, terete, not expanded at attachment to fruit;
    leaves soft; seeds plump with obtuse indistinct margins; Melon Pumpkin
    (English), Autumn & Winter Squash (American)
                                      Cucurbita maxima Duchesne
18 Fruit-stalk angular, hard; seed-margins distinct:
19 Stem, leaves and petioles setose; fruit-stalk hardly expanded at attach-
    ment to fruit; seed with raised margin; Field Pumpkin, Ornamental
    Gourds, Vegetable Marrow   *Cucurbita pepo* L.
19 Stem, leaves and petioles more or less softly hairy; fruit-stalk broadly
    expanded at attachment to fruit; seeds thin with hyaline margin; Pie
    Pumpkins   *Cucurbita moschata* Duchesne

1. FEVILLEA L. (1753)

1. F. cordifolia L., Sp. Pl. **2**: 1013 (1753).—Antidote Caccoon, Nhandiroba,
Segra Seed.

Perennial with grooved stems; leaves entire or shallowly or deeply lobed, nearly
glabrous, up to about 17 cm. long and 15 cm. broad; inflorescence puberulous;
sepals 2 mm. long; petals yellow, 4 mm. long.

Rather local in wet areas, cultivated and in thickets and on trees; 100–1950 feet;
fl. Mar–May, fr. Sept. *Cornman 599! H 8381! P 11871! 20713!* Continental
tropical Amer., West Indies.

2. CAYAPONIA Manso (1836) nom. cons.

1. C. racemosa (Mill.) Cogn. in A. & C. DC., Monogr. Phan. **3**: 768 (1881).—
Bryonia racemosa Mill. (1768). Mountain Bryony.

Perennial or annual with grooved stems up to 15 m. long; leaves subentire to
deeply 7-lobed, the lobes mostly obtuse, mucronate, scabrid adaxially, up to about
15 cm. long and broad; inflorescence glabrous; flowers greenish-white to yellow,
male about 8 mm. across, female slightly smaller; petals pubescent within; ripe
fruit smooth, orange or scarlet.

Rather common on trees and in thickets; 50–2400 feet; fl. Oct–Apr, fr. Oct–May.
A 8767! H 10107! P 16716! C. and northern S. Amer., Greater Antilles, Virgin
Is., Barbados, Trinidad.

C. attenuata (Hook. & Arn.) Cogn. and *C. laciniosa* (L.) C. Jeffrey have been
reported but further material is needed.

3. SECHIUM P. Browne (1756) nom. cons.

1. S. edule (Jacq.) Sw., Fl. Ind. Occ. **2**: 1150 (1800).—*Sicyos edulis* Jacq. (1760).
Cho Cho.

Perennial with hairy stem to 4 m. or more long from a massive fleshy root;
leaves thin, deeply cordate at base, shallowly lobed with the mid-lobe acuminate,
thinly pubescent, smooth, up to 25 cm. long and nearly as broad; flowers densely
puberulous; male flower-buds globose; female corolla 4 mm. long, petals acute,
light yellow; seed ovate, compressed, germinating within the fruit.

Commonly cultivated and established here and there; mostly at middle elevations; fl. and fr. most of the year. *Robertson UCWI 5555!* Continental tropical
Amer., West Indies; introduced into the Old World tropics.

4. CIONOSICYS Griseb. (1860)

1. C. pomiformis Griseb., Fl. Br. W.I.: 288 (1860).—Wild Melon.

Perennial with slender flexible glabrescent stem to 13 m. or more long; leaves ovate and entire to shallowly or deeply 3-lobed, the lobes acute, margin denticulate at the ends of veins, cordate with a rounded sinus, minutely scabridulous especially beneath, up to about 15 cm. long and 12 cm. broad; corolla white streaked green outside, hairy within; seeds about 12 mm. long.

Common on trees in woodlands in wet areas; 250–5000 feet; fl. Feb–May, fr. Feb–Aug. *A 6791! 9278! H 8534! Powell 1073!* Endemic.

5. SICANA Naud. (1862)

1. S. spherica Hook. f. in Curt., Bot. Mag. **116**: t. 7109 (1890).

Annual climber, glabrescent; leaves deeply 3–5-lobed, the lobes ovate, acuminate, entire or with small teeth, cordate at base with a rounded sinus, mostly minutely pustulate, up to 12 cm. long and broad; calyx pubescent or tomentose; corolla tomentose within; fruit glabrous; seeds narrowly winged.

Very rare (St. Andr.), on wooded hillsides; 3100–5000 feet; fl. Jan, Apr, fr. Apr. *Bengry! Fawcett!* Cuba.

6. MELOTHRIA L. (1753)

1. M. guadalupensis (Spreng.) Cogn. in A. & C. DC., Monogr. Phan. **3**: 580 (1881).—*Bryonia guadalupensis* Spreng. (1826). *M. fluminensis* Gardn. (1842).

Stem slender, branched, glabrescent, to 2 m. or more long; leaves entire to 3–5-lobed, the terminal lobe usually acute, the lateral and basal rounded, margin denticulate at the ends of veins, thin, more or less scabrid, up to 10 cm. long and broad; male racemes about as long as the petioles; corolla-limb spreading, sometimes as much as 1 cm. broad; fruit about 1·5 cm. long; seeds 4 mm. long.

Common in thickets and on banks; SL–5000 feet; fl. and fr. most of the year. *A 8238! H 7796! P 7666!* Bahamas, continental Amer., West Indies.

7. DOYEREA Grosourdy (1864)

1. D. emetocathartica Grosourdy, El Med. Bot. Criollo **2**: 338 (1864).— *Cissus ? cucurbitacea* Britton (1910); F. & R. Fl. Jam. **5**: 79 (1926). *Corallocarpus emetocatharticus* (Grosourdy) Cogn. (1891).

Perennial shrubby climber, the stock persistent with a short thick trunk, annual leafy growth to 6 m. or more long, internodes zig-zag, succulent, glabrescent; leaf-blade obtuse, margin sinuate to subentire, denticulate, openly cordate, scabrid on both surfaces with tubercle-based hairs, 4–9 cm. long, 3–7·5 cm. broad; racemes 2–7, in axillary clusters, up to 1·5 cm. long, crisped-pubescent; sepals and petals valvate; seeds about 3 mm. long.

Rare and local (St. Cath., Manch.), in arid coastal scrub-thicket and on rocks; SL–200 feet; fl. Nov–Feb, fr. Jan–Feb. *A 8893! H & B 10512!* Mexico to Venezuela, Puerto Rico, Guadeloupe, Margarita, Trinidad.

8. MOMORDICA L. (1753)

Bract of male peduncle closely subtending the flower; leaves lobed to about the middle, lobes coarsely lobulate-dentate   1. *balsamina*
Bract of male peduncle at or below the middle; leaves deeply lobed, lobes more or less sinuate   2. *charantia*

1. M. balsamina L., Sp. Pl. **2**: 1009 (1753).—Cerasee.

Trailing and climbing, glabrescent; stems 2 m. or more long; leaves with rhomboidal lobes (2·5–) 4–7 cm. long; bract on male peduncle roundish-cordate, toothed; fruit 3–6 cm. long, rather smooth at maturity.

Rare in the wild state (St. Cath., Manch.), on sea-beach and coastal rocks; sometimes cultivated; fl. and fr. June, Oct–Dec. *Cornman 137! 633! H 6415* Common in the drier parts of the Old World tropics.

2. M. charantia L., Sp. Pl. 2: 1009 (1753).—Wild Cerasee.
Climber to 6 m. or more, the old stem compressed and fluted, puberulous to tomentose when young; leaves cut into 5–7 narrow-based lobes, the lobes mostly obtuse and mucronulate, up to 12 cm. long; bract on male peduncle reniform to roundish-cordate, entire; fruit narrowed to both ends, 8–15 cm. long; seeds covered with red pulp.
Very common on fences, hedgerows, beaches and shrubs in disturbed areas; SL–3500 feet; fl. and fr. all the year. *A 8198! H 6779! P 11196!* In the subtropics and tropics of both hemispheres.

9. CUCUMIS L. (1753)

1. C. anguria L., Sp. Pl. 2: 1011 (1753).—West Indian Gherkin, Wild Cucumber.
Annual; stems trailing to 2 m. or more, angled, roughly setose with long white hairs; leaves deeply 3-lobed, the lateral lobes again divided, the lobes obtuse, margin denticulate, up to about 10 cm. long on a petiole as long; corolla about 1·5 cm. long; ripe fruit on long stout peduncle, ellipsoid, covered with curved prickles, 4–7 cm. long.
Locally common, in rough waste places; SL–1500 feet; fl. and fr. sporadically throughout the year. *H 9043! P 8614! Yuncker 17306!* Subtropical and tropical continental Amer., West Indies; native of Africa.

136. LYTHRACEAE

Herbs, shrubs or trees; young stems often 4-angled. Leaves usually opposite, subopposite or whorled, simple, usually entire; stipules very small or wanting. Flowers bisexual, actinomorphic or sometimes zygomorphic, solitary, cymose-fasciculate or in panicles. Perianth biseriate or uniseriate; calyx with a short or long tube (hypanthium), with 4–8 valvate lobes and sometimes as many intermediate lobes; petals as many as and alternate with sepals or absent, perigynous, crumpled in bud, imbricate. Stamens as many as or twice as many as sepals, rarely 1 or numerous, typically in 2 whorls, the outer whorl alternate with petals, hypogynous or perigynous, often unequal; anthers 2-locular, introrse, dorsifixed, opening lengthwise. Ovary superior, sessile or shortly stipitate, (1–) 2–6-locular, the septa not always complete; ovules anatropous, several to numerous, ascending on usually axile placentas; style and stigma simple. Fruit a capsule. Seeds with straight embryo; endosperm absent.
22 genera with 500 species in all regions, most numerous in the American tropics.

1 Calyx cylindrical; flowers irregular, mostly 6-merous; plants herbaceous or shrubby; stamens (5–) 11 (–12); capsule and hypanthium splitting lengthwise in fruit
1. **Cuphea**
1 Calyx campanulate or globose, about as wide as long; flowers regular:
  2 Plants herbaceous, annual; petals present or absent; stamens 4–8; lateral veins of leaves very obscure:
    3 Capsule dehiscing septicidally, the outer wall finely striate; flowers solitary in the leaf-axils; middle and upper leaves narrowed at base; stem creeping   2. **Rotala**
    3 Capsule dehiscing irregularly, smooth; flowers usually several together in the leaf-axils, fascicled or pedunculate; middle and upper leaves often cordate or auricled; stem erect   3. **Ammannia**
  2 Plants woody, shrubs or trees; petals present; stamens 10 or more; lateral veins of leaves evident:
    4 Flowers solitary in the leaf-axils; petals 5–7, yellow   4. **Heimia**
    4 Flowers in terminal panicles:
      5 Flowers 4-merous, small, very fragrant; petals creamy-white; capsule membranous, irregularly dehiscent; cultivated in gardens; native of tropical Asia; Henna
Lawsonia inermis L.

5 Flowers 6-merous, showy; petals rose-pink, mauve or white; capsule woody, loculicidally dehiscent:
  6 Shrub; leaves up to about 6 cm. long and 3 cm. broad, with 7 pairs of lateral veins;
  calyx smooth; cultivated in gardens; native of S.E. Asia and N. Australia;
  Crape Myrtle, June Rose   *Lagerstroemia indica* L.
  6 Tree; leaves up to about 15 cm. long and 8 cm. broad, with 10 pairs of lateral veins;
  calyx ribbed; planted along roadsides and in gardens; native of India to S. China,
  Indonesia, N. Australia; Queen's Flower Tree   *Lagerstroemia speciosa* (L.) Pers.

1. CUPHEA P. Browne (1756)

1 Calyx 2–2·5 cm. long, bright red, black-violet distally, spur globose; petals wanting
                                                                      1. *ignea*
1 Calyx up to 1 cm. long, green or tinged purplish, not uniformly red, gibbous or shortly
  spurred at base; petals present, purple:
  2 Flowers paired in bracteate terminal racemes; plant shrubby   2. *decandra*
  2 Flowers mostly solitary in leaf-axils; plant more or less herbaceous:
    3 Calyx 4 mm. long in flower, 5–6 mm. long in fruit; leaves 5–15 (–25) mm. long
                                                                      3. *parsonsia*
    3 Calyx 8–10 mm. long; leaves 1·5–5 cm. long:
      4 Annual viscous herb; calyx viscous-hispid, about 10 mm. long   4. *viscosissima*
      4 Perennial pubescent-stemmed herb; calyx puberulous, strongly striate, 8–9 mm.
        long   5. *melanium*

1. C. ignea A. DC. in Fl. des Serres **5**: t. 500 (1849).—*C. platycentra* of Lemaire
(1846), not Benth. (1839). Cigar Flower.
Diffusely branched perennial somewhat shrubby herb 30–50 cm. high or sometimes scrambling to 2 m.; leaf-blade elliptical, acuminate, narrowly cuneate at base,
smooth, up to 8 cm. long and 2·5 cm. broad, usually smaller; flowers solitary,
axillary; stamens 11; capsule 8–9 mm. long; seeds not winged.
Locally common (St. Andr., St. Thom.), an escape from cultivation, well established on pathside banks and in margins of open woodland; 2800–6000 feet; fl.
most of the year. *A 5696! H 11935! P 8315!* Native of Mexico, now widespread as
an ornamental.

2. C. decandra Ait. f. in Ait., Hort. Kew. ed. 2, **3**: 151 (1811).—*C. ciliata* (Sw.)
Koehne (1881), not Ruiz & Pav. (1798). *Lythrum ciliatum* Sw. (1788).
Slender-stemmed shrub 30–150 cm. high; leaves elliptic-obovate, acute to
obtuse, often cuspidate, up to 3 cm. long and 1·5 cm. broad, smooth; pedicels
slender, 2–5 mm. long; calyx green tinged reddish, about 8 mm. long; stamens
10–11; capsule about 5 mm. long.
Rare and rather local (St. Andr., Clar.), on rocks and cliffs; 300–2500 feet;
fl. Apr–Dec, fr. May–Dec. *A 8167! H 9640! P 21391!* Cuba, Hispaniola, Mexico,
Colombia.

3. C. parsonsia (L.) R. Br. ex Steud., Nomencl. Bot.: 245 (1821).—*Lythrum
parsonsia* L. (1759). Strongback.
Stem rather wiry, creeping and rooting, branched and at length ascending, up to
60 cm. long, puberulous; leaves ovate to elliptical, mostly acute, scabridulous-
margined; calyx striate, green often tinged reddish; petals about 2 mm. long;
stamens (4–) 6 (–9); capsule 3–4 mm. long; seeds narrowly winged.
Common in ditches, swamp margins and sometimes on clay pastures, swamp margins and sometimes on
strand; SL–3000 feet; fl. and fr. most of the year. *A 10051! H 12054! H & P 13315!*
Mexico, Bahamas, Greater Antilles, Dominica, Martinique.

4. C. viscosissima Jacq., Hort. Vindob. **2**: 83, t. 177 (1772).—*C. petiolata* (L.)
Koehne (1881), not Pohl ex Koehne (1877). *Lythrum petiolatum* L. (1753).
Viscous-glandular, densely pubescent annual herb; leaves ovate-lanceolate to
lanceolate, 2–5 cm. long; stamens 11.
An early casual introduction (St. Andr., St. Thom.), a weed not recently found;
reported by Macfadyen (1850). Native of E. United States.

5. C. melanium (L.) R. Br. ex Steud., Nomencl. Bot.: 245 (1821).—*Lythrum
melanium* L. (1759).
Perennial herb reputedly with an odour of onions; stems rather woody trailing

and rooting, with short curled and a few long hairs; leaves ovate, acute, scabrid, up to 4·5 cm. long and 3 cm. broad; petals up to 7 mm. long; capsule with 4–8 seeds.

Very rare, not recently reported (Macfadyen, 1850). Cuba, Hispaniola, Lesser Antilles.

Cuphea hyssopifolia Kunth, a small perennial heathlike species, native of Mexico, is occasionally grown as an ornamental pot-plant; other annual species are also sometimes cultivated.

2. ROTALA L. (1771)

1. **R. ramosior** (L.) Koehne in Mart., Fl. Bras. **13** (2): 194, t. 39, f. 1 (1877).— *Ammannia ramosior* L. (1753).

Stem creeping and spreading, rooting at nodes, tetragonal, glabrous, the flowering branches ascending and 10–15 cm. high; leaves linear-oblanceolate, 9–16 mm. long, up to 2·5 mm. broad; capsule oblong-globose, nearly 3 mm. long.

Very rare (St. Eliz.), at pond margin in wet open ground; 5–10 feet; fl. and fr. Jan. *Robertson UCWI 5413!* S. and E. United States to Brazil, West Indies, Philippines.

3. AMMANNIA L. (1753)

1 Style more than 1·5 mm. long, conspicuous, usually bent; petals present; flowers 1–3 in the axil of each leaf; leaves markedly auricled; capsule 2–3 mm. in diameter
           1. *coccinea*
1 Style very short; petals absent:
  2 Flowers 1–2 (–3) in the axil of each leaf; capsule 3 mm. or more in diameter; leaves markedly auricled, minutely punctate but smooth adaxially when dried   2. *latifolia*
  2 Flowers 4–10 in the the axil of each leaf; capsule up to 2 mm. in diameter; leaves narrow or broad at base but hardly auricled, rugulose adaxially when dried
           3a. *baccifera* ssp. *aegyptiaca*

1. **A. coccinea** Rottb., Pl. Hort. Univ. Rar. Progr.: 7 (1773).

Stem simple or more usually branched above, often spongy at base, glabrous, up to 60 cm. high; leaves decussate, linear-lanceolate, acute, glabrous but scabridulous-margined, up to 6 (–10) cm. long, mostly less than 8 mm. broad; pedicels short, slender, bracteolate; calyx about 2 mm. long in flower, much larger in fruit; petals 4 (–5), obovate, bright rose-pink to deep magenta, 1·5–2 mm. long; stamens 4 (–5).

Not uncommon in damp low pastures and thickets and in ponds, along riverbanks and in running water; SL–100 feet; fl. and fr. Aug–Apr. *A 6877! H 10377! P 18339!* E. and S. United States, Mexico to Brazil, West Indies, Pacific Is., Philippines, Japan, S.W. Europe.

2. **A. latifolia** L., Sp. Pl. **1**: 119 (1753).

Stem erect or ascending from a decumbent rooting base, branched above, glabrous, more or less tetragonal, up to 1 m. high; leaves decussate, elliptic-lanceolate, mostly rather obtuse-tipped, up to 8 cm. long and 1·8 cm. broad, usually smaller; flowers subsessile, bracteolate; calyx nearly 3 mm. long in flower, ribbed, with short spreading teeth; stamens 4 or 8, enclosed within the calyx.

Rather common in swamps, ditches, riverside mud and at lagoon margins; SL–100 (–700) feet; fl. and fr. all the year. *A 7581! H 10337! P 10193!* S.E. United States to Paraguay, West Indies; Grand Cayman. A variant with the lower leaves all narrow-based and the upper more or less oblanceolate and obtuse is represented by one collection from Jamaica (*A 8577!*; sandy roadside (Han.), 10 feet; fl. and fr. Dec.), and has been observed in herbaria from localities in Mexico and Peru.

3. **A. baccifera** L., Sp. Pl. ed. 2, **1**: 175 (1762).

3a. Ssp. **aegyptiaca** (Willd.) Koehne in Engl., Bot. Jahrb. **1**: 260 (1880).—*A. aegyptiaca* Willd. (1809).

Stem branched from near the base, more or less tetragonal, glabrous, 30–90 cm. high; leaves mostly oblanceolate, acute, 3–5 cm. long, 4–9 mm. broad, much smaller on the lateral branches; flowers very small, sessile; calyx-teeth spreading, greenish-brown; fruit red.

Rare and local (St. Andr.), in gravel of dry river-bed; 250–500 feet; fl. and fr. Oct–Feb. *Powell 744! Robertson UCWI 5454! Yuncker 17154!* S. Europe and the Old World tropics.

4. HEIMIA Link (1822)

1. **H. salicifolia** Link, Enum. Hort. Berol. 2: 3 (1822).—*Nesaea salicifolia* Kunth (1824).

Shrubby branched perennial; stems tetragonal and narrowly winged, 1–2 m. high; leaves elliptic-lanceolate, narrowed at both ends, glabrous, up to 8 cm. long and 1 cm. broad; pedicels 1–2·5 mm. long; bracteoles below the calyx leafy, obovate; calyx 5–7 mm. long, the lobes acute, bent inwards in fruit, the intermediate lobes bent outwards; petals usually 6, obovate, about 1–1·5 cm. long; capsule about 4 mm. long.

Local (St. Andr., St. Mary, St. Thom.), in riverside thickets; 500–3500 feet; fl. and fr. Feb–Sept. *A 7708! H 11972! P 24058!* Lower California, Mexico to Argentina.

137. PUNICACEAE

Shrubs or small trees, sometimes spiny. Leaves opposite, subopposite or clustered on short shoots, simple, entire, glabrous, without stipules. Flowers bisexual, actinomorphic, terminal, solitary, clustered or cymose. Perianth biseriate; calyx tubular (hypanthium), adnate to the ovary, coloured, with 5–7 valvate fleshy lobes; corolla perigynous, the petals 5–7, free, clawed, imbricate, crumpled in bud. Stamens numerous, perigynous, free; anthers 2-locular, opening lengthwise. Ovary inferior, many-locular, the loculi in 2 series at first concentric, later super-posed, the lower with axile, the upper with parietal placentation; ovules numerous, anatropous; style and stigma simple. Fruit a berry crowned by the persistent calyx-limb. Seeds with a pulpy testa; embryo straight; endosperm absent.

1 genus with 2 species, natives of S.W. Asia and Socotra.

1. PUNICA L. (1753)

1. **P. granatum** L., Sp. Pl. 1: 472 (1753).—Pomegranate.

Shrub with lax slender 4-winged branches, up to 2·5 m. high; lateral branchlets sometimes developed into spines; leaves oblanceolate (3–) 4–6 (–8) cm. long, up to 2·5 cm. broad; calyx about 2 cm. long, fleshy, light red, the lobes triangular, 5–7 mm. long; petals orange-red, caducous, 2–2·5 cm. long, 12–13 mm. broad; anthers dorsifixed, subpeltate; fruit spherical, about 5–7 cm. in diameter.

Rather common as an ornamental, rarely naturalized; fl. and fr. sporadically throughout the year. *H 11892! P 23039! Wynter UCWI 5636!* Cultivated generally in the subtropics; naturalized in S. Europe.

138. LECYTHIDACEAE

Trees. Leaves alternate, simple, without stipules. Flowers bisexual, actinomorphic or zygomorphic in axillary, terminal or cauline clusters, racemes or panicles. Perianth usually biseriate; calyx (2–) 4–6-lobed, the lobes valvate or imbricate; petals 4–8, free or connate into a campanulate ribbed tube at the base, imbricate, or absent.[1] Stamens numerous in several series, the outer sometimes staminodal and forming a false corolla[1] or a corona; filaments sometimes united and turned to one side of the flower; anthers mostly basifixed, opening by a lateral split; staminal disk present, sometimes lobed. Ovary inferior or half-inferior, 2–6-locular; ovules

---

[1] Depending on the interpretation of the petaloid part of the flower as of staminodal origin (*Napoleonoideae*).

1–many, anatropous, axile or pendulous; style usually simple. Fruit fibrous, woody or drupaceous, indehiscent or an operculate capsule. Seeds usually large, without endosperm.

24 genera with about 450 species, mostly in wet intertropical regions.

1 Petals 6, red; filaments fused and directed forwards; flowers borne in woody racemes from the trunk; fruit globose, very large, several-seeded; cultivated; native of northern S. America; Cannonball Tree   *Couroupita guianensis* Aubl.
1 Petals 4 (–5), white; filaments free, regular; fruit usually 1-seeded, crowned by the calyx-limb:
  2 Flowers showy in racemes borne on the branches; stamens spreading beyond the petals; fruit 4-sided, about 8 cm. broad; cultivated and probably naturalized in coastal parts of the parish of Portland; native of S.E. Asia, Australia, Madagascar and the Mascarene Is.   *Barringtonia asiatica* (L.) Kurz
  2 Flowers in clusters on the trunk and branches; stamens inflexed; fruit 8 (–10)-ribbed, drupaceous, 4–5 cm. broad   **1. Grias**

1. GRIAS L. (1759)

1. G. cauliflora L., Syst. Nat. ed. 10, **2**: 1075 (1759).—Anchovy, Anchovy Pear.

Tree 6–15 m. high with a slender sparingly branched trunk; leaves more or less clustered at the ends of branches, oblanceolate, acuminate, long-tapered to base, rather leathery, up to over 1 m. long and 20 cm. broad; flowers fragrant; calyx about 3 mm. long; petals 15–20 mm. long; fruit ellipsoid, pointed, 6–9 cm. long.

Rather local and gregarious, near streams and in marsh forest; SL–1000 feet; fl. Feb–May, Sept, fr. July. *A 8149! P 22127! 24775! Thompson 7250!* Probably endemic.

139. RHIZOPHORACEAE

Shrubs or trees, often with swollen nodes. Leaves usually opposite, leathery, simple, with caducous interpetiolar stipules, rarely alternate without stipules. Flowers usually bisexual, actinomorphic, solitary or cymose in axillary inflorescences. Perianth biseriate, perigynous to epigynous; calyx of 3–14 segments, more or less connate basally, adnate to the ovary or free, persistent, valvate; petals as many as sepals, usually small, free, often fleshy or leathery, often emarginate or lacerate, convolute or inflexed in bud. Stamens as many as petals, twice as many or numerous, often in pairs opposite the petals, inserted on a lobed perigynous or epigynous disk; anthers 2–many-locular, opening lengthwise. Ovary superior to inferior, 2–6-locular, rarely 1-locular by suppression of septa; ovules usually 2, pendulous, anatropous, on axile placentas; style usually solitary; stigma with as many lobes as carpels, Fruit usually indehiscent, a berry or drupe, or a tardily dehiscent septicidal capsule. Seed with a straight embryo, with or without endosperm.

17 genera with 100 or more species in the tropics.

Flowers in forked cymes; ovary inferior; sepals and petals 4, entire; 2 bracteoles united below calyx into a cup, anthers with numerous round pollen-sacs; seed germinating within the fruit while latter still attached to twig   **1. Rhizophora**
Flowers solitary or in clusters; ovary free; calyx perigynous, 4–5-lobed; petals 4–5, clawed, fringed; bracteoles absent; anthers 2- or 4- locular; seeds arillate   **2. Cassipourea**

1. RHIZOPHORA L. (1753); Graham (1964)

1. R. mangle L., Sp. Pl. **1**: 443 (1753).—Red Mangrove.

Tree with adventitious prop roots (4–) 8–16 m. high; leaves elliptical, leathery, glabrous, mostly 7–15 cm. long, 3·5–6·5 cm. broad; stipules 2·5–4 cm. long; sepals about 1 cm. long, often yellowish; petals 7–8 mm. long, villous within, white or yellow, often with a brownish balsam-scented exudate; stamens 8; fruit 2·5–3·5 cm. long; germinated radicle up to about 25 cm. long, fusiform.

Common along muddy shores and in estuarine swamps, occasional inland; SL–20 feet; fl. June–Feb, fr. July–Mar. *A 8299! H 6198! P 23841!* Coasts of continental tropical Amer., West Indies, W. Africa and Pacific Is.; Grand Cayman.

2. CASSIPOUREA Aubl. (1775); Alston (1925)

1 Flowers distinctly pedicelled:
  2 Pedicels about 5 mm. long; leaf-blade cuneate at base   1. *elliptica*
  2 Pedicels about 3 mm. long; leaf-blade obtuse at base   2. *brittoniana*
1 Flower sessile or subsessile:
  3 Leaf-blade elliptical, obtuse to rounded at base   3. *subsessilis*
  3 Leaf-blade ovate, subcordate or emarginate at base   4. *subcordata*

1. C. elliptica (Sw.) Poir. in Lam., Encycl. Méth. Bot. Suppl. **2**: 131 (1811).—
Legnotis elliptica Sw. (1788).
  Shrub 2·5 m. or tree 3–5 m. high; leaves elliptical to lanceolate, variously
acuminate, shiny, 3·5–8 cm. long, 1–3·5 cm. broad; stipules 5–6 mm. long; flowers
3–5 together; calyx-tube 3–3·5 mm. long, lobes 2–2·5 mm. long; petals white,
hairy; capsule ovoid, 3-angled, 3-valved.
  Occasional in the central and western parishes, in open woodlands on limestone;
450–2300 feet; fl. Sept–Mar, fr. Mar. *A 8654! H 10330! P 23138! 23368!* Honduras
to Panama, Cuba, Puerto Rico, Guadeloupe to Grenada, Trinidad.

2. C. brittoniana F. & R. in Journ. Bot. **64**: 14 (1926).—Coco Plum of Troy.
  Tree 2·5–10 m. high; leaves elliptical, shortly acuminate, 4–6 cm. long, 1·2–2·8
cm. broad; stipules 4 mm. long, glabrous; flowers 1–3 together; calyx-tube 2·5 mm.
long, lobes 3·5 mm. long; petals greenish or white, hairy within; capsule ridged.
  Rare and local (Trel.), in woodland on limestone hills; 1500–2000 feet; fl. Jan,
July, Nov, fr. July. *H 10670! P 18352!* Endemic.

3. C. subsessilis Britton in Bull. Torr. Bot. Club **35**: 340 (1908).
  Tree 6–8 m. high; leaves elliptical to elliptic-lanceolate, acuminate, (3–) 6–9 cm.
long, (1–) 2–3 cm. broad; stipules 5 mm. long, puberulous with appressed hairs;
flowers 1 or 2 together; calyx-tube about 3 mm. long, lobes 3–3·5 mm. long; petals
not seen; capsule oblong-conical, 1·3 cm. long.
  Very rare (Han.), in woodland on limestone; about 1300 feet; fl. and fr. Mar.
H 10307! Endemic.

4. C. subcordata Britton in Bull. Torr. Bot. Club **35**: 340 (1908).
  Shrub or tree 4 m. high; leaves ovate, shortly acuminate, 3·5–7 cm. long, 2–3 cm.
broad; stipules 8–9 mm. long; flowers 1–3 together; calyx-tube 2–2·5 mm. long,
lobes 2·5–3 mm. long; petals white; capsule as above.
  Very rare (Trel.), in marsh and along a brook in limestone hills; about 1600 feet;
fl. and fr. Sept. *H 9466!* Endemic.

140. COMBRETACEAE

Trees or shrubs, often climbing. Leaves usually alternate, less often opposite,
rarely whorled, simple, without stipules. Flowers bisexual or rarely unisexual,
usually actinomorphic, bracteate in spikes, racemes or heads, these sometimes in
panicles. Perianth biseriate or uniseriate; calyx (or receptacle) fused around the
ovary to form a hypanthium, often tubular or cupular above, 4–5 (–8)-lobed, the
lobes usually persistent, valvate; petals as many as the calyx-lobes, free and imbricate or valvate or absent. Stamens (2–) 4–5 or twice as many as the calyx-lobes and
then in 2 series, usually inserted on the calyx-tube; anthers 2-locular, opening
lengthwise. Ovary inferior, 1-locular; ovules 2–6, anatropous, pendulous from the
apex of the loculus; style slender; stigma usually simple. Fruit drupaceous and
angled and leathery or membranous and winged, 1-seeded, usually indehiscent.
Seeds without endosperm.
  18 genera with about 500 species in the subtropics and tropics.

1 Petals present; leaves opposite or subopposite:
  2 Erect shrubs or trees; petiole with 2 glands at or near apex; inflorescences axillary
    spikes and terminal panicles; bracteoles 2, adnate to calyx; petals very small
             1. Laguncularia

2 Climbing shrubs; flowers often showy:
  3 Style free from hypanthium (calyx-tube); calyx-tube up to about 2 cm. long; fruits
    broadly and thinly 4–5-winged   **2. Combretum**
  3 Style adnate to hypanthium; calyx-tube 7–8 cm. long; fruits narrowly and thickly
    winged; cultivated; native of Old World tropics; Rangoon Creeper
                                                    Quisqualis indica L.
1 Petals absent; leaves alternate or subverticillate; erect shrubs or trees; flowers small:
  4 Flowers in heads; fruiting head conelike   **3. Conocarpus**
  4 Flowers in spikes; fruit drupaceous:
    5 Calyx persistent; anthers versatile; fruit ovoid, 5-angled, 6–8 mm. long   **4. Bucida**
    5 Calyx deciduous:
      6 Anthers versatile; fruit ellipsoid, 2-edged, compressed   **5. Terminalia**
      6 Anthers fixed at the enlarged apex of filaments; fruit oblong, subterete, 1·5–2·5 cm.
      long   **6. Buchenavia**

1. LAGUNCULARIA Gaertn. f. (1807)

1. **L. racemosa** (L.) Gaertn. f. in Gaertn., Fruct. & Sem. Pl. **3**: 209, t. 217, f. 3
(1807).—*Conocarpus racemosus* L. (1759). White Mangrove.
  Shrub 3–5 m. or tree up to 20 m. high; leaves oblong or oblong-elliptical, rounded
or retuse at apex, rounded or cuneate at base, glabrous; inflorescence puberulous;
calyx 2 mm. long, not or hardly exceeded by the white or yellowish-white petals;
stamens 10, included; fruit obovoid, unequally ribbed, 1–2 cm. long.
  Common along the margins of lagoons and brackish creeks and also on the cays;
SL–10 feet; fl. May–Nov, fr. June–Oct. *A 7563! H 8275! P 15352!* Eastern tropical
Amer., West Indies; Cayman Is., W. tropical Africa.

2. COMBRETUM Loefl. (1758) nom. cons.

1 Racemes cylindrical; hypanthium cup- or bell-shaped, abruptly narrowed above the
  ovary; petals 4, clawed, spreading, limb 1 mm. long, light yellow; fruit ellipsoid,
  3–4 cm. long, 4-winged   1. *laxum*
1 Racemes secund; hypanthium tubular or funnel-shaped; petals red; fruit orbicular;
  introduced ornamental plants:
  2 Petals 4, inflexed, limb broadly ovate, 2 mm. long; fruit 4-winged; native of Mada-
    gascar   *C. coccineum* (Sonn.) Lam.
  2 Petals 5, erect, limb spathulate, 10–15 mm. long; fruit 5 (–6)-winged; native of tropical
    Africa   *C. grandiflorum* G. Don

1. **C. laxum** Jacq., Enum. Syst. Pl. Carib.: 19 (1760).—*C. marchii* F. & R. (1925).
C. robinsonii F. & R. (1925). Red Withe.
  Climbing shrub to 15 m. or more high; young twigs, petioles and veins of leaf
beneath rusty-tomentose; leaf-blade oblong-elliptical, acuminate, cuneate to
rounded or auriculate at base, up to 25 cm. long and 11 cm. broad; racemes dense,
up to about 6 cm. long in axillary or terminal panicles; fruit 3–4 cm. long.
  Rare (St. Eliz., West., Han.), on trees and in thickets in marsh or riparian forest;
25–50 feet; fl. Jan, fr. Jan, Sept. *Harris! P 24561! Skelding UCWI 3594!* Mexico to
Brazil, Cuba, Hispaniola, Trinidad. When better known *C. marchii* and *C. robinsonii* may prove to be at least varietally distinct, but in the state of our present
knowledge both species seem to fall well within the range. of *C. laxum*.

3. CONOCARPUS L. (1753)

1. **C. erectus** L., Sp. Pl. **1**: 176 (1753).—Button Mangrove.

1a. Var. erectus.
  Shrub or small tree 3–5 m. high, rarely a much larger tree to over 20 m.; sometimes, on exposed coral limestone, plants flower and fruit at 15 cm. high; leaves very
variable, linear-lanceolate and long-acute to suborbicular and rounded or emarginate,
with glands in pits along either side of the midrib beneath and on either side of the
blade at base, up to 11 cm. long and 4 cm. broad, softly leathery, glabrous or
glabrescent; flower-heads pedunculate, solitary, in clusters or more usually in
axillary and terminal racemes; fruit 7 mm. broad including the wings.

Common at the inner margins of mangrove swamps and in thickets on salinas and also on the cays; SL–20 (–100) feet; fl. and fr. most of the year. *A 7351! Campbell 6177! P 6004!* Florida, Bahamas, continental tropical Amer., West Indies; Cayman Is., W. tropical Africa.

1b. C. erectus var. **sericeus** Griseb., Fl. Br. W.I.: 277 (1860).—*C. sericeus* (Griseb.) Jiménez (1953).
Differs in having usually larger leaves covered with a soft silky indumentum.
Locally common in south coastal areas at salina margins and on limestone rocks; SL–5 feet; fl. and fr. sporadically. *A 8821! Asprey UCWI 2438! A. M. Barry!* Yucatan, Bahamas, Hispaniola.

4. BUCIDA L. (1759) nom. cons.

1. B. buceras L., Syst. Nat. ed. 10, 2: 1025 (1759).—*Buceras bucida* Crantz (1766). Black Olive.
Tree 6–25 m. high with widely spreading branches; trunk light-coloured, figured with fine striations, the old nodes marked by rings of holes; juvenile plants have spines at the nodes and oblanceolate, cuneate-based leaves with acute or subacuminate tips, older plants have obovate, round-tipped or emarginate leaves up to 9 cm. long; inflorescence yellowish-brown tomentose, up to about 8 cm. long including peduncle; flowers light yellow; stamens exserted; ovary sometimes developing into a curved hornlike gall several centimetres long.
Occasional or locally common along lagoon, swamp and river margins, mostly near the sea; SL–100 feet; fl. Feb. and sporadically throughout the year, fr. May–Aug, Nov–Dec. *A 8957! 10191! H 10816! P 11877!* Mexico to the Guianas, West Indies.

5. TERMINALIA L. (1767) nom. cons.

1 Leaves more than 12 cm. long:
  2 Leaf-blade tapered to petiole at base; drupe about 2 cm. broad with acute margins
                                                         1. *latifolia*
  2 Leaf-blade tapered but usually truncate-subcordate at base; drupe about 3 (–4) cm.
    broad with narrowly winged or unwinged margins   2. *catappa*
1 Leaves up to 8 cm. long, acute at base   3. *arbuscula*

1. T. latifolia Sw., Nov. Gen. & Sp. Pl.: 68 (1788).—Broadleaf.
Tree 20–30 m. or more high with markedly horizontal branches; trunk of young tree smooth, grey-buff, later lightly fissured and forming oblong flakes about 2 cm. broad, buttressed; leaves of saplings and regrowth acuminate, rusty-pubescent when young; leaves of mature trees glabrescent, obovate to oblanceolate, usually rounded at tip, 14–25 cm. long, usually with domatia-pits in the vein-axils, turning red when old; flowers light yellow in axillary spikes up to about 12 cm. long, often male only towards apex.
Locally common in relict woodland in gullies and depressions especially in the western parishes; (50–) 1000–2300 feet; fl. Feb–May, fr. Apr–Sept. *A 12383! P 6537! 22209! Robbins UCWI 262!* Endemic.

2. T. catappa L., Mant. Pl.: 128 (1767).—Almond, West Indian Almond.
Tree 5–16 m. high usually with low whorled widely spreading branches; leaves obovate, rounded or cuspidate at apex, up to 30 (–36) cm. long and 16 (–24) cm. broad, glabrous above, glabrescent beneath; inflorescences axillary, slender, up to 25 cm. long, with sessile bisexual flowers towards base and pedicelled male flowers towards apex.
Commonly planted and naturalized, especially near the sea in wet areas; SL–550 feet; fl. and fr. sporadically. *A 13024! A. M. Barry! H 10824! Yuncker 18412!* Native of S.E. Asia, N. Australia and the Pacific, now widespread; Grand Cayman.

3. T. arbuscula Sw., Nov. Gen. & Sp. Pl.: 68 (1788).—White Olive.
Shrub or tree to 16 m. high; leaves oblong to oblong-lanceolate, 4·5–7 cm. long,

glabrescent; spikes as long as the leaves; calyx glabrous outside, densely hairy within; male flowers and fruit not known.

Very rare (St. Eliz., Han., Trel.), in woodland on limestone hills; 900–2200 feet; fl. May. *P 24472! 31312! Swartz!* Endemic.

6. BUCHENAVIA Eichl. (1866)

1. **B. capitata** (Vahl) Eichl. in Flora **49**: 165 (1866).—*Bucida capitata* Vahl (1797). Mountain Wild Olive.

Tree 6–22 m. high with horizontally spreading branches; leaves crowded at the twig-branches, obovate, apex rounded or emarginate, leathery, shiny above, rusty-tomentose on veins beneath, 4–6·5 (–8) cm. long, ciliate-margined when young; inflorescence rusty-tomentose; spikes subglobose to oblong, 1–2 cm. long; flowers green.

Rare (St. Andr., Clar.), in woodlands and thickets on savannas in heavy soils; 200–3000 feet; fl. and fr. Feb, July. *H 6464! J.P. 2100! Proctor & Mullings 21960!* Panama to Brazil, West Indies.

141. MYRTACEAE

By G. R. Proctor

Shrubs or trees. Leaves simple, opposite (alternate in certain introduced genera), entire or rarely crenate, punctate with resinous or pellucid glands, usually pinnately-veined; principal lateral veins usually uniting distally into a more or less obscure submarginal vein extending the length of the blade; stipules absent. Flowers axillary or terminal, solitary or in bracteate inflorescences with opposite branching, usually modified in various ways, e.g. (a) elongation of the axis and reduction of lateral branches to one flower each forming a raceme; (b) suppression of the axis and reduction of lateral branches to one flower forming glomerules or umbelliform clusters; (c) reduction of lateral branches to one pair, these arising just below the flower terminating the main axis, forming a dichasium; (d) elongation of both central axis and lateral branches resulting in a panicle; and (e) transitional forms in which a panicle terminates in triads or dichasia. Flowers usually perfect, regular or essentially so. Calyx-tube (hypanthium) adnate to the ovary its whole length or sometimes prolonged beyond it; calyx-lobes usually 4 or 5, either distinct or the calyx calyptrate and circumcissile or rupturing irregularly at anthesis. Petals usually 4 or 5, rarely absent, mostly white or often tinged pinkish and readily falling. Stamens indefinitely many in 1 to many series originating around the margin of the thickened calyx-disk, usually inflexed in bud; anthers bilocular, opening (in indigenous species) by longitudinal slits. Ovary inferior, 2–many-locular; placentas axile or parietal; ovules 2 or more; style simple, elongate, with small capitate stigma. Fruit a berry, drupe or capsule. Seeds usually without endosperm.

About 60 genera, some of them poorly-defined, and nearly 3000 species, chiefly in the tropics and Southern Hemisphere.

1 Fruit a more or less fleshy berry or drupe; leaves opposite, rarely a few subopposite, never in whorls of 4:
  2 Calyx without lobes or limb unitary:
    3 Calyx closed in bud, the top circumscissile and falling off like a lid at anthesis; petals absent or minute, adherent to inside of calyx-lid; ovules usually 2 in each locule; vegetative branching dichotomous   **1. Calyptranthes**
    3 Calyx open in bud, the united petals falling off like a lid at anthesis; ovules numerous; branching not dichotomous   7.1 *Syzygium cumini*
  2 Calyx with 4–5 evident lobes, these usually persistent:
    4 Inflorescence compound, paniculate:
      5 Leaves acuminate, or if not so then sessile and subcordate at base, not or scarcely aromatic; calyx-lobes thin, often deciduous from developing fruits; seeds with relatively large twisted and folded cotyledons   **2. Myrcia**
      5 Leaves blunt or rounded at apex, petiolate, strongly aromatic; calyx-lobes thick, persistent on the fruits; seeds with spiral embryo and very short cotyledons   **3. Pimenta**

4 Inflorescence not compound, or if so, dichasially forked with a sessile terminal flower in each fork; flowers solitary, clustered or in racemes:
  6 Calyx closed or nearly so in bud, at anthesis splitting to the disk in (4–) 5 irregular or unequal lobes; stigma capitate; flowers usually solitary in leaf-axils, or rarely 3 forming a dichasium **4. Psidium**
  6 Calyx open in bud, the lobes 4, subequal; stigma not or scarcely thicker than the style:
    7 Flowers mostly 3 or 7 in a pedunculate dichasium; cotyledons 2, distinct from each other and from the radicle **5. Myrcianthes**
    7 Flowers solitary, clustered, or in contracted or elongate racemes; cotyledons, radicle and plumule not discernible as separate structures in the seed:
      8 Hypanthium abruptly contracted at base, the apex scarcely prolonged beyond the ovary; inflorescences racemose, always axillary, rarely at leafless nodes **6. Eugenia**
      8 Hypanthium attenuate downwardly at base, the apex much prolonged beyond the ovary and forming a flaring lip; inflorescences terminal, or if lateral then usually at leafless nodes on old wood **7. Syzygium**
1 Fruit a simple or compound capsule, immersed in the hardened subwoody hypanthium at maturity; adult leaves usually alternate; juvenile leaves often very different in shape and opposite or whorled; introduced species:
  9 Flowers in globose pedunculate heads, the calyces fused; leaves often in whorls of 4; large tree with aromatic elliptical leaves; planted in the mountains; native of Australia *Syncarpia procera* (Salisb.) Domin
  9 Flowers separate, distinct even when densely clustered; adult leaves mostly alternate, never whorled:
    10 Flowers in stalked axillary umbels; petals and calyx-lobes united into an operculum deciduous at anthesis; large trees with fragrant foliage, extensively planted for reforestation; natives of Australo-Malaysia *Eucalyptus* spp.
    10 Flowers densely sessile in spikes at or near tips of branches; fruits often persisting on older twigs; petals and calyx-lobes distinct, soon deciduous; shrubs or small trees planted for ornament by reason of the showy spreading stamens; natives of Australia; Bottle-brush:
      11 Stamens united in 5 bundles opposite the petals *Melaleuca* spp.
      11 Stamens not united in bundles *Callistemon* spp.

1. CALYPTRANTHES Sw. (1788) nom. cons.

1 Leaves 25–40 cm. long, the blades cordate at base; flowers numerous in sessile clusters on panicles 1. *discolor*
1 Leaves not over 15 cm. long, usually much less:
  2 Leaves sessile, subcordate-clasping at base:
    3 Inflorescence brown-tomentose; flower-buds not over 6 mm. long:
      4 Peduncles about 2 cm. long, bearing 3 heads of up to 12 sessile flowers, the lateral heads stalked; flower-buds 2·5 mm. long; leaves 1·8–2·3 cm. long 2. *maxonii*
      4 Peduncles to 5 cm. long, bearing 2 or 3 sessile flowers at apex; flower-buds 5–6 mm. long; leaves to 5 cm. long 3. *clarendonensis*
    3 Inflorescence glabrate, the buds minutely puberulous with colourless hairs; flower-buds 12–15 mm. long 4. sp. *A*
  2 Leaves usually petiolate, acute, cuneate or rounded at base:
    5 Flowers mostly solitary on axillary or terminal peduncles:
      6 Leaves ovate to elliptical, mostly widest below the middle, acuminate 5. sp. *B*
      6 Leaves obovate or elliptical, mostly widest above the middle, the apex rounded or blunt:
        7 Leaf-veins prominulous; branchlets compressed and 2-edged but not winged; peduncles 5–8 mm. long; calyx 3·5 mm. in diameter 6. *ekmanii*
        7 Leaf-veins not prominulous, not evident, except midrib; branchlets winged; peduncles 8–16 mm. long; calyx 1 mm. in diameter 7. *wilsonii*
    5 Flowers few or numerous in racemes or panicles:
      8 Leaves entirely glabrous on both surfaces:
        9 Apex of leaves obtuse, or at most bluntly acuminate:
          10 Midrib of leaf prominent adaxially; branches of inflorescence terete, or if compressed not 2-edged; flowers all pedicellate 8. *zuzygium*
          10 Midrib of leaf grooved adaxially; branches of inflorescence compressed and keeled; flowers all or mostly sessile 9. *umbelliformis*
        9 Apex of leaves long-acuminate, often sharply so:
          11 Leaves 7–15 cm. long, and up to 5·5 cm. broad; inflorescence pubescent 10. *nodosa*

R

11 Leaves up to 7 cm. long and 2·5 cm. broad; inflorescence glabrous:
  12 Branchlets 2-keeled; leaves (2–) 3–5 cm. long; berries 5–6 mm. in diameter
                                                         11. *rigida*
  12 Branchlets compressed but not keeled; leaves 5–6·5 cm. long; berries 7–10
    mm. in diameter   12. *acutissima*
8 Leaves pubescent beneath, at least when young; inflorescence more or less reddish-
  velvety:
  13 Leaves rarely more than 7 cm. long and 3 cm. broad, often smaller, the trans-
    verse veins and veinlets obscurely prominulous; flower-buds usually rounded at
    apex, rarely pointed; ripe berries 4–5 mm. in diameter   13. *pallens*
  13 Leaves 6–10 cm. long, up to 4·5 cm. broad, with fine network of prominulous
    veinlets evident on both surfaces, some of the transverse veins especially
    prominent beneath; flower-buds more or less pointed; ripe berries 5–8 mm in
    diameter   14. *chytraculia*

1. C. discolor Urb., Symb. Ant. **5**: 443 (1908).—Mountain Bay.

Shrub or small tree to 7 m. high, the young branches terete and glabrous; leaves oblong or oblong-elliptical, 7–10 cm. broad, acuminate, the veins prominulous on both surfaces, nearly or quite glabrous throughout; inflorescences densely brown-velvety, the peduncles supra-axillary and up to 20 cm. long; hypanthium 1·5 mm. long.

Very rare and local (Han.), on forested slopes; 1500–1700 feet; fl. Mar–May. *H 10279! P 20691!* Endemic.

2. C. maxonii Britton & Urb. in Urb., Symb. Ant. **7**: 296 (1912).

Shrub, the young branches compressed, 2–winged, and glabrous; leaves leathery, ovate-roundish, the midrib impressed adaxially, the veins more or less prominulous especially beneath, the margins recurved; inflorescences 2–3, terminal; peduncles about 2 cm. long.

Very rare and local (Trel.), known only from the original collection; about 1500 feet; fl. June. *Maxon 2896.* Endemic.

3. C. clarendonensis Proctor in Rhodora **60**: 323 (1958).

Shrub to 2·5 m. high, the young branches nearly terete, densely brown-tomentose; leaves rigid-leathery, deciduously brown-tomentose beneath, broadly ovate, apex blunt, the midrib impressed towards base adaxially, the veins prominulous on both surfaces; inflorescences 2–4, terminal; peduncles 2–5 cm. long; berries irregularly globose, 9–14 mm. in diameter, 2–13-seeded.

Rare (Clar.), on wooded rocky limestone hilltops; about 2500 feet; fl. Jan, fr. Dec–Jan. *P 9760! 11399!* Endemic.

A closely related but not identical variant has been found near Pike, Manchester; *P 18306!*

4. C. sp. A.

Small tree to 6 m. high, the young branches 2-keeled, glabrous; leaves leathery, glabrous, 5–7 cm. long, 3–5 cm. broad, oblong-elliptical, the apex broadly rounded, the midrib nearly flat adaxially, the veins prominulous, the tissue sparsely pellucid-dotted; inflorescences 2–4, terminal; peduncles glabrous, compressed, 5·5–7·5 cm. long, each bearing 1 (–3) sessile flower(s) at apex; flower-buds abruptly acuminate, the acumen 2–3 mm. long; flowers fragrant, white, over 2 cm. across; berries unknown.

Very rare and local (St. Ann), on wooded rocky limestone hilltops; 2100–2300 feet; fl. June–July. *P 26480!* Endemic.

A little known genus related to *Calyptranthes* but differing especially in its 3–4 or 5-locular and multiovulate ovary, is *Mitranthes* O. Berg (Linnaea **27**: 136, 316 (1856)). In Jamaica there are at least two species of this group hitherto included in *Calyptranthes*. These include *C. clarendonensis*, a species as yet undescribed (sp. A above) and probably *C. maxonii*. For further information see McVaugh, R. in Taxon **17**: 391–392 (1968).

5. C. sp. B.

Shrub 2·5 m. high, the young branches terete, glabrous; leaves thinly leathery, glabrous at maturity, minutely whitish-tomentellous when very young, 5–8·5 cm. long, 2·5–4 cm. broad, the midrib grooved adaxially, the veins obscurely prominulous; inflorescences 2–3, terminal, peduncles soon glabrous, compressed,

4·5–6·5 cm. long, each bearing 1 subsessile flower at apex; flower-buds ovoid, acuminate, about 10 mm. long, minutely whitish-tomentellous; berries unknown.

Very rare and local (St. Eliz.); 1600–1700 feet; fl. May or June. *P 20862!* Endemic.

6. **C. ekmanii** Urb. in Arkiv Bot. **22A** (10): 32 (1929).

Shrub or small tree to 4 m. high, glabrous; leaves leathery, short-petiolate, 2·5–4·5 cm. long, up to 3 cm. broad, pale beneath and glandular-punctate; peduncles solitary, supra-axillary or terminal, 1-flowered; flower-buds globose with rounded apex, 3·5 mm. in diameter; berries 7–8 mm. in diameter, the peduncles reflexed in fruit.

Rare and local (Port.), in wet woodland and thickets on limestone, known only in sterile condition as yet in Jamaica; 1500–2000 feet; *P 9820!* Hispaniola.

7. **C. wilsonii** Griseb., Fl. Br. W.I.: 233 (1860).

Shrub to 3 m. high, glabrous; leaves papery, short-petiolate, 2–3 cm. long, up to 1·8 cm. broad, dark green above, somewhat paler beneath, the tissue faintly pellucid-dotted; peduncles 1 or 2, axillary, filiform, 1-flowered; flower-buds narrowly ellipsoid, acuminate; berries unknown.

Locally frequent (Port., St. Thom.), in mossy woodlands; 1500–3000 feet. *A 9160! H & B 10747! P 9821!* Endemic.

8. **C. zuzygium** (L.) Sw., Nov. Gen. & Sp. Pl.: 79 (1788).—*Myrtus zuzygium* L. (1759).

Shrub or small tree to 10 m. high, glabrous; leaves stiff, short-petiolate, 2·5–7 cm. long, blades elliptical or narrowly obovate with obscure venation, the tissue minutely punctate-glandular on both surfaces, opaque; panicles trichotomous, few-flowered; pedicels to 8 mm. long; berries globose, black with glaucous bloom, 7–9 mm. in diameter, edible.

Common on wooded limestone hills or sometimes in thickets along streams. 500–3000 feet; fl. May–July, fr. July–Dec. *A 7276! H 11056! P 22836!* Florida, Bahamas, Cuba, Hispaniola.

9. **C. umbelliformis** Krug & Urb. in Engl., Bot. Jahrb. **19**: 596 (1895).—*C. impressa* Urb. (1908).

Small tree to 10 m. high, glabrous; leaves stiff, short-petiolate, 4–7 cm. long, the blades ovate, elliptical, obovate or rotund, paler beneath; panicles several together, terminal in an umbellate cluster; flowers sessile in clusters of 2 or 3 at ends of panicle-branches; flower-buds 3·5–4 mm. long, mucronate; berries black, 9–10 mm. in diameter.

Uncommon (St. James, Port.), on moist wooded limestone hills; 1500–3500 feet; fl. June, Aug, fr. Sept, Dec, Jan. *H 5998! P 19728!* Endemic.

The disjunct populations included under this name are not homogeneous, but not enough information is available for taxonomic segregation. The name *C. impressa* was applied to specimens (*H 7701!*) not distinguishable from typical *C. umbelliformis*, and from the same general area as the original of the latter.

10. **C. nodosa** Urb., Symb. Ant. **5**: 444 (1908).

Shrub or small tree to 10 m. high, the young branches terete, glabrous; leaves leathery, short-petiolate, the blades broadly elliptical or oblong-elliptical, the mid-rib impressed adaxially, the veins prominulous on both surfaces, the tissue densely subpellucid-glandular; inflorescences usually 4 together, terminal, the branches terete; flowers sessile in small dense clusters at apex of panicle-branches; buds obovoid, 2–3 mm. long, apex blunt or shortly apiculate; berries about 7 mm. in diameter.

Rare (Han.?, Trel., Clar.), in moist thickets especially along streams; 1500–2300 feet; fl. June, fr. June–Aug. *H 8711! P 19708!* Endemic.

11. **C. rigida** Sw., Nov. Gen. & Sp. Pl.: 80 (1788).—*C. fawcettii* Krug & Urb. (1895).

Shrub or small tree to 10 m. high, glabrous; leaves stiff and leathery, petiolate, ovate or ovate-lanceolate, the midrib impressed adaxially, the upper surface

minutely glandular-punctate; inflorescences few or several together, terminal or
subterminal; flowers in threes at ends of branches, all sessile or the lateral ones on
short pedicels; buds obovoid, acuminate, 2–3 mm. long; berries dark red.
Local (Clar., St. Andr., Port.), in mountain forests; 1500–6000 feet; fl. July–Oct,
fr. Feb–Mar. *A & S 3170! H 12901! Swartz!* Cuba.

12. C. acutissima Urb., Symb. Ant. **6**: 22 (1909).
Shrub or small tree to 7 m. high, glabrous; leaves leathery, petiolate, oblong-
lanceolate or narrowly ovate, the midrib narrowly grooved towards the base
adaxially, the tissue very minutely glandular-punctate on both surfaces, but not
pellucid, pale beneath; panicles apparently few-flowered, the branches 2-edged;
flowers unknown, evidently in sessile clusters of 3 on panicle-branches, only the
central one producing a fruit.
Very rare and local (Han.), in thickets on limestone; 1750–1786 feet; fr. Mar,
July. *H 10270! P 10431!* Endemic.

13. C. pallens (Poir.) Griseb. in Abh. Gött. Akad. **7**: 215 (1857).
Shrub or small tree to 15 m. high, the youngest branches puberulous and 2-
edged; leaves stiff, petiolate, the blades lanceolate to elliptical, acuminate, the
midrib narrowly grooved towards base adaxially, the tissue glandular punctate
with faintly pellucid dots; inflorescences axillary and terminal, the numerous small
flowers sessile in clusters at ends of panicle-branches; hypanthium about 2 mm.
long, puberulous; flowers cream tinged pink or brownish-orange, fragrant; berries
dark red.
Frequent to common on wooded limestone hillsides; near SL–2500 feet; fl.
Mar–July, fr. Apr–Oct. *A 9244! H 8943! P 21439!* Florida, Mexico, West Indies;
Cayman Is.

14. C. chytraculia (L.) Sw., Nov. Gen. & Sp. Pl.: 79 (1788).—*Myrtus chytraculia* L. (1759). *C. urbanii* F. & R. (1926). Bastard Greenheart.
Shrub or small tree to 8 (–10) m. high, the youngest branches puberulous and
slightly compressed, soon becoming glabrous and terete; leaves papery, petiolate,
the blades ovate to elliptical, acuminate, the midrib narrowly grooved adaxially,
the tissue glandular-punctate-pellucid; inflorescences axillary and terminal, the
numerous small flowers sessile in clusters at ends of panicle-branches; hypanthium
about 2 mm. long, densely reddish-tomentose; ripe berries black.
Common on wooded hillsides, in pasture thickets and beside streams, in many
types of soil; near SL–4200 feet; fl. Mar–Aug, fr. Aug–Jan. *A 11244! H 12042!
P 16480!* Florida, Cuba; a variety in C. Amer. and Colombia.

2. MYRCIA DC. ex Guill. (1827)

1 Leaves sessile or subsessile, apex blunt or rounded; flowers 4-parted   1. *skeldingii*
1 Leaves petiolate, acuminate:
  2 Leaves densely brown-tomentose beneath, likewise the young branches and inflo-
    rescence, 4–8 cm. broad; anthers 4-locular   2. *fenzliana*
  2 Leaves glabrate, if a few hairs present beneath these colourless, less than 4 cm. broad;
    anthers 2-locular:
    3 Flowers 4-parted; leaves leathery, dull; plant entirely glabrous   3. sp. *A*
    3 Flowers 5-parted; leaves papery, lustrous:
      4 Summit of ovary and interior of hypanthium densely hairy; branchlets, inflorescence
        and flower-buds pubescent with simple hairs; calyx not prolonged beyond the
        ovary; ripe fruits oblong, oval or globose   4. *splendens*
      4 Summit of ovary and interior of hypanthium glabrous; branchlets and inflorescence
        with dibrachiate hairs; flower-buds glabrous; calyx somewhat prolonged beyond
        the ovary; ripe fruits depressed-globose   5. *leptoclada*

1. M. skeldingii Proctor in Rhodora **60**: 325 (1958).
Small tree to 5 m. high, the young branches and other vascular parts minutely
strigillose; leaves leathery, 5·5–10 cm. long, broadly ovate-elliptical, veins finely
prominulous, tissue not pellucid-dotted; panicles terminal, exceeding the leaves,
many-flowered; calyx-lobes obtuse, about 0·75 mm. long; petals white, about 1 mm.
long; berries about 6 mm. in diameter, deep red.

Rare and local (Clar.–St. Ann border), in streamside thickets; 2100–2300 feet; fl. July, fr. Oct. *P 16478!* Endemic.

2. **M. fenzliana** O. Berg in Mart., Fl. Bras. **14** (1): 196 (1857).—*Gomidesia lindeniana* O. Berg (1858). *M. sintenisii* Kiaersk. (1890).

Shrub or small tree to 10 m. high; leaves thick, 6–15 cm. long, blades oblong to elliptical, strongly bullate-plicate, eventually glabrous adaxially, with midrib and veins impressed; panicles shorter than the leaves, densely-flowered; flowers sessile, fragrant; hypanthium densely velvety-tomentose; calyx-lobes pointed, 1·5 mm. long; petals white, about 3 mm. long, hairy on back; berries about 8 mm. in diameter, hairy, black.

Rare (Port.), in montane forest; 3000–3500 feet; fl. Sept. *H 7348!* S. Amer., West Indies.

3. **M. sp. A.**

Shrub 2 m. high; leaves 4·5–7·5 cm. long, blades ovate-elliptical, veins obscurely prominulous, tissue minutely punctate-glandular, dots not pellucid; panicles terminal, shorter than the leaves; flowers unknown, the subpersistent pointed calyx-lobes about 1 mm. long; berries about 5 mm. in diameter.

Rare and local (Port.), in moist woodland over limestone; 1500–2500 feet; fr. July. *P 23871!* Endemic.

4. **M. splendens** (Sw.) DC., Prodr. **3**: 244 (1828).—*Myrtus splendens* Sw. (1788). *Eugenia periplocifolia* Jacq. (1789).

Shrub or small tree to 10 m. high; leaves 2–8 cm. long, blades lanceolate, ovate or elliptical, obtuse at base, veins prominulous on both surfaces, tissue with numerous pellucid dots; panicles shorter than the leaves; flowers numerous; calyx-lobes rounded, 1–1·2 mm. long; petals white, 2–3 mm. long; berries to 10 mm. long, black.

Collected but once in Jamaica in about 1785 at an unknown locality; fl. June–Sept. Mexico, C. and S. Amer., West Indies.

5. **M. leptoclada** DC., Prodr. **3**: 244 (1828).—*Aulomyrcia leptoclada* (DC.) O. Berg (1855).

Shrub or small tree to 15 m. high; leaves 5–10 cm. long, blades elliptic-lanceolate to narrowly ovate, veins prominulous on both surfaces, a few dibrachiate hairs beneath, tissue with numerous pellucid dots; panicles equalling the leaves; calyx-lobes very unequal, 0·5–1·5 mm. long; petals white, 1·5 mm. long; berries 7 mm. long, black.

Rare (Port.), in mossy woodland; 1500–2500 feet; only sterile specimens seen. *P 19725!* C. Amer., Suriname, West Indies.

3. PIMENTA Lindl. (1821)

1 Calyx always 5-lobed, the lobes much wider than long and not clearly defined; crushed
　leaves with odour of bay rum; ovules 6–7 in each locule　　　　　　　1. *racemosa*
1 Calyx usually 4-lobed, sometimes 5-lobed in *P. jamaicensis*, the lobes nearly as long as to
　longer than wide, distinct; leaf-odour not that of bay rum:
　2 Peduncles stout, 1·5–3 mm. thick; leaves of thick leathery texture, the glandular dots
　　not pellucid, when crushed with odour of lemon, citronella or rarely turpentine;
　　calyx 4–5-lobed　　　　　　　　　　　　　　　　　　　　　　　　2. *jamaicensis*
　2 Peduncles slender, 0·5–1·2 mm. thick; leaves thin-coriaceous, the glandular dots more
　　or less pellucid; calyx always 4-lobed:
　　3 Leaves oblong to elliptical, mostly 6–20 cm. long, when crushed with odour of all-
　　　spice; panicles 4–12 cm. long, many-flowered; hypanthium smooth, puberulous;
　　　calyx-lobes rounded　　　　　　　　　　　　　　　　　　　　　　3. *dioica*
　　3 Leaves obovate, mostly 2·5–6·5 cm. long, when crushed with odour unlike allspice;
　　　panicles less than 3 cm. long, few-flowered; hypanthium densely glandular-
　　　verrucose, sparsely puberulous or glabrous; calyx-lobes acute　　　4. sp. *A*

1. **P. racemosa** (Mill.) J. W. Moore in Bernice P. Bishop Mus. Bull. **102**: 33 (1933). —*Caryophyllus racemosus* Mill. (1768). *Myrtus caryophyllata* of Jacq. (1767), not L. (1753). *Amomis caryophyllata* Krug & Urb. (1894). Bay Rum Tree.

Tree to 15 m. high with smooth whitish bark and hard heavy wood, the young branchlets flattened, 4-angled; leaves elliptical to very broadly elliptical, 4–12 cm.

long, with reticulate-prominulous venation; hypanthium obconic, nearly or quite glabrous, about 1·5 mm. long at anthesis; calyx abruptly flaring at apex; petals white, 3 mm. long; berries subglobose to ellipsoid, 8–10 mm. long.

Cultivated in Jamaica, often persisting; SL–2500 feet; fl. Jan–Mar. *H 8523! P 24427!* Native of northern S. Amer. and West Indies, introduced in Pacific Is.

2. **P. jamaicensis** (Britton & Harris) Proctor in Bull. Inst. Jam., Sci. ser. **16**: 34 (1967).—*Amomis jamaicensis* Britton & Harris (1920). *A. caryophyllata* of F. & R., partly. Wild Pimento.

Tree to 10 m. or rarely more high with hard reddish wood, the young branchlets stout, flattened; leaves elliptical to obovate, 5–20 cm. long with prominulous venation; hypanthium barrel-shaped, densely puberulous, about 2 mm. long at anthesis; calyx abruptly flaring at apex, the deltate lobes about 1 mm. long; petals pale buff, 1·5 mm. long, quickly deciduous; berries ellipsoid, 7–10 mm. long.

Local on wooded rocky limestone hillsides; 1500–2500 feet; fl. July–Aug, fr. Aug–Sept. *A 12611! H 12876! P 23872!* Endemic.

3. **P. dioica** (L.) Merr. in Contrib. Gray Herb. **165**: 37, f. 1 (1947).—*Myrtus pimenta* L. (1753). *M. dioica* L. (1759). *P. officinalis* Lindl. (1821). Allspice, Pimento.

Tree to 15 m. high with pale brown bark, the young branchlets flattened, 4-angled; hypanthium obconic, to 1·5 mm. long; calyx-lobes broadly rounded, about 1·5 mm. long, wide-spreading at anthesis; petals white, 1·5 mm. long, quickly deciduous; berries more or less globose, 4–6·5 mm. in diameter.

Common on wooded hillsides and in upland pastures; near SL–3500 feet; fl. Jan–Aug, fr. Aug–Sept. *A 6957! H 10507! P 20737!* Mexico, C. Amer., Cuba, Hispaniola, introduced in Puerto Rico, Barbados and elsewhere in the tropics.

4. **P. sp. A.**

Shrub or small tree to 8 m. high, the young branchlets somewhat flattened; hypanthium obconic, about 1 mm. long; calyx-lobes deltate, minutely ciliate, about 0·8 mm. long; petals white, 1·5 mm. long; berries ovoid, 7–8 mm. long.

Rare (St. James, St. Eliz., Trel.), on wooded limestone crags; 1600–2200 feet; fl. May, July, fr. Jan, May. *P 22536B!* Endemic.

4. PSIDIUM L. (1753)

1 Leaves sessile, rounded or subcordate at base:
  2 Hypanthium 2–3 mm. long; ovary 2-locular   1. *harrisianum*
  2 Hypanthium 5–7 mm. long; ovary 4-locular   2. *dumetorum*
1 Leaves petiolate, the blades cuneate at base:
  3 Leaves pubescent beneath; buds constricted at top of hypanthium:
    4 Lateral veins 12–20 on either side of the midrib, slightly impressed on upper side; young branches 4-angled; flowers nearly always solitary; buds completely closed at apex   3. *guajava*
    4 Lateral veins 6–10 on either side of the midrib, slightly prominulous on upper side; young branches terete, flowers solitary or 3 in a pedunculate dichasium; buds with a small aperture at apex   4. *guineense*
  3 Leaves glabrous; buds not or scarcely constricted:
    5 Leaves elliptical; branchlets 4-angled; buds closed at apex, apiculate   5. *montanum*
    5 Leaves narrowly to very broadly obovate, widest above the middle; branchlets terete or compressed:
      6 Leaves broadly acutish to short-acuminate, the midrib reaching the apex; buds open at apex   6. *cattleianum*
  3 Leaves rounded, the midrib not reaching the apex; buds closed at apex, obtuse
                                                                                7. *albescens*

1. **P. harrisianum** Urb., Symb. Ant. **7**: 294 (1912).

Shrub or small tree to 5 m. high, glabrous, the young branchlets grey, subterete; leaves roundish or ovate, 3–8 cm. long, with numerous subpellucid dots; flowers axillary or terminal, solitary or 2–6 in a short raceme, the pedicels about 3 mm. long; flower-buds globulose, not over 3 mm. long; fruit irregularly globose, to 1·3 cm. in diameter, green.

Rare in the central parishes on wooded rocky limestone hilltops; 1800–3000 feet; fl. July, fr. Dec–Jan. *H 11000! P 11396!* Endemic.

2. P. dumetorum Proctor in Bull. Inst. Jam., Sci. ser. **16**: 37, t. 14 (1967).

Shrub to 3 m. high, glabrous, the young branchlets subterete; leaves roundish, 6–9·5 cm. long, with numerous subpellucid dots and strongly recurved margins; flowers terminal, 2–4 in a short raceme, the pedicels 5–11 mm. long; flower-buds obovoid, apiculate, 13–15 mm. long; fruit unknown.

Rare and local (Clar.), in thickets along streams; 2100–2300 feet; fl. Apr, Nov. *P 19650!* Endemic.

3. P. guajava L., Sp. Pl. **1**: 470 (1753).—*P. pyriferum* L. (1762). *P. pomiferum* L. (1762). *P. guava* Griseb. (1860). Guava.

Shrub or small tree rarely over 7 m. high; leaves elliptical or oblong, mostly 7–14 cm. long, apex obtuse or acute, beneath densely coated with soft greyish mostly appressed hairs; tissue with numerous pellucid dots; mature flower-buds 13–16 mm. long, puberulous towards base; anthers 0·8–1·2 mm. long; fruit varying in size, 2–6 cm. long, globose or pyriform, usually yellow, the flesh pink or yellowish, edible.

Common in pastures and wayside thickets, sometimes cultivated; near SL–4000 feet; fl. Mar–June, or sporadically at all seasons. *H 5117! P 19714! Robertson UCWI 3177!* Florida, Mexico to S. Amer., West Indies; Grand Cayman; naturalized in the Old World tropics.

4. P. guineense Sw., Nov. Gen. & Sp. Pl.: 77 (1788).—*P. araca* Raddi (1821). *P. molle* Bertol. (1840).

Shrub to 2·5 m. or more high; leaves broadly elliptical to obovate, mostly 6–14 cm. long, apex obtuse or acute, often apiculate, beneath rather densely clothed with pale reddish erect hairs; tissue with numerous pellucid dots; mature flower-buds 10–12 mm. long, densely puberulous throughout; anthers 1·2–2·2 mm. long; fruit 1–2 cm. in diameter, globose or obovoid, yellowish-green or yellow, edible but inferior in taste.

Occasional (St. Andr.), in hillside thickets; 750–3600 feet; fl. and fr. Mar–May. *A 7027! H 6919! P 20770!* Mexico to S. Amer., Cuba. The specific epithet *guineense* is a misnomer, having been applied by Swartz in the erroneous belief that the plant originated in Africa. However, this species is truly indigenous to the Western Hemisphere tropics, perhaps including Jamaica.

5. P. montanum Sw., Nov. Gen. & Sp. Pl.: 77 (1788).—*P. wrightii* Lambert ex W. Wright (1828). Mountain Guava.

Tree to 20 m. or more high, glabrous, the young branchlets compressed and 4-angled; leaves 3–9 cm. long, acuminate, with numerous pellucid dots; flowers solitary, buds pyriform, 9–10 mm. long; calyx-lobes puberulous within; anthers about 0·8 mm. long; fruit subglobose, to 2·5 cm. in diameter, green, edible.

Widespread on wooded hillsides; 1800–4000 feet; fl. May, Aug, fr. Jan, Apr, June, Sept. *A 12523! H 9406! P 24762!* Endemic. In flavour the fruits resemble mangos rather than guavas.

6. P. cattleianum Sabine in Trans. Roy. Hort. Soc. **4**: 317, t. 11 (1821).— *P. littorale* Raddi (1823). Purple Guava, Strawberry Guava.

Shrub or small tree to 4 m. high, the young vascular parts minutely puberulous; leaves 4–12 cm. long, dark shining green, opaque; flowers solitary in axils, buds pyriform, 7–8 mm. long, glabrous, the calyx with 4–5 low rounded lobes, splitting at the sinuses into lobe-like segments; anthers about 0·5 mm. long; fruit 2·5–3·5 cm. long, obovoid or globose, reddish-purple, edible.

Naturalized (Manch., Clar., St. Ann), in pasture thickets and hill-top woodlands; 1500–3000 feet; fl. Apr, June–July, fr. Sept–Dec. *A 11677! Harris! P 19707!* Brazil; widely cultivated in C. and S. Amer. and the West Indies.

7. P. albescens Urb., Symb. Ant. **5**: 441 (1908).

Shrub 2 m. or small bushy tree to 7 m. high, glabrous; leaves 3–6 cm. long, stiff, with numerous more or less pellucid dots; flowers solitary or 2–6 in a short raceme, buds globose or pyriform, 5–7 mm. long; petals ciliate; anthers about 0·3 mm. long; fruit unknown.

Rare (Manch., St. Andr., Port.), on dry or moist wooded limestone hillsides; 900–2500 feet; fl. Feb, June, Sept, Dec. *H 9583! P 11350!* Endemic.

5. MYRCIANTHES O. Berg (1856)

1. M. fragrans (Sw.) McVaugh in Fieldiana, Bot. **29**: 485 (1963).—*Myrtus fragrans* Sw. (1788). *Eugenia fragrans* (Sw.) Willd. (1799).

Shrub or small tree to 15 m. high, glabrous or youngest branches and petioles minutely pubescent; leaves short-petiolate, 2·5–7 cm. long, 1·5–5 cm. broad, ovate, elliptical, obovate or rotund, rounded to obtusely pointed at apex, cuneate at base, densely and minutely glandular-dotted on both surfaces, some of the dots pellucid; peduncles mostly 2–4 cm. long, usually 3-flowered; hypanthium puberulous; calyx-lobes 4, rounded, in 2 unequal pairs; petals white, 4–6 mm. long; ovary 2-locular, ovules 15–20 in each locule; fruits black or nearly so, about 10 mm. in diameter, usually only 1 bean-shaped seed maturing.

Widely distributed in dry or moist woodlands; 50–7400 feet; fl. Jan, May, July, Nov, fr. Aug, Oct, Dec. *H 5023! P 9428!* Florida, Mexico, C. and northern S. Amer., West Indies; Grand Cayman.

6. EUGENIA L. (1753).—Rodwood[1]

1 Sepals, at least the larger ones 4–8 mm. long, in *E. jeremiensis* to 14 mm.:
  2 Leaves 7–9 cm. broad; pedicels 1–1·5 mm. thick; sepals 10–14 mm. long, seen in bud only, broadly rounded   **1.** *jeremiensis*
  2 Leaves usually much less than 4·5 cm. broad, occasionally much broader; pedicels much less than 1 mm. thick; sepals not over 6 mm. long:
    3 Margins of sepals ciliate at least at apex; sepals as long or longer than broad, blunt to acuminate:
      4 Flower-stalks not subtended by axillary bracts; leaves long-acuminate, finely pubescent especially beneath, hairs of upper surface soon falling; sepals pointed-acuminate   **2.** *acutisepala*
      4 Flower-stalks subtended by axillary stipule-like bracts; leaves obtuse or bluntly acuminate, glabrous; sepals blunt:
        5 Leaves oblong or narrowly elliptical with midrib grooved adaxially; fruit 6–8 mm. in diameter, not ribbed   **3.** *ligustrina*
        5 Leaves ovate, the midrib flat above or nearly so; fruit to 25 mm. in diameter or more, 8-ribbed   **4.** *uniflora*
    3 Margins of sepals glabrous throughout; sepals usually broader than long, broadly rounded:
      6 Flowers solitary or clustered chiefly in leaf-axils; leaves less than 9 cm. long:
        7 Leaves less than 3 cm. long, oblong or narrowly elliptical, noticeably paler beneath   **5.** *heterochroa*
        7 Leaves chiefly 4–8·5 cm. long, rarely more, ovate, elliptical or obovate, concolorous or slightly paler beneath:
          8 Leaves broadly rounded or slightly emarginate at apex, extremely glossy adaxially, the midrib deeply grooved towards the base   **6.** sp. *A*
          8 Leaves bluntly to sharply short-acuminate, dull green, the midrib not or but slightly impressed:
            9 Sepals 3–5 mm. long; leaves broadly ovate, leathery, the venation faintly prominulous adaxially   **7.** *mandevillensis*
            9 Sepals 5–8 mm. long; leaves elliptical, firmly papery, the venation distinctly prominulous adaxially   **8.** *sachetae*
      6 Flowers numerous in rather dense clusters chiefly at leafless nodes on old wood; leaves chiefly over 9 cm. and sometimes up to 30 cm. long:
        10 Leaves cordate at base, thick-leathery   **9.** *lamprophylla*
        10 Leaves broadly cuneate to truncate at base, firmly papery:
          11 Leaves broadly ovate or oval-oblong, usually less than 15 cm. long; chiefly in eastern and central parishes   **10.** *marchiana*
          11 Leaves elliptical or oblong-elliptical, usually more than 15 cm., often 20–30 cm. long; chiefly in western parishes   **11.** *amplifolia*

[1] This name is generally applied to most species of *Eugenia*.

1 Sepals less than 4 mm. long:
  12 Flowers solitary in leaf-axils; leaves 0·5–1·5 cm. long; confined to high mountains
    over 4000 feet   12. *alpina*
  12 Flowers several or many in a very short to elongate raceme, or clustered with the
    rachis obsolete; leaves more than 1·5 cm. long (more than 2 cm. except in *E.*
    monticola):
    13 Inflorescence puberulous or tomentose:
      14 Sepals pointed:
        15 Pedicels glabrate to sparsely short-pilose, the hairs erect; berries obovoid or pyri-
          form, asymmetrical, orange   13. *chrootrichoides*
        15 Pedicels densely tomentose, the hairs appressed-ascending; berries globose, dark
          red   14. *virgultosa*
      14 Sepals rounded:
        16 Inflorescence less than 2 cm. long, often much less:
          17 Leaf-surfaces minutely pubescent at least when young, especially on upper
            surface:
            18 Larger sepals to 2·3 mm. long; leaves densely pellucid-dotted; berries globose,
              12–13 mm. in diameter   15. *alexandri*
            18 Larger sepals 1·5–2 cm. long; leaves with relatively few pellucid dots; berries
              obovoid, mostly less than 10 mm. in diameter:
              19 Shrub with leaves obtuse or rounded at base; widely distributed in thickets
                16. *disticha*
              19 Tree with leaves narrowly cuneate at base; rare in lower montane forest
                17. *brachythrix*
          17 Leaf-surfaces always glabrous, sometimes a few hairs on midrib only:
            20 Leaf-apex obtuse or very obtusely acuminate:
              21 Leaves usually obovate or oblanceolate, or at least evidently triplinerved,
                widest above the middle; flowers sessile or subsessile, pedicels seldom over
                0·5 mm. long, often at leafless nodes   18. *maleolens*
              21 Leaves lanceolate to ovate or elliptical, widest at or below the middle; flowers
                distinctly pedicellate, nearly always axillary:
                22 Pedicels 1–3 mm. long, seldom more:
                  23 Leaves 1·5–4 cm. long, the foliage dark green, dense; sepals not over 1·3 mm.
                    long; berries 4–6 mm. in diameter   19a. *monticola* var. *monticola*
                  23 Leaves 4–8 cm. long, the foliage not especially dense:
                    24 Young branchlets minutely and densely puberulous; sepals not over 1·3 mm.
                      long; berries 4–6 mm. in diameter   19b. *monticola* var. *latifolia*
                    24 Young branchlets glabrous; sepals to 1·5 mm. long; berries to 8 mm. in
                      diameter   20. *abbreviata*
                22 Pedicels 4–8 mm. long:
                  25 Leaves mostly 6–9 cm. long; midrib grooved throughout; sepals 1·5–2 mm.
                    long   21. *isosticta*
                  25 Leaves mostly less than 5 cm. long; midrib grooved in lower half only;
                    sepals about 1·2 mm. long   22. *sloanei*
            20 Leaf-apex distinctly acuminate, sometimes sharply so:
              26 Pedicels 1–3 mm. long:
                27 Leaf-midrib grooved adaxially, glabrous   20. *abbreviata*
                27 Leaf-midrib broad and flat adaxially, minutely puberulous   23. *schulziana*
              26 Pedicels 4–8 mm. long:
                28 Leaves 6–9 cm. long, mostly over 2 cm. broad   24. *jamaicensis*
                28 Leaves less than 6 cm. long, mostly 1·5–5 cm. long:
                  29 Leaves lanceolate, widest below the middle, sharply acuminate, midrib
                    grooved its whole length; berries longer than broad, 8–10 mm. long
                      25. *wilsonella*
                  29 Leaves elliptical, widest at middle, obtusely acuminate, midrib grooved only
                    in lower half; berries globose or flattened-globose, broader than long, not
                    over 6 mm. in greatest diameter   22. *sloanei*
      16 Inflorescence usually more than 2 cm. long, rarely less:
        30 Petioles 3–8 mm. long; leaves mostly 3–7 cm. long, less than 3 cm. broad
            26. *biflora*
        30 Petioles 9–14 mm. long; leaves mostly over 7 cm. up to 14 cm. long, more than 3
          cm. broad:
          31 Lateral leaf-veins slightly prominulous adaxially; berries obovoid, to 15 mm. long
            27. *fadyenii*
          31 Lateral leaf-veins grooved adaxially; berries obliquely globulose, 5 mm. in
            diameter   28. *sulcivena*
    13 Inflorescence essentially glabrous, but bracteoles and sepals sometimes ciliate and/or
      puberulous abaxially:
      32 Inflorescence usually 1–3·5 cm. long, rarely shorter:
        33 Pedicels mostly 5–12 mm. long; berries 8–10 mm. long:

34 Longest sepals 1·5–1·8 mm. long:
  35 Leaves mostly 3–6 cm. long, the margins crenate; petals 3–3·5 mm. long, white
                                                                29. *crenata*
  35 Leaves mostly 5–8·5 cm. long, the margins entire or nearly so; petals 2·5–3 mm.
    long, usually pink or rose   30a. *harrisii* var. *harrisii*
34 Longest sepals 2–3 mm. long; petals 4 mm. long:
  36 Petioles 12–16 mm. long   30b. *harrisii* var. *grandifolia*
  36 Petioles 5–10 mm. long   31. *rendlei*
33 Pedicels 4–6 mm. long; berries 6 mm. long   32. *nicholsii*
32 Inflorescence-axis (rachis) less than 1 cm. long or absent:
  37 Rachis present, the inflorescence very shortly racemose:
    38 Pedicels less than 5 mm. long:
      39 Young branchlets minutely and rather densely puberulous; mature berries
        4–6 mm. in diameter   19. *monticola*
      39 Young branchlets glabrous; mature berries at least 7 mm. long:
        40 Sepals 0·8 mm. long; berries more or less globose, 7–10 mm in diameter
                                              33. *axillaris*
        40 Sepals 1–1·2 mm. long; berries oblong, 10–13 mm. long   34. *glabrata*
    38 Pedicels 8–16 mm. long:
      41 Leaves shortly and obtusely acuminate, 6–8 cm. long; sepals to 3·5 mm. long;
        berries 12–15 mm. in diameter   35. *polypora*
      41 Leaves long and narrowly acuminate, 4–6·5 cm. long; sepals 1·5–2 mm. long;
        berries 5–6 mm. in diameter   36. *confusa*
  37 Rachis absent, the inflorescence umbelliform:
    42 Leaves mostly less than 5 cm., rarely as much as 6 cm. long; pedicels 5–17 mm.
      long:
      43 Petioles 3–6 mm. long; sepals 1·5–2·8 mm. long; young branchlets glabrous:
        44 Leaves stiffly papery with numerous subpellucid dots; berries to 9 mm. in
          diameter   37. *rhombea*
        44 Leaves leathery, opaque; berries 5–6 mm. in diameter   38. *clarendonensis*
      43 Petioles 2–3 mm. long; sepals about 1 mm. long; young branchlets with minute
        hairs   39. *brownei*
    42 Leaves 5–10 cm. long; pedicels less than 3 mm. long:
      45 Flowers sessile or with pedicels not over 1 mm. long; leaves 5–7 cm. long;
        sepals not over 1 mm. long   40. *pycnoneura*
      45 Flowers distinctly pedicellate, the pedicels 1·5–2·5 mm. long; leaves mostly
        7–10 cm. long; sepals 1–2 mm. long   41. *eperforata*

1. E. (?) **jeremiensis** Urb. & Ekman in Arkiv Bot. **24A** (4): 29 (1932).
Small tree to 5 m. high, glabrous; leaves 10–12 cm. long, subsessile, rotund, broadly cuneate at base, rounded at apex; flowers solitary in upper leaf-axils, seen in bud only; berries unknown.
Rare and local (Han.), on wooded limestone hillsides; 1500–1700 feet; fl. buds July. *P 10416!* Hispaniola. Further information may show this to be an undescribed species distinct from the Haitian *E. jeremiensis* which was based on sterile material only.

2. E. acutisepala Proctor in Bull. Inst. Jam., Sci. ser. **16**: 29, t. 10 (1967).
Shrub or small tree to 7 m. high, the younger parts finely sericeo-lanate throughout; leaves 5–10 cm. long, narrowly to broadly ovate, midrib and veins grooved adaxially; flowers solitary or in a pedunculate cluster of 3, the lateral ones often abortive, white, fragrant; berries crimson, irregularly ellipsoid, up to 15 mm. long.
Local (St. Cath.), on wooded limestone hills; 1650–1850 feet; fl. Mar, Sept, fr. Sept. *P 25578!* Endemic.

3. E. ligustrina (Sw.) Willd. in L., Sp. Pl. ed. 4, **2**: 962 (1799).—*Myrtus ligustrina* Sw. (1788).
Shrub to 3 m. high or rarely a small tree, new growth always subtended by a series of small bractlike scales; leaves 2·5–5 cm. long, dark glossy green adaxially; flowers white, on pedicels 10–40 mm. long; berries black, edible.
Rare (St. Andr., St. Thom.), on steep wooded banks of rivers; 500–2000 feet; fl. May, fr. June–Aug. *H 12562! P 24795! Skelding & Robertson UCWI 4578!* West Indies; ascribed also to Brazil.

4. E. uniflora L., Sp. Pl. **1**: 470 (1753).—*Myrtus brasiliana* L. (1753). Suriname Cherry.
Shrub to 2·5 m. high; leaves mostly 2·5–5.5 cm. long, shortly subacuminate;

flowers white, on slender pedicels of greatly variable length (6–35 mm.); berries red, edible.

Cultivated (Han., St. Andr., St. Thom.), sometimes persisting; 250–2000 feet; fl. Jan–May, fr. Mar–May. *H 10811! P 23320!* Native probably of Brazil, cultivated in most tropical countries.

5. E. heterochroa Urb., Symb. Ant. **7**: 299 (1912).

Shrub to 3 m. high, glabrous, the young branchlets glandular-verruculose; leaves thick and leathery, the midrib deeply grooved; flowers white with rose anthers; pedicels 7–10 mm. long; berries globose, 10–12 mm. in diameter.

Rare and local (Clar.), on wooded rocky limestone hilltops; about 2500 feet; fl. July, fr. Dec, May. *H 10989! P 11400!* Endemic.

6. E. sp. A.

Shrub to 2 m. high, glabrous; leaves thick and leathery, 3–7 cm. long; flowers unkown, evidently solitary or in small clusters, chiefly at leafless nodes, subsessile or on pedicels to 4 mm. long; berries irregularly globose to ellipsoid, 8–11 mm. long, exceptionally to 15 mm. long and 5-seeded.

Local (Clar., St. Cath.), on wooded rocky limestone hilltops; 1500–2500 feet; fr. May, July. *P 10224! Webster 13644!* Endemic.

7. E. mandevillensis Urb., Symb. Ant. **7**: 306 (1912).

Occurs in two varieties as follows:

7a. Var. mandevillensis.

Shrub or small tree to 5 m. high, glabrous; leaves 4–9 cm. long, very shortly acuminate; flowers in terminal umbellate clusters, with or without a very short rachis, white, with unpleasant pungent odour; pedicels 7–20 mm. long, elongating in fruit; berries globose, warty, about 10 mm. in diameter.

Wooded limestone hillsides, the typical variety confined to central Manchester; 1250–3000 feet; fl. May, fr. Jan, Sept. *H 10600! P 19686!* Endemic.

7b. E. mandevillensis var. **perratonii** (Proctor) Proctor, unpublished.—*E. perratonii* Proctor (1958).

Differs from var. *mandevillensis* chiefly in the smaller elliptical, instead of ovate, leaves and pink flowers.

Rare (St. Cath., St. Eliz.); 1750–2600 feet; fl. June, July. *Asprey UCWI 1756! Perraton 153! P 22502!* Endemic.

8. E. sachetae Proctor in Bull. Inst. Jam., Sci. ser. **16**: 34, t. 13 (1967).

Shrub to 3 m. high, glabrous; leaves 6–10 cm. long, bluntly subacuminate, densely beset with pellucid dots; flowers solitary or 2–3 together, terminal or axillary, pedicels 14–24 mm. long; berries unknown.

Rare and local (Trel.), on wooded limestone hill; 1500–2000 feet; fl. May–July. *Fosberg 42952! H 8551!* Endemic.

9. E. lamprophylla Urb., Symb. Ant. **7**: 308 (1912).

Tree to 13 m. high, glabrous; leaves sometimes up to 13 cm. wide, the apex narrowed-obtuse; flowers pale rose, pedicels 5–13 mm. long, petals 7–8 mm. long, ciliate; berries subglobose, 8–10 mm. in diameter, reddish-black.

Rare in the central parishes, on wooded rocky hillsides along bases of limestone cliffs; 1250–2500 feet; fl. June–Sept, fr. Sept. *A 12476! H 11175! P 22748! Webster 13641!* Endemic.

10. E. marchiana Griseb., Fl. Br. W.I.: 238 (1860).

Shrub or small tree to 7 m. high, glabrous; leaves with apex obtuse or sometimes emarginate; flowers white, pedicels 5–12 mm. long, petals 7 mm. long; berries ellipsoid or globose, about 17 mm. long.

Uncommon on wooded hillsides; 1500–5550 feet; fl. July–Nov, fr. Oct, Feb. *H 5334! P 10583!* Endemic.

11. E. amplifolia Urb., Symb. Ant. **5**: 445 (1908).

Shrub or small tree to 8 m. high, glabrous; leaves with apex obtuse or very obtusely acuminate; flowers white, pedicels 4–10 mm. long, petals 4–5 mm. long; berries ellipsoid, 10–15 mm. long, bright red.

Uncommon on wooded limestone hillsides; 400–2900 feet; fl. July–Oct, Dec, fr. Dec–Mar. *A 12026! H 7050! P 10388!* Endemic.

12. E. alpina (Sw.) Willd. in L., Sp. Pl. ed. 4, **2**: 961 (1799).—*Myrtus alpina* Sw. (1788).

Dense-crowned shrub or small tree to 7 m. high, the young branchlets and leaves finely pubescent; leaves crowded, opposite, subopposite or whorled, rigid, opaque, sharply acute; flowers white; berries globose, 5–8 mm. in diameter.

Locally common (St. Andr., Port., St. Thom.), in montane forests and thickets; 4500–7400 feet; fl. May–Sept, fr. Feb, Dec. *A 10657! H 5648! P 9205!* Endemic

13. E. chrootrichoides Proctor in Bull. Inst. Jam., Sci. ser. **16**: 30, t. 11 (1967).

Straggling shrub about 2 m. high, the branchlets pendent, lax, densely short-pilose; leaves lanceolate to narrowly ovate, 2·5–7 cm. long, sharply acuminate; flowers 1–2 on a short peduncle, pedicels 2–4 mm. long; berries 9–18 mm. long.

Locally common (St. Cath., St. Ann), on wooded limestone hillsides; 1650–2600 feet; fl. Jan, July, fr. Jan. *P 20500!* Endemic.

14. E. virgultosa (Sw.) DC., Prodr. **3**: 280 (1828), in part. *Myrtus virgultosa* Sw. (1788).—*E. biflora* var. *virgultosa* (Sw.) Krug & Urb. (1895).

Occurs in two somewhat intergrading varieties as follows:

14a. Var. virgultosa.

Shrub or small tree to 7 m. high, finely pubescent; leaves lanceolate to ovate, 3·5–8 cm. long, 1·2–3 cm. broad, long-acuminate; flowers white, pedicels 2–10 mm. long; sepals to 2 mm. long; berries not seen.

In montane forests (St. Andr., Port., St. Thom.); 2750–5500 feet; fl. June–July. *P 9202!* Endemic.

14b. E. virgultosa var. **jamaicensis** (O. Berg) Proctor, unpublished.—*E. modesta* DC. var. *jamaicensis* O. Berg (1857). *E. biflora* var. *wallenii* Macf. ex Krug & Urb. (1895), illegitimate name. *E. hartii* Kiaersk. (1888).

Differs from var. *virgultosa* chiefly in the shorter petioles, much narrower leaves (0·5–1·3 cm. broad) and shorter sepals (1–1·2 mm. long).

Common (St. Andr., Port., St. Thom.), in montane forests; 4000–6500 feet; fl. July–Sept, fr. Dec. *P 9607!* Endemic.

15. E. alexandri Krug & Urb. in Engl., Bot. Jahrb. **19**: 626 (1895).

Shrub, the young branchlets puberulous; leaves oblong or narrowly ovate, 4–8 (–11) cm. long, acuminate, often acutish at base; flowers with pedicels 3–7 mm. long, sepals to 2·3 mm. long.

Rare (West., St. Eliz., St. Ann), in thickets over limestone; 1200–2400 feet; fl. Jan–July. *A 11415!* Endemic.

16. E. disticha (Sw.) DC., Prodr. **3**: 274 (1828).—*Myrtus disticha* Sw. (1788).

Shrub to 1 (–3) m. high, finely pubescent; leaves often crimson when young, oblong to ovate-elliptical, mostly 5–10 cm. long, acuminate; flowers white, pedicels 1·5–6 mm. long, petals about 5 mm. long, ciliate; ripe fruits dark purple.

In moist thickets, wooded hillsides and pastures, chiefly in the western half of the island; 400–2500 feet; fl. Feb–Aug, fr. Mar–July. *A 12429! H 9259! P 24719!* Endemic.

17. E. brachythrix Urb., Symb. Ant. **6**: 23 (1909).

Tree to 8 m. high with wide-spreading branches, puberulous; leaves lanceolate to narrowly elliptical, 4·5–7·5 cm. long, narrowly acuminate, the midrib grooved adaxially; flowers shortly racemose, pedicels 4–7 mm. long; berries 8–9 mm. long, densely glandular-granulate.

Rare (Port.), in lower montane forest; 1500–2500 feet; fl. June–Aug, fr. July Aug, Dec. *H 5306! P 19741! 23884!* Endemic.

18. E. maleolens Pers., Synops. Pl. **2**: 29 (1806).—*Myrtus buxifolia* Sw. (1788). *E. buxifolia* (Sw.) Willd. (1799), not Lam. (1789).

Shrub or small tree to 8 m. high, more or less puberulous; leaves mostly 2–5 cm. long, usually rounded at apex, rarely obtusely acuminate, cuneate at base, veins prominulous; flowers very shortly racemose, rachis not over 2 mm., sepals mostly less than 1 mm. long; berries globose, 4–6 mm. in diameter, black.

Locally common on rather dry wooded hillsides and in rocky thickets, chiefly near the south coast; SL–3000 feet; fl. June–Sept, fr. Sept–Jan. *A 12133! H 10017! P 23695!* Florida, Yucatan, British Honduras, Bahamas, Greater Antilles, Virgin Is.

19. E. monticola (Sw.) DC., Prodr. **3**: 275 (1828).—*Myrtus monticola* Sw. (1788). *E. buxifolia* of Griseb. (1860), not Lam. nor (Sw.) Willd.

Occurs in two varieties, as follows:

19a. Var. monticola.

Shrub or small tree to 8 m. high; leaves variable, narrowly lanceolate to elliptical, obtuse or very obtusely acuminate, rarely long-acuminate, midrib not grooved; flowers shortly racemose, chiefly axillary, rachis to 9 mm. long; berries purple or black, globose.

Frequent in thickets and on moist wooded hillsides; near SL–5000 feet; fl. July– Sept, fr. Jan–Sept. *A 9145! H 5380! P 22734!* West Indies.

19b. E. monticola var. **latifolia** Krug & Urb. in Engl., Bot. Jahrb. **19**: 636 (1895).

Differs from var. *monticola* chiefly in its larger leaves, mostly 4–8 cm. long and up to 4·5 cm. broad.

Wooded hillsides, chiefly on limestone; 400–3500 feet; fl. June–Oct, fr. Jan– Mar, July–Dec. *H 5051! P 10302!* Mexico, West Indies.

20. E. abbreviata Urb., Symb. Ant. **6**: 24 (1909).

Shrub; leaves ovate to elliptical, 5–7·5 cm. long, the midrib grooved adaxially; petioles 3–4 mm. long; flowers few (2–4) in very short axillary sparingly puberulous racemes; berries globulose.

Apparently rare (West., Manch.), not recently collected; about 500 feet; fl. and fr. Feb–Mar. *H 7101!* Endemic.

21. E. isosticta Urb., Symb. Ant. **7**: 305 (1912).

Shrub or small tree to 7 m. high; leaves ovate-elliptical or elliptical, densely beset with pellucid dots; petioles 7–10 mm. long; flowers few in very short axillary glabrous racemes, petals white; berries globose, 6–7 mm. in diameter.

Rare on wooded limestone hillsides; 100–1000 feet; fl. May, Sept, fr. Aug–Sept. *A 11145! H & P 13776! P 23925!* Endemic.

22. E. sloanei Urb. in Fedde, Repert. Sp. Nov. **14**: 338 (1916).

Shrub to 3 m. high, young branchlets verruculose, puberulous; leaves elliptical, obtuse or obtusely acuminate, the midrib sparsely pubescent; flowers very shortly racemose, the pedicels deciduously pubescent; berries subglobose, 1–2-seeded, 4·5–6 mm. in diameter.

Rare in thickets and on wooded hillsides; near SL–3000 feet; fl. June, Dec. fr. Sept. *H 12896! P 22504!* Endemic.

23. E. schulziana Urb., Symb. Ant. **7**: 304 (1912).

Small tree to 10 m. high, young branchlets minutely puberulous; leaves elliptical to broadly elliptical, 3·5–8 cm. long, midrib minutely puberulous; flowers very shortly racemose, sepals about 1·5 mm. long; berries globose, 5–6 mm. in diameter, black.

Rare (Han., West.), on wooded banks and hillsides; 250–1350 feet; fl. Sept. fr. Aug. *H 9765! P 23970!* Endemic.

24. E. jamaicensis O. Berg in Linnaea **27**: 237 (1856).

Shrub to 5 m. high with arching branches, young branchlets puberulous; leaves ovate or ovate-oblong, midrib grooved; inflorescence 1–1·5 cm. long, with numerous short white adpressed hairs, sepals about 1·5 mm. long; berries obliquely subglobose to ellipsoid, 5–7 mm. long.

Uncommon (West., St. Eliz., St. James), in hillside or roadside thickets; near SL–1500 feet; fl. Feb–Apr, Sept, fr. Apr–May. *A 12509! H 10320! P 22212!* Endemic.

25. E. wilsonella F. & R. in Journ. Bot. **64**: 15 (1926).

Shrub to 3·5 m. high, young branchlets puberulous or glabrate; leaves lanceolate or lanceolate-elliptical, sharply acuminate, mucronate, midrib grooved; flowers numerous in short axillary glabrate racemes, sepals to 1·5 mm. long, ciliate, mucronate, petals pinkish; berries obliquely ellipsoid or obovoid, 8–10 mm. long, blackish-red.

Rather local (Trel., Manch., Clar.), in moist thickets on limestone hillsides or open savannas; 2000–3000 feet; fl. Apr–July, Dec, fr. June–Aug, Oct. *A 8428! ACH 684! P 19827!* Endemic.

26. E. biflora (L.) DC., Prodr. **3**: 276 (1828).—*Myrtus biflora* L. (1759). *Caryophyllus fruticosus* Mill. (1768). *E. pallens* (Vahl) DC. (1828), not Poir. (1813).

Shrub or small tree to 5 (–12) m. high, puberulous; leaves variable, lanceolate to ovate, elliptical or rarely roundish; midrib grooved, often puberulous; flowers numerous, pedicels 5–15 mm. long, sepals 1·5–3 mm. long, silky-pubescent; berries obliquely globose or obovoid, 6–10 mm. in diameter.

On wooded hillsides, chiefly on limestone, commonest in the southern parishes; near SL–4000 feet; fl. Apr–Sept, fr. May–July, Dec. *A 7242! H 8951! P 19745!* Mexico, C. and S. Amer., West Indies. This is a variable species and some of the variants have been described as varieties or separate species. The validity of these taxa is uncertain and cannot be discussed here.

27. E. fadyenii Krug & Urb. in Engl., Bot. Jahrb. **19**: 622 (1895).

Shrub or small tree to 10 m. high, young branchlets with minute appressed brownish hairs; leaves more or less elliptical, midrib narrowly grooved, tissue with crowded pellucid dots; flowers racemose or rarely paniculate, white, scented; pedicels 3–12 mm. long; berries orange.

Widespread but not common, in swampy woodlands and wooded hillsides; near SL–2700 feet; fl. May–Sept, fr. Sept–Feb. *A 11794! H 5251! P 22286!* C. Amer., Cuba.

28. E. sulcivenia Krug & Urb. in Engl., Bot. Jahrb. **19**: 632 (1895).

Shrub or small tree, young branchlets puberulous; leaves leathery, ovate or ovate-elliptical, midrib etc. grooved adaxially; flowers with pedicels 3–6 mm. long, sepals to 1·5 mm. long, the shorter ones slightly apiculate.

Rare (St. Andr.), in montane forests; about 4000 feet; fl. and fr. Aug. *Hart 5049!* Endemic.

29. E. crenata O. Berg in Linnaea **27**: 226 (1856).—*E. pallens* of Griseb. (1860), partly, not DC.

Shrub or small tree to 3 m. high, young branchlets glabrous except at tips; leaves narrowly ovate or lanceolate-oblong, midrib lightly grooved, pellucid dots numerous; flowers with glandular pedicels 6–12 mm. long; berries unknown.

Rare (Manch., St. Andr., St. Thom.), on steep wooded hillsides; 2600–5500 feet; fl. May. *Cornman 603! P 23579!* Endemic. Perhaps merely a variant of *E. harrisii* but more data are needed.

30. E. harrisii Krug & Urb. in Engl., Bot. Jahrb. **19**: 632 (1895).

A variable species represented by at least two varieties as follows:

30a. Var. harrisii.

Shrub or small tree to 5 m. high, glabrous, sometimes sparsely puberulous on young branchlets and inflorescence; leaves lanceolate-oblong to ovate, obtusely acuminate, midrib grooved, the nerves slightly so; glandular dots subpellucid; flowers on glandular pedicels; bracteoles and sepals ciliate; berries obovoid or obliquely ellipsoid, 8–10 mm. long, dark red.

Rather local (St. Andr., St. Thom.), on wooded hillsides; 2800–5000 feet; fl. and fr. all the year, especially May–Sept. *A & S 3159! H 5182! Loveless UCWI 2677! P 23519!* Endemic.

30b. E. harrisii var. **grandifolia** Krug & Urb. in Engl., Bot. Jahrb. **19**: 632 (1895).

Differs from the typical variety in its larger leaves (10–12 cm. long) with longer petioles, and in having thicker pedicels and larger flowers.

Locality uncertain; fl. and fr. Sept. *H 5048!* Endemic.

31. E. rendlei Urb., Symb. Ant. **7**: 302 (1912).

Small tree to 7 m. high, glabrous; leaves ovate, 7–11 cm. long, sharply acuminate, the base rounded, midrib deeply grooved; flowers few, petals white, 4 mm. long; berries unknown.

Rare and local (St. Thom.), in moist forest on limestone; about 1700 feet; fl. Mar. *H & B 10768!* Endemic.

32. E. nicholsii F. & R. in Journ. Bot. **64**: 14 (1926).

Small tree (?), the young branchlets puberulous; leaves lanceolate to narrowly elliptical, 3–5·5 cm. long, long-acuminate, the base rounded or blunt, midrib grooved; flowers numerous, sepals to 2 mm. long, ciliate, petals 3·5–4 mm. long; berries ellipsoid, 4-seeded.

Rare and very local (St. Andr., Port.), about 4900 feet. *Nichols 23.* Endemic.

33. E. axillaris (Sw.) Willd. in L., Sp. Pl. ed. 4, **2**: 960 (1799).—*Myrtus axillaris* Sw. (1788). *E. baruensis* Jacq. (1790). *E. monticola* of Griseb. (1860), not DC. Black Cherry.

Shrub or small tree to 8 m. high, glabrous, except for ciliate sepals; leaves rather leathery, elliptical, ovate or oblong, 4–8 cm. long, obtusely acuminate, dots faintly pellucid; flowers very shortly racemose on pedicels 1–1·5 mm. long; berries black, often very numerous, persistent.

Common in thickets, wooded hillsides and upland pastures; SL–3800 feet; fl. June–Nov, fr. July–Apr. *A 8040! H 5076! P 22769!* Florida, Mexico and northern C. Amer., West Indies; Grand Cayman.

34. E. glabrata (Sw.) DC., Prodr. **3**: 274 (1828).—*Myrtus glabrata* Sw. (1788).

Shrub or small tree to 5 m. high, glabrous, except for ciliate sepals and bracteoles; leaves stiffly papery, ovate to elliptical or broadly elliptical, 3·5–9 cm. long, obtusely acuminate, with numerous faintly pellucid dots; inflorescence as in *E. axillaris* but often longer; berries dark red-purple.

Uncommon in thickets and on wooded limestone hillsides; 1500–3600 feet; fl. June, fr. Mar–Apr, July. *A 11057! H 10867! P 11812!* Cuba, Hispaniola.

35. E. polypora Urb., Symb. Ant. **6**: 24 (1909).

Tree to 20 m. high, glabrous; leaves ovate or narrowly ovate, base rounded or subtruncate, midrib prominent on both surfaces, densely beset with pellucid dots; flowers unknown; berries globose.

Very local (Han.); 1700–1780 feet; fr. Mar. *H 10271!* Endemic.

36. E. confusa DC., Prodr. **3**: 279 (1828).—*E. garberi* Sarg. (1889).

Shrub or small tree to 10 m. high glabrous; leaves leathery, lanceolate to ovate or elliptical, upper surface shining, pellucid dots dense; flowers very shortly racemose or subumbelliform, sometimes solitary, pedicels often red; berries subglobose, scarlet.

Frequent to common, on wooded ridges and limestone hillsides or in coppices in heavy clay soil; 1500–3800 feet; fl. Apr–Sept, fr. Dec, Mar–Apr. *A 11243! H 8740! P 23563!* Florida, West Indies.

37. E. rhombea (O. Berg) Krug & Urb. in Engl., Bot. Jahrb. **19**: 644 (1895).—*E. foetida* var. *rhombea* O. Berg (1856).

Shrub or small tree to 10 m. high, glabrous; leaves narrowly to broadly ovate or subrhomboid, 2·5–5·5 cm. long, short-acuminate; flowers 2–8 together, sometimes solitary; berries scarlet or black.

Rare (St. Andr., St. Cath., St. Ann), in dry rocky limestone woodland; SL–300 feet; fl. July–Aug, fr. Nov–Jan. *H 9618! P 10562!* Florida, Mexico, northern C. Amer., Bahamas, Greater Antilles, Leeward Is.

38. E. clarendonensis Urb., Symb. Ant. **7**: 305 (1912).

Shrub to 3 m. high, glabrous; leaves lanceolate to lanceolate-elliptical or narrowly ovate, 2–4 cm. long, subacuminate, midrib slightly grooved at base only, the principal nerves slightly prominulous on both surfaces; flowers mostly 1–4, petals 4 mm. long, anthers rose or pink.

On wooded rocky limestone hilltops, in west-central areas; 1650–2500 feet; fl. and fr. July. *H 10967! P 26506!* Endemic.

39. E. brownei Urb. in Fedde, Repert. Sp. Nov. **18**: 368 (1922).

Tall shrub; leaves ovate, elliptical or subrhomboid, 2·5–4 cm. long, obtusely acuminate, midrib grooved in lower half, nerves prominulous on both surfaces, pellucid dots numerous; flowers 2–4, petals 3·5 mm. long; berries unknown.

Very local (St. Eliz.), on riverbanks; 10–50 feet; fl. Sept. *H 9848!* Endemic.

40. E. pycnoneura Urb., Symb. Ant. **6**: 25 (1909).

Tree to 8 m. high, branchlets glabrous; leaves narrowly ovate, narrowly long-acuminate, midrib flat or slightly prominent, nerves prominulous on both surfaces, pellucid dots few; flowers 1–3, axillary, calyx-tube prolonged beyond ovary, the lobes deciduous; petals 1·3 mm. long, densely ciliate; berries unknown.

Rare (Port.), in montane forest; about 3500 feet; fl. Oct. *H 7448!* Endemic. The floral characters are anomalous among Jamaican species.

41. E. eperforata Urb., Symb. Ant. **6**: 25 (1909).

Small tree to 10 m. high; leaves elliptical, short-acuminate, papery, without pellucid dots, veins prominulous on both surfaces; flowers often at leafless nodes, rarely on a very short rachis, white, with conspicuously spreading stamens; berries ellipsoid, 9–11 mm. long.

Rare (St. Ann), on wooded limestone hillsides; 1400–1600 feet; fl. July, Sept, fr. Dec. *A 10037! H & P 14091!* Endemic.

Unplaced species:

Eugenia macnabiana Urb. (Symb. Ant. **6**: 104 (1909)) was described from sterile specimens and is probably a juvenile form of some other species; leaves linear, 1–1·5 cm. long, 2·5–3 mm. broad, membranous. Based on an unlocalized specimen collected by *McNab* who was in Jamaica 1838–1859.

Other species of *Eugenia* are known to occur in Jamaica, but information about most of them is thus far inadequate. *Eugenia howardiana* Proctor is described on p. 789.

7. SYZYGIUM Gaertn. (1788) nom. cons.

1 Inflorescence paniculate; flower-buds 5–6 mm. long; calyx truncate; petals united, calyptrate, falling off like a lid at anthesis **1. cumini**
1 Inflorescence racemose; flower-buds 15–30 mm. long; calyx 4-lobed, the lobes persistent; petals separate:
 2 Flowers pink to crimson, in crowded lateral clusters mostly at leafless nodes of old wood; leaves to 18 cm. wide **2. malaccensis**
 2 Flowers white or greenish-white, in mostly terminal racemes, not crowded; leaves mostly 3–5 cm. wide **3. jambos**

1. S. cumini (L.) Skeels in U.S. Dept. Agric. Bur. Pl. Indust. Bull. **248**: 25 (1912).

—*Myrtus cumini* L. (1753). *Eugenia jambolana* Lam. (1789). *S. jambolanum* DC. (1828). Damson tree, Jambolan, Java Plum.

Tree to 20 m. or more high, glabrous; leaves leathery, broadly elliptic-oblong to rotund, 7–12 (–18) cm. long, 3·5–6 (–8) cm. broad, abruptly short-acuminate, long-petiolate, with finely prominulous venation; flowers pinkish, filaments 2–6 mm. long; fruit 1-seeded, oblong to ellipsoid, 1·5–2 cm. long, purple-black, edible.

Cultivated and sometimes persisting, not common; SL–700 feet; fl. Jan–Apr, fr. May. *H 8931! Powell 534! Sullivan UCWI 9264!* Native of the Indo-Malaysian region, cultivated in most tropical countries.

2. S. malaccense (L.) Merr. & Perry in Journ. Arn. Arb. **19**: 215 (1938).— *Eugenia malaccensis* L. (1753). *Jambosa malaccensis* (L.) DC. (1828). Otaheite Apple.

Tree to 20 m. or more high, glabrous; foliage dense, leaves leathery, mostly oblong-elliptical or oblong-oblanceolate to narrowly obovate, 15–30 cm. long, short-acuminate; filaments 20–30 mm. long; fruit 1-seeded, oblong or obovoid, 4–7·5 cm. long, red or pink, rarely white, edible.

Cultivated and naturalized, common; near SL–1000 feet or over; fl. Mar–May, Nov–Dec, fr. Feb–Mar, June–July. *H 11900! P 23493! Robertson UCWI 2139!* Native of Malaya, cultivated in most tropical countries.

3. S. jambos (L.) Alston in Trimen, Handb. Fl. Ceylon, **6**: Suppl. 115 (1931).— *Eugenia jambos* L. (1753). *Jambosa vulgaris* DC. (1828). Rose Apple.

Tree to 18 m. high, glabrous; foliage dense, dark green; leaves lanceolate or very narrowly elliptical, 12–20 cm. long, narrowly acuminate; filaments to 40 mm. long; fruit 1-seeded, depressed-globose, 3–6 cm. in diameter, light yellow, rose–scented, edible.

Widely naturalized, especially in moist soil or along streams; near SL–3000 feet; fl. and fr. sporadically all the year. *A 9923! P 20788!* Native of the Indo-Malaysian and Pacific region, now pantropical.

142. MELASTOMATACEAE

By G. R. Proctor

Annual or perennial herbs, shrubs or small trees, or sometimes epiphytes or scandent vines. Leaves simple, opposite, decussate, those of a pair sometimes unequal, entire or dentate, usually with several (3–7) longitudinal nerves, these all rising at or near the base of the blade, or " -plinerved " (at least the inner nerves originating from the median one above the base of the blade), all connected transversely by smaller veins, rarely pinnate-nerved; stipules absent. Inflorescences axillary, lateral, or terminal; spicate, paniculate or cymose, or sometimes the flowers solitary or clustered. Flowers perfect; perianth regular, mostly 4–6-parted; hypanthium (calyx-tube or receptacle) subglobose to tubular, bearing sepals, petals, and stamens at its margin, free from or adherent to and often overtopping the ovary; sepals small, open or closed in bud, sometimes connate to form a circumcissile calyptra deciduous at anthesis; petals imbricate, often convolute, small or large, usually white, pink or purple. Stamens normally twice as many as the petals (rarely otherwise), often dimorphic; anthers ovoid to subulate, inflexed in bud, usually opening by a single terminal pore (rarely by more) or by longitudinal slits, the connective often enlarged at the base or prolonged below the anther-loculi and variously lobed or spurred. Ovary free or partly or wholly inferior, 2–10-celled, commonly with numerous ovules on axile placentas (rarely with few ovules on basal placentas); style solitary, simple, more or less elongated, terminated by a punctiform to peltate stigma. Fruit enclosed within the persistent hypanthium, either a berry or a loculicidally dehiscent or irregularly rupturing capsule; seeds usually numerous, minute, without endosperm, ovoid to linear, frequently cochleate.

Perhaps 200 genera and 4500 species, pantropical but most abundant in South America. The name alludes to the tongue-blackening juice in the edible fruits of certain species.

1 Leaves with 3 or more longitudinal nerves:
  2 Fruit a capsule; ovary superior, free or nearly so from calyx (fused below in *Arthro-stema fragile*); plants herbaceous or woody:
    3 Plants herbaceous, never vines; seeds cochleate:
      4 Flowers 5-parted   **1. Acisanthera**

4 Flowers 4-parted:
  5 Petals lanceolate, acute; inflorescence a panicle of numerous small white flowers
    with purple anthers   **2. Nepsera**
  5 Petals broader, not acute; inflorescence of rather few large pink flowers with light
    yellow anthers   **3. Arthrostema**
3 Plants woody:
  6 Plant a scandent or epiphytic vine; flowers paniculate in umbelliform clusters;
    petals light pink, about 1 cm. long   **5. Adelobotrys**
  6 Plant an erect shrub or small tree:
    7 Leaves hairy; cultivated ornamental shrubs:
      8 Flowers 5-parted; hypanthium without capitate-stellate hairs   6. **Tibouchina**
      8 Flowers 4-parted; hypanthium with capitate-stellate hairs; anthers 8–9 mm. long;
        native of Ceylon   *Osbeckia rubicunda* Arn.
    7 Leaves glabrous; wild shrubs or small trees of montane forests; flowers white, pink,
      or crimson   **7. Meriania**
2 Fruit a berry; ovary more or less adherent to the calyx; shrubs or small trees, never
  herbaceous:
  9 Inflorescence terminal:
    10 Calyx-limb a lidlike calyptra, deciduous at anthesis   **8. Conostegia**
    10 Calyx-limb of separate teeth or lobes, persistent after anthesis:
      11 Calyx-limb bearing long threadlike processes:
        12 Plants glabrous or nearly so   **9. Calycogonium**
        12 Plants hairy:
          13 Longer hairs glandular; flowers 6-parted; calyx-processes 5 mm. long
            10. Heterotrichum
          13 Longer hairs not glandular; flowers 7–8-parted; calyx- processes 2–3 mm. long
            18.3. *Clidemia octona*
      11 Calyx-limb without long threadlike processes or, if exterior teeth present, these
        not longer than the true sepals:
        14 Hypanthium densely covered with small ciliate scales; flowers violet or mauve,
          showy; cultivated and naturalized   **11. Melastoma**
        14 Hypanthium without ciliate scales; flowers if showy not violet or mauve; wild
          plants:
          15 Leaves, hypanthium, etc., densely rough-tuberculate, the leaves also bristly-
            scabrous beneath   17.5. *Ossaea asperifolia*
          15 Leaves, etc., not usually rough-tuberculate nor bristly-scabrous:
            16 Calyx constricted above the ovary, the limb spreading in fruit   12. **Tetrazygia**
            16 Calyx-limb erect, not spreading:
              17 Flowers (4–) 5–6-parted; petals spreading or reflexed, never crimson or cerise
                13. Miconia
              17 Flowers 4-parted; petals erect, forming a bell-shaped corolla, crimson or
                cerise   **14. Charianthus**
  9 Inflorescences lateral or axillary (rarely terminal in *Ossaea* and *Clidemia*) or flowers
    solitary or clustered at nodes:
    18 Each flower subtended by 2 pairs of large bracts more or less enclosing the hypan-
      thium; flowers solitary in the leaf-axils, large   **15. Blakea**
    18 Flowers not subtended by enclosing bracts:
      19 Flowers clustered or sometimes solitary at nodes of leafless branches:
        20 Plants entirely glabrous; calyx without long hairlike processes; berry not white-
          pellucid   **19. Mecranium**
        20 Plants bearing at least a few hairs or at least the vascular parts minutely scurfy:
          21 Flowers 5-parted   **16. Henriettea**
          21 Flowers 4-parted:
            22 Hypanthium shaggy-hairy; flowers several in a dense sessile cluster; leaves
              rough above   **17. Ossaea**
            22 Hypanthium not hairy, at most minutely scurfy; flowers distinctly stalked;
              leaves smooth, or if hairy, not rough
                18. Clidemia (especially *C. septuplinervia*)
      19 Flowers, or inflorescences, all or chiefly arising from axils of leaves:
        23 Petals acute or sharply tapering; seeds smooth   17. **Ossaea**
        23 Petals obtuse or notched:
          24 Plants with at least some hairs; anthers elongate, the connective not or scarcely
            prolonged below the loculi, and with 1 minute pore at apex; seeds smooth
                18. Clidemia
          24 Plants entirely glabrous throughout; anthers short, the connective prolonged
            below the loculi and jointed with the filament, with 1 or 2 gaping pores at
            apex; seeds granular   **19. Mecranium**
1 Leaves with a solitary longitudinal nerve, the lateral nerves pinnately arranged:
  25 Plant a more or less hairy herb with panicles of showy white or pink flowers; stamens
    8, dimorphic; anthers with 1 pore; fruit a capsule   **4. Heterocentron**

25 Plant a glabrous shrub with solitary inconspicuous axillary flowers; stamens usually 10,
uniform; anthers with 2 pores; fruit a 1–2-seeded berry   20. **Mouriri**

1. ACISANTHERA P. Browne (1756)

1. A. quadrata Pers., Synops. Pl. **1**: 477 (1805).—*Rhexia acisanthera* L. (1759).

Small annual or sub-perennial herb, usually much branched, trailing and rooting and ascending to less than 50 cm. high, glandular-hairy throughout; stems acutely quadrangular, narrowly winged on the angles; leaves 1–2 cm. long, mostly elliptical with finely denticulate margins, 3-nerved; flowers numerous, solitary in leaf-axils, subsessile; petals pink or rose, rarely white; filaments crimson upwardly, anthers yellow.

In wet boggy ground or moist pastures in clay soil, chiefly in the western half of the island; near SL–2300 feet; fl. and fr. all the year. *A 5906! H 9460! P 18446!* Mexico to S. Amer., West Indies.

2. NEPSERA Naud. (1849)

1. N. aquatica (Aubl.) Naud. in Ann. Sci. Nat. sér. 3, Bot. **13**: 28 (1849).—
Melastoma aquaticum Aubl. (1775).

Erect annual or sub-perennial herb, 25–100 cm. high in flower, usually much branched, minutely pubescent chiefly on the vascular parts or sometimes glabrous; stems dark reddish-brown, subquadrangular; leaves 2–7 cm. long, narrowly ovate or ovate, with slender petioles, chiefly 5-nerved, the margins minutely serrulate; panicles pyramidal, much branched; pedicels very slender, up to 2 cm. long.

Locally common (Clar., St. Ann, Port.), in wet boggy ground, moist pastures or edges of forest clearings; 500–2300 feet; fl. and fr. all the year. *A 11469! H 12248! P 15808!* C. and S. Amer., West Indies.

3. ARTHROSTEMA Pav. ex D. Don (1823)

1. A. fragile Lindl. in Journ. Hort. Soc. **3**: 74, fig. (1847).—*A. ciliatum* Ruiz & Pav. ex Cogn. (1891).

Weak-stemmed erect or straggling herb, apparently annual from a perennial root-system, sparsely glandular-hairy on the upper stems; stems brittle, succulent, quadrangular, irregularly branched, up to 2 m. or more long; leaves mostly 2·5– 10 cm. long, ovate, with petioles up to 2 cm. long, those of a pair unequal, the blades truncate or subcordate at base, mostly 5-nerved, the margins ciliate-serrulate; inflorescence lax, few-flowered; petals up to 2·5 cm. long, pink, sometimes bicolorous, soon falling; stamens all similar or variously dimorphic, filaments pink, anthers yellow; capsules 1–1·3 cm. long.

Common and widespread in moist thickets and swales, on moist roadside banks and along the borders of fields; 200–2600 (–4000) feet; fl. and fr. all the year. *A 6093! H 9930! P 15807!* Mexico to S. Amer., Cuba.

4. HETEROCENTRON Hook. & Arn. (1838)

1. H. subtriplinervium (Link & Otto) A. Braun & Bouché, Ind. Sem. Hort.
Berol., App. (1851); Linnaea **25**: 300 (1853).—*Melastoma subtriplinervium* Link & Otto (1821).

Erect perennial suffruticose herb or subwoody shrub, mostly 0·3–2 m. high, sparsely hairy throughout; stems quadrangular; leaves 3–10 cm. long, elliptic-oblong to rhombic-ovate, decurrent at base, the midrib giving rise to 4–6 ascending nerves on each side, the margins entire to slightly crenulate; flowers 4-parted; petals 5–10 mm. long, obovate; anthers yellow, dimorphic, the larger ones with connective elongated below into a bifid appendage; seeds cochleate.

Cultivated and naturalized locally; 1500–3500 feet; fl. Jan–Aug. *A 6558! H & P 14192!* Mexico, C. Amer., naturalized in Hawaii and cultivated in tropical Asia.

5. ADELOBOTRYS DC. (1828)

1. A. adscendens (Sw.) Triana in Journ. Bot. **5**: 210 (1867).—*Melastoma adscendens* Sw. (1798).

Subwoody vine climbing to 10 m. or more, holding to tree-trunks by adventitious roots, the young vascular parts minutely brown-hairy, the hairs deciduous; stems terete; leaves 6–13 cm. long, long-petiolate, the blades round-ovate with short-acuminate apex, chiefly 5-nerved, the margins ciliate and remotely serrulate; inflorescences terminal and lateral in upper leaf-axils the panicles made up of numerous umbels or headlike clusters, the flowers pedicellate; petals light pink or lavender, drying yellow; filaments and style rose-violet; anthers white or pale yellow, the connective with a basal spur and bearing dorsally a long antrorse appendage; capsule ribbed, shedding seeds through a terminal pore.

Rare (St. Ann, Port., St. Thom.), in moist forests; 1000–2500 feet; fl. Feb–Mar, fr. Mar–Apr. *A 9265! H & B 10729! P 23428!* Mexico to S. Amer.

6. TIBOUCHINA Aubl. (1775)

1. T. urvilleana (DC.) Cogn. in Mart., Fl. Bras. **14** (3): 358 (1885).

Shrub to 3 m. high, densely hirsute throughout; leaves of a pair unequal, petiolate, 4–12 cm. long, the blades ovate to elliptical, sharply subacuminate, 5-nerved, often red beneath; petals rich purple, about 3·5 cm. long; anthers mauve, sickle-shaped, 12–13 mm. long; capsule hairy, oblong, to 1·5 cm. long.

Cultivated chiefly in the mountains (St. Andr., Port., St. Thom.), 3000–4200 feet; fl. probably all the year, fr. Feb. *P 24257! Robertson UCWI 2985!* Native of Brazil.

7. MERIANIA Sw. (1797) nom. cons.

Veinlets of leaves thickened-prominulous, especially beneath; leaves 5–23 cm. long; bracteoles with broadly obovate tips; petals white, pink-tinged, or light crimson
          1. *leucantha*
Veinlets of leaves flat on both surfaces, not thickened; leaves mostly 3–8 cm. long; bracteoles linear or with lanceolate tips; petals light to deep crinson
          2. *purpurea*

1. M. leucantha (Sw.) Sw., Fl. Ind. Occ. **2**: 826 (1798).—*Rhexia leucantha* Sw. (1788).

Shrub or slender tree to 10 m. high; leaves petiolate, narrowly to broadly elliptical, narrowly ovate or lanceolate, obtuse or acute, rigid, 3-nerved; flowers solitary in upper leaf-axils; calyx-lobes slender or rather thick, 3–5 mm. long; petals 2·5–3·5 cm. long.

In moist thickets and forests in the eastern half of the island; (1000–) 1500–5000 feet; fl. and fr. Dec–Aug. *A 12531! H 7002! P 22069!* Endemic; a variety is reported from Cuba.

2. M. purpurea (Sw.) Sw., Fl. Ind. Occ. **2**: 829 (1798).—*Rhexia purpurea* Sw. (1788).

Shrub or slender tree to 8 m. high; leaves petiolate, elliptical or broadly elliptical, acute, papery, 3-nerved with 2 swellings on upper surface in the nerve-axils, dark green above; flowers solitary in upper leaf-axils; calyx-lobes very slender, 4–8 mm. long; petals 2·5–3·5 cm. long; anthers cream, 5·5–7 mm. long.

Locally common (St. Andr., Port., St. Thom.), in montane thickets and woodlands; 3900–6600 feet; fl. and fr. mostly Dec–July. *A 6630! Martyn 14019! P 22268!* Colombia.

Plants intermediate in character between these two species exist in areas where the more typical species are sympatric and may be natural hybrids.

8. CONOSTEGIA D. Don (1823)

1 Plants glabrous or nearly so:
2 Leaves mostly 15–30 cm. long; flower-buds 6–7 mm. long; petals white   8. *superba*
2 Leaves mostly 6–15 cm. long; flower-buds 10–17 mm. long; petals pink or rose:
3 Flower-buds obtuse or rounded, about 10 mm. long   3. *grisebachii*

3 Flower-buds acute to acuminate, 12–17 mm. long:
  4 Leaf-blades elliptical, narrowed at base, the nerves and veins yellowish beneath, the transverse veins not prominulous 1. *procera*
  4 Leaf-blades ovate to ovate-elliptical, rarely narrowly oblong, usually more or less rounded at base, the nerves and transverse veins red and prominulous beneath
2. *balbisiana*
1 Plants not glabrous:
  5 Indumentum chiefly of relatively long simple or minutely stellate-tipped abundant or sparse curved hairs:
    6 Hypanthium densely clothed in long blue hairs 7. *speciosa*
    6 Hypanthium glabrous or at most with a few stellate hairs:
      7 Leaves mostly 6–12 cm. or rarely more long; flower-buds spindle-shaped, acute to acuminate, 10 mm. or more long:
        8 Leaves 3–6 cm. broad; flower-buds about 10 mm. long; petals 7–8 mm. long
4. *pyxidata*
        8 Leaves 1·5–3 cm. broad; flower-buds 15–19 mm. long; petals 20 mm. long
5. *subprocera*
      7 Leaves mostly 12–30 cm. long; flower-buds obovoid or subglobose; 5–7 mm. long:
        9 Stems and petioles shaggy-hairy; hypanthium minutely stellate-puberulous
6. *icosandra*
        9 Stems and petioles glabrous; hypanthium glabrous 8. *superba*
  5 Indumentum not of long simple or stellate-tipped hairs:
    10 Indumentum of minute sessile stellate hairs; flower-buds 5–8 mm. long, acute
9. *montana*
    10 Indumentum of minute reddish-brown branlike ciliate scales; flower-buds 8–10 mm. long, obtuse 10. *rufescens*

1. C. procera (Sw.) DC., Prodr. **3**: 174 (1828).—*Melastoma procerum* Sw. (1788).

Shrub or tree to 15 m. high, the young branches subquadrangular, glabrous or rarely brown-tomentose; leaves petiolate, 3-nerved, the nerves often bearded in the axils beneath; petals lavender-rose, rarely white, 10–12 mm. long; anthers yellow, about 3 mm. long; style truncate at apex.

Frequent (St. Andr., Port., St. Thom.), on non-calcareous soils in montane forests; 2000–4100 (–6000) feet; fl. Mar–Aug, fr. not seen. *J.P. 1402! P 10321! Robbins UCWI 1425!* Endemic.

2. C. balbisiana Ser. in DC., Prodr. **3**: 174 (1828).—*C. procera* var. *balbisiana* (Ser.) Griseb. (1860).

Shrub or small tree up to 10 m. high, similar to *C. procera* but with smallest leaf-veins less numerous and more obscure; petals 6, unequally lobed, pink or rose, 17–22 mm. long.

Frequent in woodlands over limestone, chiefly in the western half of the island; there is an isolated population at the southern end of the John Crow Mts. (St. Thom.); 1000–3000 feet; fl. Mar–July, fr. June–Sept. *A 7792! H 8993! P 10432!* Endemic.

3. C. grisebachii Cogn. in A. & C. DC., Monogr. Phan. **7**: 700 (1891).

Small glabrous tree with subquadrangular branches; leaves 7–11 cm. long, elliptical, prominently 3-nerved with an obscure marginal pair, very rigid, usually folded and decurved; petals pink; filaments light pink; anthers yellow, about 3 mm. long; stigma subcapitate.

Very rare (St. Thom.), known only from the unlocalized type and two specimens from the southern end of the John Crow Mts.; 1500–2000 feet; fl. Mar, Aug, fr. Aug. *P 11797! Webster & Proctor 5540!* Endemic. This species seems doubtfully distinct from *C. balbisiana* except for the very different flower-buds; further study is needed.

4. C. pyxidata Proctor in Bull. Inst. Jam., Sci. ser. **16**: 38, t. 15 (1967).

Shrub to 4 m. high, the young branchlets compressed, brown-pubescent on one side; leaves with pubescent petioles, the blades glabrous, ovate-elliptical or oblong-elliptical, rounded or subcordate at base, 3-nerved with an additional obscure pair of intramarginal nerves, the margins entire or nearly so; petals white; style truncate at apex.

Rare and local (Port.), in wet mossy woodlands; 1500–3700 feet; fl. July–Aug, fr. not seen. *A 9382! ACH 990! P 10468! Webster 5606!* Endemic. A smaller-leaved variant occurs towards the northern end of the John Crow Mts. but is insufficiently known.

5. C. subprocera Proctor in Bull. Inst. Jam., Sci. ser. **16**: 38, t. 16 (1967).

Shrub or small tree to 5 m. high, the young branchlets terete, reddish-brown hirsute on one side; leaves with finely hirsute petioles, the blades glabrous, narrowly oblong-elliptical or oblanceolate, abruptly acuminate at apex with a caudate tip 1 cm. long, 3-nerved, the principal side-nerves 2 mm. or less from the entire margins; petals white or roseate, showy; style truncate at apex.

Very rare and local (Port.), in moist thickets over limestone; 1200–2000 feet; fl. Aug–Sept, fr. Dec–Jan. *P 9229! Webster & Wilson 5165!* Endemic.

6. C. icosandra (Sw. ex Wikstr.) Urb. in Fedde, Repert. Sp. Nov. **17**: 404 (1921). —*Melastoma icosandrum* Sw. ex Wikstr. (1827). *C. subhirsuta* DC. (1828).

Shrub or small tree to 10 m. high with stout subquadrangular branches; vascular parts shaggy-hirsute throughout, the hairs minutely stellate at apex; leaves to 30 cm. long, elliptical or oblong-elliptical, 3-plinerved, glabrous above, the margins wavy-dentate; flowers fragrant; petals white; filaments white; anthers yellow; stigma peltate on a bent style.

Occasional in moist thickets on limestone in the eastern half of the island; 850–2500 feet; fl. Feb–May; fr. May, Aug. *A 9336! H & B 10721! P 9975!* Mexico to S. Amer., West Indies. Material from continental areas is much less hirsute than Jamaican specimens and also tends to vary in other details.

7. C. speciosa Naud. in Ann. Sci. Nat. sér. 3, Bot. **16**: 109 (1851).

Shrub to 2 m. or more high, the branches, petioles, undersurfaces of leaves, and inflorescence densely clothed in soft stellate-tipped hairs (those of hypanthium blue); leaves petiolate, up to 25 cm. long, the blades ovate to elliptical, short-acuminate, mostly 5-plinerved, the upper surface with simple hairs; panicle many-flowered; buds about 7 mm. long, subacute; petals pink, 5–7 mm. long; anthers about 2·5 mm. long; style abruptly bent beneath the peltate stigma.

Cultivated at Castelton Gardens (St. Mary) and becoming naturalized in the vicinity; about 500 feet; fl. and fr. Sept–Nov. *Robertson UCWI 1668! Thompson 8072!* Costa Rica, Panama, northern S. Amer.

8. C. superba Naud. in Ann. Sci. Nat. sér. 3, Bot. **16**: 108 (1851).—*Melastoma superbum* Bonpl. ex D. Don (1823), name only.

Small tree to 9 m. high, with stout glabrous hollow obtusely quadrangular branches; leaves to 30 cm. long, long-petiolate, the blades thin, ovate, 3-plinerved with an additional pair of obscure submarginal nerves, sparsely hirsute on nerves beneath, glabrate, the margins subentire; flowers in umbellate clusters on the panicle-branches; petals 5 (–6), white; anthers yellow; stigma capitellate.

Rare (St. Mary, Port., St. Thom.), in moist woodlands; 1000–2500 feet; fl. July–Sept, fr. not seen. *A 11497! H & B 10562! P 23880!* Mexico, C. Amer.

9. C. montana (Sw.) DC., Prodr. **3**: 175 (1828).—*Melastoma montanum* Sw. (1788).

Shrub or tree to 6 m. high with slender spreading branches, the young branch-lets obtusely quadrangular; leaves 5–15 cm. long, petiolate, the blades narrowly ovate to elliptical or broadly elliptical, 3-plinerved with an additional pair of intra-marginal nerves, the margins subentire; petals 5 (–6), white or sometimes pink at base, about 5 mm. long; anthers 2 mm. long; stigma capitellate; ripe berries blue.

Common (St. Andr., Port., St. Thom.), in montane forests and forested ravines; 3000–6000 feet; fl. Feb, June–Sept, Nov, fr. Aug–Feb. *A 7822! H 6336! P 9443!* Endemic ? (Attributed to Martinique on the basis of *Belanger 706* (Paris)). A closely related variant occurs in the John Crow Mts. (Port.), but more material is required to establish its correct status.

10. C. rufescens Naud. in Ann. Sci. Nat. sér. 3, Bot. **16**: 108 (1851).

Shrub or tree to 12 m. high with rufescent quadrangular branches; leaves 8–19 cm. long, narrowly elliptical to elliptical, 3-nerved or 3-plinerved, glabrous above, rusty-mealy on nerves and veins beneath, the margins entire or obscurely toothed; flowers fragrant; petals (5–) 6, light pink or white tinged pink at base; anthers yellow, 3 mm. long; stigma peltate on a bent style.

Uncommon in moist forests on limestone, chiefly at the north-eastern end of the island (Port.); an isolated record from Westmoreland; 750–2250 feet; fl. Apr, Aug, Oct, fr. Oct. *A 11564! P 21476!* Colombia.

9. CALYCOGONIUM DC. (1828)

Leaves mostly 5–10 cm. long, rarely less, bearded at axils of nerves beneath; pedicels 5–10 mm. or more long   1. *glabratum*
Leaves mostly 2–5 cm. long, rarely more, fimbriate-glandular at axils of nerves beneath; pedicels less than 2 mm. long or flowers sessile   2. *rhamnoideum*

1. C. glabratum (Sw.) DC., Prodr. **3**: 168 (1828).—*Melastoma glabratum* Sw. (1788).

Shrub to 4 m. high; leaves lanceolate to elliptical, more or less acuminate, 3-plinerved; flowers 4-parted; calyx-processes 6–8 mm. long; petals dull whitish, 5 mm. long.

Rather rare (St. Cath., St. Andr., St. Thom.), on steep wooded non-calcareous hillsides; 400–2000 feet; fl. June, fr. Jan, June. *P 16447!* Cuba.

2. C. rhamnoideum Naud. in Ann. Sci. Nat. sér. 3, Bot. **16**: 85 (1851).

Shrub to 3 m. high; leaves elliptical or oblong-elliptical, acute or obtuse at apex, 3-plinerved; flowers 4-5-parted; calyx-processes 4–5 mm. long; petals white, 3–4 mm. long; anthers yellow; ripe berries black.

Uncommon (St. Cath., St. Andr., St. Thom.), in hillside thickets or on rocky banks, calcareous or not; 800–2750 feet; fl. July–Aug, fr. Apr, July–Sept. *H 6207! P 21389!* Cuba, Hispaniola.

10. HETEROTRICHUM DC. (1828)

1. H. umbellatum (Mill.) Urb. in Fedde, Repert. Sp. Nov. **15**: 14 (1917).—*Melastoma umbellatum* Mill. (1768). *M. patens* Sw. (1788). American Gooseberry.

Shrub to 3·5 m. high, the young branches, petioles and inflorescence densely clothed with long glandular hairs; leaves long-petiolate, 6–30 cm. long, ovate to very broadly ovate, more or less cordate at base, acuminate, 5–7-nerved, rarely 3-plinerved with an additional marginal pair, upper surface bearing minute scattered stellate hairs, lower surface greyish- or whitish-tomentose with crowded stellate hairs; flowers in cymose panicles, the ultimate clusters having a central subsessile flower and 2 long-pedicellate lateral ones; petals (5–) 6 (–7), white, pinkish in bud; filaments and style white; anthers yellow; ripe berries black, about 1 cm. in diameter, edible.

Rather common in moist thickets, pastures and on roadside banks, chiefly in non-calcareous areas; 200–4900 feet; fl. and fr. all the year. *A 5778! H 8652! P 20872!* Cuba, Hispaniola.

11. MELASTOMA L. (1753)

1. M. malabathricum L., Sp. Pl. **1**: 390 (1753).

Shrub to 2 m. high, the slender quadrangular branches scabrous, densely beset with minute rigid appressed ciliate scales; leaves mostly 5–12 cm. long, petiolate, the blades narrowly elliptical or oblong-elliptical, acute at both ends, 3–5-nerved, covered on both surfaces with small partially adnate bristles; flowers few (3–6), in sessile terminal corymbs, 5-parted, subtended by large deciduous bracts; anthers dimorphic.

Cultivated at Castleton Gardens (St. Mary), now naturalized locally (St. Andr., St. Mary), in hillside pastures; 500–1200 feet; fl. and fr. probably all the year. *H 10830! P 18264! Robertson UCWI 1411!* Native to tropical Asia and Polynesia.

12. TETRAZYGIA L. C. Rich. ex DC. (1828)

1 Flowers mostly 5-parted; calyx-lobes obtuse   1. *pallens*

1 Flowers always 4-parted; calyx-lobes acute to acuminate:
2 Long simple hairs present, at least on the petioles and inflorescence-branches; axils of
    transverse veins usually bearded beneath   2. *hispida*
2 Simple hairs absent; axils of veins and nerves not bearded:
    3 Leaves more than 2 cm. broad, rounded-truncate at base; calyx-tube 7–8 mm. long
                                                              3. *albicans*
    3 Leaves less than 1 cm. broad, narrowed at base; calyx-tube 1·5 mm. long or less
                                                              4. *angustifolia*

1. T. pallens (Spreng.) Cogn. in A. & C. DC., Monogr. Phan. **7**: 724 (1891).—
    Melastoma pallens Spreng. (1825). *T. ovata* Cogn. (1908). Ashes Bush, Ashes
    Wood, Black Ashes, Clover Ash.

Tree to 18 m. high, the young branches, undersurfaces of leaves, and inflorescence
densely clothed with minute stellate hairs; leaves long-petiolate, 6–22 cm. long,
the blades ovate to elliptical, acuminate, 3-nerved or 3-plinerved; flowers numerous
in panicles; petals (4–) 5, all white or white with a pink patch, 6–8 mm. long;
anthers 5–6 mm. long, yellow; style pink; ripe berry purple.

Common on wooded or partly wooded hillsides, chiefly over limestone; 300–
3000 feet; fl. and fr. most of the year. *A 10944! H 8744! P 22615!* Endemic, (attributed to Cuba and Dominica, perhaps in error). Closely related to *T. bicolor*
(Mill.) Cogn. of Florida, Cuba and Hispaniola, but seems to differ consistently in
leaf-shape and details of indumentum.

2. T. hispida (Sw.) Cogn. in A. & C. DC., Monogr. Phan. **7**: 723 (1891).—
    Melastoma hispidum Sw. (1788). *M. glandulosum* Sw. (1798). *Heterotrichum
    hispidum* (Sw.) Griseb. (1860).

Shrub or small tree to 8 m. high; leaves petiolate, 6–20 cm. long, blades narrowly
to broadly ovate, rarely elliptical, acuminate, 3–5-nerved; calyx-lobes deltate, 2–3
mm. long; petals white, pink or lavender, clawed at base, 6–8 mm. long; anthers
2–2·5 mm. long; occurs in two well marked varieties, distinguishable as follows:

Leaves scabrous-hispid on upper surface, the dense hairs of young branches, leaf-nerves
    beneath and inflorescence-branches bright red   a. var. *hispida*
Leaves nearly or quite smooth and glabrate on upper surface, the hairs elsewhere sparse
    and not red   b. var. *laevior*

2a. Var. hispida.
Chiefly occurs on wooded limestone hilltops, widely scattered but most common
in the Cockpit Country region; 1250–3000 feet; fl. June–Aug, fr. July–Aug.
H 12010! P 10579! Robertson UCWI 1989! Endemic.

2b. T. hispida var. laevior (Griseb.) Cogn. in A. & C. DC., Monogr. Phan. **7**:
    723 (1891) (as *laevis*).—*Heterotrichum hispidum* var. *laevius* Griseb. (1860).

Widely distributed on wooded limestone hillsides, more common than var.
hispida; 250–3000 feet; fl. Mar–July, fr. June–Sept. *A 10940! P 10826!* Endemic.

3. T. albicans (D. Don ex Naud.) Triana in Trans. Linn. Soc. London **28**: 100
    (1871).—*Chitonia albicans* D. Don ex Naud. (1851).

Small tree to 6 m. high, with furrowed bark; young branches, leaves beneath and
inflorescence densely clothed with minute matted whitish stellate hairs, the upper
surfaces of leaves with scattered similar hairs when young; leaves petiolate, 4–10
cm. long, blades ovate to oblong-elliptical, 3-plinerved with an additional pair of
nerves from the base; panicles few-flowered, the subsessile flowers usually in 3's at
the ends of branches; petals white, 6–8 mm. long; anthers yellow, 5–6 mm. long.

Rare (St. Andr.), on steep wooded hillsides; 3000–3200 feet; fl. and fr. Oct.
P 23565! 25591! Endemic; previously collected only once; *Wiles* about 1800.

4. T. angustifolia (Sw.) DC., Prodr. **3**: 172 (1828).—*Melastoma angustifolium* Sw.
    (1788).

Shrub or small tree; leaves petiolate, 3–8 cm. long, the blade narrowly lanceolate,
acuminate, 3-nerved, whitish-puberulous beneath with minute stellate hairs;
flowers very small in corymbiform panicles; petals yellowish or rosy, 2 mm. long;
anthers 2 mm. long.

Collected only once in Jamaica at an unknown locality in 1786; not seen here
since. The species is quite common in the islands from Puerto Rico to Trinidad.

13. MICONIA Ruiz & Pav. (1794) nom. cons.

1 Leaves glabrous on both surfaces:
  2 Leaves 3-plinerved:
    3 Flowers 4-parted; petals pink   1. *nubicola*
    3 Flowers 5-parted; petals white:
      4 Leaves not over 3·5 cm. broad, dull green, acute or subacuminate, the transverse veins somewhat obscure; hypanthium about 2 mm. long, densely powdery-stellate   2. *attenuata*
      4 Leaves mostly 4–8 cm. broad, yellow-green, acuminate, the transverse veins usually prominent beneath; hypanthium 2–2·5 mm. long, obscurely powdery-stellate
                                           3. *prasina*
  2 Leaves 3–5-nerved:
    5 Flowers on pedicels 8 mm. or more long; hypanthium 5 mm. long; branches with hairy patches above the nodes   4. *pseudorigida*
    5 Flowers sessile or on pedicels not over 1 mm. long; hypanthium not over 3 mm. long; stems not pilose:
      6 Leaf-margins ciliate; flowers secund on panicle-branches   5. *ciliata*
      6 Leaf-margins not ciliate; flowers not secund:
        7 Hypanthium glabrous, less than 2 mm. long; leaves thin:
          8 Style 2–3 mm. long; smallest areoles of leaf-veins 0·2–1 mm. across; flowers always 5-parted   6. *theaezans*
          8 Style 1·5–2 mm. long; smallest areoles of leaf-veins more than 1 mm. across; flowers 4–5-parted   7. *rubens*
        7 Hypanthium furfuraceous, 2–3 mm. long; leaves rigid   8. *quadrangularis*
1 Leaves not glabrous, indumentum at least on nerves beneath:
  9 Flowers on pedicels mostly 4–11 mm. long, 6-parted:
    10 Leaves 3-nerved, the tissue glabrous, nerves brown-stellate-furfuraceous beneath
                                         9. *rigida*
    10 Leaves 5-nerved, the undersurface tissue and nerves densely covered with matted whitish stellate tomentum   10. *dodecandra*
  9 Flowers sessile, or if rarely stalked, pedicels less than 2 mm. long:
    11 Indumentum of long simple hairs; flowers in small dense glomerules at the ends of panicle-branches   11. *matthaei*
    11 Indumentum of minute stellate hairs or scales:
      12 Leaves sessile, auriculate or subclasping, mostly 20–40 cm. or more long:
        13 Indumentum on undersurface of leaves brown-stellate; leaves 3-plinerved with an additional basal pair; inflorescence red; berries 4–5 mm. in diameter
                                     12. *impetiolaris*
        13 Indumentum on undersurface of leaves white-stellate; leaves 7-plinerved; inflorescence yellow-green; berries 7–8 mm. in diameter   13. *ampla*
      12 Leaves distinctly petiolate:
        14 Indumentum of undersurface of leaves mainly on nerves or veins, or scattered so that the underlying epidermis is clearly visible:
          15 Inflorescence spicate or racemose, the branches either absent or very short and simple:
            16 Young branches sharply quadrangular; leaves cuneate at base   14. *triplinervis*
            16 Branches terete or obtusely flattened; leaves rounded at base:
              17 Hypanthium 2–3 mm. long, whitish-stellate-tomentose; anthers white, about 1·5 mm. long; stigma peltate   15. *multispicata*
              17 Hypanthium 5–8 mm. long, brown-furfuraceous: anthers purple, 5–6 mm. long; stigma subcapitate   16. *furfuracea*
          15 Inflorescence amply paniculate, the branches variously forked:
            18 Flowers 4-parted; stigma punctate, the style narrowed at apex; anthers yellow; leaves beneath with numerous very minute brown stellate scales   17. *tetrandra*
            18 Flowers 5-parted; stigma distinctly enlarged; anthers white:
              19 Flowers secund on the panicle-branches; leaves 3-plinerved, the lateral nerves submarginal   18. *trinervia*
              19 Flowers not secund; leaves 3–5-nerved, or 3-plinerved with another submarginal pair:
                20 Leaves strongly 3-plinerved:
                  21 Leaves not over 3·5 cm. broad, dull green, acute or subacuminate, the transverse veins somewhat obscure; hypanthium about 2 mm. long, densely powdery-stellate   2. *attenuata*
                  21 Leaves mostly 4–8 cm. broad, yellow-green, acuminate, the transverse veins usually prominent beneath; hypanthium 2–2·5 mm. long, obscurely powdery-stellate   3. *prasina*
                20 Leaves 3–5 (–7)-nerved or barely 3-plinerved; hypanthium densely powdery-stellate:

22 Style 6–7 mm. long, gradually thickened to the truncate or subcapitate apex; leaves mostly 9–20 cm. long	19. *laevigata*
22 Style 4 mm. long, the stigma peltate; leaves mostly 20–30 cm. long
20. *splendens*
14 Indumentum densely covering and concealing the epidermis on undersurface of leaves:
23 Undersurface of leaves covered with golden peltate stellate scales; leaves narrowly lanceolate, less than 4 cm. broad	21. *chrysophylla*
23 Undersurface of leaves covered with matted whitish stellate hairs:
24 Flowers 5-parted, strongly secund; leaves 2·5–7 cm. broad; hypanthium 2·5–3 mm. long	22. *albicans*
24 Flowers 6-parted, not secund; leaves up to 16 cm. broad:
25 Hypanthium 6–7 mm. long; style 11–13 mm. long; leaf-blades rounded or emarginate at base	23. *serrulata*
25 Hypanthium 2 mm. long; style 2–3 mm. long; leaf-blades acute at base
24. *elata*

1. M. nubicola Proctor in Bull. Inst. Jam., Sci. ser. **16**: 42, t. 17 (1967).

Shrub about 2 m. high, glabrous throughout; leaves petiolate, 7–12 cm. long, 1·5–2·5 cm. broad, lanceolate, acuminate at both ends, 3-plinerved; flowers in clusters of 3 on panicle-branches, subsessile; hypanthium 2·5 mm. long; petals about 3 mm. long; anthers 1·5 mm. long; style 3·5 mm. long, not thickened at the truncate apex; berries purple, 3–4 mm. in diameter, not ribbed.

Very rare and local (St. Thom.), in mossy montane woodland; 6000–7000 feet; fl. and fr. Dec, fr. Jan. *H & P 14818! P 9691!* Endemic.

2. M. attenuata DC., Prodr. **3**: 186 (1828).—*M. prasina* var. *attenuata* (DC.) Cogn. (1891).

Shrub 1·5–3 m. high, the branchlets and inflorescence minutely powdery-stellate; leaves with petioles 0·5–1 cm. long, the blades oblong-elliptical or narrowly elliptical; anthers white, 2·5–3 mm. long; ripe berries dull blackish.

Frequent in low moist thickets on heavy clay soil in the western half of the island; near SL–2800 feet; fl. and fr. sporadically all the year. *A 7121! H 8703! P 23553!* Endemic?

3. M. prasina (Sw.) DC., Prodr. **3**: 188 (1828).—*Melastoma prasinum* Sw. (1788)· *Melastoma laevigatum* of Aubl. (1775), not L. (1753).

Shrub or small tree to 8 m. high, the branchlets and inflorescence glabrate or minutely powdery-stellate; leaves with petioles 0·5–2 cm. long, the blades ellipti-cal, often plicate; anthers white, about 2 mm. long; berries blue in most Jamaican material, rarely black.

Common in moist thickets; 300–2800 feet; fl. and fr. most of the year. *A 8148! H 8243! P 24813!* Mexico to Paraguay, West Indies.

It is possible that two distinguishable entities have been passing under the name *prasina* in Jamaica: one (common) with blue, rather glaucous berries, the other (rare) with somewhat larger black berries. Further study is needed. Coppice-regrowth of this plant often has the leaf-blades hairy adaxially and the margins distinctly ciliate.

4. M. pseudorigida Proctor in Bull Inst. Jam., Sci. ser. **16**: 42, t. 18 (1967).

Shrub to 2·5 m. high; leaves rigid, petiolate, 10–20 cm. long, the blades ovate-elliptical, rounded or emarginate at base, acute at apex, 3-nerved; panicle-branches with gland-tipped hairs, otherwise glabrous; hypanthium about 5 mm. long; petals white, 4 mm. long; anthers about 2 mm. long, opening by a longitudinal slit; berries reddish-purple, 6 mm. in diameter.

Very rare (Port.), in mossy woodlands on limestone; 1500–2500 feet; fl. July–Aug, fr. Feb, Aug, Dec. *A 9146! ACH 988! P 10469!* Endemic.

5. M. ciliata (L. C. Rich.) DC., Prodr. **3**: 179 (1828).—*Melastoma ciliatum* L. C. Rich. (1792). *Melastoma octandria* Mill. (1768), not *M. octandrum* L. (1753).

Shrub to 3 m. high, rarely a small tree to 7 m. high, few-branched, the younger vascular parts crimson; leaves petiolate, 8–30 cm. long, the blades oblong-lanceolate or narrowly ovate, narrowly acute, 3–5-nerved, the nerves red beneath; flowers 5-parted, sessile; hypanthium glabrous, 2–2·5 mm. long; petals pink, 2–3 mm. long; anthers 1·5 mm. long; style 2 mm. long; berries blue-black, 3–4 mm. in diameter.

Widespread in moist thickets, wooded hillsides and forest glades; 1000–2750 feet; fl. Apr–Nov, fr. all the year. *A & C 1070! A & D 13076! J. P. 2102! P 23015!* Mexico to S. Amer., West Indies.

6. M. theaezans (Bonpl.) Cogn. in Mart., Fl. Bras. **14** (4): 419 (1888).—*Melastoma theaezans* Bonpl. (1807).

Shrub or tree to 5 m. high, glabrous throughout; leaves petiolate, 5–13 cm. long, the blades narrowly elliptical, acuminate, 3-nerved, nerves often red beneath, margins entire or denticulate; flowers often short-pedicellate, fragrant; petals cream, 1–1·5 mm. long; anthers with 4 pores, about 0·8 mm. long; style 2–2·5 (–3) mm. long, stigma subpeltate; berries blue-black, 2–2·5 mm. in diameter.

Locally common (St. Andr., St. Thom.), in montane sclerophyll woodland; 3500–5800 feet; fl. Dec–Feb, June–Aug, fr. Mar, Aug–Nov. *A 11333! H 9137! P 8733!* S. Amer.

7. M. rubens (Sw.) Naud. in Ann. Sci. Nat. sér. 3, Bot. **16**: 169 (1851).—*Melastoma rubens* Sw. (1788).

Shrub or slender tree to 5 m. high, glabrous throughout; leaves petiolate, 4–10 cm. long, blades elliptical or narrowly obovate, acuminate, 3-nerved or sub-3-plinerved, margin minutely denticulate; some flowers subsessile, possibly dioecious; petals white, 0·7–0·9 mm. long; anthers with 2 pores, about 0·5 mm. long; style 1·5–2 mm. long, stigma peltate; berries blue-black, about 2 mm. in diameter.

Frequent (St. Andr., Port., St. Thom.), in montane forest; 4000–6250 feet; fl. Jan–July, fr. July–Aug. *Asprey UCWI 890! H 6340! P 23791!* Endemic.

8. M. quadrangularis (Sw.) Naud. in Ann. Sci. Nat. sér. 3, Bot. **16**: 197 (1851).—*Melastoma quadrangulare* Sw. (1788). *Pleurochaenia quadrangularis* (Sw.) Griseb (1860).

Shrub or small tree to 7 m. high, with quadrangular minutely stellate-furfuraceous branches; leaves rigid, petiolate, 6–16 cm. long, blades lanceolate, acuminate, 3-nerved, the veinlets microscopically brown-scaly beneath when very young, becoming glabrous or apparently so at maturity; panicle with ascending branches; flowers sessile, 5-parted; petals white, 2–3 mm. long; anthers white, about 1·5 mm. long; style 4 mm. long with capitate stigma; berries bluish or purple, 4–5 mm. in diameter; occurs in two varieties, as follows:

Long hairs absent   a. var. *quadrangularis*
Long gland-tipped hairs present on young branches, petioles and inflorescences; tree
   about 3 m. high   b. var. *glandulosa*

8a. Var. quadrangularis.

Frequent (St. Andr., Port., St. Thom.), in moist thickets and montane woodlands; 2000–7000 feet; fl. and fr. all the year. *A 10673! H 6434! P 23335!* Endemic.

8b. M. quadrangularis var. **glandulosa** Proctor in Bull. Inst. Jam., Sci. ser. **16**: 45 (1967).

Very rare and local (Port.), in mossy woodland on limestone; 1500–2000 feet; fl. Jan, fr. Jan–Mar. *P 5647! 9812!* Endemic.

9. M. rigida (Sw.) Triana in Trans. Linn. Soc. London **28**: 130 (1871).—*Melastoma rigidum* Sw. (1788). *Pleurochaenia rigida* (Sw.) Griseb. (1860).

Shrub or small tree up to 4 m. high, the young branches, petioles, leaf-nerves beneath and inflorescence densely and minutely brown-stellate-furfuraceous; leaves rigid, petiolate, 8–20 cm. long, the blades broadly ovate-elliptical, rounded or emarginate at base, acute at apex, 3-nerved; panicle-branches with gland-tipped hairs; hypanthium about 5 mm. long; petals white, 4–5 mm. long; anthers about 2·5 mm. long, opening by a longitudinal slit; berries brown, 5 mm. in diameter.

Locally common (St. Andr., Port., St. Thom.), in mossy montane woodlands; 4000–7400 feet; fl. and fr. sporadically all the year. *A 10647! H 6337! P 11088!* Hispaniola.

10. M. dodecandra (Desr.) Cogn. in Mart., Fl. Bras. **14** (4): 243 (1887).—*Melastoma tamonea* Sw. (1788), partly. *Melastoma dodecandrum* Desr. (1797).

Shrub or tree to 15 m. high, the young branches, petioles, etc. densely and minutely brown-stellate-furfuraceous; leaves petiolate, 7–30 cm. long, the blades narrowly ovate to elliptical, acuminate; primary panicle-branches 2 or 3 at each node; buds concealed by bracteoles 6–7 mm. long, these soon falling; hypanthium 4–5·5 mm. long at anthesis, densely whitish-stellate-tomentellous; petals translucent white with central rose-pink stripe, 7–9 mm. long; anthers yellow, turning scarlet, 6–10 mm. long; stigma peltate; berries black.

Widely distributed, on wooded hillsides and in moist thickets; 1000–4500 feet; fl. and fr. all the year. *A 11218! H 11095! P 23383!* C. and S. Amer., West Indies.

11. M. matthaei Naud. in Ann. Sci. Nat. sér. 3, Bot. **16**: 176 (1851).—*M. wilsonii* Cogn. (1912).

Shrub to 3 m. high, the young branches, petioles, leaves beneath and inflorescences densely hirsute with brown curved-ascending simple hairs; leaves petiolate, 12–25 cm. long, narrowly elliptical to elliptical, long-acuminate, 3-nerved with another submarginal pair; flowers 5-parted; hypanthium 2·5 mm. long, with minute white stellate hairs, also sparsely brown-hirsute; petals white, 3·5 mm. long; anthers about 3 mm. long; stigma capitate; berries light yellow, 5 mm. in diameter.

Rare (Port.), in wet forest thickets in only one known locality; 1750–2250 feet; fl. Feb, Apr, fr. Dec. *P 10084! 11555! 15831! Stearn 251!* Mexico to S. Amer., Cuba, Trinidad.

12. M. impetiolaris (Sw.) DC., Prodr. **3**: 183 (1828).—*Melastoma impetiolare* Sw. (1788).

Shrub or small tree to 5 m. high, the young branches, leaf-nerves beneath and inflorescences densely and minutely brown-stellate-furfuraceous; leaves oblong-obovate or broadly obovate, to 20 cm. broad; flowers 5-parted; hypanthium 2–3 mm. long; petals white, 2–3 mm. long; anthers 2–2·5 mm. long; stigma truncate; berries red, turning blue-black, 4–5 mm. in diameter.

Frequent in hillside thickets, on banks of streams and in woodland glades; 400–2500 feet; fl. Jan–Sept, fr. Apr–July. *A 10848! H 8558! P 9915!* Mexico to S. Amer., West Indies.

13. M. ampla Triana in Trans. Linn. Soc. London **28**: 101 (1871).—*M. involucrata* Donnell Sm. (1904).

Tree to 12 m. high, glabrous throughout except undersurfaces of leaves, the young branches stout, terete or obtusely flattened, hollow; leaves of a pair unequal, elliptical, subacuminate, to 22 cm. broad; flowers clustered within large deciduous bracts; petals white; fruiting panicles pentagonal in outline, long-stalked; berries ovoid, yellow-green, faintly ribbed; seeds brown, galeate, tuberculate on convex side, about 1·5 mm. long.

Rare (Clar., Port.), in wet montane forests and thickets; 1750–2250 feet; fr. Feb, Nov. *P 11552! 16492!* Guatemala, Bolivia, Trinidad.

14. M. triplinervis Ruiz & Pav., Syst. Veg. Fl. Peruv. & Chil. **1**: 105 (1798).

Straggling few-branched shrub to 2·5 m. high, the young branches, petioles, leaf-nerves beneath and inflorescences sparsely or densely and minutely stellate-furfuraceous; leaves short-petiolate, sometimes subsessile, 15–27 cm. long, the blades very narrowly elliptical, 3-nerved, the epidermis thinly and minutely stellate-pubescent, very pale beneath; flowers glomerate, sessile; hypanthium 2·5–3 mm. long; petals dull whitish, 2 mm. long; anthers 2·5–3 mm. long; style 3–4 mm. long, minutely capitate; berries bright red, 3–4 mm. in diameter.

Uncommon in moist glades and ravines at scattered localities throughout the island; 400–2000 (–4000) feet; fl. June, fr. Mar, Aug, Oct–Dec. *A 11588! H 12877! P 15832!* Mexico to S. Amer.

15. M. multispicata Naud. in Ann. Sci. Nat. sér. 3, Bot. **16**: 131 (1851).

Tree to 10 m. high, the young branches, petioles, leaf-nerves and veins beneath and inflorescences densely whitish stellate-woolly; leaves 7–17 cm. long, the blades elliptical or broadly oblanceolate, short-acuminate, 3-plinerved; flower-clusters dense, sessile or on peduncles 1–4 mm. long; petals white, 2–3 mm. long; berries black, about 5 mm. in diameter.

Uncommon in moist thickets or forests on heavy clay soil; 700–2300 feet; fl. Jan–Apr, Sept, fr. Mar, Sept. *A 9076! H & B 10540! P 9976!* Trinidad.

16. **M. furfuracea** (Vahl) Griseb., Fl. Br. W.I.: 257 (1860).—*Melastoma furfur-aceum* Vahl (1807).

Shrub or small tree to 5 m. high, the young branches, petioles, leaf-nerves beneath and inflorescences densely and minutely brownish-stellate furfuraceous; leaves mostly 15–40 cm. long, the blades elliptical or broadly elliptical, short-acuminate, 3-plinerved; inflorescence-branches 2–11 mm. long with sessile and pedicellate flowers clustered at apex; petals white, 5 mm. long; filaments white; style 8 mm. long, subcapitate; berries red, turning black, 8–10 mm. in diameter.

Not collected in Jamaica since the early 19th century, the precise locality unknown. Venezuela, Lesser Antilles.

17. **M. tetrandra** (Sw.) D. Don ex G. Don in Loudon, Hort. Brit.: 174 (1830).—*Melastoma tetrandrum* Sw. (1788).

Tree to 15 m. high, the young branches, petioles and inflorescence covered with a microscopic rusty-brown indumentum; leaves 10–20 cm. long, the blades narrowly ovate, acuminate, 3-plinerved; hypanthium 1·5 mm. long; petals white, 1·8 mm. long; anthers 1·5 mm. long; berries black, 3–4 mm. in diameter.

Uncommon (Clar., St. Andr., Port.), in moist thickets and montane forests; 1500–4300 feet; fl. Jan, May, June, fr. Apr–July. *A & S 3258! H 6408! P 22270!* West Indies.

18. **M. trinervia** (Sw.) D. Don ex G. Don in Loudon, Hort. Brit.: 174 (1830).—*Melastoma trinervium* Sw. (1788).

Shrub to 5 m. high, the young branches and petioles compressed, 2-edged, very minutely stellate-scaly, as also the inflorescence; leaves 10–25 cm. long, the blades oblong-elliptical or narrowly obovate, short-acuminate, the undersurface with numerous minute stellate scales; flowers sessile; hypanthium 2 mm. long; petals 2·5 mm. long; anthers 2 mm. long; berries blackish, 5 mm. in diameter, 10-ribbed.

Rare (Trel., St. Andr.), in hillside thickets; 2200–2500 feet; fl. Nov–Dec, fr. Mar. *H 10654! P 18430!* Mexico to S. Amer.

19. **M. laevigata** (L.) DC., Prodr. **3**: 188 (1828).—*Melastoma laevigatum* L. (1759). Johnny Berry, White Wattle.

Shrub to 3 m. high, the young branches and other vascular parts minutely stellate-puberulous; leaves narrowly ovate or elliptic-lanceolate, rarely broadly ovate, long-acuminate, the margins often sparsely ciliate; panicles many-flowered; hypanthium about 2 mm. long; petals white, 3–4 mm. long; anthers 2·7–3 mm. long; berries bluish-black, 3 mm. in diameter.

Very common weedy shrub of pastures and wayside thickets; near SL–4200 feet; fl. and fr. all the year. *A 11792! H 8781! P 22792!* Mexico to S. Amer., West Indies.

20. **M. splendens** (Sw.) Griseb., Fl. Br. W.I.: 256 (1860).—*Melastoma splendens* Sw. (1788).

Tall shrub or small tree to 8 m. high, the young branches and other vascular parts minutely brown-stellate-powdery; leaves elliptical to broadly elliptical, short-acuminate at both ends; panicles many-flowered; hypanthium 2–3 mm. long, minutely stellate-scaly; petals white, 2–3 mm. long; anthers 2–3 mm. long; berries black, 3 mm. in diameter.

Rare in moist forest glades at scattered localities; 500–2250 feet; fl. and fr. Feb–Apr. *A 9070! H 7267! P 10082!* Greater Antilles.

21. **M. chrysophylla** (L. C. Rich.) Urb., Symb. Ant. **4**: 459 (1910).—*Melastoma chrysophyllum* L. C. Rich. (1792).

Shrub or small tree to 10 m. high, the young branches 2-edged, densely brown-lepidote; leaves mostly 10–20 cm. long, acuminate, 3-nerved, glabrous and deep green on upper surface; flowers numerous, 5-parted; hypanthium 1·5 mm. long; petals white or cream, 2 mm. long; anthers 1·5 mm. long, opening by elongate slits; style 3–4 mm. long, truncate; berries 2·5–3 mm. in diameter.

Collected but once in Jamaica, near Moneague (St. Ann) in 1849–50. Mexico to S. Amer., Hispaniola, Trinidad.

22. M. albicans (Sw.) Triana in Trans. Linn. Soc. London **28**: 116 (1871).—
Melastoma albicans Sw. (1788). Georgia White Man, Whiteback.

Shrub up to 3 m. high, the young branches densely clothed with matted white appressed tomentum; leaves ovate-oblong or ovate-elliptical, 5–18 cm. long, shortly cordate at base, obtuse or acute, upper surface thinly whitish-lanate when young, soon becoming glabrous, the epidermis minutely glandular-punctate; flowers sessile; petals white, 2·5–3 mm. long; anthers 2·5 mm. long; berries blue-green, 4–5 mm. in diameter, edible.

Common in heavy clay soils in pastures and thickets and on scrubby hillsides; 500–4000 feet; fl. and fr. sporadically all the year, but especially Mar–June. *A 11789! H 11205! P 20805!* Mexico to S. Amer., West Indies.

23. M. serrulata (DC.) Naud. in Ann. Sci. Nat. sér. 3, Bot. **16**: 118 (1851).—
Diplochita serrulata DC. (1828). *Chitonia macrophylla* D. Don (1823). *Miconia macrophylla* (D. Don) Triana (1871), not Steud. (1844).

Commonly a shrub to 3 m. high, rarely a small tree; young branches and other vascular parts densely brown-stellate-puberulous; leaves 15–35 cm. long, of firm heavy texture, ovate, ovate-oblong, or broadly elliptical, acute or very short-acuminate, the margins serrulate; panicles large, heavy; flowers sessile; petals whitish, 6–8 mm. long; anthers rose-pink or purple, 6–8 mm. long; filaments and style hirtellous below; berries blue-black, 6 mm. in diameter.

Frequent in hillside thickets and along streams or borders of pastures; 350–2750 feet; fl. and fr. all the year. *A 9077! H 7639! P 10392!* Mexico to S. Amer., West Indies.

24. M. elata (Sw.) DC., Prodr. **3**: 182 (1828).—*Melastoma elatum* Sw. (1788)
Miconia eurychaenioides Griseb. (1860), partly.

Tree to 12 m. high, the 4-angled young branches and other vascular parts densely and minutely brown-stellate-scaly; leaves 15–40 cm. long, elliptical, abruptly short-acuminate, the margins denticulate; panicles pentagonal, many-flowered; flowers sessile; petals white, 1·5 mm. long; anthers 2 mm. long; stigma subpeltate; berries 10-ribbed, 4–5 mm. in diameter.

Rare and local (Port.), in lowland forest; 700–800 feet; fl. and young fr. Mar. *P 20651!* Mexico to C. Amer., Cuba.

14. CHARIANTHUS D. Don (1823)

1. C. fadyenii (Hook.) Griseb., Fl. Br. W.I.: 264 (1860).—*Tetrazygia fadyeni* Hook. (1849).

Shrub or small tree to 12 m. high, glabrous; leaves petiolate, 3–8 cm. long, the blades elliptical, blunt or short-acuminate, the apex often slightly notched, 3-plinerved, often bearded in nerve-axils and the surface minutely glandular-dotted beneath; petals 8–10 mm. long; anthers yellow, 4·5–5 mm. long; berry about 5 mm. thick, dark purplish when ripe.

Frequent on wooded limestone hills and rocky escarpments; 1200–3000 feet; fl. and fr. all the year. *A 7547! H 9234! P 16106!* Endemic.

A second species, *C. tinifolius* D. Don (1823), has been attributed to Jamaica on the basis of a *McNab* specimen which cannot now be located. Without further evidence, it is probably best to delete this doubtful record. *C. tinifolius* is said to be differentiated from *C. fadyenii* especially by the anthers having longitudinal chinks instead of a single terminal pore, and by a 4-locular instead of 2-locular ovary.

15. BLAKEA P. Browne (1756)

Corolla pink; leaves elliptical to rotundate, 5–14 cm. long   1. *trinervia*
Corolla white; leaves ovate to broadly elliptical, up to 23 cm. long   2. *urbaniana*

1. B. trinervia L., Syst. Nat. ed. 10, **2**: 1044 (1759) (incl. var. *normanii* F. & R. (1926)).—Cup and Saucer, Jamaican Rose.

A scandent or epiphytic scrambling shrub; young branches, petioles and often nerves of leaves beneath with minute branlike or elongate hairlike ciliate scales, these reddish- or yellowish-brown, the more hairlike, shaggy indumentum occurring chiefly on plants growing at higher elevations; leaves abruptly acuminate at apex, 3-nerved or 3-plinerved, the cross-veins extremely fine, parallel, close together; peduncles 3–6 cm. long, erect or somewhat recurved; bracts to 2 cm. long; filaments crimson, anthers yellow with 2 pores, the connective very thick and prolonged into a spur; style 1·5 cm. long.

Common and widely distributed on wooded rocky hillsides and in montane forests; 900–4500 feet; fl. and fr. all the year. *A 7001! H 8535! P 21938!* Endemic.

2. B. urbaniana Cogn. in Urb., Symb. Ant. **6**: 27 (1909).

Like the last but the indumentum more sparse, the peduncles shorter, up to 3 cm. long, more strongly recurved, the flowers always nodding; flowers slightly larger; filaments greenish; style 2 cm. long.

Rare and local (Han., West.), scrambling or epiphytic in woodland on limestone; 1000–1786 feet; fl. and fr. all the year. *A 8634! H 9244! P 20675!* Endemic.

16. HENRIETTEA DC. (1828)

1 Leaf-base auriculate, sessile or subsessile, blade to 40 cm. long   1. *sessilifolia*
1 Leaf-base acute or acuminate, distinctly petiolate, blade 6–20 cm. long:
  2 Leaves glabrous   2. *macfadyenii*
  2 Leaves hispid:
    3 Flowers 4-parted; calyx 2·5–3 mm. long, glabrous; petals white; anthers yellow
        3. *fascicularis*
    3 Flowers 5-parted; calyx 5–7 mm. long, finely bristly-scaly; petals pink; anthers blue
      or violet   4. *ramiflora*

1. H. sessilifolia (L.) Alain in Bull. Torr. Bot. Club **92**: 300 (1965) (as *sessiliflora*). *Melastoma sessilifolium* L. (1759).—*Henriettella sessilifolia* (L.) Triana (1871).

Shrub 3 m. or small tree to 10 m. high, the youngest parts clothed with numerous minute appressed deciduous scales, these round or deltate and densely puberulous; leaves dark green, obtuse and mucronate or acute, 5-plinerved, the tissue containing numerous minute linear cystoliths; flowers clustered at leafless nodes, on pedicels mostly 5–10 mm. long, 5-parted; petals white, obtuse to apiculate, 8–10 mm. long; anthers yellow, 4 mm. long; stigma conical, 5-ridged.

Rare and sporadic (West., St. Mary, St. Thom.), in moist forest glades; 1000–2000 feet; fl. May. *H & B 10678! P 22294!* Venezuela, Trinidad.

2. H. macfadyenii (Triana) Alain in Bull. Torr. Bot. Club **92**: 300 (1965).—*Henriettella macfadyenii* Triana (1871).

Tree 8–20 m. high; branchlets acutely 4-angled; leaves 8–14 cm. long, narrowly elliptical, apex shortly acuminate, base acute, 3-plinerved with an obscure marginal pair, glabrous, papery; petioles 1–2 cm. long; pedicels 5–8 mm. long; flower-parts in fours or fives; calyx subhemispherical, 2 mm. long, the narrow limb spreading.

Based on an unlocalized *Macfadyen* specimen from Jamaica. Puerto Rico.

3. H. fascicularis (Sw.) Gómez Maza in Anal. Hist. Nat. Madrid **23**: 68 (1894); L. O. Williams in Fieldiana, Bot. **29**: 565 (1963).—*Melastoma fasciculare* Sw. (1788). *Ossaea fascicularis* (Sw.) Griseb. (1860). *Henriettella fascicularis* (Sw.) C. Wright (1869).

Shrub or small tree to 8 m. high, the young branches densely hirsute with short broad-based hairs; leaves 7–20 cm. long, the blades elliptical, acute at both ends, 3-plinerved with another obscure submarginal pair, scabrous on upper surface, beneath densely beset with minute ciliate hairlike scales; flowers fragrant, densely clustered at nodes, on pedicels 2–6 mm. long, 4-parted; petals white, acuminate, 4 mm. long; anthers yellow, 2 mm. long; stigma minutely capitate; berries black.

Widely scattered but not common, in moist thickets or woodland clearings; 1000–2500 feet; fl. Mar–Apr, Dec, fr. Mar–Apr, Sept. *A 11030! H 11230! P 11412!* C. Amer., Greater Antilles.

4. H. ramiflora (Sw.) DC., Prodr. **3**:178 (1828).—*Melastoma ramiflorum* Sw. (1788).

Shrub or small tree to 9 m. high, the young branches densely clothed with adpressed flattened bristly hairs; leaves 7–18 cm. long, the blades elliptical to narrowly obovate, short-acuminate at apex, cuneate at base, 3-nerved or 3-plinerved, scabrous on upper surface, beneath densely covered with yellowish-brown tomentum consisting of bristly hairs each arising from a dense stellate cluster of smaller hairs; flowers clustered at leafless nodes, subsessile or on bristly pedicels to 4 mm. long; petals 8–10 mm. long; anthers 7–8 mm. long; ripe berries dull red, about 1 cm. in diameter.

Occasional (St. James, Clar., Port.), in moist thickets, chiefly in heavy clay soil; 150–2300 feet; fl. Mar, June, Sept. fr. May. *A 9069! P 16650! 23552!* Suriname, Trinidad.

17. OSSAEA DC. (1828)

1 Leaves smooth, apparently glabrous, minutely scurfy beneath under a lens; berries 8-ribbed, pellucid-white when ripe   1. *micrantha*
1 Leaves obviously hairy or rough; berries not white:
  2 Leaves 1–4 cm. long; flowers mostly solitary, long-stalked; ripe berry black
                                                                          2. *microphylla*
  2 Leaves 4–15 (–20) cm. long:
    3 Flowers stalked, or in stalked inflorescences:
      4 Stems and leaves tuberculate-scabrous; flowers rather numerous in axillary or pseudo-terminal cymose panicles, 5-parted   3. *asperifolia*
      4 Stems and leaves with hairs; flowers few or solitary, 4-parted:
        5 Pedicels about 1 mm. long; calyx 1·5 mm. long; leaves 3-nerved   4. *hirtella*
        5 Pedicels up to 20 mm. long; calyx 3 mm. long; leaves usually 3-plinerved
                                                                           5. *hirsuta*
    3 Flowers in small sessile clusters in leaf-axils or at leafless nodes:
      6 Leaves 3-plinerved with one submarginal pair; undersurface of leaves and hypanthium with simple hairs   6. *glomerata*
      6 Leaves 3-plinerved with two submarginal pairs; undersurface leaves with branched or stalked-stellate hairs, hypanthium with small ciliate scales   7. *scabrosa*

1. O. micrantha (Sw.) Cogn. in A. & C. DC., Monogr. Phan. **7**: 1066 (1891).—
Melastoma micranthum Sw. (1788). *Octopleura micrantha* (Sw.) Griseb. (1860).

Shrub or tree to 5 m. high; leaves petiolate, 8–17 (–20) cm. long, the blade narrowly elliptical to elliptical, acuminate, 3-plinerved with another submarginal pair, of thin membranous texture; panicle to 5 cm. long with deflexed or spreading branches and numerous flowers; hypanthium 2 mm. long with truncate calyx; petals 4, white, 3–4 mm. long; anthers light yellow, 2 mm. long; the juicy white berries said to be edible with flavour like black pepper.

Uncommon in moist thickets and forest glades; 1000–2100 feet; fl. and fr. all the year. *A 11573! H & B 10556! P 21475!* Mexico to S. Amer., Hispaniola.

2. O. microphylla (Sw.) Triana in Trans. Linn. Soc. London **28**: 146 (1871).—
Melastoma microphyllum Sw. (1788). *Clidemia microphylla* (Sw.) Griseb. (1860).

Small shrub rarely more than 1 m. high, the stems, leaves and hypanthium densely light-brown or reddish-hirtellous; leaves short-petiolate, elliptical to ovate, 3-plinerved; flowers long-stalked if solitary, or in pedunculate cymes of 3, the central one sessile; hypanthium 2 mm. long, bearing minute stellate hairs in addition to long simple ones; petals 4, white, turning pink or red, 4 mm. long; anthers yellow, 1 mm. long.

Frequent on rocky banks and in moist pastures; 150–2500 feet; fl. and fr. all the year. *A 9195! H 9090! P 21844!* Cuba.

3. O. asperifolia (Naud.) Triana in Trans. Linn. Soc. London **28**: 147 (1871).—
Clidemia asperifolia Naud. (1852).

Shrub to 3 m. or small tree to 8 m. high, the stems, petioles and leaf-nerves beneath densely clothed with short thick appressed bristles; leaves petiolate, 4–12 (–15) cm. long, the blades elliptical, acuminate, 3-plinerved with an obscure submarginal pair, densely tuberculate-scabrous on both surfaces; inflorescence a small cymose panicle; hypanthium densely tuberculate, 3–3·5 mm. long; petals whitish, 3–4 mm. long; anthers yellow, about 1 mm. long.

Locally frequent (St. Andr., Port., St. Thom.), in submontane forests and moist secondary thickets; 1500–4200 feet; fl. and fr. Feb–May, fr. May–Aug. *A 11945! H & B 10773! P 10192!* Cuba.

4. O. hirtella (Sw.) Triana in Trans. Linn. Soc. London **28**: 146 (1871).—
Melastoma hirtellum Sw. (1788).

Shrub to 2 m. high, hirsute on branches, petioles, and undersurfaces of leaves; leaves petiolate, 4–8 cm. long, lanceolate, acuminate, 3-nerved, the undersurface much lighter than the upper; flowers solitary or 2–3 in a small cyme; petals 1 mm. long; berry globose, deep purple, 2·5 mm. in diameter.

Rare and local (St. Thom.), in thickets on shaly banks; 500–2000 feet; fl. May–Oct. *Swartz!* Not recently collected. Endemic.

5. O. hirsuta (Sw.) Triana in Trans. Linn. Soc. London **28**: 146 (1871).—
Melastoma hirsutum Sw. (1788).

Slender straggling or diffuse shrub to 2·5 m. high, hirsute throughout, the hairs, especially of the hypanthium, usually reddish-purple; leaves petiolate, mostly 4–12 cm. long, the blades lanceolate to narrowly ovate, acuminate, 3-plinerved, rarely 3-nerved, with an obscure submarginal pair, the margins subentire or obscurely dentate; cymes few-flowered, the lateral flowers long-pedicellate; calyx-processes 5–6·5 mm. long; petals white or pink, 6–7 mm. long; anthers orange, 2 mm. long; berries blackish-purple.

Locally frequent (Port., St. Thom.), in thickets on shaly banks; 400–2500 feet; fl. and fr. all the year. *A 13231! H & B 10548! P 22953!* Endemic.

6. O. glomerata (Naud.) Triana in Trans. Linn. Soc. London **28**: 146 (1871).—
Sagraea glomerata Naud. (1852).

Shrub to 3 m. high; the stems, petioles and undersurfaces of leaves densely clothed with simple reddish-brown hairs having enlarged bases, interspersed with minute sessile stellate hairs; leaves petiolate, 4–12 cm. long, acute or blunt, the upper surface scabrid with tuberculate bristles; flower-clusters usually sessile or sometimes on a short peduncle; petals pinkish-white, 2–3 mm. long; anthers yellow, about 1 mm. long.

Occasional in thickets on shaly or limestone soils in the eastern half of the island; 750–4000 feet; fl. and fr. all the year. *A 11684! H 10028! P 22952!* Cuba.

7. O. scabrosa (L.) DC., Prodr. **3**: 169 (1828).—*Melastoma scabrosum* L. (1759).

Shrub 1–2 (–3·5) m. high, the stems densely clothed with narrow hairlike minutely pubescent scales; leaves long-petiolate, 5–17 cm. long, the blades ovate, bluntly subacuminate, scabrous with small tuberculate bristles on upper surface; flowers very densely glomerate, never pedunculate; petals pink, 2 mm. long; anthers white, 1 mm. long.

Rather uncommon (St. Cath., St. Thom.), in hillside thickets on non-calcareous soils derived from porphyry or serpentine; 500–2500 feet; fl. July–Aug, fr. not seen. *H & B 10570! P 7109!* Cuba.

18. CLIDEMIA D. Don (1823)

1 Flowers 5–8-parted:
  2 Hairs of upper leaf-surface swollen at base, the lower surface densely stellate-pubescent:
    3 Inflorescence paniculate; leaves subcordate at base   1. *strigillosa*
    3 Inflorescence spicate; leaves narrowed or blunt at base
                                          2a. *capitellata* var. *dependens*
  2 Hairs of upper leaf-surface not swollen at base:
    4 Lower leaf-surface with stellate hairs; flowers 7–8-parted; stem-hairs to 8 mm. long
                                                        3. *octona*
    4 Lower leaf-surface with simple hairs and sometimes a few stellate hairs on the veins; flowers 5–6-parted; stem-hairs less than 3 mm. long:
      5 Leaves ovate, rounded to cordate at base; inflorescence laxly-branched; calyx-hairs brownish or reddish   4. *hirta*
      5 Leaves ovate-elliptical to elliptical, inequilaterally cuneate to rounded or subcordate at base; inflorescence subsessile, densely short-branched; calyx-hairs bright red
                                                      5. *erythropogon*
1 Flowers 4-parted:
  6 Leaves with more than 3 nerves, often 10 cm. or more broad:
    7 Stems etc. minutely rusty-powdery, lacking hairs; leaf-blades inequilaterally decurrent at base   6. *septuplinervia*

7 Stems and petioles densely pilose; leaf-blades rounded to subcordate at base:
  8 Hypanthium densely hirsute, 3–4 mm. long, the calyx-processes 2–3 mm. long, hairy; leaves densely hirsute on upper surface   7. *plumosa*
  8 Hypanthium glabrous or with a few scattered hairs, 1–2 mm. long, the calyx-processes less than 1 mm. long, glabrous; leaves glabrate on upper surface, rarely with a few scattered hairs   8. *swartzii*
6 Leaves with 3 nerves, lanceolate to narrowly elliptical, less than 4 cm. broad:
  9 Leaves 3-plinerved; ripe berries nearly always glabrous   9. *grisebachii*
  9 Leaves 3-nerved:
   10 Inflorescences mostly shorter than the petioles, the pedicels not over 2 mm. long; ripe berries globose, bright glistening blue, usually bearing numerous persistent glandular hairs   10. *crossosepala*
   10 Inflorescences longer than the petioles, with long filiform branches; ripe berries oblong, purple, the hypanthium-hairs usually deciduous   11. *insularis*

1. **C. strigillosa** (Sw.) DC., Prodr. **3**: 159 (1828).—*Melastoma strigillosum* Sw. (1788).
Shrub to 1·5 m. high, the stems petioles and peduncles densely stellate-pubescent, with longer simple gland-tipped hairs; leaves thick, petiolate, 5–15 cm. long, the blades rounded to subcordate at base, acuminate, 5–7-nerved, serrulate; flowers sessile on panicle-branches, 5-parted; petals white or light green, 4–5 mm. long; anthers 2–2·5 mm. long; ripe berries bluish-black.
Locally common in moist thickets and pastures in heavy clay soil; 500–3500 feet; fl. and fr. all the year. *A 5903! H 7458! P 8066!* C. and S. Amer., Cuba, Hispaniola.

2. **C. capitellata** (Bonpl.) D. Don in Mem. Wern. Soc. **4**: 310 (1823).—*Melastoma capitellatum* Bonpl. (1806).

2a. Var. **dependens** (D. Don) Macbr. in Field Mus. Bot. **13**: 484 (1941).—*C. dependens* D. Don (1823). *C. spicata* (Aubl.) DC. (1828), not D. Don (1823). *Melastoma spicatum* Aubl. (1775).
Shrub to 2·5 m. high, the stems, petioles and peduncles stellate-pubescent and with numerous long soft simple glandular hairs; leaves of a pair unequal, petiolate, up to 20 cm. long, oblong-ovate, acuminate, 5-nerved, denticulate; flowers 5–6-parted; petals white, 6 mm. long; anthers 2 mm. long.
Attributed to Jamaica on the basis of a single specimen collected by *Wullschlaegel*, who was in this island 1847–49. However, as he also collected in other places (e.g. Nicaragua, Suriname) where this plant definitely occurs, the Jamaican record may have been caused by a mix-up of labels. Mexico to S. Amer., Cuba, Trinidad.

3. **C. octona** (Bonpl.) L. O. Williams in Fieldiana, Bot. **29**: 558 (1963).—*Melastoma octonum* Bonpl. (1806). *Heterotrichum octonum* (Bonpl.) DC. (1828).
Shrub to 2·5 m. high, the stems with 3 kinds of pubescence: short subsessile stellate hairs, longer simple gland-tipped hairs and very long eglandular ones; leaves long-petiolate, mostly 8–20 cm. long, the blades ovate or broadly ovate, cordate or subcordate at base, short-acuminate, 7-nerved; inflorescence 5 cm. or more long; petals white, 8–10 mm. long; anthers about 5 mm. long; ripe berries dark purple.
Uncommon in moist hillside thickets on rocky limestone; 1000–2250 feet; fl. and fr. all the year. *A 12417! H 8566! P 20870!* Mexico to S. Amer., Cuba.

4. **C. hirta** (L.) D. Don in Mem. Wern. Soc. **4**: 309 (1823).—*Melastoma hirtum* L. (1753). *C. hirta* var. *elegans* (Aubl.) Griseb. (1860). Soap-bush.
Shrub rarely more than 1·5 m. high, all parts densely hairy; leaves petiolate, those of a pair unequal, 4–15 cm. long, acute or short-acuminate, usually 5-nerved, the margins often crenulate and denticulate; inflorescence 3–5 cm. long; petals white, 8–11 mm. long; anthers 4·5 mm. long; ripe berries red-purple to blackish.
Common in moist pastures and thickets; 100–4000 feet; fl. and fr. all the year. *A 8112! H 12038! P 20877!* Mexico to S. Amer., West Indies, naturalized as a weed in tropical Pacific islands and Old World tropics.

5. **C. erythropogon** DC., Prodr. **3**: 157 (1828).—*C. hirta* of F. & R. (1926), partly.
Shrub to 2·5 m. high, all parts densely hairy; leaves petiolate, those of a pair

unequal, mostly 7–21 cm. long, acuminate or long-acuminate, 3–5-plinerved, the margins crenate-denticulate; inflorescence 1·5–2 cm. long, 3–7-flowered; petals white, 5–8 mm. long; anthers 2·5–3·5 mm. long; ripe berries dull violet.

Common in moist pastures, thickets and shaded ravines; 500–2300 (–3600) feet; fl. and fr. all the year. *A 8113! H 6654! P 22219!* Cuba, Hispaniola.

6. C. septuplinervia Cogn. in Mart., Fl. Bras. **14** (4): 506 (1888).

Shrub to 1·5 m. high; stems terete, very hard; leaves of a pair unequal, petiolate, 10–27 cm. long, the blades narrowly to broadly ovate, acuminate, 3–7-plinerved, undersurface pale and minutely glandular; flowers in small dense panicles at leafless nodes; petals pink or rose, 1·5–2 mm. long, acute; anthers about 1 mm. long; ripe berries indigo blue, about 1 cm. in diameter.

Rare (Port.), in moist forest thickets on limestone; 700–1750 feet; fl. and fr. probably all the year. *A 7873! ACH 962! P 19740!* S. Amer.

7. C. plumosa (Desr.) DC., Prodr. **3**: 159 (1828).—*Melastoma plumosum* Desr. (1797). *C. berteroi* (DC.) Griseb. (1860).

Shrub to 3·5 m. high, densely villous-hirsute throughout; leaves of a pair unequal, long-petiolate, 10–32 cm. long, the blades ovate, acuminate, 3-plinerved, with 2 pairs of outer basal nerves, margin denticulate; petals translucent whitish or greenish, reflexed, 2·5–4 mm. long; anthers yellow, 2 mm. long; berries purple, said to be edible.

Uncommon (Port., St. Thom.), in moist forests on limestone; 800–3000 feet; fl. and fr. all the year. *A 11563! H 9127! P 8097!* Venezuela, Hispaniola.

8. C. swartzii Griseb., Fl. Br. W. I.: 248 (1860).—*C. pilosa* (Sw.) Cogn. (1891). *Melastoma pilosum* Sw. (1788).

Straggling or arching shrub to 3 m. high, the branches, petioles, undersurface of leaves and peduncles hirsute; leaves petiolate, 8–25 cm. long, the blades oblong, elliptical, or ovate, rounded at base, acuminate, 3-plinerved with a submarginal basal pair; inflorescences 2–4 cm. long, often at leafless nodes; petals white with red spot at base, 1·5–2 mm. long; anthers 1–1·5 mm. long; berry dark purple, 2·5–3 mm. in diameter.

Rare in moist thickets and shaded ravines on limestone in the central parishes; 900–2300 feet; fl. Mar, fr. Mar–Apr. *McNab! P 16235! 22221!* Endemic.

9. C. grisebachii Cogn. in A. & C. DC., Monogr. Phan. **7**: 1009 (1891).—*Sagraea grisebachii* Triana (1871), name only.

Shrub to 3 m. high, the branches quadrangular, brown-scurfy and short-pilose; leaves of a pair unequal, petiolate, 3–18 cm. long, blades narrowly elliptical, acuminate, pale and minutely glandular beneath; flowers minute, solitary or in small clusters, pedicels to 3 mm. long; petals white, reflexed, 1 mm. long; anthers white, 0·5–0·9 mm. long; berries globose, light to dark blue, thinly setose.

Rare (Port., St. Ann), in moist thickets on limestone; 700–2250 feet; fl. and fr. Jan–Sept. *A 7878! ACH 960! P 9816! 22693!* Endemic.

10. C. crossosepala Griseb., Fl. Br. W.I.: 248 (1860).—*Sagraea crossosepala* (Griseb.) Triana (1871).

Shrub to 5 m. high, the terete branches minutely glistening-scurfy and usually pilose; leaves of a pair unequal, petiolate, 3–14 cm. long, blades narrowly elliptical or ovate-elliptical, acuminate, pale beneath; flowers minute, solitary or in small clusters; petals whitish or pale rose, erect, bristle-pointed, about 1 mm. long; anthers white, 0·9–1 mm. long; occurs in two varieties, as follows:

Stems, petioles, etc. distinctly pilose   a. var. *crossosepala*
Stems, petioles, etc. glabrous except for the minutely glistening-scurfy indumentum
                                               b. var. *adamsii*

10a. Var. crossosepala.

In scattered localities (Han., St. Andr., St. Thom.), in thickets on steep, mostly non-calcareous rocky banks and shaded cliffs; 400–3750 feet; fl. and fr. Jan–Aug. *A 10136! P 16449! 20676!* Endemic.

10b. C. crossosepala var. **adamsii** Proctor in Bull. Inst. Jam., Sci. ser. **16**: 37 (1967).

Very local (Trel), in similar habitats on limestone; about 2250 feet; fl. May, fr. Aug. *A 6762! P 8001!* Endemic.

11. C. insularis Domin in Act. Bot. Bohem. **9**: 43 (1930).—*C. capillaris* (Sw.) Griseb. (1860), not D. Don (1823). *Melastoma capillare* Sw. (1788).

Shrub to 2·5 m. high, with slender quadrangular minutely brown-scurfy branches; leaves of a pair unequal, petiolate, 4–14 cm. long, the blades narrowly lanceolate to very narrowly elliptical, long-acuminate, 3-nerved; hypanthium often with gland-tipped hairs; petals white, about 1 mm. long; anthers white, 1–1·5 mm. long; berries purple or deep violet-blue.

Widely distributed in moist thickets chiefly of limestone areas; 500–3000 feet; fl. and fr. all the year. *A 10800! H 10587! P 18323!* Cuba, Hispaniola.

19. MECRANIUM Hook. f. (1867)

1 Hypanthium less than 2 mm. long; panicles chiefly in leaf-axils, shorter than or not much
    exceeding petioles; style 2 mm. long   1. *amygdalinum*
1 Hypanthium 2·5–3 mm. long; panicles exceeding or much longer than petioles; style
    5–7 mm. long:
    2 Panicles both axillary and at leafless nodes, few-flowered; base of hypanthium obtuse;
       leaves elliptical to obovate, up to 8 cm. long   2. *purpurascens*
    2 Panicles chiefly along naked stems below the leaves, many-flowered; base of hypanthium
       acute; leaves narrowly ovate to elliptical, up to 16 cm. long   3. *virgatum*

1. M. amygdalinum (Desr.) C. Wright in Sauv. in Anal. Acad. Cienc. Habana **5**: 435 (1869).—*Melastoma amygdalinum* Desr. (1797). *Cremanium amygdalinum* (Desr.) Griseb. (1860).

Shrub or small tree to 7 m. high; leaves 5–15 cm. long, the blades narrowly elliptical to elliptical, acuminate at both ends, 3-plinerved; flowers 4-parted; hypanthium obtuse at base; petals whitish or cream, 1–1·5 mm. long; berries black, 3·5 mm. in diameter.

Frequent on wooded hillsides especially in areas of high rainfall; 950–4600 feet; fl. Mar–July, fr. June–Sept. *A 12412! H 10315! P 10433!* Greater Antilles.

Records from 5250–7375 feet (*ACH 910! A & S 3386!*) in the Blue Mountains may represent a different species intermediate between this and the next.

2. M. purpurascens (DC.) Triana in Trans. Linn. Soc. London **28**: 139 (1871).—*Ossaea purpurascens* DC. (1828). *Melastoma purpurascens* Sw. (1788), not Aubl. (1775).

Shrub or small tree to 6 m. high; leaves 3–8 cm. long, the blades long-tapered at base, abruptly acuminate at apex, 3-nerved or 3-plinerved; flowers 4-parted; petals whitish with red base, 1·5–2 mm. long; anthers white; berries globose, reddish-black.

Common (St. Andr., Port., St. Thom.), in montane forests and thickets; 3500–7000 feet; fl. July–Oct, fr. Sept–Apr. *A 10659! J. P. 1255! P 22668!* Endemic.

3. M. virgatum (Sw.) Triana in Trans. Linn. Soc. London **28**: 140 (1871).—*Melastoma virgatum* Sw. (1788). *Cremanium virgatum* (Sw.) Griseb. (1860).

Tall shrub or slender tree to 6 m. high; leaves 3-plinerved; flowers 4-parted; petals white or greenish-white, about 2 mm. long; anthers white; berries ovoid, white.

Frequent (St. Andr., Port., St. Thom.), in moist mountain thickets, (an isolated record from Clarendon); 200–5000 feet; fl. Jan–Apr, fr. Mar–Aug. *A 10825! H 6274! P 16234!* Endemic.

20. MOURIRI Aubl. (1775)

1. M. myrtilloides (Sw.) Poir. in Dict. Sci. Nat. **33**: 163 (1824).—*Petaloma myrtilloides* Sw. (1788).

Shrub or small tree to 8 m. high, glabrous throughout; leaves sessile, mostly 3–6 cm. long, lanceolate to narrowly ovate, acute or acuminate, 1-nerved, the lateral veins very obscure; flowers solitary, rarely 2, in leaf-axils on short bracteolate peduncles to 4 mm. long; hypanthium campanulate, the sepals pointeddeltate and minutely ciliolate; petals white, 5–6 mm. long; anthers 2–2·5 mm. long; berries globose, red, 6–7 mm. in diameter.

Rare (St. Ann, Clar., St. Thom.), in dry thickets or on crumbling rocky banks; 200–2500 feet; fl. Feb, Sept, fr. Feb. *H & B 10629! P 24794!* Mexico to S. Amer., Cuba, Hispaniola.

143. ONAGRACEAE

Annual or perennial herbs, sometimes shrubs or small trees. Leaves simple, opposite or alternate, without stipules or stipules caducous. Flowers mostly bisexual and actinomorphic, solitary and axillary or spicate, racemose or paniculate. Calyx forming a calyx-tube (hypanthium) adnate to the ovary, the lobes valvate, persistent or deciduous, typically tetramerous. Petals free, the same number as the sepals, contorted or imbricate, or absent. Stamens usually the same number as the sepals or twice as many, when biseriate those of the outer series alternate with the petals; anthers mostly 2-locular, opening lengthwise. Disk usually present, epigynous. Ovary inferior, (2–) 4 (–5)-locular; style solitary, slender; stigma usually capitate; ovules 1–many on axile placentas, anatropous. Fruit a capsule, berry or nut. Seeds without endosperm; embryo straight.

20 genera with about 650 species, widely distributed in temperate and subtropical regions, rarer in the tropics; especially numerous in western temperate America.

1 Calyx-tube (hypanthium) not prolonged beyond the ovary; sepals (3–) 4–6 (–7); fruit a capsule dehiscing irregularly; petals as many as the sepals or absent; seeds sometimes with a usually easily visible raphe; leaves alternate or opposite   **1. Ludwigia**
1 Calyx-tube prolonged beyond the ovary; sepals and petals 4:
  2 Fruit a 4-winged loculicidal capsule; plants (in Jamaica) herbaceous with alternate leaves   **2. Oenothera**
  2 Fruit a 4-grooved berry; plants shrubby with opposite or whorled leaves
        3. Fuchsia

1. LUDWIGIA L. (1753); Munz (1965); Raven (1963)

1 Leaves alternate:
  2 Stamens twice as many as the sepals; petals present:
    3 Sepals and petals usually 4; seeds pluriseriate in each loculus, free; pollen shed in tetrads:
      4 Capsule prominently 4-angled; raphe inconspicuous:
        5 Plants erect; more or less terrestrial shrubby herbs with leaves spread along the stems:
          6 Stems and leaves subglabrous to puberulous; petals less than 1 cm. long; capsule sessile or shortly stalked, 12–18 mm. long; seeds 0·3–0·5 mm. long   1. *erecta*
          6 Stems and leaves pubescent; petals more than 1 cm. long; capsule distinctly stalked, mostly about 2 cm. long; seeds 0·6–0·8 mm. long   2. *peruviana*
        5 Plants trailing in water; stems soft, flexible; leaves clustered in terminal rosettes, the petioles shorter towards the apex   3. *sedioides*
      4 Capsule more or less terete; raphe enlarged, equal in size to the body of the seed; erect shrubby herb with variable pubescence   4. *octovalvis*
    3 Sepals and petals usually 5 (4–7); seeds uniseriate in each loculus, each embedded in a firm piece of endocarp; raphe not enlarged; capsule terete, the seeds showing as bumps on the wall:
      7 Trailing soft-stemmed herb; leaves elliptical, acute or obtuse; capsule 1·5–2 (–2·5) cm. long; pollen grains shed singly   8. *peploides*
      7 Erect or straggling shrubby herbs, woody below:
        8 Leaves ovate to elliptical, pilose on both surfaces, minutely pellucid-dotted; sepals 3–6 mm. long, about 1·5 mm. broad; capsule 2–3·5 cm. long, subsessile; pollen shed in tetrads   5. *affinis*
        8 Leaves narrow:
          9 Leaves linear-oblanceolate, thinly pilose (in Jamaica); sepals 5–8 mm. long, 2–3 mm. broad; capsule 1·5–5 cm. long, abruptly tapered at base; pollen shed in tetrads   6. *leptocarpa*

9 Leaves linear-lanceolate, glabrous; sepals 2·5–3 mm. long; capsule 5–9 mm. long, abruptly narrowed to a slender stalk; pollen grains shed singly   7. *torulosa*

2 Stamens 4, the same number as the sepals; petals absent; trailing and rooting herb; pollen grains shed singly   9. *simpsonii*

1 Leaves opposite; stamens 4, the same number as the sepals; stems trailing and rooting; plants glabrous; pollen grains shed singly:

10 Hypanthium and capsule with 4 evident longitudinal green bands; basal bracteoles not more than 1 mm. long or not evident; petals absent; capsule 2–3 (–5) mm. long, rounded at base   10. *palustris*

10 Hypanthium and capsule lacking green bands; bracteoles above the base, 1–5 mm. long; petals present but easily shed; capsule (3–) 5–8 mm. long, narrowed at base   11. *repens*

1. L. erecta (L.) Hara in Journ. Jap. Bot. **28**: 292 (1953).—*Jussiaea erecta* L. (1753).

Annual, 60–200 cm. high; stem often red; leaves narrowly elliptical to elliptic-lanceolate, mostly rather pointed at both ends, up to 12 cm. long and 2 cm. broad; petals yellow, obovate, (2·5–) 4–6 mm. long, up to 3 mm. broad.

Locally common in swampy pastures, ditches and open savannas; near SL–2300 feet; fl. and fr. Aug, Nov–Apr. *A 10735! H 12836! P 18340!* Florida, continental tropical Amer., West Indies, Grand Cayman, tropical Africa, Madagascar.

2. L. peruviana (L.) Hara in Journ. Jap. Bot. **28**: 293 (1953).—*Jussiaea peruviana* L. (1753).

Shrubby herb, woody at the base, 60–300 cm. high; leaves elliptical, pointed at both ends, up to 15 cm. long and 3·5 cm. broad; sepals ovate, long-acuminate, up to about 1·5 cm. long and 8 mm. broad; petals yellow, up to about 2 cm. long; capsule obconic.

Rather common in boggy pastures, ditches and along streambanks and pond margins; SL–2400 feet; fl. and fr. Aug–Apr. *A 5904! H 12252! P 24778!* Florida, continental tropical Amer., Greater Antilles, Trinidad.

3. L. sedioides (Humb. & Bonpl.) Hara in Journ. Jap. Bot. **28**: 294 (1953).— *Jussiaea sedioides* Humb. & Bonpl. (1805).

Stem mostly submerged, reddish, stoloniferous; leaves mostly floating, petioles reddish increasing to about 7 cm. long, blade rhomboid, 8–15 mm. long, 5–12 mm. broad, denticulate distally; sepals ovate, very shortly acuminate, 8 mm. long, 5 mm. broad; petals yellow; capsule elongate-obconic, about 1 cm. long.

Local, in ponds, mostly in the western parishes; 25–1400 feet; fl. and fr. most of the year. *A 10341! H 11646! P 24495!* Honduras, Panama to Paraguay, Cuba.

4. L. octovalvis (Jacq.) Raven in Kew Bull. **15**: 476 (1962).—*Oenothera octovalvis* Jacq. (1760). *Jussiaea suffruticosa* L. (1753). *L. suffruticosa* (L.) Gómez Maza (1894), not Walt. (1788).

Shrubby, 60–120 (–250) cm. high; leaves linear to elliptic-lanceolate, shortly petioled, mostly pointed at both ends, up to 10 cm. long and 2·5 cm. broad; sepals ovate to broadly ovate, shortly acuminate, 8–11 mm. long, 4–7 mm. broad, sometimes tinged red; petals yellow, 1–2 cm. long; capsule (2–) 3–5 cm. long, tapered at base to a distinct slender stalk.

Common in or near water, in ponds, ditches and along riverbanks; SL–2300 feet; fl. and fr. all the year. *A 6566! 10843! H 5807! H & P 15199!* General in the subtropics and tropics; Grand Cayman.

5. L. affinis (DC.) Hara in Journ. Jap. Bot. **28**: 291 (1953).—*Jussiaea affinis* DC. (1828).

Erect or straggling undershrub up to 2·5 m. high; hairs on stems unicellular with brownish tips; leaves acute or obtuse, up to 8 (–12) cm. long and 3 (–6) cm. broad; petals yellow, 6–8 mm. long.

Very rare (St. Cath.), at riverbank; about 200 feet; fl. Apr. *P 26366!* Guatemala to Peru, the Guianas and Brazil, Lesser Antilles, Trinidad.

6. L. leptocarpa (Nutt.) Hara in Journ. Jap. Bot. **28**: 292 (1953).—*Jussiaea leptocarpa* Nutt. (1818).

Shrubby, 60–150 cm. high; leaves acute or obtuse, slender-petioled, up to 9 cm. long and 1·8 cm. broad; petals yellow, 5 (–6), 6–7 mm. long.

Rather local, in boggy pastures and pond margins in the central and western parishes; near SL–1500 feet; fl. and fr. Nov–Apr. *A 10841! 12158! P 11224!* S.E. United States, Mexico to Peru and Argentina, West Indies, tropical Africa, Madagascar. Some variants of this species in the New World are glabrous or nearly so.

7. L. torulosa (Arn.) Hara in Journ. Jap. Bot. **28**: 294 (1953).—*Jussiaea torulosa* Arn. (1835). *Oocarpon torulosum* (Arn.) Urb. (1931).

Shrubby, slender-branched herb; leaves up to 8 cm. long and 1 cm. broad; pedicels 4–6 mm. long; petals white (yellow or reddish elsewhere), about 5 mm. long; capsule torulose.

Very rare (St. Cath.), in bog; 1000 feet; fl. and fr. Jan. *P 6142!* British Honduras, Bolivia, Brazil and the Guianas, Cuba, Hispaniola.

8. L. peploides (Kunth) Raven in Reinwardtia **6**: 393 (1963).—*Jussiaea peploides* Kunth (1823). *J. repens* of F. & R. (1926), not L. (1753).

Stems trailing and rooting on mud or floating in water when sometimes swollen and bouyant; leaves tapered to base, up to 7 cm. long and 2 cm. broad, usually much smaller; sepals often red-margined, about 4·5 mm. long and 1·5 mm. broad; petals (4–) 5 (–6), yellow, falling easily.

Rather common, in boggy pastures and ponds; near SL–1500 feet; fl. and fr. most of the year. *A 10711! Proctor & Mullings 21823! Thompson 7990!* Raven (1963) recognized several subspecies of which ssp. *peploides* extends from S. United States to Argentina and through the Greater Antilles; it has also been introduced into S. Europe and some Pacific Is. Other subspecies range through the western Pacific region.

9. L. simpsonii Chapm., Fl. South. U.S. ed. 2, Suppl. 2: 685 (1892).—*L. microcarpa* of F. & R. (1926), not Michx. (1803).

Herb with trailing and rooting stems; glabrous; petioles 1–3 mm. long; leaf-blades spathulate to obovate, 5–15 mm. long, 3–7 mm. broad; flowers axillary, sessile; sepals acute, (1–) 1·5–2 mm. long; capsule obconical, obscurely angled, 2–2·5 mm. long and almost as broad.

Local (St. Eliz.), in swamps and ditches; 20–50 feet; fl. and fr. sporadically. *A 12066! H 9935! P 24836!* Bahamas, Florida, Cuba.

10. L. palustris (L.) Ell., Sketch Bot. S. Carol & Georgia **1**: 211 (1817).—*Isnardia palustris* L. (1753).

Stems trailing with ascending branches to 15 cm. high; leaves elliptical, 3–12 mm. long; flowers axillary, sessile, pinkish; sepals about 1·5 mm. long, acute; capsule corky, truncate, 2–5 mm. long.

Rare and local (Clar., Trel., St. Ann), at the muddy margins of ponds and lakes; 150–400 feet; fl. and fr. Mar, Aug. *Prior! P 24342!* Temperate Europe, western Asia and Africa, N. Amer. to Colombia, Bermuda, Greater Antilles. The description of this variable species in F. & R. (1926) refers to the larger plants found in Europe and N. Amer.

11. L. repens J. R. Forst., Fl. Amer. Sept.: 6 (1771).—*L. natans* Ell. (1821).

Stem trailing to 60 cm. or more; leaves obovate, subacute to obtuse, long-tapered to the base, up to 4·5 cm. long and 1·5 (–3) cm. broad; flowers axillary, subsessile; sepals 2–3 mm. long, acute; petals yellow or reddish, as long as sepals; capsule oblong, truncate, narrowed to the base, 3–8 mm. long.

Rather local, mostly submerged in slow-flowing streams; SL–300 feet; fl. and fr. Feb, Aug–Sept. *A 6059! H & B 10627! Proctor & Mullings 22031!* Bermuda, Bahamas, S. United States, Mexico, Cuba, Hispaniola.

2. OENOTHERA L. (1753)

1. O. rosea L'Hérit. ex Ait., Hort. Kew. **2**: 3 (1789).

Annual herb up to about 60 cm. high; stems branched erect or straggling; leaves lanceolate to oblanceolate, deeply lobed towards the base, up to 6 cm. long and 2 cm. broad; hypanthium and calyx-lobes greenish-white, ribbed or margined with reddish-purple; petals yellowish at first, later bright reddish-purple with greenish-yellow bases, about 1 cm. long; capsule obconic, ridged, opening at the top.

Rather local (St. Andr., Port., St. Thom.), a weed of roadsides and open waste places on gravelly soil; (2000–) 3500–4900 feet; fl. and fr. most of the year. *A 11206! ACH 372! H 12340! P 24839!* Bermuda, S. United States to Peru, Cuba; introduced and naturalized in some parts of Europe and the Old World tropics. Several of the yellow-flowered ornamental species of *Oenothera*, Evening Primroses, have been grown in Jamaica.

3. FUCHSIA L. (1753); Munz (1943)

Hypanthium several times longer than sepals; stamens not much exceeding sepals; stems, leaves and flowers hairy   1. *boliviana*
Hypanthium usually not longer than sepals; stamens long-exserted; plants glabrous or nearly so   2. *magellanica*

1. F. boliviana Carr. in Revue Hortic. **48**: 150, t. (1876).

Shrub up to 6 m. high, the whole plant pubescent; leaves opposite or ternate, oblong-elliptical to oblanceolate, mostly pointed at apex, shortly tapered at base, remotely denticulate, up to 25 cm. long and 7 cm. broad; flowers in racemes with persistent bracteoles; sepals dark red, 17–21 mm. long; petals light red, about 15 mm. long; berry linear-oblong, up to about 2 cm. long.

Naturalized locally (St. Andr., Port.), from a garden escape; 4000–5600 feet; fl. Mar, June, Aug, fr. Aug. *Bengry! H 5825! Lodge!* Guatemala and El Salvador to Venezuela and Bolivia.

2. F. magellanica Lam., Encycl. Méth. Bot. **2**: 565 (1788).—*F. coccinea* Ait. (1789).

Shrub 30–90 cm. high or sometimes scrambling to 3 m.; leaves opposite or ternate, rarely 4-nate, ovate-lanceolate, acute, rounded at base and there sparsely hairy beneath, denticulate, 2·5–5 cm. long, 1–2·5 cm. broad, with petioles up to 10 mm. long, usually shorter; flowers axillary, solitary on long peduncles; sepals crimson 1·5–2 cm. long; petals purple, shorter than the calyx; mature fruit not seen in Jamaica.

Naturalized locally (St. Andr., Port., St. Thom.); 4000–4500 feet; fl. Aug, Dec. *H 9130!* Native of Chile, extensively cultivated elsewhere; naturalized in W. Europe.

Plants closely resembling *F. magellanica* but slightly larger in all their parts occur naturalized in various mountainous parts of Jamaica including the summit of Blue Mountain Peak (*ACH 909! P 9436!*). They have been named *F. hybrida* Hort., a complex including many forms of garden origin.

144. HALORAGACEAE

Aquatic or terrestrial herbs or rarely undershrubs. Leaves alternate, opposite or verticillate, variable in size, often divided when submersed, with or without stipules. Flowers usually unisexual and monoecious, actinomorphic, often very small, mostly subtended by a pair of bracteoles, solitary or clustered or in corymbs or panicles. Perianth biseriate, uniseriate or altogether lacking; calyx adnate to the ovary, the limb 2–4-lobed or absent; petals 2–4, free, imbricate or valvate or absent. Stamens (1–) 2–8, free; anthers basifixed, 2-locular, opening lengthwise. Ovary inferior, 1–4-locular; ovules solitary in each loculus, anatropous, pendulous on axile placentas or when loculus solitary pendulous and parietal; styles 1–4; stigmas often plumose. Fruit a small nut or drupe, sometimes winged, indehiscent or schizocarpic. Seed endospermic, with usually straight embryo.

8 genera with about 150 species mostly in temperate and subtropical regions.

1. PROSERPINACA L. (1753)

Aquatic herbs; leaves alternate; flowers bisexual, trimerous; petals absent.

1. P. palustris L., Sp. Pl. **1**: 88 (1753).

Stems spreading to suberect, up to about 30 cm. long, rooting from the submersed parts; leaves narrowly elliptical to oblanceolate, acute or obtuse, sharply serrate, 1–3 (–6) cm. long, up to about 9 mm. or more broad, more dissected when submersed; flowers axillary; fruit trigonous, rough, 4–5 mm. long.

Rare and local (St. Eliz.), partly submerged in shallow water in swamps; 50–75 feet; fl. and fr. Feb, Aug. *P 24587! Webster & Proctor 5351!* S. E. United States.

145. ARALIACEAE

Trees, shrubs or rarely herbs, sometimes climbing. Leaves mostly alternate, simple or compound, sometimes with stellate hairs, sometimes pellucid-glandular; stipules usually present and often adnate to the dilated petiole-base. Flowers bisexual or often unisexual and polygamous or dioecious, actinomorphic, in heads, umbels or racemes in bracteate simple or compound inflorescences. Perianth biseriate, mostly 5-merous; calyx entire or toothed; corolla of (3–) 5 (–10) usually valvate often hook-tipped caducous free or rarely united petals. Glandular epigynous disk present, usually confluent with the style-bases. Stamens usually as many as the petals and alternate with them; filaments inserted on the disk; anthers 2-locular, dorsifixed, opening lengthwise. Ovary inferior with 1 or more loculi; styles as many as the carpels, free or connate or stigmas sessile; ovules solitary in each loculus, axile, pendulous, anatropous. Fruit a berry or drupe or carpels separating as distinct pyrenes. Seeds with copious endosperm and small embryo.

About 65 genera with some 800 mostly tropical species.

1 Leaves pinnately compound without pellucid glands; petals imbricate; ornamental shrub with variegated leaves; native of Pacific Is.; Aralia
          Polyscias guilfoylei (Cogn. & Marchal) L. H. Bailey
1 Leaves digitately compound or simple; petals valvate; mostly trees:
  2 Flowers sessile in heads; leaves (in our species) simple and entire, not gland-dotted
          1. Oreopanax
  2 Flowers pedicelled to subsessile in umbels:
    3 Leaves digitately compound, without pellucid glands, the petiole with a conspicuous coriaceous ligule; petals more or less connate; inflorescence a raceme of umbels
          2. Schefflera
    3 Leaves simple with more or less pellucid internal glands, the petiole without an obvious ligule; petals free   **3. Dendropanax**

1. OREOPANAX Decne. & Planch. (1854)

1. O. capitatus (Jacq.) Decne. & Planch. in Revue Hortic. sér. 4, **3**: 108 (1854). —*Aralia capitata* Jacq. (1760). Woman Wood.

Shrub 5 m. or tree to 20 m. high often with long spreading branches; leaves ovate to elliptical, acutely acuminate, cuneate to rounded at base, glabrous, up to 20 cm. long and 10 cm. broad; inflorescence terminal, paniculate, stellate-puberulous; flowers polygamo-dioecious, fragrant in globose or ovoid heads 4–6 mm. long; petals white or greenish; berry globose, 3–5 mm. in diameter, several-seeded.

Very common in secondary and marginal woodlands in moderately wet areas; 400–5000 feet; fl. Apr–Aug, fr. Apr–Dec. *A 11258! H 7684! P 22203! Stearn 976!* Continental tropical Amer., Hispaniola, Lesser Antilles, Trinidad and Tobago.

2. SCHEFFLERA J. R. & G. Forst (1776) nom. cons.

1 Flowers distinctly pedicellate; inflorescence glabrescent, at first thinly whitish-powdery-tomentose; leaflets glabrous   1. *sciadophyllum*
1 Flowers shortly pedicelled to subsessile; inflorescence pubescent to densely tomentose:
  2 Inflorescence white-powdery-tomentose; primary inflorescence-branches stout, up to 8 (–12) mm. long; calyx 3–3·5 mm. broad; leaflets with few or many whitish branched hairs beneath   2. *troyana*
  2 Inflorescence rusty-pubescent; primary inflorescence-branches slender, up to 11 mm. long; calyx less than 2 mm. broad; leaflets glabrous and dull yellowish beneath, dark green adaxially   3. *stearnii*

1. S. sciadophyllum (Sw.) Harms in Engl. & Prantl, Nat. Pflanzenf. **3** (8): 37 (1894).—*Aralia sciadophyllum* Sw. (1788) (err. *sciodaphyllum*). *Sciadophyllum brownei* Spreng. (1825). *Sciadophyllum praetermissum* Norman (1926).

Slender-trunked straggling tree, 3–6 m. or more high; leaflets 7–13, oblong, shortly and abruptly acuminate, rounded to subcordate at base, up to 30 cm. long and 11 cm. broad; inflorescences in the distal axils, pendulous, up to 60 cm. long; flowers fragrant; ripe fruits pinkish, 5-angled about 4 mm. long.

Locally common, in submontane open woodland in the central and eastern parishes; 1250–5700 feet; fl. and fr. July–Apr. *A 5776! H 7560! P 23914!* Endemic.

2. S. troyana (Urb.) A. C. Sm. in Trop. Woods **66**: 5 (1941).—*Sciadophyllum troyanum* Urb. (1908).

Sparingly or unbranched tree 3–6 m. high; leaflets 9–15, oblong-elliptical, shortly acuminate, cordate at base, up to 30 cm. long and 10·5 cm. broad; inflorescences at the ends of branches, more or less erect, up to 55 cm. long, pedicels 2·5–3 mm. long; fruit silvery-green tinged purple.

Rather rare, in woodlands and woodland margins on rocky limestone in the central and west-central parishes; 1850–3000 feet; fl. Mar–July, fr. May–Aug. *A 7559! H 9369! P 8428!* Endemic.

3. S. stearnii Howard & Proctor in Journ. Arn. Arb. **39**: 105 (1958).

Shrub or tree to about 2·5 m. high, sparingly branched; leaflets up to about 10, elliptical to broadly oblanceolate, abruptly acuminate, truncate at base, up to 18 cm. long and 8 cm. broad; inflorescences up to 22 cm. long; pedicels obsolete; fruit not seen.

Rare and local (Port.), in wet submontane thickets on limestone; 2000–2500 feet; fl. Jan–Mar. *A 9156! HPS 14761! P 16255!* Endemic.

3. DENDROPANAX Decne. & Planch. (1854); A. C. Smith (1944)

1 Inflorescence a raceme of umbels   1. *arboreus*
1 Inflorescence a simple umbel:
  2 Peduncle jointed (1·5–) 2 cm. or more above the base, the lower section like the upper and not corky:
    3 Leaf-blade narrowly cuneate to rounded at base, pinnately veined   2. *pendulus*
    3 Leaf-blade cordulate at base, subtriplinerved   3. *ovalifolius*
  2 Peduncle not obviously jointed or joint within 1 cm. of base and proximal section corky:
    4 Leaf-tip distinctly acuminate, the acumen mostly rather abrupt and sharply pointed; blade acuminate at base, elliptical   5a. *nutans* var. *nutans*
    4 Leaf-tip rounded to obtuse or acute, rarely broadly and gradually acuminate:
      5 Leaf-base cordate or subcordate; blade broadly ovate   4. *cordifolius*
      5 Leaf-base cuneate to obtusely rounded:
        6 Leaf-blade ovate to lanceolate or elliptical, broadest at or below the middle:
          7 Leaf-blade rounded at base; peduncle about 4 cm. long   7. *grandis*
          7 Leaf-blade cuneate at base:
            8 Peduncle 4–10 cm. long; leaf-margin more or less reflexed
                                    5b. *nutans* var. *obtusifolius*
            8 Peduncle 10–25 cm. long; leaf-margin usually flat   6. *swartzii*
        6 Leaf-blade obovate to oblanceolate, broadest above the middle:
          9 Leaf-blade broadly obovate, twice as long as broad or broader; pedicels up to about 2 cm. long:
            10 Peduncle at least 10 cm. long   8. *portlandianus*
            10 Peduncle short and stout, 2–5 cm. long   9. *grandiflorus*
          9 Leaf-blade oblanceolate, mostly 2·5–3 times as long as broad:
            11 Peduncle less than 1 mm. thick; pedicels few, 10–15 mm. long   10. *filipes*
            11 Peduncle at least 2 mm. thick; pedicels (12–) 15–25 mm. long   11. *blakeanus*

1. D. arboreus (L.) Decne. & Planch. in Revue Hortic. sér. 4, **3**: 107 (1854).—*Aralia arborea* L. (1759). *Gilibertia arborea* (L.) Marchal (1891). Angelica Tree, Galipee.

Tree 3–16 m. high; bark smooth, buff-grey with white slash exuding small drops of brownish-orange sap; leaves softly fleshy and faintly aromatic when crushed, elliptical to ovate-elliptical or obovate, acuminate, cuneate to rounded at base, very shiny above, up to 15 cm. or more long and about 10 cm. broad; inflorescence terminal; peduncle 1·5–8 cm. long; pedicels 6–8 mm. long; flowers at first green turning yellow; petals 4–6; fruit at first green, then white, finally blackish-purple, 5–6 angled when dry, 5–7 mm. long.

Common in damp sheltered areas, particularly along hedgerows and roadsides in areas of high rainfall; 250–4500 (–6000) feet; fl. Dec–July, fr. Dec–Sept. *A 5887! J. P. 976! 1092! P 22284!* Mexico to Bolivia and Peru, Greater Antilles, St. Vincent, Grenada, Trinidad.

2. D. pendulus (Sw.) Decne. & Planch. in Revue Hortic. sér. 4, **3**: 107 (1854).
—*Hedera pendula* Sw. (1788). *Gilibertia elongata* (Britton) F. & R. (1926). *G. pendula* (Sw.) Marchal ex Urb. (1899).

Slender-trunked tree 5–10 m. high or shrub 2·5–4 m.; branches weak and spreading; leaves very variable in size and petiole-length, mostly ovate-lanceolate, apex acute to less frequently obtuse, 3–18 cm. long, 1–7 cm. broad; inflorescence pendulous, up to 30 cm. long; pedicels 1·5–2 cm. long; flowers green; berry 5-seeded.

Rather local, in submontane woodlands on limestone or shale; 2000–5000 feet; fl. Sept–May, fr. Feb. *A 8174! 11597! H 10874! P 23104!* Endemic.

3. D. ovalifolius (F. & R.) Adams in Phytologia **21** (2): 66 (1971).—*Gilibertia ovalifolia* F. & R. (1926).

Shrub 4 m. or slender-trunked tree to 8 m. high; leaves ovate, acute to shortly acuminate, up to 13 cm. long and about 7 cm. broad; peduncle 20–26 cm. long; pedicels 15–18 mm. long; petals 4 mm. long, green; fruit not known.

Rare and local (St. Eliz., St. James), in woodlands on limestone; 1300–2000 feet; fl. Feb, June. *H 9188! 12378!* Endemic.

4. D. cordifolius Britton in Bull. Torr. Bot. Club **39**: 4 (1912).—*Gilibertia cordifolia* (Britton) F. & R. (1926).

Tree about 6 m. high; leaves broadly ovate, obtuse to shortly acuminate, up to 20 cm. long and 12 cm. broad; peduncle 10–15 cm. long, stout; pedicels 2–2·5 cm. long; petals greenish to white, about 5 mm. long; fruit not known.

Very rare (Han.), on wooded limestone hillside; 1500–1700 feet; fl. Mar. *P 20684!* Endemic.

5. D. nutans (Sw.) Decne. & Planch. in Revue Hortic. sér. 4, **3**: 107 (1854).
—*Hedera nutans* Sw. (1788). *Gilibertia nutans* (Sw.) Marchal ex Urb. (1899).

5a. Var. nutans.
Shrub 2 m. or gnarled tree to 6 m. high; leaves broadly elliptical, very variable in size and petiole length, up to 11 cm. long and 9 cm. broad; peduncle 5–20 cm. long; pedicels 2–4 cm. long; petals 4·5 mm. long, green.

Local (Port., St. Thom.), in montane and submontane woodlands and thickets; (2500–) 6200–7400 feet; fl. Dec–Sept, fr. Dec–Aug. *A 10661! P 9694! Royer UCWI 3373!* Endemic.

5b. D. nutans var. obtusifolius Adams in Phytologia **21** (2): 66 (1971).
Differs from var. *nutans* in the leaves being mostly smaller, 4–8 cm. long and 2–4·5 cm. broad, mostly rounded-obtuse at tip and cuneate at base, rather leathery with the margins usually reflexed; peduncle 4–10 cm. long, rather stiff; pedicels 10–13 mm. long; petals about 3 mm. long.

Local (Port., St. Thom.), with or near populations of var. *nutans*; 2200–6800 feet; fl. Feb, July–Aug, fr. Aug, Dec. *A 10693! P 19736! Robbins UCWI 93! Webster & Proctor 5518!* Endemic.

6. D. swartzii (F. & R.) A. C. Sm. in Trop. Woods **66**: 3 (1941).—*Gilibertia swartzii* F. & R. (1926).

Tree about 6 m. high; leaves elliptical, acute to obtuse at tip, cuneate at base, the blade up to 10 cm. long and 4 cm. broad; peduncle mostly 15–20 cm. long; pedicels about 1·5 cm. long, numerous.

Rare (Port., St. Thom.), in woodlands and thickets in wet areas; 2000–3000 feet; fl. Mar–June, fr. Jan. *A. M. Barry! J. P. 983!* partly. Endemic.

7. D. grandis Britton in Bull. Torr. Bot. Club **39**: 4 (1912).—*Gilibertia grandis* (Britton) F. & R. (1926).
Tree 12–16 m. high; leaves ovate to ovate-elliptical, obtuse at both ends, leathery, up to 15 cm. long; peduncle stout, erect; pedicels 18–20 mm. long; petals 4–5 mm. long; fruit not seen.
Very rare (St. Ann), in forest. *Prior.* Endemic.

8. D. portlandianus Proctor in Bull. Inst. Jam., Sci. ser. **16**: 48, t. 20 (1967).
Shrub about 2·5 m. high; leaves obovate-elliptical, obtuse, leathery, up to 18 cm. long and 9·5 cm. broad, pinnately veined; peduncle up to about 17 cm. long; pedicels about 2 cm. long; flowers green; petals about 4 mm. long; fruit not known.
Rare and local (Port.), in moist mossy limestone forest; 1500–2000 feet; fl. Apr–May. *duQuesnay 342! P 9987!* Endemic.

9. D. grandiflorus Britton in Bull. Torr. Bot. Club **39**: 3 (1912).—*Gilibertia grandiflora* (Britton) F. & R. (1926).
Shrub 3 m. or tree to 12 m. high; leaves, rounded or bluntly pointed at tip, leathery, up to 14 cm. long and 7 cm. broad; peduncle erect; pedicels few, up to 17, up to 2·5 cm. long in fruit; flowers greenish-yellow, slightly fragrant; petals about 7 mm. long; berry 10–12 mm. long.
Very local (Clar.), on limestone rocks in woodland; about 2500 feet; fl. Dec–Jan, May–July, fr. May–July. *H 10994! P 11402! 23175!* Endemic.

10. D. filipes Britton in Bull. Torr. Bot. Club **41**: 9 (1914).—*Gilibertia filipes* (Britton) F. & R. (1926).
Slender straggling shrub to 3 m. high or small tree; leaves rounded to acute at tip, thin or leathery with rather prominent venation, 7–12 cm. long, 2–3·5 cm. broad; peduncle 7–11 cm. long; pedicels up to about 15; petals 1·5–2 mm. long; fruit not known.
Rare and local (Clar., Trel.), on limestone rocks in woodland; 2000–2500 feet; fl. Mar–May. *A 12846! H 11057! P 20758! Robertson UCWI 967!* Endemic.

11. D. blakeanus Britton in Bull. Torr. Bot. Club **39**: 4 (1912).—*Gilibertia blakeana* (Britton) F. & R. (1926). *D. oblanceatus* Proctor (1967).
Tree with slender trunk, 7–8 m. high; leaves mostly acute at tip, leathery, 4–20 cm. long, 2–8 cm. broad; peduncle 11–15 cm. long; pedicels up to about 25; berry subglobose, about 5 mm. in diameter.
Local (Trel. Port., St. Thom.), in thickets and moist woodland on limestone; 1500–2500 feet; fl. Jan, June, fr. Jan. *H & B 10761! P 9810!* Endemic.

146. UMBELLIFERAE (AMMIACEAE)

Herbs or rarely shrubs with hollow stems. Leaves alternate, rarely opposite, usually compound and often with sheathing petioles; stipules absent. Flowers usually bisexual, actinomorphic, in simple or compound indeterminate umbels or heads, with or without bracts and bracteoles. Perianth typically biseriate; calyx adnate to ovary, the free teeth 5 or obsolete; petals 5, free, the tips usually inflexed. Stamens 5, free, inflexed in bud, alternating with the petals, inserted on an epigynous disk; anthers 2-locular, opening lengthwise. Ovary inferior, 2-locular; ovules anatropous, solitary in each loculus, pendulous; styles 2. Fruit a schizocarp of 2 mericarps united and separating by their faces (commissure), flattened dorsally (parallel to commissure) or laterally (at right angles to commissure) or terete; each mericarp with 5 primary ribs (1 dorsal rib, 2 lateral ribs nearer the commissure and 2 intermediate ribs), with or without oil-tubes in the intercostal spaces; mericarps 1-seeded, usually suspended after separation by a carpophore. Seed with endosperm and small embryo.
Some 200–300 genera with about 3000 species occurring mainly in north temperate regions.

1 Inflorescence involucrate-capitate with obsolete rays and pedicels; basal leaves simple, serrate-dentate, upper leaves opposite, deeply toothed   1. **Eryngium**
1 Inflorescence distinctly umbellate with evident rays and/or pedicels; leaves simple or compound:
  2 Umbels simple or proliferous (superposed); leaves simple:
    3 Leaves peltate; involucre wanting or inconspicuous   2. **Hydrocotyle**
    3 Leaves not peltate; involucre of 2 conspicuous ovate to suborbicular bracts
                                            3. **Centella**
  2 Umbels variously compound; leaves compound:
    4 Fruit armed with glochidiate bristles, dorsally flattened; involucel of ciliate bracteoles; involucre of 3-fid or more compound bracts   4. **Daucus**
    4 Fruit not bristly:
      5 Fruit dorsally flattened with some or all of the wings ribbed; involucel wanting; flowers yellow:
        6 Dorsal wings of fruit well developed; cultivated glaucous annual with a strong anise odour; native of Europe; Dill   *Anethum graveolens* L.
        6 Dorsal wings of fruit absent, lateral wings prominent; plant biennial, puberulous
                                            5. **Pastinaca**
      5 Fruit terete or laterally flattened with ribs not prominently winged; plants glabrous:
        7 Annuals; involucre and involucels absent; flowers white   6. **Apium**
        7 Biennials or perennials, aromatic; flowers not white:
          8 Involucels absent; plant glaucous; petals yellow; naturalized   7. **Foeniculum**
          8 Involucels present; cultivated plants rarely escaping:
            9 Involucel shorter than the flowers; leaves decompound; petals yellowish; native of S. Europe; Parsley   *Petroselinum crispum* (Mill.) A. W. Hill
            9 Involucel longer than the flowers; leaves pinnate; petals greenish to brown; robust herb with large edible taproot; described as from Jamaica but native in northern S. America; Apio, Arracacha   *Arracacia xanthorrhiza* Bancroft

1. ERYNGIUM L. (1753)

1. E. foetidum L., Sp. Pl. **1**: 232 (1753).—Fit Weed, Spirit Weed.

Taprooted biennial unpleasantly scented herb; shoots divaricately branched in flower, 15–40 cm. high; basal leaves oblanceolate, obtuse, up to 30 cm. long and 4 cm. broad; flower-heads cylindrical, up to about 1·5 cm. long, green, subtended by a whorl of 5–6 unequal bracts resembling upper leaves; sepals 0·5–1 mm. long; petals 0·6–0·7 mm. long; fruit scaly, about 2 mm. long.

Locally common, along damp shaded roadsides and cultivation margins mostly on heavy soils; 200–2300 feet; fl. and fr. all the year. *A 6532! H 11888! P 24124!* Continental tropical Amer., West Indies; introduced and naturalized in W. Africa and Uganda.

2. HYDROCOTYLE L. (1753)

1 Inflorescence an interrupted sometimes bifurcated spike; fruits in tight clusters on pedicels usually less than 2 mm. long; petioles glabrous; leaf-blades up to 6 cm. in diameter   1. *verticillata*
1 Inflorescence a simple umbel or occasionally proliferous:
  2 Pedicels up to 12 (–25) mm. long in fruit; petioles glabrous, up to 40 cm. long; leaf-blade 0·5–8 cm. in diameter; peduncles up to 35 cm. long but usually longer than petioles; umbels many-flowered   2. *umbellata*
  2 Pedicels very short, the flowers and fruits subsessile; leaf-blades not exceeding 2 cm. in diameter:
    3 Petioles glabrous to densely retrorse-villous, 5–35 mm. long; peduncles 5–15 mm. long; fruit glabrous to sparingly hirsute, oil-bearing glands absent
                                          3. *pusilla*
    3 Petioles glabrous, 3–14 mm. long; peduncles longer than the petioles 3–16 mm. long; fruit glabrous, oil-bearing glands present   4. *brittonii*

1. H. verticillata Thunb., Diss. Hydroc.: 5 (1798).

Stem trailing and rooting; petioles up to 25 cm. long; leaf-blades orbicular, shallowly lobed and crenate; inflorescence up to 17 cm. long with 2–7 few-flowered verticils; fruit ellipsoidal, cuneate at base, 1–3 mm. long, 2–4 mm. broad with distinct dorsal and lateral ribs, oil-bearing glands conspicuous.

Rather rare, in ditches and marshes; 50–700 feet; fl. and fr. Jan–June. *H 7094!*

10373! Powell 256! E. & S. United States, Bermuda, Bahamas, Mexico, Greater Antilles, Guadeloupe, tropical Africa and Australia.

2. H. umbellata L., Sp. Pl. **1**: 234 (1753).
Habit like the last; umbels terminal or two superposed; petals greenish-white; fruit globose to ellipsoidal, 1–2 mm. long, 2–3 mm. broad, the dorsal surface acute with the dorsal and lateral ribs obtuse.
Frequent, in swamps and pond margins; SL–4000 feet; *A 12350! P 6144! Thompson 6168! Yuncker 18043!* From Canada to temperate S. Amer., Bermuda, Greater Antilles, Guadeloupe, tropical and S..Afric .

3. H. pusilla A. Rich. in Ann. Sci. Phys. **4**: 167, t. 52, f. 2 (1820).
Stems filiform, rooting at the nodes; leaf-blades suborbicular, 3–20 mm. in diameter, shallowly 5–8-lobed, crenate, thin, glabrous or villous above, glabrous beneath; umbels 2–6-flowered; fruit 0·5 mm. long, 1 mm. broad, ribs obsolete, orange when mature.
Locally common (St. Andr., Port.), on roadside and pathside banks in damp shady places; 2500–4900 feet; fl. and fr. July–Aug. *A & S 3487! H & B 10542! J. P. 2092! P 23760!* Mexico to Uruguay, Greater Antilles.

4. H. brittonii Mathias in Brittonia **2**: 239 (1936).
Like the last; leaf-blade 4–11 mm. in diameter, glabrous or sparsely setulose above; umbels 1- or 2-flowered; fruit ellipsoidal, about 1 mm. long, 1·5 mm. broad.
Very rare (St. Andr.); about 5000 feet; known only from the type, *Britton 177.* Endemic.

3. CENTELLA L. (1763)

1. C. asiatica (L.) Urb. in Mart., Fl. Bras. **11** (1): 287, t. 78, f. 1 (1879).—*C. erecta* (L. f.) Fernald (1940). *Hydrocotyle asiatica* L. (1753).
Stoloniferous perennial, thinly villous; petioles very short to 20 (–50) cm. long; leaf-blade up to 5 (–10) cm. long and 6 (–9) cm. broad, rounded at tip, cordate at base, margin repand-dentate or crenate; peduncles slender, up to 8 (–20) cm. long; pedicels 2–5, up to 4 mm. long; petals white edged with rose; fruit ellipsoid,3–4 mm. long, 3–5 mm. broad.
Locally common in moist open savannas, pastures and banks mostly on heavy clay soils; 1500–2750 feet; fl. Mar–July, Dec, fr. May–Sept, Dec–Jan. *A 7116! H 11091! P 15933!* General in the subtropics and tropics; Grand Cayman.

4. DAUCUS L. (1753)

1. D. carota L., Sp. Pl. **1**: 242 (1753).—Wild Carrot.
Taprooted biennial branched herb 30–90 cm. high; leaves 2–3-pinnate, hairy, the leaflets toothed; calyx obsolete; petals white or tinged purplish; in fruit the outer rays arch over the inner; fruit oblong-ovoid, 3–4 mm. long.
Locally frequent (St. Andr., Port., St. Thom.), on open grassy banks, roadsides and in waste or cultivated places; (1500–) 3000–5250 feet; fl. Dec–Aug, fr. Mar–Aug, Dec. *A 7273! H 11917! P 24573!* Native of Eurasia; introduced and naturalized from Canada to Mexico and C. Amer., Greater Antilles, Barbados, mostly only where cultivated in tropical regions.

5. PASTINACA L. (1753)

1. P. sativa L., Sp. Pl. **1**: 262 (1753).—Wild Parsnip.
Taprooted biennial herb 60–90 cm. high; leaves pinnate; leaflets toothed; petals yellow; mature fruits 6 mm. long and 5 mm. broad.
Locally common (St. Andr., St. Thom.), a weed along roadsides, on banks and in pastures and waste places; 3300–5000 feet; fl. and fr. Dec–Aug. *A 10561! H 10116! P 6210!* Native of Europe; Hispaniola.

6. APIUM L. (1753)

1. **A. leptophyllum** (Pers.) F. von Muell. ex Benth., Fl. Austral. 3: 372 (1867).—
Pimpinella leptophylla Pers. (1805). *Cyclospermum leptophyllum* (Pers.)
Sprague (1923).
Glabrous annual herb 15–40 cm. high; leaves divided into filiform segments;
peduncles 1–2 cm. long; pedicels (0–) 2–7 mm. long; calyx-teeth lacking; petals
ovate; mericarps 1·2–1·3 mm. long, semi-terete with 5 equal rounded ribs.
Occasional as a weed of sandy roadsides, ditches and cultivated ground; (300–)
1500–3500 feet; fl. and fr. Dec–May, Aug. *A 5977! H 11894! Powell 215!* General
in the subtropics and tropics, casual in S.W. Europe.
A. graveolens L., Celery, is cultivated in hill gardens in Jamaica.

7. FOENICULUM Mill. (1754)

1. **F. vulgare** Mill., Gard. Dict. ed. 8 (1768).—Fennel.
Perennial erect branched herb up to 2 m. high; leaves finely divided into filiform
segements; peduncles 1·5–6·5 cm. long; primary rays (15–) 17–21 (–40) somewhat
unequal, 1–6·5 cm. long; pedicels 2–10 mm. long; calyx-teeth lacking; mature fruit
oblong, 3·5–4·5 mm. long, 1·5–2 mm. broad, with acute ribs.
Locally common (St. Andr., Manch., Trel.), along roadsides and in open waste
places; 2500–3900 feet; fl. and fr. most of the year. *A 5796! H 12383! Powell 1581!*
Yuncker 18035! Native of Europe and the Mediterranean region, introduced into
the New World and naturalized in Greater Antilles, Virgin Is., Barbados.

147. GARRYACEAE

Shrubs or trees, evergreen. Leaves opposite, simple, entire, without stipules but
petiole-bases united by a ridge. Flowers unisexual, dioecious, in terminal and
axillary simple or branched inflorescences. Male flowers pedicelled, decussate in
pairs; perianth uniseriate of 4 valvate segments often coherent apically; stamens 4,
alternate with sepals and protruding between them; anthers 2-locular, opening
lengthwise. Femal flowers sessile,' bracteate in opposite pairs, without perianth;
ovary 1-locular; ovules 2, anatropous, pendulous from near the apex; styles 2, free
subulate, spreading. Fruit a 1–2-seeded berry, globose to ovoid, the styles persistent.
Seeds with copious endosperm; embryo small.
1 genus with 15 species in S.W. United States, Mexico, Cuba, Jamaica,
Hispaniola.

1. GARRYA Douglas ex Lindl. (1834)

1. **G. fadyenii** Hook., Ic. Pl. 4: t. 333 (1840).—*Fadyenia hookeri* Endl. (1848), not
F. hookeri (Sweet) Maxon (1908) which is a fern.
Shrub 2–5 m. or slender-branched bushy tree to 8 m.; young twigs, leaves
beneath and inflorescences hoary pubescent, mature leaves very shiny above; leaf-
blades narrowly elliptical, apex acute to obtuse, mucronate, base narrowly cuneate
to rounded, up to 10 (–11) cm. long and about 3·5 (–4) cm. broad; male flowers
about 3 mm. long; fruit globose, ripening deep purple or black, up to about 8 mm.
in diameter.
Locally common (St. Andr., Port., St. Thom.), in thickets, woodland margins
and on exposed slopes; 1800–6000 feet; fl. Dec–May, fr. Jan–Sept. *A 8778! 11198!
H 12414! P 9608! 23561!* Cuba, Hispaniola.

148. CLETHRACEAE

Shrubs or trees, pubescent or tomentose. Leaves alternate, simple entire or serrate, without stipules. Flowers bisexual, actinomorphic, in terminal bracteate racemes or panicles; bracts lanceolate, deciduous. Perianth biseriate; calyx 5-merous, persistent the lobes obtuse, imbricate; petals 5, free, imbricate, caducous. Stamens 10 (–12) in two series, hypogynous; anthers sagittate, inflexed in bud, 2-locular, opening by long apical pores; pollen simple. Ovary superior, 3-locular; ovules numerous, anatropous, borne on axile intrusive placentas; style simple; stigma (2–) 3 (–4)-lobed. Fruit a 3-valved loculicidal capsule, the valves separating from the central axis. Seeds angled or flattened, often winged; embryo cylindrical; endosperm fleshy.

1 genus with about 35 species, mostly in the tropics of America and Asia.

1. CLETHRA L. (1753); Sleumer (1967)

Calyx 5–6 mm. long, with short tomentum and numerous long brown hairs on the midline; inflorescence-branches and leaves beneath densely beset with long brown hairs
                                                                                            1. *alexandri*
Calyx 3–4·5 mm. long, with short pale tomentum only or sometimes a few longer whitish hairs; inflorescence-branches and leaves at least on the veins beneath with a short buff to rusty tomentum; lamina beneath with or without a shorter whitish felt   2. *occidentalis*

1. **C. alexandri** Griseb., Fl. Br. W.I.: 142 (1860).

Gnarled shrub or tree to 6 m. high; leaves obovate-elliptical, broadly rounded at base, obtuse to shortly acuminate at tip, margin subentire or distally with short teeth, sparsely hairy above, densely pale-felted beneath on the lamina, up to about 18 cm. long and 10 cm. broad; racemes numerous, in upper leaf-axils and a terminal panicle, up to about 20 cm. long; flowers fragrant; petals white, eroded, ciliate, hairy within, 7 mm. long; capsule densely villous 6–7 mm. broad.

Locally common (St. Andr., St. Thom.), in thickets and montane woodland; 5500–7400 feet; fl. July–Nov, fr. Dec–Mar. *A 10692! ACH 897! J.P. 1279! P 24024! Stearn 108!* Endemic.

2. **C. occidentalis** (L.) Kuntze, Revis. Gen. Pl. 2: 389 (1891).—*Tinus occidentalis* L. (1759). *C. tinifolia* Sw. (1788). *C. bracteata* Griseb. (1860). *C. jamaicensis* Britton (1914) Soapwood.

Straggling shrub to 3 m. or tree 4–16 m. high, often deciduous; trunk shortly buttressed; bark reddish-brown, underbark streaked crimson; leaves entire to obscurely serrate-dentate in distal half, more deeply serrate in some variants and in juveniles, elliptical to oblong-obovate, shortly rounded at base, acute to obtuse or rounded at apex, 3–20 cm. long, 1·5–7 cm. broad, glabrescent above, usually rusty-pubescent on the midrib and veins beneath, shortly white-tomentose between the veins but less so in juveniles and old leaves to almost glabrous; flowers fragrant in a panicle of racemes up to 30 cm. long; petals white, eroded, ciliate, about 4 mm. long; capsule densely short-tomentose, 3–4 mm. broad; seeds striate-winged, about 1 mm. long.

Rather common in the eastern parishes at middle and higher elevations, occasional in the central parishes, absent from the west, in well-lit woodlands and sub-montane thickets; (400–) 2000–5300 feet; fl. Jan–Apr, July–Nov. fr. Aug–Dec and sporadically. *A 7935! H 9454! J.P. 652! P 20665!* Mexico to El Salvador and Honduras.

Clethra in Jamaica requires at least biometrical study in order to distinguish ontogenetic, ecological and genetical characteristics before any further taxonomic treatment is contemplated. Plants intermediate to these two species occur between about 5000 and 6500 feet in the Blue Mountains and may be hybrids.

149. ERICACEAE

Trees or erect or scrambling shrubs, sometimes small but always more or less woody. Leaves alternate or rarely opposite or whorled, simple, entire or serrate, without stipules. Flowers bisexual, actinomorphic or slightly zygomorphic in terminal or axillary racemes or axillary clusters. Perianth biseriate; calyx of 4–8 sepals, free or adnate to the ovary, the lobes valvate; corolla gamopetalous with 4–7 lobes or teeth. Stamens usually twice as many as corolla-lobes, free; anthers often appendaged, 2-locular, introrse, usually opening by terminal pores; pollen compound. Ovary superior or inferior, 4–5-locular with numerous anatropous or amphitropous ovules on axile placentas; style simple; stigma capitate. Fruit a capsule, berry or drupe. Seeds with short embryo and fleshy endosperm.

70–80 genera with over 2000 species, mostly in cold or temperate regions and on tropical mountains.

1 Fruit a capsule; ovary superior; racemes short, subumbellate; shrubs or trees:
  2 Capsule septicidal; corolla showy, crimson, the limb mostly widely campanulate and somewhat zygomorphic; inflorescences terminal; plants cultivated mostly in the mountains; *Rhododendron arboreum* Sm., naturalized locally; (*Azalea* spp. in gardens.)
  2 Capsule loculicidal; corolla small, white, urceolate, regular; inflorescences terminal and axillary; wild plants   **1. Lyonia**
1 Fruit a berry; ovary inferior; raceme-axis distinct:
  3 Leaves crenate-serrate, pinnately veined; erect shrub or tree; bracts subpersistent; corolla urceolate, thin, the lobes simple, imbricate   **2. Vaccinium**
  3 Leaves entire, subplamately veined; climbing or scrambling shrub; bracts caducous; corolla campanulate, fleshy, the lobes flanged, valvate-induplicate   **3. Symphysia**

1. LYONIA Nutt. (1818) nom. cons.

Corolla 4·5–6·5 mm. long; pedicels up to 7 mm. long; all young parts including flowers scaly, axes and petioles whitish puberulous; flowers mostly 5-merous   1. *jamaicensis*
Corolla 3–3·5 mm. long; pedicels up to 4 mm. long; indumentum as above but scales less persistent, axes glabrescent; flowers mostly 4-merous   2. *octandra*

1. L. jamaicensis (Sw.) D. Don in G. Don, Gen. Syst. **3**: 831 (1834).—*Andromeda jamaicensis* Sw. (1788).

Shrub or tree 3–12 m. high with red bark; young vegetative growth rusty-scaly; leaves narrowly elliptical to oblanceolate, obtuse to shortly and bluntly acuminate at tip, acute at base, the margin entire or irregularly sinuate, up to 10·5 cm. long and 5·5 cm. broad, usually much smaller on flowering branches, the venation slightly prominent; flowers fragrant; stamens 10; ovary setulose-hispid; capsule up to 6 mm. long, with persistent style and septa; seeds linear.

Locally common, particularly in St. Andrew and St. Thomas in thickets and woodlands on non-calcareous soils; 1000–4000 (–5000) feet; fl. May–Nov, Jan, fr. July–Jan. *A 8267! 11191! H 12206! P 23832!* Endemic.

2. L. octandra (Sw.) Griseb., Fl. Br. W.I.: 142 (1860).—*Andromeda octandra* Sw. (1788).

Shrub or tree like the last, 1·2–9 m. high; all flower-parts and fruit smaller; stamens 8.

Rare and local (St. Andr., St. Thom.), in montane woodlands and thickets; 4750–7400 feet; fl. June–July, fr. Aug. *ACH 893! H 5812! Weaver 860!* Endemic.

2. VACCINIUM L. (1753)

1. V. meridionale Sw., Nov. Gen. & Sp. Pl.: 63 (1788).—Bilberry.

Shrub 1–4 m. or tree to 13 m. high; bark reddish, flaking; branchlets pubescent; leaves elliptical to ovate, acute, mucronate, rounded at base, margin crenate, 2–4 (–4·5) cm. long, 0·8–2 (–2·5) cm. broad, glabrous except petiole and a few short hairs on midrib adaxially, rather leathery; racemes pubescent, up to about 4 cm.

long; pedicels 1–4 mm. long; calyx-lobes 4, distinct, shortly triangular; corolla white or tinged pink or sometimes brilliant carmine, 5–7 mm. long; stamens 8; anther-sacs each with a hornlike bristle introrsely; berries black when ripe, subspherical, about 8 mm. in diameter.

Locally common (St. Andr., Port., St. Thom.), in thickets and woodland margins on steep rocky banks; 3000–7400 feet; fl. Feb–May, Sept–Dec, fr. Feb–May Aug, Nov. *A 6344! 10683! H 9133! P 22277!* Colombia, Venezuela, Peru.

3. SYMPHYSIA C. Presl (1827)

1. S. racemosa (Vahl) Stearn in Taxon **21**: (1972).—*Hornemannia racemosa* Vahl (1810).

Shrub scrambling and climbing to 3 m.; leaves broadly ovate, shortly narrowed from a rounded base, acuminate, up to 15 cm. long and 7 cm. broad, the main lateral veins with about 3 pairs radiating from near the base, rather leathery, aromatic; racemes glabrous, terminal or subterminal, 3–4 cm. long; pedicels about 1·5 cm. long in flower; calyx cupular, shortly and broadly 6-lobed, the lobes minutely apiculate; corolla green, 5–6 mm. long, the tube nearly as broad; anthers without bristles; berries purple, obovoid, about 12 mm. long.

Very rare (Port.), in wet thickets on limestone; about 2500 feet; fr. Mar, Aug. *A 9142! P 16289!* Venezuela(?), Guyana(?), Greater Antilles, Guadeloupe, Dominica, Martinique.

150. THEOPHRASTACEAE

Trees or shrubs. Leaves alternate, simple, entire or toothed, punctate, without stipules. Flowers usually bisexual, actinomorphic, in axillary or terminal racemes, rarely solitary. Perianth biseriate, (4–) 5-merous; calyx persistent, segments free or rarely connate, imbricate; corolla gamopetalous, rotate to funnel-shaped, the lobes imbricate. Stamens (4–) 5, epipetalous low on corolla-tube opposite the lobes; filaments usually free; anthers introrse or extrorse, 2-locular, opening lengthwise; staminodes 5, opposite the sepals, globose or petaloid. Ovary superior, 1-locular; ovules numerous, anatropous, embedded in mucilage on a free central placenta; style simple, usually long but sometimes obsolete; stigma discoid or capitate. Fruit a drupe or berry; seeds usually several, rarely solitary; endosperm present.

4 genera with about 100 species in the American subtropics and tropics.

1. JACQUINIA L. (1759)

Leaves leathery, alternate or opposite; flowers white or yellowish, fragrant; anthers extrorse; staminodes petaloid:

1 Inflorescence umbelliform, the axis very short; calyx 3·5–6 mm. long; corolla-tube 6–9 mm. long; plants of rocky limestone in the interior:
  2 Calyx 5–6 mm. long; corolla-tube 8–9 mm. long   1a. *macrantha* var. *macrantha*
  2 Calyx 3·5–4 mm. long; corolla-tube 6–7 mm. long   1b. *macrantha* var. *clarendonensis*
1 Inflorescence racemose; calyx 1·5–3 mm. long; corolla-tube 2·5–5 mm. long; coastal plants:
  3 Leaves large, mostly more than 5 cm. long and 2 cm. broad   2. *arborea*
  3 Leaves small, up to 4·5 (–5·5) cm. long and 2 (–2·5) cm. broad:
    4 Corolla about 2·5 mm. long; calyx about 1·5–2 mm. long; lower surface of leaf when dried striate with fine longitudinal wavy lines   3. *proctorii*
    4 Corolla 3·5–5 mm. long; calyx about 2·5 mm. long; lower surface of leaf without fine lines as above   4. *keyensis*

1. J. macrantha Urb., Symb. Ant. **5**: 455 (1908).

1a. Var. macrantha.

Shrub or tree 1·8–3·6 m. high; leaves obovate, oblanceolate or elliptical, cuneate at base, obtuse or rounded, 4–7 cm. long, 1·5–4 cm. broad; petiole 3–6 mm. long; corolla creamy-white or buff, the lobes 4 mm. long; fruit not known.

Rare (Trel.), in crevices of precipitous rocks; 1500–2200 feet; fl. Sept–Nov. *H 9064! 9480!* Endemic.

1b. J. macrantha var. **clarendonensis** Stearn, unpublished.
Small tree to 4 m. high; leaves obovate to broadly obovate, narrowly cuneate at base, margins reflexed, 1·5–4 cm. long, 1–2·5 cm. broad; petiole about 2 mm. long; corolla-tube pinkish, the lobes yellowish-white, about 3 mm. long; fruit not known.
Rare (Clar.), on rocky limestone hilltop; about 2500 feet; fl. Dec. *P 11409! Stearn 8!* Endemic.

2. J. arborea Vahl, Eclog. Amer. **1**: 26 (1797).
Shrub or tree 1–6 m. high, branches and subtending leaves spiral or subverticillate; leaves obovate, cuneate at base, rounded at tip, 3–11 cm. long, 1–5·5 cm. broad; raceme up to about 8 cm. long; flowers fragrant; corolla white; ripe fruit red, 7–8 mm. in diameter.
Common in thickets, on dunes and near swamps, mostly on limestone near the sea; SL–50 feet; fl. Nov–Apr, fr. Dec–Feb, July–Sept. *A 8819! 11947! P 23478!* West Indies.

3. J. proctorii Stearn, unpublished.
Dense-crowned shrub about 2·5 m. high; leaves oblanceolate to narrowly obovate, tapered at base to short petiole, rounded at tip, margin revolute, 1·5–5·5 cm. long, 4–23 mm. broad; raceme 2–3·5 cm. long; flowers odorous; corolla light yellow turning brownish-orange.
Rare and local (St. Eliz.), on dry coastal limestone; 50–100 ft; fl. May. *P 15379! Stearn 1032!* Endemic.

4. J. keyensis Mez in Urb., Symb. Ant. **2**: 444 (1901).
Shrub or tree (0·5–) 1–6 m. high; leaves obovate to cuneate-spathulate, more or less emarginate, the margins deeply revolute, 1–3 cm. long, up to 1 cm. broad; racemes terminal, short; flowers fragrant; corolla white or yellowish; fruit obovoid-globose, apiculate, about 11 mm. long and 8 mm. broad.
Local (St. Ann, St. Mary, St. Thom.), on exposed coral limestone near the sea; SL–10 feet; fl. and fr. Nov–Mar. *A 6137! H 11678! T. L. Lewis 1298! P 11252!* Florida, Bahamas, Cuba, Hispaniola.

151. MYRSINACEAE

Trees or shrubs. Leaves mostly alternate, rarely whorled, simple, often leathery with linear or punctate resin-glands, without stipules. Flowers bisexual or unisexual, actinomorphic, often glandular, in clusters, cymes, spikes, racemes or panicles. Perianth biseriate; sepals 4–5 (–6), free or connate, imbricate or valvate, persistent; petals 4–5 (–6), fused or rarely free, corolla rotate or tubular, aestivation contorted, imbricate or valvate. Stamens as many as corolla-lobes and on the same radii; filaments usually adnate to the corolla and distinct; anthers 2-locular, introrse, opening lengthwise or by apical slits or pores, usually longer than the filaments. Ovary superior or more or less inferior, 1-locular, with few to numerous ovules on a basal or free-central placenta; the placenta often overcapping or surrounding the ovules; ovules semianatropous to semicampylotropous; style short or none; stigma simple or lobed. Fruit a drupe with stony endocarp or a berry. Seed with cylindrical embryo and copious endosperm.
32 genera and about 1000 species, in the subtropics and tropics.

1 Flowers in umbellate clusters on very short scaly axillary spurs; style short or wanting; stigma conspicuous **1. Myrsine**
1 Flowers usually in terminal racemes or panicles, sometimes inflorescences axillary and then long-pedunculate:

2 Flowers bisexual with stamens and style both well developed in the same flower; corolla mostly 6–12 mm. long; ovules numerous, scattered in several series over the placenta   2. **Ardisia**
2 Flowers unisexual, dioecious, the male flowers with well developed exserted stamens, the female flowers with abortive included stamens but exserted styles; corolla 1·5–6 mm. long; ovules 3–4 in a single horizontal row on the placenta   3. **Wallenia**

1. MYRSINE L. (1753); Stearn (1969)[a]

Young shoots and petioles minutely pubescent; leaf-blades mostly narrowly elliptical and less than 2·5 cm. broad; corolla-lobes joined for the lower 0·5 mm. or more
                                                                                                    1. *coriacea*
Young shoots and petioles glabrous; leaf-blades mostly obovate or narrowly obovate and more than 2·5 cm. broad; corolla-lobes free to the base   2. *acrantha*

1. M. coriacea (Sw.) R. Br. ex Roem. & Schult. in L., Syst. Veg. ed. nov. **4**: 511 (1819).—*Samara coriacea* Sw. (1788). *Rapanea ferruginea* (Ruiz & Pav.) Mez (1901). Colic Wood.

Shrub about 2 m. or tree to 6 m. high; leaves cuneate at base, acute or obtuse at tip, 2–13 (–15) cm. long, 1·3–3 (–4) cm. broad, with brownish-red gland-dots; pedicels very short; flowers greenish, mostly if not all, functionally unisexual, the male flowers with ample pollen and reduced gynoecium, 2·5–3 mm. long, the female flowers with well developed gynoecium bearing a large fleshy green stigma and anthers without pollen, 1·5–2 mm. long, probably also dioecious; mature fruit black, about 4 mm. in diameter.

Common in woodlands and thickets on limestone, shale and serpentine and also on open poorly drained savannas; 2200–7400 feet; fl. Dec–Mar, July–Sept, fr. Mar–Apr, Aug. *A 12207! A. von der Porten! P 23118!* Mexico to Argentina, West Indies.

2. M. acrantha Krug & Urb. in Notizbl. Bot. Gart. Berlin **1**: 79 (1895).

Tree up to 6 m. high, glabrous except the bud-scales minutely ciliate; leaves cuneate at base, obtuse and often more or less emarginate at tip, 4·5–8 (–14) cm. long, 2–3 (–6) cm. broad, the larger leaves on non-flowering branches regrown from coppicing, punctate but not gland-lined; flowers as above, male about 2·5 mm. long, female about 2 mm. long, sometimes 4-merous; fruit 4–4·5 mm. in diameter when dry.

Occasional in open woodlands on limestone; 2400–4500 feet; fl. Oct–Nov, fr. Dec–Jan, June. *A 10114! H 8295! P 23130!* Probably also in Cuba and Hispaniola.

The sexuality and distribution of the flowers of the two Jamaican species of *Myrsine* require further study.

2. ARDISIA Sw. (1788) nom. cons.; Stearn (1969)[a]

1 Inflorescences all axillary; pedicels 1–4 cm. long, clustered almost umbellately at the end of the peduncle; anthers 5–6 mm. long   1. *solanacea*
1 Inflorescences terminal and sometimes also axillary; pedicels mostly 1–7 mm. long, sometimes to 12 mm. long but then racemosely arranged along the lateral peduncle; anthers 1·5–5 mm. long:
2 Lateral peduncles compound, i.e. bearing 3–4 secondary peduncles each umbellately 2–6-flowered; leaves 7–19 cm. long, the blades narrowly obovate to oblanceolate; filaments of stamens 0·5 mm. long   2. *jamaicensis*
2 Lateral peduncles mostly simple, i.e. directly bearing 3–28 pedicelled flowers, the secondary branches if rarely present racemose; leaves mostly less than 12 cm. long; filaments 1–5 mm. long:
3 Leaves closely clustered at tips of shoots, less than 5·5 cm. long; inflorescence contracted, less than 4 cm. long and broad, the branches 3–6-flowered; pedicels 1–4 mm. long   3. *byrsonimae*
3 Leaves more widely spaced, some or all more than 5·5 cm. long; inflorescence more than 4 cm. long and broad, the branches usually many-flowered; pedicels up to 15 mm. long:
4 Inflorescence somewhat fastigiate, the densely many-flowered branches ascending at about 40°; corolla about 5 mm. long; anthers about 1·5 mm. long, much shorter than the filaments   4. *densiflora*

4 Inflorescence looser, the branches spreading at about 50–60°; corolla 7–12 mm. long;
  anthers 2·5–5 mm. long, equalling or longer than the filaments:
5 Pedicels 1–4 (–5) mm. long; leaf-blades with well-marked reticulate venation
                                         5. *dictyoneura*
5 Pedicels 4–15 mm. long; leaf-blades not markedly reticulate:
6 Sepals 1–2 mm. broad; anthers 3–5 mm. long, apically dehiscent   6. *tinifolia*
6 Sepals 2·5–4 mm. broad:
7 Anthers 3–3·5 mm. long, laterally dehiscent almost to the base; sepals ciliate
                                        7. *urbanii*
7 Anthers 4·5–5 mm. long, apically dehiscent; sepals almost or quite glabrous
                                      8. *brittonii*

1. **A. solanacea** Roxb., Pl. Coast Coromand. **1**: 27, t. 27 (1795).—*A. humilis* of Griseb. (1861), not Vahl (1794). Blackberry, Craing Craing.
Shrub up to 3 m. or tree 4 m. high; leaves elliptical, cuneate at base, shortly acuminate, fleshy, more or less obviously reticulate-veined and pellucid-dotted when dry, up to 17 cm. long and 7 cm. broad; peduncle up to 3 (–6·5) cm. long; sepals 4 mm. long; corolla light mauve to rose; fruit oblate, pink at first, finally black, the juice purple, the flesh white, about 8 mm. high and 11 mm. broad.
Locally common, naturalized on streambanks and forming secondary thickets in moderately wet areas; 40–1500 (–4100) feet; fl. Jan–Sept, fr. July–Feb. *A 10094! 11482! H 11637! P 16546!* Native of S.E. Asia, now widespread in cultivation.

2. **A. jamaicensis** Lundell in Wrightia **4** (3): 120 (1969).—*A. compressa* of Stearn (1969), not Kunth (1819).
Shrub up to 3 m. or tree to 6 m. high; leaves oblong-elliptical to oblanceolate, cuneate at base, acuminate; corolla white or pale rose with light brown flecks.
Uncommon and local (West., Han., St. James), in thickets on rocky limestone and in clay beside streams; 200–2250 feet; fl. Feb–Apr. *A 13208! H 10253! P 16166!* Endemic.

3. **A. byrsonimae** Stearn in Bull. Brit. Mus. (Nat. Hist.) Bot. **4** (4): 162, f. 23 A–E, t. 8b (1969).
Shrub 1·5 m. or tree up to 5 m. high; young twigs rusty-puberulent; leaves obovate, cuneate at base, rounded at tip and sometimes emarginate, 3–5·5 cm. long, 1·3–3·5 cm. broad, leathery, with obscure venation; calyx about 3 mm. long: corolla pink, 7–8 mm. long; fruit broadly ellipsoidal, 4–5 mm. high, 5–6 mm. broad.
Rare and local (Clar.), on rocky limestone hilltop; about 2500 feet; fl. Jan, Aug, fr. Dec–Jan, Aug. *H 12799! P 11408! Stearn 9!* Endemic.

4. **A. densiflora** Krug & Urb. in Notizbl. Bot. Gart. Berlin **1**: 79 (1895).
Tree 6–8 m. high; leaves obovate to oblong-elliptical, cuneate at base, obtuse or very shortly and bluntly acuminate, smooth, 5·5–10·5 cm. long, 2–4·5 cm. broad; pedicels 1–6 mm. long; calyx 2·5 mm. long; corolla white, 5 mm. long; flowers fragrant; fruit 5–6 mm. broad.
Uncommon and local (St. Andr.), in montane woodlands and thickets; 3000–5000 feet; fl. Nov–Dec, fr. Jan–Mar, July. *H 6578! 10029! P 24511!* Mexico, British Honduras.

5. **A. dictyoneura** Urb., Symb. Ant. **6**: 28 (1909).
Shrub 3–5 m. or tree up to 15 m. high; young twigs densely pilose; leaves obovate to elliptical, obtuse or rounded at tip, rigid, conspicuously veined, 5–10 cm. long, 2–4·5 cm. broad; calyx about 2·5 mm. long; corolla pink, about 12 mm. long; fruit 6–6·5 mm. high, 6·5–8 mm. broad.
Occasional in the central and western parishes, in woodland on limestone; 500–3000 feet, fl. Aug, Nov, fr. Oct–Mar. *H 10345! P 11316! 22979! Stearn 351!* Endemic.

6. **A. tinifolia** Sw., Nov. Gen. & Sp. Pl.: 48 (1788).—*A. coriacea* Sw. (1788). *A. harrisiana* Mez (1901). *A. troyana* Urb. (1908).
Shrub to about 3 m. or tree up to 10 m. high; leaves obovate to elliptical, acute to obtuse or rounded at tip, venation obscure or evident, 3–14 cm. long, 1·5–7 cm. broad, leathery; sepals 1·5–3 mm. long; corolla mauve to rose-pink, 7–8 mm. long; ripe fruit black or purplish-black.

Common in thickets and woodlands mainly on limestone, in all except the wettest and driest areas; SL–3500 feet; fl. June–Dec, fr. most of the year. *A 8424! 12579! H 8741! Loveless UCWI 1198! P 11062!* Endemic. Stearn (1969)ᵃ has discussed the variability of this species and has recognized tentatively four unnamed varieties representing more or less distinct geographical populations distinguished by differences of leaf-size, shape and texture, number of flowers, shape of sepals and length of anthers.

7. A. urbanii Stearn in Bull. Brit. Mus. (Nat. Hist.) Bot. **4** (4): 160, f. 22 (1969).—
A. rosea Urb. (1908), not King & Gamble (1905).

Glabrous tree 6–8 m. high; leaves elliptical to oblong, shortly and obtusely acuminate at tip, 7·5–15 cm. long, 3·5–6 cm. broad; inflorescence about 12 cm. broad; pedicels 5–11 mm. long; sepals 3–5 mm. long; petals rosy-pink, 12 mm. long, revolute.

Rare and local (Trel.), in woodland on limestone; about 1600 feet; fl. July–Sept, fr. Sept. *H 9418! 9419!* Endemic.

8. A. brittonii Stearn in Bull. Brit. Mus. (Nat. Hist.) Bot. **4** (4): 159, f. 21, t. 8a (1969).

Tree up to 4·5 m. high; young twigs rusty-pulverulent; leaves oblong-elliptical, cuneate at base, obtuse at tip, glabrous, leathery, 8–17 cm. long, 3–8 cm. broad; inflorescences about 15 cm. long and 10 cm. broad with branches up to 9 cm. long; pedicels up to 15 mm. long; sepals 4–4·5 mm. long; petals rosy-pink, about 9 mm. long, revolute.

Rare and local (St. Thom.); fl. Sept. *H & B 10558!* Endemic.

3. WALLENIA Sw. (1788) nom. cons.; Stearn (1969)ᵃ

1 Leaves with base broadly rounded and slightly cordate, paired or in whorls of three, fairly large; petiole less than 6 mm. long or lacking:
  2 Leaves narrowed towards the acute or obtuse apex, 3–5 (–7) cm. broad
          1. *subverticillata*
  2 Leaves rounded at apex, 6–17 cm. broad:
    3 Leaves sessile   2. *clusioides*
    3 Leaves with 3–6 mm. long petiole   3. *elliptica*
1 Leaves with base cuneate or attenuate, or if rounded the petiole then 1 cm. or more long, usually alternate, only occasionally paired or whorled:
  4 Leaves pinnately veined beneath, i.e. without raised cross-veins between the main lateral veins:
    5 Lamina usually abruptly contracted at base into petiole; inflorescence usually minutely pubescent   4. *grisebachii*
    5 Lamina gradually narrowed into petiole; inflorescence glabrous:
      6 Leaf-blades mostly less than 12 cm. long and 6 cm. broad; petiole up to 1 cm. long; fruits vertically ribbed   5. *laurifolia*
      6 Leaf-blades up to 25 cm. long and 11 cm. broad, mostly more than 12 cm. long and 6 cm. broad; petiole 1–2 cm. long:
        7 Fruits vertically ribbed   6. *sylvestris*
        7 Fruits not ribbed   7. *discolor*
  4 Leaves reticulate-veined beneath, with slightly raised cross-veins between the main lateral lines:
    8 Petiole up to about 8 mm. long; leaf-blades mostly less than 11 cm. long:
      9 Leaf-blade mostly broadest at the middle and gradually narrowed from there to apex:
        10 Petiole 1–4 mm. long; blade inconspicuously punctate beneath   8. *fawcettii*
        10 Petiole 5–8 (–10) mm. long; blade conspicuously punctate beneath   9. *venosa*
      9 Leaf-blade mostly broadest beyond the middle with apex rounded or obtuse:
        11 Leaf-blades 2–8 cm. long, less than 3·5 cm. broad   10. *crassifolia*
        11 Leaf-blades mostly larger, 5–11 cm. long:
          12 Leaves coriaceous, firm, greyish or glaucescent; female calyx about 1·4 mm. long   11. *corymbosa*
          12 Leaves chartaceous, thinner:
            13 Leaves drying light green; female calyx about 1·5 mm. long   12. *purdieana*
            13 Leaves drying brownish; female calyx 1 mm. long   13. *xylosteoides*
    8 Petiole 8–30 mm. long; leaf-blades mostly more than 11 cm. long:
      14 Leaf-blades nearly three times as long as broad, mostly narrowly elliptical:
        15 Base of blade narrowly cuneate or attenuate   14. *erythrocarpa*
        15 Base of blade rounded or abruptly contracted   4. *grisebachii*

14 Leaf-blades about twice as long as broad, mostly elliptical or narrowly obovate:
  16 Lateral veins spreading outwards from the midrib at an angle of about 40°
                                      15. *punctulata*
  16 Lateral veins spreading outwards from the midrib at an angle of about 50–70°:
    17 Male flower with calyx about 3 mm. long, corolla 5–6 mm. long; persistent
      fruiting calyx about 2·5 mm. long   16. *calyptrata*
    17 Male flower with calyx 1·5–2·5 mm. long, corolla 2·5–4 mm. long; persistent
      fruiting calyx not more than 1·5 mm. long:
      18 Leaves usually densely blackish-punctate and somewhat mottled beneath,
        strongly veined, the base cuneate   9. *venosa*
      18 Leaves inconspicuously punctate and not mottled beneath, lightly veined, the
        base cuneate or rounded or abruptly contracted:
        19 Leaves drying light green; inflorescence glabrous, loose; male flower with
          corolla yellowish, not punctate; style about 0·8 mm. long   12. *purdieana*
        19 Leaves drying brownish; inflorescence minutely pubescent, compact; male
          flower with corolla reddish, distinctly punctate; style about 1·5 mm. long
                                      4. *grisebachii*

1. **W. subverticillata** (Britton) Ekman ex Urb., Symb. Ant. **9**: 412 (1925).—
Petesioides subverticillata Britton (1910).
Shrub 1–3 m. high; leaves very variable, mostly oblong-lanceolate, acute, sub-sessile, 5–20 cm. long; bracts in inflorescence oblong, light green, 5–7 mm. long; calyx of male flowers 2–2·5 mm. long, corolla white, about 2·5 mm. long; calyx of female flowers about 1 mm. long, corolla a little longer; fruit bright purplish-red, 5·5 mm. broad.
Local (Port., St. Thom.), in woodlands on limestone; 1000–2200 feet; fl. Dec–Mar, fr. Mar. *A 12966! H & B 10694! P 22102!* Cuba, Hispaniola.

2. **W. clusioides** (Griseb.) Mez in Urb., Symb. Ant. **2**: 411 (1901).—*Ardisia clusioides* Griseb. (1861), partly.
Glabrous shrub or tree 2–6 m. high; leaves obovate to obovate-elliptical, auriculate at base, leathery, 15–31 cm. long; pedicels very short, the flowers densely clustered and umbellate to capitulate; calyx of male flowers 1·5–2 mm. long, corolla about 2·5 mm. long, greenish-yellow; female flowers unknown.
Uncommon, in western parishes, in woodland on rocky limestone; 1100–2500 feet; fl. Jan–Mar. *H 9177! 10210! P 10028!* Endemic.

3. **W. elliptica** Urb., Symb. Ant. **6**: 30 (1909).
Plant glabrous; leaves elliptical or ovate-elliptical, broadest at the middle, thinly leathery, 12–20 cm. long, 6–8·5 cm. broad; calyx and corolla of female flowers about 1·5 mm. long, 4-merous; fruiting pedicels up to 1·5 mm. long; fruit globose, about 3·5–4 mm. in diameter.
Rare (St. Ann). *Britton & Hollick 2776! Prior.* Endemic.

4. **W. grisebachii** Mez in Urb., Symb. Ant. **2**: 411 (1901).
Shrub 2–3 m. or tree to 5 m. high; leaves elliptical, shortly acute to obtuse at tip, up to 18 cm. long and 7·5 cm. broad; inflorescence compact, many-flowered, 3–4-pinnate; pedicels hardly 2 mm. long; flowers 4-merous, brownish-red; calyx of male flowers 2 mm. long, corolla 4 mm. long; calyx of female flowers 1·5–2 mm. long, corolla about 1·5 mm. long.
Occasional in the central parishes, in woodland on rough limestone; 2000–2900 feet; fl. Oct–Apr. *A 11766! Britton & Hollick 2776! P 8247! Robertson UCW1 5470!* Endemic.

5. **W. laurifolia** Sw., Nov. Gen. & Sp. Pl.: 31 (1788).—*W. clusiifolia* Griseb. (1861).
Shrub or tree 1–5 m. high; leaves elliptical to obovate-elliptical, acute or obtuse; petiole 3–10 mm. long; flowers yellowish, fragrant; calyx of male flowers 1·5–2·5 mm. long, corolla 2·5–3 mm. long; calyx and corolla of female flowers about 1·5 mm. long; fruit bright red, about 3 mm. broad when dry.
Widespread but not common, on limestone cliffs and in thickets on exposed rocky banks; 50–2600 feet; fl. Sept–Apr, fr. Sept–May. *A 12519! H 6880! H & B 10613! P 24151! 24175! Stearn 226!* Cuba, Hispaniola.

6. W. sylvestris Urb. in Fedde, Repert. Sp. Nov. **13**: 468 (1915).

Shrub 4 m. or tree up to 13 m. high; leaves elliptical to narrowly elliptical, rarely obovate, obtuse or rounded at apex, up to 22 cm. long and 9 cm. broad; petioles 8–15 mm. long; calyx and corolla of female flowers about 2 mm. long; fruit subglobose, 4–4·5 mm. broad when dry.

Rare and local (Port., St. Thom.), in swampy parts of forest on limestone; 1500–2500 feet; fl. and fr. Mar. *H & B 10690! 10720! P 10477!* Endemic.

7. W. discolor Urb., Symb. Ant. **6**: 29 (1909).

Leaves elliptical to obovate-elliptical, obtuse at apex, leathery, 18–25 cm. long, 7–11 cm. broad; pedicels up to 3 mm. long; calyx of female flower 1·3 mm. long; fruit globose, 3 mm. broad.

Very rare (? St. Ann), known only from the type, *Prior.* Endemic.

8. W. fawcettii Mez in Urb., Symb. Ant. **2**: 408 (1901).

Glabrous shrub about 3 m. high; leaves elliptic-lanceolate, cuneate at base, acute at apex, chartaceous, 3·5–9 cm. long, 1·5–3·5 cm. broad; flowers 4-merous on pedicels hardly 1·5 mm. long; calyx of male flowers 1·5 mm. long, corolla 2·5 mm. long; calyx of female flowers about 1 mm. long.

Rare and local (Port., St. Thom.), in montane woodland; 5000–5500 feet; fl. Nov. *H 5422!* Endemic.

9. W. venosa Griseb., Fl. Br. W.I.: 394 (1861).

Glabrous straggling shrub 2–5 m. or tree to 6 m. high; leaves elliptical to obovate, cuneate at base, obtuse or rounded at apex, rather leathery, 7–11 (–14) cm. long, 3–5 cm. broad; flowers white or greenish, fragrant, on pedicels about 3 mm. long; calyx of male flowers 1·5–2 mm. long, corolla about 3 mm. long; calyx of female flowers 0·8 mm. long, corolla about 1 mm. long; fruit globose, 5–5·5 mm. broad when dry.

Occasional, in woodlands and thickets on limestone; SL–3000 feet; fl. Sept–Mar, fr. Aug, Nov–Mar. *A 11845! H 8768! 10032! P 18400!* Endemic.

10. W. crassifolia Mez in Urb., Symb. Ant. **2**: 409 (1901).

Shrub 1·5–2·5 m. high; branches thick, glabrous; leaves elliptical to narrowly elliptical, broadly cuneate at base, rigid, prominently veined; petioles about 5 mm. long; pedicels 3–4 mm. long; calyx of male flowers about 2 mm. long, corolla 4 mm. long; calyx of female flowers about 1 mm. long; fruit globose, about 4 mm. broad when dry.

Locally frequent (St. Andr., St. Thom.), in montane thickets and woodlands; 4750–7000 feet; fl. Jan–Feb, fr. Jan. *A 10666! J.P. 1035! P 9524! Weaver 1853!* Endemic.

11. W. corymbosa Urb., Symb. Ant. **5**: 457 (1908).

Glabrous shrub to 3 m. or slender tree up to 13 m. high; leaves narrowly obovate to obovate-elliptical, cuneate at base, acute, obtuse or rounded at apex, 6–11 cm. long, 3·3–5·5 cm. broad; petioles 4–8 mm. long; inflorescence corymbose, about 5 cm. broad; pedicels up to 7 mm. long; flowers light yellow or white; calyx of male flowers 1·5–2 mm. long, corolla 3–4 mm. long; calyx of female flowers about 1·4 mm. long, corolla about 2·5 mm. long; fruit about 3·5 mm. broad.

Locally common (Clar., Trel., St. Ann), in woodlands on limestone; 2000–2800 feet; fl. Dec, fr. July, Nov–Mar. *H 12781! 12819! H & P 15170!* Endemic.

12. W. purdieana Mez in Urb., Symb. Ant. **2**: 408 (1901).

Glabrous shrub 3 m. or slender tree up to 8 m. high; leaves elliptical, shortly cuneate at base, rounded at apex, up to about 9 cm. long and 4·5 cm. broad; inflorescence shortly paniculate; pedicels up to 1·5 mm. long; flowers 4-merous, light green; calyx of male flowers about 2 mm. long, corolla 3 mm. long; calyx of female flowers about 1·5 mm. long, corolla 1·5–2 mm. long; fruit pink, globose, vertically ribbed, about 4 mm. broad when dry.

Occasional in west-central and western parishes, in woodland on limestone; 500–3000 feet; fl. Aug–Jan, fr. Sept–Apr. *A 9866! H 9096! 10305! HPS 14663! P 22992!* Endemic.

13. W. xylosteoides (Griseb.) Mez in Urb., Symb. Ant. 2: 409 (1901).—*Ardisia xylosteoides* Griseb. (1861), partly (Jamaican plant only).

Leaves elliptical, rounded or broadly cuneate at base, obtuse at apex, 5–11 cm. long, 2–5 cm. broad; inflorescences few-flowered; pedicels up to 6 mm. long; female flowers 5-merous with rounded sepals, corolla a little longer.

Rare (St. Ann), on wooded limestone slopes; about 1500 feet. *HPS 14608! J.P. 1118! Prior.* Endemic.

14. W. erythrocarpa Urb., Symb. Ant. 6: 29 (1909).

Glabrous shrub 4 m. or tree up to 7 m. high; leaves elliptical to oblong-elliptical, acute or rather obtuse at apex, leathery, prominently reticulate-veined, 11–21 cm. long, 4·5–7·5 cm. broad; petioles 1–3 cm. long; inflorescences corymbose, many-flowered; pedicels up to 3·5 mm. long; female flowers 4-merous, calyx scarcely 1 mm. long; ripe fruit red, globose, 4–4·5 mm. in diameter.

Rare and local (West., Han.), in woodlands on limestone; 500–1800 feet; fl. Feb, fr. Mar. *H 7080! 10343!* Endemic.

15. W. punctulata Urb., Symb. Ant. 7: 322 (1912).

Leaves obovate to obovate-elliptical, cuneate at base, obtuse at apex, leathery, densely prominently reticulate, 11–15 cm. long, 5·5–7 cm. broad; petioles 8–12 mm. long; pedicels 1·5–4 mm. long; female flowers 4-merous, sepals 1·5 mm. long, corolla 1·7 mm. long; fruit globose, 3·5 mm. in diameter.

Known only from the type (Manch.), *Britton 3281.* Endemic.

16. W. calyptrata Urb., Symb. Ant. 5: 458 (1908).

Shrub 1·5–2·5 m. or tree up to 7 m. high; leaves obovate to narrowly obovate, cuneate at base, obtuse or rounded at apex, thick, prominently veined, 8–14 cm. long, 3·5–6 cm. broad; petioles 8–12 mm. long; pedicels up to 8 mm. long; calyx of male flowers yellow, 3–3·5 mm. long, corolla 5–6 mm. long; calyx of female flowers 2·5 mm. long, corolla 3 mm. long; ripe fruit red, about 3 mm. in diameter.

Local (St. Andr., Port., St. Thom.), in montane woodland on sheltered slopes; 4250–7000 feet; fl. July, Oct–Nov, fr. July. *A 7440! Cornman 837! P 9683!* Endemic.

Stearn (1969)[a] has discussed the problems of defining taxa in *Wallenia* in Jamaica. The greatest need is for the accurate correlation of male and female flowers of the same taxon. Collectors should search carefully for plants of both sexes in the same area and those with appropriate facilities should attempt to bring to flowering maturity a population of plants derived from fruits of the same tree.

152. PRIMULACEAE

Annual or perennial herbs. Leaves alternate, opposite or whorled, without stipules. Flowers bisexual, actinomorphic or rarely slightly zygomorphic, solitary, umbellate, racemose or paniculate. Perianth mostly biseriate, typically 5-merous; calyx gamosepalous, persistent; corolla gamopetalous, rotate, campanulate or funnel-shaped, (4–) 5 (–7)-lobed, rarely wanting; aestivation imbricate. Stamens epipetalous, inserted on the same radii as the corolla-lobes; anthers 2-locular, opening lengthwise, introrse. Ovary superior or rarely half-inferior by fusion with calyx-tube, 1-locular; ovules numerous, semianatropous, on a free central placenta; style solitary; stigma capitate. Fruit a 5-valved or transversely dehiscent capsule; seeds angular with a small straight embryo and copious endosperm.

28 genera and about 800 species, most numerous in north temperate regions.

1. ANAGALLIS L. (1753)

Leaves opposite or alternate; capsule globose, circumscissile.

Corolla bright blue or red, fringed with glands; leaves all opposite; pedicels recurved in fruit; flowers axillary   **1.** *arvensis*

Corolla white, not fringed with glands; leaves alternate; pedicels erect in fruit; flowers in terminal racemes   2. *pumila*

1. A. arvensis L., Sp. Pl. **1**: 148 (1753).—Blue or Scarlet Pimpernel.

Annual weedy herbs with slender quadrangular winged erect or trailing branches, up to about 40 cm. high; leaves broadly ovate, sessile, cordate at base, obtuse, punctate, up to 2·5 (–3) cm. long and 1·5 cm. broad; pedicels to 2·5 cm. long in fruit; sepals lanceolate, scabrid-keeled, hyaline-margined, 3–4 mm. long; corolla (4–) 5-merous, scarlet, salmon-pink or deep blue with crimson around a white eye; capsule about 4 mm. in diameter.

Locally common, the blue variant more frequent than the red, in cultivations, pastures and on roadsides and open waste ground; 2600–5000 feet; fl. and fr. Feb–July. *A 6423! 6619! H 9201! P 23854! 24765!* Native of W. Europe, now widespread in both hemispheres; Hispaniola. The taxonomic distinction of the blue, red and other corolla-colour variants within this species has been made but the exact relationships of Jamaican plants to those in their native areas remains to be determined.

2. A. pumila Sw., Nov. Gen. & Sp. Pl.: 40 (1788).

Slender-stemmed much-branched glabrous procumbent annual herb, up to 9–15 cm. high; some lower leaves opposite or subopposite, sessile or shortly petioled, elliptical or broadly elliptical, acute at both ends, entire, up to 8 mm. long and 4 mm. broad; pedicels up to 6 mm. long in fruit; sepals lanceolate, aristate, 2–3 mm. long; capsule 2·5 mm. in diameter; seeds 3-angled, brown, minutely rugose.

Very rare (Clar.), in marshy places; known in Jamaica only from the type, *Swartz*. General in the subtropics and tropics.

153. PLUMBAGINACEAE

Perennial herbs or shrubs, sometimes climbers. Leaves alternate, without true stipules. Flowers bisexual, actinomorphic, bracteate, in heads, spikes, racemes or panicles of cymules or cincinni. Perianth biseriate, calyx tubular, often ribbed and folded in bud, 5-lobed, sometimes with smaller secondary lobes; corolla gamopetalous, often deeply divided into 5 imbricate and contorted lobes. Stamens 5, hypogynous or perigynous, opposite the corolla-lobes; anthers 2-locular, opening lengthwise. Disk absent. Ovary superior, 1-locular; ovule solitary, anatropous, pendulous from a basal funicle; styles 5, opposite the sepals, free or connate, sometimes heterostyle. Fruit an indehiscent utricle or a tardily dehiscent circumscissile capsule. Seed with granular endosperm or endosperm lacking; embryo straight.

10 genera with 350 species mostly of semi-arid regions of the Old World.

1. PLUMBAGO L. (1753)

Shrubs with broad non-rosulate leaves; inflorescence racemose (of cymules reduced to 1 flower); pedicels short; calyx with stalked glands; filaments free from corolla but shortly connate at base; pollen monomorphic.

1 Calyx without glands on lower half or one-third, about 13 mm. long; inflorescence minutely pubescent, short, the flowers crowded on 6 cm. or less of the axis; corolla light blue or rarely white, tube about 3·5 cm. long, lobes obovate, about 13 mm. long; widely cultivated; native of S. tropical Africa; Blue Plumbago (*P. capensis* Thunb.)
                                                                   P. auriculata Lam.
1 Calyx uniformly glandular almost to base, about 8–10 mm. long; inflorescence glabrous, elongated; corolla smaller, not blue:
  2 Corolla rose-coloured, the tube 2–3 cm. long, the lobes broadly elliptical, 10–12 mm. long; cultivated; native of S. Asia; Red Plumbago   *P. indica* L.
  2 Corolla white, the tube 1·5–2 cm. long, the lobes obovate, 5–7 mm. long; wild in Jamaica   **1.** *scandens*

1. P. scandens L., Sp. Pl. ed. 2, **1**: 215 (1762).—Wild Plumbago.

Straggling shrubby herb to 1·2 m. or more high; leaves broadly ovate to elliptical or lanceolate, more or less acuminate, tapered to petiole, up to 10 (–12) cm. long and 6 cm. broad, usually smaller; corolla white, sometimes drying reddish, the lobes mucronate; anthers exserted.

Rather common in hedgebanks, on fences and in thickets and waste places on sandy or gravelly soil; SL–1250 feet; fl. and fr. most of the year. *A 6165! P 23047! Robertson UCWI 9148!* Texas, Arizona, Florida, Mexico to Argentina, Bahamas, West Indies.

154. SAPOTACEAE

Trees or shrubs, often with milky sap. Leaves usually alternate, rarely opposite, simple, entire or rarely toothed; stipules absent or early caducous. Flowers usually bisexual, sometimes female only in polygamous specimens, actinomorphic, solitary, clustered or in racemes in the axils or at leafless nodes. Perianth 2–several-seriate; sepals 5, spirally arranged in one whorl or 4–8 in 2 whorls, free or shortly united; corolla gamopetalous with a long or short tube, with 4–10 lobes in 1–3 series, each lobe sometimes with 2 dorsal appendages or lateral lobules, usually imbricate. Stamens epipetalous, free, fertile opposite corolla-lobes; anthers 2-locular, opening lengthwise; staminodes often present, alternating with corolla-lobes and fertile stamens, often petaloid. Ovary superior, 4–12-locular; ovules 1 in each loculus, anatropous, axile in basal-ascending, middle or apical positions; style simple; stigma small, sometimes lobed. Fruit a berry. Seeds 1–few, with hard shiny testa and often a large hilum; embryo straight; endosperm oily when present.

About 800 species in some 40 genera in the subtropics and tropics.

1 Sepals in 2 distinct series, 6 (3 + 3) or 8 (4 + 4), 4–10 mm. long; corolla 5–10 mm. long, the lobes with or without appendages; staminodes present, petaloid; seeds with endosperm and a long lateral hilum; leaves with obscure secondary veins, glabrous or glabrescent; fruit brown, scurfy   1. **Manilkara**
1 Sepals usually imbricate or spiral, (4–) 5 (–9), not in 2 distinct series, except sometimes in *Pouteria* decussate (2 + 2):
  2 Staminodes absent; corolla-lobes without appendages; secondary leaf-veins parallel to the lateral primaries; ovary pubescent; hilum rather short; endosperm present   2. **Chrysophyllum**
  2 Staminodes present:
    3 Corolla-lobes with lateral appendages; hilum small, basilateral; endosperm present or absent   3. **Bumelia**
    3 Corolla-lobes without appendages:
      4 Primary lateral veins of leaves straight, the secondary hardly distinguishable from them and very numerous and close together; flowers up to about 14 in axillary clusters; fruit 1 (–3)-seeded, about 5 cm. long and 4 cm. broad; hilum long, lateral; endosperm present   4. **Micropholis**
      4 Primary lateral veins of leaves distinctly curved, the secondary more or less transverse and reticulate:
        5 Flowers numerous in axillary clusters; sepals about 2 mm. long; ovary glabrous or nearly so; hilum small; endosperm present; petiole pouched at junction with leaf-blade due to approximation of margins   5. **Mastichodendron**
        5 Flowers solitary or few together; sepals 5–6 mm. long; ovary hairy; hilum long, lateral; endosperm absent; petiole not pouched as above   6. **Pouteria**

1. MANILKARA Adans. (1763) nom. cons.

1 Leaf-blades mostly more than 6 cm. broad; flowers 1 or 2 together on pedicels 2–3 cm. long; appendages of corolla-lobes usually equalling or exceeding the lobes; fruit about 4 cm. in diameter   1. *excisa*
1 Leaf-blades rarely as much as 6 cm. broad; pedicels 1–2·5 cm. long:
  2 Flowers solitary; corolla-lobes without appendages; fruits mostly 5–8 cm. in diameter; tree mostly in cultivation   2. *zapota*
  2 Flowers in clusters of 3–6; corolla-lobes with appendages shorter than the lobes; fruits 1·5–2 cm. in diameter; wild tree   3. *sideroxylon*

1. **M. excisa** (Urb.) Gilly ex Cronquist in Bull. Torr. Bot. Club **72**: 555 (1945).— *Mimusops excisa* Urb. (1908). Sapodilla, Sappa.

Tree 10–30 m. or more high, with thick minutely pubescent branchlets; leaves narrowly elliptical to obovate, rounded and emarginate at tip, drying silvery-grey or greenish-yellow, 9–20 cm. long, 5·5–10 cm. broad; petioles 2·5–5 cm. long; calyx 8–9 mm. long, the 3 outer sepals rusty-pubescent, the inner greyish; corolla white, the tube about 4 mm. long, lobes about 5 mm. long; staminodes petaloid.

Local (St. James, Trel.), on wooded limestone hills; 1750–2500 feet; fl. Jan, fr. Nov. *A 12446! H 8813! 8961! P 22935! 23176!* Endemic.

2. M. zapota (L.) P. van Royen in Blumea **7**: 410 (1953).—*Achras zapota* L. (1753), partly. *Sapota achras* Mill. (1768). Naseberry.

Tree up to 15 m. or rarely more high; branchlets and young leaves beneath brownish-pubescent; leaves elliptical to oblanceolate, cuneate at base, rounded acute or shortly acuminate at tip, dark green, 3–13 cm. long, 1·5–5 cm. broad; petioles 6–30 mm. long; calyx 6–10 mm. long, the outer rusty-pubescent, the inner light green; corolla white, the tube 5–6 mm. long, lobes 3–4 mm. long; staminodes petaloid.

Cultivated, relict and escaped generally; 50–1250 feet; fl. and fr. all the year. *A 10915! H 8634! P 10387! 24295!* Native of tropical Amer., throughout the West Indies, introduced into the Old World tropics; Grand Cayman.

3. M. sideroxylon (Griseb.) Dubard in Ann. Mus. Colon. Marseille **23**: 15, f. 6 (1915).—*Sapota sideroxylon* Griseb. (1861). Naseberry Bullet.

Tree up to 25 m. high, the young branchlets thick, rusty-pubescent; leaves elliptical to oblanceolate, rounded or obtuse at tip, 7–16 cm. long, 2·5–5 cm. broad; petioles 1·5–3 cm. long; calyx 6–7 mm. long; corolla-tube 2·5 mm. long, lobes 3–4 mm. long; staminodes petaloid.

Frequent in woodlands on limestone in wetter areas; 40–2300 feet; fl. Mar, June–Oct, fr. June–Jan. *A 12571! Cornman 820! H 5379! P 22784! 24733!* Cuba.

2. CHRYSOPHYLLUM L. (1753)

1 Leaves silvery-greyish pubescent beneath, glabrescent; fruit ovoid or obovoid, 1·5–2 cm. long, 1-seeded, dark blue   **1. argenteum**
1 Leaves densely rusty-reddish or silky-pubescent beneath:
  2 Pedicels 7–12 mm. long; flowers 10–30 in a cluster; corolla-lobes as long as or a little longer than the tube; stigma-lobes 7–12; fruit subglobose, purple or green, mostly 8–10 seeded, 3–8 (–10) cm. in diameter; tree mostly in cultivation   **2. cainito**
  2 Pedicels 4–8 mm. long; flowers 3–10 together; corolla-lobes shorter than the tube; stigma-lobes 5; fruit oblong-ovoid, blue or blackish, 1 (–2)-seeded, 1–2 cm. long, up to about 1 cm. broad; wild tree   **3. oliviforme**

1. C. argenteum Jacq., Enum. Syst. Pl. Carib.: 15 (1760).

Shrub or tree to 13 (–20) m. high; leaves elliptical to narrowly oblong-elliptical, cuneate at base, acuminate at tip, 4·5–12 cm. long, 2–4·5 cm. broad; petioles 8–12 mm. long; calyx about 2 mm. long; corolla 3–4 mm. long, with short rounded lobes.

Rare (Trel., St. Thom.), in secondary thickets; 250–600 feet; fr. Jan. *A 10424! P 10514!* Venezuela, West Indies.

2. C. cainito L., Sp. Pl. **1**: 192 (1753).—Star Apple.

Tree up to 15 (–20) m. high; leaves elliptical to oblong-elliptical, cuneate at base, mostly shortly acuminate, (3–) 7–16 cm. long, (1·5–) 4·5–8 cm. broad, glabrous and glossy adaxially; petioles 1·5–3 cm. long; calyx 1–1·5 mm. long; corolla yellow, greenish-yellow or purplish, fragrant, tube 1·5–1·8 mm. long, lobes spreading, 1·8–2 mm. long.

Common, mostly along roadsides and in pastures and yards where planted; 100–1400 feet; fl. Aug–Sept, Dec, fr. Nov–Mar. *A 7941! H 12128! P 9627!* Native in the Greater Antilles, cultivated elsewhere throughout the New World tropics and introduced into W. Africa.

3. C. oliviforme L., Syst. Nat. ed. 10, **2**: 937 (1759).—Wild Star Apple.

Shrub 3 m. or tree to 15 m. high; leaves ovate to elliptical, shortly acuminate, 1·5–9 cm. long, 1–4 cm. broad, becoming glabrous and glossy adaxially; petioles 8–12 mm. long; calyx 1·5 mm. long; corolla light green to greenish-yellow, fragrant, tube about 2·5 mm. long, lobes reflexed, 1·5 mm. long.

Locally common, in woodlands and thickets on limestone; 30–3000 feet; fl. July–
Feb, fr. Dec–Mar. *A 9947! H 5363! 9955! P 22766! 23151!* Florida, British
Honduras, Bahamas, Greater Antilles, cultivated elsewhere.

3. BUMELIA Sw. (1788) nom. cons.; Stearn (1968)

1 Calyx 2–2·5 (–3) mm. long; flowers mostly 3–15 in a cluster:
  2 Petiole 15–32 mm. long; leaf-blades often more than 8 cm. long and 4 cm. broad, the
    lateral veins clearly evident; fruits subglobose, 12–18 mm. or more in diameter
                                                             1. *nigra*
  2 Petiole 2–15 mm. long; leaf-blades rarely as much as 8 cm. long and 4 cm. broad, the
    lateral veins inconspicuous; fruits 4–10 mm. in diameter:
    3 Leaf-blades mostly narrowly elliptical, acute or shortly acuminate; fruits ellipsoid
      6–10 mm. long   2. *salicifolia*
    3 Leaf-blades elliptical to rotundate or obovate or oblanceolate, obtuse, rounded or
      emarginate at tip; fruits 8–10 mm. long:
      4 Young branchlets, pedicels and sepals glabrous; leaves green beneath when young;
        fruits oblong-ellipsoid, mostly 4–5 mm. broad   3. *rotundifolia*
      4 Young branchlets, pedicels and sepals minutely hairy; leaves brown-tomentose
        beneath when young; fruits subglobose to broadly ellipsoid, 9–10 mm. broad
                                                      4. *americana*
1 Calyx (3–) 3·5–4 mm. long; flowers mostly up to 8 in a cluster:
  5 Sepals 7–9, the outer as well as the inner brown-tomentose; pedicels persistently hairy
                                                        5. *octosepala*
  5 Sepals 5–8, the two outer glabrous or sparsely hairy, the inner tomentose or sericeous;
    pedicels glabrous or sparingly hairy:
    6 Leaf-blades not more than 5 cm. broad:
      7 Leaves without a raised network of veins beneath; calyx 3·5–4 mm. long 6. *montana*
      7 Leaves with a raised network of veins beneath; calyx about 3 mm. long   7. sp. *A*
    6 Leaf-blades some or all more than 5 cm. broad, a raised network of veins immediately
      visible with a lens:
      8 Petiole 1–10 mm. long; leaf-blades obovate to broadly obovate   8. *bullata*
      8 Petiole 10–20 mm. long; leaf-blades elliptic-oblong   9. sp. *B*

1. B. nigra Sw., Nov. Gen. & Sp. Pl.: 49 (1788).—*Dipholis nigra* (Sw.) Griseb.
    (1861). Black or Red Bullet or Bully.
Tree 8–20 m. high, with curving branches; leaves oblong-elliptical to lanceolate,
cuneate at base, acute or shortly acuminate, 3–18 cm. long, 1–7 cm. broad, glabrous
or nearly so, venation slightly prominent beneath; petiole not pouched at apex;
flowers white, 3–10 at axils and leafless nodes, fragrant; pedicels 5–10 mm. long;
corolla 4–4·5 mm. long, the lateral lobes pointed; fruit 1-seeded, yellow or purplish.
    Rather common in central and western parishes, in pastures and woodland
margins on limestone; 25–3000 feet; fl. July–Sept, fr. Nov–May. *A 11744! 12693!
H 5777! P 22265! 26625!* Endemic.

2. B. salicifolia (L.) Sw., Nov. Gen. & Sp. Pl.: 50 (1788).—*Achras salicifolia* L.
    (1762). White Bullet or Bully.
Tree 4–10 (–22) m. high, the branchlets at first rusty-pubescent; leaves ovate,
narrowly elliptical or lanceolate, acuminate at base and apex, 3–12 cm. long, 1·5–
3·5 cm. broad; flowers cream or yellow, mostly 5–7 in a cluster, fragrant; pedicels
3–5 mm. long; corolla 4–5 mm. long; fruits finally shiny black, 5–6 mm. broad.
    Common in woodlands and thickets on limestone in moderately arid areas and at
salina margins; SL–1500 (–3000) feet; fl. Dec–June, fr. Mar–July. *A 6829! 12336!
H 9616! 10170! P 23227!* Florida, C. Amer., Bahamas, Greater Antilles, Virgin Is.,
Antigua, Guadeloupe, Barbados; Cayman Brac.

3. B. rotundifolia Sw., Nov. Gen. & Sp. Pl.: 50 (1788).—*B. clarendonensis* Urb.
    (1915). *B. peckhamensis* Urb. (1925).
Shrub 2–2·5 m. or tree up to 10 m. high, the stem sometimes spiny at base; bark
rough, rectangular-scaly; leaf-blade glabrous, 1·5–7 cm. long, 0·6–4 cm. broad;
flowers white, 3–15 in a cluster, fragrant; pedicels 4–10 mm. long, glabrous;
corolla 4·5–5 mm. long.
    Rather common in central and western parishes, in thickets and woodlands in
arid limestone areas; 50–3000 feet; fl. Nov–May, fr. Dec, May–July. *A 8764!
H 11040! 11111! P 23113! 26878!* Endemic.

4. B. americana (Mill.) Stearn in Journ. Arn. Arb. **49**: 282 (1968).—*Maurocenia americana* Mill. (1768). *B. retusa* Sw. (1788).

Shrub 2–2·5 m. or tree to 13 m. high; occasional short branches develop into spines; leaf-blades dark glossy green adaxially; flowers greenish-white to light yellow, 3–12 in a cluster, fragrant; pedicels 3–9 mm. long, pubescent; corolla about 4·5 mm. long.

Frequent in central and western parishes, mostly in coastal areas on arid honey-combed limestone; SL–200 (–750) feet; fl. Sept–Mar, June, fr. Sept–May. *H 9729! P 17427! Stearn 314!* Bahamas, Cuba.

5. B. octosepala (Urb.) Stearn in Journ. Arn. Arb. **49**: 284 (1968).—*Dipholis octosepala* Urb. (1912).

Shrub 5 m. or tree up to 10 m. high, the branchlets rusty-tomentose, becoming grey and glabrous; leaves lanceolate to narrowly obovate, long-acuminate, acute or rarely obtuse, 7–18 cm. long, 2–9 cm. broad; flowers 2–7 in axillary clusters; pedicels 4–8 mm. long, tomentose; corolla white, the tube 4·5–5 mm. long; fruit narrowly ellipsoid, 10–15 mm. long.

Uncommon in central parishes, on well-drained rocky limestone; 1500–2500 feet; fl. May–Sept, Jan, fr. Sept–Jan. *H 11049! 12798! P 22742!* Endemic.

6. B. montana Sw., Nov. Gen. & Sp. Pl.: 49 (1788).—*Dipholis montana* (Sw.) Griseb. (1861). Mountain Bullet or Bully.

Tree 3–15 m. high, the branchlets brown-sericeous, becoming grey and glabrous; leaves elliptical to obovate, obtuse to shortly acuminate, 3–9 cm. long, 1·5–4·5 cm. broad, glabrous; flowers 1–3 (–5) in axillary clusters; pedicels 3–10 mm. long, stout, sparsely hairy; corolla white, the tube 4·5 mm. long; fruit ellipsoid, bluish-black to brownish-purple, about 15 mm. long and 8–10 mm. broad.

Locally common (St. Andr., Port., St. Thom.), in montane woodlands; 3000–750 feet; fl. July–Nov, fr. Jan–Aug. *A 12552! H 5370! 6691! P 9538!* Endemic.

7. B. sp. A.

Tree up to 16 m. high, the branchlets brown-sericeous becoming glabrous; leaves broadly to narrowly elliptical, acute or shortly acuminate, 4–9 cm. long, 2·5–5 cm. broad, glabrous, coriaceous; flowers solitary; pedicels about 5 mm. long, sparsely appressed-hairy.

Rare and local (St. Eliz.); 2100–2200 feet; fl. Sept. *H 9742! 9803!* Endemic.

8. B. bullata (Howard & Proctor) Stearn in Journ. Arn. Arb. **49**: 284 (1968).—*Dipholis bullata* Howard & Proctor (1958).

Tree up to 8 m. high, with glabrous branchlets; leaves obovate, broadly cuneate at base, rounded or obtuse at tip, 3·5–13 cm. long, 2–7 cm. broad, glabrous, cori-aceous; flowers 2–4 in axillary clusters; pedicels 5–15 mm. long; corolla yellowish, the tube 2·5–3 mm. long; fruit narrowly ellipsoid, 15 mm. or more long, about 5 mm. broad.

Rare and local (Port.), in woodland on limestone in wet areas; 1500–2500 feet; fl. Oct, fr. Jan. *HPS 14579!* Endemic.

9. B sp. B.

Tree about 8 m. high, the branchlets tomentose becoming glabrous; leaves elliptical, acute, 5–18 cm. long, 2·5–8 cm. broad, glabrous; flowers 3–8 in axillary clusters; pedicels 7–9 mm. long, sparsely appressed-hairy.

Rare and local (Han.), in forest on limestone; 1500–1700 feet; fl. Apr. *P 10039!* Endemic.

4. MICROPHOLIS (Griseb.) Pierre (1890)

1. M. rugosa (Sw.) Pierre, Not. Bot. Sapot.: 41 (1891).—*Chrysophyllum rugosum* Sw. (1788). Beef Apple, Bull Apple.

Tree up to 16 m. high, the branchlets rusty-sericeous when young; leaves oblong-elliptical, mostly acuminate, 5–19 cm. long, 2–8·5 cm. broad; flowers on pedicels up to 7 mm. long; sepals rounded, rusty-pubescent, 2–2·5 mm. long; corolla 3·5–4 mm. long; ripe fruit yellow or orange-yellow; seeds black or dark brown, 2·5 cm. long and about 1 cm. broad, with a white hilum nearly as long as seed.

Occasional, in central and western parishes, mostly in open situations on limestone hills; 1500–3000 feet; fl. Mar-May, fr. Mar–July. *H 11016! 11043! P 9955! 16471!* Endemic.

5. MASTICHODENDRON (Engl.) H. J. Lam (1939)

1. M. foetidissimum (Jacq.) H. J. Lam in Med. Bot. Mus. Utrecht **65**: 521 (1939).—*Sideroxylon foetidissimum* Jacq. (1760). *S. jamaicense* Urb. (1904). *M. floribundum* (Griseb.) Cronquist (1946). Mastic.

Tree 6–16 m. or more high, with scaly reddish bark; leaves oblong-ovate to elliptical, cuneate at base, obtuse, acute or acuminate at tip, up to 16 (–20) cm. long and 6·5 (–8) cm. broad; petioles slender, up to 7 cm. long; pedicels 4–10 mm. long, pubescent or glabrous; corolla greenish-yellow, 3·5–5·5 mm. long, the lobes oblong, obtuse; ripe fruit ovoid or pyriform, yellow, 1–2·5 cm. long, 1–3-seeded.

Occasional, in central and eastern parishes, in open hillside woodland; 20–3000 feet; fl. Jan, July, fr. Jan–Sept. *A 11365! H 11217! H & B 10715! C. B. Lewis!* Florida, Bahamas, Antilles south to Martinique, Grand Cayman.

6. POUTERIA Aubl. (1775)

Terminal buds and young growth with inconspicuous appressed hairs; pedicels mostly exceeding 1 cm. long; sepals 4 (–5), subequal; fruit 2–6 cm. long   **1. multiflora**
Terminal buds shaggy with somewhat spreading hairs; pedicels up to about 4 mm. long; sepals 8–12, the outermost much shorter than the innermost; fruit usually 8–20 cm. long; cultivated; probably native in C. Amer.; Mammee Sapota
     P. sapota (Jacq.) H. E. Moore & Stearn

1. P. multiflora (A.DC.) Eyma in Rec. Trav. Bot. Néerl. **33**: 164 (1936).—*Lucuma multiflora* A.DC. (1844). Galimenta, Bully Tree.

Slender-trunked tree (3–) 6–25 m. or more high; branchlets rufous-pubescent when young; leaves oblong to obovate-elliptical, cuneate at base, obtuse to shortly acuminate at tip, 6–36 cm. long, 2·5–15 cm. broad, dark green adaxially and glabrous at maturity; flowers solitary or few together, up to 8; pedicels 8–15 (–20) mm. long; corolla greenish-white, 7–11 mm. long; fruit ovoid to subglobose, 1 (–4)-seeded.

Common in woodlands on limestone; 50–3000 feet; fl. Apr–Dec, fr. Sept–May. *A 10120! H 9099! 11221! H & P 13608! P 23107!* Bermuda, West Indies.

Other species of *Pouteria* otherwise sometimes called *Lucuma* and including the Canistal or Egg Fruit (*P. campechiana* (Kunth) Baehni) are cultivated in Jamaica.

155. EBENACEAE

Shrubs and trees. Leaves alternate, simple, entire, coriaceous, without stipules. Flowers usually unisexual, dioecious or rarely monoecious, actinomorphic, in axillary inflorescences. Perianth biseriate; calyx 3–7-lobed, persistent and often enlarging in fruit; corolla 3–7-lobed, urceolate, the lobes imbricate. Male flowers usually in small cymes; pistillode present. Stamens as many as corolla-lobes or 2–3 times as many, hypogynous or epipetalous, free or united in pairs; anthers 2-locular, introrse, opening lengthwise. Female flowers usually solitary; staminodes often present. Ovary superior, 2–16 locular; ovules (1–) 2 in each loculus, anatropous, pendulous on axile placentas; styles and stigmas 2–8, styles free or basally connate. Fruit a soft or leathery berry. Seed with straight embryo and hard endosperm.

2 genera with about 450 species; *Diospyros*, the largest genus, has representatives in the subtropics and tropics of both hemispheres, the other genera mainly African.

1. DIOSPYROS L. (1753)

Flowers unisexual, dioecious; calyx accrescent; corolla-lobes imbricate and contorted.

1. D. tetrasperma Sw., Nov. Gen. & Sp. Pl.: 62 (1788).—Clamberry.

Shrub 2–4 m. or tree to 16 m. high; trunk usually slender; wood heavy, fine-grained, sapwood pinkish-brown, old heartwood jet-black; leaves narrowly obovate to oblanceolate, rounded and retuse at apex, base cuneate, 3–8 (–10) cm. long, 1–3·5 (–5) cm. broad, glabrescent, often with a few large glands beneath; inflorescences and flowers densely covered with short appressed hairs; male flowers 5–6 mm. long with 8 stamens in 2 unequal series; female calyx cup-shaped, about 4 mm. long, expanding to 6 mm. long in fruit with 4–5 rounded lobes; fruit sub-globose, 4-seeded, 10–14 mm. in diameter, glabrescent.

Common except in the extreme eastern parishes, in woodlands and thickets on limestone; SL–2700 feet; fl. May–Dec, fr. June–Mar. *A 6316! 9774! H 9922! 10011! H & P 13917! 14518!* British Honduras, Cuba, Hispaniola.

An undescribed species of *Diospyros* has been discovered recently in the parish of Hanover (on wooded limestone hilltop; 1700 feet; fl. May, fr. Apr. *P 28636! 31304!*). It is a small tree distinguished from *D. tetrasperma* by its larger leaves and fruits, the latter 2–2·5 cm. in diameter.

Several introduced species cultivated in Jamaica include *D. discolor* Willd., the Philippine Persimmon or Velvet Apple.

156. SYMPLOCACEAE

Trees or shrubs. Leaves alternate, simple, without stipules. Flowers bisexual or rarely unisexual, actinomorphic, in racemes, panicles or axillary clusters. Perianth biseriate; calyx (4–) 5-lobed, the lobes imbricate or valvate, persistent; corolla gamopetalous, divided nearly to the base into (3–) 5 (–11) imbricate lobes. Stamens (4–) 15 or more, mostly in 2–several series, epipetalous, free or filaments connate; anthers short, 2-locular, opening lengthwise. Ovary inferior to half-inferior, 2 (–5-) locular[1]; ovules 2 (–4) in each loculus, anatropous, pendulous on axile placentas; style simple; stigma capitate or up to 5-lobed. Fruit a berry or drupe crowned by the persistent calyx, 1-5-locular with a seed in each loculus. Seeds with copious endosperm; embryo straight or curved.

1 genus with about 300 species in tropical and subtropical Asia and tropical America.

1. SYMPLOCOS Jacq. (1760)

1 Corolla-lobes (5–) 7–10, 5–6 mm. broad; corolla-tube[2] 11–13 mm. long; calyx 3·5 mm. long, sepals 2·5–3 mm. broad; leaf-margin entire to obscurely crenulate
             1. *octopetala*
1 Corolla-lobes 5 (–6), 3–4 mm. broad:
  2 Corolla-tube 10–12 mm. long; sepals 1·5–2 mm. broad; leaf-margin entire to crenulate, glandular-serrulate at least when young   2. *martinicensis*
  2 Corolla-tube about 15 mm. long; sepals 2·2 mm. broad; leaf-margin serrulate to crenulate with up to 40 teeth on each side   3. *tubulifera*

1. S. octopetala Sw., Nov. Gen. & Sp. Pl.: 109 (1788).—*Ternstroemia crenata* Macf. (1837).

Tree 4–12 m. high; young buds grey-sericeous; petioles 6–15 mm. long; leaf-blade elliptical, shortly acuminate, cuneate to rounded at base, margin when young with peglike glands, shiny on both surfaces, up to 8 (–10) cm. long, 1·3–4·5 (–5·5) cm. broad; inflorescences 1–2 (–3)-flowered; flowers white, fragrant; sepals ciliate; fruit ellipsoid, about 2 cm. long and 7 mm. broad.

Local (St. Andr., Port.), in montane woodland on steep hillsides; 3800–4600 feet; fl. Dec–Feb, May–June, fr. July. *A 12542! P 22275! Robbins UCWI 2540!* Endemic.

---

[1] The flowers of *S. octopetala* have, around the upper part of the ovary and surrounding the base of the style, a continuous ring of tissue which is distinct. It bears long unicellular hairs of the same type as those on the style-base but apart from these this tissue resembles glandular disc and is conspicuous in sections of living flowers by its dark orange colour.
[2] Corolla-tube refers to the connate and free portion proximal to the spreading lobes.

2. S. martinicensis Jacq., Enum. Syst. Pl. Carib.: 24 (1760).—*S. jamaicensis* Krug & Urb. (1892).

Tree 12–18 m. high; young buds grey-buff-sericeous; petioles 6–10 mm. long; leaf-blade elliptical, acuminate, cuneate at base, minutely appressed-puberulous on midrib and proximal lamina beneath, margin glandular when young, 5–14 cm. long, 2–6 cm. broad; inflorescences (1–) 3 (–5)-flowered; flowers white, fragrant.

Rare and local (Port.), in wet woodlands on limestone; 2000–4500 feet; fl. Mar, July. *A 9130! P 22121! 23790!* Honduras, Guatemala, Guyanas, Puerto Rico, Lesser Antilles, Trinidad. The calyx of Jamaican representatives is generally a little larger than that of Lesser Antillean plants and the inflorescence is fewer-flowered.

3. S. tubulifera Krug. & Urb. in Engl., Bot. Jahrb. **15**: 331 (1892).

Tree; leaf-blade elliptical, glabrous above, pilose on midrib beneath with ascending hairs, 3·5–13 cm. long, 1·5–5·5 cm. broad; calyx-lobes rounded, imbricate, ciliate, about 1·3 mm. long; corolla-lobes 5–6, connate part of tube very short, white.

Rare and local (St. Cath., St. Ann). *H 8899. Prior.* Endemic.

157. OLEACEAE

Trees or erect or climbing shrubs. Leaves opposite or subopposite, rarely alternate, simple or pinnate, without stipules. Flowers bisexual or unisexual, actinomorphic, often fragrant, in axillary or terminal racemes, cymes or panicles. Perianth biseriate, uniseriate or absent; calyx 4–many-lobed, the lobes valvate; corolla typically gamopetalous, but often deeply lobed or rarely the segments quite free; lobes 4 (–12), mostly imbricate. Stamens hypogynous or epipetalous, 2 or rarely 4, free; anthers 2-locular, often apiculate with the produced connective, opening lengthwise. Disk absent. Ovary superior, 2-locular; ovules usually 2 in each loculus, anatropous, pendulous or ascending on axile placentas; style simple or absent; stigma capitate or bifid. Fruit a berry, drupe, loculicidal or circumscissile capsule or a samara. Seeds usually with fleshy endosperm; embryo straight.

27 genera with 600 species, widely distributed in tropical and temperate regions, with a high concentration in eastern Asia.

1 Corolla-tube conspicuous, often longer than the (4–) 5–9 corolla-lobes; fruit often 2-lobed, berrylike, black; plants often scrambling or climbing; leaves pinnate, 3-foliolate or simple **1. Jasminum**
1 Corolla-tube shorter than the 4 corolla-lobes or petals free or wanting; fruit entire, drupaceous; erect shrubs or trees; leaves simple:
 2 Corolla wanting or very small and early caducous; anthers much shorter than the slender filaments; flowers usually unisexual, small and inconspicuous in short axillary bracteate racemes; drupe 1–2-seeded **2. Forestiera**
 2 Corolla present, white; anthers as long as or longer than the inconspicuous filaments; flowers mostly bisexual in corymbs or panicles; drupe 1-seeded:
  3 Flowers in axillary corymbs, fragrant; corolla 4 mm. long with broad imbricate petals; leaves narrowly elliptical, 4–8 cm. long; evergreen tree cultivated in the mountains; native of Himalaya, China and Japan *Osmanthus fragrans* Lour.
  3 Flowers in terminal or axillary panicles; petals narrow:
   4 Petals linear, thin, free or nearly so and joined by the filaments; adaxial surface of leaf not lepidote; endocarp thick, woody **3. Linociera**
   4 Petals clavate, fleshy, joined at the base into a tube about 2 mm. long and longer than the calyx; adaxial surface of leaf lepidote; endocarp thin, fragile **4. Haenianthus**

1. JASMINUM L. (1753)

1 Leaves compound:
 2 Leaflets usually 7, up to 3 cm. long and 1·5 cm. broad but mostly much smaller, glabrescent; calyx-lobes linear-filiform, 4–7 mm. long; corolla-tube very slender about 2 cm. long *1. grandiflorum*
 2 Leaflets 3, up to 5 cm. long and 4 cm. broad, thinly pubescent beneath, with domatia; calyx-teeth very short; corolla-tube about 1·5 cm. long *2. fluminense*

T

1 Leaves simple:
3 Stems, leaves and calyx glabrous; calyx-lobes triangular, less than 1 mm. long; flowers
  in terminal cymes; corolla white, the tube 1 cm. long; rare in cultivation; native of
  Australia; Wax Jasmine   *J. volubile* Jacq.
3 Stems, leaves and calyx more or less hairy; calyx-lobes linear, 5–10 mm. long:
4 Cymules several, clustered and subsessile in heads usually terminating short lateral
  branches; stems densely pubescent; leaves ovate, subcordate at base   3. *multiflorum*
4 Cymules laxly stalked, mostly 3-flowered; stems thinly pubescent; leaves ovate-
  elliptical to rotundate, rounded to cuneate at base   4. *sambac*

1. J. grandiflorum L., Sp. Pl. ed. 2, **1**: 9 (1762).—Poet's Jasmine.
Climbing or straggling shrub to 5 m. or more high, sometimes deciduous;
leaves imparipinnate, the terminal leaflet larger, narrowly ovate, acute, narrowed to
base; cymes on slender branches of a terminal panicle; corolla 5-lobed, white
tinged crimson outside; ovary with 2 ovules in each cell.
Common in gardens and naturalized occasionally, 600–4250 feet; fl. Feb, May–
Aug. *A 10491! H 10822!* Native of eastern China, now cultivated for ornament in
many warm countries.

2. J. fluminense Vell., Fl. Flumin.: 10 (1825).—Azores Jasmine.
Evergreen scrambling shrub or climbing to 12 m. high; stems pubescent;
leaflets broadly ovate, acute or acuminate, truncate-subcordate at base; cymes
lateral and terminal; corolla 5–9-lobed, white; ovary with 1 ovule in each cell.
Locally common (St. Andr., Clar.), in shady waste places, hedgerows and thickets;
50–800 feet; fl. and fr. Aug–Mar. *A 9786! Porter & Simpson 976! Stearn 31!*
Native of tropical Africa, introduced and naturalized in the American tropics.

3. J. multiflorum (Burm. f.) Andr., Bot. Repos. **8**: t. 496 (1807).—*Nyctanthes
multiflora* Burm. f. (1768). *J. pubescens* (Retz.) Willd. (1797). Hairy Jasmine.
Shrub 60–120 cm. high or branches long and trailing or climbing to 6 m.;
leaves mostly acute, up to 6 cm. long and 4 cm. broad; lateral flowering shoots with
the lower pair of leaves much smaller than the upper; calyx densely pubescent, the
lobes 7–9 mm. long; corolla white, the tube 1·5–2 cm. long, the lobes 6–9, narrowly
oblong-elliptical, nearly as long; ovary with 1 ovule in each cell.
Common in cultivation and occasionally naturalized (St. Andr., West., Port.),
in waste places; 10–3000 feet; fl. most of the year. *A 12298! H 11924! P 16165!*
Native of tropical Asia, introduced elsewhere.

4. J. sambac (L.) Ait., Hort. Kew. **1**: 8 (1789).—*Nyctanthes sambac* L. (1753).
Arabian Jasmine.
Habit like the last; leaves acute to rounded at tip, mucronate, with domatia
beneath, up to 11 cm. long and 7 cm. broad; calyx thinly pubescent, the lobes
linear, 5–10 mm. long; corolla white turning pink or brownish-pink, the tube 12–
15 mm. long; lobes 5–9; flowers with partly aborted corollas with the lobes developed into sterile anthers sometimes occur.
Cultivated and sometimes escaping (St. Andr., St. Cath., St. Thom.), on to
roadside banks; 500–1500 feet; fl. most of the year. *A 12569! Thornton UCWI 986!*
Native of East Indies, cultivated and occurring in many varieties elsewhere.

2. FORESTIERA Poir. (1810)

Leaves narrowly elliptical, oblanceolate or obovate, obtuse, narrowed to base, glabrous and
  pitted beneath, inflorescences less than 1 cm. long; mature fruit slightly curved
                                                                    1. *segregata*
Leaves elliptical to narrowly ovate, more or less acuminate, pubescent but not pitted
  beneath; inflorescences 1–2 cm. long; mature fruit regularly ellipsoid   2. *rhamnifolia*

1. F. segregata (Jacq.) Krug & Urb. in Engl., Bot. Jahrb. **15**: 339 (1892), partly.—
Myrica segregata Jacq. (1787).
Diffusely slender-branched shrub 1–3 m. or tree to 6 m. high; young twigs
shortly puberulous; leaves mostly opposite, thin, 2–7 cm. long, 0·5–3 cm. broad;
racemes 3–7-flowered; bracteoles decussate, white-ciliate; male flowers with a
distinct pedicel about 2·5 mm. long, bearing 2–4 stamens with filaments 4–6 mm.

long; ripe fruits obliquely fusiform, acute, bluish-purple (also reported as red), about 7 mm. long.

Locally common in arid areas along the south coast on sand or limestone rocks; 5–500 (–1500) feet; fl. Sept–Mar, fr. almost all the year. *A 11812! H 9015! 9537! P 20566! 23621!* Bermuda, Florida, Greater Antilles, Virgin Is., Antigua; Grand Cayman.

2. F. rhamnifolia Griseb., Cat. Pl. Cub.: 169 (1866).

Shrub or tree 4·5–6 m. high; twigs and leaves especially beneath with short erect hairs; leaves subcoriaceous, 3–9 cm. long; 1·5–5 cm. broad; racemes clustered, axillary or below the leaves; bracteoles greenish-yellow; pedicels about 1·5 mm. long; fruit ellipsoid to narrowly obovoid, purplish, about 1 cm. long.

Rare (St. Andr., Port., St. Thom.), in thickets on shaly hillsides; 600–1500 feet; fl. May–June, Oct, fr. June–Nov. *A 7260! H 8526! 9017!* Cuba, Hispaniola, St. Croix, Guadeloupe, Martinique and Grenada [see note p. 789].

3. LINOCIERA Sw. ex Schreb. (1791) nom. cons

1 Leaves gland-dotted beneath, the dots prominent adaxially at least when dry; domatia
  absent; stamens as long as petals, about 5–7 mm. long; anthers linear, 5–6 mm. long
                                                                  1. *ligustrina*
1 Leaves not gland-dotted; domatia present; stamens much shorter than petals, 1·5–2 mm.
  long including the oblong anthers:
  2 Petals 12–15 (–20) mm. long:
    3 Petiole slender 1–2 cm. or more long; petals 1–1·3 mm. broad; style 0·5 mm. long;
      leaves rather leathery   2. *domingensis*
    3 Petiole 4–5 mm. long; petals less than 1 mm. broad; style 1 mm. long; leaves thin
                                         3. *sp. A*
  2 Petals 4–7 mm. long; petioles 7–12 mm. long:
    4 Primary panicle-branches thyrsoid, up to about 5 cm. long; petals 5–7 mm. long
                                   4. *jamaicensis*
    4 Primary panicle-branches racemose, up to about 2·5 cm. long; petals 4 mm. long
                                   5. *sp. B*

1. L. ligustrina (Sw.) Sw., Fl. Ind. Occ. **1**: 50, t. 2 (1797).—*Thouinia ligustrina* Sw. (1788).

Straggling-branched tree up to 6 m. high; leaves elliptical, gradually obtusely acuminate, coriaceous, up to 6 cm. long and 2 cm. broad; petioles up to 1 cm. long; panicles 3–5 cm. long; flowers white, fragrant; ripe fruits dark purple, about 1 cm. long.

Rare in the southern parishes, in woodlands on limestone; 50–900 feet; fl. July, fr. Mar, July–Sept. *H 11736! C. B. Lewis! Tulloch 208!* Greater Antilles.

2. L. domingensis (Lam.) Knobl. in Bot. Centralbl. **61**: 87 (1895).—*Chionanthus domingensis* Lam. (1791). *L. latifolia* Vahl (1805). *Mayepea domingensis* (Lam.) Krug & Urb. (1892). Ironwood, White Rosewood.

Tree with whitish bark 8–16 m. high; leaves elliptical, acuminate, narrow at base, subcoriaceous, up to 15 cm. long and 6 cm. broad, domatia-pits in vein-axils hairy; fruit oblong, 12–15 mm. long.

Widely distributed but not common, in woodlands on limestone or shale, mainly in wet areas; 1000–4000 feet; fl. Feb–Apr, fr. Apr. *H 7095! 10293! P 23279!* British Honduras, Guatemala, Greater Antilles.

3. L. sp. A.

Tree 10 m. high; young stems, inflorescence-branches and petioles golden-puberulous; leaves oblanceolate-elliptical, rather abruptly acuminate, cuneate at base, thin, glabrescent adaxially, with extensive domatia-pits abaxially, up to 12·5 cm. long and 4·5 cm. broad; inflorescence-branches very slender.

Very rare (West.), on wooded limestone hillside; 1200–1400 feet; fl. Apr. *P 22206!* Endemic.

4. L. jamaicensis Urb., Symb. Ant. **2**: 456 (1901).

Shrub 3 m. or tree to 12 m. high; leaves elliptical or narrowly elliptical, bluntly acute, obtuse or bluntly acuminate, coriaceous, with hairy domatia abaxially, 6–10 cm. long, 2·5–4 cm. broad.

Rare (St. Andr.), in woodlands; 1500–3500 feet; fl. Nov–Dec. *H 6076! 6643!* Endemic.

5. **L. sp. B.**

Shrub 3 m. high; leaves elliptical, shortly and bluntly acuminate, broadly cuneate at base, coriaceous, domatia-pits large and filled with yellowish hairs, up to 16 cm. long and 6·5 cm. broad.

Very rare (Trel.), on wooded limestone hilltops; 2000–2200 feet; fl. Sept, Dec. *P 21372! 21856!* Endemic.

In spite of their rarity, species of *Linociera* show great diversity in Jamaica; yet another species, without gland-dots or domatia, from woodlands on arid limestone (St. Cath.; fl. and fr. Aug. *Tulloch 144!*), remains to be described.

4. HAENIANTHUS Griseb. (1861)

1. **H. incrassatus** (Sw.) Griseb., Fl. Br. W.I.: 405 (1861).—*Chionanthus incrassatus* Sw. (1788).

Tree to 15 m. high; young twigs brownish, with prominent lenticels in vertical rows; glabrous except for minute rounded scales on youngest branches and adaxial leaf-surfaces; leaves elliptical, acute or shortly acuminate, narrow at base, paler beneath and pitted, long-petioled, 4–11 cm. long, 1·5–4·5 cm. broad; panicle pedunculate, lepidote; corolla 5–6 mm. long; ripe fruits dark blue, 14–15 mm. long.

Locally common (St. Andr., Port.), in montane forest; 3900–5500 feet; fl. May–June, fr. Mar, May, Dec. *H 6359! P 19666!* Endemic.

158. LOGANIACEAE

Trees, shrubs or herbs with or without intraxylary phloem in the wood. Leaves simple, opposite, verticillate or rarely alternate, entire to dentate; stipules absent or represented by thickened ridges or sheathing flanges between the leaf-bases. Flowers bisexual, actinomorphic or nearly so, solitary or in cymose, racemose, paniculate or capitate inflorescences. Perianth biseriate; calyx (2–) 4–5-lobed; corolla gamopetalous with 4–5 (–7) or rarely more contorted, imbricate or valvate lobes. Stamens as many as the corolla-lobes and alternate with them, rarely twice as many or fewer, epipetalous; anthers 2-locular, opening lengthwise. Ovary superior or partly inferior, 2 (–3–4)-locular, rarely 1-locular; ovules amphitropous or anatropous, numerous on axile placentas; styles simple, rarely 2; stigma capitate or 2-lobed. Fruit usually a septicidally dehiscent 2-valved capsule, rarely a berry or drupe. Seeds small with straight embryo and fleshy or bony endosperm, often winged.

About 30 genera with some 800 species in the subtropics and tropics.

1 Shrubs:
  2 Flowers 4-merous with imbricate or subvalvate corolla-lobes; fruit a septicidal 2-valved capsule; leaves entire to dentate, usually thin, with glandular, stellate or lepidote indumentum   **1. Buddleja**
  2 Flowers 5-merous with contorted corolla-lobes; corolla showy, creamy-white, waxy; sepals rounded, strongly imbricate; fruit a berry; leaves entire, obovate, leathery, glabrous; cultivated ornamentals; native of S.E. Asia:
    3 Corolla about 14 cm. long   *Fagraea ridleyi* King & Gamble
    3 Corolla about 6 cm. long   *Fagraea obovata* Wall.
1 Herbs, mostly annual:
  4 Corolla-lobes imbricate; flowers solitary in the forks of branches; ovary partly inferior; styles permanently united; capsule loculicidal   **2. Polypremum**
  4 Corolla-lobes valvate; flowers secund in terminal inflorescences; ovary superior; capsule septicidal:
    5 Inflorescence dichotomous with racemose branches; corolla white, 1·5 mm. long; ovary apocarpous with a common style, the halves soon separating   **3. Mitreola**

5 Inflorescence spicate; corolla pink, 1 cm. long; styles permanently united
4. **Spigelia**

1. BUDDLEJA L. (1753); Norman (1966)

1 Inflorescence paniculate, with narrow leafy bracts in lower part, the branches 3–25 cm. long; calyx and capsule tomentose; corolla yellow, pilose inside, tomentose outside, 3·5–4 mm. long, slightly exceeding or shorter than the calyx, the lobes subvalvate, erect, the tube about 2 mm. long   1. *americana*
1 Inflorescence usually unbranched or consisting of a long terminal thyrse with or without basal branches; capsule not tomentose; corolla white, mauve, lilac or purple with broadly imbricate then spreading lobes, the tube longer than the lobes and calyx; introduced, natives of China:
   2 Calyx and capsule scaly; corolla-tube slightly curved, scaly outside; shrub to 2 m. high; naturalized locally   *B. lindleyana* Fortune ex Lindl.
   2 Calyx pubescent, not scaly; capsule glabrous; corolla-tube straight, glabrous or slightly pubescent; cultivated ornamental   *B. davidii* Franchet

1. B. americana L., Sp. Pl. **1**: 112 (1753).
Shrub with tetragonal branches, with intraxylary phloem in the wood, 1·5–4 m. high, aromatic-glandular and also with a soft pale stellate-tomentose indumentum; leaves elliptic-lanceolate, long-acuminate, long-tapered to base, serrulate, paler beneath, 9–30 cm. long, 1·5–8 cm. broad; inflorescence a terminal thyrsoid panicle, the primary branches opposite; style persistent, exserted; stigma oblong, 2-lobed.
Locally common on well drained loose rocky slopes, open rock faces, river gravel and in thickets at pasture margins; 350–3500 feet; fl. and fr. Dec–May. *A 6007! 7105! H 10146! P 24510!* Mexico to Venezuela and Bolivia, Cuba.

2. POLYPREMUM L. (1753)

1. P. procumbens L., Sp. Pl. **1**: 111 (1753).
Diffuse annual or perennial herb, 6–30 cm. high; leaves linear, acute, 8–22 mm. long, the margin minutely denticulate-ciliate; sepals 4 (–5), almost distinct; corolla white, tube about 2 mm. long, lobes rounded and about 2·5 mm. long, shorter than the calyx; capsule rounded in outline, compressed, glabrous.
Very rare, a weed of damp places, not reported since Macfadyen. S. United States, Bahamas, Mexico, C. and S. Amer., Cuba, Hawaii.

3. MITREOLA L. (1758)

1. M. petiolata (J. F. Gmel.) Torr. & A. Gray, Fl. N. Amer. **2**: 45.—*Ophiorrhiza mitreola* L. (1753). *Cynoctonum mitreola* (L.) Britton (1894). Mitrewort.
Erect slender annual herb 5–40 (–90) cm. high; leaves ovate-elliptic, acute, cuneate at base, entire, remotely puberulous, 2–5 cm. long, 1–2·5 cm. broad; inflorescence-branches up to 5 cm. long; calyx (4–) 5-lobed, 1 mm. long in flower, 1·5 mm. long in fruit; fruiting carpels opening adaxially.
Locally common, in alluvial limestone clay, coral sand and muddy swamp margins and ditches; SL–700 feet; fl. and fr. most of the year. *A 8559! H 9759! P 7649!* S-E. United States, Mexico, Greater Antilles, Trinidad; introduced into tropical Asia and also in W. Africa.

4. SPIGELIA L. (1753)

1. S. anthelmia L., Sp. Pl. **1**: 149 (1753).—Pink Weed, Worm Grass.
Erect annual herb 5–40 cm. high; stem glabrous; leaves broadly lanceolate, long-acute, tapered to base, scabridulous above, puberulous-setulose on veins beneath, up to 12 cm. long and 4 cm. broad, subsessile, the upper in two pairs at stem-apex; sepals acute, 2 mm. long; corolla-tube white with magneta stripes on either side of midline of lobes; fruit 2-lobed, scaly-warted, separating from persistent base, 5 mm. in diameter.
Common, a weed of waste open stony or cultivated ground; SL–1600 feet; fl. and fr. all the year. *A 5622! H 8542! P 15430!* Native of tropical Amer. now generally distributed in the tropics of both hemispheres.

159. GENTIANACEAE

Herbs, shrubs or small trees. Leaves opposite, entire, without stipules but often with an interpetiolar ridge or membrane. Flowers bisexual, actinomorphic or rarely subregular, showy in cymose inflorescences. Perianth biseriate; calyx tubular or sepals nearly free, the lobes often keeled or winged, usually imbricate; corolla gamopetalous, the lobes contorted in bud. Stamens epipetalous, alternating with corolla-lobes; filaments free; anthers usually versatile, 2-locular, opening length-wise or rarely by pores. Ovary superior, 1-locular with 2 parietal placentas or 2-locular by connation of the intrusive placentas; ovules mostly numerous, anatro-pous; style simple or absent; stigma entire or 2-lobed. Fruit usually a septicidal capsule. Seeds small with small embryo and fleshy endosperm.

About 70 genera with some 1000 species of world-wide distribution.

1 Saprophytic herbs lacking green colour; stem simple; leaves reduced to scales; inflorescence terminal with erect flowers   **1. Leiphaimos**
1 Holophytic plants with green fully differentiated leaves:
  2 Annual or therophytic herbs; inflorescence terminal:
  3 Flowers 4-merous; calyx broadly winged; corolla pink, mauve or purple
                                                                             2. Schultesia
  3 Flowers mostly 5-merous; calyx very narrowly keeled:
    4 Flowers small, not exceeding 1·5 cm. long, numerous in a much-branched cyme,
      sessile or shortly pedicelled; corolla pink, tube exceeding calyx   **3. Centaurium**
    4 Flowers large, 2–3 cm. long, solitary to few on pedicels up to 5 cm. long; corolla
      bluish-purple, tube shorter than calyx   **4. Eustoma**
  2 Perennial shrubby herbs, shrubs or small trees; corolla yellow or greenish-yellow:
    5 Corolla salver- or funnel-shaped, more or less persistent; stigma subentire; calyx-
      lobes narrow; inflorescences mostly axillary; leaves thin; capsule not beaked or
      beak usually short and fragile   **5. Lisianthius**
    5 Corolla campanulate, slightly zygomorphic, deciduous; stigma 2-lipped; calyx-lobes
      broadly deltate, overlapping; inflorescence terminal; lower leaves rather fleshy,
      more than 25 cm. long and 10 cm. broad; capsule with a strong persistent beak
                                                                           6. Macrocarpaea

1. LEIPHAIMOS Cham. & Schlecht. (1831)

Flowers solitary, rich yellow to orange   1. *aphylla*
Flowers in a dichasial cyme; whole plant light buff   2. *parasitica*

1. L. aphylla (Jacq.) Gilg in Engl. & Prantl, Nat. Pflanzenf. **4** (2): 104 (1895).—
  Gentiana aphylla Jacq. (1760).
  Slender erect perennial herb 5–18 cm. high; base of stem white, yellow or light orange above; scale-leaves opposite; perianth 5-merous; corolla-tube 2–2·5 cm. long, persistent in fruit.
  Rare (Clar., St. Mary, Port., St. Thom.), on rotten logs and leaves in forest or in fern tussocks in open swamp; 1750–2300 feet; fl. Jan–Mar, July–Sept, Dec, fr. Mar. *A 9372! Skelding UCWI 3360! P 15888!* Costa Rico to Paraguay, West Indies.

2. L. parasitica Cham. & Schlecht. in Linnaea **6**: 387 (1831).—*Voyria mexicana* Griseb. (1838).
  Erect glabrous herb 10–15 cm. high; first flower terminal, later flowers secund on inflorescence-branches; flowers 4–5 mm. long; capsule 2·5–3 mm. broad.
  Very rare (Han., St. James, Trel.), in decaying leaves and amongst tree roots in forest on limestone; 750–2100 feet; fl. and fr. Nov–Feb. *H 9176! P 11294!* S. Florida, Bahamas, Mexico, Honduras, Cuba, Hispaniola.

2. SCHULTESIA Mart. (1827) nom. cons.

Corolla 1·5–2·3 cm. long; calyx-tube prominently reticulate-veined   1. *guyanensis*
Corolla 3–4·7 cm. long; calyx-tube with rather inconspicuous veins, not reticulate
                                                                             2. *brachyptera*

1. S. guyanensis (Aubl.) Malme in Arkiv Bot. **3** (12): 9 (1904).—*Exacum guyanense* Aubl. (1775). *S. stenophylla* Mart. (1827).
Erect herb 15–30 cm. high with 4-angled stem; leaves ovate to oblong, sessile, 2–3 cm. long; calyx 1·5 cm. long, the lobes subulate 4–7 mm. long; corolla-lobes obovate.
Very rare, only once reported, *Macfadyen!* Mexico to Paraguay, Cuba, Hispaniola, Trinidad; variety in W. Africa.

2. S. brachyptera Cham. in Linnaea **8**: 8 (1833).—*S. heterophylla* Miq. (1847).
Erect herb 15–45 cm. high, simple-stemmed or sparingly branched above; leaves lanceolate to linear-lanceolate, sessile, 1–5 cm. long; calyx 2–3 cm. long, the lobes attenuated to fine points 8–10 mm. long, veins of the wing pectinate but finely branched and anastomosing marginally; corolla-lobes shorter than the tube.
Very rare, reported only by *Macfadyen!* Mexico to Paraguay, Greater Antilles.

3. CENTAURIUM Hill (1756)

With Wm. T. Stearn

Plant usually tall but very variable, 10–90 cm. high, branched above; basal rosette of leaves usually present at flowering time; flowers sessile or subsessile   **1.** *erythraea*
Plant low-growing, 3–15 cm. high, usually branched from near the base, the basal rosette of leaves usually absent at flowering time; flowers mostly short-pedicelled
          2. *pulchellum*

1. C. erythraea Rafn, Danm. Holsteens Fl. **2**: 75 (1800).—*Erythraea centaurium*, *C. minus* and *C. umbellatum* of various authors.
Simple or bushily branched glabrous herb; leaves of basal rosette oblanceolate, up to 5 cm. long and 1·5 cm. broad, longitudinally nerved, of the flowering stem smaller, ovate to lanceolate, sessile; flowers (4–) 5-merous, with 2 bracteoles immediately below calyx; calyx 3–5 mm. long, green; corolla-tube up to 9 mm. long, greenish, lobes pink to whitish at base, spreading.
Very locally common (St. Andr.), a weed of pathsides and open waste ground; 3400–5500 feet; fl. and fr. Dec–Mar, but mostly May–Aug. *A 7429! H 9545!* Native of Europe, N. Africa, W. and C. Asia, introduced and naturalized in Amer.

2. C. pulchellum (Sw.) Druce, Fl. Berks.: 342 (1898).—*Gentiana pulchella* Sw. (1783).
Herb like the last; leaves of flowering stem elliptical, up to 2·5 cm. long and 6 mm. broad, 1–3-nerved; pedicels 2–10 mm. long; calyx 5–6 mm. long, green; corolla-tube 6–7 mm. long, lobes pink, spreading.
Very rare (St. Ann), not recently collected. *McNab! March!* Native of Europe, N.W. Africa and C. Asia, naturalized in Bermuda, Barbados and W. tropical Africa.

4. EUSTOMA Salisb. (1806)

1. E. exaltatum (L.) G. Don, Gen. Syst. **4**: 211 (1837).—*Gentiana exaltata* L. (1762). *E. silenifolium* Salisb. (1806).
Glabrous erect herb 10–50 (–90) cm. high; stems and leaves glaucous; leaves oblong to elliptical, obtuse or acute, sessile, up to 8 cm. long and 3 cm. broad; calyx-lobes filiform-subulate, 10–12 mm. long; corolla-lobes nearly 2 cm. long; capsule ellipsoid, 1·5–2·5 cm. long, valves about 8 mm. broad.
Local (St. Ann, St. Thom.), in coastal sands and at margins of marshes and in ditches near the sea; near SL; fl. Nov–June, fr. Jan–June. *H 11605! Loveless UCWI 516! P 24210!* S. United States, Mexico to Venezuela, Cuba, Hispaniola.

5. LISIANTHIUS P. Browne (1756)

By R. E. Weaver

1 Inflorescence-branches greatly contracted, the inflorescences appearing umbellate or capitate and subtended by 2 pairs of opposite involucral bracts; peduncles compressed:

2 Pedicels in fruit 9–20 mm. long; corolla-lobes one-tenth to one-eighth as long as tube
1. umbellatus

2 Pedicels in fruit less than 7·5 mm. long, the fruits often sessile; corolla-lobes one
quarter to one-third as long as tube
2. capitatus

1 Inflorescence-branches elongate, the inflorescences open, dichasioid or thyrsiform,
occasionally reduced to a single flower, and subtended by a pair of opposite bracts,
but these not involucrate; peduncles terete:

3 Plants puberulous-scabridulous, especially the inflorescence:

4 Leaves on flowering branches cuneate or at most rounded at base   3. longifolius

4 Leaves on flowering branches, at least the upper ones, distinctly cordate at base
4. cordifolius

3 Plants glabrous:

5 Corolla persistent on mature fruits:

6 Calyx-lobes conspicuously winged abaxially; leaves membranaceous, elliptical or
lanceolate
3. longifolius

6 Calyx-lobes not winged abaxially; leaves coriaceous, oblanceolate to obovate
5. adamsii

5 Corolla not persistent on mature fruits:

7 Stamens and style greatly exserted, exceeding the corolla by at least 3 cm. in mature
flowers; corolla-tube constricted for at least half its length   6. exsertus

7 Stamens and style included, or if exserted, not exceeding the corolla by more than
1·5 cm.; corolla-tube constricted for less than half, usually about one-third its
length:

8 Leaves up to 22 cm. long and 7 cm. broad, usually less than 3 times as long as
broad; capsule including beak 14–20 mm. long, the beak 3–6 mm. long
7. latifolius

8 Leaves up to 17 cm. long and 5 cm. broad, but usually less than 11 cm. long and
2·5 cm. broad, usually more than 3 times as long as broad; capsules including
beak 8–13 mm. long, the beak less than 1·5 mm. long   8. troyanus

1. L. umbellatus Sw., Nov. Gen. & Sp. Pl.: 40 (1788).

Glabrous sparsely branched shrubs 1·5–6 m. high, occasionally arborescent;
leaves oblanceolate to obovate, rarely elliptical, attenuate at base, acuminate, 7–29
cm. long, 2–9·5 cm. broad; inflorescences thyrsiform, but the branches greatly
contracted and appearing umbellate; peduncles ascending, 7–19 cm. long; calyx
scarious, 4–6·5 mm. long, persistent; corolla tubular-funnelform, the tube bright
yellow, 22–31 mm. long, 4–6 mm. broad, the lobes greenish, orbicular, as broad
as long, erect, 2–4 mm. long; styles and stamens exserted, exceeding the corolla
by 5–15 mm.; capsules 10–15 mm. long.

Very local (Han., Trel.), in moist thickets on limestone hills; 1000–2000 feet;
fl. Jan–May, fr. May–Sept. *A 8005! H 9268! Weaver 1832!* Endemic.

2. L. capitatus Urb., Symb. Ant. **6**: 33 (1909).

Similar to the last but leaves 6–31 cm. long and 1·5–6·5 cm. broad; peduncles
7–23 cm. long; calyx 7·5–9 mm. long; corolla tubular-campanulate, the tube 16·5–
22 mm. long and 7·5–10 mm. broad and the lobes ovate, longer than broad, 4·5–
6·5 mm. long; style and stamens exceeding the corolla by 15–30 mm.; capsules
14–22 mm. long.

Local in central and eastern parishes, in moist thickets on limestone hills and rare
on rocky headlands by the sea; SL–2700 feet; fl. and fr. Jan–Oct. *A 10833! H 6683!
Weaver 940! 1002! 1291!* Endemic. *L. umbellatus* and *L. capitatus* are unique in the
genus in having apparently umbellate or capitate inflorescences. They are closely
related and have often been confused.

3. L. longifolius L., Mant. Pl.: 43 (1767).—*Leianthus longifolius* (L.) Griseb.
(1862). *Leianthus longifolius* var. *gracilis* Griseb. (1862). *Lisianthus gracilis*
(Griseb.) Perkins (1902). Jamaican Fuchsia.

Erect or ascending sparsely to profusely branched subshrubs 0·5–2 m. high,
completely glabrous to moderately puberulous-scabridulous; leaves elliptical to
lanceolate, cuneate to long-attenuate at base, short to long acuminate or rarely
rounded at apex, glabrous or ciliolate, 1·5–13·5 cm. long, 0·4–4·0 cm. broad, the
leaves on axillary or flowering branches but slightly different in shape from those on
main axes; inflorescences variable from long-peduncled many-flowered thyrses to
simple dichasia or nearly sessile single flowers, but always on determinate axillary
branches; calyx 7–22 mm. long, glabrous to scabridulous, the lobes winged or at

least keeled abaxially, the wing 0·3–4 mm. broad; corolla glutinous, tubular-funnelform, persistent on mature fruit, 2·9–6·5 cm. long, the tube bright yellow, the lobes greenish, ovate-lanceolate to oblong, acuminate, recurved, 8–22 mm. long; stamens and styles included or slightly exserted; capsule 8–17 mm. long.

Rather common, on exposed banks or thickets on limestone or serpentine soils; 25–5000 feet; fl. and fr. all the year. *A 5457! 8947! H 9030! 12031! P 23239! Weaver 894! 941!* Endemic. *L. longifolius* presents a rather perplexing taxonomic problem. The species is extremely variable. It is made up of a large number of local populations recognized by different combinations of intergrading characters (e.g. length of calyx, width of calyx-wing, length of corolla, pubescence, etc.). None of the populations or groups of populations are clearly distinct, and none are here given taxonomic status of any sort.

4. L. cordifolius L., Mant. Pl.: 43 (1767).—*Leianthus longifolius* (L.) Griseb. var. *cordifolius* (L.) Griseb. (1862).

Profusely branched subshrubs 20–90 cm. high, puberulous-scabridulous in vegetative parts; leaves 0·5–5 cm. long, 0·4–2·5 cm. broad, those on main axes ovate-lanceolate to lanceolate, cuneate or rounded at base, acuminate, those on axillary branches ovate, rounded or, at least the upper ones, cordate at base, acute or acuminate, ciliolate; inflorescences terminating axillary shoots, few-flowered, dichasioid or reduced to a single flower; calyx 7–10 mm. long, glabrous to scabridulous, the lobes slightly keeled abaxially; corolla glutinous, tubular-funnelform, persistent on mature fruit, 3–4 cm. long, the tube bright yellow, the lobes greenish, ovate-lanceolate, acuminate, recurved, 9–13 mm. long; stamens included or nearly so; style usually exserted, exceeding the corolla by no more than 7 mm.; capsule 10–13 mm. long.

Local (St. Andr.), in dry thickets and in crevices of steep limestone cliffs; 200–1300 feet; fl. and fr. Nov–June, fr. June–Aug. *A 7095! H 10055! Webster & Wilson 4860! Weaver 1821!* Endemic. Closely related to and perhaps only varietally distinct from *L. longifolius.*

5. L. adamsii Weaver in Brittonia **22**: 11, f. 1 (1970).

Weak-stemmed spindly shrub 0·2–2 m. high, glabrous throughout; leaves coriaceous, oblanceolate to obovate, rarely elliptical, attenuate at base, abruptly short-acuminate, 6–18 cm. long and 2–6·5 cm. broad; inflorescences axillary, 4–18-flowered, dischasioid or thyrsiform, the peduncles 8–20 cm. long; calyx unkeeled, 7–11 mm. long; corolla tubular-campanulate, slightly ventricose, 4–5·5 cm. long, persistent on mature fruit, the tube constricted in lower third, bright yellow, the lobes ovate-lanceolate, acuminate, recurved, 12–20 mm. long; stamens and style as long as corolla or exceeding it by no more than 13 mm.; capsules 14–17 mm. long.

Rare (Clar., Han., Manch.), on rocky limestone hills; 1000–2300 feet; fl. and fr. Apr–Oct, fr. Feb. *A 13205! P 10037! 16182! Weaver 1293!* Endemic.

6. L. exsertus Sw., Nov. Gen. & Sp. Pl.: 40 (1788).

Glabrous shrub, 1–6 m. high, rarely a slender tree; stems unbranched in lower portions but usually profusely branched above, brittle; leaves elliptical or rarely elliptic-oblong, the bases attenuate, acuminate or rarely acute, 3–13·4 cm. long and 1–5·5 cm. broad; inflorescences axillary, nearly always opposite, 1–34-flowered, dichasioid or thyrsiform, but sometimes reduced to a single flower, the peduncles to 6 cm. long; calyx unkeeled, 6–9 mm. long; corolla tubular-campanulate, not persistent on mature fruit, 2–3·3 cm. long, light yellow, the lobes greenish, broadly ovate, erect, 3–5 mm. long; capsule ellipsoid, 10–16 mm. long.

Occasional in the eastern and central parishes, in thickets and on well drained banks and in wet elfin woodland; 600–3500 feet; fl. and fr. nearly all the year. *A 9151! H 7452! Weaver 951! 1824!* Endemic and quite distinct from all other members of the genus.

7. L. latifolius Sw., Nov. Gen. & Sp. Pl.: 40 (1788).

Glabrous shrub 0·3–2 m. high; stems sparsely branched, brittle; leaves elliptical, 3·6–22 cm. long, 1·5–7 cm. broad, usually less than 3 times as long as broad; inflorescences axillary, 1–18-flowered, dichasioid or thyrsiform, but occasionally reduced to a single flower, the peduncles naked or with 1–2 pairs of leaflike bracts; calyx unkeeled, 7–11 mm. long; corolla funnelform, not persistent on mature fruits, 3·5–5·5 cm. long, bright yellow, the lobes greenish, ovate-lanceolate to oblong, acute or acuminate, recurved, 11–18 mm. long; stamens and style included or nearly so.

Local and quite rare (St. Andr., St. Thom., Port.), in montane forests or elfin woodland; 1500–7000 feet; fl. and fr. most of the year. *H 7003! P 9990! Weaver 1827! 1953!* Endemic and similar to the following.

8. L. troyanus Urb., Symb. Ant. **6**: 32 (1909).

Sparsely branched subshrubs 0·5–1·5 m. high, glabrous throughout; leaves narrowly elliptical or tending to oblanceolate, long-attenuate at base, acuminate or rarely acute, up to 17 cm. long and 5 cm. broad but usually much smaller and more than 3 times as long as broad; inflorescences axillary, 3–15-flowered, dichasioid or thyrsiform; calyx unkeeled, 6·5–12 mm. long; corolla tubular-funnelform, not persistent on mature fruits, 3·5–4·5 mm. long, the tube bright yellow, the lobes greenish, ovate-lanceolate, acuminate, recurved, 9–14 mm. long; style and stamens included or barely exceeding corolla.

Uncommon in the western parishes, in thickets or on dry banks on limestone; 200–2000 feet; fl. and fr. July–Mar. *H 8789! P 11295! Weaver 1272!* Endemic and similar to *L. latifolius* which is larger and more robust in all ways. Very similar also to *L. laxiflorus* Urb. in Puerto Rico and *L. domingensis* Urb. in Hispaniola.

6. MACROCARPAEA (Griseb.) Gilg (1895)

1. M. hartii Krug & Urb. in Notizbl. Bot. Gart. Berlin **1**: 80 (1895).—*Lisianthus thamnoides* Griseb. (1862), partly.

Shrub 4 m. or tree 5–6 m. high, glabrous or nearly so, the leaves sometimes puberulous on midrib beneath; leaves obovate to oblanceolate, cuspidate-acuminate, long-tapered to and connate-amplexicaul at base, up to 40 cm. long and 15 cm. broad; calyx 8–9 mm. long; corolla light yellow or yellowish-green, about 2·5 cm. long; stamens included; capsule 1·5–2·5 cm. long, conical, sometimes curved or beak oblique.

Local (St. Andr., Port.), in wet montane and submontane woodlands; (2500–) 3500–4500 feet; fl. mostly May–Sept, fr. July–Sept. *A 7671! H 7718! J.P. 1417! P 23868! Weaver 952!* Endemic.

160. MENYANTHACEAE

Perennial herbs. Leaves alternate, simple or trifoliolate, entire or toothed, without stipules. Flowers bisexual, actinomorphic, solitary or in subumbellate cymes. Perianth biseriate, 5 (–6)-merous; calyx deeply cleft, persistent; corolla gamo-petalous, rotate or funnel-shaped, the lobes induplicate-valvate in bud. Stamens equal in number to corolla-lobes and alternate with them, epipetalous, uniform, filaments free; anthers 2-locular, opening lengthwise. Ovary superior to partly inferior, 1-locular, with usually 2 parietal placentas; ovules numerous, anatropous; style simple; stigma 2–3-lobed. Fruit a capsule or indehiscent, often bursting irregularly, sometimes 2-valved. Seeds with hard testa and copious endosperm.

5 genera with 40 species of swamp or aquatic plants with worldwide distribution.

1. NYMPHOIDES Séguier(1754); Ornduff (1969)

Leaves simple, cordate; flowers in clusters closely subtended by a leaf-blade and thus appearing to arise from a petiole, 5-merous: corolla-tube short, the lobes fimbriate-margined, often pilose within; corolla appendaged within between the stamens; capsule opening irregularly.

1. N. indica (L.) Kuntze, Revis. Gen. Pl. **2**: 429 (1891).—*Menyanthes indica* L. (1753). *Villarsia humboldtiana* Kunth (1818). *Limnanthemum humboldtianum* (Kunth) Griseb. (1838). *N. humboldtiana* (Kunth) Kuntze (1891).

Perennial aquatic herb with trailing stem and floating leaves; petioles variable; leaf-blade suborbicular to broadly ovate, entire, glabrous, softly fleshy, sometimes purplish beneath, up to about 10 cm. broad; pedicels 2–5 cm. long, deflexed and longer in fruit; calyx-lobes acute, about 6 mm. long; corolla rotate, about 2·5 cm. across, the lobes white, yellow at the base; flowers heterostylous; ovary and style 6 or 10 mm. long; filaments yellow, 4 or 1 mm. long; anthers greenish-yellow, turning black; capsule ovoid, enclosed by calyx, about 6 mm. long, many-seeded; seeds lenticular, about 1·5 mm. in diameter.

Rather local, more common in the central and western parishes, in ponds and swamps; SL–1500 feet; fl. May–Feb, fr. June–Feb. *A 9452! H 11642! P 10395!* Continental tropical Amer., Greater Antilles, Guadeloupe, Trinidad; Grand Cayman and Old World tropics.

161. APOCYNACEAE

Trees, erect or climbing shrubs or perennial herbs, often with milky or viscid sap. Leaves opposite, verticillate or alternate, simple, entire, sometimes with minute stipules. Flowers bisexual, actinomorphic, solitary, racemose or cymose. Perianth biseriate, typically 5-merous; calyx often glandular within, the lobes imbricate; corolla gamopetalous, the lobes contorted or rarely valvate in bud, the tube often appendaged within (corona). Stamens (4–) 5, alternate with corolla-lobes, epipetalous, distinct; anthers free or coherent around the style, 2-locular, opening lengthwise, often sagittate and apiculate with the connective produced; pollen granular or in tetrads. Disk usually present, annular, cupular or of separate glands. Ovary mostly superior, 1-locular with 2 parietal placentas or 2-locular with axile placentas or carpels 2 and more or less free with adaxial placentas in each carpel; style 1, split at the base or entire, thickened below the apex; ovules 2 or more in each carpel, anatropous or campylotropous. Fruit of solitary or paired follicles or a capsule, berry or drupe. Seeds comose, winged or arillate, with large embryo and endosperm.

200 genera with about 2000 species in all regions.

1 Leaves spiral or alternate; erect shrubs or trees:
  2 Flowers white, less than 2 cm. long; branchlets drooping, slender, with alternate leaves; fruit drupaceous, oblong-obovoid, 10–13 mm. long, 9–10 mm. broad
    2. Vallesia
  2 Flowers showy, if white then also with a yellow eye, 3·5–7 cm. long; branchlets thick and rather fleshy; leaves spiral:
    3 Ultimate branchlets up to 3 cm. thick; leaves oblong, elliptical or oblanceolate; corolla-limb rotate; inflorescence long-pedunculate, usually many-flowered; flowers fragrant; fruit follicular
      1. Plumeria
    3 Ultimate branchlets 4–5 mm. thick; leaves linear; corolla-limb funnel-shaped; inflorescence shortly pedunculate, the branches 1–3-flowered; flowers yellow or pinkish-orange, not fragrant, about 7 cm. long; cultivated ornamental, sometimes relict; native of tropical America; Lucky Nuts, Lucky Seeds, Milk Bush
      Thevetia peruviana (Pers.) K. Schum.
1 Leaves opposite or whorled:
  4 Leaves normally in whorls of 3 or 4, rarely opposite:
    5 Flowers less than 1·5 cm. long; glands present in leaf-axils or on base of petiole as well; fruit a fleshy drupe finally black; wild erect shrubs or small trees
      3. Rauvolfia
    5 Flowers 4 cm. or more long; cultivated or relict ornamental shrubs:
      6 Corolla often pink (rarely white or yellow), with laciniate appendages at mouth of tube, the lobes overlapping in bud to the right; anthers with long hairy apical appendages; fruit follicular, smooth; erect shrub with dark green leaves elongated and pointed at both ends; native of Eurasia; Oleander   *Nerium oleander* L.
      6 Corolla without appendages, the lobes overlapping in bud to the left; anthers without apical appendages; fruit a spiny capsule; natives of Brazil:
        7 Corolla yellow; erect or climbing more or less glabrous shrubs; Yellow Allamanda
        Allamanda cathartica L.

7 Corolla purplish-mauve; erect or scrambling hirsute shrub; Purple Allamanda
Allamanda violacea Gardn.
4 Leaves opposite:
  8 Erect herbs, shrubs or trees; seeds, as far as known, not comose:
    9 Plant more or less herbaceous; flowers and fruits subsessile in leaf-axils; corolla-
      tube 2·5 cm. long   4. **Catharanthus**
    9 Shrubs or trees; flowers and fruits pedicellate in pedunculate corymbose in-
      florescences; corolla-tube not exceeding 2 cm. long (nearly 2·5 cm. long in
      Tabernaemontana divaricata):
      10 Lateral veins of leaf numerous and almost parallel; anthers with threadlike apical
        appendages; fruit samaroid   5. **Cameraria**
      10 Lateral veins of leaf distant, curved; anthers without threadlike apical appendages;
        fruits follicular:
        11 Inflorescence erect, much-branched and many-flowered, long-pedunculate;
          calyx without squamellae within; follicles straight; seeds flattened, angled,
          winged at each end   6. **Strempeliopsis**
        11 Inflorescences pendulous to ascending, mostly few-flowered and shortly ped-
          unculate; calyx with squamellae within; follicles gibbous; seeds rather
          rounded, with fleshy coloured arils (as far as known)   7. **Tabernaemontana**
  8 Twining shrubs; seeds, as far as known, comose:
    12 Corolla-lobes of erect bud normally overlapping or twisted to the left (when seen
      from side), white or greenish-white:
      13 Inflorescences alternate-racemose; calyx without squamellae within; corolla-tube
        6–12 mm. long; stamens not adhering to stigma; fruit a slender 1-seeded follicle
        produced into a beak barbed with hairs   8. **Anechites**
      13 Inflorescences dichasial to thyrsoid; calyx with squamellae within; corolla-tube
        1–3 mm. long; follicles several-seeded, linear   9. **Forsteronia**
    12 Corolla-lobes of erect bud normally overlapping or twisted to the right (when seen
      from side); stamens adhering to stigma:
      14 Corolla funnel-shaped, expanded above the stamens and open-mouthed:
        15 Calyx without squamellae within; corolla white with yellow eye
                                          10. **Rhabdadenia**
        15 Calyx with squamellae within; corolla yellow:
        16 Anthers with linear apical appendages; calyx 10–18 mm. long
                                        11. **Urechites**
        16 Anthers without linear apical appendages; calyx up to 5 mm. long
                                        12. **Angadenia**
      14 Corolla salver-shaped, the tube slightly expanded at insertion of stamens, then
        cylindrical or constricted distally; calyx with squamellae within:
        17 Corolla-tube 3–4 (–5) cm. long; follicles straight, smooth, not torulose
                                        13. **Echites**
        17 Corolla-tube less than 2 cm. long:
          18 Corolla-tube 4–6 mm. long; follicles constricted at intervals   14. **Mandevilla**
          18 Corolla-tube about 15 mm. long; fruit not known   15. **Secondatia**

1. PLUMERIA L. (1753); Woodson (1938)

1 Corolla-tube gradually broadening upwards; leaves pandurate, the tip usually long-
  acuminate, marginal vein obscure, up to 30 cm. long and 15 cm. broad; flowers
  numerous in a congested inflorescence, white with yellow eye; cultivated ornamental,
  originally described from Curaçao   *P. pudica* Jacq.
1 Corolla-tube cylindrical or nearly so; leaves with convex-rounded to nearly straight
  lateral margins, the tip not long-acuminate:
  2 Inflorescence lax, corymbose; leaves mostly obovate to oblong-lanceolate, obtuse to
    shortly acuminate, the marginal vein distinct, up to 50 cm. long and 15 cm. broad;
    corolla white, yellow or suffused rose; widespread cultivated ornamental; originally
    described from Jamaica but not native there and more probably indigenous to
    continental America; Frangipani   *P. rubra* L.
  2 Inflorescence compact, the secondary peduncles unequal; leaves very variable from
    linear-oblanceolate to broadly elliptical, rounded to acute and emarginate to mucro-
    nate at tip, the marginal vein obscure; corolla white with yellow eye; wild plants
                                                         1. *obtusa*

1. P. obtusa L., Sp. Pl. **1**: 210 (1753).—*P. marchii* Urb. (1902). *P. jamaicensis*
Britton (1910). *P. confusa* Britton (1915). Wild Frangipani.
  Shrubs or trees up to 6 (–12) m. high; leaves long-petioled, the blade oblong-
elliptical, oblanceolate to obovate, usually narrowed to base, with numerous parallel
lateral veins, up to 24 (–33) cm. long and 12 (–17) cm. broad; pedicels slender, up

to about 10 mm. long; corolla-tube up to 20 mm. long, the lobes about twice as long; follicles up to 22 cm. long and 2·5 cm. broad.

Locally common in thickets and woodlands on limestone, often in exposed situations near the sea; SL–3000 feet; fl. mostly Mar–Sept, fr. Mar–Oct. *A 7304! P 8636! 23423! 24015!* Woodson was unable to resolve the complex of variants in Jamaica taxonomically and little further information has been gathered which suggests how this could be done. There is great diversity of form and ecological differentiation in this group in Jamaica but a thorough biometric investigation is required to state it clearly. In the the broad sense in which Woodson construed this species the distribution extends through Bahamas, Greater Antilles, Cayman Is. and Honduras.

2. VALLESIA Ruiz & Pav. (1794)

1. V. antillana Woodson in Ann. Miss. Bot. Gard. **24**: 13 (1937).

Shrub or tree 1–5 m. high, with drooping branches; bark yellowish-buff, twisted-fissured; leaves ovate-elliptical to elliptic-lanceolate, acute or mostly shortly acuminate, up to 9 cm. long and 3 cm. broad, light green, rather fleshy; inflorescence simple or dichotomous-cymose; flowers fragrant; corolla 5 (–6)-merous, the tube 5–7 mm. long, the lobes about 4 mm. long; fruit opalescent, on pedicel up to 7 mm. long.

Uncommon and rather local (St. Cath., Trel., St. Ann), in woodland and thicket on coral sand or arid limestone near the sea; SL–100 feet; fl. Feb–Mar, fr. June, Oct. *A 12364! H 10157! P 17428!* Florida, Bahamas, Cuba, Hispaniola.

3. RAUVOLFIA L. (1753)

Leaves shiny adaxially, glabrous; peduncles rather stout, these and pedicels glabrous; corolla-tube 4–6 mm. long, lobes 3–4 mm. long; fruit 7–18 mm. broad   1. *nitida*
Leaves dull adaxially, softly pubescent abaxially; peduncles and pedicels slender, pubescent; corolla-tube 3–4 mm. long, lobes 1–1·5 mm. long; fruit 5–8 mm. broad
      2. *tetraphylla*

1. R. nitida Jacq., Enum. Syst. Pl. Carib.: 14 (1760).—Glasswood.

Glabrous shrub to 4 m. or tree to 10 m. high; trunk slender with pale bark; leaves unequal in whorls of (2–) 4 (–7), ovate-elliptical, to elliptic-lanceolate, mostly acutely acuminate, (2–) 5–18 cm. long, (1–) 3–6·5 cm. broad, glossy; lateral branchlets often terminated by a whorl of leaves; peduncles axillary or terminal, up to 4 cm. long; calyx 1·5–2 mm. long, the lobes rounded; corolla greenish-white to yellowish; fruit 1–2-seeded.

Occasional in woodland margins and thickets on limestone in heavy soils and in sublittoral swamp forest; SL–3500 feet; fl. May–July, fr. Aug–Oct. *A 12522! H 9743! J.P. 1413! Powell 548!* Bahamas, Greater Antilles, Virgin Is., St. Kitts.

2. R. tetraphylla L., Sp. Pl. **1**: 208 (1753).

Pubescent shrub up to 2·5 m. high; twigs with yellowish bark; leaves often very unequal, (3–) 4 (–5) in a whorl, broadly elliptical to oblanceolate, acute or acuminate, 1–12 cm. long, 4–45 mm. broad, not glossy, the smallest leaves often subsessile; peduncles axillary and terminal, obsolete to about 1 cm. long; calyx 1·5–2 mm. long, the lobes ovate; corolla greenish-white or white speckled pink; fruit 1-2-seeded.

Rather uncommon, in the eastern parishes and along the north coast, in sandy thickets and on roadside and ditch banks where ground water available; SL–500 feet; fl. and fr. June–Dec. *A 11508! 12623! P 23919!* Mexico to Venezuela, Greater Antilles, St. Thomas, Barbados, Trinidad and Tobago.

4. CATHARANTHUS G. Don (1837)

1. C. roseus (L.) G. Don, Gen. Syst. **4**: 95 (1837).—*Vinca rosea* L. (1759). Old Maid, Periwinkle, Ram-Goat Rose.

Bushy herb up to about 80 cm. high; leaves oblong-elliptical, obtuse, rounded or mucronate-acuminate at tip, 2–7 cm. long, 1·5–3 cm. broad; flowers solitary or up to 3 in short axillary cymes; pedicels very short; calyx-lobes narrow, up to 7 mm.

long; corolla-tube inflated immediately below the limb, the lobes obovate, 1·5–2·5 cm. long; follicles subterete, 1·5–3·5 cm. long, puberulous, subtended on either side by a subulate yellowish gland.

Common in gardens and waste places, mostly at low elevations; fl. and fr. all the year. *H 7006! H & P 14481! J.P. 626!* Originally described from Madagascar, now widespread in the subtropics and tropics. Most plants of this species have one of the three following colour expressions in the corolla-limb, other colours being extremely rare; rose-pink with reddish-purple eye (var. *roseus*), white with reddish-purple eye (var. *ocellatus* G. Don) and white with greenish-yellow eye (var. *albus* G. Don).

5. CAMERARIA L. (1753)

1. C. latifolia L., Sp. Pl. **1**: 210 (1753).—Bastard Manchineel.

Glabrous tree up to 13 m. high; trunk smooth, with light brown bark; leaves broadly ovate to suborbicular, cuneate at base, acuminate at tip, the acumen acute to caudate-obtuse, with numerous almost straight lateral veins, 1·5–5 (–6) cm. long, 0·7–2·8 cm. broad; inflorescence a terminal few-flowered cyme; calyx-lobes 1–1·5 mm. long; corolla-tube 5–9 mm. long, lobes obovate, 6–15 mm. long, white; fruit 4–5 cm. long, 1·5–2·5 cm. broad.

Very rare (West., Han., St. James), in open woodland on rocky limestone hills; 300–450 feet; fl. Dec. *A 8652! Britton & Hollick 2045!* British Honduras, Cuba, Hispaniola.

C. angustifolia L. was reported from Jamaica by Lunan, but the occurrence of this species outside Hispaniola has not been confirmed; it may be distinguished by the linear-lanceolate leaves, up to 8 cm. long but not exceeding 3 mm. broad.

6. STREMPELIOPSIS Benth. (1876)

1. S. arborea Urb., Symb. Ant. **5**: 461 (1908).—Gutterwood.

Shrub 1–2 m. or tree to 8 (–12) m. high; stipules semicircular, about 2·5 mm. broad, caducous; leaves oblong-oblanceolate, cuneate at base, obtuse or abruptly and bluntly acuminate, up to 16 cm. long and 7 cm. broad, rather leathery, shiny; inflorescences terminal; peduncles up to 11 cm. long, whitish; calyx-lobes triangular, 1–1·5 mm. long; corolla-tube 4–6 mm. long, lobes about 4 mm. long, white with yellow eye; follicles 5–8 cm. long, ribbed; seed-body 1·5–1·8 cm. long.

Rare and local (Han., Trel.), in woodlands on craggy limestone; 1300–1750 feet; fl. Aug–Apr, fr. Apr, Oct. *A 6778! H 8675! 10265! P 21430!* Endemic.

7. TABERNAEMONTANA L. (1753)

1 Corolla-tube 14–18 (–24) mm. long:
  2 Leaf-blade up to about 10 cm. long with 4–6 lateral veins on each side of midrib, the secondary veins faint; corolla swollen at insertion of stamens in middle of tube

                                                             1. *discolor*
  2 Leaf-blades mostly more than 10 cm. long with 7–11 lateral veins on each side of midrib, the secondary veins evident:
    3 Flowers usually " double ", the corolla-tube up to 24 mm. long, swollen at the middle, yellow within, greenish without, the limb white with crimped lobe-margins; leaves oblong-elliptical, up to about 8 cm. broad, distinctly acutely acuminate; common cultivated ornamental; origin unknown; Coffee Rose (*Ervatamia divaricata* (L.) Burkill)   *T. divaricata* (L.) R. Br. ex Roem. & Schult.
    3 Flowers " single ", the corolla-tube 15–18 mm. long, swollen distally, yellowish-green, the limb yellow with entire lobe-margins; leaves ovate to broadly elliptical, up to 10 cm. broad, rounded, obtuse or cuspidate at tip; wild in coastal communities

                                                             2. *laurifolia*
1 Corolla-tube 5–11 mm. long:
  4 Corolla-tube swollen at insertion of stamens in distal one-third; leaf-blade up to 9 (–14) cm. long with 5–9 lateral veins on each side of midrib; calyx 1·5–2 mm. long, the lobes acute or rather obtuse   3. *ochroleuca*
  4 Corolla-tube swollen at about the middle; leaf-blades mostly longer than 7 lateral veins on each side of midrib; calyx 2–4 mm. long, the lobes obtuse or rounded:
    5 Corolla-lobes 14–16 mm. long   4. *ovalifolia*

5 Corolla-lobes 4–8 mm. long:
  6 Corolla-tube 5–6·5 mm. long, the lobes 4–5 mm. long; follicles semiorbicular to
    subspherical   5. *wullschlaegelii*
  6 Corolla-tube 7–8 mm. long, the lobes 7–8 mm. long:
    7 Leaves glaucescent; follicles almost lanceolate in outline; in north-western parishes
                                     6. *glaucescens*
    7 Leaves yellowish-green; follicle unknown; eastern parishes   7. *rendlei*

1. T. discolor Sw., Nov. Gen. & Sp. Pl.: 52 (1788).
    Glabrous shrub about 2 m. high; leaves unequal, elliptical to narrowly elliptical, cuneate at base, acuminate, 3–8 cm. long, 1–3·5 cm. broad; cymes loosely few-flowered; pedicels 10–12 mm. long; calyx about 2 mm. long, the lobes rounded; corolla-tube 14–15 mm. long, the lobes 6–10 mm. long, white or yellowish.
    Very rare and not known from recent collections. *Swartz*. Endemic.

2. T. laurifolia L., Sp. Pl. **1**: 210 (1753).
    Glabrous shrub 1·5–2·5 m. or tree up to 10 m. high; leaves of a pair often unequal, rounded to broadly cuneate at base, 4–20 cm. long, 2·5–10 cm. broad; cymes small, compact; pedicels 3–8 mm. long; calyx 2·5–3 mm. long, the lobes obtuse or rounded; corolla-lobes 10–12 mm. long; follicles up to 5 cm. long and 1·5 cm. thick; seeds embedded in soft orange pulp, about 5 mm. long.
    Common in coastal thickets and at mangrove margins on limestone; SL–50 feet; fl. and fr. most of the year. *A 10975! Loveless UCWI 2507! Stearn 645!* Also in Grand Cayman.

3. T. ochroleuca Urb., Symb. Ant. **6**: 34 (1909).—*T. discolor* of Woodson (1938), not Sw.
    Glabrous shrub from 2 m. or tree up to 6 m. high; leaves unequal, elliptical, narrowly attenuate-cuneate at base, acuminate, 3–11 (–14) cm. long, 1–4 (–5) cm. broad; cymes lax; pedicels very slender, 6–10 mm. long; corolla-tube yellow, 9–10 mm. long, the lobes white then yellow, 7–10 mm. long, 2·5–3·5 mm. broad; follicles narrowly ellipsoid up to 5 cm. long and 1 cm. broad.
    Local (Han.), in woodland on limestone; 900–1700 feet; fl. July, Dec. *A 8591! H 10297! Webster & Wilson 5081!* Endemic.

4. T. ovalifolia Urb., Symb. Ant. **5**: 462 (1908).
    Glabrous tree about 8 m. high; leaves unequal, elliptical, cuneate at base, obtuse or acute, 3–22 cm. long, 1·5–9 cm. broad, rather leathery, with 5–6 veins on each side of midrib on smaller leaves and up to 10 or 11 on the larger; cymes repeatedly branched, lax, corymbose, many-flowered; pedicels 3–15 mm. long; corolla white, the tube about 11 mm. long, the lobes 8 mm. broad; follicles lanceolate.
    Rare (Han.), on limestone hillside; 900–1200 feet; fl. Mar–May. *H 9239! 10276!* Endemic.

5. T. wullschlaegelii Griseb., Fl. Br. W.I.: 409 (1861).—*T. lactea* Urb. (1909).
    T. calcicola Urb. (1915).
    Glabrous shrub 2–3 m. or tree up to 10 m. high; leaves equal or unequal, elliptical to narrowly lanceolate, cuneate to rounded at base, acute or shortly acuminate, 3–18 cm. long, 1·5–8 cm. broad, with 8–10 prominent lateral veins on each side of midrib; cymes lax, many-flowered; pedicels 2–9 mm. long; corolla at first greenish, later cream-coloured or yellow, the lobes 2·5 mm. broad; follicles 2·5–4 cm. long, up to 2·5 cm. broad.
    Occasional, in the central parishes, on wooded limestone hills; 1500–3000 feet; fl. Dec–Apr, fr. Oct–Dec. *A 10123! 12491! H 8688! 8815! H & P 15079! P 16200!* Endemic.

6. T. glaucescens Urb., Symb. Ant. **6**: 35 (1909).
    Glabrous tree about 8 m. high; leaves unequal, narrowly elliptic-oblanceolate, cuneate at base, acute or shortly acuminate, 4–12 cm. long, 2–5 cm. broad; cymes lax, many-flowered, almost sessile; pedicels about 2 mm. long; corolla-tube greenish, the lobes 3–4 mm. broad, white; follicles lanceolate, up to about 4·5 cm. long.
    Rare (St. James, St. Ann), in woods at mostly low elevations; fl. and fr. Mar. *H 10342! 10358!* Endemic.

7. **T. rendlei** Stearn in Journ. Arn. Arb. **52**: 618 (1971).

Glabrous shrub or tree up to 8 m. high; leaves unequal, oblong-elliptical to narrowly lanceolate, cuneate at base, shortly acuminate, 4·5–20 cm. long; 2–8 cm. broad, with 9–12 prominent lateral veins on each side of midrib; cymes many-flowered; pedicels 5–10 mm. long; corolla-lobes about 3 mm. broad.

Rare and local (Port., St. Thom.), in wet secondary forest on limestone; 1000–2250 feet; fl. Feb–Apr. *H & B 10728! P 10112! Stearn 509!* Endemic.

8. ANECHITES Griseb. (1861)

1. **A. nerium** (Aubl.) Urb. in Fedde, Repert. Sp. Nov. **16**: 150 (1919).—*Apocynum nerium* Aubl. (1775). *Anechites asperuginis* (Sw.) Griseb. (1861).

Stem slender, pubescent, up to 10 m. long; leaves elliptical to ovate or lanceolate, cuneate to rounded or subcordate at base, acuminate, 4–12 cm. long, 1·5–5 cm. broad, with short curved hairs on main veins beneath; petiole up to 2 cm. long; inflorescences loosely few-flowered; pedicels 7–11 mm. long; calyx about 2 mm. long, the lobes acuminate; corolla-lobes obovate, spreading, 10–13 mm. long; follicles almost terete, 6–8 cm. long.

Very rare (St. Thom.), in dry pastures. Only once reported, *Wilson*. Panama, Colombia, Ecuador, Greater Antilles.

9. FORSTERONIA G. F. W. Meyer (1818)

Anther-tips barely exserted; filaments free and distinct; calyx-lobes much longer than corolla-tube; leaves glandular at base of midrib adaxially, the lateral veins making an angle of about 45° with midrib   **1. wilsonii**

Anthers wholly exserted; filaments adherent to style, free from one another; calyx-lobes about as long as corolla-tube; leaves without glands, the lateral veins making an angle of about 70° with midrib   **2. floribunda**

1. **F. wilsonii** (Griseb.) Woodson in Ann. Miss. Bot. Gard. **22**: 174 (1935).— *Thyrsanthus wilsonii* Griseb. (1862).

Glabrous woody vine; leaves elliptical, obtuse at base, acutely acuminate, 4–11 cm. long, 1·5–4 cm. broad, shortly petiolate; inflorescences much shorter than leaves, with many small white flowers; calyx-lobes linear-lanceolate, 3–4 mm. long; corolla subrotate, the tube 1·2–1·5 mm. long, the lobes 2–2·5 mm. long; fruits unknown.

Rare (St. Andr., Manch.), in thickets; 750–2000 feet; fl. Apr, Aug. *P 22624!* Endemic.

2. **F. floribunda** (Sw.) A. DC. in DC., Prodr. **8**: 437 (1844).—*Echites floribunda* Sw. (1788). Milk Withe.

Glabrous trailing and climbing shrubby vine up to 10 m. high, temporarily an erect shrub after coppicing; sap yellow, acrid, very adhesive; leaves oblong-elliptical or elliptical, obtuse at base, obtuse or shortly acuminate at tip, 4–11 cm. long, 1–5 cm. broad, shortly petiolate; inflorescence a dense terminal panicle with numerous small greenish-white fragrant flowers; calyx-lobes ovate, 2–3 mm. long; corolla-tube 2–3 mm. long, the lobes 4–4·5 mm. long; follicles linear, slightly curled, 15–30 cm. long; seed 7–9 mm. long; pappus tawny 15–21 mm. long.

Frequent in the central and western parishes, in thickets and woodlands on limestone; 250–2900 feet; fl. May–Jan, fr. Sept–June. *A 8937! 11793! P 10279!* Cuba.

10. RHABDADENIA Müll. Arg. (1860)

1. **R. biflora** (Jacq.) Müll. Arg. in Mart., Fl. Bras. **6** (1): 175 (1860).—*Echites biflora* Jacq. (1760). *Echites paludosa* Vahl (1798).

Glabrous shrubby vine with soft flexible stems; leaves obovate to oblong-lanceolate, mostly obtuse at base, cuspidate, 5–12 cm. long, 1·5–5 cm. broad; petioles up to 2 cm. long; inflorescence a small dichasial cyme, 1–5-flowered;

calyx-lobes ovate-oblong, leafy, nearly 1 cm. long; corolla mostly white, but often pinkish outside, 4·5–7·5 cm. long; follicles slightly curved, slender, 9–14 cm. long; seed about 2·5 cm. long, the pappus about as long.

Common at the margins of mangrove swamps and in thickets near the sea; SL–20 feet; fl. all the year, fr. May–Dec. *A 5670! H 8186! Loveless UCWI 2708! P 11260!* Florida, Mexico to northern S. Amer., West Indies; Cayman Is.

11. URECHITES Müll. Arg. (1860)

1. **U. lutea** (L.) Britton in Bull. New York Bot. Gard. **5**: 316 (1907).—*Vinca lutea* L. (1756). Nightsage, Nightshade.

Glabrous or pubescent shrubby vine up to 3 m. or more high; sap milky or merely turbid; leaves oblong, obovate to subrotundate, obtuse or shortly cordate at base, obtuse or cuspidate at tip, 3–9 cm. long, 0·5–6 cm. broad; inflorescences axillary or terminal, cymose, with usually few large showy flowers; calyx-lobes linear-lanceolate; corolla yellow, often marked red at level of anthers, 3·5–6·5 cm. long; follicles rather woody, slightly curved but not reflexed, 8–20 cm. long; pappus white.

Common in fields, dunes and mangrove margins and also on fences, hedges and walls; SL–3000 feet; fl. and fr. all the year. *A 9840! 12891! 12892! H 9346! H & P 13888! Stearn 209!* Florida, Bahamas, Greater Antilles, Lesser Antilles south to St. Vincent; Cayman Is.

12. ANGADENIA Miers (1878)

1. **A. lindeniana** (Müll. Arg.) Miers, Apocyn. S. Amer.: 180 (1878).—*Rhabdadenia lindeniana* Müll. Arg. (1860). *Echites jamaicensis* Griseb. (1862).

Slender-stemmed climber up to 6 m. high; stems at first hispid with spreading or retrorse hairs, glabrescent; leaves elliptical to oblong-lanceolate, obtuse at base, acuminate, 2–6·5 cm. long, 1·5–3 cm. broad, glossy adaxially, more or less hispid on midrib beneath and margins when young; inflorescence several-flowered, longer than leaves; pedicels 10–12 mm. long; calyx-lobes ovate-lanceolate; corolla about 5 cm. long; follicles slightly torulose, 19–22 cm. long; pappus light buff.

Frequent, in the central and western parishes, in thickets and woodland margins and on banks in limestone areas; 500–3000 feet; fl. and fr. most of the year. *A 6758! 7316! H 9006! P 6478! Stearn 908!* Cuba.

13. ECHITES P. Browne (1756)

1. **E. umbellata** Jacq., Enum. Syst. Pl. Carib.: 13 (1760).—Deadly Nightshade.

Shrubby twining climber with clear viscid sap; leaves broadly ovate to oblong-elliptical, cuneate or rounded to subcordate at base, obtuse or shortly cuspidate at tip, 4–12 cm. long, 2–8 (–12) cm. broad, glabrous or minutely puberulous on the veins beneath; inflorescences shorter than the leaves, few-flowered, clearly peduncled; calyx-lobes ovate, up to 5 mm. long; distal part of corolla-tube twisted, the lobes spreading up to 2·5 (–3) cm. long; follicles thick, rigid, reflexed at maturity, 9–26 cm. long; seeds about 5 mm. long.

Common in thickets and at woodland margins, also on fences and in waste grassy places on the ground; SL–1000 (–2600) feet; fl. and fr. most of the year. *A 5420! H 12109! H & P 14473!* Florida, Mexico, British Honduras, Colombia, Bahamas, Cuba, Hispaniola, Grand Cayman.

14. MANDEVILLA Lindl. (1840)

1. **M. torosa** (Jacq.) Woodson in Ann. Miss. Bot. Gard. **19**: 64 (1932).—*Echites torosa* Jacq. (1760).

Slender-stemmed twining shrub, the young parts mostly thinly pubescent; leaves elliptical, mostly shortly cordate at base, acute to acuminate, 2–7 cm. long, 0·7–3 cm. broad, discolorous, with glands at base of lamina adaxially; inflorescences axillary, racemose, about as long as the leaves; pedicels 7–10 mm. long; calyx-lobes lanceolate, 1·5–2 mm. long; corolla yellow, the lobes spreading, 4–5 mm. long; follicles torulose, somewhat falcate, light reddish-brown and smooth, 9–20 cm. long.

Common in thickets and woodland margins and on limestone banks; SL–5000 feet; fl. and fr. most of the year. *A 5495! P 19817! Stearn 795!* Mexico.

15. SECONDATIA A.DC. (1844); Woodson (1935)

1. **S. macnabii** (Urb.) Woodson in Ann. Miss. Bot. Gard. **19**: 385 (1932).—
Orthechites macnabii Urb. (1909).

Shrub; leaves lanceolate to oblong-lanceolate, acute to obtuse at base, acuminate at tip, 5–7 cm. long, 1·3–2·3 cm. broad, glabrous; inflorescence terminal, sub-corymbose, 3–6-flowered; pedicels 3–4 mm. long; calyx-lobes ovate, 2 mm. long; corolla-lobes obliquely elliptical, acute, 7–8 mm. long; fruit unknown.

Very rare and known only from the type, *McNab.* Endemic.

162. ASCLEPIADACEAE

Perennial herbs, shrubs or rarely small trees, often climbing, generally with milky sap. Leaves usually opposite, less often whorled, alternate or vestigial in succulent forms, simple, generally entire; minute stipules sometimes present. Flowers bisexual, actinomorphic, in cymose, racemose or umbelliform inflorescences. Perianth biseriate, 5-merous; calyx-tube short, the segments imbricate, valvate or open; corolla sympetalous, the lobes contorted or valvate. Stamens 5, distinct or more usually connate around the gynoecium and adherent to the stigma; corona if present formed either wholly or partly by appendages to the corolla or filaments; anthers 2-locular, rarely 4-locular, often with scaly terminal or subterminal appendages; pollen granular in tetrads, associated with 5 spoon-shaped translators (Periplocoideae) or agglutinated in each anther-sac as a pollinium and these linked laterally in pairs between adjacent anthers by translators joined at a gland (Ascle-piadoideae), each gland on the radius of a corolla-lobe. Disk absent. Ovary superior of 2 free carpels; styles 2, linked by a single massive 5-lobed stigma, the receptive surfaces coincident with the translator-glands; ovules numerous on adaxial placentas in each carpel. Fruit a pair of follicles, commonly one only developing, dehiscing adaxially, the placenta becoming free as a replum. Seeds mostly rather flat with a micropylar coma of silky hairs; endosperm small.

About 250 genera with 2000 species, predominantly tropical.

1 Corolla tubular or funnel-shaped, 4 cm. or more long, the tube evident, the lobes over-lapping to the right:
  2 Corona-scales attached to corolla; stamens with distinct filaments; pollen granular; corolla showy, light mauve to deep reddish-purple   **1. Cryptostegia**
  2 Corona-scales attached to staminal column; pollen agglutinated; corolla white, about 4 cm. long; flowers fragrant in stalked axillary umbels; cultivated ornamental twiner; native of Madagascar   *Stephanotis floribunda* Brongn.
1 Corolla rotate or campanulate, not exceeding 3 cm. long, the tube shorter than the lobes or usually so, the lobes valvate or contorted in aestivation; stamens with united filaments; pollen in each anther-loculus agglutinated into a waxy pollinium:
  3 Erect herbs, undershrubs or shrubs, occasionally arborescent; corolla-lobes valvate:
    4 Corona-lobes flattened laterally each with a short curved basal spur; petals erect; leaves mealy-glaucous, broad at base with a patch of erect dark bristles adaxially   **2. Calotropis**
    4 Corona-lobes folded, without a basal spur; petals reflexed; leaves not glaucous, without bristles   **3. Asclepias**
  3 Twiners:
    5 Corolla-lobes crisped and deeply indented, about 7 mm. long; flowers racemose; pollinia horizontal; follicle large, obliquely ovoid-ellipsoid, smooth, hairy, blunt, thick-walled; seed-margin crenate-dentate   **4. Fischeria**
    5 Corolla-lobes not crisped, entire or at most shallowly emarginate; other characters not combined as above:

6 Stigma deeply 2-fid; corona-scales 5, swollen, erect; pollinia pendulous; corolla rotate, the lobes linear, acute, up to 2 cm. long; leaves deeply cordate; follicles long-pointed **5. Oxypetalum**
6 Stigma not 2-fid; corolla-lobes if linear, much smaller than above:
  7 Inflorescences long-stalked (in Jamaican species) with peduncles up to 11 cm. long; corona double, the inner of 5 short fleshy scales, the outer annular; corolla rotate, the lobes spreading, about 5 mm. long; pollinia pendulous; seed-margin broken **6. Sarcostemma**
  7 Inflorescences mostly shortly stalked or subsessile, the whole shorter than the leaf subtending; seed-margin (as far as known) entire:
    8 Corolla rotate, 8 mm. or more in diameter; anthers with spreading dorsal appendages; pollinia horizontal; corona single or double **7. Gonolobus**
    8 Corolla cup- or bell-shaped or if rotate much smaller than above; anthers without spreading dorsal appendages; corona single:
      9 Follicles prominently 5-angled, oblong, acuminate; corolla-lobes reflexed beyond the middle; corona of 5 shallow lobed thickenings below the anthers; anther-appendages 5 broad thickened flaps crimping the margin of the massive stigma; pollinia horizontal **8. Jacaima**
      9 Follicles (as far as known in Jamaican species) smooth or warted, not angled or winged; other characters not combined as above:
        10 Pollinia pendulous; corolla campanulate or rotate; except in *C. jamaicense*, in which leaves cordate and inflorescence racemose, flowers less than 3 mm. long in umbelliform inflorescence and leaves mostly narrow or absent at flowering time **9. Cynanchum**
        10 Pollinia erect; corolla urceolate, the lobes erect or recurved, obliquely emarginate; flowers 4·5–8 mm. long in compact or lax cymes; leaves present at flowering time, cordate only in *M. fusca* **10. Marsdenia**

1. CRYPTOSTEGIA R. Br. (1820)

Corona-scales 5 and 2-fid or 10; corolla light mauve outside, whitish within; rampant twiner naturalized in coastal swamps **1. grandiflora**
Corona-scales 5, entire; corolla reddish-purple; suberect or climbing ornamental shrub; native of Madagascar; Purple Allamanda, Rubber Vine
C. madagascariensis Boj. ex Decne.

1. C. grandiflora R. Br. in Edw., Bot. Regist. **5**: t. 435 (1820).—India-rubber Vine.

Woody twiner to 6 m. or more high, with copious latex; leaves elliptical, up to 11 cm. long and 6 cm. broad, rather leathery; flowers in terminal forked cymes; corolla mostly 5–6 (–7) cm. long; corona-lobes up to 1·5 cm. long; follicles usually paired, divergent, 10–12 cm. long, woody with longitudinal ribs or wings.

Locally common (St. Andr., St. Cath., St. Thom.), in thickets near the sea; SL–600 feet; fl. and fr. sporadically throughout the year. *A 6462! P 23934! Skelding & Robertson UCWI 5562!* Native of Madagascar; naturalized in Greater Antilles, Barbados, Grand Cayman.

2. CALOTROPIS R. Br. (1810)

1. C. procera (Ait.) Ait. f. in Ait., Hort. Kew. ed. 2, **2**: 78 (1811).—*Asclepias procera* Ait. (1789). Auricula Tree, Dumb Cotton, French Cotton.

Shrub to 4 m. or small tree to 5·5 m. high; leaves subsessile, oblong-obovate, cordate at base, shortly acuminate at tip, up to 25 (–30) cm. long and 15 (–18) cm. broad; cymes extra-axillary, pedunculate, umbelliform; calyx purplish within, the lobes 5–6 mm. long; corolla about 2 cm. broad, greenish-white outside, purple within; follicles inflated, obliquely ovoid, up to about 10 cm. long.

Locally common (St. Andr.), in arid sandy or gravelly waste places; SL–1200 feet; fl. and fr. most of the year. *A 8209! A. M. Barry! H 10052!* Native of dry parts of tropical Asia, introduced and naturalized elsewhere.

3. ASCLEPIAS L. (1753)

1 Corona-lobes without an inner basally attached horn; follicles subglobose, inflated, covered with long soft bristles; leaves linear-lanceolate **1. physocarpa**

1 Corona-lobes with a horn arising from the base of the inner face; follicles fusiform, smooth:
2 Leaves lanceolate; corolla scarlet; corona yellow　　　　　2. *curassavica*
2 Leaves ovate-lanceolate; corolla light green; corona white with or without blackish-purple marks near the base　　　　　3. *nivea*

1. A. physocarpa (E. Meyer) Schltr. in Engl., Bot. Jahrb. **21**, Beibl. 54: 8 (1896).
—*Gomphocarpus physocarpus* E. Meyer (1837).
Shrubby branched herb 0·6–1·5 m. high, whitish-pubescent in the upper parts; leaves up to 11 (–15) cm. long and 1·2 cm. broad; umbels racemose, lateral, up to about 10-flowered, peduncles 2–3 cm. long; calyx reddish; corolla greenish at first, becoming white, the lobes ovate, ciliate on one margin, about 7 mm. long; corona-lobes 3 mm. long, shorter than gynostegium.
Local (St. Andr., St. Ann, St. Thom.), in fields and on pathside banks; (1500–) 2800–5000 feet; fl. Mar–Nov, fr. Apr–Dec. *A 7691! 11275! Loveless UCWI 2124! P 23675!* Native of tropical and S. Africa, introduced and naturalized elsewhere.

2. A. curassavica L., Sp. Pl. **1**: 215 (1753).—Redhead, Red Top.
Sparingly branched erect or decumbent herb up to about 1 m. high; leaves acute or acuminate, 4–11 (–15) cm. long, 1–3 (–5) cm. broad; flowers in extra-axillary and terminal umbels; pedicels 1–2 cm. long; corolla-lobes 6–8 mm. long; staminal column about 2 mm. long; follicles usually solitary, erect, pointed, about 8 cm. long; seeds reddish-brown, about 7 mm. long.
A rather common weed of pastures, limestone banks and waste places generally; SL–3600 feet; fl. and fr. all the year. *H 6915! H & P 13347!* Throughout the American subtropics and tropics, introduced into the Old World as an ornamental but rarely fully naturalized there; Grand Cayman.

3. A. nivea L., Sp. Pl. **1**: 215 (1753).—Whitehead.
Like the last and sometimes regarded as not more than varietally distinct; flowers a little smaller; corolla-lobes about 5 mm. long; staminal column 1 mm. long; seeds 5 mm. long.
Locally common in pastures at middle elevations, uncommon elsewhere; (50–) 1200–4200 feet; fl. and fr. all the year. *A 6338! P 24118! Wood!* Greater Antilles, Virgin Is., Martinique. A number of plants with characters intermediate between *A. curassavica* and *A. nivea* have been described either as subordinate taxa or hybrids. Observations in Jamaica, wherever these two species occur naturally together and of plants in cultivation, suggest very strongly that all variants are completely interfertile and that this fact accounts for the existence of a wide and inconsistent range of combinations of leaf-shapes, flower-sizes and corona and corolla colours (*A 10128! H 5276!*).

4. FISCHERIA DC. (1813)

1. F. crispiflora (Sw.) K. Schum. in Engl. & Prantl, Nat. Pflanzenf. **4** (2): 228 (1895).—*Cynanchum crispiflorum* Sw. (1788).
Twiner with strong stems, up to 6 m. or more high; stems and leaves pubescent with short and long brownish hairs, odorous when bruised; leaves ovate to ovate-elliptical, cordate at base, cuspidate-apiculate at tip, 5–15 (–20) cm. long, 4–10 (–14) cm. broad; peduncle and inflorescence-axis eventually up to about 30 cm. long, bearing successively clusters of flowers; flowers fragrant; pedicels about 2 cm. long; calyx-lobes 4 mm. long, abruptly acuminate; corolla greenish-white, reticulate-marked within, the lobes about 7 mm. long; follicle up to 16 cm. long and 7 cm. broad; seeds blackish-brown, broadly clawed.
Local in sheltered thickets and woodlands on limestone in wet areas; 500–1750 feet; fl. Oct–May, fr. all the year. *A 12812! H 9154!* Cuba.

5. OXYPETALUM R. Br. (1810) nom. cons.

1. O. cordifolium (Vent.) Schltr. in Urb., Symb. Ant. **1**: 269 (1899).—*Gothofreda cordifolium* Vent. (1808). *O. riparium* Kunth (1819).
Slender-stemmed high-climbing twiner, puberulous on young parts; leaves

ovate, deeply cordate, acuminate, 5–10 cm. long, 3–5 cm. broad; cymes few-flowered, shorter than leaves; pedicels filiform; calyx-lobes linear-lanceolate, 4–5 mm. long; follicles about 8 cm. long.

Very rare and without any precise locality. Continental tropical Amer., Cuba, Puerto Rico.

6. SARCOSTEMMA R. Br. (1810)

1. S. clausum (Jacq.) Schult. in L., Syst. Veg. ed. nov., **6**: 116 (1820).—*Asclepias clausa* Jacq. (1760). Milk Withe.

Rather succulent twiner, herbaceous, up to 3 m. or more long; leaves oblong or ovate or narrower to linear, rounded to subcordate at base, acute or acuminate at apex, 3–8 cm. long, up to about 3 cm. broad, glabrous, sometimes absent at flowering time; peduncles longer than leaves; pedicels slender, 7–20 mm. long; flowers fragrant; calyx-lobes puberulous, 2·5–3 mm. long; corolla greenish-white turning yellowish, the lobes spreading and about 5 mm. long; follicles puberulous, 5–8 cm. long.

Locally common in the southern parishes, in thickets near rivers or mangrove swamps; SL–1650 feet; fl. Oct–Mar, fr. Nov–Apr. *A 8311! H 5414! H & P 15189!* Florida, continental tropical Amer., Bahamas, Greater Antilles, Grenada, Trinidad; Grand Cayman.

7. GONOLOBUS Michx. (1803)

1 Stems and leaves glabrous; corolla 3–4 cm. broad   1. *stellatus*
1 Stems and leaves hispid, pubescent or glabrescent; corolla mostly smaller:
  2 Corolla 2–3 cm. broad:
    3 Outer corona ringlike   2. *jamaicensis*
    3 Outer corona wanting   3. *stapelioides*
  2 Corolla barely 1 cm. broad, glabrous:
    4 Leaves pubescent   4. *pubescens*
    4 Leaves glabrescent   5. *rhamnifolius*

1. G. stellatus Griseb., Fl. Br. W.I.: 420 (1862).

Glabrous twiner; leaves elliptical, cuneate at base, shortly acuminate, 3–8 cm. long, 1–4·5 cm. broad; peduncles 1·5–2 cm. long, shorter than petioles; corolla green, the lobes 1·5–2 cm. long, glabrous except for a puberulous fleshy ring inside at base; corona double, tawny yellow.

Rare (Manch., St. Thom.), in woodlands on limestone; fl. Mar. *H & B 10731! Purdie!* Endemic.

2. G. jamaicensis Rendle in Journ. Bot. **74**: 345 (1936).

Woody twiner to 6 m. high; branches hispid; leaves oblong to ovate-elliptical, rounded to openly cordate at base, shortly acuminate, 5·5–8·5 cm. long, 3–4·5 cm. broad, papery when dry; inflorescences about 5-flowered; flower-buds conical, beaked; corolla greenish outside, purplish-red and puberulous inside at base, the lobes tapering, acuminate, 1·2 cm. long, 4–5 mm. broad at base, glabrous.

Rare (Port.), in montane woodland; 3900–5000 feet; fl. June. *H 7687! J.P. 968! P 6803!* Endemic.

3. G. stapelioides Desv. in Ham., Prodr. Pl. Ind. Occ.: 32 (1825).—*Fischeria cincta* Griseb. (1862).

Woody twiner up to 3 m. or more high; hairs on young shoots crimson, those at nodes recurved; leaves narrowly oblong to elliptical, rounded at base, shortly acuminate, 5–9·5 cm. long, 1·5–4·5 cm. broad; pedicels about 12 mm. long; flower-buds subglobose, bluntly beaked; corolla green, the lobes broadly ovate, thick, undulate, 10–12 mm. long, with arcs of stiff hairs and whitish at base within; corona-lobes erect and forked at tip.

Rare (Port., St. Thom.), in montane woodland; 3950–5000 feet; fl. Feb, June. *A 12705! H 5561! J.P. 608!* Endemic.

4. G. pubescens Griseb., Fl. Br. W.I.: 420 (1862).

Twiner, densely hispidulous with soft recurved hairs; leaves ovate-oblong, rounded at base, mucronate, 1·5–3·5 cm. long, puberulous on both surfaces; fascicles few-flowered; calyx-lobes pilose; corolla 7 mm. broad, the lobes ovate-deltate, acute, papillose within the tip; outer corona 5-lobed.

Known only from the type, *Macfadyen*. Endemic.

5. G. rhamnifolius Griseb., Fl. Br. W.I.: 420 (1862).

Twiner; stem puberulous, glabrescent; leaves ovate to oblong-elliptical, acuminate, 4–9 cm. long, 1·5–4 cm. broad; flowers in shortly stalked umbels; pedicels 8 mm. long, puberulous; corolla about 8 mm. broad, the lobes ovate-oblong, obtuse; outer corona with serrulate margin, inner of 5 flat fleshy lobes.

Rare (St. Ann), in thickets. *Prior! P 11825! Stearn 529!* Endemic.

8. JACAIMA Rendle (1936)

1. J. costata (Urb.) Rendle in Journ. Bot. **74**: 340, f. 1 (1936).—*Poicilla ? costata* Urb. (1909).

Twiner up to 8 m. high, with slender puberulous glabrescent stems and branchlets; leaves oblong to oblong-elliptical, rounded to subcordate at base, acute, 5–10 cm. long, 1·5–3 cm. broad, sparsely pubescent on veins beneath, papery when dry; peduncle 2–6 mm. long; pedicels stiff, recurved, about 5 mm. long; flowers greenish-yellow; calyx pilose, nearly 2·5 mm. long; corolla thinly pilose, about 4 mm. long, lobes 2·5 mm. long, reflexed distally, hollowed and keeled proximally; corona green; follicle glaucous, blue-green, about 6·5 cm. long.

Rare and very local (St. Andr., St. Cath.), on trees and shrubs in thicket on limestone, near SL–800 feet; fl. Aug, Nov–Feb, fr. Jan, June. *duQuesnay 400! H 9590! 10006! P 18335! 21944!* Endemic.

9. CYNANCHUM L. (1753)

1 Inflorescence racemelike with evident internodes between the pedicels; corolla about
　6 mm. long; leaf-blades cordate, more than 3 cm. broad 　　　　　　**1. *jamaicense***
1 Inflorescence umbellike, the pedicels arising close together; corolla 1·5–3 mm. long,
　less than 1 cm. broad; leaf-blades smaller, mostly narrowed at base:
　2 Plants usually or mostly leafless at flowering time; corolla glabrous within; corona-
　　lobes shorter than gynostegium, broadly ovate or semiorbicular, purple; plants
　　glabrous:
　　3 Corolla dark purple 　　　　　　　　　　　　　　　　　　**2. *atrorubens***
　　3 Corolla greenish-white to yellowish 　　　　　　　　　　　**3. *leptocladum***
　2 Plants leafy at flowering time:
　　4 Lobes of corona about half as long as gynostegium:
　　　5 Corolla-lobes glabrous within; plant glabrous 　　　　　　　**4. *fawcettii***
　　　5 Corolla-lobes villous within; plant puberulous 　　　　　　　　**5. *rendlei***
　　4 Lobes of corona equalling or exceeding gynostegium; corolla white or tinged pink;
　　　plants glabrous:
　　　6 Corona-lobes linear; corolla 2·5–3 mm. long 　　　　　　　　**6. *albiflorum***
　　　6 Corona-lobes broader; corolla up to nearly 2·5 mm. long:
　　　　7 Corona-lobes narrowly triangular 　　　　　　　　　　　　　**7. *priorii***
　　　　7 Corona-lobes ovate 　　　　　　　　　　　　　　　　　　　**8. *harrisii***

1. C. jamaicense (Griseb.) Woodson in Ann. Miss. Bot. Gard. **28**: 210 (1941).—*Enslenia jamaicensis* Griseb. (1862).

Slender glabrous twiner; leaves unequal in each pair, ovate, deeply cordate at base, shortly and abruptly acuminate, 5–9 cm. long, 3–5·5 cm. broad; petiole slender, 2–4·5 cm. long; pedicels up to 1 cm. long, puberulous in a line on one side; corolla divided nearly to base, the lobes oblong-lanceolate, acute, glabrous; corona slightly exceeding gynostegium, the tips shortly 2-lobed.

Known only from the unlocalized type, *Wilson*. Endemic.

2. C. atrorubens (Schltr.) Alain in Mem. Soc. Cub. Hist. Nat. **22**: 120 (1955).—*Metastelma atrorubens* Schltr. (1899).

Much branched slender climber often forming a tangled network of green stems; leaves (mostly absent at flowering time) elliptical, apiculate, up to 2 cm. long and 3·5 mm. broad; cymes subsessile, 4–6-flowered; pedicels puberulous, filiform, 3–4 mm. long; corona-lobes ovate, acuminate; follicles 6–7 cm. long, puberulous, glabrescent.

Rather common, on shrubs and trees in thickets and woodland margins; 1500–4000 feet; fl. Oct–Mar, fr. Jan–Feb. *A 8225! 8452! H 12805! P 6216!* Cuba.

3. **C. leptocladum** (Decne.) Jiménez in Rhodora **62**: 238 (1960).—*Vincetoxicum leptocladum* Decne. (1844). *Amphistelma filiforme* Griseb. (1862).
Like the last; leaves linear-oblong to oblanceolate, 1–2·5 cm. long, less than 5 mm. broad; cymes up to 8-flowered; pedicels about 5 mm. long; corolla-lobes oblong, obtuse, 1·6–2 mm. long; follicle 5–6 cm. long, very slender.
Rather common on shrubs, trees and rocks, mostly in limestone areas; 1250–2900 (–5000) feet; fl. Aug–Mar, fr. Feb–Mar. *A 8222! H 12856! H & P 14947!* Greater Antilles; Cayman Is.

4. **C. fawcettii** (Schltr.) Stearn in Phytologia **21** (3): 138 (1971).—*Metastelma fawcettii* Schltr. (1899).
Glabrous or glabrescent twiner; leaves lanceolate to elliptic-lanceolate, cuneate at base, acute or shortly acuminate at tip, 1·5–3 cm. long, 5–10 mm. broad; cymes 10–12-flowered, the pedicels puberulous, 3–4 mm. long; flowers fragrant; corolla rotate, yellowish-white tinged brown, the lobes 2 mm. long, oblong.
Rather local (St. Andr., Port., St. Thom.), on shrubs and trees in montane thickets; 2500–6500 feet; fl. Sept–Jan. *H 7404! P 8332!* Endemic. This species resembles *C. leptocladum* closely and is probably only a variant in which the leaves are retained at flowering.

5. **C. rendlei** Stearn in Phytologia **21** (3): 138 (1971).—*Metastelma jamaicense* Schltr. (1908), not *C. jamaicense* (Griseb.) Woodson.
Slender-stemmed twiner to 3 m. or more high; young stems curled-puberulous, older stems with yellowish bark; leaves opposite, clustered on the reduced shoots, oblong to rotundate-elliptical, obtuse and sometimes emarginate, mucronate, 12–16 mm. long, 7–10 mm. broad or smaller on the branchlets, thinly puberulous beneath; pedicels up to 1 mm. long; flowers fragrant, 4·5–5 mm. long, often solitary; corolla campanulate, white or yellow, the lobes 2·5–3·5 mm. long, linear-oblong, glabrous outside, whitish-hairy within; corona-lobes shortly ovate; follicle about 5 cm. long, warted proximally.
Locally frequent, on shrubs and trees in arid areas on limestone mostly near the sea; 10–1000 feet; fl. Oct–Mar, fr. Oct, Apr. *A 6841! 10068! H 8866! H & P 15129!* Endemic.

6. **C. albiflorum** (Griseb.) Stearn in Phytologia **21** (3): 138 (1971).—*Metastelma albiflorum* Griseb. (1862). *M. hartii* Schltr. (1899).
Slender-stemmed glabrous twiner; leaves elliptical to elliptic-lanceolate, acute or obtuse, minutely mucronate, 17–25 mm. long, 7–10 mm. broad or smaller on the branchlets; cymes few-flowered; pedicels about 2 mm. long; flowers fragrant; corolla campanulate, glabrous outside, hairy within.
Occasional (St. Cath., Clar., Manch.), on shrubs and trees in thickets; SL–1750 feet; fl. Nov–Mar. *A 13058! Cornman 908! H 10180!* Endemic.

7. **C. priorii** (Rendle) Stearn in Phytologia **21** (3): 137 (1971).—*Metastelma priorii* Rendle (1936).
Slender glabrous twiner, the internodes generally 3–4 cm. long; leaves lanceolate, acute, shortly mucronate, up to 3–4 cm. long, 12–15 mm. broad; cymes several-flowered; pedicels 2–3 mm. long; flowers very small; corolla-lobes ovate, acute, 1·7 mm. long, puberulous within.
Rare and local (Clar., Manch., St. Ann), in open woodland on rocks and trees and in thickets in exposed limestone areas; 2300–2600 feet; fl. Mar, Dec, fr. Dec. *A 8377! P 23367!* Endemic.

8. C. harrisii (Schltr.) Stearn in Phytologia **21** (3): 137 (1971).—*Metastelma harrisii* Schltr. (1899). *M. parviflorum* of Griseb. (1862), not R. Br. (1810).

Glabrous twiner; leaves lanceolate to oblong, acute, mucronate, 1·5–4·3 cm. long, 5–15 mm. broad; pedicels glabrescent, 2–4 mm. long; corolla openly campanulate, 3 mm. long, the lobes lanceolate, acute, 2 mm. long, glabrous outside, the margins within densely puberulous.

Occasional, in the eastern parishes, on shrubs and trees in open submontane thickets; 2000–5000 feet; fl. most of the year. *A 7834! 8247! H 8195! P 20807!* Endemic.

10. MARSDENIA R. Br. (1810)

1 Leaves cordate, up to about 14·5 cm. long and 10·5 cm. broad; cyme compact, with 20 or more greenish flowers about 7·5 mm. long; corolla-lobes spreading-recurved, ciliate, hyaline-margined; throat of corolla not closed by hairs but with short hairs in 5 vertical bands; corona with 5 distinct subulate incurved horns   1. *fusca*
1 Leaves not cordate, smaller than above; cyme rather open; flowers not over 6 mm. long; corolla-lobes erect, not ciliate; throat of corolla closed by long hairs; corona inconspicuous, without distinct curved horns:
   2 Leaves smooth between the main lateral veins beneath; corolla about 3·5 mm. broad at widest part, pinkish; corona of blunt inverted Y-shaped thickenings below each anther with a central ridge having a minute free tip   2. *troyana*
   2 Leaves finely reticulated all over the abaxial surface; corolla hardly 3 mm. broad at widest part, light green or cream-coloured; corona of flat more or less triangular thickenings below each anther without a central ridge:
      3 Corona-lobes lanceolate, subacute, the tips free and reaching nearly to the anther-sacs   3. *clausa*
      3 Corona-lobes triangular, the tips blunt, hardly free and ending well below the anther-sacs   4. *macfadyenii*

1. M. fusca C. Wright ex Griseb., Cat. Pl. Cub.: 178 (1866).

Robust perennial vine with copious milky sap, glabrous, up to 6 m. or more high; leaves broadly ovate-elliptical, truncate-subcordate at base, cuspidate at tip, paler beneath; peduncle extra-axillary, thick, 10–12 mm. long; pedicels short; calyx 4·5 mm. long, the lobes oblong, hyaline-margined, ciliate; corolla-lobes 4 mm. long, glandular and tinged purplish on inner surface; ripe follicle spongy-woody, glabrous, about 13 cm. long and 8 cm. broad when dehisced; seeds flat, entire and thin-margined, body 10 mm. long and 7·5 mm. broad, coma about 5 cm. long.

Very rare (Clar.), twining on shrubs and trees in thicket on limestone; 5–10 feet; fl. Apr–July, fr. May, Nov. *A 13004! Stearn 765!* Cuba, Guadeloupe, Martinique.

2. M. troyana Urb. in Fedde, Repert. Sp. Nov. **16**: 36 (1919).

Strong twiner to about 5 m. high; leaves oblong-elliptical, cuneate or rounded at base, rounded or shortly acuminate at apex, up to 10 cm. long and 4·5 cm. broad, glabrous, paler beneath; pedicels broader distally, 5–6 mm. long; sepals suborbicular, ciliate, 2–2·5 mm. long, pink; corolla urceolate, the tube 3 mm. long, the lobes 2 mm. long and broadly ovate and obtuse, maroon inside; follicle narrowly lanceolate, acuminate, about 9 cm. long, smooth; coma of seed about 3·5 cm. long.

Rare and local (Trel.), on rocks and trees in woodland on limestone; 1600–2000 feet; fl. Apr–May, fr. Sept. *A 12831! H 9515! R 1871!* Endemic.

3. M. clausa R. Br. in Mem. Wern. Soc. **1**: 30 (1810).

Rather woody vine with milky sap, glabrescent; leaves narrowly oblong to oblanceolate, long-cuneate at base, acute to obtuse and mucronate at tip, up to 9 cm. long and 2 cm. broad, rather leathery, paler beneath; sepals broadly ovate, ciliate, concave, about 2 mm. long; corolla campanulate, about 5 mm. long, the lobes about 3 mm. long, light yellow-green.

Not adequately confirmed for Jamaica, the record based on an unlocalized specimen, *Swartz!*

4. M. macfadyenii Rendle in Journ. Bot. **75**: 349, f. 2 (1937).

Like the last; young shoots sparsely puberulous; leaves up to 7 cm. long and 3 (–4) cm. broad, glabrous except the petiole and a few small white marginal hairs;

flowers with a musky odour, cream-coloured; sepals orbicular to obovate, ciliate, about 1·5 mm. long; corolla campanulate, about 4·5 mm. long, the lobes a little shorter than tube; fruit not known.

Uncommon, in thickets at salina margin and in woodland on arid coastal limestone; 10–50 feet; fl. Oct–Jan. *A 13057! H & P 15109!* Endemic.

The Wax Plant, *Hoya carnosa* (L. f.) R. Br., which has thick leathery dark green leaves and umbels of fragrant pink waxy flowers, is a native of eastern Asia and is sometimes cultivated as an ornamental.

163. CONVOLVULACEAE

Herbs or shrubs, erect or more usually climbing, rarely small trees, often with milky sap. Leaves alternate, mostly simple, rarely absent (*Cuscuta*), without stipules. Flowers bisexual, actinomorphic, axillary, solitary or in cymes, bracteate, pedicels jointed. Perianth biseriate, (4–) 5-merous; sepals usually free, imbricate, often unequal; corolla gamopetalous, entire or lobed, induplicate-contorted or imbricate. Stamens epipetalous, alternate with corolla-lobes, inserted towards base of corolla, often unequal; anthers 2-locular, opening lengthwise. Disk usually present. Ovary superior, 1–4-locular; ovules 1 or 2 in each loculus, axile, erect, anatropous; style(s) terminal or gynobasic; stigma 2-lobed or capitate. Fruit a loculicidal capsule or indehiscent. Seeds smooth or hairy; embryo large; endosperm usually cartilaginous, scanty.

About 50 genera with some 1500 species, generally distributed.

1 Parasitic twiners with viscid sap; leaves absent or reduced to scales; flowers small, clustered; corolla with 5 fimbriate scales inside; styles 2, unequal   **1. Cuscuta**
1 Erect or climbing herbs or shrubs with green leaves; flowers often showy; corolla not appendaged:
  2 Ovary lobed; styles 2, gynobasic; small creeping herbs with reniform or suborbicular leaves   **2. Dichondra**
  2 Ovary not lobed; style or styles terminal:
    3 Styles 2; stigmas 4; erect or trailing herbs or undershrubs, not twining
                                                     3. Evolvulus
    3 Style 1:
      4 Flowers numerous, small, white, fragrant, in racemes or panicles; stigma slightly 2-lobed; vigorous woody twiner; cultivated; native of East Indies; Christmas Vine, White Coralilla or Coralita   *Porana paniculata* Roxb.
      4 Flowers solitary or in clusters or cymes; if corolla small, stigma distinctly 2-lobed or stigmas 2:
        5 Stigmas flattened above, elliptical, oblong or linear; pollen smooth or nearly so:
          6 Sepals equal or nearly so; corolla not yellow with a purple eye; indumentum when present often of trifid stellate hairs; leaves often gland-dotted
                                                     4. Jacquemontia
          6 Sepals unequal; corolla light yellow with deep purple eye; indumentum of simple thick-based hairs; leaves minutely pustulate   **5. Hewittia**
        5 Stigmas subglobose:
          7 Outer sepals markedly broader and longer than inner   **6. Aniseia**
          7 Outer sepals if at all not markedly larger than inner:
            8 Pollen grains smooth; anthers twisted after dehiscence   **7. Merremia**
            8 Pollen grains spinose; anthers straight:
              9 Fruit dehiscent; capsule usually ovoid to globose, (2–) 4-valved, (2–) 4-seeded, clasped by persistent sepals   **8. Ipomoea**
              9 Fruit indehiscent, dry with sepals spreading at base when mature:
              10 Capsule usually 1-seeded; corolla mostly white, funnel-shaped, 2–3 cm. long
                                                   9. Turbina
              10 Capsule usually 4-seeded; corolla mostly purple, campanulate; robust vine with large cordate leaves densely white-tomentose beneath; cultivated; native of India; Elephant Ear Vine   *Argyreia nervosa* (Burm. f.) Boj.

1. CUSCUTA L. (1753); Yuncker (1932), (1965)

1 Filaments shorter than the anthers: calyx-lobes rounded, overlapping, much shorter than the tube; corolla-lobes short, obtuse; capsule circumscissile   1. *americana*

1 Filaments equal to or longer than the anthers; calyx-lobes equal to or longer than the tube; capsule membranous and inflated, hollowed between the styles, opening irregularly or indehiscent or sometimes circumscissile in *C. umbellata*:
2 Calyx divided nearly to the base, the lobes deltate-triangular, acute or acuminate; corolla-lobes narrowly triangular, reflexed at anthesis; filaments 1–2 mm. long
2. *umbellata*
2 Calyx divided one-half to two-thirds to the base; corolla-lobes broadly ovate to deltate, ascending or spreading; filaments 0·5–0·9 mm. long:
3 Calyx much shorter than the corolla-tube, the lobes not overlapping at base; corolla fleshy; capsule thickened at base of style 3. *indecora*
3 Calyx about as long as the corolla-tube, the lobes overlapping slightly at base; corolla membranous; capsule not especially thickened at top:
4 Corolla-lobes acute; anthers longer than broad; interstylar aperture scarcely gaping 4. *campestris*
4 Corolla-lobes obtuse; anthers about as broad as long; interstylar aperture large 5a. *obtusiflora* var. *glandulosa*

1. C. americana L., Sp. Pl. **1**: 124 (1753).—Dodder, Love Vine.
Stems twining, deep yellow or orange, smooth, up to 2 mm. thick; flowers tightly clustered in cymose inflorescences; base of calyx light green; corolla greenish or yellowish-white; capsule ovoid-globose; seeds about 1.5 mm. long.
A common parasite on herbs, shrubs and low trees; near SL–1200 feet; fl. and fr. most of the year. *P 9844! Yuncker 17008!* Throughout tropical and southern subtropical Amer.

2. C. umbellata Kunth, Nov. Gen. **3**: 121 (1819).
Stems light greenish-yellow to orange, slender; flowers on thick tapered pedicels in loose cymes, light green to white; corolla about twice as long as calyx; capsule depressed-globose with a thickened collar around the interstylar aperture; seeds dark brown, 1–2 mm. long.
Locally common, especially on fleshy-leaved herbs near the sea; SL–50 (–600) feet; fl. and fr. July–Jan. *A 12882! Campbell 5974! P 16497!* S. United States, Mexico to northern S. Amer., West Indies.

3. C. indecora Choisy in Mém. Soc. Phys. & Hist. Nat. Genève **9**: 278, t. 3, f. 3 (1841).
Stems orange, very slender; flowers greenish-white; capsule globose; seeds about 1·7 mm. long.
Frequent on herbs and shrubs in waste places; 50–2100 feet; fl. and fr. most of the year. *H & P 14525! Yuncker 17957!* S. United States, Mexico, Venezuela, West Indies.

4. C. campestris Yuncker in Mem. Torr. Bot. Club **18**: 138, f. 14 (1932).
Stems yellow-orange; flowers light green; calyx with fine glandular lines towards base; capsule depressed-globose, finely striate, with flat linear glands near the apex, splitting between the styles and elsewhere vertically.
Occasional (St. Andr., Clar., St. Eliz.), mostly on herbaceous weeds; 50–2250 feet; fl. and fr. Mar–Sept. *A 7555! H & B 10637! Robertson UCWI 9238!* Native of tropical Amer., now established in many warm countries.

5. C. obtusiflora Kunth, Nov. Gen. **3**: 122 (1819).

5a. Var. glandulosa Engelm. in Trans. Acad. Sci. St. Louis **1**: 492 (1859).
Stems light yellow to brownish-orange, slender; flowers in cymose clusters; pedicels green; calyx with raised blisterlike glands; corolla white; capsule depressed-globose, about 3 mm. in diameter, with raised glands like the calyx.
Rather rare (St. Cath., Han., Trel.), on herbs in low lying damp places; 400–1000 feet; fl. and fr. Jan–Mar, Aug. *A 10846! 12160! P 10531!* United States, Mexico, Cuba, Puerto Rico.

2. DICHONDRA J. R. & G. Forst. (1776); Tharp & Johnston (1961)

1. D. repens J. R. & G. Forst., Charact. Gen. Pl.: 39, t. 20 (1776).
Small herbs with slender branched trailing stems rooting frequently; leaves soft, reniform to suborbicular, deeply cordate, slender-petioled; flowers solitary on

slender pedicels, white or light green; fruit dicoccous, nodding. Widespread in warm countries.

This variable species has two main variants in Jamaica:

a. Leaves up to about 1 cm. broad, dark green, sparsely villose to glabrescent; common in grassy waste places, lawns and on roadside banks; SL–2000 feet; fl. Dec–Jan, fr. Jan–Mar. *A 6087! P 6027! Skelding UCWI 3094!* (*D. micrantha* Urb.)

b. Leaves mostly very small but some up to 2 cm. or more broad, densely appressed silky-tomentose especially beneath, mostly light green; common at roadsides amongst loose stones and on rock or earth banks; 3000–5000 feet; fl. and fr. most of of the year. *A 5847! J.P. 1130! P 7018!* (*D. sericea* Sw.)

3. EVOLVULUS L. (1762)

1 Erect ascending-branched undershrub 20–45 cm. high; leaves narrowly elliptical, mostly less than 10 mm. long and 4 mm. broad, the uppermost almost scalelike; sepals about 2·5 mm. long   1. *arbuscula*
1 Plant branched near the ground the shoots ascending spreading or radiating:
   2 Flower-stalk (peduncle and pedicel together) shorter than calyx; leaves linear-lanceolate, silky beneath   2. *sericeus*
   2 Flower-stalk as long as or longer than calyx:
     3 Shoots creeping and rooting at nodes; leaves oblong-elliptical, broadly obovate or almost orbicular, less than twice as long as broad, often cordate:
       4 Stems villose with minute spreading hairs; peduncle very short, shorter than pedicel (bracteoles near base of flower-stalk); corolla deeply 5-lobed   3. *nummularius*
       4 Stems with appressed hairs or nearly glabrous; peduncle as long as or longer than pedicel (bracteoles at or above middle of flower-stalk); corolla shallowly 5-lobed   4. *glaber*
     3 Shoots prostrate, ascending or erect, not rooting; leaves usually more than twice as long as broad:
       5 Leaves oblong-elliptical to broadly obovate, glabrous or with sparse appressed hairs; sepals 3–4 mm. long   4. *glaber*
       5 Leaves ovate to lanceolate, narrowly elliptical or linear, usually very hairy; stems mostly with both spreading long hairs and appressed hairs; sepals 2·5–3 mm. long:
         6 Stems mostly spreading; corolla 4·5–12 mm. across the limb   5. *alsinoides*
         6 Stems mostly erect or ascending; corolla 3–4·5 mm. across   6. *filipes*

1. **E. arbuscula** Poir. in Lam., Encycl. Méth. Bot. Suppl. 3: 459 (1814).—Sea Thyme.

Stem and leaves villous; peduncle short or obsolete; pedicel as long as or longer than calyx; corolla 5–8 mm. across limb, purple, blue or white.

Rare (Manch., St. Eliz.), on coastal limestone; SL–750 feet; fl. Sept–Dec, fr. Nov–Jan. *H 9720! Robertson UCWI 9221!* Cuba, Hispaniola, Bahamas, Little Cayman.

2. **E. sericeus** Sw., Nov. Gen. & Sp. Pl.: 55 (1788).

Perennial herb with lax slender branches 10–40 cm. long; leaves oblanceolate to narrowly elliptical, up to 3 cm. long and 5 mm. broad; plants often galled to form heads of densely hairy overlapping leaves; calyx 5–6 mm. long; corolla white or light blue, 6–7 mm. across limb; capsule about 3·5 mm. in diameter; seeds black.

Occasional (St. Andr., Clar., Manch., St. Eliz.), along roadsides and in arid waste places on limestone; 200–1500 feet; fl. Aug–Dec, fr. Aug–Feb. *A 7737! H 11690!* S.E. United States, continental tropical Amer., West Indies.

3. **E. nummularius** (L.) L., Sp. Pl. ed. 2, 1: 391 (1762).—*Convolvulus nummularius* L. (1753).

Perennial prostrate-branched herb with regular internodes; leaves 3–10 (–15) mm. long, 3–10 mm. broad; pedicels short; calyx about 2·5 mm. long; corolla white, rarely light blue, 6–7 mm. across limb; capsule nearly 3 mm. in diameter, 1-locular.

Locally common as a weed of waste ground, lawns and grazed pastures in damp or heavy soils; SL–800 feet; fl. and fr. Nov–Feb. *A 8345! H 10037! P 24348!* General in the tropics; Grand Cayman.

4. E. glaber Spreng. in L., Syst. Veg. ed. 16, **1**: 862 (1824) [see note p. 789].

Herb with ascending or prostrate and rooting branches; stems and leaves silky-pubescent, at least when young; leaves oblong to obovate, 1–3 cm. long, 5–14 (–18) mm. broad; peduncles at least 5 mm. long, 1–2-flowered; corolla rotate, 7–12 mm. in diameter, white or light blue with a white centre; capsule subglobose, about 2·5 mm. in diameter.

Occasional on flat open sandy ground, ditchbanks, salina margins and limestone rocks near the sea; SL–30 feet; fl. and fr. Nov–Mar. *A 8959! P 7609! Robertson UCWI 5353!* Florida, northern S. Amer., West Indies.

5. E. alsinoides (L.) L., Sp. Pl. ed. 2, **1**: 392 (1762).—*Convolvulus alsinoides* L. (1753). Speedwell.

Herbs with slender spreading branched stems; stems and leaves variously sparsely to densely pilose; several varieties differing in hairiness and corolla-size have been reported in Jamaica:

5a. Var. alsinoides.

Leaves 2–4 cm. long, 3–7 mm. broad, thinly pubescent; corolla light bright blue, 9–12 mm. in diameter. Frequent in cultivation as an ornamental pot-plant. *Wood 7534!*

5b. E. alsinoides var. **debilis** (Kunth) Ooststr. in Meded. Bot. Mus. & Herb. Utrecht **14**: 33 (1934).

Leaves pilose or villous with mostly rather long spreading hairs; corolla blue or white, 5–8 mm. in diameter. Rare weed (St. Eliz.); 400 feet; fl. Feb. *H 7241!* S. United States, Mexico to Bolivia, Cuba, Hispaniola.

5c. E. alsinoides var. **grisebachianus** Meisn. in Mart., Fl. Bras. **7**: 344 (1869).

Leaves sparsely or densely pilose; corolla 4·5–5·5 mm. in diameter. *Houston.* Florida, Mexico to Brazil, Bahamas, Cuba, Hispaniola, Lesser Antilles.

6. E. filipes Mart. in Flora 24 (2), Beibl.: 100 (1847).

Annual diffusely ascending-branched herb to 60 cm. high; leaves up to 2 cm. long and 3 mm. broad; peduncles very slender, ascending, 1·5–3·5 cm. long; corolla light blue; capsule subglobose, membranous, about 2·5 mm. in diameter.

Rather rare (St. Andr., St. Eliz.), a weed of open grassy fields and sandy waste places; 20–800 feet; fl. and fr. Nov–Jan. *A 8293! H 12444! P 16095!* Mexico to Brazil.

4. JACQUEMONTIA Choisy (1834)

1 Sepals 5–9 mm. long, longer than valves of capsule; peduncle 1–14 cm. long, often longer than leaves; corolla blue, 1·5–2 cm. long   1. *pentantha*
1 Sepals 2·5–4 mm. long, shorter than valves of capsule; peduncle up to 3 cm. long, never longer than leaves; corolla white or tinged pink or if light blue less than 1 cm. long:
  2 Leaf-blade oblong-elliptical with deeply emarginate tip; inflorescence 1–2-flowered; pedicels recurved in fruit; corolla light blue, 8–10 mm. long; plant prostrate
       2. *obcordata*
  2 Leaf-blade narrowly lanceolate or narrowly oblong to ovate, the tip acuminate, acute, obtuse or occasionally shallowly emarginate; inflorescence mostly several–many-flowered; pedicels not recurved in fruit; corolla white or tinged pink; plant twining:
    3 Inflorescence subsessile; pedicels 2–3 mm. long; sepals acuminate; corolla about 6 mm. long   3. *verticillata*
    3 Inflorescence distinctly pedunculate, peduncle 5–25 mm. long; some pedicels at least 4 mm. long; sepals acute, obtuse or rounded; corolla 1·2–1·4 cm. long:
      4 Leaf-blade densely pubescent or tomentose, cordate or subcordate; outer sepals rounded; corolla shallowly lobed; stigma-lobes filiform, 1·5–2 mm. long
         4. *nodiflora*
      4 Leaf-blade minutely pubescent or glabrous, cuneate, rounded or almost truncate; outer sepals acute; corolla deeply lobed to about halfway; stigma-lobes elliptical, about 0·5 mm. long   5. *havanensis*

1. J. pentantha (Jacq.) G. Don, Gen. Syst. **4**: 283 (1837).—*Convolvulus pentanthos* Jacq. (1791).

Strong-stemmed slender twiner, probably mostly annual; stems and leaves glabrous or pubescent; leaves ovate, acuminate, cordate at base, 2–7 (–9) cm. long, 1–5 cm. broad; cymes contracted, with bracts about 5 mm. long; outer sepals ovate, acuminate, the inner smaller; capsule subglobose, 4-valved, 3–4 mm. in diameter; seeds glabrous.

Locally common in the central and western parishes in thickets and rough waste places; near SL–2200 feet; fl. Aug–Mar, fr. Oct–Feb. *A 8530! H 11817! P 11192!* Florida, Mexico to Peru, West Indies, introduced into the Old World tropics as an ornamental where it behaves as a perennial.

2. J. obcordata (Millsp.) House in New York State Mus. Bull. 233–234: 63 (1921). —*Convolvulus obcordatus* Millsp. (1900). *J. subsalina* Britton (1925).

Stems prostrate, rooting at nodes, slender, strigose, up to 50 cm. long, numerous and radiating from a woody stock; leaves glabrous, 1–3 cm. long; outer sepals ovate, rounded at tip, about 3 mm. long; capsule subglobose, about 4 mm. in diameter.

Very rare (St. Eliz.), in low-lying coastal area; fl. Sept. *H 9812!* Mexico, Cuba, Puerto Rico, Antigua, Guadeloupe.

3. J. verticillata (L.) Urb., Symb. Ant. **3**: 339 (1902).—*Ipomoea verticillata* L. (1759). *Convolvulus micranthus* Roem. & Schult. (1819).

Slender-stemmed twiner, more or less appressed pubescent; stems trailing or climbing to about 2 m.; leaves oblong to lanceolate, mucronate, cordate at base, 1–3 (–4) cm. long, 5–14 mm. broad; cymes several-flowered; sepals acute, 2·5–3 mm. long; corolla mauve to light pink, darkening on drying, 3–4 mm. across the limb; capsule globose, 2–3 mm. in diameter; seeds rugulose, brown.

Occasional in exposed thickets and rough open places in limestone areas; near SL–700 (–2600) feet; fl. and fr. Nov–Feb. *A 6194! H 12750! P 17433!* Greater Antilles.

4. J. nodiflora (Desr.) G. Don, Gen. Syst. **4**: 283 (1837).—*Convolvulus nodiflorus* Desr. (1792) [see note p. 789].

Strong-stemmed climber to 6 m. high; stems and leaves densely pubescent; leaves ovate, obtuse or acute, mucronate, 2–6 cm. long, 1·5–3 (–4) cm. broad; cymes several-flowered; pedicels 3–6 mm. long; sepals rounded, 2–3 mm. long; corolla white; capsule globose, 3–4 mm. in diameter; seeds glabrous.

Locally common (St. Andr., St. Cath., Trel., St. Thom.), in open thickets and on banks in rather arid limestone areas; 50–2500 feet; fl. Oct–May, fr. Dec–Mar. *A 5499! H 9528! Powell 1032!* Continental tropical Amer., West Indies.

5. J. havanensis (Jacq.) Urb., Symb. Ant. **3**: 342 (1902).—*Convolvulus havanensis* Jacq. (1767). *C. jamaicensis* Jacq. (1768). *J. jamaicensis* (Jacq.) Hallier f. (1899).

Twining herb with slender pubescent glabrescent stems 1–2 m. long, probably annual; leaves oblong-ovate to linear, obtuse or acute, mucronate, cuneate to rounded at base, 1·5–4 (–5·5) cm. long, 3–20 mm. broad; cymes 1–several-flowered, often branched, the pedicels ascending; sepals broadly ovate, acute or shortly acuminate, about 2 mm. long; corolla white with usually a band of pink or brownish-purple along the outer surface or each lobe; capsule subglobose, about 4 mm. long; seed rough.

Common in thickets and open woodlands on limestone, mostly near the sea; SL–1500 feet; fl. and fr. Nov–Mar. *A 6141! H 9999! P 5994!* Florida, Mexico, Bahamas, Greater Antilles, Antigua, Barbuda, Virgin Is.; Cayman Is.

5. HEWITTIA Wight & Arn. (1837)

1. H. sublobata (L.f.) Kuntze, Revis. Gen. Pl. **1**: 441 (1891).—*Convolvulus sublobatus* L.f. (1781).

Tough-stemmed climber or scrambler; whole plant hispid with thinly dispersed long thick-based hairs; leaves broadly ovate, entire or 3–lobed, abruptly acuminate, mucronate, cordate at base with a deep rounded sinus, 4–10 cm. long along the midrib up to 12 cm. across the basal lobes; peduncle axillary, 4–10 cm. long, bibracteolate below the solitary or paired flowers; outer sepals ovate, 1·5 cm. long, 1 cm. broad; corolla about 2·5 cm. long; capsule about 1 cm. in diameter; seeds 5–6 mm. long.

Locally abundant in the eastern parishes as a weed of rough pastures and on fences or in thickets; near SL–750 feet; fl. Nov–June, fr. Mar–June. *A 9431! P 20835! Stearn 559!* Widespread in the Old World tropics, apparently a recent introduction into Jamaica.

6. ANISEIA Choisy (1834)

1. **A. martinicensis** (Jacq.) Choisy, Conv. Diss. Sec.: 144 (1837).—*Convolvulus martinicensis* Jacq. (1763). *Ipomoea martinicensis* (Jacq.) G. F. W. Meyer (1818).

Glabrescent slender-stemmed twiner; leaves elliptical to oblong-lanceolate, rounded and long-mucronate at apex, up to 9 cm. long and 3 cm. broad; flowers mostly solitary on long axillary peduncles; bracteoles subulate; outer sepals 1·5–3 cm. long, leafy, decurrent on the pedicel; corolla about 2·5 cm. long, white, pilose outside; capsule 2 cm. long, 4-valved, 4-seeded.

Rare (St. Cath., St. Eliz.), climbing on shrubs and herbs at swamp and pond margins; 20–1000 feet; fl. and fr. Dec–Jan. *A 10396! 12163! P 21845!* Widespread in tropical countries.

7. MERREMIA Dennst. ex Endl. (1841) nom. cons.

1 Leaves not divided, ovate, deeply cordate; inflorescence umbelliform, usually many-flowered; corolla yellow, 2–2·5 cm. long; capsule dark brown; seeds velvety-brown-tomentose　　　　　　　　　1. *umbellata*
1 Leaves deeply divided or digitately compound; inflorescence loose or few-flowered:
　2 Leaves digitately compound into (3–) 5 leaflets; corolla less than 3 cm. long:
　　3 Leaflet-margin serrate-dentate; peduncles glandular; calyx 5–8 mm. long; corolla creamy-white to light yellow　　　　2. *quinquefolia*
　　3 Leaflet-margin entire or nearly so; peduncles not glandular; calyx 15–23 mm. long; corolla white　　　　　　　3. *aegyptia*
　2 Leaves palmately or pedately divided into 5–7 segments; corolla 3·5 cm. or more long; seeds glabrous:
　　4 Leaf-segments coarsely and bluntly toothed or lobed; petioles hairy; sepals up to 3 cm. long in fruit; corolla white with purple or pink eye; capsule 1–1·5 cm. long　　　　　　　　　4. *dissecta*
　　4 Leaf-segments entire; petioles glabrous; sepals 4–6 cm. long in fruit; corolla yellow; capsule about 3 cm. long　　　　　　5. *tuberosa*

1. **M. umbellata** (L.) Hallier f. in Engl., Bot. Jahrb. **16**: 552 (1893).—*Convolvulus umbellatus* L. (1753). *Ipomoea umbellata* (L.) G. F. W. Meyer (1818), not L. (1759). *I. polyanthes* Roem. & Schult. (1819).

Glabrescent twiner or stems sometimes trailing and rooting; leaves obtuse, acute or acuminate, 2–9 cm. along the midrib, 2–8 cm. across the basal lobes; petiole often stipuliform at base; peduncle variable, terete or narrowly raggedly winged; sepals about 8 mm. long; capsule slightly flattened-globose, about 10 mm. broad.

Common on fences and in thickets and waste places; SL–1000 feet; fl. and fr. Nov–Mar. *A 10066! Allwood UCWI 762! A. von der Porten!* Widespread in the tropics; Grand Cayman.

2. **M. quinquefolia** (L.) Hallier f. in Engl., Bot. Jahrb. **16**: 552 (1893).—*Ipomoea quinquefolia* L. (1753). Rock Rosemary.

Trailing or twining to 4 m. or more high; stems and petioles slender, sparingly pilose, glabrescent; leaflets elliptical to lanceolate, pointed at both ends, 2–6 (–7·5) cm. long and 7–20 (–25) mm. broad; inflorescence (2–) 3–6-flowered; corolla 1·5–2 cm. long; capsule 8–10 mm. in diameter; seeds with whitish curled hairs.

Locally common, mostly in rather exposed well drained thickets and waste places; SL–1250 feet; fl. Nov–May, fr. Dec–Mar. *A 6154! H 8263! P 8309!* Mexico to Peru, West Indies.

3. **M. aegyptia** (L.) Urb., Symb. Ant. **4**: 505 (1910).—*Ipomoea aegyptia* L. (1753).

Trailing or twining, the stems petioles, peduncles and calyx usually covered with pungent hairs up to 5 mm. long, rarely glabrous; leaflets thin, elliptical, acuminate, narrowed at sessile base, sparingly pilose, 3–9 (–12) cm. long, up to 3·5 cm. broad; peduncles stiff, 7–12 cm. long, few-flowered; corolla 2·5–3 cm. long; capsule about 12 mm. broad, enclosed by persistent pointed calyx; seeds smooth.

Rather common in the south-eastern parishes, very rare elsewhere, in thickets and waste places; near SL–900 feet; fl. Oct–Feb, fr. Dec–Feb. *A 6199! (A 8284!* glabrous variant) *H 9997! P 8160!* Widespread in the tropics.

4. M. dissecta (Jacq.) Hallier f. in Engl., Bot. Jahrb. **16**: 552 (1893).—*Convolvulus dissectus* Jacq. (1767). *Ipomoea dissecta* of Griseb. (1862), not Willd. Know You.

Twiner mostly with rather long stiff yellowish hairs, glabrescent; leaf-segments ovate to lanceolate in outline, up to about 6 cm. long and 2 cm. broad; peduncles variable, 1–2-flowered; sepals glabrous; capsule 1–1·5 cm. long.

Cultivated and widely escaped on to fences and in thickets and waste ground; near SL–600 feet; fl. Nov–Mar, June–Aug, fr. Jan–Mar, June. *Cornman! P 20730! Robertson UCWI 1090! 9191!* Native of tropical Amer., now widespread; Cayman Is.

5. M. tuberosa (L.) Rendle in Dyer, Fl. Trop. Afr. **4**: 104 (1905).—*Ipomoea tuberosa* L. (1753). Wood Rose.

Robust climber, glabrous or glabrescent; leaf-segments elliptical, acuminate, up to 12 cm. long and 4 cm. broad; peduncles 12 cm. or more long, branched above, several-flowered; sepals becoming woody and spreading in fruit, glabrous; seeds 1·5–2·5 cm. long and broad.

Occasional, probably originally a garden escape, in roadside thickets and on banks and trees; SL–3200 feet; fl. Oct–Dec, fr. Nov–Mar. *A 9801! 9833! P 9573! Robertson UCWI 5651!* Native of tropical Amer., now widely grown as an ornamental throughout the tropics.

8. IPOMOEA L. (1753)

1 Leaves pinnate with linear segments up to 1 mm. broad; corolla salver-shaped, red or white, tube about 3 cm. long, limb less than 1·5 cm. across; seeds narrowly ovoid
12. quamoclit

1 Leaves entire, lobed or digitately compound:
  2 Leaf-blade digitately compound into 3–7 leaflets, cut to junction with petiole or nearly so:
    3 Leaflets not exceeding 6 cm. long; calyx up to 7 mm. long; inflorescences 1–few-flowered:
      4 Peduncle filiform, spirally coiled, 1-flowered; calyx 4–5 (–6) mm. long; corolla funnel-shaped, 1·7–1·8 cm. long   1. *spiralis*
      4 Peduncle rigid, (1–) 2–3-flowered; calyx 6–7 mm. long:
        5 Corolla funnel-shaped, the limb spreading, up to 7 cm. long and 5–6 cm. across; leaflets ovate to lanceolate; peduncle at least 1 cm. long   2. *cairica*
        5 Corolla salver-shaped, the limb reflexed, tube 2·5–3 cm. long, limb about 2 cm. across; leaflets narrowly elliptical to linear; peduncle rarely as much as 5 mm. long   3. *tenuifolia*
    3 Leaflets larger, at least some of them 6 cm. long, others often up to 12 cm. long; calyx 8–20 mm. long; inflorescences mostly several-flowered:
      6 Leaflets 5, rarely 3; corolla usually crimson, purple or pink:
        7 Sepals of opened flowers almost equal in length, about 1 cm. long; corolla about 5 cm. long   4. *horsfalliae*
        7 Sepals of opened flowers markedly unequal, the outermost 8–9 mm. long, inner 10–13 mm. long; corolla about 6 cm. long   5. *rubella*
      6 Leaflets 3:
        8 Sepals (8–) 12–18 mm. long; corolla white or white tinged pink; leaflets with 5–11 pairs of lateral veins prominent beneath   6. *ternata*
        8 Sepals 7–10 mm. long; corolla purplish-pink; leaflets with 10–20 pairs of fine lateral veins   7. *lineolata*
  2 Leaf-blade entire or lobed but not divided nearly to the junction with petiole:
    9 Corolla-tube 7–12 cm. long, linear or almost so; limb 8–14 cm. across, white; capsule 2–2·5 cm. broad; seeds about 1 cm. long; flowers nocturnal:
      10 Outer sepals with an awnlike appendage 2–12 mm. long, inner mucronate; capsule ovoid, pointed; seeds glabrous, usually white   8. *alba*
      10 Outer sepals mucronate but not awned; capsule subglobose; seeds densely puberulous, long-hairy on the ridges, usually brown   9. *macrantha*

9 Corolla-tube shorter, up to about 6 cm. long; capsule and seeds smaller; flowers diurnal:
　11 Sepals with 3–5 prominent longitudinal ribs, 18–22 mm. long; bracts usually leafy; corolla-tube 4–5 cm. long, limb up to 9 cm. across, purple　　10. *setifera*
　11 Sepals without longitudinal ribs:
　　12 Sepals with an oblong marginate blade about 3 mm. long with a sub-apical linear appendage 2·5–4 mm. long; corolla salver-shaped, scarlet or light crimson, tube 2·5–3·5 cm. long　　11. *hederifolia*
　　12 Sepals not as above:
　　　13 Sepals narrowly lanceolate to linear, 5–10 times as long as broad, conspicuously pilose or ciliate below with tubercle-based bristles 3–4 mm. long:
　　　　14 Peduncles 2·5–16 cm. long; sepals 1·5–3·5 cm. long; corolla 5·5–6 cm. long, light blue　　13. *nil*
　　　　14 Peduncles 5–30 mm. long; sepals 1·3–1·6 cm. long; corolla 2·5–3 cm. long, deep bluish-purple　　14. *meyeri*
　　　13 Sepals otherwise, not more than 3 times as long as broad, if bristly less than 1·3 long:
　　　　15 Leaf-blades always distinctly 2-lobed or emarginate at apex, leathery or succulent; glabrous sea-shore plants with long trailing and rooting stems:
　　　　　16 Leaves broadly oblong to suborbicular, rarely twice as long as broad; corolla purple or very rarely white　　15. *pes-caprae*
　　　　　16 Leaves mostly oblong, at least 3 times as long as broad; corolla white or light yellow with a purple eye　　16. *stolonifera*
　　　　15 Leaf-blades not or very rarely emarginate, rarely oblong; rarely coastal:
　　　　　17 Outer sepals suborbicular; robust climbing or erect shrubs with loose many-flowered inflorescences; corolla pink:
　　　　　　18 Calyx usually glabrous; corolla normally 5–6 cm. long; leaves up to 20 cm. broad　　17. *phyllomega*
　　　　　　18 Calyx pubescent; corolla normally 8–9 (–10) cm. long; leaves up to 15 cm. broad:
　　　　　　　19 Twiner; leaf-blades broadly ovate, shortly pointed to acuminate　18. *carnea*
　　　　　　　19 Erect shrub; leaf-blades ovate, long-tapered acuminate　　19. *fistulosa*
　　　　　17 Outer sepals much longer than broad:
　　　　　　20 Corolla yellow with purple base; sepals almost equal, narrowly ovate, acute, scarious-margined, about 6 mm. long; capsule ovoid, about twice as long as sepals, usually tipped with persistent style-base　　31. *ochracea*
　　　　　　20 Corolla not yellow; sepals and capsule otherwise:
　　　　　　　21 Sepals conspicuously ciliate with bristly hairs up to 1 mm. long; corolla 1·5–2·5 (–4) cm. long; ovary and capsule pilose:
　　　　　　　　22 Sepals 5–7 mm. long; peduncles as long as or longer than the petioles; corolla 1·5–2 cm. long, reddish-purple　　20. *triloba*
　　　　　　　　22 Sepals 9–12 mm. long:
　　　　　　　　　23 Peduncles longer than petioles; corolla 2·5–3 (–4) cm. long, pink or purple　　21. *trichocarpa*
　　　　　　　　　23 Peduncles shorter than petioles; corolla 1·5–2 cm. long, mostly white　　22. *lacunosa*
　　　　　　　21 Sepals not conspicuously ciliate; corolla 2·5–8 cm. long:
　　　　　　　　24 Leaf-blades reniform to suborbicular, apex sometimes emarginate, basal lobes rounded, rather thick; sepals markedly unequal, 5–10 mm. long; trailing plant of low pastures　　23. *asarifolia*
　　　　　　　　24 Leaf-blades otherwise, apex acute or acuminate:
　　　　　　　　　25 Leaf-blades sagittate, basal lobes from half to as long as the median lobe; sepals unequal, 5–10 mm. long, red-margined　　24. *sagittata*
　　　　　　　　　25 Leaf-blades otherwise or if sagittate the basal lobes short:
　　　　　　　　　　26 Sepals with bristlelike tips:
　　　　　　　　　　　27 Stem trailing; leaf-blades ovate to deeply lobed; ovary hairy; plants cultivated for edible tubers　　25. *batatas*
　　　　　　　　　　　27 Stem twining; leaf-blades mostly ovate to broadly ovate; ovary glabrous; plants wild:
　　　　　　　　　　　　28 Sepals ovate, shortly mucronate, verrucose, 4–6 mm. long; corolla 2·5–3 cm. long, white　　32. *obscura*
　　　　　　　　　　　　28 Sepals, at least the outer, lanceolate, smooth or minutely punctate, 7–12 mm. long; corolla 4 cm. or more long, pink or purple　　26. *tiliacea*
　　　　　　　　　　26 Sepals without bristlelike tips:
　　　　　　　　　　　29 Stems trailing on mud or floating in water, mostly thick, hollow or spongy, rooting at nodes; leaves often sagittate; sepals obtuse　　27. *aquatica*
　　　　　　　　　　　29 Stems usually twining, slender and rather woody; plants of well drained places; leaves rarely sagittate:

30 Sepals about equal in length, the inner either slightly longer or slightly shorter than the outer, attenuated to the acute or acuminate tip:
  31 Corolla salver-shaped, tube narrow and about 8 mm. across at the mouth, limb 4·5–5 cm. across; anthers exserted; leaves often grey-silky   28. *jamaicensis*
  31 Corolla funnel-shaped, tube gradually expanded to 1–2 cm. across at mouth, limb to about 6 cm. across; anthers included; leaves usually less evidently hairy   29. *acuminata*
30 Sepals markedly unequal in length, the inner much longer than the outer, obtuse or acute:
  32 Corolla funnel-shaped, blue or white, gradually expanded; anthers included   30. *cyanantha*
  32 Corolla salver-shaped, pinkish purple, limb 5–7 cm. across; anthers exserted; naturalized after cultivation at high altitudes in St. Andrew; native of C. America   *I. purga* (Wender.) Hayne

1. I. spiralis House in Muhlenbergia **3**: 40 (1907).—*I. pulchella* of Griseb. (1862), not Roth (1821).
    Annual herbaceous slender-stemmed twiner, glabrous except here and there on peduncles and axillary branches; leaflets (4–) 5–7, lanceolate-elliptical, tapered to a narrow tip, 2–4·5 cm. long, 3–9 mm. broad; peduncles 2–3·5 cm. long; corolla mostly purple; capsule 2-locular, smooth, globose, about 9 mm. in diameter; seeds black, pubescent, with longer white hairs on inner margin, 5–6 mm. long.
    Rare and local (St. Cath.), in muddy ditch; about 30 feet; fl. and fr. Nov. *A 11896!* C. Amer., West Indies.

2. I. cairica (L.) Sweet, Hort. Brit.: 287 (1827).—*Convolvulus cairicus* L. (1759).
    Slender-stemmed woody twiner to 8 m. high; leaflets mostly 5, obtuse, mucronate, thinly pubescent on veins and margin, 1–4 cm. long, 6–15 mm. broad; corolla mauve.
    Uncommon in cultivation (Manch., Trel.), naturalized on walls, fences and trees; 1750-2300 feet; fl. Mar–May, Sept. *H & B 10615! Stearn 390! P 17470! I. cairica* is a native in the Old World tropics and is now widespread in cultivation as an ornamental. It has been reported from Cuba, Barbados, Trinidad and Cayman Is.

3. I. tenuifolia (Vahl) Urb., Symb. Ant. **5**: 472 (1908).—*Convolvulus tenuifolius* Vahl (1794). *I. fawcettii* Urb. ex House (1908).
    Tough-stemmed high climbing woody vine; leaflets mostly 5, shortly mucronate, glabrous, up to 4 cm. long, 1–10 mm. broad; corolla-tube light green, lobes rosy-mauve with fimbriate margins.
    Rare (St. Andr., St. Cath.), in thickets in arid limestone areas; 100–1000 feet; fl. Nov–Apr, fr. Feb–Apr. *A 10775! H 10010! P 17412!* Endemic.

4. I. horsfalliae Hook. in Curt., Bot. Mag. **61**: t. 3315 (1834).
    Twiner to 10 m. high; stem to 1·5 cm. or more thick; leaflets obovate to oblong-elliptical or oblanceolate, acuminate, mostly cuneate at base, rather thick in texture, up to 12 cm. long and 5·5 cm. broad; inflorescence corymbose or paniculate with stout peduncles; corolla typically brilliant crimson but plants with pinkish-purple corollas are otherwise indistinguishable; stamens exserted; capsule abruptly pointed, 13–16 mm. long; seeds dark brown, 6–7 mm. long, with tawny-silky marginal hairs up to 10 mm. long.
    Frequent in thickets and woodlands on limestone or shale; 200–3000 feet; fl. June–Dec, fr. June–Jan. *A 7262! P 19807! Robbins UCWI 780!* Also in Puerto Rico, and widely introduced into other tropical countries.

5. I. rubella House in Bot. Gaz. **43**: 414 (1907).—*I. grisebachii* Urb. (1903), not Prain (1894). *I. plumeriana* House (1907).
    Robust twiner like the last, climbing to 7 m. or more; leaflets often not so obviously acuminate, rarely obtuse and emarginate; inflorescences often larger with more numerous flowers; corolla larger, up to 7 cm. across the limb, pink to reddish-purple or sometimes crimson, usually darker in the tube; stamens usually included; capsule and seeds as above.
    Rather common in thickets and woodlands, mainly in coastal areas but occasional inland; SL–3000 feet; fl. Apr–Jan, fr. May–Jan. *A 8042! H 8727! P 16746!*

Endemic. Not all specimens fall clearly into one or other of the above two species and further study is required to determine the status of intermediates.

6. I. ternata Jacq., Pl. Hort. Schoenbr. **1**: 16, t. 37 (1797).
Robust woody twiner climbing to 16 m. or more; old stems up to 5 cm. or more thick, the bark shedding in rectangular flakes; younger stems smooth or warted; leaflets oblong-elliptical, shortly acuminate, rounded to cuneate at base, thick and fleshy, up to 18 cm. long and 8 cm. broad; inflorescences 1–few-flowered, axillary or terminal on pendulous lateral branches, or corymbose and often cauliflorous on old plants; corolla campanulate, tube about 5 cm. long; capsule 2 cm. long; seeds 7–8 mm. long, surrounded by a fringe of tawny silky hairs up to 13 mm. long, minutely puberulous on the surface.
Frequent in hilly wooded areas; 1300–5000 feet; fl. July–Sept, fr. Sept–Mar. *A 11747! H 6598! P 10577!* Endemic.

7. I. lineolata Urb., Symb. Ant. **3**: 355 (1903).
Slender-stemmed twiner climbing to 5 m. or more; bark thin, yellowish; leaflets elliptical, acuminate, thin, soft, up to 10 cm. long and 4·5 cm. broad; inflorescences usually few-flowered; corolla funnel-shaped, tube 3–4 cm. long; capsule 1 cm. long; seeds 5 mm. long with tawny silky marginal hairs 5–6 mm. long.
Occasional in open woodland and on rocky banks in limestone areas; 500–2500 feet; fl. May–Dec, fr. June–Dec. *Cornman A3G1! Harris A! P 10429!* Endemic.

8. I. alba L., Sp. Pl. **1**: 161 (1753).—*Convolvulus aculeatus* L. (1753). *I. bona-nox* L. (1762). Moonflower, Night Ipomoea.
Strong-stemmed twiner climbing to about 5 m., probably mostly annual; leaves broadly ovate, acuminate to a fine hairlike tip, deeply cordate, entire or shallowly lobed, thin, up to about 15 cm. long and broad; flowers solitary or few in an axillary corymb, fragrant at night; corolla-tube and outer flanges of limb light green.
Frequent on fences, roadside thickets and along riverbanks; 50–750 feet; fl. and fr. sporadically throughout the year. *A 10421! H 8458! P 20703!* Native of tropical Amer., now widely cultivated and naturalized in the tropics.

9. I. macrantha Roem. & Schult. in L., Syst. Veg. ed. nov. **4**: 251 (1819).—*I. longiflora* R. Br. (1810), not Willd. (1809). *I. tuba* (Schlecht.) G. Don (1837).
Perennial with thick-barked stems climbing and scrambling to 8 m. long; leaves rather fleshy, entire or 3-lobed; capsule truncate or shortly and abruptly pointed.
Locally common on rocks and in thickets and cactus scrub near the sea, also on most of the cays; SL–20 feet; fl. and fr. Nov–July. *A 11445! H 9543! P 11488!* Mexico to the Guianas, Bahamas, West Indies; Cayman Is.; introduced into the Old World tropics but native in the Pacific.

10. I. setifera Poir. in Lam., Encycl. Méth. Bot. **6**: 17 (1804).—*I. rubra* (Vahl) Millsp. (1900), not (L.) L. (1774).
Stems twining and trailing, hirsute or glabrous, rooting at nodes; young vegetative parts and leaf-veins often purplish; leaves broadly ovate, rounded to shortly cuspidate and minutely emarginate at tip, deeply cordate, thin, up to 9 cm. long along midrib, 8 cm. broad across the basal lobes; flowers solitary to subumbellate-corymbose; stamens unequal and deeply included; capsule subglobose, about 1 cm. in diameter; seeds tomentose.
Very local (St. Mary, Port.), on fences, trees and trailing at swamp margins and in ditches; 10–200 feet; fl. Dec–Mar, fr. Feb–Mar. *A 12295! 12297! H 12468! Larter 137!* Continental tropical Amer., West Indies; naturalized in W. Africa.

11. I. hederifolia L., Syst. Nat. ed. 10, **2**: 925 (1759).—*Quamoclit coccinea* of authors, not (L.) Moench (1794). *I. coccinea* of Griseb. (1862), not L. (1753).
Annual slender-stemmed twiner; leaves entire or occasionally shortly lobed, acuminate, openly cordate, thin, up to about 8 cm. long and broad; peduncles often very long, the flowers in dichasial cymes; corolla-tube often slightly curved; capsule 6–7 mm. in diameter; seeds irregularly warted, shortly velvety, dark brown.
Rather local (St. Andr., Manch., St. Thom.), on walls and fences and in road-

side thickets and waste places; near SL–2500 feet; fl. Oct–June, fr. Nov–Apr.
P 15885! Robertson UCWI 3498! Yuncker 17210! Native or tropical Amer., now
widely cultivated and naturalized in warm countries.

12. I. quamoclit L., Sp. Pl. **1**: 159 (1753).—*Quamoclit pennata* Boj. (1837).
Slender-stemmed twiner; leaves in outline about 7 cm. long and 4 cm. broad,
the basal pinnae often again divided; flowers solitary or few together on short or
long peduncles; capsule ovoid, about 1 cm. long.
Mostly cultivated, escaping occasionally; near SL–1600 feet; fl. July, Nov–Jan,
fr. Oct–Jan. *Hinchcliffe UCWI 756! H & P 14255!* Native of tropical Amer.,
now widespread.

13. I. nil (L.) Roth, Catalecta Bot. **1**: 36 (1797).—*Convolvulus nil* L. (1762).
Kaladana.
Slender strong-stemmed annual twiner, generally pilose; sap only very thinly
milky; leaves usually 3-lobed, the lobes shortly acuminate, openly cordate at base,
up to 16 cm. along midrib and 20 cm. across the lobes; flowers (1–) 2–6 together in a
subumbellate cyme on a usually long stout peduncle; corolla drying pinkish; ovary
3-locular developing (3–) 4–5 (–6) seeds in a glabrous capsule 1 cm. in diameter;
seeds black, subglabrous.
Occasional on fences and in thickets and rough waste places; SL–700 feet;
fl. Nov–Feb, fr. Dec–Feb. *A 8507! H 9155! P 18429!* General in the tropics.

14. I. meyeri (Spreng.) G. Don, Gen. Syst. **4**: 275 (1837).—*Convolvulus meyeri*
Spreng. (1825).
Annual twiner with milky sap, usually rather low-growing; leaves mostly entire,
sometimes 3-lobed, acuminate, openly cordate, thinly pilose adaxially glabrous or
with minute white scales beneath, up to 8 cm. along midrib, and 7 cm. across the
lobes; flowers arranged as above; peduncles 0·5–2 (–10) cm. long; corolla whitish at
base; capsule puberulous, 6–8 mm. in diameter, beaked; seeds finely pubescent.
Rather local (St. Andr., St. Eliz.), on fences and in thickets and rough waste
places; 50–700 feet; fl. Nov–Feb, fr. Dec–Feb. *A 8509! P 8280! Robertson UCWI
768!* Mexico to Venezuela, Greater Antilles, Trinidad.

15. I. pes-caprae (L.) R. Br. in Tuckey, Narrat. Exped. Zaire: 477 (1818).—
Convolvulus pes-caprae L. (1753).

15a. Ssp. brasiliensis (L.) Ooststr. in Blumea **3**: 533 (1940).—*Convolvulus
brasiliensis* L. (1753). *I. brasiliensis* (L.) Sweet (1818). Beach Morning Glory.
Stems trailing, glabrous, to 6 m. or more long; leaves entire-margined, up to
10 cm. long and broad; cymes 1–few-flowered; peduncle stout; pedicels variable in
length up to about 5 cm. long; sepals ovate-oblong, outer 6–10 mm. long, inner
8–15 mm. long; corolla 3–5 cm. long; anthers included; capsule 2-locular, subglobose, about 15 mm. in diameter; seeds brownish-pubescent.
Common on beaches and sandy waste places near the sea; SL–20 feet; fl. and fr.
most of the year. *Allwood UCWI 773! 9159! P 11504!* General on tropical shores;
Cayman Is. The subspecies *pes-caprae* is restricted to certain Old World localities.

16. I. stolonifera (Cyrillo) J. F. Gmel. in L., Syst. Nat. ed. 13, **2**: 345 (1791).—
Convolvulus stoloniferous Cyrillo (1788). *I. acetosifolia* (Vahl) Vahl ex Roem.
& Schult. (1819).
Plant with numerous underground stolons; leaves up to 10 cm. long and about
2·5 cm. broad; flowers solitary, erect; peduncles very short to about 6 cm. long;
sepals oblong, outer obtuse, mucronulate, 12–15 mm. long, inner acute, 14–18
mm. long; corolla 3·5–5 cm. long; anthers included; capsule globose, 10–15 mm.
long; seeds smooth.
Very rare (St. Andr.), in open sand at top of beach; 5 feet; fl. Oct–Feb. *Chapman
UCWI 22629! P 21393! Robertson UCWI 26635! 27087!* Widespread on the shores
of warm countries.

17. I. phyllomega (Vell.) House in Ann. New York Acad. Sci. **18**: 246 (1908).—
Convolvulus phyllomega Vell. (1825) (as *philomega*).

Robust glabrous woody twiner; leaves broadly ovate, acuminate, entire-margined, cordate, appressed-pilose beneath, 8–25 cm. long, 5–20 cm. broad; peduncles many-flowered; calyx 10–15 mm. long; sepals slightly cordate at base; ovary 2-locular; capsule 4-seeded.

Rare (St. Ann, St. Thom.), at margin of forest clearing; 2100 feet; fl. July. *Bengry & Lewis!* British Honduras to Brazil, Hispaniola, Guadeloupe, Dominica, Martinique, Trinidad & Tobago.

18. I. carnea Jacq., Enum. Syst. Pl. Carib.: 13 (1760).

Robust climber, velvety puberulous at least on young parts; leaves broadly ovate, entire-margined, cordate, thin, 5–15 cm. long; sepals suborbicular, about 5 mm. long; corolla pubescent; capsule ovoid, about 15 mm. long; seeds woolly.

Mostly cultivated and probably not naturalized (St. Andr., St. Thom.), in moist ravines; 400–700 feet; fl. Jan–Mar. *J.P. 2089! P 9828!* Nicaragua to Venezuela, Peru, Puerto Rico, St. Vincent, Margarita, Aruba.

19. I. fistulosa Mart. ex Choisy in DC., Prodr. **9**: 349 (1845).

Shrub up to 3 m. high, with lax branches; leaves truncate-subcordate at base, very long-tapered acuminate, puberulous, up to about 15 (–30) cm. long and 8 (–17) cm. broad; inflorescence cymose-paniculate; sepals suborbicular, outer about 5 mm. long, inner up to 7 mm. long, puberulous; corolla puberulous; fruit not seen.

Mostly cultivated and common in gardens; SL–750 feet; fl. sporadically through-out the year. *A. M. Barry! Larter 161! P 23073!* Native of Brazil, now widespread in cultivation.

20. I. triloba L., Sp. Pl. **1**: 161 (1753).

Variably pubescent slender-stemmed annual twiner; stems mostly trailing up to 1 m. long; leaves shallowly repand to 3 (–5)-lobed, terminal lobe acuminate, base cordate-subhastate with rounded sinus, 1–5 (–10) cm. long, up to 4·5 cm. broad across the basal lobes; flowers solitary or few together in a subumbellate cyme on a short or long peduncle; capsule subglobose, about 4 mm. in diameter, crowned by the persistent style-base; seeds dark brown or black, smooth.

Common, especially in the southern parishes near the sea, a weed of sandy ground and grassy swamp margins and also on hedges and in thickets; SL–1000 feet; fl. and fr. all the year. *A 6193! H 9840! P 16094!* S. United States to S. Amer., Bahamas, Cuba, Puerto Rico, Lesser Antilles, Trinidad, Grand Cayman, naturalized in the Old World tropics. This species is very variable in the size and shape of its leaves; some specimens approach closely Jamaican examples of *I. trichocarpa*.

21. I. trichocarpa Ell., Sketch Bot. S. Carol. & Georgia **1**: 258 (1817).

Slender-stemmed thinly pilose annual twiner; leaves ovate, entire or deeply 3-lobed, cordate with rounded sinus, up to 5 cm. along the midrib and 6 cm. across the basal lobes; peduncles few-flowered, stout; corolla much shorter in the Jamaican specimens than typically in S. United States; seeds glabrous.

Very rare (St. Andr.), possibly a casual introduction; 600 feet; fl. and fr. Nov–Dec. *Allwood UCWI 772!* S. United States, C. Amer., ? Cuba.

22. I. lacunosa L., Sp. Pl. **1**: 161 (1753).

Like the last but with white corolla.

Very rare (St. Eliz.); 50–100 feet. *P 16096!* S. United States. The interrelation-ships of *I. triloba*, *I. trichocarpa* and *I. lacunosa* in Jamaica require further study.

23. I. asarifolia (Desr.) Roem. & Schult. in L., Syst. Veg. ed. nov. **4**: 251 (1819). —*Convolvulus asarifolius* Desr. (1792).

Tough-stemmed perennial herb with trailing usually glabrous branches often rooting at nodes; leaf-blades up to 8 (–12) cm. in diameter, the proximal lateral veins con-spicuously radiating from the base; flowers few, cymose-subumbellate, on usually short stout peduncles; corolla deep reddish-purple, 5–7 cm. long; capsule de-pressed-globose, 10–15 mm. in diameter; seeds puberulous.

Locally common in the western parishes, in low-lying damp open pastures and wet waste ground; 10–1500 feet; fl. Sept–Apr, fr. Sept–Apr. *A 12011! H 9826! Proctor & Mullings 21824!* Mexico, Cuba, tropical Africa and Asia.

24. I. sagittata Poir., Voy. Barbarie **2**: 122 (1789).
Slender-stemmed, probably annual, glabrous twiner; mid-lobe of leaf up to 5 cm. long and 1 cm. broad near its base; flowers solitary; corolla (5–) 6–7·5 cm. long, light purple, paler outside and in tube; capsule globose, about 1 cm. in diameter, glabrous, beaked; seeds glabrous or minutely puberulous with long marginal hairs.
Rare and local (St. Eliz., Han.), in swamp margins; near SL–50 feet; fl. June–Sept, Dec, fr. July–Sept, Dec. *A 8046! 8581! H & P 14529! Robertson UCWI 9216!* S. Europe, N. Africa, S.E. United States, Bermuda, Bahamas, Cuba.

25. I. batatas (L.) Lam. in Tabl. Encycl. & Méth., Bot. **1**: 465 (1793).—*Convolvulus batatas* L. (1753). Sweet Potato.
Perennial from large tuberous roots; stems trailing, glabrous or nearly so; leaves entire or lobed, acuminate, cordate, 5–15 cm. long; flowers few together on short or long peduncles; sepals 7–11 mm. long; corolla trumpet-shaped, pinkish-mauve, rose or white with purple marks, 3–4·5 (–5) cm. long; capsule very rarely formed 2-locular; seeds glabrous.
Common in cultivation and relict or escaping; SL–4500 feet; fl. Nov–Feb. *A 10517! P 18380!* Native probably in Mexico, now cultivated as numerous varieties in all tropical countries; Cayman Is.

26. I. tiliacea (Willd.) Choisy in DC., Prodr. **9**: 375 (1845).—*Convolvulus tiliaceus* Willd. (1809). *I. fastigiata* (Roxb.) Sweet (1827). Wild Potato, Wild Slip.
Glabrous or sparingly pubescent twiner to 5 m. or more high; leaves usually entire and broadly ovate, rarely shortly 3-lobed, acuminate, cordate, 4–12 cm. long, 3–8 cm. broad; cymes few to many-flowered; corolla light mauve to purple with paler tube and darker eye, 3·5–6 cm. long; capsule depressed-globose, 2-locular, about 8 mm. broad and 6 mm. long; seeds dark brown or black, minutely puberulous or glabrous.
Very common in woodland and thicket margins and rough grassy places; SL–3000 feet; fl. July–Aug, Nov–Apr, fr. Aug, Dec–Mar. *A 6321! 7841! H 8430! P 8281!* Florida, Bahamas, Mexico to Brazil, West Indies; Grand Cayman, also in the Old World tropics.

27. I. aquatica Forsk., Fl. Aegypt.-arab.: CVI, 44 (1775).
Stems glabrous, trailing extensively; leaves broadly triangular to lanceolate, more or less hastate at base, tapered to a small obtuse tip, up to 10 cm. long and 5 cm. broad; flowers solitary or few in subumbellate cymes; sepals 6–8 mm. long; corolla light mauve, limb almost white with purple rays, darker within, about 5 cm. long and about 6 cm. across the limb; fruit not seen in Jamaica.
Rather local (St. Cath., West.), on muddy riverbanks and in roadside ditches; near SL–75 feet; fl. Nov–Feb. *A 10152! P 18342! Stearn 391!* Cuba, Trinidad the Guianas, widespread in the Old World tropics. The Jamaica plants are smaller in all their dimensions than typical Old World examples.

28. I. jamaicensis G. Don, Gen. Syst. **4**: 278 (1837).—*Convolvulus jamaicensis* Spreng. (1824), not Jacq. (1768). *Convolvulus tomentosus* L. (1753).
Strong-stemmed twiner to 6 m. or more high; leaves shallowly or deeply 3-lobed, rarely the basal lobes again divided, mid-lobe acuminate, base cordate, up to 12 cm. long and broad, very variably hairy from glabrous to densely silky tomentose adaxially; flowers solitary to numerous in compound bracteate racemes; sepals 10–14 mm. long; corolla bright crimson to magenta, tube 5·5–6 cm. long; capsule subglobose, about 1 cm. in diameter, beaked.
Rather common in the southern parishes, in thickets in arid limestone areas; 50–1800 feet; fl. Oct–Apr, fr. Nov–Feb. *A 9242! H 11688! P 17443!* Endemic. A very variable species having close affinity with *I. acuminata* and possibly hybridizing with that species (*A 8516!*).

29. I. acuminata (Vahl) Roem. & Schult. in L., Syst. Veg. ed. nov. **4**: 228 (1819).
—*I. cathartica* Poir. (1816). *Convolvulus acuminatus* Vahl (1794).
Glabrous or appressed-hairy twiner with slender stems to 5 m. or more long,

often trailing and rooting; leaves entire and broadly ovate or 3–sub-5-lobed, mid-lobe acuminate, base cordate with an open or rounded sinus, 4–10 cm. along the midrib, up to 9 cm. broad; flowers solitary to several, in bracteate subumbelliform cymes; sepals lanceolate, 1–2 cm. long; corolla usually bluish-purple, rarely white, mostly 5·5–7 cm. long; capsule subglobose, 10–12 mm. in diameter, shortly beaked; seeds dark brown or blackish, puberulous.

Common in open waste places and thickets, often near the sea in sandy ground, rarer in the drier areas; SL–2000 feet; fl. Aug–Apr, fr. Nov–Dec. *A 8575! H & B 10781! P 11272!* Generally but not uniformly spread throughout the tropics; Cayman Is.

30. I. cyanantha Griseb., Fl. Br. W.I.: 469 (1862).—*I. plicata* Urb. ex House (1908).

Slender-stemmed woody twiner to 5 m. or more high; stems and petioles pubescent, glabrescent; leaves ovate, acuminate, mostly openly cordate to truncate, up to 8 cm. long and 6 cm. broad on the main stems, smaller on the lateral flowering branches; flowers solitary, pedicels exceeding the short axillary peduncles; outer sepals 6–7 mm. long, inner 10–13 mm. long; corolla purplish-blue or white, 4·5–7 cm. long, fragrant; capsule long-acuminately beaked, about 17 mm. long.

Occasional, mainly in the central parishes, in woodland margins and hillside thickets, mostly in limestone areas; 1200–3000 feet; fl. Aug–Nov, fr. Aug–Oct. *A 11692! 11724! H 8997! P 18279!* Endemic.

31. I. ochracea (Lindl.) G. Don, Gen. Syst. **4**: 270 (1837).—*Convolvulus ochraceus* Lindl. (1827).

31a. Var. curtissii (House) Stearn in Proc. Linn. Soc. **170**: 145 (1959).—*I. curtissii* House (1908).

Perennial; stems twining or trailing and rooting; stems, petioles, inflorescence-branches and leaf-veins beneath densely and shortly pubescent; leaves rounded-ovate, shortly acuminate, mucronate, deeply cordate, up to 9 cm. along the midrib and 10 cm. broad; flowers 1–several in compact or lax inflorescences; corolla widely trumpet-shaped, about 4 cm. long; capsule 11–14 mm. long, beaked; seeds blackish, minutely pitted, glabrous except at the hilum.

Locally common (St. Andr., St. Cath., Manch.), on banks, shrubs and fences, weedy and mostly near habitations; 75–3500 feet; fl. and fr. Dec–Mar. *A 8769! H 8447! A. von der Porten!* Panama, Cuba. The variety *ochracea* occurs throughout tropical Africa.

32. I. obscura (L.) Ker-Gawl. in Edw., Bot. Regist. **3**: t. 239 (1817).—*Convolvulus obscurus* L. (1762).

Slender twiner like the last; sepals ovate-elliptical, about 5 mm. long, pubescent; corolla white, funnel-shaped, 2·5–3 cm. long; anthers included; capsule globose, pointed, 7–8 mm. in diameter; seeds black, finely pubescent.

Very rare (St. Eliz.), on river bank; 50 feet; fl. and fr. Jan. *Stearn 164!* Antigua, Barbados and part of a complex development of closely related species in tropical Africa which includes *I. ochracea*.

9. TURBINA Raf. (1838)

1. T. corymbosa (L.) Raf., Fl. Tellur. **4**: 81 (1838).—*Convolvulus corymbosus* L. (1759). *Rivea corymbosa* (L.) Hallier f. (1893). Christmas Pops.

Scrambling twiner with slender thinly pubescent stems to 8 m. or more high; leaves ovate, acute or shortly acuminate, openly cordate or sinus rounded, up to 6 (–7) cm. long and broad; inflorescence repeatedly branched and many-flowered, corymbose; sepals lanceolate, acute or obtuse, hyaline-margined, 5–9 mm. long; corolla white with pewter-grey coloration at base outside, dark purplish-red inside, 2–3 cm. long, fragrant; capsule ellipsoid, beaked, 8–10 mm. long; seed hairy.

Common on walls, fences and shrubs and trees in open thickets and woodlands, mostly in limestone areas; 50–2000 feet; fl. June–Jan, fr. Dec–Jan. *A 8671! 8228! H 8161! P 8270! 24371!* Continental tropical Amer., West Indies.

164. COBAEACEAE

Climbing shrubs. Leaves alternate, pinnately compound, the terminal pair of leaflets modified into tendrils, the basal pair of leaflets stipuliform. Flowers bisexual, actinomorphic, solitary or a few together on long axillary peduncles. Perianth biseriate; calyx gamosepalous, 5-lobed; corolla campanulate, the lobes contorted in bud. Stamens 5, epipetalous; anthers exserted, versatile, 2-locular, opening lengthwise. Disk large, fleshy, deeply lobed. Ovary superior, 3-locular; ovules anatropous, 2 or more in each loculus on axile placentas; style filiform, 3-fid. Fruit a septicidal 3-valved capsule. Seeds winged; embryo white, fleshy with large cotyledons, without endosperm.

1 genus with about 20 species native from Central to southern subtropical America.

1. COBAEA Cav. (1791)

1. C. scandens Cav., Ic. Descr. Pl. **1**: 11, tt. 16, 17 (1791).

Glabrescent climber up to 10 m. or more high; expanded leaflets 2 pairs, petiolulate, their blades elliptical, acute at both ends, 3–11 cm. long, 1·5–4·5 cm. broad; tendrils repeatedly forked; peduncles stout, 15–25 cm. long; calyx-lobes ovate-orbicular, 2·5–3·5 cm. long; corolla mostly purple, whitish at base, 5–6 cm. long; flowers protandrous, filaments coiling after pollen shedding followed by elongation of style and separation of 3 stigmas; capsule ellipsoidal, 5 cm. long.

Cultivated and naturalized locally (St. Andr., Port., St. Thom.), on trees and shrubs along roadsides and woodland margins; 3500–4900 feet; fl. Feb, fl. and fr. June–Aug. *A 7751! A. M. Barry! Robertson UCWI 1733!* Native of Mexico and elsewhere in C. and S. Amer., otherwise widely introduced and cultivated as an ornamental.

165. HYDROPHYLLACEAE

Annual or perennial herbs, rarely shrubs, often hairy or scabrid, sometimes spiny. Leaves alternate, rarely opposite, often in basal rosettes, entire to pinnately or palmately lobed, without stipules. Flowers bisexual, actinomorphic, in dichasial, umbellate or helicoid cymes, sometimes axillary and solitary. Perianth biseriate, usually 5-merous; sepals connate or almost free, imbricate, often with appendages between; corolla gamopetalous, the lobes imbricate or contorted. Stamens mostly 5, alternate with corolla-lobes, epipetalous towards the base of corolla-tube; anthers 2-locular, opening lengthwise. Ovary superior, 1-locular with 2 parietal or intrusive placentas or 2-locular; ovules usually numerous, pendulous, anatropous or amphitropous; styles 1 or 2. Fruit usually a loculicidal capsule, rarely septicidal or indehiscent. Seeds usually sculptured; endosperm copious or thin; embryo straight.

20 genera with 270 species; widespread, except Australia, mostly in N. America.

Plants erect, spiny; flowers several together in cymes, rarely solitary; corolla deeply lobed
to more than halfway; overy 2-locular   **1. Hydrolea**
Plants with prostrate branches, not spiny; flowers solitary or paired; corolla shortly lobed;
ovary 1-locular, the intrusive placentas contiguous but not fused   **2. Nama**

1. HYDROLEA L. (1762) nom. cons.

Stem and leaves with short and long glandular hairs; leaves narrowly elliptical, up to 8 cm.
long and 3·5 cm. broad; sepals up to 7 mm. long; corolla blue to violet, about 1·5 cm. long
                             1. spinosa
Stem and leaves minutely hairy, glabrescent; leaves oblanceolate, up to 2·5 cm. long and
7 mm. broad; sepals about 2·5 mm. long; corolla white, about 3 mm. long
                             2. nigricaulis

1. H. spinosa L., Sp. Pl. ed. 2, **1**: 328 (1762).

Shrubby herb, 30–200 cm. high, the main stem often creeping, the branches ascending and spreading; spines on stem mostly more than 1 cm. long, purplish; leaves acute, narrowly cuneate at base; flowers fragrant; sepals lanceolate, glandular hairy; corolla campanulate; filaments slender, blue; capsule ovoid, up to about 8 mm. long.

Rare (St. Eliz., Trel.), in swamps and ponds; 1500–2000 feet; fl. Jan–Apr, fr. Apr. *H 8680! 10676! P 20593!* Mexico to northern S. Amer., Cuba, Trinidad, introduced into Malaysia.

2. H. nigricaulis C. Wright ex Griseb., Cat. Pl. Cub.: 207 (1866).

Shrubby herb, 20–75 cm. high, the main stem erect with flexuous ascending branches; spines slender, 3–8 mm. long; leaves acute, narrowly cuneate at base; cymes short, few-flowered, mostly axillary; sepals lanceolate, glabrous; corolla funnel-shaped; capsule broadly ovoid, 3 mm. long.

Rare (Clar., St. Eliz.), in shallow water at pond margins; 25–300 feet; fl. and fr. Sept–Jan. *H 12553! P 24338! Yuncker 18049!* Cuba.

2. NAMA L. (1759) nom. cons.

1. N. jamaicense L., Syst. Nat. ed. 10, **2**: 950 (1759).—*Marilaunidium jamaicense* (L.) Kuntze (1891).

Annual herb with numerous spreading prostrate branches; branches pilose, to 40 cm. long, more or less winged by decurrent petiole-margins; leaves obovate to spathulate, up to 7 cm. long and 3 cm. broad, usually much smaller, apex obtuse; flowers subsessile to distinctly pedicelled; sepals linear, about 7 mm. long in flower, elongating in fruit; corolla 5–6 mm. long, white or tinged mauve, greenish towards the base, narrowly funnel-shaped, lobes 1–1·5 mm. long, rounded; capsule 2-valved, about 5 mm. long.

Rather common as a weed of lawns, roadsides and waste places; 100–2000 feet; fl. and fr. Oct–May. *A 6010! H 10325! P 17424!* Florida, Texas, Bermuda, Mexico to Venezuela, West Indies.

166. BORAGINACEAE

Herbs, shrubs or trees. Leaves usually alternate, simple, entire or toothed, without stipules. Flowers mostly bisexual, usually actinomorphic, sometimes zygomorphic, in cymose spikes, racemes, heads or panicles. Perianth biseriate, 5 (–7)-merous; sepals free or basally connate, imbricate or valvate, sometimes irregular; corolla gamopetalous, rotate, salver-shaped, funnel-shaped or campanulate, the lobes imbricate or contorted in bud, sometimes with pleats or appendages in the tube or partly closing the mouth. Stamens 5 (–7), epipetalous, usually uniform, alternating with corolla-lobes; filaments free; anthers 2-locular, opening lengthwise. Disk present or absent. Ovary superior, 2-locular but often becoming falsely 4-locular at maturity, entire or deeply 4-lobed; ovules usually 4, 2 in each carpel, anatropous, axile, erect, horizontal or pendulous; style 1, gynobasic or terminal, simple, 2-lobed or 4-lobed. Fruit of 4 nutlets or a 1–4-seeded nut or drupe. Seeds with or without endosperm.

100 genera and about 2000 species with cosmopolitan distribution.

1 Style gynobasic, arising between the 4 lobes of the ovary; corolla with invaginated saccate scales at mouth; herb with long basal leaves; fruit glochidiate; flowers unilateral   **1. Cynoglossum**
1 Style(s) or stigma terminal; ovary entire or hardly lobed; corolla without scales at mouth; fruit not prickly:
  2 Style single with solitary simple or 2-lobed stigma or stigma sessile and usually large and peltate or conical:
    3 Inflorescences of simple or compound unilateral spikes or racemes uncoiling in development:

4 Leaves narrowly oblanceolate, entirely covered with silky tomentum, numerous, crowded; fruit hollowed at base, conical-ovoid, corky; coastal shrub

2. Mallotonia

4 Leaves and fruit not as above:
  5 Fruit dry, breaking up into 2 or 4 nutlets; annual herbs or low much branched small-leaved undershrubs **3. Heliotropium**
  5 Fruit drupaceous with fleshy mesocarp at least at first, sometimes lobed and later drying into 2–4 nutlets; erect or climbing shrubs, sometimes small trees

4. Tournefortia

3 Inflorescence otherwise, cymose-paniculate, the flowers not obviously unilateral nor uncoiling in development; shrubs or trees, see 7* below:
2 Styles 2, free or united below, simple (2 stigmas) or 2-fid (4 stigmas); fruit drupaceous; shrubs or trees; flowers not usually obviously or regularly unilateral on the inflorescence-branches:
  6 Plants armed with axillary spines: leaves mostly clustered on short shoots, usually rather small, entire; flowers in few-flowered cymes or solitary; stigmas 2

5. Rochefortia

  6 Plants not armed; leaves not clustered:
    7 Stigmas 4 **6. Cordia**
    7* Stigmas 2:
      8 Corolla almost rotate, the tube shorter than the lobes; inflorescence a many-flowered dichotomous-cymose panicle; calyx imbricate in bud **7. Ehretia**
      8 Corolla salver-shaped, the tube longer than the lobes; inflorescence corymbose-cymose; calyx valvate in bud, the lobes separating irregularly **8. Bourreria**

1. CYNOGLOSSUM L. (1753)

1. C. amabile Stapf & Drummond in Kew Bull. **1906**: 202 (1906).
Short-lived perennial herb forming at first a rosette of leaves, 25–90 cm. high in flower; basal leaves long-petiolate with elliptical blades up to 15 cm. long and 4 cm. broad, upper leaves sessile, lanceolate, smaller, puberulous above, whitish-pubescent beneath; sepals ovate-elliptical, 2·5–4 mm. long, spreading in fruit, yellowish-hispid; corolla-limb up to 10 (–12) mm. across, bright blue or mauve; nutlets 4, separating, about 2·5 mm. broad.
Naturalized (St. Andr., Clar., Manch.), in open waste places and roadsides, sometimes cultivated at lower elevations; 2500–4000 feet; fl. and fr. Dec–Apr, July–Aug. *A 8411! H & P 14254! Robertson UCWI 1094!* Native of W. China and N. India, introduced into Mexico and Puerto Rico.

2. MALLOTONIA (Griseb.) Britton (1915)

1. M. gnaphalodes (L.) Britton in Ann. Miss. Bot. Gard. **2**: 47 (1915).—*Heliotropium gnaphalodes* L. (1759). *Tournefortia gnaphalodes* (L.) Kunth (1818). Seaside Lavender.
Rather succulent bushy shrub 0·6–2 m. high; leaves up to about 8 (–10) cm. long and 8–9 mm. broad behind the tip, obtuse, greyish-green; inflorescences peduncu-late, simple or branched, flowering portion short and dense; calyx hispid-tomentose, the lobes 2–3 mm. long; corolla a little longer than the calyx, white tinged pink in throat, base of tube green, hairy outside, fragrant; fruit 5–6 mm. long.
Widespread but not common on coastal limestone and amongst beach boulders; SL–20 feet; fl. and fr. most of the year. *A 8029! H 9522! P 20664!* Bermuda, Florida, C. Amer., Bahamas, Greater Antilles, Lesser Antilles (drier islands), Pedro and Morant Cays, Grand Cayman.

3. HELIOTROPIUM L. (1753)

1 Plant glabrous and glaucous with fleshy oblanceolate leaves   1. *curassavicum*
1 Plant hairy; leaves not fleshy:
  2 Leaves about 2 mm. long, tightly overlapping along prostrate spreading branches; plant deeply taprooted forming mats; corolla white with yellow eye; on beaches and coastal limestone; Grand Cayman   *H. humifusum* Kunth
  2 Leaves larger; plants not forming cushions or mats:
    3 Leaves usually less than 1·5 cm. broad, if as much or more then rounded at the tip; fruit separating into 4 nutlets:

4　Leaf-blade obovate; cymes very slender, without bracts; stigma subsessile
　　　　　　　　　　　　　　　　　　　　　　　　　　　　　2. procumbens
4　Leaf-blade narrowly ovate to linear-lanceolate; cymes with bracts; style evident:
　5　Leaves beneath densely covered with an indumentum of fine ascending hairs;
　　　cymes 1–3 (–4) cm. long, few-flowered with inconspicuous linear bracts; corolla-
　　　limb up to 7 mm. across　　　　　　　　　　　　　　　　**3. ternatum**
　5　Leaves beneath visible between appressed bristly hairs; cymes eventually elongated,
　　　4–15 cm. long with scattered leafy bracts; corolla-limb 2–3 mm. across
　　　　　　　　　　　　　　　　　　　　　　　　　　　　　4. fruticosum
3　Leaves at least 1·5 cm. broad or mostly so, acute or acuminate:
　6　Cymes short, much branched forming a rounded corymb; leaves elliptical, pointed;
　　　anthers minutely ciliate near tip; fruit separating into 4 nutlets; corolla violet,
　　　purple or white, fragrant; cultivated in gardens; native of Peru; Cherry Pie,
　　　Heliotrope　　　　　　　　　　　　　　　　　　　　**H. arborescens** L.
　6　Cymes elongated, unbranched, slender; anthers glabrous; nutlets united in pairs:
　　7　Leaves elliptical, pointed at both ends, mostly less than 4 cm. broad; corolla white,
　　　　usually with a yellow eye, hairy in throat; nutlets rounded, bluntly ribbed
　　　　　　　　　　　　　　　　　　　　　　　　　　　　5. angiospermum
　　7　Leaves broadly ovate, more or less truncate at base, mostly more than 4 cm. broad;
　　　　corolla light violet with yellow eye, rarely white, glabrous in throat; nutlets
　　　　pointed, sharply ribbed　　　　　　　　　　　　　　　　**6. indicum**

1. H. curassavicum L., Sp. Pl. **1**: 130 (1753).
Branches spreading from a central woody stock, up to about 50 cm. or more
long, the flowering shoots ascending; leaves oblanceolate to linear, up to 5 cm. long,
often much smaller; inflorescences simple or forked, up to about 5 cm. long; calyx
about 1·5 mm. long; corolla white with greenish-yellow eye.
Locally common in low-lying sandy or gravelly places and on limestone rocks
near the sea; SL–20 (–50) feet; fl. and fr. most of the year. *A 6029! H 8191!*
C. B. Lewis! Florida, Texas, Bahamas, tropical Amer., West Indies; Grand Cayman
and on some Old World tropical shores.

2. H. procumbens Mill., Gard. Dict. ed. 8 (1768).—*H. inundatum* Sw. (1788).
Annual branched subwoody hispid herb 5–30 cm. high; leaves slender-petioled,
blade up to 3 cm. long and 1·3 cm. broad, obovate; cymes 1–3; corolla white, tube
about 1·2 mm. long, limb about 1 mm. across; nutlets brown.
Rare and sporadic (St. Cath., Clar., Han.), a weed in low-lying wet places and at
pond margins; 200–900 feet; fl. and fr. Apr, Sept. *H 8509!* S. United States to
Panama and Paraguay, Bahamas, Greater Antilles, Antigua, Trinidad.

3. H. ternatum Vahl, Symb. Bot. **3**: 21 (1794).—*H. fruticosum* of Griseb. (1862),
　　not L. Wild Marjoram.
Low trailing or erect bushy shrub up to 1·5 m. high, aromatic; leaves rigid,
shortly petioled, up to 2 cm. long and about 6 mm. broad, often much narrower
with the margins recurved; cymes mostly simple; corolla white with yellow eye,
tube 4 mm. long, limb 3–7 mm. across; nutlets black.
Locally common (St. Andr., St. Cath., St. Thom.), among limestone rocks on
open arid hillsides and in coastal sand (prostrate variant); SL–600 feet; fl. and fr.
all the year. *A 10001! H 10002! P 7508!* Continental tropical Amer., Bahamas,
West Indies; Grand Cayman.

4. H. fruticosum L., Syst. Nat. ed. 10, **2**: 913 (1759).
Low annual subwoody herb, 5–30 cm. high, with ascending branches; leaves
shortly petioled, elliptic-lanceolate, pointed at both ends, up to 2 cm. long and
5 mm. broad, the margins somewhat recurved; cymes simple; corolla white, tube
about 2 mm. long; nutlets brown.
Rare (St. Cath., Clar., St. Eliz.), a casual weed of open grassy places; about 300
feet; fl. and fr. Aug, Dec. *H 12753!* Guatemala to Colombia, Greater Antilles,
Curaçao, Margarita.

5. H. angiospermum Murr., Prodr. Stirp. Gotting.: 217 (1770).—*H. parvi-*
florum L. (1771). Dog's Tail.
Annual thinly hispid bushy herb to 1 m. high; leaves up to 8 (–10) cm. long and
3·5 (–5) cm. broad, usually smaller, thin, flat, slender-petioled; cymes simple or

forked, up to about 12 cm. long in fruit; stigma sessile; fruit glabrous but covered with minute scales; nutlets 1·5–2 mm. long.

Common as a field and garden weed mostly in rather shady places; SL–700 (–4000) feet; fl. and fr. all the year. *A 5482! H 8214! P 9473!* S. United States, Mexico to N. Chile, Bolivia and Guyana, West Indies, Trinidad; Grand Cayman.

6. H. indicum L., Sp. Pl. **1**: 130 (1753).—Scorpion Weed, Wild Clary.

Annual erect taprooted hispid herb up to 1 m. or more high; leaves broad at base and often subcordate at junction with the long winged petiole, up to 15 cm. long and 6 (–10) cm. broad, irregularly sinuate-margined; cymes simple, up to 15 cm. or more long in fruit; stigma borne on an evident style; fruit glabrous; nutlets nearly 3 mm. long.

Common as a weed of pastures, cultivated ground and waste places; SL–1300 feet; fl. Jan–June, Sept, fr. Feb–June. *A 6932! H 8516! P 6560!* Pantropical; Grand Cayman.

4. TOURNEFORTIA L. (1753)

1 Leaf-blades on flowering shoots rarely as much as 6 cm. long and 3 cm. broad; corolla-lobes much longer than broad, acuminate; style evident; fruit 2–4-lobed; climbing shrubs:
  2 Leaves densely grey-pubescent beneath, the hairs so close that their bases are not visible between them, lanceolate to narrowly lanceolate; inflorescence at least twice-forked   1. *poliochros*
  2 Leaves glabrous to pubescent beneath with individual hairs visible:
    3 Leaf-blades narrowly lanceolate; inflorescence slender, simple or usually once-forked   2. *minuta*
    3 Leaf-blades broadly ovate to lanceolate; inflorescence at least twice-forked:
      4 Leaves mostly pubescent beneath; corolla-tube 2 mm. long; ripe fruit white, usually with black spots   3. *volubilis*
      4 Leaves almost glabrous beneath; corolla-tube 4–5 mm. long; ripe fruit yellow, usually marked with black spots   4. *maculata*
1 Leaf-blades on flowering shoots larger, some at least 8 cm. long and 3·5 cm. broad:
  5 Leaves conspicuously hairy especially beneath:
    6 Scrambling or climbing shrub; inflorescence with spreading coarse hairs; leaves about twice as long as broad   5. *hirsutissima*
    6 Erect shrub; inflorescence with appressed fine hairs; leaves about three times as long as broad   6a. *astrotricha* var. *astrotricha*
  5 Leaves glabrous or with only a few scattered hairs:
    7 Corolla-lobes rounded, blunt:
      8 Calyx-lobes long-acute, about twice to three times as long as tube; anthers included; erect or straggling shrub   6b. *astrotricha* var. *subglabra*
      8 Calyx-lobes acute or obtuse, about as long as tube; anthers exserted, about 2·5 mm. long; erect shrub or small tree   7. *staminea*
    7 Corolla-lobes acute, acuminate or mucronate:
      9 Inflorescence compact, the flowers subcontiguous on short branches; corolla-lobes rounded, mucronate; straggling or scandent shrub; stigma sessile   8. *bicolor*
      9 Inflorescence lax, the flowers at length clearly separated; corolla-lobes long-acuminate; stigma on a slender style:
        10 Leaf-blades elliptical, long-tapered at base; erect shrub or small tree   9. *glabra*
        10 Leaf-blades ovate to oblong-elliptical, rounded to broadly cuneate at base; straggling or climbing shrub   4. *maculata*

1. T. poliochros Spreng. in L., Syst. Veg. ed. 16, **1**: 644 (1824).

Climbing shrub with slender twining branchlets; leaf-blades up to 6 cm. long and 1·8 cm. broad; inflorescence-branches slender, up to about 5 cm. long; flowers narrow, green, 3·5–4 mm. long; fruit green turning white with black spots.

Occasional in woodlands and thickets on limestone; 100–1600 feet; fl. Sept–Mar, fr. Oct–Mar. *A 9239! H 10355!* Bahamas, Turks and Caicos Is., Cuba. This species comes very close to some variants of *T. volubilis*.

2. T. minuta Bert. ex Spreng. in L., Syst. Veg. ed. 16, **1**: 644 (1824).

Slender-stemmed scrambling and twining shrub to 1 m. or more high; leaves up to 5·5 cm. long, 2–10 mm. broad; inflorescence-branches 2–4 cm. long; calyx-lobes subulate, about 1 mm. long.

Rare and local, in arid thickets and woodlands on limestone; 20–400 feet; fl. and fr. Nov–Feb. *A 5664! H & B 10787! P 7522!* Cuba (?), Hispaniola.

3. T. volubilis L., Sp. Pl. **1**: 140 (1753).—Chigger Nut (Jigger Nit).
Trailing or twining to 6 m.; leaves mostly ovate, acutely acuminate, broadly cuneate to rounded at base, up to 9 cm. long and 4 cm. broad; inflorescence-branches up to 6 cm. long in fruit; calyx-lobes lanceolate-subulate, 1·5–2 mm. long; corolla green turning deep yellow.
Common on limestone banks and rocks and in thickets; SL–800 feet; fl. and fr. all the year. *A 9240! A. M. Barry! H 10356!* Florida, Bahamas, C. and S. Amer., West Indies; Grand Cayman. An extremely variable species within which some authors include both the foregoing. A more vigorous, more hairy variant with cordate-based leaves up to 13 cm. long and 7 cm. broad occurs on dry limestone hills near Kingston (*A 9238*).

4. T. maculata Jacq., Enum. Syst. Pl. Carib.: 14 (1760).—*T. laurifolia* of Griseb. (1862), not Vent.
Straggling or climbing shrub to 6 m. or more; leaves mostly broadly ovate, acute or acuminate, broadly cuneate to rounded at base, up to 12 cm. long and 7 cm. broad, soft and thin, often minutely white-punctate above; inflorescence-branches slender, up to about 5 cm. long; flowers about 5 mm. long; corolla green or yellowish; fruit about 5 mm. broad.
Occasional in the central and western parishes, in woodland and thickets on rocky limestone; 50–2500 feet; fl. Apr–May, fr. July. *A 6789! H 10973!* Mexico to Peru and Brazil, West Indies.

5. T. hirsutissima L., Sp. Pl. **1**: 140 (1753).—Cold Withe.
Erect shrub to 2 m. or more usually a robust scrambler or vine to 6 m., hispid with brownish hairs; leaves broadly elliptical to ovate, acutely acuminate, broadly cuneate to rounded at base, blades up to 15 (–26) cm. long and 8 (–13·5) cm. broad; inflorescence-branches rather compact, up to about 4 cm. long in fruit; flowers about 7 mm. long, fragrant; corolla white; stigma sessile; fruit a soft white berry 8–10 mm. in diameter.
Very common as a weed on banks and waste ground and secondary in thickets and woodland margins; SL–4200 feet; fl. Feb–Oct, fr. Apr–Oct. *A 6754! A. M. Barry! Stearn 786!* Florida, continental tropical Amer., West Indies.

6. T. astrotricha DC., Prodr. **9**: 520 (1845).

6a. Var. astrotricha.
Shrub with stout branches 1·2–3 m. high; leaves mostly elliptical, obtuse to acute or shortly acuminate, mostly broadly cuneate at base, blades up to 25 cm. long and 14 cm. broad, scabrid; inflorescence-branches up to about 12 cm. long in fruit; flowers 5 mm. long, fragrant; corolla green turning white then brownish-pink, the lobes rounded; style evident, persistent; fruit fleshy, shiny, white, subglobose, up to about 1 cm. in diameter.
Frequent in southern coastal areas in woodlands, thickets and on rocks at mangrove margins; SL–1500 feet; fl. and fr. Mar–July, Oct–Jan. *A 5493! H 8930! P 27539!* Grand Cayman.

6b. T. astrotricha var. **subglabra** Stearn in Journ. Arn. Arb. **52**: 633 (1971).
Straggling shrub 2·5–5 m. high; leaves elliptical, pointed at both ends, blades up to 18 cm. long and 7 cm. broad; inflorescence-branches up to 20 cm. long in fruit; flowers about 6 mm. long; corolla white turning reddish; style 1–1·5 mm. long; fruit subglobose, white, fleshy, 6–9 mm. in diameter.
Rare (St. Eliz., Han., St. James), in waste places; SL–100 feet; fl. and fr. Mar–May. *H 10351! P 23475! Stearn 1029!* Grand Cayman.

7. T. staminea Griseb., Fl. Br. W.I.: 484 (1862).
Shrub or tree 1·5–5 m. high with lax brittle branches; bark deeply fissured with broad furrows; leaves obovate to oblong-elliptical, acuminate, cuneate at base, blades up to 25 cm. long and 10 cm. broad; inflorescence-branches up to about 20 cm. long in fruit; flowers 4–5 mm. long, very fragrant; corolla white turning

yellow, greenish outside; stigma sessile; fruit broadly ovoid, blunt with 2 grooves when dry, 4·5–6 mm. long, yellow.

Occasional in the central and western parishes in thickets and woodland margins on limestone and on cliffs; 450–2750 feet; fl. and fr. sporadically. *A 7997! 8611! H 8663! P 7180!* Endemic.

8. T. bicolor Sw., Nov. Gen. & Sp. Pl.: 40 (1788).

Low bush to 1 m. high or scandent shrub; leaves elliptical, acuminate, broadly cuneate at base, blades up to 20 cm. long and 9 cm. broad; inflorescence-branches up to 4 cm. long; flowers 5–6 mm. long, fragrant; corolla white, the tube greenish outside, drying yellow; fruit subglobose, 3–4 mm. long.

Occasional, on roadside banks and in thickets; SL–3500 feet; fl. Apr–Sept, fr. Sept. *A 12391! H 9596!* Continental tropical Amer., West Indies.

9. T. glabra L., Sp. Pl. **1**: 141 (1753).—*T. cymosa* L. (1762).

Shrub 2–3 m. or tree to 6 m. high; stem weak; leaves oblanceolate to elliptical, long-acuminate, long-tapered to base, blades up to 35 cm. long and 9 cm. broad; inflorescence-branches up to 25 cm. long in fruit; flowers 6 mm. long; corolla green to greenish-white; fruit conical-ovoid, ripening fleshy and white, ribbed when dry, about 3 mm. long and 4 mm. broad.

Rather common in damp thickets and woodland margins and on roadside banks; 600–5600 feet; fl. June–Nov, fr. June–Feb. *A 8137! H 9427! P 6891!* C. Amer., Cuba, Hispaniola.

5. ROCHEFORTIA Sw. (1788)

1 Leaves up to 8 mm. broad, obovate to oblanceolate, glabrous; flowers solitary, subsessile; spines conspicuous on short lateral shoots; calyx-segments ovate   1. *acanthophora*
1 Leaves 10–18 (–25) mm. broad, obovate, at least some slightly hairy; flowers usually in cymose clusters, pedicelled; spines on main shoots only or lacking; calyx-segments suborbicular, ciliate:
  2 Spines lacking or sparse; hairs on adaxial leaf-surface minute without thickened bases; leaf-margin smooth; pedicels pubescent   2. *cuneata*
  2 Spines numerous; hairs on adaxial leaf-surface bristlelike, usually broadened at base; leaf-margin more or less ciliate; pedicels glabrescent   3. *acrantha*

1. R. acanthophora (DC.) Griseb., Fl. Br. W.I.: 482 (1862).—*Ehretia acanthophora* DC. (1845). Greenheart Ebony.

Shrub or tree 2–6 m. high with profuse multiple branching, the distal branches drooping; leaves rigid, blade 2–15 mm. long, shortly petioled to subsessile; fruit red when ripe, subglobose, about 3·5 mm. in diameter, 4-seeded.

Local (St. Cath., St. Ann, St. Thom.), in scrub woodland and on limestone rocks and at salina margins in arid areas near the sea; SL–200 feet; fr. Jan, Oct. *H & P 15091! Loveless UCWI 2455! P 19747!* Greater Antilles, Antigua.

2. R. cuneata Sw., Nov. Gen. & Sp. Pl.: 54 (1788).

Shrub or tree to 5 m. high, sparingly spiny; leaves 1–5 cm. long, 7–25 mm. broad, distinctly petioled; cymes axillary or terminal.

Rare (Han., Trel.), on wooded limestone hills; 1300–**1500** feet; fl. Mar. *H 10302!* Hispaniola, Antigua.

3. R. acrantha Urb., Symb. Ant. **5**: 479 (1908).—Green Ebony.

Shrub or tree 1–5 m. high with flexuous branches; leaves 1–3 cm. long, 5–12 mm. broad, shortly petioled; cymes terminal and axillary; corolla-lobes spreading, about 5 mm. across; stigmas exserted; drupe subglobose, 8–10 mm. in diameter.

Rare and local (West., Trel.), in woodland on rocky limestone; 30–2000 feet; fl. Dec–Mar. *H 8821! 10232!* Endemic.

6. CORDIA L. (1753)

1 Inflorescence lax, branched, corymbose or paniculate, not spicate or globose-capitate:
  2 Calyx 7–15 mm. long in flower, enlarging and enclosing the fruit:
    3 Leaves rough; calyx setose, not or only shallowly ribbed; corolla red or orange, 4–5·5 cm. long, deciduous, lobes 6–7, rounded   1. *sebestena*

3 Leaves smooth; calyx puberulous, with a few longer hairs, ribbed; corolla white, 1·5–2 cm. long, persistent, turning brown, lobes 5–6, oblong   2. *gerascanthus*
2 Calyx 2–6 mm. long, not permanently enclosing the fruit, but persistent at its base; corolla 2–10 mm. long, usually 5-lobed:
   4 Leaves mostly very large, up to 50 cm. long and 36 cm. broad, hairy on both surfaces; inflorescence shaggy-tomentose; calyx densely hairy, not ribbed   3. *macrophylla*
   4 Leaves smaller, up to 20 cm. long and 10 cm. broad:
     5 Leaf-margin toothed, blade as broad as long to twice as long as broad, adaxial surface pustulate with the broadened bases of minute bristly hairs; calyx ribbed, membranous distally and fragmenting irregularly at anthesis; corolla-limb broadly funnel-shaped with crisped margins   4. *alba*
     5 Leaf-margin entire, blade distinctly longer than broad; calyx not ribbed, 2–5-lobed or toothed:
       6 Leaves cuneate at base, usually more than twice as long as broad; adaxial surface minutely punctulate with pustules not associated with hairs; inflorescence pubescent:
         7 Leaves ovate to lanceolate, acutely long-acuminate, broadly cuneate at base, thin, scabridulous adaxially, up to 9 cm. long and 4 cm. broad; corolla campanulate, about 2·5 mm. in diameter   5. *bifurcata*
         7 Leaves elliptical to oblanceolate, mostly narrowly cuneate at base, obtuse, acute, rarely shortly acuminate or cuspidate, usually rather leathery, smooth and shiny adaxially, larger than above; inflorescence lax, 5–15 cm. broad; corolla rotate to funnel-shaped:
           8 Leaves usually broadest above the middle; corolla 6–8 mm. in diameter, the lobes about 1·5–2 mm. broad   6. *collococca*
           8 Leaves usually broadest at the middle; corolla 12 mm. in diameter, the lobes 3·5–5 mm. broad   7. *laevigata*
       6 Leaves rounded or subcordate at base, usually less than twice as long as broad, leathery; adaxial surface very inconspicuously or not punctulate, puberulous or glabrous:
         9 Leaves broadest below the middle, the blade ovate or narrowly ovate, acutely acuminate, usually subcordate at base; inflorescence and calyx densely puberulous with short appressed brownish hairs   8. *elliptica*
         9 Leaves broadest at the middle, the blade oblong to elliptical and rounded at base:
           10 Leaf-tip abruptly long-acuminate, the point 5–18 mm. long; inflorescence and calyx puberulous with short spreading brown hairs   9. *troyana*
           10 Leaf-tip acute, shortly acuminate or cuspidate, the point 1–8 mm. long; inflorescence puberulous with short appressed brownish hairs; calyx glabrous   10. *harrisii*
1 Inflorescence dense, unbranched; leaves serrate or dentate:
  11 Inflorescence cylindrical, spicate, 1·5–4 cm. long, excluding peduncle:
    12 Leaves densely hairy beneath; calyx with evident short stiff hairs in upper part   11. *brownei*
    12 Leaves almost glabrous or with very short inconspicuous hairs beneath; calyx minutely hairy appearing glabrous to the unaided eye   12. *jamaicensis*
  11 Inflorescence globose-capitate:
    13 Hairs on adaxial surface of mature leaf without conspicuously broadened bases; peduncles mostly axillary, slender, up to 4 cm. long, with spreading hairs; calyx-lobes at most mucronate   13. *linnaei*
    13 Hairs on adaxial surface of mature leaf with broadened and often swollen bases; peduncles terminal or internodal; calyx-lobes attenuated into linear appendages 1·5–6 mm. long; corolla white:
      14 Calyx (excluding appendages) about 5 mm. long; leaves ovate to broadly ovate; peduncle up to 8 cm. long, with ascending rusty-brown hairs; corolla up to about 9 mm. long   14. *clarendonensis*
      14 Calyx (excluding appendages) 2·5–4 mm. long; leaves lanceolate to ovate; peduncle with whitish hairs; corolla 4–6 mm. long:
        15 Mature leaves rigid, thick, the abaxial surface usually deeply depressed between the veins, hairs on adaxial surface with raised swollen bases; peduncles 3–7 cm. long   15. *bullata*
        15 Mature leaves thinner, the abaxial surface normally flat, hairs on adaxial surface without raised swollen bases; peduncles up to 2 (–2·5) cm. long   16. *globosa* var. *humilis*

1. C. sebestena L., Sp. Pl. **1**: 190 (1753).—Red or Scarlet Cordia.
Shrub (usually when coppiced) or tree to 6 m. high; leaves ovate, mostly obtusely pointed, shortly cordate at base, up to 20 cm. long and 15 cm. broad or longer; inflorescence corymbose; calyx tubular, 10–15 (–18) mm. long, shortly lobed; corolla salver-shaped; stamens 5–7; ripe fruit white, pointed, up to 4 cm. long.

Locally common in sandy thickets and on limestone rocks in arid coastal areas;
SL–580 feet; fl. and fr. most of the year. *A 8892! H 6766! P 24211!* Florida,
continental tropical Amer., West Indies; Grand Cayman; introduced and cultivated for ornament in the Old World tropics.

2. C. gerascanthus L., Syst. Nat. ed. 10, **2**: 936 (1759).—Panchallon, Spanish
Elm.

Deciduous tree to 6–20 m. or more high; leaves ovate, lanceolate or elliptical,
more or less acuminate, up to 18 cm. long and 6 cm. broad, of juvenile plants
pellucid-dotted and with squarrose hairs on midrib beneath; inflorescence rounded-
corymbose; flowers fragrant; calyx 7–10 mm. long, shortly lobed; corolla salver-
shaped, marcescent and forming a parachute for the combined calyx and ovary in
fruit; fruit oblong, with persistent style-base at tip.

Common on limestone hills mostly in rather dry areas; SL–2000 (–2500) feet;
fl. Dec–June, fr. Mar–May. *A 6330! H 8954! P 8386!* Mexico to Nicaragua,
Greater Antilles; Cayman Is.

3. C. macrophylla L., Sp. Pl. ed. 2, **1**: 274 (1762).—Fish-leaf, Manjack.

Tree 8–15 m. high with low heavy spreading branches; leaves ovate to lanceolate,
acute or acuminate, rounded to truncate or subcordate at base; inflorescence a
large dichotomous panicle; calyx about 4 mm. long; corolla white, lobes spreading
and reflexed; flower fragrant; fruit ovoid, pointed, about 1 cm. long.

Occasional, rather widespread, in thickets and secondary woodlands; 500–2300
feet; fl. Dec–Mar, fr. Mar–May. *A 8980! H & P 13660! Stearn 447!* Endemic.

4. C. alba (Jacq.) Roem. & Schult. in L., Syst. Veg. ed. nov. **4**: 466 (1819).—
Varronia alba Jacq. (1760). *C. dentata* Poir. (1806). Duppy Cherry.

Shrub or tree to 10 m. high, sometimes branches long and scrambling; leaves
broadly ovate, obtuse or shortly pointed, broadly cuneate to rounded at base, up
to 10 cm. long and 8 cm. broad; inflorescence dichotomously much branched,
compact, rounded; calyx campanulate, 3–4 mm. long, whitish distally; corolla
yellowish-white, about 15 mm. across, fragrant; ripe fruit white.

Locally abundant (St. Andr., St. Cath., St. Mary), on gravelly alluvial plains;
5–700 feet; *A 6834! P 23505! Stearn 307!* Continental tropical Amer., West
Indies.

5. C. bifurcata Roem. & Schult. in L., Syst. Veg. ed. nov. **4**: 466 (1819).

Shrub 1–1·2 m. high; stem slender, erect, brittle; leaves with short white hairs
on midrib and veins beneath; petioles sharply bent near base, the base persistent
as a peg after leaf-fall; inflorescence shortly divaricately branched; calyx white-
strigose with 5 short deltate teeth; stamens 5, shortly exserted; fruit light red,
about 2·5 mm. in diameter.

Rare and local (St. Ann), on banks and at thicket margin; 1650–1750 feet; fl.
and fr. Oct. *A 11796!* Costa Rica to Brazil and northern Argentina.

6. C. collococca L., Fl. Jam.: 14 (1759) (err. *callococca*).—Clammy Cherry.

Deciduous or semi-deciduous tree 8–16 m. high with grey bark and spreading
branches; leaves obovate to elliptic-oblanceolate, acute to bluntly acuminate, up
to 18 cm. long and 7 cm. broad; calyx splitting irregularly, about 3 mm. long;
corolla creamy-white, the lobes often reflexed; ripe fruit red, about 1 cm. in diameter.

Common in thickets and along roadsides and pasture margins; 10–1300 feet;
fl. Feb–Apr, fr. Mar–June. *A 6835! A. M. Barry! Stearn 335!* Mexico to northern
S. Amer., West Indies.

7. C. laevigata Lam. in Tabl. Encycl. & Méth., Bot. **1**: 422 (1792).—*C. nitida*
Vahl (1793).

Shrub or tree like the last, trunk with dark grey rather smooth bark, branches
spreading and drooping; leaves elliptical, more or less acuminate, up to 18 cm.
long and 7 cm. broad; calyx about 4 mm. long; corolla white or light purple;
ripe fruit red, up to 2 cm. in diameter.

Frequent in woodlands on limestone; 100–3000 feet; fl. Mar–Dec, fr. June–Feb.

A 7907! 12560! H 10281! P 7265! C. Amer., Greater Antilles, Virgin Is., Guadeloupe.

8. C. elliptica Sw., Nov. Gen. & Sp. Pl.: 47 (1788).—*C. fawcettii* Krug & Urb. (1895).
Tree 6–10 m. high with spreading branches; leaves up to 20 cm. long and 10 (–12) cm. broad; inflorescence-branches several times forked, the flowers markedly unilateral, fragrant; calyx campanulate, 6–7 mm. long; corolla white; ripe fruit white.
Rare in central and eastern parishes, in wet secondary forest on limestone hills; 1500–3500 feet; fl. July, Oct–Feb, fr. Nov–Apr. *A 12963! H 6718! P 7638!* Endemic.

9. C. troyana Urb., Symb. Ant. **5**: 475 (1908).
Shrub or tree 3–8 m. high, with spreading zig-zag branches; leaves up to 18 cm. long and 9 cm. broad; inflorescence and calyx as above; corolla creamy-white.
Rare (Clar., Trel., St. Ann), in woodlands on rocky limestone; 1500–2800 feet; fl. Mar–Apr, July, fr. Mar, Aug, Dec. *A 11053! H 8819! P 8416!* Endemic.

10. C. harrisii Urb., Symb. Ant. **5**: 474 (1908).
Tree like the last, up to 8 m. high; corolla white.
Very rare (Trel.), in woodland on limestone hill; 2000 feet; fl. Sept. *H 9421!* Endemic, and known only from the type.

11. C. brownei (Friesen) I. M. Johnston in Journ. Arn. Arb. **31**: 177 (1950).—*Mountjolya brownei* Friesen (1933). Black Sage.
Shrub 1·2–3 m. high; leaves elliptic-lanceolate, up to 10 cm. long and 3 cm. broad; flowers develop mostly basipetally; calyx 2·5–3 mm. long; corolla white, lobes 5 with small intermediate lobes, faintly fragrant; ripe fruit subglobose, red.
Common in thickets and open woodlands and on roadside banks; SL–3750 (–5000) feet; fl. and fr. most of the year. *A 5478! P 7215! Stearn 667!* Grand Cayman.

12. C. jamaicensis I. M. Johnston in Journ. Arn. Arb. **31**: 179 (1950).—*C. cylindristachya* of Griseb. (1862), partly.
Shrub, erect or scrambling, 1·2–3 m. or up to 6 m. high; leaves ovate-lanceolate to elliptic-lanceolate, up to 10 cm. long and 4 cm. broad; calyx 3–4 mm. long; corolla white; ripe fruit red.
Widespread, especially in the central and western parishes, but nowhere common, in old pastures, thickets and woodlands on limestone; 50–3000 feet; fl. most of the year, fr. Aug–Nov (–Jan). *A 12722! H 8732! P 7790!* Endemic.

13. C. linnaei Stearn in Journ. Arn. Arb. **52**: 627 (1971).—*Varronia lineata* L. (1759), illegitimate name.
Straggling shrub, 1·5–3 m. or more high with very tough stems, sometimes a small tree; leaves lanceolate, acutely long-acuminate, repand-dentate, up to 10 cm. long and 2·5 cm. broad; inflorescence about 1 cm. in diameter; calyx 2·5–3 mm. long; corolla greenish-white; ripe fruit red.
Locally common on roadside banks and in thickets on gravelly soil; 650–2500 feet; fl. Sept–Jan, fr. Sept–Mar. *A 5479! H 9434! P 29205!* Mexico to Venezuela, Cuba, Hispaniola.

14. C. clarendonensis (Britton) Stearn in Journ. Arn. Arb. **52**: 631 (1971).—*Varronia clarendonensis* Britton (1914).
Shrub with weak straggling branches up to about 2 m. high; leaves broadly ovate, obtuse, broad at base, dentate, up to 10 cm. long and 5·5 cm. broad; inflorescence about 1·5 cm. in diameter.
Rare (Clar., Trel.), in open woodland on arid limestone; 1300–2500 feet; fl. May–Sept, fr. Sept. *H 10995! H & P 14124! R 1927! Weaver 1007!* Endemic.

15. C. bullata (L.) Roem. & Schult. in L., Syst. Veg. ed. nov. **4**: 462 (1819).—*Varronia bullata* L. (1759).
Shrub 1–2·5 m. high, often with straggling branches; leaves broadly ovate to

elliptic-lanceolate, mostly obtuse, up to 10 cm. long and 6 cm. broad, very harshly scabrid; inflorescence about 1 cm. in diameter; fruit red.

Widespread on limestone rocks in secondary open woodlands and thickets; SL–3000 feet; fl. most of the year, fr. July–Nov. *A 5582! 12558! H 5915! P 23643!* Endemic.

16. C. globosa (Jacq.) Kunth, Nov. Gen. **3**: 76 (1818).—*Varronia globosa* Jacq. (1760).

16a. Var. **humilis** (Jacq.) I. M. Johnston in Journ. Arn. Arb. **30**: 98, 117 (1949).— *V. humilis* Jacq. (1760). Gout Tea, Wild Sage.

Shrub, usually erect, 2–6 m. high or sometimes scrambling to 6 m.; leaves rhomboid-ovate to elliptic-lanceolate, acute or obtuse, up to 6 cm. long and 3 cm. broad; inflorescence 8–12 mm. in diameter; ripe fruit oblong, red.

Very common on limestone or shale rocks in thickets and open woodlands, mainly in rather arid areas; SL–2750 feet; fl. and fr. all the year. *A 7259! 7969! H 6567! P 8498! C. globosa* in the strict sense occurs in the Lesser Antilles and northern S. Amer.; var. *humilis* occurs in Florida, Bahamas, Mexico to Panama, Greater Antilles and Grand Cayman.

7. EHRETIA P. Browne (1756)

1. E. tinifolia L., Syst. Nat. ed. 10, **2**: 936 (1759).—Bastard Cherry.

Tree 6–15 m. high, rarely a shrub (coppiced); leaves and twigs glabrous; leaves broadly elliptical to elliptic-oblanceolate, mostly obtuse, cuneate to rounded at base, up to 20 cm. long and 8 cm. broad; inflorescence glabrescent; calyx-lobes ciliate, otherwise glabrous; corolla white to cream-coloured, fragrant, tube 1 mm. long, lobes about 2·5 mm. long; stamens free nearly to base of corolla; fruit yellow to red, finally black, drupaceous with 2 pyrenes, about 5 mm. in diameter.

Fairly common, in secondary woodlands, sometimes on coastal limestone; 20–1800 feet; fl. and fr. Dec–Sept. *A 11513! H 6838! P 15044!* Mexico, Cuba, Hispaniola, Grand Cayman.

8. BOURRERIA P. Browne (1756) nom. cons.

1 Inflorescence and outside of calyx closely and densely hairy; style 2-fid   1. *baccata*
1 Inflorescence etc. glabrous:
  2 Style divided into 2 branches 1–2·5 mm. long, each stigma about 1 mm. broad
          2. *venosa*
  2 Style grooved but not divided; stigma bilobed, about 1 mm. broad   3. *succulenta*

1. B. baccata Raf., Sylva Tellur.: 42 (1838).—*Cordia bourreria* L. (1759).

Tree 3–8 m. high; leaves elliptical, obtuse to rounded at tip, cuneate at base, up to 9 cm. long and 4·5 cm. broad; inflorescence corymbose; calyx 5–7·5 mm. long; corolla 5 (–6)-lobed, white, fragrant; ripe fruits yellow or red, 7–11 mm. in diameter.

Locally common, in thickets and woodlands on limestone, mostly in coastal rather arid or exposed areas; 15–3000 feet; fl. and fr. Nov–Mar, June–Aug. *A 8966! H 6074! Porter 1029!* Endemic [see note on p. 790].

2. B. venosa (Miers) Stearn in Journ. Arn. Arb. **52**: 625 (1971).—*Crematomia venosa* Miers (1869).

Shrub or slender tree 2–9 m. high, often with drooping branches; leaves obovate to elliptic-oblanceolate, obtuse or rounded sometimes emarginate at tip, up to 9 cm. long and 4·5 cm. broad; calyx 5–6 mm. long; corolla 5 (–6)-lobed, white; ripe fruit red, 5–8 mm. in diameter.

Widespread in thickets, rocky pastures and on limestone cliffs in exposed or arid areas; SL–3000 feet; fl. most of the year, fr. June–Dec. *A 9843! H 8981! P 23075!* Cayman Is.

3. B. succulenta Jacq., Enum. Syst. Pl. Carib.: 14 (1760).
Shrub or tree 5–6 m. high; branches often pendulous; leaves, inflorescence and fruit as above.
Occasional on rocky hillsides; 20–800 feet; fl. Mar–Oct, fr. Mar–Dec. *A 8161! H 12084!* Florida, Mexico to Venezuela, West Indies. Without flowers it is impossible to distinguish *B. succulenta* and *B. venosa*; in Jamaica the degree of division of the style is inconsistent even on the same plant, thus suggesting that these two species may not be really distinct.

167. VERBENACEAE

Herbs, shrubs or trees, often with quadrangular twigs. Leaves opposite or whorled, rarely subopposite or alternate, mostly simple, sometimes compound, without stipules. Flowers usually bisexual, zygomorphic or less often actinomorphic or nearly so, in racemes, spikes, heads or variously cymose and often paniculate. Perianth biseriate, 4–5-merous; calyx persistent, up to 8-toothed; corolla sympetalous, usually with as many lobes as calyx, the lobes imbricate, commonly salverform, sometimes 2-lipped. Stamens (2–) 4 (–5), with or without staminodes, epipetalous; anthers 2-locular, opening lengthwise. Ovary superior, 2–9-locular, often 4-locular; ovules usually solitary in each loculus or paired, anatropous, erect or rarely pendulous, on axile or sometimes during development parietal or rarely free central placentas; style terminal, solitary. Fruit usually drupaceous with as many pyrenes as ovules or these sometimes paired, or of nutlets. Seed with straight embryo and endosperm usually wanting.
About 100 genera with some 2600 species in the subtropics and tropics and elsewhere mostly in S. temperate regions.

1 Leaves digitately compound with up to 9 leaflets; erect shrubs or trees; fertile stamens 4; drupe with 1 4-locular pyrene   **13. Vitex**
1 Leaves simple:
  2 Flowers sessile:
    3 Flowers in small heads subtended by an involucre of 3 large pink tomentose spreading bracts; corolla small, strongly 2-lipped; robust hairy climbing ornamental shrub; native of tropical Asia   *Congea tomentosa* Roxb.
    3 Flowers in heads or spikes with involucres if present of several imbricate bracts; corolla-limb subregular to obscurely 2-lipped:
      4 Calyx conspicuous, distinctly longer than broad; erect herbs or undershrubs:
        5 Stamens 2; anther-loculi widely divergent; pollen grains lobed; axis of spike long and slender with depressions into which calyx fits; fruit with 2 nutlets   **5. Stachytarpheta**
        5 Stamens 4; anther-loculi parallel; axis of spike without depressions:
          6 Fruit with 2 nutlets; spike slender, elongated; calyx 8–11 mm. long; corolla-tube about 9 mm. long   **4. Bouchea**
          6 Fruit with 4 nutlets; spikes short and compact or, if elongated and slender, flowers much smaller than above   **1. Verbena**
      4 Calyx obscure and more or less hidden by bracts, not much longer than broad, up to 2·5 mm. long; plants shrubby or if herbaceous creeping or trailing; stamens 4; fruit with usually 2 pyrenes:
        7 Fruit fleshy, drupaceous; calyx-margin truncate or very shallowly toothed; corolla-tube 4·5 mm. or more long; erect or scrambling shrubs or undershrubs   **2. Lantana**
        7 Fruit dry; calyx distinctly 2- or 4-toothed or cleft; corolla-tube up to 3·5 mm. long; undershrubs or herbs   **3. Lippia**
  2 Flowers distinctly stalked:
    8 Inflorescence racemose, elongated, the flowers maturing from below upwards:
      9 Herb with toothed leaves; fruiting calyx inflated, subglobose, covered with hooked hairs; stamens 4; nutlets 2-seeded   **6. Priva**
      9 Shrubs or trees with entire leaves:
        10 Woody climber; calyx showy, bluish, the lobes exceeding the tube and the dark purplish-blue corolla; cultivated ornamental; native of tropical S. America   *Petrea volubilis* L.
        10 Erect shrubs or trees; calyx shorter than corolla, truncate or only shortly toothed; fruits smooth, fleshy:

11 Calyx in flower grooved, in fruit fleshy, yellow, enclosing 4 2-seeded nutlets; plant sometimes spiny; corolla white or mauve; leaf-margin not glandular
7. **Duranta**

11 Calyx in flower smooth, in fruit cup-shaped, not fleshy; drupe with 2 2-seeded nutlets; plants not spiny; corolla white or yellowish; leaf-margin often glandular near the base
8. **Citharexylum**

8 Inflorescence cymose, the middle flowers of the clusters opening first, rarely flowers solitary and axillary, sometimes in axillary clusters or heads, commonly in panicles; shrubs or trees:

12 Corolla regular; stamens 4–6, equal:

13 Leaves very large, up to 15 cm. broad on flowering shoots, much larger on others; calyx inflated and bladderlike in fruit; flowers rather small, white, in large panicles; stamens exserted; drupe with 1 4-locular pyrene; deciduous cultivated timber tree; native of tropical Asia; Teak *Tectona grandis* L.f.

13 Leaves not more than 12 cm. broad; calyx perhaps accrescent but not inflated in fruit:

14 Hairs on stems and leaves stellate; stamens inserted at or near base of corolla-tube; stigma peltate or capitate; drupe with 1-seeded pyrenes
9. **Callicarpa**

14 Hairs simple; stamens inserted at or above middle of corolla-tube; stigma 2-lobed or 2-fid:

15 Stigma-branches filamentous, 1–6 mm. long, either stamens or style usually exserted; calyx accrescent; drupe with 1-seeded pyrenes; heterostylous erect or climbing shrubs; flowers mostly 4-merous
10. **Aegiphila**

15 Stigma-lobes short, about 1 mm. long; stamens and style hardly exserted, the anthers almost sessile; calyx not accrescent; drupe with 1 4-locular pyrene; tree
11. **Petitia**

12 Corolla irregular:

16 Fertile stamens 2, more or less included; staminodes 2; drupe with 1 4-locular pyrene; flowers small, numerous, purple, in terminal thyrsoid panicles
12. **Cornutia**

16 Fertile stamens 4, exserted; flowers usually showy:

17 Calyx-limb spreading, saucer-shaped, subentire, 1·5–3 cm. across; corolla shortly 2-lipped, lemon-yellow or red; lax cultivated ornamental shrub with shallowly serrated leaves; native of India; Chinese Hat
Holmskioldia sanguinea Retz.

17 Calyx not as above:

18 Scrambling, twining or erect shrubs; leaf-margins various; corolla 5-lobed without an elongated lower lip, not yellow; stamens somewhat unequal; drupe with 4 1-seeded pyrenes
14. **Clerodendrum**

18 Tree; corolla 4-lobed with lower lip longer than upper, about 4·5 cm. long, yellow and brown, hairy; stamens markedly didynamous; leaves long-stalked with broadly elliptical or ovate acute entire blades; calyx rusty-brown, glandular, 5 mm. long; planted for ornament; native of tropical Asia
Gmelina asiatica L.

1. VERBENA L. (1753)

1 Leaves long-petiolate, the blades ovate, the veins hardly impressed nor prominent beneath; spikes up to 20 cm. or more long, the flowers not contiguous; corolla-tube 2–2·5 mm. long
1. *scabra*

1 Leaves sessile, half-amplexicaul, the blades linear-lanceolate to oblong, the veins impressed and prominent beneath; spikes compact, up to about 7 cm. long; corolla larger:

2 Leaf-blade linear-lanceolate, deeply serrate with antrorse teeth; bracts as long as or shorter than calyx; corolla-tube about 6 mm. long
2. *bonariensis*

2 Leaf-blade oblong-elliptical, serrate with spreading teeth; bracts much longer than calyx; corolla-tube about 10 mm. long
3. *rigida*

1. V. scabra Vahl, Eclog. Amer. **2**: 2 (1798).

Erect annual bushy slender-branched herb up to 1·2 m. high; leaves opposite or whorled, the blades serrate-crenate, up to 12 cm. long and 5 cm. broad; spikes paniculate; bracts shorter than calyx; corolla white, mauve or light rose with a darker eye, drying bluish, funnel-shaped; nutlets 1–1·5 mm. long, ribbed on outer side.

Local in swamp margins and wet pastures; 10–2300 feet; fl. and fr. most of the year. *A 10967! H 9937! H & P 14546!* Bermuda, S.E. United States, Mexico, Greater Antilles.

2. V. bonariensis L., Sp. Pl. **1**: 20 (1753).

Erect perennial quadrangular-stemmed herb with long internodes, up to 1·5 m. high; leaves acute, scabrid, up to 11 cm. long and 1·5 cm. broad; corymbs of spikes 2–5 cm. across at the end of long peduncles; bracts 2–3 mm. long; calyx about 3 mm. long; corolla-limb about 4 mm. broad, bright purple drying blue; nutlets 1·5–2 mm. long, longitudinally ribbed on outer side.

Locally common (St. Andr., Port., St. Thom.), at roadsides and in old cultivations or gravelly waste open places; 1700–5200 feet; fl. and fr. most of the year. *A 6431! H 9132! J.P. 1169! P 23518!* Native of subtropical S. Amer., introduced into Bermuda, United States, Puerto Rico and elsewhere and escaping from cultivation.

3. V. rigida Spreng. in L., Syst. Veg. ed. 16, **4** (2): 230 (1827).

Herb, spreading and rooting, with flowering shoots ascending to 60 cm. high; leaves acute, ascending, scabridulous, up to 7 cm. long and 1·5 cm. broad; spikes 3–5 together, terminal; bracts 5–6 mm. long, ciliate; calyx 4 mm. long; corolla-limb about 8 mm. broad, violet to bright bluish-purple; nutlets as above.

Cultivated ornamental escaping occasionally (St. Andr., Manch., St. Eliz.), on roadsides and in rough pastures; 2000–3500 feet; fl. and fr. May–Sept. *A 11205! H 11969! P 23557!* Native of subtropical S. Amer., introduced into Bermuda, United States, Cuba and elsewhere.

2. LANTANA L. (1753)

1 Leaves in whorls of 3 or 4, long-cuneate at base; outer or lower bracts acuminate, up to about 15 mm. long; spikes becoming cylindrical and ultimately as much as 4 cm. long; corolla bright purple to mauve   1. *trifolia*
1 Leaves mostly opposite, occasionally in whorls of 3, mostly rather shortly cuneate at junction of petiole and blade, otherwise blade truncate or rounded; outer or lower bracts elliptical, ovate or lanceolate, less than 10 mm. long; spikes capitate or sub-capitate:
   2 Stem clothed with long white spreading hairs and short glandular hairs; corolla yellow-orange becoming reddish; submontane species   2. *insularis*
   2 Stem pubescent with short hairs sometimes mixed with glandular hairs, glabrescent:
     3 Corolla yellow, orange or red; outer bracts ovate or lanceolate, up to 2 mm. broad:
       4 Lower surface of leaves thinly pubescent or almost glabrous; calyx as broad as long   3. *camara*
       4 Lower surface of leaves with numerous short hairs; calyx longer than broad   4. *urticifolia*
     3 Corolla white, pink or purplish, usually with a yellow eye; outer bracts usually ovate, 2–5 mm. broad, forming an involucre:
       5 Leaves less than twice as long as broad; stamens inserted at and above middle of corolla-tube:
         6 Leaf-blade broadly obovate to suborbicular, with irregular tertiary venation adaxially, margin minutely dentate or subentire, base broadly cuneate, apex rounded   5. *involucrata*
         6 Leaf-blade broadly ovate, with tertiary venation demarcating regular oblong areoles adaxially, margin distinctly dentate, base truncate and broadly cuneate, apex obtuse to acute   6. *reticulata*
       5 Leaves more than twice as long as broad; stamens inserted at and below middle of corolla-tube:
         7 Leaf-blade ovate, shallowly crenulate with up to about 25 teeth on each side or almost entire, smooth or nearly so adaxially; western Jamaica   7. *jamaicensis*
         7 Leaf-blade lanceolate, crenulate with 30 or more distinct teeth on each side, the adaxial surface rugose or bullate; eastern Jamaica   8. *angustifolia*

1. L. trifolia L., Sp. Pl. **2**: 626 (1753).

Undershrub 60–120 cm. high; stems pilose with white bristly hairs; leaves ovate-lanceolate, serrate-margined, acute at tip, up to 12 cm. long and 4·5 cm. broad; peduncle up to 7 cm. long; corolla with a white or yellow eye, the tube 7–8 mm. long, the limb 5–6 mm. across; drupes bright purple.

Common in rough pastures and waste places; 500–3000 (–3600) feet; fl. and fr.

all the year. *A 6641! J.P. 1305! P 6162! 24179!* Throughout tropical Amer.; W. Africa.

2. L. insularis Moldenke in Carib. For. **2**: 16 (1940).

Diffuse erect or scrambling shrub to 1·5 m. or more high; stem sometimes prickly on the angles; leaves broadly ovate, truncate-subcordate at base, acute, up to 10 cm. long and 6 cm. broad; peduncles 4–7 cm. long; head about 2 cm. in diameter; bracts elliptical, 5–7 mm. long; corolla-tube 4·5–6 mm. long, limb 4 mm. across; drupe bluish-black, about 7 mm. in diameter.

Rather local (St. Andr., St. Thom.), on pathside banks and in thickets; 3000–4700 feet; fl. most of the year, fr. Oct–Apr. *A 12687! Larter 234! A. von der Porten!* Cuba.

3. L. camara L., Sp. Pl. **2**: 627 (1753), partly.—*L. aculeata* L. (1753). *L. scabrida* Ait. (1789). *L. brittonii* Moldenke (1941). White Sage, Wild Sage.

Suberect, scrambling or climbing aromatic shrub 1–6 m. high; stem sometimes prickly; leaves ovate, cuneate to subcordate at base, acuminate, crenate-serrate, up to 12·5 cm. long and 5·5 cm. broad; peduncles up to 6 cm. long; corolla-tube 6–9 mm. long, the limb 5–7 mm. across; drupe slate-blue to black, about 4 mm. in diameter.

Very common in rough pastures, waste places and thickets; SL–5000 feet; fl. and fr. all the year. *A 9228! H 12263! P 24178!* General in the tropics.

4. L. urticifolia Mill., Gard. Dict. ed. 8 (1768).—Black Sage.

Erect or subscandent shrub 1–3 m. high; leaves broadly ovate, truncate and cuneate at base, acute or acuminate, crenate, up to 8 cm. long and 5 cm. broad, aromatic; peduncles up to 6·5 cm. long; bracts lanceolate; corolla-tube 5–7 mm. long, the limb 4–6 mm. across; drupe blue-black.

Locally common in the drier parts of the southern parishes, in thickets on limestone gravel or sand; 50–3000 feet; fl. and fr. almost all the year. *A 5647! H 10103! P 18266!* West Indies; Grand Cayman.

5. L. involucrata L., Cent. Pl. **2**: 22 (1756).—Wild Mint.

Shrub up to 2 m. high, with slender branches, minty-aromatic; leaf-blade up to 3 cm. long and 2·5 cm. broad, often much smaller, slender-petioled; peduncles up to 6 cm. long; corolla-tube about 3 mm. long, the limb 3–4 mm. across; drupe light mauve to reddish-purple.

Common on rocky limestone and in thickets or pastures mostly near the sea; SL–600 feet; fl. and fr. Nov–May. *A 9814! H 10003! P 7714!* Florida and generally throughout the Caribbean area, Galapagos Is.

6. L. reticulata Pers., Synops. Pl. **2**: 141 (1806).

Shrub like the last; stem and undersurface of leaf often densely puberulous and whitish; plant pungent-aromatic; leaf-blade up to 5 cm. long and 2·6 cm. broad, occasionally slightly larger, often much smaller; peduncles up to 7·5 cm. long.

Rather common in rocky thickets and woodland margins on limestone; SL–2700 feet; fl. and fr. all the year. *A 6012! H 9647! H & P 14527!* Mexico, Cuba, Hispaniola, Grenadines, Barbados, Margarita.

7. L. jamaicensis Britton in Bull. Torr. Bot. Club **37**: 357 (1910).

Straggling shrub up to 2·5 m. or scandent to 6 m. high; leaves sometimes in threes, the blade rounded at base to a short wedge, acuminate with a blunt acumen, 2·5–9 cm. long, 1–4 cm. broad, finely tomentellous beneath; peduncles flexuous, up to 5·5 cm. long; heads up to about 1·5 cm. broad; corolla-tube 4·5–6 mm. long, the limb 5·5–6 mm. across; drupe white tinged mauve.

Local (Manch., St. Eliz., Trel., St. Ann), in thickets and woodland margins on limestone hills; 1500–2600 feet; fl. Jan–Apr, Sept, fr. Apr–Sept. *A 12483! H 10671! HPS 14670! P 22196!* Endemic.

8. L. angustifolia Mill., Gard. Dict. ed. 8 (1768).—*L. stricta* Sw. (1788).

Straggling shrub about 1 m. or subscandent up to 3 m. high; leaves sometimes in threes, shortly petiolate, broadly cuneate to rounded at base, long-acute at tip, 1·5–8 cm. long, 0·5–3 cm. broad, densely pale-pubescent beneath; peduncles slender, often curved, 3–12 cm. long; head about 1·5 cm. broad; corolla-tube about 6 mm. long, the limb about 6 mm. across; drupe dark rose-purple.

Locally rather common (St. Andr., St. Cath., St. Thom.), on steep more or less sheltered shale or serpentine banks and in pastures and thickets; 1600–5000 feet; fl. and fr. most of the year. *A 10075! H 10798! P 7338!* Endemic.

Plants of *Lantana* have been found in Jamaica which are putative hybrids between described species. *L. fucata* Lindl. var. *antillana* Moldenke (1952) may be a hybrid between *L. jamaicensis* and *L. reticulata*. Other intermediates exist which could be: *L. involucrata* × *L. reticulata*, *L. camara* × *L. insularis*, *L. angustifolia* × *L. camara* and *L. angustifolia* × *L. trifolia*.

3. LIPPIA L. (1753)

1 Spikes or heads on peduncles not exceeding 2 cm. long and much shorter than the leaves; aromatic shrub   1. *alba*
1 Spikes or heads on elongating peduncles as long as or eventually longer than the leaves; herbs or undershrubs:
  2 Leaves linear-lanceolate, toothed almost to the base, the veins deeply impressed adaxially and prominent abaxially; aromatic woody herb, erect or decumbent at base but not long-creeping   2. *stoechadifolia*
  2 Leaves spathulate, obovate or elliptical, toothed only in the distal half or two-thirds, the veins not deeply impressed nor as prominent abaxially; herbs with prostrate rooting stems; plants not markedly aromatic:
    3 Leaves mostly obovate to spathulate, rounded at tip, with teeth around distal margin only, venation rather obscure; hairs on margin of calyx hardly 0·1 mm. long
                                                      3. *nodiflora*
    3 Leaves elliptical to rhomboidal, obtuse at tip, with few large teeth in distal two-thirds, venation distinct; hairs on margin of calyx 0·3–0·4 mm. long   4. *strigulosa*

1. L. alba (Mill.) N. E. Br. in Britton & Wilson, Sci. Surv. Porto Rico & Virg. Is. **6**: 141 (1925).—*Lantana alba* Mill. (1768). *Lippia geminata* Kunth (1818). Colic Mint, Cullen Mint, Guinea Mint.

Much branched shrub with long straggling slender branches, up to 1·5 m. high; leaves opposite or in threes, oblong-elliptical, cuneate at base, obtuse to rounded at tip, 1–3 cm. long, 0·9–1·5 (–2) cm. broad, tomentellous on both surfaces and veins prominent beneath; peduncles rarely as much as 1·5 cm. long; spikes elongating to 2 cm. long; corolla white, pink or light bluish-purple, the tube 3–3·5 mm. long, the limb 1·5 mm. across.

Local and sporadic, in the southern parishes, sometimes cultivated, mostly in thickets and gravelly waste places near the sea; fl. and fr. Mar–Oct. *H & B 10593! P 9421! Robertson UCWI 3355!* Texas, Mexico to Argentina and general throughout the West Indies; Grand Cayman.

2. L. stoechadifolia (L.) Kunth, Nov. Gen. **2**: 265 (1818).—*Verbena stoechadifolia* L. (1753). *Phyla stoechadifolia* (L.) Small (1909).

Herb, woody and rooting from decumbent base, up to about 60 cm. high; leaves opposite, cuneate at base, acute at tip, regularly serrate-dentate, up to 5 cm. long and 1 cm. broad; veins impressed above, prominent beneath, the adaxial hairs setose and diverging from the costae; peduncles up to 6·5 cm. long.

Very local in swamps, rice fields and damp low-lying places; 10–30 feet; fl. and fr. July–Jan. *A 10161! H 11748! H & P 14072! Skelding UCWI 3642!* Florida, Mexico, Bahamas, Greater Antilles, Antigua, Guadeloupe.

3. L. nodiflora (L.) Michx., Fl. Bor. Amer. **2**: 15 (1803).—*Verbena nodiflora* L. (1753). *Phyla nodiflora* (L.) Greene (1899).

Low creeping perennial herb, with branches sometimes up to 3 m. or more long on open sandy beaches; leaves long-cuneate at base, up to 4 cm. long and 2 cm. broad; peduncles up to 10 cm. long; bracts purple or black-purple; spike up to 12 mm. long; corolla white or mauve with yellow eye, the tube about 2 mm. long, the limb about 1·5 mm. across; fruit obovoid, about 1·5 mm. long, minutely pubescent.

Common in damp low-lying grassland, coastal thickets and on upper beaches; SL–1200 (–2300) feet; fl. and fr. all the year. *A 5539! 6878! H 8192! Powell 257!* Throughout the tropics; Grand Cayman.

4. L. strigulosa Martens & Gal. in Bull. Acad. Roy. Brux. **11** (2): 319 (1844).— *L. reptans* of authors, not Kunth (1818). *Phyla strigulosa* (Martens & Gal.) Moldenke (1947).

Trailing and rooting perennial herb like the last; leaves stiff; peduncles 4·5–8 cm. long; bracts reddish-purple; spike globose to ovoid, 5–14 mm. long; corolla white, mauve or pink, the tube about 2·5 mm. long, the lower lip 1·5 mm. long; fruit 1·8–2 mm. long, glabrous.

Occasional in ditches and damp pastures and at pond margins mainly where sandy; SL–100 feet; fl. and fr. most of the year. *A 6051! H 9724! P 16087! Stearn 206!* Mexico to Argentina, West Indies.

4. BOUCHEA Cham. (1832) nom. cons.

1. B. prismatica (L.) Kuntze, Revis. Gen. Pl. **2**: 502 (1891).—*Verbena prismatica* L. (1753). *B. ehrenbergii* Cham. (1832). Wild Vervine.

Erect annual herb 30–60 cm. high; stem quadrangular, pubescent; leaves broadly ovate, broadly cuneate at base, obtuse, the marginal teeth large and obliquely ovate, up to 7 cm. long and 5 cm. broad; spike simple, up to over 20 cm. long, nearly 1 mm. thick; corolla light blue, pinkish or light purple; nutlets linear, beaked, 9–12 mm. long.

Rather local in the southern parishes, a weed of open ground in thin pastures on limestone and dry alluvial gravel; 20–1000 feet; fl. and fr. June–Jan. *A 5511! 5655! H 11792! Powell 1020!* S. United States, Mexico to northern S. Amer., Bahamas, Greater Antilles, Virgin Is., Antigua, Barbados.

5. STACHYTARPHETA Vahl (1804) nom. cons.

1 Plant, at least on stems, petioles and inflorescence, villous or tomentose; inflorescence-axis 5–7 mm. thick; corolla bright rose-pink, the limb exceeding 12 mm. across
                 1. *mutabilis*

1 Plant at most sparsely hairy, often glabrescent, or glabrous; inflorescence-axis narrower; corolla light to deep blue or violet, the limb not exceeding 10 mm. across:
  2 Leaves linear to oblong-lanceolate or oblanceolate, remotely serrate; calyx unequally toothed   **5.** *angustifolia*
  2 Leaves ovate to oblong-elliptical, crenate-serrate or dentate; calyx equally 4-toothed:
    3 Bracts 7–9 mm. long; inflorescence-axis 3–4·5 mm. thick   **2.** *adulterina*
    3 Bracts 3–6 mm. long; inflorescence-axis narrower:
      4 Inflorescence-axis 2·5–3 mm. thick for most of its length; fruiting calyx in a furrow distinctly narrower than the axis; bracts lanceolate, about 2 mm. broad; corolla usually deep violet-blue, the limb up to about 9 mm. across   **3.** *jamaicensis*
      4 Inflorescence-axis 1·5–2 mm. thick for most of its length; fruiting calyx in a furrow about as wide as the axis and projecting slightly from it; bracts linear-subulate, less than 1 mm. broad; corolla usually light mauve-blue or nearly white, the limb 5–6 mm. across   **4.** *cayennensis*

1. S. mutabilis (Jacq.) Vahl, Enum. Pl. **1**: 209 (1804).—*Verbena mutabilis* Jacq. (1789).

Shrub with green ascending leafy branches, up to about 2·5 m. high; leaves ovate-elliptical, cuneate with lamina long-decurrent on petiole, acute or obtuse at tip, crenate, up to 14 cm. long and 10 cm. broad; spike simple, up to 50 cm. or more long; bracts 8–10 mm. long; calyx 2–fid with 2 points on each lobe; corolla-tube 15–20 mm. long, the limb 12–20 mm. across; fruit 11–12 mm. long.

Generally dispersed on banks, pathsides and in open areas in hilly districts; (500–) 1250–4000 feet; fl. and fr. most of the year. *A 5470! H 6789! J.P. 1459! P 22536!* Native of C. and northern S. Amer., also in Hispaniola, Trinidad and Tobago and reported from Cuba, introduced and naturalized in the Old World tropics and elsewhere.

2. S. adulterina Urb. & Ekman in Arkiv Bot. **22A** (17): 105 (1929).

Robust herb up to about 1 m. high; leaves ovate-elliptical, broadly cuneate at

base, acute at tip, coarsely crenate, hispid on veins beneath, up to 13 cm. long and 8 cm. broad; corolla deep purplish-blue, the tube 8–9 mm. long; fruit not known.
Rather local (Manch., Trel.), on roadside banks and in thickets; 1000–2000 feet; fl. Apr–Aug. *A 6774! P 10600! Thornton UCWI 1179!* Hispaniola. This plant is intermediate in character between *S. mutabilis* and *S. jamaicensis* and is strongly suspected to have originated as a hybrid between these two species.

3. S. jamaicensis (L.) Vahl, Enum. Pl. **1**: 206 (1804).—*Verbena jamaicensis* L. (1753). Porter Weed, Vervine.
Herb 30–100 cm. high; leaves oblong-elliptical, long-cuneate at base, obtuse at tip, crenate-dentate, more or less glabrous, up to 9 cm. long and 4·5 cm. broad; spike to 55 cm. long; bracts ovate-lanceolate, about 6 mm. long; corolla 8–10 mm. long; fruit 7 mm. long.
Very common as a weed of waste places at low elevations, especially in pastures and sandy thickets near the sea and on the cays; SL–2800 feet; fl. and fr. all the year. *A 5532! H 41! P 11507! Stearn 503!* Florida, Bahamas, Mexico to northern S. Amer., West Indies; Grand Cayman; introduced into the Pacific Is.

4. S. cayennensis (L. C. Rich.) Vahl, Enum. Pl. **1**: 208 (1804).—*Verbena cayennensis* L. C. Rich. (1792).
Herb up to about 1 m. high; stem glabrescent or with short appressed hairs; leaves oblong-elliptical, long-cuneate at base, acute or obtuse at tip, serrate-crenate, up to 7 cm. long and 4 cm. broad; spike to 25 cm. long; bracts about 4 mm. long; corolla 6–7 mm. long; fruit 6 mm. long.
Common, especially in rough pastures and damp waste places and also in cultivations; 20–4100 feet; fl. and fr. all the year. *A 5547! 6372! H 6592! P 8723! Stearn 500!* Mexico to Peru and Argentina, West Indies, introduced into the Old World.

5. S. angustifolia (Mill.) Vahl, Enum. Pl. **1**: 205 (1804).—*Verbena angustifolia* Mill. (1768).
Erect herb with numerous adventitious roots at base, 60–100 cm. high; leaves 3–9 cm. long, up to 2 cm. broad, rarely larger; spike to 25 cm. long; corolla light blue or violet.
Very rare (St. Eliz.), in swamp; 20–40 feet; fl. and fr. Jan. *P 23149!* Tropical Amer. and Africa, Cuba.

6. PRIVA Adans. (1763)

1. P. lappulacea (L.) Pers., Synops. Pl. **2**: 139 (1806).—*Verbena lappulacea* L. (1753). Clammy Bur, Fasten-'pon-Coat, Styptic Bur, Velvet Bur.
Annual or short-lived perennial taprooted herb to about 1 m. high; stem quadrangular; leaves ovate to triangular, truncate-subcordate at base, acute or acuminate at tip, crenate-dentate, up to 9 cm. long and 5·5 cm. broad; racemes up to about 15 cm. long; fruiting calyx 5–6 mm. long; corolla light mauve or whitish, 4–5 mm. long; pyrenes echinate, 3–3·5 mm. long.
A common weed of cultivations, roadsides and waste places; 30–1750 feet; fl. and fr. all the year. *A 5429! H 6785! P 23897!* Subtropics and tropics of Amer., recently introduced into West Africa and Asia.

7. DURANTA L. (1753)

1. D. repens L., Sp. Pl. **2**: 637 (1753).—Angel's Whisper, Poison Macca.
Shrub up to about 4 m. or rarely a small tree to 6 m. high; branches often drooping and trailing, with or without short axillary spines; leaves opposite or subopposite, elliptical, long-cuneate at base, acuminate at tip, the margin entire or more or less serrate-crenate distally, up to 7 cm. long and 3·5 cm. broad, thinly puberulous; racemes axillary and unbranched up to 20 cm. long or terminal and then branched; corolla mauve, light bluish-purple or white, the tube about 6 mm. long, the limb 6–12 mm. across; fruiting calyx pear-shaped, 9 mm. long and 6–7 mm. broad.

Common on roadside banks and in thickets, also cultivated for ornament; 450–5500 feet; fl. and fr. most of they ear. *A 9869! H 11880! J. P. 1095! P 6604!* Florida, Bermuda, Mexico to Argentina, West Indies; Grand Cayman; introduced into the Old World and the Pacific area.

8. CITHAREXYLUM L. (1753)

1 Calyx at anthesis 2–2·5 mm. long; corolla-tube about 2·5 mm. long; leaves ovate to lanceolate, acute or acuminate   1. *tristachyum*
1 Calyx at anthesis 3–4 mm. long; corolla-tube 4–6 mm. long; leaves mostly elliptical to narrowly obovate, acute, obtuse or sometimes emarginate:
   2 Pedicels at anthesis up to about 1 mm. long, in fruit 1–2 mm. long; lateral veins of leaves rather straight, diverging from the midrib at an angle of 30–40°; veinlets adaxially forming a prominent network   2. *fruticosum*
   2 Pedicels at anthesis 1–2 mm. long, in fruit 2–3·5 mm. long; lateral veins of leaves curved, diverging from the midrib at a basal angle of 50–60°; veinlets adaxially rather obscure
     3. *caudatum*

1. C. tristachyum Turcz. in Bull. Soc. Nat. Moscou **36** (2): 209 (1863).—*C. urbanii* O. E. Schulz (1908).
Shrub 2–3 m. or tree to 5 m. high; leaves ovate-lanceolate or elliptical, cuneate at base, if acuminate the acumen pointed, up to 10 cm. long and 3·5 (–4) cm. broad, thinly spreading-pubescent or puberulous on midrib and veins beneath; racemes slender, up to 14 cm. long, sometimes branched; pedicels 0·5–1·5 mm. long; corolla yellow; fruits subglobose, black, about 5 mm. in diameter.
Local and uncommon (St. Andr., St. Thom.), on banks, mostly of loose shale; 1000–3500 feet; fl. Jan, July–Sept, fr. July, Oct. *A 7729! H 6724! J.P. 1433! P 27582!* Cuba.

2. C. fruticosum L., Syst. Nat. ed. 10, **2**: 1115 (1759).—Yellow Fiddlewood.
Shrub 2–3 m. or tree to 15 m. high, glabrous; bark stringy, lightly and closely longitudinally fissured, dull grey; leaves ovate to oblong-elliptical, cuneate at base, rounded to obtuse and emarginate at tip or rarely acute, up to about 14 cm. long and 5 (–6) cm. broad; racemes up to 15 cm. long, rarely branched; flowers fragrant; corolla white, funnel–shaped, the tube 4–6 mm. long, the lobes spreading, 2·5–3 mm. long; fruit red at first, finally black, subglobose, 8–12 mm. long.
Common in thickets and woodland on limestone especially near the sea, also on the cays; SL–1100 (–2500) feet; fl. June–Nov, fr. July–Mar. *A 6026! 11868! H 8608! J.P. 1076! P 10384!* Florida, Venezuela, Suriname, West Indies.

3. C. caudatum L., Sp. Pl. ed. 2, **2**: 872 (1763).—Fiddlewood, Juniper Berry.
Shrub or tree like the last, up to 12 m. high; leaves obtuse to shortly acuminate at tip, rarely emarginate, up to 17 cm. long and 6 (–9) cm. broad; racemes up to 20 cm. long, simple; corolla white, cream or pinkish, the tube about 4·5 mm. long, the lobes about 2 mm. long; fruit finally purplish-black, 6–12 mm. long.
Common in thickets and woodlands in the limestone and shale hills; (700–) 1500–5600 feet; fl. and fr. most of the year. *A 10812! H 8546! H & P 13683!* Bahamas, Mexico to Colombia, Greater Antilles, Dominica.
The criteria for distinguishing *C. fruticosum* and *C. caudatum* usually hold and are to a great extent correlated with a fairly distinct ecological separation of the two taxa.

9. CALLICARPA L. (1753)

Leaves almost glabrous beneath, with numerous small cupular golden glands; stellate hairs present mostly along the scarcely raised veins; leaf-margin toothed; calyx glabrous but with glands; fruits 3–4 mm. broad   1. *ferruginea*
Leaves densely covered beneath with stellate hairs; veins beneath prominent; leaf-margin subentire; calyx densely hairy; fruits 4–5 mm. broad   2. *reticulata*

1. C. ferruginea Sw., Nov. Gen. & Sp. Pl.: 31 (1788).
Shrub to 2·5 m. or small tree to 5 m. high; branches slender, densely covered with yellowish stellate hairs, straggling; leaves elliptic-lanceolate, cuneate at base,

long-acute, crenate-serrate, at least at maturity with a few scattered stellate hairs beneath, up to 14 cm. long and 4 cm. broad; peduncles axillary; inflorescence more or less divaricate, up to 6 cm. long; calyx lightly 4-ribbed; corolla funnel-shaped, pink or white, 2·5–3 mm. long; drupe subglobose, rose-purple, 3–4 mm. in diameter.

Occasional in sheltered submontane and montane woodlands; 1500–6500 feet; fl. Apr–Aug, fr. June–Dec. *A 11456! 12527! H & B 10554! P 6802! 16530!* Cuba.

2. C. reticulata Sw., Nov. Gen. & Sp. Pl.: 31 (1788).

Shrub with slender obtusely 4-angled branches, densely covered when young with stellate hairs and minute prickles; leaves narrowly ovate to elliptical, acute at apex, 6–10·5 cm. long, 2–4·5 cm. broad; drupe subglobose, covered with golden glands.

Very rare; known only from the unlocalized type, *W. Wright!*

10. AEGIPHILA Jacq. (1763)

1 Inflorescence loosely paniculate with more than 12 flowers:
  2 Leaves oblong-ovate, rounded-subcordate at base, up to about 19 cm. long and 9 cm.
    broad; calyx 3–4 mm. long   1. *elata*
  2 Leaves ovate to lanceolate, cuneate at base, up to 7 (–10) cm. long and 2 (–3·5) cm.
    broad; calyx about 2 mm. long   2. *oligoneura*
1 Inflorescences small and more or less congested, up to 9-flowered:
  3 Petioles 2–3 (–4) mm. long; leaves rounded to subcordate at base; inflorescences about
    3-flowered; calyx 3 mm. long at anthesis:
    4 Young branches and calyx densely pilose with spreading yellowish hairs; corolla-tube
      18 mm. long, lobes about 7 mm. long   3. *foetida*
    4 Young branches and calyx shortly appressed-pilose with yellowish hairs; corolla-tube
      12 mm. long, lobes 8 mm. long   4. *plicata*
  3 Petioles 4–10 mm. long; leaves rounded, obtuse or broadly cuneate at base; calyx 4–8
    mm. long:
    5 Flowers and fruits clustered in the leaf-axils; peduncle obsolete; pedicels not over 1
      mm. long   5. *obtusa*
    5 Flowers solitary or in evident cymes or subpaniculate, the peduncles and/or pedicels
      over 3 mm. long:
      6 Young branches and calyx with long spreading yellowish hairs; calyx at anthesis
        7 mm. long; corolla-tube 9–10 mm. long, lobes 8–9 mm. long   6. *uniflora*
      6 Young branches and calyx with short patent hairs, puberulous or glabrous:
        7 Flowers in threes; peduncles up to 20 mm. long; calyx 4–6 mm. long; corolla-tube
          16 mm. long, lobes about 7 mm. long   7. *trifida*
        7 Flowers 3–9 in subpaniculate racemes; peduncles 1–3 mm. long; calyx 4 mm. long;
          corolla-tube 12 mm. long, lobes 4–6 mm. long   8. *swartziana*

1. A. elata Sw., Nov. Gen. & Sp. Pl.: 31 (1788).

Straggling-branched shrub to 2 m. or climber to 6 m. high, puberulous at least when young; leaves shortly acuminate, glabrescent; petioles up to 10 mm. long; inflorescence a terminal panicle up to 18 cm. long, the ultimate branches cymose; corolla light yellow, tube about 7 mm. long; stamens 4, exserted in short-styled flowers, included in long-styled flowers; fruit yellow, subglobose, half-enclosed by the cup-shaped calyx, 9–15 mm. long, with 4 fusiform pyrenes.

Rather common, on banks and in hillside thickets; SL–4000 feet; fl. most of the year, fr. Apr–Sept. *A 6655! 11200! J.P. 1155! P 23913!* Mexico to Venezuela, Greater Antilles, Martinique, Barbados; Grand Cayman.

2. A. oligoneura Urb. in Fedde, Repert. Sp. Nov. **16**: 40 (1919).

Shrub with glabrous slender branches, up to 3 m. high; leaves long-acuminate with a blunt acumen, glabrous; petioles 3–4 mm. long; pedicels 2·5–3·5 mm. long; calyx glabrous; corolla sulphur-yellow, tube about 8 mm. long, lobes 3–4 mm. long.

Rare (Manch., St. Eliz.), in crevices of limestone rocks; 300–800 feet; fl. Nov. *H 11716!* Probably endemic.

3. A. foetida Sw., Nov. Gen. & Sp. Pl.: 32 (1788).

Shrub with arching or straggling slender hairy branches, 2·5–5 m. high; leaves

oblong-elliptical, shortly acuminate, 3–12 cm. long, 1–4·5 cm. broad; corolla light yellow, 4-lobed, tube 18 mm. long, lobes 7 mm. long; fruit sugblobose, yellow or red, half-covered by persistent calyx.
Rare (Manch., Han., St. Ann), in woodland on limestone; 1500–2000 feet; fr. Jan. *H & P 15047! Prior! P 23967!* Endemic.

4. A. plicata Urb., Symb. Ant. **3**: 365 (1903).
Leaves ovate to ovate-lanceolate, acuminate, 5–7·5 cm. long, 3–4 cm. broad, glabrous, minutely glandular abaxially; axillary peduncles up to 3 mm. long; pedicels less than 1 mm. long; corolla-tube 12 mm. long, lobes 8 mm. long and 2 mm. broad.
A very obscure species known only from the type, *Bertero!* Endemic.

5. A. obtusa Urb., Symb. Ant. **5**: 486 (1908).
Straggling shrub to 3 m. or arborescent to 5 m. high; young branches with dense ascending or appressed yellowish hairs, these on petioles and inflorescence-branches sometimes spreading; leaves ovate to oblong-elliptical, rounded at base, obtuse or bluntly acuminate, 4–11 cm. long, 2–5·5 cm. broad, hairy on veins abaxially; calyx 4 mm. long; corolla white or light yellow, tube 12–15 mm. long, lobes 4–6 mm. long; ripe fruit broadly ellipsoid, orange, about 6 mm. long, half-covered by the cupular calyx.
Rare and local (St. Cath., Manch.), in woodland and woodland margins on limestone; 1500–2800 feet; fl. Jan, Aug–Sept, fr. Sept. *A 11720! 12629! H 8996! Robertson UCWI 5411!* Endemic.

6. A. uniflora Urb., Symb. Ant. **3**: 365 (1903).
Shrub with slender hairy branches; leaves narrowly ovate to elliptical, obtuse or rounded at base, acuminate, 6–8·5 cm. long, 2·5–3·5 cm. broad, shortly pilose on veins abaxially; fruit 8–9 mm. long, 7–8 mm. broad.
Very rare (Port.); 3800 feet; fl. and fr. Jan. *H 5533!* Endemic.

7. A. trifida Sw., Nov. Gen. & Sp. Pl.: 32 (1788).
Shrub with glabrescent branches, up to 6 m. high; leaves ovate to elliptic-lanceolate, obtuse or broadly cuneate at base, acuminate, rarely rounded at base, up to 10 cm. long and 3·5 cm. broad; petiole 5–10 mm. long; pedicels 1–4 mm. long; calyx with very short appressed hairs; fruit ellipsoid.
Local (St. Andr., Port., St. Thom.), on limestone, shale or serpentine; 1500–4100 feet; fl. Mar–Aug, fr. Mar, Aug, Dec. *H 9372! P 7683! 8746!* Endemic.

8. A. swartziana Urb., Symb. Ant. **3**: 364 (1903).
Shrub; leaves ovate-elliptical to narrowly ovate-elliptical, rounded at base, shortly acuminate, 6–7 cm. long, 3–4 cm. broad, glabrous adaxially, shortly pilose on veins abaxially; petiole 4–5 mm. long; pedicels 3–6 mm. long.
Very rare, known only from the type, *Swartz!*
The only really well known species of *Aegiphila* in Jamaica is *A. elata*. When other more complete material has been examined and compared in the living state, and differences of floral structure inherent in the heterostylous or incipient dioecious conditions are understood, besides the full ranges of hairiness, inflorescence-branching and leaf-shape, a drastic taxonomic revision may be required.

11. PETITIA Jacq. (1760)

1. P. domingensis Jacq., Enum. Syst. Pl. Carib.: 12 (1760).
Tree 4–13 m. high, with flaking bark; leaves long-petiolate, ovate to oblong-elliptical or lanceolate, broadly cuneate at base, acuminate, up to 24 cm. long and 12 cm. broad, glabrescent adaxially, puberulous beneath and yellowish with numerous minute glands, prominently veined; inflorescences amply paniculate, the small flowers in clusters; calyx funnel-shaped, 2 mm. long; corolla greenish-white, short; fruit ripening red, subglobose, about 4 mm. long.
Common in secondary thickets, pastures and woodland on limestone; 10–2300 feet; fl. and fr. all the year. *A 6302! H 8776! P 8665!* Bahamas, Greater Antilles; Cayman Is., cultivated elsewhere.

12. CORNUTIA L. (1753)

1. C. thyrsoidea Moldenke in Fedde, Repert. Sp. Nov. 40: 160, 193 (1936).—
C. jamaicensis Moldenke (1936).
   Shrub 1·5 m. or tree to 16 m. high; leaves broadly ovate, broadly cuneate to
rounded at base, acuminate, up to 12 cm. long and 8·5 cm. broad, thinly puberulous, glabrescent; inflorescence up to about 25 cm. long, the axis 4-angled; calyx
campanulate, 2 mm. long, puberulous; corolla purple to blue-violet, about 8 mm.
long, the tube pubescent or glabrous.
   Uncommon, in woodland margins on limestone hills; 1000–3000 feet; fl. May–
Sept, fr. June. H 5199! 9252! Powell 554! P 10259! Endemic.

13. VITEX L. (1753)

Cultivated shrub or small tree; leaflets mostly 7, thin, discolorous, densely tomentellous
   beneath, pointed at both ends; inflorescence terminal, compactly cylindrical with many
   mauve flowers; fruit oblong-globose, about 3 mm. long; native of the Mediterranean
   region; Tree of Chastity   V. agnus-castus L.
Wild tree; leaflets mostly 5, rather leathery, glabrescent except on veins beneath, rather
   blunt at base and apex; inflorescence open; fruit ovoid-subglobose, about 15 mm. in
   diameter   1. umbrosa

1. V. umbrosa Sw., Nov. Gen. & Sp. Pl.: 93 (1788).—Box Wood.
   Tree 8–15 m. high; bark flaky; trunk up to 1 m. in diameter, fluted at base;
leaves with (4–) 5 (–6) leaflets; leaflets elliptic-lanceolate, broadly cuneate to rounded
at base, obtuse to shortly and bluntly acuminate at tip, up to 20 cm. long and 8·5
cm. broad; petioles up to 13 cm. long; petiolules 0·6–4·5 cm.; inflorescences
axillary; calyx 3·5 mm. long; corolla purple or blue-violet, hairy, about 14 mm.
long; drupe yellow.
   Occasional in pasture margins and on wooded hillsides; 500–1800 feet; fl. May–
Sept, fr. June–Sept. H 10579! 11975! P 19783! Probably endemic, although
reported from Hispaniola.

14. CLERODENDRUM L. (1753)

1 Leaves narrow and cuneate at base, paired or in threes; petioles less than 12 mm. long;
   shrubs with erect or lax and trailing branches:
   2 Petiole-bases forming persistent spines 2–4 mm. long; leaf-blade entire; indigenous
      species of arid coastal situations   1. aculeatum
   2 Petiole-bases not forming spines; leaves oblanceolate, more or less serrate; introduced
      ornamental shrubs, natives of E. Africa:
      3 Corolla-tube 6–12 cm. long, straight, longer than the lobes, all white; flowers in
         compact terminal cymes or heads; leaves coarsely dentate, up to about 9 cm. long
         and 3 cm. broad; Musical Notes   C. incisum Klotzsch
      3 Corolla-tube 6–7 mm. long, bent, shorter than the lobes, the anterior lobe violet-blue
         within, the others lighter blue; flowers in small axillary dichasial cymes; leaves
         shallowly serrate, up to 12 cm. long and 4·5 cm. broad   C. ugandense Prain
1 Leaves broad and truncate-subcordate to cordate at base, paired:
   4 Erect shrubs with long-petiolate broadly ovate pubescent leaves:
      5 Flowers crowded in a dense corymbose head; calyx 1–2 cm. long; corolla white or
         pink, usually double   2. philippinum
      5 Flowers in a large loose panicle, the primary branches ascending, racemose; calyx
         3–10 mm. long; corolla scarlet, the tube much longer than calyx; leaves up to over
         30 cm. long and 20 cm. broad, dark green; cultivated ornamental; native of Java
                                                         C. speciosissimum Van Geert ex Morren
   4 Climbing shrubs with short-petiolate broadly to narrowly ovate glabrous or puberulous
      leaves; flowers in forking axillary lax cymes; calyx angled; corolla red, the tube about
      25 mm. long:
      6 Calyx white, 13–30 mm. long; stamens about 18 mm. long; cultivated ornamental;
         native of West tropical Africa; Rice and Peas   C. thomsoniae Balfour f.
      6 Calyx purplish, mostly 15–18 mm. long; stamens about 25 mm. long; ornamental of
         hybrid origin, a cross between C. thomsoniae and C. splendens G. Don also of West
         tropical Africa   C. × speciosum D'Ombrain

1. C. aculeatum (L.) Schlecht. in Linnaea **6**: 750 (1831).—*Volkameria aculeata* L. (1753).

Shrub up to 3 m. high, with slender straggling spiny branches; leaves paired, in threes or sometimes, on reduced lateral branches, in clusters, elliptical to lanceolate, acute or acuminate, up to 7·5 cm. long and 2·3 cm. broad, puberulous or glabrous adaxially, puberulous abaxially; flowers few on axillary peduncles up to 2·5 cm. long; corolla white, the tube 1·5–2·2 cm. long, the lobes 6–8 mm. long; stamens long-exserted, the filaments purple, unequal, 2·5–3·5 cm. long; fruit 5 mm. long, splitting into 2 parts at maturity.

Frequent on limestone rocks, sand dunes and gravelly wastes, mostly near the sea; SL– 800 feet; fl. June–Feb. *A 9441! H 10792! P 11255!* Bermuda, Bahamas, Mexico to Venezuela and the Guianas, West Indies mostly in the drier islands; Grand Cayman; introduced into Senegal, Guinea and Gambia.

2. C. philippinum Schauer in DC., Prodr. **11**: 667 (1847).—*Volkameria fragrans* Vent. (1804). *C. fragrans* Willd. (1809), illegitimate name. Julius Plague, Lady Nugent's Rose.

Shrub 1–2·5 m. high, almost everywhere pubescent, branches mostly stout; leaves broadly ovate, truncate-subcordate and shortly cuneate at base, acute or acuminate, shallowly repand-dentate, up to over 25 cm. long and broad; petioles 2–24 cm. long; flowers fragrant; fruit only very rarely developed.

Occasional on shady roadsides and in clearings; 400–4200 feet; fl. almost all the year. *A 10259! H 8704! A. von der Porten!* Native of China, now general in the tropics.

168. AVICENNIACEAE

Shrubs or trees. Leaves opposite, simple, without stipules. Flowers bisexual, subactinomorphic, sessile in leaf-axils or in terminal spikes or heads, each subtended by 3 imbricate bracteoles; perianth biseriate; sepals 5, imbricate, almost free, persistent; corolla almost regular, the tube short, lobes 4, imbricate. Stamens 4, epipetalous; anthers 2-locular, opening lengthwise, introrse. Ovary superior, incompletely 4-locular, the placenta free central; ovules 4, apical, pendulous, orthotropous; style solitary; stigma-lobes 2. Fruit a compressed asymmetrical 2-valved capsule, deciduous with the calyx; pericarp leathery, 1–seeded. Seed with folded cotyledons, the testa not evident, germinating within the pericarp.

One genus with about 11 species on subtropical and tropical shores.

1. AVICENNIA L. (1753); Stearn (1958)

1. A. germinans (L.) L., Sp. Pl. ed. 3, **2**: 891 (1764).—*Bontia germinans* L. (1759). *A. nitida* Jacq. (1760). Black Mangrove.

Shrub to 3 m. or tree to 10 m. or more high; roots characteristically developing numerous straight erect blunt branches up to about 30 cm. long (pneumatophores); leaves pubescent when young, oblong to oblong-lanceolate, rounded to cuneate at base, mostly obtuse at tip, 3–10 (–16) cm. long, 1–4·5 cm. broad, pitted adaxially greyish-green; panicles 2–5 cm. long; corolla about 1 cm. across, the lobes rounded, white with yellow towards base within; capsule apiculate, mostly 2·5–4 cm. long.

Common in all saline and brackish communities around the coast and on the cays; near SL; fl. Nov–Aug, fr. Nov–Dec, July–Aug. *A 8824! H 8190! P 24214! Stearn 287!* Coasts of the American subtropics and tropics from Florida to Brazil, Ecuador, Peru and W. Africa.

169. LABIATAE (LAMIACEAE)

Herbs, rarely shrubs or trees, with usually 4-angled stems. Leaves opposite or whorled, simple, mostly toothed or crenate, sometimes lobed, often aromatic and glandular; stipules absent. Flowers bisexual, rarely unisexual, usually zygomorphic, solitary and axillary or more usually in compact cymes often forming a false whorl (verticillaster) and these in terminal or axillary often bracteate racemes or panicles. Perianth biseriate; calyx tubular or campanulate, regular or 2-lipped, basically 5-merous but the upper 3 teeth often united, often accrescent; corolla tubular at the base with the limb usually 2-lipped, the upper 2 lobes usually united to form an entire or emarginate hood or rarely absent, the lower 3 lobes forming the lower lip, with imbricate aestivation. Stamens 4 and didynamous or 2; filaments epipetalous; anthers typically 2-locular but 1 loculus sometimes aborted, opening lengthwise, introrse. Annular or unilateral disk or gland present. Ovary superior, of 2 bilobed carpels, becoming 4-parted or deeply 4-lobed; ovules 4, basal, erect, anatropous; style solitary, central and usually gynobasic; stigma 2-fid or entire. Fruit of 4 free or paired 1-seeded nutlets enclosed within the calyx. Seed with usually a straight embryo with flat cotyledons; endosperm absent or scanty.

About 3200 species in 200 genera; cosmopolitan but concentrated in the Mediterranean region.

1 Leaves, including those of inflorescence, deeply 3-partite and lobed, usually long-petiolate   **5. Leonurus**
1 Leaves entire, crenate, toothed or at most narrowly incised, not deeply lobed:
  2 Upper lip of corolla evidently absent, lower lip 5-lobed; calyx swollen, 5-toothed, villous, not ribbed or nerved   **1. Teucrium**
  2 Upper lip of corolla well developed, lower lip 3-lobed or entire:
    3 Calyx equally 2-lipped; the lips closed after flowering:
      4 Calyx-lips truncate, minutely mucronate; corolla blue, about 1·5 cm. long
                                            7. Salvia
      4 Calyx-lips rounded, entire, the upper crested on the back; corolla scarlet, 2–2·5 cm. long   **2. Scutellaria**
    3 Calyx regular or unequally 2-lipped, the lobes toothed, the upper not crested on the back:
      5 Calyx regular or nearly so, the teeth more or less equal:
        6 Stamens protruding beyond upper lip of corolla; corolla about 15 mm. long; anthers of posterior stamens 1-locular, of anterior stamens 2-locular
                                            3. Epimeredi
        6 Stamens not protruding beyond upper lip of corolla; corolla 2·5–8 mm. long; anthers all alike:
          7 Lower lip of corolla concave (scoop-shaped), abruptly bent downwards; stamens declinate; flowers mostly very small, in cymes forming panicles, spikes or heads; mostly coarse weedy herbs   **10. Hyptis**
          7 Lower lip of corolla convex or flat; flowers solitary, in small axillary or terminal cymes or verticillate, not in panicles, spikes or heads:
            8 Calyx-tube lightly reticulate-veined, with few long hairs; upper lip of corolla hooded; stamens ascending, exserted from corolla-tube but covered by hood; erect annual weedy herb   **4. Stachys**
            8 Calyx-tube with strong straight ribs, puberulous or glabrous; upper lip of corolla flat; stamens directed forwards, unequal and two included in corolla-tube; creeping or tufted perennial herb or erect shrub; leaves small
                                            6. Satureja
      5 Calyx irregular, usually 2-lipped, the upper lip often concave:
        9 Stamens 2, the connective elongated and produced; anthers 1-locular   **7. Salvia**
        9 Stamens 4, the connective small:
          10 Inflorescence a dense terminal oblong head with broad sessile rounded entire cuspidate bracts; calyx 2-lipped, the upper with 3 short teeth and closed after flowering; trailing and rooting herb   **8. Prunella**
          10 Inflorescence interrupted, the verticils or axillary cymes distinct; calyx tubular, not closed after flowering; erect herbs:
            11 Calyx with 8–10 spinescent teeth, the posterior greatly elongated, the anterior eventually incurved; corolla very hairy, orange; stamens ascending under corolla-hood   **9. Leonotis**
            11 Calyx with up to 5 teeth; corolla blue, mauve or white tinged purple; stamens resting on lower lip of corolla:

12 Lower lip of corolla flat or slightly concave, not much longer than upper; filaments free from one another **11. Ocimum**
12 Lower lip of corolla strongly concave, much longer than upper; filaments shortly joined at base into a sheath **12. Plectranthus**

1. TEUCRIUM L. (1753)

1. **T. vesicarium** Mill., Gard. Dict. ed. 8 (1768).—*T. inflatum* Sw. (1788).
Faintly aromatic erect perennial herb up to about 1 m. high; leaves triangular-ovate, truncate-subcordate at base, acute, deeply dentate, up to about 9 cm. long and 4·5 cm. broad, glabrescent adaxially; inflorescence terminal, compact, sub-racemose, up to about 12 cm. long; calyx up to 7 mm. long; corolla-tube greenish-white, limb mauve-pink with darker lines and spots in throat.
Occasional to frequent in the north-central parishes, at roadsides and in damp thickets and rough pastures on limestone; (SL–) 500–2700 feet; fl. and fr. all the year. *A 8983! H 11998! P 22659!* Mexico to Argentina, Cuba, Hispaniola.

2. SCUTELLARIA L. (1753)

1. **S. ventenatii** Hook. in Curt., Bot. Mag. **72**: t. 4271 (1846).
Straggling perennial herb up to about 1 m. high; leaves broadly ovate, openly cordate, acute or acuminate, with crenations broad and shallow, up to about 10 cm. long and 6 cm. broad, puberulous, paler beneath, long-petiolate; inflorescence a terminal raceme, short and corymbose at first, elongating to about 15 cm. long in fruit; nutlets discoid, black, tuberculate, about 1·5 mm. in diameter, released by caducous upper calyx-lip.
Frequent on roadside limestone rocks and walls and in field margins and thicket undergrowth, an escape from cultivation; 1000–27000 feet; fl. and fr. all the year. *A 6570! H 12387! H & P 14970! Stearn 434!* Native of Colombia and Venezuela, introduced and naturalized in Trinidad and elsewhere.

3. EPIMEREDI Adans. (1763)

1. **E. indicus** (L.) Rothm. in Fedde, Repert. Sp. Nov. **53**: 12 (1944).—*Nepeta indica* L. (1753). *Anisomeles ovata* Ait. f. (1811).
Robust annual herb up to about 1 m. high; leaves broadly ovate, truncate-rounded at base and then shortly and broadly cuneate, mostly acute at tip, coarsely crenate, up to about 9 cm. long, and 6 cm. broad, glandular and pubescent; whorls of flowers cymose, ultimately unilateral; calyx 5-lobed, about 7–8 mm. long at anthesis, the lobes pinnately veined; corolla-tube white, limb at least mid-lobe of lip deep bright purple.
Locally naturalized and common (St. Andr., Port., St. Thom.), on shaded road-side banks and in moist pastures; 30–1000 feet; fl. and fr. Sept–Mar. *A 8089! H 8298! P 11380! 22076!* Native of E. Indies, introduced in Trinidad.

4. STACHYS L. (1753)

1. **S. arvensis** (L.) L., Sp. Pl. ed. 2, **2**: 814 (1763).—*Glechoma arvensis* L. (1753).
Annual herb up to about 30 cm. high, sometimes decumbent and rooting from lower nodes; leaf-blades oblong-ovate, rounded to truncate-subcordate at base, obtuse, crenate with rounded crenations, up to 3·5 cm. long and 2·5 cm. broad, light green, thinly hairy, the lower long-petiolate, the upper floral leaves very shortly petiolate; calyx up to about 7 mm. long in fruit; corolla-tube white, limb rose-pink to mauve with a few purple spots and streaks.
Locally common (St. Andr., Port., St. Thom.), a weed of cultivations, clearings and roadsides; 2000–7400 feet; fl. and fr. Nov–June. *A 5716! H 8590! P 11417!* Native of Europe, introduced and naturalized elsewhere.

5. LEONURUS L. (1753)

1. L. sibiricus L., Sp. Pl. **2**: 584 (1753).—Greasy Bush, Honey Weed.
Erect annual herb up to about 1 m. high; leaf-segments linear to lanceolate and again cleft or lobed, puberulous; petioles of lower leaves up to 10 cm. or more long; inflorescence verticillate, each whorl subtended by two leaflike bracts of which the uppermost are entire or nearly so; calyx-tube 3·5–4 mm. long, the teeth aristate; corolla pinkish-mauve, the lower lip streaked purple, 11–12 mm. long; nutlets rugose, brown.
Frequent, a weed of open fields, waste places and gravelly riverbanks; SL–4000 feet; fl. and fr. all the year. *A 6569! H 6976! P 23340!* Native of E. Asia, introduced and widespread in the American subtropics and tropics.

6. SATUREJA L. (1753); Epling & Játiva (1966)

Creeping or tufted herb, rooting freely; flowers solitary in axils on pedicels up to 5 mm. long; mouth of calyx with a fringe of white hairs   1. *brownei*
Erect or scrambling shrub; branchlets scabrid; flowers few in abbreviated axillary cymes, rarely solitary; mouth of calyx without a fringe of hairs   2. *viminea*

1. S. brownei (Sw.) Briq. in Engl. & Prantl, Nat. Pflanzenf. **4** (3a): 300 (1896).—
Thymus brownei Sw. (1788). *Micromeria brownei* (Sw.) Benth. (1834). Penny Royal.
Slender-branched glabrescent herb, with erect shoots sometimes up to 20 cm. high; leaves broadly ovate to rhomboid-orbicular, truncate at base, rounded at tip, very shallowly crenate with few teeth, less than 1 cm. long and broad; calyx about 4·5 mm. long; corolla light violet with dark dots in throat.
Occasional in damp places; 50–4000 feet; fl. most of the year, fr. Feb–Mar, July–Aug. *H 12887! P 9964! Stearn 170! Weaver 1174!* Tropical C. and S. Amer., Cuba, Hispaniola.

2. S. viminea L., Syst. Nat. ed. 10, **2**: 1096 (1759).—*Micromeria obovata* Benth. (1834). *M. viminea* (L.) Urb. (1919). All Heal, Savory, Wild Mint.
Aromatic shrub with long slender branches, up to 3 m. high; leaves very variable, suborbicular to oblanceolate, scabrid or smooth adaxially, more or less whitish-pubescent abaxially, 5–20 mm. long, 4–9 mm. broad; calyx more or less scabrid-puverulous, also with yellow glands, 2·5–3 mm. long; corolla light blue, lavender or white, very variable in length, 3–10 mm. long.
Locally common (St. Andr., Port., St. Thom.), in thickets on steep shale or sepentine banks and gravelly streambeds; 500–4500 feet; fl. Oct–July, fr. Nov–Apr. *A 11942! 12144! H 6817! J.P. 1157! P 20810! 23307!* Cuba, Hispaniola.

7. SALVIA L. (1753); Epling (1938–39)

1 Calyx truncate, very obscurely 2-lobed, densely white-woolly, 5–6 mm. long in fruit; corolla violet-blue, about 15 mm. long; leaves lanceolate, long-cuneate at base, remotely serrate, subacute, puberulous on veins, up to about 7 cm. long and 2 cm. broad; cultivated ornamental perennial herb; native of tropical America; Blue Salvia
                                                                                  S. farinacea Benth.
1 Calyx distinctly lobed:
  2 Corolla bright scarlet, rarely white or yellow:
    3 Calyx scarlet, inflated, about 15 mm. long in flower; leaves cuneate at base, glabrous; corolla about 25 mm. long; cultivated ornamental; native of Brazil; Red Salvia
                                                                                  S. splendens Ker-Gawl.
    3 Calyx green or tinged purple, not inflated, about 7 mm. long in flower; leaves broadly cuneate to subcordate at base, pubescent; corolla-tube 13–19 mm. long, overall up to 23 mm. long   1. *coccinea*
  2 Corolla not bright scarlet or yellow, very rarely all white, less than 20 mm. long:
    4 Inflorescence dense and spikelike with large persistent bracts; corolla mauve or blue; introduced annual herb to 1 m. high, not recently collected; native of C. Mexico, naturalized in S.W. Europe   *S. hispanica* L.
    4 Inflorescence of distinct whorls of flowers; bracts small and/or deciduous; wild species:

5 Corolla purple; calyx with sessile or inconspicuous glands; shrubs; bracts deciduous:
  6 Calyx minutely puberulous, 6–8 mm. long at anthesis, up to 10 mm. long in fruit,
    the upper lip acuminate; racemes up to 30 cm. long   *2. clarendonensis*
  6 Calyx pubescent or lanate, up to 7 mm. long in fruit, the upper lip obtuse or
    cuspidate; racemes up to about 8 cm. long:
    7 Indumentum canescent   *3. jamaicensis*
    7 Indumentum tomentose   *4. eriocalyx*
5 Corolla blue or blue and white; calyx with stalked glands; herbs:
  8 Bracts eventually deciduous; erect annual herb, rarely decumbent and rooting at
    base; fruiting calyx 5·5–7 mm. long; leaf-blade truncate at base   *5. serotina*
  8 Bracts more or less persistent; creeping or straggling herbs with stems often or
    always rooting near base; other characters not combined as above:
    9 Stem slender, prostrate and trailing and rooting regularly; leaves truncate-
      subcordate at base   *6. tenella*
    9 Stems straggling and ascending, often bushily branched; leaves cuneate at base:
      10 Mature calyces 2·8–3·5 mm. long, not gaping in fruit, upper lip obscurely
        3-mucronate or rounded-truncate, lower curved upwards, the teeth obscure,
        acute but not spiny   *7. occidentalis*
      10 Mature calyces up to 7 mm. long, gaping in fruit, upper lip and teeth of lower
        clearly mucronate or spiny; lower lip straight:
        11 Mature calyces 3·5–5 mm. long   *8. misella*
        11 Mature calyces 5·5–7 mm. long   *9. riparia*

1. S. coccinea Buc'hoz ex Etlinger, Salvia: 23 (1777).—Scarlet Sage.
Herb up to 120 cm. high, aromatic; stem with short curled pubescence and mostly also with long hairs; leaves broadly triangular-ovate, crenate-serrate, obtuse or acute, up to about 6 cm. long and 4 cm. broad; bracts subpersistent; calyx 8–10 mm. long in fruit, puberulous, with sessile glands; stamens exserted.
Rather common on shaded or open rocky banks; 600–3500 feet; fl. and fr. all the year. *A 6891! H 8795! P 6504!* Throughout subtropical and tropical Amer., introduced into West Africa and Polynesia.

2. S. clarendonensis Britton in Bull. Torr. Bot. Club **48**: 340 (1922).
Shrub with slender spreading puberulous to tomentellous branches, up to 1·5 (–3) m. high; leaves lanceolate, cuneate or rounded at base, serrate, long-acuminate, up to 15 cm. long and 4 cm. broad; corolla 14–15 mm. long.
Rare and local (Clar.), in open woodland on steep sheltered limestone slope; 2250–2350 feet; fl. Dec–Apr, Aug, fr. Jan, Aug. *A 12400! A & C 1243! H 12787! P 29955!* Endemic.

3. S. jamaicensis Fawcett ex Urb., Symb. Ant. **1**: 396 (1899).
Shrub up to 2·5 m. high; leaves oblong-lanceolate, broadly cuneate to rounded at base, serrate, long-acute, 5–10 cm. long, 2–4 cm. broad; calyx 5–6 mm. long at anthesis; corolla about 15 mm. long.
Rare and local (St. Andr.), in montane forest on limestone; 5500–5750 feet; fl. Nov. *H 7327! Hart 1415! P 9539!* Endemic. Perhaps not more than varietally distinct from the next.

4. S. eriocalyx Bert. ex J. A. & J. H. Schult., Mant. **3**, Addit. 2: 246 (1827).
Straggling-branched shrub up to about 1·5 m. high; leaves lanceolate, broadly cuneate at base, serrate, long-acute to gradually acuminate, up to 12 cm. long and 4 cm. broad; calyx bright purple, 5 mm. long at anthesis; corolla 15 mm. long.
Rare (St. Andr.), in thickets on hillsides; 1000–3500 feet; fl. May–June, Nov, fr. June. *H 6645! P 8714! 23731! Skelding UCWI 25099!* Endemic.

5. S. serotina L., Mant. Pl.: 25 (1767).—Chicken Weed, Little Woman.
Aromatic herb up to about 45 cm. high; leaf-blade broadly ovate, truncate-subcordate at base, apex obtuse or rounded, margin crenate, 1·5–4·5 cm. long, 1–2·5 cm. broad, variably hairy, commonly glabrescent between the veins beneath; petioles up to 2 (–3) cm. long; racemes 5–12 cm. long; calyx at anthesis 3–4 mm. long; pedicels on old inflorescences ascending, about 3 mm. long.
Common as a weed of cultivated ground and pastures on shallow limestone soils; 20–2700 (–3500) feet; fl. and fr. all the year. *A 6982! 10363! H & P 15200! Stearn 147!* Bermuda, Florida, Bahamas, Mexico to Panama, Greater Antilles and south to

Grenada and Barbados; Grand Cayman. A variable species from the broader concept of which plants identified as *S. micrantha* Vahl. and *S. caymanensis* Millsp. & Uline are not obviously distinct.

6. S. tenella Sw., Nov. Gen. & Sp. Pl.: 14 (1788).

Stems slender, puberulous, the flowering shoots up to about 20 cm. high; leaves broadly ovate, blunt-tipped, crenate-dentate, mostly about 1 cm. long and broad, puberulous on both surfaces; inflorescence-axis glandular; calyx at anthesis about 2·2 mm. long; corolla mottled violet-blue with deep violet lines in throat, 3 mm. long.

Local (St. Andr., Port., St. Thom.), on mossy rocks and stones in moist shady places; 3000–4900 feet; fl. and fr. most of the year. *A 7702! P 22677!* Reported from Dominican Republic by Epling (1938) and from Bahamas by Hitchcock (1893), the latter almost certainly erroneously.

7. S. occidentalis Sw., Nov. Gen. & Sp. Pl.: 14 (1788).—Field Basil.

Diffusely branched aromatic herb with shoots up to about 120 cm. high; leaf-blades rhomboid-ovate, cuneate and decurrent at base, acute, coarsely serrate, with usually evenly spaced large appressed hairs on adaxial surface, up to 5 (–9) cm. long, and 3·5 (–5) cm. broad; raceme up to about 16 (–19) cm. long with broadly ovate acuminate bracts; calyx at anthesis about 2 mm. long; corolla-tube about 2·5 mm. long.

Common on roadside banks and in rocky thickets, pebbly riverbeds and waste places; 10–4000 feet; fl. and fr. mostly Nov–Apr. *A 12097! P 21933! Swartz!* Throughout tropical Amer.; Grand Cayman.

8. S. misella Kunth, Nov. Gen. 2: 290 (1818).

Like the last but distinguished by the spine-tipped calyx gaping in fruit; leaf-blades 1·5–3·5 cm. long, 1–2 cm. broad.

Rare (St. Andr., Manch.), a weed; 700–1750 feet; fl. and fr. Nov–Jan. *H 8213! P 22877!* Reported from Lower California, Mexico, Cuba, Puerto Rico. *S. misella* has been identified in some herbaria with the previous species.

9. S. riparia Kunth, Nov. Gen. 2: 300 (1818).

Diffusely branched herb up to about 1 m. high; leaf-blade broadly ovate to subrhomboidal, long-cuneate and decurrent nearly to base of petiole, acute, finely serrate; otherwise like the last but with calyx larger in fruit; bracts ovate to lanceolate, generally narrower than in *S. occidentalis*.

Frequent, in stony waste places on limestone; 100–2700 (–4000) feet; fl. and fr. Nov–Feb, May–July. *A 11956! Yuncker 17455!* Mexico to Peru, Cuba, Hispaniola.

8. PRUNELLA L. (1753)

1. P. vulgaris L., Sp. Pl. 2: 600 (1753).

Trailing and rooting herb with flowering shoots ascending to about 25 cm. high; leaves ovate, truncate to rounded and cuneate at base, obtuse at tip, subentire, up to about 4 cm. long and 2 cm. broad; petiole slender, up to about 1 cm. long; stem and leaves with long sparse hairs; bracts ciliate; corolla bright bluish-purple.

Rare and local (St. Andr., Port.), along damp sheltered grassy pathsides; 4450–5000 feet; fl. Dec–Apr, July, fr. Feb–Apr. *A 8703! 12677! Stearn 187! Watt!* Native of Europe, now widely naturalized; Hispaniola.

9. LEONOTIS (Pers.) R. Br. (1810)

1. L. nepetifolia (L.) Ait. f. in Ait., Hort. Kew. ed. 2, 3: 409 (1811).—*Phlomis nepetifolia* L. (1753). Bald Bush, Christmas Candlestick.

Erect annual herb 1–2 m. high; leaves broadly ovate, truncate-subcordate and cuneate at base, deeply crenate-dentate, up to 12 cm. long and 10 cm. broad, puberulous; flowers in dense remote globose glomerules up to 6 cm. in diameter at maturity; calyx at anthesis about 12 mm. long, elongating to 20–25 mm.; corolla about 20 mm. long; nutlets black, smooth.

Rather common, a weed of fields, roadsides and waste ground; 100–1000 (–3000) feet; fl. and fr. all the year, but mainly Oct–Mar. *A 5487! H 6990! H & P 14483!* Native of tropical Africa, now general in the tropics.

10. HYPTIS Jacq. (1787) nom. cons.

1 Inflorescences paniculate of cymes, either verticillate or of numerous small bracteate heads, or open or unilaterally pectinate:
  2 Calyx-teeth deltate, about 1 mm. long, the calyx-tube barrel-shaped in fruit and about 2 mm. long; flowers in numerous superposed clusters subtended by reduced leaves or bracts; leaves cuneate at base   1. *verticillata*
  2 Calyx-teeth aristate or subulate, with a more or less obvious tuft of whitish hairs between the lobes; leaves mostly rounded to truncate-subcordate and then shortly cuneate at base:
    3 Flowers in numerous subglobose glomerules subtended by appressed ovate or elliptical erect bracts, on peduncles 2–5 mm. long; fruiting calyx-tube 4–6 mm. long; corolla usually blue   2. *mutabilis*
    3 Flowers in open or lax cymules without appressed subtending bracts:
      4 Calyx-tube in fruit 5–7 mm. long, glandular, the teeth triangular at base, aristate; corolla mauve to purplish-blue; plant strongly aromatic   3. *suaveolens*
      4 Calyx-tube in fruit about 2·5 mm. long, pubescent, the teeth subulate; corolla white or tinged pink; plant not strongly aromatic   4. *pectinata*
1 Inflorescences uninterrupted spikes or globose heads usually solitary on terminal or axillary peduncles; leaf-blades cuneate at base, short- or long-decurrent on petiole:
  5 Inflorescences compact spikes up to 5 (–9) cm. long; calyx with conspicuous shortly stalked glands; corolla blue or violet   5. *spicigera*
  5 Inflorescences capitate, involucrate, up to 2 cm. in diameter in fruit; calyx without conspicuous glands:
    6 Stems prostrate, trailing and rooting; leaves ovate to lanceolate; narrowly cuneate from a broad base; involucral bracts broadly elliptical to obovate; corolla white with mauve marks on lip; nutlets rough   6. *atrorubens*
    6 Stems erect; leaves elliptical to lanceolate, long-cuneate from a narrow base; corolla white; nutlets smooth:
      7 Calyx-tube in fruit 7–8 mm. long; flower-heads 1–2·5 cm. broad; peduncles up to 6 cm. long; involucral bracts lanceolate to spathulate   7. *capitata*
      7 Calyx-tube in fruit 2·5–3 mm. long; flower-heads up to 1·5 cm. broad; peduncles 3–15 mm. long; involucral bracts linear-subulate   8. *brevipes*

1. H. verticillata Jacq., Collect. **1**: 101 (1787).—John Charles.
Erect perennial shrubby herb with many branches, 1–3 m. high; leaf-blade elliptic-lanceolate, acute, serrate except towards base, up to 10 cm. long and 3·5 cm. broad, thin, glabrous except on midrib beneath; glomerules about 1 cm. in diameter; corolla usually white; nutlets minutely reticulate.
Frequent as a weed of thickets and rough pastures; 20–1600 feet; fl. and fr. Oct–Jan. *A 8219! H 6864! P 22854!* Florida, Mexico to Colombia, Greater Antilles, Dominica, Martinique.

2. H. mutabilis (L. C. Rich.) Briq. in Bull. Herb. Boiss. **4**: 788 (1896).—*Nepeta mutabilis* L. C. Rich. (1792). *H. spicata* Poit. (1806).
Coarse erect annual herb up to about 2 m. high; leaf-blade ovate to rhomboid-ovate, shortly acuminate, 3–6 cm. long, variably hairy from hirtellous only on the veins beneath to pubescent throughout; petioles 1–4 cm. long; corolla about 6 mm. long; nutlets oblong, about 1 mm. long, black, smooth.
Known in Jamaica only from an unlocalized report by Fawcett (1893). General in the American subtropics and tropics and introduced into the Old World and Polynesia.

3. H. suaveolens (L.) Poit. in Ann. Mus. Hist. Nat. Paris **7**: 472, t. 29, f. 2 (1806).
—*Ballota suaveolens* L. (1759). Pignut, Spikenard.
Annual erect herb with glandular indumentum, up to 1·5 (–3) m. high, woody at base; leaf-blade broadly ovate, acute or obtuse, doubly serrate-dentate, 3–10 cm. long, 2–6 cm. broad, adaxially with rather long hairs, abaxially pubescent; axillary cymes pedunculate, about 2 cm. long and broad; corolla-tube 4–6 mm. long; nutlets light brown, ribbed, nearly 4 mm. long.

Common in rough pastures and gravelly waste places on shale or limestone; 10–2900 feet; fl. Oct–Apr, fr. Oct–June. *A 6221! H 11859! P 22873!* Native of the American tropics now general in the Old World and Polynesia; Grand Cayman.

4. H. pectinata (L.) Poit. in Ann. Mus. Hist. Nat. Paris **7**: 474, t. 30 (1806).— *Nepeta pectinata* L. (1759). Piaba.

Erect or straggling herb 1–3 m. high; leaf-blade broadly ovate, acute, serrate-dentate, up to 11 cm. long and 6·5 cm. broad, shortly pubescent abaxially; petioles slender, up to 5 cm. long; cymes subsessile, commonly forked; corolla-tube about 1·5 mm. long; nutlets black, smooth.

Common along roadsides and on waste rocky ground; SL–4400 feet; fl. and fr. Nov–Mar, June. *A 6387! H & P 13411! Yuncker 17579!* General in the tropics; Grand Cayman.

5. H. spicigera Lam., Encycl. Méth. Bot. **3**: 185 (1789).—*H. americana* (Aubl.) Urb. (1918), not Briq. (1897). *Nepeta americana* Aubl. (1775).

Annual erect aromatic herb up to 2 m. high, the angles of the stem scabrid; leaf-blade lanceolate, acute, crenate-serrate, up to 6 (–8) cm. long and 2 (–3) cm. broad, scabridulous adaxially; calyx 10-ribbed, about 5 mm. long at maturity; corolla about 4 mm. long; nutlets blackish-brown, minutely reticulate.

Rare and local (St. Cath., Clar.), in damp open grassy places; 100–2300 feet; fl. and fr. Dec–Jan. *P 15900! 16135! Yuncker 17985!* Mexico to Brazil and Peru, Greater Antilles, naturalized in Africa and Asia.

6. H. atrorubens Poit. in Ann. Mus. Hist. Nat. Paris **7**: 466, t. 27, f. 3 (1806).

Perennial creeping herb with sparingly pubescent long slender stems, with ascending flowering branches up to 25 cm. high; leaves obtuse, crenate, 1·5–4·5 cm. long, up to 2 cm. broad; peduncles slender, 5–10 mm. long; heads 8–10 mm. in diameter; bracts 4–6 mm. long, ciliate; calyx about 4 mm. long at anthesis; corolla-tube 4–5 mm. long; nutlets ellipsoidal, bright brown.

Very rare and known only from one unlocalized gathering, *March 1321.* General throughout the tropics in moist places.

7. H. capitata Jacq., Collect. **1**: 102 (1787).—Ironwort, Wild Caesar Obeah.

Erect annual herb up to 1·2 (–2) m. high; stem glabrescent; leaves ovate-elliptical, acute or obtuse, serrate or crenate-serrate, up to 15 (–20) cm. long and 7·5 (–11) cm. broad; bracts 5–7 mm. long, obscured at maturity by reflexed lower calyces; corolla-tube 2·5–3 mm. long; nutlets oblong, dark brown, smooth, about 1 mm. long.

Common in damp low-lying places and cultivated ground; 50–4000 feet; fl. and fr. Dec–Apr. *A 6081! 10409! H 12280! P 6178!* Throughout tropical Amer., Asia and in Polynesia.

8. H. brevipes Poit. in Ann. Mus. Hist. Nat. Paris **7**: 465 (1806).

Slender annual erect herb 30–90 cm. high; slender roots with fusiform tubers; stem appressed-hirtellous; leaves lanceolate, acute, coarsely irregularly crenate, 4–6 (–10) cm. long, 1–2 (–4) cm. broad, thinly pubescent on both surfaces; bracts spreading or reflexed, 4–6 mm. long; corolla-tube 2·5–3 mm. long; nutlets oblong, blackish, smooth, about 0·5 mm. long.

Rare and local (St. Cath.), in pond margins and swamps; 700–900 feet; fl. and fr. Jan. *A 10355! 12944!* Native of Brazil, widespread elsewhere in continental tropical Amer., Trinidad and Asia.

11. OCIMUM L. (1753)

1 Leaves coarsely serrate-dentate; perennial undershrub; fruiting calyx saccate at base, the
 lower lip inflexed at maturity, about 6 mm. long   1. *gratissimum*
1 Leaves entire or at most shallowly serrate; annuals; fruiting calyx not saccate at base,
 the lower lip directed forwards:
 2 Pedicels 4–7 mm. long; filaments unappendaged; calyx 7–9 mm. long at maturity
                    2. *micranthum*

2 Pedicels 2–4 mm. long; upper filaments with a toothlike appendage at base; calyx 5–6 mm. long at maturity; glabrescent herb 30–60 cm. high; leaves narrowly ovate, acute, up to about 4 cm. long and 2 cm. broad; commonly cultivated as a culinary herb; native of E. Indies; Basil   *O. basilicum* L.

1. O. gratissimum L., Sp. Pl. **2**: 1197 (1753).—African Tea Bush.
   Shrubby herb up to 120 cm. high, thinly puberulous on stem, petioles, leaf-veins, pedicels and calyx; leaves broadly elliptical, acuminate at both ends, 6–12 cm. long, 3–6 cm. broad; racemes rather lax, up to about 20 cm. long; flowering calyx 2·5–3 mm. long; fruiting calyx more or less galeate, the upper lip rather flat, the upper lobe of lateral tooth well developed and obtuse or truncate; corolla light blue.
   Very rare (Clar.), at roadside; about 50 feet; fl. and fr. Jan. *Robertson UCWI 23369!* Native probably of tropical Asia, now widespread in tropical Africa but only occasional in the West Indies; Cuba, Hispaniola, Martinique.

2. O. micranthum Willd., Enum. Hort. Berol.: 630 (1809).—Barsley.
   Low-branched bushy strongly aromatic herb up to about 50 cm. high, pubescent on stem, leaf-veins, pedicels and calyx; leaves broadly ovate-elliptical, cuneate at base, acute or acuminate at tip, up to 7 cm. long and 5 cm. broad; racemes compact, up to about 10 cm. long; flowering calyx about 4 mm. long; fruiting calyx cam-panulate, the upper lip concave, the upper lobe of lateral tooth hardly developed; corolla white or mauve blotched purple or violet.
   Common in open ground; SL–3700 feet; fl. and fr. all the year. *A 6984! A. M. Barry! H 11903!* General in the American subtropics and tropics; Cayman Is.

12. PLECTRANTHUS L'Hérit. (1788) nom. cons.

Leaf-blades thick and fleshy with prominent veins beneath, hairy, mostly less than 10 cm. long with petioles less than 2 cm. long; 4 lower calyx-teeth subequal   1. *amboinicus*
Leaf-blades thin, almost glabrous, mostly more than 10 cm. long with longer petioles; 2 lateral calyx-teeth obtuse, 2 lower acute   2. *blumei*

1. P. amboinicus (Lour.) Launert in Mitt. Bot. Staatss. München **7**: 298 (1968).—
   Coleus amboinicus Lour. (1790). French Thyme, Soup Mint.
   Lax straggling thick-stemmed herb, the flowering branches ascending to about 60 cm. high, strongly aromatic; leaves broadly ovate, truncate-subcordate and shortly cuneate, acute, obtuse or rounded at tip, crenate, up to about 10 cm. long and 8 cm. broad, rarely larger with longer petioles; inflorescence-branches up to about 40 cm. long; upper-lip of calyx mostly oblong; corolla bright lilac, darker within, 7–8 mm. long.
   Cultivated and locally naturalized (St. Andr., St. Thom.), on rocky limestone banks; SL–2600 feet; fl. Apr–Aug, fr. May–July. *A 6951! H 9595!* Widespread in tropical countries and grown as a pot-herb; origin uncertain but possibly southern tropical Africa.

2. P. blumei (Benth.) Launert in Mitt. Bot. Staatss. München **7**: 301 (1968).—
   Coleus blumei Benth. (1832). Joseph's Coat.
   Branched bushy herb up to 1 (–2) m. high; leaves broadly ovate, rounded and shortly cuneate at base, acuminate, shallowly crenate, commonly variegated with light green and purple in patterns affecting the centre, veins or margins inde-pendently; inflorescence-branches up to 50 cm. long with many tiers of short cymes; upper lip of calyx oblate; corolla blue-violet, about 12 mm. long.
   Cultivated and occasionally relict or escaping especially in damp shady places; 500–3500 feet; fl. most of the year. *A 5559! H 11893! P 10143!* Well known ornamental plants possibly of garden origin; first described from the East Indies.
   Besides the foregoing a number of introduced culinary herbs are grown in Jamaica, particularly in hill gardens. These include various species and hybrids of *Mentha*, such as *M. spicata* L. (*M. viridis* (L.) L.), Spearmint and *M.* × *smithiana* R. A. Graham, as well as *Thymus vulgaris* L., Thyme, *Rosmarinus officinalis* L., Rosemary, *Salvia officinalis* L., Sage and *Origanum majorana* L., Marjoram.

170. SOLANACEAE

Herbs, erect or climbing shrubs or trees. Leaves alternate, simple or compound by dissection, without stipules. Flowers bisexual, actinomorphic or slightly zygomorphic, solitary or in usually cymose inflorescences. Perianth biseriate; calyx 4–6-lobed, tubular or deeply divided, mostly persistent and often accrescent; corolla sympetalous, tubular or rotate, (4–) 5 (–6) or rarely more -lobed, the lobes valvate, valvate-induplicate or convolute and often folded. Stamens 2, 4 or 5, with or without staminodes, epipetalous; anthers 2-locular or 1 loculus undeveloped, opening lengthwise or by apical pores. Disk usually present. Ovary superior, typically 2-locular; ovules mostly numerous, anatropous or amphitropous, on axile placentas; style solitary; stigma simple or lobed. Fruit a berry or septicidal capsule. Seed with curved embryo and fleshy endosperm.

85 genera with about 2300 species; cosmopolitan but with a high concentration in south subtropical and tropical America.

1 Fruit a capsule; corolla tubular:
2 Perfect stamens 4:
3 Flowers in panicles; corolla 2·5–3 cm. long, the tube twisted and yellow, the limb scarlet; ornamental shrub with small strongly veined leaves, in hill gardens; native of Colombia   *Streptosolen jamesonii* Miers
3 Flowers solitary or few in short racemes; corolla not yellow and red, the tube not twisted:
4 Annual herb with violet-blue, mauve or rarely white flowers   1. **Browallia**
4 Shrubs with white or yellowish fragrant showy flowers   2. **Brunfelsia**
2 Perfect stamens 5:
5 Flowers in terminal racemes or panicles; leaves sessile or nearly so, clothed with glandular hairs; corolla funnel-shaped   3. **Nicotiana**
5 Flowers usually solitary; leaves petiolate:
6 Stamens uniform; plants shrubby even if short-lived or arborescent, not obviously glandular-pubescent; calyx at least 3 cm. long; corolla trumpet-shaped, white or tinged purple, showy   4. **Datura**
6 Stamens 4 in 2 pairs and one smaller; plant herbaceous, somewhat glandular-pubescent; calyx much smaller than above; corolla funnel-shaped, variously coloured but often pink; cultivated ornamentals mostly of garden origin; native of southern S. Amer.   *Petunia* spp.
1 Fruit a berry; stamens all fertile, usually 5:
7 Corolla-tube distinctly longer than lobes; shrubs or trees:
8 Woody climbers with solitary flowers; corolla funnel-shaped, (15–) 18 cm. or more long, the lobes imbricate   5. **Solandra**
8 Erect shrubs or trees; flowers not solitary; corolla-lobes valvate or plicate:
9 Flowers in racemes or panicles of racemes; corolla-tube subcylindrical, 10–22 mm. long; hard-wooded shrubs or trees   6. **Cestrum**
9 Flowers long-pedicelled in umbelliform clusters:
10 Stamens included; corolla narrowly funnel-shaped, 3·5–4 cm. long, deep violet, the lobes very short; leaves ovate-elliptical, softly tomentose, up to 10 cm. long and 4·5 cm. broad; soft-wooded shrub up to about 2 m. high, cultivated as an ornamental in hill gardens; native of Colombia and Ecuador
                                                   Iochroma cyaneum (Lindl.) M. L. Green
10 Stamens exserted; corolla broadly funnel-shaped, under 1 cm. long, greenish-white to cream, the lobes recurved; small tree   7. **Dunalia**
7 Corolla rotate or campanulate with tube mostly much shorter than lobes:
11 Berry enclosed in the inflated calyx; plants herbaceous; flowers solitary; corolla usually openly campanulate, yellow with or without purple spots at base
                                                                              8. **Physalis**
11 Berry not enclosed in the inflated calyx:
12 Anther-loculi separated by a thickened connective; corolla light pink, 1·5–2·5 cm. in diameter; fruit ovoid, smooth, orange tinged red, 6–7 cm. long; small tree with ovate cordate leaves; cultivated for the edible fruit; native of tropical continental Amer.; Tree Tomato   *Cyphomandra betacea* (Cav.) Sendtn.
12 Anther-loculi not separated by a broad thickened connective:
13 Anthers mostly opening by terminal pores and typically but not invariably connivent in a cone around the style, yellow; herbs, erect shrubs, climbers or trees, with entire or divided leaves, sometimes with prickles on stems and leaves; corolla usually white, mauve or pink; fruit often globose   9. **Solanum**
13 Anthers opening longitudinally; unarmed herbs or undershrubs:

14 Anthers connivent in a cone around the style, produced into a sterile tip, yellow; leaves pinnate; corolla clear yellow; calyx deeply divided 10. **Lycopersicon**
14 Anthers not connivent in a cone around the style, often greyish-blue; leaves entire or at most repand-dentate; corolla white or green:
  15 Calyx 5-lobed, spreading and papery in fruit; flowers in stalked umbels; leaves repand-dentate 11. **Saracha**
  15 Calyx subentire, cupular-appressed on the fruit and usually fleshy; flowers in sessile umbels, subsolitary or solitary; leaves entire or nearly so
      12. **Capsicum**

1. BROWALLIA L. (1753)

1. B. americana L., Sp. Pl. **2**: 631 (1753).—Jamaican Forget-me-not.

Annual herb from as small as 2–3 cm. on poor soil to 60 cm. high; leaves ovate to lanceolate, rounded to broadly cuneate at base, acuminate, up to 8 cm. long and 4 cm. broad, thin; pedicels up to 10 (–15) mm. long; calyx up to 8 mm. long; corolla-tube up to 15 cm. long; capsule about 7 mm. long.

Common in the eastern parishes on roadsides and banks in damp shady areas; 150–5000 feet; fl. and fr. most of the year. *A 5402! H 6916! P 20618!* General in tropical Amer.

2. BRUNFELSIA L. (1753)

1 Leaves thickly leathery with recurved margins, tip obtuse or rounded, usually less than 8 cm. long, shiny; fruit smooth, shiny; calyx about 8 mm. long in flower; corolla glabrous 1. *splendida*
1 Leaves not or only softly leathery, margins flat, tip usually acute or acuminate:
  2 Calyx 15 mm. or more long, lobes about 3 mm. long; corolla-tube puberulous:
    3 Leaf-blades elliptic-lanceolate, acute, 5–11 cm. long, thin 2. *jamaicensis*
    3 Leaf-blade oblanceolate, usually acuminate, up to 17 (–20) cm. long, softly leathery
      3. *plicata*
  2 Calyx up to 10 mm. long, lobes 1–1·5 mm. long; corolla-tube glabrous or thinly puberulous:
    4 Fruit spherical, up to 5 cm. in diameter, reticulate-sculptured on the surface (at least at maturity and when dry) with double-margined areoles; calyx 7–10 mm. long; corolla-tube glabrous; petioles usually slender, 5–8 mm. long; leaf-blade glabrous
      4. *maliformis*
    4 Fruit subspherical to ovoid, 1·5–3 cm. in diameter, smooth; calyx 5–7 mm. long:
      5 Leaves puberulous at least on midrib beneath; petioles slender, 8–15 mm. long; corolla-tube glabrous 5. *membranacea*
      5 Leaves glabrous; petioles 6–7 mm. long; corolla-tube thinly puberulous
        6. *undulata*

1. B. splendida Urb., Symb. Ant. **5**: 491 (1908).

Shrub 2–3 m. or tree to 6 m. high, the young shoots puberulous; leaves obovate to elliptical, cuneate at base, 3–8 (–10) cm. long 1·5–3·5 (–4) cm. broad, midrib prominent beneath, lateral veins obscure; petiole 2–12 mm. long; pedicels 5–15 mm. long; calyx up to 13 mm. long in fruit; corolla-tube 4·5–7·5 cm. long, limb about 4 cm. across; fruit subspherical, about 2·5 cm. in diameter.

Uncommon (St. Cath., Clar., Trel.), in woodlands on rocky limestone; 1500–2500 feet; fl. Apr–June, fr. Mar–July. *H 8508! 10977! P 24822! Proctor & Alain 24898!* Endemic.

2. B. jamaicensis (Benth.) Griseb., Fl. Br. W.I.: 432 (1862).—*B. nitida* Benth. var. *jamaicensis* Benth. (1846). *B. harrisii* Urb. (1903).

Shrub 2 m. or tree to 8 m. high; leaves acute at base and apex or slightly acuminate at tip, 5–11 cm. long, 1·5–2·5 (–3·5) cm. broad; petiole 3–8 mm. long; pedicels 4–8 mm. long in flower, up to 18 mm. long in fruit; calyx 15–18 mm. long, puberulous; corolla-tube 6·5–10 cm. long, limb up to about 5 cm. across; fruit 2·5–4 cm. in diameter, smooth or minutely verrucose.

Rather local (St. Andr., Port., St. Thom.), in montane woodland; 4900–5700 feet; fl. Aug, Nov, fr. Feb–Mar, Aug, Nov. *A 10581! H 7472! J.P. 891! P 9520!* Endemic.

3. B. plicata Urb., Symb. Ant. **6**: 39 (1909).

Shrub 3–4 m. high; leaves shortly and acutely acuminate, up to 17 (–20) cm. long and 5 (–6) cm. broad, smooth, leathery; petiole 10–12 mm. long; pedicels 7–10 mm. long; calyx up to 20 mm. long; corolla-tube up to 11 cm. long, limb up to 6·5 cm. across; fruit 2·5–4 cm. in diameter, smooth.

Uncommon, in the central parishes, in woodland on limestone; 2000–3000 feet; fl. Aug, fr. Jan, Apr, Sept–Oct. *A 11032! Britton & Hollick 2748! H 8970! P 11022!* Endemic.

4. B. maliformis Urb., Symb. Ant. **3**: 372 (1903).—? *B. fawcettii* Urb. (1903).

Shrub 2–5 m. or tree to 13 m. high; leaves oblong-elliptical, acute or acuminate at base and apex, 7–15 cm. long, 2–4 cm. broad; pedicels 5–9 (–18) mm. long in flower, 10–20 mm. long in fruit; calyx-lobes ciliolate, otherwise glabrous; corolla-tube 5–11 cm. long, limb up to about 5 cm. across, ageing primrose-yellow.

Frequent, in woodland and on limestone cliffs; 700–3800 feet; fl. Mar, Sept, fr. Sept–Apr. *A 6742! H 7398! Jekyll 8125! HPS 14723!* Endemic.

5. B. membranacea Urb., Symb. Ant. **5**: 492 (1908).

Shrub 3–4 m. high; young shoots and leaves abaxially yellowish-puberulous; leaves oblong-elliptical, broadly or narrowly cuneate at base, obtuse, broadly acute or shortly acuminate at tip, 4–10 cm. long, 1·5–4·5 cm. broad; pedicels 5–13 mm. long; corolla-tube 5–8 cm. long, limb 3–5 cm. across, at first white turning yellow or orange-yellow.

Local (St. Andr., St. Cath.), in thickets on limestone rocks; 200–750 feet; fl. Feb–July, fr. Apr, Oct–Dec. *A 6964! Fawcett 8527! H 8929! P 22067!* Endemic.

6. B. undulata Sw., Nov. Gen. & Sp. Pl.: 90 (1788).

Shrub 1·5–2 m. high; leaves narrowly oblanceolate, abruptly cuneate at base, shortly acuminate, 9–15 cm. long, 2·5–5 cm. broad, slightly leathery; corolla-tube about 10 cm. long, 3 mm. broad, limb about 4·5 cm. across; ripe fruit orange.

Uncommon, mostly in dry limestone woodland in north coastal areas; about 100 feet; fl. May, fr. Mar–May. *P 20699! 22260!* Endemic.

3. NICOTIANA L. (1753)

1. N. tabacum L., Sp. Pl. **1**: 180 (1753).—Tobacco.

Annual glandular-pubescent erect herb 1–2 m. high; leaves elliptical to oblong-lanceolate, narrowed at amplexicaul base, short- or long-acuminate, up to 30 cm. or more long and 15 cm. or more broad; calyx 10–12 mm. long in flower, up to 18 mm. long in fruit; corolla-tube dull green, about 4·5 cm. long, limb pink with acute lobes; capsule ovoid-ellipsoid, about 2 cm. long.

Cultivated and sometimes escaping, an important crop in some lowland areas; 50–1500 (–3000) feet; fl. and fr. Mar, June, Sept. *A 9434! H 11930! A. von der Porten!* Native of S. Amer., now widespread in all warm countries.

N. plumbaginifolia Viv. was reported for Jamaica in Alain, Fl. de Cuba (1957), but no specimen has been seen. It is distinguished by the capsules being about 9 mm. long.

4. DATURA L. (1753)

1 Shrubs or small trees; flowers pendulous; corolla 25–30 cm. or more long:
  2 Calyx equally 5-lobed; free part of filaments about 3·5 cm. long; stigma oblong-linear
    with 2 narrow faces, about 7 mm. long   1. *suaveolens*
  2 Calyx spathaceous, split down one side; free part of filaments about 6 cm. long; stigma
    triangular with 2 broad elliptical faces, about 5 mm. long   2. *candida*
1 Annual herbs or undershrubs; flowers erect; corolla up to about 18 cm. long:
  3 Spines of capsule short, deltoid, 2–4 mm. long   3. *metel*
  3 Spines of capsule subulate or capsule smooth:
    4 Calyx 6–8·5 cm. long; corolla 10–16 cm. long; fruit pendulous; seeds brown
                             4. *innoxia*
    4 Calyx 3·5–5 cm. long; corolla 7–10 cm. long; fruit erect; seeds black   5. *stramonium*

1. D. suaveolens Humb. & Bonpl. ex Willd., Enum. Hort. Berol.: 227 (1809).—
    Brugmansia suaveolens (Humb. & Bonpl. ex Willd.) Bercht. & Presl (1823).
    Angel's Trumpet.

Shrub or tree to 6 m. high; twigs and leaves usually puberulous; leaves ovate to elliptical, equal or unequal at base, acute or acuminate, up to 30 cm. or more long and about 20 cm. broad; petioles up to 12 cm. long; pedicels 2–6 cm. long; calyx-tube about 8 cm. long, lobes about 3·5 cm. long, triangular, acute; flowers fragrant at night; corolla greenish-white, sometimes orange-yellow; anthers 3 cm. long; capsule fusiform, 6–12 cm. long.

Occasional, mostly in cultivation in hill gardens; near SL–3500 feet; fl. May, Sept–Dec, fr. Feb. *H 9599! P 20815!* Native of Brazil, now widespread as a planted ornamental.

2. **D. candida** (Pers.) Safford in Journ. Acad. Sci. Washington **11**: 182 (1921).—
   Brugmansia candida Pers. (1805). Angel's Trumpet.
   Shrub or small tree like the last, calyx gradually expanded, 15 cm. long including lobe, the split about 6 cm. long; corolla creamy-white fading light pinkish-orange.
   Common in cultivation and naturalized in sheltered places; 500–3600 feet; fl. June, Sept–Feb, fr. June, Oct–Nov. *P 15445! 24152!* Native of Peru, now in all warm countries.

3. **D. metel** L., Sp. Pl. **1**: 179 (1753).—Thorn Apple[1]
   Glabrous or very thinly puberulous erect herb up to 2 m. high; leaves ovate-lanceolate to elliptical, unequal at base, acute or acuminate, 5–17 cm. long; petioles slender, 3–7 cm. long; calyx about 6 cm. long, the base forming a wing-collar about 3 cm. broad in fruit; corolla white tinged violet, 14–18 cm. long; capsule ovoid, 4–6 cm. long, nodding, with short blunt spines.
   Rare (St. Eliz.), in sandy soil behind beach; near SL; fl. and fr. May. *P 15362!* Native of tropical Asia, now scattered throughout the subtropics and tropics of both hemispheres.

4. **D. innoxia** Mill., Gard. Dict. ed. 8 (1768).—Prickly Bur.
   Odorous straggling-branched herb up to 2·5 m. high; plant finely glandular-pubescent; leaves broadly ovate, unequal at base, acute, up to 25 cm. long; corolla white; capsule globose or ovoid-globose, 2·5–4 cm. in diameter, subtended by a wing-collar up to about 7 cm. broad.
   Rare (St. Andr.), in arid waste places; about 50 feet; fl. and fr. Nov–Jan. *P 8170!* Native of tropical S. Amer., now widespread.

5. **D. stramonium** L., Sp. Pl. **1**: 179 (1753).—Devil's Trumpet, Jimson Weed,
   Trimona.
   Pungent-odorous weedy herb like the last with robust puberulous or glabrescent stems; leaves ovate, sinuate-lobed, up to about 20 cm. long; corolla white or tinged purplish; capsule ovoid, up to about 5 cm. long, prickly or smooth.
   Occasional in well drained sandy or gravelly waste places; 200–4000 feet; fl. Mar–July, fr. Apr–Oct. *A 11808! Larter 300! P 10215! Thornton UCWI 2506!* Native in the American subtropics and tropics, now widespread. The following rather arbitrary varieties have been named:

1 Stem green; corolla white:
  2 Fruit prickly   *D. stramonium* var. *stramonium*
  2 Fruit smooth   *D. stramonium* var. *inermis* (Jacq.) Timmerman
1 Stem and corolla purplish:
  3 Fruit prickly   *D. stramonium* var. *tatula* (L.) Torr.
  3 Fruit smooth   *D. stramonium* var. *godronii* Danert

5. SOLANDRA Sw. (1787) nom. cons.

Leaves and calyx glabrous; berry with red pulp   1. *grandiflora*
Leaves and calyx hairy; berry with white pulp   2. *hirsuta*

1. **S. grandiflora** Sw. in Vet. Akad. Handl. Stockh. **8**: 300, t. 11 (1787).—Chalice
   Vine.
   Climber to 10 m. with thick woody stems; leaves elliptical or oblong-elliptical,

---

[1] Name applied to all herbaceous species of *Datura* with spiny fruits.

cuneate at base, cuspidate at tip, up to 14 cm. long and 6 (–7) cm. broad; petiole slender; pedicels terminal often on short lateral branches; calyx tubular, 2–5-cleft, 7–9·5 cm. long; corolla up to 25 cm. long, the tube narrow at base, the lobes rounded, greenish-white to cream-coloured with brownish-purple lines; fruit ovoid, up to about 5 cm. long and broad, pointed to persistent style.

Frequent on trees, sometimes cultivated; 1500–5000 feet; fl. Dec–Apr, fr. May, Aug. *A 12292! H 10895! Powell 1056!* Continental tropical Amer., West Indies.

2. **S. hirsuta** Dunal in DC., Prodr. **13** (1): 535 (1852).

Climber like the last; leaves elliptical to obovate-elliptical; calyx 8–10 cm. long; corolla 21–23 cm. long; fruit ovoid, 5·5 cm. long, 3·7 cm. broad, with thickened persistent style-base, 4-locular.

Locally frequent (St. Andr., Port., St. Thom.), on trees in montane woodland, mostly near clearings; 3500–4500 feet; fl. Mar–May, fr. June. *A 12372! Powell 1065! Skelding UCWI 4440!* Endemic.

6. CESTRUM L. (1753)

1 Petioles pubescent; peduncles stellate-tomentose; filaments of stamens not appendaged:
  2 Corolla 22–25 mm. long including the linear-lanceolate to subulate lobes 4–7 mm. long; calyx about 1 mm. long   1. *latifolium*
  2 Corolla 11–18 mm. long including the ovate-lanceolate lobes 2·5–3·5 mm. long; calyx 2–3 mm. long   2. *hirtum*
1 Petioles glabrous or nearly so; peduncles usually glabrous but sometimes thinly tomentose; filaments straight or appendaged:
  3 Corolla 16–24 mm. long; filaments with a toothlike appendage; berry white
                                             3. *nocturnum*
  3 Corolla 13–18 mm. long; filaments not appendaged; berry purple-black   4. *diurnum*

1. **C. latifolium** Lam. in Tabl. Encycl. & Méth., Bot. **2**: 5 (1794).

Shrub about 2 m. high; stellate-pubescent on the younger parts; leaves broadly elliptical or ovate, cuneate or rounded at base, acute or acuminate, 4–13 cm. long, 2·5–5·5 (–10) cm. broad; inflorescences short, axillary with racemose branches; calyx shortly 5-lobed; corolla-tube sublinear; fruit about 8 mm. long.

Rare and possibly only cultivated, not recently collected; about 700 feet; fl. Sept. *J. P. 720! Thompson 8063!* Northern S. Amer., Martinique, St. Lucia, St. Vincent, Barbados, Trinidad.

2. **C. hirtum** Sw., Nov. Gen. & Sp. Pl.: 49 (1788).

Shrub 2–4 m. or tree to 8 m. high; young stems, leaves and inflorescences densely stellate-tomentose; leaves ovate to oblong-elliptical, rounded to cordate at base, acuminate, up to 18 cm. long and 7·5 (–9·5) cm. broad, glabrescent; inflorescences very short; flowers fragrant; calyx purplish; corolla greenish or yellow-green often tinged purple especially the lobes; filaments of stamens slightly swollen and bent at insertion on corolla-tube but not appendaged; fruit blackish-purple, ellipsoid, 8 mm. long.

Common in mature woodlands in gullies and on sheltered hillsides; 1000–6000 feet; fl. Dec–Mar, fr. Jan–July. *A 6523! H 10201! P 22109! 22296!* Cuba.

3. **C. nocturnum** L., Sp. Pl. **1**: 191 (1753).—Jasmine, Lady-of-the-Night.

Shrub 2–4 m. high with lax drooping branches; leaves ovate to elliptic-lanceolate, rounded to broadly cuneate at base, acutely acuminate, up to 15 cm. long and 6 cm. broad, glabrous at maturity; pedicels up to 9 mm. long; flowers very numerous developing simultaneously, with very heavy fragrance at night; calyx about 3 mm. long; corolla greenish-white to cream; fruit about 1 cm. long.

Common in cultivation and occasionally relict or escaping into secondary thickets; 100–1700 feet; fl. and fr. Jan–Aug. *A 6106! H 7055! P 7585!* West Indies, now widespread in the tropics as an ornamental

4. **C. diurnum** L., Sp. Pl. **1**: 191 (1753).—Wild Jasmine.

Shrubs 2–5 m. or trees to 10 m. high; branches often drooping; leaves oblong-elliptical, cuneate at base, acuminate, up to 15 cm. long and 6·5 cm. broad, petiole up to 2·5 cm. long, sometimes puberulous at first; flowers fragrant at night; corolla greenish-white to cream-coloured; fruit purplish-blue to black.

Common in thickets and woodland margins; SL–5100 feet; fl. mostly Sept–May, fr. Sept–June. A number of varieties have been named based mainly on differences of calyx size. Those reported for Jamaica are:

4a. Var. **portoricense** O. E. Schulz in Urb., Symb. Ant. **5**: 490 (1908).

Calyx less than 3 mm. long in flower; leaves thin, smaller and more rounded-based than in other varieties. Rare; SL–1800 feet; *P 23582!* Puerto Rico.

4b. C. **diurnum** var. **venenatum** (Mill.) O. E. Schulz in Urb., Symb. Ant. **6**: 263 (1909).—*C. venenatum* Mill. (1768).

Calyx 3–4·5 mm. long. Common in thickets; 20–2700 feet; *A 5424! H 12822! J.P. 1362! P 17452!* Bahamas, Cayman Is.

4c. C. **diurnum** var. **odontospermum** (Jacq.) O. E. Schulz in Urb., Symb. Ant. **6**: 264 (1909).—*C. odontospermum* Jacq. (1798).

Calyx 5–8 mm. long. Frequent in woodlands; 250–5100 feet; *A 6422! H 12343! P 23132!* Continental S. Amer.

7. DUNALIA Kunth (1818) nom. cons.

1. D. **arborescens** (L.) Sleumer in Lilloa **23**: 124 (1950).—*Atropa arborescens* L. (1756). *Acnistus arborescens* (L.) Schlecht. (1832).

Shrub 3 m. or tree up to 6 (–12) m. high, with spreading and straggling branches; young stems and leaves rusty-pubescent; leaves elliptical, long-cuneate, acute or acuminate, up to 30 cm. long and 14 cm. broad, glabrescent on the lamina; flowers in clusters of 30 or more, fragrant; pedicels 1–2 (–3·5) cm. long; calyx campanulate, subentire, splitting in fruit, about 3 mm. long; ripe fruit orange, smooth, globose, 8–11 mm. in diameter.

Locally common (St. Andr., Port., St. Thom.), rare in St. Ann, mostly in submontane secondary thickets and clearings; 1200–5000 feet; fl. sporadically throughout the year, fr. Mar–July. *A 11203! 11291! H 9556! P 8046!* Continental tropical Amer., West Indies.

8. PHYSALIS L. (1753); Waterfall (1967)

1 Fruiting calyx 10-ribbed or -angled or more or less terete:
  2 Flowering pedicels 5–10 mm. long; calyx at anthesis (3–) 4–7 mm. long; plant thinly
    pubescent to subglabrous; leaf-base broadly or narrowly cuneate   1. *angulata*
  2 Flowering pedicels 6–8 mm. long; calyx at anthesis 8–9 mm. long; plant tomentose;
    leaf-base openly cordate   2. *peruviana*
1 Fruiting calyx 5-ribbed:
  3 Fruiting calyx softly hairy, 1·5–3 cm. long; corolla yellow with at most 5 brown lines at
    base within; plant variably hairy, sometimes viscid-glandular   3. *pubescens*
  3 Fruiting calyx glabrous or nearly so, 2·5–4·5 cm. long; corolla light yellow with dark
    purple spots at base; plant subglabrous   4. *cordata*

1. P. **angulata** L., Sp. Pl. **1**: 183 (1753).—Wild Gouma, Winter Cherry.

Herb up to about 120 cm. high; leaves ovate to lanceolate, broadly to narrowly cuneate at base, margin incised or sinuate-toothed, 5–11 cm. long, 3·5–8 cm. broad; petioles 4–8 cm. long; corolla dull yellow and usually darker olive at base without-spots or spots indistinct, 6–12 mm. long, 7–12 mm. broad; anthers blue or violet, 2–2·5 mm. long; fruiting calyx 2–3·5 cm. long; berry 10–12 mm. in diameter.

Common as a weed of cultivations, alluvial gravel and pasture margins; SL–5000 feet; fl. and fr. most of the year. *A 8305! H 10780! P 9462! 10503!* General in the subtropics and tropics; Grand Cayman.

2. P. **peruviana** L., Sp. Pl. ed. 2, **2**: 1670 (1763).—Cape Gooseberry.[1]

Straggling-branched herb to 1 m. or shrubby up to 2 m. high; leaves ovate, more

---

[1] Name applied to several species of *Physalis*.

or less cordate at base, acuminate, margin sparingly toothed or entire, 5–10 cm. long, 4–8 cm. broad; petioles 1–4 cm. long; corolla yellow with blackish-purple spots near base within, 10–14 mm. long, 12–15 mm. broad; anthers bluish, 3·5–4 mm. long; fruiting calyx 3–4 cm. long; berry oblong, 12–20 mm. long, 10–15 mm. broad.

Uncommon and local (St. Andr., St. Thom.), on banks and in hillside thickets; 2850–5200 feet; fl. and fr. June–Dec. *A 5842! H & B 10546! P 9603!* Native tropical S. Amer., introduced and naturalized elsewhere in the New and Old Worlds.

3. P. pubescens L., Sp. Pl. **1**: 183 (1753).

Herb 10–80 cm. high; leaves ovate, truncate to broadly and sometimes unequally cuneate at base, acuminate, irregularly shallowly repand-dentate, 4–9 cm. long, 2–4 cm. broad; petioles 2–7 cm. long; pedicels 3–6 mm. long; calyx at anthesis 4–10 mm. long; corolla 7–10 (–12) mm. long, 10–15 mm. broad; anthers purple, 1·5–3 mm. long; berry 10–18 mm. in diameter.

Occasional, a weed of pastures and waste ground; 400–2000 feet; fl. Feb, Aug–Sept, fr. Feb–Mar, Aug–Sept. *Britton 1602! P 15691! Yuncker 18146!* General in the American tropics, introduced into West Africa.

4. P. cordata Mill., Gard. Dict. ed. 8 (1768).—*P. turbinata* Medic. (1780).

Like the last and perhaps not really distinct; slightly larger in most parts; leaves broadly cuneate to rounded-subcordate at base.

Frequent, a weed of cultivated ground and waste places; 100–600 feet; fl. and fr. sporadically throughout the year. *A 8280! 11263! H 10212! P 15689!* General in the American subtropics and tropics.

9. SOLANUM L. (1753)

1 Climbers or wide-spreading scramblers; flowers usually bluish or mauve, often pendulous:
  2 Calyx entire or nearly so:
    3 Leaves simple, glabrous or puberulous; flowers solitary to umbellate; calyx truncate; corolla rotate; ripe fruits yellow   **1.** *stellatum*
    3 Leaves pinnate with 4–5 pairs of glabrous lanceolate leaflets; flowers paniculate; calyx-lobes represented by minute mucros; corolla deeply lobed; ripe fruits red; plant climbing by petiole-tendrils; native of tropical Amer., now widespread in cultivation   *S. seaforthianum* Andr.
  2 Calyx distinctly lobed; corolla rotate; leaves glabrous, simple or compound; cultivated ornamentals:
    4 Leaves mostly simple and ovate-lanceolate, entire, up to about 5 cm. long and 2 cm. broad, or 3-foliolate; corolla about 2·5 cm. across, white tinged with blue; native of S. Amer.; Potato Vine   *S. jasminoides* Paxt.
    4 Leaves simple or pinnately compound, larger than above; plant sometimes prickly; corolla 2·5–5 cm. across, light purple; native of C. Amer.; Privy Vine   *S. wendlandii* Hook. f.
1 Erect herbs, shrubs or trees:
  5 Calyx entire; shrub with often zig-zag branches; leaves entire; corolla bluish mauve   **1.** *stellatum*
  5 Calyx lobed:
    6 Leaves compound; unarmed cultivated herb with underground stem-tubers; flowers white or mauve, 2–3 cm. in diameter; native of the Andes; Irish Potato   *S. tuberosum* L.
    6 Leaves simple, entire or variously dissected; plants without stem-tubers:
      7 Leaf-blade at maturity glabrous or if slightly hairy then the hairs simple and short, entire or only shallowly toothed; plants unarmed or prickles on stem very few and small:
        8 Corolla deeply divided, the lobes longer than the tube and more or less reflexed at maturity; inflorescences extra-axillary or often internodal; pedicels slender:
          9 Annual herb, usually with sinuate or shallowly toothed leaves; inflorescence stalked, umbelliform, the pedicels spreading in all directions; corolla-lobes white, 2 mm. long; ripe berry black, about 6 mm. in diameter   **2.** *americanum*
          9 Perennial undershrub or arborescent shrub with entire leaves:
            10 Inflorescence umbelliform; corolla-lobes light mauve or white, 5–7 mm. long; ripe berry black, about 6 mm. in diameter   **3.** *antillarum*
            10 Inflorescence racemose, ultimately many-flowered, the pedicels pendulous in 2 rows; corolla-lobes greenish-white, 5–6 mm. long; ripe berry yellow or brown, about 1 cm. in diameter   **4.** *parcebarbatum*

8 Corolla lobed to about halfway, subrotate; inflorescences racemose; pedicels more or less tapered:
  11 Pedicels winged, broader distally, up to about 2 cm. long in fruit; corolla white, about 1 cm. long; leaves elliptical   **5. *acropterum***
  11 Pedicels not winged; corolla bluish-mauve, larger than above:
    12 Leaves elliptical; pedicels up to about 1 cm. long; corolla 2–3 cm. across   **6. *havanense***
    12 Leaves obovate to oblanceolate; pedicels 1·5–4 cm. long; corolla 3·5–6·5 cm. across   **7. *troyanum***
7 Leaf-blade distinctly hairy or lepidote:
  13 Indumentum of stems, leaves, flowers and fruits lepidote; juvenile twigs prickly, of mature tree mostly not; leaves entire; corolla deeply divided, 4–5-merous; pedicels tapered, recurved in fruit   **8. *punctulatum***
  13 Indumentum of simple or stellate hairs:
    14 Inflorescence terminal, corymbose, with long erect dichotomously branched peduncles; unarmed shrubs with entire leaves having a pungent tarry odour when bruised; indumentum stellate; calyx-lobes deltate; corolla white or greenish, 5–7 mm. long; fruit sparsely stellate-hairy:
      15 Leaf-blade shortly cuneate to subtruncate at base, broadest below the middle; calyx and corolla densely beset with short-armed stellate hairs   **9. *erianthum***
      15 Leaf-blade long-tapered at base, broadest at or beyond the middle; calyx and corolla shaggy with long-armed stellate hairs   **10. *umbellatum***
    14 Inflorescences lateral or axillary, flowers sometimes solitary; plants often armed with prickles on stems and leaves; leaves entire or often angled or variously lobed and dissected:
      16 Hairs on leaves simple; weedy undershrubs with numerous straight or slightly curved prickles; leaves angular-lobed; flowers solitary or few in short inflorescences; fruit glabrous:
        17 Calyx not prickly; fruit yellow, conical or broadly ovoid with a conical tip, up to about 6 cm. long; corolla mauve or bluish-purple with lobes up to 2 cm. long   **11. *mammosum***
        17 Calyx prickly; fruit white turning orange or red, globose, 2–2·5 cm. in diameter; corolla white, about 1 cm. long   **12. *ciliatum***
      16 Hairs on leaves stellate:
        18 Flowers large and showy, 3 cm. or more in diameter; corolla mauve or purple at least at anthesis, not deeply lobed; cultivated plants:
          19 Inflorescence-branches racemose, elongated; corolla fading from blue-violet to nearly white, up to 7 cm. in diameter; leaves up to 40 cm. or more long, usually deeply cut, small prickly tree planted for ornament; native of Brazil; Potato Tree   ***S. macranthum*** Dunal
          19 Inflorescence few-flowered; corolla purplish, 3–6 cm. in diameter; leaves flaccid, lobed, up to about 20 cm. long; calyx large and baggy; short-lived shrub grown for its edible fruits; native of S. Asia; Aubergine, Egg Plant, Garden Egg   ***S. melongena*** L.
        18 Flowers smaller, not or little over 2 cm. in diameter:
          20 Leaves deeply pinnately lobed and dentate; calyx prickly, 6 mm. long; corolla lavender or white, about 2 cm. across; fruit 1–1·5 cm. in diameter; weedy annual herb   **13. *campechiense***
          20 Leaves entire or shallowly lobed; shrubs:
            21 Flowers 4-merous, solitary; calyx zygomorphic, the 2 lateral lobes much smaller than the anterio-posterior; corolla white, the lobes lanceolate, 8–9 mm. long; leaves entire; plant armed with acicular prickles   **14. *aquartia***
            21 Flowers 5-merous, not solitary; calyx regular:
              22 Inflorescence an elongating unilateral coiled raceme, rarely branched; leaves narrowly oblong-elliptical, entire; prickles acicular or lacking; corolla bluish-mauve to violet   **15. *bahamense***
              22 Inflorescence not as above, often internodal, with short unilateral racemose branches; leaves broadly ovate to rhomboid-ovate, entire or lobed; prickles broad-based; corolla white:
                23 Leaves subsessile, cuneate at base; prickles recurved; inflorescence subsessile; ripe berry orange-red, mostly 7–8 mm. in diameter   **16. *jamaicense***
                23 Leaves distinctly petiolate, retuse on one side at least at base; prickles few, more or less straight; inflorescence pedunculate with 2–3 short branches; ripe berry usually greenish-yellow, about 15 mm. in diameter   **17. *torvum***

1. S. stellatum Jacq., Collect. **3**: 254 (1790).—*Lycianthes stellata* (Jacq.) Bitter (1920).

Trailing or straggling much branched shrub to 6 m. high, occasionally erect 1–2 m. high; leaves ovate to elliptical, cuneate at base, acute or acuminate, up to 11 cm. long and 5 cm. broad; petioles 8–10 mm. long; pedicels slender, up to 3 cm. long; calyx cupular, 3 mm. long, accrescent; corolla 2–2·5 cm. across, often green-banded outside; ripe fruits about 1 cm. in diameter.

Frequent, in thickets and woodland margins on limestone; 500–3000 feet; fl. Jan–June, Sept–Oct, fr. Apr–Oct. *A 11005! H 12510! P 22264!* Endemic. Typical *S. stellatum* has glabrous leaves or not more than a few hairs on the midrib beneath. Occupying similar habitats in the central and western parishes, particularly in Clarendon and St. Ann, plants with puberulous leaves are not uncommon. These have been named var. *puberulum* O. E. Schulz and they approach *S. virgatum* Lam. of Hispaniola, which is distinguished by having pubescent leaves and more numerous flowers on pedicels rarely reaching 15 mm. long even in fruit.

2. S. americanum Mill., Gard. Dict. ed. 8 (1768).—Black Nightshade, Gouma.

Thinly pubescent spreading-branched herb up to 60 (–80) cm. high; stems usually smooth but sometimes with shallow soft prickles on the upper internode ridges; leaves ovate, the blade decurrent on the petiole, acute, often sinuate-margined or dentate, up to 14 cm. long and 7 cm. broad; peduncles up to 2 cm. long in fruit; anthers 1–1·5 mm. long.

Common weed of cultivations and disturbed ground; 50–3000 feet; fl. and fr. all the year. *A 6145! H 8538! P 24759!* Throughout the American subtropics and tropics. Taxonomic distinctions in the group of *S. nigrum* L. and *S. nodiflorum* Jacq. are obscure and authors vary widely in the application of names to the tropical representatives of this group. Jamaican plants have been assigned to *S. americanum* var. *nodiflorum* (Jacq.) Edmonds (1971).

3. S. antillarum O. E. Schulz in Urb., Symb. Ant. **6**: 164 (1909).

Undershrub resembling the last in most characters, up to 1·5 m. high; leaves lanceolate, cuneate at base, gradually long-acuminate, up to 15 cm. long and 4 cm. broad; peduncles up to 4·5 cm. long in fruit; pedicels 11–13 mm. long; anthers 3·5 mm. long.

Occasional (St. Andr., Port., St. Thom.), along sheltered rocky pathsides and on banks; 3600–7250 feet; fl. and fr. sporadically throughout the year. *A 11219! R. D. Henry! Herbert UCWI 26337!* Greater Antilles.

4. S. parcebarbatum Bitter in Fedde, Repert. Sp. Nov. **18**: 51 (1922).

Shrub 1·2–2 m. or small tree to 3 m. high; leaves pungent odorous, elliptic-lanceolate or elliptical, cuneate at base, acute or acuminate, thin, up to 10 (–20) cm. long and 3 (–4·5) cm. broad, often accompanied by a smaller subopposite leaf on distal branches; calyx 2 mm. long, the lobes 1 mm. long and abruptly acuminate.

Rare and local (St. Ann), in secondary thicket; about 1800 feet; fl. (in cult.) all the year, fr. Nov–May. *A 12743! 12881! 13005!* Guatemala, Costa Rica, Panama.

5. S. acropterum Griseb., Fl. Br. W.I.: 437 (1862).

Shrub 1·5–4 m. or tree to 10 m. high; leaves cuneate at base, shortly acuminate, up to 13 cm. long and 6 cm. broad; flowers fragrant, mostly pendulous; calyx-lobes linear-oblong beyond a triangular base, 4 mm. long; fruit ellipsoidal about 12 mm. long.

Occasional, in woodland on rocky limestone; 1200–3000 feet; fl. Jan–Apr, fr. Mar, Aug. *H 10298! H & P 15024! Stearn 499!* Endemic.

6. S. havanense Jacq., Enum. Syst. Pl. Carib.: 15 (1760).

Shrub with straggling branches 1–3 m. high; young shoots and petioles puberulous; leaves cuneate at base, obtuse, acute or very shortly acuminate at tip, shiny adaxially, 2·5–11 cm. long, 1–4·5 cm. broad; flowers fragrant; calyx unequally divided the lobes oblong-elliptical; anthers broadly elliptical, 4·5 mm. long; fruit ovoid, 12–23 mm. long, dark purple.

Rather common, in thickets and woodlands on limestone mostly in dry areas;

SL–750 (–2250) feet; fl. and fr. most of the year. *A 6333! H 10036! H & P 13765!* Cuba.

7. S. troyanum Urb., Symb. Ant. **5**: 487 (1908).
Arborescent shrub 1·2–2·5 m. high; leaves long-cuneate at base, obtuse or shortly acuminate at tip, up to 15 cm. long and 6 cm. broad; flowers 5 (–6)-merous, fragrant; calyx-lobes broadly ovate, 4–5 mm. long; anthers elliptical, 5·5 mm. long; fruit 10–13 mm. in diameter.
Occasional in the central parishes, in woodland on craggy limestone rocks and on cliffs; 1500–3000 feet; fl. Mar–Sept, fr. June–Oct. *A 11230! H 9000! P 21339! Stearn 922!* Endemic.

8. S. punctulatum Dunal in DC., Prodr. **13** (1): 122 (1852).
Tree 5–18 m. high; leaves oblong-elliptical, unequally rounded at base, acute, obtuse or shortly acuminate at tip, 5–18 cm. long, 2–8·5 cm. broad; flowers few, fragrant in leaf-opposed dichotomous cymes; pedicels very stiff, up to 2·5 (–3·5) cm. long; calyx cupular, 2·5 mm. long, with very short rounded lobes; corolla-lobes light mauve, about 12 mm. long; anthers 5 mm. long; fruit ripening light orange, 15–16 mm. in diameter.
Locally common (St. Andr., Port., St. Thom.), in montane woodland, rare (Clar., St. Ann) in woodland on limestone; (2000–) 4400–6500 feet; fl. and fr. sporadically throughout the year. *A 10616! J.P. 2115! P 23787!* Endemic.

9. S. erianthum D. Don, Prodr. Fl. Nepal.: 96 (1825).—Wild Susumber.
Weak-stemmed arborescent shrub 2–5 m. high, covered in all overground parts with a pale indumentum of minute stellate hairs; leaves ovate, long-petioled, acuminate, up to 30 cm. long and 15 cm. broad; peduncle up to about 10 cm. long; corolla mostly white, green in centre or green-veined, about 1·5 cm. across; ripe fruit about 1 cm. in diameter.
Frequent, in thickets and steep banks on limestone; SL–700 feet; fl. and fr. all the year. *A 5425! H 6998! P 23222!* Native of tropical Asia, now widespread in the American subtropics and tropics.

10. S. umbellatum Mill., Gard. Dict. ed. 8 (1768).
Arborescent shrub like the last, 1–3 m. high; indumentum of larger more sparsely spread hairs; leaves oblong-elliptical, acuminate at both ends, much paler beneath, up to 18 cm. long and 5 cm. broad; peduncle 7–8 cm. long; flowers somewhat nodding; corolla white; ripe fruit 11–12 mm. in diameter.
Local (Clar.), in thickets on limestone; about 2300 feet; fl. June–Aug, fr. Aug. *A 12598! P 19711!* Cuba, Hispaniola.

11. S. mammosum L., Sp. Pl. **1**: 187 (1753).—Bachelor's Pear.
Short-lived diffusely branched undershrub up to 1·5 m. high, rather densely tomentose; leaves broadly ovate, cordate, the marginal teeth and tip acute, up to about 11 cm. long and broad, sometimes up to 20 cm. long; petiole 2–10 cm. long; calyx about 5 mm. long.
Occasional as a weed of pastures and roadsides; 400–1700 feet; fl. Aug–Jan, fr. Aug–Apr. *A 10371! H 6356! P 10625!* Subtropical and tropical Amer.

12. S. ciliatum Lam. in Tabl. Encycl & Méth., Bot. **2**: 21 (1794).—*S. aculeatissimum* of authors, not Jacq. Cockroach Poison.
Diffusely branched shrubby herb up to about 1 m. high; prickles on stem mostly slightly deflexed; leaves broadly ovate, truncate-subcordate at base, coarsely dentate or lobed with 3–4 segments on each margin, up to 15 cm. long and 13 cm. broad, thinly hairy at maturity, the margin ciliate; petiole 1–7 cm. long; calyx 3–4 mm. long; seeds thin, light yellow, about 4 mm. in diameter with a broad hyaline margin.
Common in pastures and open ruinate on limestone; SL–3000 feet; fl. and fr. Nov–July. *A 6783! 10339! H 11090! P 23625!* Tropical Amer., E. Asia, W. Africa. *S. aculeatissimum* Jacq. is an Old World plant with brown fruits and smaller seeds.

13. S. campechiense L., Sp. Pl. **1**: 187 (1753).—*S. guanicense* Urb. (1899).
Prickly herb about 50 cm. high; leaves broadly ovate in outline, cordate at base,

tips of lobes acute or obtuse, pilose with long-stalked stellate hairs; cymes 1–2-flowered; pedicels 5–10 mm. long, recurved in fruit.

Very rare (St. Cath.), in open field; about 700 feet; fl. and fr. Jan. *H & P 15187!* Costa Rica, Greater Antilles.

14. S. aquartia Dunal, Hist. Sol.: 187 (1813).—*S. aculeatum* (Jacq.) O. E. Schultz (1909), not St.-Lag. (1880).

Diffusely slender-branched prickly shrub up to 2 m. high; leaves oblong to oblong-elliptical or sometimes ovate, unequally subcordate at base, rounded to obtuse at tip, densely stellate-tomentose on both surfaces, 1–5 cm. long, 1–3·5 cm. broad; larger calyx-lobes 2–4 mm. long, rounded; petals 2–2·5 mm. broad at base; filaments slender, 2·5 mm. long; anthers 9 mm. long; style declinate, 16–17 mm. long; ripe fruit globose, light orange-red, 3·5–6 mm. in diameter.

Rare and local (St. Cath.), in thicket on exposed arid limestone hillside; about 700 feet; fl. and fr. all the year in cultivation, during rainy weather in the natural environment. *A 12798! 12870!* Cuba, Hispaniola.

15. S. bahamense L., Sp. Pl. 1: 188 (1753).—Canker Berry.

Shrub often with drooping branches, up to 2·5 m. high, variably prickly and sometimes lacking prickles except on the lower part of the stem or altogether; leaves lanceolate, usually acute and scabrid, up to 14 cm. long and 5 cm. broad; racemes up to 15 cm. long; corolla-lobes linear, 7–12 mm. long; ripe fruits orange-red to scarlet, 6–7 mm. in diameter.

Common on arid limestone and sandy waste places near the sea; SL–1600 feet; fl. and fr. all the year. *A 6315! H 7226! P 6106!* Florida, Bahamas, Cuba, Hispaniola, Grand Cayman, but not obviously distinct from *S. racemosum* Jacq. and *S. persicifolium* Dunal of Puerto Rico and the Lesser Antilles.

16. S. jamaicense Mill., Gard. Dict. ed. 8 (1768).

Straggling often low-branched undershrub 0·6–2 m. high; leaves sinuate-margined or shallowly lobed, acute or obtuse at tip, densely stellate-pubescent, up to 12 (–25) cm. long and nearly as broad; petiole very short; pedicels 4–12 mm. long; calyx-lobes linear, about 4 mm. long; corolla-lobes acute, about 5 mm. long; anthers oblong, about 5 mm. long.

Locally common in rough pastures and damp waste places; SL–2500 feet; fl. and fr. all the year. *A 7322! 10340! Hart 1461! P 16245!* Throughout tropical Amer.

17. S. torvum Sw., Nov. Gen. & Sp. Pl.: 47 (1788).—*S. ficifolium* Ort. (1800). Gully Bean, Susumber, Turkey Berry.

Shrub 1–4 m. high; leaves broadly ovate, truncate-subcordate at base, margin entire or lobed, the lobes and tip acute or acuminate, stellate-pubescent, up to 18 (–25) cm. long and 15 (–18) cm. broad; petioles 1·5–5 cm. long; pedicels 5–8 mm. long; calyx about 4 mm. long, the lobes ovate, acute; anthers linear, 6·5 mm. long.

Common in woodland clearings, thickets and waste places; 50–4200 feet; fl. and fr. all the year. *A 10542! H 8773! P 8141!* General in the tropics.

10. LYCOPERSICON Mill. (1754)

1. L. esculentum Mill., Gard. Dict. ed. 8 (1768).—*Solanum lycopersicum* L. (1753). Tomato.

Annual coarse straggling herb up to 1 m. or more high; plant viscid-pubescent and odorous; leaves petiolate; leaflets small and large, ovate or ovate-lanceolate, serrate-dentate, usually acute; fruit subspherical, slightly flattened, orange-red or red when ripe.

Commonly cultivated in many varieties. A naturalized variant has been identified as:

1a. Ssp. galeni (Mill.) Luckwill in Aberdeen Univ. Stud. 121: 23 (1943).—*L. galeni* Mill. (1768). Wild Tomato.

Leaflets in about 3 pairs of the larger interspersed with smaller, the terminal up to 5·5 cm. long and 3 cm. broad; inflorescences up to about 8 cm. long, 3–10-flowered; calyx-lobes lanceolate, about 12 mm. long; corolla about 16 mm. across; fruit 11–18 mm. in diameter.

Occasional (St. Andr., St. Mary), on rough stone walls and in sheltered waste places; 700–3500 feet; fl. and fr. Mar–Aug. *A 11204! H 11896! P 23682!* Native of western S. Amer., introduced into other parts of the American subtropics and tropics and into W. Africa; cultivated and naturalized.

11. SARACHA Ruiz & Pav. (1794)

1. S. antillana Krug & Urb. in Notizbl. Bot. Gart. Berlin 1 (2): 80 (1895).

Annual rather robust pilose herb up to about 1 m. high; leaves ovate, long-cuneate at base, shallowly repand-dentate with few teeth or subentire, acute at tip, up to 15 cm. long and 5 cm. broad; peduncle up to 13 mm. long; pedicels 7–12 mm. long; calyx-lobes broadly ovate, the whole calyx accrescent to about 14 mm. broad in fruit; corolla white, broadly campanulate, folded between the bases of the lobes; ripe fruit red, 7–8 mm. in diameter.

Local and not common (St. Andr., Port., St. Thom.), on shale banks and in damp shady waste places; 3500–5000 feet; fl. and fr. Nov–June. *A 10146! H 5109! P 6825!* Greater Antilles.

12. CAPSICUM L. (1753)

1 Flowers numerous in each umbel, on slender pedicels, 4-merous   1. *macrophyllum*
1 Flowers solitary or 2–3 together, 5-merous:
   2 Plant annual or short-lived; fruits often wrinkled, of various shapes, sizes and colours and also varying in degree of acridity; numerous cultivated varieties known as Paprika, Piment, Red or Sweet Peppers, etc.; not known in the wild state but obtained originally from tropical American plants   *C. annuum* L.
   2 Perennial shrubs:
     3 Fruit conical-ellipsoid, 12–20 mm. long or longer in cultivars   2. *frutescens*
     3 Fruit globose-ellipsoid, about 8 mm. long   3. *baccatum*

1. C. macrophyllum (Humb. & Bonpl.) Standl. in Journ. Acad. Sci. Washington 17: 16 (1927).—*Witheringia macrophylla* Humb. & Bonpl. (1816).[1]

Weak-stemmed undershrub with thinly puberulous shoots, 1·2–2·5 m. high; leaves ovate-elliptical, bluntly rounded and unequal at base, subentire, acuminate, up to 21 cm. long and 11 cm. broad, often accompanied by subopposite much smaller leaves; calyx cupular, about 1 mm. long; corolla yellowish-green to light yellow, the tube about 2 mm. long, the lobes recurved, 5–7 mm. long; anthers white; fruit green turning orange then light red, subglobose, 5–6 mm. in diameter.

Rare and local (Port., St. Thom.), in secondary thickets and along pathsides in damp shady places; 1000–2000 feet; fl. most of the year, fr. Jan–Mar, July. *A 9087! 12958! H & B 10779! P 9818! 19724!* Venezuela.

2. C. frutescens L., Sp. Pl. 1: 189 (1753).—Bird Pepper, Cayenne Pepper[2], Chilli.[2]

Shrub 1–2 m. high with erect or spreading branches, the branchlets often shallowly zig-zag; leaves ovate, shortly cuneate, acuminate, up to 9 (–15) cm. long and 4·5 (–7) cm. broad, the midrib patchily hairy otherwise subglabrous; pedicels up to 3 cm. long; calyx 2·5–3 mm. long; corolla light green, about 3·5 mm. long; fruit yellow, bright orange or red, very acrid.

Occasional in thickets and waste places; 30–700 feet; fl. and fr. sporadically. *A 9832! H 11864! Stearn 769!* Probably native of tropical Amer., now widespread in warm countries.

3. C. baccatum L., Mant. Pl.: 47 (1767).—Bird Pepper.

Shrub like the last with often markedly divaricate branching and sometimes climbing to 6 m. high; leaves up to 7·5 cm. long and 3·5 cm. broad, with domatial hairs.

---

[1] A recent monograph (Hunziker, 1969), which has not been studied, attributes Jamaican plants to *W. solanacea* Kunth.

[2] The taxonomy of cultivated peppers is extremely difficult; some authors regard these as varieties of *C. annuum*.

Common in rough pastures, thickets and woodland margins; SL–2300 feet; fl. and fr. most of the year. *A 6167! H 9327! H & P 14049!* Probably widespread at least in the American tropics. This species has so often been combined with the last that it is quite impossible to work out accurately the distribution of either of them from published data.

171. SCROPHULARIACEAE

Herbs, shrubs or climbers, rarely trees. Leaves alternate, opposite or whorled, simple, entire, toothed or deeply divided, without stipules. Flowers bisexual, zygomorphic or subactinomorphic, solitary or in various types of usually bracteate inflorescences. Perianth biseriate; calyx 4–5-toothed or deeply divided, the segments imbricate or valvate, persistent; corolla sympetalous, rotate, campanulate or cylindrical, often 2-lipped, lobes 4–5 (–8), imbricate. Stamens epipetalous, mostly 4 and didynamous, sometimes 2 or 5 or the fifth represented by a staminode; anthers usually 2-locular and opening introrsely and lengthwise. Disk usually present. Ovary superior, typically 2-locular; ovules numerous, anatropous, on large axile placentas; style solitary, often persistent; stigma 2-lobed. Fruit usually a capsule opening septicidally, loculicidally or by pores, rarely a berry. Seeds smooth, angled or winged, with straight or slightly curved embryo and fleshy endosperm.

About 200 genera with some 3000 species; cosmopolitan.

1 Fertile stamens 5; flowers in dense terminal and subterminal racemes; corolla rotate, with 5 nearly equal lobes and very short tube, yellow; large herbs with alternate leaves
　　　　　　　　　　　　　　　　　　　　　　　　　　　　1. Verbascum
1 Fertile stamens 2 or 4, very rarely 5:
　2 Perfect stamens 2:
　　3 Posterior lip of corolla much smaller than anterior or obsolete; leaves opposite:
　　　4 Erect herbs with pinnately divided leaves; anterior lip of corolla entire, inflated, yellow, posterior lip small　　　　　　　　　　　　**2. Calceolaria**
　　　4 Stems creeping, forming mats; leaves very small, entire; anterior lip of corolla 3-lobed, white, posterior lip obsolete　　　　　　**19. Hemianthus**
　　3 Posterior lip of corolla well developed or corolla subregular:
　　　5 Corolla rotate, 4-cleft, the tube very short; staminodes wanting　　23. **Veronica**
　　　5 Corolla tubular or campanulate, 5-cleft with upper lip only very shortly divided:
　　　　6 Staminodes wanting; sepals very unequal, the outer broadly ovate
　　　　　　　　　　　　　　　　　　　　　　　17.4. *Bacopa innominata*
　　　　6 Staminodes porrect; sepals subequal, linear-lanceolate
　　　　　　　　　　　　　　　　　　　　　　　18.3. *Lindernia rotundifolia*
　2 Perfect stamens 4 (–5):
　　7 Plants twining or trailing and more or less scandent; leaves mostly alternate:
　　　8 Lower lip of corolla with a prominent palate closing the mouth; sepals glabrous:
　　　　9 Corolla with curved spur; capsule with both loculi dehiscent; sepals oblong-lanceolate　　　　　　　　　　　　　　　　　　**5. Cymbalaria**
　　　　9 Corolla without a spur; adaxial loculus of capsule indehiscent; sepals linear-lanceolate, striate　　　　　　　　　　　　　**7. Maurandella**
　　　8 Lower lip of corolla without a palate, the corolla-mouth open:
　　　　10 Connective with a retrorse appendage; sepals broadly ovate to ovate-lanceolate, pubescent; seeds broadly winged　　　　　　　　**8. Lophospermum**
　　　　10 Connective not appendaged; sepals linear-lanceolate, with stalked glands; seeds not winged　　　　　　　　　　　　　　　　**9. Maurandya**
　　7 Plants erect, lax or creeping but not scandent; herbs, undershrubs or shrubs:
　　　11 Leaves all alternate:
　　　　12 Flowers solitary, paired or up to 4 in leaf-axils; corolla subregular, campanulate, white; stamens 4 or 5　　　　　　　　　　　**21. Capraria**
　　　　12 Flowers in dense terminal racemes; corolla irregular; stamens 4:
　　　　　13 Corolla with spur and orange palate, otherwise yellow; leaves narrow　**4. Linaria**
　　　　　13 Corolla campanulate without spur of palate, purpl'sh or white, usually spotted; leaves broad　　　　　　　　　　　　　　**22. Digitalis**
　　　11 Leaves mostly opposite, whorled or obsolete, but the upper often alternate:
　　　　14 Anterior part of corolla-tube saccate; corolla 2-lipped-subrotate, blue or purplish, showy　　　　　　　　　　　　　　　　**3. Angelonia**
　　　　14 Anterior part of corolla-tube not saccate:

15 Mouth of corolla closed by a raised palate; base of corolla not spurred:
16 Calyx-lobes ovate, shorter than capsule in fruit; showy annual cultivated ornamental; native of Mediterranean region; Snapdragon

Antirrhinum majus L.
16 Calyx-lobes linear, posterior at least much longer than capsule in fruit

6. **Misopates**
15 Mouth of corolla not closed by a raised palate:
17 Posterior lobe(s) of corolla internal in bud; plants mostly semiparasites with short roots and often drying blackish:
18 Corolla salver-shaped, purple or mauve; flowers in spikes with bracts shorter than calyx   24. **Buchnera**
18 Corolla campanulate or rounded:
19 Flowers in spikes with large leafy bracts; corolla yellow; leaves lanceolate, coarsely dentate   25. **Alectra**
19 Flowers in lax racemes; corolla white or tinged rose or mauve; leaves linear:
20 Anther-sacs of same stamen equal; pedicels up to 3 mm. long, without bracteoles; stem glabrous; corolla usually white   26. **Agalinis**
20 Anther-sacs of same stamen unequal; pedicels 25–45 mm. long, with 1 or 2 bracteoles about the middle; stem scabrid-hispidulous; corolla pinkish

27. **Anisantherina**
17 Posterior lobe(s) of corolla external in bud; plants not parasitic:
21 Stigma capitate; inflorescence usually compound; fertile stamens 4; staminode present:
22 Corolla red, tube cylindrical, slightly saccate at base, 11–25 mm. long; leaves distinctly glandular-punctate beneath, ovate or obsolete   10. **Russelia**
22 Corolla green and white, obliquely campanulate, about 3 mm. long; leaves with minute glands beneath   11. **Scrophularia**
21 Stigma 2-lipped; inflorescence racemose, spicate or flowers solitary in leaf-axils; leaves often distinctly gland-dotted:
23 Calyx divided to about halfway; flowers subsecund in a terminal raceme; basal leaves long-stalked, subrosulate   12. **Mazus**
23 Calyx usually divided beyond the middle, if otherwise then flowers not in racemes:
24 Corolla rotate, the 4 lobes spreading and longer than tube; erect branched herb   20. **Scoparia**
24 Corolla evidently tubular, the lobes shorter than tube:
25 Sepals free:
26 Sepals equal, linear-lanceolate; pedicels without bracteoles; branches 4-angled and narrowly winged, creeping and rooting freely; petioles very short; seeds reticulate   15. **Cheilophyllum**
26 Sepals unequal either in length or breadth:
27 Bracteoles 2 at base of pedicel; corolla yellow   16. **Mecardonia**
27 Bracteoles 2 under calyx or wanting; corolla white, pink or mauve

17. **Bacopa**
25 Sepals fused at base, mostly equal or subequal:
28 Bracteoles 2 under calyx; glandular-aromatic herbs or under-shrubs with sessile subamplexicaul elongated leaves; pedicels very short, anther-sacs stipitate   13. **Stemodia**
28 Bracteoles wanting; low herbs with spreading or creeping and rooting branches, not aromatic; leaves mostly petiolate and rather small:
29 Anther-sacs stipitate; branches terete, spreading diffusely, not rooting, glandular-pubescent; seeds longitudinally furrowed, otherwise smooth   14. **Lendneria**
29 Anther-sacs at most divergent; branches often creeping and rooting, mostly angled and sometimes winged, not glandular-pubescent; seeds transversely lined or verrucose   18. **Lindernia**

1. VERBASCUM L. (1753)

Plant densely covered with woolly stellate indumentum, not glandular; raceme compact the pedicels very short: staminal hairs pale yellowish   1. *thapsus*
Plant thinly clothed with simple or sparingly branched hairs, the upper all gland-tipped; raceme open, the pedicels up to 5 mm. long in fruit; staminal hairs purple   2. *virgatum*

1. V. thapsus L., Sp. Pl. **1**: 177 (1753).

Erect herb 1–2 m. high; leaves elliptical to obovate-lanceolate, the lower long-stalked, the upper winged-decurrent, acute, crenulate, up to 30 cm. or more long; corolla 15–30 mm. across; capsule ovoid, septicidal, 7–8 mm. long.

Cultivated and rarely naturalized (St. Andr.), from hill gardens; 4800–5000 feet; fl. Jan, May, fr. May. *Larter 247! P 25687!* Native of Europe and W. Asia, naturalized in N. Amer., Bermuda.

2. V. virgatum Stokes in With., Bot. Arr. Brit. Pl. ed. 2, 1: 227 (1787).

Erect taprooted herb up to 1·5 m. high; basal leaves oblanceolate, long-stalked, coarsely crenate, up to 25 cm. long and 7 cm. broad, the upper linear-lanceolate to ovate, shorter upwards, sessile, auriculate; corolla up to 30 mm. across; capsule globose, septicidal, 8–9 mm. in diameter.

Sparingly naturalized (St. Andr.), along pathsides and on cleared slopes; 3000–5000 feet; fl. and fr. Mar, July. *H 12118! Robertson UCWI 26502!* Native of S.W. Europe, N. Africa, Azores, naturalized in S.W. United States and Bermuda.

2. CALCEOLARIA L. (1770) nom. cons.

Calyx-lobes broadly ovate, acute or shortly acuminate, about 6 mm. long and 5 mm. broad; corolla about 1 cm. broad   1. *chelidonioides*
Calyx-lobes triangular-ovate, broadly acute, about 4·5 mm. long and broad; corolla about 6 mm. broad   2. *pinnata*

1. C. chelidonioides Kunth, Nov. Gen. 2: 378 (1818).

Annual erect or lax herb 6–60 cm. high or scrambling to 3 m.; stem reddish, often rooting at base; leaves simple and toothed to bipinnatifid, the lobes ovate to lanceolate, thinly hairy, the margins glandular-ciliate, up to about 10 cm. long and 6 cm. broad; inflorescence with gland-tipped hairs; pedicels up to 25 mm. long in fruit; capsule conical-ovoid, 7–8 mm. long, the valve-tips rounded; seeds brown, lightly grooved and transversely rugose.

Locally common (St. Andr., Port., St. Thom.), on pathside and roadside banks, mostly in damp shady places; 1850–5600 feet; fl. and fr. Dec–Sept. *A 5986! H 9136! 10069! P 23349!* Native Mexico to Ecuador and Venezuela.

2. C. pinnata L. in Vet. Acad. Handl. Stockh. **31**: 286 (1770).

Weak-stemmed herb up to about 70 cm. high; leaves bipinnatisect; like the last only smaller in all parts.

Rare (St. Andr., Port.), on moist gravelly roadside bank; 3800–4000 feet; fl. and fr. Mar–Apr. *J.P. 952! Larter 209! P 22248!* Native of Peru.

3. ANGELONIA Humb. & Bonpl. (1812)

1. A. angustifolia Benth. in DC., Prodr. **10**: 254 (1846).

Erect laxly branched thinly hairy perennial herb up to about 60 cm. high; leaves opposite or in threes, linear-lanceolate, hardly clasping at base, serrulate or subentire up to about 6 cm. long and 8 mm. broad, sparsely ciliate towards base with some of the hairs gland-tipped; pedicels slender, mostly a little over 1 cm. long; sepals ovate, acute, acuminate or cuspidate, about 3 mm. long; corolla about 1·5 cm. long; capsule globose, about 6 mm. in diameter.

Cultivated and naturalized locally in open damp places; 25–900 feet; fl. and fr. sporadically throughout the year. *Fawcett 8401! H 11665! Proctor & Mullings 21840!* Generally distributed in tropical Amer. *A. pilosella* Kickx (*A. cubensis* Robins.) was reported for Jamaica by Alain, Fl. Cub. **4**: 408 (1957). Older exsiccata from Jamaica had been named *A. cubensis* but were re-named *A. angustifolia* by Pennell; the distinction of these two species, maintained by Alain, is not clear.

4. LINARIA Mill. (1754)

1. L. vulgaris Mill., Gard. Dict. ed. 8 (1768).

Glaucous perennial herb up to 80 cm. high, usually much branched; leaves lanceolate to linear-lanceolate, up to 8 cm. long; pedicels longer than calyx; sepals ovate or lanceolate; corolla 15–25 mm. long; capsule ovoid, more than twice as long as calyx.

Very rare; only once reported (Manch.), *Wullschlaegel*. Native of Europe and W. Asia, naturalized in N. Amer.

5. CYMBALARIA Hill (1756)

1. C. muralis Gaertn., Meyer & Scherb., Fl. Wett. **2**: 397 (1800).
Glabrous perennial herb with slender trailing stems up to 80 cm. long; leaves reniform, openly cordate, shortly 3–7-lobed, up to about 2·5 cm. long; pedicels about 2 cm. long in flower, greatly elongating in fruit; sepals about 2 mm. long; corolla 8–10 mm. long, including spur; capsule globose, 4 mm. long; seeds with thick flexuous ridges.
Introduced and naturalized locally (Manch.), on stone walls; 2000–2300 feet; fl. and fr. Oct–Nov. *P 22924! Robertson UCWI 5675!* Native of S. Europe, naturalized elsewhere.

6. MISOPATES Raf. (1840)

1. M. orontium (L.) Raf., Aut. Bot.: 158 (1840).—*Antirrhinum orontium* L. (1753).
Erect annual weedy herb up to 60 cm. high; stem glabrous; leaves linear to linear-oblanceolate, sparsely hairy, entire, up to 5 cm. long and 1 cm. broad, the upper much smaller; raceme open, with leafy bracts; pedicels 1–5 mm. long; calyx and capsule with long hairs; corolla about 10 mm. long, white or mauve with purple lines and lip yellow laterally, glandular in throat; capsule obliquely ovoid, 5–7 mm. long.
Rather local (St. Andr.), on open sandy or gravelly banks; 2500–5000 feet; fl. and fr. Nov–May. *A 6437! J.P. 1168! P 23558!* Native Europe, W. Asia, N. Africa and Canary Is.; Hispaniola.

7. MAURANDELLA (A. Gray) Rothm. (1943)

1. M. antirrhiniflora (Humb. & Bonpl. ex Willd.) Rothm. in Fedde, Repert. Sp. Nov. **52**: 27 (1943).—*Maurandya antirrhiniflora* Humb. & Bonpl. ex Willd. (1806).
Slender twiner with glabrous stems to 3 m. or more long; leaves petiolate, triangular-hastate, cordate, the basal lobes and tip acute, otherwise entire, 1–5 cm. long; flowers solitary; peduncles up to about 3 cm. long, longer than the petioles; calyx-lobes 7–12 mm. long; corolla about 1·5 cm. long, violet-blue with white throat and tube; capsule depressed-globose, about 1 cm. in diameter.
Uncommon (Manch., St. Eliz., Trel.), naturalized on walls, barbecues and limestone ledges; 1000–2300 feet; fl. and fr. Feb–May, Sept. *V. J. Foote! H 9660! P 26880!* Native of southern United States and Mexico, naturalized elsewhere in Bermuda, Bahamas, Cuba, St. Thomas and Barbados.

8. LOPHOSPERMUM D. Don (1827)

1. L. erubescens D. Don in Sweet, Brit. Flow. Gard. ser. 2, **1**: under t. 75 (1830).
—*Maurandya erubescens* (D. Don) A. Gray (1868).
Twining shrub with long slender branches; leaves broadly triangular-ovate, cordate, acute, coarsely dentate, pubescent, up to about 8 cm. long and 9 cm. broad; petioles tendriliform; peduncles solitary, up to 4 cm. long; calyx 2–3 cm. long; corolla-tube 4·5 cm. long, pink outside, white with deep pink blotches inside, lobes rounded; capsule subglobose, hispid, about 1·5 cm. in diameter.
Local (St. Andr., Manch., St. Thom.), naturalized on walls, rocky banks and trees; 1500–5700 feet; fl. Nov–Aug, fr. Dec–Aug. *A 5764! H 6920! Stearn 407!* Native of Mexico, introduced and cultivated in Bermuda, Puerto Rico, Hawaii and elsewhere.

9. MAURANDYA Ort. (1797)

1. M. barclaiana Lindl. in Edw., Bot. Regist. **13**: t. 1108 (1827).
Slender twiner with glabrous stems and leaves; leaves hastate-triangular, cordate or truncate at base, sometimes sparingly lobulate, acute, up to about 4·5 cm. long

and broad; peduncles up to 3 cm. long; calyx about 1 cm. long; corolla about 4 cm. long, light to bright rose-purple; capsule globose, glabrous, about 8 mm. in diameter.

Uncommon (St. Andr., Manch., St. Ann), naturalized on walls and rocky banks; 2000–3500 feet; fl. and fr. Dec–June. *Harris! P 26879! Stearn 598!* Native of Mexico, introduced elsewhere.

10. RUSSELIA Jacq. (1760)

Leaves often poorly developed; stem many-ribbed; sepals broadly ovate, cuspidate, glabrous; corolla up to about 25 mm. long, glabrous within   1. *equisetiformis*
Leaves opposite or in threes, joined across the base by a line of hairs; stem 4–6-ribbed; sepals ovate-lanceolate, acuminate, thinly pubescent; corolla 11–14 mm. long, pubescent within with yellow hairs; cultivated ornamental, native of C. and S. Amer.
                                                        R. sarmentosa Jacq.

1. **R. equisetiformis** Cham. & Schlecht. in Linnaea **6**: 377 (1831).
Lax much branched shrub with shoots up to 1·5 m. long, sometimes arching and rooting; leaves up to seven in a whorl or obsolete, shortly petiolate, ovate, serrate-dentate, up to about 2 cm. long; pedicels about 1 cm. long; sepals 2 mm. long; capsule broadly ovoid, about 5 mm. long, beaked.

Common in cultivation and in some areas as an escape on roadside banks; 300–2600 feet; fl. most of the year, fr. Jan–Feb, May. *H 9884! Stearn 264! 1021!* Native of Mexico now widely planted as an ornamental in the tropics; Grand Cayman.

11. SCROPHULARIA L. (1753)

1. **S. minutiflora** Pennell in Proc. Acad. Nat. Sci. Philad. **75**: 18 (1923).—*S. micrantha* Desv. (1825), not Urville (1822).
Weak-stemmed erect or decumbent herb up to 30 (–50) cm. high, almost glabrous but with glandular hairs in the inflorescence; leaves petiolate, ovate, broadly cuneate at base, coarsely dentate, up to about 4 (–8) cm. long and 2 cm. broad; pedicels 3–10 mm. long, filiform; sepals linear-lanceolate, about 2·5 mm. long; capsule ovoid, acute, about 4 mm. long.

Rare and local (St. Thom.), on pathside rocky banks; 3500–4300 feet; fl. and fr. Jan–Feb. *A 10487! P 24536!* Greater Antilles.

12. MAZUS Lour. (1790)

1. **M. pumilus** (Burm. f.) Steenis in Nova Guinea n. s. **9**: 31 (1958).—*Lobelia pumila* Burm. f. (1768). *M. japonicus* (Thunb.) Kuntze (1891).
Erect thinly pubescent annual herb 5–10 cm. high; basal leaves obovate, rounded at tip, subentire, up to 3 cm. long and 1 cm. broad; upper leaves smaller, cuneate at base; pedicels ascending, 5–6 mm. long and calyx 6–7 mm. long in fruit; corolla about 7 mm. long, tube violet; lip paler with 2 yellow crests and white hairs, 3-lobed; capsule subglobose, a little longer than calyx-tube.

Very local (St. Andr.), a weed of lawns; 600–700 feet; fl. and fr. Jan–Apr. *H 8206! 9203! P 20796!* Native of S.E. Asia.

13. STEMODIA L. (1759) nom. cons.

Corolla glabrous; bracteoles lanceolate to ovate, leafy, longer than calyx; leaves 1–2 (–2·5) cm. long   1. *maritima*
Corolla glandular-pubescent; bracteoles linear, shorter than calyx; leaves up to 9 cm. long
                                                        2. *durantifolia*

1. **S. maritima** L., Syst. Nat., ed. 10, **2**: 1118 (1759).
Much branched perennial herb or undershrub up to about 1 m. high, the stems often decumbent with numerous ascending shoots, tetragonal when young and often sinuate-winged, viscous-glandular; leaves lanceolate, cordate-amplexicaul, acute, serrate the length of both margins; flowers subsessile; calyx 2–3 mm. long; corolla

5·5 mm. long, blue, light mauve or white, the upper lip subentire; capsule elongate-ovoid, 2·5 mm. long; seeds minutely punctate, apiculate.

Common at salina and mangrove margins and in wet brackish areas generally; SL–20 feet; fl. and fr. all the year. *A 5667! H 9567! P 23692!* Northern S. Amer. to Brazil, Bahamas, Greater Antilles, Curaçao; Grand Cayman.

2. S. durantifolia (L.) Sw., Obs. Bot.: 240 (1791).—*Capraria durantifolia* L. (1759).

Erect weak-stemmed glandular-aromatic herb 30–100 cm. high; leaves mostly linear-oblanceolate, shortly amplexicaul, long-acute, serrate-dentate distally, entire towards base, mostly less than 1 cm. broad, the upper much smaller; pedicels 1–2·5 mm. long; calyx 5 mm. long; corolla 7 mm. long, blue or dark or light mauve, white or light buff in tube, the lobes with darker lines, upper lip emarginate; capsule conical, as long as calyx; seeds oblong, dark brown, almost smooth.

Frequent in ditches, along riverbanks and in swamps and swampy grassland; SL–1000 feet; fl. and fr. most of the year. *A 8834! 12164! H 11907! P 24347!* Continental tropical Amer., Bahamas, West Indies.

14. LENDNERIA Minod (1918)

1. L. verticillata (Mill.) Britton in Britton & Wilson, Sci. Surv. Porto Rico & Virg. Is. 6: 184 (1925).—*Erinus verticillatus* Mill. (1768).

Low annual weedy herb with branches 5–15 cm. long; leaves opposite or verticillate, ovate, cuneate at base, acute or obtuse at tip, crenate-serrate, 6–18 mm. long, 3–9 mm. broad, almost glabrous; petioles slender, 4–8 mm. long; flowers solitary or paired in the axils on peduncles 1–2 mm. long; calyx 3 (–5) mm. long, segments lanceolate; corolla 4–5 mm. long, tube bearded within, lobes purplish and finely lined; capsule subglobose, about 2·5 mm. long.

Uncommon (St. Andr., St. Mary), a weed of cultivated land, lawns, nursery beds and pots; 500–600 feet; fl. and fr. Apr–May. *H 12061! P 20798!* Throughout tropical Amer.

15. CHEILOPHYLLUM Pennell (1920)

1. C. jamaicense Pennell in Bull. Torr. Bot. Club 62: 256 (1935).

Low herb with glandular-puberulous glabrescent branches 15–25 cm. long; leaves ovate to oblong-elliptical, cuneate at base, acute, few-toothed, the margins reflexed and thickened, 4–7 mm. long, 2–5 mm. broad, shortly petioled, thinly puberulous; peduncles solitary, 2–3 (–10) mm. long; sepals lanceolate, 2·5 mm. long; corolla white, about 2·5 mm. long; capsule ovoid, acute, 2 mm. long.

Very rare (Clar.), in low-lying savanna in shade of grasses; 100 feet; fl. and fr. Dec. Apparently known only from the type *H 12737!*

16. MECARDONIA Ruiz & Pav. (1798)

1. M. procumbens (Mill.) Small, Fl. Southeast. U. S.: 1065, 1338 (1903).— *Erinus procumbens* Mill. (1768). *Lindernia dianthera* Sw. (1788).

Spreading-branched mostly annual glabrous herb; stems sometimes rooting, winged-angled, up to about 25 cm. long; leaves subsessile, elliptical, cuneate at base, subacute, serrulate, 1–2·5 cm. long, up to 1 cm. broad; peduncles solitary, up to about 15 mm. long; bracteoles oblanceolate, 4–5 mm. long; outer sepals at flowering 5·5 mm. long, the corolla a little longer, in fruit 9 mm. long, the capsule ellipsoid and much shorter; upper lip of corolla emarginate with fine blackish-purple lines, hairy within; seeds blackish-brown, longitudinally grooved and minutely verrucose.

Occasional in central and eastern parishes, a weed of shaded lawns, low pastures, sandy or gravelly riverbanks and marshy places; 100–3900 feet; fl. and fr. most of the year. *A 7648! H 12053! P 24372!* Generally distributed through subtropical and tropical continental Amer. and the West Indies.

17. BACOPA Aubl. (1775) nom. cons.

1 Stems erect or usually so; leaves toothed; pedicels up to about 1 mm. long; capsule glandular-punctate   **1. sessiliflora**

1 Stems creeping or floating; leaves entire; pedicels 3 mm. or more long; capsule not
    glandular:
  2 Bracteoles 2 under calyx; leaves mostly oblanceolate, with very obscure venation;
      corolla 7–10 mm. long   2. *monnieri*
  2 Bracteoles wanting; leaves mostly ovate or obovate, distinctly palmately veined from
      base; corolla 3–4 mm. long:
    3 Leaves broadest beyond the middle; outer sepal oblong-ovate to elliptical, parallel-
        veined, up to about 2 mm. broad   3. *repens*
    3 Leaves broadest below the middle; outer sepal ovate-orbicular, cordate, reticulate-
        veined, about 3 mm. broad   4. *innominata*

1. B. sessiliflora (Benth.) Edwall, Fl. Paulista 2: 175 (1897).—*Herpestis sessili-flora* Benth. (1836).

Erect annual much branched herb 20–40 cm. high, with few scattered hairs;
leaves sessile, narrowly oblanceolate, narrowed to base, acute, margin serrate
beyond middle, up to 5·5 cm. long and 5 mm. broad, 1-nerved, punctate on both
surfaces; bracteoles filiform, shorter than calyx; calyx 3·5–4 mm. long, the corolla a
little longer and white; capsule ovoid, 3–4 mm. long.

Very rare (St. Eliz.), in wet open sandy ground; 10–50 feet; fl. and fr. Dec.
P 18445! Mexico to Brazil, Cuba, Hispaniola.

2. B. monnieri (L.) Pennell in Proc. Acad. Nat. Sci. Philad. 98: 94 (1946).—*Lysimachia monnieri* L. (1756).

Herb with diffuse trailing and rooting branches up to 50 cm. long, glabrous,
sometimes forming mats and loose cushions on open ground; leaves sessile, cuneate
at base, rounded at tip, very rarely shallowly toothed, 5–15 mm. long, up to 5 mm.
broad; peduncles solitary, slender, up to 25 (–35) mm. long; bracteoles oblong,
shorter than calyx; calyx 6 mm. long, 2 inner sepals keeled and partly enclosing
fruit; corolla pink, light mauve or white; stamens 4 or 5; capsule ovoid.

Common in wet places, especially in short-grass pastures near the sea; SL–100
(–1200) feet; fl. and fr. all the year. *A 8073! 12055! H 9566! 8194! P 24589!* Subtropics and tropics generally.

3. B. repens (Sw.) Wettst. in Engl. & Prantl, Nat. Pflanzenf. 4 (3b): 77 (1891).—*Gratiola repens* Sw. (1788). *Macuillamia repens* (Sw.) Pennell (1923).

Herb with creeping and rooting thinly pubescent or glabrous branches up to
40 cm. long; leaves sessile, obovate-elliptical, narrowed at base, rounded at tip,
5–7-veined, 8–20 (–30) mm. long, 3–10 (–15) mm. broad; peduncles solitary or 2
together in leaf-axils, 4–10 mm. long, thinly pubescent; sepals 3–4 mm. long;
corolla white, about 4 mm. long; capsule broadly ellipsoid, 3–4 mm. long; seeds
reticulate, pale.

Rare (Clar., Trel.), in shallow water of muddy ponds; 150–400 feet; fl. and fr.
Aug, Dec. *P 10509! 24340!* Tropical S. Amer., Greater Antilles, Trinidad.

4. B. innominata (Gómez Maza) Alain in Rev. Soc. Cub. Bot. 13: 61 (1956).—*Conobea innominata* Gómez Maza (1894). *Herpestis rotundifolia* Gaertn. f. (1807), not *Bacopa rotundifolia* (Pursh) Wettst. (1891).

Much branched herb with weak trailing and rooting stems up to 25 cm. long;
leaves sessile, broadly ovate to oblong, obtuse or rounded at tip, palmately 5–7-
veined, 5–12 mm. long, 4–7 mm. broad; peduncles solitary, 2–10 mm. long, slender;
outer sepals 4·5 mm. long; corolla white, 4 mm. long; capsule ellipsoid, about 3
mm. long; seeds reticulate, brown.

Rare (St. Eliz., St. James, Trel.), in wet places; 100–500 feet; fl. Aug–Oct, fr.
Aug. *P 10550!* S. United States, Greater Antilles, Guadeloupe.

18. LINDERNIA All. (1766)

1 Calyx (3–) 6–10 mm. long, divided to a little over half way, lobes acute; pedicels much shorter
    than calyx, 2–3 mm. long; capsule about 10 mm. long, fusiform-ellipsoid, long-acumi-
    nate; leaf-blades oblong-ovate to suborbicular, broadly cuneate at base   1. *diffusa*
1 Calyx 2–4 mm. long; pedicels much longer than calyx, up to 12 mm. long; capsule 3–4
    mm. long, rounded at apex; leaf-blades broadly ovate, more or less truncate at base:
  2 Leaves petiolate; calyx 2·5–4 mm. long, divided less than halfway   2. *crustacea*
  2 Leaves sessile; calyx 2–2·5 mm. long, divided more than halfway   3. *rotundifolia*

1. L. diffusa (L.) Wettst. in Engl. & Prantl, Nat. Pflanzenf. **4** (3b): 79 (1891).—
Vandellia diffusa L. (1767).

Low branched herb the stems creeping and often rooting, with lines of short
white hairs; leaves obtuse to rounded or subacute, serrate, ciliate, 1–3·5 cm. long,
1–2 cm. broad, often purple beneath; calyx-segments white-pubescent on midribs;
corolla about 5 mm. long, tube white, lip 3-lobed, white with yellow spot centrally
in throat, hood purplish-brown or light purple; seeds verrucose.

Occasional weed in damp places on heavy soils; 500–2300 feet; fl. and fr. almost
all the year. *A 9079! H 12057! P 16571!* General in the tropics.

2. L. crustacea (L.) F. von Muell., Syst. Census Austral. Pl. **1**: 97 (1882).—
Capraria crustacea L. (1767).

Erect or diffuse slender-branched sparsely pubescent herb up to 20 cm. high;
leaves ovate to elliptical, rounded to subcordate at base, obtuse at tip, serrate, blade
5–15 (–30) mm. long, 3–10 (–15) mm. broad, obscurely gland-dotted; petiole 1–6
mm. long; calyx-segments ovate, acute or obtuse; corolla 5–7 (–10) mm. long,
violet, whitish within with 2 light yellow crests.

Local (St. Andr., St. Mary), a weed of cultivations and lawns in low-lying places;
590–700 feet; fl. and fr. Oct–Apr. *H 11872! P 20797!* Widespread in the subtropics and tropics.

3. L. rotundifolia (L.) Alston in Trimen, Handb. Fl. Ceylon **6**, Suppl: 214 (1931).
—*Gratiola rotundifolia* L. (1771). *Ilysanthes rotundifolia* (L.) Benth. (1846).
L. microcalyx Pennell & Stehlé (1938).

Low herb with rooting branches, glabrous or nearly so; leaves sessile or subsessile, suborbicular, rounded-subcordate at base, rounded at tip, entire or occasionally shallowly crenate-serrate, 2–12 mm. long, 2–7 mm. broad, about 7-veined
from base; pedicels 5–13 mm. long; calyx-segments linear-lanceolate; corolla 7–12
mm. long, white with violet or blue spots or all white; staminodes clavate, light
blue.

Uncommon and local (Port., St. Thom.), in wet grassy places; 50–2200 feet;
fl. Jan–Apr, fr. Apr. *Britton 2581! P 20645! Yuncker 18079!* Scattered thinly
through the tropics, rare in the West Indies being reported only from Guadeloupe,
Martinique and French Guiana and doubtfully from Cuba.

19. HEMIANTHUS Nutt. (1817)

1. H. callitrichoides Griseb. in Mem. Amer. Acad. n. s. **8**: 522 (1862).

Small plant with numerous prostrate branches forming tight mats, glabrous;
leaves sessile, ovate to elliptical, cuneate, 1·5–3 mm. long, 0·8–2·3 mm. broad,
light green; peduncles 0·2–0·5 mm. long; calyx shortly lobed, 0·8 mm. long;
corolla white, about 1 mm. long; anthers yellow; capsule globose, about 1 mm.
long.

Occasional (St. Cath., Trel., St. Ann), on rocks in streams or riverside mud;
150–250 feet; fl. and fr. Feb, June. *A 9471! Britton 3027! Stearn 182! Yuncker
18293!* Bahamas, Greater Antilles.

20. SCOPARIA L. (1753)

1. S. dulcis L., Sp. Pl. **1**: 116 (1753).—Sweet Broom.

Much branched erect annual taprooted herb 30–60 (–100) cm. high, glabrous;
leaves opposite or verticillate, linear-oblanceolate to elliptical, cuneate at base,
acute, entire, serrate or doubly serrate, up to 5 cm. long and 1 (–2) cm. broad;
pedicels commonly paired; calyx-lobes oblong, up to 2 mm. long and 1 mm. broad;
corolla white, 3–4 mm. across, lobes reflexed, densely bearded at throat; capsule
ovoid-globose, a little longer than sepals.

Frequent in damp sandy waste places and salina margins among grasses or in
thickets; SL–1000 feet; fl. and fr. all the year. *A 11915! H 6573! H & B 10628!
P 15702!* General in the subtropics and tropics.

21. CAPRARIA L. (1753)

1. C. biflora L., Sp. Pl. **2**: 628 (1753).—Goatweed.

Herb or undershrub up to 1·5 m. high, very variable in hairiness from densely villous to almost glabrous; leaves oblanceolate to elliptic-lanceolate, narrowed at base, acute, serrate at least in the distal half to entire, up to 9 cm. long and 1·5 (–2) cm. broad; pedicels slender, flexuous, up to about 15 mm. long, rarely as many as 4 together, without bracteoles; sepals linear-lanceolate, 4–6 mm. long; corolla about 1 cm. long; capsule oblong-ovate to ellipsoid, 2-sulcate, 4–6 mm. long.

Common in disturbed ground, along roadsides, at salina margins and in ditches; SL–1500 feet; fl. and fr. all the year. *A 6013! 11826! Fawcett 8245! P 6127! 21442! Stearn 872!* Throughout subtropical and tropical Amer., sparingly introduced and established in the Old World.

22. DIGITALIS L. (1753)

1. D. purpurea L., Sp. Pl. **2**: 621 (1753).

Erect perennial herb up to about 1 m. high; basal leaves long-stalked, cauline sessile running on to bracts; stem glabrescent below; leaf-blades elliptical to lanceolate, crenate to entire, pubescent, the lower up to 15 cm. long and 7 cm. broad; pedicels slender; outer sepals ovate-elliptical, acute, about 12 mm. long and 7 mm. broad; corolla about 4 cm. long; capsule ovoid.

Very local (St. Thom.), in clearing in thicket; 7400 feet; fl. and fr. July. *ACH 921! 923!* Native of S.W. Europe and Morocco.

23. VERONICA L. (1753)

1 Flowers subsessile, pedicels less than 1·5 mm. long; seeds flattened, smooth:
  2 Leaves ovate to narrowly ovate, rounded at base; plant hairy   **1.** *arvensis*
  2 Leaves narrowly oblanceolate to oblong, cuneate at base; plant almost glabrous
                                               2. *peregrina*
1 Flowers long-pedicelled, pedicels 2–20 mm. long:
  3 Plant perennial; stems creeping and rooting; leaves entire or shallowly crenate; pedicels up to 5 mm. long; seeds flattened, smooth   **3.** *serpyllifolia*
  3 Plant annual; stems decumbent, rooting only near base; leaves deeply toothed; pedicels up to about 20 mm. long; seeds hollowed out on one side, rugose on the other
                                               4. *persica*

1. V. arvensis L., Sp. Pl. **1**: 13 (1753).

Annual herb 5–25 cm. high, with ascending branches from the base; stems with spreading hairs in 2 bands; lower leaves stalked, the upper sessile, crenate, pubescent, 5–15 mm. long, 3–10 mm. broad; flowers in long loose erect racemes; bracts longer than flowers; calyx 4–5 mm. long, segments lanceolate, obtuse, unequal; corolla blue; capsule about 3 mm. broad, lobes rounded, ciliate.

Locally frequent (St. Andr., St. Thom.), in cultivated ground and open waste places; 3200–6000 feet; fl. and fr. Feb–Mar. *A 10533! H 9546! P 23267!* Native of Europe, temperate Asia and N. Africa; naturalized in N. Amer., Bermuda, Hispaniola.

2. V. peregrina L., Sp. Pl. **1**: 14 (1753).

Annual herb 5–30 cm. high, with several ascending branches from the base; leaves shortly petiolate, remotely serrate-crenate distally or entire, up to 30 mm. long and 8 mm. broad; flowers in erect racemes; bracts longer than flowers; calyx 3–5 mm. long, segments lanceolate, obtuse, subequal; corolla white or very light mauve, shorter than calyx; capsule about 3 mm. broad, shallowly notched, glabrous.

Locally common (St. Andr., St. Thom.), a weed of cultivated ground and shady waste places; 1900–5000 feet; fl. and fr. Jan–May. *A 6615! 10534! H 10799! P 19664!* E. United States, Bermuda, Mexico, Hispaniola, naturalized in Europe.

3. V. serpyllifolia L., Sp. Pl. **1**: 12 (1753).

Herb with spreading and ascending branches 5–15 cm. high; stem puberulous with short upcurved hairs; leaves subsessile on the lower shortly stalked, blade ovate or broadly elliptical, broadly cuneate at base, obtuse at apex, 8–15 mm. long, 4–10 mm. broad; racemes loose; bracts longer than pedicels; calyx 3 mm. long, segments oblong; corolla light blue or mauve with darker lines, rarely all white, about 2 mm. long; capsule about 3 mm. long and 4 mm. broad, ciliate with gland-tipped hairs.

Locally common (St. Andr., Port., St. Thom.), a weed of grassy damp sheltered places; 3800–7400 feet; fl. and fr. all the year. *A 7746! 10554! H 9518! Powell 1075!* Europe, temperate Asia, N. Africa, N. and S. Amer.

4. V. persica Poir. in Lam., Encycl. Méth. Bot. **8**: 542 (1808).

Laxly-branched pubescent annual herb with stem up to 45 cm. long; leaves stalked, blade ovate, subcordate or rounded at base, obtuse at tip, 5–30 mm. long, 5–20 mm. broad; pedicels 10–25 mm. long, recurved in fruit; bracts leaflike; calyx 3–4 mm. long in flower enlarging to 6·5 mm. in fruit, segments ovate; corolla light blue or purplish-blue with darker lines, 4 mm. or more long; capsule nearly twice as broad as long, 6–8 mm. broad, thinly ciliate with gland-tipped hairs.

Occasional (St. Andr., St. Thom.), a weed of sheltered pathside and roadside banks; 3200–4900 feet; fl. and fr. Feb–Aug. *A 6428! 6699! H 12397! P 24679!* Native of W. Asia; naturalized in Europe and Amer.

24. BUCHNERA L. (1753); Philcox (1965)

Calyx hispid or scabrid only on nerves, sometimes glabrescent; corolla mostly glabrous externally above calyx, lightly pubescent within; capsule oblong, abruptly apiculate, 6–7 mm. long, hardly exceeding calyx   1. *longifolia*
Calyx totally pubescent to pubescent in upper one-third only with short white crisped callus-based hairs; corolla pilose externally; capsule ovoid to spherical, tapered-apiculate, 5–6 mm. long, exceeding calyx   2. *floridana*

1. B. longifolia Kunth, Nov. Gen. **2**: 340 (1818).

Erect herb (5–) 30–60 (–110) cm. high; leaves linear-oblanceolate to linear, scabridulous, up to 13 cm. long, 1·5–5 (–8) mm. broad; calyx tubular, the lobes short triangular, up to 9 mm. long in fruit; corolla purple or violet, the lobes spreading regularly, with white hairs in throat, 9–13 mm. long.

Occasional in damp grassy places on heavy soils; 1400–2800 feet; fl. Apr–May, Aug–Dec, fr. Apr–May, Sept–Oct. *A 7144! H 6192! 11104! P 19657! Stearn 986!* Florida, C. Amer., Bahamas, Greater Antilles, Trinidad.

2. B. floridana Gandog. in Bull. Soc. Bot. France **66**: 217 (1919).

Erect pilose-scabrid to subglabrous herb 20–60 cm. high; leaves elliptic-lanceolate to linear, the lower obtuse at tip, 3–8 cm. long, 3–13 mm. broad; calyx tubular, 3–4 mm. long, the lobes very short; corolla violet-blue or white, 6–10 mm. long.

Uncommon in sandy pastures and waste places; SL–2300 feet; fl. and fr. Jan–Sept. *A 8075! H 9771! H & P 14524!* Mexico, British Honduras, S.E. United States, Bahamas, Greater Antilles, introduced in Trinidad.

25. ALECTRA Thunb. (1784)

1. A. fluminensis (Vell.) Stearn in Journ. Arn. Arb. **52**: 635 (1971).—*Scrophularia fluminensis* Vell. (1825). *Melasma melampyroides* (L. C. Rich.) Pennell (1925).

Coarse erect little-branched taprooted herb, 30–80 cm. high; stem brittle, setulose; leaves ovate to lanceolate, subsessile, rounded at base, long-acute, scabrid, 2–5 cm. long, 5–15 mm. broad; pedicels up to about 2 mm. long; calyx about 8 mm. long; corolla about 9 mm. long; capsule subglobose, almost enclosed in calyx, about 9 mm. long; seeds conical to linear, truncate at both ends, a little over 1 mm. long.

Rare (St. Eliz., West., Han.), at swamp margins and in wet grassy places on clay soils; 40–900 feet; fl. and fr. Nov–Dec. *A 8641! 11986! P 27703!* C. and S. Amer., West Indies but apparently absent from Cuba.

26. AGALINIS Raf. (1837) nom. cons.

1. A. albida Britton & Pennell in Bull. Torr. Bot. Club **42**: 391 (1915).—*Gerardia albida* (Britton & Pennell) Pennell (1935).

Slender annual subglabrous herb up to about 60 cm. high; leaves linear-subulate, scabridulous, often recurved, up to about 2·5 cm. long; racemes lax, few- to many-flowered; pedicels ascending, up to 3 mm. long in fruit; calyx-tube about 2·5 mm. long, the lobes triangular-ovate, acuminate; corolla 10–15 mm. long, campanulate, lobes ciliate, white sometimes tinged with rose; capsule globose, 4–5 mm. long.

Rare and local (Clar.), in damp grassy savannas; 2000–2500 feet; fl. and fr. sporadically throughout the year. *A 5916! 7134! H 12222! P 15822!* Cuba, Grand Cayman.

27. ANISANTHERINA Pennell (1920)

1. A. hispidula (Mart.) Pennell in Mem. Torr. Bot. Club **16**: 106 (1920).—*Gerardia hispidula* Mart. (1829).

Annual branched hispid and scabrid herb 30–45 cm. high; leaves sessile, linear, obtuse or rounded at base, acute at tip, 2·5–5 cm. long; raceme very lax; calyx 4 mm. long, with short subulate teeth; corolla 15–25 mm. long; capsule ovoid-globose, glabrous.

Very rare (Manch.), only once reported, *Wullschlaegel*. Panama, Guyana, Brazil, Cuba.

172. BIGNONIACEAE

Trees, climbing or erect shrubs or rarely herbs. Leaves opposite or rarely alternate, simple or pinnately compound, if trifoliolate the terminal leaflet often modified to a tendril, sometimes digitate, without stipules (pseudostipules formed from arrested axillary branches frequently occur). Inflorescences mostly racemose or cymose. Flowers bisexual, zygomorphic. Perianth biseriate; calyx campanulate, usually with 5 teeth or lobes, sometimes truncate, sometimes spathaceous; corolla gamo-petalous, 5-lobed, the lobes usually imbricate, when bilabiate the upper lip of 2 lobes the lower of 3. Stamens 4 or 2, epipetalous, alternating with corolla-lobes; anthers coherent or distinct, 2-locular, opening lengthwise, the loculi often widely divergent. Staminodes 1, 3 or 0. Hypogynous disk present. Ovary superior, 2-locular with 2 axile placentas in each loculus or 1-locular with 2 bifid parietal placentas; style terminal simple; stigma 2-lobed; ovules numerous, anatropous. Fruit a capsule or indehiscent. Seeds from capsular fruits often winged, without endosperm; embryo straight.

120 genera with about 800 species mostly in tropical America but some in the subtropics and tropics of the Old World.

1 Trees or shrubs, if climbing never by tendrils or rootlets; leaves simple or compound but never regularly bifoliolate:
  2 Fruit indehiscent, gourdlike; leaves simple; seeds not winged:
    3 Leaves mostly clustered or spiral on reduced shoots; fruit hard, unilocular, spherical
        1. Crescentia
    3 Leaves alternate; fruit leathery, bilocular, broadly oval to ellipsoid   **2. Enallagma**
  2 Fruit a loculicidal capsule with the valves at right angles to the septum:
    4 Leaves simple, opposite or verticillate; fertile stamens 2; staminodes 3; fruit very long and slender; seeds hairy   **3. Catalpa**
    4 Leaves compound, opposite; stamens 4; seeds winged:
      5 Leaves digitately 3–5 foliolate (sometimes simple or 1-foliolate); leaflets at least beneath with minute peltate scales; corolla white, yellow or pink; often leafless at flowering; capsule elongated   **4. Tabebuia**
      5 Leaves pinnate or bipinnate (rarely uppermost 3-foliolate or simple):
        6 Leaves bipinnate; staminode much longer than the stamens, clothed with gland-tipped hairs; corolla mauve or violet; capsule short and broad, woody   **5. Jacaranda**
        6 Leaves simply pinnate; staminode inconspicuous; capsule elongated; seeds winged:
          7 Leaflets entire; trees; valves of fruit woody; calyx spathaceous, very large; corolla scarlet, scoopshaped; stamens exserted   **6. Spathodea**

7 Leaflets serrate; shrubs; valves of fruit thin; calyx regular, campanulate, 5-toothed:
  8 Stamens exserted; lax much branched erect or scrambling shrub; corolla orange-scarlet, laterally compressed; calyx 4–6 mm. long   **7. Tecomaria**
  8 Stamens included; corolla more or less salver-shaped:
    9 Erect shrub or small tree; corolla yellow; calyx about 5 mm. long   **8. Tecoma**
    9 Scrambling climber; corolla mostly pink with darker lines in throat; calyx about 15 mm. long   **9. Podranea**
1 Climbers by tendrils or rootlets; leaves often bifoliolate:
  10 Leaves all simple; climber by rootlets; fruit subglobose, indehiscent   **10. Schlegelia**
  10 Leaves 2–3-foliolate, rarely simple; climbers by tendrils; fruit capsular, septicidally dehiscent:
    11 Tendrils trifid-clawed; corolla yellow   **11. Doxantha**
    11 Tendrils simple or branched, twining:
      12 Corolla elongate-tubular; pseudostipules small or absent:
        13 Tendrils simple; corolla-tube straight, white; valves of capsule smooth   **12. Tanaecium**
        13 Tendrils often branched; corolla-tube curved, rich orange-red; fruit not known in Jamaica   **13. Pyrostegia**
      12 Corolla campanulate or salver-shaped; pseudostipules usually evident, foliaceous:
        14 Corolla light yellow; tendrils 3-branched at tip; fruit ellipsoid, flattened, woody, strongly and densely spiny on the surface   **14. Pithecoctenium**
        14 Corolla purple; tendrils simple; fruit not known in Jamaica   **15. Saritaea**

1. CRESCENTIA L. (1753)

1. C. cujete L., Sp. Pl. **2**: 626 (1753).—Calabash Tree.
Tree 6–10 m. high; leaves oblanceolate to elliptical tapering to the base, shortly acuminate at the tip, up to 20 cm. long and 6 cm. broad, usually smaller; flowers borne on the old wood; calyx up to about 2 cm. long, deeply split; corolla 4–6 cm. long, broadly campanulate, greenish-white to greenish-yellow tinged rose and purple-veined; fruit up to about 25 cm. in diameter.
Common along roadsides and in old pastures, thickets and woodland margins; SL–1400 feet; fl. and fr. May–Jan. *A 8356! H 9382! H & P 14490!* Widespread in the tropics, frequently in semi-cultivation; Grand Cayman.

2. ENALLAGMA (Miers) Baill. (1888) nom. cons.

1. E. latifolia (Mill.) Small, Fl. Miami: 171 1913).—*Crescentia latifolia* Mill. (1768). Wild Calabash.
Tree 6–10 m. high; leaves oblanceolate to obovate, shortly tapered to the base, cuspidate, up to 22 cm. long and 6 cm. broad; flowers solitary; calyx 2·5–4 cm. long, spathaceous, irregularly (2–) 3-lobed; corolla cream to yellowish-green tinged purplish or dull rose, 5–6 cm. long; fruit 5–12 cm. long, somewhat pointed apically, fragrant.
Rather local in coastal woodland on limestone and in marsh forest; near SL (–600 feet); fl. and fr. sporadically. *Asprey UCWI 2538! H & P 15057! P 24516!* Florida, Mexico to Colombia and Guianas, Greater Antilles, Martinique, St. Vincent Trinidad (cult.).

3. CATALPA Scop. (1777); Paclt (1952)

1. C. longissima (Jacq.) Dum.-Cours., Bot. Cult. **2**: 190 (1802).—*Bignonia longissima* Jacq. (1760). French Oak, Mast Wood, Yoke Wood.
Tree 5–25 (–30) m. high with fissured, sometimes reticulated, light grey bark, occasionally deciduous; petioles slender, 1·5–2·5 cm. long; leaf-blade ovate-lanceolate, rounded at base, mostly long-acute at tip, 5–11 (–14) cm. long, 2–4 (–5·5) cm. broad; inflorescence a small panicle; corolla mostly white, pinkish on lobes, yellow in throat with purple lines and yellow bands in tube, 25–30 mm. long, 30–34 mm. broad, tube about 12 mm. long; stamens included; fruit 35–77 cm. long, 4 mm. broad; seeds plumed at both ends.
Locally common, especially in the southern parishes, in open plains and in gullies and thickets mostly on gravelly soil; SL–700 (–2300) feet; fl. and fr. most of the year. *A 7955! H 9229! P 19780!* Hispaniola, Martinique, introduced into the Pacific.

4. TABEBUIA DC. (1838)

1 Corolla clear rich yellow, the limb with scattered branched hairs outside; leaflets obovate, acuminate, rusty-stellate pubescent beneath at least when young; cultivated tree; native of Trinidad and continental S. Amer.; Yellow Poui (Jamaica)[1]

T. rufescens J. R. Johnston

1 Corolla usually pink or white, glabrous outside, ciliate-margined, yellow in throat; leaflets not pubescent:

2 Leaflets obtuse to rounded at tip, the middle 3 mostly oblanceolate; calyx 9–14 mm. long　　　　　　　　　　　　　　　　　　　　　　　　　1. *riparia*

2 Leaflets acute to acuminate, the middle 3 mostly lanceolate to elliptical or oblong:

3 Calyx 11–16 mm. long; corolla 5–8·5 cm. long, light pink; larger leaflets up to 20 cm. long and 4·5 (–6) cm. broad; fruit to 23 cm. long, 8–10 mm. broad, sparsely lepidote, glabrescent　　　　　　　　　　　　　　　　2. *angustata*

3 Calyx 15–20 (–28) mm. long; fruit 20–35 cm. long, densely lepidote at least when young:

4 Leaflet-tip gradually acuminate, the acumen obtuse; corolla 5–6·5 (–7) cm. long, white, often light yellow in bud　　　　　　　　　　　　3. *platyantha*

4 Leaflet-tip caudate-acuminate, the acumen sharply acute; leaflets 6–25 (–35) cm. long, 2·5–12 (–18) cm. broad; corolla (6–) 7–8 cm. long, purplish-pink to almost white; cultivated; native Mexico to Ecuador and Venezuela; Pink Poui

T. rosea (Bertol.) DC.

1. T. riparia (Raf.) Sandwith in Taxon **4**: 44 (1955).—*Bignonia leucoxylon* L. (1753), not *T. leucoxylon* DC. (1838). *Leucoxylon riparium* Raf. (1838). White Cedar.

Slender tree 6–20 m. high or a shrub from 1 m.; bark pale, flaking irregularly; branches slender, ascending; leaflets variable, mostly about 7 cm. long and 2 cm. broad, but often larger or smaller; corolla 5–7 cm. long, white or light rose, streaked with light pink or crimson within; fruit lepidote up to about 25 cm. long and 7 mm. broad, beaked.

Rather common, especially in the southern parishes in thickets and open wood-lands on arid limestone and on sea-cliffs; 25–300 (–1000) feet; fl. Feb–May, Sept, fr. May–Sept. *A 10923! H 9214! P 16249! Stearn 640!* Endemic. A closely related plant with fruits 10–12 cm. long (*Fawcett!*) occurs in Grand Cayman.

2. T. angustata Britton in Bull. Torr. Bot. Club **42**: 376 (1915).

Tree 6–12 (–25) m. high or a shrub from 3 m., sometimes with a thick trunk; bark grey; leaflets glossy, shortly narrowed or more or less rounded at base, acumen indefinite, obtuse; valves of fruit to 17 mm. broad when open.

Rather common in fields, hedgerows and woodland margins, occasionally on limestone rocks or swamp margins near the sea, mostly in the central and western parishes, probably sometimes planted; SL–3000 feet; fl. Mar–Sept, fr. May–July. *A 10974! H 9253! H & P 14209! Loveless UCWI 2561! P 20837! Stearn 781!* Cuba.

3. T. platyantha (Griseb.) Britton in Bull. Torr. Bot. Club **42**: 379 (1915).—*Tecoma platyantha* Griseb. (1861). *Tabebuia jamaicensis* Britton (1915).

Tree 5–13 m. high; bark grey; leaflets mostly rather large, up to 25 cm. long, (4·5–) 6–12 cm. broad; ripe fruit 20–27 cm. long, 8–9 mm. broad.

Sparsely scattered through the central and western parishes in woodland on limestone; (250–) 1500–2750 feet; fl. and fr. Jan–May. *A 10936! Purdie! Stearn 356!* Endemic.

5. JACARANDA Juss. (1789)

1. J. mimosifolia D. Don in Edw., Bot. Regist. **8**: t. 631 (1822).—Jacaranda.

Tree up to 10 (–15) m. high; leaves with 12 or more pairs of pinnae; pinnules about 16 pairs, linear to narrowly elliptical, aristate, up to 23 mm. long and 5 mm. broad; inflorescence a much branched panicle; calyx about 2·5 mm. long; corolla obliquely campanulate widening from a narrow tube, densely pubescent outside, 3–4·5 (–5) cm. long; ripe capsule 4·5–6 cm. broad.

---

[1] This is the Black Poui of Trinidad; the Yellow Poui, *T. serratifolia* (Vahl) Nicholson, has leaves finally glabrous.

Frequently cultivated; 600–2500 feet; fl. and fr. Apr.–June *Loveless UCWI 2133!* *Wynter UCWI 3267!* Native of N.W. Argentina and Uruguay, extensively planted in the subtropics and tropics.

6. SPATHODEA Beauv. (1805)

1. S. campanulata Beauv., Fl. Oware & Benin **1**: 47, tt. 27, 28 (1805).—African Tulip Tree.
Tree to 16 m. high, deciduous; bark grey; leaves often unequal in each pair or one not developing; leaflets (4–) 6–7 (–8) pairs with an odd terminal leaflet, ovate to oblong-elliptical, acuminate, unequal at base with a large gland or a few smaller glands beneath, 4–11 (–14) cm. long, 1·5–5 (–6) cm. broad, subglabrous adaxially, puberulous to thinly hispid on the veins beneath; racemes terminal; calyx 5 (–7) cm. long, shortly tomentose, distinctly ribbed; corolla scarlet, yellow at base, within and marginally, margin crisped, 10–12 cm. long, about 10 cm. deep; follicles (14–) 18–23 (–30) cm. long, 3·5–4 cm. broad, replum leathery; seeds laterally winged, 2–3 cm. broad, wing transparent.
Commonly planted; SL–1500 feet; fl. and fr. most of the year. *Thompson!* Native of W. tropical Africa, now widespread in the tropics; Grand Cayman.
A closely related species, *S. nilotica* Seem., with the leaflets tomentose beneath, is from N.E. tropical Africa and has been grown in Jamaica.

7. TECOMARIA Spach (1840)

1. T. capensis (Thunb.) Spach, Hist. Vég. Phan. **9**: 137 (1840).—*Bignonia capensis* Thunb. (1800).
Lax-branched shrub 1–2 m. high, the branches trailing and rooting; leaves with 7 crenate-serrate leaflets, the terminal leaflet larger up to 4·5 cm. long and 2·5 cm. broad, glabrous except petiole and near the proximal lateral veins beneath; rachis narrowly winged; inflorescence compound subracemose; corolla curved, funnel-shaped, 3·5–5 cm. long; fruit (rarely formed), about 8 cm. long, the valves 5–6 mm. broad.
Naturalized locally (St. Andr., Manch., St. Eliz.) and cultivated; 600–2400 feet; fl. mostly Sept–May, fr. Mar. *A 12355! H 9678! H & P 13743!* Native of S. and E. tropical Africa, now widely cultivated and escaped.

8. TECOMA Juss. (1789)

1. T. stans (L.) Kunth, Nov. Gen. **3**: 144 (1819).—*Bignonia stans* L. (1763).
Shrub 1–5 m. high, rarely a small tree; leaflets (3–) 7 (–9), lanceolate, serrate, the terminal larger, up to 11 cm. long and 2·5 cm. broad, pubescent or glabrous except domatia in the proximal vein-axils beneath; flowers fragrant, borne in small clusters or singly in a usually unbranched terminal raceme; corolla campanulate, sub-regular, up to 5 cm. long; fruit 10–20 cm. long, 6–7 mm. broad, drying light brown; seeds including wing about 18 mm. long.
Locally abundant on cut-over limestone hillsides and waste sandy places; SL–1000 (–2400) feet; fl. and fr. most of the year. *A 9245! H 8230! H & P 13590!* Native of the New World tropics, now widespread; Grand Cayman.

9. PODRANEA Sprague (1904)

1. P. ricasoliana (Tanfani) Sprague in Dyer, Fl. Cap. **4** (2): 450 (1904).—*Tecoma ricasoliana* Tanfani (1887). Pandorea, Zimbabwe Creeper.
Scrambling shrub, sometimes climbing to 10 m. or more; bark reticulate-fissured, light grey, soft; leaflets mostly 9 or 11, petiolulate, coarsely serrate-dentate, glabrous, the terminal larger, up to 7 cm. long and 3 cm. broad; inflorescence a terminal panicle; flowers with a fruity odour; calyx campanulate, the lobes reflexed, aristate, 15–18 mm. long, whitish; corolla 5–7 cm. long, the lobes spreading, hairy within the tube abaxially; fruit (rarely formed) up to 28 cm. long and about 8 mm. broad.

Common in cultivation up to about 2500 feet; fl. most of the year. *A 11226!*
Native of S.E. and S. tropical Africa, now widely cultivated.

10. SCHLEGELIA Miq. (1844)

Calyx shallowly lobed, about 4 mm. long; corolla white, tube 8–10 mm. long, lobes 3–4
mm. long; inflorescences racemose, fascicled, 3–5 cm. long   1. *axillaris*
Calyx truncate, campanulate, 6–9 mm. long; corolla white at base, rosy-purple to crimson
distally and in throat, tube 2·5–3·5 cm. long, lobes 8–10 mm. long; inflorescence very
short, the flowers clustered or two from a common peduncle   2. *parasitica*

1. **S. axillaris** Griseb., Fl. Br. W.I.: 445 (1862).—*S. urbaniana* K. Schum. ex
Duss (1897).
Climbing shrub to 10 m. or more high; leaves elliptical, narrowed to both ends,
7–13 (–15) cm. long, 3·5–6·5 (–10·5) cm. broad, rather leathery; inflorescence
puberulous; bracts and bracteoles 1–1·5 mm. long; fruit subspherical, about 8 mm.
in diameter.
Rare and local (Port.), trailing and climbing on trees in secondary forest; 1000–
3000 feet; fl. Apr–May, fr. Aug. *Bot. Dept. 10311! P 10100! Robbins!* Dominica,
Guadeloupe.

2. **S. parasitica** (Sw.) Miers ex Griseb., Fl. Br. W.I.: 445 (1862).—*Tanaecium
parasiticum* Sw. (1788).
Climbing shrub to 10 m. or more high; bark soft and fleshy, grey; leaves elliptical,
shortly acuminate, tapered to base, up to 22 cm. long and 12 cm. broad, rigidly
leathery; calyx white tinged pink or purple; corolla-lobes recurved; fruit spherical,
–33·5 cm. in diameter, hard, smooth; seeds angular with purple testa.
Frequent in woodlands and on rocky slopes in moderately wet limestone or shale
areas; 50–3400 feet; fl. and fr. most of the year. *A 7762! H 9350! P 6400! Stearn
530!* Cuba.

11. DOXANTHA Miers (1863)

1. **D. unguis-cati** (L.) Miers in Proc. Roy. Hort. Soc. London 3: 190 (1863)
(as *unguis*).—*Bignonia unguis-cati* L. (1753).
High-climbing vine to 20 m. or more high, rooting frequently from the nodes;
leaflets 2, variable in shape from lanceolate to broadly ovate, 3–7 (–10) cm. long,
glabrous or pubescent beneath; flowers mostly solitary, axillary or on short axillary
branches; calyx openly campanulate, about 1 cm. long; corolla (4–) 7–9 cm. long,
tube funnel-shaped, yellow, rest of limb light orange with orange lines in throat,
lobes spreading; fruit up to 35 (–70) cm. long, 1 (–1·6) cm. broad; seeds with
membranous wings.
Cultivated and occasionally naturalized. *A 9233! Karling UCWI 4825!* Native of
continental tropical Amer., now widespread in cultivation.

12. TANAECIUM Sw. (1788)

1. **T. jaroba** Sw., Nov. Gen. & Sp. Pl.: 92 (1788).—*T. exsertum* Griseb. (1862).
High-climbing vine to 20 m., often with pendulous branches; leaflets 3 or 2 with
a long tendril, ovate, more or less rounded at base, acuminate, the acumen blunt,
up to 15 cm. long and 9 cm. broad, with several small gland-areas abaxially;
inflorescence a short corymb, terminal on a short lateral branch; calyx tubular,
truncate, glabrescent, 1–1·5 cm. long; corolla-tube gradually expanded, densely
puberulous outside, 12–15 cm. long, the lobes spreading, white; fruit oblong,
woody, smooth, 15–30 cm. long, about 7 cm. broad; seeds corky.
Cultivated and naturalized locally (St. Eliz., Han.), in secondary woodlands and
riparian forest; 50–1000 feet; fl. Feb–June, fr. Mar–Sept, Dec. *H 12092! Palmer
UCWI 2064! Proctor & Stearn 15383! Stearn 1040!* Native of Brazil and the
Guianas.

13. PYROSTEGIA C. Presl (1845)

1. P. venusta (Ker-Gawl.) Miers in Proc. Roy. Hort. Soc. London **3**: 188 (1863).
—*Bignonia venusta* Ker-Gawl. (1818). Flame Vine.
High-climbing vine to 15 m., with pendulous branches; leaflets 2 or 3 or 2 with a long tendril 3-branched at tip, ovate, acuminate, broadly cuneate to unequally rounded or subcordate at base, up to 9 cm. long and 6 cm. broad; inflorescences terminal on short lateral branches, corymbose; pedicels slender, 1–2 cm. long; calyx openly campanulate, 4–5 mm. long, shortly toothed; corolla-tube 6–7 cm. long, slightly bilaterally flattened, gradually expanded, the lobes spreading.
Common in cultivation, occcasionally relict; 300–2400 feet; fl. Nov–Feb. Native of Brazil, now widespread as an ornamental plant in tropical gardens.

14. PITHECOCTENIUM Mart. ex Meisn. (1840)

1. P. echinatum (Jacq.) Baill., Hist. Pl. **10**: 8, f. 17–20 (1888).—*Bignonia echinata* Jacq. (1760). Monkey Comb.
High-climbing vine to 25 m., with pendulous branches; stem ridged; leaflets 3 or 2 with a long tendril mostly 3-branched at tip with each of the branches 3-clawed, broadly rounded-ovate, acuminate, openly cordate at base, up to 15 cm. long and 10 cm. broad, minutely lepidote between the veins; inflorescences terminal on lateral leafy branches, shortly corymbose; calyx campanulate, truncate, 7–8 mm. long, densely tomentellous; corolla curved, 4·5–5 cm. long, densely tomentellous outside except at base of tube, lobes spreading, creamy-white with light yellow throat changing to primrose-yellow; fruit about 15 cm. long and 6 cm. broad; seeds flimsy-winged, 5–8 cm. broad.
Locally common (St. Andr., St. Cath., St. Mary), on trees near streams and in gullies; 50–1200 feet; fl. and fr. May–June. *A 8332! 9433! H 10938! Stearn 792!* Mexico to Bolivia and Argentina, Cuba, Trinidad.

15. SARITAEA Dugand (1945)

1. S. magnifica (Sprague ex Steenis) Dugand in Caldasia **3**: 263 (1945).— *Arrabidaea magnifica* Sprague ex Steenis (1927).
Scrambling vine to about 10 m.; leaflets 3 or 2 with a tendril, lowest leaves simple, obovate, shortly and bluntly cuspidate, long-tapered to the base, up to about 15 cm. long and 8 cm. broad; pseudostipules leafy; inflorescence terminal or subterminal on lateral leafy branches, short, corymbose; calyx truncate, light green, narrowed at both ends, about 1 cm. long; corolla-tube about 6 cm. long, gradually expanded, the basal 1 cm. pale, the rest bright mauve-purple, white within, lobes broadly rounded, spreading, bright mauve-purple, subequal.
Commonly cultivated as an ornamental; 700–1200 feet; fl. mostly Aug–Jan. Native of Colombia, now widely grown in the tropics.
Other members of Bignoniaceae sometimes cultivated in Jamaica include *Kigelia africana* (Lam.) Benth. (Sausage Tree), *Parmentiera cereifera* Seem. (Candle Tree) and *Pseudocalymma alliaceum* (Lam.) Sandwith (Garlic Vine).

173. PEDALIACEAE

Annual or perennial herbs, rarely shrubs. Leaves opposite or the upper alternate, simple, entire or lobed, or rarely digitately compound, without stipules. Flowers bisexual, zygomorphic, solitary or in simple dichasia in the leaf-axils. Perianth biseriate; calyx (4–) 5-merous; corolla gamopetalous, usually bilabiate with 5 imbricate lobes. Stamens epipetalous, (2 or) 4, didynamous, the posterior fifth represented by a staminode; anthers often contiguous in pairs, 2-locular, opening lengthwise. Disk present, fleshy. Ovary usually superior, 2- or falsely 4-locular; ovules anatropous, 1–many on axile placentas; style solitary, slender; stigmas 2.

Y

Fruit a loculicidal capsule or nut, often spiny or with horns, hooks or wings. Seeds smooth or reticulated, with straight embryo, with or without endosperm.

16 genera with about 50 species mostly in drier parts of the Old World tropics.

1. SESAMUM L. (1753)

Leaves entire or lobed to digitately compound; flowers solitary, with glands at bases of pedicels; corolla 2-lipped, not spurred; stamens included; ovary divided into 4 loculi by false septa; capsule with equal loculi without horns or spines.

1. S. indicum L., Sp. Pl. 2: 634 (1753).—Beniseed (Oil of Sesame).

Annual herb with simple or branched tetragonal stem 30–100 (–200) cm. high; lower leaves divided, mostly 3-partite, sparsely pubescent, up to 20 cm. long and 10 cm. broad; pedicel 2–5 mm. long; calyx-lobes linear; corolla white, pink or purple-tinged or spotted, pubescent outside, about 3 cm. long; capsule narrowly oblong, beaked, pubescent and lepidote, 2·5–3 cm. long; seeds obovoid, compressed, 2·5–3 mm. long, 1·5 mm. broad.

Sparsely cultivated and occasionally escaping; fl. Sept–Dec, fr. Sept–Jan. *Harris!* Native of the Old World tropics, now widespread.

174. GESNERIACEAE

Herbs, erect, climbing or epiphytic shrubs or rarely trees. Leaves opposite and often unequal in the pair, whorled or alternate, simple, often with large septate hairs and toothed, without stipules. Flowers bisexual, almost all zygomorphic, solitary, fascicled, umbellate or cymose-paniculate from the upper axils. Perianth biseriate, (4–) 5-merous; calyx-segments free or connate, valvate or imbricate; corolla gamopetalous, rotate or distinctly tubular, often bilabiate, the lobes imbricate. Stamens epipetalous, 2, 4 or rarely 5, if 4 didynamous and a staminode often present; anthers often coherent or connate, 2-locular, opening lengthwise or rarely by pores. Glandular disk present. Ovary superior to inferior, 1-locular; ovules numerous, anatropous, on 2 parietal often intrusive and sometimes confluent placentas; style solitary; stigma often bilobed. Fruit a loculicidal capsule or a berry. Seeds small, usually with copious endosperm.

140 genera with about 1800 species, mostly subtropical and tropical in Asia and C. and S. America.

1 Fertile stamens 2, exserted; anthers with porose dehiscence; corolla-limb usually blue, violet, pink or white, flat, much longer than tube; stem short with approximate or rosette leaves; small decorative herbaceous pot-plants with many cultivars; native of E. Africa; African Violet *Saintpaulia ionantha* Wendl.
1 Fertile stamens 4; corolla-lobes usually shorter than tube; natives of the American tropics:
  2 Ovary superior; leaves opposite, sometimes subopposite or whorled; fruit a berry:
   3 Disk annular; erect terrestrial shrubs; calyx-lobes shorter than tube; pedicels 1–4 in the leaf-axils, not pedunculate **1. Besleria**
   3 Disk a large posterior gland or irregular separate glands; herbs or undershrubs, often trailing, climbing or epiphytic:
    4 Calyx-lobes much shorter than tube; inflorescence several-flowered, umbellate, on axillary peduncles up to 4 cm. long; calyx red, 5-angled, about 1 cm. long; corolla yellow blotched red, about 1·5 cm. long; sparingly cultivated soft-stemmed herb to 60 cm. high, sometimes escaping; native of Trinidad and Panama *Chrysothemis pulchella* (Donn ex Sims) Decne.
    4 Calyx-lobes much longer than tube or sepals almost free:
     5 Stoloniferous cultivated ornamental herbs; flowers solitary or few on very short axillary peduncles; corolla often with fringed lobes, tube more or less swollen and spurred at base adaxially; leaves often variously coloured:
      6 Limb of corolla lavender with a light yellow patch, tube 2·5–3 cm. long; native of C. Amer. *Episcia lilacina* Hanstein
      6 Limb of corolla red:
       7 Corolla-tube about 2·5 cm. long, yellow with red spots within; calyx-lobes usually entire; native of Colombia and Venezuela *Episcia cupreata* (Hook.) Hanstein

7 Corolla-tube 3·5–4 cm. long, not spotted within; calyx-lobes usually toothed; native of Colombia, Brazil, the Guianas   *Episcia reptans* Mart.
5 Epipetric, epiphytic or scandent herbs or undershrubs; flowers solitary or fascicled in the leaf-axils; corolla-lobes not fringed, tube gibbous but not spurred at base; leaves sometimes purplish beneath but not variegated:
8 Corolla-tube cylindrical, lobes more or less equal and uniformly spreading   **2. Alloplectus**
8 Corolla-tube gradually expanded, lobes unequal, the 2 posterior connate, the anterior and lateral more or less reflexed   **3. Columnea**
2 Ovary wholly or partly inferior; disk annular or of 5 similar lobes; fruit a capsule:
9 Leaves alternate; undershrubs, shrubs or small trees, some rosette herbs:
10 Filaments inserted at base of corolla-tube; anthers often exserted; corolla often clear yellow, green, white or red; inflorescences various but rarely paniculate   **4. Gesneria**
10 Filaments inserted on corolla-tube well above the base; anthers included; corolla usually greenish-yellow tinged or spotted brownish-red or purple; inflorescences cymose-paniculate on long simple peduncles   **5. Rytidophyllum**
9 Leaves opposite; anthers included; herbs or shrubs:
11 Inflorescences cymose-paniculate, several times branched, terminating long axillary peduncles, up to about 30-flowered; leaves scabrid; shrub   **6. Heppiella**
11 Inflorescences otherwise, flowers usually solitary or few from leaf-axils, peduncles if present short; plants herbaceous, with scaly rhizomes:
12 Leaf-blades cordate; calyx leafy; corolla saccate at base, campanulate, violet-blue, the lower lobe fringed; ovary 10-ribbed; disk annular   **7. Gloxinia**
12 Characters not as above:
13 Disk entire or nearly so; leaves uniformly green, sometimes purplish beneath   **8. Achimenes**
13 Disk of 5 separate glands; leaves variegated; cultivated ornamentals occasionally escaping; natives of Colombia:
14 Veins of leaf purplish-brown in light green lamina; corolla rose-pink with darker spots   *Kohleria amabilis* (Planch. & Linden) Fritsch
14 Veins of leaf light green or white in dark green lamina; corolla red distally outside, yellow spotted red within
    Kohleria bogotensis (Nicholson) Fritsch

1. BESLERIA L. (1753)

1. B. lutea L., Sp. Pl. **2**: 619 (1753).
Shrub 1–2·5 m. or tree up to about 3 m. high; young leaves minutely pubescent, glabrescent; leaf-blades ovate-elliptical, cuneate, acuminate, remotely serrate, thin, up to 27 cm. long and 9 (–12) cm. broad; pedicels slender, up to 2 (–3) cm. long; calyx yellow tinged reddish, 8–10 mm. long, the teeth about 3 mm. long; corolla creamy-yellow or white, about 1·5 cm. long, the lobes rounded; ripe fruit red or reddish-orange, about 8 mm. in diameter.
Common in damp shady places; 100–5000 feet; fl. and fr. all the year. *A 5783! H 6313! J.P. 892! P 9617!* Cuba, Hispaniola, wetter islands of the Lesser Antilles south to Grenada.

2. ALLOPLECTUS Mart. (1829) nom. cons.; Stearn (1969)[b]

Calyx-segments entire, with colourless glandular hairs   1. *pubescens*
Calyx-segments laciniate with 2–5 prominent teeth on each side, without gland-tipped hairs   2. *grisebachianus*

1. A. pubescens (Griseb.) Fawcett, Prov. List Jam.: 28 (1893).—*Pterygoloma pubescens* Griseb. (1862).
Small creeping undershrub; stems 2–3 mm. thick; leaves often unequal in each pair, the blade elliptical to narrowly obovate, cuneate, acuminate, denticulate, 4·5–9 cm. long, 2–3 cm. broad, densely hirsute with hairs of 6–9 cells, some of them glandular; pedicels 7–15 mm. long; sepals narrowly elliptical, about 1 cm. long; corolla orange-red, about 2 cm. long, the lobes 1–2 mm. long; fruit subglobose, reddish, 4–5 mm. in diameter.
Rare and local (Port.), on trees and rocks in wet woodlands on limestone; 1000–2000 feet; fl. Aug–Dec, fr. Jan–Feb. *HPS 14792! Robbins UCWI 605! Weaver 1919!* Endemic.

2. A. grisebachianus (Kuntze) Urb., Symb. Ant. **2**: 357 (1901).—*Pterygoloma cristatum* Griseb. (1862). *Columnea grisebachiana* Kuntze (1891).

Straggling climber to 5 m. high or epiphytic, stems slender, much branched, when young clothed densely with crimson or colourless hairs; leaves mostly unequal in each pair, the blade elliptical or ovate, rounded at base, acuminate, margin shallowly denticulate, 2·5–6 cm. long, 1·5–3 cm. broad, rather thick and fleshy; pedicels 2–10 mm. long; sepals elliptic-lanceolate, 10–12 mm. long; corolla scarlet or crimson with paler or white lines running from base to each lobe, about 2 cm. long, the lobes 1–1·5 mm. long; fruit subglobose, white, about 7 mm. in diameter.

Occasional in the central parishes, on banks, rocks, trees and decaying logs; 1500–3000 feet; fl. Aug–Apr, fr. Aug, Dec, Mar. *A 7797! H 6488! 8413! H & P 13629! P 22648!* Endemic.

3. COLUMNEA L. (1753); Stearn (1969)[b]

By B. D. Morley

1 Corolla weakly 2-lipped, the anterior lobe 5–8 mm. long; larger leaves less than 6 cm. long; branches 1–3 mm. thick   1. *jamaicensis*
1 Corolla strongly 2-lipped, the anterior lobe more than 1 cm. long; at least larger leaves more than 6 cm. long; branches mostly 5–10 mm. thick:
   2 Calyx-lobes pinnatifid; leaves purplish-red abaxially   2. *rutilans*
   2 Calyx-lobes toothed or entire; leaves green abaxially:
     3 Calyx-lobes strongly toothed with teeth more than 1 mm. long, usually red or tinged red:
       4 Mature leaves pilose with erect 5–6-celled hairs   3. *hirsuta*
       4 Mature leaves scabridulous with appressed 2–3-celled hairs   4. *fawcettii*
     3 Calyx-lobes entire, mostly green:
       5 Mature leaves and calyx sericeous or appressed-tomentose:
         6 Corolla yellow; leaves appressed-sericeous with 8–10 celled hairs   5. *argentea*
         6 Corolla red and yellow; leaves tomentose with 5–7-celled hairs   6. *harrisii*
       5 Mature leaves and calyx pilose, hispid or scabrid:
         7 Posterior sepal free, lateral and anterior fused at base; corolla usually light yellow   7. *subcordata*
         7 All sepals free:
           8 Mature leaf strigose-scabridulous with sparse appressed 2–3-celled hairs; corolla usually yellow   8. *brevipila*
           8 Mature leaf hispid or pilose; hairs erect, 4–10-celled:
             9 Leaves not markedly unequal in each pair obovate-elliptical, with colourless hairs; corolla yellow   9. *urbanii*
             9 Leaves markedly unequal in each pair, the larger at least twice as long as the smaller, oblong-elliptical to obovate, with colourless or reddish hairs:
               10 Petioles of larger leaves 2–8 mm. long; leaves with colourless hairs 1·5–2·5 mm. long; corolla 3·5–5 cm. long, usually red and yellow   10. *proctorii*
               10 Petioles up to about 20 mm. long; leaves hispid and ciliate with red hairs up to 4 mm. long; corolla 2·5–3·5 cm. long, yellow with red hairs   11. *hispida*

1. C. jamaicensis Urb., Symb. Ant. **2**: 359 (1901).

Stems branched diffusely and spreading to 2 m. long; leaves unequal in the pair, ovate to elliptical, cuneate, acute or shortly acuminate, subentire, sparingly setose, 2·5–5·5 cm. long, 1–2·5 cm. broad; pedicels 1·5–3·5 cm. long; calyx-lobes ovate or lanceolate, 8–16 mm. long, red or green, entire or minutely toothed; corolla 2·5–3 cm. long, upper lip 5–10 mm. long, yellow to orange streaked with brownish-red, with long red multicellular hairs; fruit ovoid, white or pink, 6–8 mm. long.

Rare and local (West., Trel.), scrambling on stone walls, rocky banks, trees and rotting logs; 1200–2250 feet; fl. July–Mar, fr. Dec. *A 12794! H 10200! P 21518!* Endemic.

2. C. rutilans Sw., Nov. Gen. & Sp. Pl.: 94 (1788).

Trailing and occasionally rooting shrub with branches to 3 m. or more long; leaves unequal in the pair, the larger lanceolate, unequal at base, acute or shortly acuminate at tip, dark green usually with short hairs adaxially, crimson beneath with appressed longer hairs, up to 18 cm. long and 6 cm. broad; pedicels 1–2 cm. long; calyx-lobes with 3–6 pairs of lateral teeth, 2–3·5 cm. long; corolla 4–5·5 cm. long, upper lip to 3 cm. long, yellow or light orange streaked with scarlet; fruit globose, white, about 12 mm. broad.

Locally common in the western parishes, epiphytic or on limestone rocks; 1500–3000 feet; fl. Mar–July, fr. Apr, July. *A 12419! H 9370! P 10036!* Endemic. In certain parts of its range this species is sympatric with *C. urbanii* and hybrids having characters of each of the species combined in different ways have been described.

3. **C. hirsuta** Sw., Nov. Gen. & Sp. Pl.: 94 (1788).
Shrub with scrambling and rooting hairy branches up to 1 m. or more long; leaves unequal in the pair, the larger oblong-elliptical to obovate, rounded or cuneate at base, acute or shortly acuminate, shallowly serrate, 4·5–12·5 cm. long, 2–5·5 cm. broad; pedicels 1–2 cm. long, 1–3 together; calyx-lobes 2–4 cm. long; corolla 4–5·5 cm. long, upper lip about 2·5 cm. long, red and yellow; fruit white, about 10 mm. broad.
Common (St. Andr., Port., St. Thom.), epiphytic or on rocks in montane and submontane woodlands and thickets; 1250–5700 feet; fl. all the year, fr. Feb–Apr, July. *A 6357! 11922! H 9123! P 23255!* Endemic.

4. **C. fawcettii** (Urb.) Morton in Contrib. U.S. Nat. Herb. **29**: 8 (1944).
Scrambling shrub with branches to 2 m. or more long; leaves equal or unequal, elliptical, cuneate, shortly acuminate, remotely toothed, rather fleshy, 5–10 (–16) cm. long, 2·5–4·5 cm. broad; pedicels 8–20 mm. long, 1–3 together; calyx-lobes narrowly lanceolate, 1·6–3·3 cm. long; corolla 4–5·5 cm. long, upper lip about 2·8 cm. long, crimson to pinkish-red streaked light yellow or white; fruit ovoid, white, about 10 mm. broad.
Frequent in the eastern parishes, on trees and rocks in woodland; 400–2750 feet; fl. all the year, fr. Mar–Apr, July–Aug. *A 6505! 7611! Fawcett 6587! P 8603!* Endemic.

5. **C. argentea** Griseb., Fl. Br. W.I.: 465 (1862).
Scrambling shrub with ascending or trailing branches up to 1 m. or more long; leaves of a pair subequal, narrowly elliptical to oblanceolate, rounded or cuneate, at base, shortly acuminate, remotely toothed to subentire, greyish-silky on both surfaces, 7·5–17 cm. long, 2–4 cm. broad; pedicels 4–25 mm. long, 1–4 together; calyx-lobes linear-lanceolate, 1·4–2·4 cm. long; corolla 4–5·5 cm. long, upper lip 2–2·5 cm. long, yellow; fruit ellipsoidal, pink, 8–10 mm. broad.
Local (Manch.), on trees or on limestone rocks in thickets and woodlands; 2000–2900 feet; fl. Nov–Apr, fr. Feb. *A 11018! H 8201! P 11615!* Endemic.

6. **C. harrisii** (Urb.) Britton ex Morton in Contrib. U.S. Nat. Herb. **29**: 10 (1944).
Straggling branched shrub with rather lax slender stems up to 1 m. or more long; leaves equal or unequal, oblong-elliptical, unequally rounded at base, shortly acuminate, remotely serrate, 7–14 cm. long, 2–5 cm. broad; pedicels 5–25 mm. long, 1–4 together; calyx-lobes lanceolate, 1·7–2·5 cm. long; corolla 4–5 cm. long, upper lip about 2·3 cm. long, yellow variably striped with red; fruit red, about 10 mm. broad.
Rare and local (West., Han., St. James), epiphytic and on limestone banks; 300–1200 feet; fl. Sept–Apr. *A 12796! Fawcett 8480! H 7522! Proctor & Mullings 22010!* Endemic.

7. **C. subcordata** Morton in Contrib. U.S. Nat. Herb. **29**: 6 (1944).
Sparsely branched lax shrub with rather succulent stems up to 2 m. long; leaves unequal in the pair, the larger oblong-elliptical, obliquely subcordate at base, abruptly and shortly acuminate at apex, 8–14 cm. long, 4·5–6·5 cm. broad; pedicels 15–20 mm. long, 1–4 together; free segment of calyx linear-lanceolate, entire, 1·5–2·2 cm. long; corolla 4–5 cm. long, the upper lip about 2·5 cm. long, light yellow; fruit not known.
Rare and local (Trel.), epiphytic on trees and on rocks; 400–1500 feet; fl. Mar. *Morley 53c! Proctor & Stearn 11767! Stearn 472!* Endemic.

8. **C. brevipila** Urb., Symb. Ant. **6**: 41 (1909).
Stem scandent, rooting, up to 1 m. or more long, with minute appressed hairs; leaves subequal, oblong-elliptical, cuneate, shortly acuminate, obscurely toothed, 5–13·5 cm. long, 2–5 cm. broad; pedicels 1–3 cm. long, 1–5 together; calyx-lobes

lanceolate, entire, 2–3 cm. long; corolla 4·5–5·5 cm. long, the upper lip 2–2·5 cm. long, yellow; fruit white tinged red, about 10 mm. broad.

Local (St. Eliz., West.), epiphytic on trees; 500–2500 feet; fl. Feb–Apr. *H 10199! Proctor & Mullings 22062! Stearn 433!* Endemic.

9. C. urbanii Stearn in Bull. Brit. Mus. (Nat. Hist.) Bot. **4** (5): 220 (1969).

Scrambling shrub with stems ascending up to about 1 m. high; leaves subequal, subsessile or shortly petiolate, subrotundate to broadly obovate, broadly cuneate, obtuse or rounded at tip, subentire, coarsely hairy, 4–12 cm. long, 2–2·5 cm. broad; pedicels 6–16 mm. long, 1–3 together; calyx-lobes lanceolate, entire, 1·2–2·5 cm. long; corolla 3·5–5 cm. long, the upper lip 1·8–2·5 cm. long, light yellow; fruit pink, about 10 mm. broad.

Occasional (Clar., Manch., St. Ann), on stone walls and rocks, less often on trees; 2400–2900 feet; fl. Dec–Mar, fr. Dec. *A 10127! P 16201! Robertson UCWI 2720! Stearn 361!* Endemic.

10. C. proctorii Stearn in Bull. Brit. Mus. (Nat. Hist.) Bot. **4** (5): 222 (1969).

Scrambling shrub; stems hispid; leaves markedly unequal in the pair, the larger oblong-elliptical, unequally rounded at base, shortly acuminate, subentire, hispid on both surfaces, 5–13 cm. long, 2·5–6·5 cm. broad; pedicels 10–23 mm. long, 1–3 together; calyx-lobes narrowly lanceolate, acute, 2–3 cm. long; corolla 3·5–5 cm. long, the upper lip about 1·8 cm. long, light yellow, streaked red; fruit white or pink, about 10 mm. broad.

Occasional in the north-western parishes, epiphytic and on rocks; 400–3000 feet; fl. Dec–Apr, fr. May. *A 12420! Fawcett 8916! P 23009! Stearn 451!* Endemic.

11. C. hispida Sw., Nov. Gen. & Sp. Pl.: 94 (1788).

Lax shrub with trailing branches to 2 m. or more long, young parts with stiff red hairs; leaves markedly unequal in the pair, the large elliptical to oblong-elliptical, unequally cuneate at base, obtuse, obscurely serrate or subentire, 6–16 cm. long, 3·5–7 cm. broad; pedicels 5–10 mm. long, 2–3 together; calyx-lobes oblong-lanceolate, entire, 1·5–2 cm. long; corolla 2·5–3·5 cm. long, the upper lip about 1·5 cm. long, dull yellow; fruit white or pink.

Rare and local (St. James, Trel.), in crevices in limestone cliffs; 1600–2000 feet; fl. Feb–Mar. *Morley & Read 287! P 22577! Stearn 449!* Endemic.

4. GESNERIA L. (1753)

1 Herbs, sometimes woody, or undershrubs not over 1 m. high, often much smaller:
  2 Corolla cylindrical, red, up to 3·5 cm. long:
    3 Calyx-lobes much longer than tube, linear-lanceolate to lanceolate, prominently 3 (–5)-nerved; peduncle shorter than branched subumbellate part of inflorescence, sometimes obsolete　　　　　　　　　　　　　6. *acaulis*
    3 Calyx-lobes 2–3 times longer than tube, ovate, not prominently nerved; peduncle longer than branched umbellate to bisumbellate part of inflorescence　　　7. sp. *A*
  2 Corolla campanulate, white, white marked pink, or salmon-pink; flowers mostly on solitary axillary pedicels, sometimes paired or few together on short common peduncles:
    4 Acaulescent rosette-herb; corolla not usually over 2 cm. long, pink or red at base, the lobes toothed　　　　　　　　　　　　　　　　1. *pumila*
    4 Stem evident, sparingly or not branched, at least 15 cm. long; corolla usually more than 2 cm. long, the lobes entire or nearly so; pedicels deflexed or declinate in fruit:
      5 Calyx 4-lobed, bent upwards in fruit, lobes oblong, about 8 mm. long; leaves obovate; corolla salmon-pink, about 4 cm. long　　　　　　2. *proctorii*
      5 Calyx 5-lobed; leaves mostly oblanceolate; corolla white or white tinged rose, 2–2·5 cm. long:
        6 Leaves glabrous; pedicels 10–15 mm. long; calyx-lobes lanceolate to oblanceolate, about 8 mm. long, glabrous　　　　　　　　3. *neglecta*
        6 Leaves hairy on midrib beneath; pedicels up to 25 mm. long; calyx-lobes ovate to ovate-elliptical, about 6 mm. long, hairy　　　　　4. *mimuloides*
1 Shrubs or small trees:
  7 Calyx-limb divided less than halfway to base, openly cupular, the lobes deltate and about 6 mm. long　　　　　　　　　　　　　　　　5. *calycina*
  7 Calyx-limb divided almost or quite to the top of the ovary into free lobes:

8 Corolla red, more or less cylindrical:
  9 Pedicels 1-flowered, solitary; corolla glabrous; leaves glabrous, narrowed to a short
    rounded base; sepals lanceolate-subulate, 8–9 mm. long   8. *jamaicensis*
  9 Pedicels several together, subumbellate on a common peduncle; corolla hairy;
    leaves pubescent at least beneath:
    10 Leaves broadly rounded at base; sepals ovate-lanceolate to elliptical, up to 8 mm.
      long   9. *scabra*
    10 Leaves narrowed to a small rounded base; sepals lanceolate-subulate, 5–6 mm.
      long   10. *sphaerocarpa*
8 Corolla yellow, greenish-yellow or green, sometimes tinged pink or orange:
  11 Corolla cylindrical or narrowly campanulate:
    12 Pedicels 1-flowered, solitary; sepals filiform; corolla glabrous, 12–14 mm. long
                      11. *harrisii*
    12 Pedicels several together on a common peduncle; sepals oblong-elliptical; corolla
      hairy, 15–17 mm. long   12. *fawcettii*
  11 Corolla openly campanulate, glabrous:
    13 Leaves cuneate at base, margins remotely dentate, caudate-acuminate at tip;
      petioles 1–2 cm. long; flowers solitary:
      14 Sepals entire; pedicels about 2 cm. long; vegetative buds tomentose, apparently
        not glutinous   13. *clandestina*
      14 Sepals with one or two large teeth on the proximal margins; pedicels 2·5–4 cm.
        long; vegetative buds glutinous especially on drying   14. sp. *B*
    13 Leaves broadly cuneate to equally or unequally rounded at base, margins sub-
      entire to crenate-serrate, acute or shortly acuminate at tip; vegetative buds
      essentially glabrous but glutinous especially on drying:
      15 Flowers subumbellate, 1–4 together on axillary peduncles; sepals subulate,
        6–8 mm. long; corolla 2–2·5 cm. long; petioles up to about 3 cm. long
                      15. *exserta*
      15 Flowers solitary, axillary; sepals 1 cm. or more long; corolla showy, 3–4 cm. long:
        16 Petioles 7–12 mm. long; pedicels up to 4·5 cm. long; sepals narrowly lanceolate,
          1–2 cm. long   16. *alpina*
        16 Petioles slender, up to 3 cm. long; pedicels up to 9 cm. long; sepals sublinear,
          about 3·5 cm. long   17. *calycosa*

1. G. pumila Sw., Nov. Gen. & Sp. Pl.: 90 (1788).

Rosette herb with fibrous roots or a weak taproot; leaves oblanceolate, obtuse, coarsely crenate or dentate, almost glabrous, up to 10 cm. long and 3 cm. broad; petiole very short, pedicels slender, up to 2·5 cm. or more long, sometimes branched, sometimes deflexed; calyx-lobes lanceolate, 5–9 mm. long; corolla 1·5–1·8 (–2·5) cm. long, the abaxial lobes fimbriate; capsule ovoid, 5 mm. long.

Locally common in damp shady places on limestone rocks; 1000–3000 feet; fl. July, Oct–Mar, fr. Mar, June–Oct. *A 9091! H 10667! P 22991!* Endemic. This is a very variable species which may be susceptible to taxonomic subdivision when better known. Some variants (e.g. *H 10667*) approach *G. neglecta* closely.

2. G. proctorii Stearn, unpublished.

Erect unbranched somewhat succulent herb up to about 30 cm. high, forming colonies; stem, petiole and midrib of leaf scurfy; leaves obovate, cuneate at base, obtuse, doubly repand-denticulate, up to 10 cm. long and 4 cm. broad, with 8–10 pairs of lateral veins, paler beneath; pedicels up to 6 cm. long; fruit not known.

Rare and local (Port.), in thickets on limestone in wet area; 2000–3000 feet; fl. Apr. *Osmaston 5176! P 5732!* Endemic.

3. G. neglecta (Hook.) Kuntze, Revis. Gen. Pl. **2**: 473 (1891).—*Conradia neglecta* Hook. (1852). *G. leiocarpa* Urb. & Britton (1912).

Undershrub up to about 1 m. high; leaves mostly oblanceolate, long-cuneate at base, obtuse, coarsely crenate, up to 10 cm. long and 3 (–4) cm. broad, warted on midrib abaxially; filaments crimson, exserted; capsule obovoid, about 5 mm. long.

Rare and local (Clar.), in crevices and on ledges of limestone rocks; 2500–2800 feet; fl. Nov–Jan, fr. Mar. *H 10880! 12763! P 8213!* Endemic.

4. G. mimuloides (Griseb.) Urb., Symb. Ant. **2**: 377 (1901).—*Conradia mimuloides* Griseb. (1862).

Unbranched undershrub up to about 30 cm. high; leaves oblanceolate to obovate, long-cuneate to a very narrowly rounded base, obtuse, serrate, up to 11 cm. long and 3 (–3·5) cm. broad; filaments pink, exserted.

Rare and local (St. Andr., Port.), on moist shaded limestone ledges; 4900–5750 feet; fl. Feb, fr. July. *Harris! J. P. 1125! P 9527! 23762!* Endemic.

5. G. calycina Sw., Nov. Gen. & Sp. Pl.: 90 (1788).—*Pentaraphia calycina* (Sw.) Hanstein (1865).

Shrub 2·5–3 m. or tree to 8 m. high; leaves oblong-oblanceolate, cuneate at base, acuminate, shallowly dentate, up to 24 cm. long and 5·5 cm. broad, essentially glabrous; peduncles 6–12 cm. long; flowers subumbellate, 2–5 together; calyx about 1·5 cm. long; corolla light green, campanulate, about 2·5 cm. long; stamens exserted; capsule linear, 2-valved, splitting from apex, about 1·5 cm. long.

Local (Port.), in moist shady woodland on limestone; 1200–1500 feet; fl. Jan–Mar, July, fr. Feb–Apr. *A 9343! ACH 957! P 5682! 7635!* Endemic.

6. G. acaulis L., Syst. Nat. ed. 10, 2: 1110 (1759).

Lax herb or undershrub with very short or long stem up to 60 cm. or more long; leaves oblanceolate, long-tapered to base, usually acute at tip, margins finely crenulate to deeply doubly serrate, up to 15 (–21) cm. long and 5·5 (–6) cm. broad, glabrous or scabridulous adaxially, clustered towards the ends of branches; inflorescences compact, axillary, shorter than the leaves; pedicels up to 24 mm. long; calyx very variable, the lobes including filiform or long-tapered acumen up to 15 mm. long; corolla slightly bent and narrower distally, crimson to light scarlet or vermilion, hairy; stamens included.

Common, on cliffs, rocks and banks in shade; 25–2500 feet; fl. and fr. all the year. *A 8451! 12644! H 9003! P 22145!* Endemic. Several more or less distinct populations require biometric study.

7. G. sp. A.

Undershrub up to about 1 m. high; stem pubescent with the leaves spread uniformly along it; leaves oblanceolate, long-tapered to base, shortly acuminate, crenate-dentate, 8–15 cm. long, 1.5–3·2 cm. broad, shortly setulose-hispid adaxially, with 8–12 pairs of ascending lateral veins; peduncles up to 7 cm. long; inflorescence paniculate with slender pedicels up to 23 mm. long; calyx 5–6 mm. long, the tube 1·5–2·5 mm. long, the lobes acuminate and subulate-tipped; corolla light scarlet, 18–24 mm. long, the lobes 2–3 mm. long and denticulate, hairy; stamens exserted.

Rare and local (Manch., St. Ann), on limestone banks; 2100–2300 feet; fl. Aug, Nov. *A 12643! P 29384!* Endemic.

8. G. jamaicensis Britton in Bull. Torr. Bot. Club **48**: 341 (1922).

Shrub with straggling branches 1·2–3 m. high; leaves elliptic-lanceolate to oblanceolate, cuneate to a shortly rounded base, acute or acuminate at tip, serrate, up to 15 cm. long and 4·2 cm. broad, leathery; pedicels filiform, up to 8·5 cm. long; corolla rose-pink, 12–14 mm. long; stamens exserted; capsule obovoid, 7–8 mm. long.

Very rare (St. Eliz.); 500–1300 feet; fl. and fr. Mar, June. *H 12374! 12509!* Endemic.

9. G. scabra Sw., Nov. Gen. & Sp. Pl.: 89 (1788).

Shrub up to about 2·5 m. high; leaves oblong-elliptical, broadly rounded at base, obtuse, shallowly dentate, up to 10·5 cm. long and 4·5 cm. broad; peduncles up to 8 cm. long, pubescent, few to many-flowered; capsule obconic, 5–8 mm. long, about 7 mm. broad at top.

Uncommon (Han., St. James), in open woodlands and thickets on limestone; 500–1750 feet; fl. and fr. sporadically throughout the year. *H 9249! Osmaston 5004! P 16641! 23137!* Endemic.

10. G. sphaerocarpa Urb., Symb. Ant. **5**: 499 (1908).

Shrub up to about 2·5 m. high with straggling and drooping branches; leaves oblanceolate, long-tapered to a shortly rounded base, shortly acutely acuminate, serrate, up to 12 cm. long and 3·5 cm. broad, scabrid adaxially; peduncles up to 11 cm. long, pubescent, several flowered; capsule subspherical, 4–5 mm. long.

Local (St. Eliz.), on rocky limestone; 1000–1250 feet; fl. and fr. May–June, Sept. *H 9956! 12361! P 20845!* Endemic.

11. G. harrisii Urb., Symb. Ant. **5**: 497 (1908).

Straggling-branched undershrub or shrub 1–3 m. high; leaves clustered at ends of branches, oblong-elliptical, cuneate at base, obtuse or shortly acute, crenate-serrate distally with rather blunt teeth, glabrous, up to 9 cm. long and 3·5 cm. broad; pedicels very slender, about 3 cm. long; sepals 11–14 mm. long; corolla sulphur-yellow; capsule obconic, 5 mm. long.

Uncommon (Clar., Manch., Trel.), in crevices of limestone rocks; 1400–2500 feet; fl. Nov–June, fr. Nov–Apr. *H 8670! 9066! P 9770!* Endemic.

12. G. fawcettii Urb., Symb. Ant. **5**: 500 (1908).

Shrub 1–3 m. or small tree; leaves oblanceolate, narrowed to a short rounded base, acutely acuminate at tip, irregularly repand-dentate with acute teeth, up to 14 cm. long and 4·5 cm. broad, dark green adaxially, paler beneath; peduncle up to 5 cm. long; pedicels up to 1·5 cm. long; sepals 6–8 mm. long, toothed distally; corolla bright orange-yellow.

Rather local (West., Trel.), in damp shady places on limestone; 1400–1700 feet; fl. Mar, Aug–Sept, fr. Mar, Aug. *H 9883! Poulter in A 10893! P 15756!* Endemic.

13. G. clandestina (Griseb.) Urb., Symb. Ant. **2**: 377 (1901).—*Conradia clandestina* Griseb. (1862). *Pentaraphia clandestina* (Griseb.) Fawcett (1893).

Shrub about 2 m. high; leaves oblanceolate, cuneate at base, caudate-acuminate, shallowly remotely dentate, up to 15 cm. long and 3·5 cm. broad, thinly pilose; sepals nearly 3 cm. long; corolla yellowish-green, about 4 cm. long.

Rare and very local (St. Thom.), in damp ravine; about 2000 feet; fl. Jun–Mar, fr. Mar, Aug. *H & B 10679! P 27785!* Endemic.

14. G. sp. B.

Shrub up to about 2·5 m. high; leaves as of the last, the marginal teeth more numerous, up to 12 cm. long and 3·5 cm. broad, not pilose but scurfy with minute resinous flakes; sepals 2–3 cm. long; corolla greenish or dull rose.

Rare and very local (Port.), in woodland on limestone in wet area; 1500–2000 feet; fl. Jan–Mar, fr. Jan. *A 9108! P 9799!* Endemic.

15. G. exserta Sw., Nov. Gen. & Sp. Pl.: 90 (1788).

Shrub or tree 1–6 m. high; leaves oblong-lanceolate, broadly and unequally rounded at base, shortly acuminate, shallowly crenate-dentate to subentire, up to 15 (–22) cm. long and 6 (–7) cm. broad, usually much smaller, darker adaxially; petioles reddish; peduncles up to 5 (–6) cm. long; pedicels rather brittle, 1–4 cm. long; corolla greenish-white turning light yellow; stamens and style exserted; capsule obovoid, 10 mm. long; seeds dark reddish-brown, fusiform, about 0·5 mm. long.

Common in central and eastern parishes in thickets and on rocky banks; 600–4000 feet; fl. June–Aug, Nov–Mar, fr. July–Aug, Nov–Apr. *A 5821! H 5547! P 9841!* Endemic.

16. G. alpina (Urb.) Urb., Symb. Ant. **5**: 498 (1908).—*G. calycosa* var. *alpina* Urb. (1901).

Shrub 2–6 m. high; leaves ovate, broadly cuneate to rounded at base, shortly and broadly acuminate, serrate, up to 9·5 cm. long and 4·5 cm. broad, papery, dark green; flowers horizontal or pendulous; corolla light dull primrose-yellow, about 3 cm. long; capsule lightly 10-ribbed, about 1 cm. long.

Local (St. Thom.), in montane woodland; 5250–6500 feet; fl. Nov–Jan, fr. Nov–Feb. *A 10700! H 7547! P 9616! Stearn 100!* Endemic.

17. G. calycosa (Hook.) Kuntze, Revis. Gen. Pl. **2**: 473 (1891).—*Conradia calycosa* Hook. (1844).

Shrub or tree 2–5 m. high; leaves oblong-elliptical, broadly cuneate to rounded at base, shortly acuminate, shallowly crenate-dentate, up to 19·5 cm. long and 8 cm. broad, with 10–12 pairs of lateral veins; corolla light greenish-yellow, about 4 cm.

long; stamens and style long-exserted; capsule 16–21 mm. long excluding persistent sepals.

Occasional in western parishes in shaded gullies and on sheltered banks; 400–2000 feet; fl. most of the year, fr. May, Aug, Nov–Dec. *A 10197! H 12004! P 15684!* Endemic.

5. RYTIDOPHYLLUM Mart. (1829)

1 Leaves subsessile or with petioles not over 5 mm. long, tomentose; peduncles up to about 18 cm. long, slender; inflorescence cymose-paniculate, 3–4 times branched, the branches often unilaterally ascending   1. *tomentosum*
1 Leaves with distinct petioles mostly over 2 cm. long, not tomentose; peduncles up to over 30 cm. long, stout; inflorescence as above but more or less corymbose:
  2 Leaves scabrid adaxially; inflorescence usually at least three times branched
          2a. *grande* var. *grande*
  2 Leaves smooth adaxially; inflorescence usually only twice branched
          2b. *grande* var. *laevigatum*

1. R. tomentosum (L.) Mart. ex G. Don, Gen. Syst. **4**: 650 (1838).—*Gesneria tomentosa* L. (1753), partly. Search-me-Heart.

Undershrub or shrub up to 3 m. high, somewhat aromatic and glandular; leaves elliptic-lanceolate, long-tapered at both ends, serrate-denticulate, up to 45 cm. long and 9 cm. broad, usually light green and often with rounded false stipules developed at base; pedicels slender; calyx usually green but often tinged red, lobes 2–3 mm. long; corolla essentially greenish-yellow but often spotted or tinged purplish-red within, about 18 mm. long; capsule broadly obovoid, 8–9 mm. long.

Common on open banks and in pastures; 25–3500 feet; fl. and fr. all the year. *A 5556! 6889! J.P. 1135! P 6479!* Endemic but very closely related to *R. villosulum* (Urb.) Morton, originally described as *R. tomentosum* forma *villosulum* Urb., endemic to Cuba.

2. R. grande (Sw.) Mart. ex G. Don, Gen. Syst. **4**:650 (1838).—*Gesneria grandis* Sw. (1788). *R. grande* var. *phaeanthum* Urb. (1901). Cow's Tongue.

2a. Var. grande.

Shrub or tree 2–5 m. high, erect, sparingly branched, brittle; leaves often clustered near the branch-tips, oblanceolate, long-cuneate, acuminate, serrulate, up to 40 cm. long and 9 cm. broad, very rough adaxially with short hairs; peduncles up to 50 cm. or more long; calyx-teeth narrowly deltate, 2–5 mm. long; corolla essentially yellowish-green but often reddish-purple at base within or whole corolla brownish-red and rimmed with green, puberulous, 12–18 mm. long; capsule ovoid, about 8 mm. long.

Locally common in clearings and glades in forest or sheltered thickets on limestone; 950–3800 feet; fl. July–Sept, Jan, fr. July–Dec. *A 7912! 11569! H 8779! H & P 14167!* Endemic.

2b. R. grande var. laevigatum Adams in Phytologia **21** (2): 71 (1971).

Like the last; leaves elliptic-oblanceolate, cuneate at base, serrulate, acute or acuminate, up to 27 cm. long and 6 cm. broad; petioles up to 3 cm. long; young buds very sticky; corolla yellow or greenish-yellow, 14–16 mm. long; flowers faintly lemon-scented.

Local (St. James, Trel.), on rocky limestone in thickets and on open banks 1300–1600 feet; fl. and fr. sporadically throughout the year. *A 6786! H & P 14418 HPS 14656!* Endemic.

6. HEPPIELLA Regel (1853)

1. H. corymbosa (Sw.) Urb., Symb. Ant. **2**: 368 (1901).—*Gesneria corymbosa* Sw. (1788).

Shrub up to about 60 cm. high; leaves ovate-elliptical, rounded at base, acute, serrate-dentate, up to 9 cm. long and 4·5 cm. broad; peduncles up to about 18 cm. long; pedicels and calyx often dark red; calyx about 3 mm. long; corolla yellow to brownish-orange with red hairs, more or less blotched red within, 11–21 mm. long; anthers at mouth of corolla, dark red; filaments yellow; capsule hemispherical, about 4 mm. broad.

Very local (St. Andr.), in crevices of limestone rocks; 100–400 feet; fl. Nov–July, fr. Dec–July. *A 6967! ACH 323! H 9621! P 10205!* Endemic.

7. GLOXINIA L'Hérit. (1785)

1. G. perennis (L.) Druce in Rep. Bot. Exch. Club Br. Is. **3**: 418 (1914).— *Martynia perennis* L. (1753).

Low spreading or erect herb up to about 1·2 m. high; stem and petioles streaked crimson; leaves broadly ovate, cordate at base, rounded or obtuse at tip, sometimes cuspidate, crenate-serrate, up to about 15 cm. long and 11 cm. broad; petioles up to 8 cm. long; inflorescence subracemose with leafy bracts; sepals elliptical, up to nearly 2 cm. long and 8–10 mm. broad; corolla about 3 cm. long; flower mint-scented.

Occasional naturalized on stone walls and roadside banks, cultivated in gardens at middle elevations; 1000–2600 feet; fl. Sept, Dec–Mar. *A 12375! H 9829! H & P 13514!* Native Colombia to Brazil and Peru, introduced elsewhere; Puerto Rico.

8. ACHIMENES Pers. (1806) nom. cons.

Corolla-tube up to about 5 cm. long, bluish-violet, rarely rose or white; leaves ovate to lanceolate, purple beneath, up to 8 cm. long and 2·5 cm. broad; cultivated pot-plant and occasionally established on garden walls; native Mexico to Panama   *A. longiflora* DC.
Corolla-tube 10–15 mm. long, bright crimson to light rose-red; leaves ovate-elliptical to lanceolate, often reddish beneath; naturalized herb   1. *erecta*

1. A. erecta (Lam.) H. P. Fuchs in Act. Bot. Neerl. **12**: 15 (1963).—*Columnea erecta* Lam. (1786). *A. coccinea* (Scop.) Pers. (1806).

Slender-stemmed herb trailing at base; erect flowering shoots 10–20 cm. or more high; stems and leaves beneath often reddish, especially on the veins; leaves cuneate at base, acute or obtuse at tip, the margin with rather few large anticous teeth, 1·5–4 (–6) cm. long, 1–2 (–3) cm. broad; pedicels solitary or paired, slender, reddish, up to 3 cm. long; sepals lanceolate, 3–4 mm. long; corolla yellow in throat, the limb flat; capsule obovoid, 6–7 mm. long.

Locally common in the eastern parishes, on damp shaded rocky banks and moist ledges; 250–3750 feet; fl. Aug–Apr, fr. Nov–Apr. *A 9236! 12671! H 12866! P 18377!* Native Mexico to Panama, introduced elsewhere; Hispaniola.

175. LENTIBULARIACEAE

Herbs of damp or wet places or epiphytes, mostly insectivorous. Leaves alternate on stolons or in a basal rosette, entire, finely dissected, reduced or absent, often bearing complex traps or glands. Flowers bisexual, zygomorphic, bracteolate, in scapose racemes or solitary. Perianth biseriate; calyx 2–5-merous, the lobes open or imbricate, persistent; corolla gamopetalous, 5-merous, lobes imbricate, 2-lipped, the lower lip saccate or spurred. Stamens 2, epipetalous, inserted at base of corolla-tube, 2 staminodes often present; anthers 2-locular, opening lengthwise by a common slit. Disk absent. Ovary superior, 1-locular; ovules anatropous, 2–many on a free central placenta; style simple or 2-lobed stigma sessile. Fruit a capsule, 2- or 4-valved or circumscissile, rarely 1-seeded and indehiscent. Seeds small, without endosperm.

5 genera with about 250 species; cosmopolitan.

1. UTRICULARIA L. (1753); P. Taylor (1955)

Calyx 2-lobed, the lobes entire; insectivorous bladders on branches of finely dissected submerged leaves or on rhizoidal outgrowths; true roots lacking.

1 Terrestrial or epiphytic herbs; leaves usually entire; traps on rhizoids, leaves or stolons;
2 Habitat mossy banks or trees, terrestrial or epiphytic; flowers more than 2 cm. long:
    calyx more than 1 cm. long in fruit; bracteoles present; corolla white or light mauve;
    leaves 3 cm. or more long; traps on rhizoids only   **1.** *alpina*
2 Habitat damp open savanna, terrestrial; flowers and leaves much smaller than above;
    bracteoles absent; corolla yellow:
    3 Lower lip of corolla as long as spur or a little longer; calyx not exceeding capsule
                                                                     2. *subulata*
    3 Lower lip of corolla about half as long as spur; calyx exceeding capsule   **3.** *pusilla*
1 Floating aquatic herbs; leaves divided; traps on leaves only; corolla yellow:
    4 Pedicels recurved in fruit; leaves repeatedly divided   **4.** *foliosa*
    4 Pedicels erect in fruit; leaves up to 1 cm. long, once or twice divided   **5.** *gibba*

1. U. alpina Jacq., Enum. Syst. Pl. Carib.: 11 (1760).—*U. montana* Jacq. (1763).
Perennial herb, with tuber at base of scape; leaves ovate-lanceolate, 3–12 (–20) cm. long, 1–3 cm. broad, rather leathery, rosulate; scape to 30 cm. long; raceme 1–6-flowered, up to 15 cm. long; corolla 3–4·5 cm. long, usually white with a yellow patch on the gibbous palate, lower lip up to 5 cm. broad; spur slender; capsule subglobose, about 8 mm. in diameter; seeds fusiform.
Very rare (Port.), on mossy tree-trunks and banks; 1500 feet; fr. Mar. *Robertson UCWI 3440!* Continental tropical Amer., Lesser Antilles, Trinidad.

2. U. subulata L., Sp. Pl. **1**: 18 (1753).—*Setiscapella subulata* (L.) Barnh. (1913).
Erect herb, 5–25 cm. high; leaves linear, up to 2 cm. long; inflorescence slender, 1–several-flowered, up to 10 cm. long; pedicels erect or spreading, 2–8 mm. long; calyx-lobes orbicular; corolla with upper lip emarginate, lower lip 3-lobed; spur subulate; capsule globose, about 1·5 mm. in diameter; seeds ovoid, striate.
Rare and local (Clar., St. Eliz.), in wet open savannas; SL–2200 feet; fl. Mar, Dec. *A 5929! Cornman!* E. United States, continental tropical Amer., Greater Antilles, Trinidad, tropical Africa.

3. U. pusilla Vahl, Enum. Pl. **1**: 202 (1804).—*Setiscapella pusilla* (Vahl) Barnh. (1925).
Erect herb up to 12 cm. high; leaves few, linear-oblanceolate, scattered on the stolons; inflorescence slender, usually several-flowered, up to 7 cm. long; pedicels spreading, up to 4 mm. long; calyx-lobes ovate; corolla with upper lip entire, lower lip 3-lobed; spur subulate; capsule globose, about 1·5 mm. in diameter; seeds ovoid, striate.
Very local (Clar.), in puddles and pond margins in open savanna; 50–2300 feet; fl. and fr. Jan–Mar, May, Aug, Nov–Dec. *A 5933! A & C 1136! P 24602!* Continental tropical Amer., West Indies.

4. U. foliosa L., Sp. Pl. **1**: 18 (1753).
Stolons long, branched, about 2 mm. thick, often pinkish; leaves usually either with many traps or with very few or none; scape up to 20 cm. long, several-many-flowered; pedicels 5–15 mm. long; corolla 6–11 mm. long, upper lip orbicular, lower lip suborbicular, entire or emarginate; spur conical, shorter than the lower lip; capsule globose, 4–6 mm. in diameter; seeds lenticular with a membranous wing, 2–2·5 mm. in diameter.
Locally common (St. Eliz., West.), in swamps, ditches, ponds and sluggish streams; SL–50 feet; fl. and fr. almost all the year. *A 10166! 12052! P 24526! Yuncker 18001!* S. United States, continental tropical Amer., Cuba, Hispaniola, Trinidad, tropical Africa.

5. U. gibba L., Sp. Pl. **1**: 18 (1753).—*U. obtusa* Sw. (1788).
Stolons slender, numerous from base of scape; scape up to 10 cm. long, flexuous few-flowered; pedicels 8–18 mm. long; corolla 6–10 mm. long, upper lip broadly ovate, lower lip orbicular, entire; spur conical, slightly longer than the lower lip; capsule globose, about 2 mm. in diameter; seeds lenticular with a narrow wing, about 1 mm. in diameter.
Occasional in ponds and marshes; 25–2850 feet; fl. and fr. Dec–May. *A 10837! 12051! P 15886! Stearn 143!* S. United States, continental tropical Amer., West Indies, tropical Africa.
Pinguicula elongata Benjamin was reported for Jamaica by Grisebach. This record was based on a specimen collected by Purdie in Colombia.

176. ACANTHACEAE

Herbs, shrubs or trees; shoots often angled and swollen above the nodes. Leaves opposite rarely in whorls of 3, simple, usually entire, often with conspicuous cystoliths, without stipules. Flowers bisexual, zygomorphic or subregular, solitary, fascicled or cymose in the leaf-axils or in terminal or axillary cymes, racemes, spikes or panicles, often with large bracts. Perianth biseriate; calyx with 4 or 5 segments or lobes, imbricate or valvate, or reduced to a ring; corolla gamopetalous, usually 2-lipped, sometimes subregular or 1-lipped, the lobes imbricate or contorted. Stamens 4, didynamous, with or without a staminode, or 2 with or without 2 (–3) staminodes, very rarely 5, epipetalous and alternate with the corolla-lobes; filaments free or connate in lateral pairs; anthers 2-locular, sometimes with the loculi unequal, or 1-locular, opening lengthwise; pollen variously grooved, honeycombed or otherwise sculptured. Disk present. Ovary superior, 2-locular; ovules anatropous or amphitropous, 2 or more in each loculus, usually superposed on axile placentas; style simple, filiform; stigma capitate or lobed. Fruit very rarely a berry, more usually a capsule, often narrowed to both ends, mostly elastically dehiscent, the valves recurving and leaving the central axis on which the seeds were borne on curved retinacula (indurated funicles). Seeds usually flat; endosperm absent or very thin.

250 genera with about 2500 species in the tropics.

1 Flowers solitary or paired in leaf-axils or in terminal racemes with inconspicuous bracts, long-peduncled; short inconspicuous calyx and base of corolla enclosed by a pair of large spathaceous bracteoles; ovules collateral; capsule abruptly beaked; seeds hemispherical with a hollow on the lower surface; funicle not hooklike; plants often twining
                                                              1. Thunbergia
1 Flowers in bracteate heads or spikes, or loose racemes, cymes or panicles, or solitary or fascicled in leaf-axils, never both long-peduncled and conspicuously bracteolate; ovules mostly superposed; capsule mostly linear-oblong, ellipsoid or clavate, often dorsi-ventrally compressed, not long-beaked; seeds usually compressed; funicle hooklike; plants sometimes trailing but not climbing:
2 Inflorescence of terminal, or rarely axillary, simple or compound spikes or heads with conspicuous imbricate bracts:
3 Bracts mostly more than 2 cm. broad, red; corolla tubular with very short equal revolute lobes, yellow; fertile stamens 2, exserted   **2. Sanchezia**
3 Bracts less than 2 cm. broad; corolla-lobes well-developed:
4 Leaves of a pair markedly unequal in size, serrate; stamens 4   **3. Goldfussia**
4 Leaves of a pair more or less equal:
5 Bracts less than 3 mm. broad, bristle-tipped; calyx-segments 5; corolla more or less 2-lipped, about 5 mm. long, the adaxial lobe shorter and entire; stamens 4
                                           4. Teliostachya
5 Bracts up to 15 mm. broad:
6 Corolla subregular to zygomorphic, the lobes more or less equal and spreading, or if 2-lipped stamens strongly didynamous:
7 Stamens 2, exserted; staminodes 2 arising from bases of filaments of fertile stamens; sepals 5, half-connate; corolla-lobes equally spreading, contorted in bud, deep blue; bracts with white patches between the veins; ornamental shrub sometimes escaping; native of N. India and S. China
                                      Eranthemum pulchellum Andr.
7 Stamens 4:
8 Sepals 4, very unequal; stamens unequal, 2 exserted, 2 very short and included with or without a staminode between; corolla with anterior lobe longer than the others, yellow; aestivation imbricate; shrubs with supra-axillary spines
                                           5. Barleria
8 Sepals or calyx-lobes 5; stamens equal or only slightly unequal; anthers included or at the corolla-throat; corolla-lobes equal or nearly so, not yellow:
9 Corolla 1-lipped, the 5 lobes directed sideways and downwards, light vermilion; aestivation rolled-imbricate, the anterior corolla-lobe completely enclosed; sepals free, the adaxial 2-toothed at apex; cultivated shrub, native of tropical Asia   *Crossandra undulifolia* Salisb.
9 Corolla-lobes spreading in all directions, white or mauve; sepals shortly connate, all entire:
10 Herbs; anthers 2-locular, in the corolla-throat; calyx-lobes valvate; staminode absent; bracteoles lanceolate, larger than sepals; corolla-lobes contorted   **6. Blechum**

10 Shrub; anthers 1-locular, included; calyx-lobes imbricate; staminode
    filamentous; bracteoles triangular-ovate, much smaller than sepals;
    corolla-lobes imbricate   **7. Neriacanthus**
6 Corolla distinctly 2-lipped; stamens 2:
  11 Anther-loculi one above the other; upper lip of corolla straight; corolla-lobes
     imbricate; calyx-segments 5   **8. Drejerella**
  11 Anther-loculi side by side or nearly so:
    12 Stems creeping and rooting; plant herbaceous; leaves variegated with white
      veins; corolla light yellow; ornamental pot-plant, occasionally naturalized
      (St. Mary); native of continental tropical America
                           Fittonia argyroneura E. Coem.
    12 Stems erect; plant shrubby; leaves all green:
      13 Corolla white, lined purple, lower lip longer than the tube, upper lip curved;
        inflorescences mostly axillary   **9. Justicia**
      13 Corolla red, lower lip shorter than the tube, upper lip straight; inflorescence
        terminal   **10. Pachystachys**
2 Inflorescence not spicate or capitate or if spicate bracts small and not overlapping:
  14 Flowers solitary, fascicled or in contracted cymes in the leaf-axils:
    15 Flowers pedicelled; stamens 2; staminodes 2; shrubs or small trees, often with
      opposite spines   **11. Oplonia**
    15 Flowers sessile or subsessile; stamens 4, didynamous:
      16 Calyx-segments 4, the anterior-posterior larger; corolla 3 cm. or more long,
        subregular   **5. Barleria**
      16 Calyx-segments 5, equal or almost so:
        17 Corolla 2-lipped, less than 2 cm. long   **12. Hygrophila**
        17 Corolla more or less regular, 3–4 cm. long   **15. Ruellia**
  14 Flowers in axillary or more usually terminal peduncled cymes, spikes, racemes,
    thyrses or panicles:
    18 Corolla more or less regular with 5 spreading lobes; stamens 4:
      19 Leaves of a pair very unequal in size, serrate; corolla resupinate with 2 vertical
        rows of hairs within, aestivation contorted; filaments united in a band,
        unequal   **3. Goldfussia**
      19 Leaves of a pair equal; corolla not resupinate, without rows of hairs inside:
        20 Corolla narrowly funnel-shaped, almost cylindrical, the lobes very short,
          imbricate; stamens equal, included; shrub   **13. Salpixantha**
        20 Corolla broadly funnel-shaped with conspicuous lobes; stamens didynamous
          and with the filaments connate in pairs; herbs:
          21 Inflorescence racemose, one-sided; corolla-lobes imbricate in bud; seeds
            rugose or tuberculate, without mucous hairs   **14. Asystasia**
          21 Inflorescence cymose; corolla-lobes contorted in bud; seeds with mucous
            hairs swelling conspicuously in water   **15. Ruellia**
    18 Corolla distinctly 2-lipped:
      22 Fertile stamens 4   **15. Ruellia**
      22 Fertile stamens 2:
        23 Stems 6-ribbed; each flower subtended by 2 bracts and 4 bracteoles
                           16. Dicliptera
        23 Stems 4-ribbed or -angled; each flower subtended by 1 bract and 2 bracteoles:
          24 Shrubs with large leaves, mostly more than 10 cm. long and 4 cm. broad;
            flowers 1·5–5 cm. long; staminodes present:
            25 Inflorescence less than 6 cm. long; corolla 3–3·5 cm. long
                           17. Graptophyllum
            25 Inflorescence more than 8 cm. long, interrupted, the flowers clustered at the
              lower nodes; corolla about 2·5 cm. long, purple, or white; introduced but
              not recently found; native of Lesser Antilles and Trinidad
                        Odontonema nitidum (Jacq.) Kuntze
        24 Herbs with smaller narrow leaves; flowers less than 1·5 cm. long:
          26 Capsule up to 4-seeded; flowers sessile or nearly so; corolla up to 8 mm. long;
            anther-loculi not at the same height   **9. Justicia**
          26 Capsule up to 12-seeded; flowers pedicelled; corolla 11–13 mm. long; anther-
            loculi at the same height   **18. Andrographis**

1. THUNBERGIA Retz. (1780) nom. cons.

1 Erect or straggling-branched shrubs with brittle tetragonal stems; corolla-aestivation
  imbricate; cultivated ornamentals:
2 Leaf-blade obtuse or rounded at base, subentire, up to 11 cm. long; bracteoles about
  3 cm. long; corolla-limb about 6 cm. across, deep purple with yellow throat; native
  of east tropical Africa   *T. affinis* S. Moore

2 Leaf-blade cuneate at base, angular-toothed, smaller than above; bracteoles 1·5–2 cm.
  long; corolla-limb up to 4 cm. across, dark or light purple, light blue or white with
  yellow throat; native of west tropical Africa   *T. erecta* (Benth.) T. Anders.
1 Twining vines; corolla-aestivation contorted:
3 Petiole winged; calyx-lobes about 10, subulate; corolla orange, yellow or rarely white
  with or without a blackish-purple eye, tube about 2 cm. long   1. *alata*
3 Petiole not winged:
4 Calyx with about 12–15 slender subulate teeth; corolla white with greenish eye, up to
  5 cm. long (tube 2·5–3 cm.)   2. *fragrans*
4 Calyx entire, ringlike; corolla light bluish-mauve or rarely white, 8 cm. long (tube
  4–4·5 cm.)   3. *grandiflora*

1. T. alata Boj. ex Sims in Curt., Bot. Mag. **52**: t. 2591 (1825).—Black-eyed Susan.
  Stems slender; leaf-blade pubescent on both surfaces, about 5 cm. long and
broad; flowers usually solitary in the leaf-axils; bracteoles green, keeled, lightly
fused, about 1·8 cm. long; capsule with beak 2–2·5 cm. long.
  Common in cultivation and as an escape trailing on the ground at roadsides and
on shrubs and trees; 500–2250 (–3500) feet; fl. and fr. most of the year. *A 12119!*
H & P 13469! P 10188! Native of E. and S. Africa, introduced into many warm
countries.

2. T. fragrans Roxb., Pl. Coast Coromand. **1**: 47, t. 67 (1796).—*T. laevis* Nees
(1832). White Nightshade.
  Stem slender; leaf-blade lanceolate to ovate, subentire to repand-dentate, long-
acute, truncate to subcordate at base, up to 12 cm. long and 6 cm. broad, puberulous on both surfaces; flowers mostly solitary in the leaf-axils; bracteoles green,
keeled, free, 1·5–2 cm. long; capsule with compressed beak, about 2·5 cm. long,
pubescent.
  Widely naturalized on fences and in thickets and waste places; 50–2800 feet;
fl. and fr. most of the year. *A 5513! A. M. Barry! H & P 14268!* Native of tropical
Asia, now widespread.

3. T. grandiflora Roxb. in Edw., Bot. Regist. **6**: t. 495 (1820).
  Robust climber from large tuberous woody stock, the flowering branches often
pendulous; leaf-blade broadly ovate, angular, cordate, up to about 15 cm. long and
broad, rather scabrid; flowers axillary or decussate in one-sided terminal racemes;
bracteoles yellowish-green, gland-dotted, not keeled, lightly fused adaxially, 4 cm.
long; capsule not formed in Jamaica.
  Cultivated and relict on fences, shrubs and trees; mostly at low elevations; fl.
nearly all the year. *H 11895! A. von der Porten!* Native of India, now widespread in
cultivation in warm countries.

2. SANCHEZIA Ruiz & Pav. (1794)

1. S. nobilis Hook. f. in Curt., Bot. Mag. **92**: t. 5594 (1866).
  Robust herb or shrub to 2 m. or more high; leaf-blade elliptical, acuminate,
narrowed to winged petiole, up to 25 cm. or more long, midrib white; inflorescence
usually paniculate, the flowers in compact cymes subtended by showy bracts.
  Occasional on sheltered banks and sometimes forming thickets near streams,
originally a garden escape; 400–2700 feet; fl. Jan–June. *P 20714! Robertson UCWI
5512!* Native of Ecuador, now widely distributed in cultivation.

3. GOLDFUSSIA Nees (1832)

1 Leaves narrowed at base into a petiole; calyx-segments 5, more or less equal:
2 Stems and leaves hairy; base of leaf-blade markedly asymmetric; flowers crowded in
  short axillary spikes with large leafy bracts   1. *glomerata*
2 Stems and leaves glabrous; base of leaf-blade more or less symmetrical; flowers in
  loose panicles with small bracts   2. *colorata*
1 Leaves sessile with a broad clasping base; calyx 2-lipped, the adaxial lip 3-fid, the
  abaxial 2-fid; flowers in elongated spikes with conspicuous bracts; corolla purple;
  cultivated in gardens; native of Burma   *Perilepta dyerana* (Mast.) Bremek.

1. G. glomerata Nees in Wall., Pl. Asiat. Rar. **3**: 88 (1832).—*Strobilanthes glomerata* (Nees) T. Anders. (1867).
  Bushy herb or scrambling shrub to 2 m. high; leaves unequally elliptical,

acuminate, rounded at base on one side, cuneate on the other, 1·5–15 cm. long, 1–6·5 cm. broad; corolla purplish-blue, 5–5·5 cm. long.

Local (St. Andr., Port., St. Thom.), on pathside and streamside banks, forming thickets in sheltered places; (2000–) 3500–5050 feet; fl. most of the year. *A 5799! H 10027! P 6184!* Native of N.E. India and Burma.

2. G. colorata Nees in Wall., Pl. Asiat. Rar. **3**: 89 (1832).—*Strobilanthes colorata* (Nees) T. Anders. (1867).

Shrubby herb to over 2 m. high; leaves elliptical, caudate-acuminate, cuneate to rounded at base, up to 20 cm. long and 8 cm. broad, dark green above, reddish-purple beneath; bracteoles ovate, 3 mm. long; calyx-lobes oblong, obtuse, 5–7 mm. long; corolla light mauve to white with fine bluish veins, 3·2 cm. long.

Local (St. Andr., Port.), on moist shaded banks and forming thickets on bouldered streambanks; 2400–3500 feet; fl. Jan–Mar, Aug. *A 6340! J.P. 49! P 23209!* Native of eastern Himalayas.

4. TELIOSTACHYA Nees (1847)

1. T. alopecuroidea (Vahl) Nees in Mart., Fl. Bras. **9**: 72 (1847).—*Ruellia alopecuroidea* Vahl (1798). *Lepidagathis alopecuroidea* (Vahl) R. Br. ex Griseb. (1862).

Branched herb with slender 4-angled trailing and ascending stems to 50 cm. long; leaves ovate to elliptical, up to 5 cm. long and 2 cm. broad; spikes 1·5–4 (–8) cm. long; bracts strongly nerved, ciliate; corolla light mauve or almost white; capsule shorter than bracts.

Rather local (St. Andr., Clar., Port.), in open damp savannas and shaded pathside banks; 2000–2300 feet; fl. Dec–Apr, fr. Dec–July. *A 5910! P 10107! Stearn 231!* Widespread in the tropics.

5. BARLERIA L. (1753)

1 Plant spineless; corolla violet, pink or white, 6–7 cm. long, the anterior lobe free only near the top of the tube; anterior-posterior sepals about 2 cm. long, pectinate-ciliate; flowers solitary or in small cymes in the axils   1. *cristata*
1 Plant with paired supra-axillary branched spines; corolla yellow, 3–4·5 cm. long, the anterior lobe free from about the middle of the tube; all 4 sepals entire:
   2 Leaves mostly linear-elliptical, 4–10 times as long as broad, usually with a red midrib; inflorescence spicate, terminal; bracts suborbicular or broadly elliptical, closely imbricate, hiding the calyx   2. *lupulina*
   2 Leaves elliptical, 2–3 times as long as broad; flowers in terminal and axillary spikes; bracts spreading, narrowly elliptical, not hiding the calyx   3. *prionitis*

1. B. cristata L., Sp. Pl. **2**: 636 (1753).

Bushy leafy shrub 60–150 cm. high; leaves elliptical, scabridulous with short stiff hairs, up to 8 cm. long and 3 cm. broad; bracteoles linear-subulate, about half as long as the larger sepals; stamens abaxial in the corolla; fruit not formed in Jamaica.

Common in cultivation, very rarely relict or naturalized; mostly at low elevations; fl. almost all the year. *P 23101! Yuncker 17857!* Native of India, introduced into many warm countries.

2. B. lupulina Lindl. in Edw., Bot. Regist. **18**: t. 1483 (1832).

Erect or straggling shrub 60–240 cm. high, much branched; leaves up to 12 cm. long and 2 (–2·5) cm. broad; spikes 5–7 cm. long; outer sepals ovate, 11 mm. long, 5 mm. broad; inner sepals lanceolate, 9–10 mm. long, 2 mm. broad; capsule ovate-lanceolate, 1·3–1·5 cm. long with one seed in each loculus; seed about 8 mm. long densely covered with matted hairs.

Naturalized in waste places on banks and in hedgerows, mostly on sandy soil; 250–1000 feet; fl. Dec–Sept, fr. Feb–Sept. *A 7970! H 6790! P 16170!* Native of Madagascar, introduced elsewhere; Hispaniola, Grenada, Barbados, Trinidad.

3. B. prionitis L., Sp. Pl. **2**: 636 (1753).

Erect shrub 60–180 cm. high; leaves up to 10 cm. long and 4 cm. broad; spikes indefinite, the lower bracts not clearly distinguishable from leaves; sepals acuminate, spine-tipped, about 1·5 cm. long; capsule beaked, 1·5 cm. long; seed as above.

Locally common and gregarious mostly in open low-lying waste ground; SL–2250 feet; fl. and fr. Jan–June. *A 6215! P 20700! Robertson UCWI 5378!* Native of India, introduced elsewhere; Antigua, Grenada, Barbados.

6. BLECHUM P. Browne (1756)

1 Corolla about 3 cm. long, the tube about 1·7–2 cm. long; bracts usually glabrous, sometimes minutely ciliate with very short hairs; bracteoles ciliate with hairs scarcely 1 mm. long; leaves acuminate   1. *blechoides*
1 Corolla up to 1·8 cm. long, the tube to about 1·5 cm. long; bracts ciliate with some hairs at least 1 mm. long; bracteoles ciliate with hairs sometimes 2 mm. long:
  2 Leaves acute; bracts ciliate with hairs mostly less than 1 mm. long; corolla-tube slender, curved, gradually expanded above the base; tube 8–14 mm. long; limb 3–11 mm. across, usually light mauve   2. *pyramidatum*
  2 Leaves acuminate; bracts ciliate with hairs mostly 1–2 mm. long; corolla-tube broader, abruptly expanded at about 3 mm. above the base into the funnel-shaped mouth, limb 11–15 mm. across, usually white   3. *killipii*

1. **B. blechoides** (Sw.) Hitchc. in Rep. Miss. Bot. Gard. **4**: 115 (1893).—*Ruellia blechioides* Sw. (1788). *B. laxiflorum* Juss. (1807).

Shrubby herb 90–120 cm. high; leaves ovate to lanceolate, broadly cuneate at base, up to 10 cm. long and 4 cm. broad, slender-petioled; spikes 2–6 cm. long; corolla light violet or white; capsule shortly beaked, 7–8 mm. long.

Rather local (St. Andr., and western parishes), in sheltered woodlands on limestone or shale; (300–) 900–2000 (–4000) feet; fl. and fr. Sept–Apr. *A 10199! H 12506 P 22204!* Cuba.

2. **B. pyramidatum** (Lam.) Urb., in Fedde, Repert. Sp. Nov. **15**: 323 (1918).—*Barleria pyramidata* Lam. (1785). *Blechum brownei* Juss. (1807).

Low trailing branched herb with ascending flowering shoots 5–60 cm. high, rooting at lower nodes; leaves elliptical or ovate to lanceolate, 1–6 (–9) cm. long, 0·8–2·5 (–4) cm. broad; spikes 2–5 cm. long, sometimes flowers solitary in the axils of leaves; corolla light mauve, the tube white at base, rarely all white; capsule ovoid, about 12–16 seeded, the septum incomplete above, about 6 mm. long, puberulous.

Common as a weed of roadsides, field margins and waste places; SL–2700 (–3500) feet; fl. and fr. Nov–Aug. *A 6213! H 6407! Proctor & Mullings 22034!* Throughout tropical Amer.; Grand Cayman; introduced into the Pacific.

3. **B. killipii** Leonard in Journ. Acad. Sci. Washington **32**: 184 (1942).

Lax herb 30–90 (–200) cm. high; stems obtusely 4-angled, pubescent; leaves ovate to lanceolate, 2·5–6 (–9) cm. long, 1·5–2 (–4) cm. broad; spikes short and rather open; corolla usually white, rarely very light mauve, fragrant; capsule 6 mm. long, puberulous.

Rare and local (Port., St. Thom.), on shaded rocky banks and ledges and in ravines in woodland on limestone or serpentine; 400–2100 feet; fl. Dec–July, fr. Jan–July. *A 7605! A& S 3134! P 16253! 23793!* Endemic. Plants intermediate between *B. killipii* and *B. pyramidatum* occur in intermediate localities and habitats; in shady woodlands in Portland and St. Thomas, at elevations from about 2000 to 3500 feet, plants with the small but pure white corolla of otherwise typical *B. pyramidatum* have the more erect habit and narrower acuminate leaves of *B. killipii*.

7. NERIACANTHUS Benth. (1876)

1. **N. purdieanus** Benth. in Benth. & Hook. f., Gen. Pl. **2**: 1102 (1876).—*Salpixantha purdieana* (Benth.) S. Moore (1927).

Shrub 2–3·6 m. high; branches terete, brittle; leaves elliptical to obovate, obtuse or rounded at tip, cuneate at base, margins sometimes recurved, punctate beneath, up to 8·5 cm. long and 4·5 cm. broad; spikes few-flowered, up to about 6 cm. long, including the stout peduncle; bracts obovate, leafy; corolla-tube slender, greenish-white, 1·6–2·2 cm. long, limb white, fragrant; capsule 1·5–1·7 cm. long, oblanceolate, blunt, shiny, glabrous; seeds 1 or 2, broadly spurred, papillose.

Rare and local (Manch., Trel), on cliffs and wooded limestone hilltops; 1300–2200 feet; fl. Jan–Apr, fr. Apr–May. *A 6801! H 10669! P 20760!* Endemic.

8. DREJERELLA Lindau (1900) [see note on p. 790]

1 Spikes drooping or horizontal, 3–10 cm. long; bracts light green turning brownish-orange or red; corolla white with red markings; cultivated ornamental; native of Mexico; Shrimp Plant   *D. guttata* (Brandegee) Bremek.
1 Spikes erect or ascending, 2–4 (–6) cm. long; bracts mostly green or tinged purplish; corolla pink, crimson or purple; wild plants:
  2 Shoots densely covered with soft spreading hairs to 2 mm. long; petioles 2–6 (–10) mm. long; corolla 3–3·5 cm. long, rose-pink   1. *jamaicensis*
  2 Shoots sparsely hairy with short inconspicuous hairs or almost glabrous; petioles mostly 1–7 cm. long:
    3 Leaf-tip acuminate; bracts lanceolate to narrowly ovate, acute; corolla up to 3 cm. long, rose-pink to crimson   2. *nemorosa*
    3 Leaf-tip obtuse or acute; bracts rotundate to elliptical, rounded or abruptly acute; corolla 3·5–4 cm. long, dull crimson or rosy purple   3. *blechoides*

1. D. jamaicensis (Britton) S. Moore in Journ. Bot. **65**: 221 (1927).—*Jacobinia ? jamaicensis* Britton (1914). *Justicia jamaicensis* (Britton) Stearn (1971).
Straggling shrub 60–120 cm. high; leaves ovate to lanceolate, mostly long-attenuate-acuminate, broadly cuneate and shortly petioled or rounded-subcordate and subsessile, purplish beneath when young, up to 15 cm. long and 6 cm. broad; spikes up to 6 cm. long, the lower bracts up to 2 cm. long and nearly 1 cm. broad.

Rare and local (Clar., Trel.), on cliffs and in thickets on rocky limestone; 1700–2500 feet; fl. May–Jan, fr. Jan. *A 12822! H 10978! P 11051!* Endemic.

2. D. nemorosa (Sw.) Lindau in Urb., Symb. Ant. **2**: 223 (1900).—*Justicia nemorosa* Sw. (1788). *Beloperone nemorosa* (Sw.) Nees (1862).
Straggling shrub to 1·5 m. high; leaves ovate to ovate-lanceolate, gradually acuminate, cuneate at base, up to 12 cm. long and 6 cm. broad; spikes 1·5–3 cm. long, the lower bracts about 12 mm. long; capsule pubescent, 9 mm. long.

Occasional on rocky ledges in shaded ravines in limestone or shale areas; 50–3100 feet; fl. Mar, Sept, fr. Mar. *H 5650! H & B 10569! P 23334!* Endemic.

3. D. blechoides Lindau in Urb., Symb. Ant. **6**: 43 (1909).—*Justicia blechoides* (Lindau) Stearn (1971).
Bushy shrub 40–90 cm. high; leaves elliptical to ovate, tapered to an obtuse or acute tip, hardly acuminate, broadly cuneate at base, up to 10 cm. long and 4·5 cm. broad; spikes 2–4 cm. long, the lower bracts about 12 mm. long.

Rare and local (Trel.), on sheltered limestone rocks and cliffs; 1250–2800 feet; fl. Dec–Apr. *A 6735! H 8958! Powell 775! P 15755!* Endemic. Further study may show that *D. blechoides* and *D. nemorosa* are not really distinct.

9. JUSTICIA L. (1753)

1 Erect shrub; leaves up to 9 cm. broad; spikes dense with conspicuous overlapping bracts; corolla about 3 cm. long   1. *adhatoda*
1 Trailing or decumbent-branched herbs; leaves hardly exceeding 3 cm. broad; inflorescence with small subulate bracts often shorter than the calyx of the remote flowers:
  2 Inflorescences axillary and terminal with branches 3–7 in whorl, with a few glan-tipped hairs; corolla 3–5 mm. long   2. *comata*
  2 Inflorescences terminal, simple or with alternate branches, with numerous gland-tipped hairs; corolla 7–8 mm. long   3. *pectoralis*

1. J. adhatoda L., Sp. Pl. **1**: 15 (1753).—*Adhatoda zeylanica* Medic. (1790).
Shrub with ascending appressed-pubescent branches, 1–2·5 m. high; leaves lanceolate, acuminate, cuneate at base, up to 25 cm. long and 9 cm. broad; bracts elliptical, 1·5–2 cm. long; bracteoles about 12–14 mm. long; corolla white lined purple; filaments thicker than the style.

Rare and local (St. Thom.), an escape from cultivation on shaded waste ground; 1800–2000 feet; fl. Jan–Mar. *A 10258! H 7642! P 23321!* Native of East Indies, introduced and often naturalized elsewhere; Cuba, Guadeloupe, Martinique, Barbados.

2. J. comata (L.) Lam., Encycl. Méth. Bot. **1**: 632 (1785).—*Dianthera comata* L. (1759).

Stems slender, decumbent and often rooting from the lower nodes, rarely erect, up to 60 cm. high; leaves lanceolate to elongate-elliptical, long-tapered to both ends, almost glabrous, 2–10 cm. long, up to 3 cm. broad; common peduncles 1·5–4 cm. long, the terminal inflorescences with several superposed whorls of branches; flowers unilateral in the ultimate spikes; calyx about 1·5 mm. long; corolla white or tinged mauve; capsule 3·5 mm. long.

Frequent in ditches and at muddy pond margins; SL–1100 feet; fl. and fr. Aug–May. *A 10346! H 11941! P 21846!* British Honduras and Guatemala to Colombia, Greater Antilles, Trinidad and Tobago.

3. J. pectoralis Jacq., Enum. Syst. Pl. Carib.: 11 (1760).—*Dianthera pectoralis* (Jacq.) Murr. (1784). Fresh Cut.

Stems trailing and rooting sparingly from the nodes; internodes slender with 1 or 2 lines of curved hairs; flowering branches somewhat ascending to 60 cm. high; leaves linear to ovate-lanceolate, mostly long-acute, cuneate to broadly rounded at base, 3–8 cm. long, 0·5–3 cm. broad; inflorescence 4–10 (–12·5) cm. long; flowers more or less unilateral on the branches; calyx about 2 mm. long; corolla purple with white markings; fruit not seen.

Rare (Trel. St. Mary), by pathsides and in secondary thickets, occasionally cultivated; 40–1500 feet; fl. Jan–May. *A 12245! Morley 114!* Mexico to northern S. Amer., West Indies.

10. PACHYSTACHYS Nees (1847)

1. P. coccinea (Aubl.) Nees in DC., Prodr. **11**: 319 (1847).—*Justicia coccinea* Aubl. (1775).

Weak-stemmed erect or sometimes scrambling shrubby herb 1–3 m. high; stems glabrescent, swollen above the nodes; leaves broadly elliptical to obovate, acuminate, mostly cuneate or shortly and often unequally rounded at base, very thin when dry, up to 25 cm. long and 13 cm. broad; spikes 6–15 cm. long; bracts ovate to elliptical, up to 2·5 cm. long; corolla 5–6 cm. long; fruit not seen.

Locally common, a weed in shady waste places near streams, mostly in alluvial soils; 20–2000 feet; fl. most of the year. *A 5515! H 11933! P 8031!* Native of Guyana, probably naturalized in the West Indies; introduced into cultivation as an ornamental in the Old World tropics.

11. OPLONIA Raf. (1838); Stearn (1971)

1 Corolla white; stamens always included (homostyled), filaments about 0·8 mm. long and shorter than the anthers; low shrub up to 90 cm. high; leaves mostly ovate or broadly ovate, 1–2·5 cm. long, up to about 1·5 cm. broad, shortly rounded and minutely apiculate at tip, rounded to subtruncate at base; plant with slender axillary spines
1. *acicularis*
1 Corolla crimson, purple or mauve; stamens exserted or included (heterostyled, with each type of flower on separate plants), filaments 1·5–7 mm. long, longer than the anthers when exserted; shrubs (60 cm.–) 1–2 m. or small trees; leaves cuneate at base:
2 Leaves mostly 3–10 cm. long, 1·8–4·5 cm. broad, rounded to broadly acute and apiculate at tip; flowers 4–15 in each fascicle; corolla bluish-purple; spines absent
2. *jamaicensis*
2 Leaves mostly less than 2·5 cm. long and 2 cm. broad, emarginate at tip; flowers 1–4 in each fascicle; spines often present:
3 Leaves not more than 12 mm. long and 6 mm. broad, oblanceolate or obovate; corolla light mauve with purple spots
3. *microphylla*
3 Leaves mostly 1–2·5 cm. long and 0·5–2 cm. broad, mostly ovate to elliptical, occasionally oblanceolate:

4 Corolla light crimson to rich purple; plant mostly of limestone woodlands in the interior hills   4. *armata* var. *armata*
4 Corolla light purple, mauve or almost white; plant usually coastal
4. *armata* var. *pallidior*

1. O. acicularis (Sw.) Stearn in Bull. Brit. Mus. (Nat. Hist.) Bot. **4** (7): 312 (1971).—*Justicia acicularis* Sw. (1788). *Anthacanthus acicularis* (Sw.) Nees (1847). *A. jamaicensis* Griseb. (1862).

Undershrub 10–90 cm. high; twigs puberulous; flowers solitary in the axils; corolla-tube about 7 mm. long; capsule 15–16 mm. long, 4–seeded; seeds flat, rugose, notched at one end, 2·5–3 mm. long.

Rare (West., Port., St. Thom.), in damp sheltered thickets and woodlands on limestone rocks; SL–1500 feet; fl. Mar, June, Aug, fr. Mar, June. *A 9258! H 8609! Proctor & Stearn 11818!* Endemic.

2. O. jamaicensis (Lindau) Stearn in Bull. Brit. Mus. (Nat. Hist.) Bot. **4** (7): 300 (1971).—*Psilanthele jamaicensis* Lindau (1908).

Shrub 2–4 m. high with straggling and arching branches; bark fissured, greyish; twigs nearly glabrous; corolla-tube 7–8 mm. long; capsule 20 mm. long.

Occasional, in the central parishes, in shady woodland on limestone; 1600–2700 feet; fl. Feb–May, Aug, fr. Mar–May, Aug. *A 11034! H 9362! P 8433!* Endemic.

3. O. microphylla (Lam.) Stearn in Bull. Brit. Mus. (Nat. Hist.) Bot. **4** (7): 307 (1971).—*Justicia microphylla* Lam. (1791).

Shrub 1·2–2·4 m. high with twigs densely pubescent or two sides of tetragonal stems glabrous; corolla-tube about 8 mm. long; capsule pointed, 12–20 mm. long; seeds subspherical, reticulate-marked.

Rather local (St. Andr., St. Cath., St. Thom.), in thickets and woodlands and on cliffs in arid limestone areas; fl. sporadically throughout the year, fr. Jan. *A 5665! P 18266! Stearn 66!* Greater Antilles, Lesser Antilles south to Grenadines.

4. O. armata (Sw.) Stearn in Bull. Brit. Mus. (Nat. Hist.) Bot. **4** (7): 301 (1971).—*Justicia armata* Sw. (1788). *Anthacanthus armatus* (Sw.) Nees (1847).

4a. Var. armata.
Shrub or rarely a small tree with soft fissured bark, up to 5 m. high; branches slender erect or arching, glabrous or puberulous; spines 5–10 mm. long or absent or sometimes only on lower spreading vegetative shoots; leaves sometimes clustered on short shoots; corolla-tube about 8 mm. long; capsule 15 mm. long, 4-seeded; seeds rugulose.

Occasional in woodlands on limestone or serpentine rocks; 500–2800 feet; fl. Sept–May, fr. Jan–July. *A 11062! H 9478! P 10236!* Endemic.

4b. O. armata var. **pallidior** Stearn in Bull. Brit. Mus. (Nat. Hist.) Bot. **4** (7): 302 (1971).—*Psilanthele minor* Lindau (1912).

Shrub with slender spreading usually puberulous branches 1–3 m. high, rarely a small tree, often spiny especially below; corolla-tube 7–9 mm. long; capsule 16–17 mm. long.

Locally common, in the western and central parishes in rocky coastal woodlands on limestone, rarer inland; SL–800 (–1700) feet; fl. Sept–May, fr. Jan–May, Sept. *A 11143! 11147! H 10257! P 23236! Stearn 863!* Endemic.

12. HYGROPHILA R. Br. (1810)

Leaves linear to narrowly elliptical, rarely as much as 12 mm. broad, with about 7 pairs of lateral veins; stems essentially glabrous, striate with cystoliths; cymes few-flowered
1. *guianensis*
Leaves elliptic-lanceolate, up to 5 cm. broad, with about 15 pairs of lateral veins; stems softly hispid with long pluricellular hairs; cymes many-flowered   2. *costata*

1. H. guianensis Nees in Hook., Lond. Journ. Bot. **4**: 634 (1845).

Herb with erect or ascending branches 30–90 cm. high; leaves linear to elliptic-lanceolate, long-tapered at both ends, 2–10 cm. long, 3–10 mm. broad, thinly scabridulous beneath; flowers clustered in the leaf-axils; calyx 5–6 mm. long in flower, up to 9 mm. long in fruit, the lobes subulate and paler margined; corolla puberulous, white or light mauve, 8–9 mm. long; capsule linear, acute, 9–14 mm. long, 10–16-seeded; seeds flat, smooth, about 1 mm. long.

Very local (St. Cath.), on silt and gravel banks and rocks in river; 200–250 feet; fl. and fr. Dec–Mar. *A 6474! P 8293! Yuncker 18276!* Mexico to Argentina, West Indies.

2. H. costata C. & T. Nees, Pl. Hort. Med. Bonn. Ic. **1**: 7, t. 3 (1824).

Erect herb up to 1 m. or more high; leaves oblong to elliptic-lanceolate, narrowed at base to a very short petiole, subacute at tip, up to 15 cm. long and 5 cm. broad, thinly pubescent on veins; calyx-teeth linear, about 8 mm. long; corolla puberulous, white or light rose, 9–11 mm. long; capsule (12–) 14–17 mm. long, 12–18-seeded; seeds as above.

Rare and local (St. Eliz.), at muddy river bank; about 25 feet; fl. and fr. Jan. *P 24518!* Continental tropical Amer., Greater Antilles.

13. SALPIXANTHA Hook. (1845)

1. S. coccinea Hook. in Curt., Bot. Mag. **71**: t. 4158 (1845).—*Geissomeria coccinea* (Hook.) T. Anders. ex Griseb. (1862).

Shrub 1·5–3 m. high or small tree; leaves oblanceolate to ovate-elliptical, obtuse or shortly acuminate, cuneate at base, with 6–9 pairs of curved lateral veins, glabrous, up to 16·5 cm. long and 5·5 cm. broad; spikes axillary and terminal, the flowers decussately arranged, the rachis puberulous, up to 25 cm. long in fruit; calyx-lobes minutely ciliolate; corolla 2–3·8 cm. long, scarlet or crimson; capsule smooth, shiny, about 17 (–21) mm. long and 7–9 mm. broad, 4-seeded; seeds 5–6 mm. long, papillose-warted.

Uncommon in the central and western parishes, in rocky limestone woodlands; 1300–3000 feet; fl. Sept–Apr, fr. Dec–July. *A 10887! H 8163! P 8630!* Endemic.

14. ASYSTASIA Blume (1826)

1. A. gangetica (L.) T. Anders. in Thwaites, Enum. Pl. Zeyl.: 235 (1860).—*Justicia gangetica* L. (1756).

Straggling herb with flowering branches up to about 60 cm. or scrambling to 4 m. high; leaves ovate, rounded to a winged petiole at base, acuminate, blade up to 7·5 cm. long and 4·5 (–6) cm. broad; sepals free, valvate, about 8 mm. long; corolla white or light yellow or variously suffused with violet or dull purple, 3·5–4 cm. long; capsule clavate, acuminate, about 2·5 cm. long, 4-seeded; seeds double-margined, denticulate, 3–5 mm. long, 2–4 mm. broad.

Rather common in gardens, occasionally escaping; 50–600 feet; fl. and fr. most of the year. *Asprey UCWI 2039!* Native of the Old World tropics.

15. RUELLIA L. (1753)

1 Corolla red or pink, the tube compressed from side to side; anthers exserted or at throat of corolla; inflorescences axillary with long peduncles; capsule clavate or ellipsoid (14–) 17–18 (–20) mm. long:
  2 Calyx with minute gland-tipped hairs, the lobes tapered, shorter than tube, ciliolate; capsule clavate, glabrous except at tip of beak; leaves broadly ovate, acuminate, broad at base; corolla (4·5–) 5–6 cm. long, bright red   1. *macrophylla*
  2 Calyx with long gland-tipped hairs, the lobes linear-oblong to spathulate, much longer than tube or tube obsolete:
    3 Leaves elliptic-lanceolate, acuminate, long-cuneate at base, glabrous; corolla rose-pink, 5–6 cm. long; capsule ellipsoid, puberulous   2. *costata*
    3 Leaves ovate to elliptical, acute or obtuse, broad at base of blade, pubescent; corolla scarlet, 3·5–5 cm. long; capsule narrowly ellipsoid or clavate, glabrous except for a few gland-tipped hairs distally   3. *elegans*

1 Corolla blue, mauve or white, the tube terete or compressed anterio-posteriorly; calyx-lobes linear-lanceolate to subulate; capsule oblong-fusiform:
    4 Leaves linear to narrowly lanceolate, up to about 20 cm. long and 2 cm. broad; panicles mostly axillary with long peduncles; calyx-lobes tapered, with a few short gland-tipped hairs; capsule oblong-ellipsoid, 2·2–2·5 cm. long, glabrous except for a patch of hairs on each side of beak   **4. brittoniana**
    4 Leaves ovate to broadly lanceolate or elliptical:
        5 Flowers subsessile in the leaf-axils; plant hairy but not glandular; capsule pubescent, about 8 mm. long   **5. geminiflora**
        5 Flowers in pedunculate axillary cymes or if solitary also distinctly stalked; capsule glabrous:
            6 Stem, leaves, inflorescence-branches and calyx with gland-tipped hairs; limb of corolla 2 cm. across; anthers at throat of corolla; capsule 10–15 mm. long   **6. paniculata**
            6 Stem, leaves, etc. without gland-tipped hairs; limb of corolla 3–4·5 cm. across; anthers included; capsule 2–2·5 cm. long   **7. tuberosa**

1. R. macrophylla Vahl, Symb. Bot. **2**: 72, t. 39 (1791).—*Stemonacanthus macrophyllus* (Vahl) Nees (1845).
Erect little-branched shrubby herb 1·2–2 m. high with tetragonal stems at first thinly tomentose, later glabrous; leaves tomentose on midrib and veins, glabrescent, 6–18 (–26) cm. long, 4–8 (–16) cm. broad; capsule 12–16-seeded; seeds orbicular, 2–3 mm. broad.
Uncommon (St. Andr., St. Eliz.), mostly in gardens and probably not native; 400–1500 feet; fl. and fr. Jan–May. *H 7064! P 22259! Wynter UCWI 27222!* Colombia, Venezuela, Cuba, St. Vincent.

2. R. costata (Nees) Hiern in Vid. Medd. Nat. For. Kjøbenh. **1877–78**: 76 (1878). —*Arrhostoxylon costatum* Nees (1847). *R. acuminata* Griseb. (1862).
Herb up to 45 cm. high; stems tetragonal, glabrous except for lines of minute hairs; leaves 7–16 cm. long, 2–5 cm. broad; capsule about 12-seeded.
Uncommon (Manch., St. Mary), mostly in cultivation; 700–2000 feet; fl. and fr. Jan, Sept. *H & B 10599! A. von der Porten 2456!* Native of Brazil.

3. R. elegans Poir. in Lam., Encycl. Méth. Bot. Suppl. **4**: 727 (1816).—*R. formosa* Andr. (1810), not Humb. & Bonpl. (1807).
Herb with spreading and ascending branches up to 80 cm. high; stems tetragonal, thinly pilose; leaves 5–11 cm. long, 2–5 cm. broad, thinly pilose; capsule 8–10-seeded.
Rather local (St. Andr., St. Thom.), naturalized on pathside banks, old walls and stony waste ground; (600–) 3400–5000 feet; fl. and fr. most of the year. *A 6693! H 9597! A. von der Porten!* Native of Brazil.

4. R. brittoniana Leonard in Journ. Acad. Sci. Washington **31**: 96, f. 1 (1941).
Erect herb with purplish tetragonal stems up to about 1 m. high, glabrescent; leaves with a few long hairs on midrib beneath; distal inflorescence-branches with glandular hairs; corolla deep mauve to light purple, 4–5 cm. long, about 4 cm. across the limb; capsule about 20-seeded; seeds about 2·5 mm. in diameter.
Rare and local (St. Cath., St. Mary), along riverbanks and at pond margins; 200–1950 feet; fl. and fr. sporadically throughout the year. *A 9442! P 20717!* Native of Mexico, introduced elsewhere; Hawaii.

5. R. geminiflora Kunth, Nov. Gen. **2**: 240 (1818).
Perennial herb with erect or ascending tetragonal branches, up to 30 cm. high; leaves obtuse or rounded at tip, cuneate at base, thinly pubescent, up to about 4 cm. long and 1·5 cm. broad; corolla bluish-mauve, 3–4 cm. long.
Rare (St. Andr., West.), probably a garden escape; 1300–1500 feet; fl. Apr–May, fr. May. *J.P. 724! Skelding UCWI 3249!* C. and S. Amer., Cuba, Hispaniola, Guadeloupe.

6. R. paniculata L., Sp. Pl. **2**: 635 (1753).
Bushy herb 60–90 cm. high; leaves ovate to lanceolate, obtuse, acute or acuminate, long-tapered at base, up to 15 cm. long including petiole and 5 cm. broad,

usually much smaller; inflorescence-branches divergent; bracts oblanceolate; corolla light mauve-blue or rarely white, 3–3·5 cm. long.

Rather local (St. Cath., Clar., St. Thom.), in open low-lying ground and thickets on heavy soils; SL–50 feet; fl. and fr. Oct–Mar. *A 8809! H 12474! P 7608!* Texas, Mexico to Colombia and Venezuela, Cuba.

7. R. tuberosa L., Sp. Pl. **2**: 635 (1753).—Duppy Gun, Menow Weed.

Usually perennial herb with thick elongated fusiform roots; stems erect, 5–60 cm. high, pilose above, glabrescent; leaves elliptical, obtuse or acute, tapered at base, up to 12 cm. long and 4·5 cm. broad; inflorescences few-flowered; pedicels accrescent and sepals reflexed in fruit; corolla mauve to bluish-purple, rarely white, the tube about 4·5 cm. long; shorter stamens sometimes sterile.

Very common in pastures and waste places and on roadside banks; SL–700 feet; fl. and fr. most of the year but especially during rainy weather. *A 5444! H 9012! P 9471!* S. United States to Guyana, West Indies, tropical Asia, E. Africa, Ghana; Grand Cayman.

16. DICLIPTERA Juss. (1807) nom. cons.

1. D. sexangularis (L.) Juss. in Ann. Mus. Hist. Nat. Paris **9**: 267 (1807).— *Justicia sexangularis* L. (1753).

Annual much branched almost glabrous herb up to about 120 cm. high; leaves ovate to lanceolate, long-petioled, acuminate, thinly pilose, blade up to 10 cm. long and 5 cm. broad; branches of inflorescence more or less unilateral racemes; bracteoles spathulate to subulate; corolla 2·4–3 cm. long, scarlet; anthers exserted; capsule about 5 mm. long, opening widely on dehiscence, 2-seeded; seeds minutely spinulose, about 2 mm. in diameter.

Occasional beside streams, ditches and swamps, usually in partial shade; 20–2700 feet; fl. and fr. Dec–Apr. *A 6305! H 9156! P 8380!* Bahamas, Florida, Mexico to Colombia, West Indies.

17. GRAPTOPHYLLUM Nees (1832)

1. G. pictum (L.) Griff., Notulae **4**: 139 (1854).—*Justicia picta* L. (1762). *G. hortense* Nees (1832). Caricature Plant, Match-me-not.

Shrub 3–4 m. or tree to 6 m. high; leaves oblong-elliptical, shortly acuminate, broadly cuneate at base, glabrous, up to 15 cm. long and 8 cm. broad, dark green and plain or variegated; flowers 1–3 together subtended by small ciliolate bracts and bracteoles; corolla reddish-purple, the 3 segments of the lower lip strongly recurved.

Uncommon in gardens and occasionally escaped; 600–2000 feet; fl. Jan, Apr, July. *A 7620! H 11968!* Native of Malaya now widely distributed in the tropics.

18. ANDROGRAPHIS Wall. ex Nees (1832)

1. A. paniculata (Burm. f.) Wall. ex Nees in Wall., Pl. Asiat. Rar. **3**: 116 (1832).— *Justicia paniculata* Burm. f. (1768). Rice Bitters.

Annual taprooted herb, often bushily branched, up to 1 m. high; leaves lanceolate, long-tapered to both ends, up to 9 cm. long and 2·5 cm. broad, glabrous; main panicle-branches racemose, the pedicels erect; pedicels, calyx and capsule with gland-tipped hairs; capsule oblong-fusiform, (13–) 16–19 (–21) mm. long.

Locally common (St. Andr., St. Thom.), in waste places and gravelly exposed wayside banks; SL–1750 feet; fl. and fr. all the year. *A 5476! H 6566! P 19770!* Native of India and Ceylon, introduced elsewhere; Cuba, Hispaniola, Barbados.

177. MYOPORACEAE

Shrubs or trees. Leaves alternate or opposite, entire, without stipules. Flowers bisexual, zygomorphic, solitary or in axillary cymose clusters. Perianth biseriate; calyx 5-lobed, persistent; corolla gamopetalous, usually 5-lobed, sometimes bilabiate, the aestivation imbricate. Stamens 4 (–5), epipetalous, mostly didynamous; staminode sometimes present; anthers 2-locular, opening lengthwise. Ovary superior, 2– (3–10)-locular; ovules superposed in pairs, 2–8 in each loculus, anatropous, pendulous on axile placentas; style solitary; stigma simple. Fruit a drupe; seeds small with little or no endosperm.

5 genera with about 180 species, mostly in the Australasian region; 1 in the West Indies.

1. BONTIA L. (1753)

Leaves alternate; flowers solitary, axillary; corolla-limb 2-lipped, the upper 2-lobed, the lower 3-lobed and recurved, the mid-lobe densely bearded; ovary 2-locular with 4 ovules in each loculus.

1. B. daphnoides L., Sp. Pl. 2: 638 (1753).

Shrub or low bushy tree up to 6 m. or more high; leaves oblong to linear-lanceolate, narrowed at base, acuminate or acute at apex, mid-vein prominent beneath but lateral veins obscure, 3–11 cm. long, 8–20 mm. broad, glandular-punctate; peduncles 1–3 cm. long; calyx-lobes broadly ovate, imbricate, about 3 mm. long, acuminate, ciliate; corolla about 2 cm. long, tawny-yellow blotched purple; drupe ovoid, tapered upwards to a pointed persistent style-base, 10–16 mm. long.

Very rare (St. Cath.), in thickets on limestone; near SL; fr. July. *Tulloch 71!* Northern S. Amer., West Indies; Grand Cayman. Until the single cited collection was made in 1970, this widespread Caribbean species was known in Jamaica only in cultivation (*H 9277!*).

178. PLANTAGINACEAE

Herbs or undershrubs. Leaves all radical or nearly so, alternate or opposite, sometimes sheathing at the base, mostly longitudinally veined, without stipules. Flowers usually bisexual, small, actinomorphic, bracteate but without bracteoles, in scapose capitate or spicate inflorescences. Perianth biseriate; calyx tubular (3–) 4-lobed or -toothed; corolla scarious, (3–) 4-lobed or -toothed, lobes imbricate. Stamens (1, 2 or) 4, epipetalous, alternate with corolla-lobes, exserted; anthers 2-locular, opening lengthwise. Ovary superior, (1–) 2 (–4)-locular; ovules semi-anatropous, 1 or more in each loculus on axile or basal placentas; style simple, 2-fid. Fruit a circumscissile capsule or a bony nut. Seeds with fleshy endosperm; embryo usually straight.

3 genera with about 250 species mostly in the cosmopolitan genus *Plantago*.

1. PLANTAGO L. (1753)

Weedy rosette herbs; stamens 4; ovary 2-locular; ovules 2 or more; capsule normally circumscissile.

1 Leaf-blade less than twice as long as broad and more or less distinct from the petiole; sepals 4; corolla-lobes about 1 mm. long; seeds several, finely ridged from the hilum on one side, longitudinally ridged on the other   1. *major*
1 Leaf-blade at least three times as long as broad, narrowly elliptical and decurrent into the petiole; corolla-lobes 1·8–2·5 mm. long; seeds 2, smooth:
  2 Margin of bract broad; sepals 3; spikes dense with flowers hiding the axis; corolla-lobes spreading; seeds 2–3·5 mm. long, 1·5 mm. broad, deeply hollowed on the hilum side   2. *lanceolata*

2 Margin of bract narrow; sepals 4; spikes slender, looser, the axis visible between the flowers; corolla-lobes of fertile flowers erect, of male flowers spreading; seeds 1·3–2 mm. long, 0·7 mm. broad, nearly flat on the hilum side   3. *virginica*

1. P. major L., Sp. Pl. **1**: 112 (1753).—English Plantain.

Annual or short-lived perennial herb with a taproot producing numerous lateral fibrous roots; leaf-blade broadly ovate-elliptical, obtuse, up to 17 cm. long and 12·5 cm. broad; inflorescence 15–60 cm. long, the scape often 20 cm. or more long, the spike 20 cm. or more long; flowers perfect, green; capsule broadly ellipsoid, 3–4 mm. long, with up to 30 seeds; seeds nearly 1 mm. long.

Frequent as a weed of damp sandy or gravelly roadsides and banks; 1500–6500 feet; fl. and fr. most of the year. *A 7185! 11409! H 6302! P 23676!* Native of the Old World, now cosmopolitan.

2. P. lanceolata L., Sp. Pl. **1**: 113 (1753).—Ribwort Plantain.

Perennial herb; leaves elliptic-lanceolate, usually acute, entire, about 5-ribbed, 5–25 cm. long, up to nearly 2 cm. broad; scape ribbed, much longer than the cylindrical head, the latter blackish when young, very variable; flowers perfect, protogynous; corolla white with dark brown stripes; capsule ellipsoid, 3–5 mm. long.

Rather local, a weed of roadsides, banks and grassy places; (2000–) 2800–7400 feet; fl. and fr. most of the year. *A 5721! H 11978! J.P. 1485!* Native of Europe, now widespread.

3. P. virginica L., Sp. Pl. **1**: 113 (1753).

Annual or probably mostly annual villous herb proliferating sparingly by means of taprooted basal innovations; leaves oblanceolate, rounded at tip, remotely dentate, up to 8 (–15) cm. long and 2 cm. broad; scape terete, usually slightly longer than the spike, up to 15 cm. long; spike up to 12 cm. long; flowers unisexual probably dioecious; corolla yellowish or tinged pink; capsule ovoid, about 2 mm. long; seeds orange-brown.

Locally common (St. Andr., Port., St. Thom.), in open or shaded well drained places by paths and roadsides; 3200–5600 feet; fl. and fr. Feb–May. *A 6620! H 11979! P 24695!* Native of S. and E. United States.

179. CAPRIFOLIACEAE

Erect or climbing shrubs or small trees, rarely herbs. Leaves opposite, simple or pinnate; stipules usually lacking. Flowers mostly bisexual, actinomorphic or zygomorphic, in cymose inflorescences. Perianth biseriate; calyx usually small, 5-lobed or -toothed; corolla gamopetalous, rotate or bilabiate, typically 5-lobed, the lobes imbricate. Stamens (4–) 5, epipetalous, alternating with corolla-lobes; anthers 2-locular, opening lengthwise. Ovary inferior, (1–) 2–5-locular; ovules 1–many in each loculus, pendulous, usually on axile placentas; style slender or obsolete; stigmas 3–5. Fruit a berry or drupe. Seeds with a usually small straight embryo; endosperm fleshy.

About 15 genera and 400 species mostly in the northern Hemisphere.

1 Leaves pinnate with an interpetiolar stipular ridge; erect soft-wooded shrub; corolla regular; fruit a berry   **1. Sambucus**
1 Leaves simple without stipules:
  2 Vine; corolla 2-lipped, at least 20 mm. long; ovary with several pendulous ovules in each loculus; style long; stigma capitate; fruit a berry   **2. Lonicera**
  2 Erect shrubs or trees; corolla rotate, less than 10 mm. long; ovary with one pendulous ovule in each loculus; style short; stigmas 3; fruit a drupe; plant rank-scented when dry   **3. Viburnum**

1. SAMBUCUS L. (1753)

1. S. simpsonii Rehder in Sarg., Trees & Shrubs **2**: 187, t. 175 (1911).—Elder.

Shrub 2·5–4 m. or tree to 6 m. high; leaves and inflorescence sparsely pubescent;

lower leaflets mostly 3-foliolate; leaflets elliptic-lanceolate, long-acuminate, serrate, 3-11 cm. long, up to 3·5 cm. broad; inflorescence a broad compound 4–5-rayed corymb; corolla white, 5–7 mm. across the limb; fruit (not seen in Jamaica) globose, purplish-black or black, 5–6 mm. in diameter.

Occasional on open roadside banks and in thickets; 50–3200 feet; fl. most of the year especially during rainy periods. *A 11422! H 11932! Powell 127!* S.E. United States, C. Amer., sporadic and mostly escaped from cultivation in the West Indies south to St. Vincent and Barbados; Grand Cayman.

2. LONICERA L. (1753)

1. L. japonica Thunb., Fl. Jap.: 89 (1784).—*L. confusa* of Rendle (1936). Honeysuckle.

Twining woody vine to 5 m. high or trailing on the ground, pubescent; leaves ovate to oblong, acute, rounded at base, 4–7 cm. long, 2–3 cm. broad; flowers fragrant, mostly in shortly pedunculate pairs, bracteate in a terminal panicle; corolla at first white changing to yellow and light orange, about 4 cm. long, the tube very narrow, glandular-pubescent outside; stamens and style exserted.

Cultivated in hill gardens and commonly escaping on to roadside banks and into secondary thickets; (600–1500 cult.) –4500 feet; fl. Mar–Sept, fr. Dec. *A 7490! H 10897! P 9602!* Native of E. Asia, now widely distributed; a weed in parts of the United States, mostly cultivated for ornament in the West Indies.

3. VIBURNUM L. (1753)

1 Leaves and inflorescence-branches glabrous or nearly so; lamina between veins glabrous
                     1. *alpinum*
1 Leaves, at least on main veins beneath, and inflorescences-branches stellate-pubescent:
  2 Lamina between veins glabrous; leaf-margin sinuate to shallowly dentate
               2c. *villosum* var. *subdentatum*
  2 Lamina between veins stellate-pubescent; leaf-margin entire or nearly so:
    3 Drupe 5–6 (–7) mm. long, ovoid:
      4 Indumentum on abaxial surface of leaf of dense contiguous stellate hairs
               2a. *villosum* var. *villosum*
      4 Indumentum on abaxial surface of leaf of sparse non-contiguous stellate-hairs
               2b. *villosum* var. *glabrescens*
    3 Drupe 8–10 mm. long, oblong-ellipsoid
                   3. *arboreum*

1. V. alpinum Macf. ex Britton in Bull. Torr. Bot. Club **37**: 352 (1910).

Shrub or tree 2–8 m. high; leaves ovate to elliptical, acuminate, rounded to obtuse at base, with or without domatia in the vein-axils beneath, 4–12 cm. long, 2–6 cm. broad; inflorescence (4–) 5–7-rayed; corolla about 5 mm. in diameter across the limb, white; stamens exserted; drupe black, ovoid, 5–6 mm. long.

Locally common in thickets and woodland margins in central and eatern parishes; 1300–4600 (–7000) feet; fl. Dec–Aug, fr. Mar–Oct. *A 10482! H 9197! P 19715 !* Endemic.

2. V. villosum Sw., Nov. Gen. & Sp. Pl.: 54 (1788).

2a. Var. villosum.

Shrub or tree, 2–6 m. high, often with straggling branches; leaves broadly ovate to elliptical, usually with a few scattered stellate hairs adaxially; flowers fragrant; corolla white, 4–5 mm. across; ripe drupes purple or black.

Rather common in thickets; 800–5000 (–7000) feet; fl. Dec–Sept, fr. Dec–Aug. *A 6008! H 9255! C. B. Lewis!* Cuba.

2b. V. villosum var. **glabrescens** Griseb., Fl. Br. W.I.: 315 (1861).

Shrub; leaves broadly rounded to subcordate at base, adaxially with stellate hairs only on the main veins.

Rare and local (St. Andr., St. Ann); 2000–4500 feet; fl. Apr, fr. July–Aug. *Asprey UCWI 888! Robbins UCWI 215! Weaver 847!* Endemic.

2c. V. villosum var. **subdentatum** Griseb., Fl. Br. W.I.: 315 (1861).

Shrub or tree to 5 m. high; leaves broadly ovate, sharply acuminate, shortly cordate at base, thick-textured, up to 10 cm. long and 6·5 cm. broad, usually smaller; drupe ovoid, 6 mm. long, 5 mm. broad, purplish.

Local (St. Thom.), in thickets; 7200–7400 feet; fl. Feb–Aug, fr. June–Sept. *A 10684! Asprey UCWI 1030! Proctor & Cooley 7073!* Endemic.

3. V. arboreum Britton in Bull. Torr. Bot. Club **37**: 351 (1910).

Shrub 3–4 m. or tree to 14 m.; leaves elliptical, shortly acuminate, obtuse at base, glabrescent adaxially, loosely stellate-pubescent adaxially, (6–) 8–12 cm. long, (4–) 5–6·5 cm. broad.

Rare and local (St. Cath., St. James, Trel.), in woodland on limestone hills; 1500–2000 feet; fl. Dec, fr. July–Dec. *A 13013! H 9475! ? P 8660! 10390!* Endemic.

The taxonomy of *Viburnum* in Jamaica is extremely difficult. The divisions se out above are in most respects arbitrary and further intermediatevariants exist.

180. RUBIACEAE

Trees, shrubs or herbs. Leaves opposite or whorled, simple, entire or rarely toothed; stipules interpetiolar or intrapetiolar, sometimes foliaceous and indistinguishable from leaves, sometimes reduced to glandular setae. Flowers mostly bisexual and actinomorphic, in essentially cymose inflorescences, the cymules reduced to a solitary flower or in panicles, heads or rarely spikes, with or without bracts. Perianth biseriate; calyx-lobes 4–5, with open aestivation; corolla gamopetalous, 4–5 (–10)-lobed, with valvate, imbricate or contorted aestivation. Stamens epipetalous, as many as the corolla-lobes and alternate with them; anthers mostly dorsifixed 2-locular, opening lengthwise. Ovary inferior or rarely half-inferior, crowned by a more or less well developed disk, rarely 1-locular with parietal placentation, usually 2-locular with axile, basal or apical placentation, rarely loculi several; ovules 1–several in each loculus, anatropous; style usually slender, mostly 2-lobed with 2 stigmas or with one 2-lobed stigma. Fruit a capsule, berry or drupe. Seeds sometimes winged; endosperm usually present and copious; embryo straight or curved.

Over 450 genera with more than 5000 species; generally distributed but most numerous in the tropics.

1 Ovules few to many on axile placentas in each ovary-loculus:
  2 Fruit a capsule:
    3 Seeds broadly winged or plumose-appendaged; ovules many, ascending; stipules interpetiolar, entire:
      4 Seeds plumose at one end, pointed at the other; corolla white, the lobes contorted; capsule septicidal; flowers showy, solitary or few together, fragrant   1. **Hillia**
      4 Seeds winged:
        5 Climbing shrubs; capsule septicidal; corolla-lobes valvate; seeds winged all round   2. **Manettia**
        5 Erect shrubs or trees:
          6 Capsule septicidal, opening from the base upwards; corolla-lobes valvate; seeds irregularly winged all round; anthers included   3. **Cinchona**
          6 Capsule loculicidal:
            7 Corolla-lobes valvate; seeds with wing attenuated in opposite directions; flowers in panicles; filaments declinate; anthers included   4. **Macrocnemum**
            7 Corolla-lobes imbricate; seeds winged all round; flowers in corymbose panicles to solitary:
              8 Anthers and stigma long-exserted; corolla-lobes linear   5. **Exostema**
              8 Anthers included; corolla-lobes mostly rounded   9. **Rondeletia**
    3 Seeds not broadly winged or plumose-appendaged:
      9 Herbs; stipules usually toothed; corolla-lobes valvate:
        10 Inflorescences terminal, showy; capsule septicidal; calyx without intermediate teeth, the lobes often unequal; corolla white, pink, purple or crimson, the tube 1·5–2 cm. long; ovules many; mostly cultivated; native of tropical Africa
                                          Pentas spp.
        10 Inflorescences axillary; flowers small, white; capsule loculicidal:

11 Ovary inferior; ovules many; roots fibrous; calyx without intermediate teeth
37. **Oldenlandia**

11 Ovary half-inferior; ovules few; roots tuberous; calyx with minute intermediate teeth
38. **Lucya**

9 Shrubs or trees; stipules entire; ovules numerous:

12 Corolla-lobes valvate or reduplicate-valvate:

13 Flowers small, numerous, in stout-peduncled bracteate cymose panicles; corolla widely open, 5–6 mm. long; calyx-limb subtruncate; capsule septicidal; seeds compressed, marginate
6. **Chimarrhis**

13 Flowers showy, solitary or few together in the leaf-axils (apparently terminal and umbellate in *P. harrisii*); corolla funnel-shaped, at least 4 cm. long; calyx-lobes evident; capsule loculicidal; seeds angular, appendaged at base
7. **Portlandia**

12 Corolla-lobes imbricate; stipules usually silky-hairy adaxially:

14 Leaves thickly fleshy, less than 1 cm. long; capsule septicidal, half-superior; flowers solitary, 4-merous; stipules adnate to leaf-bases, persistent; coastal plant
8. **Rachicallis**

14 Leaves not fleshy or very small, more than 1 cm. long; capsule loculicidal or rarely septicidal, inferior; flowers in cymes, corymbs or panicles, rarely solitary; not coastal plants:

15 Stamens inserted in the throat of the corolla; flowers (4–) 5-merous, regular; leaves elliptical or broader, not resinous, opposite or rarely 3 in a whorl; stipules free, or rarely connate, caducous early or late
9. **Rondeletia**

15 Stamens inserted in the lower part of the corolla-tube; flowers 4 (–6)-merous, the lobes of the corolla unequal; leaves narrow with revolute margins, resinous; stipules connate between the petioles, persistent; capsule loculicidal
10. **Acrosynanthus**

2 Fruit a soft or leathery berry:

16 Corolla-lobes valvate:

17 Inflorescence axillary; berry globose:

18 Climbing shrub; stipules suborbicular; corolla 5-lobed; style 4–5-armed; fruit white
11. **Sabicea**

18 Creeping herbs; stipules linear; corolla 4-lobed; style 2-armed; fruit blue
12. **Coccocypselum**

17 Inflorescence terminal:

19 Climbing shrub; inflorescence involucrate-capitate; calyx truncate; style 4-armed
31. **Schradera**

19 Erect or scrambling shrubs or trees; calyx toothed or lobed; corolla densely hairy in throat:

20 Inflorescence corymbose; style 2-armed; berry oblong-ellipsoid; cultivated shrubs
13. **Mussaenda**

20 Inflorescence spiciform; style 4-armed; berry globose; wild plant
14. **Gonzalagunia**

16 Corolla-lobes contorted or imbricate; shrubs or trees:

21 Corolla-lobes contorted; seeds immersed in pulp, large, in subspherical fruits:

22 Inflorescence axillary; flowers 1–few in clusters
15. **Randia**

22 Inflorescence terminal, branched; flowers numerous:

23 Corolla slightly irregular, curved in bud, up to 11 cm. long when open; ovary 1–2-locular; anterior anthers separating explosively; cultivated; native of tropical S. Amer.
Posoquiera longiflora Aubl.

23 Corolla regular, not curved, the tube up to about 2·5 cm. long; ovary 1-locular; anthers not explosive
16. **Casasia**

21 Corolla-lobes imbricate; seeds small:

24 Ovary 4–5-locular; inflorescence terminal; fruit ovoid or oblong; leaves in whorls of 3 or opposite
27. **Hamelia**

24 Ovary 2-locular; inflorescences or solitary flowers axillary; fruit ovoid or sub-globose:

25 Undershrub with herbaceous branches without spines; leaves opposite, rather large; stamens inserted at throat of corolla
28. **Hoffmannia**

25 Shrubs with spines; leaves opposite but the pairs often clustered, small; flowers solitary; stamens inserted at base of corolla-tube
17. **Catesbaea**

1 Ovules solitary in each ovary-loculus or 2 in a 1-locular ovary:

26 Ovary 1-locular; ovules 2; shrub or small tree; corolla-lobes valvate, 4; fruit globose with a persistent tubular calyx-limb
29. **Faramea**

26 Ovary 2- or more- locular with as many ovules:

27 Ovules pendulous from the top of the loculi; trees or shrubs:

28 Stamens inserted at or near the throat of the corolla; corolla-lobes imbricate; endosperm absent or scanty:

29 Inflorescence a terminal panicle with regularly cymose shortly stalked flowers; calyx-lobes ovate or triangular, persistent; corolla-lobes about as long as tube; fruit coccoid, 2-seeded
24. **Machaonia**

29 Inflorescences axillary, the forked main branches with sessile or subsessile flowers along the upper side; calyx-limb tubular, truncate or irregularly shortly toothed; corolla-lobes shorter than the tube; fruit drupaceous with hard endocarp:

    30 Calyx-limb persistent; fruit oblong, 2-seeded (in Jamaican species) 25. **Antirhea**

    30 Calyx-limb deciduous; fruit globose, 3–9-seeded   26. **Guettarda**

28 Stamens inserted at the base of the corolla-tube; endosperm fleshy; inflorescences mostly axillary:

  31 Ovary 5–22-locular; inflorescence paniculate; corolla-lobes imbricate, 5–10, much longer than the tube; ripe fruit drupaceous, black   20. **Erithalis**

  31 Ovary 2-locular:

    32 Corolla-lobes imbricate, much shorter than the tube; stipules fringed; flowers in short axillary fasciculate cymes; drupe white   21. **Scolosanthus**

    32 Corolla-lobes valvate:

      33 Erect resinous shrubs or trees; stipules truncate, ciliate; inflorescence fasciculate, each cluster subtended by a cup-shaped bract; corolla-lobes longer than the tube; fruit ovoid   22. **Phialanthus**

      33 Climbers; stipules apiculate, eciliate; inflorescence subracemose; corolla-lobes shorter than the tube; fruit compressed, white   23. **Chiococca**

27 Ovules axile or basal; herbs, shrubs or trees:

  34 Corolla-lobes contorted or imbricate; ovules axile:

    35 Leaves in whorls of 3, small and narrow with revolute margins; stipules fringed, persistent; internodes very short; corolla pink; drupe white; coastal plant

                      18. **Strumpfia**

    35 Leaves mostly opposite, large, flat; stipules entire, eventually deciduous; internodes long; cultivated plants:

      36 Bracteoles connate into an epicalyx; flowers in axillary clusters; corolla white; ripe berry red   19. **Coffea**

      36 Bracteoles not connate; flowers in large terminal or axillary corymbs; corolla red, yellow or white, the tube slender and markedly longer than the lobes; ripe berry usually black; (the commonest species is *I. coccinea* L.; corolla orange-red with acute lobes; native of tropical Asia)   *Ixora* spp.

  34 Corolla-lobes valvate:

    37 Stipules indistinguishable from leaves, thus plant appearing to have leaves in whorls of 4; diffusely trailing herb with tetragonal stems; ovules basal; fruit red, superior   46. **Relbunium**

    37 Stipules not resembling the leaves:

      38 Fruit a fleshy compound berry derived from a tight head of flowers without an involucre; ovules axile; wood of roots yellow; corolla and ripe fruit white

                      30. **Morinda**

      38 Fruits not compound:

        39 Fruits drupaceous; ovules basal; stipules various:

          40 Flowers in involucrate heads:

            41 Creeping herb; bracts green   32. **Geophila**

            41 Erect shrub; bracts red   33. **Cephaelis**

          40 Flowers not in involucrate heads:

            42 Ovary-loculi and style-arms 6–7 (–12); flowers in axillary fascicles

                      34. **Lasianthus**

            42 Ovary-loculi and style-arms 2; flowers mostly in terminal corymbose panicles:

              43 Corolla-tube straight, not or hardly swollen at base; flowers small

                      35. **Psychotria**

              43 Corolla-tube more or less curved and swollen at base; flowers often showy

                      36. **Palicourea**

        39 Fruits dry; ovules axile; stipules mostly setose-toothed:

          44 Ovary 3-locular; flowers small in a dense terminal involucrate head; corolla usually 6-lobed; herb with prostrate hairy branches   39. **Richardia**

          44 Ovary 2-locular; flowers in axillary or terminal clusters; corolla 4-lobed:

            45 Fruit indehiscent, a dry berry with 2 pyrenes; trailing or diffuse mainly coastal undershrub; corolla-tube longer than the strongly revolute lobes, up to 10 mm. long, greenish-white or tinged pink   40. **Ernodea**

            45 Fruit dehiscent or separating into 2 cocci; corolla small, white or tinged pink or mauve:

              46 Fruit separating into 2 cocci:

                47 Cocci indehiscent; trailing herbs or climbing undershrubs   41. **Diodia**

                47 Cocci dehiscent at the base; erect slender herb   42. **Hemidiodia**

              46 Fruit dehiscent, the cocci not separating; erect or diffuse weedy herbs:

                48 One valve of fruit remaining closed   43. **Spermacoce**

                48 Both valves of fruit opening:

                  49 Dehiscence septicidal at apex   44. **Borreria**

                  49 Dehiscence transverse   45. **Mitracarpus**

1. HILLIA Jacq. (1760)

Corolla 6-lobed, the lobes linear-lanceolate; leaf-tip cuspidate-acuminate   1. *parasitica*
Corolla 4-lobed, the lobes ovate; leaf-tip obtuse   2. *tetrandra*

1. H. parasitica Jacq., Select. Stirp. Amer. Hist.: 96, t. 66 (1763).
Straggling epiphytic shrub, rooting from the branches; leaves ovate, obtuse at base, thick, up to 10 cm. long and 5 cm. broad; corolla-tube 8–10 cm. long; capsule tetragonal, 6–8 (–9·5) cm. long.
Rather local (Port.), on trees in submontane forest; 2000–4250 feet; fl. May–Sept, fr. Sept–Jan. *A 10280! H 5773!* Tropical S. Amer., West Indies.

2. H. tetrandra Sw., Nov. Gen. & Sp. Pl.: 58 (1788).
Erect or scrambling shrub 1·2–3 m. high or sometimes epiphytic; leaves obovate or oblong-obovate, cuneate at base, succulent, brittle, 4–6 cm. long, 2–3 cm. broad; corolla-tube greenish, (1·8–) 3–6 cm. long; capsule 3·5–6 cm. long.
Local in the south-eastern parishes and St. Ann and Trelawny, in woodlands on limestone; 1500–2650 feet; fl. July–Oct, fr. July–Dec. *A 11717! H 9448! H & P 14290!* Cuba.

2. MANETTIA Mutis ex L. (1771) nom. cons.

Corolla deep violet distally to pinkish at base; calyx usually 4-lobed with 4 interposed teeth; flowers few in axillary cymes   1. *lygistum*
Corolla red; calyx 8-lobed; flowers mostly solitary in the axils   2. *coccinea*

1. M. lygistum (L.) Sw., Nov. Gen. & Sp. Pl.: 37 (1788).—*Petesia lygistum* L. (1759).
Woody twiner to 3 m. high, with tetragonal stems; leaves broadly ovate, acute or acuminate, rounded at base, with lateral veins impressed adaxially and prominent beneath, leathery, 1–5 cm. long, 1–3·5 cm. broad, puberulous; corolla-tube 8–10 mm. long, puberulous outside, shaggy-pubescent in distal half within; capsule obovoid, about 5 mm. long.
Locally common (St. Andr., Port., St. Thom.), in montane thickets; (2500–) 4250–7400 feet; fl. and fr. most of the year. *A 10645! P 4313! Stearn 106!* Probably endemic; old records from Hispaniola have not been confirmed recently.

2. M. coccinea (Aubl.) Willd. in L., Sp. Pl. ed. 4, **1** (2): 625 (1798).—*Nacibea coccinea* Aubl. (1775). *M. uniflora* Kunth (1818).
Nearly glabrous climbing shrub; leaves ovate, acuminate, thin, the veins not prominent, 4–6 cm. long, 2–3·5 cm. broad; pedicels slender; corolla about 15 mm. long; capsule obovoid, 7 mm. long.
Very rare (Manch.), not seen since collected by *Purdie* in 1843. Mexico to Venezuela, Cuba, St. Vincent, Trinidad.

3. CINCHONA L. (1753)

Leaves broadly ovate-elliptical, obtuse, with about 10 pairs of lateral veins, glandular beneath; petiole, midrib and lateral veins puberulous beneath; corolla-tube 14–15 mm. long   1. *pubescens*
Leaves ovate-elliptical to elliptical, acute or shallowly acuminate, with about 8 pairs of lateral veins, not glandular, glabrous beneath except in domatia-pits; corolla-tube 10–13 mm. long   2. *officinalis*

1. C. pubescens Vahl in Skrivt. Nat. Selsk. Kjøbenh. **1**: 19 (1790).—Quinine Tree.
Tree 6–9 m. high with broad crown; young stems puberulous; leaves broad at base, shiny adaxially, up to 22 cm. long and 14 cm. broad or larger; inflorescence paniculate; flowers pink, fragrant; inner margins of corolla-lobes clothed with long hairs; capsule oblong-fusiform, 21–26 mm. long; seeds including wing about 6 mm. long
Formerly cultivated and now naturalized along roadsides and pasture margins in the mountains (St Andr., Port.); 3500–5000 feet; fl. Mar, Aug, Nov, fr. Dec–Mar. *A 9871! Proctor & Steyermark 7718!* Native of Peru.

2. C. officinalis L., Sp. Pl. **1**: 172 (1753).

Shrub 2–3 m. high or a small tree; plant glabrous or nearly so; leaves cuneate at base, 4–15 cm. long, 2–7 cm. broad, with red midrib; inflorescence as above; corolla mauve or dull crimson; capsule 12–22 mm. long; seeds 3·5–5 mm. long.

Occasional (St. Andr., St. Thom.), on grassy mountain slopes, relict from former cultivation; 5300–6500 feet; fl. July, Nov–Feb, fr. July, Nov–Apr. *A 10588! 10632! P 9440!* Native of Peru.

4. MACROCNEMUM P. Browne (1756)

1. M. jamaicense L., Fl. Jam.: 14 (1759).—White Thorn.

Shrub or tree to 13 m. high; leaves ovate to oblong-obovate, cuspidate to shortly acuminate, rounded or obtuse at base, appressed-hairy on midrib and veins when young, 10–18 cm. long, 6–11 cm. broad; inflorescence terminal or axillary, long-peduncled; corolla light green to greenish-yellow, the tube 9–12 mm. long; capsule cylindrical, 2–2·5 cm. long; seeds about 2 mm. long.

Local in wet rocky forest and in pasture margins on heavy soils, mostly in the north-central and eastern parishes; 400–1750 feet; fl. Nov–Apr, fr. Apr–Sept. *A 9250! H 12025! Stearn 579!* Endemic.

5. EXOSTEMA (Pers.) L. C. Rich. (1807)

1 Flowers solitary, axillary; corolla usually white, tube 2·5–3 cm. long; stipules acuminate; capsule 8–15 mm. long, 6–8 mm. broad   1. *caribaeum*
1 Flowers in terminal cymes; corolla usually pink:
   2 Corolla-tube 5–7·5 cm. long; capsule 2–2·5 cm. long, 1–1·5 cm. broad; stipules ovate
           2. *brachycarpum*
   2 Corolla-tube 2–2·5 cm. long; capsule 1–2 cm. long, up to 1 cm. broad; stipules deltate
           3. *triflorum*

1. E. caribaeum (Jacq.) Schult. in L., Syst. Veg. ed. nov. **5**: 18 (1819).—*Cinchona caribaea* Jacq. (1760). Caribee Bark Tree, Jesuit Bark.

Shrub to 5 m. or tree to 10 m. high; leaves lanceolate, acute, 3–5 cm. long, 1–2·5 cm. broad, hairy or glabrous; pedicels 4–8 mm. long; corolla at first green then white ageing yellow, rarely light pink; seeds suborbicular, 2·5 mm. in diameter.

Fairly common, in arid coastal thickets and woodlands on limestone; SL–800 feet; fl. May–Dec, fr. Sept–Feb. *A 9845! H 9314! P 7527!* Florida, Bahamas, Mexico to Costa Rica and throughout the northern drier parts of the Caribbean region to St. Lucia and the Grenadines; Grand Cayman.

2. E. brachycarpum (Sw.) Schult. in L., Syst. Veg. ed. nov. **5**: 19 (1819).—*Cinchona brachycarpa* Sw. (1788). Maroon Lance.

Tree (4–) 10–12 m. high; leaves oblong-elliptical, obtuse to shortly and bluntly acuminate, 12–15 cm. long, 4–9·5 cm. broad; stipules ovate; panicle corymbose, many-flowered; corolla finally dark crimson; style clavate, exserted; seeds flat, subulate.

Local in the central and western parishes, in woodland on limestone; 1500–3000 feet; fl. July–Sept, fr. Aug–Mar. *A 12723! H 8787! P 11309!* Endemic.

3. E. triflorum (W. Wright) G. Don, Gen. Syst. **3**: 481 (1834).—*Cinchona triflora* W. Wright (1787).

Tree 12 m. high; leaves lanceolate, acuminate, with 7–9 pairs of lateral veins, 6–14 cm. long, 2·5–4·5 cm. broad; stipules broadly triangular; panicles slender-branched with the pedicelled flowers in threes at the ends of the branches; calyx-tube 2·5 mm. long, lobes about 1 mm. long; corolla at first light green, later pink, ageing crimson; seeds flat, irregularly winged, about 6 mm. long and 2·5 mm. broad.

Rare and local (St. Ann, Port.), in sheltered moist woodlands on limestone; about 1500 feet; fl. June, fr. Dec. *A 9449! 10057!* Endemic.

6. CHIMARRHIS Jacq. (1763)

1. C. cymosa Jacq., Select. Stirp. Amer. Hist.: 61 (1763).

1a. Ssp. **jamaicensis** Urb., Symb. Ant. **1**: 411 (1899).—Wild Fiddlewood.

Tree up to 15 m. high; leaves elliptical to obovate, cuspidate-acuminate, narrowed to base, usually with conspicuous domatia in the vein-axils beneath, up to 20 cm. long and 10 cm. broad; stipules deciduous; peduncle up to 12 cm. long; corolla white, (4–) 5-merous, fragrant; filaments about as long as corolla-lobes and radiating between them; capsule 5 mm. long; seeds about 0·5 mm. broad with a narrow jagged wing.

Occasional in woodlands on limestone in areas of high rainfall; 900–3800 feet; fl. Mar, June–Oct, fr. July–Oct. *A 9289! H 9449! P 19742!* Endemic. Urban recognized typical ssp. *cymosa* in the Lesser Antilles and ssp. *microcarpa* Urb. in Cuba.

7. PORTLANDIA P. Browne (1756)

1 Calyx-lobes 1–6 mm. long; corolla 6 cm. long, light crimson 　　　　　 1. *microsepala*
1 Calyx-lobes 1 cm. or more long:
　2 Leaf-blade suborbicular, subsessile, cordate at base; flowers clustered subterminally in the uppermost axils; corolla white tinged rose 　　　　　　　　　　 2. *harrisii*
　2 Leaf-blade ovate, elliptical or obovate, distinctly petiolate, cuneate to rounded but hardly cordate at base; flowers solitary, paired or few in the axils:
　　3 Corolla crimson; leaves obtuse at tip, rounded at base 　　　　　　　 3. *coccinea*
　　3 Corolla white, often lined or tinged rose; leaves acute or acutely cuspidate:
　　　4 Calyx-lobes ovate-lanceolate, leafy; capsule ribbed; corolla-tube 10–20 cm. long
　　　　　　　　　　　　　　　　　　　　　　　　　　　　 4. *grandiflora*
　　　4 Calyx-lobes lanceolate to linear-lanceolate; capsule round-sided to obscurely angled; corolla-tube 4–10 cm. long:
　　　　5 Leaves rounded to broadly cuneate at base, 7·5–16 cm. long, 4–10 cm. broad
　　　　　　　　　　　　　　　　　　　　　　　　　　　　 5. *latifolia*
　　　　5 Leaves narrowly cuneate at base, 6·5–12 cm. long, 2·5–4·5 cm. broad
　　　　　　　　　　　　　　　　　　　　　　　　　　　　 6. *albiflora*

1. P. microsepala Urb. in Fedde, Repert. Sp. Nov. **13**: 478 (1915).

Shrub; stipules acute; leaves ovate, acute, rounded at base, 10–13 cm. long, 5·5–7·5 cm. broad; pedicels probably solitary, 5–15 mm. long; capsule obovoid, about 2 cm. long.

Rare and local (St. Ann); about 2000 feet; fl. Jan–Apr, fr. Mar–Aug. *HPS 14598! Prior! P 9329!* Endemic.

2. P. harrisii Britton in Bull. Torr. Bot. Club **39**: 8 (1912).

Shrub 1·5–5 m. high or slender little-branched tree to 8 m.; stipules obtuse, about 10 mm. long; leaves 12–14 cm. long and broad; pedicels 2–5 in the upper axils, about 10 mm. long; calyx-lobes about 15 mm. long, 3·5 mm. broad; corolla white tinged rose, 8–9 cm. long; capsule obovoid.

Rare and local (Clar., St. Ann), in thickets and open woodlands on limestone; 2200–2500 feet; fl. Sept, fr. Dec–July. *A 12488! H 10975! P 9781!* Endemic.

3. P. coccinea Sw., Nov. Gen. & Sp. Pl.: 42 (1788).

Shrub 0·5–1·5 (–3) m. high; stipules acute, 6–7 mm. long; leaves ovate to oblong-ovate, up to 12 cm. long and 7·5 cm. broad; pedicels solitary or paired, about 1 cm. long; calyx-lobes 15–18 mm. long; corolla light or dark crimson, often with white lines in the folds within, about 6 cm. long; capsule obovoid, 5-ribbed, about 1·7 cm. long.

Rare (St. Cath., St. Eliz., Trel.), on limestone crags and cliffs; 500–1900 feet; fl. Jan–Apr, Oct, fr. sporadically. *A 6773! H 12515! P 11056!* Endemic.

4. P. grandiflora L., Syst. Nat. ed. 10, **2**: 928 (1759) (incl. var. *parviflora* S. Moore (1930)).—Bell Flower.

Shrub 2–3 m. or tree to 6 m. high; stipules acute to rounded at tip, 5–10 mm. long; leaves up to 16 cm. long and 10 cm. broad; pedicels solitary, stout, 10–15 mm. long; calyx-lobes 1·5–3·5 cm. long; corolla greenish-white tinged pink in bud, opening white, fading cream; capsule obovoid, 2–2·5 cm. long.

Fairly common, on limestone rocks and cliffs in thickets and open woodlands; SL–1800 feet; fl. and fr. all the year. *A 6326! 6810! H 8662! P 16657!* Endemic; cultivated in gardens in E. Africa and elsewhere.

5. P. latifolia Britton & Harris ex S. Moore in Journ. Bot. **68**: 108 (1930).

Shrub 1–2·5 m. high; stipules more or less acuminate, 7–10 mm. long; leaves ovate to elliptical; pedicels solitary, 10–18 mm. long; calyx-lobes 1·5–2·5 cm. long, 3–6 mm. broad; corolla light golden-buff often tinged crimson in bud, opening white; capsule obovoid, 1·5–2·5 cm. long.

Locally common in the eastern and east-central parishes, on limestone cliffs and in rocky thickets and woodlands; SL–900 feet; fl. May–Jan. fr. May–Feb. *A 9914! H 12669! P 15536!* Endemic. *P. latifolia* occupies a more or less central position in a cline between *P. grandiflora* and *P. albiflora*. It resembles closely the Mexican *P. platantha* Hook.

6. P. albiflora Britton & Harris ex Standl. in N. Amer. Fl. **32**: 12 (1918).

Shrub or tree 1·5–5 m. high; stipules acute or shortly acuminate, 6 mm. long; leaves obovate to narrowly elliptical; pedicels solitary or a few together, 10–15 mm. long; calyx-lobes up to 15 mm. long and 2·5 mm. broad; corolla white, tinged crimson and greenish at base in bud; capsule subglobose, 1·3–1·8 cm. long; seeds 2·5–3 mm. long, up to 2 mm. broad, minutely muriculate.

Very local (St. Andr.), on limestone rocks and cliffs; 300–600 feet; fl. Mar–Nov, fr. Apr–Oct. *A 6961! ACH 322! H 12670!* Endemic. Plants very near to *P. albiflora* have been so-named from Manchester and Portland parishes but typical specimens are known only from one locality in south St. Andrew.

8. RACHICALLIS DC. (1830)

1. R. americana (Jacq.) Kuntze, Revis. Gen. Pl. **1**: 281 (1891).—*Hedyotis americana* Jacq. (1760).

Flexible-branched shrub 30–120 cm. high; leaves sessile, oblong, obtuse, imbricate; flowers axillary, sessile; corolla deep yellow, the tube 5–6 mm. long, silky outside; style included or exserted; capsule 3 mm. long; seeds angular, pitted.

Locally common (St. Ann, St. Mary), on coastal limestone rocks; SL–10 feet; fl. Jan–May, fr. Mar. *A 6131! H 10176! P 11827!* Mexico, Bahamas, Cuba, Hispaniola, Cayman Is.

9. RONDELETIA L. (1753)

With G. R. Proctor

1 Inflorescence several- to many-flowered, paniculate or loosely cymose, or if merely 3-branched, the branches usually branched again; bracts usually small and awl-shaped, rarely foliose, (lanceolate in *R. subsessilifolia*):
  2 Corolla glabrous outside or nearly so:
    3 Leaves sessile or nearly so, rounded or subcordate at base:
      4 Leaves large, up to 30 cm. long, often purple beneath; Cockpit country
           1. *amplexicaulis*
      4 Leaves smaller, not often over 16 cm. long, green on both surfaces although veins sometimes red:
        5 Corolla tawny yellow or rarely pink, tube 6–7 mm. long; western Jamaica 2. *harrisii*
        5 Corolla bright coral pink, tube 11–13 mm. long; eastern mountains
           3. *subsessilifolia*
    3 Leaves distinctly petiolate, mostly cuneate at base:
      6 Leaves less than 5 cm. long, glabrous; central Jamaica   4. *daphnoides*
      6 Leaves usually much more than 5 cm. long, minutely hairy on nerves beneath; eastern mountains:
        7 Corolla-tube 8–10 mm. long, twice as long as lobes; leaves narrowly cuneate at base
           5. *pallida*
        7 Corolla-tube 12–14 mm. long, three times as long as lobes; leaves broadly cuneate to rounded at base   6. *elegans*
  2 Corolla pubescent outside:
    8 Calyx glabrous; Blue Mts.   7. *racemosa*
    8 Calyx pubescent:
      9 Calyx-lobes minute, not over 0·7 mm. long, as broad as or broader than long:
        10 Corolla-lobes almost as long as tube, tube less than 3 mm. long; inflorescence elongate-paniculate; central and western Jamaica   8. *laurifolia*
        10 Corolla-lobes much shorter than tube; inflorescence cymose-paniculate, not elongated:

11 Leaves minutely pubescent, especially beneath:
   12 Leaves in pairs, with petioles 2–4·5 cm. long; capsules deeply bisulcate;
      northern central Jamaica   9. *petiolata*
   12 Leaves usually in threes, with petioles 1–2 cm. long; capsules depressed-
      globose; north central, central and south coast Jamaica   10. *stipularis*
11 Leaves glabrous or nearly so:
   13 Corolla-tube 6–7 mm. long; leaves 12–20 cm. long; extreme north-western
      Jamaica   11. *nemoralis*
   13 Corolla-tube 3–5 mm. long; leaves up to 10 (–15) cm. long:
      14 Leaves elliptical, broadest at middle, apex blunt to acute, corolla-tube 4–5 mm.
         long; central and western Jamaica   12. *polita*
      14 Leaves obovate, broadest beyond middle, apex rounded; corolla-tube 3 mm.
         long; central Jamaica   13. *adamsii*
9 Calyx-lobes longer than broad, mostly 1–2 mm. long:
   15 Leaves softly hirsute with hairs 1–2 mm. or more long; calyx-lobes 5–7 mm. long;
      mostly south-eastern Jamaica   14. *hirsuta*
   15 Leaves glabrate or minutely pubescent chiefly on nerves with hairs 0·1–0·5 mm.
      long; calyx-lobes 1–2 mm. long:
      16 Stipules deltate, merely acute; petioles mostly 1–2 cm. long; inflorescence more
         than 3-branched, often amply paniculate; Port Royal Mts.   15. *impressa*
      16 Stipules long-acuminate; petioles less than 0·5 cm. long; inflorescence only
         3-branched; central Jamaica   23. *jamaicensis*
1 Inflorescence small, loosely or densely capitate to subumbellate with few or rarely
   solitary flowers; bracts small to large and foliose:
17 Leaves evidently pubescent on both surfaces, often distinctly hirsute:
   18 Bracts subtending inflorescence broadly foliaceous, up to 2·5 cm. long and 1·5 cm.
      broad; eastern mountains   16. *portlandensis*
   18 Bracts much smaller and narrower, less than 1·5 cm. long:
      19 Peduncles much shorter than or barely equalling the flowering heads:
      20 Flowers solitary or rarely 2 or 3 together; calyx-lobes densely spreading-hirsute,
         5–6 mm. long; north-western Jamaica   17. *lingulata*
      20 Flowers several in a dense nodding head; calyx-lobes appressed-pubescent, about
         3 mm. long; west-central Jamaica   18. *saxicola*
      19 Peduncles equalling to much longer than the heads:
      21 Calyx-lobes 5–7 mm. long; corolla-tube 6–8 mm. long; hairs on midrib beneath
         up to 2 mm. or more long; mostly south-eastern Jamaica   14. *hirsuta*
      21 Calyx-lobes 3–4 mm. long; corolla-tube (6–) 8–12 mm. long; hairs on midrib
         beneath up to about 1 mm. long:
         22 Leaves mostly 5–15 cm. long, cuneate to rounded at base, distinctly petiolate;
            Port Royal Mts. and western St. Thomas   19. *hirta*
         22 Leaves mostly 2–4 cm. long, cordate at base, sessile or nearly so; western St.
            Thomas   20. *brachyphylla*
17 Leaves glabrous, glabrate or pubescent on one surface only with small or minute hairs:
23 Inflorescences subsessile, the peduncle less than 4 mm. long; Cockpit country
                                            21. *cymulosa*
23 Inflorescences distinctly stalked with peduncles equalling or exceeding flowering
   heads:
   24 Leaves of thin or papery texture, dull green adaxially, if paler beneath not whitish:
      25 Flowering heads loosely cymose, distinctly 3-branched:
      26 Calyx-lobes 3 mm. long, blunt, broadest beyond middle; corolla-tube 8–11 mm.
         long; hairs on midrib beneath 0·5–1 mm. long; north-western Jamaica
                                         22. *umbellulata*
      26 Calyx-lobes 2 mm. long, acute, broadest below middle; corolla-tube about 6 mm.
         long; hairs on midrib beneath 0·2–0·4 mm. long; central Jamaica
                                         23. *jamaicensis*
      25 Flowering heads densely capitate, not distinctly branched:
      27 Leaves elliptical, broadest at middle, acuminate; calyx-lobes 2 mm. long;
         corolla-tube 1 mm. in diameter; north-western Jamaica   24. *dolphinensis*
      27 Leaves obovate, broadest beyond middle, blunt; calyx-lobes 4 mm. long;
         corolla-tube 2·5 mm. in diameter; central Jamaica   25. *glauca*
   24 Leaves of stiff leathery texture, glossy green adaxially, markedly paler or whitish
      beneath:
      28 Leaves ovate, cordate at base; central Jamaica   26. *clarendonensis*
      28 Leaves oblong to elliptical, cuneate at base:
      29 Leaves densely velvety-pubescent beneath; north-central Jamaica 27. *sylvestris*
      29 Leaves minutely and sparsely appressed-pubescent beneath:
      30 Stipules 10–16 mm. long; calyx-lobes 6–7 mm. long; north-western Jamaica
                                          28. *cincta*
      30 Stipules 7–9 mm. long; calyx-lobes about 5 mm. long; central Jamaica
                                          29. *incana*

1. R. amplexicaulis Urb., Symb. Ant. **5**: 502 (1908).

Shrub or tree 1·5–6 m. high; leaves opposite or in threes, obovate to oblong-obovate, subcordate at base, acuminate, margin recurved, crimson beneath, 12–36 cm. long, 5–14 cm. broad, with 9–12 pairs of lateral veins; stipules about 10 mm. long; bracts linear-lanceolate; calyx-tube campanulate, 2 mm. long, lobes subulate, 2 mm. long; corolla-tube crimson, 10–13 mm. long, lobes suborbicular, 3 mm. long, creamy-pink; capsule about 4 mm. broad.

Rare and local (Trel.), in open glades of wooded area and on clay banks; 1500–2250 feet; fl. and fr. Apr–May, Aug. *A 6771! 12418! H 8567! P 15730!* Endemic.

2. R. harrisii Urb., Symb. Ant. **6**: 43 (1909).

Shrub 2–2·5 m. or tree to 4 m. high; leaves oblong-obovate, rounded to shortly cordate at base, acute or acuminate, thin, glabrous, 10–21 cm. long, 4–9 cm. broad, with 5–7 pairs of lateral veins; stipules deltate, acute, 3 mm. long; peduncles 1·5–4 cm. long; bracts small; calyx-tube subglobose, 1 mm. long, lobes triangular, 3 mm. long; corolla-lobes suborbicular, 1·5 mm. long, pink; capsule depressed-globose, up to about 6 mm. broad.

Uncommon (Han., Trel., St. Ann), in woodland on limestone; 500–2500 feet; fl. Mar, Aug, fr. Mar–Apr, Aug. *H 10310! P 15641! 16197!* Endemic.

3. R. subsessilifolia Proctor in Bull. Inst. Jam., Sci. ser. **16**: 66, t. 29 (1967).

Shrub or bushy tree 2–3 m. high; petioles up to 3 mm. long; leaves oblong-obovate, obtuse to subcordate at base, subacute at tip, thin, firm, glabrous adaxially, appressed-pubescent abaxially, 6–13 cm. long, 3–6 cm. broad, with 6–7 pairs of lateral veins; stipules deltate, acuminate, 5–6 mm. long; peduncles 2·5–5 cm. long; bracts lanceolate, about 3 mm. long; calyx-tube subglobose, about 1·5 mm. long, lobes triangular, 0·5 mm. long; corolla-lobes rounded, 2–3 mm. long; capsule depressed-globose, up to 7 mm. broad.

Very local (Port.), in mossy thickets in wet area on limestone; 2300–2500 feet; fl. Jan–Apr, fr. July–Aug. *A 9138! ACH 985! P 5695!* Endemic.

4. R. daphnoides Griseb., Fl. Br. W.I.: 327 (1861).

Slender glabrous shrub; leaves obovate to oblong-obovate, bluntly acute or emarginate, thinly papery, mostly 3·5–5·5 cm. long, 1·5–2·5 cm. broad, with 4–5 pairs of lateral veins; petioles about 5 mm. long; stipules deltate, acute, 2 mm. long; peduncles up to 1 cm. long, few-flowered; pedicels 2–5 mm. long; bracts subulate, up to 5 mm. long; calyx-tube subglobose, barely 1·5 mm. long, lobes triangular, acute, 0·3 mm. long; corolla apparently reddish, tube cylindrical, 7 mm. long, lobes suborbicular, 2 mm. long; capsule globose, 4 mm. broad.

Very rare (St. Ann), known only from the type, *Prior 469*. Endemic.

5. R. pallida Britton in Bull. Torr. Bot. Club **37**: 358 (1910).

Shrub or tree 2–10 m. high; leaves oblong-elliptical, acute or acuminate, thin, firm, mostly 6–12 (–20) cm. long, 1·8–4 (–5·5) cm. broad, with 4–6 (–8) pairs of lateral veins; petioles 1–2 cm. long; stipules deltate, acuminate, 3–4 mm. long, silky within; peduncles 1–1·5 cm. long, few-flowered; pedicels 2–5 mm. long; bracts linear-subulate, 2–3 mm. long; calyx-tube campanulate, 1–1·5 mm. long, lobes 1 mm. long; corolla tawny yellow or white, lobes suborbicular, 4 mm. long; capsule depressed-globose, about 5 mm. broad.

Uncommon and local (Port., St. Thom.), in woodland on limestone in wet area; 1200–2000 feet; fl. Feb–Mar, fr. Mar, Aug. *A 11579! H & B 10680! 10724! P 7644!* Endemic.

6. R. elegans Britton in Bull. Torr. Bot. Club **37**: 358 (1910).

Shrub or tree 2·5–10 m.; leaves elliptical, mostly acuminate, thin, firm, 7–15 cm. long, 3–6·5 cm. broad, with 4–7 pairs of lateral veins; stipules deltate, 3–4 mm. long, silky within; inflorescence as above; calyx-tube campanulate, 2 mm. long, lobes triangular, a little over 1 mm. long; corolla crimson, lobes tawny yellow, obovate, 4 mm. long; capsule depressed-globose, strongly ribbed, about 5·5 mm. broad.

Uncommon and local (Port., St. Thom.), in moist secondary forest on limestone hills; 1200–2250 feet; fl. Mar, fr. Mar, Aug. *H & B 10744! P 11806!* Endemic. Very close to and possibly not really distinct from the last.

7. R. racemosa Sw., Fl. Ind. Occ. **1**: 360 (1797).

Tree 6–9 m. high; leaves broadly to narrowly elliptical, cuneate at base, thin, glabrous, mostly 7–13 cm. long, 3·5–5 cm. broad, with 6–7 pairs of lateral veins; petioles 1·5–3 cm. long; stipules broad, cuspidate-acuminate, 3 mm. long, silky within; peduncles up to about 3 cm. long; bracts small, subulate; calyx-tube top-shaped, 2 mm. long, lobes very short, rotundate; corolla-tube funnel-shaped, 2 mm. long, pink, lobes rotundate, barely 1·5 mm. long, cream-coloured; capsules ovoid, about 4 mm. long and 3·5 mm. broad.

Rather local (St. Andr., Port., St. Thom.), in submontane woodlands; 3000–5500 feet; fl. July–Dec, fr. Sept–Jan. *H 5262! J.P. 1101! P 9649! Skelding UCWI 24945!* Endemic.

8. R. laurifolia Sw., Fl. Ind. Occ. **1**: 363 (1797).

Tree with slender trunk 4–15 m. high, shrublike when coppiced; leaves elliptical, cuneate at base, acute or acuminate, thin, firm glabrous, 5–14 cm. long, 2·5–6 cm. broad, with about 5 pairs of lateral veins; petioles up to 2 cm. long; stipules deltate, 3 mm. long, silky within; panicles 5–15 cm. long, axillary, many-flowered; calyx-tube depressed-globose, 1 mm. long, lobes triangular, 0·5 mm. long; corolla white or creamy-pink, tube 1·5–2·5 mm. long, lobes suborbicular, 1·3 mm. long; capsule depressed-globose, about 2 mm. long.

Generally distributed in woodlands on limestone in areas of moderate rainfall; 50–2300 feet; fl. Dec–Apr, fr. Feb–June. *A 10932! 13091! H 10359! P 16085!* Endemic.

9. R. petiolata Proctor in Bull. Inst. Jam., Sci. ser. **16**: 62, t. 27 (1967).

Shrub about 3 m. high; leaves broadly elliptical, cuneate-decurrent at base, shortly acuminate, minutely tomentose on both surfaces, 13–17 cm. long, 6–8 cm. broad, with 5–6 pairs of lateral veins; stipules deltate, acuminate, 4–5 mm. long; panicles 4–6 cm. long; peduncles 1·5–2 cm. long; bracts 3–6 mm. long; calyx-tube subglobose, woolly-tomentellous, about 1 mm. long, lobes broadly deltate, 0·5 mm. long; corolla light yellow, tube 4–6 mm. long, broadening apically, lobes 1–1·5 mm. long; capsule tomentellous, about 2 mm. long.

Rare (St. James, St. Ann), on wooded limestone hillsides; about 500 feet; fl. May–July, fr. July. *Asprey UCWI 500! P 16462!* Endemic.

10. R. stipularis (L.) Druce in Rep. Bot. Exch. Club Br. Is. **3**: 423 (1914).— *Petesia stipularis* L. (1759). *R. trifolia* Jacq. (1763). *R. tomentosa* Sw. (1788).

Shrub 0·6–3 m. or tree 4–10 m. high; leaves oblong-obovate to elliptical, cuneate at base, acute or acuminate, thin, smooth or rough adaxially, pubescent abaxially, glabrescent, 5–10 cm. long, 2–4 cm. broad, with 4–5 pairs of lateral veins; stipules deltate, acute, 3·5 mm. long, silky within; peduncles mostly 1–2 cm. long; bracts subulate, 2–4 mm. long; flowers few together, sessile or subsessile; calyx-tube subglobose, 1 mm. long, lobes 0·2–0·5 mm. long; corolla yellow to salmon-red, greenish at base, tube about 5 mm. long, lobes suborbicular, about 1·5 mm. long; capsule 3–4 mm. broad.

Frequent in the east-central and central parishes, on limestone cliffs and in thickets on rocky hillsides; 150–1200 feet; fl. most of the year, fr. Mar–May. *A 8972! H 8933! P 20565! Stearn 701!* Endemic.

11. R. nemoralis Proctor in Bull, Inst. Jam., Sci. ser. **16**: 61, t. 26 (1967).

Slender erect shrub 3 m. high; leaves broadly elliptical to rotundate, cuneate at base, acute or obtusely acuminate, coriaceous, 6–9 cm. broad, with 6–8 pairs of lateral veins, with a few small hairs along margins and on midrib beneath; petioles 2–3 cm. long; stipules broadly deltate, acuminate, 3–4 mm. long; peduncles 1–2 cm. long; bracts about 1·5 mm. long; calyx-tube about 1 mm. long, ribbed, white-hirtellous, lobes deltate, 0·7 mm. long; corolla creamy-white, lobes about 2·5 mm. long; capsule 4–6 mm. long.

Rare and local (Han.), on wooded limestone hill; 250–750 feet; fl. and fr. Nov. *P 11286!* Endemic.

12. R. polita Griseb., Fl. Br. W.I.: 326 (1861).

Shrub up to 3 m. or tree to 7 m. high; leaves obovate to oblong-elliptical, broadly or narrowly cuneate at base, firm-textured, 6–12 (–14) cm. long, 2·5–5 (–6·5) cm. broad, with 4–6 pairs of lateral veins; petioles about 1 cm. long; stipules broadly deltate, acuminate, 2·5 mm. long; peduncles about 1 cm. long; pedicels up to 3 mm. long; bracts minute; calyx-tube top-shaped, 1·5 mm. long, lobes deltate, 0·5 mm. long; corolla light pink or white tinged pink or light yellow, lobes suborbicular, about 2 mm. long; capsule 3–4 mm. long.

Frequent in central and western parishes, on rough exposed limestone and in thickets; 250–2600 feet; fl. and fr. most of the year. *A 8375! 12045! H 10337A! P 11318!* Endemic.

13. R. adamsii Proctor in Bull. Inst. Jam., Sci. ser. **16**: 53, t. 22 (1967).

Shrub or tree 3–4 m. high; leaves obovate, narrowly cuneate at base, subcoriaceous, 4·5–8·5 cm. long, 1·8–3·5 cm. broad, with about 4 pairs of lateral veins; petioles 4–9 mm. long; stipules broadly deltate, acuminate, 3–3·5 mm. long; peduncles 3–5 mm. long; flowers sessile; bracts deltate, acuminate; calyx as above; corolla salmon-pink, lobes about 1·5 mm. long; capsule 2–2·5 mm. long.

Local (Clar., Trel.), on rocky limestone; 1800–2500 feet; fl. May, fr. Aug. *P 10222! Webster & Proctor 5418!* Endemic.

14. R. hirsuta Sw., Nov. Gen. & Sp. Pl.: 41 (1788).

Shrub about 2 m. or tree up to 5 m. high; leaves elliptical, broadly cuneate to rounded at base, acute or shortly acuminate, mostly 7–11 cm. long, 3–5 cm. broad, with 5–6 pairs of lateral veins; petioles 3–9 mm. long; stipules triangular, acuminate, hairy outside, silky within, 7–13 mm. long; inflorescence sparingly branched; peduncles 4–8 cm. long; pedicels about 5 mm. long; bracts narrow, up to 5 mm. long; calyx-tube globose, hairy, about 2 mm. long; corolla yellow, tube up to 8–12 mm. long, lobes obovate, 2 mm. long; capsule depressed-globose, bisulcate, finally glabrous, about 7 mm. broad.

Local (Port., St. Thom.), in ravines and on cliffs in forest; 400–1800 feet; fl. Aug–Jan, fr. Aug–Dec. *H & B 10573! P 7430! 27784!* Endemic.

15. R. impressa Krug & Urb. in Urb., Symb. Ant. **1**: 412 (1899).

Shrub 1·5–4 m. high with pale terete pubescent twigs; leaves sometimes in threes, ovate to oblong-ovate, cuneate at base, acute or shortly acuminate, thin, 5–10 cm. long, 2–4·5 cm. broad, with 4–6 pairs of lateral veins, sparsely pubescent beneath; inflorescence few-flowered, pubescent; calyx-tube subglobose, about 1·5 mm. long, lobes barely 1 mm. long; corolla tawny-yellow, tube about 6 mm. long, lobes suborbicular, 2 mm. long; capsule globose, 4–5 mm. broad.

Rare (St. Andr.), in thickets; 600–2250 feet; fl. and fr. Feb. *H 11870! Robertson UCWI 3224!* Endemic.

16. R. portlandensis Proctor in Bull. Inst. Jam., Sci. ser. **16**: 65, t. 28 (1967).

Shrub or tree 2–6 m. high; young stems and leaves bright crimson; leaves oblong-ovate to elliptical, cuneate at base, acute or acuminate, thin, firm, 6–15 cm. long, 2–5 cm. broad, with 5–6 pairs of lateral veins; petioles 1–2 (–3) cm. long; stipules ovate-deltate, caudate-acuminate, up to 12 mm. long, pubescent in midline; peduncles mostly 1·5–2·5 cm. long; bracts hirsute; calyx-tube densely pubescent, 1·5 mm. long, lobes oblong-lanceolate, 7·5 mm. long; corolla bright yellow, tube cylindrical, glabrous outside except distally, 10 mm. long, lobes suborbicular, 4 mm. long; capsule globose, 6·5 mm. broad.

Rare and local (Port.), in damp thickets and mossy forest on limestone; 1500–2500 feet; fl. Dec–Apr, fr. Dec–Mar, July–Aug. *A 9129! ACH 950! P 9981!* Endemic.

17. R. ligulata Urb., Symb. Ant. **5**: 503 (1908).

Shrub 1–6 m. high; leaves oblong-oblanceolate to elliptical, cuneate at base, acute or obtuse at tip, firm-textured, 4–10 cm. long, 2–3·5 cm. broad, with 3–4 pairs of lateral veins; petioles 5–10 mm. long; stipules ovate-lanceolate, acuminate, densely hirsute outside, silky within, 12–18 mm. long; peduncles 2–8 mm. long; bracts ovate, acuminate, about 12 mm. long; calyx-tube top-shaped, 2 mm. long;

corolla-tube crimson, 12 mm. long, lobes orbicular, lemon-yellow, 4·5 mm. long; capsule depressed-globose, 5 mm. long.

Local (Han., Trel.), in moist woodland on limestone; 1500–2500 feet; fl. sporadically, fr. Jan. *H 10668! P 11310!* Endemic.

18. R. saxicola Britton in Bull. Torr. Bot. Club **39**: 8 (1912).

Shrub 4 m. high; twigs at first 4-angled and hirsute becoming terete and glabrous; leaves oblong-oblanceolate, cuneate at base, acute or obtuse at tip, papery, 3·5–9 cm. long, 1·5–3 cm. broad, with 5–6 pairs of lateral veins; petioles 5–10 mm. long; stipules ovate-lanceolate, acuminate, appressed-hirsute outside, 6–10 mm. long; peduncles 4–8 mm. long; bracts ovate to linear-lanceolate, 3–4 mm. long; flowers not known.

Rare and local (Manch.), on rocky cliff; 2200 feet; young fr. Sept. *H & B 10609!* Endemic.

19. R. hirta Sw., Nov. Gen. & Sp. Pl.: 41 (1788).

Shrub 0·3–4 m. or tree up to 5 m. high; leaves ovate or oblong-ovate, rounded to cuneate at base, acute or acuminate, papery, 5–16 cm. long, 2–6·5 cm. broad, with 6–7 pairs of lateral veins; petioles 4–9 mm. long; stipules triangular, acute, 4–6 mm. long, appressed-hirsute outside, silky within; peduncles up to about 7 cm. long; outer bracts foliaceous, 5–10 mm. long, the others subulate or lanceolate; calyx-tube 1·5 mm. long; corolla-tube green to dull yellow or streaked pink, hairy 6–8 (–10) mm. long, lobes yellow to orange, suborbicular, 2 mm. long; capsule puberulous, about 6 mm. broad.

Locally frequent (St. Andr., St. Thom.), in thickets and open woodlands on steep rocky slopes and on cliffs; 800–3000 feet; fl. and fr. most of the year. *A 5828! 8273! 9970! H 10591! P 9840!* Endemic. It is probable that heterostylous variants exist in this and related species and that differences of corolla-size are correlated with style types.

20. R. brachyphylla Proctor ex Adams in Phytologia **21** (2): 70 (1971).

Shrub 2 m. or more high or small tree; leaves broadly ovate, cordate, shortly acuminate to a sharply pointed tip, papery to firm, glabrous adaxially except on midrib, hairy on veins beneath, 2–8 cm. long, 1·5–4·5 cm. broad; petiole 0–3 mm. long; stipules deltate-acuminate, appressed-pilose, 5 mm. long; peduncle up to 4 cm. long; pedicels 0·5–4 mm. long; bracts subulate; calyx-tube ovoid, 2 mm. long; corolla-tube crimson, 12 mm. long, thinly pilose, lobes orbicular, deep yellow, glabrous distally, 4 mm. long; capsule bisulcate, thinly pilose, 5 mm. long, 6 mm. broad.

Rare and local (St. Thom.), on rocky serpentine hillsides; 2500–3000 feet; fl. and fr. Jan–Mar, July. *A 12139! 13236! P 23304!* Endemic.

21. R. cymulosa Proctor in Bull. Inst. Jam., Sci. ser. **16**: 55, t. 23 (1967).

Shrub 1–2·5 m. high with glabrous twigs; leaves broadly elliptical, cuneate to obtuse at base, obtuse at tip, glabrous except for a few minute hairs beneath, 3–6 (–7) cm. long, 1·5–3·5 cm. broad, with about 4 pairs of lateral veins; petioles 2–4 mm. long; stipules deltate, keeled, acuminate, 2 mm. long; inflorescence capitate, 3–5-flowered; pedicels slender; bracts about 1 mm. long; calyx-tube subglobose, 1 mm. long, lobes oblanceolate, obtuse, 2 mm. long; corolla puberulous, cream tinged red, tube 5–6 mm. long, lobes about 1 mm. long; capsule globose, 2–3 mm. long.

Local (Trel.), on limestone cliff and in thickets; 1250–1400 feet; fl. Feb–Apr, fr. Feb–Apr, Aug. *A 6795! P 20585!* Endemic. In the type locality *R. cymulosa* is sympatric with *R. sylvestris* and plants intermediate between the two species (e.g. *P 20586!*) are putative hybrids.

22. R. umbellulata Sw., Nov. Gen. & Sp. Pl.: 41 (1788).

Shrub about 3 m. high; leaves elliptic-lanceolate, narrowly cuneate at base, acuminate, papery, nearly glabrous adaxially, hirsute on veins beneath, 6–9·5 cm. long, 2–3 cm. broad, with 4–6 pairs of lateral veins; petioles hirsute, about 1 cm. long; stipules ovate, caudate-acuminate, appressed-hirsute on both surfaces, up to

13 mm. long; peduncles mostly 2·5–5 cm. long; outer bracts foliaceous, others lanceolate to linear-lanceolate, up to 3 cm. long; flowers sessile or subsessile; calyx-tube broadly top-shaped, hairy, 1·5 mm. long; corolla light tawny yellow, tube dilated at throat, lobes obovate, 2–2·5 mm. long; capsule globose, 5 mm. broad.

Rare (West., St. James), on shaded banks; 650–800 feet; fl. Feb. *H 9189! P 16163!* Endemic.

23. R. jamaicensis Proctor in Bull. Inst. Jam., Sci. ser. **16**: 56, t. 25 (1967).

Shrub 1·5 m. high; leaves oblong-elliptical, shortly cuneate at base, acute, subcoriaceous, minutely pubescent on both surfaces, 4–11 cm long, 1·5–3·5 cm. broad, with 4–6 pairs of lateral veins; petioles 2–4 mm. long; stipules deltate, acuminate, hirsute, about 5 mm. long; peduncles 1–1·3 cm. long; bracts deltate, acuminate, 3–3·5 mm. long; flowers few; calyx-tube globose, about 1·1 mm. long; corolla-tube light tawny yellow, lobes orange-yellow, 2 mm. long; capsule glabrate, 2–5 mm. long.

Rare and local (St. Ann), on wooded limestone hillside; about 2500 feet; fl. and fr. Mar. *P 16198!* Endemic.

24. R. dolphinensis Proctor in Bull. Inst. Jam., Sci. ser. **16**: 56, t. 24 (1967).

Tree 5–10 m. high; leaves elliptical, cuneate, sharply acuminate, membranous, sparsely appressed-hirtellous on both surfaces, 3·5–6·5 cm. long, 1–2 cm. broad, with 3–4 pairs of lateral veins; petioles 4–7 mm. long; stipules oblong, abruptly acuminate, sparsely appressed-hirtellous, 5–7 mm. long; bracts 4–6 mm. long; corolla cream-coloured, pubescent, tube about 6 mm. long, lobes 1·5–2 mm. long; capsule globose, bisulcate, about 4 mm. broad.

Rare and local (Han.), in moist forest on limestone; 1500–1700 feet; fl. Mar–Apr, fr. Mar. *P 20686!* Endemic.

25. R. glauca Griseb., Fl. Br. W.I.: 329 (1861).

Low or scrambling shrub 0·6–3 m. or tree up to 5 m. high; leaves obovate, cuneate at base, obtuse to rounded at tip, up to 9 cm. long and 5 cm. broad, puberulous or glabrous and shiny adaxially, appressed-puberulous abaxially; petioles 6–13 mm. long; stipules ovate to ovate-lanceolate, acuminate, about 8 mm. long, densely sericeous, deciduous; inflorescence few-flowered; peduncles 3–8 (–10) mm. long; outer bracts 5–7 mm. long; calyx-tube obovoid, 4 mm. long; corolla light yellow, tube sericeous, 8 mm. long, lobes suborbicular, silky, 4–4·5 mm. long; capsule depressed-globose, 4–4·5 mm. long.

Rather local (St. Cath., Trel., St. Ann), on sheltered limestone banks and on jagged rocks in woodland; 2000–3000 feet; fl. and fr. Dec–Aug. *A 12632! 12847! H 8881! P 16196!* Endemic.

26. R. clarendonensis Britton ex S. Moore in Journ. Bot. **68**: 111 (1930).

Shrub up to 1·5 m. high; leaves subsessile, obtuse to rounded at tip, margins reflexed, 3–7 cm. long, 2–5 cm. broad, with about 6 pairs of lateral veins, appressed-pilose on midrib abaxially; stipules deltate, acuminate, 5–6 mm. long, appressed-sericeous, keeled; inflorescence subcapitate; peduncles up to 2 cm. long; bracts linear, 5–7 mm. long; calyx-tube hemispherical, silky, 1·5 mm. long, lobes linear, 6 mm. long; corolla white or pinkish, tube silky, 7 mm. long, lobes suborbicular, 2·5 mm. long; capsule depressed-globose, grey-hairy, 5 mm. broad.

Rare and local (Clar., St. Ann), on exposed craggy limestone; 2300–2500 feet; fl. Dec–Jan, fr. sporadically; *A 11049! H 12774! Robertson UCWI 1999! P 11390!* Endemic.

27. R. sylvestris S. Moore in Journ. Bot. **68**: 111 (1930).

Shrub 1–3 m. high; leaves oblong-obovate, obtuse to rotundate at tip, leathery, at first silky-pubescent adaxially, tomentose abaxially, 2·5–8 (–10) cm. long, 1·5–3·5 (–6) cm. broad, with 5–6 pairs of lateral veins; petioles 4–16 (–22) mm. long; stipules ovate, acute, appressed-hairy outside, silky within, 6–20 mm. long, deciduous; peduncle short; flowers 1–3, fragrant; bracts ovate to ovate-lanceolate, acute, 5–6 mm. long; calyx-tube ovoid, about 3 mm. long, lobes lanceolate, about 7 mm. long; corolla white, tube 8 mm. long, lobes ovate, 2·5 mm. long; capsule subglobose, 5 mm. broad.

Rather local (Clar., Trel., St. Ann), on cliffs and in thickets on limestone; 1000–2500 feet; fl. Dec–June, fr. sporadically. *A 6799! H 11030! P 11917!* Endemic.

28. R. cincta Griseb., Fl. Br. W.I.: 329 (1861).

Shrub up to 2 m. or tree to 6 m. high; leaves oblong-obovate to elliptical, obtuse or acute at tip, leathery, glabrescent, 5–9 cm. long, 2–3 (–5·5) cm. broad, with 3–5 pairs of lateral veins; petioles about 1 cm. long; stipules ovate or oblong, acuminate, hairy outside, silky within, deciduous; flowers few in cymose heads; peduncles 5–10 mm. long; outer bracts ovate, 8–10 mm. long; calyx-tube subglobose, 3 mm. long, lobes ovate-lanceolate, acute; corolla white, tube broad, 9 mm. long, lobes suborbicular, 3 mm. long; capsule depressed-globose, sericeous, about 6 mm. broad.

Rare and local (Han.), in exposed places on steep rocky hillside; 1750–1780 feet; fl. Mar–Apr, fr. July. *H 10272! P 10049!* Endemic.

29. R. incana Sw., Nov. Gen. & Sp. Pl.: 41 (1788).

Shrub 1–2 m. high; leaves elliptical, obtuse at tip, pubescent on midrib abaxially, 5–10 cm. long, 2·5–4 cm. broad, with 4–5 pairs of lateral veins; petioles 5–15 mm. long; stipules ovate-lanceolate, acuminate, finely silky on both surfaces, deciduous; flowers in clusters of three, fragrant; peduncles 1–2 cm. long; pedicels up to 7 mm. long; bracts ovate-lanceolate, acuminate, about 6 mm. long; calyx-tube ovoid, 3 mm. long, lobes ovate, obtuse; corolla white, tube narrowly funnel-shaped, 10 mm. long, lobes ovate, 4 mm. long; capsule globose, tomentose, 7 mm. broad.

Very rare (Manch.), on limestone rocks; about 2300 feet; fr. Feb. *P 16178!* Endemic.

10. ACROSYNANTHUS Urb. (1913)

1. A. jamaicensis Howard & Proctor in Journ. Arn. Arb. **39**: 101 (1958).

Shrub 0·6–4 m. high, often with drooping flexible branches; stipules triangular, ciliate, up to 2 mm. long; leaves linear-lanceolate, 2·5–4·5 cm. long, 3–8 mm. broad, dark green adaxially, whitish beneath; inflorescence cymose with 3-flowered cymules; flowers fragrant; corolla-tube white with orange base, pubescent outside, 2·5–3 mm. long; capsule about 3 mm. long; seeds ellipsoidal, compressed, about 1 mm. long.

Rare and local (Trel.), on limestone cliffs; 1200–1500 feet; fl. Apr–Aug, fr. July–Aug. *A 6796! A & S 3241! H & P 14391!* Endemic.

11. SABICEA Aubl. (1775)

1. S. hirta Sw., Nov. Gen. & Sp. Pl.: 46 (1788).

Woody climber up to 2 m. or more high, with slender hairy branches; stipules up to about 12 mm. broad; leaves elliptical, acuminate, up to 12 cm. long and 7 mm. broad, hairy especially on the veins; cymes umbellate, bracteate, 3-flowered; corolla white, sparsely hirsute, the tube about 5 mm. long; berry globose, sparsely hairy, white, about 7 mm. in diameter.

Rare and rather local in the western parishes and Portland, in moist secondary thickets; 900–1500 feet; fl. and fr. Feb–May, Sept. *A 9110! H 9246! P 22123!* Endemic.

12. COCCOCYPSELUM P. Browne (1756) nom. cons.

Flower-heads usually sessile or subsessile, rarely peduncles up to 20 mm. long; corolla-lobes blue or mauve, the tube whitish and about 6 mm. long; berry 6–9 mm. in diameter
    1. *herbaceum*

Flower-heads mostly stalked; corolla white, the tube about 4·5 mm. long; berry about 4 mm. in diameter
    2. *pseudotontanea*

1. C. herbaceum Aubl., Hist. Pl. Guiane **1**: 68 (1775).

Herb with usually hairy trailing and rooting branches, the stem and petioles often tinged mauve; leaves ovate to oblong-ovate, obtuse, 2–4·5 (–6) cm. long and 1–2 (–2·5) cm. broad; heads mostly 2–3-flowered, rarely 1-flowered.

Common, along pathside banks and in damp grassy savannas in wet areas; 700–5050 feet; fl. and fr. all the year. *A 5737! H 12275! P 19751!* Greater Antilles. A glabrate variant is known from the parish of Clarendon (*J.P. 1027! P 15818!*).

2. C. pseudotontanea Griseb., Fl. Br. W.I.: 322 (1861).
Like the last but smaller in all parts and apparently not rooting so freely; peduncles 4–40 mm. long, sometimes shorter if flowers solitary.
Rare and rather local (St. Andr., Port.), on pathside banks; 3900–5000 feet; fl. and fr. sporadically throughout the year. *A 5874! J.P. 1482! P 19752!* Endemic.

13. MUSSAENDA L. (1753)

1 Enlarged sepal red; leaves densely tomentose; lax scrambling shrub; native of tropical Africa; Ashanti Blood   *M. erythrophylla* Schumach.
1 Enlarged sepals not red; leaves at most thinly pubescent adaxially at maturity; erect shrubs:
  2 Leaves up to about 20 cm. long, distinctly acuminate, with 12–15 pairs of lateral veins; enlarged sepal white; corolla orange-yellow; tropical Asia   *M. treutleri* Stapf
  2 Leaves up to about 8 (–10) cm. long, acute or only very slightly acuminate, with 5–7 pairs of lateral veins; enlarged sepal greenish-cream adaxially, white beneath; corolla light yellow; tropical Africa   *Pseudomussaenda flava* Verdcourt

14. GONZALAGUNIA Ruiz & Pav. (1794)

1. G. brachyantha (A. Rich.) Urb., Symb. Ant. **7**: 400 (1912).—*Gonzalea brachyantha* A. Rich. (1850).
Shrub or tree 2–10 m. high; leaves elliptic-lanceolate, acuminate, cuneate at base, 4–15 cm. long, 1·5–4 (–5) cm. broad; stipules caudate, silky within, 4–8 mm. long; cymes small and few-flowered on a common axis 3–18 cm. long; corolla greenish-white to light yellow, 4–5-merous, the tube about 2 mm. long; berry globose, white, 3 mm. in diameter.
Generally distributed and rather common in woodlands on limestone or shale; 1200–2750 feet; fl. June–Mar, fr. Aug–Mar. *A 11690! H 12381! P 18404!* Cuba, Hispaniola.

15. RANDIA L. (1753)

1. R. aculeata L., Sp. Pl. **2**: 1192 (1753).—Box Briar, Indigo Berry, Ink Berry.
Deciduous shrub or tree 1·2–6 m. high; leaves obovate to suborbicular, mostly obtuse; flowers sessile or subsessile, mostly solitary, fragrant; corolla-tube green, 4–7 mm. long, hairy in throat, lobes 4–6, white, ovate, acute, 3·5–4 mm. long; berry bluish-black, up to about 12 mm. in diameter; occurs in three more or less distinct varieties as follows:

1a. Var. aculeata.
Branches armed with paired patent or ascending spines about 1 cm. long; twigs, leaves and flowers externally glabrous; leaves thick, rigid, shiny, 1–4 cm. long, 7–25 mm. broad, narrowed at base.
Rather common in thickets and woodlands on rocky limestone or serpentine, mostly in areas of rather low rainfall; 50–2900 feet; fl. Apr–May, fr. Jan, Apr, Oct. *A 8215! H 11776! P 19655!* Florida, Bermuda, Bahamas, Mexico to Venezuela, West Indies; Grand Cayman.

1b. R. aculeata var. **jamaicensis** (Spreng.) Adams in Phytologia **21** (2): 70 (1971). —*Gardenia jamaicensis* Spreng. (1824). *R. jamaicensis* (Spreng.) Krug & Urb. (1899).
Like the last but young stems, leaves and corolla shortly and densely or sparsely hairy; leaves typically truncate to subcordate at base.
Rare and local (St. Andr., Clar.), in thickets and open woodlands on rough limestone or shale; 700–2600 feet; fl. Feb, June, fr. Apr. *A 6389! ACH 128!* Endemic.

1c. R. aculeata var. **mitis** (L.) Griseb., Fl. Br. W.I.: 318 (1861).—*R. mitis* L. (1753).

Branches usually unarmed; leaves thin, obtuse or acute, mostly 3–8 cm. long and 1·5–4 cm. broad, narrowed at base.

Frequent at swamp margins and in woodlands on limestone in areas of moderate to high rainfall; near SL–3000 feet; fl. ? July, fr. July–Nov. *A 11366! H 10711! P 21359!* Probably endemic.

16. CASASIA A. Rich. (1850)

1. C. longipes Urb., Symb. Ant. **5**: 506 (1908).—*C. piricarpa* Urb. (1908).

Slender-trunked tree up to 11 m. high; leaves mostly broadly elliptical, shortly acuminate at base and apex, shiny on both surfaces, with 7–10 pairs of conspicuous lateral veins, up to 28 cm. long and 16 cm. broad; petiole stout up to 4·5 cm. long; stipules early deciduous, deltate, 10–12 mm. long, 8–9 mm. broad at base; cymes rather short and compact; flowers greenish at first, fragrant; corolla white, the lobes a little shorter than the tube which shrinks on drying; fruit hard, black when ripe, up to 9 cm. in diameter; seeds flat, irregularly discoid, about 1 cm. in diameter.

Uncommon in the central parishes, in woodland on limestone; 100–3000 feet; fl. Apr–Sept, fr. most of the year. *A 6890! 13141! H 11199! P 21365!* Endemic.

17. CATESBAEA L. (1753)

Leaves mostly 1–2·5 cm. long; flowers pendulous; corolla 6–14 cm. long, green at first, finally yellow; cultivated ornamental shrub; native of Bahamas and Cuba; Lily Thorn
                                                                                        C. spinosa L.
Leaves rarely as much as 1 cm. long; flowers patent; corolla 8–10 mm. long
                                                                                        1. *parviflora*

1. C. parviflora Sw., Nov. Gen. & Sp. Pl.: 30 (1788).

Shrub up to 1·5 m. high, with numerous erect or drooping branches; leaves sometimes clustered on short lateral shoots, orbicular to obovate, rigidly leathery, glabrous, 5–8 mm. long, 3–5 mm. broad; corolla white, 5–8 (–10) mm. long, the lobes ovate, acute; berry ovoid, white or black, about 4 mm. long.

Rare (Clar., Han., Port.), on coastal rocks and at salina margins; near SL; fl. Nov, fr. Mar. *A 12998! H 10193!* Cuba, Puerto Rico, Antigua.

18. STRUMPFIA Jacq. (1760)

1. S. maritima Jacq., Enum. Syst. Pl. Carib.: 28 (1760).

Much branched shrub up to about 1 m. high; leaves sessile, linear, pubescent beneath, 1–2 cm. long, 1–2 mm. broad; racemes few-flowered, shorter than the leaves; corolla-tube about 1·5 mm. long; drupe globose, about 3·5 mm. in diameter.

Local, on coastal limestone; SL–10 feet; fl. Dec–Mar, June, fr. Sept–Mar. *A 6130! H 9726! P 15200!* Florida, Mexico, Bahamas, Greater Antilles, drier islands of the Lesser Antilles – Anguilla, Antigua, Barbuda, Guadeloupe, Marie Galante, Barbados; Grand Cayman.

19. COFFEA L. (1753)

1 Stipules obtuse to subacute; leaves rounded to shortly acuminate at tip, obovate, oblanceolate or broadly elliptical, 10–40 cm. long, 5–15 cm. broad; bracteoles without leafy appendages; corolla-tube mostly 10–15 mm. long; berries 12–25 mm. long; native of tropical Africa; Liberian Coffee   *C. liberica* Bull ex Hiern
1 Stipules cuspidate to acuminate; leaves distinctly acuminate at tip:
    2 Appendages to bracteoles up to 20 mm. long and 9 mm. broad, the calyx hidden at anthesis; leaves elliptical to obovate, 12–35 cm. long, 5–12 cm. broad; domatia puberulous; berries 9–17 mm. long; native of tropical Africa; Robusta Coffee
                                                                        C. canephora Pierre ex Fröhner
    2 Appendages to bracteoles up to 5 mm. long and 1·5 mm. broad, the calyx exposed at anthesis; leaves ovate, elliptical or oblong, 7–18 cm. long, 3–8 cm. broad; domatia usually glabrous; berries 13–17 mm. long; native of Arabia and N.E. tropical Africa; Arabian Coffee   *C. arabica* L.

20. ERITHALIS P. Browne (1756)

1 Ovary 15–22 locular; branches stout, more or less quadrangular; flowers more than 1 cm.
  long and broad; stipules truncate; fruit 5–8 mm. in diameter; inflorescence rounded-
  corymbose   1. *quadrangularis*
1 Ovary 5–12-locular; branches slender, more or less terete; flowers less than 1 cm. long
  and broad; stipules apiculate; fruit 2–4 mm. in diameter:
  2 Petioles stout, up to 5 mm. long; inflorescence rounded-corymbose; corolla-lobes 4–8,
    4–8 mm. long; leaves often obovate   2. *fruticosa*
  2 Petioles slender, up to 15 mm. long; inflorescence pyramidal; corolla-lobes usually 5,
    4 mm. long; leaves mostly oblong   3. *harrisii*

1. **E. quadrangularis** Krug & Urb. in Notizbl. Bot. Gart. Berlin 1: 320 (1897).
   Tree 5–10 m. high; leaves thinly leathery, broadly ovate or rounded-elliptical,
up to 12 cm. long and 10 cm. broad; corolla usually 6–7-lobed, white fading yellow;
ripening fruit dull pink, fleshy when fresh, 7–9 mm. long, 6–8 mm. broad, about
20-ribbed when dry, flattened globose, 5–7 mm. in diameter.
   Local (Clar., Manch., St. Ann), in woodland on limestone; 1500–2500 feet;
fl. Mar–May, fr. May–Aug. *A 12521! H 11048! P 19678!* Endemic.

2. **E. fruticosa** L., Syst. Nat. ed. 10, 2: 930 (1759).—*E. odorifera* Jacq. (1763).
   E. harrisii var. *angusta* S. Moore (1935).
   Shrub 0·3–5 m. high, usually glabrous but branches sometimes minutely
puberulous; leaves leathery, shiny, obtuse, very variable in shape, the margins
usually reflexed and the veins rather obscure, up to 11 cm. long and 6 cm. broad
but mostly much smaller; corolla white fading yellow or pinkish; fruit finally shiny,
black.
   Common on coastal rocks, occasional on limestone cliffs inland; SL–800 (–2000)
feet; fl. and fr. most of the year, but more frequently Dec–June. *A 8545! H 12114!
P 6103! Stearn 323!* Florida, Bahamas, C. Amer., West Indies southwards to
Grenada; Grand Cayman.

3. **E. harrisii** Urb., Symb. Ant. 5: 514 (1908).
   Shrub 3 m. or tree to 8 m. high, with slender straggling sometimes puberulous
branches; leaves thinly leathery, not very shiny and with the veins usually clearly
apparent, sometimes puberulous, up to 10 cm. long and 5 (–7) cm. broad; corolla
white; fruit dull pinkish, finally black.
   Locally common in the central parishes, in woodland margins on rough limestone; 1200–2500 feet; fl. Apr–Oct, fr. Aug–Sept. *A 12652! H 8974! P 21441!*
Endemic. Plants with characters intermediate between *E. fruticosa* and *E. harrisii*
occur in both coastal and inland localities.

21. SCOLOSANTHUS Vahl (1797)

1. **S. multiflorus** (Sw.) Krug & Urb. in Urb., Symb. Ant. 1: 443 (1899).—*Ixora* ?
   multiflora Sw. (1788).
   Erect shrub 2–4·5 m. high, sometimes spiny; young branchlets and calyx setulose-
puberulous; leaves ovate, acute or obtuse, leathery, 2–4·5 cm. long, 1–2·5 cm. broad;
corolla tawny yellow, the tube about 7 mm. long; drupe ellipsoidal, 5 mm. long,
3·5 mm. broad.
   Rare and local (St. Cath., Clar., St. Eliz., St. Ann), in coastal woodland on
limestone; SL–100 (–500) feet; fl. and fr. Mar, July, Oct. *Asprey UCWI 4822!
H 10374!* Endemic.

22. PHIALANTHUS Griseb. (1861)

1 Leaf-margin distinctly revolute, blade broadest at middle, shortly cuneate at base;
  inflorescence up to 12-flowered; calyx-tube obovoid   1. *revolutus*
1 Leaf-margin flat or only slightly or very narrowly revolute; inflorescences with 4–8
  flowers; calyx-tube turbinate-clavate:
  2 Leaf-blade lanceolate to obovate, broadest at middle, acute at base   2. *myrtilloides*
  2 Leaf-blade oblanceolate, broadest towards tip, long-cuneate at base   3. *jamaicensis*

1. P. revolutus Urb. in Fedde, Repert. Sp. Nov. **17**: 407 (1921).

Shrub or tree 3–6 m. high, the terminal buds and leaves resinous; leaves elliptical, obtuse, up to 6 (–9) cm. long and 2 (–3) cm. broad, paler beneath; stipules truncate; corolla 1·5 mm. long, yellowish.

Very rare (St. Cath., Clar., Manch.), in coastal thickets; 50–500 feet; fl. Sept. *A 8847! H 10525!* Endemic.

2. P. myrtilloides Griseb., Fl. Br. W.I.: 335 (1861).

Shrub 1–3 m. high, with arching branches; resinous; leaves obtuse, 1–5 cm. long, 5–18 mm. broad; calyx-lobes enlarging in fruit; corolla 2 mm. long including lobes; fruit cylindrical, 1·5 mm. in diameter.

Very rare (Trel.), on steep slopes of limestone ravine; about 1200 feet; fl. Aug. *A & C 1204! Howard 14134!* Bahamas, S. Caicos, Cuba, Puerto Rico.

3. P. jamaicensis Urb., Symb. Ant. **5**: 515 (1908).

Shrub or tree to 6 m. high; young twigs curved-ascending; leaves rounded at tip, 1·5–4 cm. long, 4–10 mm. broad; stipules fringed at first; calyx minutely scurfy; corolla 1·5 mm. long, greenish.

Very rare (St. Andr.), in thickets on arid limestone; 800–900 feet; fl. Oct. *H 9023!* Endemic.

23. CHIOCOCCA P. Browne (1756)

Leaves ovate to lanceolate, long-acute to narrowly acuminate at tip; racemes about as long as the leaves, branched; pedicels up to 5 mm. long   1. *alba*
Leaves ovate to elliptical, obtuse to shortly and bluntly acuminate at tip; racemes about as long as the leaves in flower, slightly longer in fruit, simple or very sparingly branched; pedicels up to 2 mm. long   2. *parvifolia*

1. C. alba (L.) Hitchc. in Rep. Miss. Bot. Gard. **4**: 94 (1893).—*Lonicera alba* L. (1753). David's Root, Snowberry.

Trailing and twining shrub climbing to 5 m. or more; leaves thin, acute at base, 3–9 cm. long, 1·5–3 (–3·5) cm. broad; petioles often twisted; flowers (4–) 5 (–6)-merous; calyx small; corolla campanulate, cream-coloured to yellow, tube 2·5–5 mm. long, lobes reflexed, hyaline-margined; fruit white, compressed, 4–5 mm. broad.

Common, in thickets and on limestone banks and cliffs; SL–2700 feet; fl. and fr. all the year. *A 5572! H 5866! P 19683!* Florida, Bahamas, Mexico, C. Amer., West Indies.

2. C. parvifolia Wullschl. ex Griseb., Fl. Br. W.I.: 337 (1861).

Climbing shrub to 3 m. or more high; leaves broadly cuneate at base, 1–3·5 cm. long, 5–15 (–20) mm. broad; corolla usually light yellow, tube about 3 mm. long; fruit like the last.

Rather common, in thickets and thin open woodlands on limestone or serpentine; SL–3000 feet; fl. Apr–Nov, fr. June–Jan. *A 7281! H 9648! P 19684!* Florida, Bahamas, Cuba, Hispaniola, Antigua, St. Vincent, Trinidad and Tobago.

24. MACHAONIA Humb. & Bonpl. (1806)

Flowers normally 5-merous; leaves 4–7 cm. long, 3–4·5 cm. broad; fruit papillose
                                                                      1. *rotundata*
Flowers usually 4-merous; leaves 1·5–2·5 cm. long, 8–13 mm. broad; fruit typically roughly hairy but sometimes smooth   2. *cymosa*

1. M. rotundata Griseb., Fl. Br. W.I.: 348 (1861).

Unarmed or very sparingly spiny shrub 2–3 m. high; branches glabrescent; leaves ovate, obtuse or shortly acuminate; inflorescence-branches pubescent; calyx-tube narrowly obovoid, nearly glabrous, 2·5 mm. long, lobes white; corolla greenish-white, tube 2 mm. long, fragrant; fruit obovoid, obscurely ribbed, 5 mm. long.

Local, at inner mangrove margins and in brackish coastal thickets; near SL; fl. Aug–Oct, fr. Aug–Sept. *A 8079! H 10640! M. Jeffrey-Smith!* Endemic.

2. M. cymosa (Sw.) Griseb., Fl. Br. W.I.: 348 (1861).—*Lippia cymosa* Sw. (1788).
   M. cymosa var. *glabrescens* S. Moore (1930).
   Sparingly spiny shrub 2–3 m. high, with slender puberulous branches; leaves ovate to oblong-ovate, obtuse to shortly acute; panicle pubescent; calyx-tube obovoid, hispidulous, 1·5 mm. long, the lobes as long as the tube; corolla white, tube 2 mm. long, fragrant; fruit obconical, strongly 4-ribbed, 4·5–5 mm. long, smooth or sparsely papillose in var. *glabrescens* S. Moore, otherwise roughly hairy.
   Rare and local, in rocky thickets; 200–1250 feet; fl. June–July, Oct–Nov, fr. Oct–Dec. *A 13229! H 11694! P 19704! Yuncker 17106!* Endemic.

25. ANTIRHEA Commers. (1789)

1 Leaves velvety-tomentose beneath, without obvious domatia; corolla white, about 5 mm.
      long   1. *tomentosa*
1 Leaves glabrous or nearly so:
   2 Corolla tomentose, tube 6 mm. long; leaves with domatia in the vein-axils beneath
                                                                          2. *jamaicensis*
   2 Corolla glabrous or minutely puberulous; leaves without domatia:
      3 Leaves thin, shiny; corolla-tube 3·5–5 mm. long   3. *lucida*
      3 Leaves thick, leathery, opaque; corolla-tube 8–9 mm. long   4. *coriacea*

1. A. tomentosa (Sw.) Fawcett, Prov. List Jam.: 19 (1893).—*Laugeria tomentosa* Sw. (1788).
   Tall shrub or tree to 6 m. high; leaves oblong-ovate, obtuse at tip and base, 6–14 cm. long, 3·5–7 cm. broad; stipules triangular, about 4 mm. long; inflorescence twice-forked.
   Very rare, in thickets in the western parts of the island. Known only from the type, *Swartz*. Endemic.

2. A. jamaicensis Urb., Symb. Ant. **1**: 435 (1899).—Gold Spoon, May Day Mahogany, Pigeon Wood, Susan Wood.
   Shrub 2·5 m. or tree 6–20 m. high, with spreading branches; leaves thin, ovate to obovate, rounded to obtusely cuspidate at tip, 6·5–15 cm. long, 3–9 cm. broad; stipules about 7 mm. long; inflorescence once-forked; corolla greenish-yellow tinged pink; fruit 7–12 mm. long.
   Occasional, in woodlands on limestone in moderately wet areas; 1200–4000 feet; fl. Feb–Aug, fr. Mar–Nov. *H 11989! P 22207!* Endemic.

3. A. lucida (Sw.) Hook. f. in Benth. & Hook. f., Gen. Pl. **2**: 100 (1873).—*Laugeria lucida* Sw. (1788).
   Tree 6–13 m. high; leaves oblong-ovate to elliptical, obtuse at tip and base, 6–12 cm. long, 3·5–6 cm. broad, glabrous, with very faint veins; stipules acuminate, 6–8 mm. long; inflorescence once-forked; corolla white, campanulate; fruit about 1 cm. long.
   Very local (St. Thom.), on limestone rocks in woodland; 20–800 feet; fl. July, Nov, fr. Nov. *A 11363! H 11679!* Bahamas, Greater Antilles, St. Croix. A specimen from Trelawny parish, 1500 feet, *H & P 14417!*, has larger obovate leaves and may be a distinct species.

4. A. coriacea (Vahl) Urb., Symb. Ant. **1**: 436 (1899).—*Laugeria coriacea* Vahl (1797). Peg Wood.
   Shrub 3 m. or tree to 11 m. high; leaves ovate to elliptical, obtuse at tip and base, 5–10 cm. long, 3–5 cm. broad, with prominent lateral veins, golden-brown when young; stipules obtuse, 5–7 mm. long, red when young; inflorescence twice-forked; corolla greenish-white to white or yellow tinged pink; fruit 7–8 mm. long.
   Uncommon in the central and eastern parishes, in wet woodland on limestone; 1400–2500 feet; fl. Mar–Nov, fr. May–June. *A 12470! H 8730! P 18401!* Puerto Rico, Lesser Antilles.

26. GUETTARDA L. (1753)

1 Leaf-blade scabrid adaxially, tomentose abaxially; stipules lanceolate, acuminate and
      subulate-tipped   1. *scabra*

1 Leaf-blade glabrous or puberulous adaxially, smooth:
  2 Flowers solitary or inflorescence few-flowered and subcapitate, the peduncle and primary branches very slender; leaves mostly less than 5 cm. long and 3 cm. broad; stipules ovate to lanceolate; corolla-tube 6–9 mm. long   5. *elliptica*
  2 Flowers several to numerous in distinctly forked cymes; peduncle stout, erect in flower: leaves mostly more than 5 cm. long; corolla-tube at least 10 mm. long:
    3 Stipules mostly subulate-tipped; corolla-tube 12–23 mm. long; leaf-blade 5–17 cm. long, 3–9 cm. broad, densely appressed silky-pubescent beneath or glabrescent; common variable species   2. *argentea*
    3 Stipules ovate to ovate-lanceolate, at most acute; leaf-blade thinly appressed-pubescent to glabrescent abaxially; rare species:
      4 Leaf-blade up to 12 cm. long and 5 cm. broad; corolla probably not as long as 2·5 cm; lateral veins ascending at an angle of 20°–40° from the midrib   3. *frangulifolia*
      4 Leaf-blade 12–20 cm. long, 9–13 cm. broad; corolla 2·5–3·8 cm. long; lateral veins ascending at an angle of 50° or more from the midrib   4. *longiflora*

1. G. scabra (L.) Vent., Choix Pl.: t. 1 (1803).—*Matthiola scabra* L. (1753).

Shrub or tree to 10 m. high; leaves ovate to oblong-ovate, cuspidate or mucronate at tip, rounded to shortly cordate at base, 3–15 cm. long, 2–12 cm. broad, rather leathery; peduncles hairy, 2–12 cm. long; flowers compact, white; corolla-tube 10–21 mm. long; fruit globose, 5–8 mm. in diameter when dry, 3–6-seeded.

Reported by Grisebach on the basis of three early nineteenth Century collections, but not confirmed. Otherwise throughout the area from Florida to Brazil and in most of the West Indian islands.

2. G. argentea Lam., Encycl. Méth. Bot. **3**: 54 (1789).—*G. argentea* var. *glabrata* Urb. (1909). *G. potamophila* Urb. (1909). *G. constricta* Britton (1910).

Tree up to 10 m. high; bark rather smooth, mottled, staining yellow; leaves variable in shape, size and hairiness, mostly broadly ovate, shortly acuminate or cuspidate, rounded or shortly cordate at base, sometimes repand-margined, typically silky beneath but less hairy in juveniles and some low-elevation forms; peduncles up to 12 cm. long, but usually shorter than the leaves; corolla usually pinkish, 6–7-lobed; ripe fruit purple, soft, about 1 cm. in diameter, 4–6-seeded.

Common, in open woodlands on limestone or serpentine; near SL–3000 (–5000) feet; fl. May–Dec, fr. June–Jan. *A 12586! H 10993! P 10226!* Hispaniola.

3. G. frangulifolia Urb., Symb. Ant. **6**: 47 (1909).

Very like the last and probably only a minor variant of it; shrub 1·3–3 m. or tree to 12 m. high; leaf-blade ovate-lanceolate to pandurate-lanceolate or lanceolate, acuminate, glossy adaxially; stipules ovate-lanceolate; peduncle 2·5–8·5 cm. long; inflorescence-branches short, twice-forked.

Uncommon and local (St. Cath., Han., Trel.), in thickets on exposed limestone cliffs and crags; 700–2650 feet; fl. not known, fr. July–Sept. *A 11707! H 9472! P 15760!* Endemic.

4. G. longiflora Griseb., Fl. Br. W.I.: 332 (1861), excl. ref. Cuba.

Tree 10 m. high; leaves ovate, cuspidate at tip, rounded or subcordate at base, glabrous adaxially; peduncles shorter than the leaves; corolla-lobes 5–6, 5–6 mm. long; fruit not known.

Very rare (St. Ann, St. Thom.), not recently collected. *Prior 459!* Endemic. The *March* specimen, cited by Grisebach, lacks flowers and resembles juvenile *G. argentea*. The *Prior* specimen fits Grisebach's description perfectly and should be regarded as typical. The Cuban references in Grisebach and Fawcett & Rendle are to a distinct species *G. combsii* Urb.

5. G. elliptica Sw., Nov. Gen. & Sp. Pl.: 59 (1788).—Velvet Seed.

Shrub or tree 2–7 m. high; leaves ovate to elliptical, obtuse, acute, cuspidate or rarely acuminate at tip, acute to truncate at base, 1–7·5 (–15) cm. long, 1–4·5 (–7) cm. broad, puberulous; peduncles subfiliform, up to 3 cm. long; corolla greenish-white to dull yellow; fruit dark blood-red turning blackish-purple, 4–8 mm. in diameter when dry, 3–6-seeded.

Common in dry pastures, alluvial gravel and on coastal limestone in thickets; SL–3000 feet; fl. June–Nov, fr. July–Jan. *A 5617! 10206! H 7709! P 11338!* Florida, Bahamas, Mexico to Venezuela, Greater Antilles, Virgin Is.; Grand Cayman.

27. HAMELIA Jacq. (1760)

1 Flowers mostly sessile or very shortly stalked on the inflorescence-branches; corolla
   1–2 cm. long:
2 Leaves scabrid   6. *papillosa*
2 Leaves not scabrid:
3 Leaves opposite, at most puberulous; corolla campanulate, yellow sometimes fading
   red, about 10 mm. long   1. *axillaris*
3 Leaves whorled in threes or fours, generally pubescent; corolla more or less cylindrical, orange-red fading dark crimson, 13–20 mm. long   2. *patens*
1 Flowers mostly pedicellate; corolla 1·5–5 cm. long:
4 Leaves mostly opposite, glabrous; corolla-tube narrowly funnel-shaped, 1·5–2 cm.
   long, at first green maturing through yellow and orange to salmon-red
            3. *chrysantha*
4 Leaves mostly in whorls of three, more or less hairy or scabrid; corolla-tube salver-
   shaped to campanulate, yellow or sometimes streaked orange:
5 Stipules subulate-hornlike, curved inwards; corolla-tube 4–5 cm. long; leaves smooth,
   with domatia in the vein-axils beneath   4. *ventricosa*
5 Stipules flat, inconspicuous; corolla-tube 1·5–3·5 (–4) cm. long:
6 Leaf-blade smooth, pubescent   5. *cuprea*
6 Leaf-blade scabrid   6. *papillosa*

1. H. axillaris Sw. in Upfostr.–Sälsk. Tidn. **1785** (19): 148 (1785).

Shrub 60 cm. to 2·5 m., rarely a tree to 5 m. high; leaves broadly ovate to obovate, acuminate, 5–10 cm. long, 3–5 cm. broad, puberulous and with domatia
beneath; cymes puberulous; fruit oblong-ovoid, about 6 mm. long, black when ripe.

Common in thickets and pasture margins in wet areas; SL–2000 feet; fl. May–
Nov, fr. June–Jan. *A 5564! H 12390! P 10307!* Honduras to Peru and Brazil,
Greater Antilles, Virgin Is., Trinidad.

2. H. patens Jacq., Enum. Syst. Pl. Carib.: 16 (1760).

Shrub 1–2 m. or tree to 6 m. high, the plant including corolla and fruit pubescent;
leaves obovate to lanceolate, acuminate, 5–15 cm. long, 2·5–6·5 cm. broad; ripe
fruit obovoid, up to 7 mm. long, at first yellow then red, finally black with red disk.

Locally common in the eastern parishes, in thickets and on roadside banks;
50–1250 feet; fl. and fr. Jan–Mar, June–Sept. *A 6153! H 12403! P 10633!* Florida,
Bermuda, Mexico to Paraguay, Greater Antilles, Martinique, Trinidad.

3. H. chrysantha Sw., Nov. Gen. & Sp. Pl.: 46 (1788).

Shrub or tree 2–8 m. high; leaves ovate to lanceolate, acute or acuminate, 4–10
cm. long, 2–3·5 cm. broad; pedicels slender, up to 7 mm. long; ripe fruit subspherical, 4·5–5 mm. long, with persistent subulate calyx-teeth.

Uncommon in several western and central parishes, in sheltered wooded hollows
and gullies on limestone; 700–2000 feet; fl. and fr. sporadically throughout the
year. *A 10198! H 9365! P 15767!* Endemic.

4. H. ventricosa Sw. in Upfostr.–Sälsk. Tidn. **1785** (19): 148 (1785).—Prince
Wood.

Shrub 3–8 m. or tree to 10 m. high; leaves elliptic-lanceolate, acuminate, puberulous on the veins beneath, 6–15 cm. long, 2–4·5 cm. broad; stipules about 5 mm.
long; corolla markedly constricted above the base, bright yellow; ripe fruit ovoid,
9–12 mm. long, soft, deep red.

Locally common in wet woodlands on limestone; 800–2300 feet; fl. Apr–Oct,
fr. Aug–Dec. *A 7789! H 9355! P 19691!* Endemic.

5. H. cuprea Griseb., Fl. Br. W.I.: 320 (1861).

Shrub 1·5–4 m. or tree 5–10 m. high; leaves ovate, acute or acuminate, 4–9 cm.
long, 1·5–3·5 cm. broad, glabrous; stipules triangular, acuminate, about 1·5 mm.
long; pedicels rather stout, up to 7 mm. long, older flowers subsessile; corolla
constricted above the base, bright yellow or orange-yellow, 2–3·5 cm. long; fruits
ovoid, 5–7 mm. long, ripening through red to purplish-black.

Common in the central and eastern parishes, in coastal thickets and on wooded
limestone hillsides; SL–2600 feet; fl. and fr. most of the year. *A 6805! H 9343!
Powell 985!* Cuba, Hispaniola, Grand Cayman. Plants with short corollas have been
identified with *H. chrysantha* but **that** species has a much narrower corolla-tube and

opposite leaves; plants with longer corollas resemble *H. ventricosa* but may be distinguished by the stipules.

6. H. papillosa Urb., Symb. Ant. **5**: 508 (1908).—*H. scabrida* Britton (1912).
Shrub or tree 2–5 m. high; leaves broadly ovate, cuspidate-acuminate, 5·5–10 cm. long, 2·5–6 cm. broad, scabrid adaxially, more or less pubescent abaxially; pedicels wanting or up to 6 mm. long; corolla as above, clear yellow or sometimes streaked with orange, 1·5–4 cm. long; fruit oblong-ovoid, about 10 mm. long.
Rather rare in the west-central parishes, on exposed craggy limestone cliffs and at woodland margins; 1300–3000 feet; fl. Feb–July, fr. Feb–Aug. *A 11047! ACH 1258! H 10968! Webster & Proctor 5398!* Endemic.

28. HOFFMANNIA Sw. (1788)

1. H. pedunculata Sw., Nov. Gen. & Sp. Pl.: 30 (1788).
Shrubby herb up to 1·5 m. high; leaves obovate-lanceolate, acuminate, long-cuneate at base, thin, rough adaxially, pubescent on the nerves beneath, mostly 7–15 cm. long, 3–4 cm. broad; cymes pedunculate; pedicels 3–4 mm. long; corolla yellow with red streaks towards the base, the tube 2 mm. long, the lobes about 4 mm. long; fruit ovoid, red, 6 mm. long.
Very rare (St. Andr., Port.), in moist submontane woodlands; 2500–3900 (–5000) feet; fl. and fr. July. *P 6917! 10171!* Endemic.

29. FARAMEA Aubl. (1775)

1. F. occidentalis (L.) A. Rich. in Mém. Soc. Hist. Nat. Paris **5**: 176 (1834).—
Ixora occidentalis L. (1759). Wild Coffee, Wild Jessamine.
Shrub or tree 1·2–6 m. high, glabrous; leaves elliptical, acuminate, cuneate at base, shiny, 8–16 cm. long, 2·5–7 cm. broad; stipules awned, broad at base; corymbs axillary or terminal, mostly shorter than the leaves; flowers very fragrant; corolla white drying black, the tube 10–18 mm. long, the lobes lanceolate, up to about 10 mm. long; fruit globose, purplish-black, about 7 mm. in diameter, the persistent calyx-tube variable in length being up to 2·5 mm. long in some populations in Hanover.
Common in shady woodlands on limestone; 400–4000 feet; fl. and fr. most of the year. *A 9362! H 10815! P 11045!* Mexico to Peru and Brazil, West Indies.

30. MORINDA L. (1753)

Lax trailing or climbing shrub; stipules acuminate; leaves linear-lanceolate to oblong-lanceolate, up to about 10 cm. long; flower-heads subsessile; compound fruit sub-spherical, about 2 cm. in diameter   1. *royoc*
Erect shrub or tree; stipules rounded; leaves broadly elliptical to oblong-ovate, up to 30 (–45) cm. long; flower-heads pedunculate; compound fruit ovoid-ellipsoid, up to 10 cm. long and 6 cm. broad   2. *citrifolia*

1. M. royoc L., Sp. Pl. **1**: 176 (1753).—Red Gal, Strongback.
Shrub with slender flexible stems, 0·6–2 (–6) m. high; leaves pointed at both ends, 4–11 cm. long, 1–3·5 cm. broad; corolla-tube about 5 mm. long.
Very common, in pastures and thickets on limestone, also on some of the cays in the coral sand; SL–1000 feet; fl. and fr. all the year. *A 5500! H 6606! P 7713!* Florida, Mexico to Venezuela, Bahamas, Cuba, Hispaniola, Aruba, Curaçao, Grand Cayman.

2. M. citrifolia L., Sp. Pl. **1**: 176 (1753).—Hog Apple.
Shrub or tree 3–6 m. or more high; leaves acute, cuspidate or obtuse at tip, cuneate to rounded at base, shiny, up to 45 cm. long and 24 cm. broad; corolla-tube about 1 cm. long; ripe fruit foetid.
Locally common in open places near the sea, cultivated inland; SL–100 (–600) feet; fl. and fr. most of the year. *A 7565! P 11487! Robertson UCWI 2794!* Native of tropical Asia and Australia, widely naturalized in the New World tropics.

31. SCHRADERA Vahl (1797) nom. cons.

1. **S. involucrata** (Sw.) K. Schum. in Mart., Fl. Bras. **6** (6): 295 (1889).—*Fuchsia involucrata* Sw. (1788).

Robust climbing or epiphytic shrub to 8 m. or more high; juvenile branches slender with thin acutely acuminate smaller leaves; stipules obovate to spathulate, up to about 2 cm. long; leaves ovate to oblong, obtuse or acute, cuneate at base, leathery, 5–10 cm. long, 2–5·5 cm. broad; peduncles 2–4 cm. long; flowers 6–8-merous, up to 4 cm. in diameter, fragrant; corolla white, tube 5–20 mm. long, lobes thick and wedge-shaped in section, up to 18 mm. long; fruit subglobose, about 12 mm. in diameter, many-seeded.

Locally very common, in submontane and montane woodlands on trees and rocky banks; 1500–5200 feet; fl. most of the year, fr. sporadically. *A 11016! H 8894! H & P 15059!* Endemic.

32. GEOPHILA D. Don. (1825) nom. cons.

1. **G. repens** (L.) I. M. Johnston in Sargentia **8**: 281 (1949).—*Rondeletia repens* L. (1759). *G. herbacea* (Jacq.) K. Schum. (1891).

Stem prostrate and rooting; petioles lined-pubescent, up to 5 cm. long; leaf-blade ovate to reniform or suborbicular, deeply cordate, 1–5 cm. long, paler beneath; peduncle up to 3 cm. long; flowers 1–3 or rarely more in an umbel subtended by small bracts; corolla white, tube 6–10 mm. long; fruit scarlet, about 5 mm. in diameter.

Rare in the eastern parishes, in damp shady places or in pastures on heavy soils; 500–1750 feet; fl. June, Dec, fr. Dec. *A 7906! H 12291!* General in the tropics.

33. CEPHAELIS Sw. (1788) nom. cons.

1. **C. elata** Sw., Nov. Gen. & Sp. Pl.: 45 (1788).

Straggling shrub or tree 1·5–6 m. high; branches stout but lax and brittle; leaves elliptical to obovate-elliptical, cuspidate-acuminate, cuneate at base, shiny, 10–23 cm. long, 3·5–7 cm. broad; stipules persistent at swollen nodes; bracts 2, waxy, scarlet or crimson, forming an involucre up to 7 cm. across; corolla white, about 15 mm. long; fruit about 6 mm. long.

Locally common (St. Andr., Port., St. Thom.), in sheltered facies of montane woodland; (850–) 2000–4500 feet; fl. and fr. Dec–Aug. *A 7898! H 5191! P 7824!* Endemic.

34. LASIANTHUS Jack (1823) nom. cons.

1. **L. lanceolatus** (Griseb.) Gómez Maza, Noc. Bot. Sist., Habana: 86 (1893).— *Hoffmannia ? lanceolata* Griseb. (1862). *L. moralesii* (Griseb.) C. Wright (1869).

Straggling shrub 1–4 m. high; leaves elliptical to obovate or oblanceolate, usually acuminate at tip, cuneate or rounded at base, 5–14 cm. long, 1·5–5 cm. broad flowers sessile, mostly (3–) 4-merous; corolla white, 4–6 mm. long; fruit subglobose, blackish, about 10 mm. in diameter.

Very rare (Clar.), in thickets on clay soil; 2300 feet; fl. and fr. Dec, July. *A & D 13075! P 16491!* Greater Antilles.

35. PSYCHOTRIA L. (1759) nom. cons.

1 Herbs or undershrubs with soft often decumbent brittle stems; pyrenes 3-ribbed in a fleshy fruit about 1 cm. long and broad:
  2 Leaves acuminate; panicles distinctly stalked; fruits ripening red, finally black
        1. *uliginosa*
  2 Leaves rounded at tip; panicles sessile or subsessile; fruits white   2. *discolor*
1 Erect shrubs or trees; pyrenes 4–5-ribbed, or warted or smooth in a usually ellipsoid fruit less than 1 cm. broad:

3 Leaves subsessile or petiole very short (less than 5% of length of blade):
  4 Leaves at least 6 cm. long; petioles up to 5 mm. long:
    5 Leaf-blade oblong-oblanceolate, mostly 8–10 cm. long, cordate at base   3. *subcordata*
    5 Leaf-blade obovate, up to 30 cm. long, long-cuneate at base   4. *grandis*
  4 Leaves up to 3 cm. long and 1·5 cm. broad, mostly oblong, obtuse; stipules 1·5
    (−3·5) mm. long; ripe fruits red:
    6 Flowers sessile; young branches pubescent; lateral veins obscure beneath 5. *manna*
    6 Flowers pedunculate; young branches at most slightly scabrid, not pubescent:
      7 Lateral veins (in dried leaf) obscure beneath   6. *myrstiphyllum*
      7 Lateral veins prominent beneath   7. *wullschlaegelii*
3 Leaves distinctly petiolate (the petiole mostly at least 10% of length of blade):
  8 Stipules with a persistent basal ring of tissue having two teeth on each side of the node,
    withering in situ and creating a buff smooth scar at the thickened node; ripe fruits
    blue, purple or black, never orange or red:
    9 Leaves with domatia (hair-filled pits) in the vein-axils beneath:
      10 Corolla, calyx and inflorescence-branches bright mauve, the corolla fading white
        or paler mauve; leaves oblanceolate to narrowly elliptical, long-cuneate at base
                          8. *corymbosa*
      10 Corolla white; calyx and inflorescence-branches green or brownish-purple;
        leaves elliptical, broadly cuneate at base   9. *domatiata*
    9 Leaves without domatia; corolla often yellow or cream-coloured in bud, opening
      yellow or white; inflorescence-branches usually green:
      11 Primary branches of inflorescence without bracts; leaves softly fleshy with about
        5 pairs of widely spaced lateral veins; corolla drying orange   10. *patens*
      11 Primary branches of inflorescence usually subtended by subulate or rarely
        foliaceous bracts; leaves thin or firmly leathery, with usually 7 or more pairs of
        regular closely spaced lateral veins, acuminate:
        12 Lateral branches of inflorescence at rightangles to the main axis; flowers in tight
          bracteolate clusters; ripe fruits china-blue; leaves with a pale ciliolate margin
                          11. *brachiata*
        12 Lateral branches of inflorescence more or less ascending, not distinctly rectan-
          gular; ripe fruits purple or black:
          13 Leaves large, 10–24 by 5–20 cm., with 11–15 pairs of lateral veins; corolla
            3–4 mm. long; dry fruit uniformly ribbed   13. *berteroana*
          13 Leaves smaller, not exceeding 15 (−20) cm. long and 7 cm. broad, with up to
            11 (−14) pairs of lateral veins; corolla (5−) 6 mm. or more long; dry fruit
            didymous with a broad commissural groove, wrinkled or warty:
            14 Branched portion of inflorescence as long as or longer than peduncle:
              15 Leaves glabrous, rather leathery, obovate; corolla 10–15 mm. long, glabrous
                outside   14. *dolichantha*
              15 Leaves generally pubescent, thin, ovate to oblong-lanceolate; corolla 5–6 mm.
                long, pubescent outside   15. *pubescens*
            14 Branched portion of inflorescence usually less than half as long as peduncle:
              16 Peduncle and inflorescence-branches glabrous
                          16a. *pedunculata* var. *pedunculata*
              16 Peduncle, inflorescence-branches, calyx and corolla pubescent
                          16b. *pedunculata* var. *caudata*
  8 Stipules deciduous wholly or in parts, any basal ring becoming detached eventually,
    leaving an often rusty-hairy brownish scar; corolla white; ripe fruits usually
    orange or red, when ripe, longitudinally ribbed when dry:
    17 Stipules fused, sheathing the bud, usually falling together in one piece having split
      along one side only:
      18 Panicles sessile:
        19 All young parts of plant covered with long reddish hairs   17. *coeloneura*
        19 Plant glabrous to thinly pubescent   18. *nervosa*
      18 Panicles stalked:
        20 Stems and leaves hairy; leaves cuneate to broadly rounded or even subcordate at
          base   19. *hirsuta*
        20 Stems and leaves glabrous:
          21 Stipules rusty-hairy, up to 20 mm. long; leaves more or less lanceolate
                          20. *dasyophthalma*
          21 Stipules glabrous, less than 12 mm. long; leaves elliptical:
            22 Leaf-tip acute
                              21. *congesta*
            22 Leaf-tip sharply acuminate   22. *lunanii*
    17 Stipules free or nearly so, deciduous separately or breaking up or withering to a
      basal ring separating later:
      23 Bracts and bracteoles conspicuous, white, ciliate, mostly close under the flowers;
        stipules lobed or lacerate; fruits black   12. *amplifolia*
      23 Bracts and bracteoles inconspicuous or obsolete:
        24 Leaf-blade with narrow white ciliated margins, drying silvery-grey 23. *marginata*

24 Leaf-blade with undifferentiated glabrous margins:
  25 Stipules 2-lobed, the lobes long, subulate-acuminate   24. *tenuifolia*
  25 Stipules entire or merely fimbriate:
    26 Leaves rigidly leathery with revolute margins   25. *clarendonensis*
    26 Leaves softly leathery to thinly herbaceous:
      27 Leaf-blade usually folded longitudinally   26. *plicata*
      27 Leaf-blade flat:
        28 Stipules on vegetative twigs (5–) 7–15 mm. long:
          29 Panicle very lax, twice- or more-branched, the long branches terminated by spheroidal clusters of flowers; stipules acuminate; leaf-blade broadly rounded to slightly cordate at base, up to 20 cm. long and 11 cm. broad; corolla 3 mm. long   27. *foetida*
          29 Panicle more or less compact, the branches ascending, the flowers not in tight spheroidal clusters; stipules blunt; leaf-base rounded to cuneate:
            30 Corolla-tube 8 mm. long; leaves obovate, tapered to the base, stiffly leathery   28. *clusioides*
            30 Corolla-tube not exceeding 6·5 mm. long; leaves at most thinly leathery:
              31 Calyx truncate; corolla-tube 6·5 mm. long; leaves obovate; stipules oblong; inflorescence glabrous   29. *siphonophora*
              31 Calyx shortly toothed; corolla-tube 3–5 mm. long; stipules broadly ovate; inflorescence usually puberulous   30. *dura*
        28 Stipules 2–6 (–8) mm. long; calyx lobed or toothed:
          32 Stipules acuminate; leaves thin:
            33 Corolla 3·5 mm. long; stipules abruptly acuminate, 4 mm. long; leaves obtuse or acute   35. *foetens*
            33 Corolla 1·5 mm. long; stipules ovate, broadening above the base and becoming membranous, caudate-acuminate, 6–8 mm. long; leaves acuminate   36. *dolphiniana*
          32 Stipules blunt, at first rather fleshy, the apical portions often separating from a basal ring:
            34 Pairs of lateral veins mostly 7–11; inflorescence whitish-puberulous, glabrescent, sessile or pedunculate; leaves rather variable in shape; corolla-tube 3–6·5 mm. long   31. *glabrata*
            34 Pairs of lateral veins mostly 5–8; inflorescence pedunculate:
              35 Inflorescence distinctly puberulous with greenish-golden hairs corolla-tube 3·5–4 mm. long   32. *sloanei*
              35 Inflorescence glabrous; leaf-base mostly narrowly cuneate:
                36 Leaf-blade ovate to elliptical, thin; corolla-tube 3·5 mm. long   33. *purdiaei*
                36 Leaf-blade oblong to oblong-lanceolate, rather firm; corolla-tube 5 mm. long   34. *balbisiana*

1. P. uliginosa Sw., Nov. Gen. & Sp. Pl.: 43 (1788).

Weak-stemmed undershrub 0·6–2 m. high, often decumbent and rooting; leaves obovate, up to 25 cm. long and 12 cm. broad; flowers 4–5-merous; corolla-tube white, 4 mm. long, lobes pink, reflexed, horned at tip.

Locally common, in damp shady woodlands; 1000–6600 feet; fl. and fr. most of the year. *A 8612! H 8431! P 8484!* Continental tropical Amer., West Indies.

2. P. discolor (Griseb.) Rolfe in Kew Bull. **1893**: 258 (1893).—*P. uliginosa* var. *discolor* Griseb. (1861).

Low bushy herb 30–60 cm. high; leaves like the last, narrowed into a long broad petiole; corolla greenish-white.

Uncommon and rather local, in wet forest on limestone; 900–2250 feet; fl. Mar–Sept, fr. Mar–Nov. *A 9105! H & B 10571! P 9977!* Endemic.

3. P. subcordata Britton in Bull. Torr. Bot. Club **37**: 358 (1910).

Shrub 2–2·5 m. high; leaves subsessile, 2–3·5 cm. broad; corolla white, about 2 mm. long; fruit 6–7 mm. long.

Very local (St. Thom.), in wet woodlands on limestone; 1500–2500 feet; fl. Mar, fr. Aug. *P 11805! Webster & Proctor 5519!* Endemic.

4. P. grandis Sw., Nov. Gen. & Sp. Pl.: 43 (1788).

Shrub or tree 3–5 m. high; leaves obovate-elliptical, shortly acuminate or cuspidate, up to 12 cm. broad; corolla-tube 3 mm. long; fruit scarlet, globose, obscurely ribbed, 5–6 mm. in diameter when dry.

Very local (St. Thom.), gregarious in moist shaded ravines and forested hillsides; 400–1500 feet; fl. Apr–June, fr. Aug–Jan. *Cornman 641! H & B 10549! P 16576! 24876!* Costa Rica, Greater Antilles.

5. P. manna Urb., Symb. Ant. **5**: 518 (1908).
    Much branched shrub 1·5–2·5 m. high; stems and petioles densely rough-puberulous; leaf-blade glabrous or with a few hairs on midrib beneath; stipules acute or obtuse, 1·5 mm. long; flowers in small cymules; corolla white, 5–6 mm. long; fruit ovoid, 4·5 mm. long.
    Rather local (Clar., Trel., Port.), in wet thickets and woodlands on rocky lime-stone; 1750–2500 feet; fl. May–Sept, fr. Aug, Dec–Mar. *A 9119! H 10666! P 11361!* Endemic.

6. P. myrstiphyllum Sw., Nov. Gen. & Sp. Pl.: 44 (1788).
    Shrub 1–2 m. high; leaves 1·5–3 cm. long, 5–13 mm. broad; stipules 1·5 mm. long; corolla white, 5 mm. long; fruit red, ellipsoidal, 5–6 mm. long.
    In scattered central and western localities, in woodlands and thickets on rocky limestone; 100–2600 feet; fl. and fr. July–Nov. *H 12684! P 9251!* Endemic.

7. P. wullschlaegelii Urb., Symb. Ant. **6**: 50 (1909).
    Shrub, probably not really distinct from the last; stipules 2·5–3·5 mm. long.
    Only twice collected (St. Ann). *Prior!* Endemic.

8. P. corymbosa Sw., Nov. Gen. & Sp. Pl.: 44 (1788).
    Shrub 1–3 m. or tree to 6 m. high; leaves 5–12 cm. long, 1·5–4 cm. broad, with 6–9 pairs of prominent lateral veins; corolla-tube glabrous or thinly puberulous, 4–4·5 mm. long; fruit purple, 5 mm. in diameter.
    Locally abundant (St. Andr., Port., St. Thom.), in montane thickets and wood-lands; 3500–7400 feet; fl. and fr. all the year. *A 5884! H 9144! P 6082!* Endemic.

9. P. domatiata Adams in Phytologia **21** (2): 69 (1971).
    Shrub 2–2·5 m. or tree to 6 m. high; leaves 4–17 cm. long, 2–7 cm. broad with 7–11 pairs of lateral veins; corolla-tube tomentose, 3–4 mm. long; fruit blackish-purple, 5 mm. long, 6 mm. broad.
    Local (Port.), in wet forest and mossy thickets on limestone; 1750–3000 feet; fl. Jan–Mar, Aug, fr. Jan, Aug. *A 9375! HPS 14757! P 10464!* Endemic.

10. P. patens Sw., Nov. Gen. & Sp. Pl.: 45 (1788).
    Shrub 1–3 m. high; leaves broadly elliptical, acuminate, mostly 7–13 cm. long, 3–5 cm. broad, rather brittle, shiny adaxially, the petiole and midrib often becoming reddish on drying; corolla white, the tube 4 mm. long; immature fruit greyish-blue, ripening black, 2–4 mm. broad.
    Rather local (St. Andr., St. Ann, Port.), in submontane woodland; 2500–4800 feet; fl. July–Aug, fr. Oct–Mar. *A 6254! 11918! HPS 14616! Robbins UCWI 2591!* C. and S. Amer., Greater Antilles, Trinidad.

11. P. brachiata Sw., Nov. Gen. & Sp. Pl.: 45 (1788).
    Shrub 1–3·5 m. high; leaves elliptical, acuminate, up to 15 cm. long and 6 cm. broad; corolla greenish-white turning yellow, with a tuft of blue hairs at the tip of each lobe, the tube 4·5 mm. long; fruits strongly ribbed when dry, 4 mm. long.
    Rather common in the central and eastern parishes, in rocky woodland and on riverside rocks; 400–4000 feet; fl. Mar, July–Sept, fr. July–Sept, Dec–Mar. *A 9262! H 8977! P 8473!* Guatemala to Colombia and Peru, Greater Antilles, Trinidad.

12. P. amplifolia Räusch., Nomencl. ed. 3: 56 (1797).—*Naletonia violacea* (Aubl.) Bremek. (1934), not *P. violacea* Aubl. (1775). *Nonatelia violacea* Aubl. (1775). *P. inundata* of Proctor (1967), not Benth. (1841).
    Shrub about 1·5 m. high; stipules 1–2 cm. long; leaves elliptic-lanceolate, cuneate at base, caudate-acuminate, the lateral veins numerous, curved, alternating with finer veins, up to 20 cm. long and 5 (–7) cm. broad; branches of inflorescence

puberulous; corolla white, the tube about 7 mm. long; ripe fruit ovoid, 6–7 mm. long.

Very rare and local (Clar.), in shady thickets on wet clay soil in savanna; about 2300 feet; fl. June–July, fr. Oct. *P 16493! 16711! W. Wright!* Nicaragua, Guianas.

13. P. berteroana DC., Prodr. **4**: 515 (1830).

Shrub or tree 2–6 m. high; branches stout, puberulous; leaves ovate to oblong-obovate, shortly acuminate; corolla white or light yellow; ripe fruit bluish-black, subspherical, 3 mm. long when dry. Easily confused with *Palicourea barbinervia* q.v.

Very local (Port., St. Thom.), in damp shady woodlands on limestone; 1250–3000 feet; fl. Jan–May. fr. Feb–Aug. *A 9263! H & B 10749! P 7640! Weaver 1213!* C. and S. tropical Amer., West Indies.

14. P. dolichantha Urb., Symb. Ant. **5**: 517 (1908).

Shrub or tree 2–7 m. high; leaves cuspidate-acuminate, 5–10 cm. long, 2–5·5 cm. broad; corolla white to cream-coloured; ripe fruit blackish-purple, about 4 mm. broad.

Rather common in the central parishes, in thickets and woodland on rocky limestone; 1200–3000 feet; fl. June–Sept, fr. Aug–Jan. *A 10097! H 9399! HPS 14687!* Endemic.

15. P. pubescens Sw., Nov. Gen. & Sp. Pl.: 44 (1788).

Shrub or tree 1–4 m. high, sometimes straggling; leaves acutely acuminate, 7–16 cm. long, 3–7 cm. broad; corolla white to light yellow or tinged pink; ripe fruit blackish-purple, 3–4 mm. broad.

Very common, in thickets and open woodland on limestone; SL–3000 feet; fl. Apr–Jan, fr. June–Oct. *A 6997! H & B 10586! P 23659!* C. Amer., Bahamas, Greater Antilles, St. Thomas, St. Kitts.

16. P. pedunculata Sw., Nov. Gen. & Sp. Pl.: 44 (1788).

16a. Var. pedunculata.

Shrub 2–4 m. or tree to 8 m. high; leaves obovate to elliptical, cuspidate-acuminate, narrowed to base, essentially glabrous, 7–14 (–20) cm. long, 3–6 (–7) cm. broad; peduncle (5–) 10–14 cm. long; corolla-tube about 8 mm. long, glabrous outside, yellow and sometimes tinged red in bud, light yellow or white when open; ripe fruit bluish-purple to black, 4–5 mm. broad.

Common mostly in central and western parishes, in woodland margins and sheltered ravines on limestone; 150–3000 feet; fl. and fr. all the year. *A 12815! H 8979! P 10412!* Endemic.

16b. P. pedunculata var. **caudata** Adams in Phytologia **21** (2): 69 (1971).

Shrub 2 m. or tree to 5 m. high, like the last but pubescent; leaves caudate-acuminate, up to 15 cm. long and 7 cm. broad; corolla-tube about 5 mm. long, pubescent outside.

Uncommon, in the central parishes, in margins of woodlands on rocky limestone; 1400–2000 feet; fl. June–July, fr. ? Aug. *A 7296! Osmaston 5017!* Endemic.

17. P. coeloneura Urb., Symb. Ant. **5**: 519 (1908).

Shrub 1–3 m. high; leaves obovate narrowed to both ends, acuminate, up to 10 cm. long and 4·5 cm. broad; corolla-tube tinged greenish, 2·5–3 mm. long; ripe fruit red, oblong, 8–10 mm. long; a strong odour of licorice is emitted when the stems or leaves are crushed.

Locally abundant in a small area (St. Cath., St. Ann), in woodland and woodland margins on rocky limestone hills; 1700–3000 feet; fl. July–Aug, fr. July–Aug, Dec–Feb. *A 7801! H 6455! Webster & Proctor 5626!* Apparently endemic but very close to *P. rufescens* Kunth (S. tropical Amer.) with which it shares an extreme position in a range of variants, differing in the density of rufous indumentum, from the completely glabrous *P. fadyenii* Urb. (Jamaica) through *P. nervosa* Sw. (Jamaica), *P. undata* Jacq. (Florida, Bahamas, Cuba) to *P. portoricensis* DC. (Puerto Rico) and others extending throughout the American tropics.

18. P. nervosa Sw., Nov. Gen. & Sp. Pl.: 43 (1788).—*P. undata* Jacq. (1798). *P. fadyenii* Urb. (1913).

Shrub 1–2 (–3) m. high, like the last but glabrous or nearly so; leaves ovate to narrowly lanceolate, acuminate at apex and base, up to 11 cm. long and 4·5 cm. broad, if at all hairy having domatia in the vein-axils beneath, the margins often ciliate; corolla-tube about 4 mm. long; fruit ellipsoid, red, 6–7 mm. long.

Very common, in well drained thickets and woodland margins; SL–4100 feet; fl. and fr. most of the year. *A 7083! H 8836! P 8749!* Grand Cayman. When construed in a broad sense this species extends throughout the Caribbean area.

19. P. hirsuta Sw., Nov. Gen. & Sp. Pl.: 42 (1788).

Shrub 1·5–3 m. high; leaves ovate to ovate-lanceolate, acuminate, 5–12 cm. long, 2·5–7 cm. broad, with 7–10 pairs of lateral veins; stipules 7–12 (–22) mm. long; corolla-tube 5 mm. long; fruit red, 6 mm. long.

Rare and local (St. Ann, Port., St. Thom.), in sheltered woodlands on limestone; 400–3300 feet; fl. Mar–Sept, fr. Aug–Nov. *A 9252! H & B 10584! P 10099!* Endemic.

20. P. dasyophthalma Griseb., Fl. Br. W.I.: 341 (1861).

Shrub or bushy tree 2·5–3 m. high; leaves acuminate, up to 10 cm. long and 3·5 cm. broad, with 6–8 pairs of lateral veins; corolla-tube 3 mm. long; fruit ellipsoidal, 6 mm. long.

Rare and local (St. Andr., Manch.), in dense thickets on limestone hills; 1100–2900 feet; fl. Jan, July, fr. Aug. *ACH 664! H 11130! P 21918!* Endemic.

21. P. congesta Spreng. ex DC., Prodr. **4**: 515 (1830).

Shrub or small tree; leaves elliptic-lanceolate, up to 10 cm. long and 4 cm. broad, with 5–6 pairs of lateral veins; corolla 3–4 mm. long; fruit ellipsoidal, 4 mm. long.

An obscure species not confirmed by recent collections; apparently endemic. It is possible that when the stipules are better known, this species will be aligned with *P. dolphiniana* q.v.

22. P. lunanii Urb., Symb. Ant. **7**: 468 (1913).

Shrub 2 m. high; leaves rhomboid-elliptical, narrowed to both ends, up to 7 cm. long and 3 cm. broad; corolla tetramerous, 3 mm. long including lobes, creamy-white; fruit oblong-ovoid, 6–7 mm. long.

Very rare and local (St. Cath.), in woodland on limestone; 1700–2600 feet; fl. Dec., fr. Dec–Feb. *A 10043! H 8889!* Endemic.

23. P. marginata Sw., Nov. Gen. & Sp. Pl.: 43 (1788).

Shrub 1–2·5 m. high; leaves lanceolate, acute or acuminate, long-cuneate at base, up to 15 cm. long and 4 cm. broad, with 10–12 pairs of lateral veins; stipules 9–13 mm. long; corolla white or yellow, the tube 2·5 mm. long; fruit subspherical, red, 3–4 mm. in diameter. That black berries occur in this species, was reported by S. Moore (1936) and Yuncker (1958).

Frequent in the north-central and eastern parishes, in woodland on limestone, particularly in sheltered areas or near rivers; 400–1750 feet; fl. and fr. Jan–Apr, Aug, *A 9048! H 11638! P 11537! Yuncker 18411!* Guatemala to Colombia, Cuba. Trinidad.

24. P. tenuifolia Sw., Nov. Gen. & Sp. Pl.: 43 (1788).

Shrub 1–3 (–6) m. high; leaves broadly elliptical, usually abruptly acutely acuminate, broadly or narrowly cuneate at base, up to 15 cm. long and 6 cm. broad, very pale beneath; stipules 7–9 mm. long; inflorescence thinly puberulous, shortly stalked or sessile; corolla-tube 1·5–2 mm. long; fruit red, oblong, 6–8 mm. long.

Occasional, in woodlands and woodland margins on limestone; 550–2500 feet; fl. July–Jan, fr. July–Mar. *A 9287! ACH 758! P 21283!* Guatemala, Venezuela, Antilles south to Barbados and Grenada.

25. P. clarendonensis Urb., Symb. Ant. **7**: 443 (1913).

Shrub or tree 2·5–8 m. high; leaves ovate to oblong-ovate, obtuse or minutely cuspidate, 3·5–5·5 cm. long, 2·5–3·5 cm. broad, with 6–7 pairs of lateral veins; corolla-tube broad, 4 mm. long; fruit unknown.

Rare and local (Clar.), on rocky limestone hilltops; 2500 feet; fl. Jan, July *H 10988! P 9756!* Endemic.

26. P. plicata Urb., Symb. Ant. **7**: 438 (1913).
Shrub 3 m. or tree 5–12 m. high, with gnarled branches and close-noded twigs; leaves oblong-ovate, broadly acuminate, up to 7 cm. long and 3 cm. broad; peduncle up to 3·5 cm. long; inflorescence glabrous; corolla-tube about 3·5 mm. long; fruit red, ellipsoidal, 7–8 mm. long.
Rare and local (Trel., St. Ann), in woodland on rocky limestone; 1200–2200 feet; fl. Apr, Aug, fr. Aug–Oct. *A & C 1215! Cornman 800! H 9479! P 22838!* Endemic. This interesting species has many features suggestive of aged plants in exposed situations; when better known it may prove to be more appropriately placed in the *P. glabrata* complex.

27. P. foetida Griseb., Fl. Br. W.I. 342 (1861).
Shrub 2–3 m. high; leaves broadly ovate to oblong-ovate, obtuse or cuspidate-acuminate, often mottled and lustrous beneath when dry; peduncle up to 12·5 cm. long; fruit ellipsoidal, 8 mm. long.
Rather rare, mainly in the eastern parishes, in wet shady woodland on limestone; 200–3000 feet; fl. Apr, July–Aug, fr. Mar, Nov. *A 7925! H 10687! P 7365!* Endemic.

28. P. clusioides Proctor in Bull. Inst. Jam., Sci. ser. **16**: 51, t. 21 (1967).
Tree 5–10 m. high; leaves obtuse or very shortly cuspidate up to about 20 cm. long and 10 cm. broad, becoming reddish on drying, with 11–12 pairs of lateral veins; stipules oblong, rounded, about 8 mm. long; fruit tapering into the thickened pedicel.
Very local (Port.), in wet forest on limestone; 1750–2000 feet; fl. Aug, young fr. Aug, fr. Feb. *A 9386! P 10460!* Endemic.

29. P. siphonophora Urb., Symb. Ant. **5**: 516 (1908).
Shrub or tree to 4 m. high; leaves cuspidate at tip, long-cuneate at base into a stout petiole, up to 14 cm. long and 5·5 cm. broad, with 10–12 pairs of lateral veins; stipules 10–15 mm. long; fruit ovoid, 5 mm. long.
Rare and very local (Trel.), in woodland on limestone; about 2000 feet; fl. Aug, fr. Sept. *H 8760! 9412!* Endemic.

30. P. dura Griseb., Fl. Br. W.I.: 340 (1861).—*P. troyana* Urb. (1908). *P. danceri* Urb. (1913). *P. jenmanii* Urb. (1913).
Shrub or tree 3–10 m. high; petioles 4–22 mm. long; leaves mostly oblong-elliptical, acuminate, broadly rounded to cuneate at base, 10–18 (–20) cm. long, 4–6·5 (–8) cm. broad, softly leathery, shiny and dark green adaxially, with (8–) 10–12 (–14) pairs of lateral veins; peduncle stout, mostly 5–7·5 cm. long, very rarely with basal subsidiary branches; corolla-lobes 5 (–7); fruit ovoid, 5–7 mm. long.
Locally common, in thickets and woodlands on limestone; SL–2200 feet; fl May–Nov, fr. sporadically. *A 11192! 11834! H 8561! P 11288! Webster & Proctor 5520!* Endemic.

31. P. glabrata Sw., Nov. Gen. & Sp. Pl.: 43 (1788).—*P. brownei* of S. Moore (1936), not Spreng. (1824). *P. swartzii* Urb. (1913).
Shrub 1·5–3 m. or tree to 12 m. high; petioles up to 15 mm. long; leaves ovate to elliptic-lanceolate, acute or acuminate, usually abruptly narrowed at base, 5–14 cm. long, 2·5–6·5 cm. broad; peduncle up to 4 cm. long; corolla-lobes 5; ripe fruit ellipsoid, 6–7 mm. long.
Very common except in the eastern parishes in thickets and woodlands mostly on rocky limestone; SL–3000 feet; fl. and fr. all the year. *A 10254! H 12594! P 11112!* Endemic.

32. P. sloanei Urb., Symb. Ant. **7**: 445 (1913).—*P. harrisiana* Urb. (1913).
Shrub 2–5 m. or tree to 10 m. high; leaves ovate to oblanceolate, acute or shortly acuminate, 4–11 cm. long, 2–4·5 cm. broad, drying dark above and pale beneath; dry fruit oblong-ovoid, about 7 mm. long.

Locally frequent (St. Andr., Port., St. Thom.), in gullies and sheltered places in submontane and montane woodland; 2000–5500 feet; fl. Aug–Jan, fr. Nov–Feb. *A 10272! Hart J. P. 1436! H 10111! P 21955!* Endemic.

The taxonomy of the components here included in species 30–32 is extremely difficult. The characters of leaf-shape, vein-number, stipule-length, corolla-length and presence or absence of a simple peduncle are all unreliable. Consistent groupings have been possible only by accepting quantitative characters within much broader limits than hitherto.

33. P. purdiaei Urb., Symb. Ant. **7**: 439 (1913).—*P. platoensis* Urb. (1913).

Shrub 3–4 m. high; leaves obtuse or shortly acuminate, up to 10 cm. long and 4·5 cm. broad; peduncle 2·5–3·5 cm. long; fruit obovoid, 5–6 mm. long.

Rare and local (St. Andr., St. James, Port.), in woodlands on limestone; 1500–3500 feet; fl. July–Aug, fr. Jan, July–Aug. *H 5534! Loveless UCWI 1774! Osmaston 5006! Purdie!* Endemic.

34. P. balbisiana DC., Prodr. **4**: 517 (1830).

Climbing or erect shrub 1–4 m. or tree to 5 m. high; young vegetative shoots rather glutinous; leaves obtuse or shortly acuminate, mostly 4·5–9 cm. long and 2·5–4 cm. broad, often drying copper-coloured or bluish-grey, very glossy adaxially; stipules 2–5 mm. long; corolla white or tinged yellow; fruit narrowly ellipsoid with conspicuous ribs, 4·5–5·5 mm. long.

Very common, in thickets on rocky limestone hills, mostly in rather arid areas; 100–2500 feet; fl. and fr. all the year. *A 10290! H 5237! P 20829!* Endemic. Some variants of this species occupying more moist habitats approach *P. dura*.

35. P. foetens Sw., Nov. Gen. & Sp. Pl.: 43 (1788).

Tree 6–7 m. high; leaves ovate-elliptical, up to 11 cm. long and 5 cm. broad, drying grey, with 8–10 pairs of lateral veins; fruit oblong, scarlet, imperfectly known.

Rare (St. Eliz., St. Mary); 900–1000 feet; fl. Apr, July *P 22216! Thompson 8028!* Endemic.

36. P. dolphiniana Urb., Symb. Ant. **7**: 440 (1913).—*P. pusilliflora* S. Moore (1931). *P. congesta* of S. Moore (1936), partly.

Diffusely branched shrub 1–3 m. high; leaves ovate to broadly elliptical, 5–12 cm. long, 3–6 cm. broad, with 7–9 (–15) pairs of lateral veins; panicle sessile or with peduncle up to 7 cm. long; fruit ellipsoid, 4–5 mm. long.

Rather local in the extreme eastern and western parishes, in woodlands and woodland margins on limestone in wet areas; 400–2400 feet; fl. Mar–Aug, fr. sporadically. *A 7890! H 6046! Webster & Wilson 5047!* Endemic.

36. PALICOUREA Aubl. (1775)

1 Flowers sessile or subsessile in threes at the ends of spreading panicle-branches; corolla strongly curved, white or mauve with the lobes tinged pinkish outside; leaves glabrous, paler beneath; stipules breaking up　　　　　　　　　　1. *domingensis*
1 Flowers all stalked; corolla weakly curved or straight; stipules more or less persistent:
  2 Stipules 5–8 mm. long; midrib hairy beneath:
    3 Corolla up to 10 mm. long, densely puberulous outside, white; lateral veins of leaf straight in their proximal half; midrib shortly and regularly fringed with hairs; cystoliths in adaxial epidermis absent; dry fruit 8–10 ribbed, 3–4 mm. long
　　　　　　　　　　　　　　　　　　　　　　　　　2. *barbinervia*
    3 Corolla 15–20 mm. long, glabrous or nearly so, yellow, white or turning mauve or brownish-red; lateral veins of leaf arching from the origin; midrib pubescent; cystoliths conspicuous; fruit ovoid, faintly ribbed, 4 mm. long　　3. *alpina*
  2 Stipules up to 4 mm. long; midrib hairy or glabrous:
    4 Leaves usually hairy on midrib beneath, rarely glabrous; cystoliths conspicuous; panicle-branches ascending in flower; corolla orange to red, about 8 mm. long; dry fruit more or less distinctly 8-ribbed, ovoid, pointed apically, 4 mm. long　4. *crocea*
    4 Leaves glabrous or only thinly puberulous on the midrib beneath; cystoliths obscure or absent; panicle-branches spreading; dry fruit distinctly dicoccous:
      5 Corolla including lobes 11–19 mm. long, white or mauve to magenta　　5. *wilesii*
      5 Corolla including lobes 22–32 mm. long, rose-violet to magenta　　6. *pulchra*

1. P. domingensis (Jacq.) DC., Prodr. **4**: 529 (1830).—*Psychotria domingensis* Jacq. (1760).

Laxly branched shrub 0·8–2·5 m. high; stipules with teeth about 2 mm. long, withering in situ; leaves elliptic-lanceolate, obtusely acuminate, cuneate at base, up to 12 (–16) cm. long and 5·5 cm. broad; corolla-tube about 12 mm. long, lobes 5–6 mm. long; ripe fruit black, glossy, very plump and juicy, about 7 mm. broad.

Common among craggy limestone rocks in shade and in sheltered woodlands on steep slopes; 50 1750 (–3000) feet; fl. and fr. most of the year. *A 7861! H 5347! P 7869!* Greater Antilles, Virgin Is., northern Leeward Is. to Antigua.

2. P. barbinervia DC., Prodr. **4**: 530 (1830).

Shrub or tree 5–7 m. high; leaves ovate, shortly acuminate, cuneate at base, up to 25 cm. long and 10 cm. broad; pedicels bright yellow or tinged purplish; ripe fruit purplish-black.

Very local (Port.), in moist secondary thickets on heavy clay soil; about 1000 feet; fl. ?Nov–Dec, fr. Jan–Feb. *A 9071! Mitchell 10071! P 22091!* Greater Antilles, Trinidad. Without flowers this species bears a striking resemblance to *Psychotria berteroana* but may be distinguished from that species by the dried fruit which has bluntly rounded rather than sharply angled ribs.

3. P. alpina (Sw.) DC., Prodr. **4**: 528 (1830).—*Psychotria alpina* Sw. (1788).

Shrub or more usually a tree 2–6 m. high; upper internodes and petioles sometimes shaggily hairy; leaves elliptical, shortly acuminate, rather obtuse at base, up to 15 cm. long and 6 cm. broad, with 9–15 (–20) pairs of lateral veins prominent beneath; inflorescence usually shortly stalked, pyramidal, the branches often bright yellow or reddish-brown; ripe fruit deep purple.

Locally common (St. Andr., Port., St. Thom.), in submontane and montane woodland; (2000–) 2500–7000 feet; fl. Mar–Nov, fr. July–Mar. *A 6443! H 10563! P 15554!* Mexico to Panama, Greater Antilles.

4. P. crocea (Sw.) Schult. in L., Syst. Veg. ed. nov. **5**: 193 (1819).—*Psychotria crocea* Sw. (1788). *Palicourea riparia* Benth. (1841).[1]

Shrub 1·2–3 m. high; leaves ovate-lanceolate, acutely acuminate, mostly acute at base, up to 12 cm. long and 6 cm. broad; inflorescence-branches brownish-red and ascending in flower, more spreading and purplish in fruit; ripe fruit slightly laterally flattened, brownish-purple to dark blue.

Common on damp roadside banks and in sheltered woodland margins; 50–2250 feet; fl. and fr. almost all the year. *A 9284! H 9435! P 15847!* When construed in a broad sense this species ranges from Honduras to Peru and Paraguay and throughout the Antilles. The Antillean races tend to have the leaves glabrous or with hairs only in the vein-axils, bright red inflorescence-branches and yellow corollas. The mainland variants resemble more the plants occurring in Jamaica.

5. P. wilesii Adams in Phytologia **21** (2): 68 (1971).—*P. riparia* of S. Moore (1936), not Benth. (1841).

Shrub 1·2–4 m. or tree 5 m. high, with terete woody twigs; leaves broadly lanceolate, acuminate with the tip acute, usually cuneate at base, up to 20 cm. long and 6 cm. broad; inflorescence-branches mauve to light purple, rarely yellowish; ripe fruit smooth, black, 5–6 mm. long.

Locally common (St. Andr., Port., St. Thom.), in shady woodland on limestone or shale; 950–3750 feet; fl. May–Aug, fr. sporadically. *A 7475! 11926! H 5203! P 23278!* Endemic. This species falls between *P. pulchra* and the non-Jamaican glabrous races of *P. crocea*; it replaces *P. pulchra* in eastern Jamaica.

6. P. pulchra Griseb., Fl. Br. W.I.: 345 (1861).

Erect or scrambling shrub 1·5–2·5 m. or tree to 4 m; leaves elliptic-lanceolate, acuminate, usually obtuse at base, up to 16 cm. long and 7 cm. broad; inflorescence-branches and younger parts of stem often reddish-purple; flowers drooping but fruiting pedicels mostly erect; ripe fruit black, compressed, broader than long.

---

[1] In this description Bentham remarked on sexual dimorphism in *Palicourea* and *Psychotria*.

Frequent in central and western parishes, in woodlands and woodland margins on limseone hills; 750–2900 feet; fl. most of the year, fr. Aug–Dec. *A 11017!* *H 12368! P 21352!* Endemic. The shape of the corolla-lobes varies considerably in this species; a striking variant from Trelawny (*R 1904!*) has triangular lobes to a broader than usual corolla of deep magenta colour, combined with a leaf with purple abaxial surface.

37. OLDENLANDIA L. (1753)

1 Flowers mostly in pedunculate umbelliform cymes, rarely one or a pair without peduncle and then pedicels shorter than the leaves; leaves linear or linear-lanceolate, mostly 1–5 cm. long, 2–7 mm. broad   **1.** *corymbosa*
1 Flowers solitary or in pairs at the nodes:
  2 Pedicels much shorter than the leaves, less than 2 mm. long; internodes much longer than the leaves; leaves ovate-elliptical to broadly lanceolate; calyx-lobes ovate-lanceolate   **2.** *uniflora*
  2 Pedicels at least half as long as the leaves:
    3 Leaves linear, up to 5 (–7) cm. long; flowers often in pairs; calyx-lobes triangular
                                              3. *lancifolia*
    3 Leaves lanceolate to elliptic-lanceolate, 1–1·5 cm. long; flowers nearly all solitary or rarely 2 on a common peduncle; calyx-lobes lanceolate   **4.** *pumila*

1. O. corymbosa L., Sp. Pl. **1**: 119 (1753).
Erect up to about 20 cm. high or prostrate, annual or becoming slightly woody at the base; stipules minute, toothed; leaves rough on the margins, paler beneath; corolla about as long as the calyx, becoming white, rarely pink or mauve; capsule 2–3 mm. in diameter.
Locally common, a weed of pastures, grassy waste places and lawns; 50–700 feet; fl. and fr. all the year. *A 8362! H 11887! P 10493!* Widespread in the tropics.

2. O. uniflora L., Sp. Pl. **1**: 119 (1753).
Erect or trailing herb with rather rough stems and leaves; leaves 5–15 mm. long, 2–10 mm. broad; corolla white, shorter than the calyx; capsule globose, rough, 2 mm. in diameter.
Rare and very local (St. Eliz.), in damp low-lying places; about 300 feet; fl. and fr. Sept. *Britton 1495!* S.W. United States, Cuba, Puerto Rico.

3. O. lancifolia (Schumach.) DC., Prodr. **4**: 425 (1830).—*Hedyotis lancifolia* Schumach. (1827). *O. herbacea* of S. Moore (1936), not (L.) Roxb. (1814).
Straggling glabrous weak-stemmed herb, more or less rhizomatous and rooting at the lower nodes; leaves sessile, 1–8 mm. broad; stipules with small bristly teeth; corolla white; capsule 3–3·5 mm. broad, broader than long.
Rare (St. Eliz.), in moist places; 25–300 feet; fl. and fr. Jan–Feb, Sept. *H 9877! P 21930! Yuncker 18046!* Native of tropical Africa and Madagascar, naturalized in the Greater Antilles and several scattered areas in the American tropics.

4. O. pumila (L.f.) DC., Prodr. **4**: 425 (1830).—*Hedyotis pumila* L.f. (1781). *O. crystallina* of S. Moore (1936), not Roxb. (1820).
Low annual branched from the base, rooting sparingly from spreading prostrate branches; corolla white, shorter than the calyx; capsule 2 mm. in diameter, longer than broad.
Rare and very local (St. Andr.), a weed of lawns; 600–700 feet; fl. and fr. Dec. *H 11844! P 20794!* Native of S.E. Asia, rare in tropical Africa and Amer.

38. LUCYA DC. (1830) nom. cons.

1. L. tetrandra (L.) K. Schum. in Engl. & Prantl, Nat. Pflanzenf. **4** (4): 27 (1891).—*Peplis tetrandra* L. (1759).
Slender-stemmed perennial herb, tufted from a tuber, up to 12 (–15) cm. high; leaves ovate to oblong-ovate, obtuse, cuneate at base, scabridulous, 1–2 (–2·5) cm. long, 5–8 (–15) mm. broad; pedicels solitary, 3–6 mm. long; flowers usually 4-merous; calyx 8-toothed; corolla white, tube about 1 mm. long; fruit about 2·5 mm. long and 4 mm. broad; seeds ovoid, black, 1·5 mm. long.

Uncommon and local, in pockets and crevices of limestone rocks and among tree roots; 300–2300 feet; fl. and fr. most of the year. *A 5599! H 9626! P 11034!* Greater Antilles.

39. RICHARDIA L. (1753)

1. R. brasiliensis Gomez, Mem. Ipecac.: 31, t. 2 (1801).

Annual herb with roughly pubescent prostrate branches; leaves oblong-lanceolate to elliptical, acute, greyish-green, 2–4 cm. long, up to 2 cm. broad; involucre 4-leaved; flower-heads up to about 1·5 cm. in diameter; corolla (5–) 6-lobed, white but often with the lobes purple-margined in bud, tube 4 mm. long; fruit obovoid, separating into indehiscent cocci 2·5 mm. long; seeds ovoid, about 2 mm. long.

Rather rare (St. Andr., Manch., St. Thom.), a weed of damp shady places; 400–5000 feet; fl. and fr. Jan–Sept. *A 7403! 12683! H 8583! P 27536!* S.E. United States, C. and S. Amer., Cuba, Hawaii.

40. ERNODEA Sw. (1788)

1. E. littoralis Sw., Nov. Gen. & Sp. Pl.: 29 (1788).

Shrub with 4-angled trailing and scrambling branches up to 1·5 m. long; leaves lanceolate to linear-lanceolate, sessile, mucronate, spine-tipped, leathery, 2–4 cm. long, 2–10 mm. broad, continued at the base by a line of hairs on each side; flowers solitary; anthers exserted; fruit yellow, crowned by the persistent calyx, about 5 mm. long and 3 mm. broad.

Frequent in rocky coastal thickets and on limestone cliffs, very rare inland; SL–200 (–3000) feet; fl. and fr. Dec–Apr. *A 8544! H 10219! P 20663!* Florida, Mexico, Honduras, Greater Antilles and in the northern islands of the Lesser Antilles to Guadeloupe; Grand Cayman.

41. DIODIA L. (1753)

1 Scrambling undershrub or weak climber; flowers up to about 5 together in the axils; calyx-lobes 4, unequal; fruit smooth, 3·5–5 mm. long   1. *sarmentosa*
1 Weak slender-stemmed herbs; flowers usually solitary:
   2 Calyx-lobes 4; leaves linear to linear-lanceolate; stem scabrid; fruit hispid, 2·5 mm. long   2. *teres*
   2 Calyx-lobes 2; leaves oblong; stem and fruit glabrous; fruit 4–5 mm. long   3. *simplex*

1. D. sarmentosa Sw., Nov. Gen. & Sp. Pl.: 30 (1788).

Branches 4-angled, rough when young, scrambling, up to 2·4 m. long; leaves shortly petiolate, oblong-lanceolate to elliptical, acute, scabrid adaxially, 2·5–5 cm. long, 1–2 cm. broad; corolla light green or white, the tube about 2 mm. long; fruit oblong-ovoid, brown, usually about 4 mm. long.

Local, in wet secondary thickets and ditches; 700–2500 feet; fl. and fr. Jan–May, Aug–Sept. *A 8101! 10847! H 12040! P 7142!* General throughout the tropics.

2. D. teres Walt., Fl. Carol.: 87 (1788).—*D. prostrata* Sw. (1788).

Annual; stems much branched, slender, hardly angled, pubescent, up to 30 cm. long; leaves sessile, narrowly revolute-margined, acute, 5–40 mm. long, 1–3 mm. broad; corolla white or pink, 4–5 mm. long; fruit obovoid, 4–4·5 mm. long, the cocci ribbed on the back.

Very rare (St. Andr.), in stony waste places; about 400 feet; fl. and fr. Sept. *H 12145!* S.E. United States, C. Amer., Cuba.

3. D. simplex Sw., Nov. Gen. & Sp. Pl.: 29 (1788).

Stems ascending or decumbent, the main stem often rhizomatous and rooting freely from the nodes; upright stems 4-angled, up to 30 cm. long; leaves sessile, acute at both ends, ciliate, 1–4 cm. long, 5–20 mm. broad; corolla white, the tube very slender, 3·5–7 mm. long, lobes ovate, acute, 3·5 mm. long, 1·5 mm. broad; fruit ovoid, keeled on the back, 4–5 mm. long.

Uncommon in the central parishes, in boggy pastures and pond margins among grasses; 25–1000 feet; fl. and fr. Aug–Jan. *A 10405! H 10633! P 15709!* Cuba.

42. HEMIDIODIA K. Schum. (1888)

1. H. ocymifolia (Willd.) K. Schum. in Mart., Fl. Bras. **6** (6): 29, t. 72 (1888).—
Spermacoce ocymifolia Willd. (1818).
Erect herb to 1 m. or straggling to 3 m. high; stems slender, puberulous on the 4 angles when young; leaves lanceolate, narrowed to both ends, scabrid on the margins and veins, 3–7 cm. long, 1–2 cm. broad; corolla white with the lobe-margins pink, 2–3 mm. long; fruit setulose on the distal half, 3·5 mm. long.
Uncommon, a weed of roadsides and open rough pastures in wet areas; SL–2300 feet; fl. and fr. most of the year. *A 7128! Hart 1445! P 16241!* Throughout tropical Amer. and also in Malaysia.

43. SPERMACOCE L. (1753)

Capsule shortly setulose; leaves lanceolate to linear-lanceolate, scabridulous
                                                                        1. confusa
Capsule glabrous; leaves lanceolate to ovate-lanceolate, glabrous, smooth   **2. tenuior**

1. S. confusa Rendle in Journ. Bot. **74**: 12, ff. D–F (1936).
Annual erect herb 30–90 cm. high; stem scabridulous on the angles; leaves hardly petiolate, acuminate, narrowed to the base, 2·5–5 cm. long, 2–5 (–10) mm. broad; corolla light mauve at base and tips of the lobes in bud, otherwise white or all white or pinkish, 2 mm. long; fruit subglobose, 2·5 mm. long, readily deciduous.
Common, in waste places; 10–2400 feet; fl. and fr. all the year. *A 5624! H 12749! P 9466!* Throughout subtropical and tropical Amer.
S. tetraquetra A. Rich. is recorded by Alain (1962) as occurring in Jamaica but no specimen has been seen. It is distinguished from *S. confusa* by having the leaves ovate to lanceolate and the plant being hirsute.

2. S. tenuior L., Sp. Pl. **1**: 102 (1753).
Perennial trailing and straggling herb to 2 m. high or suberect to 1·2 m. high; branches few, 4-angled; leaves shortly petiolate, acute at apex and base, 2–5 (–8·5) cm. long, 8–14 (–22) mm. broad; corolla white, tube 0·5 mm. long, lobes 1·5 mm. long; fruit 2·5–3 mm. long.
Rather uncommon, in swamps and at pond margins; 30–2100 feet; fl. and fr. July, Dec–Mar, *A 9317! Campbell 6135! P 10398!* S.E. United States, tropical S. Amer., Bahamas, Greater Antilles, Barbados; Grand Cayman.

44. BORRERIA G. F. W. Meyer (1818) nom. cons.

1 Leaves linear to linear-lanceolate, with 2–4 pairs of ascending lateral veins; stipule-sheath baggy, up to 4 mm. long; flower-heads about 15 mm. broad; leafy inflorescence-bracts and arrested leafy axillary branches well developed; calyx-teeth unequal:
  2 Bracts of at least the upper inflorescences spreading or deflexed; stipule-teeth 3–5; capsule glabrous or nearly so   **1. verticillata**
  2 Bracts of the inflorescences ascending from their bases; stipule-teeth up to 9; capsule pilose distally   **2. spinosa**
1 Leaves ovate-elliptical to elliptic-oblanceolate with (4–) 5 (–7) pairs of curved lateral veins; stipule-sheath short, tight, with 5–10 teeth on each side; leafy inflorescence-bracts 0–2 at each node; arrested axillary branches poorly or not developed:
  3 Flower-heads 3–7 mm. broad; calyx-lobes unequal, 2 much smaller than the other 2; capsule glabrous or thinly pilose distally; leaves subsessile, rarely as much as 3 cm. long   **3. ocymoides**
  3 Flower-heads 8–14 mm. broad; calyx-lobes subequal; capsule setose in the distal half; leaves more or less distinctly petioled, usually more than 3 cm. long   **4. laevis**

1. B. verticillata (L.) G. F. W. Meyer, Prim. Fl. Esseq.: 83 (1818).—*Spermacoce verticillata* L. (1753). Wild Scabious.
Perennial bushy herb or undershrub 30–120 cm. high; branches glabrescent, the

young twigs 4-angled and shortly setulose on the angles at first; leaves shortly petiolate, acute to acuminate at apex and base, rough on the margins and midrib beneath, 1·5–4·5 cm. long, 1·5–5 mm. broad, lateral veins obscure; flower-heads mostly terminal; corolla all white or tinged red, 2–3 mm. long; fruit subglobose, 1·5–2 mm. long.

Common at lowest and higher elevations, rarer at middle elevations, on open waste ground and stony banks; SL–5000 feet; fl. and fr. all the year. *A 6121! H 11744! P 7337!* Throughout tropical Amer. and also in tropical Africa.

2. **B. spinosa** (Jacq.) DC., Prodr. **4**: 542 (1830).—*Spermacoce spinosa* Jacq. (1760).

Stiffly erect shortly branched annual taprooted herb up to 1 m. high; leaves subsessile, acuminate, rough-margined, up to 10 cm. long and 6 mm. broad; corolla white, 5 mm. long; fruit oblong, 3–4 mm. long.

Rare (St. Andr., St. Eliz., St. Thom.), a weed of dry areas; 10–600 feet; fl. and fr. Nov–Dec. *H 6945! P 18450!* Thinly scattered through tropical Amer.

3. **B. ocymoides** (Burm. f.) DC., Prodr. **4**: 544 (1830).—*Spermacoce ocymoides* Burm. f. (1768).

Weak slender-stemmed low bushy herb up to 25 cm. high, glabrous or hairy on the stem-angles; leaves acute, rough-margined, 1–3 cm. long, 2–10 mm. broad; corolla white or tinged pink, about 1 mm. long; fruit 1 mm. long.

Rather local and restricted to the eastern parishes, a weed of roadsides and open sandy pastures; 20–2250 feet; fl. and fr. Mar–Apr, Aug–Dec. *A 6533! H 11922! P 16572!* General in the tropics.

4. **B. laevis** (Lam.) Griseb., Fl. Br. W.I.: 349 (1861).—*Spermacoce laevis* Lam. (1791). Button Weed.

Lax trailing or rarely erect annual or long-lived herb, with angular branches up to 1 m. or more long, variably pubescent; leaves acute or acuminate, 2–5·5 cm. long, 5–25 mm. broad; corolla white, the lobes hairy within and often tipped pink, 4 mm. long; fruit 2·5–3 mm. long.

Very common, in pastures and waste places and on roadside banks; SL–5000 feet; fl. and fr. all the year. *A 6858! H 6712! P 9411!* Throughout subtropical and tropical Amer., Hawaii.

45. MITRACARPUS Zucc. (1827)

1. **M. villosus** (Sw.) DC., Prodr. **4**: 572 (1830).—*Spermacoce villosa* Sw. (1788).

Erect annual herb 10–35 cm. high, the lower branches often spreading; stem often blackish, puberulous to villous; leaves subsessile to shortly petiolate, oblong to lanceolate, acute or obtuse, scabridulous, 2–5 cm. long, 5–20 mm. broad; flower-heads about 1 cm. in diameter, mostly axillary; shorter calyx-teeth hyaline; corolla white, 2·5 mm. long; fruit membranous, nearly 1 mm. long, the upper deciduous portion hairy.

Occasional to locally common, a weed of roadsides and waste places; 150–5000 feet; fl. and fr. Dec–July. *A 7240! H 11862! P 11418!* Tropical Amer.

46. RELBUNIUM (Endl.) Hook. f. (1873)

1. **R. hypocarpium** (L.) Hemsl., Biol. Centr. Amer. Bot. **2**: 63 (1881).—*Valantia hypocarpia* L. (1759).

Diffusely scrambling branched herb with a musty odour; stems 4-angled, rough, up to about 1 m. long; leaves and leafy stipules oblong-ovate, 1-nerved, hispid, up to 10 mm. long and 5 mm. broad; flowers solitary or few in the leaf-axils; pedicels slender; corolla greenish-white, to greenish-yellow, about 2·5 mm. in diameter; fruit orange-red, about 3 mm. broad.

Locally common (St. Andr., Port., St. Thom.), in montane thickets and on rocky banks; 2700–7400 feet; fl. and fr. most of the year. *A 5758! H 8588! P 9434!* Mexico to Argentina, Hispaniola.

181. CAMPANULACEAE

Annual or perennial herbs or shrubs, sometimes arborescent, often with milky sap. Leaves alternate, rarely opposite or whorled, simple, without stipules. Flowers bisexual, often showy, in bracteate cymes, racemes or heads or rarely solitary. Perianth biseriate; calyx-tube (hypanthium) adnate to the ovary, often free above, with (3–) 5 (–10) lobes; corolla actinomorphic and campanulate or tubular with usually 5 free or fused imbricate or valvate petals, or zygomorphic and bilabiate with the tube often split down one side. Stamens as many as the corolla-lobes or petals and alternate with them, epipetalous or free from the corolla, free from each other or variously connate; anthers free or united, introrse, 2-locular, opening lengthwise. Ovary inferior or half-inferior, 2–, 3–, 5– or 10-locular; style simple, slender, sometimes with 2–5 branches and stigmatic surfaces; ovules numerous, anatropous, on usually axile placentas. Fruit a capsule with slit, porose or circumscissile dehiscence or a berry. Seeds small with straight embryo and fleshy endosperm.

Some 2000 species in 60–70 genera, mostly temperate and subtropical.

1 Corolla zygomorphic, 2-lipped, tube split down one side and often curved; anthers
   connate; filaments connate distally   **1. Lobelia**
1 Corolla actinomorphic, straight-tubed:
  2 Tube of corolla much longer than the lobes; anthers connate   **2. Hippobroma**
  2 Tube of corolla shorter than the lobes; anthers and filaments free   **3. Legousia**

1. LOBELIA L. (1753); McVaugh (1943)[1]

1 Annual herb; leaves ovate, sinuate-crenate, with slender petioles, rather small; corolla
   about 4 mm. long, mauve and white   *1. cliffortiana*
1 Perennial herbs or shrubs; leaves elongated, serrate or crenate-dentate, with usually
   broad petioles; corolla 1·5–3·5 cm. long, mostly yellowish or light green, often tinged
   crimson or brown:
  2 Raceme contracted within a terminal leaf-rosette; leaf-margin doubly dentate with teeth
    filiform-appendaged   *2. harrisii*
  2 Raceme more or less elongated:
   3 Bracts leaflike:
    4 Stem usually branched, 1–2 m. high; upper stem pubescent; corolla usually brownish-
     purple   *3. martagon*
    4 Stem unbranched, 5–6 m. high; upper stem strongly pubescent; corolla green
     tinged rose   *4. alticaulis*
   3 Bracts slender, usually inconspicuous:
    5 Corolla glabrous or nearly so; inflorescence usually shorter than the leaves:
     6 Calyx-lobes entire or with a few small teeth; leaves narrowed to base, margin
      serrulate, lateral veins ascending; plants glabrous or glabrescent:
      7 Sepals 1·5–2·5 cm. long; inflorescence more or less radial   *5. acuminata*
      7 Sepals up to 1 cm. long; inflorescence unilateral   *6. innominata*
     6 Calyx-lobes distinctly serrate; lateral leaf-veins spreading; plants pubescent or
      glabrescent; inflorescence unilateral:
      8 Leaves oblong-elliptical, shortly cuneate to obtuse at base, denticulate
           7. grandifolia
      8 Leaves oblanceolate, long-tapered to base, crenate-serrate *8. fawcettii*
    5 Corolla pubescent; inflorescence usually overtopping the leaves:
     9 Leaves elliptical to oblong-lanceolate; inflorescence markedly unilateral:
      10 Margin of lamina entire near the base, crenate above; capsules borne on straight
       thickened pedicels   *9. caledoniana*
      10 Margin of lamina near the base produced into filiform appendages about 1 cm.
       long, upper margin denticulate; stem winged by decurrent leaf-bases; capsules
       nodding on slender recurved pedicels   *10. assurgens*

[1] McVaugh (N. Amer. Fl. **32A** (1): 1–134) distinguishes the genera *Lobelia* and *Pratia* on the basis of the latter having fleshy indehiscent fruits. The distinction is not clear in Jamaican species where all the endemic species seem to form a natural group. Although the fruits are fleshy at first they tend to dehisce if sufficiently dried, the extent to which they may not do so probably depending on other environmental factors. None has been observed to produce a true berry.

9 Leaves linear-lanceolate, serrate or serrulate:
  11 Leaf-tip acute, margin serrulate; inflorescence radial   11. *viridiflora*
  11 Leaf-tip long-caudate, margin deeply and widely-spaced serrate; inflorescence
    radial or unilateral   12. *caudata*

1. L. cliffortiana L., Sp. Pl. **2**: 931 (1753).
  Glabrous annual herb 15–35 (–50) cm. high; stem and leaves light green; leaves thinly papery when dry, blades 2–4 cm. long; corolla-lobes white, tube mauve with 2 green spots at throat; androecium purple; pedicels elongating in fruit; capsule about 6 mm. long.
  A weed of shallow roadside ditches and damp banks, mostly in the eastern half of the island; 150–2000 feet; fl. and fr. Jan–May. *A 8939! H 10845! P 17425!* Florida, Mexico to tropical S. Amer., Greater Antilles, Dominica, Martinique, Trinidad; introduced into Java and Mauritius.

2. L. harrisii Urb., Symb. Ant. **5**: 520 (1908).—*Pratia harrisii* (Urb.) McVaugh (1943).
  Shrublet branched from the base, up to 20 cm. high; leaves broadly oblanceolate, glabrous, up to 15 cm. long and 3 cm. broad; pedicels 1–1·5 cm. long; calyx-tube about 5 mm. long; corolla curved, orange-yellow, 2–2·5 cm. long, glabrous; capsule campanulate, about 6 mm. long.
  Very rare and local (Trel.) in crevices of honeycombed limestone rocks in woodland; 2000–2300 feet; fl. Jan, June, fr. June–July. *H 8695!* Endemic.

3. L. martagon (Griseb.) Hitchc. in Rep. Miss. Bot. Gard. **4**: 103 (1893).—*Tupa martagon* Griseb. (1861).
  Low-branched thick-stemmed shrub or sparsely branched, 1–2 (–3) m. high, the lower stem leafless and covered with scars; lower leaves elliptic-lanceolate, acute, serrulate, up to 15 cm. long and 2·5 cm. broad; upper leaves elliptical, 5–6 cm. long and 2 cm. broad, merging to bracts; flowers about 4 cm. long, the corolla puberulous; pedicels up to 7 cm. long in fruit; capsule nodding about 1 cm. long and broad.
  Locally common (St. Andr., St. Thom., St. Ann), in open thickets; (2500–) 5200–7400 feet; fl. and fr. Feb–Apr, July–Oct. *A 10633! H 12672! P 7063!* Endemic.

4. L. alticaulis Proctor in Bull. Inst. Jam., Sci. ser. **16**: 69, t. 30 (1967).
  Shrub like the last with a thick simple stem covered with leaf-scars; lower leaves linear-oblanceolate, acute, serrulate, up to 30 cm. long and over 3 cm. broad; upper leaves elliptical, 8–10 cm. long, merging to bracts; flowers about 3 cm. long, the corolla lined with short erect hairs; pedicels up to 4·5 cm. long in fruit; capsule nodding, about 1 cm. long and broad.
  Rare and local (Port.), in wet mossy thicket and woodland over limestone; 2000–2500 feet; fl. and fr. Mar–Aug. *ACH 1006! P 5734! 16270!* Endemic.

5. L. acuminata Sw., Nov. Gen. & Sp. Pl.: 117 (1788).—*Pratia acuminata* (Sw.) McVaugh (1943). *L. alexia* Wimmer (1935).
  Robust shrubby herb 60–120 cm. high; whole plant glabrous except the staminal tube distally; leaves lanceolate or oblanceolate, acute or acuminate, up to 38 cm. long and 9·5 cm. broad; corolla 3–3·5 cm. long, white to light green sometimes tinged pinkish near the base; capsule obconical, about 1·5 cm. long and 1 cm. broad.
  Frequent on banks and in thickets on limestone in partial shade in the western parishes; 300–2700 feet; fl. July, Dec–Apr, fr. Feb–Apr. *A 10959! H 9162! P 17472!* Endemic.

6. L. innominata Rendle in Journ. Bot. **73**: 274, f. 1b (1935).—*Pratia innominata* (Rendle) McVaugh (1943).
  Coarse shrubby herb like the last; corolla 2–2·5 cm. long, deeply curved, light green to greenish-yellow, with purple marks.
  Rather local (St. Mary, Port., St. Thom.), in moist thickets and woodland on limestone; 1000–2500 feet; fl. and fr. Jan–June. *A 10870! H & B 10703! P 7642!* Endemic.

7. L. grandifolia Britton in Bull. Torr. Bot. Club **37**: 359 (1910).—*Pratia grandifolia* (Britton) McVaugh (1943).

Erect shrubby herb 1–2 m. high; leaf-blades up to 35 cm. long and 12·5 cm. broad; calyx-lobes 1·2–1·5 cm. long; corolla 2·5–3 cm. long, light green or greenish-yellow; anthers buff-pink to light brown, the two anterior conspicuously bearded; capsule about 8 mm. long and broad.

Rather local (Port., St. Thom.), in damp thickets and woodland on limestone rocks; 700–2000 feet; fl. and fr. Feb–Aug. *A 7872! H & B 10725! P 22095!* Endemic.

8. L. fawcettii Urb., Symb. Ant. **1**: 452 (1899).—*Pratia fawcettii* (Urb.) McVaugh (1943).

Robust herb like the last 1–2·4 m. high; leaf-blades up to 35 cm. long and 10 cm. broad; calyx-lobes 1–2 cm. long; corolla 3–3·5 cm. long, light green or green tinged dusky purple, sometimes light yellow; capsule as above.

Rather common in the central and western parishes, very rare in the east, on sheltered rocky banks; 1000–3000 feet; fl. Jan–July, fr. Jan–Aug. *A 10797! P 11890! Stearn 961!* Endemic.

9. L. caledoniana Adams in Phytologia **21** (2): 67 (1971).

Soft-stemmed shrubby herb like the last, 1–1·5 m. high; sap copious, milky turning pink on exposure to air; leaf-blades broadly oblanceolate, acuminate, narrowed to a distinct petiole, up to 32 cm. long and 7·5 cm. broad; calyx-lobes with a few small teeth, 10–13 mm. long; corolla about 4 cm. long, light green, deeply curved; capsule as above, on a rigid fruiting pedicel about 2 cm. long, apparently circumscissile.

Rare and local (Port.), in montane woodland on eroded limestone rocks; 4600 feet; fl. May, fr. May, Sept. *A 11629! 12547!* Endemic.

10. L. assurgens L., Syst. Nat. ed. 10, **2**: 1237 (1759).—*L. jamaicensis* (Urb.) Urb. (1930). Red Heart.

Shrubby herb, probably short-lived or sometimes annual, lower stem and leaves nearly glabrous, 1–2 (–2·5) m. high, branched above; leaf-blades elliptic-oblanceolate, long-acuminate, hardly petiolate, up to 45 cm. long and 9·5 cm. broad; whole inflorescence tomentose; calyx-lobes with a few very small teeth, up to about 1·5 cm. long; corolla 2·5 cm. long, dull reddish-purple; capsule variable in size up to about 9 mm. long and broad.

Frequent locally (St. Andr., Port., St. Thom.), on banks at woodland margins and along sheltered roadsides and pathsides; 2800–5300 feet; fl. and fr. Dec–Sept. *A 7749! H 12453! P 21949!* Cuba, Hispaniola, with a variety in Puerto Rico.

11. L. viridiflora McVaugh in N. Amer. Fl. **32A** (1): 91 (1943).—*L. salicina* of Rendle (1936), not Lam. (1792).

Simple or branched slender-stemmed herb 1–2·5 m. high, lower stem and leaves glabrous or nearly so; leaf-blades long-acute, shortly petiolate, up to 28 cm. long and 3·5 cm. broad, dark green and shiny adaxially, inflorescence pubescent; calyx-lobes entire, 3–5 mm. long; corolla 2·5–3 cm. long, nearly straight, light green at base, the lobes olive-green or reddish-brown within; capsules 8–9 mm. in diameter on slender curved pedicels.

Locally common, particularly in the northern and eastern parishes, on roadside banks and in pasture thickets on limestone; SL–2500 feet; fl. and fr. most of the year. *A 9766! H 11996! P 20725!* Endemic.

12. L. caudata (Griseb.) Urb., Symb. Ant. **1**: 454 (1899).—*Tupa caudata* Griseb. (1861).

Shrub to 1·2 m. high the lower leafless stem covered with abscission scars; leaf-blades up to 16 cm. long and 9 mm. broad from a slender petiole, hispid on the midrib abaxially; inflorescence puberulous; calyx-lobes toothed, 3–4 mm. long; corolla 1·5–3 cm. long, curved distally, light green or yellowish; capsule 6–8 mm. long on curved pedicels.

Rare and very local (St. Andr., Port.), on open rocky slopes; 4400–4900 feet; fl. and fr. May–July. *H 11778! Lodge! P 23759!* Endemic.

2. HIPPOBROMA G. Don (1834)

1. H. longiflora (L.) G. Don, Gen. Syst. **3**: 717 (1834).—*Lobelia longiflora* L. (1753). *Isotoma longiflora* (L.) C. Presl (1836). Horse Poison, Madam Fate.

Thinly pubescent herb 25–60 cm. high with acrid milky sap; roots thick and fleshy, whitish; stem simple or sparingly branched; leaves oblanceolate, sessile, sinuate-dentate and denticulate, mostly 10–13 cm. long, 2–4 cm. broad; flowers solitary in axils of upper leaves; calyx-lobes narrow, denticulate, about 1 cm. long; corolla-tube slender, white or greenish, puberulous, up to about 11 cm. long; corolla-lobes elliptic-lanceolate, white with a green midvein, up to about 2·5 cm. long; capsule obovoid-obconic, about 1·5 cm. long, nodding on a slender pedicel.

Widely distributed on damp sheltered banks and along paths and roadsides in clay over limestone; 200–2700 feet; fl. and fr. nearly all the year. *A 5681! H 6782! P 20738!* Continental tropical Amer. and throughout the West Indies; introduced into some Pacific Is.

3. LEGOUSIA Durande (1782)

1. L. perfoliata (L.) Britton in Mem. Torr. Bot. Club **5**: 309 (1894).—*Campanula perfoliata* L. (1753). *Specularia perfoliata* (L.) A. DC. (1830).

Erect or straggling usually unbranched annual herb with stems in flower 20–40 (–90) cm. long, thinly hispid on angles of stem, leaf-margins and veins beneath; leaves sessile, broadly ovate to suborbicular, crenate, up to 1·6 cm. long and 1·5 cm. broad, usually smaller; flowers solitary in the axils of upper leaves; calyx-tube and lobes each about 4 mm. long; corolla broadly campanulate, bluish-violet, about 7 mm. long; stamens white; stigma violet; capsule about 5 mm. long.

Very local (St. Andr.), on pathside and roadside banks; 3300–5200 feet; fl. and fr. Mar–Aug. *A 7441! H 12087! Powell 331!* Temperate N. Amer. to Mexico and S. Amer. in the mountains; Hispaniola.

182. GOODENIACEAE

Perennial herbs or shrubs without milky sap. Leaves alternate or rarely opposite or basal, simple, without stipules. Flowers bisexual, usually zygomorphic, axillary in cymes, racemes, heads or solitary. Perianth biseriate; calyx-limb short, 5-lobed; corolla bilabiate or 1-lipped, split down one side, the lobes usually 5, valvate or induplicate. Stamens 5, alternate with corolla-lobes, free or shortly adnate to base of corolla; anthers free or coherent around the style, introrse, 2-locular, opening lengthwise. Ovary inferior, rarely half-inferior or superior, 1–2 (–4)-locular; ovules anatropous, erect or ascending, 1 or more in each loculus on basal or axile placentas; style slender; stigma simple or 2–3-branched, surrounded by a cup. Fruit a capsule, drupe or nut. Seeds with straight embryo; endosperm present.

About 12 genera and 300 species mostly in the Australasian region.

1. SCAEVOLA L. (1771) nom. cons.

Leaves entire; inflorescence cymose; corolla-lobes winged; anthers free; ovary inferior with 1 ovule in each loculus; fruit drupaceous with a 2-seeded endocarp.

1. S. plumieri (L.) Vahl, Symb. Bot. **2**: 36 (1791).—*Lobelia plumieri* L. (1753).

Shrub 60–200 cm. high, much branched; leaves obovate, 4–8 cm. long, up to 3·5 cm. broad, fleshy, with a tuft of silky hairs in the axil; petiole narrowly winged; corolla about 2·5 cm. long, green and glabrous outside, lobes white within, tube filled with short woolly hairs; drupe ovoid to subglobose, soft, dark purple to bluish-black, 10–20 mm. long; endocarp rugose.

Occasional in sandy places near the sea; SL–10 feet; fl. and fr. most of the year. *A 12256! H 9533! P 15214!* Widespread on tropical shores; Grand Cayman.

183. COMPOSITAE (ASTERACEAE)

Herbs, shrubs, climbers or small trees. Leaves alternate or opposite, simple or variously divided, without stipules. Flowers (florets) crowded into heads (capitula) surrounded by a calyxlike involucre of one or more series of free or connate bracts (involucral bracts or phyllaries); sometimes the heads compound with the capitula few- or single-flowered; receptacle paleate, setose, pitted or naked, usually convex, sometimes elongated or concave. Florets of one or two kinds in each capitulum, hermaphrodite, unisexual or neuter, rarely dioecious, the outer ones often ligulate (ray-florets), the inner ones tubular (disk-florets), or all tubular, or all ligulate. Calyx epigynous, reduced to a pappus of persistent or caducous hairs, bristles or scales, or absent. Corolla sympetalous, 4–5-fid (actinomorphic disk-florets), filiform, salverform or campanulate, or ligulate or rarely bilabiate (zygomorphic ray-florets). Stamens 5, rarely 4, epipetalous; filaments free; anthers connate into a tube, rarely free, 2-locular, opening lengthwise, often appendaged at apex and tailed at base. Ovary inferior, 1-locular; ovule solitary, basal, anatropous; style of hermaphrodite or female florets usually 2-fid, the style-arms smooth, papillose or hairy, tapered, rounded, deltate or truncate, with or without a terminal appendage. Fruit (achene) sessile, sometimes beaked. Seed without endosperm; embryo straight.

A cosmopolitan family with an estimated 950 genera and 20,000 species.

1 Florets all ligulate (each with a single anterior corolla-limb) and hermaphrodite; sap usually milky:
  2 Achenes without a pappus   54. **Lapsana**
  2 Achenes with a pappus:
    3 Pappus-hairs, at least the inner, plumose; inner achenes beaked; receptacle-scales present; rosette-herbs   55. **Hypochoeris**
    3 Pappus-hairs at most somewhat barbellate; receptacle-scales wanting:
      4 Capitulum solitary, scapose; scape hollow; achenes tuberculate in upper part, long slender-beaked; rosette-herb   56. **Taraxacum**
      4 Capitula several in each inflorescence; herbs with leafy stems:
        5 Achene distinctly beaked, compressed, with minute ascending marginal hairs, black; pappus monomorphic   57. **Lactuca**
        5 Achene narrowed above but not beaked, brown or grey:
          6 Achene compressed, unequally ribbed, smooth or muricate, glabrous or with minute descending hairs, light brown; pappus dimorphic, of long setae and fine hairs intermixed; ligules narrow   58. **Sonchus**
          6 Achene columnar and more or less terete, equally ribbed, the ribs muricate or verrucose; ligules broad:
            7 Pappus dimorphic; achenes grey, glabrous, muricate on the ribs
                                                                         59. **Launaea**
            7 Pappus monomorphic; achenes reddish-brown, with ascending hairs on the upper part, verruculose on the ribs   60. **Crepis**
1 Florets not all ligulate; sap usually clear:
  8 Florets, at least in centre of capitulum, unequally 2-lipped; anther-base long-tailed:
    9 Shrubs or undershrubs with numerous cauline leaves; inflorescences corymbose-paniculate, the capitula pedunculate; receptacle minutely scaly; corolla yellow, anterior lip distinctly longer than posterior   52. **Trixis**
    9 Rosette-herbs with long-scaped solitary capitula:
      10 Style-arms evident, linear; ligules of outer florets small, shorter than involucre, white, pink or reddish-purple; achenes more or less beaked   53. **Chaptalia**
      10 Style-arms very short, broadly ovate, obtuse; ligules showy, longer than involucre, usually yellow, red or pink; achenes narrowed but not beaked; cultivated ornamental; native of Transvaal; Barberton Daisy, Gerbera
                                              Gerbera jamesonii Bolus ex Hook. f.
  8 Florets not 2-lipped, either ligulate (ray) and tubular (disk), or all tubular:
    11 Florets all tubular and regular (capitula discoid):
      12 Capitula unisexual:
        13 Leaves opposite, at least some bipinnatifid; plants monoecious; female capitula 1-flowered; receptacle scaly   24. **Ambrosia**
        13 Leaves alternate or wanting, at most toothed or lobed:
          14 Plants monoecious; female capitula 2-flowered, the involucre with hooked spines; receptacle scaly; leaves broad, toothed or lobed   25. **Xanthium**

14 Plants dioecious; female capitula several- to many-flowered, the involucre or spineless bracts in several series; receptacle naked; leaves small and rotundate, reduced or wanting   38. **Baccharis**

12 Capitula bisexual:

15 Capitula 1-flowered, very rarely 2-flowered; involucres tubular, 5-toothed, in globose or campanulate bracteate glomerules; pappus a fringed ring sometimes shortly awned   23. **Lagascea**

15 Capitula 3- or more-flowered; involucre of separate bracts, if united then only at the base:

16 Involucral bracts in one main series, valvate or very narrowly imbricate:

17 Leaves opposite; receptacle naked; involucral bracts 4; pappus bristly; mostly climbers   46. **Mikania**

17 Leaves alternate:

18 Receptacle scaly; achenes compressed, with or without wings, 2-awned   15. **Verbesina**

18 Receptacle naked; pappus of numerous rough or silky bristles:

19 Involucral bracts and sometimes the leaves glandular, aromatic, glaucous; capitula homogamous   27. **Porophyllum**

19 Involucral bracts not glandular:

20 Capitula heterogamous, disk florets perfect, marginal florets pistillate with filiform tubular corolla; style-arms crowned with divergent hairs surrounding an appendage of fused papillae   32. **Erechtites**

20 Capitula homogamous, all florets perfect:

21 Terminal appendages of style-arms greatly prolonged, frequently recurved; leaves and stems with bright purple hairs; florets orange; ornamental climber, escaping; native of Java   *Gynura aurantiaca* (Blume) DC.

21 Terminal appendages not exceeding style-arms in length or altogether wanting:

22 Tips of style-arms truncate or low-domed, not appendaged, crown of diverging hairs usually well developed; florets yellow; calyculus often present; herbs, shrubs or trees   30. **Senecio**

22 Tips of style-arms acute, appendaged, crown of diverging hairs imperfectly formed; florets mostly mauve or crimson; calyculus wanting; rather glaucous weedy herbs   33. **Emilia**

16 Involucral bracts in 2- to several series, if very few then usually distinctly imbricate:

23 Leaves alternate:

24 Receptacle-scales present; capitula homogamous:

25 Involucral bracts spine-tipped; leaves distichous; achenes obovoid-oblong; pappus scaly; florets mauve   51. **Acanthodesmos**

25 Involucral bracts not spine-tipped; leaves spiral:

26 Achenes compressed, with or without wings, 2-awned; florets orange or white   15. **Verbesina**

26 Achenes oblong; pappus bristly; florets yellow   28. **Neurolaena**

24 Receptacle-scales wanting:

27 Capitula heterogamous; pappus bristly:

28 Involucral bracts scarious   39. **Gnaphalium**

28 Involucral bracts herbaceous:

29 Capitula in glomerules   40. **Pterocaulon**

29 Capitula in corymbs   41. **Pluchea**

27 Capitula homogamous:

30 Capitula few-flowered, aggregated into headlike glomerules; pappus of few bristles:

31 Pappus-bristles 5 or more, all straight   47. **Elephantopus**

31 Pappus-bristles few, the 2 lateral longer and vertically folded   48. **Pseudelephantopus**

30 Capitula many-flowered, not usually clustered:

32 Pappus tubular, shortly toothed; flower-heads in axils of cauline leaves   49. **Struchium**

32 Pappus 2-seriate, the inner of bristles, the outer of shorter bristles or scales; flower-heads in corymbs or in unilateral bracteate cymes   50. **Vernonia**

23 Leaves opposite:

33 Receptacle without scales:

34 Achenes 10-ribbed; pappus bristly; annual herb   42. **Brickellia**

34 Achenes 5-ribbed:

35 Pappus of 3 or 4 stalked clavate glands; anthers without a terminal appendage   43. **Adenostemma**

35 Pappus of scales or bristles; anthers appendaged:

36 Pappus of 5 awned or awnless scales   44. **Ageratum**

36 Pappus of numerous bristles:
  37 Involucral bracts several to many in (1-) 2-several series, imbricate;
    florets (2-) 3–50   45. **Eupatorium**
  37 Involucral bracts 4 (–5) in 1–2 series, subvalvate or imbricate; florets
    usually 4   46. **Mikania**
33 Receptacle scaly:
  38 Pappus of about 16 linear-tapering scales; florets about 15, yellow   3. **Calea**
  38 Pappus not of scales:
    39 Receptacle columnar or conical:
      40 Achenes oblong, 5-angled, without pappus; erect pubescent undershrub
                                          17. **Isocarpha**
      40 Achenes compressed, 2-awned:
        41 Erect or trailing herbs; capitula solitary or few   7. **Spilanthes**
        41 Climbing or scrambling shrubs; capitula in cymose panicles   8. **Salmea**
    39 Receptacle convex or flat:
      42 Achenes with 2–3 stiff erect retrorsely barbellate awns   6. **Bidens**
      42 Achenes not awned or awns smooth, winged or caducous:
        43 Capitula heterogamous, the outer florets fertile, the inner staminate,
          white; coarse erect shrubs with broad leaves; achenes at length dru-
          paceous   19. **Clibadium**
        43 Capitula homogamous:
          44 Climbing shrub with lanceolate leaves; capitula in cymose panicles;
            florets white; awns of achene unequal, the longer broadly winged
                                      16. **Notoptera**
          44 Erect or lax herbs or undershrubs; achenes short and thick; awns if
            present short:
            45 Florets yellow in small axillary short-peduncled often nodding heads;
              pappus a ciliate cupule; weedy herb   9. **Eleutheranthera**
            45 Florets white in heads on erect peduncles; pappus of short caducous
              awns; coarse undershrub   12. **Melanthera**
11 At least some outer florets ligulate, the ligules sometimes very small:
  46 Receptacles-scales wanting:
    47 Involucral bracts connate for most of their length; leaves opposite or alternate,
      pinnatisect; plants glandular and strongly aromatic; cultivated annual orna-
      mentals, natives of Mexico:
      48 Florets of one colour, light sulphur-yellow to deep orange, in heads up to about
        7 cm. across; erect herb with open growth up to about 70 cm. high; African
        Marigold   *Tagetes erecta* L.
      48 Florets usually yellow or orange and red, in heads up to about 4 cm. across;
        herb with low bushy growth up to about 35 cm. high; French Marigold
                                        Tagetes patula L.
    47 Involucral bracts free or connate only at base:
      49 Leaves opposite:
        50 Involucral bracts in one main series; small or slender-branched glandular-
          aromatic herbs; leaves entire, mostly linear, often bristle-margined 26. **Pectis**
        50 Involucral bracts in 4–5 series; coarse perennial herb; leaves toothed, broadly
          ovate, whitish-tomentose beneath   29. **Liabum**
      49 Leaves alternate:
        51 Involucral bracts in one main series:
          52 Style-arms of disk-florets truncate   30. **Senecio**
           52 Style-arms of disk-florets subulate-appendaged   31. **Gynoxis**
        51 Involucral bracts in (1–) 2–4 series:
          53 Pappus cupular   34. **Egletes**
          53 Pappus of bristles:
            54 Ligule of ray-florets very short; involucre 2–3-seriate; erect annual herbs
                                        35. **Conyza**
            54 Ligule of ray-florets clearly evident, spreading:
              55 Involucral bracts in (1–) 2 (–3) series; small or spreading-branched herbs
                with oblanceolate or spathulate often toothed leaves   36. **Erigeron**
              55 Involucral bracts in 3 (–4) series; erect herb with linear or lanceolate sub-
                entire leaves   37. **Aster**
  46 Receptacle-scales present:
    56 Receptacle-scales flat or very narrow, not folded around the achenes:
      57 Leaves alternate on the upright shoots, otherwise in a basal rosette; pappus
      absent:
        58 Leaves finely 2–3 times pinnatisect, lanceolate in outline; ray- and disk-florets
          usually white, rarely pinkish, never yellow; strongly aromatic herb with
          numerous small capitula in a broad corymb; rare casual introduction in the
          mountains; native of Europe and W. Asia; Milfoil, Yarrow
                                        Achillea millefolium L.

58 Leaves lobed or pinnatifid; ray- and disk-florets yellow or orange; ray-florets fertile; inner achenes compressed, narrowly winged **4. Chrysanthellum**
57 Leaves opposite:
  59 Achenes of 2 kinds, the outer flat or concave and pectinate-winged, the inner angular and crowned with 2–3 bristles **5. Synedrella**
  59 Achenes all more or less alike:
    60 Ray-florets 30 or more, the ligules very narrow, white; receptacle-scales very narrow, few; pappus of small teeth **11. Eclipta**
    60 Ray-florets rarely as many as 10; receptacle-scales broad, flat; pappus conspicuous:
      61 Pappus of scales or plumose bristles; ray-florets (3–) 4–6; leaves simple:
        62 Pappus of plumose bristles; style-branches elongate, linear with subulate tips; ray-corollas with 1 or 2 inner lobes; receptacle-scales persistent **1. Tridax**
        62 Pappus of laciniate or ciliate scales; style-branches very short without terminal appendages; ray-corollas without inner lobules; receptacle-scales tardily deciduous **2. Galinsoga**
      61 Pappus of retrorsely barbellate or smooth awns; ray-florets up to about 9; inner involucral bracts connate at base:
        63 Achenes not beaked; annual herbs or shrubby climbers; leaves simple or pinnate **6. Bidens**
        63 Achenes beaked; erect annual herbs with pinnatisect leaves; cultivated and escaping; natives of tropical America:
          64 Ligule of ray-florets 1–1·5 cm. long, oblong-oblanceolate, pink *Cosmos caudatus* Kunth
          64 Ligule of ray-florets 2–3 cm. long, obovate, yellow or orange *Cosmos sulphureus* Cav.

56 Receptacle-scales folded around the achenes:
  65 Leaves alternate:
    66 Ray-florets neuter with large showy yellow or orange to red ligules; achenes more or less compressed-tetragonal without wings; awns mostly early deciduous:
      67 Flower-heads up to 30 cm. or more across, often drooping especially in fruit; at least the lower leaves cordate at base; achenes 7 mm. or more long; cultivated for oil-yielding fruits and ornament; many cultivars originating probably from a Mexican plant; Sunflower *Helianthus annuus* L.
      67 Flower-heads mostly 5–15 cm. across; leaf-blade decurrent on the petiole; achenes 5–6 mm. long **14. Tithonia**
    66 Ray-florets usually fertile, not yellow; flower-heads not over 1 cm. across; achenes compressed; awns persistent:
      68 Awns erect; achenes mostly with winged margins; leaves toothed to pinnatifid **15. Verbesina**
      68 Awns reflexed laterally; achenes of ray-florets keeled; leaves bipinnatifid **21. Parthenium**

  65 Leaves opposite:
    69 Flower-heads axillary, sessile or subsessile; pappus wanting:
      70 Achenes (or ray-florets) enclosed by hook- and spine-bearing inner involucral bracts, in a single series; disk-florets functionally male only **22. Acanthospermum**
      70 Achenes enclosed by smooth inner involucral bracts and receptacle-scales, in several series; disk-florets bisexual or male only **20. Enydra**
    69 Flower-heads mostly pedunculate; pappus usually evident:
      71 Receptacle columnar-conical; pappus of awns **7. Spilanthes**
      71 Receptacle flat or convex, not elongated:
        72 Achenes more or less compressed, usually awned; plants annual:
          73 Leaves entire, all opposite, usually with longitudinal lateral veins; ray-florets fertile, of various colours; involucral bracts in 3 or more series; cultivated ornamentals, sometimes escaping; native continental tropical America *Zinnia* sp.
          73 Leaves toothed, the upper alternate; ray-florets neuter, yellow; involucral bracts in 2 (–3) series **13. Simsia**
        72 Achenes angular; pappus cupular, toothed; plants perennial; florets yellow:
          74 Erect coastal shrub; leaves oblanceolate to linear, entire, often silvery-tomentellous and smooth **18. Borrichia**
          74 Creeping and rooting stoloniferous herbs; leaves toothed or 3-lobed, rough **10. Wedelia**

The foregoing key has been constructed in the knowledge that accurate placing of genera in their tribes often requires careful examination of style-arms and anther-bases with optical magnifying equipment that is not always available. The need to establish these characters has been reduced to a minimum resulting in a key in which the genera do not appear in the sequence of their tribes. The tribes represented by indigenous and naturalized *Compositae* in Jamaica are: I. *Heliantheae* (genera 1–25), II. *Helenieae* (genera 26 and 27), III. *Senecioneae* (genera 28–33), IV. *Astereae* (genera 34–38), V. *Inuleae* (genera 39–41), VI. *Eupatorieae* (genera 42–46), VII. *Vernonieae* (genera 47–51), VIII *Mutisieae* (genera 52–53) and IX. *Cichorieae* (genera 54–60). The tribe *Anthemideae* is represented by the rare early introduction *Achillea millefolium* L.

1. TRIDAX L. (1753); Powell (1965)

1. T. procumbens L., Sp. Pl. **2**: 900 (1753).

Perennial herb with straggling branches up to 40 cm. long; leaves ovate to lanceolate, serrate or dentate, rough-hairy, 2–7 (–12) cm. long, 1–4 (–6) cm. broad; capitula campanulate, 7–10 mm. in diameter, solitary on axillary peduncles 7–23 cm. long; ray-florets 3–6, ligules 2·5–5 mm. long, light yellow to cream-coloured; disk light yellow; achenes 2–2·5 mm. long, densely pilose.

Rare and rather local, along the south coast, in disturbed ground and along roadsides in arid limestone areas; 10–700 feet; fl. and fr. July, Nov–Feb. *A 6017! 6308! Howard, Proctor & Wagenknecht 20524!* Native of C. Amer., now general in the subtropics and tropics; Cayman Brac.

2. GALINSOGA Ruiz & Pav. (1794)

Stems glabrous or nearly so; hairs on peduncle short, appressed-ascending, mixed with shortly stalked spreading glands; leaves ovate, subentire; receptacle-scales 3-fid; pappus-scales about as long as body of achene, longer than corolla of disk-florets
1. *parviflora*
Stems and peduncles pilose with long spreading hairs and long-stalked glands; leaves broadly ovate, serrate-crenate; receptacle-scales entire or shortly laciniate; pappus-scales about half the length of body of achene, shorter than corolla of disk-florets
2. *ciliata*

1. G. parviflora Cav., Ic. Descr. Pl. **3**: 41, t. 281 (1796).

Lax-branched annual herb up to about 75 cm. high; leaves 2–5 cm. long, 1–3 cm. broad; capitula about 4–5 mm. in diameter; ray-florets white; disk-florets yellow; achenes 1·2–1·3 mm. long.

Rather rare (St. Andr., St. Thom.), in disturbed ground and sandy or gravelly waste places; 1600–4300 feet; fl. and fr. Jan–July. *A 7696! 11179! H 8586! Mrs. K. L. Hart!* Native subtropical S. Amer., Hispaniola, introduced into Europe, tropical Africa and elsewhere.

2. G. ciliata (Raf.) Blake in Rhodora **24**: 35 (1922).—*Adventina ciliata* Raf. (1836). *G. parviflora* of S. Moore in F. & R. (1936), partly not Cav.

Annual weedy herb like the last, (3–) 8–35 cm. high; leaves 0·5–6 cm. long, 0·3–4 cm. broad; ray-florets white or pinkish; some pappus-scales usually awned.

Frequent in the eastern and east-central parishes, a weed of sandy and gravelly waste places, roadsides and cultivations; 1500–5200 feet; fl. and fr. most of the year. *A 10330! 10796! A. M. Barry! H 10115!* Native Mexico to Chile, Puerto Rico, introduced in temperate N. Amer., Europe and tropical Africa.

3. CALEA L. (1763)

1. C. jamaicensis (L.) L., Sp. Pl., ed. 2, **2**: 1179 (1763).—*Santolina jamaicensis* L. (1759). Camphor Bush, Halbert Weed.

Erect and bushy or lax scrambling shrub 0·6–3 m. high; leaves ovate to ovate-lanceolate, rounded to obtuse at base, acute or acuminate, entire or sparsely toothed, glandular-punctate, 2–5 cm. long, 0·8–2·8 cm. broad; capitula discoid, about 4–5 mm. broad; involucral bracts in 3–4 series; receptacle-scales ovate, laciniate, about 6 mm. long; achenes about 2 mm. long; pappus-scales about 3 mm. long.

Widespread and locally common on exposed limestone or shale banks and cliffs and in well drained pastures; SL–3500 feet; fl. and fr. most of the year. *A 8245! H 11653! P 18384!* Endemic. A variant with small blunt leaves has been described as var. *parvifolia* S. Moore (1929) and is known from a small area in the parish of St. Andrew (*H 10064! P 15881!*)

4. CHRYSANTHELLUM L. C. Rich. (1807)

1. C. americanum (L.) Vatke in Abh. Nat. Bremen **9**: 122 (1885).—*Anthemis americana* L. (1753).

Rosette herb with erect or prostrate branches up to 35 cm. long; leaf-blades more or less ovate in outline, up to about 2 cm. long and 1·5 cm. broad, the segments oblong-lanceolate; petioles up to 3 cm. long, amplexicaul; peduncles up to 3–6 cm. long; involucral bracts in 1–2 series, about 4 mm. long; ray-florets 8–12; receptacle-scales linear, 3 mm. long; achenes 3 mm. long.

Locally common in lawns, grazed pastures and along roadsides; SL–2000 feet; fl. and fr. most of the year. *A 8022! 11426! H 12107! P 6431!* General in the tropics.

5. SYNEDRELLA Gaertn. (1791) nom. cons.

1. S. nodiflora (L.) Gaertn., Fruct. & Sem. Pl. **2**: 456, t. 171, f. 7 (1791).—*Verbesina nodiflora* L. (1755). Fatten Barrow.

Erect herb up to about 75 cm. high, sometimes flowering when very small in grazed pastures or lawns; stems appressed-pubescent; leaves ovate, cuneate at base, subentire to crenate-serrate, 3–9 cm. long, 1·5–5 cm. broad; petiole winged, bristly-ciliate; flower-heads crowded and subsessile in the leaf-axils; outer green involucral-bracts 2, 7·5–10 mm. long, inner glumaceous; ray-florets 3–5 (–9), yellow; disk-florets 6–10 (–13); achenes 3·5–4 mm. long.

Common weed; SL–2300 feet; fl. and fr. all the year. *A 9225! H 6874! P 7358!* General in the tropics; Grand Cayman.

6. BIDENS L. (1753)

1 Erect or diffusely branched annual weedy herbs:
  2 Leaves bipinnate; outer involucral bracts linear-subulate, early reflexed, not ciliate; achenes mostly 4-awned, few outer setose-pubescent, inner smooth, curved at maturity; ray-florets small, yellow, not or hardly exceeding involucre
               1. *cynapiifolia*
  2 Leaves pinnate or simple; outer involucral bracts spathulate, ciliate; achenes mostly 2 (–3)-awned, uniform, the body with sparse ascending hairs, straight at maturity:
    3 Ray-florets very small or wanting, cream or yellow; involucral bracts glabrescent except ciliate margin, erect   2a. *pilosa* var. *dubia*
    3 Ray-florets conspicuous with large white limb; involucral bracts puberulous at least near tip, outer spreading and recurved   2b. *pilosa* var. *radiata*
1 Climbing, trailing or lax shrubs; ray-florets present, yellow, orange-yellow or reddish-orange:
  4 Leaves compound; achenes setose, the awns 1·5–2·5 mm. long, spreading, yellow:
    5 Leaves bipinnatisect to bipinnate, almost glabrous   3. *dissecta*
    5 Leaves pinnate, the leaflets deeply toothed, pubescent:
      6 Leaves thinly pubescent on both surfaces   4a. *reptans* var. *reptans*
      6 Leaves densely tomentose beneath   4b. *reptans* var. *tomentosa*
  4 Leaves simple, glabrous or nearly so:
    7 Twining climber; leaves broadly rounded to truncate at base; achenes long-setose, curved, up to 16 mm. long including barbed awns up to 6 mm. long   5. *shrevei*
    7 Scrambling, trailing or laxly erect shrubs; leaves cuneate at base; achenes smooth, straight, 10–12 mm. long, with awns barely 1 mm. long:
      8 Leaf-blade deeply serrate-margined   6. *clarendonensis*
      8 Leaf-blade typically entire   7. *trelawniensis*

1. B. cynapiifolia Kunth, Nov. Gen. **4**: 235 (1820).—Spanish Needle.

Herb with slender glabrescent spreading branches 15–120 cm. high or rarely to 2 m.; leaflets mostly serrate, the lateral ovate, the terminal lanceolate, up to 5 cm. long and 2 cm. broad, cuneate at base, caudate-acuminate at tip; peduncles up to

8 cm. long; capitula about 25-flowered; involucral bracts 5–6 mm. long; ray-florets 4 or 5 with usually yellow ligules about 4 mm. long; achenes curved outwards, 10–14 mm. long.

Frequent on partly shaded waste stony ground, along roadsides and in open woodlands; 350–1250 feet; fl. and fr. all the year. *A 5486! H 8182! Webster & Wilson 5060!* Continental tropical Amer., Bahamas, Greater Antilles, Antigua, Trinidad and Tobago; Grand Cayman.

2. B. pilosa L., Sp. Pl. **2**: 832 (1753).—Spanish Needle.

Erect variably hairy herbs, mostly 60–100 cm. high with spreading branches; lowest leaves often simple, upper pinnate with 1–3 pairs of lateral leaflets; leaflets ovate to lanceolate, usually truncate at base, acuminate at tip, coarsely serrate, up to 7 (–10) cm. long and 3 (–4·5) cm. broad but usually smaller. Variation mainly affects the presence, size and colour of ray-florets; typical *B. pilosa* does not seem to have been confirmed for Jamaica but the two following varieties have been recognised:

2a. Var. dubia O. E. Schulz in Urb., Symb. Ant. **7**: 135 (1911).

Ray-florets are almost always wanting in this variant in Jamaica.

Locally common in damp shady places, along pasture margins and stream banks on heavy soils; (1400–) 2900–5000 feet; fl. and fr. Sept–Apr. *A 6373! 10026! Britton 2631! H 12115!* Cuba.

2b. B. pilosa var. radiata Sch. Bip. in Hist. Canar. **3** (2, 2): 242 (1844).

Ray-florets usually 5 or 6 with an 8-veined limb up to 1 cm. broad; disk-florets light yellow; achenes straight, black, thinly hispid, 9–13 mm. long, awns 2–3, slightly divergent, up to 3 mm. long.

A common weed of roadsides and waste places; SL–4500 feet; fl. and fr. all the year. *A 10315! 13157! P 7749! Thompson 7927!* Generally distributed in the tropics of both hemispheres; Grand Cayman.

3. B. dissecta (O. E. Schulz) Sherff in Bot. Gaz. **56**: 493 (1913).

Erect or scrambling shrub 1–2·5 m. high; internodes pubescent distally, glabrescent; leaves with 3–4 pairs of pinnatisect leaflets, the ultimate segments 1–1·5 mm. broad, overall outline up to 10 cm. long and 6 cm. broad; involucre about 5 mm. long; ligules of ray-florets golden-yellow, 1 cm. long and 5–6 mm. broad; achenes 5–10 mm. long, brown.

Local (St. Thom.), in thickets and woodland margins on serpentine; 1000–2900 feet; fl. and fr. Nov–Mar. *A 12128! H 12302! J.P. 1212! P 7443!* Endemic.

4. B. reptans (L.) G. Don in Sweet, Hort. Brit. ed. **3**: 360 (1839).—McKatty Weed.

4a. Var. reptans.

Twining climber up to 6 m. high; leaves mostly with 3 or 5 leaflets; leaflets ovate-lanceolate, to lanceolate, serrate-dentate, acuminate, the terminal up to 6 cm. long and 2·5 cm. broad; involucre 5–6 mm. long; ray-florets 5–8, yellow, the ligules about 1 cm. long; achenes up to 12 mm. long, curved, brown.

Common on open or shady banks and in woodland margins and thickets; 500–2800 feet; fl. and fr. Oct–May. *A 5449! 12384! H 5617! P 7657!* Mexico to Colombia and Venezuela, Cuba, Puerto Rico, Lesser Antilles south to St. Vincent.

4b. B. reptans var. tomentosa O. E. Schulz in Urb., Symb. Ant. **7**: 141 (1911).

Like the last with pubescent branchlets and leaves whitish beneath.

Locally common (St. Andr., Port., St. Thom.), in thickets and on open ground on steep hillsides; 3500–5000 feet; fl. and fr. Dec–Feb, June–Sept. *A 7756! 10496! H 12327! P 6815!* Endemic.

5. B. shrevei Britton in Bull. Torr. Bot. Club **37**: 359 (1910).

Climbing shrub up to 16 m. high; old stems woody, up to 7 cm. thick, with pale corky bark; leaves ovate-lanceolate, crenate-serrate with gland-tipped teeth, acuminate, up to 11 cm. long and 4 cm. broad; involucre about 8 mm. long; ray-florets about 8, yellow, the ligules oblong-obovate, about 15 mm. long.

Locally common (St. Andr., Port., St. Thom.), in thickets and woodland margins; 3000–7000 feet; fl. and fr. Mar–July, Nov–Dec. *A 7032! H 6735! P 20817!* Endemic.

6. B. clarendonensis Britton in Bull. Torr. Bot. Club **39**: 9 (1912).

Trailing and scrambling shrub up to 4 m. high; leaves ovate, acute or obtuse at tip, marginal serratures gland-tipped, 5–7·5 cm. long, 3–4 cm. broad; petioles up to 2 cm. long; involucral bracts 8–12 mm. long; capitula with about 5 ray-florets and 20 disk-florets, all orange or yellow-orange.

Very rare (Clar.); about 2500 feet; fl. and fr. July–Sept. *H 10987!* Endemic.

7. B. trelawniensis Proctor in Bull. Inst. Jam., Sci. ser. **16**: 69 (1967).

Scrambling shrub with resinous sap to 5 m. high; leaves ovate to ovate-lanceolate, acute to subacuminate at tip, sometimes serrate or even pinnatifid, 5–8 cm. long, 1·5–3 cm. broad; petioles 1–1·5 cm. long; involucral bracts 7–9 mm. long; capitula with 3–6 ray-florets, orange drying reddish; disk-florets 9–15; ligules of ray-florets mostly 10–12 mm. long and 5 mm. broad.

Local (Trel.), among limestone crags and on wooded limestone hillside; 1600–2200 feet; fl. Dec–July, fr. Jan, June. *A 12818! P 21337! R 1919! Weaver 1001!* Endemic.

7. SPILANTHES Jacq. (1760)

1 Leaves entire, mostly elliptic-lanceolate, subsessile; flower-heads solitary or rarely paired, on stout peduncles up to 18 cm. long, subhemispherical, discoid; awns rigid
                                                               1. *urens*
1 Leaves shallowly or remotely serrate with petioles up to 1·5 cm. long; flower-heads few to several at branch-ends or upper axils, on slender peduncles up to 6 cm. long, ovoid, radiate; awns hairlike:
  2 Leaves ovate-lanceolate; florets golden-yellow or orange   2. *uliginosa*
  2 Leaves ovate; florets white   3. *radicans*

1. S. urens Jacq., Enum. Syst. Pl. Carib.: 28 (1760).—Pigeon Coop.

Herb with trailing and rooting branches, ascending in flower to about 15 cm. high, nearly glabrous; leaves very variable from elliptic-orbicular to linear-lanceolate, triplinerved, rather fleshy, obtuse, 3–8 cm. long, 1–1·8 (–2·3) cm. broad; involucre of 2 series of bracts up to about 5 mm. long; florets very numerous, white; achenes buff-brown, shiny, costate, thinly pubescent and ciliate, 2·5 mm. long, with 2 incurved broad-based awns.

Common in damp pastures, at swamp margins and on rocks near the sea; SL–150 (–500) feet; fl. and fr. most of the year. *A 6875! 11647! H 9756! P 11096!* C. and S. Amer., Cuba, Hispaniola, Martinique, St. Vincent, Curaçao, Grand Cayman.

2. S. uliginosa Sw., Nov. Gen. & Sp. Pl.: 110 (1788).

Erect or more rarely a prostrate-branched herb, 15–75 cm. high; leaves mostly subacute, 2–6 cm. long, 0·5–2 cm. broad; involucral bracts about 2 mm. long; ligules of ray-florets about 1 mm. long; achenes black, compressed, ciliate on both margins, 1·5–1·7 mm. long.

Common in damp places particularly on heavy soils; 200–1750 feet; fl. and fr. most of the year. *A 9231! 11468! H 7457! 9960!* partly (BM!) *Proctor & Mullings 22026!* C. and S. Amer., Hispaniola, Lesser Antilles, Trinidad and Tobago, West Africa.

3. S. radicans Jacq., Collect. **3**: 229 (1790).

Erect annual herb up to 1 m. high; leaves acute at tip, up to 6 cm. long and 3 cm. broad; capitula, florets and achenes as in the last.

Rare (St. Eliz., West.), a weed of roadsides and cultivations; 1150–1500 feet; fl. and fr. Jan, Sept, Dec. *A 6079! 12793! H 9960,* partly (UCWI!). Mexico to Venezuela.

8. SALMEA DC. (1813) nom. cons.

1 Leaves obovate to elliptic-obovate, entire, glabrous, up to 5 cm. long and 2·5 cm. broad; capitula numerous in terminal corymbs; erect shrub; Bahamas, Cuba, Grand Cayman
S. petrobioides Griseb.
1 Leaves ovate to ovate-lanceolate:
  2 Leaves distinctly stalked, glabrescent and subentire at least on flowering branches
1. *scandens*
  2 Leaves subsessile, pubescent beneath, repand-dentate   2. *sessilifolia*

1. **S. scandens** (L.) DC., Cat. Hort. Monsp.: 141 (1813).—*Bidens scandens* L. (1753).
Shrub, climbing to 10 m. high or low and scrambling, variably hairy and more so when young; leaves ovate to ovate-lanceolate, narrowly to broadly cuneate, entire or subentire on flowering branches, repand-dentate on vegetative shoots, acute to acuminate at tip, 4–10 (–13) cm. long, 1·5–4 (–5·5) cm. broad; peduncles 5–15 mm. long; involucral bracts in about 3 series, oblanceolate, about 4 mm. long; capitula up to about 7 mm. broad; florets numerous, white or greenish-white; achenes oblong, ciliate, 2–3 mm. long.
Common in thickets and woodlands; 50–5250 feet; fl. and fr. July, Nov–Jan. *A 8431! 8590! H 6660! P 9612!* Mexico to Paraguay, Greater Antilles, Trinidad.

2. **S. sessilifolia** Griseb., Fl. Br. W.I.: 375 (1861).
Stems densely rufous-hirsute; leaves ovate, acuminate, 5–10 cm. long, 3–5 cm. broad; corymbs open; involucral bracts in about 5 series, ovate, submembranous, about 4 mm. long.
Very rare (West.), known only from the type, *Purdie!* Endemic. The principal vegetative characters by which *S. sessilifolia* is distinguished from *S. scandens* are all displayed in some degree by juvenile plants of the latter.

9. ELEUTHERANTHERA Poit. ex Bosc (1803)

1. **E. ruderalis** (Sw.) Sch. Bip. in Bot. Zeit. **24**: 165 (1866).—*Melampodium* ? *ruderale* Sw. (1806).
Erect branched annual faintly aromatic herb up to about 50 cm. high, much resembling *Synedrella nodiflora* in habit; leaves ovate, serrate-dentate, pubescent and with scattered yellow glands beneath, up to 6 cm. long and 3·5 cm. broad; peduncles up to 1·5 cm. long; involucral bracts leafy, hispid, 5 mm. long; receptacle-scales folded around achenes, persistent; about 4 mm. long; florets yellow, mostly hermaphrodite and pentamerous, occasionally small sterile ray-florets developed marginally; achenes obovoid, puberulous, 3 mm. long.
Apparently uncommon but reported from seven parishes and most likely over-looked on account of its resemblance to *Synedrella*, a weed of shady undisturbed ground, cultivations and pastures; 150–1600 feet; fl. and fr. Nov–July. *A 5481! 7091! 9190!* General in the wetter parts of the tropics.

10. WEDELIA Jacq. (1760) nom. cons.

Stems, leaves and peduncles with long white spreading hairs; involucral bracts 6–8 mm. long; achenes 3·5–4 mm. long   1. *gracilis*
Stems, leaves and peduncles with few short setulose or appressed hairs, glabrescent but variable; involucral bracts 8–12 mm. long; achenes 4–5 mm. long   2. *trilobata*

1. **W. gracilis** L. C. Rich. in Pers., Synops. Pl. **2**: 490 (1807).
Herb with spreading and rooting stoloniferous branches, the flowering shoots ascending to about 20 cm. high; leaves obovate, narrowed to base and more or less petiolate, coarsely dentate-lobulate, 1–3·5 cm. long, up to 2·5 cm. broad; peduncles up to 12 cm. long; flower-heads solitary, 1·5–2 cm. broad; ray-florets 9–13, golden-yellow, often darker proximally.
Rather common, especially in rough damp low-lying pastures; SL–2700 feet; fl. and fr. all the year. *A 7639! 9463! 12377! H 8667! Powell 840!* Greater Antilles, Antigua.

2. **W. trilobata** (L.) Hitchc. in Rep. Miss. Bot. Gard. **4**: 99 (1893).—*Silphium trilobatum* L. (1759). Creeping Ox-eye, Marigold.

Herb like the last but less hairy, larger and more robust in all parts; on beaches creeping branches may extend 2 m. or more; leaves obovate or oblong-obovate, usually strongly 3-lobed but rarely weakly lobed to subentire, mostly serrate-dentate, up to 7 (–13) cm. long and 5 cm. broad; peduncles up to 15 cm. long; flower-heads 2 cm. or more broad; florets yellow or orange-yellow.

Common in damp pastures, on roadside banks and in waste places, trailing on beaches in wet areas and on some of the cays; SL–4000 feet; fl. and fr. all the year. *A 5516! 6707! H 12125! P 6753!* Florida, C. and S. Amer., Bahamas, Greater Antilles, Virgin Is., Barbados, Trinidad and Tobago, Grand Cayman, West Africa, Hawaii.

11. ECLIPTA L. (1771) nom. cons.

1. **E. alba** (L.) Hassk., Pl. Jav. Rar.: 528 (1848).—*Verbesina alba* L. (1753).

Creeping and rooting or more usually an erect herb up to 60 (–100) cm. high; leaves sessile, elliptic-lanceolate, mostly long-cuneate at base, serrulate to sub-entire, acute, scabridulous, 2·5–7 (–10) cm. long, 0·5–2 (–2·5) cm. broad; peduncles slender, up to 4·5 cm. long; involucral bracts ovate, acute, 3·5–4 mm. long; ligules of ray-florets linear, 2 mm. long; achenes rugose, black, 2 mm. long.

Common in wet places; SL–3800 feet; fl. and fr. most of the year. *A 6033! H 8560! P 6758!* General in the subtropics and tropics.

12. MELANTHERA Rohr (1792)

1. **M. aspera** (Jacq.) L. C. Rich. ex Spreng., Neue Entdeck. **3**: 40 (1822).—*Calea aspera* Jacq. (1789). *M. deltoidea* Michx. (1803), illegitimate name.

Herb or undershrub, erect 50–100 cm. high or laxly scrambling to 1·5 m.; stem and leaves rough; leaves ovate-rhombic to deltate, cuneate at base, serrate-dentate, acuminate, up to 10 cm. long and 7 cm. broad; capitula mostly solitary on ped-uncles up to 10 cm. long; involucral bracts in 2–3 series, the inner about 4 mm. long, obtuse or acute; achenes 3 mm. long, the bristles 2–4, 1–3 mm. long.

Frequent in open sandy or gravelly waste places mostly near the sea; SL–2600 feet; fl. and fr. all the year. *A 5538! 7586! H 11739! P 11502!* Florida, Mexico, Bahamas, Cuba.

13. SIMSIA Pers. (1807)

1. **S. jamaicensis** Blake in Proc. Amer. Acad. **49**: 388 (1913).

Shrubby annual herb 1–2·5 m. high; stems and leaves hispid and glandular-aromatic; petioles 5–6 cm. long; leaf-blades broadly ovate, truncate-subcordate at base, serrate-dentate, broadly acuminate, up to 13 cm. long and 11 cm. broad; capitula stalked in erect panicles, about 1·5 cm. across; involucral bracts about 1·5 cm. long; receptacle-scales about 1 cm. long; ray-florets 0–10, often 8; achenes 6 mm. long with 2 awns about 4 mm. long.

Very local (St. Andr., St. Cath.), in river gravel and on well drained exposed limestone rubble; 300–650 feet; fl. and fr. Nov–Feb. *A 8766! H 6953! P 21945!* Endemic.

14. TITHONIA Desf. ex Juss. (1789)

Leaves usually undivided or rarely 3-lobed, crenate or serrate; flower-heads 4–8 cm. broad on peduncles up to 30 cm. long; ligulate florets usually orange-scarlet; annual herb up to 2 m. high casually escaping from gardens; native of C. Amer.

T. rotundifolia (Mill.) Blake

Leaves usually 3- or 5-lobed, crenate; flower-heads 10–15 cm. broad; florets golden-yellow; perennial shrub   *1. diversifolia*

1. **T. diversifolia** (Hemsl.) A. Gray in Proc. Amer. Acad. **19**: 5 (1883).—*Mirasolia diversifolia* Hemsl. (1881). Mexican Sunflower.

Straggling shrub up to 5 m. high; young branches and petioles tomentose; leaves ovate-deltate in outline, the lobes acuminate, up to about 30 cm. long and

18 cm. broad, aromatic; flower-heads solitary or few together, long-pedunculate; involucral bracts very unequal; receptacle-scales aristate-acuminate, 12 mm. long.

Locally common, naturalized on roadside banks and in cultivation; 560–2250 feet; fl. and fr. Nov–Apr. *H 11877! P 6562!* Native of C. Amer.

15. VERBESINA L. (1753); S. F. Blake (1925)

1 Herbs; capitula heterogamous with usually both ray- and disk-florets; involucral bracts in 2–several series; leaves sessile:
  2 Florets deep reddish-orange; longer awn hooked; leaves toothed; capitula solitary or few on peduncles 5 cm. or more long   1. *alata*
  2 Florets white drying cream or yellow; both awns straight; leaves lobed, pinnatifid; capitula numerous in broad dense corymbs; peduncles up to about 1·5 cm. long
                                                      2. *pinnatifida*
1 Shrubs or small trees with viscid gummy aromatic sap; capitula homogamous, discoid; involucral bracts in 1–2 (–3) series; florets white:
  3 Leaves petiolate, petiole 2–10 (–25) mm. long:
    4 Leaves scabrid, serrulate   3. *rupestris*
    4 Leaves smooth, glabrous except midrib puberulous:
      5 Leaf-blade serrate, linear-oblanceolate to elliptic-lanceolate, acute at tip, the lateral veins making an angle of 30–50° with midrib   4. *nervosa*
      5 Leaf-blade entire or minutely callus-toothed, obovate, abruptly acuminate to narrowly obtuse, the lateral veins making an angle of 60–80° with midrib
                                                      5. *karsticola*
  3 Leaves at least below inflorescence sessile:
    6 Internodes of upper stem, and occasionally of inflorescence-branches, winged; leaves smooth, glabrescent:
      7 Leaves at least on stem serrate-dentate distally, uppermost subentire; involucral bracts about 5 mm. long, mostly shorter than adjacent receptacle-scales
                                                   6. *petrobioides*
      7 Leaves remotely and shortly callus-toothed; involucral bracts 8–11 mm. long, very few, mostly longer than adjacent receptacle-scales   7. *portlandiana*
    6 Internodes of stem and inflorescence without wings:
      8 Leaves smooth, appressed-puberulous at first, glabrescent, at least of stem serrate-margined   6. *petrobioides*
      8 Leaves not smooth:
        9 Leaf-blade harshly scabrid with tubercle-based hairs on both surfaces, lobed-dentate   8. *aspera*
        9 Leaf-blade thinly scabridulous adaxially, shortly hispid abaxially, serrate distally
                                                       9. *propinqua*

1. V. alata L., Sp. Pl. **2**: 901 (1753).

Low straggling-branched perennial herb with winged stems up to about 40 cm. high; leaves ovate to narrowly elliptic-lanceolate, repand-dentate, acute to obtuse, sessile, up to about 12 cm. long and 4 cm. broad, scabridulous; peduncles up to 15 (–20) cm. long; capitula about 1 cm. broad; ligulate florets fertile; achenes puberulous, 4 mm. long, the wings pale and ciliate.

Locally common, particularly in the northern parts of central parishes, in pastures and along roadsides; near SL–3000 feet; fl. and fr. all the year. *A 9909! 13154! H 12021! P 10313!* Greater Antilles, Virgin Is., Guadeloupe, Curaçao.

2. V. pinnatifida Sw., Nov. Gen. & Sp. Pl.: 114 (1788).

Erect robust annual herb 1·5–4 m. high; stem smooth, glaucous, very occasionally winged, striate when dry; petiole-wing linear, amplexicaul; leaves mostly deeply pinnatifid with about 4 pairs of segments, the upper less divided to subentire, the lower up to 50 cm. long and 30 cm. broad, scabridulous adaxially, hispidulous abaxially; capitula about 7–8 mm. broad; involucral bracts acute, about 5 mm. long; ray-florets (1–) 3–4 (–5), the ligule 3 (–4)-toothed; disk-florets 25–35 in each head; achenes black, 6 mm. long, with or without a broad paler wing.

Locally common in rocky waste places, hillside pastures and thickets; 700–2600 (–5000) feet; fl. and fr. Nov–Jan. *A 5584! H 8180! P 7434!* Cuba. A variant with lower leaves shortly lobed and upper leaves entire (*Sinha 102!* from the parish of St. Thomas) requires further study.

3. V. rupestris (Urb.) Blake in Amer. Journ. Bot. **12**: 631 (1925).—*Chaeno-cephalus rupestris* Urb. (1903).

Shrub or tree 1–4 m. high; leaves oblanceolate, long-cuneate at base, mostly acute at tip, 6–12 (–18) cm. long, 1·5–4 (–6) cm. broad; flowers very fragrant; heads numerous in panicles 10–20 cm. broad; outer involucral bracts ovate, subacute, inner oblong, rounded at tip, up to about 3 mm. long; florets (6–) 10–15; receptacle-scales obovate, 5 mm. long, puberulous; achenes black, 5 mm. long, narrowly marginate to broadly winged.

Local (St. Andr., St. Thom.), on rocky shale banks and on serpentine, sometimes forming thickets; 900–3000 feet; fl. and fr. Nov–Feb (–Mar). *A 10078! 12129! H 7813! P 9589!* Endemic.

4. V. nervosa Blake in Amer. Journ. Bot. **12**: 631 (1925).—*Chaenocephalus venosus* Urb. (1908), not *V. venosa* Greene (1882).

Shrub 2–2·5 m. high; leaves cuneate at base, 5–11 (–16) cm. long, 1·5–2·5 (–3) cm. broad; panicles about 10 cm. broad; involucral bracts oblong, obtuse, 6 mm. long; florets about 16; receptacle-scales obovate, 5·5 mm. long; achenes 3 mm. long (? immature).

Rare and local (St. Andr., Port., St. Thom.), in montane thickets; 5500–7000 feet: fl. and fr. Nov. *H 9109! P 9439!* Endemic.

5. V. karsticola Proctor in Bull. Inst. Jam., Sci. ser. **16**: 76, t. 34 (1967).

Shrub 2–5 m. high, arborescent with branches clustered at the top; leaves cuneate at base, 4–9 (–20) cm. long, 2–3 (–5) cm. broad; panicles compact, 4–6 cm. broad; involucral bracts oblong-linear, blunt, 6–8 mm. long; florets about 10; receptacle-scales obovate; achenes oblong-obovate, 5–6 mm. long, narrowly marginate or winged.

Rare and local (St. James, Trel.), on rocky ledges and in woodland on limestone; 2000–2200 feet; fl. and fr. Nov–Dec. *P 20763! 23002! 25628!* Endemic.

6. V. petrobioides (Griseb.) Blake in Amer. Journ. Bot. **12**: 631 (1925).—*Chaenocephalus petrobioides* Griseb. (1861).

Shrub or tree 2–6 m. high with branches mostly clustered terminally; twigs and inflorescence-branches with or without narrow wings not decurrent from leaf-bases; leaves obovate to oblanceolate, the lower sessile and up to 40 cm. long and 10 cm. broad, upper often shortly stalked and much smaller; leaf-margin serrate-dentate to subentire; panicles up to 20 cm. broad; heads 12–16-flowered; receptacle-scales obovate, about 7 mm. long; achenes about 4 mm. long, narrowly or broadly winged.

General in the central parishes in woodlands and thickets on limestone crags and cliffs; 500–3000 feet; fl. Sept–Mar, fr. Dec–Apr. *A 7319! 10203! H 8806! 12778! P 11393! 22978!* Endemic. The presence or absence of stem-wings and the degree of serration of leaf-margin at first suggest a taxonomic division within this species, but other features as well as distribution and flowering times do not support this; plants with winged and unwinged twigs may be found growing together (*P 11392! 11393!*).

7. V. portlandiana Proctor in Bull. Inst. Jam., Sci. ser. **16**: 79, t. 35 (1967).

Arborescent shrub about 3 m. high with the habit of the last; leaves oblanceolate, acute or acuminate, the lower up to 20 cm. long and 4 cm. broad, the upper 7–12 cm. long and 1·5–3 cm. broad; panicles about 15 cm. broad; heads 10–14-flowered; receptacle-scales obovate, 6–7 mm. long; achenes about 4 mm. long.

Rare and local (Port.), in woodland on limestone; 1500–2500 feet; fl. and fr. Dec–Apr. *A 9152! P 9986! 11372!* Endemic.

8. V. aspera Blake in Amer. Journ. Bot. **12**: 631 (1925).—*Chaenocephalus lobatus* Urb. (1908), incl. var. *brachyphyllus* Urb., not *V. lobata* Gaudich. (1826).

Shrub 1–1·5 m. high; leaves oblanceolate to obovate, coarsely toothed or lobed, cuneate at base, 2·5–10 (–17) cm. long; 1–3 (–8) cm. broad; panicles up to 10 cm. broad; involucral bracts reflexed, 7 mm. long; florets about 20; receptacle-scales acute, 5–6 mm. long; achenes about 5 mm. long.

Local (St. Andr., St. Thom.), on limestone cliffs; 100–900 feet; fl. and fr. Oct–Dec. *H 9510! P 11382!* Endemic.

9. V. propinqua (Britton) Blake in Amer. Journ. Bot. **12**: 632 (1925).—*Chaenocephalus propinquus* Britton (1910).

Erect shrub 2–3 m. high; leaves oblanceolate, coarsely serrate distally, long-cuneate at base, 5–11 cm. long, 1·5–3 cm. broad; panicles about 5 cm. broad; involucral bracts obtuse, 5 mm. long; florets about 18; receptacle-scales obovate, 4 mm. long; achenes obovate, 4 mm. long, broadly or narrowly winged.

Very locally common (St. Eliz.), in rocky thickets; 1500–1600 feet; fl. and fr. July–Nov. *H 9672! P 11333! Robertson UCWI 9205!* Endemic.

16. NOTOPTERA Urb. (1901)

1. N. hirsuta (Sw.) Urb., Symb. Ant. **2**: 466 (1901).—*Bidens hirsuta* Sw. (1788). *Salmea hirsuta* (Sw.) DC. (1813).

Erect shrub 1–2 m. or woody climber to 10 m. high; plant pubescent or tomentose and also, even the corolla, covered with minute stalked and sessile glands; leaves lanceolate to ovate-lanceolate, rounded to truncate-subcordate at base, long-acute, subentire or rarely repand-dentate, 6·5–15 cm. long, 2·5–6·5 cm. broad; petioles slender, up to 14 mm. long; heads 2–5 in a cluster, very shortly stalked; involucral-bracts up to 3 mm. long; receptacle-scales acute, up to about 7 mm. long; achenes including winged awn about 6 mm. long.

Locally common especially in the south-eastern parishes, in woodlands and secondary thickets on limestone; SL–2900 feet; fl. and fr. Nov–June. *A 8385! 8849! H 10158! Webster & Wilson 4888!* Endemic.

17. ISOCARPHA R. Br. (1817)

1. I. oppositifolia (L.) Cass. in Dict. Sci. Nat. **24**: 19 (1822).—*Santolina oppositifolia* L. (1759).

Brittle-stemmed taprooted thinly pubescent shrubby herb up to about 75 cm. high; leaves oblanceolate to elliptic-oblanceolate, tapered to a very narrow sessile base, entire or shallowly toothed, acute to rounded at tip, light green, 3–5 (–10) cm. long, 0·5–1 (–2·5) cm broad, minutely glandular and aromatic when crushed; florets white, fragrant, in heads 6–10 mm. long; capitula in clusters at the tip of a peduncle 3–18 cm. long; involucral bracts and receptacle-scales 3–4 mm. long; achenes about 2 mm. long.

Frequent in rocky places, especially the drier aspects of coastal woodland and thickets, also at lagoon margins and sandy waste ground; SL–400 (–1200) feet; fl. and fr. Nov–July. *A 5654! 6974! H 5448! P 9465!* Texas to Venezuela and Colombia, Bahamas, Cuba, Trinidad and Tobago, Cayman Is.

18. BORRICHIA Adans. (1763)

1. B. arborescens (L.) DC., Prodr. **5**: 489 (1836).—*Buphthalmum arborescens* L. (1759). Seaside Ox-eye.

Silvery-tomentose to glabrous shrub up to about 1 m. high; resinous-aromatic; leaves rather succulent, sessile, mucronate, 2–5·5 (–8) cm. long, 5–8 (–13) mm. broad; peduncles terminal, erect, up to 4·5 cm. long; capitula about 2·5 cm. across; ray-florets about 14, fertile, the ligules about 7 mm. long; achenes 3–4-angled, 3·5–4 mm. long.

Rather common on limestone sea-cliffs, gravelly beaches and in pastures or on bare coral rocks near the sea; SL–20 feet; fl. and fr. most of the year. *A 6138! H 9237! P 6648!* Florida, Mexico, Bermuda, Bahamas, Greater Antilles, Antigua, Barbuda, Barbados; Grand Cayman.

19. CLIBADIUM Allam. ex L. (1771)

Pistillate florets 3–6, about half as many as hermaphrodite florets (6–12) in each head; receptacle-scales few to 0; sterile achenes hairy all over; leaves lanceolate to ovate
1. surinamense

Pistillate florets 7–12, more numerous than hermaphrodite florets (5–10) in each head; receptacle scaly throughout; sterile achenes hairy at top; leaves mostly broadly ovate
2. terebinthinaceum

1. C. surinamense L., Mant. Pl. Alt.: 294 (1771).—Jackass Breadnut.
Shrub 1–4 m. high, shortly and coarsely pubescent; leaves cuneate at base, crenate-serrulate, gradually acuminate, scabridulous on both surfaces, up to 13 cm. long and 5 cm. broad; capitula globose, about 5 mm. broad, clustered on the panicle-branches; involucral bracts ovate, about 4 mm. long; achenes obovoid, about 3 mm. long.
Occasional in central and western parishes in moist thickets and open savannas mostly on heavy soils; 300–2300 feet; fl. and fr. Mar–Apr, July–Dec. *A 7517! 8674! H 9886! P 15925! 21495!* C. and S. Amer., Hispaniola, sporadically in Lesser Antilles to Trinidad and Tobago.

2. C. terebinthinaceum (Sw.) DC., Prodr. **5**: 506 (1836).—*Trixis terebinthinacea* Sw. (1788).
Shrub 1–2 m. high with roughly pubescent branches; leaves broadly ovate to ovate-elliptical, rather abruptly acuminate at both ends, serrulate, scabridulous, up to 22 cm. long and 14 cm. broad; petioles up to 5·5 cm. long; capitula subglobose to broadly ovoid, loosely clustered; involucral bracts ovate, 2–3 mm. long; achenes black, 1·5 mm. long.
Frequent mainly in the eastern or western parishes and at higher elevations, a distribution more or less complementary to that of the last, in thickets and woodland margins and on banks; (500–) 1250–5000 feet; fl. and fr. all the year. *A 6894! 7160! H 7356! P 8313!* Costa Rica, Colombia, Cuba.

20. ENYDRA Lour. (1790)

1. E. sessilis (Sw.) DC., Prodr. **5**: 637 (1836).—*Eclipta sessilis* Sw. (1788).
Stem trailing and rooting and ascending to about 60 cm. high in flower, puberulous especially below nodes; leaves oblong-obovate, narrowed to somewhat amplexicaul base, coarsely toothed, obtuse, 2·5–4 cm. long, 10–18 mm. broad; capitula mostly solitary, 7–10 mm. across; involucral bracts 4 in 2 pairs, outer about 7 mm. long and 5 mm. broad; receptacle-scales 3·5–4 mm. long; florets light green, limb of ray-florets very small, 3-lobed, disk-florets 5-merous; achenes compressed, glabrous, 3 mm. long.
Rare and local (St. Eliz., St. James), in ponds and swamps; about 30 feet; fl. and fr. Mar, July. *A 11375! H 10348!* Tropical S. Amer., Greater Antilles.

21. PARTHENIUM L. (1753)

1. P. hysterophorus L., Sp. Pl. **2**: 988 (1753).—Dog-flea Weed, Wild Wormwood.
Annual diffusely branched taprooted aromatic herb up to about 120 cm. high; leaves in outline up to 12 (–15) cm. long and 7 (–12) cm. broad, more or less pubescent, the ultimate segments about 2–4 mm. broad, obtuse; capitula on slender peduncles in loose panicles; involucral bracts in 2 series, ovate, obtuse, about 2 mm. long; florets white; achenes broadly obovoid, black, 2 mm. long.
Common along roadsides and in shady or open waste places; 100–2300 feet; fl. and fr. all the year. *A 5443! H 8212! H & P 14487!* American subtropics and tropics and introduced into the Old World; Grand Cayman.

22. ACANTHOSPERMUM Schrank (1820)

1. A. humile (Sw.) DC., Prodr. **5**: 522 (1836).—*Melampodium humile* Sw. (1788). Sheep Bur.
Annual herb with spreading branches up to about 60 cm. high; stem roughly hairy; leaf-blade broadly ovate with few coarse teeth then abruptly narrowed to a cuneate toothed sessile base, up to 4 cm. long and 2·5 cm. broad; capitula in fruit 1–1·5 cm. broad to tips of spines; outer involucral bracts 5, obovate, ciliate, 3·5 mm. long; inner involucral bracts enclosing achenes about 4 mm. long, the spines also about 4 mm. long, marginal hooks much smaller; receptacle-scales subtending disk-florets about 1·5 mm. long; florets greenish-yellow.

Occasional at roadsides, on waste ground and in pastures; SL–3000 feet; fl. and fr. most of the year. *A 6985! 13147! H 9964! P 10596!* Florida, Panama, Cuba, Hispaniola, Virgin Is.

23. LAGASCEA Cav. (1803) nom. cons.

1. L. mollis Cav. in Anal. Cienc. Nat. **6**: 332, t. 44 (1803).—*Nocca mollis* Jacq. (1805).

Straggly-branched hairy annual herb up to about 1 m. high; leaves opposite or alternate, petiolate, ovate, broadly cuneate to rounded-truncate at base, acute or acuminate, serrate to subentire, 2·5–7 (–9) cm. long, 1–6 cm. broad; glomerules 1–1·5 (–2) cm. broad on short or long stalks, subtended by about 6 ovate-oblong foliose bracts 5–12 mm. long; capitula numerous; involucre ribbed, the lobes ciliate, acute, about 6 mm. long; florets white or mauve; achenes black, 3 mm. long.

Frequent to locally common, a weed of waste places; 50–2300 feet; fl. and fr. all the year. *A 5442! 8310! H 11798! P 10248!* General throughout tropical Amer.

24. AMBROSIA L. (1753)

Stems creeping; lower leaves tripinnatifid; male capitula in compact racemes　1. *hispida*
Stems erect or ascending, sometimes the lower branches rooting; lower leaves bipinnatifid; male capitula in lax racemes　2. *peruviana*

1. A. hispida Pursh, Fl. Amer. Sept. **2**: 743 (1814).—Bay Tansy.

Perennial herb with straggling branches up to about 50 cm. long; leaves ovate in outline, the blade up to about 5 cm. long and 3·5 cm. broad, canescent; racemes up to about 10 cm. long; male capitula up to about 20-flowered, 3–6 mm. broad; female capitula up to about 10 in a cluster; achenes 2·5–3 mm. long, black and shiny.

Rare (St. Cath., St. James), in sandy places mostly near the sea, or cultivated; SL–600 feet; fl. and fr. Jan–Feb, May–June. *A von der Porten! Thompson! Wynter UCWI 27406!* Florida, C. Amer., Bahamas, Greater Antilles, Saba, St. Kitts, Barbados; Grand Cayman.

2. A. peruviana Willd. in L., Sp. Pl. ed. 4, **4**: 377 (1805).—*A. paniculata* Michx. (1803), illegitimate name. Wild Tansy, Wormwood.

Slightly aromatic shrubby herb 0·6–1·2 (–2·5) m. high; leaves deltate to narrowly ovate in outline, up to about 18 cm. long and 12 cm. broad, the ultimate segments lanceolate, thinly pubescent; panicle-branches slender, up to about 12 cm. long; male capitula about 12-flowered, 2–4 mm. broad; female capitula 3–8 in a cluster; achenes keeled, 2–2·5 mm. long.

Locally common, mostly in the eastern parishes and St. Ann, on open roadside banks and pasture margins; 750–4000 feet; fl. and fr. Dec–Aug. *A 6883! 7710! H 7259! H & P 13369!* Florida, Bahamas, Cuba, Hispaniola. The taxonomic affinities of Jamaican plants are not clear; the variation and relationships of local populations require further study. Plants with larger male capitula and more numerous female heads have been distinguished as a variety, but no recent collections of this variant have come to hand.

25. XANTHIUM L. (1753)

1. X. occidentale Bertol., Lucubr. Re Herb.: 38 (1822).—*X. chinense* of S. Moore in F. & R. (1936), not Mill. (1768).

Coarse annual taprooted herb up to 1 (–2) m. high; stem rough; leaves broadly ovate, 3–5-lobed, cordate or cordate-truncate at base, 3-nerved, coarsely toothed, up to 14 cm. long and 18 cm. broad; petioles up to 11 cm. long; male capitula many-flowered, 8 mm. broad, with involucral bracts lanceolate, 1·5 mm. long; female involucre in fruit ovoid, 10–15 mm. long, 7–9 mm. broad; achenes obovoid, without pappus.

Widespread but uncommon and casual in open waste places; 10–1800 feet; fl. and fr. Jan, May, Sept, Dec. *A 8804! H 8537! P 16696!* Bahamas, Greater Antilles, Saba, Antigua, Martinique.

26. PECTIS L. (1759)

1 Involucral bracts (7–) 8; ray-florets 6–8; pappus of up to 10 laciniate scales sometimes
  with bristle-tips; peduncles up to 25 mm. long   **1. *swartziana***
1 Involucral bracts 5; ray-florets up to 5:
  2 Peduncles filiform, 12–38 mm. long, solitary:
    3 Pappus of 2–4 spreading subulate glabrous bristles; glands on undersurface of leaf
      small, scattered   **2. *linifolia***
    3 Pappus of 2–5 erect antrorsely setulose aristate scales; glands on undersurface of leaf
      large in two rows   **3. *febrifuga***
  2 Peduncles short or obsolete, 0–10 mm. long, 1–3 or a few together:
    4 Heads sessile or subsessile; ray-florets 2–4; glands on undersurface of leaf small in
      several rows; pappus of aristate scales; leaf-tip mucronate or shortly aristate
                                       4. *ciliaris*
    4 Heads on peduncles up to about 9 mm. long; ray-florets 5; glands on undersurface of
      leaf large in 2–4 rows:
      5 Herb with erect branches; stem-angles rough below nodes; leaf-tips mostly mucro-
        nate; pappus-bristles hardly broadened at base   **5. *floribunda***
      5 Herb with diffusely ascending or prostrate branches; stem puberulous in 2 bands;
        leaf-tips mostly aristate; pappus of aristate scales   **6. *linearifolia***

1. P. swartziana Less. in Linnaea **6**: 711 (1831).
Erect annual branched herb 6–25 cm. high; leaves linear-lanceolate, obtuse or acute and mucronate at apex, 10–35 mm. long, 1·5–4 mm. broad, with 1–3 pairs of basal bristles, glands scattered; capitula 17–20-flowered; involucral bracts 3·5–4 mm. long; achenes 2·5 mm. long.
Reported once only by Macfadyen. Mexico to Bolivia, Cuba, Hispaniola.

2. P. linifolia L., Syst. Nat. ed. 10, **2**: 1221 (1759).
Erect diffusely branched annual glabrous herb 30–120 cm. high; leaves linear to linear-lanceolate, acute, 2–8 cm. long, 1–6 mm. broad, with 1–2 pairs of basal bristles; capitula 6–9-flowered; ligules of ray-florets yellow striped purple or all deep crimson or dark purple outside and yellow within; involucral bracts linear, blackish-purple, 5–6 mm. long; achenes 5 mm. long.
Frequent on exposed limestone rocks, river gravel and in arid waste places; 20–2500 feet; fl. and fr. most of the year. *A 7951! 9853! H 12127! P 11200!* S.W. United States, Mexico to Colombia and Venezuela, West Indies, Galapagos Is.

3. P. febrifuga Van Hall in Fl. Jard. Pays-Bas **4**: 33 (1861).
Aromatic annual bushily slender-branched erect taprooted herb up to 20 cm. high; leaves linear, mucronulate, 5–20 mm. long, 1–1·5 mm. broad, with 3–5 pairs of basal bristles; capitula 11–18-flowered; ray- and disk-corollas yellow; involucral bracts acute, 5 mm. long; achenes nearly 3 mm. long.
Locally common (St. Andr.), in stony waste places of gritty alluvium; 400–600 feet; fl. and fr. Feb, May–Sept. *A 7949! ACH 141! Bengry! H 12139!* Costa Rica to Venezuela, Puerto Rico, Virgin Is., Aruba, Bonaire, Curaçao.

4. P. ciliaris L., Syst. Nat. ed. 10, **2**: 1221 (1759).—Donkey Weed.
Annual herb with spreading branches 5–15 (–30) cm. high; leaves linear-oblong, 1·5–2·5 cm. long, 1·5–4 mm. broad, with 3–4 pairs of marginal bristles; capitula 9–17-flowered; florets yellow or ligules reddish outside; involucral bracts obtuse or rounded, 6 mm. long; achenes 3 mm. long.
Locally common (St. Andr., St. Cath., Han.), arid pastures, waste places and coral strand; SL–2000 feet; fl. and fr. all the year. *A 5469! H 12067! P 15793!* Greater Antilles.

5. P. floribunda A. Rich in Sagra, Hist. Cuba, Parte 2, **11**: 36 (1850).—Verbena Grass.
Branched annual herb or undershrub up to 60 cm. or more high, fragrantly

aromatic; leaves linear, acuminate, 1–4 (–6) cm. long, 1–3 mm. broad, with 1–6 pairs of marginal bristles; capitula 11–12-flowered; florets purple or yellow; involucral bracts acute, 5–6 mm. long; achenes 2·5 mm. long.

Occasional in pastures and waste places; SL–2500 feet; fl. and fr. Oct–Dec. *A 7849! 9939! Fawcett 7453! P 11179!* C. Amer., Greater Antilles, St. Lucia.

6. P. linearifolia Urb., Symb. Ant. **5**: 276 (1907).

Annual aromatic taprooted herb 10–30 cm. high; leaves linear, obtuse, 1–2·5 cm. long, mostly about 2 mm. broad, with 3–4 pairs of long marginal bristles; capitula 11–12-flowered; florets yellow; involucral bracts oblong, acute or obtuse, about 4·5 mm. long; achenes black, thinly pilose, 3 mm. long.

Rare (St. Eliz.), in thickets near swamp; near SL; fl. and fr. Nov. *P 27689!* Florida.

P. prostrata Cav., like the last but with larger leaves having 5–9 pairs of marginal bristles and scattered glands, was reported for Jamaica by Alain, Fl. Cub. **5**: 225 (1962), but no specimen has been seen.

27. POROPHYLLUM Guett. (1754)

1. P. ruderale (Jacq.) Cass. in Dict. Sci. Nat. **43**: 56 (1826).—*Cacalia porophyllum* L. (1753). *Kleinia ruderalis* Jacq. (1760). *P. ellipticum* Cass. (1826).

Aromatic glaucous annual herb up to 120 cm. high; leaves mostly elliptical or oblong, petiolate, with or without marginal and superficial glands, the blades very thin, up to 6 cm. long and 2·5 cm. broad; peduncles swollen at the top, 2–4 cm. long, solitary or few together; involucral bracts 5, lightly imbricated by hyaline margins, 17–20 mm. long, about 2 mm. broad; florets about 30; corolla-tube slender, 12·5 mm. long, green but purplish-red distally, puberulous; achenes 8–8·5 mm. long, hispidulous; pappus of numerous slender barbellate setae.

Occasional and generally scattered on waste ground and strand; SL–2300 feet; fl. and fr. sporadically throughout the year. *A 7294! 10050! H 12631! P 16028!* Bahamas, throughout tropical Amer.; Grand Cayman.

28. NEUROLAENA R. Br. (1817)

1. N. lobata (L.) Cass in Dict. Sci. Nat. **34**: 502 (1825).—*Conyza lobata* L. (1753). American Goldenrod, Cow-gall Bitter.

Erect shrubby coarse herb up to 2·5 (–4) m. high; stems and leaves impart a yellow colour to the skin when handled; leaves ovate to narrowly elliptical, usually long-pointed at both ends, entire, toothed or 3-lobed, 5–24 (–30) cm. long, 2·5–8 cm. broad; peduncles slender; capitula about 6 mm. broad; involucral bracts in 4 series, innermost about 5·5 mm. long; florets deep yellow; achenes glabrous, about 2 mm. long.

Frequent, especially on heavy soils in damp areas; 25–2800 feet; fl. and fr. Jan–Aug. *A 6639! 9049! H 8503! P 20679!* Throughout tropical Amer.

29. LIABUM Adans. (1763)

1. L. umbellatum (L.) Sch. Bip. in Journ. Bot. **1**: 236 (1863).—*Amellus umbellatus* L. (1759).

Perennial often shortly rhizomatous herb 20–120 cm. high; young stem and undersurface of leaves whitish-tomentose; lower leaves with winged petioles, upper amplexicaul, up to 18 cm. long and 8 cm. broad; capitula 10–13 mm. broad, in corymbose umbels; outer involucral bracts ovate, inner linear-lanceolate, up to 5·5 mm. long; ray-florets in several series, yellow; achenes minutely hairy, 2 mm. long; pappus-bristles in 2 series, about 6 mm. long.

Locally common on shaded moist rocky and mossy banks; 1300–6500 feet; fl. and fr. Feb–Aug. *A 6622! 10803! H 10844! P 3924!* Cuba, Hispaniola.

30. SENECIO L. (1753)

1 Annual herb; ray-florets wanting; leaves deeply pinnatifid   **1.** *vulgaris*
1 Perennial erect or climbing shrubs or trees; ray-florets present; leaves shortly pinnately
   divided, toothed or entire:
   2 Leaves divided to about halfway to midrib in proximal part, the lobes more or less
      triangular, up to 40 cm. long and 12 cm. broad, equally coarsely pubescent on both
      surfaces; small tree   **2.** *swartzianus*
   2 Leaves at least on flowering branches entire or repand-dentate, smaller:
      3 Florets white; leaves white-tomentose beneath; shrub or undershrub   **3.** *discolor*
      3 Florets yellow, orange or reddish-orange; leaves glabrous or nearly so on both surfaces:
         4 Ligules of ray-florets about 2 cm. long, reddish-orange; cultivated ornamental
            twiner; native of Philippines; Smuggler's Joy   *S. confusus* Elmer
         4 Ligules of ray-florets 2–6 mm. long, yellow or light orange:
            5 Leaves on flowering branches distinctly toothed, each tooth-tip a thick gland, with
               7–12 pairs of conspicuous lateral veins prominent beneath when dry   **4.** *fadyenii*
            5 Leaves on flowering branches entire or if denticulate the tooth-tips not obviously
               thickened; lateral veins up to about 7 pairs, obscure:
               6 Involucral bracts 5; florets 5–6; erect shrub with obtusely tipped leaves
                                                                        5. *tercentenariae*
               6 Involucral bracts 6–8 (–10); florets 8–14; leaf-tips mostly acute or acuminate;
                  leaves on non-flowering branches narrower and more distinctly toothed:
                  7 Erect or straggly-branched shrub or small tree; leaves on flowering branches
                     cuneate at base; lateral veins making an angle of 70–80° with midrib
                                                                        6. *swartzii*
                  7 Climbing shrub; leaves broadly cuneate to rounded at base; lateral veins making
                     an angle of 30–60° with midrib   **7.** *hollickii*

1. S. vulgaris L., Sp. Pl. **2**: 867 (1753).—Groundsel.
   Erect sparingly pubescent herb up to about 40 cm. high; leaves oblanceolate to
obovate in outline, the lower petiolate, the upper amplexicaul, up to 8 cm. long and
3 cm. broad; capitula about 5 mm. broad; involucral bracts linear, many, the inner
longer, 5–6 mm. long; florets light yellow; achenes puberulous; pappus white.
   Local (St. Andr.), a weed of cultivated and open ground; 3800–4900 feet; fl. and
fr. Mar–Apr. *A 6424! H 12339!* Native in the temperate Old World, introduced
into United States, Bermuda, Cuba, Hispaniola.

2. S. swartzianus Bueck, Index ad DC. Prodr. **2**: vi (1840).—*Cineraria laciniata*
Sw. (1806). *S. laciniatus* (Sw.) DC. (1838), not Bertol. (1813).
   Shrub or tree up to about 5 m. high; leaves obovate in outline; petioles swollen
at base, 2–3 cm. long; capitula numerous, compactly clustered; involucres about
6 mm. broad; ray-florets about 6, yellow, the ligules about 4 mm. long; disk-
florets about 20; achenes glabrous, 1·5 mm. long; pappus-bristles white, rough,
7·5 mm. long.
   Local (Port., St. Thom.), in montane mossy forest; 7000–7300 feet; fl. and fr.
Nov, Apr–May. *Asprey UCWI 1029! Harris! P 8193!* Endemic.

3. S. discolor (Sw.) DC., Prodr. **6**: 412 (1838).—*Cineraria discolor* Sw. (1788).
Whiteback.
   Erect or scrambling shrub 1–4 m. high; leaves lanceolate to oblong-lanceolate,
rounded to subcordate at base, acute or obtuse at tip, like the stem cottony at first
adaxially and glabrescent, softly leathery, 5–13 cm. long, 1·5–4 cm. broad; capitula
numerous in open corymbs, fragrant; involucre about 4 mm. broad; involucral
bracts 8–9, about 4·5 mm. long; ray-florets 3–6, white, the ligules 2 mm. long; disk-
florets about 12, cream-coloured; achenes hairy, 2 mm. long; pappus-bristles white,
rough, 3·5 mm. long.
   Common on roadside banks and in thickets on well drained slopes; 100–3000
feet; fl. and fr. Dec–Aug. *A 6380! H 8425! P 6070!* Endemic but closely related to
S. almironcillo Gómez Maza of Cuba.

4. S. fadyenii Griseb., Fl. Br. W.I.: 382 (1861).—*S. dolichanthus* (Krug & Urb.)
S. Moore (1929).
   Glabrous shrub 2–3 m. high; leaves oblong to oblanceolate, cuneate to rounded
at base, obtuse or acute at tip, thick, brittle, 10–14 cm. long, 4–6·5 cm. broad;

petioles 2·5–4·5 cm. long; capitula numerous, fragrant in flower; involucre 4–6 mm. broad; involucral bracts 5–6, 6·5–9 mm. long; ray-florets 2–3, orange–yellow, the ligules 3·5–4·5 mm. long; disk-florets 3–5; achenes ribbed, 2 mm. long; pappus-bristles white, rough, about 8 mm. long.

Occasional (St. Andr., Port., St. Thom.), in moist shady submontane woodlands; 2100–5500 feet; fl. and fr. Mar–June. *A 12964! H 5113! J.P. 1359!* Endemic.

5. S. tercentenariae Proctor in Bull. Inst. Jam., Sci. ser. **16**: 75, t. 33 (1967).

Erect shrub 2–3 m. high; leaves glabrous obovate to obovate-elliptical, cuneate at base, thinly leathery, 5–8 (–10) cm. long, up to about 4 cm. broad; petioles 1–2 cm. long; capitula in corymbs 3–6 cm. across, fragrant in flower; involucral bracts 5, 7 mm. long; ray-florets 2–3, bright yellow, the ligules about 4 mm. long; disk-florets usually 3; achenes ribbed, about 2 mm. long; pappus-bristles white, rough, up to about 4·5 mm. long.

Local (Port.), in wet mossy woodland on limestone; 1500–2000 feet; fl. and fr. Mar–Apr. *P 9983!* Endemic.

6. S. swartzii DC., Prodr. **6**: 411 (1838).

Glabrous shrub or tree up to 4 m. high; branches weak and straggling, the younger somewhat fleshy; leaves oblong-elliptical to obovate, obtuse to acuminate at tip, softly fleshy but not thick, up to 18 cm. long and 6·5 cm. broad; petioles 1–2·5 cm. long; involucral bracts 6–10, 7–8 mm. long; ray-florets 3–6, yellow or golden-yellow, the ligules about 5 mm. long; disk-florets 5–8; achenes ribbed, nearly 3 mm. long; pappus-bristles white, rough, 7 mm. long.

Occasional in woodland and woodland margins on limestone; 1000–2500 feet; fl. and fr. Dec–Apr. *A 8621! 8880! H 8892! Proctor & Mullings 22003!* Endemic.

7. S. hollickii Britton ex Greenm. in Ann. Miss. Bot. Gard. **3**: 201 (1916).

Trailing or climbing shrub up to 6 m. or more high; juvenile plants with linear-lanceolate repand-dentate leaves bright purple beneath, hairy at first; leaves of mature shoots ovate to oblong or elliptic-lanceolate, often thick and succulent, acute or apiculate at tip, 4–7 (–10) cm. long, 2·5–4 cm. broad; petioles 5–10 mm. long; involucral bracts about 8, 7·5 mm. long; ray-florets 4, golden-yellow, the ligules about 4 mm. long; disk-florets 6–7; achenes ribbed, about 3 mm. long; pappus-bristles white, rather rough, 6 mm. long.

Frequent in central parishes (), in woodland margins and thickets on craggy limestone; 1250–2500 feet; fl. and fr. Dec–May. *A 6737! 8995! H 11983! P 8426!* Endemic.

The criteria for distinguishing species in the group represented by species 5–7 above require further study; some plants with affinity of *S. swartzii* but climbing habit and others with different combinations of involucre and floret characters exist.

31. GYNOXYS Cass. (1827)

1. G. incana (Sw.) Less., Syn. Gen. Comp.: 390 (1832).—*Cineraria incana* Sw. (1806).

Shrub 2–4 m. high with thin cobwebby indumentum; leaves lanceolate, sessile or subsessile, acute or acuminate, pinnatisect basally with segments up to 15 mm. long, 10–25 (–35) cm. long, 5–7 (–13) cm. broad, light green; corymbs stalked with many capitula; involucral bracts 10–12, oblong, acute, about 7 mm. long; ray-florets 4–6, yellow, the ligules 6 mm. long; disk-florets 8–14, the corollas leathery at base, 10 mm. long; achenes slightly hairy, 1·5 mm. long; pappus-bristles white, smooth, 8 mm. long.

Local (St. Ann, St. Mary, Port., St. Thom.), in thickets and hillside woodlands on limestone; 100–2000 feet; fl. and fr. Nov, Feb–June. *duQuesnay 210! H 5539! P 11849!* Endemic.

32. ERECHTITES Raf. (1817); Belcher (1956)

1. E. hieraciifolia (L.) Raf. ex DC., Prodr. **6**: 294 (1838).—*Senecio hieracifolius* L. (1753).

Erect more or less hairy weedy herb up to about 1 m. high; leaves sessile, oblong to linear-lanceolate, often auricled at base, obtuse or acuminate, margin coarsely toothed or pinnatifid with acute segments, 4–12 cm. long, 1–4 cm. broad; capitula clustered or in open corymbs; involucral bracts nearly 1 cm. long; florets cream-coloured to greenish-yellow; achenes ribbed, slightly hairy, about 3 mm. long; pappus-bristles white, smooth, 10 mm. long.

Two varieties are recognized:

Bracteoles of calyculus extending less than one-quarter of length of involucre; bracteoles of peduncle much shorter than involucre; all bracteoles glabrous or beset with uni-cellular hairs only   1a. var. *hieraciifolia*
Bracteoles of calyculus extending more than one-quarter of length of involucre; bracteoles of peduncle approximately as long as involucre; all bracteoles ciliolate with multicellular hairs   1b. var. *cacalioides*

1a. Var. hieraciifolia.

Peduncles usually rather long.

Frequent on roadside banks and pasture margins on clay or sandy soil; 20–2500 feet; fl. and fr. most of the year. *A 6497! 8662! H 12637! H & P 13946!* Canada, United States, the Guianas, Greater Antilles; Grand Cayman; also in Hawaii and sporadic in C. Europe.

1b. E. hieraciifolia var. **cacalioides** (Fischer ex Spreng.) Griseb., Fl. Br. W.I.: 381 (1861).—*Senecio cacalioides* Fischer ex Spreng. (1819).

Occasional along damp roadsides, in boggy ground and among rocks by rivers; 450–2000 feet; fl. and fr. Mar–July, Dec. *A 6567! H 10890! P 7855! Webster 5615!* Mexico to Uruguay, Bahamas, Antilles sporadically; also Malaya, Indonesia, China.

33. EMILIA Cass. (1817)

Involucre (7–) 9–11 mm. long, conical at base; florets mauve or rarely whitish, exserted from involucre by 0·5–2 mm.; style-arms truncate, with a tuft of hairs; pollen colourless, grains 24–28μ in diameter   1. *sonchifolia*
Involucre (9–) 10–14 mm. long, rounded or truncate at base; florets crimson or bright scarlet, rarely orange, exserted from involucre by 2–5 mm.; style-arms with a tapered hairy appendage; pollen yellow, grains 30–40μ in diameter   2. *javanica*

1. E. sonchifolia (L.) DC. in Wight, Contrib. Bot. Ind.: 24 (1834).—*Cacalia sonchifolia* L. (1753). *E. sagittata* of S. Moore (1936).

Slender-branched annual herb up to about 50 cm. high, somewhat glaucous, thinly pubescent; lower leaves lyrate with broadly winged petioles; upper leaves lanceolate, amplexicaul, irregularly serrate-dentate, acute, up to 10 cm. long and 4 cm. broad; capitula in open corymbs, on peduncles up to 10 cm. or more long; florets 8–12 mm. long, mostly 30–45 in each head; achenes about 3 mm. long.

Frequent as a weed of bare stony ground and rough pastures; SL–2500 feet; fl. and fr. all the year. *A 7099! 8566! H 6648! H & P 13337!* General in the tropics.

2. E. javanica (Burm. f.) Robins. in Philipp. Journ. Sci. Bot. **3**: 217 (1908).—*Hieracium javanicum* Burm. f. (1768). *E. sonchifolia* of S. Moore (1936), partly *E. sagittata* DC. (1838). *E. coccinea* of authors, not (Sims) G. Don. Cupid's Shaving Brush.

Short-lived herb like the last but slightly more robust in all vegetative parts, up to about 1 m. high; florets 11–14·5 mm. long, mostly 40–65 in each head.

Common as a weed of open ground and well drained waste places; SL–4000 feet; fl. and fr. all the year. *A 10320! 11916! H 6727! H & P 13338! P 15934!* General in the American and Asian tropics; replaced mostly by yellow- or light red-flowered *E. coccinea* in tropical Africa.

E. sonchifolia and *E. javanica* are sometimes found growing together in Jamaica. Lee, B. in Sci. Notes & News **2** (4): 14–15 (1966) has shown that *E. sonchifolia* is diploid (2n=10), has cleistogamous flowers and is obligately inbreeding; *E. javanica* is tetraploid (2n=20) and is outbreeding. See also Vuilleumier, B. S. in Journ. Arn. Arb. **50**: 122–123 (1969).

34. EGLETES Cass. (1817)

1. E. prostrata (Sw.) Kuntze, Revis. Gen. Pl. **1**: 334 (1891).

Herb with prostrate branches ascending in flower, wispily hairy at first; leaves obovate, sessile, cuneate at base, rounded and serrate at tip, 1·5–2·5 cm. long, 7–12 mm. broad; peduncles usually solitary, extra-axillary, 1–3 cm. long; involucral bracts in 2–3 series, the outer lanceolate, acute, 3 mm. long, the inner longer; ray-florets white, the ligules about 4 mm. long: disk-florets numerous, yellow; achenes obovoid, with glandular hairs.

Very rare (St. Ann); not recently collected. Venezuela, West Indies.

35. CONYZA Less. (1832) nom. cons.

Leaves linear-oblanceolate to linear, entire or subentire, very thinly pubescent and often with remote marginal and midrib hairs only; capitula up to about 4 mm. long in flower
                                                                        1. *canadensis*
Leaves oblanceolate to linear-elliptical, crenate-serrate, densely greyish-pubescent with ascending hairs; capitula up to about 6·4 mm. long in flower   2. *bonariensis*

1. C. canadensis (L.) Cronquist in Bull. Torr. Bot. Club **70**: 632 (1943).— *Erigeron canadensis* L. (1753). *E. pusillus* Nutt. (1818). Canada Fleabane.

Erect annual taprooted herb up to about 1 m. high; leaves numerous, in a basal rosette at first, the lower long-tapered to base, mucronate, up to 5 (–7) cm. long and 5 (–9) mm. broad; panicle-branches racemose; capitula numerous, up to about 5 mm. across in flower; ligules of ray-florets white, very short; disk-florets yellow; achenes 1 mm. long.

Common on roadside banks and in rough pastures; SL–3800 feet; fl. and fr. all the year. *A 5581! H 11772! P 11658!* Native of temperate N. Amer., now widely naturalized; Bahamas, Greater Antilles, Antigua, Trinidad.

2. C. bonariensis (L.) Cronquist in Bull. Torr. Bot. Club **70**: 632 (1943).— *Erigeron bonariensis* L. (1753). *Leptilon bonariense* (L.) Small (1903). Asthma Weed.

Erect herb like the last up to about 120 cm. high, but coarser with larger fewer leaves; lower leaves long-tapered to base, acute, up to 10 (–12) cm. long and 2 (–3) cm. broad; peduncles slender; capitula up to nearly 1 cm. across in flower; ray-florets very small, white; disk-florets light or greenish-yellow; achenes 1 mm. long, minutely hairy.

Locally common along roadsides and in sheltered rocky waste places; 550–5200 feet; fl. and fr. Dec–July. *A 6438! J.P. 938! A. von der Porten!* Subtropical and tropical continental Amer., West Indies, West Africa, Pacific Is.

36. ERIGERON L. (1753)

1 Stems woody, diffusely branched, slender, rooting, leafy with numerous partly developed axillary buds; capitula solitary on naked or nearly naked terminal or axillary peduncles; ray-florets more than twice as long as involucre   1. *karvinskianus*
1 Rosette-herbs with short main stem and more or less erect or ascending flowering branches rooting only at the base; rosette-leaves usually present at flowering time; capitula solitary or few in small panicles; ray-florets spreading but not greatly exceeding involucre:
  2 Rosette-leaves entire or shallowly crenate, scabridulous; inflorescence subscapose, lax, not or sparingly branched, with at least the upper scape-leaves linear-subulate; capitula about 6–7 mm. broad in flower   2. *cuneifolius*
  2 Rosette-leaves serrate-dentate, pubescent; inflorescence branched, leafy; capitula 7–10 mm. broad in flower   3. *jamaicensis*

1. E. karvinskianus DC., Prodr. **5**: 285 (1836).—Rockside Daisy.

Perennial herb or undershrub with trailing branches up to 1 m. or more long; leaves elliptic-lanceolate, entire or with 1 (–2) pairs of lateral teeth, the tips mucronate, rarely over 3 cm. long and 1 cm. broad, thinly pubescent; peduncles up to about 7 cm. long or rarely much longer; capitula about 2 cm. broad in flower;

ray-florets numerous in 2 series, the ligules at first white, fading pink then reddish-purple; disk-florets yellow; achenes about 1 mm. long.

Widespread and locally common in hilly districts, a weed of roadside and trailside banks; 2000–7400 feet; fl. and fr. all the year. *A 5698! H 11881! H & P 13516!* Mexico to Venezuela, Hispaniola.

2. E. cuneifolius DC., Prodr. **5**: 288 (1836).

Annual or short-lived herb, 5–30 cm. high in flower; basal leaves oblong-obovate, obtuse, up to 4 cm. long and 1·3 cm. broad; scape-leaves 5–10 mm. long; capitula solitary or very few on slender peduncles; involucral bracts 2·5–4·5 mm. long, green; ray-florets 4–5 mm. long, white; disk-florets light yellow; achenes hairy, 1 mm. long.

General in damp short grass pastures on clay or rocky soils; 20–3000 feet; fl. and fr. all the year. *A 6588! 7114! H 11970! P 18302!* Greater Antilles, Virgin Is.

3. E. jamaicensis L., Syst. Nat. ed. 10, **2**: 1213 (1759).

Annual or short-lived herb regenerating by basal innovations; flowering shoots branched above, up to 30 cm. high; basal leaves oblanceolate, obtuse, toothed or shortly lobed with mucronate tips, up to 9 (–12) cm. long and 1·5 (–2) cm. broad; scape-leaves like the basal leaves only smaller or the uppermost entire and linear-oblong; capitula in a small corymb; involucral bracts 2–4 mm. long; ray-florets numerous, about 5 mm. long, white; disk-florets light yellow; achenes hairy, 1·5 mm. long.

Occasional on wet ledges, rocks and roadside banks and beside streams, mostly in shaded or sheltered places; 200–2500 feet; fl. and fr. all the year. *A 6092! 7173! H 6263! P 21399!* Greater Antilles.

37. ASTER L. (1753)

1. A. exilis Ell., Sketch Bot. S. Carol. & Georgia **2**: 344 (1823).

Annual or short-lived taprooted herb 15–200 cm. high, minutely puberulous on stem and leaves; leaves long-tapered to apex and to half-amplexicaul narrow base, serrulate or more usually entire, up to 15 (–20) cm. long and 1·5 (–3·5) cm. broad; panicle large with spreading linear-bracteate branches; capitula 7–9 mm. broad in flower; involucral bracts hyaline-margined, 2·5–5 mm. long; ray-florets numerous, very light mauve; disk-florets light yellow; achenes thinly pubescent, 1–1·5 mm. long.

Frequent as a weed of sandy roadsides, excavations, limestone pastures and salina margins; SL–3500 feet; fl. and fr. Sept–Mar. *A 5978! 8412! 9834! H 12154! H & P 13937!* S.E. United States, Mexico, Bahamas, Cuba.

38. BACCHARIS L. (1753)

1 Leaves fully developed on sterile shoots and juveniles only, oblanceolate, coarsely toothed, reduced to short scales on flowering branches; ultimate branches fastigiate, very numerous, with a capitulum terminating each   1. *scoparia*
1 Leaves fully developed on all shoots, entire or sparingly toothed; ultimate branches not fastigiate:
   2 Leaves obovate, tapered to base, the petiole obsolete, the tip rounded and subcuspidate; capitula numerous in dense usually terminal clusters   2. *dioica*
   2 Leaves elliptical, acuminate at base to a more or less distinct but winged petiole, the tip obtuse; capitula solitary or few together on terminal and axillary peduncles; cultivated ornamental; native of east coast N. America   *B. halimifolia* L.

1. B. scoparia (L.) Sw., Fl. Ind. Occ. **3**: 1339 (1806).—*Chrysocoma scoparia* L. (1759). Bitter Broom, Mountain Broom.

Much branched shrub 1·2–3 m. high, viscid in the younger parts; leaves on seedlings and sterile branches oblanceolate, with 1 or 2 pairs of deltate teeth, the main lateral veins intramarginal, up to 1·5 (–2·5) cm. long and about 5 mm. broad; male capitula about 5 mm. long with 14–16 (–20) florets, the corollas (4–) 5-merous, the stigma undivided; female capitula about 4 mm. long with about 10 florets, the corollas 5-merous, the stigma bifid; florets white or cream-coloured; achenes glabrous, 1·5 mm. long.

Common at middle elevations in eastern parishes especially on steep open hill-sides, occasional in central parishes and absent from the west; 500–7400 feet; fl. and fr. Aug–Feb. *A 5712! 8274! 9966! H 6628! P 18431!* Cuba.

2. B. dioica Vahl, Symb. Bot. **3**: 98, t. 74 (1794).

Shrub 0·6–2·5 m. high, the young twigs angled and resinous; leaves crowded on the branches, obscurely triplinerved, leathery, gland-dotted, up to 4·5 cm. long and 1·8 cm. broad; capitula sessile, the male about 5 mm. long and 20-flowered; female capitula about 7 mm. long and 50-flowered; florets white; achenes glabrous, nearly 2 mm. long.

Local (St. Andr., St. Ann, St. Mary), in thickets and rough patures on limestone; SL–3000 feet; fl. and fr. July–Aug, Nov. *A 9912! Asprey UCWI 2259! H 9009! P 15530!* Florida, Bahamas, Greater Antilles, St. Croix, Montserrat.

39. GNAPHALIUM L. (1753)

1 Inflorescence as a whole corymbose, usually broader than long or at least broadly rounded; receptacle rather flat, foveolate:
  2 Inner involucral bracts mostly less than 4 mm. long; receptacle 1·3–1·5 mm. broad; leaves linear-oblanceolate, long-acute, more or less smooth-margined   **1. albescens**
  2 Inner involucral bracts up to 5·5 mm. long; receptacle about 2·5 mm. broad; leaves elliptic-oblanceolate, acute sinuate-undulate-margined   **2. domingense**
1 Inflorescence as a whole spiciform, longer than broad:
  3 Leaves in lower part of inflorescence shorter than subtended flowering branches; receptacle 2 mm. broad, muricate, convex-umbonate at post-maturity; inner involucral bracts 4 mm. long, pinkish   **3. purpureum**
  3 Leaves in lower part of inflorescence longer than subtended flowering branches; receptacle 1·5–1·9 mm. broad, almost smooth or obscurely foveolate, crateriform at post-maturity:
    4 Inner involucral bracts 4 mm. long, dark brown at tip   **4. americanum**
    4 Inner involucral bracts less than 3 mm. long, creamy-buff   **5. luteo-album**

1. G. albescens Sw., Nov. Gen. & Sp. Pl. 112 (1788).

Erect taprooted herb up to 1 m. high, the whole vegetative plant densely white-woolly at least at first; leaves numerous, the basal rosette not persistent at flowering, glabrescent and variably scabridulous adaxially, up to 7·5 cm. long and 1 cm. broad; capitula numerous, ovoid in flower, about 4 mm. long; involucral bracts in up to about 5 series, the outer ovate and hyaline, silvery-white or light lemon-yellow, the inner oblong and thick towards base; florets numerous, light yellow; achenes oblong, 0·5 mm. long.

Locally frequent (St. Andr., Clar., St. Thom.), mostly in exposed well drained submontane situations; 1800–5000 feet; fl. and fr. Dec–Mar. *A 5845! H 6991! P 9647!* Endemic.

2. G. domingense Lam., Encycl. Méth. Bot. **2**: 743 (1788).

Herb like the last, up to 45 (–60) cm. high, woolly-tomentose and aromatic-glandular; leaves glabrescent and strongly discolorous, up to 5 (–7) cm. long and about 12 mm. broad; capitula in compact clusters, campanulate in flower, 5–6 mm. long; involucral bracts silvery-white, the outer pointed, the inner obtuse; florets yellow; achenes about 0·5 mm. long.

Very local (St. Thom.), on exposed pathside banks; 7200–7400 feet; fl. and fr. Jan–Apr. *A 10675! Asprey UCWI 1037! Weaver!* Hispaniola.

3. G. purpureum L., Sp. Pl. **2**: 854 (1753).—*G. americanum* of S. Moore (1936), partly.

Erect or laxly ascending herb with a few branches from a more or less leafy basal rosette, up to about 40 cm. high; leaves spathulate, woolly, up to 5 cm. long and 1 cm. broad, the cauline smaller; capitula in compact clusters, about 5 mm. long and 4 mm. broad; achenes oblong, light brown, glandular, 0·5 mm. long.

Rare (St. Andr., St. Thom.), on exposed banks and in open pastures; 3000–4100 feet; fl. and fr. Feb. *A 6351! 10559! J.P. 2083!* Subtropical N. and S. Amer. and sporadically in the West Indies and elsewhere.

4. G. americanum Mill., Gard. Dict. ed. 8 (1768).—Cottonweed.

Erect sparingly branched white-woolly annual herb up to about 30 cm. high; leaves linear-oblanceolate to linear, obtuse or acute, glabrescent adaxially and strongly discolorous, smooth, up to 10 cm. long and 12 mm. broad; leaves in inflorescence linear; capitula narrowly ovoid or urceolate, 4–5 mm. long; involucral bracts acute; florets creamy-white; achenes 0·5 mm. long.

Common (St. Andr., Port., St. Thom.), a weed of open ground in montane communities; (3000–) 4000–7400 feet; fl. and fr. Dec–July. *A 6448! 10635! ACH 517! Nicholls! P 9891!* C. and S. Amer., Greater Antilles.

5. G. luteo-album L., Sp. Pl. **2**: 851 (1753).—Cudweed.

Herb like the last, rosette-leaves spathulate, obtuse, apiculate, more or less cottony on both surfaces, up to 7 (–11) cm. long and 1·5 (–2·5) cm. broad; cauline leaves oblanceolate; capitula 3·5–4 mm. long.

Frequent (St. Andr., Port., St. Thom.), a weed of shady pathsides and pastures in submontane communities; 550–3500 feet; fl. and fr. Sept–May. *A 6597! 12807! P 23580! Robertson UCWI 2426!* Native of the Old World subtropics and tropics, sporadically naturalized in America.

G. indicum L., a more diffusely low-branched plant with receptacle 1 mm. broad, has been reported for other islands of the Greater Antilles. *G. pensylvanicum* Willd. has been reported but material is thus far inadequate for confirmation.

40. PTEROCAULON Ell. (1823)

Flower-head clusters aggregated into a more or less continuous globose or ellipsoidal terminal mass; stem broadly winged; capitula about 1 cm. long   1. *alopecuroides*
Flower-head clusters arranged in an interrupted spike; stem narrowly winged; capitula about 7 mm. long   2. *virgatum*

1. P. alopecuroides (Lam.) DC., Prodr. **5**: 454 (1836).

Erect herb up to 60 cm. high; leaves oblong to oblong-ovate, serrulate, obtuse or acute, tomentose beneath, 3–9·5 cm. long, 1–3·5 cm. broad; involucral bracts in 3 series, the inner 5–6 mm. long; achenes about 1 mm. long, hairy; pappus buff or brownish.

Rare (St. Andr., Clar., St. Mary), a weed of poorly drained pastures; 600–2300 feet; fl. and fr. Sept, Dec. *A 5937! P 15998! Thompson 7942!* S. Amer., West Indies.

2. P. virgatum (L.) DC., Prodr. **5**: 454 (1836).—*Gnaphalium virgatum* L. (1759). Golden Cudweed.

Erect herb like the last but more slender with smaller sometimes solitary capitula; leaves linear-lanceolate, entire or nearly so, obtuse, tomentose beneath, mostly 3–7 cm. long, 3–6 (–15) mm. broad; involucral bracts in 3 series, the inner about 5 mm. long; achenes as above.

Very rare (St. Cath.), a weed of pastures at low elevations; fl. and fr. May. *H 12048!* Texas, Mexico to Paraguay, Greater Antilles, Virgin Is.

41. PLUCHEA Cass. (1817)

1 Leaves sessile or subsessile, truncate to cordate-amplexicaul at base, serrate-denticulate; perennial herb   1. *rosea*
1 Leaves, at least the lower cauline, distinctly petiolate and cuneate at base:
   2 Annual herb; leaves shallowly serrate-dentate; young branches grey-buff pubescent; capitula hemispherical, pubescent   2. *odorata*
   2 Shrub; leaves entire or nearly so; young branches pale rusty-tomentose, remaining velutinous; capitula campanulate, tomentose   3. *carolinensis*

1. P. rosea Godfrey in Journ. Elisha Mitchell Sci. Soc. **68**: 266 (1952).

Erect herb up to about 1 m. high; stem pubescent; leaves oblong, obtuse or acute, glandular-punctate, up to 8 cm. long and 2·5 cm. broad; capitula more or less clustered in small corymbs, sessile or shortly peduncled, 5–6 mm. long, about 8 mm. broad; outer involucral bracts oblong, obtuse, densely puberulous distally, ciliate, 5 mm. long; florets rose-purple; achenes 1 mm. long, thinly pubescent, ribbed.

Rare (West., Han.), on strand and grassy roadside; near SL; fl. and fr. Mar–Apr. *A 13198! Wedderburn 127!* Florida, Bahamas, Mexico, Cuba.

2. P. odorata (L.) Cass. in Dict. Sci. Nat. **42**: 3 (1826).—*Conyza odorata* L. (1759). *P. purpurascens* (Sw.) DC. (1836). Bitter Tobacco.

Erect aromatic herb 60–120 cm. high; leaves lanceolate to ovate-lanceolate, mostly obtuse, up to 9 cm. long and 3 cm. broad; pubescent beneath; corymbs mostly 2–4 cm. across; capitula about 5 mm. long; outer involucral bracts ovate, 2 mm. long, inner oblong-lanceolate, acute, 4 mm. long; florets bright purple; achenes 1 mm. long.

Frequent in swamps and damp sandy places; SL–250 feet; fl. and fr. most of the year. *A 6049! H 7223! P 20559!* S.E. United States, C. Amer., West Indies; Grand Cayman.

3. P. carolinensis (Jacq.) G. Don in Sweet, Hort. Brit. ed. 3: 350 (1839).— *Conyza carolinensis* Jacq. (1787). *P. odorata* of S. Moore (1936), not (L.) Cass. Wild Tobacco.

Coarse faintly aromatic undershrub or shrub 1–3 m. high; leaves elliptical, sometimes with a few minute teeth, acute, softly pubescent beneath, coarsely puberulous above, up to 15 cm. long and 5 cm. broad; corymbs 5–10 cm. or more across; capitula about 6 mm. long; outer involucral bracts ovate, obtuse, 2 mm. long, inner oblong-lanceolate, acute, 5 mm. long; florets light pink or mauve; achenes 0·5 mm. long.

Common in saline thickets and open waste places on limestone; SL–2800 feet; fl. and fr. Dec–June, Sept. *A 6329! 6464! H 5619! P 7857!* Florida, Bermuda, Mexico to Venezuela, Bahamas, West Indies, Pacific Is.

42. BRICKELLIA Ell. (1822) nom. cons.

1. B. diffusa (Vahl) A. Gray, Pl. Wright. **1**: 86 (1852).—*Eupatorium diffusum* Vahl (1794).

Annual glabrescent herb 60–240 cm. high; leaves opposite or alternate, the blade broadly ovate or deltate, cordate or truncate !at base, crenate-serrate, 3-nerved, 2·5–8 cm. long, 2–6·5 cm. broad; petioles slender; panicles axillary or terminal, the branches slender; peduncles up to 2 cm. long; involucres oblong, 7·5 mm. long; achenes hairy, nearly 2 mm. long; pappus white, 4 mm. long.

Uncommon and local (St. Andr., West., St. Thom.), a weedy herb in sandy sheltered places, often near streams; 700–1100 feet; fl. and fr. Oct, Jan. *A 10238! A. M. Barry! H 6987!* Mexico to Argentina, Greater Antilles, Trinidad.

43. ADENOSTEMMA J. R. & G. Forst. (1776)

Stem trailing and rooting from and between the nodes; leaves less than 5 cm. long, entire or shallowly serrate-dentate; capitula about 6 mm. broad   1. *verbesina*
Stem erect; lower leaves mostly more than 6 cm. long, distinctly toothed or lobed; capitula about 8 mm. broad   2. *brasilianum*

1. A. verbesina (L.) Sch. Bip. in Journ. Bot. **1**: 235 (1863).—*Cotula verbesina* L. (1759).

Diffusely branched herb with flowering shoots ascending to about 25 cm. high, thinly glandular-pubescent; leaves ovate, truncate at base, obtuse, up to 4·5 cm. long and 3·3 cm. broad; petiole slender; capitula solitary or 2–4 in small corymbs; involucral bracts in 2 series, obtuse, glabrescent; florets about 20, the tube green, the limb white; achenes 2–2·5 mm. long, smooth, minutely glandular; pappus of 2 anterior longer stalked glands and 1 posterior shorter gland, sometimes 4, 2 large and 2 small.

Uncommon in the eastern parishes, rare elsewhere, along moist pathsides and clearings in forest; 1000–4000 feet; fl. and fr. all the year. *A 5856! H 5745! 8809! P 8604!* C. and S. Amer., West Indies.

2. A. brasilianum (Pers.) Cass. in Dict. Sci. Nat. **25**: 363 (1822).—*Verbesina brasiliana* Pers. (1807).

Erect stout-stemmed branched glandular herb up to about 1 m. high; leaves ovate, narrowed at base into a winged petiole, obtuse, up to 15 cm. long and 12 cm.

broad; corymbs with several capitula on densely glandular-pubescent peduncles; involucral bracts pubescent; florets about 50, white; achenes 2–2·5 mm. long, warted and coarsely glandular on the body; pappus of usually 2 clavate glands.

Very rare (Trel.), on shaded rocky slopes; 2000 feet; fl. and fr. Oct. *H 12622!* Subtropical and tropical S. Amer., Cuba, Hispaniola.

44. AGERATUM L. (1753)

1 Pappus of 5 short oblong awnless scales 0·5 mm. long   1. *latifolium*
1 Pappus of 5 awned scales or some shorter and awnless:
   2 Involucral bracts oblong-lanceolate with subacute tips, glabrous or thinly pilose; involucre 4–6 mm. broad   2. *conyzoides*
   2 Involucral bracts linear-lanceolate with long acuminate tips, hairy with short and long spreading hairs; involucre mostly 5–7 mm. broad   3. *houstonianum*

1. A. latifolium Cav., Ic. Descr. Pl. **4**: 33, t. 357 (1797).

Annual sparsely hairy herb up to about 1 m. high; leaves ovate or oblong-ovate, truncate at base, bluntly dentate, 2–4·5 cm. long, 2·5 cm. broad; capitula lax or crowded in terminal corymbs; involucral bracts oblong-lanceolate, acute, about 4 mm. long; florets 30–40, blue ,mauve or white; achenes 2 mm. long, slightly hairy.

Rare (Clar., St. Ann); about 2000 feet; fl. and fr. Dec. *H 12825!* C. and S. Amer., Bahamas, Cuba, Hispaniola.

2. A. conyzoides L., Sp. Pl. **2**: 839 (1753).

Annual erect pubescent herb up to 60 (–90) cm. high; leaves broadly ovate, crenate or serrate, acute or obtuse, up to 7 cm. long and 4·5 cm. broad; capitula stalked in compact corymbs; involucral bracts 4–5 mm. long; florets about 50, mauve or white; achenes 2 mm. long; pappus 2·5 mm. long.

Common as a weed of pastures and waste places; 100–5200 feet; fl. and fr. all the year. *A 5870! 6975! 10851! H 11973! P 6211!* General in warm countries. The variety *inaequipaleaceum* Hieron. in Engl., Bot. Jahrb. **19**: 44 (1894), differs in having some of the pappus-scales short and without awns; it is rare, reported from Portland parish (*P 11565!*).

3. A. houstonianum Mill., Gard. Dict. ed. 8 (1768).

Annual erect herb like the last but usually larger in all parts and with longer whitish hairs; involucral bracts 5 mm. long; florets about 70, mauve, white or yellow, but usually light bright blue; achenes and pappus as in *A. conyzoides*.

Locally rather common (St. Andr., Port., St. Thom.), along roadsides, on banks and in waste places especially in submontane communities; 300–4900 feet; fl. and fr. all the year. *A 7689! 8250! H 11971! P 6212!* Native of Mexico now widespread in the subtropics and tropics, cultivated and sometimes escaping.

45. EUPATORIUM L. (1753)

1 Annual erect weak-stemmed herb with thin ovate-deltate leaves; involucral bracts more or less hyaline, loosely imbricate in several series   1. *microstemon*
1 Perennial undershrubs, shrubs, trees or climbers:
   2 Vines climbing to 5 m. or more; leaves triplinerved or 3-nerved from base; inflorescence broadly paniculate of numerous capitula in small rounded corymbs; involucral bracts ovate to oblong, mostly obtuse:
     3 Leaves ovate to oblong-lanceolate, sometimes in threes; petiole up to 5 mm. long; inflorescence glabrescent; capitula 4-flowered, with 4–5 series of deeply imbricate involucral bracts up to 2·5 mm. long   2. *tetranthum*
     3 Leaves broadly ovate, paired; petioles mostly over 1 cm. long; inflorescences densely puberulous:
       4 Capitula 5–6-flowered, with 2–3 series of loosely imbricate involucral bracts up to about 2 mm. long   17. *simile*
       4 Capitula about 10-flowered, with about 4 series of very unequal tightly imbricate involucral bracts up to 4·5 mm. long   3. *hardwarense*
   2 Erect or lax undershrubs, shrubs or trees:
     5 Receptacle densely hairy; tall undershrub with large broadly ovate cordate 3-nerved minutely velvety leaves   4. *macrophyllum*

5 Receptacle glabrous:
  6 Capitula cylindrical or campanulate with usually (3–) 4 or more series of more or less
    deciduous very unequal deeply imbricate involucral bracts; whole[1] receptacle
    oblong to linear:
    7 Leaves 3-nerved or triplinerved, mostly coarsely serrate-dentate, with numerous
      small yellow glands in pits beneath; florets often bluish, more than 10 in each
      capitulum:
      8 Leaves lanceolate, long-tapered to subsessile base; inflorescences corymbose,
        several-headed   5. *ivifolium*
      8 Leaves ovate, rhombic-ovate or ovate-lanceolate, abruptly narrowed to slender
        petiole:
        9 Capitula numerous in broad open or compact corymbs; florets 15–25, white or
          light bluish-mauve   6. *odoratum*
        9 Capitula solitary or up to 3 together at branch-ends:
          10 Leaves hairy, membranous; florets bright bluish-mauve, about 50
                                                                             7. *heteroclinium*
          10 Leaves glabrous, leathery; florets violet-blue, 25–30   8. *rigidum*
    7 Leaves pinnately nerved, mostly subentire but sometimes shallowly serrate-dentate,
      without yellow glands in pits; florets usually white, up to 5 in each capitulum:
      11 Leaf-blade with translucent lines accompanying the veinlets only; leaf-base
        and tip long-attenuated; corolla cylindrical with very short lobes:
        12 Inflorescence glabrous   9a. *critoniforme* var. *critoniforme*
        12 Inflorescence pubescent   9b. *critoniforme* var. *pubescens*
      11 Leaf-blade with translucent dots and lines in the areoles; leaf-base cuneate,
        leaf-tip obtuse to acutely acuminate; corolla salver-shaped:
        13 Inflorescence-branches densely to thinly pubescent; petiole 1–2 (–3) cm. long;
          involucre 3·5–4·5 (–5) mm. long, 2–3 (–5)-flowered; leaf-blade ovate to
          elliptic-lanceolate; pappus-bristles fine   10. *parviflorum*
        13 Inflorescence-branches glabrous; petiole up to 1 cm. long; involucre 5 (–6) mm.
          long, 3–5-flowered; leaf-blade elliptical to lanceolate:
          14 Pappus-bristles coarse; leaf-tips acute to obtuse   11. *platychaetum*
          14 Pappus-bristles fine; leaf-tips acute to acuminate   12. *dalea*
  6 Capitula campanulate with 1–3 (–4) series of persistent usually narrowly or loosely
    imbricate involucral bracts; receptacle short, flat or broadly convex:
    15 Leaves subsessile, amplexicaul:
      16 Branchlets hirsute   13. *cordifolium*
      16 Branchlets thinly hispid   14. *dolphini*
    15 Leaves distinctly petiolate:
      17 Leaves pinnately veined, the lateral veins in about 4 equally disposed pairs;
        branches rusty-pubescent:
        18 Leaf-blade broadly ovate-elliptical to obovate, repand, broadly cuneate to
          rounded at base, bluntly rounded at tip, coriaceous, glabrescent between the
          veins   15. *hammatocladum*
        18 Leaf-blade elliptical, crenate, cuneate at base, acute to obtuse at tip, perga-
          maceous, pubescent on both surfaces   16. *schizanthum*
      17 Leaves 3-nerved, triplinerved or at least the lateral veins unequally spaced along
        midrib:
        19 Inner involucral bracts up to 2·5 mm. long, as long as to a little longer than
          achenes at maturity; capitula in rounded clusters, small, on filiform peduncles;
          leaf-blades mostly subentire:
          20 Leaves broadly ovate, truncate to openly cordate at base, gradually acuminate,
            densely tomentellous beneath, 3-nerved or triplinerved; petioles up to 4 cm.
            long; capitula 5–6-flowered   17. *simile*
          20 Leaves ovate, oblong-ovate or ovate-lanceolate; petioles 5–15 mm. long;
            capitula 8–18-flowered:
            21 Leaves mostly 3-nerved from base and rounded to truncate-subcordate,
              gradually tapered to a small rounded tip, pubescent on both surfaces
                                                                           18. *villosum*
            21 Leaves triplinerved, acuminate at base, long-acuminate at tip, essentially
              glabrous   19. *gracilipes*
        19 Inner involucral bracts 3–6 mm. long, one and a half times to twice as long as
          achenes at maturity; leaf-blades mostly toothed:
          22 Leaves 3-nerved from base:
            23 Stems, petioles and inflorescence with numerous stalked glands; florets
              about 50 (–65) in each capitulum; achenes 1·2–1·5 mm. long
                                                                           20. *adenophorum*

---

[1] As the involucral bracts are deciduous, the shape referred to is that at post-maturity
after the bracts as well as the achenes have been shed.

23 Stems, petioles and inflorescence rusty-hirsute, glabrescent, without stalked glands; florets about 25 in each capitulum; achenes 2 mm. long
21. *corylifolium*
22 Leaves triplinerved or irregularly pinnately veined, the main pair of ascending lateral veins arising distinctly above the base; branchlets pubescent:
24 Leaf-blades lanceolate, serrate distally, eglandular, the main lateral veins and midrib subparallel 22. *riparium*
24 Leaf-blades broadly to narrowly ovate, crenate to dentate, with numerous sessile or pitted yellow glands at least beneath, the main lateral veins curved-ascending:
25 Leaf-bases cordate or rounded-subcordate; leaf-blade broadly ovate, coarsely dentate to serrate-dentate; involucral bracts straight, oblong, obtuse, prominently 2–3-nerved, inner 3 mm. long; achenes glabrous
23. *montanum*
25 Leaf-bases cuneate to truncate; leaf-blade ovate, crenate or serrate-crenate; involucral bracts obscurely nerved, inner 5–6 mm. long; achenes puberulous:
26 Involucral bracts lanceolate, acuminate, often turned to one side; inflorescences open; leaf-bases cuneate and decurrent on petiole; capitula about 10-flowered 24. *contortum*
26 Involucral bracts oblong or linear-oblong, obtuse or acute, straight; leaf-bases broadly cuneate to truncate; capitula 10–20-flowered:
27 Leaf-lamina densely puberulous between the veins beneath, with numerous small glands; inflorescences broadly corymbose, up to about 12 cm. across 25. *triste*
27 Leaf-lamina glabrescent between the veins beneath, with few large glands; inflorescences rounded, compact, up to about 8 cm. across 26. *hartii*

1. E. microstemon Cass. in Dict. Sci. Nat. **25**: 432 (1822).

Herb rooting at and near base of glabrescent striate stem, up to about 1 m. high; leaf-blade cuneate at base, crenate, acuminate at tip, 2·5–5 (–6) cm. long, 1·5–4 (–5) cm. broad, 3-nerved; petiole up to 3 cm. long; panicles several-headed; capitula 4 mm. broad; outer involucral bracts acute, inner obtuse or rounded, striate, 4 mm. long; florets 15–30, white or very light mauve; achenes 1·5 mm. long.

Rather common as a weed of shaded roadside banks and waste places in moderately wet areas; 250–2800 feet; fl. and fr. most of the year. *A 6118! 8589! H 12639! P 21509!* Continental tropical Amer., West Indies, Ivory Coast, Ghana.

2. E. tetranthum Griseb., Fl. Br. W.I.: 360 (1861).

Trailing or high-climbing shrub; leaves broad and mostly rounded at base, very shallowly repand-dentate, acute at tip, the veinlets forming prominent reticulations on both surfaces, triplinerved, up to 8 cm. long and 3·5 cm. broad, glabrous except puberulous on midrib beneath; involucre about 2 mm. broad; florets light green, the corolla 2 mm. long; achenes top-shaped, 1–1·3 mm. long.

Rare and local (Manch.), on wooded limestone hills; about 2300 feet; fl. and fr. Jan. *P 21878!* Endemic.

3. E. hardwarense Proctor ex Adams in Phytologia **21** (6): 409 (1971).

High-climbing shrub; leaves truncate-subcordate at base, dentate, shortly acuminate, up to 9 cm. long and 8 cm. broad, with numerous small yellow glands on both surfaces; capitula light greenish-brown; involucral bracts about 16, very unequal in 3–4 series; achenes glandular, 1·7–2·3 mm. long.

Rare and local (Port.), in montane woodland; about 3900 feet; fl. and fr. Nov. *P 22970!* Endemic.

4. E. macrophyllum L., Sp. Pl. ed. 2, **2**: 1175 (1763).

Woody herb up to 2 m. high; branches strongly striate, velvety; leaves broadly cordate at base, crenate, acute or broadly acuminate, up to 13 cm. long and 12 cm. broad; petiole up to about 5 cm. long; capitula numerous in rounded compact panicles, whitish, 35–50-flowered, about 5 mm. broad; involucral bracts in about 6 series, the inner up to 5 mm. long; achenes 1·2–1·3 mm. long.

Rather local and uncommon, mostly in rather wet parts of northern parishes; 400–1550 feet; fl. and fr. Mar–Sept. *H 10647! P 8655!* Mexico to Paraguay, West Indies.

5. E. ivifolium L., Syst. Nat. ed. 10, **2**: 1205 (1759).—*Osmia ivifolia* (L.) Sch. Bip. (1866).

Bushy undershrub up to about 1 m. high; stems and veins of leaves slightly rough, glabrescent; leaves distantly toothed, 3–5 (–8) cm. long, 0·5–1·5 (–2) cm. broad; corymbs trichotomous or bifurcately branched, thinly puberulous; involucral bracts in about 5 series, up to 6 mm. long, 3-nerved with green deltate tips; florets about 20, lavender-blue; achenes glabrous or nearly so, 3 mm. long.

Rare (St. Eliz.), in old pasture; about 100 feet; fl. and fr. Oct. *P 21461!* S. United States to Paraguay, Greater Antilles, Guadeloupe, Martinique, Trinidad and Tobago.

6. E. odoratum L., Syst. Nat. ed. 10, **2**: 1205 (1759).—*Osmia odorata* (L.) Sch. Bip. (1866). Christmas Bush, Jack-in-the-Bush.

Erect or scrambling hairy or glabrate shrub up to 2 (–3) m. high; leaves truncate or cuneate at base, mostly coarsely toothed, acute or acuminate, up to about 7 cm. long and 3 cm. broad; corymbs many-headed; involucral bracts in 4–5 series, up to 7 mm. long, 3-nerved with obtuse or subacute tips; florets 14–25, white to light mauve; achenes 4–5 mm. long.

Very common as a weed of pastures and clearings on limestone and waste places generally; SL–2300 feet; fl. and fr. Nov–Apr. *A 6096! 6720! H 8216! P 21849!* Florida to Paraguay, West Indies, West Africa, Malaya; Grand Cayman.

7. E. heteroclinium Griseb., Fl. Br. W.I.: 358 (1861).—*E. maxwelliae* S. Moore (1928).

Shrub with erect main stem and ascending more or less herbaceous flowering branches, 45–120 cm. high, rather coarsely hairy; leaves broadly cuneate at base, serrate-dentate at least distally, acute or obtuse, 1·5–5·5 cm. long, 1–2·5 cm. broad, fragrant on drying; involucral bracts in about 7 series, striate, the inner narrower and 7–8 mm. long; achenes 4–5 mm. long.

Locally frequent (St. Andr., St. Thom.), on open roadside banks, especially shale or gravel screes, and cliffs; 500–5000 feet; fl. and fr. most of the year. *A 5452! 10251! J.P. 1014! P 20816!* Endemic.

8. E. rigidum Sw., Nov. Gen. & Sp. Pl.: 111 (1788).

Erect brittle-stemmed herb or undershrub up to about 60 cm. high; leaves truncate to shallowly cordate at base, coarsely crenate, acuminate, 2–5 cm. long, 1–3 cm. broad, puberulous on veins adaxially otherwise glabrous and shiny; involucral bracts in 3 series, dark-striate, the inner about 7 mm. long; achenes 5 mm. long.

Rare and local (Trel.), on exposed gravelly limestone on wooded hillsides and on cliffs; 1250–2200 feet; fl. and fr. Nov–Feb. *Henry! P 18350! Swartz!* Endemic.

9. E. critoniforme Urb., Symb. Ant. **1**: 458 (1900).

Shrub or tree with brittle branches, 2–8 m. high; leaves elliptic-lanceolate, cuneate at base, entire, sinuate or serrate, acutely acuminate, thin, up to 11 (–16) cm. long and 4 (–6) cm. broad; petioles 1–2 (–3·5) cm. long; panicle of lax corymbs. Two varieties are distinguished:

9a. Var. critoniforme.

Inflorescence glabrous; capitula sessile in small clusters; involucre 5·5 mm. long, the bracts in 4–5 series; corolla-tube greenish-white; stigmas reddish-brown; achenes prominently ribbed, 3 mm. long.

Rather local and uncommon (St. Andr., Port., St. Thom.), in montane forest; 4000–6500 feet; fl. and fr. June–Aug. *ACH 304! Harris! J.P. 1043! P 15550!* Endemic.

9b. E. critoniforme var. pubescens Adams in Phytologia **21** (6): 408 (1971).

Inflorescence pubescent; capitula slightly larger than above.

Rare and local (Port.), in damp rocky thickets; 1700–1900 feet; fl. and fr. July. *J.P. 1360! P 28401!* Endemic.

10. E. parviflorum Sw., Nov. Gen. & Sp. Pl.: 111 (1788).

Shrub 2–3 m. or tree up to 8 m. high; leaves elliptical, cuneate at base, serrate, acute or acuminate; panicles large or small, the branches always more or less hairy; achenes puberulous, about 2 mm. long.

Common in sheltered gullies and woodland margins; 700–7000 feet; fl. and fr. July–Feb. *A 8423! H 9069! P 9654!* Endemic. There are several vicariants within the broad circumscription of this species:

10a. A. The " typical " variant has leaves ovate to broadly elliptical, broadly cuneate at base, serrate-dentate, acuminate, up to 27 cm. long and 14 cm. broad; inflorescence-branches rather densely pubescent; involucre 4–5 mm. long; florets 2 or 3 in each head. This is common mainly at lower altitudes up to about 4000 feet.

10b. B. The " submontane " variant has leaves elliptical to elliptic-lanceolate, cuneate at base, serrate, acute to acuminate, 7–15 cm. long, 2–6 cm. broad; inflorescence thinly pubescent, glabrescent; involucre 4–4·5 mm. long; florets 3–5. This variant occurs in St. Andrew and St. Thomas between 4500 and 6150 feet and flowers mainly between September and December. *A 10644! H 9142!*

10c. C. The " John Crow Mts." variant has leaves like the last but broader with less evident pellucid dots and lines; inflorescence thinly pubescent, glabrescent; involucre 3·5–4 mm. long; florets mostly 3. This occurs in wet woodlands on limestone in eastern Portland between 1500 and 2500 feet and flowers in July and August. *Osmaston 5157! P 10478! Wilson & Murray 595!*

10d. D. The " montane " variant has linear-lanceolate leaves 4–5 times longer than broad, shallowly serrate and acute at both ends, up to about 13 cm. long and 2·5 cm. broad; inflorescence thinly pubescent; involucre 4–5 mm. long; florets up to 5. This occurs in St. Thomas between 6000 and 7000 feet and flowers from September to November. *P 7224! 24027!*

11. E. platychaetum Urb. in Notizbl. Bot. Gart. Berlin **8**: 23 (1921).

Shrub 4–6 m. high; leaves elliptic-lanceolate, cuneate at base, shallowly serrate, up to 11 cm. long and 2·7 cm. broad; capitula sessile; involucre 5 mm. long, the bracts in about 4 series, rotundate; florets white, 3 or 4; achenes glabrous, 2 mm. long.

Rare and local (St. Cath., St. Ann), in woodland on limestone; 2700–3000 feet; fl. and fr. Aug, Dec. *H 8985! H & P 13631!* Endemic.

12. E. dalea L., Syst. Nat. ed. 10, **2**: 1204 (1759).—Cigar Bush.

Shrub 2–5 m. or tree up to 6 m. high, erect or rather long-branched and scrambling; leaves elliptical or lanceolate, serrulate, up to 16 cm. long and 5 cm. broad, coumarin-scented on drying; capitula sessile in small clusters; involucre 5 mm. long, the bracts in about 4 series; florets white, 3–5, fragrant; achenes hairy between the ribs, 3 mm. long.

Locally common in thickets and woodland on limestone, often in rather arid or rugged exposed areas; 25–2000 feet; fl. and fr. Sept–Feb. *A 5595! H 9059! P 15882!* Cuba.

13. E. cordifolium Sw., Nov. Gen. & Sp. Pl. 111 (1788).

Shrub about 1 m. high; leaves ovate, cordate, acute, triplinerved, rough adaxially, mostly 4·5–7 cm. long and 3·5–5 cm. broad; capitula few in a compact sub-sessile inflorescence, about 7 mm. broad with hairy involucral bracts in 4 series; inner involucral bracts obtuse, about 5 mm. long; florets about 45; achenes about 2 mm. long.

Very rare, unlocalized, known only from the type, *Swartz!* Endemic.

14. E. dolphini Urb., Symb. Ant. **5**: 522 (1908).

Shrub 1·5–2 m. high, like the last but much less hairy; leaves coarsely and sharply serrate, acutely acuminate, glabrescent except on nerves, minutely gland-pitted beneath, 4–10 cm. long, 2–6 cm. broad; capitula about 5 mm. long; involucral

bracts recurved latterly; florets about 50, lavender-blue; achenes nearly 3 mm. long.
Very local (Han.), in thickets on exposed limestone hilltop; 1600–1790 feet;
fl. May–July, fr. June–Oct. *H 9250! P 7279!* Endemic.

15. E. hammatocladum Robins. & Britton in Proc. Amer. Acad. **55**: 246 (1918).
Shrub 1·2–3 m. high; leaves opposite or sometimes in threes or fours, 5–6·5
(–8) cm. long, 4–5 cm. broad; panicle terminal; capitula sessile or on filiform
peduncles; involucral bracts 2–3-seriate, oblong, obtuse, 2·5–3 mm. long; florets
5–7, whitish; achenes 2 mm. long, glandular.
Uncommon (Clar., St. James, Trel.), among rocks on wooded limestone hills;
1800–2500 feet; fl. and fr. Dec–May. *H 12795! P 16157! Robertson UCWI 1998!*
Endemic.

16. E. schizanthum Griseb., Fl. Br. W.I.: 361 (1861).
Shrub with rusty-tomentose branches; leaves mostly 3–7 cm. long, 1·5–2·5 cm.
broad; panicle terminal, shorter than the larger leaves; capitula peduncled, about 5
mm. broad; involucral bracts up to 5 mm. long; achenes glandular, 3–3·5 mm. long.
Very rare, unlocalized, known only from the type, *McNab*. Endemic.

17. E. simile Proctor in Bull. Inst. Jam., Sci. ser. **16**: 71, t. 31 (1967).
Climbing or scrambling shrub 2·5 m. or more high; leaves with glands sunken
in deep pits beneath, blades up to 15 cm. long and 10 cm. broad; petioles up to
4 cm. long; capitula in rounded clusters in broad terminal rectangular panicles;
involucral bracts 8 in 2–3 series, 1–2 mm. long; achenes 1·3–1·7 mm. long.
Widely scattered but uncommon, in woodland on rugged limestone; 1250–3800
feet; fl. and fr. Apr–May, Nov. *A 12601! P 8543!* Endemic.

18. E. villosum Sw., Nov. Gen. & Sp. Pl.: 111 (1788).—Bitter Bush.
Much branched velvety-pubescent shrub, woody below, 1–4 m. high; leaves
slightly aromatic, bitter to taste, 2·5–8 (–13) cm. long, 1–4 (–7) cm. broad, often
obscurely toothed; petioles 5–10 mm. long; capitula numerous on peduncles 3–5
mm. long; involucral bracts 2–2·5 mm. long; florets white or lavender, fragrant;
achenes sparsely glandular, 1·5–2 mm. long.
Common on banks and in open thickets on limestone; SL–2400 (–5000) feet;
fl. and fr. Sept–May. *A 5620! 6884! H 12242! P 18399!* Florida, Bahamas, Cuba,
Barbados, Grand Cayman.

19. E. gracilipes Urb., Symb. Ant. **5**: 522 (1908).
Laxly branched shrub 2–4 m. high; branches finely puberulous; leaves minutely
glandular-punctate beneath, subentire, up to 14 cm. long and 15 cm. broad;
petioles 10–15 mm. long; capitula about 3·5 mm. broad; involucral bracts 1·5–2
mm. long; florets white or bluish; achenes obscurely glandular, about 1·5 mm. long.
Very locally common (St. Andr., St. Thom.), in thickets on steep hillsides;
600–2200 feet; fl. and fr. Nov–Mar. *H 7773! P 7433! Yuncker 17973!* Endemic.

20. E. adenophorum Spreng. in L., Syst. Veg. ed. 16, **3**: 420 (1826).—*E.
glandulosum* Kunth (1820), not Michx. (1803).
Herb or bushy undershrub 60–120 cm. high; leaves rhombic-ovate, broadly
cuneate at base, crenate-serrate, acutely acuminate, 3–8·5 cm. long, 1·5–5·5 cm.
broad; capitula stalked in corymbs, about 6 mm. broad; involucral bracts up to
about 3 mm. long; florets white; achenes black, glabrous.
Locally common (St. Andr., St. Thom.), along pathsides in shady montane
woodlands; 3250–7000 feet; fl. and fr. Feb–July. *A 6712! 7411! Asprey UCWI 306!
P 8744!* Mexico, Madeira, Pacific Is. including Hawaii.

21. E. corylifolium Griseb., Fl. Br. W.I.: 361 (1861).
Undershrub; leaves ovate, rounded or obtuse at base, serrate or crenate-serrate,
acute or subacuminate at tip, 2·5–4·5 (–8) cm. long, 1·5–2·5 (–5·5) cm. broad;
capitula few in stalked corymbs, about 6 mm. broad; involucral bracts up to 5 mm.
long; corolla barely 3 mm. long; achenes slightly hairy.
Very rare (St. Andr.), in montane area; about 4000 feet. *Masson!* Cuba.

22. E. riparium Regel, Gartenfl. **15**: 324, t. 525 (1866).

Slender-branched perennial shrubby herb up to 1 m. high, rather straggling with flexuous purplish stems; leaves narrowly cuneate at base, entire proximally, acuminate, 3–7 (–11) cm. long, 1–2 (–2·5) cm. broad; capitula few in open corymbs, 5 mm. broad; inner involucral bracts 4–4·5 mm. long; florets 25–28, white; achenes black, thinly puberulous, 1·5–1·8 mm. long; pappus pink.

Common in the eastern parishes, rather rare in the central parishes and not reported from the west, a weed of stony banks, cleared slopes and rough pastures; 1500–7000 feet; fl. and fr. Jan–July, Oct. *A 6400! H 7812! P 20718!* Mexico, Cuba, Hawaii.

23. E. montanum Sw., Nov. Gen. & Sp. Pl.: 111 (1788).

Erect or scrambling shrub 1–2·5 m. high; branches rough with short hairs; leaves shiny and rough adaxially, hairy especially on the nerves abaxially, 7–12 (–15) cm. long, 3–7 (–8·5) cm. broad; capitula in open corymbs, about 4 mm. broad; involucral bracts in 2–3 series, the inner (2·5) 3 (–3·5) mm. long; florets about 25; corolla dirty white tinged pink; achenes 1·5–1·8 mm. long.

Locally common (St. Andr., Port., St. Thom.), on roadside banks and in open thickets; 1000–1700 (–3500) feet; fl. and fr. sporadically throughout the year. *A 5619! 9885! J.P. 8157! P 19769!* Endemic.

A similar but narrower-leaved species is the very obscure *E. nervosum* Sw. (1788), unlocalized and represented at Kew by incomplete specimens only.

24. E. contortum Adams in Phytologia **21** (6): 408 (1971).

Shrub about 1·5 m. high, with lax densely glandular branches; leaves ovate to ovate-lanceolate, long-acute at tip, puberulous on midrib adaxially and generally abaxially, up to 5·8 cm. long and 2·3 cm. broad; capitula numerous in open corymbs; involucral bracts 8–10, outer 2·5 mm. long, inner about 5·5 mm. long; achenes mostly 5-angled, black, 3 mm. long; pappus 1·5–2·5 mm. long.

Very rare (St. Ann), on wooded limestone hilltop; about 2000 feet; fl. and fr. Jan. Known so far only from the type collection, *Howard, Proctor & Wagenknecht 20512!* Endemic.

25. E. triste DC., Prodr. **5**: 166 (1836).—Old Woman's Bitter Bush.

Shrub 1–3 m. high with rusty-velvety branches; leaves ovate or oblong-ovate, broadly truncate-rounded at base, sometimes lobulate, crenate, acute or obtuse, up to 8 cm. long and 4·5 cm. broad; corymbs rather open with few heads; capitula about 7 mm. broad; involucral bracts linear-oblong, acute or obtuse, the inner 5–6 mm. long; florets white tinged pink, fragrant; achenes black, shortly hairy, 3·5 mm. long.

Common in thickets or pastures and open rocky hillsides; 1800–5000 feet; fl. and fr. Dec–June. *A 8371! 8393! H 9135! P 15954!* Endemic.

26. E. hartii Urb., Symb. Ant. **3**: 395 (1903).

Shrub 1·2–2 m. high with tomentellous branches; leaves oblong-ovate, rounded at base, serrate, acute or obtuse at tip, up to 4·5 cm. long and 2·5 cm. broad; corymbs few, compact; capitula about 8 mm. broad; involucral bracts oblong, obtuse, often tinged reddish-purple, the inner about 5·5 mm. long; florets white; anthers light purple; achenes shortly hairy, 3 mm. long.

Very local (St. Thom.), in and forming low thickets; 7400 feet; fl. Dec, fl. and fr. Jan–Feb. *A 10677! Harris! H & P 14821! Weaver!* Endemic. This species is probably no more than an ecad of *E. triste*; examples of intermediate hairiness and gland-size occur in other montane situations.

46. MIKANIA Willd. (1803) nom. cons.

1 Inflorescence-branches spicate or racemose; capitula less than 4 mm. long:
  2 Flower-heads clustered, sessile or subsessile in spikes 5–8 cm. long; leaves deltate,
    openly cordate   1. *hastata*
  2 Flower-heads solitary or in pairs, shortly pedunculate in racemes; leaves ovate,
    rounded at base   2. *swartziana*

1 Inflorescence of rounded clusters or corymbs of heads:
  3 Capitula subsessile and sessile in compact rounded clusters:
    4 Leaves 3-nerved, shallowly dentate   8. *micrantha*
    4 Leaves triplinerved, entire or subentire:
      5 Involucral bracts 4 mm. long   3a. *brachycarpa* var. *brachycarpa*
      5 Involucral bracts 2 mm. long   3b. *brachycarpa* var. *purdieana*
  3 Capitula mostly stalked, in broad or rounded corymbs:
    6 Leaves densely tomentellous beneath with veinlets prominent around the smallest areoles; involucral bracts tomentellous, obtuse to rounded at tip; achenes glandular-puberulous:
      7 Involucral bracts about 8 mm. long, broadly oblong; leaves ovate   4. *jamaicensis*
      7 Involucral bracts 5–7 mm. long:
        8 Leaves ovate; petioles thick; involucral bracts oblong-elliptical   5. *maxonii*
        8 Leaves deltate-ovate; petioles slender; involucral bracts oblong   6. *troyana*
    6 Leaves pubescent to thinly scabridulous or glabrous, the smallest veinlets not prominent; involucral bracts acute, acuminate or mucronate:
      9 Outer involucral bracts and leaves beneath uniformly pubescent; involucral bracts acuminate, 6–8 mm. long; leaves openly cordate and rather angular at base
                                                     7. *cordifolia*
      9 Outer involucral bracts patchily pubescent, glabrescent or glabrous:
        10 Leaves shallowly dentate, angular-hastate at base, glabrous or nearly so; involucral bracts mostly acute, 3–4 mm. long; inflorescence-bracts lanceolate to subulate, inconspicuous   8. *micrantha*
        10 Leaves sinuate-dentate to subentire, the basal halves rounded, thinly scabrid-hirtellous on both surfaces; involucral bracts mucronate, 5–6 mm. long; inflorescence-bracts obovate to elliptical, conspicuous, up to 9 mm. long
                                                   9. *montverdensis*

1. M. hastata (L.) Willd. in L., Sp. Pl. ed. 4, **3**: 1742 (1803).—*Eupatorium hastatum* L. (1759).

Shrub with strongly striate branches soon becoming glabrous; leaves 3-nerved, denticulate or sinuate, acute or acuminate, mostly 3–7 cm. long and 2–5 cm. broad; petioles up to 3 cm. long; panicles axillary or terminal, up to 30 cm. long; involucral bracts oblong, obtuse, 3 mm. long; corolla white, fragrant, 2 mm. long; achenes rough, 1·5 mm. long.

Very rare and not collected for over 100 years. Venezuela, Cuba, Trinidad.

2. M. swartziana Griseb., Fl. Br. W.I.: 363 (1861).

Glabrous shrub with smooth striate branches; leaves entire or sinuate-margined, triplinerved, acute, mostly 5–8 cm. long and 2·5–4·5 cm. broad; petioles 5–12 mm. long; racemes mostly 2–5 cm. long; involucral bracts oblong, obtuse, 3 mm. long; corolla 2·3 mm. long; achenes with short hairs, nearly 1·5 mm. long.

Very rare (Manch.), only twice collected. *Purdie!* Hispaniola.

3. M. brachycarpa Urb., Symb. Ant. **5**: 220 (1907).

3a. Var. brachycarpa.

High-climbing woody vine; leaves ovate, shallowly cordate and produced on to petiole, entire or denticulate, caudate-acuminate, glabrous except on nerves, 5–9·5 cm. long, 2·5–6·5 cm. broad; petioles 1–2 cm. long; involucral bracts oblong, rounded at tip, nearly glabrous; corolla greenish-white; achenes minutely glandular, 1·5 mm. long.

Rare (Port.), in montane forest; 3800–4500 feet; fl. and fr. Dec–Feb. *H 9126! P 22071!* Endemic.

3b. M. brachycarpa var. **purdieana** Urb., Symb. Ant. **5**: 220 (1907).

Like the last but leaves markedly cordate at base and blade not produced on to petiole.

Very rare (West.), known only from the type, *Purdie!* Endemic.

4. M. jamaicensis Robins. in Contrib. Gray Herb. **64**: 12 (1922).

Woody climber with tomentose branches; leaves cordate, entire, obtuse, slightly rough adaxially; petioles tomentose, 10–15 mm. long; corolla 7 mm. long; achenes with minute glandular hairs, 6 mm. long.

Rare and local (Port.), in forest; about 2000 feet; fl. and fr. Oct. *J.P. 984!* Endemic.

5. M. maxonii Proctor in Bull. Inst. Jam., Sci. ser. **16**: 72 (1967).
Like the last, the branches densely rusty-tomentose; leaves subcordate at base, entire or sinuate, subacuminate, up to 8 cm. long and 6 cm. broad; petioles up to 3 cm. long; corolla pink, 5·5 mm. long; achenes glandular-puberulous, 4 mm. long.
Rare and local (St. Thom.), in montane forest; about 5000 feet; fl. and fr. Dec. *P 9639!* Endemic.

6. M. troyana Urb., Symb. Ant. **5**: 226 (1907).
Climber like the last, up to 8 m. high; leaves broadly subcordate at base, subentire, acute to subacuminate, up to 10 cm. long and 9 cm. broad; petioles up to about 2 cm. long; corolla cream-coloured, fragrant; achenes glandular, 3·5 mm. long.
Uncommon (Han., Trel.), climbing over shrubs in thickets; 1000–2000 feet; fl. and fr. Dec. *A 8601! H 8822!* Endemic.

7. M. cordifolia (L.f.) Willd. in L., Sp. Pl. ed. 4, **3**: 1746 (1803).—*Cacalia cordifolia* L. f. (1781).
Wide-ranging scrambler and climber; leaves broadly ovate, entire or toothed, obtuse or acute, 3–8 (–15) cm. long, 2–6 (–14) cm. broad, 3-nerved, faintly aromatic; petioles up to 8 cm. long; corolla white or greenish-white, 6 mm. long; achenes glabrous, 3–4 mm. long.
Rather common in the eastern hills (St. Andr., St. Thom.), much less so elsewhere, in thickets and woodlands; 150–3500 feet; fl. and fr. Dec–Mar. *A 6715! 10263! H 7794! P 16120!* S. United States to Paraguay, Greater Antilles, Virgin Is., Guadeloupe, Trinidad.

8. M. micrantha Kunth, Nov. Gen. **4**: 134 (1820).—*M. congesta* DC. (1836).
Guaco.
Much branched extensively scrambling and twining slender-stemmed vine, thinly hairy or glabrous; leaves broadly ovate, shallowly or coarsely toothed, acute or acuminate, 4–10 (–12) cm. long, 2·5–6 (–8) cm. broad, 3-nerved; petioles tendriliform, 2–9 cm. long; florets white or greenish, fragrant; corolla 2·5–3 mm. long; achenes glandular, 1·2–1·8 mm. long.
Common, especially in wet places; SL–2300 feet; fl. and fr. Sept–Mar, June–July. *A 6025! 8888! H 9934! P 16019!* Throughout tropical Amer.
Both *M. cordifolia* and *M. micrantha* possess at the nodes of young vegetative shoots, membranous semi-translucent interpetiolar enations resembling stipules. Such structures are so unusual in Compositae that special attention has been drawn to them; they wither on older shoots and have not been observed on flowering branches.

9. M. montverdensis Proctor in Bull. Inst. Jam., Sci. ser. **16**: 75, t. 32 (1967).
High-climbing vine with herbaceous sparsely pubescent branches, up to 12 m. high; leaves broadly ovate to rotundate, plicate-acuminate at tip, up to 9 cm. long and 8 cm. broad, 5-nerved; petioles tendriliform, up to 5 cm. long; corolla white, about 5 mm. long; achenes thinly puberulous, glabrescent, about 3 mm. long.
Very local (Port.), in thickets and disturbed montane forest; about 3800 feet; fl. and fr. Jan–Mar. *A 12261! Proctor & Powell 22282!* Endemic.

47. ELEPHANTOPUS L. (1753)

Leaves obovate or elliptic-obovate; inflorescence unequally branched with spreading peduncles; glomerules subtended by broad leafy bracts; pappus of 5 bristles
1. *mollis*
Leaves linear to linear-oblanceolate; inflorescence elongate-spicate; glomerules clustered on the main axis or short basal branches, with inconspicuous bracts; pappus of numerous bristles
2. *angustifolius*

1. E. mollis Kunth, Nov. Gen. **4**: 26 (1820).—Elephant Foot.
Erect fibrous-rooted bushy herb up to about 120 cm. high; leaves tapered to a narrowly winged petiole and shortly tubular sheathing base, crenate-serrate, acute or obtuse, pubescent, 6–9 (–18) cm. long, 3–5 (–7·5) cm. broad; glomerules up to

about 2 cm. long; capitula 4-flowered; involucral bracts acuminate, the inner up to 7 mm. long; corolla white, 6·5 mm. long; achenes 10-ribbed, 2–3 mm. long.

Rather common along roadsides, in pastures and among rocks in thickets and marginal woodland; 850–4200 feet; fl. and fr. Nov–Feb. *A 8398! Fawcett & Harris 7008! P 11214!* Throughout the American subtropics and tropics and generally in the Old World and Pacific.

2. E. angustifolius Sw., Nov. Gen. & Sp. Pl.: 115 (1788).

Erect pubescent herb up to about 1 m. high; stock short with numerous spreading roots; leaves arising mostly near the stem-base, thinly pubescent and scabridulous, remotely shallowly crenate or shortly toothed, acute or obtuse, up to 35 cm. long and 3·5 cm. broad; glomerules about 1 cm. long; capitula 3–5-flowered; inner involucral bracts oblong-lanceolate, acute, 8 mm. long; corolla white, 6 mm. long; achenes 10-ribbed, 2 mm. long.

Rare (St. Andr., Clar.), on open clay banks and in exposed rough pastures; 800–2700 feet; fl. and fr. sporadically throughout the year. *A 5938! H 9025! P 9734!* Continental tropical Amer., West Indies.

48. PSEUDELEPHANTOPUS Rohr (1792)

1. P. spicatus (B. Juss. ex Aubl.) C. F. Baker in Trans. Acad. Sci. St. Louis **12**: 45, 55 (1902).—*Elephantopus spicatus* B. Juss. ex Aubl. (1775). Dog's Tongue, Packy Weed.

Erect perennial herb up to 70 (–100) cm. high, from a strongly rooted stock; lower leaves obovate to oblanceolate, narrowed at base to short sheathing petiole, crenulate-sinuate, obtuse, scabrid, up to 10 cm. long and 5 cm. broad; upper leaves linear; inflorescence-branches subspicate; glomerules sessile, 1–4-headed; capitula 4-flowered; inner involucral bracts 8 mm. long; florets white or mauve to light purple; achenes 10-ribbed, 5 mm. long; longer pappus bristles doubly bent, 6 mm. long.

Common weed in pastures, on banks and along roadsides; SL–3500 feet; fl. and fr. Nov–June. *A 5577! H 6936! P 7457!* Continental tropical Amer. and West Indies, introduced into W. Africa, E. Asia and Guam.

49. STRUCHIUM P. Browne (1756)

1. S. sparganophora (L.) Kuntze, Revis. Gen. Pl. **1**: 366 (1891).—*Ethulia sparganophora* L. (1763). *Sparganophorus vaillantii* Crantz (1766).

Trailing, ascending or erect-branched herb up to about 1 m. high; branches angled, glabrescent; leaves elliptic-lanceolate, cuneate, serrate, acuminate, puberulous, 6–9 (–12) cm. long, 1·5–5 (–6) cm. broad; capitula subglobose, up to about 8 mm. broad; involucral bracts ovate or ovate-lanceolate, acuminate, inner 3–4 mm. long; florets numerous, white or pink; achenes prominently 2–4-angled, 1·5 mm. long; pappus white, 1 mm. long.

Rather uncommon, in wet places; near SL–900 feet; fl. and fr. Jan–Aug. *A 10353! H 11905! P 8523!* Throughout tropical Amer., introduced into the Old World tropics.

50. VERNONIA Schreb. (1791) nom. cons.

1 Erect annual herbs:
  2 Capitula in flower about 1 cm. long; florets rose-coloured; involucral bracts terminated by a distinct green appendage narrower than the base; leaves serrate; introduced from India and formerly cultivated but not recently seen   *V. anthelmintica* Willd.
  2 Capitula in flower 5–6 mm. long; florets mauve to purple; involucral bracts integral, not appendaged; leaves sinuate-crenate to subentire; common weed   **1.** *cinerea*
1 Lax, scrambling or erect shrubs; capitula cymose, often spread along one side of divaricate inflorescence-branches or, if branches short, clustered; leaves entire:
  3 Pappus white; cymes elongated; leaves usually acute or acuminate and densely sericeous beneath   **2.** *divaricata*
  3 Pappus light or dull brown; cymes elongated or contracted; leaves puberulous beneath, often very thinly so:

4 Leaf-tips obtuse or rounded:
  5 Involucres 10–11 mm. long, turbinate at base, of about 7 (–10) series of involucral bracts, subsolitary in the axils of broad leafy bracts   3. *rigida*
  5 Involucres 7–8 mm. long, narrowly rounded at base, of 4–5 series of involucral bracts, solitary or more usually clustered in association with elliptical leafy bracts
    4. *harrisii*
4 Leaf-tips acute or acuminate:
  6 Leaves on main stem mostly in whorls of three, sometimes four; capitula in small clusters in the axils of large leaflike bracts, sometimes solitary; involucres turbinate with 6–7 series of bracts   5. *verticillata*
  6 Leaves on main stem not whorled; involucres campanulate with 4–5 series of bracts:
    7 Leaf-blade acuminate; capitula more or less spread along divaricate short or long bracteate branches   6. *acuminata*
    7 Leaf-blade acute; capitula more or less clustered   7. *pluvialis*

1. V. cinerea (L.) Less. in Linnaea **4**: 291 (1829).—*Conyza cinerea* L. (1753). *Senecioides cinerea* (L.) Kuntze (1903).

Erect taprooted pubescent herb up to 60 (–150) cm. high; leaves ovate to lanceolate, cuneate at base, sinuate-dentate to subentire, acute or obtuse, mostly 1·5–5 cm. long and 1–3 cm. broad; inflorescence-branches spreading; capitula pedunculate, 14–20-flowered; involucral bracts in 5 series, linear-lanceolate, acuminate; achenes brown, shortly bristly, 1·5 mm. long; pappus white, outer scales extremely short, inner bristles about 3 mm. long.

Very common, a weed of pastures and waste places; SL–2500 feet; fl. and fr. all the year. *A 9204! H 11793! C. B. Lewis!* General in the tropics; Grand Cayman.

2. V. divaricata Sw., Fl. Ind. Occ. **3**: 1319 (1806).—Fleabane.

Erect or lax shrub up to 2 m. high or scrambling to 3 m.; branches usually tomentose but indumentum very variable; leaves ovate-lanceolate, more or less rounded at base, acute or acuminate, 2·5–6·5 (–10) cm. long, 1·5–3·5 (–5) cm. broad, variable in texture and hairiness but usually scabrid and bullate adaxially; inflorescence mostly open with scorpioid branches; involucres 5–6 mm. long and broad with 5 rows of involucral bracts; florets 13–17; corolla white, mauve or reddish-purple, about 6 mm. long; achenes appressed-silky, about 1 mm. long; pappus 3 mm. long.

Common in many open situations; SL–5000 feet; fl. and fr. all the year. *A 6341! 8234! H 6993! P 6041!* Also in Grand Cayman.

3. V. rigida Sw., Fl. Ind. Occ. **3**: 1322 (1806).

Shrub with slender flexuous tomentose branches, up to 1·5 m. high; leaves broadly ovate to orbicular, 2–4·5 cm. long, 2–4 cm. broad, at first pubescent; inflorescence-branches typically elongated and leafy; florets about 12–15; corollas bright purple; achenes bristly, 2·5 mm. long; pappus-scales about 1 mm. long, setae 5–6 mm. long;

Rare and local (Trel.), in thickets on limestone; 1250–1400 feet; fl. and fr. Jan–Feb. *P 20579! R 1749!* Endemic.

4. V. harrisii S. Moore in Journ. Bot. **66**: 164 (1928).

Shrub with straggling brownish-tomentose branches up to 3 m. high; leaves ovate or oblong-ovate, rounded at both ends, glabrescent and shiny adaxially, up to 5·5 cm. long and 4 cm. broad; inflorescence-branches typically rather short; florets 9–10; corollas white, mauve or rosy-purple; achenes bristly, nearly 2 mm. long; pappus 4 mm. long.

Rare and local (Clar., Trel.), on exposed crags on wooded limestone hills; 2200–2500 feet; fl. and fr. mostly Dec–Jan. *A 11045! A & C 1260! H 12776! P 9766! Stearn 6!* Endemic.

5. V. verticillata Proctor ex Adams in Phytologia **21** (6): 409 (1971).

Shrub up to 2·5 m. high with spreading arching branches, sparsely puberulous; lower and inflorescence-leaves alternate; leaf-blades ovate to oblong-ovate, rounded at base, acuminate at tip, 3–7 cm. long, 1·5–5 cm. broad, darker and shiny adaxially; involucres about 8 mm. long, the outer bracts very short, the inner up to

about 6 mm. long; florets 7; corolla bright purple, 5 mm. long; achenes bristly, 1·5 mm. long; outer pappus-scales pale, 1–1·5 mm. long, inner setae 5–6 mm. long.
Rare and local (St. James), on rocky limestone hills; 1400–2200 feet; fl. and fr. Dec–Mar. *P 22987! 24715!* Endemic.

6. V. acuminata Less. in Linnaea **6**: 663 (1831).—*V. expansa* Gleason (1906).
Erect shrub 1–2 m. high or scrambling and climbing to 3–6 m., pubescent at least on young branches; leaves narrowly to broadly elliptical, cuneate at base, 3–8 cm. long, 1·5–2·5 (–3·5) cm. broad; petioles up to 5 mm. long, usually much shorter; involucres about 7 mm. long, broadly campanulate, the inner bracts oblong and about 5 mm. long; florets 11–15; corolla usually white, about 6 mm. long; achenes bristly, 1–1·5 mm. long; outer pappus-scales pale, nearly 1 mm. long, inner setae 3–4·5 mm. long.
Common in thickets and woodland margins on limestone; SL–5000 feet; fl. mainly Dec. but recorded fl. and fr. almost all the year. *A 8606! H 8205! P 7339!* Endemic.

7. V. pluvialis Gleason in Bull. Torr. Bot. Club **40**: 312 (1913).—*V. reducta* Gleason (1913).
Weak-stemmed erect or trailing shrub 0·5–3·5 m. high, shortly appressed-pubescent at least on young branches; leaves narrowly elliptical, cuneate at base, 3–7 cm. long, 1–2·5 cm. broad; petioles up to 3 mm. long; involucres 6–7 mm. long, narrowly campanulate, the inner bracts 4–5 mm. long; florets 5–12; corolla mauve fading white, 6–7 mm. long; achenes bristly 1·5 mm. long; outer pappus-scales 1 mm. long, inner setae 3–4 mm. long.
Local but not rare (St. Andr., Port., St. Thom.), in montane thickets; 4800–7400 feet; fl. and fr. June–Feb. *A 10682! H 11633! P 9206!* Endemic.

51. ACANTHODESMOS Adams & duQuesnay (1971)

1. A. distichus Adams & duQuesnay in Phytologia **21** (6): 405 (1971).
Shrub with spreading cottony-hairy glandular branches up to 1 m. long from near the base; leaves and axillary branches coplanar on each shoot; internodes about 2 cm. long; leaves linear-oblong, unequal at shortly petiolate base, remotely denticulate, long-acute, whitish-cottony beneath, 4–8 cm. long, 8–14 mm. broad, with a cluster of (1–) 3–5 spine-tipped bracteoles leaf-opposed at the base; capitula leaf-opposed, sessile, solitary or approximate on reduced axillary shoots; involucral bracts 15–16, spine-tipped, the outer up to 12 mm. long, in about 5 series; florets 12–13 accompanied by receptacle-scales resembling the inner involucral bracts; corolla 7·5 mm. long; achenes smooth, 1·7 mm. long; pappus-scales unequal, 0·7–1·3 mm. long.
Rare and local (Clar.), in thickets at salina margin; near SL; fl. and fr. Oct–May. *A 12999! 13056! P 31173!* Endemic.

52. TRIXIS P. Browne (1756)

1. T. inula Crantz, Inst. Rei Herb. **1**: 329 (1766).—*Inula trixis* L. (1759). *T. radialis* Kuntze (1891). *Perdicium radiale* L. (1763), illegitimate name.
Shrub with suberect, scrambling or climbing branches, pubescent at first, up to 2·5 m. high; leaves oblong-ovate to elliptic-lanceolate, obtuse or rounded at base, acute, mucronate, pubescent beneath, 4–10 cm. long, 2–3·5 cm. broad, remotely toothed; capitula about 1 cm. broad; involucral bracts 8, 8–10 mm. long; florets 10–12, unequally radiate; receptacle shortly scaly; achenes hairy, 4–6 mm. long; pappus-bristles 9 mm. long.
Local and uncommon (St. Andr., Clar.), in thickets on gravelly or rocky ground in rather arid areas; 50–900 feet; fl. and fr. Nov–June. *A 6962! 9949! H 6942! P 9457!* Mexico to Venezuela, Cuba, Hispaniola, Grenadines, Trinidad, Margarita.

53. CHAPTALIA Vent. (1802) nom. cons.

1 Leaves oblanceolate, gradually tapered to base; involucral bracts persistently silky-tomentose   **1.** *dentata*

1 Leaves lyrate, more or less abruptly narrowed above the base:

  2 Terminal segment of leaf 4–10 cm. long; involucral bracts persistently silky-tomentose   **2.** *nutans*

  2 Terminal segment of leaf 1–4 cm. long; involucral bracts glabrescent   **3.** *pumila*

1. C. dentata (L.) Cass. in Dict. Sci. Nat. **26**: 104 (1823).—*Tussilago dentata* L. (1763).

Rosette-herb; leaves subentire to shallowly runcinate, acute, white-tomentose beneath, 2–10 (–12) cm. long, 8–17 mm. broad; scape cottony-hairy at first, glabrescent, up to 35 cm. long; involucral bracts in 3 series, linear-lanceolate, innermost 9 mm. long; ligulate florets white or greenish-white turning light pink; achenes fusiform, puberulous, 3–4 mm. long, with beak 4–6 mm. long; pappus-bristles rough, reddish-brown, 9 mm. long.

Occasional on open clay banks and on shallow bauxitic soil on limestone; 250–3000 feet; fl. and fr. Oct–June. *A 7110! H 8856! P 18304!* Bahamas, Greater Antilles.

2. C. nutans (L.) Polak. in Linnaea **41**: 582 (1878).—Heal-and-Draw.

Rosette-herb; leaves repand-dentate, the terminal segment oblong-ovate, obtuse, white beneath, 5–14 (–20) cm. long, 2·5–5 cm. broad; scape tomentose, up to 70 cm. long; capitula nodding; involucral bracts in 3–4 series, linear-lanceolate, acuminate, innermost 15 mm. long; ligulate florets at first white turning pink then dark crimson; achenes fusiform, puberulous, 3 mm. long, long-beaked, spreading in a globose head at maturity; pappus-bristles rough, light brown, 11 mm. long.

Common on roadside banks, a weed of pastures and gardens on bauxitic and clay soils and in open grassy savannas; 500–4500 feet; fl. and fr. Aug–May. *A 7107! H 8752! P 18303!* S. United States to Argentina, West Indies.

3. C. pumila (Sw.) Fawcett, Prov. List Jam.: 22 (1893).

Rosette-herb; leaves runcinate to crenate or lobulate proximally, white beneath, 2–5 (–8) cm. long 7–12 (–14) mm. broad; scape glabrescent, filiform, 5–18 (–25) cm. long; involucral bracts in 3 series, linear, innermost 7 mm. long; florets white, becoming purplish; achenes fusiform, puberulous, 2·5 mm. long, the beak about 3·5 mm. long; pappus-bristles rough, light brown 4 mm. long.

Locally common (St. Andr., St. Thom.), on open or shaded earthy banks; 3000–6500 feet; fl. and fr. Dec–Apr. *A 6663! H 10033!* Cuba. Some plants (e.g. *A 10250!* from St. Thomas) show characters intermediate to *C. dentata*; others (e.g. *A 10641! P 23485!* from 6000–6500 feet; in St. Thomas) seem to be a montane vicariant of *C. pumila* and approach *C. crispata* Urb. & Ekman of Haiti. The latter have larger capitula with maroon-coloured phyllaries.

54. LAPSANA L. (1753)

1. L. communis L., Sp. Pl. **2**: 811 (1753).—Nipplewort.

Annual glabrous or hairy herb up to about 1 m. high; lower leaves lyrate-pinnatifid with a rotundate terminal lobe the proximal lobes much smaller, repand-toothed, 5–13 cm. long including petiole, 1–5 cm. broad; upper leaves ovate; capitula few, about 1 cm. long; involucral bracts 8; florets 8–12, yellow; achenes smooth, glabrous, slightly curved, about 4-ribbed with 3–4 minor ribs between each major pair, 3–3·5 mm. long.

Locally common (St. Andr., Port., St. Thom.), a weed of pathsides and roadsides; 2800–7000 feet; fl. and fr. most of the year. *A 7379! H 8587! P 21948!* Native of Europe, Asia and N. Africa, introduced and naturalized in Canada, United States and Hispaniola.

55. HYPOCHOERIS L. (1753)

Leaves glabrous or nearly so; involucral bracts glabrous, smooth; ligules hardly exceeding involucre in flower; involucre 12–15 mm. long   1. *glabra*
Leaves hispid; involucral bracts glabrous or minutely ciliolate, with at least the middle series having a short row of ascending teeth along the upper part of the midrib; ligules exceeding involucre; involucre 18–25 mm. long   2. *radicata*

1. H. glabra L., Sp. Pl. **2**: 811 (1753).—Smooth Cat's Ear.

Rosette-herb with a long-lived stock, probably not behaving as an annual in Jamaica; leaves linear, runcinate-pinnatifid, up to 10 cm. long and about 1 cm. broad; inflorescence 15–35 cm. high; involucral bracts green, tinged red at tips; florets yellow distally, whitish at base.

Local (St. Andr., St. Thom.), on pathsides, open banks and in grassy clearings; 3300–7400 feet; fl. and fr. Dec–Feb, June–Aug. *A 7393! 7444! P 23486!* Native of Europe, W. Asia and N. Africa, where reported as calciphobe; also in E. tropical Africa.

2. H. radicata L., Sp. Pl. **2**: 811 (1753).—*H. glabra* of S. Moore (1936), partly. Cat's Ear.

Rosette-herb like the last; leaves oblanceolate, sinuate-pinnatifid, up to 16 cm. long and 3 cm. broad; inflorescence 20–45 cm. high; involucral bracts dull green; florets yellow, the outer rays purplish abaxially.

Very local (St. Andr.), on moist roadside banks and a weed of lawns; 4000–5000 feet; fl. and fr. Dec–July. *A 12684! H 8581! P 15559!* Native of Europe, W. Asia and N. Africa. Other exsiccata, cited under *H. glabra* by S. Moore in Fawcett & Rendle, but which are also this species, are *J.P. 1221!* and *H 10927!*

56. TARAXACUM Weber (1780) nom. cons.

1. T. officinale Weber in Wiggers, Prim. Fl. Holsat.: 56 (1780).—Dandelion.

Perennial rosette-herb; leaves runcinate-pinnatifid, thin, up to 30 cm. long and 5 cm. broad; scape simple, up to 27 cm. long in flower, longer in fruit; capitula up to about 5 cm. in diameter in flower; florets yellow; achenes 3·5–4 mm. long, the beak 3–4 times longer, greenish-buff.

Rather uncommon, a weed of paths and roadsides on higher ground; 2600–5250 feet; fl. and fr. Dec–Sept. *A 5844! Bengry IJ 2336! H 10926!* Native in north temperate regions, now widely distributed; Greater Antilles.

57. LACTUCA L. (1753)

1. L. jamaicensis Griseb., Fl. Br. W.I.: 384 (1861).

Herb, usually erect, up to 2 m. high; roots swollen, fusiform; lower leaves usually coarsely toothed, up to 12 cm. long and 4 cm. broad; upper leaves subentire, smaller, amplexicaul; inflorescence ample; peduncles with small bracts; capitula about 1 cm. broad; inner involucral bracts oblong, outer much shorter, ovate, imbricate; florets light yellow or cream-coloured; achenes broadly elliptical with a prominent midrib on each face, about 4 mm. long excluding beak.

Occasional to locally common at higher elevations, mostly on disturbed ground; 300–4500 feet; fl. and fr. all the year. *A 8154! H 11010! P 21375!* Endemic.

L. sativa L., the cultivated Garden Lettuce, is grown extensively in Jamaica in several varieties.

58. SONCHUS L. (1753)

Achenes smooth, obovate-elliptical; leaf-base auricles more or less rounded   1. *asper*
Achenes rugulose, grooved, oblanceolate, narrowed to base; leaf-base auricles more or less pointed   2. *oleraceus*

1. S. asper (L.) Hill, Herb. Brit. **1**: 47, t. 34, f. 2 (1769).—*S. oleraceus* var. *asper* L. (1753).

Annual herb 30–90 cm. high; leaves strongly spine-toothed, undulate, glossy adaxially, the upper mostly subentire, the lower with a narrow terminal lobe, up to about 15 cm. long and 4 cm. broad; capitula about 1 cm. broad; involucral bracts glabrescent; florets light yellow; achenes with a few downward pointing hairs, about 2 mm. long.

Uncommon (St. Andr., St. Thom.), a weed of open waste places in the mountains; 3900–7400 feet; fl. and fr. Jan, Apr, July. *A 8781! SCH 913! J.P. 934!* Native of Eurasia, now cosmopolitan.

2. S. oleraceus L., Sp. Pl. **2**: 794 (1753).—Sow-thistle.

Annual herb like the last, 60–150 cm. high; leaves entire to deeply pinnatifid, the lower with a broad terminal lobe, sharply dentate, up to 20 cm. long and 7 cm. broad; florets light yellow or outer florets tinged mauve in midline; achenes 2·5–3 mm. long.

Rather common, a weed of disturbed ground and waste places; 10–5500 feet; fl. and fr. Sept–May. *A 5790! H 11974! H & P 13309!* Native of Eurasia and N. Africa, now cosmopolitan.

59. LAUNAEA Cass. (1822)

1. L. intybacea (Jacq.) Beauverd in Bull. Soc. Bot. Genève, sér. 2, **2**: 114 (1910).—
Lactuca intybacea Jacq. (1784). Wild Lettuce.

Glabrous erect annual herb up to 2 m. high; lower cauline leaves coarsely toothed, oblanceolate, up to 20 cm. long and 7 cm. broad; upper leaves smaller; inflorescence with many ascending branches; peduncles bracteate; outer involucral bracts ovate, acute, inner linear-lanceolate, about 10 mm. long; florets light yellow; achenes 3·5–4 mm. long.

Occasional, a weed of sandy or gravelly waste places in drier areas; SL–800 feet; fl. and fr. Dec–Feb. *A 8700! 8765! H 12900! A. M. Barry!* Florida, Mexico, Venezuela, Bahamas, Greater Antilles, Virgin Is., Antigua, Barbados, Trinidad, introduced and naturalized throughout subtopical and tropical Africa and extending to India.

60. CREPIS L. (1753)

1. C. japonica (L.) Benth., Fl. Hongk.: 194 (1861).

Annual herb, 5–50 cm. high in flower; leaves mostly basal, lyrate-pinnatifid, long-petioled, oblanceolate with a large terminal lobe, up to 18 cm. long and 5 cm. broad; inflorescence scapose, much branched above; capitula about 5 mm. broad; florets light yellow, often tinged red and fading to orange; achenes 10-ribbed, about 1·5 mm. long.

A common weed of shady paths, roadsides and waste places; 350–4500 feet; fl. and fr. most of the year. *A 5756! H 12909! P 16007!* Native of eastern Asia; Cuba, Hispaniola.

APPENDIX I

Aids to the Determination of Families

1. KEY TO THE FAMILIES OF MONOCOTYLEDONS

THE sequence of orders and families of monocotyledons in this Flora is based on that of John Hutchinson as set out in his *Families of Flowering Plants*, 2nd edition, 1959. It is appropriate that adapted versions of his keys should be used for the identification of families in Jamaica.

Key to the artificial groups of monocotyledon families

Ovary superior:
  Carpels free or only slightly united at the base, or gynoecium reduced
    to 1 carpel with 1 stigma; all aquatic (except some palms)   Group 1
  Carpels more or less completely united with usually more than 1 stigma:
    Perianth composed of dissimilar calyx and corolla, the former often
      green, the latter usually larger and white or coloured, but never
      united into a single tube, or both series dry and hyaline   Group 2
    Perianth composed of similar segments in one or two series, usually
      conspicuous and all petaloid, free or if united then connate in the
      proximal part into a single tube, or small and inconspicuous when
      inflorescence a spadix (*Araceae*)   Group 3
    Perianth sepaloid, dry, glumaceous or represented by " hypogynous
      setae ", " scales " or " lodicules ", or absent; flowers minute and
      arranged in the axils of scaly bracts (glumes) in spikelets, or in
      spadices or panicles, often subtended by spathes or leafy bracts   Group 4
Ovary inferior or semi-inferior:
  Perianth composed of separate calyx and corolla remaining in two
    distinct series   Group 5
  Perianth composed of similar or mostly similar usually petaloid seg-
    ments free or united at the base into a single tube   Group 6

Group 1

Flowers bracteate:
  Aquatic or semi-aquatic herbs:
    Carpels several, often whorled; tufted herbs rooted in soil or mud
          2. **Alismataceae** (p. 32)
    Carpel solitary, fused to bract (spathe); floating rosette-herb
          21. **Araceae** (p. 68)
    Trees; carpels 3; leaves folded in bud   25. **Palmae** (p. 73)
  Flowers ebracteate; aquatic herbs:
    Marine or brackish-water plants:
      Flowers spicate, bisexual   4. **Ruppiaceae** (p. 34)
      Flowers axillary or cymose, monoecious or dioecious:
        Ovule apical, pendulous; marine perennials   5. **Zannichelliaceae** (p. 34)
        Ovule basal, erect; brackish-water annual   6. **Najadaceae** (p. 35)
    Freshwater plants:
      Plant-body minute without distinction of stem and leaf; stamen 1–2; carpel 1
          22. **Lemnaceae** (p. 71)

Plant-body of distinct stem and leaves:
Flowers clustered in the branch-axils or solitary; stamen 1; carpel 1
6. **Najadaceae** (p. 35)
Flowers in terminal or axillary spikes:
Spikes axillary; flowers bisexual; stamens 4; carpels 4
3. **Potamogetonaceae** (p. 33)
Spikes terminal, densely cylindrical; flowers unisexual; stamens 2–5; carpel 1
23. **Typhaceae** (p. 72)

Group 2

Flowers very small in heads:
Flowers bisexual, involucrate; stamens 3; sepals unequal; ovary 1-locular with 3
parietal placentas   9. **Xyridaceae** (p. 40)
Flowers unisexual, monoecious; ovary 2–3-locular with 1 pendulous ovule in
each loculus   10. **Eriocaulaceae** (p. 41)
(*Eriocaulaceae* have not been reported from Jamaica)
Flowers in cymes, panicles, racemes or spikes, rarely solitary, not capitate:
Leaves folded in bud; flowers usually in panicles with large spathaceous bracts
25. **Palmae** (p. 73)
Leaves not folded in bud:
Ovary 1-locular; placentation parietal; stamens 3; aquatic herbs with very
narrow leaves bidentate at apex   8. **Mayacaceae** (p. 40)
Ovary 2–3-locular; placentation axile or basal; leaf-sheaths closed:
Ovules 1 to few; fruit usually a capsule; flowers cymose or solitary, often
enclosed in a spathe in bud and fruit   7. **Commelinaceae** (p. 36)
Ovules numerous; fruit fleshy and indehiscent or a capsule with appendaged
seeds; mostly epiphytes   11. **Bromeliaceae** (p. 41)

Group 3

Flowers arranged in a scapose umbel, subtended by spathaceous bracts; plant
usually bulbous; leaves radical   27. **Amaryllidaceae** (p. 77)
Flowers otherwise or if subumbellate or capitate bracts not spathaceous:
Flowers in a spadix subtended by or enclosed in a spathe, very small; ovary
1–2- or more-locular   21. **Araceae** (p. 68)
Flowers not in a spadix; ovary usually of 3 carpels, rarely of 1 carpel:
Aquatic herbs; inflorescence subtended by a spathelike leaf-sheath; flowering
branches bearing only 1 leaf   18. **Pontederiaceae** (p. 63)
Terrestrial herbs or shrubs, sometimes climbing:
Flowers unisexual, dioecious, small; stems climbing or straggling, often prickly;
leaves with reticulate venation; fruit a berry   19. **Smilacaceae** (p. 65)
Flowers bisexual; leaves with parallel veins; fruit a capsule or berry:
Stamens 6   17. **Liliaceae** (p. 61)
Stamens 3   30. **Haemodoraceae** (p. 83)

Group 4

Leaves folded in bud, with strong parallel nerves, often pinnately or flabellately
divided or nerved:
Perianth-segments 4 (–0); ovary free or sunken in the axis; flowers monoecious
in a spadix   24. **Cyclanthaceae** (p. 72)
Perianth-segments 6 in two distinct series; mostly tall plants with simple or
little-branched stems; flowers small in panicles or spikes   25. **Palmae** (p. 73)
Leaves not folded in bud:
Leaves not grasslike; inflorescence subtended by a spathaceous bract; flowers
bisexual or monoecious:
Terrestrial or rarely aquatic plants with well developed leaves
21. **Araceae** (p. 68)
Floating plants with a minute plant-body not differentiated into stem and leaves;
flowers very small; perianth wanting   22. **Lemnaceae** (p. 71)

Leaves linear (oblong-lanceolate or ovate in some grasses):
Perianth present; flowers bisexual; stamens 3 or 6   34. **Juncaceae** (p. 132)
Perianth absent or represented by " hypogynous setae ", " scales " or
" lodicules ":
Trees or shrubs often with aerial or prop roots; flowers dioecious, capitate or
spicate; leaves keeled with spiny margins   26. **Pandanaceae** (p. 76)
Herbs (including bamboos in *Gramineae*):
Flowers in a terminal spadix, monoecious; leaves smooth
    23. **Typhaceae** (p. 72)
Flowers in the axils of scaly bracts (glumes) in spikelets; fruit indehiscent:
Flowers in the axil of a single bract and collected into spikelets, the latter
variously arranged, from solitary to umbellate, paniculate or capitate and
often with leafy bracts subtending; leaves usually with closed sheaths;
stems mostly solid and triquetrous; embryo erect, free from the pericarp
    35. **Cyperaceae** (p. 133)
Flowers enclosed in a bract and bracteole (lemma and palea); arranged in
spikelets without leafy bracts; leaves usually with open sheaths; stems
mostly terete with hollow internodes; embryo usually adnate to the pericarp
    36. **Gramineae** (p. 158)

Group 5

Inner perianth zygomorphic:
Stamens 1 (2, or very rarely 3) inserted on a prolongation of the floral axis
(column), often with agglutinated pollen; ovary sometimes spirally twisted;
seeds without endosperm, very small   33. **Orchidaceae** (p. 87)
Stamens 3 or more; ovary not twisted; seeds with endosperm; pollen granular:
Stamens 3   31. **Iridaceae** (p. 83)
Stamens 5 or 6:
Leaves and bracts spiral   12. **Musaceae** (p. 54)
Leaves and bracts distichous   13. **Strelitziaceae** (p. 54)
Inner perianth actinomorphic:
Stamens 3 or more; petaloid staminodes absent:
Calyx tubular, soon split down one side, 3–5-dentate; bracts large and
spathaceous, deciduous, coloured; flowers unisexual   12. **Musaceae** (p. 54)
Calyx actinomorphic, lobes imbricate; flowers bisexual:
Stamens 6; bracts often coloured   11. **Bromeliaceae** (p. 41)
Stamens 3; bracts usually membranous, green   31. **Iridaceae** (p. 83)
Stamen 1, the remainder transformed into petaloid staminodes often more
conspicuous than the inner perianth and often asymmetrical:
Anthers 2-locular; sepals united into a sometimes spathaceous tube; ovules
numerous; embryo straight   14. **Zingiberaceae** (p. 55)
Anthers 1-locular; sepals free or connivent:
Ovules numerous in each loculus; embryo straight; petiole without a callus
    15. **Cannaceae** (p. 58)
Ovules solitary in each loculus; embryo curved; petiole with a pulvinate callus
at junction with blade   16. **Marantaceae** (p. 60)

Group 6

Ovules spread all over the inner walls of the carpels or on intrusive septa; flowers
borne in spathaceous bracts; seeds without endosperm; submarine aquatic (or
cultivated in fresh water)   1. **Hydrocharitaceae** (p. 31)
Ovules borne on placentas or at base or apex of the ovary; land plants:
Stamens 1 (or 2); perianth zygomorphic; ovary sometimes twisted; seeds without
endosperm; epiphytic or terrestrial (rarely saprophytic) herbs
    33. **Orchidaceae** (p. 87)
Stamens 3, 6 or more:
Stems climbing; leaves usually broad; stems or roots more or less tuberous;
flowers small, unisexual; fruit 1- or 3-winged   20. **Dioscoreaceae** (p. 66)

Stems not climbing; flowers bisexual:
Inflorescence scapose, umbellate or rarely flower solitary, subtended by an involucre of one or more spathaceous bracts; stamens 6
27. **Amaryllidaceae** (p. 77)
Inflorescence not as above, if appearing umbellate then not involucrate:
Stamens 3; perianth more or less actinomorphic:
Small saprophytic herbs; ovary and fruit often winged
32. **Burmanniaceae** (p. 86)
Herbs with well developed green leaves and thick rhizomes or corms
31. **Iridaceae** (p. 83)
Stamens 6:
Inflorescence scapose, radical, with flowers solitary to spicate or racemose; perianth without a tube (but sometimes the ovary long-beaked and resembling a narrow tube); leaves strongly parallel-veined, often plicate
29. **Hypoxidaceae** (p. 82)
Inflorescence terminal, long-racemose, spicate or paniculate, sometimes very large; perianth usually with a distinct tube; leaves stiff, sometimes with prickly margins and rather indistinct veins   28. **Agavaceae** (p. 80)

2. REFERENCES TO KEYS TO THE FAMILIES OF DICOTYLEDONS AND NOTES ON VEGETATIVE CHARACTERS

THE problems of preparing a key to the families of dicotyledons have received much thought during the preparation of this Flora. There were two main alternatives; one was to reproduce an existing key, perhaps in a modified form, such as has been done for the monocotyledons; the other was to prepare a completely new key based on local representatives of the families. Although the usefulness of some sort of key for the identification of dicotyledon families is undeniable, the allocation of space for it in what is already a large book, is a cogent objection. Moreover, in regard to the first alternative, published keys to families are available in several standard horticultural and botanical works and some of these are listed below; the two most recent of the listed publications were written specially for the purpose of meeting this need and both are quite modestly priced.

In regard to the second alternative of constructing a special key to the families of dicotyledonous flowering plants as they are represented in Jamaica, it has been found that, to be really useful to the field worker or the main body of more or less inexperienced botanists, the arrangement would have to be an almost totally artificial one. This entails assemblage of facts in ways largely inconsistent with formal taxonomic treatments (for example as in the key to the genera of *Compositae* used in this work), and therefore requiring much more trial and testing than could be sensibly devoted to this purpose at this time. In brief, the construction of such a key is a major undertaking for the future.

Study of the flora of Jamaica as a whole, and particularly in the field where representatives of many different species are encountered in quite small areas, reveals the outstanding characteristic that a very small proportion of the plants can be identified immediately by applying formal methods. Most of the species seen in one area at one time are neither in flower nor fruit and of those that have any reproductive parts present, only a few will have both flower and fruit. This means that, however adequate facilities for identification may be, getting to know the flora is a piecemeal process which may take a long time.

Apart from asking someone who already knows the botanical name or the family of a particular plant, and apart from a process of cross-referring from common names, a great deal can be achieved by learning the vegetative features by means of which many of the larger or commoner families may be recognized. These features include the following main characters of which the expression of any one, or the association of several, may greatly assist the recognition of a taxonomic group: leaves simple or compound; leaves opposite or alternate; leaves present or absent (allowing for deciduousness); leaf-blade entire or divided or toothed;

leaves with or without superficial or pellucid glands; stipules present or absent (allowing for early fall); presence or absence of spines, milky sap, aromatic principals, caustic or urticating principals; types of hairs, scales or other forms of indumentum. Some or all of these characters can be useful for identification at all taxonomic levels as study of the keys to the genera and species testify, and a few simple examples will show how important families can also be recognized or suspected:

Leaves simple, opposite, entire, with interpetiolar stipules   *Rubiaceae* or *Rhizophoraceae*
Leaves simple, opposite, entire or toothed, with subparallel longitudinal veins and
    stipules   *Urticaceae*
Leaves simple, opposite, entire or toothed, with subparallel longitudinal veins, without
    stipules   *Melastomataceae*
Leaves simple, opposite with pellucid glands
                                                    Myrtaceae or *Compositae* (*Eupatorium* in part)
Leaves simple, opposite, with cystoliths
                                    Acanthaceae (part), *Urticaceae* (part), *Rubiaceae* (part)

Where flowers and fruits are also available the following books may be consulted.

Bailey, L. H., 1947. The Standard Cyclopedia of Horticulture. Macmillan, New York.
——, 1949. Manual of Cultivated Plants. Macmillan, New York.
Davis, P. H. and Cullen, J., 1965. The Identification of Flowering Plant Families. Oliver & Boyd, Edinburgh and London.
Hutchinson, J., 1959. The Families of Flowering Plants, I. Dicotyledons ed. 2. Oxford.
——, 1967. Key to the Families of Flowering Plants of the World. Oxford.
Lawrence, G. H. M., 1951. Taxonomy of Vascular Plants. Macmillan, New York.
Little, E. L. and Wadsworth, F. H., 1964. Common Trees of Puerto Rico and the Virgin Is. U.S.D.A., Forest Service, Washington, D.C.

Very useful modern family keys for tropical Floras have also been published in:

Flora Zambesiaca 1 (1): 55–76, 1960. [Key to families and higher groups by J. E. Dandy] Crown Agents, London.
Flora of West Tropical Africa 2: 474–495, 1963. [Key to the families of dicotyledons by J. E. Dandy] Crown Agents, London.

APPENDIX II

Principal Taxonomic References

1. REFERENCES TO FLORAS AND CONSULTED WORKS OF A GENERAL NATURE CONCERNING THE FLORA AND VEGETATION OF JAMAICA AND SOME OF THE OTHER WEST INDIAN ISLANDS PUBLISHED SINCE 1936

THE volumes of the Fawcett & Rendle, Flora of Jamaica contain a progressive bibliography up to and including 1936, the date of publication of Vol. 7 of that work.

Adams, C. D., Magnus, K. E. and Seaforth, C. E., 1963. Poisonous plants in Jamaica. Department of Extra-mural Studies, University of the West Indies, Mona, Jamaica.
——, 1966. A checklist of the orchids of Jamaica. Amer. Orch. Soc. Bull. 35 (12): 995–999.
——, 1969. A botanical description of Big Pelican Cay. Atoll Res. Bull. 130.
——, Kasasian, L. and Seeyave, J., 1970. Common weeds of the West Indies. University of the West Indies, St. Augustine, Trinidad.
——, and duQuesnay, M. C. in Woodley, J. D. (editor), 1971. Vegetation, pp. 48–119. Hellshire Hills Scientific Survey 1970. University of the West Indies, Mona, Jamaica and Institute of Jamaica.
Alain, Hno., 1953–57. Flora de Cuba 3–4. Habana, Cuba.
——, 1963. Flora de Cuba 5. Rio Piedras, Puerto Rico.
Arnoldo, Fr. M., 1954. Zakflora (Pocket Flora of Curaçao, Aruba and Bonaire). Willemstad, Curaçao.
——, 1954, Gekweekte en nuttige planten van de Nederlandse Antillen. Willemstad, Curaçao.
Asprey, G. F. and Robbins, R. G., 1953. The vegetation of Jamaica. Ecol. Monogr. 23: 359–412.
—— and Thornton, P., 1953–55. Medicinal plants of Jamaica (4 parts). West Indian Medic. Journ. 2 (4), 3 (1), 4 (2–3).
—— and Loveless, A. R., 1958. The dry evergreen formations of Jamaica II. The raised coral beaches of the north coast. Journ. Ecol. 46: 547–570.
Beard, J. S., 1949. The natural vegetation of the Windward and Leeward Is. Oxford For. Mem. 21.
——, 1949. Forest trees of the Windward and Leeward Is. Unpubl. tss. (Herb. UCWI).
Carroll, E. and Sutton, S., 1965. A cumulative index to the 9 vols. of Symbolae Antillanae. Arnold Arboretum of Harvard University, Jamaica Plain, Mass.
Chapman, V. J., 1944. The 1939 Cambridge University Expedition to Jamaica, Part I. A study of the botanical processes concerned in the development of the Jamaican shore-line. Journ. Linn. Soc. London 52: 407–447.
D'Arcy, W. G., 1967. Annotated checklist of the dicotyledons of Tortola, Virgin Is. Rhodora 69 (No. 780): 385–450.
Eyre, A., 1966. The botanic gardens of Jamaica. Andre Deutsch, London.
Gooding, E. G. B., Loveless, A. R. and Proctor, G. R., 1965. Flora of Barbados. H.M.S.O., London.
Hitchcock, A. S., 1936. Manual of the grasses of the West Indies. U.S.D.A., Washington, D.C.

Hodge, W. H., 1954. Flora of Dominica, B.W.I., Part I. Lloydia 17 (1–3).

Howard, R. A. and Proctor, G. R., 1957. The vegetation on bauxite soils in Jamaica. Journ. Arn. Arb. 38: 1–41, 151–169.

Jiménez, J. de J., 1966. Catalogus Florae Domingensis, Supl. I. Arch. Bot. Biogeog. Ital. 39–43 (1963–67).

León, Hno., 1947. Flora de Cuba 1. Habana, Cuba.

—— and Alain, Hno., 1951. Flora de Cuba 2. Habana, Cuba.

Little, E. L. and Wadsworth, F. H., 1964. Common trees of Puerto Rico and the Virgin Is. U.S.D.A., Forest Service, Washington, D.C.

Loveless, A. R. and Asprey, G. F., 1957. The dry evergreen formations of Jamaica I. The limestone hills of the south coast. Journ. Ecol. 45: 799–822.

Moscoso, R. M., 1943. Catalogus Florae Domingensis, Part I. Spermatophyta. New York.

Proctor, G. R., 1961. Practical problems confronting the writers of Floras; our knowledge of the flora of the West Indies. Rec. Adv. Bot. 9: 929–932.

——, 1964. The vegetation of the Black River morass. Black River morasses reclamation project, Appendix H. Ministry of Agriculture and Lands, Jamaica.

Questel, A., 1941. The Flora of St. Bartholomew. Basse-Terre, Guadeloupe.

Shreve, F., 1942. The vegetation of Jamaica. Chron. Bot. 7 (4): 164–166.

Stearn, W. T., 1959. A botanist's random impressions of Jamaica. Proc. Linn. Soc. London 170 (2): 134–147.

——, 1965. Grisebach's Flora of the British West Indian Islands: A biographical and bibliographical introduction. Journ. Arn. Arb. 46 (3): 243–285.

Steers, J. A., 1940. The cays and the Palisadoes, Port Royal, Jamaica. Geog. Rev. 30 (2): 279–296.

——, 1940. The coral cays of Jamaica. Geog. Journ. 95 (1): 30–42.

—— and Chapman, V. J. et al., 1940. Sand cays and mangroves in Jamaica. Geog. Journ. 96 (5): 305–328.

Stehlé, H. and M. and Quentin, L., 1937. Flore de la Guadeloupe et Dépendences 2 (1). Basse-Terre, Guadeloupe; ibid., 1948–49, 2 (2–3). Montpellier, France.

——, 1939. Flore descriptive des Antilles françaises. 1. Les Orchidales. Fort-de-France, Martinique.

Storer, D. P., 1958. Familiar trees and cultivated plants of Jamaica. London and Institute of Jamaica, Kingston.

Swabey, C., 1941. The principal timbers of Jamaica. Dept. Sci. & Agric. Bull. 29, new series. Kingston, Jamaica.

Trinidad and Tobago, Flora of, 1928–. 3 Vols. in progress. Department of Agriculture (later Ministry of Agriculture), Port of Spain, Trinidad.

Vélez, I. and Overbeek, J. van, 1950. Plantas indeseables en los cultivos tropicales. Rio Piedras, Puerto Rico.

Walter, M., 1934. Some Jamaica wild flowers. Kingston, Jamaica.

——, 1938. Some more Jamaica wild flowers.

Williams, R. O. and Williams, R. O., 1951. The useful and ornamental plants of Trinidad and Tobago, 4th ed. Port of Spain, Trinidad.

2. REFERENCES TO MONOGRAPHS AND REVISIONS

THIS list of references to publications is not exhaustive. Additional references to revisions may be sought in the citations of species in the text. Although these works have been consulted and are hereby acknowledged, this author has not necessarily agreed in all respects with the revisions nor always accepted a classification. Numbers and names in parentheses refer to the families and genera in this volume.

Adams, C. D., 1970. Mitt. Bot. Staatss. München 8: 99–110 (48.1 Pilea).

Airy Shaw, A. K. and Forman, L. L., 1967. Kew Bull. 21 (1): 1–19 (112.3 Spondias).

Alston, A. H. G., 1925. Kew Bull. 1925: 241–276 (139.2 Cassipourea).

Benson, L., 1941. Amer. Journ. Bot. 28: 748–754 (94.13 Prosopis).

Blake, S. F., 1925. Amer. Journ. Bot. 12: 625–640 (183.15 Verbesina).

Borssum Waalkes, J. van, 1966. Blumea 14 (1): 1–251 (122.7 Sida, etc.).

Bunting, G. S., 1968. Gent. Herb. 10 (2): 136–168 (21.2 Philodendron).

Burch, D., 1966. Ann. Miss. Bot. Gard. 53: 90–99 (109.32 Euphorbia-Chamaesyce).

Cheesman, E. E., 1947. Kew Bull. **1947**: 106–117 (12.1 *Musa*).
——, 1948. Kew Bull. **1948**: 145–153 (12.1 *Musa*).
Clayton, W. D., 1965. Kew Bull. **19** (2): 287–296 (36.21 *Sporobolus*).
Clement, I. D., 1957. Contrib. Gray Herb. **180**: 1–91 (122.7 *Sida*).
Coursey, D. G., 1967. Yams. Longmans, Green & Co., London (20.1 *Dioscorea*).
Cowan, R. S., 1968. Fl. Neotrop. Monogr. **1**: 1–228 (93.1 *Swartzia*).
Cristobal, C. L., 1960. Opera Lilloana **4**: 1–230 (124.3 *Ayenia*).
Cronquist, A., 1944. Brittonia **5**: 133–137 (104.4 *Alvaradoa*).
Den Hartog, C., 1964. Blumea **12** (2): 289–312 (5.1 *Halodule*).
——, 1970. Verhand. K. Nederl. Akad. Wet., Natuurk. ser. 2, 59 (1): 1–275 (1.4–5 *Thalassia, Halophila*; 5.1–2 *Halodule, Syringodium*).
Dressler, R. L., 1957. Contrib. Gray Herb. **182**: 1–188 (109.33 *Pedilanthus*).
——, 1962. Ann. Miss. Bot. Gard. **48**: 329–341 (109.32 *Euphorbia–Poinsettia*).
Drummond, J. R., 1907. Rep. Miss. Bot. Gard. **18**: 25–75 (28.2 *Furcraea*).
Epling, C., 1938–39. Fedde, Repert. Sp. Nov., Beih. **110**: 1–160, **112**: 161–380 (169.7 *Salvia*).
—— and Játiva, C., 1966. Brittonia **18** (3): 244–248 (169.6 *Satureja*).
Fries, R. E., 1908. Kungl. Svensk. Vet. Akad. Handl. **43** (4): 1–114 (122.4 *Wissadula*).
Gale, S., 1944. Rhodora **46**: 89–134, 159–197, 207–249, 255–278 (35.1 *Rhynchospora*).
Goldberg, A., 1967. Contrib. U.S. Nat. Herb. **34**: 191–362 (124.4 *Melochia*).
Graham, S. A., 1964. Journ. Arn. Arb. **45** (3): 286–293 (139.1 *Rhizophora*).
Harling, G., 1958. Acta Hort. Berg. **18**: 1–428 (24.1 *Carludovica*).
Henrard, J. T., 1950. Monogr. Digitaria. Leiden (36.43 *Digitaria*).
Hermann, F. J., 1957. Baileya **5** (4): 149–151 (7.2 *Setcreasea*).
Hespenheide, H. A., 1968. Proc. Acad. Nat. Sci. Philad. **120** (1): 1–23 (33.39 *Lepanthes*).
Holttum, R. E., 1955. Malayan Agri.-Hort. Assoc. Mag. **12** (2): 6–7 (58.4 *Bougainvillea*).
Howard, R. A., 1949. Bull. Torr. Bot. Club **76**: 89–100 (122.13 *Thespesia*).
——, 1957. Journ. Arn. Arb. **38**: 81–106 (54.4 *Coccoloba*).
Hunziker, A. T., 1969. Kurtziana **5**: 101–179 (170.12 *Witheringia*).
Hutchinson, J., 1920. Kew Bull. **1920**: 275–281 (82.2 *Bocconia*).
Hutchinson, J. B., 1943. Trop. Agric. **20** (3): 56–58 (122.14 *Gossypium*).
Iltis, H. H., 1960. Brittonia **12**: 279–294 (83.1 *Cleome*).
——, 1967. Amer. Journ. Bot. **54**: 953–962 (83.1 *Cleome*).
Ingram, J., 1966. Gent. Herb. **10** (1): 1–38 (109.10 *Argythamnia*).
Jablonski, E., 1968. Phytologia **16**: 393–434 (109.20 *Sapium*).
Jarrett, F. M., 1959. Journ. Arn. Arb. **40**: 113–155, 298–326, 327–368 (47.7 *Artocarpus*).
Johnston, M. C., 1964. Amer. Journ. Bot. **51**: 1113–1118 (119.6 *Ziziphus*).
Kearney, T. H., 1954a. Leafl. West. Bot. **7** (6): 138–150 (122.7 *Sida*).
——, 1954b. Leafl. West. Bot. **7** (5): 122–130 (122.11 *Pavonia*).
Killip, E. P., 1938. Field Mus. Bot. **19**: 1–613 (130.1 *Passiflora*).
Kimnach, M., 1961. Cact. and Succ. Journ. Amer. **33**: 11–16 (64.4 *Disocactus*).
Kobuski, C. E., 1941a. Journ. Arn. Arb. **22**: 457–496 (77.2 *Freziera*).
——, 1941b. Journ. Arn. Arb. **22**: 395–416 (77.4 *Cleyera*).
——, 1943. Journ. Arn. Arb. **24**: 60–76 (77.3 *Ternstroemia*).
——, 1949. Journ. Arn. Arb. **30**: 166–186 (77.1 *Laplacea*).
Kopp, L. E., 1966. Mem. New York Bot. Gard. **14** (1): 1–117 (68.6 *Persea*).
Kostermans, A. J. G. H., 1957. Reinwardtia **4**: 227 (68.7 *Cinnamomum*).
——, 1961. Reinwardtia **6**: 17–24 (68.7 *Cinnamomum*).
Krukoff, B. A., 1939. Brittonia **3**: 205–337 (95.26 *Erythrina*).
——, 1969. Phytologia **19** (3): 113–175 (95.26 *Erythrina*).
Kuijt, J., 1961. Wentia **6**: 1–145 (51.6 *Dendrophthora*).
Kükenthal, G., 1949. Engl., Bot. Jahrb. **74** (3): 375–509 (35.1 *Rhynchospora*).
Lee, B., 1966. Sci. Notes and News (Jamaica) **2** (4): 14–15 (183.33 *Emilia*).
Lourteig, A., 1952. Not. Syst. **14**: 234–248 (8.1 *Mayaca*).
McLaughlin, A. D., 1944. Catholic Univ. Amer., Biol. Stud. **5**: 1–108 (35.4 *Cyperus*).
McVaugh, R., 1943. N. Amer. Fl. **32A**: 1–134 (181.1 *Lobelia*, etc.).

Merrill, E. B., 1934. Contrib. Arn. Arb. **8**: 47 (47.5 *Poikilospermum*).
Munz, P. A., 1943. Proc. Calif. Acad. Sci. ser. 4, **25**: 1–138 (143.3 *Fuchsia*).
——, 1965. N. Amer. Fl. ser. 2, **5**: 1–278 (143.1 *Ludwigia*, etc.).
Norman, E. M., 1966. Gent. Herb. **10** (1): 47–114 (158.1 *Buddleja*).
Ogden, E. C., 1943. Rhodora **45**: 57–105, 119–163, 171–214 (3.1 *Potamogeton*).
Ornduff, R., 1969. Brittonia **21** (4): 346–352 (160.1 *Nymphoides*).
Paclt, J., 1952. Candollea **13**: 241–285 (172.3 Catalpa).
Pfeifer, H. W., 1966. Ann. Miss. Bot. Gard. **53**: 115–196 (53.1 *Aristolochia*).
Philcox, D., 1965. Kew Bull. **18** (2): 275–315 (171.24 *Buchnera*).
Porter, D. M., 1969. Contrib. Gray Herb. **198**: 1–153 (102.3 *Kallstroemia*).
Powell, A. M., 1965. Brittonia **17** (1): 47–96 (183.1 *Tridax*).
Raven, P. H., 1963. Reinwardtia **6**: 327–427 (143.1 *Ludwigia*).
Read, R. W., 1969. Brittonia **21** (1): 83–90 (11.5 *Pitcairnia*).
Rechinger, K. H. Jr., 1937. Field Mus. Bot. **17**: 1–151 (54.3 *Rumex*).
Rogers, C. M., 1963. Brittonia **15**: 97–122 (98.1 *Linum*).
Rogers, D. J., 1949. Ann. Miss. Bot. Gard. **36**: 475–477 (59.1 *Stegnosperma*).
Rominger, J. M., 1962. Univ. Ill. Biol. Monogr. **29**: 1–132 (36.39 *Setaria*).
Sauer, J., 1964. Brittonia **16** (2): 106–181 (95.23 *Canavalia*).
Schery, R. W., 1942. Ann. Miss. Bot. Gard. **29**: 183–244 (122.12 *Malvaviscus*).
Schneider, C., 1918. Bot. Gaz. **65**: 1–41 (43.1 *Salix*).
Schulz, O. E., 1904. Symb. Ant. **5**: 17–47 (19.1 *Smilax*).
Sealy, J. R., 1954. Kew Bull. **1954**: 201–240 (27.6 *Hymenocallis*).
Seifriz, W., 1920. Amer. Journ. Bot. **7**: 83–94 (36.2 *Chusquea*).
Simmonds, N. W., 1959. Bananas. Longmans, Green & Co., London (12.1 *Musa*).
——, 1962. The Evolution of the Bananas. Longmans, Green & Co., London (12.1 *Musa*).
Sleumer, H., 1967. Engl., Bot. Jahrb. **87**: 36–175 (148.1 *Clethra*).
Smith, A. C., 1944. N. Amer. Fl. **28B**: 3–41 (145.3 *Dendropanax*).
Smith, C. E., 1960. Fieldiana Bot. **29** (5): 292–341 (106.2 *Cedrela*).
Smith, L. B., 1960. Phytologia **7** (5): 249–257 (11.4 *Hohenbergia*).
Stearn, W. T., 1958. Kew Bull. **1958**: 35 (168.1 *Avicennia*).
——, 1968. Journ. Arn. Arb. **49**: 280–289 (154.3 *Bumelia*).
——, 1969a. Bull. Brit. Mus. (Nat. Hist.) Bot. **4** (4): 143–178 (151.1–3 *Myrsine, Ardisia, Wallenia*).
——, 1969b. Bull. Brit. Mus. (Nat. Hist.) Bot. **4** (5): 179–236 (174.2–3 *Alloplectus, Columnea*).
——, 1971. Bull. Brit. Mus. (Nat. Hist.) Bot. **4** (7): 259–323 (176.11 *Oplonia*).
Storey, W. B., 1938. Amer. Soc. Hort. Sci. Proc. **35**: 83–85 (132.1 *Carica*).
Swingle, W. T., 1943. The Botany of Citrus. Univ. of Calif. (103.6 *Citrus*).
Taylor, P., 1955. Fl. Trin. and Tobago **2** (5): 288–300 (175.1 *Utricularia*).
Tharp, B. C. and Johnston, M. C., 1961. Brittonia **13**: 346–360 (163.2 *Dichondra*).
Thomas, J. L., 1960. Contrib. Gray Herb. **186**: 1–114 (114.1 *Cyrilla*).
Trelease, W., 1913. Mem. Nat. Acad. Sci. Washington **11**: 1–299 (28.1 *Agave*).
——, 1916. Monogr. Phorad. Urbana. (51.5 *Phoradendron*).
Waterfall, U. T., 1967. Rhodora **69**: 82–120, 203–239, 319–329 (170.8 *Physalis*).
Webster, G. L., 1956–58. Journ. Arn. Arb. **37**: 91–122, 217–268, 340–359; **38**: 51–80, 170–198, 295–373; **39**: 49–100, 111–212 (109.1 *Phyllanthus*).
Weimarck, H., 1940. Lunds Univ. Årsskr. N.F. Avd. 2, **36** (1): 1–141 (31.1 *Aristea*).
Windler, D. R., 1966. Austral. Journ. Bot. **14** (3): 379–420 (94.10 *Neptunia*).
Withner, C. L. and Stevenson, J. C., 1968. Amer. Orch. Soc. Bull. **37** (1): 21–32 (33.57 *Oncidium*).
Woodson, R. E. Jr., 1935. Ann. Miss. Bot. Gard. **22**: 153–306 (161.15 *Secondatia*, etc.).
——, 1938. N. Amer. Fl. **29**: 103–192 (161.1 *Plumeria*, etc.).
Yuncker, T. G., 1932. Mem. Torr. Bot. Club **18**: 113–331 (163.1 *Cuscuta*).
——, 1960. Bull. Inst. Jam., Sci. ser. **11**: 1–56 (38.1–3 *Peperomia, Piper, Pothomorphe*).
——, 1965. N. Amer. Fl. ser. 2, **4**: 1–51 (163.1 *Cuscuta*).

Revision Notes

p. 172 Add:

Eragrostis tenuifolia Hochst. ex Steud., Synops. Pl. Glum. **1**: 268 (1854).
Perennial with culms 20–40 cm. high; panicle open; spikelets 7–10-flowered; glumes minute.
Very rare (St. Andr.), on waste ground; about 550 feet; fl. Dec. *A13255!*
Native of the Old World tropics. [Specimen kindly determined by Dr. N. L. Bor.]

p. 528 Add:

Eugenia howardiana Proctor in Bull. Inst. Jam., Sci. ser. **16**: 33, t. 12 (1967).
Shrub up to 3·5 m. high; young twigs silky-puberulous, glabrescent; leaves leathery, narrowly ovate to elliptical, cuneate at base, sharply acuminate at tip, with numerous pellucid dots, undersurface densely appressed-puberulous, 2·5–6 cm. long, 1–2 (–2·5) cm. broad; racemes 3–5-flowered, short, densely puberulous; peduncle 4–10 mm. long; pedicels 4–8 mm. long; bracts sharply deltate, 1·5–2 mm. long; calyx-lobes rounded, 1·5–2 mm. long; berry irregularly globose, 8–11 mm. in diameter, minutely glandular-verrucose, red when ripe.
Rare (Manch., St. Ann), in hilltop scrub forest; 300–2300 feet; fl. July. *H & P 14355!* Endemic. Closest to *E. acutisepala* Proctor but leaves and calyx-lobes smaller and inflorescence racemose.

p. 579

The Jamaican representative of *Forestiera rhamnifolia* is now distinguished as var. **pilosa** Stearn in Journ. Arn. Arb. **52**: 615 (1971) and is also in Cuba; the glabrous variant, var. *rhamnifolia*, comprises the Antillean distribution listed.

p. 604

Evolvulus glaber Spreng. is **E. convolvuloides** (Willd.) Stearn; the combination will be validly published in 1972.

p. 605

Recent palynological investigations by W. T. Stearn have shown that the pollen of the Jamaican species hitherto included in *Jacquemontia* is of two types illustrated in Rep. on Brit. Mus. (Nat. Hist.) **1966–68**: 45 (1969), i.e. dodecocolpate in typical *Jacquemontia* but tricolpate in *J. nodiflora* as it is also in *Convolvulus arvensis* L. Thus on the basis of pollen structure (and stigma criteria), *J. nodiflora* (Desr.) G. Don is more properly placed in *Convolvulus* where it was originally described.

p. 625

Stearn, W. T. in Journ. Arn. Arb. **52**: (1971) has further appraised the representation of this genus in Jamaica and now recognizes two species in the affinity of *B. baccata*, distinguished as follows:

Hairs of inflorescence and calyx appressed *B. baccata*
Hairs of calyx erect or spreading (*Ehretia velutina* DC.) *B. velutina* (DC.) Gürke

Bourreria velutina seems to be restricted to the area of Port Henderson Hill in St. Catherine (*Powell 558, Webster & Wilson 4930, Yuncker 17477*), but a full re-examination of plants from neighbouring areas has yet to be carried out.

p. 690

Stearn, W. T. in Journ. Arn. Arb. **52**: (1971) has placed the Jamaican species of *Drejerella* in *Justicia* on the ground that pollen differences are not adequate for the separation of these two genera. *Drejerella guttata* becomes *Justicia brandegeana* L. B. Sm. & Wassh.

Index of Common Names

Index of Botanical Names

H5